In this work, 60 specialists come together to discuss the regional occurrences of Jurassic rocks around the paleo-Pacific. Their tectonic settings, stratigraphic sequences, and fossil assemblages are covered in detail; regional biozones based on palynomorphs, protistans, plants, and invertebrates are defined; and superregional standard zones based on ammonites are established. Numerous tables are used to document and illustrate intracontinental and intercontinental circum-Pacific correlations, and a large atlas illustrates more than 1,000 circum-Pacific index and guide fossils. Biogeographic distributions of different fossil groups are plotted on Jurassic plate-tectonic reconstructions of the paleo-Pacific; biorealms, -provinces and -subprovinces are distinguished for the different groups.

Not only is this the first comprehensive synthesis of Jurassic geology and paleontology of the Pacific area, but it appears to be the only one of its kind for any geological system. This book will prove indispensable to academic geologists, paleontologists, and stratigraphers interested in the Jurassic period.

# The Jurassic of the Circum-Pacific

# The Jurassic of the Circum-Pacific

EDITED BY

G. E. G. WESTERMANN
*McMaster University*

International Geological Correlation Programme Project 171: Jurassic of the Circum-Pacific

CAMBRIDGE UNIVERSITY PRESS
Cambridge, New York, Melbourne, Madrid, Cape Town, Singapore, São Paulo

Cambridge University Press
The Edinburgh Building, Cambridge CB2 2RU, UK

Published in the United States of America by Cambridge University Press, New York

www.cambridge.org
Information on this title: www.cambridge.org/9780521351539



First published 1992
This digitally printed first paperback version 2005

*A catalogue record for this publication is available from the British Library*

*Library of Congress Cataloguing in Publication data*

The Jurassic of the circum-Pacific / edited by G. E. G. Westermann.

p. cm.

Includes bibliographical references and index.

ISBN 0–521–35153–7

1. Geology, Stratigraphic – Jurassic. 2. Geology – Pacific Area.
3. Paleontology – Pacific Area. I. Westermann, Gerd Ernst Gerold,
1927–
QE681.J875 1992
551.7′66′091823–dc20 91–35703
CIP

ISBN-13 978-0-521-35153-9 hardback
ISBN-10 0-521-35153-7 hardback

ISBN-13 978-0-521-01992-7 paperback
ISBN-10 0-521-01992-3 paperback

# Contents

# Contributors

**L. Beauvais**
Laboratoire Paleontologique Invertebrat, Université Pierre-et-Marie Curie, Tour 24, 4 Place Jussieu, 75230 Paris, Cedex 5, France

**M. Bradshaw**
Bureau of Mineral Resources, Geology and Geophysics, GPO Box 378, Canberra 2601, Australia

**W. K. Braun**
Department of Geology, University of Saskatchewan, Saskatoon, Saskatchewan S7N OWO, Canada

**M. M. Brooke**
Department of Geological Sciences, University of Saskatchewan, Saskatoon, Saskatchewan S7N OWO, Canada

**J. H. Callomon**
Department of Chemistry, University College London, 20 Gordon Street, London WC1H 0A1, England

**M. Z. Cao**
Nanjing Institute of Geology and Paleontology, Academia Sinica, 39 Eastern Beijing Road, Nanjing, Jiangsu 210008, People's Republic of China

**A. B. Challinor**
25 Bailey Avenue, Hamilton, New Zealand

**A. L. Cione**
Division Paleontología Vertebrados, Museo de Ciencias Naturales, Paseo del Bosque, 1900 La Plata, Argentina

**A. S. Dagys**
Institute of Geology and Geophysics, Akademgorokok University Avenue 3, 630090 Novosibirsk, Russia (USSR)

**S. E. Damborenea**
Division Paleontología Vertebrados, Museo de Ciencias Naturales, Paseo del Bosque, 1900 La Plata, Argentina

**R. L. Detterman**[†]
Branch of Alaskan Geology, U.S. Geological Survey, 345 Middlefield Road, Menlo Park, CA 94025, USA

[†]Deceased.

**P. Doyle**
British Antarctic Survey, Natural Environmental Research Council, High Cross, Madingley Road, Cambridge CB3 0ET, England

**Z. Gasparini**
Division Paleontología Vertebrados, Museo de Ciencias Naturales, Paseo del Bosque, 1900 La Plata, Argentina

**M. B. Gordon**
Department of Geological Sciences, The University of Texas at Austin, Austin, TX 78713–7909, USA

**C. A. Gulisano**
Casilla de Correo 25, 8300 Neuquén, Argentina

**H.-J. Gursky**
Institut für Geologie und Paläontologie, Phillip Universität, Hans-Meerwein Strasse, D 3550 Marburg, Germany

**R. L. Hall**
Department of Geology, University of Calgary, Calgary, Alberta T2N 1N4, Canada

**A. von Hillebrandt**
Institut für Geologie und Paläontologie, Hardenbergstrasse 42, 1 Berlin 12, Germany

**P. J. Howlett**
British Antarctic Survey, Natural Environmental Research Council, High Cross, Madingley Rd., Cambridge CB3 0ET, England

**D. L. Jones**
Department of Geology and Geophysics, University of California, Berkeley, CA 94720, USA

**E. D. Kalacheva**
VSEGEI, Srednii Prospekt 74, 199026 St. Petersburg (Leningrad) V-26, Russia (USSR)

**T. Kimura**
Department of Astronomy and Earth Sciences, Tokyo Gakugei University, Koganei, Tokyo 184, Japan

**E. L. Lebedev**
Geological Institute, Academy of Sciences, Moscow, Russia (USSR)

**B. P. Liu**
Department of Paleontology and Stratigraphy, Chinese University of Geosciences, 29 Chengfu Road, Haidian, Beijing 100083, People's Republic of China

**M. O. Manceñido**
División Paleozoología Invertebrados, Museo de Ciencias Naturales, Paseo del Bosque, 1900 La Plata, Argentina

**E. M. Markovich**
VSEGEI, Srednii Prospekt 74, 199026 St. Petersburg (Leningrad) V-26, Russia (USSR)

**S. Mizutani**
Department of Earth Sciences, Nagoya University, Chikusa, Nagoya, 464, Japan

**J. Mojica**
Department of Geoscience, Universidad Nacional, Apt. 14490, Bogotá, Colombia

**T. I. Nal'nyaeva**
IgiG SO ANSSR, Universitetskiy Prospekt 3, Novosibirsk 90, Russia (USSR)

**G. S. Odin**
Department de Geologie Sédimentaire, Université Pierre-et-Marie Curie, 4 Place Jussieu, case 119, 75252 Paris, Cedex 5, France

**J. G. Ogg**
Department of Earth and Atmospheric Sciences, Purdue University, West Lafayette, IN 47907, USA

**T. M. Okuneva**
USEGEI, Srednii Prospekt 74, 199026 St. Petersburg (Leningrad) V-26, Russia (USSR)

**F. Olóriz**
Facultad de Ciencias, Departamento de Geología, Universidad de Granada, Granada, Spain

**O. Palacios**
INGEOMIN, Calle Pablo Bermudes, Lima, Peru

**K. V. Paraketzov**
State Geological Survey, Magadan 6, Russia (USSR)

**Judith Totman Parrish**
Department of Geology, University of Arizona, Tucson, AZ 85721, USA

**E. Pessagno, Jr.**
Programs in Geosciences, University of Texas at Dallas, P.O. Box 830688, Richardson, TX 75083–0688, USA

**J. A. Peterson**
Department of Geology, University of Montana, Missoula, MT 59812, USA

**I. V. Polubotko**
VSEGEI, Srednii Prospekt 72-B, St. Petersburg (Leningrad) V-26, Russia (USSR)

**T. P. Poulton**
Institute of Sedimentary and Petroleum Geology, Geological Survey of Canada, 3303 33rd Street N.W., Calgary, Alberta T2L 2A7, Canada

**Y. S. Repin**
VSEGEI, Liteinyi Prospekt 39, 196104 St. Petersburg (Leningrad) D-104, Russia (USSR)

**A. C. Riccardi**
División Paleozoología Invertebrados, Museo de Ciencias Naturales, Paseo del Bosque, 1900 La Plata, Argentina

**D. B. Rowley**
Department of Geophysical Sciences, University of Chicago, 5734 South Ellis Avenue, Chicago, IL 60637, USA

**A. Salvador**
Department of Geological Sciences, University of Texas, Austin, TX 78712, USA

**V. A. Samylina**
Komarov Botanical Institute, Academy of Sciences, St. Petersburg (Leningrad) 22, Russia (USSR)

**W. A. S. Sarjeant**
Department of Geology, University of Saskatchewan, Saskatoon, Saskatchewan S7N OWO, Canada

**T. Sato**
Institute of Geoscience, University of Tsukuba, Sakura-mura, Niihari-gun, Ibaraki 305, Japan

**C. Schubert**
Centro Ecología, Instituto Venezolano de Investigaciones Científicas, Apartado 1827, Caracas 1010A, Venezuela

**I. I. Sey**
VSEGEI, Srednii Prospekt 74, 199026 St. Petersburg (Leningrad) V-26, Russia (USSR)

**P. Smith**
Department of Geological Sciences, University of British Columbia, Vancouver, British Columbia, V6T 1W5, Canada

**R. Sukamto**
Geological Survey of Indonesia, Jalan Diponegoreo 57, Bandung, Indonesia

**D. G. Taylor**
Department of Earth Sciences, Portland State University, Portland, OR 97207, USA

**M. R. A. Thomson**
British Antarctic Survey, High Cross, Madingley Road, Cambridge, CB3 OET, England

**H. W. Tipper**
Geological Survey of Canada, 100 West Pender Street, Vancouver, British Columbia V6B 1R6, Canada

**W. Volkheimer**
Museo "B. Rivadavia," Avenida Angel Gallordo 470, 1404 Buenos Aires, Argentina

**S. E. Wang**
Institute of Geology, Chinese Academy of Geological Sciences, Beiwanzhuang Road, Beijing 100037, People's Republic of China

**Y. G. Wang**
Nanjing Institute of Geology and Paleontology, Academia Sinica, 39 Eastern Beijing Road, Nanjing, Jiangsu 210008, People's Republic of China

**Zhang Wang-Ping**
Institute of Geology, Chinese Academy of Geological Sciences, Beiwanzhuang Road, Beijing 100037, People's Republic of China

**G. E. G. Westermann**
Department of Geology, McMaster University, 1280 Main Street West, Hamilton, Ontario L8S 4M1, Canada

**J. S. Yü**
Institute of Geology, Chinese Academy of Geological Sciences, 26 Beiwanzhuang Road, Beijing 100036, People's Republic of China

# Preface

In contrast to the "classical" Jurassic outcrops of Europe, the Jurassic of the much larger Pacific rim has been studied intensively only in the past few decades, and often by Europeans. Much of the circum-Pacific chronostratigraphy was created by Europeans using European standards, even European zonal names. Scientific interchange between Pacific-rim countries often remained minimal, and the creation of independent stratigraphic scales and standards, as usually required by the record, lagged. Furthermore, the Subcommission on Jurassic Stratigraphy (International Stratigraphic Commission) has consistently neglected the extra-European globe.

In 1979 we formed the Circum-Pacific Research Group, which in 1980 became affiliated with the International Geologic Correlation Program (UNESCO and IUGS) as Project 171 – Circum-Pacific Jurassic. The project was extended in 1985 and ended officially in 1987. Its aim was to assemble all workers (approximately 150) involved in Jurassic-related projects in the circum-Pacific area into an informal body for the following purposes: (1) to bring together workers from developing, sometimes politically isolated countries, as well as the often unknown foreign researchers working in those countries; (2) to exchange the latest research results in a newsletter; (3) to hold several field meetings, with UNESCO/IUGS financial support (held in Canada, Mexico, Argentina, and Japan); (4) to circulate special papers (eventually 17) on topical themes for immediate news dispersal; (5) to publish range and correlation charts in the *Newsletters on Stratigraphy* (parts 1–4 now published: Soviet Union, People's Republic of China, South America, Southeast Asia); and (6) to publish this synthesis in book form at the end of the project.

This volume presents the first synthesis of any geological system in the entire circum-Pacific area. I requested the various specialists in the regions and disciplines to contribute original chapters to this volume and have attempted to shape these disparate contributions into a single, comprehensive work.

GERD E. G. WESTERMANN

# Acknowledgments

I greatly thank the many authors for their contributions concerning many topics and countries, especially Dr. Terry P. Poulton, Dr. Alberto C. Riccardi, Dr. Irena I. Sey, and Dr. Wang Yigang, who organized major chapters. Without all of them this work could not have been accomplished. I am also grateful to the patient editors of Cambridge University Press, New York, especially Mrs. Katharita Lamoza, and to BookMasters for excellent typesetting. Mrs. Elisabeth Laforme retyped many of the foreign manuscripts and Mr. David Hodson redrafted many of the illustrations. I am especially indebted to my research associate, Mr. Cameron Tsujita, for undertaking the tremendous task of producing the two indexes for this book. Many years of my own scientific work that preceded this work, as well as many of the expenses incured during the production of this work, were generously funded by the Natural Sciences and Engineering Research Council of Canada under the grant ''Jurassic Ammonoids and Cephalopod Functional Morphology.''

# Introduction

G. E. G. WESTERMANN

The Jurassic period is of particular significance for a variety of geological and biological disciplines: Not only has it served as the "classical" example for biostratigraphy and chronostratigraphy, but also it marked the beginning of the last megacycle of plate movements, that is, the "final" breakup of Pangea. The history of the Pacific Ocean (with its ancestor, Panthalassa) and the motions of the multitudinous suspect/tectonostratigraphic terranes now surrounding it can be traced back to the Jurassic, which formed the oldest extant ocean floor. Palinspastic reconstructions by geophysicists are required for the Jurassic paleogeography of the Pacific rim, because most of the current continental margins were, at that time, islands or submarine plateaus (terranes) within the single ocean (Panthalassa). Paleobiogeography provides limits to the geophysical placements of the terranes, especially for Northern Hemisphere versus Southern Hemisphere alternatives and narrowly spaced blocks and terranes forming isthmuses or seaways for faunal migration.

The essential condition for reconstructing a detailed Jurassic history of the Pacific rim is, of course, a precise time correlation between the distant autochthonous (craton) and allochthonous (terranes) Jurassic sequences. In addition to the development of radiometric and magnetic time scales, the construction of refined biostratigraphic regional zonations is essential. In general, ammonite zones remain by far the most practical, precise means for defining and identifying coeval chronostratigraphic units, not only between Pacific regions but also globally. Other plant and animal taxa are becoming increasingly important for biostratigraphy and biogeography, and a multitaxial, integrated approach is beginning to develop. Yet the synthesis of the data presented here has only begun and will take decades to complete. Basic biostratigraphic data continue to be published, for example, in the *Newsletters on Stratigraphy,* in a series entitled "Jurassic Taxa Ranges and Correlations Charts for the Circum Pacific," edited by G. E. G. Westermann and A. C. Riccardi.

The Jurassic time scale used here (see Chapter 1) is necessarily of a provisional nature, because radiochronometry for the Jurassic remains exceptionally coarse and controversial. A unique "compromise scale" (Table I.1) has been developed, combining Odin's revised radiometric data with Westermann's biochronological estimates for the relative durations of stages (averaged and scaled subzones).

Table I.1. *Compromise Jurassic time scale*

| Period | Stage | Age (Ma) | Duration (m.y.) |
|---|---|---|---|
| Late Jurassic | (top) | 135 | |
| | Tithonian | | 6 |
| | | 141 | |
| | Kimmeridgian | | 5 |
| | | 146 | |
| | Oxfordian | | 8 |
| | | 154 | |
| Middle Jurassic | Callovian | | 6 |
| | | 160 | |
| | Bathonian | | 7 |
| | | 167 | |
| | Bajocian | | 9 |
| | | 176 | |
| | Aalenian | | 4 |
| | | 180 | |
| Early Jurassic | Toarcian | | 7 |
| | | 187 | |
| | Pliensbachian | | 7 |
| | | 194 | |
| | Sinemurian | | 7 |
| | | 201 | |
| | Hettangian | | 4 |
| | (base) | 205 | |

The spelling of the names of zones follows that currently in wide use in Northwest Europe and North America, developed essentially by John Callomon in order to clearly distinguish between biostratigraphic and chronostratigraphic units.

Standard zones (chronozones), defined by their bases in stratotypes, are capitalized in roman type (e.g., Ovalis Zone). Biozones (range zones, assemblage zones, etc.), defined by taxa, are printed in italics, like the index taxon [e.g., *Duashnoceras floresi* (Assemblage) Zone].

The same spelling applies for subzones. Lowercase spelling is used for "zone/subzone" if the unit is informal or if used without its specific name attached [e.g., ". . . all palynozones/subzones are . . ." (contrasting with Callomon's spelling)].

The radiochronological scale has recently been complemented by the magnetic-reversal scale, which, in its latest form for the Jurassic, is here presented by J. Ogg (Chapter 2). Although still in its infancy for the Jurassic Period, this scale promises great advances as a technique for precise correlations.

The palinspastic maps for three Jurassic intervals have been developed by D. Rowley. They appear to be the first reconstructions for the entire Pacific area to include the positions of major terranes, and they have been used as the basis for biogeographic and oceanographic descriptions.

The backbone of this book, of course, is composed of the regional geological and biostratigraphic descriptions presented by numerous specialists. The order of description proceeds clockwise from Alaska to North-East Russia. Included are zone-by-zone listings of the important faunas and floras and regional correlation charts.

In the chapters on biozones and chronozones (Part IV), several authors summarize the regional biostratigraphic data on floras and faunas, in an attempt to construct a superregional biochronology for the Jurassic of the Pacific rim. This has been partially successful for the ammonites, which therefore receive extended treatment. The more recent significant developments in palynomorph stratigraphy are also reflected in a major contribution, whereas the other taxa are treated in less detail. Part IV is supplemented by an Appendix: Index and Guide Fossils, which forms an essential part of this volume.

The Jurassic biogeography of several animal and plant classes has also been given ample space here, and the ammonite biogeography is applied to terrane and plate reconstructions. Specialists working on single classes have, unfortunately, used different names and categories for the biogeographic provinces, realms, and so forth. These biogeographic units are thus based on single higher-animal or plant taxa, not on faunal and/or floral communities (biomes). A solution for the problem of combining the different overlapping sets of biogeographic classifications into a single set awaits a future thorough synthesis of the data presented here.

Paleobiogeographic units, as we understand them, are highly dynamic. They come and go, wax and wane with geological time by spatial expansion and contraction. The units also change in their degrees of endemism, in the proportions as well as the rankings of the geographically restricted taxa. Biogeographic ranking, from realm to subprovince, thus depends on the spatial extent *and* the degree of endemism within the units. It is to some degree subjective and is expected to change through geological time. For example, Subprovince 1 of Province I and Subrealm A may later become Province 1 of Subrealm A or a new Subrealm B. At the level of faunal realms, only the Boreal Realm and Tethyan Realm are retained here for the Jurassic, throughout which they persisted. The former East Pacific Realm is here lowered to a subrealm. The biogeographic term "cosmopolitan" usually refers to an extremely wide, circumglobal ("pantropical") distribution, which nevertheless is restricted to the Tethyan Realm and its overlap with the Boreal Realm (Subboreal Province), but excludes the Boreal Province. "Cosmopolitan" is thus almost synonymous with "pan-Tethyan."

# Part I: Time scales

# 1 Numerical time scale in 1988

G. S. ODIN

### Introduction

The numerical time scale is an important tool for geologists, but often such scales are based on a minimal amount of data. The problem is particularly acute for Jurassic time; this problem results from the unavoidable constraint that a realistic numerical time scale must be derived from appropriate radiometric measurements. These measurements can be undertaken only with adequate *geochronometers*, in which the Jurassic sequence is particularly lacking. Adequate geochronometers consist of stratigraphically well controlled (within a stage) potassium- or uranium-bearing minerals or rocks. This stratigraphic constraint eliminates most plutonic rocks, leaving only volcanics or glauconitic sediments as sources for precise geochronological tie points. However, where few data are available, plutonic rocks are useful because they are easily dateable and their mineral fractions frequently are better preserved than those of volcanic rocks.

Because of the limited number of adequate control points along the sequence, usually it is necessary to interpolate between these geochronologically founded tie points in order to suggest numbers for stage boundaries. An abundant recent literature proposes many new suggestions for interpolation; but the difficult task of establishing and qualifying valid tie points is rarely undertaken. Often, the examination of the geochronological data is given a minor treatment in papers dealing with time-scale calibration, when it should be the foundation. This chapter is intended, first, to detail some of the Jurassic scale proposals in the literature and clarify the manner in which they were constructed and, second, to summarize the recently obtained (1983–1988) radiometric data in comparison with previous data, with the objective of constructing an improved numerical time scale.

### Foundation of some representative Jurassic time scales

There are two syntheses that discuss the calibration of the Jurassic time scale and summarize a large collection of radiometric studies from which it was calibrated. The first was published in 1964 following "A Symposium on the Phanerozoic Time Scale" organized by the Geological Society of London and Glasgow (GSLG 1964). It was followed, 18 years later, by the work entitled *Numerical Dating in Stratigraphy* (Odin 1982). In the meantime, a synthesis by Afanass'Yev and Zykov (1975) was published in Russian, but the details on the geochronology are not sufficient for adequate utilization, because field locations and decay constants are not shown; another compilation essentially recombined previous data for which geological or analytical details were documented elsewhere (Armstrong 1978).

The GSLG time scale cited 11 radiometric ages within the Jurassic system (Howarth 1964); half of them were from batholiths with poor stratigraphic control (with more than one full stage involved in the stratigraphic definition). The resulting scale placed the beginning of the period between 190 and 195 Ma, and the end at about 135 Ma. Because most of its ages were obtained using the K-Ar dating method, and the decay constants have been recalibrated since then, the 1964 estimates should be "corrected." The correction results in an increase in age estimates by about 2.3%, or 3–4 Ma (Ma is the conventional unit for million years, i.e., "Mega Annum").

Because of imprecision and the small number of tie points available within the Jurassic period, the scale recommended "as a basis for discussion" in 1964 was based on an interpolation between 192 Ma and 135 Ma (57 Ma for 11 stages), assuming that stages were of approximately equal lengths (5–6 Ma). This is shown in Figure 1.1, column 1.

The same numbers of 135 and 192 Ma were used in the time scale constructed by Van Hinte (1976, 1978). That author added new constraints, including (1) an age for the Middle Jurassic (Bajocian–Bathonian boundary at 165 Ma), (2) assignment of a duration of 1.0 Ma to each ammonite zone below that Bajocian–Bathonian boundary (26 zones for 27 Ma), and (3) use of a new stage subdivision, following the recommendation of the "Colloquium on the Jurassic," held in Luxembourg in 1962.

For the Late Jurassic stages, Van Hinte (1978) considered a combination of new radiometric ages and an equal, standard

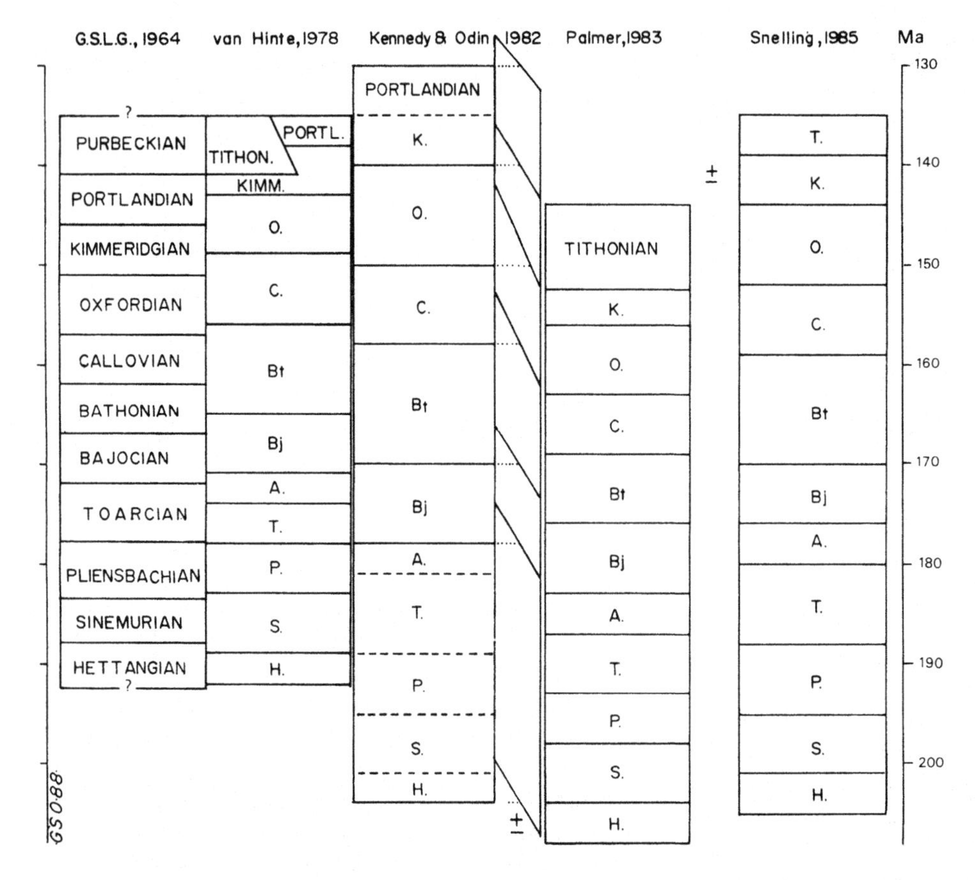


**Figure 1.1. Evolution of the Jurassic numerical time scale. The equal-stage proposal of 1964 (G.S.L.G.) was improved by Van Hinte (1978) using mainly the equal-standard-zone interpolation system and was replaced in 1982 by a scale using more geochronological ages. The scale of Palmer (1983) essentially differs by assuming a much older Jurassic/Cretaceous boundary (144 Ma); large error bars (up to ±10–30 Ma) were suggested by Palmer. The scale of Snelling (1985) resulted from a cooperative effort based on geochronological dates and does not essentially differ from that of Kennedy and Odin (1982), but considers the necessity for combining different opinions.**

system for zone durations and proposed ages of 138 Ma, 141 Ma, and 143 Ma for the bases of the Portlandian, Tithonian, and Kimmeridgian stages, respectively. The durations of Oxfordian and Callovian stages were then estimated, assuming a mean ammonite-zone duration of 1.0 Ma. Being subdivided into a large number of ammonite zones, the Bathonian became a longer stage than most of the others in the Jurassic (column 2 in Figure 1.1).

The new, "modern" synthesis of time-scale calibration edited by Odin in 1982 gathered supplementary radiometric ages obtained during the previous 18 years. That allowed a geochronological calibration for the base of the Jurassic period and stage boundaries from Bajocian to Portlandian time (Kennedy and Odin 1982). In the absence of precise radiometric ages, the assumption of equal durations of ammonite standard zones was used to estimate the ages of stage boundaries within the Liassic portion of the period (26 zones for 26 Ma). It was shown that those interpolated ages did not contradict the 11 (stratigraphically imprecise) geochronological results considered as reliable to document the Liassic epoch (Armstrong 1982; Kennedy and Odin 1982). Eleven more radiometric ages were considered useful for calibration of the Bajocian to Callovian stages, and 10 for the Oxfordian to Portlandian stages.

The two Jurassic scales by Van Hinte (1978) and Kennedy and Odin (1982) took into account the available geochronological studies (and number of ammonite standard zones defined within each stage where few data were available), and that resulted in unequal durations for the stages. That was in contrast to another subsequent proposal made by Harland et al. (1982), who surprisingly returned to the equal-duration hypothesis for the 20 stages (of 6 or 7 Ma duration) between the middle Triassic (Anisian) and the middle Cretaceous (Albian). That poorly constrained (and intellectually naive) exercise resulted in an interpolated age estimate of 144 Ma for the Jurassic–Cretaceous boundary, with no new precise radiometric data to support that proposal. The same number was again accepted, after interpolation, by Kent and Gradstein (1985) and the derived time scale proposed as the standard for the North American geological community (Palmer 1983).

This age of 144 Ma makes the latter scales incompatible with existing (in 1982) geochronological data and consequently with the scale of Kennedy and Odin (1982) for the last two stages of the sequence (Kimmeridgian and Tithonian). Taking into account the margins of uncertainties based on examination of the radiometric data, Figure 1.1 shows that columns 3 and 4 are compatible for the stage boundaries from the Hettangian to Callovian.

In 1984, a new effort was made by the Geological Society of London to gather contributions dealing with this question. No new radiometric ages were published, and the Jurassic scale suggested (Hallam et al. 1985; Snelling 1985) was reasonably similar to the one shown by Odin (1982), with the Jurassic period accepted to be 70 m. y. long, between 205 and 135 Ma (column 5 in Figure 1.1).

### The Jurassic–Cretaceous boundary age

The main difference among the modern scales currently used in the literature concern the age estimates for the Jurassic–

Cretaceous boundary. A poorly documented age of 135 Ma was used in 1964. In his revision of the GSLG (1964) scale, Lambert (1971) commented on that age as follows: "the boundary may fall in the interval 125–145 Ma and there is no sound reason for taking the mean figure."

Kennedy and Odin (1982) and Odin (1985) commented on the Jurassic–Cretaceous boundary age using all available radiometric data. For those authors there were essentially four sets of data available to calibrate the limit at that time: Those were all justifiably selected (buried at less than a few hundred meters) glauconitic samples from which K-rich and homogeneous glauconitic minerals were separated. All samples were well calibrated with good fossil assemblages in three different basins (i.e., Paris Basin, South England, and Moscow Area). Those data were supported by a small number of other glaucony ages from Switzerland and southeastern France (Table 11 in Kennedy and Odin 1982). All of those ages were quite consistent and indicated apparent ages between 135 Ma and 130 Ma for Portlandian or Volgian sediments. Two other ages of 134 Ma were calculated from Early Cretaceous (Ryazanian) K-poor glauconitic minerals that were not considered reliable because of the possibility of argon inheritance (Odin and Dodson 1982). The glaucony ages considered reliable cannot be suspected to have been significantly rejuvenated in the absence of important burial or tectonism in the areas of collection. However, it cannot be claimed that the minor tectonic flexure in some outcrops or postdepositional early diagenetic recrystallization has not modified the apparent age by a few million years.

According to these data, the Jurassic–Cretaceous boundary can be proposed to lie within the interval of time 130–135 Ma. Two remarks should be added: On the one hand, it is unlikely that any disturbances would have *similarly modified* the radiometric clocks in all samples dated from five different areas; on the other hand, the opinion favored now is that some of the Portlandian (and Volgian) sediments previously included in the Jurassic are earliest Cretaceous in age, with the Jurassic–Cretaceous limit (i.e., the Tithonian–Berriasian boundary) older than the top of the Volgian or the top of the Portlandian (*sensu* gallico) and the base of the Ryazanian (Westermann 1984; Wimbledon 1984; cf. Jeletzky 1984). More recently, Zeiss (1986) has suggested that the late Volgian was a time equivalent to the lower Berriasian; therefore, the interval of time of 130–135 Ma quoted earlier might be regarded as earliest Cretaceous in age.

Because glauconies are commonly problematic as geochronometers when inappropriately selected (and it is a difficult task to be certain that they are appropriate), it is necessary to reconsider stratigraphically less precise data from plutonic or volcanic rocks that have been used in the past for estimating the age of the Jurassic–Cretaceous boundary.

In his computation, Armstrong (1978, Fig. 2) showed the Berriasian–Tithonian boundary at about 142 Ma; the single nonglaucony age used by Armstrong is discussed later. Lanphere and Jones (1978) reported details on two geochronological studies, including the one used by Armstrong. The Shasta Bally batholith from California was cited as being overlain by late or possibly middle Hauterivian sediments. No maximum stratigraphic age was assumed; calculated ages of 134 Ma (pooled K-Ar ages of biotite and hornblende) and 136 Ma (zircon U-Pb) were reported. That was a maximum age for the Hauterivian sediments. Armstrong (1978) quoted an age of 134 Ma for the same batholith, but showed the batholith as Kimmeridgian or younger. No stratigraphic details were available to us concerning the latter attribution. In short, the Shasta Bally batholith radiometric age [135 Ma for a Kimmeridgian(?) to Hauterivian pluton] does not contradict (nor confirm) a Jurassic–Cretaceous boundary at 130 Ma or 135 Ma.

The second geochronological result discussed by Lanphere and Jones (1978) concerns a tuff of Late Valanginian age from central Alaska; with a biotite K-Ar age of about 136 Ma being reported. Those authors emphasized that a single mineral age is not sufficient for estimating with confidence the time of sedimentation of the tuff, especially when the mineral used is a biotite with a potassium content slightly lower than 2.2wt% K. The reliability of that age is highly questionable, because alteration of the geochronometer was evident: In general, biotite is considered suspect if its potassium content is lower than 6–6.5%, and it is worth noting that the altered biotites may give ages that are too old as well as ages that are too young (Obradovich and Cobban 1976; Odin et al. 1991). Lanphere and Jones (1978) concluded that "no [really definitive] new data on the age of the Jurassic–Cretaceous boundary [have] been reported since Lambert (1971) reviewed the situation" (see the beginning of this section).

The current situation is not greatly improved in regard to the data from plutonic or volcanic rocks. But the earliest glaucony ages and our knowledge of their significance have been improved during the past decade; in the view that a rejuvenation of all glaucony ages of more than 10% would be needed to confirm an age of 144 Ma for the Jurassic–Cretaceous boundary, the latter appears older than is acceptable from any geochronological study.

### New radiometric data published since 1982

Among the data considered by Webb (1982) and Odin and Létolle (1982) for calibration of the Triassic–Jurassic boundary, the early Liassic volcanics from the Hartford Basin (eastern United States) were heavily weighted. They indicated an age of 190–200 Ma (Webb 1982). However, Webb (1982, abstracts NDS 202 and 203) commented on these results as follows: "stratigraphic control on the dated volcanics is excellent but geochemical uncertainties are high." Recently, Seidemann (1988) published a synthesis of his studies on the volcanics of that basin. He noted that some of these volcanics contain "small and variable amounts of excess argon," and he concluded that the age of 187 ± 3 Ma ($2\sigma$) should replace the older ones (190–200 Ma). If this conclusion is accepted, the Triassic–Jurassic boundary could be at or even younger than 200 Ma. The interval of time 204 ± 4 Ma includes this younger number and the significantly different estimate of 208 Ma (Armstrong 1982; Palmer 1983); this latter number seems less probable today.

Hess, Lippolt, and Borsuk (1987) dated Liassic Caucasian volcanics that were stratigraphically located between middle Pliensbachian and upper Toarcian mollusk- and brachiopod-bearing sediments. The volcanics were not metamorphosed, but were affected by spilitic alteration. However, Hess et al. (1987) were able to obtain some fresh plagioclase (carefully identified as high-temperature plagioclase using x-ray diffractometry) and unaltered biotite (potassium contents 6.4% and 7.2% K for two separates). From 14$^{40}$AR-$^{39}$Ar step-heating age measurements the authors found an age of 185 ± 6 Ma (2$\sigma$) for the emplacement of the volcanics. Hess et al. (1987) concluded that that result confirmed the data from Armstrong (1982); it showed that the Toarcian–Aalenian boundary age of 187 Ma calculated by Palmer (1983) and similar scales was too old, and the estimate of 181 Ma by Kennedy and Odin (1982) seemed to be the best approximation. An even younger interpolated estimate (178 Ma) would not be unacceptable according to the few definitive radiometric data available (Kennedy and Odin 1982, pp. 574–7) if we favor a modest duration for the Bathonian stage as supported by Westermann (1984; 1988).

Bellon et al. (1986) repeated the dating of the basalt of "Les Vignes" (French Massif Central) previously studied by Baubron et al. (1978); see also abstract NDS 214 in Odin (1982). The basalt was stratigraphically located above Upper Bajocian sediments and below accepted Upper Bathonian sediments. The rocks were weathered; two lava flows and one crosscutting dike were dated as whole rocks. They gave consistent apparent K-Ar ages, allowing the authors to suggest an age of 168 ± 8 Ma (2$\sigma$) for their time of eruption, with a later dike event at 163 ± 8 Ma. The age of emplacement is in good agreement with both the estimate of Kennedy and Odin (170–158 Ma) and the proposal of Palmer (176–168 Ma) for the lower and upper boundaries of the Bathonian stage, taking into account the large analytical errors and the stratigraphic imprecision.

Another supplementary study was undertaken concerning the Oxfordian Swiss glauconies previously considered by Odin (1982, NDS 141). Three glauconitic horizons that correlated individually with the Cordatum and Densiplicatum Standard Zones and the Parandieri Standard Subzone (Lower Oxfordian) were collected by Fischer and Gygi (1987). Those authors indicated that the sediments had not been deeply buried nor tectonically affected. The separated Oxfordian glauconitic minerals were potassium-rich. K-Ar ages of 149.2, 148.5, and 145.9 ± 1.8 Ma (2$\sigma$) were obtained from the bottom to the top of the sequence. These ages allow us to exclude the youngest portion of the interval of time suggested by Kennedy and Odin (1982) for the Callovian–Oxfordian boundary: 150 (+3 or −8) Ma. Therefore, an age of 150 Ma can be suggested as the glaucony age, and minimum estimate, for that boundary. However, this age of 150 Ma still remains incompatible with the scales of the Palmer type, which locate the same boundary at 163 Ma.

The latter old age was based on studies of the Klamath Mountains in California and Oregon (Saleeby et al. 1982; Harper, Saleeby, and Cashman 1986). In the Smith River area (northern California), the stratigraphic sequence comprises the Josephine ophiolite, overlain by the Galice Formation. The latter comprises a thin hemipelagic sequence, with cherts showing ghosts of radiolaria that "range tentatively from Callovian to Oxfordian in age" (Saleeby et al. 1982). This hemipelagic sequence is overlain by a thick flysch facies in which the bivalve *Buchia concentrica* is found. This fossil is considered to indicate a middle Oxfordian to late Kimmeridgian age (Imlay 1984, Fig. 6). The Josephine ophiolite was dated using zircons from two plagiogranites at 157 ± 2 Ma (U-Pb concordant ages). Slightly discordant U-Pb ages on zircons from a porphyritic quartz keratophyre at 150 ± 2 Ma constrained the minimum age of the intruded Galice flysch (Saleeby et al. 1982, p. 3837). Elsewhere, in the Rogue River area, the Galice Formation depositionally overlies the azoic-type Rogue Formation; the latter has been dated at 157 ± 1.5 Ma (U-Pb zircon age from an interlayered tuff-breccia) and postdated at 150 ± 2 Ma using zircons from dacite dikes that cut the upper Rogue Formation (Saleeby 1984).

In short, U-Pb zircon ages define a range between 157 Ma and 150 Ma for a portion of the time range of *B. concentrica* (middle Oxfordian–late Kimmeridgian), and the older age of 157 Ma is from rocks located below sediments of tentative Callovian to Oxfordian age. Therefore, according to these data, the base of the Oxfordian stage should be older than 150 Ma, and possibly much older, depending on the precise time when the Galice flysch was deposited with the *B. concentrica* Zone. The Oxfordian–Kimmeridgian boundary could be older or younger than this age for a similar reason.

Among the new possibilities for time-scale calibration, a major event has been the recent development of the research undertaken in China. The geochronological laboratory of Yichang has obtained data useful for an estimate of the Jurassic–Cretaceous boundary (Ye Bodan 1986, 1988). According to that author, the Jiande Group comprises three formations: The Laocun Formation is now considered to represent the basal part of the Cretaceous system in South China ("Berriasian"); it is overlain by the Huanjian Formation (still considered "Berriasian") and the Shouchang Formation ("Valanginian"). In these three formations, the Jiande biota comprises essentially lagoonal to freshwater fossils: plants, fishes, bivalves, charophyta, insects, ostracods, and estherids (Ye Bodan 1988).

The volcanosedimentary Laocun Formation has been dated, with the following results: a Rb-Sr whole-rock isochron age of 129 ± 3 Ma; two confirmatory K-Ar sanidine and biotite ages of 127 Ma and 126.3 Ma. The Huanjian Formation provided U-Pb zircon ages with a lower intercept at 129.5 ± 2 Ma. The overlying Valanginian volcanics (Shouchang Formation) provided a whole-rock acid volcanic Rb-Sr isochron age of 122.6 ± 2.5 Ma, a volcanic feldspar $^{39}$Ar-$^{40}$Ar step-heating age of 121.8 Ma, and a U-Pb zircon age (lower intercept) of 124 ± 3 Ma. According to Ye Bodan (1986, 1988), these ages "suggest that the Cretaceous–Jurassic boundary age may be about 130 Ma old in South China." Additional biostratigraphic information is still needed [the ostracods, for instance, are freshwater and endemic (R. Damotte, personal communication)] to give appropriate stratigraphic weight to these Chinese geochronological results; for example, although the Laocun Formation is now regarded as Berriasian in age, earlier studies considered the overlying Shouchang Formation as latest Jurassic in age.

Still, from Early Cretaceous sediments, preliminary results

have recently been presented (although in abstract form only) by Bralower et al. (1987), who studied two bentonite horizons found in north California (Grindstone Creek section). A low-uranium (*a priori* a favorable criterion) zircon separate yielded concordant U-Pb results (no inheritance) indicating an age of 135.2 ± 1.5 Ma for one horizon. Additional measurements using fission-track dating applied to apatite and zircon gave younger ages, around 115–125 Ma, suggesting a geochemical diagenetic disturbance; therefore, the preliminary fission-track ages are unreliable (C. W. Naeser, personal communication). The biostratigraphic study undertaken in parallel makes that study a good future reference for time-scale calibration (Bralower, Monechi, and Thierstein 1989). The stratigraphic location suggested according to calcareous nannofossils (earliest Late Berriasian) is of good quality in itself and is generally confirmed by *Buchia* as well as by indirect magnetostratigraphic correlation with deep-sea and land sections including the stratotype (Galbrun 1986).

The single-age, single-geochronometer, single-dating-method age available will influence our proposal for the age for the Jurassic–Cretaceous boundary, as discussed later.

### Biozonal interpolation hypothesis

The use of numbers of standard zones to subdivide a period of time has long been tested by various authors. This seems to assume constant evolutionary rates, a hypothesis that may appear of low reliability, but in fact requires only that Jurassic zones (and subzones) vary randomly; their durations appear to be unrelated to rates of evolution (Westermann 1988). It is useful to discuss here whether or not the result from such an interpolation hypothesis is reasonable as a function of the available radiometric data. The number of northwestern European ammonite standard zones per stage was the basis for the interpolation made by Van Hinte (1978). The principle used by Westermann (1984, 1988) was similar, but the application was more complex, because some portions of the Jurassic period were subdivided using zones, and others using more or less numerous subzones; Westermann assumed that the more numerous subzones probably represented more confidently the time involved. He suggested that nonsubdivided zones should be weighted more heavily than subzones, but less heavily than subdivided zones, and proposed this ratio: subzone/zone = 0.75. As a result, a nonsubdivided zone represents 1 *unit of time*, a zone subdivided in two subzones represents 2 × 0.75 units, a zone subdivided into three subzones represents 3 × 0.75 units, and so on. That interpolation system also considered several ammonite zonations from different bioprovinces and averaged between them.

Westermann (1984, 1988) used anchors at 144 Ma and 208 Ma for the end and beginning of the period, respectively, according to Kent and Gradstein (1985). The numbers of zonal units for the stages calculated by Westermann have been reused here to calculate a "new" time scale using the limits geochronologically defined at 135 Ma and 204 Ma (Figure 1.2, column "Westermann," with the number of units reported). A number of 9 units was chosen for the Tithonian stage because this stage is defined in

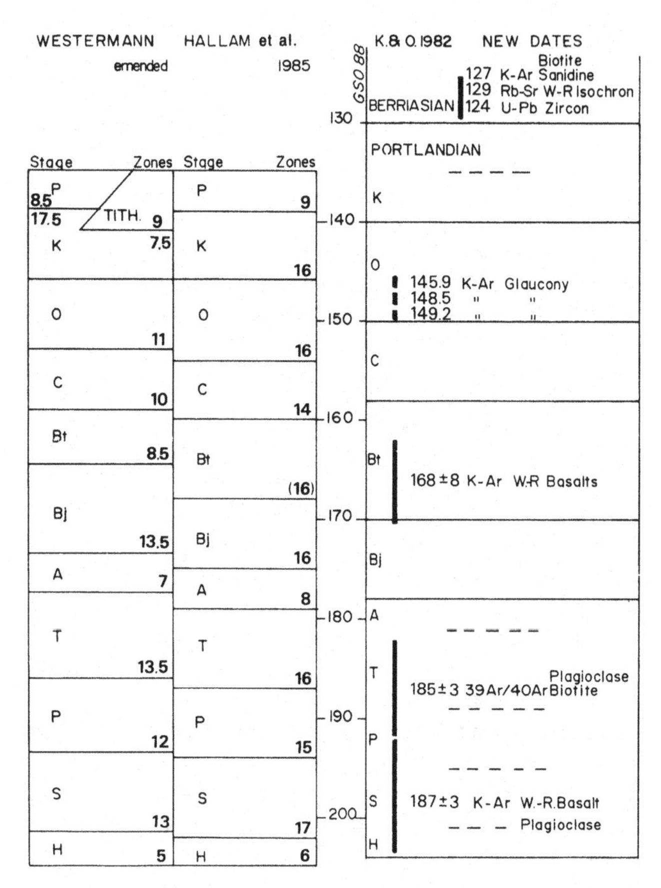


**Figure 1.2. Jurassic interpolated time scales. The two scales use biozonal interpolations between 205 Ma and 135 Ma: rounded numbers are preferred as the tie points for the limits of the period. To the left, the scaled subzone interpolation system of Westermann (1984) is used; to the right, the proposal by Hallam et al. (1985) is shown. Boldface numbers indicate the number of zonal units considered.**

the Tethyan Realm for which Westermann calculated this number of units.

A similar, simpler subzonal interpolation hypothesis has been suggested by Hallam et al. (1985), who used the ammonite subzones recognized in England. However, because the English Bathonian zones are not subdivided into subzones, and because the two Bajocian and Bathonian stages, of similar facies, have similar thicknesses, the number of zones of the Bathonian was arbitrarily increased to make it similar to the Bajocian stage (Figure 1.2, column to the right). The two columns are very similar as far as durations are concerned (except for the Bathonian); this shows that the more or less sophisticated biozonal interpolation hypothesis is reasonably well reproducible when ammonites are concerned.

Now, if we compare the columns in Figure 1.2 with the columns shown in Figure 1.1, we observe that there is no crucial disagreement. (In this test comparison we exclude the Liassic portion of the scales, because they are all based on the same system of interpolation.) A conclusion of that exercise is that the rate of ammonite standard zone (subzone) change is not widely different from one to another portion of the Jurassic period. The single apparent disagreement is in the Bathonian versus Bajocian stage-duration ratio, which is variously larger than 1, according to

Kennedy and Odin (1982) and Snelling (1985), equal to 1, according to the scale of Palmer (1983) or Hallam et al. (1985), or smaller than 1, in the scale of Westermann (1984, 1988); this is linked to the latter author's opinion that the English Bathonian was "over-split" and has only four or five "good" zones, as in other bioprovinces (G. Westermann, personal communication). The error margins in geochronological calibration indicated by Kennedy and Odin (1982) and Palmer (1983) are large enough to include the Westermann (emended) and Hallam proposals, which are therefore generally compatible with geochronological results.

### Magnetostratigraphic interpolation hypothesis

A number of time scales proposed for the Late Jurassic–Early Cretaceous portion of the stratigraphic sequence are based, at least in part, on the following assumptions: (1) the sea-floor spreading rate is constant over long periods of time in a given ocean, (2) the magnetic sequence in deep seas is well established, (3) the correlation between the sea-floor anomaly sequence and the usual definition of stages is well known, (4) it is enough to know the ages of the two extreme points to interpolate the ages of all other points in between, and (5) the numerical ages of these two extreme points are well enough known to make them absolute tie points. Usually, the model is applied to Hawaiian lineations in the northwestern Pacific Ocean for Late Jurassic–Early Cretaceous time (Larson and Hilde 1975; Kent and Gradstein 1985, Lowrie and Ogg 1986).

This method of interpolation is potentially useful. However, the first assumption remains hypothetical as far as long periods are concerned. This has been shown by comparison of spreading rates in different oceans for similar periods (Heirtzler et al. 1968; Curry 1985; Odin and Curry 1985). The correlation between sea-floor anomaly sequence and stage definitions is still in a state of flux, with sometimes difficult interpretations (Channell, Ogg, and Lowrie 1982; Steiner et al. 1985; Galbrun 1986; Ogg and Steiner 1988). The numerical ages of the two extreme points of the sequence usually considered (i.e., Oxfordian and Aptian) are still debated and are far from being established with acceptable precision. Currently, we still have to wait for a better connection between magnetic sequences known from deep seas and the available records on land, where good fossil control is available; there still is no accepted correlation of deep-sea nannofossil and microfossil stage "definitions" and onshore macrofossil definitions, and boundaries in the two situations usually are at different levels; more precise radiometric data are needed to calibrate the sequence in sufficient tie points to allow correction of the possible changes in the rate of sea-floor spreading, as accepted in some magnetostratigraphic interpolations (Lowrie and Alvarez 1981).

The important efforts being made in this domain by many specialists will soon improve the validity of the interpolation method. However, the magnetostratigraphic interpolation hypothesis is still based on too much assumption and great imprecision; application of a potentially good method in unsatisfactory conditions might well conceal the need for more work and discourage future research in both geochronology and magnetostratigraphy. Currently, it appears premature to derive stage durations from the Jurassic magnetic sequence, and application seems difficult for the older Jurassic portion, where fundamental data gathering is still in progress (Galbrun, Gabilly, and Rasplus 1988); we should prefer to wait for better use and discussion of that potentiality in the future.

### Discussion and conclusions

The numerical time scale for the Jurassic period is based on an insufficient number of geochronological studies. Up to 1982, the Upper Jurassic stratigraphic sequence was calibrated mainly using glaucony K-Ar ages; Early and Middle Jurassic time was calibrated mainly using plutonic or volcanic rocks, frequently imprecisely located within the stratigraphic sequence. The age of the beginning of the period was comparatively well known and was located with confidence in the interval 200–208 Ma. The age of the end of the period was a debated question and was shown to be around either 130 Ma or 145 Ma, depending on the authors cited.

New data are now available; they are collated in Figure 1.3. Volcanic rocks collected from above the Jurassic–Cretaceous boundary were dated in China and in the United States. The Chinese data are geochronologically well founded, with three dating methods applied to at least four different geochronometers. The consistent results suggest an age of about 130 Ma for basal Cretaceous sediments.

In contrast, the North American datum is geochronologically based on a still preliminary single U-Pb zircon age; but the biostratigraphic control is well established, leaving aside the question of stage definition itself. Considering the fact that the age of 135.2 ± 1.5 Ma (Bralower et al. 1987) may be located in the middle of the Berriasian stage, and the fact that this stage is accepted to be 5 ± 1 Ma long (Figure 1.1), a tentative extrapolation by addition of the calculated age to half the probable stage duration (2.5 ± 0.5 Ma) is suggested. An estimate of 138 ± 2 Ma results for the Jurassic–Cretaceous boundary. The maximum age of the Jurassic–Cretaceous boundary is therefore 140 Ma (it might equally be as young as 136 Ma). If this result is combined with the age suggested by the Chinese results (similar to ages deduced from glauconite dating), the Jurassic–Cretaceous numerical age must fall in the interval of time 135 ± 5 Ma (i.e., 130–140 Ma). This interval appears to be the best compromise and might be made more precise at 135 ± 3 Ma without exclusion of fundamental results. The radiometric data indicate that an age at or above 140 Ma is not acceptable for the Jurassic–Cretaceous boundary.

New Late Jurassic data include new glaucony ages indicating that the Callovian–Oxfordian boundary must be older than 150 Ma, and this is an improvement over the previous proposal of Kennedy and Odin (1982). In addition, we must consider the information published from the Galice Formation in the Klamath Mountains in the United States. There, a middle Oxfordian and late Kimmeridgian bivalve is postdated at 150 Ma, which means a minimum age of roughly 155 Ma for the basal Oxfordian. The maximum age of 157 Ma can be accepted as not much older than both the previously cited fossils and a possibly Callovian–Oxfordian radiolarian fauna. All radiometric data can be recon-

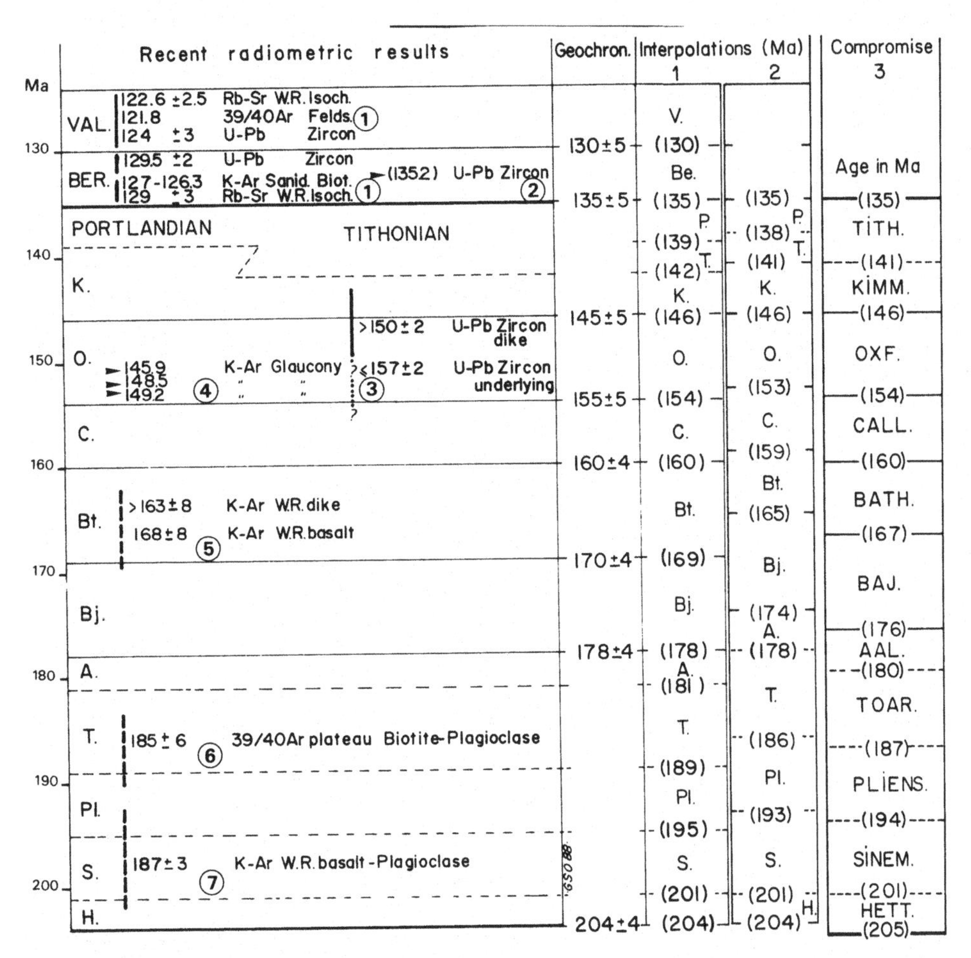


**Figure 1.3. Recent radiometric results and the improved numerical time scale of the Jurassic. Sources: 1, Ye Bodan (1986); 2, Bralower et al. (1987); 3, Saleeby et al. (1982); 4, Fischer and Gygi (1987); 5, Bellon et al. (1986); 6, Hess et al. (1987); 7, Seidemann (1988). Geochronological data allow definition of time intervals within which boundaries must lie; this is the data basis for time-scale synthesis. The numbers listed here (column geochron.) are fully documented by Kennedy and Odin (1982); a few modifications have been made in agreement with the new radiometric data shown. From this data basis, subjective interpolation systems allow proposal of estimates (in parentheses) for which no ± can be rigorously defined. Col. 1 gives a conservative picture influenced by Kennedy and Odin (1982) and Hallam et al. (1985) for the Bajocian-Bathonian relative lengths; col. 2 gives another solution (short Bathonian) in agreement to Westermann's relative scale (1984); col. 3 gives "means" between the central value resulting from the radiometric data and col. 2; the latter is a compromise suggested by the editor as the best present interim evaluation for the Jurassic time scale, used in this volume.**

ciled by using the interval 150–160 Ma (155 ± 5 Ma) to locate the Callovian–Oxfordian boundary, keeping in mind that the two extreme values are geochronologically documented.

Concerning the Middle Jurassic portion of the scale, only one new datum has been obtained, at 168 ± 8 Ma, for Bathonian basalts. This age is consistent with previous geochronological estimates of 160 ± 4 Ma and 170 ± 4 Ma for the upper and lower boundaries of the Bathonian, respectively. The age of 170 Ma appears more appropriate than a younger age, according to the geochronological results, and this conservative number was used to draw the interpolated scale in column 1 (Figure 1.3). However, Westermann (1988) believes that the Bathonian stage was of average duration only, based on the number of ammonite zonal units in different bioprovinces. In this circumstance, an interpolated age as young as 167 Ma results for the Bajocian–Bathonian boundary; column 2 in Figure 1.3 was drawn according to this opinion.

Concerning Early Jurassic time, the new data from the northern Caucasus provided by Hess et al. (1987) generally agree with previous geochronological estimates for the ages of the Pliensbachian and Toarcian stages and data from British Columbia (Armstrong 1982). But a new study by Seidemann (1988) would suggest that the younger side of the previous estimate of 204 ± 4 Ma by Kennedy and Odin (1982) for basal Jurassic is more probable. The interval of time 204 ± 4 Ma is kept unchanged in order to reconcile all geochronological estimates for location of the Triassic–Jurassic boundary: around 208 Ma by Armstrong (1982), and 200 Ma by Seidemann (1988).

The relative durations of stages are difficult to estimate using the preceding radiometric data alone. The equal-subzone interpolation hypothesis is an interim approach that currently proves to be consistent with radiometric data, when key points are themselves geochronologically well founded. It is reasonable to assume that the hypothesis will give a better approximation (1) when the number of zonal units increases, as has been the case for the Cretaceous and Paleogene zonation; (2) when short stages are considered, and therefore a small number of zonal units for each stage, groups of stages should be considered for more realistic interpolation (Odin 1985); (3) the key points between which the interpolation is undertaken should be as near to each other as possible and should be considered with their complete levels of uncertainty, in addition to the stratigraphic (correlation to the sequence), geochemical (initial inheritance, and late alteration), and analytical problems (multimineral and multimethod datings). Modern research in magnetostratigraphy (Chapter 2) is a promising possibility to supplement and improve the subjective biozonal approach for interpolation between geochronologically based tie points.

The relative durations of stages suggested by diverse authors result in the following main conclusions: There are at least three comparatively long stages within the Jurassic (i.e., Toarcian, Bajocian, and Oxfordian); there are three at least 50% shorter stages, (i.e., Hettangian, Aalenian, and Portlandian). The Tithonian stage often is preferred to the Portlandian and has a longer duration.

On the basis of the intervals of time defined from radiometric data for some stage boundaries and the relative durations suggested by interpolation systems, it is possible to propose some sort of "relative numerical scales," as shown on the extreme right in Figure 1.3; these ages, in parentheses, are given without error bars because it is difficult to evaluate with some rigor and objectivity the quality of the interpolation. Moreover, the numbers of standard zones accepted for the different stages depend on the basin where they are identified, among other problems. Figure 1.3 shows two interpolations; column 1 depends mostly on a single basin and is closer to the central geochronological values; column 2 is a picture more closely related to the interpolation by Westermann (1984, 1988). Finally, a "mean" between geochronological central values and biozone interpolation estimates in column 2 was suggested by the editor (column 3) as the recommended scale for this volume. The resulting numbers appear to be the best compromise, taking into account the necessity to avoid elimination of certain sets of data, considering the low number of geochronological results available and the difficulty of evaluating some of them. It remains necessary both to narrow the still important "plus-or-minus" intervals of time geochronologically defined for most boundaries and to document the paleomagnetic sequence and its precise connection to stage definition in continuous sedimentary sequences on land. Clarification of the stratigraphic definitions of some of these stage boundaries is also an important mortar for the edifice to which this volume brings some stones.

Field geologists are invited to search for dateable material in volcanosedimentary sequences for further improvement of the Jurassic numerical scale; preferably, volcanic rocks should be considered if unaltered minerals can be separated from them and if they can be directly correlated to marine biostratigraphic controls.

### Acknowledgments

I thank J. D. Obradovich, who reviewed a previous draft of this chapter, improved the English usage, and drew my attention to additional references. An anonymous reviewer made informative comments and encouraged consideration of one of the magnetostratigraphic interpolations in the literature; this application appears somehow premature if unuseful critique is to be avoided. T. J. Bralower and K. R. Ludwig were kind enough to answer my requests for more details on the data published in their abstract from the 1987 GSA meeting, for which a paper is now in preparation; discussion with C. W. Naeser concerning fission-track dating of the same formation was helpful. Professor J. B. Saleeby provided useful information concerning the stratigraphy and geochronology of the Klamath Mountains. The proposed scale obviously was influenced by G. E. G. Westermann; I thank him for his suggestions and confidence. Final review for English language and stratigraphic and more general purposes by T. J. Bralower and W. J. Kennedy is much appreciated.

This chapter is a contribution to the IUGS-UNESCO International Geological Correlation Programme, Project 196, "Numerical Calibration of the Phanerozoic Time Scale." It summarizes the efforts made during the past five years by the authors cited. The latter have our deep appreciation for their help in the time-consuming search for new tie points.

### References

Afanass'Yev, G. D., & Zykov, J. I. (1975). Echelle géochronologique du Phanérozoïque (in Russian). *Doklady Akad. Nauk SSSR*, pp. 1–106.

Armstrong, R. L. (1978). Pre-Cenozoic Phanerozoic time scale. In G. V. Cohee et al. (eds.), *The Geologic Time Scale* (pp. 73–92). Tulsa: American Association of Petroleum Geologists, Studies in Geology 6.

(1982). Late Triassic–Early Jurassic time scale calibration in British Columbia, Canada. In G. S. Odin (ed.), *Numerical Dating in Stratigraphy* (pp. 509–14). New York: Wiley.

Baubron, J. C., Defaut, B., Demange, J., & Maury, R. C. (1978) Une coulée sous-marine d'âge jurassique moyen dans les Causses. *C. R. Acad. Sci. (Paris), 287*, 225–7.

Bellon, H., Fabre, A., Sichler, B. & Bonhomme, M. G. (1986). Contribution to the numerical calibration of the Bajocian–Bathonian boundary: $^{40}Ar/^{40}K$ and paleomagnetic data from Les Vignes basaltic complex (France). In G. S. Odin (ed.), *Calibration of the Phanerozoic Time Scale, Chem. Geol. (Isotope Geosci. Sect.), 59*, 155–62.

Bralower, T. J., Ludwig, K. R., Obradovich, J. D. & Jones, D. L. (1987). Berriasian (Early Cretaceous) radiometric ages from the Great Valley sequence, California, and their biostratigraphic and magnetic correlation. *Geol. Soc. Am. Abstr. Progr., 19*, 598.

Bralower, T. J., Monechi, S., & Thierstein, H. R. (1989). Calcareous nannofossil zonation of the Jurassic–Cretaceous boundary interval and correlation with the geomagnetic polarity time scale. *Mar. Microphal, 14*, 153–235.

Channell, J. E. T., Ogg, J. G., & Lowrie, W. (1982). Geomagnetic polarity in the Early Cretaceous and Jurassic. *Philos. Trans. R. Soc. Lon., 306*, 137–46.

Curry, D. (1985). Oceanic magnetic lineaments and the calibration of the Late Mesozoic–Cenozoic time scale. In N. J. Snelling (ed.), *The Chronology of the Geological Record* (pp. 269–72). Oxford: Blackwell Scientific.

Fischer, H. & Gygi, R. A. (1987). Glauconite K-Ar dating of ammonite subzones in an Oxfordian succession, northern Switzerland (abstract). *Terra Cognita, 7*, 329.

Galbrun, B. (1986). La séquence de polarité magnétique au passage Jurassique Crétacé: corrélations entre les données magnétobiostratigraphiques et la succession des anomalies magnétiques océaniques. *C. R. Acad. Sci. (Paris), 303*, 495–8.

Galbrun, B., Gabilly, J., & Rasplus, L. (1988). Magnetostratigraphy of the Toarcian stratotype sections at Thouars and Airvault (Deux-Sèvres, France). *Earth Planet. Sci. Lett., 87*, 453–62.

GSLG (1964). Geological Society Phanerozoic time scale 1964. *Quart. J. Geol. Soc. London, 120*, 260–2.

Hallam, A., Hancock, J. M., La Brecque, J. L., Lowrie, W., & Channell, J. E. T. (1985). Jurassic and Cretaceous geochronology, Jurassic to Palaeogene magnetostratigraphy. In N. J. Snelling, (ed.), *The Chronology of the Geological Record* (pp. 118–40). Oxford: Blackwell Scientific.

Harland, W. B., Cox, A. V., Llewellyn, P. G., Pickton, C. A. G., Smith, A. G., & Walters, R. (1982). *A Geologic Time Scale*. Cambridge University Press.

Harper, G. D., Saleeby, J. B., Cashman, S. (1986). Isotopic ages of the

Nevadan Orogeny in the Western Klamath Mountains, California and Oregon. *Geol. Soc. Am. Abstr. Progr., 18,* 114.

Heirtzler, J. L., Dickson, G. D., Herron, E. M., Pittman, W. C., & Le Pichon, X. (1968). Marine magnetic anomalies, geomagnetic field reversals and motions of the ocean floor and continents. *J. Geophys. Res., 73,* 2119–36.

Hess, J. C., Lippolt, H. J., & Borsuk, A. M. (1987). Constraints on the Jurassic time scale by $^{40}Ar/^{39}Ar$ dating of North Caucasian volcanic rocks. *J. Geol., 95,* 563–71.

Howarth, M. K. (1964). The Jurassic period. In W. B. Harland et al. (eds.), *The Phanerozoic Time Scale, Quart. J. Geol. Soc. London, 120,* 203–5.

Imlay, R. W. (1984). Jurassic ammonite succession in North America and biogeographic implications. In G. E. G. Westermann (ed.), *Jurassic–Cretaceous Biochronology and Paleogeography of North America* (pp. 1–12). Geological Association of Canada, Special Paper 27.

Jeletzky, J. A. (1984). Jurassic–Cretaceous boundary beds of western and arctic Canada and the problem of the Tithonian-Berriasian stages in the Boreal Realm. In G. E. G. Westermann (ed.), *Jurassic–Cretaceous Biochronology and Paleogeography of North America* (pp. 174–250). Geological Association of Canada, Special Paper 27.

Kennedy, W. J., & Odin, G. S. (1982). The Jurassic and Cretaceous time scale in 1981. In G. S. Odin (ed.), *Numerical Dating in Stratigraphy* (pp. 557–92). New York: Wiley.

Kent, D. V., & Gradstein, F. M. (1985). A Cretaceous and Jurassic time scale. *Geol. Soc. Am. Bull. 96,* 1419–27.

Lambert, R. St. J. (1971). The Pre-Pleistocene Phanerozoic time scale. In *The Phanerozoic Time-Scale, a Supplement* (Special publication), (pp. 9–31). London: Geological Society of London.

Lanphere, M. A., & Jones, D. L. (1978). Cretaceous time scale from North America. In G. V. Cohee et al. (eds.), *The Geological Time Scale* (pp. 259–68). Tulsa: American Association of Petroleum Geologists, Studies in Geology 6.

Larson, R. L., & Hilde, T. W. C. (1975). A revised time scale of magnetic reversals for the Early Cretaceous, Late Jurassic. *J. Geophys. Res., 80,* 2586–94.

Lowrie, W., & Alvarez, W. (1981). One hundred million years of geomagnetic polarity history. *Geology, 9,* 392–7.

Lowrie, W., & Ogg, J. G. (1986). A magnetic polarity time scale for the Early Cretaceous and Late Jurassic. *Earth Planet. Sci. Lett., 76,* 341–9.

Obradovich, J. D., & Cobban, W. A. (1976). A time scale of the western interior of North America. *Geological Association of Canada, Special Papers 13,* 31–54.

Odin, G. S. (ed.) (1982). *Numerical Dating in Stratigraphy.* New York: Wiley.

(1985). Concerning the numerical ages proposed for the Jurassic and Cretaceous geochronology. In N. J. Snelling (ed.), *The Geochronology of the Geological Record* (pp. 196–8). Oxford: Blackwell Scientific.

Odin, G. S., & Curry, D. (1985). The Palaeogene time scale: radiometic dating versus magnetostratigraphic approach. *J. Geol. Soc. London, 142,* 1179–88.

Odin, G. S., & Dodson, M. H. (1982). Zero isotopic age of glauconies. In G. S. Odin (ed.), *Numerical Dating in Stratigraphy* (pp. 277–305). New York: Wiley.

Odin, G. S., & Létolle, R. (1982). The Triassic time scale in 1981. In G. S. Odin (ed.), *Numerical Dating in Stratigraphy* (pp. 523–35). New York: Wiley.

Odin, G. S., Montanari, A., Deino, A., Drake, R., Guise, P. G., Kreuzer, H. & Rex, D. C. (1991). Reliability of volcano-sedimentary biotite ages across the Eocene–Oligocene boundary. *Chem. Geol. (Isot. Geosc. Sect.), 86,* 203–24.

Ogg, J. G., & Steiner, M. B. (1988). Late Jurassic and Early Cretaceous magnetic polarity time scale. In R. Rocha & A. F. Soares (eds.), *Proceedings of the Second International Symposium on Jurassic Stratigraphy,* Lisbon, September 1987 (pp. 1125–38).

Palmer, A. R. (1983). 1983 Geologic Time Scale. *Geology, 11,* 503–4.

Saleeby, J. B. (1984). Pb/U zircon ages from the Rogue River area, Western Jurassic Belt, Klamath Mountains, Oregon. *Geol. Soc. Am. Abstr. Progr. 16,* 331.

Saleeby, J. B., Harper, G. D., Snoke, A. W., & Sharp, W. D. (1982). Time relations and structural-stratigraphic patterns in ophiolite accretion, West Central Klamath Mountains, California. *J. Geophys. Res., 87,* 3831–48.

Seidemann, D. E. (1988). The hydrothermal edition of excess $^{40}Ar$ to the lava flows from the Early Jurassic in the Hartford Basin (northeastern U.S.A.): implications of the time scale. *Chem. Geol. (Isot. Geosci. Sect.), 72,* 37–46.

Snelling, N. J. (1985). An interim time scale. In N. J. Snelling (ed.), *The Chronology of the Geological Record* (pp. 261–5). Oxford: Blackwell Scientific.

Steiner, M. B., Ogg, J. G., Melendez, G., & Sequeiros, L. (1985). Jurassic magnetostratigraphy. 2: Middle–Late Oxfordian of Aguilon, Iberian Cordillera, northern Spain. *Earth Planet. Sci. Lett., 76,* 151–66.

Van Hinte, J. E. (1976). A Jurassic time scale. *Am. Assoc. Petrol. Geol. Bull., 60,* 489–97.

(1978). A Jurassic time scale. In G. V. Cohee et al. (eds.), *The Geologic Time Scale* (pp. 289–98). Tulsa: American Association of Petroleum Geologists, Studies in Geology 6.

Webb, J. A. (1982). Triassic radiometric dates from eastern Australia. In G. S. Odin (ed.), *Numerical Dating in Stratigraphy* (pp. 515–22, 876–9). New York: Wiley.

Westermann, G. E. G. (1984). Gauging the duration of stages: a new approach for the Jurassic. *Episodes, 7,* 26–8.

(1988). Duration of Jurassic stages based on averaged and scaled subzones. In F. P. Agterberg & C. N. Rao (eds.), *Recent Advances in Quantitative Stratigraphic Correlation* (pp. 90–100). Delhi: Hindustan Publishing, Recent Research in Geology 12.

Wimbledon, W. A. (1984). The Portlandian, the terminal Jurassic stage in the Boreal Realm. In O. Michelson & A. Zeiss (eds.), *International Symposium on Jurassic Stratigraphy, Erlangen, 1984* (pp. 531–50). Copenhagen: International Commission on Jurassic Stratigraphy.

Ye Bodan (1986). The boundary age of Cretaceous–Jurassic period in South China (abstract). *Terra Cognita, 6,* 221.

(1988). Radiometric age of the Jurassic/Cretaceous boundary in South China. In G. S. Odin, *Calibration of the Phanerozoic Time Scale, Bull. Liais. Inf. I.G.C.P. Proj. 196 (offset Paris), 7,* 31–8.

Zeiss, A. (1986). Comments on a tentative correlation chart for the most important marine provinces at the Jurassic/Cretaceous boundary. *Acta Geol. Hungary, 29,* 27–30.

# 2 Jurassic magnetic-polarity time scale

J. G. OGG

The magnetic-polarity time scale for the Jurassic (Figure 2.1) has been compiled from several magnetostratigraphic studies. Correlation of the marine magnetic-anomaly M sequence with Lower Cretaceous and Upper Jurassic sediment sections has enabled assignment of precise biostratigraphic ages to the corresponding polarity chrons M0 through M25, as reviewed in Ogg (1988) and Ogg and Steiner (1988).

It has been suggested that the Cretaceous–Jurassic boundary (Berriasian–Tithonian boundary) be defined as the base of reversed-polarity chron M18r (Ogg and Lowrie 1986). The Tithonian and Kimmeridgian stages have been correlated to Tethyan ammonite zones in the Spanish Subbetic (Ogg et al. 1984), enabling the assignment of the Tithonian–Kimmeridgian boundary to approximately polarity chron M22A. The Kimmeridgian–Oxfordian boundary is near polarity chron M25. Therefore, polarity chrons corresponding to marine magnetic anomalies M26 through M28 should have a Late Oxfordian age. There have not yet been any magnetostratigraphic studies of Upper Oxfordian sections to verify these chrons.

The Middle Oxfordian polarity sequence is derived from ammonite-rich sections in northern Spain (Steiner et al. 1985). Lower Oxfordian through Callovian ammonite-zone sequences in southern Poland have yielded the polarity pattern shown in Figure 2.1 (Ogg et al. 1990); the condensed nature of these sediments and the recent reports of close-spaced magnetic anomalies from Pacific crust of presumed Early Oxfordian–Callovian age (Handschumacher and Gettrust 1985; Handschumacher et al. 1988) suggest that the actual polarity pattern in this interval is much more complex. There have been no magnetostratigraphic studies from earliest Callovian and late Bathonian – the reported Pacific marine magnetic anomalies would indicate the continuation of a high frequency of reversals into this interval.

Bajocian through Middle Bathonian ammonite-zone sediments in southern Spain have yielded a very high frequency of magnetic reversals (Steiner, Ogg, and Sandoval 1987). The polarity pattern shown in Figure 2.1 is a correlation of the main features in these sections; there may be more short-duration polarity zones.

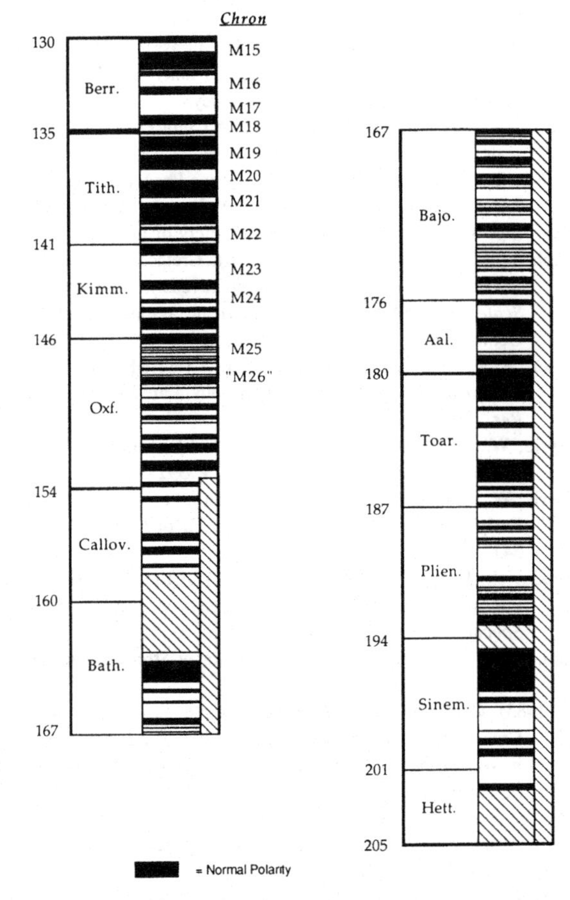


**Figure 2.1. Magnetic-polarity time scale for the Jurassic.**

The Aalenian, Toarcian, and Pliensbachian polarity pattern is from a single ammonite-zone pelagic sediment section in southern Switzerland (Horner and Heller 1983). This pattern lacks verification in other sections; therefore a higher degree of uncertainty is indicated. The Toarcian stratotype in the Paris Basin has yielded a polarity pattern having a frequency of reversals similar to that of the Switzerland Toarcian (Galburn, Gabilly, and Rasplus 1988); however, the sedimentation rates for this shallow-marine stratotype were much more variable than for the Switzerland section, and its ammonite zones cannot be directly correlated to the Tethyan zonation; therefore the Switzerland pattern is drawn in Figure 2.1.

Sinemurian and latest Hettangian sections in Austria have yielded the tentative polarity patterns (Steiner and Ogg 1988) shown in Figure 2.1. These sections lack definite age controls. Magnetostratigraphic results from the eastern United States (McIntosh, Hargraves, and West 1985) and southern Germany (J. Ogg unpublished data) suggest that this interval may instead be dominated by normal polarity.

The polarity patterns in Figure 2.1 are scaled to the individual stage durations of the compromise scale of Odin (see column 3 of Figure 1.3). That scale within the Jurassic is a compromise between the absolute geochronological data and the relative biochronological data from the ammonite-zone/subzone scaling method of Westermann; the resulting boundary ages are well within the error margins of the radiometric scale. The age of the Triassic–Jurassic boundary has been rounded off. In most cases, the pattern *within* the stage is scaled according to the magnetostratigraphic correlations with ammonite zones and using the subzonal method of Westermann (see the discussion by Odin, Chapter 1). The only exception is for the M-sequence chrons, which are scaled within individual stages according to their patterns in the Hawaiian lineations of the Pacific, which are commonly assumed to have formed during a period of constant spreading rate; however, if a subzonal scaling method was used, these patterns would not change significantly.

## References


Galbrun, B., Gabilly, J., & Rasplus, L. (1988). Magnetostratigraphy of the Toarcian stratotype sections at Thouars and Airvault (Deux-Sèvres, France). *Earth Planet. Sci. Lett., 87,* 453–62.

Handschumacher, D. W., & Gettrust, J. F. (1985). Mixed polarity model for the Jurassic "quiet zones": new oceanic evidence of frequent pre-M25 reversals. *Eos, 66,* 867.

Handschumacher, D. W., Sager, W. W., Hilde, T. W. C., & Bracey, D. R. (1988). Pre-Cretaceous tectonic evolution of the Pacific plate and extension of the geomagnetic polarity reversal time scale with implications for the origin of the Jurassic "Quiet Zone." *Tectonophysics, 155,* 365–80.

Horner, F., & Heller, F. (1983). Lower Jurassic magnetostratigraphy at the Breggia Gorge (Ticino, Switzerland) and Alpi Turati (Como, Italy). *Geophys. J., Royal Astron. Soc., 73,* 705–18.

McIntosh, W. C., Hargraves, R. B., & West, C. L. (1985). Paleomagnetism and oxide mineralogy of the Upper Triassic to Lower Jurassic red beds and basalts in the Newark Basin. *Geol. Soc. Am. Bull., 96,* 463–80.

Ogg, J. G. (1988). Early Cretaceous and Tithonian magnetostratigraphy of the Galicia Margin Ocean Drilling Program, Leg 103. *Proc. Ocean Drilling Program, Scientific Results, 103,* 659–82.

Ogg, J. G., & Lowrie, W. (1986). Magnetostratigraphy of the Jurassic/Cretaceous boundary. *Geology, 14,* 547–50.

Ogg, J. G., & Steiner, M. B. (1988). Late Jurassic and Early Cretaceous magnetic polarity time scale. In R. Rocha (ed.), *Proceedings of the Second International Symposium on Jurassic Stratigraphy* (pp. 1125–38). Lisbon, September 1987.

Ogg, J. G., Steiner, M. B., Oloriz, F., & Tavera, J. M. (1984). Jurassic magnetostratigraphy. 1. Kimmeridgian–Tithonian of Sierra Gorda and Carcabuey, southern Spain. *Earth Planet. Sci. Lett., 71,* 147–62.

Ogg, J. G., Wieczorek, J., Steiner, M. B., Hoffman, M. (1990). Jurassic magnetostratigraphy. 4. Early Callovian through Middle Oxfordian of the Krakow Uplands (Poland). *Earth Planet. Sci. Lett., 104,* 289–303.

Steiner, M. B., & Ogg, J. G. (1988). Early and Middle Jurassic magnetic polarity time scale. In R. Rocha (ed.), *Proceedings of the Second International Symposium on Jurassic Stratigraphy* (pp. 1097–111). Lisbon, September 1987.

Steiner, M. B., Ogg, J. G., Melendez, G., & Sequeiros, L. (1985). Jurassic magnetostratigraphy. 2: Middle–Late Oxfordian of Aguilon, Iberian Cordillera, northern Spain. *Earth Planet. Sci. Lett., 76,* 151–66.

Steiner, M. B., Ogg, J. G., & Sandoval, J. (1987). Jurassic magnetostratigraphy. 3. Bathonian–Bajocian of Carcabuey, Sierra Harana and Campillo de Arenas (Subbetic Cordillera, southern Spain). *Earth Planet. Sci. Lett., 82,* 357–72.

# Part II: Circum-Pacific base map

## 3 Reconstructions of the circum-Pacific region

D. B. ROWLEY

### Introduction

This chapter attempts a simplified paleogeographic assessment of the major allochtonous elements important to the Jurassic of the circum-Pacific region. Sketch reconstructions of the Pliensbachian (~190 Ma), Bathonian (~165 Ma), and Tithonian (~140 Ma) stages are presented. The reconstructions are extremely simplified, in part reflecting the scale of the maps, and in part reflecting a limited understanding of the detailed relationships between the various tectonic elements, the so-called terranes (Coney, Jones, and Monger 1980), that comprise the tectonic collage that characterizes circum-Pacific and Asian regions. A conservative approach has been followed, and on the average only limited motion (on the order of hundreds to thousands of kilometers) between the larger tectonic elements (sometimes referred to as superterranes) has been inferred. No attempt has been made to reconstruct the central Pacific Ocean basin or the plate motions associated with the evolution of the Pacific, Phoenix, Farallon, Kula, Izanagi I, and Izanagi II plates, which have been summarized elsewhere by Hilde, Uyeda, and Kronke (1977), Engebretson, Cox, and Gordon (1984, 1985), and Henderson, Gordon, and Engebretson (1984), among others. The summary presented here depends heavily on our compilation of a Mesozoic and Cenozoic tectonic map of the world (Rowley et al. 1985) as part of the Paleogeographic Atlas Project, as well as on syntheses by Davis, Monger, and Burchfiel (1978), Dickinson (1981), and Hamilton (1978) for the western United States, Monger, Price, and Templeman-Kluit (1982), Monger et al. (1985), and Armstrong (1988) for the Canadian Cordillera, Kosygin and Parfenov (1981) and Parfenov and Natal'in (1985, 1986) for northeast Asia, Kimura, Mihashita, and Miyasaka (1983), and Okada (1982) for Hokaido, Taira and Ogawa (1988), Taira, Saito, and Hashimoto (1983), Hattori (1982), and Tanaka and Nozawa (1977) for Japan, and Sengör (1984, 1987) for Asia. Data for other portions of the circum-Pacific, including Alaska, Asia, Southeast Asia–Indonesia, New Zealand, Antarctica, and South America, are based on a survey of the available literature that is too long to cite here.

Figure 3.1 shows the present-day distribution of the continental blocks and tectonic elements that are incorporated in the Jurassic reconstructions. Figure 3.1 also shows the age distribution of Jurassic versus younger oceanic and/or oceanic-arc-founded lithosphere. The very limited distribution of Jurassic oceanic lithosphere emphasizes the problems in inferring Pacific oceanic spreading geometries and plate kinematics, as well as the vast magnitudes of lithosphere that have been subducted around and/or within the Pacific since the Jurassic. All of the regions underlain by post-Jurassic oceanic lithosphere are excluded from any further discussion in this chapter. Figure 3.1 also emphasizes those regions of the present continents and adjacent oceanic-arc systems that have been accreted after the Jurassic and therefore can be excluded from the reconstructions. Included with these are regions of post-Jurassic accretionary wedges and oceanic arcs, such as the Shimanto belt of Honshu (Taira and Ogawa 1988), the Philippines, Palawan, northern Kalimantan, eastern Indonesia, and northern New Guinea arc and accretionary assemblages (Hamilton 1979), parts of New Zealand (Korsch and Wellman 1988), most of the Scotia Sea (Barker and Hill 1981), most of the Caribbean (Pindell 1985; Rowley and Pindell 1989; Pindell and Barrett 1990), parts of the Franciscan and equivalent assemblages (Blake et al. 1985), the Yukatat, Chugach-Kodiak, and Prince William complexes of southern Alaska (Plafker, Jones, and Pessango, 1977; Plafker 1987), and the Koryak and eastern Kamchatka complexes of northeast Asia (Kosygin and Parfenov 1981).

The discussion begins with an extremely circumscribed review of the tectonic elements that have been incorporated into the reconstructions, beginning with the derivation of the reconstructions of the circum-Atlantic and circum-Antarctic plate motions, followed by a discussion of the rationale upon which this set of reconstructions is based. The reconstructions are discussed by region, beginning with Asia and proceeding clockwise around the circum-Pacific.

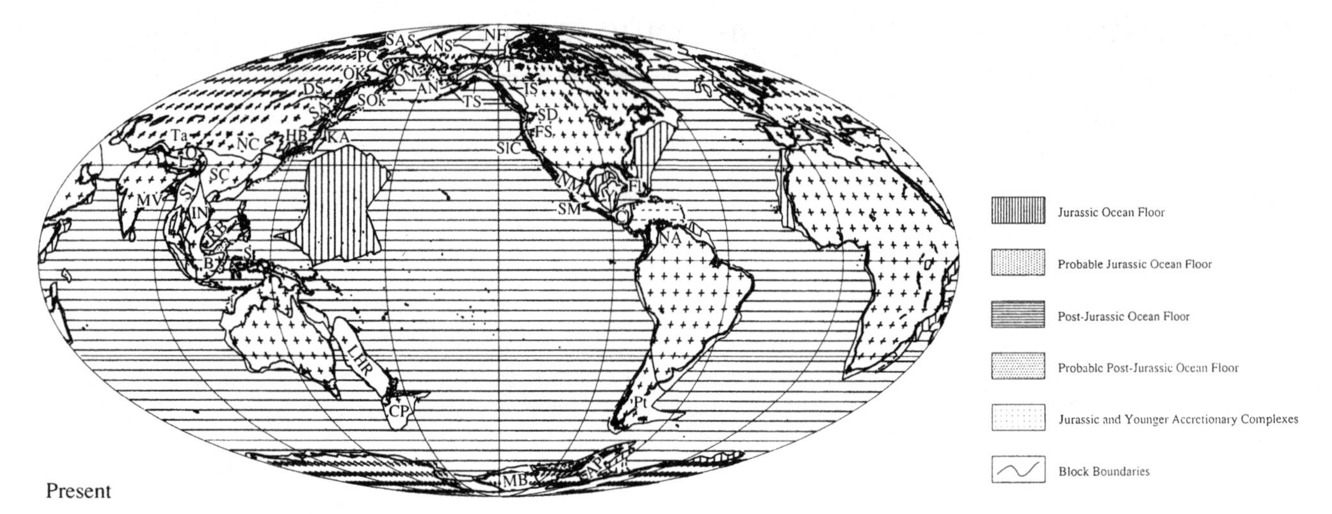


**Figure 3.1. Present-day map showing distribution of Jurassic versus post-Jurassic oceanic lithosphere, probable Jurassic and present-day locations of blocks (abbreviations) discussed in text and shown in Figures 3.2–3.4, and accretionary wedges and oceanic arcs accreted during and subsequent to the Jurassic. A, Alexander Terrane; AN, Angayucham terrane (includes Togiak, Nyack, and Goodnews Terranes); AP, Antarctic Peninsula; B, Borneo Block; C, Chortis Block; CC, Cache Creek suture; CP, Cambell Plateau; DS, Dzhagdi suture; Fl, South Florida Block; FS, Foothills suture; HB, Hida Block; IN, Indochina Block; IS, Intermontane Superterrane (includes Slide Mountain, Quesnellia, Cache Creek, Stikinia Terranes); KA, Kitikami-Abukuma Block; L, Lhasa Block; LHR, Lord Howe Rise–Norfolk Ridge; MB, Marie Byrd Land Block; MV, Mount Victoria Block; NA, Northern Andes Blocks; NC, North China (+ Sino-Korea) Block; NF, Nixon Fork (includes Ruby, Kilbuck, and Dillinger Terranes); NM, North Mexico Block; NS, North Slope–Chukotka Block (includes Brooks Range and Seward Peninsula); NW, North Wrangellia Terrane; OK, Okhotsk Block; OM, Omolon Block; OS, Owen Stanley metamorphics; P, Peninsular Terrane; PC, Poulosney-Chersky Ranges; Pt, Patagonia Block; Q, Qiangtang Block; QS, Quesnellia Terrane; RB, Reed Bank Block; S, Sula Platform; SA, Sikhote Alin; SAS, South Anyuy suture; SC, South China (= Yangtze) Block; SD, Seven Devils Terrane; SI, Sibumasu (= Shan Thai or Shan Thai–Malaya) Block; Sk, Sakhalin Block; SIC, Slate Creek Terrane; SOk, Sea of Okhotsk Block; ST, Stikinia Terrane; SW, South Wrangellia Terrane; Ta, Tarim Block; TS, Talkeetna Superterrane (includes Peninsular, North and South Wrangellia, and Alexander Terranes); Y, Yucatan Block; YT, Yukon-Tanana Assemblage.**

### Reconstructions: Pangean framework

Because of the very limited distribution of Jurassic oceanic lithosphere in the Pacific, and the enormous changes in paleogeography resulting from the breakup of Pangea, the starting point for a discussion of the Jurassic plate-kinematic evolution of the circum-Pacific region must be with an analysis of the breakup and dispersal history of Pangea. A reappraisal of the geological and geometric constraints on the pre-Jurassic fits of the continents surrounding the Arctic and North Atlantic (Rowley and Lottes 1988), Central and South Atlantic (Pindell 1985; Klitgord and Schouten 1986; Pindell et al. 1988; Rowley and Pindell 1989), and Indian and southwest Pacific oceans (Lawver and Scotese 1987; Lottes and Rowley 1990) has recently been undertaken. For the purposes of this chapter the continental fits of Pangea recently presented by Lottes and Rowley (1990) and the reconstructions of the Jurassic breakup of Pangea (D. Rowley, unpublished data) are used as the starting point for an analysis of the circum-Pacific region. In addition, a preliminary palinspastic base map is used for the Asian domain, based on an analysis of the deformation within Asia resulting from the India–Eurasia collision (Rowley 1989). These reconstructions provide the continental framework within which the migrations of continental and oceanic elements that now compose the circum-Pacific have evolved.

The reconstructions are oriented with respect to the paleomagnetic pole. For the purposes of this chapter, we have used the mean global poles derived from well-constrained paleomagnetic studies of North America, Europe, Africa, South America, India, and Australia that have been summed in reconstructed coordinates. The global means used here are not substantially different from those used by Ziegler, Barrett, and Scotese (1981).

### Asia

All of the major continental blocks of traditional Pangea, including Gondwana and Laurasia, as well as several of the smaller Asian blocks, North China (NC),[1] and Tarim (Ta), were already assembled by the end of the Paleozoic. In addition, most of the continental blocks of China and southeast Asia, including South China (SC, also referred to as the Yangtze Block), Indochina (IN), Qiangtang (Q), and Sibumasu (SI), had been sutured together and had also collided with the southern margin of Eurasia by the Carnian to Norian stages of the Late Triassic (Sengör 1984; Sengör et al. 1988; Chang, Pan, and Sun 1989; Nie, Rowley, and Ziegler 1990. In the south, only the Lhasa Block (L), India, and possibly the Mount Victoria (MV) Block (Mitchell 1989) were not already part of Asia by the Jurassic.

The history of collision among these various blocks is now generally agreed upon, although some controversy surrounds the history of collision of the Sibumasu and Indochina Blocks along the

[1] Abbreviations of tectonic elements as shown in Figures 3.1–3.4.

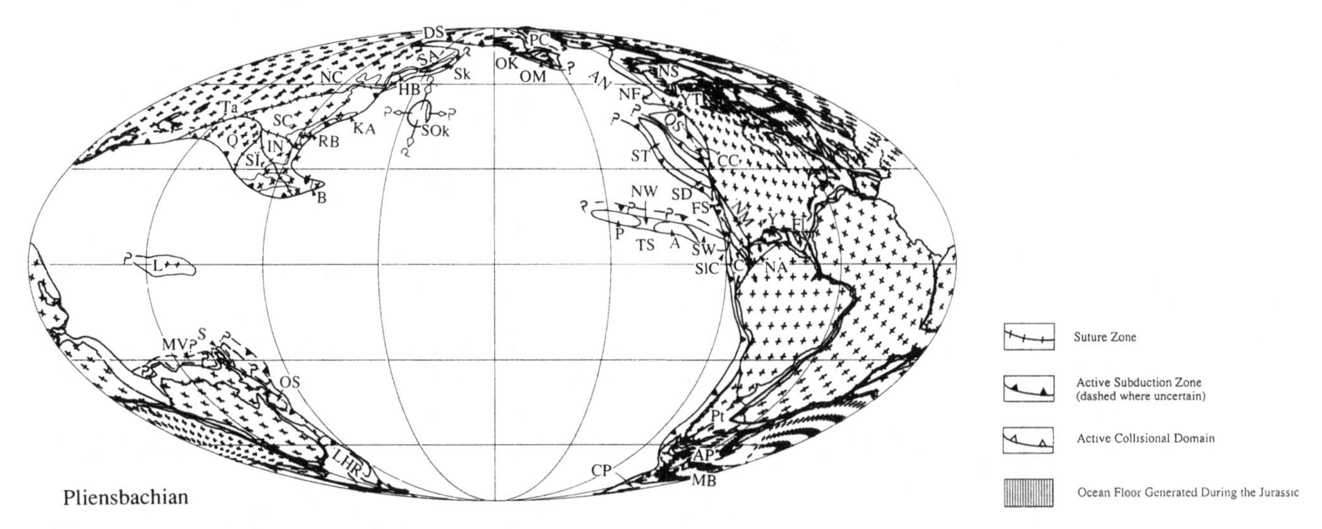


**Figure 3.2. Pliensbachian reconstruction of the circum-Pacific region. Abbreviations and blocks discussed in text. Lines with filled triangles, subduction zones; unadorned solid lines, block boundaries including passive continental margins; lines with hachures, sutures.**

Nan–Uttradit–Sra Kaeo suture, and particularly along its southerly continuation, the Bentong–Raub line in eastern Malaysia. This issue is briefly examined. Audley-Charles (1988) and Audley-Charles, Ballantyne, and Hall (1988) have argued that Sibumasu collided during the Late Cretaceous, much later than previously thought by Stauffer (1983), Bunopas (1981), Sengör (1984), Sengör and Hsü (1984), and Sengör et al. (1988), among others. Audley-Charles et al. (1988) supported this timing by noting that the Sibumasu Block experienced significant Late Cretaceous deformation. In addition, they noted the long-recognized similarity between Sibumasu and northwest Australia and assumed that the Sibumasu Block represented the continental block that left the northwest shelf of Australia in the Late Jurassic–Early Cretaceous, thereby requiring a time of suturing that postdates this rifting event. However, the stratigraphic and structural relationships along the Sibumasu–Indochina suture appear quite incompatible with this interpretation. For example, in the north in Thailand and extending into Yunnan, Norian to Lower Jurassic nonmarine redbeds of the Tembeling and equivalent units extend across the Nan–Uttradit suture zone, precluding a post-Norian age of suturing in this area (Bunopas 1981). Farther south, along the Bentong–Raub suture, stratigraphic relationships adjacent to the west of the suture, summarized by Metcalfe (1988), display a history of shallow-marine shelf sedimentation, followed by Middle to Late Triassic rapid subsidence and drowning, associated with deposition of a thick flysch sequence (Semanggol Formation) characterized by an easterly provenance. This sequence is unconformably covered by Jurassic redbeds. To the east of the Bentong–Raub suture, marine volcanics and volcaniclastics of Triassic age are also covered by nonmarine redbeds of Jurassic age (Metcalfe 1988). The straightforward interpretation of these relationships is that the collision of Sibumasu and the Malaysian Peninsula occurred in the Late Triassic (Hutchison 1975, 1982; Sengör 1984; Metcalfe 1988). It therefore seems highly improbable that this suture could record a Late Cretaceous collision, and Figure 3.2 shows the Sibumasu Block juxtaposed with Indochina in the Pliensbachian.

The Lhasa Block, a continental fragment of Gondwana derivation (Chang and Cheng 1973; Chang et al. 1989) is situated between the Indus–Yarlung Zangbo[2] and Banggong Co–Nujiang sutures. The Lhasa Block represents the western continuation of a loosely associated group of fragments that Sengör (1979, 1984) referred to as the Cimmerian continent. This block apparently left Gondwana in the Early Mesozoic, and it was faunally distinct from India during the Mesozoic (Westermann 1988; Chang et al. 1989). Closure of the Banggong Co–Nujiang suture is constrained by the presence of ophiolitic clasts in, and an unconformable overlap of ophiolites by, the Zigetang Formation, a Jurassic to Cretaceous clastic unit (Chang et al. 1989). This is true only if the ophiolites represent the fore arc of the Qiangtang Block (Q) to the north. The further unconformable overlap by Early Cretaceous *Orbitolina*-bearing limestones implies, however, that a Jurassic closure is likely. The reconstructions (Figures 3.2–3.4) show a progressive northward transit of the Lhasa Block across Paleo-Tethys, and its arrival adjacent to the Qiangtang Block by the Tithonian.

The only other block to collide with Asia during the Mesozoic was the little-known Mount Victoria Block of Burma (Mitchell 1989). This block is known from metasediments, associated Carnian *Halobia*-bearing flysch, that are unconformably overlain by Albian, which Mitchell interprets as the age of collision. The outcrop of this block is quite limited, and its actual extent and derivation are not known. However, Mitchell has suggested a Gondwana derivation, and in Figures 3.2–3.4 a possible derivation from the west margin of Australia (Sengör et al. 1988) is implied, but this is not constrained by any independent data.

Asia is schematically palinspastically restored to at least a pre-Tertiary configuration (Rowley 1989), including closure of the Sea of Japan and the South China Sea in Figures 3.2–3.4.

[2] This suture often is referred to as the Indus–Yarlung Zangbo River or simply the Zangbo suture. Such usage is either redundant or misleading, as *Zangbo* (also transliterated as Zangpo or Tsangpo) means *river* in Tibetan.

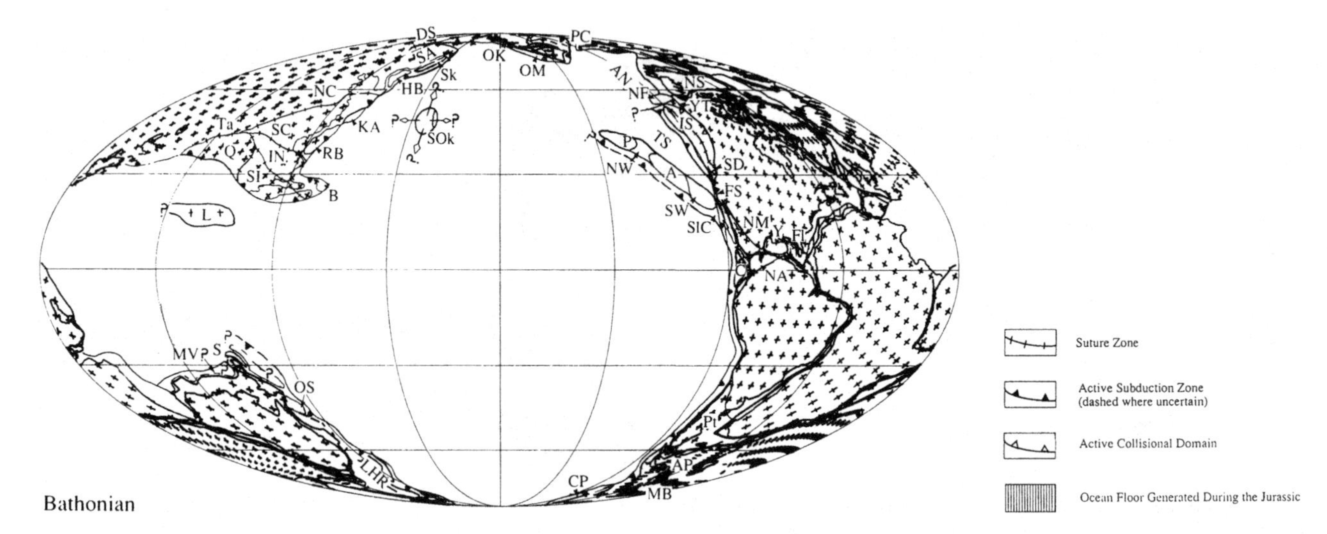


**Figure 3.3. Bathonian reconstruction of the circum-Pacific region. Abbreviations and blocks discussed in text. Lines with filled triangles, subduction zones; lines with open triangles, active collision zones; unadorned solid lines, block boundaries including passive continental margins; lines with hachures, sutures; vertical ruling, Jurassic oceanic lithosphere.**

Seafloor spreading in the South China Sea occurred during the interval of approximately 30–18 Ma, associated with the separation of the Reed Bank (RB) Block (including the North Palawan–Calamian Islands and Mindoro Blocks) from the South China Block in a generally north–south direction (Holloway 1982; Taylor and Hayes 1983), which has been reconstructed. The Reed Bank Block collided with the Palawan arc in the Miocene (Hamilton 1979), but it apparently did not fully collide farther south with the northwestern margin of Kalimantan (B, Borneo). Borneo has undergone substantial (>60° anticlockwise) post-Cretaceous rotation with respect to Indochina, and this rotation is also restored on the reconstructions. The Sea of Japan opened in the Miocene, with a quite complex geometry of opening (Otofuji, Matsuda, and Nohda 1985). Reconstruction of this opening places the Hida Block (HB) adjacent to Korea, suggesting that it and adjacent Mesozoic accretionary wedges of the Tamba and Mino Terranes (Hattori 1982) represent the marginal facies of the North China Block (NC). Equivalents of the Tamba and Mino Terrane accretionary wedge extend northward in the reconstructed positions into Sikhote-Alin (Mel'nikov and Golozubov 1980). The Abukuma-Kitikami Block (KA) of northern Japan is composed of displaced fragments characterized by sequences that extend back to at least the late Early Paleozoic, with continental and/or continental-margin affinities (Tanaka and Nozawa 1977). Permo-Carboniferous floras from coals are typical of the Cathaysian Realm, indicating a low-paleolatitude derivation of these blocks in the Permo-Carboniferous, and potential affinities with North or South China, or Indochina (Nie et al. 1990). The presence of blocks of late Early and Middle Paleozoic sediments that may have been derived from the Abukuma-Kitikami Block enclosed in Cretaceous Mélanges of the Chichibu belt (Tanaka and Nozawa 1977) is suggestive of Early Cretaceous transform-related collision of the Abukuma-Kitikami Block with more interior parts of Japan and Sikhote-Alin (SA). In Figures 3.2–3.4, I infer that the Abukuma-Kitikami Block was derived from the eastern margin of the South China Block and place it in approximately Taiwan's present position with respect to South China. Much of what currently underlies Taiwan probably was not accreted to the margin of South China until the Cretaceous or more recently, and therefore this position is plausible. The actual placement of the Abukuma-Kitikami Block within the reconstructions is quite uncertain.

The Dzhagdi suture (DS) (Kosygin and Parfenov 1981), north of the Amur River in Transbaikalia, also referred to as the Mongolo–Okhotsk suture, is an extremely important element in the amalgamation history of Asia, as it was the only oceanic basin extending into Asia that remained open through the Jurassic. The Dzhagdi ocean was the remnant of a once much larger ocean basin that in Late Paleozoic times extended into central Mongolia, with a configuration similar to that of the Sea of Japan, except that it apparently was open to the east (Rowley et al. 1985; Nie et al. 1990). The Kerulen Range and the Bureya Massif delimited this ocean's southern margin, and these blocks were in turn attached in the Jurassic to the already assembled elements of Asia to the south. The Dzhagdi ocean closed diachronously from the Permian in the west to the latest Jurassic in the east, which accounts for the otherwise unaccounted-for lack of superposition of paleomagnetic poles of Permian through Jurassic from areas of Asia south of this suture. The ocean is therefore seen to be progressively closing during the successive reconstructions.

The Sea of Okhotsk Block, which includes at least the western margin of Kamchatka and the "Academy of Sciences Rise" (Gnibidenko 1985), and the Hidaka complex and Kurile Islands (Kimura et al. 1983) did not collide with the Hokaido–Sakhalin (SA) margin and the margin offshore the parautochthonous Okhotsk and Omolon Massifs until the Late Cretaceous (Parfenov et al. 1979; Kosygin and Parfenov 1981; Takahashi 1983), when the extensive Andean Okhotsk–Chukotsk volcanic-plutonic belt ceased its activity. The nature of the basement of the Sea of Okhotsk Block is controversial and beyond the scope of this discussion. It is assumed to contain continental crust of Precambrian or Paleozoic age (Gnibidenko 1985), but its derivation and paleolatitude are

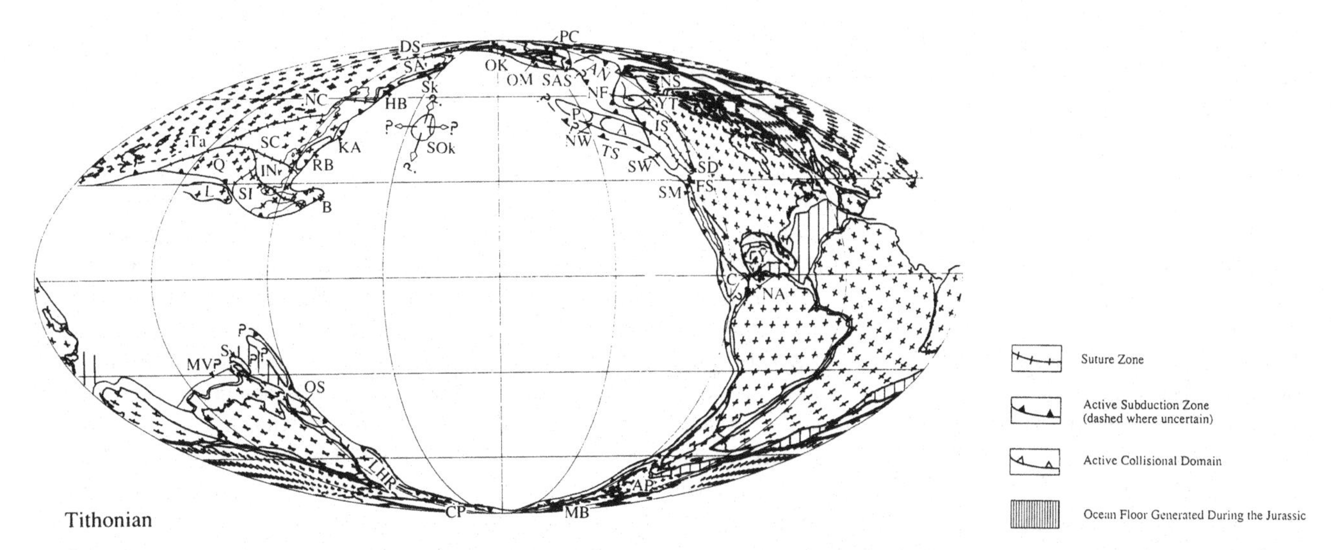


**Figure 3.4. Tithonian reconstruction of the circum-Pacific region. Abbreviations and blocks discussed in text. Lines with filled triangles, subduction zones; lines with open triangles, active collisional zones; unadorned solid lines, block boundaries including passive continental margins; lines with hachures, sutures; vertical ruling, Jurassic oceanic lithosphere.**

not known for any of the reconstructions, and it is placed in the western Pacific solely for convenience. The question marks surrounding it in Figures 3.2–3.4 emphasize the uncertainty of its reconstruction. It is possible that it was once contiguous with the Abukuma-Kitikami Block of Japan and was separated from it along sinistral strike-slip faults after the collision of the Kitikami-Abukuma Block with the Sikhote-Alin margin of Asia.

### Northeast Asia

The reconstruction for northeast Asia, and in particular the circum-Verkhoyansk domain to the south of the South Anyuy suture (SAS) (Seslavinskii 1979), and including the Okhotsk (OK), Omolon (OM), and Pri-Kolymsk Blocks, as well as the Paleozoic sequences within the Chersky and Poulosney Ranges (PC) remains poorly understood (Kosygin and Parfenov 1981; Fujita and Newberry 1983; Parfenov and Natal'in 1985). The Okhotsk Block is believed to be a parautochthonous element of the Aldan shield, based on the mapped continuity of Archean basement around the southern end of the Sette Daban foldbelt that separates it from the Aldan shield, as well as on the continuity of Paleozoic and Early Mesozoic shelf sediments within the Verkhoyansk and northern Sette Daban foldbelts to the north and west of the Okhotsk Block. McElhinny (1973) utilized Soviet paleomagnetic results for the Permian to Jurassic units from the Omolon Block[3] to suggest that it was displaced with respect to the Aldan shield. These results are in general accord with Jurassic floras from the Omolon (Samylina and Yefimova 1968) that suggest a potentially more southerly position with respect to coeval floras of the Verkhoyansk region (A. Ziegler personal communication 1989). However, Permian floras of the Omolon are distinctly Angaran (Meyen 1976, personal communication 1986) and suggest a northerly position adjacent to the Aldan shield during the Permian. The lack of a well-defined suture between the Omolon Block and the Pri-Kolymsk Block along the Sugoy foldbelt hinders the interpretation of the relations in this area (Merzlyakov et al. 1982). We prefer an interpretation of the Lower Paleozoic sequences of the Pri-Kolymsk Block and the Chersky and Poulosney (PC) Ranges as the upthrust basement of the Verkhoyansk passive margin, with the Pri-Kolymsk Block having undergone substantial Middle Jurassic to earliest Cretaceous counterclockwise rotation with respect to the Aldan shield (Fujita and Newberry 1983; Natal'in and Parfenov 1983). The Omolon Block is interpreted to have been emplaced adjacent to the Pri-Kolymsk Block along a transpressional sinistral transform coeval with the Middle Jurassic to earliest Cretaceous orogenic collapse of the Verkhoyansk, Chersky and Poulosney, and Alazea belts, which was also synchronous with the closure of the South Anyuy suture to the north of the Omolon Block in Early Cretaceous times (Kosygin and Parfenov 1981; Parfenov and Natal'in 1985; Rowley and Lottes 1988).

[3] Much confusion has resulted from of the use of the term "Kolyma Block" as used by McElhinny (1973), among many others, in the interpretation of the tectonic evolution of northeast Asia. The geographic region to which the name Kolyma Block has been applied does not encompass a single entity, but instead comprises an amalgam of blocks with markedly different stratigraphic and structural histories. The Omolon Block (Kosygin and Parfenov, 1981) is one such block.

### Alaska and Western Canada

Alaska has been interpreted to comprise many dozens of allochthonous terranes that may have moved relative to North America (Coney et al. 1980; Jones et al. 1983). In this treatment, a very simplified interpretation is presented. Areally the largest block is the North Slope–Chukotka Block (NS). This block extends from near the present Arctic-basin shelf edge to the front of the Endicott allochthons of the Brook Range (but which for simplicity and because of the scale of the maps are here included in the North Slope–Chukotka Block) and their apparent equivalents farther west in Chukotka (Churkin and Trexler 1981), then reappearing in parautochthonous windows, such as the Doonerack window, farther south (Oldow et al. 1987). The shelf and offshelf equivalents of the North Slope–Chukotka Block in the Brooks

Range were shortened by perhaps 500 km during Late Jurassic to Cretaceous times (Oldow et al. 1987), and our reconstructions assume similar shortening along the length of this orogen. The southern boundary of this block is marked by a series of dismembered ophiolitic and oceanic volcanic assemblages that surround the Yukon-Koyukuk Basin (Patton et al. 1977) and broadly comprise the Angayucham Terrane (Jones et al. 1981). Angayucham equivalents appear to continue westward into the Chukotka (Fujita and Newberry 1983) and still farther west to at least the South Anyuy suture of Seslavinskii (1979). The North Slope–Chukotka Block rotated away from the Canadian Arctic margin to open the Amerasian basin (Grantz and May 1982) starting in the Hauterivian and completed by the late Aptian to earliest Albian (Rowley and Lottes 1988). Therefore, for the purposes of our reconstructions, the North Slope–Chukotka Blocks is portrayed in its pre-Hauterivian position adjacent to the Canadian Arctic islands. The southern edge of the Endicott Terrane is palinspastically reconstructed approximately 500 km to the south (in present coordinates).

The Angayucham oceanic assemblage includes rocks of Devonian to Jurassic age (Patton et al. 1977), which correlates well with the age span of the Endicott passive margin and is inferred to represent the basement offshore of the Endicott Terrane prior to its obduction. Obduction and collision of the Angayucham Terrane (AN) with the North Slope–Chukotka Block passive margin may have occurred diachronously, beginning initially in the vicinity of the Seward Peninsula, where Paleozoic shelfal sequences were subjected to blueschist-facies conditions at approximately 160 Ma (Armstrong et al. 1986), and the western Brooks Range in Oxfordian(?) (Balkwill et al. 1983) or earliest Cretaceous (Detterman 1973), apparently propagating both east and west during the Early Cretaceous. Coeval suturing of the Angayucham Terrane occurred along the Ramparts ophiolite belt of Patton et al. (1977) with the Ruby and Nixon Fork Terranes (NF), which along with at least the Dillinger Terrane (included in NF) (Jones et al. 1981, 1982; Churkin, Foster, and Chapman 1982) are continental-shelf and shelf-margin sequences that can be interpreted to represent a Florida-like prong extending away from the Cordilleran margin of North America in Paleozoic and early Mesozoic times (Churkin et al. 1982). The Angayucham–Ruby/Nixon Fork suture extends at least as far as southwestern Alaska, where it has been referred to as the Kanektok suture (Box 1982), which juxtaposes Togiak, Nyack, and Goodnews arc/subduction-related assemblages (= Angayucham) with Precambrian basement (Kilbuck Terrane, included in NF) and Cretaceous foreland-basin sediments (Kuskokwim Formation) (Box 1985). No further extension of this Late Jurassic(?)–Early Cretaceous suture zone is recognized to the east in Alaska. Oceanward-directed subduction is interpreted to have begun between the Bathonian (Figure 3.3) and Tithonian (Figure 3.4) reconstructions. The Tithonian reconstruction shows a locus of collision in the vicinity of the Seward peninsula, and that collision had not yet begun along adjacent segments to both the north (present west) and the south (present east), reflecting the inferred diachroneity along this margin.

A younger collision of probable Albian age in Alaska is responsible for the juxtaposition of the Talkeetna Superterrane [TS comprising Peninsular (P), Wrangellia (NW and SW), and Alexander (A) Terranes] (Csejtey, Cox, and Evarts 1982) with the complex assemblage adjacent to the north and east of the Denali Fault, at least some (e.g., Dillinger and Nixon Fork Terranes, NF) of which appears to represent displaced segments of the North American passive continental margin (Jones et al. 1981). The complexly deformed Jura-Cretaceous flysch belt of southern Alaska marks the suture zone between these terranes (Csejtey et al. 1982). The Peninsular Terrane appears to extend to the west to the Bering Sea shelf edge, but the location of the westward continuation of the Jura-Cretaceous flysch-belt suture remains unknown (Scholl, Grantz, and Vedder 1987). Closure between the Talkeetna Superterrane and fragments to the north is generally inferred to have occurred as a result of a subduction zone dipping northeastward (present orientation) beneath the already accreted Intermontane Superterrane (IS) in the south, which presumably continued to the north along the southern margin of the NF. Initiation of subduction along the present Pacific side of the Talkeetna Superterrane is still poorly dated, but judging from the oldest blueschists within the Uyak–McHugh complex (Moore and Connelly 1977) and the age of Talkeetna-arc magmatism, an Early Jurassic initiation seems likely. Similar conclusions are supported by sequences farther south on Vancouver Island, where the coeval Bonanza Volcanics are preserved (Armstrong 1988). Abundant accretion of sediments in the form of the Chugach, Prince William, and Yukatat complexes did not begin until the Late Cretaceous (Moore 1973; Winkler and Plafker 1981), after the Talkeetna Superterrane had already docked against the North American continent.

The Talkeetna Superterrane was amalgamated at least prior to the Late Jurassic, when the Gravina-Nutzotin and equivalents constituted an overlap assemblage linking all three of the terranes (Coney et al. 1980). Additional support is provided by the Barnard Glacier pluton that stitches across the Northern Wrangellia (NW)–Alexander contact and has been dated at 309 $\pm$ 5 Ma, demonstrating a much earlier linkage of these terranes (Gardner et al. 1988).

Farther south in Western Canada, the Talkeetna Superterrane, which has also been referred to as Terrane II (Monger, Price, and Templeman-Kluit 1982) or Insular Superterrane (Price, Monger, and Roddick 1985), lies outboard of a complex assemblage of island-arc [Quesnellia (QS), Stikinia (ST), and much of the Yukon-Tanana (YT) Terranes], subduction-accretion complexes with incorporated seamount complexes (Cache Creek, Bridge River, Cadwallader, Chilliwack-Nooksack Terranes), and back-arc-basin/subduction-accretion wedge sequences (Slide Mountain Terrane, included in QS) that comprise Terrane I (Monger et al. 1982) or the Intermontane Superterrane of Price et al. (1985). The Intermontane Superterrane (IS) has been interpreted to have been amalgamated before being sutured to North America (Monger et al. 1982). Suturing between Stikinia and Quesnellia to close the Cache Creek ocean has been interpreted to have occurred already in the Late Triassic (Monger et al. 1982); however, recent findings of Early or Middle Jurassic (Pliensbachian to Bajocian) Radiolaria in cherts of the western Cache Creek, correlative with cherts also found in the Bridge River and Chilliwack-Nooksack among other Western Cordilleran accretionary complexes, suggest a Middle Jurassic age of suturing (Cordey et al. 1987). Quesnellia and Stikinia are therefore shown as separate during the Pliensbachian (Figure 3.2) and together by the Bathonian (Figure 3.3) in the reconstructions. Their locations prior to amalgamation are quite uncertain, but faunal affinities with Late Paleozoic and Early Mesozoic sequences of western North America suggest a position not too distant off the western margin in the Early and Middle Jurassic (Price et al. 1985; Smith and Tipper 1986). Suturing of the Intermontane Superterrane with the Paleozoic–Early Mesozoic passive-type margin of Western Canada by westward subduction and consequent collision and obduction of the Slide Mountain back-arc basin may also have begun as early as the Late Triassic (Templeman-Kluit 1979a) or earliest Jurassic (before 194 Ma) (Hansen 1988) in the north and may have progressed diachronously to the south, where collision is believed to have begun in the early Middle Jurassic (Monger et al. 1982; Price et al. 1985), at approximately 180 Ma as plutons stitched the Quesnellia–North America suture (Armstrong 1988). Some support for diachroneity is provided by the ages of the oldest westerly derived clastics on the shelf, which are Late Triassic to Early Jurassic(?) in the north (Templeman-Kluit 1979a,b), and by the late Oxfordian turbidites of the Fernie Group (Price et al. 1985) in the south. Initiation of the east-facing Quesnellia/Yukon-Tanana arc that resulted in the closure of the Slide Mountain back-arc basin appears to have begun near the Permo-Triassic boundary (~246 Ma)(Erdmer and Armstrong 1988), at least in the north, and thus to have predated the suturing of Stikinia and Quesnellia in the Late Triassic to the pre–Middle Jurassic (~176 Ma). In terms of the reconstructions, the closure of the Slide Mountain back-arc basin along the Teslin suture by the Early Jurassic would require Quesnellia to have been adjacent to North America at that time, and therefore it provides a tie point for reconstruction of the Jurassic evolution of this region. That arc-related magmatism overstepped eastward across the North America–Quesnellia suture only in southern British Columbia, and not farther north (Armstrong 1988), may suggest that the southward tapering of Stikinia seen today reflects its original geography. An interesting point that should be made is that Hansen (1988) has argued for an oblique dextral collision along the Teslin suture. If collision occurred earliest along the Teslin part of the margin, one might have predicted a sinistral, rather than a dextral, sense of obliquity.

Thus, the Pliensbachian reconstruction (Figure 3.2) portrays the initial collision of Quesnellia and Slide Mountain only in the north, but with Quesnellia lying not far offshore. A southward-widening remnant of the Cache Creek ocean is shown separating Quesnellia and Stikinia; it had closed by eastward subduction by the Bathonian (Figure 3.3). Eastward subduction beneath Stikinia is inferred from the geographic distribution of approximately coeval accretionary-wedge assemblages (Cadwallader, Bridge River) and arc-related volcanism (Armstrong 1988). Coalescence of the Intermontane Superterrane (IS) and collision with North America proceeded through the Bathonian (Figure 3.3) and Tithonian (Figure 3.4), such that by the Bathonian, east-dipping subduction-related magmatism first affected the North American basement, but only in the south.

Collision of the outboard Talkeetna Superterrane with the already-assembled Intermontane Superterrane and its associated accretionary wedges (Bridge River, Tyaughton, Hozameen, and Nooksack) in the Canadian Cordillera and San Juan thrust belt appears to have begun in the late Middle to Late Jurassic (Tipper 1984; Potter 1986; Brandon, Cowan, and Vance 1988). In the San Juan foldbelt/thrustbelt, suturing predated the depositional overlap of Jura-Cretaceous clastics across the various terrane boundaries (Brandon et al. 1988). These more recent data require an older age of suturing than the late Early Cretaceous age previously suggested based on westerly provenance of sediments in the Methow–Tyaughton trough (Monger et al. 1982). In the reconstructions, the southward extension of the Talkeetna Superterrane is portrayed as colliding quite obliquely with the western North American margin, but to have arrived adjacent to the western Intermontane Superterrane margin by the Late Jurassic (Figure 3.4). Note that a northward-propagating closure of this ocean is implied.

**Western Cordillera of the United States**

Still farther south, in Washington State, south of the southern extension of Stikinia (Monger et al. 1982), structural and stratigraphic relationships become still more complicated and uncertain and are beyond the scope of this chapter (Davis et al. 1978; Vance et al. 1980), involving a complex arrangement of terranes, in part reflecting the crosscutting nature of post-Jurassic dextral strike-slip faults (including the Early Tertiary Straight Creek and Fraser River faults and the Middle Cretaceous Ross Lake, Hozameen, Passayten, and Yalakom faults) and substantial Late Cretaceous west-vergent thrusting (Vance et al. 1980; Brandon et al. 1988) that have rearranged elements within this region. In addition, the relationships of the various Jurassic ophiolitic rocks that are arrayed within west-central Washington, including Ingalls and various other units, such as the Cretaceous blueschists (Shuksan), are unclear.

Margin sequences are next exposed from beneath extensive Tertiary volcanics and sediments in the Riggins area of western Idaho and then again in the almost east–west-striking Blue Mountain belt across Oregon, from southeasternmost Washington and westernmost Idaho to the Klamaths region. Three general sequences are exposed in this area, including (1) the Seven Devils Terrane (SD), a volcanic-arc sequence, (2) an intermediate Paleozoic(?) to Jurassic oceanic accretionary sequence, including ophiolitic rocks, blueschists, and mélange that incorporate Tethyan fusulinids, and (3) a southern Triassic continental-margin-arc sequence (Huntington arc) developed on top of the already-emplaced Sonomia Terrane (Davis et al. 1978; Dickinson and Thayer 1978; Hamilton 1978). Collision of the Seven Devils Terrane with the Huntington arc to close the central-mélange-belt ocean is believed to have occurred prior to the end of the Jurassic (Brooks and Vallier 1978). Lund, Snee, and Sutter (1985) have suggested a late

Early Cretaceous age for this collision, but this is difficult to accept, as restoration of subsequent dextral strike-slip faults (mentioned earlier) places both the Intermontane and southern Talkeetna Superterranes to the west of the Seven Devils Terrane and therefore makes it virtually impossible to close this basin later than the early Late Jurassic.

In southern Oregon and California, important Jurassic elements include the western Klamaths and their probable southerly extensions in central California, the Slate Creek (SlC) arc and Coast Range ophiolite, and of course the Jurassic components of the Franciscan subduction-accretion complex (Davis et al. 1978; Hamilton 1978). Relationships among these various elements are still quite controversial (Davis et al. 1978). Closure between the ensimatic Middle to Late Jurassic Slate Creek arc (Edelman et al. 1989) and the Jurassic arc of the Sierra Nevada resulted in termination of the Callaveras subduction zone along the Foothills suture (FS) (Saleeby et al. 1978). Whether subduction was both east- and west-dipping beneath the respective arcs (Schweickert 1978) or only east-dipping remains uncertain. Plutonic stitching across the Slate Creek–Foothills suture zone occurred by 165 Ma, thereby constraining the timing in this region (Edelman et al. 1989). Suturing of the Slate Creek arc with the Jurassic Sierran arc often is inferred to have initiated development of the Late Jurassic Nevadan orogeny of the Western Cordillera. The overlap of the Middle to Late Jurassic Smartville arc across this belt documents the continuation or initiation of Franciscan subduction along the western margin of the just-accreted Slate Creek Terrane immediately after the formation of the Western Klamath–Coast Ranges ophiolitic basement in a strongly oblique subduction system to the west of the Slate Creek arc (Harper, Saleeby, and Norman 1985).

#### Central America

The geology of Mexico, including Baja California, is complicated by the obliquely transecting sinistral strike-slip faults associated with the Mohave–Sonora megashear (Silver and Anderson 1974; Anderson and Schmidt 1983), by the pre-Trans-Mexican volcanic belt (Walper 1980; Pindell and Dewey 1982; Rowley and Pindell 1989), and by the obscuring cover of the extensive Tertiary volcanics (Lopez-Ramos 1976). Like California to the north, two Jurassic arc sequences are preserved within Mexico, a Middle to Late Jurassic Andean arc in Sinaloa and Sonora (Rangin 1978), presumably the southern extension of the eastern Sierra Nevada arc, and an ensimatic arc preserved along the west coast of Baja and extending northward into southern California (Gastil, Philips, and Allison 1975; Gastil, Morgan, and Krummenacher 1978). Closure between these two arcs predated the Albian and may have been roughly synchronous with closure of the Foothills suture farther north in California.

The placement of the Chortis Block (C) south of what is the present-day Middle America Trench is based on analysis of the kinematic evolution of the Caribbean region by Pindell and Dewey (1982) and Pindell and Barrett (1990), which places it along this margin in pre–Late Cretaceous times. Upon closure of the Atlantic, and accounting for motion along the various faults in Mexico, the Chortis Block fits between northwestern South America and southern Mexico (Pindell et al. 1988; Rowley and Pindell 1989). The Jurassic history of this block is restricted to limited exposures of the El Plan Formation, a Triassic(?)–Jurassic sequence of terrestrial-to-shallow-marine sandstones and shales, with minor conglomerates and coals, and the later Jurassic Todos Santos red continental sandstones, conglomerates, and shales that very locally contain andesites (Weyl 1980). These units appear, at least in some regions, to have been deposited in extensional basins, presumably associated with the opening of the proto-Caribbean.

#### South America

Along the northern and northwestern margins of South America, extensional basins developed in the Jurassic associated with the opening of the Central Atlantic and proto-Caribbean (Maze 1984). Farther south along the western margin of South America there is clear evidence of subduction-related arc magmatism (Cobbing 1985). As pointed out by Ziegler et al. (1981), the locus of Jurassic arc magmatism was situated along the present-day coast and migrated to the east in the Cretaceous and Tertiary, which they interpreted as being due to progressive subduction erosion of crust from the western margin. In southern South America, extensional deformation and widespread volcanism affected most of this area from the Late Triassic through the Early Cretaceous (Uliana and Biddle 1987), with the development of the Rocas Verde back-arc basin in the Late Jurassic (Dalziel 1981). Dalziel argues for a quite limited width of the Rocas Verde trough, which terminated just south of 50°S, too small to be portrayed on these reconstructions. Arc magmatism continued west of the Rocas Verde trough along Cordillera Darwin and farther north, and on the reconstructions (Figures 3.2–3.4) a continuous Andean-type margin is shown.

#### Antarctica

The Antarctic peninsula (AP) includes the further continuation of the southern Andean belt, as well as continental and continental-margin assemblages associated with its separation from the Falkland Plateau to open the Weddell Sea, starting in the late Middle Jurassic (LeBrecque and Barker 1981). The Antarctic Peninsula is rotated counterclockwise in the reconstructions to fit adjacent to the Falkland Plateau, where it too experienced at least Jurassic extensional deformation similar to that observed in southern South America (Uliana and Biddle 1987) and South Africa (Dingle, Siesser, and Newton 1983), and with the same overall orientation (Dalziel et al. 1987). Our reconstructions treat Marie Byrd Land (MB) as a separate block, separated from East Antarctica and the Antarctic Peninsula by extensional basins underneath the Ross Ice Shelf. The reconstructed geometry allows the Marie Byrd Block and the Antarctic Peninsula to maintain continuity of their present-day Pacific margin, which is suggested by the apparent continuity of the active margin assemblages.

### Australia and New Zealand

The Cambell Plateau (CP) fits against the Marie Byrd Block margin, with closure of the southwest Pacific (Weissel, Hayes, and Herron 1977). The northern margin of the Cambell Plateau is inferred to be the westward continuation of the Antarctic Peninsula active margin. Schistose rocks interpreted to represent the along-strike continuation of the Otago and/or Torlesse belt (of accretionary-prism origin) of South Island, New Zealand (Spörli 1978), crop out on the Chatham Islands (Adams and Robinson 1977). Jurassic isotopic ages from metamorphic minerals within the schists are compatible with the inferred continuation of the continentward-dipping subduction margin, which is also suggested by the presence of Jurassic plutonic rocks on the Bounty Islands (Adams and Robinson 1977). Jurassic arc and accretionary-wedge sequences continue in New Zealand, where they belong to the eastern Torlesse and Waipapa Terranes (Spörli 1978), and presumably offshore to the northwest along the Lord Howe Rise and Norfolk Ridge (LHR), along which Jurassic arc and/or fore-arc sequences reappear on the surface in New Caledonia beneath the subsequently obducted ophiolites (Brothers and Lillie 1988). The Lord Howe Rise, Norfolk Ridge, and New Zealand are reconstructed to positions along the eastern Australian margin by removing the effects of Tasman Sea opening.

The Paleocene to Eocene opening of the Coral Sea (Weissel and Watts 1979) presumably was responsible for detachment of a block off the Queensland continental margin. That block is not exposed, but is generally inferred to underlie the Owen Stanley metamorphics (OS) along the Papuan peninsula (Davies, Symonds, and Ripper 1985). Immediately to the west, Late Triassic–Early Jurassic arc-related volcanics are exposed immediately north of the Kubor Ranges (Milsom 1985) and apparently mark the last place where this active margin is preserved along the western margin of Gondwana. Jurassic rifting and subsequent seafloor spreading (Hamilton 1979; Pigram and Panggabean 1984) along the northern margin of central New Guinea detached a fragment of crust (shown as queries in Figures 3.2–3.4) whose present location is completely unknown. On the reconstructions, this fragment is depicted as pulling away from the New Guinea passive margin by the Tithonian, as a result of back-arc spreading. Farther west, the Sula Platform (S) is another fragment of Australian crust (Hamilton 1979; Pigram and Panggabean 1984; Audley-Charles 1988). The Sula Platform was displaced from the northern Vogelkopf (Bird's Head) margin of New Guinea along the complex South Sula–Sorong and North Sula–Sorong fault system during the Miocene (Hamilton 1979). The Sula Platform is shown at a pre-Miocene position north of the Vogelkopf for all of the reconstructions.

The final segment is the northwestern and western Australian margin. This margin is a passive margin that underwent rifting and drifting in the Jurassic (von Rad and Exon 1982). The margin facies are exposed around the Banda Sea, where the Pliocene Banda-arc–Australian-continental-margin collision has uplifted and exposed offshelf sequences in Timor, Tanimbar, Kai, and Seram islands (Audley-Charles, Barber, and Carter 1979; Hamilton 1979; Audley-Charles 1988). The present locations of the block or blocks that left the northwestern and western margin of Australia are unknown, although, as mentioned earlier, two suggestions have been put forth. The first, by Audley-Charles et al. (1988) and Audley-Charles (1988), is that the Sibumasu Block represents the departed fragment. This suggestion was discounted because of problems with the timing of collision between Sibumasu and Indochina. A second alternative (the one depicted on the reconstructions) is that the Mount Victoria (MV) Block is the crustal fragment that departed (Sengör et al. 1988). This requires that the Mount Victoria Block be much larger than its presently very limited exposure (Mitchell 1989) would suggest.

### Discussion and conclusions

Three reconstructions of the circum-Pacific during the Jurassic are presented to suggest possible sites of derivation and histories of motion for the various tectonically significant elements that are now dispersed in the circum-Pacific orogenic collage. The maps are schematic, but attempt to incorporate constraints imposed by geological relations. The maps reflect a quite conservative approach, with relatively limited motions implied between the various blocks. In some cases this probably is quite reasonable, but in others the motions may have been at least an order of magnitude larger than here inferred. An important aspect of these reconstructions is that the geometry of plate boundaries within and along the margins of the Pacific must have been complex, and in that way not significantly different from that observed in the present western Pacific. The difference that is evident between the northern and southern Pacific is interesting. The south is depicted as a quite simple, continuous, Pacific Ocean–facing Andean margin extending from northern South America to northern New Guinea. Although seamounts and other oceanic materials have been accreted to the margin, and at least one back-arc basin evolved (Rocas Verde), there is no obvious evidence of large-scale accretion of arcs or continental blocks along this entire length of margin. In contrast, the northern Pacific is characterized by complex arrangements of subducting boundaries associated with progressive assembly and accretion of oceanic arcs and continental fragments to the surrounding North American and Eurasian continents. There is an approximate correlation with the history of large-scale accretion and the opening of the Central Atlantic, but in detail the correlation is not good. For example, the initiation of westward subduction and consequent collision of the Quesnellia–Slide Mountain assemblage with the western passive margin of Canada predates the opening of the Central Atlantic by approximately 25 m.y. The time of stitching of the Quesnellia–North America suture by plutons coincides well with the initiation of spreading in the Central Atlantic, so one might argue that there is some correlation, but this is tenuous at best. As best as one can determine, there was little change in the evolution of the southern margins that was controlled by the breakup of western and eastern Gondwana. The opening of the Rocas Verde trough correlates in part, but really postdates the breakup in this region.

The term "Pangea," meaning "all land," has been used to imply that at some point during the Late Paleozoic or Early Mesozoic all of the continental crust of the world was amalgamated into a single supercontinent. Subsequently that supercontinent progressively disintegrated to arrive at the modern, quite widely dispersed continental configuration. A point that can be made from these maps is that a true Pangea may never have existed. The window of time during which Pangea could have existed was limited by the time when southern Asia welded itself to Laurasia in the Late Triassic, and by the time when northern Gondwana disintegrated and gave rise to the Lhasa and associated blocks of the Cimmerian continent. Northern Gondwana had already begun to disintegrate during the Late Permian and came into being no later than the Middle to Late Triassic (Sengör et al. 1988). Thus, the Pliensbachian reconstruction shown in Figure 3.2 comes closest to the time when Pangea was most closely approached, but apparently never quite achieved.

The maps are schematic, but it is hoped that they will provide a useful base upon which to examine biogeographic questions, or upon which to infer the pre- and syn-Jurassic age distributions of oceanic lithosphere within the paleo-Pacific from the distributions and ages of oceanic sequences, including seamounts and pelagic sediments, accreted to the margins of the various terranes. This may be the only way to place any constraints on pre-Cretaceous plate motions in this enormous segment of the earth's outer lithospheric shell.

## References


Adams, C. J. D., & Robinson, P. (1977). Potassium-argon ages of schists from Chatham Island, New Zealand Plateau, southwest Pacific. *New Zealand J. Geol. Geophys.*, *20*, 287–301.

Anderson, T. H., & Schmidt, V. A. (1983). The evolution of Middle America and the Gulf of Mexico–Caribbean Sea region during Mesozoic time. *Geol. Soc. Am. Bull.*, *94*, 941–66.

Armstrong, R. L. (1988). Mesozoic and early Cenozoic magmatic evolution of the Canadian Cordillera. In S. P. Clark, B. C. Burchfiel, & J. Suppe (eds.), *Processes in Continental Lithosphere Deformation* (pp. 55–92). Geological Society of America, Special Paper 218.

Armstrong, R. L., Harakal, J. E., Forbes, R. B., Evans, B. W. & Thruston, S. P. (1986). Rb-Sr and K-Ar study of metamorphic rocks of the Seward Peninsula and southern Brooks Range, Alaska. In B. W. Evans & E. H. Brown (eds.), *Blueschists and Eclogites*, (pp. 185–203). Geological Society of America, Memoir 164.

Audley-Charles, M. G. (1988). Evolution of the southern margin of Tethys (North Australian region) from early Permian to late Cretaceous. In M. G. Audley-Charles & A. Hallam (eds.), *Gondwana and Tethys* (pp. 79–100). Geological Society of London, Special Publication 37.

Audley-Charles, M. G., Ballantyne, P. D., & Hall, R. (1988). Mesozoic-Cenozoic rift-drift of Asian fragments from Gondwanaland. *Tectonophysics*, *155*, 317–30.

Audley-Charles, M. G., Barber, A. J., & Carter, D. J. (1979). Geosynclines and plate tectonics in Banda arcs, eastern Indonesia. *Am. Assoc. Petrol. Geol. Bull.*, *63*, 249–52.

Balkwill, H. R., Cook, D. G., Detterman, R. L., Embry, A. F., Hakansson, E., & Mail, A. D. (1983). Arctic North America and northern Greenland. In B. M. Moullade & A. E. Nairn (eds.), *The Phanerozoic Geology of the World. II: The Mesozoic* (pp. 1–31). Amsterdam: Elsevier.

Barker, P. F., & Hill, I. A. (1981). Back arc extension in the Scotia sea. *Philos. Trans. R. Soc. Lond.*, *A300*, 249–62.

Blake, M. C., Engebretson, D. C., Jayko, A. S., & Jones, D. L. (1985). Tectonostratigraphic terranes in southwest Oregon. In D. G. Howell (ed.), *Tectonostratigraphic Terranes of the Circum-Pacific Region* (vol. 1, pp. 147–57). Houston: Circum-Pacific Council for Energy and Mineral Resources.

Box, S. E. (1982). Kanektok suture, SW Alaska: geometry, age and relevance. *Trans. Am. Geophys. Union*, *63*, 915.

(1985). Early Cretaceous orogenic belt in northwestern Alaska: internal organization, lateral extent and tectonic interpretation. In D. G. Howell (ed.), *Tectonostratigraphic Terranes of the Circum-Pacific Region* (vol. 1, pp. 137–45). Houston: Circum-Pacific Council for Energy and Mineral Resources.

Brandon, M. T., Cowan, D. S., & Vance, J. A. (1988). *The Late Cretaceous San Juan Thrust System, San Juan Islands, Washington*. Geological Society of America, Special Paper 221.

Brooks, H. C., & Vallier, T. L. (1978). Mesozoic rocks and tectonic evolution of eastern Oregon and western Idaho. In D. G. Howell & A. McDougall (eds.), *Mesozoic Paleogeography of the Western United States* (pp. 133–45). Los Angeles: Pacific Section SEPM.

Brothers, R. N., & Lillie, A. R. (1988). Regional geology of New Caledonia. In A. E. M. Nairn, F. G. Stehli, & Uyeda, S. (eds.), *The Ocean Basins and Margins. Vol. 7B: The Pacific Ocean* (pp. 325–74). New York: Plenum.

Bunopas, S. (1981). *Paleogeographic History of Western Thailand and Adjacent Parts of South-East Asia – a Plate Tectonic Interpretation*. PhD thesis, Victoria University, Wellington, New Zealand. Thailand: Geological Survey, Department of Mineral Resources.

Chang, C. F., & Cheng, H. L. (1973). Some tectonic features of the Mt. Jolmo Lungma area, southern Tibet, China. *Sci. Sinica*, *14*, 247–65.

Chang, C. F., Pan, Y. S., & Sun, Y. Y. (1989). The tectonic evolution of the Qinghai-Tibet Plateau: a review. In A. M. C. Sengör (ed.), *Tectonic Evolution of the Tethyan Region* (pp. 415–76). Norwell, Mass.: Kluwer, NATO ASI Series 259.

Churkin, M., Foster, H. L., Chapman, R. M. (1982). Terranes and suture zones in east central Alaska. *J. Geophys. Res.*, *87*, 3718–30.

Churkin, M., & Trexler, J. H. (1981). Continental plates and accreted oceanic terranes in the arctic. In A. E. M. Nairn, M. Churkin, & F. G. Stehli (eds.), *The Ocean Basins and Margins, Vol. 5: The Arctic Ocean* (pp. 439–92). New York: Plenum.

Cobbing, E. J. (1985). The central Andes: Peru and Bolivia. In A. E. M. Nairn, F. G. Stehli, & S. Uyeda (eds.), *The Ocean Basins and Margins. Vol. 7A: The Pacific Ocean* (pp. 219–64). New York: Plenum.

Coney, P. J., Jones, D. L., & Monger, J. W. H. (1980). Cordilleran suspect terranes. *Nature*, *288*, 329–33.

Cordey, F., Mortimer, N., DeWever, P., & Monger, J. W. H. (1987). Significance of Jurassic radiolarians from the Cache Creek terrane, British Columbia. *Geology*, *15*, 1151–4.

Csejtey, B., Cox, D. P., & Evarts, R. C. (1982). The Cenozoic Denali fault system and the Cretaceous accretionary development of Southern Alaska. *J. Geophys. Res.*, *87*, 3741–54.

Dalziel, I. W. D. (1981). Back-arc extension in the southern Andes: a review and critical appraisal. *Philos. Trans. R. Soc. Lond.*, *A300*, 319–35.

Dalziel, I. W. D., Garrett, S. W., Grunow, A. M., Pankhurst, R. J., Storey, B. C., & Vennum, W. R. (1987). The Ellsworth-Whitmore Mountains crustal block: its role in the tectonic evolution of West Antarctica. In G. D. McKenzie, (ed.), *Gondwana Six: Structure, Tectonics and Geophysics* (pp. 173–82). Washington, D.C.: American Geophysical Union, Geophysical Monograph 40.

Davies, H. L., Symonds, P. A., & Ripper, I. D. (1985). Structure and evolution of the southern Solomon Sea region. *BMR J. Austral. Geol. Geophys.*, *9*, 49–68.

Davis, G. A., Monger, J. W. H., & Burchfiel, B. C. (1978). Mesozoic construction of the cordilleran "collage," central British Columbia to central California. In D. G. Howell & K. A. McDougall (eds.), *Mesozoic Paleogeography of the Western United States* (pp. 1–32). Los Angeles: Pacific Section SEPM.

Day, H. W., Moores, E. M., & Tuminas, A. C. (1985). Structure and tectonics of the northern Sierra Nevada. *Geol. Soc. Am. Bull., 96,* 436–50.

Detterman, R. L. (1973). Mesozoic sequence in arctic Alaska. In M. G. Pitcher (ed.), *Arctic Geology* (pp. 376–87). Tulsa: American Association of Petroleum Geologists, Memoir 19.

Dickinson, W. R. (1981). Plate tectonics and the continental margin of California. In W. G. Ernst, (ed.), *The Geotectonic Development of California* (pp. 1–28). Englewood Cliffs, N.J.: Prentice-Hall.

Dickinson, W. R., & Thayer, T. P. (1978). Paleogeographic and paleotectonic implications of Mesozoic stratigraphy and structure in the John Day inlier of central Oregon. In D. G. Howell & K. A. McDougall (eds.), *Mesozoic Paleogeography of the Western United States* (pp. 147–61). Los Angeles: Pacific Section SEPM, Pacific Coast Paleogeography Symposium 2.

Dingle, R. V., Siesser, W. G., & Newton, A. R. (1983). *Mesozoic and Tertiary Geology of Southern Africa.* Rotterdam: A. A. Balkema.

Edelman, S. H., Day, H. W., Moores, E. M., Zigan, S. M., Murphy, T. P., & Hacker, B. R. (1989). *Structure across a Mesozoic Ocean–Continent Outline Zone in the Northern Sierra Nevada, California.* Geological Society of America, Special Paper 224.

Engebreston, D. C., Cox, A., & Gordon, R. G. (1984). Relative motions between oceanic plates of the Pacific Basin. *J. Geophys. Res., 89.* 10291–310.

(1985). *Relative Motions between Oceanic and Continental Plates in the Pacific Basin.* Geological Society of America, Special Paper 206.

Erdmer, P., & Armstrong, R. L. (1988). Permo-Triassic isotopic dates for blueschists, Ross River area, Yukon. In *Yukon Geology and Exploration.* Whitehorse, Yukon Territory: Department of Indian and Northern Affairs.

Fujita, K., & Newberry, J. T. (1983). Accretionary terranes and tectonic evolution of northeast Siberia. In M. Hashimoto & S. Uyeda (eds.), *Accretion Tectonics in the Circum-Pacific Regions* (pp. 43–57). Tokyo: Terra Scientific Publishing.

Gardner, M. C., Bergman, S. C., Cushing, G. W., MacKevett, E. M., Plafker, G., Campbell, R. B., Dodds, C. J., McClelland, W. C., & Mueller, P. A. (1988). Pennsylvanian pluton stitching of Wrangellia and the Alexander terrane, Wrangell Mountains, Alaska. *Geology, 16,* 967–71.

Gastil, R. G., Morgan, G. J., & Krummenacher, D. (1978). Mesozoic history of peninsular California and related areas east of the Gulf of California. In D. G. Howell & K. A. McDougall (eds.), *Mesozoic Paleogeography of the Western United States* (pp. 107–15). Los Angeles: Pacific Section SEPM, Pacific Coast Paleogeography Symposium 2.

Gastil, R. G., Philips, R. P., & Allison, E. C. (1975). *Reconnaissance Geology of the State of Baja California.* Geological Society of America, Memoir 140.

Gnibidenko, H. (1985). The Sea of Okhotsk–Kurile Islands ridge and Kurile-Kamchatka Trench. In A. E. M. Nairn, F. G. Stehli, & S. Uyeda (eds.), *The Ocean Basins and Margins, Vol. 7A: The Pacific Ocean* (pp. 377–418). New York: Plenum.

Grantz, A., & May, S. D. (1982). Rifting history and structural development of the continental margin north of Alaska. In J. S. Watkins & C. L. Drake (eds.), *Studies in Continental Margin Geology* (pp. 77–100). Tulsa: American Association of Petroleum Geologists, Memoir 34.

Hamilton, W. (1978). Mesozoic tectonics of the western United States. In D. G. Howell, & K. A. McDougall (eds.), *Mesozoic Paleogeography of the Western United States* (pp. 33–70). Los Angeles: Pacific Section SEPM, Pacific Coast Paleogeography Symposium 2.

(1979). *Tectonics of the Indonesian Region.* U.S.G.S. Geological Survey Professional Paper 1078.

Hansen, V. L. (1988). A model for tectonic accretion – Tukon-Tanana and Slide Mountain terranes, northwest North America. *Tectonics, 7,* 1167–77.

Harper, G. D., Saleeby, J. B., & Norman, E. A. S. (1985). Geometry and tectonic setting of sea-floor spreading for the Josephine ophiolite, and implications for Jurassic accretionary events along the California margin. In D. G. Howell (ed.), *Tectonostratigraphic terranes of the Circum-Pacific Region* (vol. 1, pp. 239–57). Houston: Circum-Pacific Council for Energy and Mineral Resources.

Hattori, I. (1982). The Mesozoic evolution of the Mino terrane, Central Japan: a geologic and paleomagnetic synthesis. *Tectonophysics, 85,* 313–40.

Henderson, L. J., Gordon, R. G., & Engebretson, D. C. (1984). Mesozoic aseismic ridges on the Farallon plate and southward migration of shallow subduction during the Larimide orogeny. *Tectonics, 3,* 121–32.

Hilde, T. W. C., Uyeda, S., & Kronke, L. (1977). Evolution of the western Pacific and its margin. *Tectonophysics, 38,* 145–65.

Holloway, N. H. (1982). North Palawan block, Philippines – its relation to Asian mainland and role in evolution of South China Sea. *Am. Assoc. Petrol. Geol. Bull., 66,* 1355–83.

Hutchison, C. S. (1975). Ophiolite in southeast Asia. *Geol. Soc. Am. Bull.,* 86, 797–806.

(1982). Southeast Asia. In A. E. M. Nairn & R. G. Stehli (eds), *The Ocean Basins and Margins, Vol. 6, The Indian Ocean* (pp. 451–512). New York: Plenum.

Jones, D. L., Howell, D. G., Coney, P., & Monger, J. W. H. (1983). Recognition, character, and analysis of tectonostratigraphic terranes in western North America. In M. Hashimoto & S. Uyeda (eds.), *Accretion Tectonics in the Circum-Pacific Regions* (pp. 21–35). Tokyo: Terra Scientific Publishing.

Jones, D. L., Silberling, N. J., Berg, H. C., & Plafker, G. (1981). *Map Showing Tectonostratigraphic Terranes of Alaska, Columnar Sections, and Summary Description of Terranes.* U.S. Geological Survey, open-file report 81-792.

Jones, D. L., Silberling, N. J., Gilbert, W., & Coney, P. (1982). Character, distribution, and tectonic significance of accretionary terranes in the central Alaska Range. *J. Geophys. Res., 87,* 3709–17.

Kimura, G., Mihashita, S., & Miyasaka, S. (1983). Collision tectonics in Hokkaido and Sakhalin. In M. Hashimoto & S. Uyeda (eds.), *Accretion Tectonics in the Circum-Pacific Regions* (pp. 123–34). Tokyo: Terra Scientific Publishing.

Klitgord, K. D., & Schouten, H. (1986). Plate kinematics of the central Atlantic. In P. R. Vogt, & B. E. Tucholke (eds.), *The Geology of North America: The Western North Atlantic Region: M* (pp. 351–78). Geological Society of America.

Korsch, R. J., & Wellman, H. W. (1988). The geological evolution of New Zealand and the New Zealand region. In A. E. M. Nairn, F. G. Stehli, & S. Uyeda (eds.), *The Ocean Basins and Margins, Vol. 7B: The Pacific Ocean* (pp.411–82). New York: Plenum.

Kosygin, Y. A., & Parfenov, L. M. (1981). Tectonics of the Soviet Far East. In A. E. M. Nairn, M. Churkin, & F. G. Stehli (eds.), *The Ocean Basins and Margins, Vol. 5: The Arctic Ocean* (pp. 377–412). New York: Plenum.

Lawver, L. A., & Scotese, C. R. (1987). A revised reconstruction of Gondwanaland. In G. D. McKenzie (ed.), *Gondwana Six: Structure, Tectonics and Geophysics* (pp. 17–23). Geophysical Monograph 40. Washington, D.C.: American Geophysical Union.

LeBrecque, J. L., & Barker, P. (1981). The age of the Weddell Sea. *Nature, 290,* 489–92.

Lopez-Ramos, E. (1976). *Carta Geologica de la Republica Mexicana.* Comite de la carta geologica de Mexico.

Lottes, A. L., & Rowley, D. B. (1990). Reconstruction of the Laurasia and Gondwana segments of Permian Pangaea. In W. S. McKerrow

& C. R. Scotese (eds.), *Paleozoic Paleogeography and Biogeography* (pp. 383–95). Geological Society of London, Memoir 12.

Lund, K., Snee, L. W., & Sutter, J. F. (1985). Style and timing of suture-related deformation in island arc rocks of western Idaho. *Geol. Soc. Am. Abstr. Progr., 17,* 367.

McElhinny, M. W. (1973). *Paleomagnetism and Plate Tectonics.* Cambridge University Press.

Maze, W. B. (1984). Jurassic La Quinta Formation in the Sierra de Pierja, northwestern Venezuela: geology and tectonic environments of red beds and volcanic rocks. In W. E. Bonini, R. B. Hargraves, & R. Shagam (eds.), *The Caribbean–South American Plate Boundary and Regional Tectonics* (pp. 263–82). Geological Society of America Memoir 162.

Mel'nikov, I. G. & Golozubov, V. V. (1980). Olistostrome sequences and consedimentational tectonic nappes in Sikhote-Alin. *Geotectonics, 14,* 310–17.

Merzlyakov, V. M., Terekhov, M. I., Lychagin, P. P., & Dylevskiy, Ye. F. (1982). Tectonics of the Omolon Massiv. *Geotectonics, 16,* 52–60.

Metcalfe, I. (1988). Origin and assembly of south-east Asian continental terranes. In M. G. Audley-Charles & A. Hallam (eds.), *Gondwana and Tethys* (pp. 101–18). Geological Society of London, Special Publication 37.

Meyen, S. V. (1976). Carboniferous and Permian lepidophytes of Angaraland. *Paleont. Abteil. B, 157,* 112–57.

Milsom, J. (1985). New Guinea and the western Melanesian arcs. In A. E. M. Nairn, F. G. Stehli, & S. Uyeda (eds.), *The Ocean Basins and Margins, Vol. 7B: The Pacific Ocean* (pp. 551–606). New York: Plenum.

Mitchell, A. H. G. (1989). The Shan Plateau and western Burma: Mesozoic–Cenozoic plate boundaries and correlations with Tibet. In A. M. C. Sengör (ed.), *Tectonic Evolution of the Tethyan Region,* (pp. 567–83). NATO ASI Series 259. Norwell, Mass: Kluwer.

Monger, J. W. H., Clowes, R. M., Price, R. A., Simony, P. S., Riddihough, R. P., & Woodsworth, G. J. (1985). *Continent–Ocean Transect B2: Juan de Fuca Plate to Alberta Plains,* Geological Society of America.

Monger, J. W. H., Price, R. P., Templeman-Kluit, D. J. (1982). Tectonic accretion and the origin of the two major metamorphic and plutonic welts in the Canadian Cordillera. *Geology, 10,* 70–5.

Moore, J. C. (1973). Cretaceous continental margin sedimentation, southwestern Alaska. *Geol. Soc. Am. Bull., 874,* 595–614.

Moore, J. C., & Connelly, W. (1977). Tectonic history of the continental margin of southwestern Alaska: late Triassic to earliest Tertiary. *Proc. Alaska Geol. Soc. Symp.*

Natal'in, B. A., & Parfenov, L. M. (1983). Accretional and collisional eugeosynclinal folded systems of the northwest Pacific rim. In M. Hashimoto, & S. Uyeda (eds.), *Accretion Tectonics in the Circum-Pacific Regions* (pp. 59–68). Tokyo: Terra Scientific Publications.

Nie, S. Y., Rowley, D. B., & Ziegler, A. M. (1990). Constraints on the locations of Asian microcontinents in Palaeo-Tethys during the Late Palaeozoic. In W. S. McKerrow & C. R. Scotese (eds.), *Palaeozoic Palaeogeography and Biogeography* (pp. 397–409). Geological Society of London, Memoir 12.

Okada, H. (1982). Geological evolution of Hokkaido, Japan: an example of collision orogenesis. *Proc. Geol. Assoc. Lond., 93,* 201–12.

Oldow, J. S., Seidensticker, C. M., Phelps, J. C., Julian, F. E., Gottschalk, R. R., & Boler, K. W., (1987). Balanced cross sections through the central Brooks range, and North Slope, Arctic Alaska. *Am. Assoc. Petrol. Geol. Bull.*

Otofuji, Y., Matsuda, T., & Nohda, S. (1985). Paleomagnetic evidence for the Miocene counterclockwise rotation of northeast Japan – rifting processes of the Japan Arc. *Earth Planet. Sci. Lett., 75,* 265–77.

Parfenov, L. M., Karsakov, L. P., Natal'in, B. A., Popeko, V. A., & Popeko, L. I. (1979). Ancient sialic blocks in the folded structures of the Far East. *Geol. Geofiz., 20,* 21–33.

Parfenov, L. M., & Natal'in, B. A. (1985). Mesozoic accretion and collision tectonics of northeastern Asia. In D. G. Howell (ed.), *Tectonostratigraphic Terranes of the Circum-Pacific Region* (vol. 1, pp. 363–73). Houston: Circum-Pacific Council for Energy and Mineral Resources.

(1986). Mesozoic tectonic evolution of northeastern Asia. *Tectonophysics, 127,* 291–304.

Patton, W. W., Tailleur, I. L., Broage, W. P., Lanphere, M. A. (1977). Preliminary report on the ophiolites of northern and western Alaska. In R. G. Coleman & W. P. Irwin (eds.), *North American Ophiolites* (pp. 51–7). Oregon Department of Geology and Mineral Industries Bulletin 95.

Pigram, C. J., & Panggabean, H. (1984). Rifting of the northern margin of the Australian continent and the origin of some microcontinents in eastern Indonesia. *Tectonophysics, 107,* 331–53.

Pindell, J. L. (1985). Alleghenian reconstruction and subsequent evolution of the Gulf of Mexico, Bahamas and Proto-Caribbean. *Tectonics,4,* 1–39.

Pindell, J. L., & Barrett, S. F. (1990). Geological evolution of the Caribbean region: a plate-tectonic perspective. In G. Dengo & J. E. Case (eds.), *The Geology of North America, The Caribbean Region* (pp. 405–32). Boulder: Geological Society of America.

Pindell, J. L., Cande, S. C., Pitman, W. C., Rowley, D. B., Dewey, J. F., LaBreque, J., & Haxby, W. (1988). A plate-kinematic framework for models of Caribbean evolution. *Tectonophysics, 155,* 121–38.

Pindell, J. L., & Dewey, J. F. (1982). Permo-Triassic reconstruction of western Pangea and the evolution of the Gulf of Mexico/Caribbean region. *Tectonics, 1,* 179–211.

Plafker, G. (1987). Regional geology and petroleum potential of the northern Gulf of Alaska continental margin. In D. W. Scholl, A. Grantz, & J. G. Vedder, (eds.), *Geology and Resource Potential of the Continental Margin of Western North America and Adjacent Ocean Basins – Beaufort Sea to Baja California* (pp. 229–68). Earth Science Series 6. Houston: Circum-Pacific Council for Energy and Mineral Resources.

Plafker, G., Jones, D. L., & Pessango, E. A. (1977). *A Cretaceous Accretionary Flysch and Melange Terrane along the Gulf of Alaska Margin.* U.S. Geological Survey Circular 751-B, B41-b-B43.

Potter, C. J. (1986). Origin, accretion, and postaccretionary evolution of the Bridge River terrane, southwest British Columbia. *Tectonics, 5,* 1027–41.

Price, R. A., Monger, J. W. H., & Roddick, J. A. (1985). Cordilleran cross-section, Calgary to Vancouver: Trip 3. In D. Templeman-Kluit (ed.), *Field Guides to Geology and Mineral Deposits in the Southern Canadian Cordillera.* Geological Society of America, cord. sect. mtg., Vancouver, B.C.

Rangin, C. (1978). Speculative model of Mesozoic geodynamics, central Baja California to northeastern (Mexico). In D. G. Howell & K. A. McDougall (eds.), *Mesozoic Paleogeography of the Western United States* (pp. 85–106). Pacific Coast Paleogeography Symposium 2. Los Angeles: Pacific Section SEPM.

Rowley, D. B. (1989). Tectonic evolution of Asia with an emphasis on the Mesozoic and Cenozoic, *28th International Geological Congress Abstracts, 2,* 723.

Rowley, D. B., & Lottes, A. L. (1988). Plate kinematic reconstructions of the North Atlantic and Arctic: Late Jurassic to Present. *Tectonophysics, 155,* 73–120.

Rowley, D. B., & Pindell, J. L. (1989). End Paleozoic–Early Mesozoic western Pangean reconstruction and its implications for the distribution of Precambrian and Paleozoic rocks around Meso-America. *Precamb. Res., 42,* 411–44.

Rowley, D. B., Raymond, A., Parrish, J. T., Lottes, A. L., Scotes, C. R., & Ziegler, A. M. (1985). Carboniferous paleogeographic, phytogeographic and paleoclimate reconstructions. *Int. J. Coal Geol., 5,* 7–42.

Saleeby, J. B., Goodin, S. E., Sharp, W. D., & Busby, C. J. (1978). Early Mesozoic paleotectonic-paleogeographic reconstructions of the southern Sierra Nevada region. In D. G. Howell & K. A. McDougall (eds.), *Mesozoic Paleogeography of the Western United States* (pp. 311–36). Pacific Coast Paleogeography Symposium 2. Los Angeles: Pacific Section SEPM.

Samylina, V. A., & Yefimova, A. F. (1968). First finds of an early Jurassic flora in the Kolyma River basin. *Doklady Akad. Nauk SSSR, 179,* 28–30.

Scholl, D. W., Grantz, A., & Vedder, J. G. (eds.). (1987). *Geology and Resource Potential of the Continental Margin of Western North America and Adjacent Ocean Basins – Beaufort Sea to Baja California.* Earth Science Series 6. Houston: Circum-Pacific Council for Energy and Mineral Resources.

Schweickert, R. A. (1978). Triassic and Jurassic Paleogeography of the Sierra Nevada and adjacent regions, California and Western Nevada. In D. G. Howell & K. A. McDougall (eds.), *Mesozoic Paleogeography of the Western United States* (pp. 361–84). Pacific Coast Paleogeography Symposium 2. Los Angeles: Pacific Section SEPM.

Sengör, A. M. C. (1979). Mid-Mesozoic closure of Permo-Triassic Tethys and its implications. *Nature, 279,* 590–3.

(1984). *The Cimmeride Orogenic System and the Tectonics of Eurasia.* Geological Society of America, Special Paper 195.

(1987). *Tectonic Subdivisions and Evolution of Asia.* Technical University of Istanbul publication 40.

Sengör, A. M. C., Altiner, D., Cin, A., Ustaomer, T., & Hsu, K. J. (1988). Origin and assembly of the Tethyside orogenic collage at the expense of Gondwana Land. In M. G. Audley-Charles & A. Hallam (eds.), *Gondwana and Tethys* (pp. 119–81). Geological Society of London, Special Publication 37.

Sengör, A. M. C., & Hsü, K. J. (1984). The Cimmerides of eastern Asia: history of the eastern end of Paleo-Tethys. *Mem. Soc. Geol. France, N.S., 147,* 139–67.

Seslavinskii, K. B. (1979). South Anyui Sutura (West Chukota). *Dolkady Akad. Nauk SSSR, 249,* 1181–6.

Silver, L. T., & Anderson, T. H. (1974). Possible left-lateral early to middle Mesozoic disruption of the southwestern North American craton. *Geol. Soc. Am. Abstr. Progr. 6,* 955.

Smith, P. L., Tipper, H. W. (1986). Plate tectonics and paleobiogeography: Early Jurassic (Pliensbachian) endemism and diversity. *Palaios, 1,* 399–412.

Spörli, K. B. (1978). Mesozoic tectonic, North Island, New Zealand. *Bull. Geol. Soc. Am., 89,* 415–25.

Stauffer, P. H. (1983). Unravelling the mosaic of Paleozoic crustal blocks in southeast Asia. *Geol. Rund., 72,* 1061–80.

Taira, A., & Ogawa, Y. (eds.). (1988). The Shimanto belt, southwest Japan – studies on the evolution of an accretionary prism. *Mod. Geol., 12,* 542.

Taira, A., Saito, Y., & Hashimoto, M. (1983). The role of oblique subduction and strike-slip tectonics in the evoluton of Japan. In T. W. C. Hilde & S. Uyeda (eds.), *Geodynamics of the Western Pacific-Indonesian Region* (pp. 303–16). Geodynamics Series 11, American Geophysical Union.

Takahashi, M. (1983). Space-time distribution of late Mesozoic to early Cenozoic magmatism in East Asia and its tectonic implications. In M. Hashimoto & S. Uyeda (eds.), *Accretion Tectonics in the Circum-Pacific Regions* (pp. 69–88). Tokyo: Terra Scientific Publishing.

Tanaka, K., & Nozawa, T. (eds.). (1977). *Geology and Mineral Resources of Japan, Vol. 1.* Geological Survey of Japan.

Taylor, B., & Hayes, D. E. (1983). Origin and history of the South China Sea basin. In D. E. Hayes (ed.), *The Tectonic and Geologic Evolution of Southeast Asian Seas and Islands: Part 2* (pp. 23–56). Washington, D.C.: American Geophysical Union.

Templeman-Kluit, D. J. (1979). *Five Occurrences of Transported Synorogenic Clastic Rocks in Yukon Territory. Current Research, Part A.* Geological Survey of Canada, paper 79-1A.

(1979b). *Transported cataclasite, ophiolite and granodiorite in Yukon: Evidence of Arc–Continent Collision.* Geological Survey of Canada, Paper 79-14.

Tipper, H. W. (1984). The allochthonous Jurassic–Lower Cretaceous terranes of the Canadian Cordillera and their relation to correlative strata of the North American craton. In G. E. G. Westermann (ed.), *Jurassic–Cretaceous Biochronology and Paleogeography of North America* (pp. 113–20). Geological Association of Canada, Special Paper 27.

Uliana, M. A., & Biddle, K. T. (1987). Permian to Late Cenozoic evolution of northern Patagonia: main tectonic events, magmatic activity and depositional trends. In G. D. McKenzie (ed.), *Gondwana Six: Structure, Tectonics and Geophysics* (pp. 271–86). Geophysical Monograph 40. Washington D.C.: American Geophysical Union.

Vance, J. A., Dugan, M. A., Blanchard, D. P., & Rhodes, J. M. (1980). Tectonic setting and trace element geochemistry of Mesozoic ophiolitic rocks in western Washington. *Am. J. Sci., 280-A,* 359–88.

von Rad, U., & Exon, N. F. (1982). Mesozoic–Cenozoic sedimentary and volcanic evolution of the starved passive continental margin off northwest Australia. In J. S. Watkins & C. L. Drake (eds.), *Studies in Continental Margin Geology* (pp. 253–81). Memoir 34. Tulsa: American Association of Petroleum Geologists.

Walper, J. L. (1980). The tectonic-sedimentary history of Caribbean basins and their hydrocarbon potential. *Mem. Canadian Soc. Petrol. Geol., 6,* 887–911.

Weissel, J. K., Hayes, D. E., & Herron, E. M. (1977). Plate tectonics synthesis: the displacements between Australia, New Zealand and Antarctica since the Late Cretaceous. *Marine Geology, 25,* 231–77.

Weissel, J. K., & Watts, A. B. (1979). Tectonic evolution of the Coral Sea Basin. *J. Geophys. Res., 84,* 4572–82.

Westermann, G. E. G. (1988). Middle Jurassic ammonite biogeography supports ambi-Tethyn origin of Tibet. In M. G. Audley-Charles & A. Hallam (eds.), *Gondwana and Tethys* (pp. 235–9). Geological Society Special Publication 37.

Weyl, R. (1980). *Geology of Central America.* Berlin: Gebrüder Borntraeger.

Winkler, G. R., & Plafker, G. (1981). *Geologic Map and Cross Sections of the Cordova and Middleton Island Quadrangles, Southern Alaska.* U.S. Geological Survey open-file report 81-1164.

Ziegler, A. M., Barrett, S. F., & Scotese, C. R. (1981). Paleoclimate, sedimentation, and continental accretion. *Philos. Trans. R. Soc. Lond., A301,* 253–64.

# Part III: Regional geology and stratigraphy

## 4 Western Canada and United States

T. P. POULTON, R. L. DETTERMAN, R. L. HALL, D. L. JONES, J. A. PETERSON, P. SMITH, D. G. TAYLOR, H. W. TIPPER, and G. E. G. WESTERMANN

### INTRODUCTION[1]

Prior to the general acceptance of the theory of plate tectonics, western North America was both a classic model and an enigma for structural geologists captivated by the eugeosyncline–miogeosyncline paradigm. Soon after the theory of plate tectonics settled in, accreted terranes followed, pioneered by one of the leaders of the plate-tectonic movement (Wilson 1968). Western North America, unique among the various boundary regions of the Pacific Ocean in being dominated by thin slivers of accreted terranes separated by transcurrent faults (Howell and Jones 1989), has played a leading role in the development of the concepts of terrane accretion. Insofar as the Jurassic was a primary period of accretion, the Jurassic rocks in western North American have played a major role in these developments.

The North American continent that split away from Europe during the Jurassic is estimated to have been 20–30% smaller than at present and to have grown by 300–500 km along its Pacific Coast by the accretion of about 100 terranes between 200 and 50 m.y. ago (Figure 4.1A). The western margin of North America prior to about Middle Jurassic time was a passive margin (miogeocline), with mainly cratonic sediment sources, persisting 450 m.y. since a major Late Proterozoic rifting event (Sloss 1982; Stott and Aitken 1982). Depending on latitude along the North American margin, the Jurassic and more recent history of the margin is one of (1) accretion of exotic Pacific terranes, (2) jostling, uplift, and magmatism related to crustal thickening, and (3) sedimentation of orogenic detritus in successor basins and in adjacent foreland basins on the craton.

Lithofacies and biofacies trends and characteristics along the ancient margin of cratonic North America are critical for interpreting the accretion history of the circum-Pacific terranes. Cratonic paleomagnetic and paleontologic data provide the standards for interpretation of the affinities and original locations of the allochthonous terranes. In most of the North American Cordillera north of Mexico the major change from a passive to an active continental margin took place by the Late Jurassic (Coney 1989). The change is indicated by newly developed sediment sources in the Late Jurassic and by Middle and Late Jurassic radiometric dates on metamorphic and igneous rocks within the collision belts, both indicating uplifts associated with the accretionary process. The cratonic rocks also contain a record of epeirogenic and eustatic events, presumably partly related to the early stages of the opening of the North Atlantic and the Central Atlantic Ocean in the east.

The material that was accreted to western North America was diverse: slices of North America previously rifted away from or transported along the western continental margin; volcanic-arc and related materials formed adjacent to western North America; and oceanic sea-floor material, oceanic island arcs, and oceanic plateaus. Ben-Avraham et al. (1981) have compared Mesozoic and later western North America with Recent southeastern Asia, where trenches, arcs, and oceanic plateaus lie off the continental margins. Additionally, small amounts of Lower Jurassic and older continental fragments may be pieces of a rifted mid-Pacific continent. Plate-motion studies (Cox, Debiche, and Engebretson 1989) indicate Middle Jurassic spreading and more recent spreading from a complicated, ever-changing series of mid-Pacific spreading centers, so that the great bulk of the material accreted to North America should be of general northeastern Pacific affinity. This suggestion is supported by recent interpretations of North American affinities for many Jurassic and more recent fossils (Taylor et al. 1984; Tipper 1984a). All the North Pacific plates, as well as the spreading ridges and triple junctions, were convergent with respect to the relatively westward moving North American plate. After accretion, northerly transport of many exotic terranes along transcurrent faults has resulted from the important northerly component of the relative motions of Pacific plates at various times, especially in the Tertiary and Quaternary.

Western North American Jurassic geology is further complicated by interactions with the opening of the Gulf of Mexico in the south and the Arctic Ocean basins in the north. These areas

[1] By T. P. Poulton; Geological Survey of Canada Contribution No. 38989.

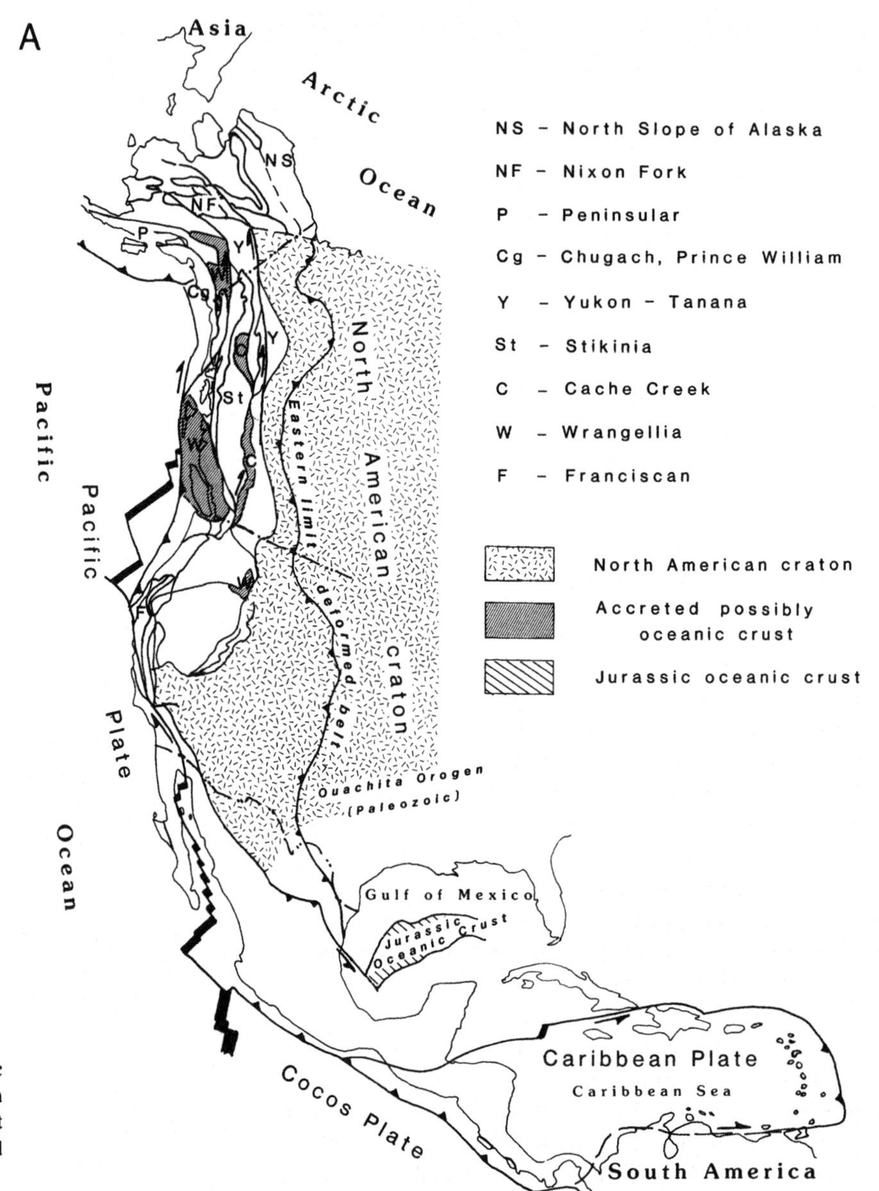


**Figure 4.1A. Generalized tectonic map of western North America showing only the most important terranes for the Jurassic. (Adapted from Coney 1989.)**

remain the foci of many unresolved problems. The opening of the Arctic Ocean and Gulf of Mexico influenced the marine organisms, and presumably the climates, that pertained, as a result of redistribution of marine currents through connections between the Pacific, "Tethyan," and Arctic seas.

Unraveling the history of cordilleran geology in the light of plate tectonics has been a stimulus for exciting advances since the early 1970s, linking many disparate field and theoretical studies. Paleontology [the first suggestion that cratonic and cordilleran fossils indicate different paleolatitudes was by Tozer (1970)], lithostratigraphic analysis (Tipper 1984a), and paleomagnetism (Stone and McWilliams 1989) have played the leading roles in terrane analysis. They demonstrate the different characters and affinities of contemporaneous rocks and fossils across terrane and terrane–continent boundaries and contribute to (and limit) speculations about where exotic terranes may have come from and when.

The contributions in this chapter are organized by area, from north to south on the stable craton, and then from north to south in the cordilleran orogenic belt (Figure 4.1B).

## NORTHWESTERN CANADA AND NORTHERN ALASKA[2]

### Introduction

Jurassic sedimentary rocks were deposited unconformably above a variety of Precambrian, Paleozoic, and Triassic units, at the northwestern corner of the ancient continent, on a shelf called the Brooks-Mackenzie Basin (Figure 4.2) (Balkwill et al. 1983). This interpretation assumes palinspastic continuity within one sin-

[2] By T. P. Poulton; Geological Survey of Canada Contribution No. 28589.

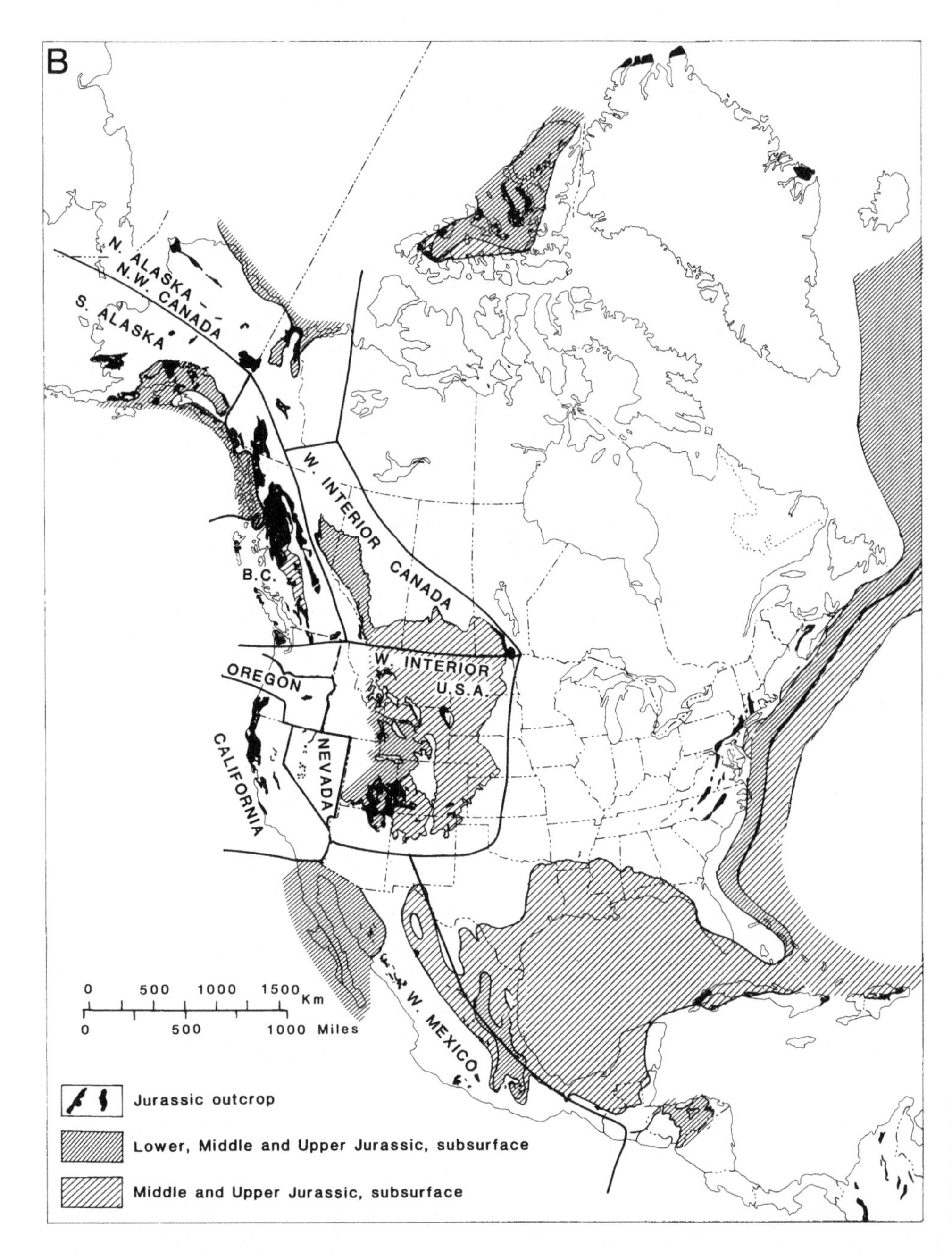


**Figure 4.1B. Index map of North America showing the Jurassic in outcrops and subsurface, as well as the areas of responsibility covered by the authors of this chapter.**

gle sedimentary basin. However, the part that includes northern Alaska and the adjacent northwestern Yukon has been suggested as being allochthonous with respect to adjacent parts of the northeastern Yukon. Most interpretations of this kind favor an original site adjacent to the Canadian Arctic islands. Various models have been proposed, each of which treats the origin of the Arctic Ocean basins of today, but thus far none has been universally accepted (Balkwill et al. 1983). Two primary interpretations involve (1) counterclockwise rotation and (2) left-lateral strike-slip faulting along an extension of the Kaltag fault of northern Alaska (Figure 4.2). In the majority of interpretations, the strata of the Brooks-Mackenzie Basin are the northernmost of the rocks that originated along the western North American border and its possible extension into the Arctic during the Jurassic.

Small uplifts along the northwestern margin of the continent (Dixon 1982; Poulton, Leskiw, and Audretsch 1982) may be high blocks along an ancient rift margin, if the hypothesis of rotation of northern Alaska away from Arctic Canada is correct, or they may represent jostled blocks in a strike-slip regime.

The southeastern margin of the Brooks-Mackenzie Basin coincided with a fault zone where Lower Jurassic to Oxfordian rocks were deposited against high-standing blocks of Upper Paleozoic rocks (Young, Myhr, and Yorath 1976; Dixon 1982; Poulton 1982; Poulton et al. 1982). Upper Jurassic rocks were deposited across this fault zone as a consequence of a regional transgressive episode, and they are preserved sporadically on the craton far to the east.

Both local and distant sediment sources are indicated by the petrology of the Jurassic rocks; the sources were to the southeast,

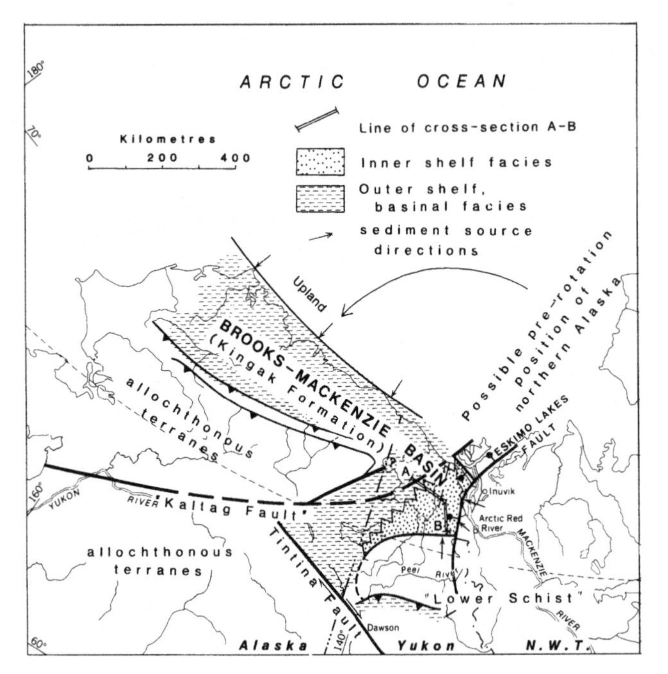


**Figure 4.2. Index map of northeastern Alaska and northwestern Canada showing major Jurassic facies belts and simplified tectonic elements.**

consistent with facies changes that indicate more-proximal deposits in that direction. The Jurassic sedimentary wedge becomes coarser-grained, and generally shallow in aspect, as it thins and becomes less complete southeastward.

During the Jurassic, fine clastic rocks, now preserved in a belt running east–west across the central Yukon (''Lower Schist Division'') probably were deposited in a southern extension of the Brooks-Mackenzie Basin, on the south side of a promontory of the ancient continental margin that jutted westward into what is now northern Alaska (Poulton and Tempelman-Kluit 1982).

The northern and northeastern margins of the Brooks-Mackenzie Basin in Canada are poorly known. In northern Alaska, the northern margin in Early and Middle Jurassic time was a highland called Barrow Arch (Detterman 1970, 1973; Rickwood 1970; Balkwill et al. 1983). In Canada, some sort of discontinuous northern highland may be indicated by a series of enigmatic uplifts. They apparently were not sources of any significant amount of sediment (Poulton 1982). It is not clear whether these uplifts were part of a continuous highland area or were simply individual elements of a more extensive series of *en echelon* uplifts that characterized the unstable cratonic margins in much of Arctic Canada. Neither is it clear to what extent any northern highlands that may have existed were allochthonous and, if so, where they originated.

The southwestern basin margin during Early and Middle Jurassic time is completely unknown, apparently having been displaced by subsequent large-scale movement along Tintina and other transcurrent faults. The areas to the southwest are presently occupied by allochthonous terranes.

## Stratigraphy and fossil occurrences

The Jurassic rocks on the eastern basin margin represent a series of marine cycles of easterly derived sandstone and shale complexes (Bug Creek Group and the Husky, Porcupine River, and North Branch Formations) (Jeletzky 1967, 1977; Dixon 1982; Poulton et al. 1982; Poulton 1984, 1989a; Braman 1985). These sandstones grade to outer-shelf shale and siltstone facies (Kingak Formation) (Poulton 1982) to the north, west, and southwest (Figure 4.3) and become thicker, perhaps exceeding 1,200 m in some places. They are also more complete in terms of the preserved record in those directions (Young et al. 1976; Poulton 1982). In northern Alaska, the Jurassic is mainly in a shale facies, the original Kingak Formation (Leffingwell 1919; Detterman et al. 1975). There are minor thin sandstones in the lower part, apparently derived from the north (Detterman 1970, 1973; Rickwood 1970; Detterman et al. 1975). In northwestern Alaska, Upper Jurassic turbidites derived from the south demonstrate uplift of the ancestral Brooks Geanticline in that direction (Campbell 1967), perhaps during early stages of the Brookian Orogeny and the initial subsidence of the Colville Trough foredeep (Balkwill et al. 1983). No similar tectonic events were recorded in adjacent parts of Canada until well into the Cretaceous.

The succession along the southeastern margin of the Brooks-Mackenzie Basin records five major regressive episodes, each preceded by transgressions that are remarkable for the consistency with which progressive onlap over the craton can be seen in the preserved record (Poulton et al. 1982). Thus, the basal Jurassic rocks onlapping the craton to the southeast become progressively younger eastward and southeastward, ranging in age from Hettangian in the northwest to Late Jurassic in the southeast. The depocenters in the Lower, Middle, and Upper Jurassic show a progressive southerly shift associated with transgression in that direction.

The Jurassic fossils and paleontological literature have been summarized by Poulton (1978, 1984), Detterman et al. (1975), and Fensome (1987).

### *Lower Jurassic*

The oldest beds known, early Hettangian shelf sandstones basal to the Kingak Formation, contain *Psiloceras* sp. at their base and *Psiloceras* (*Caloceras*) cf. *P.* (*C.*) *johnstoni* (Sowerby) and associated bivalves in their upper part (Frebold and Poulton 1977; Poulton 1991). They represent the Planorbis Standard Zone.

Except for localized Hettangian and Sinemurian basin-margin sandstones and conglomerates, the lowest strata regionally are late Sinemurian open-shelf shales and siltstones with articulated crinoid remains and *Chondrites*, attesting to poorly oxygenated bottom conditions and lack of currents. They grade upward into glauconitic, shallow-shelf sandstones that are evidence for storm and/or tidal depositional agents, and are late Sinemurian to early Toarcian. The overlying unit is a blanket of Toarcian black shale. The Lower Jurassic in northwestern Yukon and northern Alaska is represented by the lower Kingak Formation. This consists largely

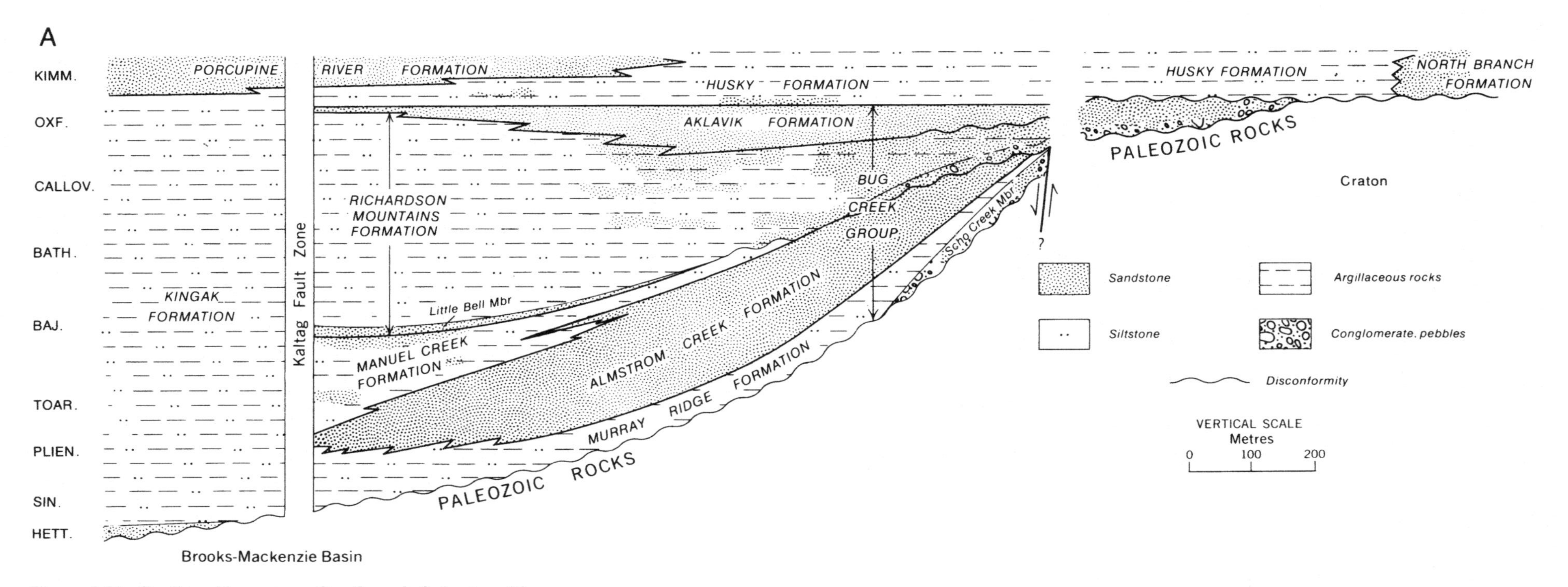


**Figure 4.3A. Stratigraphic cross section through facies transitions from outer-shelf Kingak Shale (NW) in northern Yukon to inner-shelf Bug Creek Group, Husky, Porcupine, and North Branch Formations (SE). Line of section is shown in Figure 4.4.**

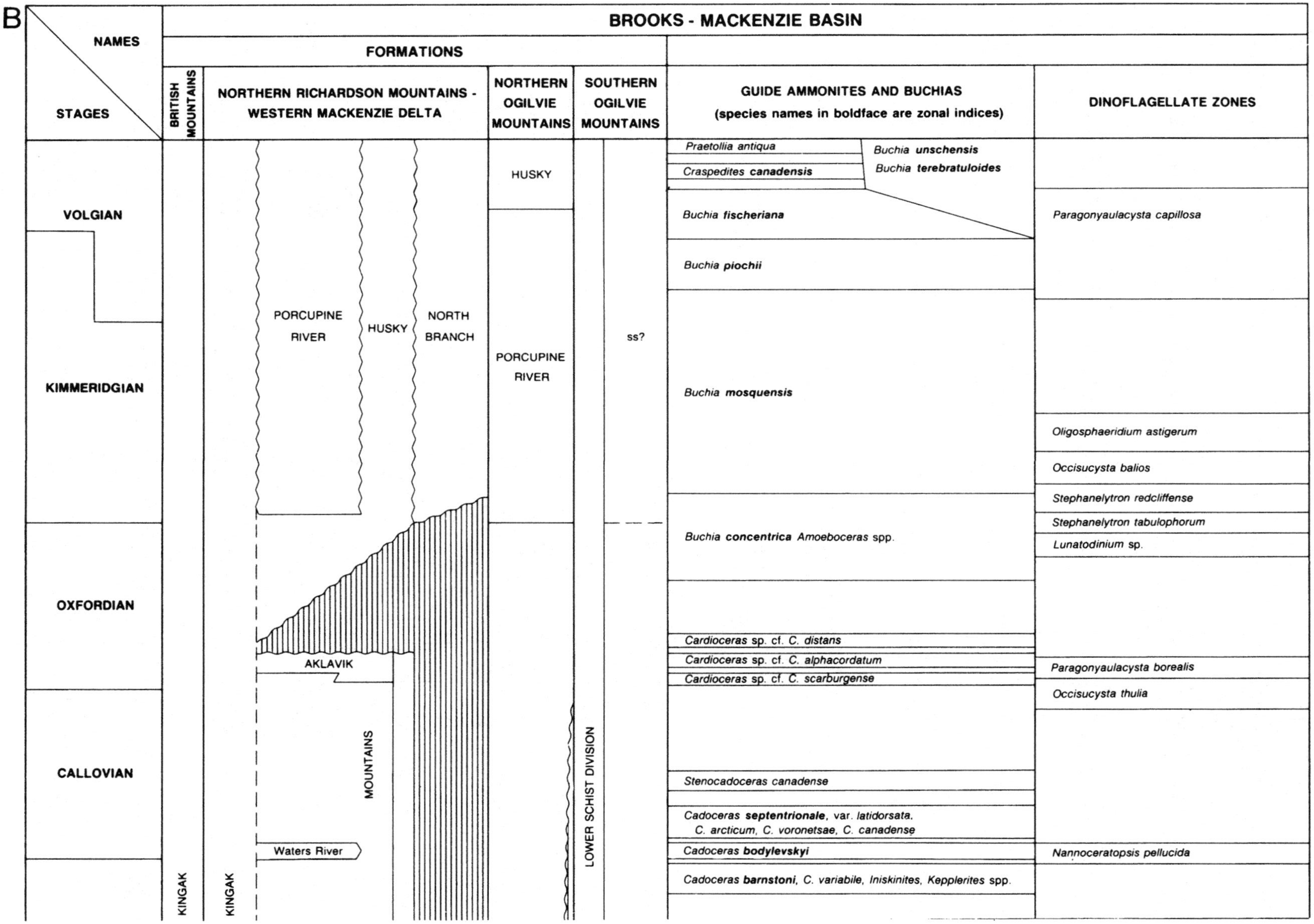

B

| STAGES \ NAMES | BROOKS - MACKENZIE BASIN: FORMATIONS: BRITISH MOUNTAINS | NORTHERN RICHARDSON MOUNTAINS - WESTERN MACKENZIE DELTA | NORTHERN OGILVIE MOUNTAINS | SOUTHERN OGILVIE MOUNTAINS | GUIDE AMMONITES AND BUCHIAS (species names in boldface are zonal indices) | DINOFLAGELLATE ZONES |
|---|---|---|---|---|---|---|
| VOLGIAN | KINGAK | PORCUPINE RIVER; HUSKY; NORTH BRANCH | HUSKY | ss? | *Praetollia antiqua*; *Craspedites* ***canadensis***; *Buchia* ***unschensis***; *Buchia* ***terebratuloides***; *Buchia* ***fischeriana*** | *Paragonyaulacysta capillosa* |
| KIMMERIDGIAN | KINGAK | PORCUPINE RIVER; HUSKY; NORTH BRANCH | PORCUPINE RIVER | ss? | *Buchia* ***piochii***; *Buchia* ***mosquensis*** | *Oligosphaeridium astigerum*; *Occisucysta balios*; *Stephanelytron redcliffense* |
| OXFORDIAN | KINGAK | KINGAK; HUSKY; NORTH BRANCH; AKLAVIK | | LOWER SCHIST DIVISION | *Buchia* ***concentrica*** *Amoeboceras* spp.; *Cardioceras* sp. cf. *C. distans*; *Cardioceras* sp. cf. *C. alphacordatum*; *Cardioceras* sp. cf. *C. scarburgense* | *Stephanelytron tabulophorum*; *Lunatodinium* sp.; *Paragonyaulacysta borealis* |
| CALLOVIAN | KINGAK | KINGAK; MOUNTAINS; Waters River | | LOWER SCHIST DIVISION | *Stenocadoceras canadense*; *Cadoceras* ***septentrionale***, var. *latidorsata*. *C. arcticum*, *C. voronetsae*, *C. canadense*; *Cadoceras* ***bodylevskyi***; *Cadoceras* ***barnstoni***, *C. variabile*, *Iniskinites*, *Kepplerites* spp. | *Occisucysta thulia*; *Nannoceratopsis pellucida* |

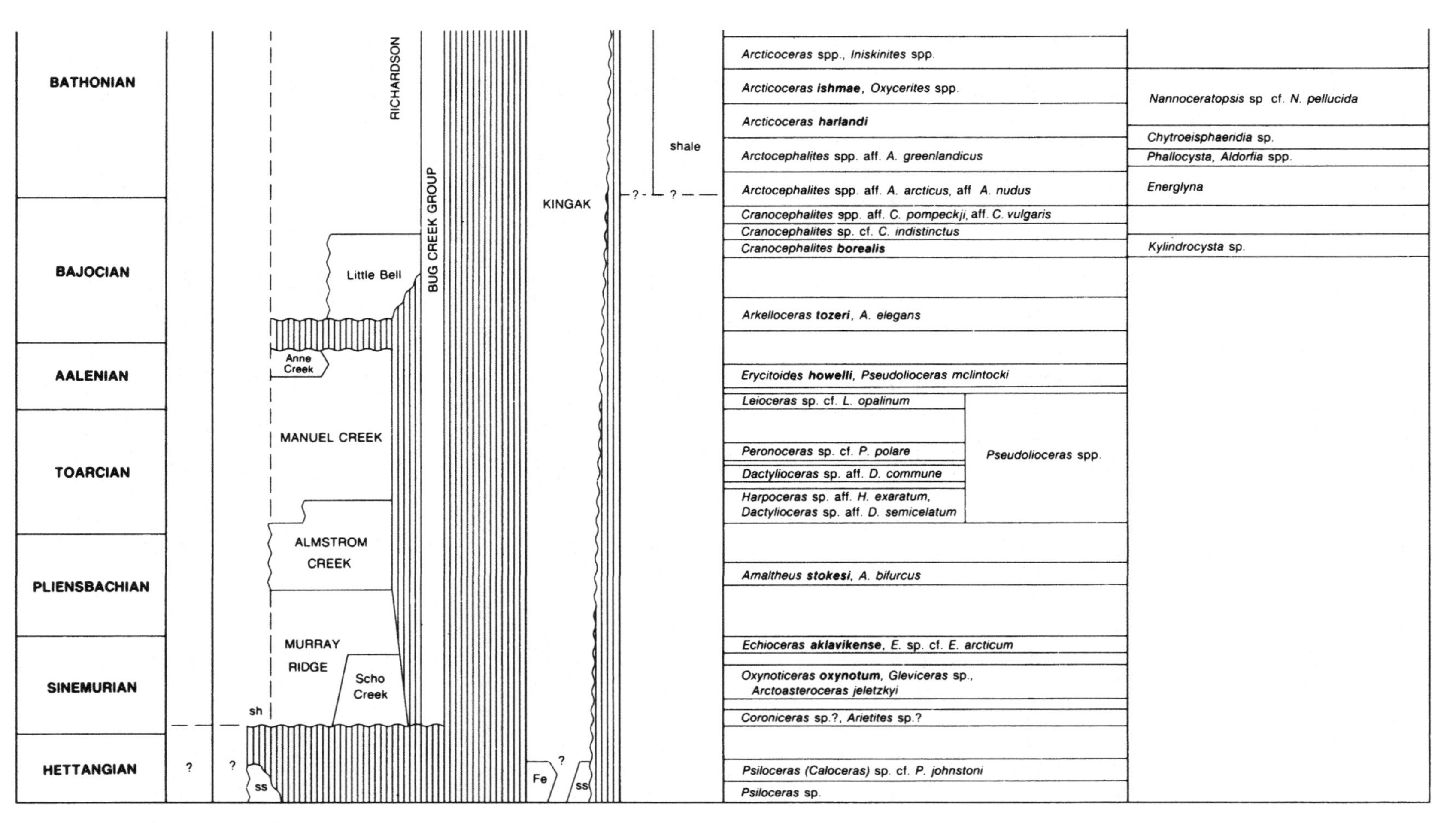


**Figure 4.3B.** Chart showing lithologic units and ammonite and dinoflagellate assemblages, the latter compiled by E. H. Davies. Species names in bold print indicate assemblage zones; stages not drawn to scale.

of finely laminated black shales with few fossils, with basal sandstone and glauconitic sandstone interbeds locally deposited near the northern margin of the Brooks-Mackenzie Basin along the present Arctic Coast area. Late Pliensbachian turbidite sandstones are present in northwesternmost Yukon.

Lower Sinemurian ammonites, occurring in shales of the Murray Ridge Formation and assigned to the Bucklandi and Semicostatum Standard Zones, respectively, include *Coroniceras*(?) and *Arnioceras* cf. *A. douvillei* (Bayle). The Upper Sinemurian Oxynotum Standard Zone is represented by *Oxynoticeras oxynotum* (Quenstedt), *Gleviceras* sp. *Aegasteroceras* (*Arctoasteroceras*) *jeletzkyi* Frebold, and *Aegasteroceras* sp. *Echioceras aklavikense* Frebold is assigned to the Raricostatum Standard Zone (Frebold 1960; Jeletzky 1967; Poulton 1991).

Upper Pliensbachian sandstones of the Almstrom Creek Formation and shales of the Kingak Formation (Frebold, Mountjoy, and Tempelman-Kluit 1967) contain *Amaltheus stokesi* (Sowerby) and *A. bifurcus* Howarth, representing the Margaritatus Standard Zone.

*Ovaticeras* cf. *O. ovatum* (Young & Bird) in association with *Collina*(?) aff. *C.*(?) *simplex* Fucini in the Kingak Shale are not precisely dated but appear to be early, perhaps earliest, Toarcian (Poulton 1991). The uppermost Almstrom Creek and lower Manuel Creek Formations contain *Dactylioceras commune* (Simpson), which may indicate the lower Bifrons Standard Zone. Middle and Upper Toarcian beds in the Kingak and Manuel Creek Formations contain *Pseudolioceras lectum* (Simpson) that may indicate the Thouarsense Standard Zone.

The locally abundant and varied bivalves, and the less prominent crinoids and brachiopods, are similar to their counterparts in northeastern Siberia and northwestern Europe. For the most part, the Lower Jurassic faunas are thus readily correlatable with the standard Northwest European zonation (Frebold 1975; Poulton 1991), and no new terminology had to be introduced. A few Lower Jurassic microfossils and palynomorphs have been described (Poulton et al. 1982).

### *Middle Jurassic*

The Aalenian is represented by a sandstone gradationally overlying the Toarcian shale, in what appears to be an offshore bar sequence near the eastern basin margin. Aalenian beds are also present farther west in the basinal shales, as indicated by ammonites. The earliest Bajocian in northern Yukon is represented by a transgressive sandstone, overlying a disconformity that coincides with a major regressive event (Poulton et al. 1982; Poulton 1988). A heterogeneous sequence of Bajocian through early Oxfordian shelf shales, siltstones, and sandstones follows. In a local depocenter in northern Yukon, some thick Lower Callovian sandstone units may have represented shoals or barrier islands. In northern Alaska, a few bentonite beds interlayered in the Kingak Shale may be products of the extensive Jurassic basic igneous activity that occurred contemporaneously in the landmass south of the Brooks-Mackenzie Basin (Reiser, Lanphere, and Brosgé 1965). Ammonites from the Kingak in northern Alaska have been described by Imlay (1955, 1976, 1981c), and the stratigraphy has been outlined by Detterman et al. (1975).

The Brooks-Mackenzie Basin represents a region of mixed Boreal and northeast Pacific faunas in the Middle Jurassic [the Bering Province of Taylor et al. (1984)]. The Lower Aalenian is indicated by *Leioceras* aff. *L. opalinum* (Reinecke), and perhaps by some of the *Pseudolioceras* occurrences. Higher beds are characterized by northern Pacific or Boreal faunas for which regional zones have been erected (Westermann 1964b; Sey, Kalacheva, and Westermann 1986). Upper Aalenian beds, assigned to the *Erycitoides howelli* Zone, are characterized by *E. howelli* (White) in association with *Pseudolioceras* aff. *P. whiteavesi* (White) in northern Yukon. *Tmetoceras* in the Kingak of northern Alaska (Imlay 1955) represents some part of the Aalenian as well. The Aalenian beds are in the Kingak Shale and the upper part of the Manuel Creek Formation.

*Arkelloceras elegans* Frebold (Frebold et al. 1967), *A. tozeri* Frebold, and *A. mclearni* Frebold (Poulton 1978) in the basal sandstone of the Richardson Mountains Formation probably are early Bajocian, as suggested by Westermann (1964a) for the genus in Alberta. This genus is characteristic of the Bering Province of the northern Pacific area (Taylor et al. 1984) and of the Canadian Arctic islands Boreal Bajocian. *Cranocephalites borealis* (Spath) and *C. warreni* Frebold occur higher in the same unit. They indicate the Borealis Zone of Callomon (1959, 1984), which is probably a Boreal provincial equivalent of part of the upper Bajocian. Overlying sandstones with *Cranocephalites* cf. *C. indistinctus* Callomon, *C.* spp. aff. *C. pompeckji* (Madsen), *C. maculatus* Spath, and *C. vulgaris* Spath represent the Indistinctus and Pompeckji Zones, respectively, which may be equivalent to lower Bathonian in the standard European classification.

The Kingak Shale and the sandstones and shales of the Richardson Mountains Formation contain ammonite faunas dominated by a Boreal sequence of cardioceratids ranging from middle Bathonian through lower Callovian (Frebold 1961, 1978; Poulton 1987). In contrast, certain minor elements of the Bathonian ammonite fauna, such as the genus *Iniskinites*, are of eastern Pacific affinity, and others, such as *Phylloceras* and *Cadomites*(?), are cosmopolitan. The mixing is explained by the location of the basin at the northwestern corner of the continent in the Jurassic (Poulton 1982), in the region of faunal overlaps between the Boreal Realm and East Pacific Subrealm of the Tethyan Realm.

The lowest zone in this sequence, the *Arctocephalites spathi* Zone, contains *A. spathi* Poulton, and *A.* cf. *ellipticus* Spath. The overlying *Arctocephalites porcupinensis* Zone contains *A. porcupinensis* Poulton, *A. callomoni* Frebold and *A.* spp. aff. *nudus* Spath. These zones are approximately equivalent with the Boreal Arcticus Standard Zone, although the former may fill a gap below the Arcticus Zone in Greenland. Younger beds, assigned to the *Arctocephalites amundseni* Zone, contain *A. amundseni* Poulton, *A. arcticus* (Whitfield), *A. praeishmae* Poulton, and *A. kigilakhensis* Voronetz. The youngest probable middle Bathonian rocks, the *A. frami* Zone, contain *A. frami* Poulton, *A. belli* Poulton, and the earliest *Arcticoceras* and *Kepplerites*. The *A.* (?) *amundseni* and *A. frami* Zones are correlated with the Greenlandicus Zone of East Greenland.

The lowest beds of probably late Bathonian age are characterized by *Arcticoceras harlandi* Rawson. They were assigned to the *A. harlandi* Zone locally (Poulton 1987) and are here included within the lower Boreal Ishmae Standard Zone, which contains, in its upper part in Canada, *Arcticoceras ishmae* (Keyserling), other late *Arctocephalites* species, *Loucheuxia bartletti* Poulton, several species of *Kepplerites* and *Iniskinites*, *Cadomites* sp., *Oxycerites birkelundi* Poulton, *Parareineckeia* sp., *Choffatia*(?) sp., and various phylloceratids. The youngest late Bathonian zone preserved, the *Cadoceras barnstoni* Zone, is characterized by *Cadoceras barnstoni* (Meek), *C. variabile* Spath, *Loucheuxia bartletti* Poulton, *Kepplerites* spp., including *K.* aff. *K. rosenkrantzi* Spath, *Iniskinites yukonensis* Frebold, *I. variocostatus* (Imlay), *Cadomites*(?) sp., and various phylloceratids.

The lowest Callovian is represented by *Cadoceras bodylevskyi* Frebold, the basis for the *C. bodylevskyi* Zone (Poulton 1987). Younger species present in the area include *C. septentrionale* Frebold, *C. arcticum* Frebold, *C. voronetsae* Frebold, and *C.* (*Stenocadoceras*) *canadense* Frebold (Frebold 1964a). These species are not dated in detail, but suggestions as to their correlations with other Arctic areas are given by Callomon (1984). The *Cadoceras* and associated *Kepplerites* faunas are essentially Boreal or Subboreal in affinity (Callomon 1984; Taylor et al. 1984; Poulton 1987). Throughout most of northern North America, fossils diagnostic of the middle and upper Callovian are either unknown or unproven, suggesting the presence of a regional disconformity. However, *Longaeviceras* cf. *L. stenolobum* (Keyserling), representing the Athleta Standard Zone of the Upper Callovian, is present in the Kingak Shale of northern Alaska (Callomon 1984).

Middle Jurassic microfossils and palynomorphs have been illustrated by Poulton et al. (1982), and the latter have been summarized and zoned by Davies (Poulton 1989a). Bivalves and belemnites are common but not diverse, being represented mainly by *Retroceramus*, *Oxytoma*, *Cylindroteuthis*, *Pachyteuthis*, and *Belemnoteuthis*. They are similar to those described from other Boreal regions. Scaphopods occur here and there.

### *Upper Jurassic*

At the top of the heterogeneous Middle Jurassic clastic succession of the eastern basin margin is a prograded shelf sandstone of early Oxfordian age (upper Richardson Mountains and Aklavik Formations). Late Oxfordian and younger Jurassic rocks in the eastern Brooks-Mackenzie Basin are sandstones of the Porcupine River and North Branch Formations and shales of the Kingak and Husky Formations. The sandstones extend basinward in shallow-marine facies, grading into the Husky Shale to their north and Kingak Shale to the west and south. As much as 3,000 m of southerly derived, Upper Jurassic sandstone turbidites in northwestern Alaska (Campbell 1967) are the first indication of orogenic activity there (Balkwill et al. 1983).

The Lower Oxfordian is represented at several localities by species of *Cardioceras* whose correlations have been suggested by Callomon (1984). *Cardioceras* aff. *scarburgense* (Young & Bird) and aff. *C. alphacordatum* Spath, representing about the Mariae and Cordatum Standard Zones, occur in the Kingak Shale (Frebold et al. 1967) and in the uppermost beds of the Richardson Mountains Formation (Poulton et al. 1982). *C.* aff. *C. distans* (Whitfield) from the Lower Schist Division of central Yukon (Poulton and Tempelman-Kluit 1982) represents the Cordatum Standard Zone, according to Callomon (1984), as does *C.* cf. *hyatti* Reeside from the Kingak Shale of northern Alaska. Callomon (1984) assigned other occurrences of *C.* cf. *hyatti* Reeside, from the Kingak of northern Yukon, to the Middle Oxfordian Densiplicatum Standard Zone.

*Amoeboceras* spp. indet. and *Buchia concentrica* (Sowerby) indicate some part of the Upper Oxfordian or Lower Kimmeridgian in the lower Husky Shale. Only *A.* (*Prionodoceras*) has been identified in any detail. One species was assigned by Callomon (1984) to the Serratum Standard Zone, another to the Glosense Zone, and yet another, in association with *B. concentrica*, to the Rosenkrantzi Standard Zone. Post–early Oxfordian Jurassic rocks are subdivided primarily on the basis of a series of *Buchia* species, described by Jeletzky (1965, 1966, 1973, 1984), who erected a zonation for these characteristically Boreal bivalves in western and Arctic Canada. Ammonites are rare, almost entirely phylloceratids, and other fossils uncommon in the *Buchia*-bearing interval. Foraminifera from the Kimmeridgian and younger Jurassic have been described by Hedinger (in press), and palynomorphs by Fensome (1987). The latter were summarized and zoned by Davies (Poulton 1989a). The Jurassic–Cretaceous boundary falls within the Kingak and Husky Formations, where it is identified primarily by *Buchia* species.

## Ammonite and Buchia zones

### *Hettangian*

1. Planorbis (Standard) Zone – *Psiloceras* cf. *johnstoni* (Sowerby), *Psiloceras* spp.

### *Sinemurian*

1. Bucklandi Zone – *Coroniceras* sp.
2. Semicostatum Zone – *Arnioceras* cf. *douvillei* (Bayle)
3. Oxynotum Zone – *Oxynoticeras oxynotum* (Quenstedt), *Gleviceras* sp., *Aegasteroceras* (*Arctoasteroceras*) *jeletzkyi* Frebold, *Aegasteroceras* sp.
4. Raricostatum Zone – *Echioceras aklavikense* Frebold

### *Pliensbachian*

1. Margaritatus Zone – *Amaltheus stokesi* (Sowerby), *A. bifurcus* Howarth

*Toarcian*

1. Tenuicostatum Zone(?) – *Ovaticeras* cf. *ovatum* (Young & Bird), *Collina*(?) aff. *simplex* Fucini
2. Bifrons Zone(?) – *Dactylioceras commune* (Simpson)
3. Thouarsense Zone(?) – *Pseudolioceras lectum* (Simpson), *Pseudolioceras* sp.

*Aalenian*

1. Opalinum Zone(?) – *Leioceras* aff. *opalinum* (Reinecke), *Tmetoceras* sp.
2. *Erycitoides howelli* Zone – *E. howelli* (White), *Pseudolioceras* aff. *whiteavesi* (White), *Planammatoceras* sp.

*Bajocian*

1. *Arkelloceras tozeri* Zone – *Arkelloceras tozeri* Frebold, *A. mclearni* Frebold, *A. elegans* Frebold
2. Borealis Zone – *Cranocephalites borealis* (Spath), *C. warreni* Frebold

*Bajocian or Bathonian*

1. Indistinctus Zone – *Cranocephalites indistinctus* Callomon
2. Pompeckji Zone – *Cranocephalites* spp. aff. *pompeckji* (Madsen), *C. maculatus* Spath, and *C. vulgaris* Spath

*Bathonian*

1. *Arctocephalites spathi* Zone – *A. spathi* Poulton, *A.* cf. *ellipticus* Spath
2. *Arctocephalites porcupinensis* Zone – *A. porcupinensis* Poulton, *A.* aff. *sphaericus* Spath, *A. callomoni* Frebold, *A.* aff. *nudus* Spath
3. *Arctocephalites amundseni* Zone – *A. amundseni* Poulton, *A.* aff. *sphaericus*, *A. porcupinensis* Poulton, *A. callomoni* Frebold, *A.* aff. *nudus*, *A. praeishmae* Poulton, *A. arcticus* (Whitfield), *A. kigilakhensis* Voronetz, *Iniskinites* sp.
4. *Arctocephalites frami* Zone – *A. frami* Poulton, *A.*(?) *belli* Poulton, *Arcticoceras* sp., *Kepplerites* sp., *Iniskinites* sp.
5. Ishmae Zone – *Arcticoceras ishmae* (Keyserling), *A. harlandi* Rawson, *Arctocephalites*(?) *belli* Poulton, *Loucheuxia bartletti* Poulton, *Kepplerites* sp., *Iniskinites* sp., *Cadomites* sp., *Oxycerites birkelundi* Poulton, *Parareineckeia* sp., *Choffatia*(?) sp.
6. *Cadoceras barnstoni* Zone – *C. barnstoni* (Meek), *C. variabile* Spath, *Paracadoceras* sp., *Loucheuxia bartletti* Poulton, *Kepplerites* aff. *rosenkrantzi* Spath, *Iniskinites variocostatus* (Imlay), *I. yukonensis* Frebold, *Cadomites* sp.

*Callovian*

1. *Cadoceras bodylevskyi* Zone – *C. bodylevskyi* Frebold
2. Athleta Zone – *Longaeviceras* cf. *stenolobum* (Keyserling), *Grossouvria* sp.

*Oxfordian*

1. Mariae Zone – *Cardioceras* spp. aff. *scarburgense* (Young & Bird)
2. Cordatum Zone – *Cardioceras alphacordatum* Spath, *C.* aff. *distans* (Whitfield), *C.* cf. *hyatti* Reeside
3. Densiplicatum Zone – *Cardioceras* (*Maltoniceras*) sp.
4. Glosense Zone – *Amoeboceras*(?) sp.
5. Serratum Zone – *Amoeboceras* (*Prionodoceras*) sp.
6. Rosenkrantzi Zone – *Amoeboceras* sp., *Buchia concentrica* (Sowerby)

*Oxfordian–Kimmeridgian*

1. *Buchia concentrica* Zone – *B. concentrica* (Sowerby)

*Kimmeridgian–Tithonian*

1. *Buchia mosquensis* Zone – *B. mosquensis* (von Buch)

*Tithonian*

1. *Buchia piochii* Zone – *B. piochii* (Gabb)
2. *Buchia fischeriana* Zone – *B. fischeriana* (d'Orbigny)

## WESTERN INTERIOR CANADA[3]

### Introduction

The Jurassic system in Western Interior Canada (Springer, MacDonald, and Crockford 1964; Stott 1970; Hall 1984; Poulton 1984) records clearly the changes in tectonic and depositional regimes of the Western Canada Basin caused by accretion of allochthonous terranes to the western edge of North America (Figure 4.4). Jurassic manifestations of these events comprise the earliest pulses of the Columbian Orogeny and are clearly seen in the cratonic Jurassic record beginning in the early Late Jurassic. What had been a tectonically quiet region, with relatively stable environments of deposition (Figure 4.5A), was abruptly transformed into a highly mobile region with orogenic uplift in the west and an associated narrow, arcuate foredeep, the Alberta Trough, to its east (Figure 4.5B). The Early and Middle Jurassic conditions are recorded in a thin sequence of craton-derived clastics, with units of chemical deposits, punctuated by many disconformities. The Late Jurassic conditions are recorded by a thick, apparently continuous sequence of westerly derived clastics. Both of these major sequences thicken and become more complete westward, that is, toward the cratonic margin in the lower sequence, and into the

[3] By T. P. Poulton and R. L. Hall; Geological Survey of Canada Contribution No. 32689.

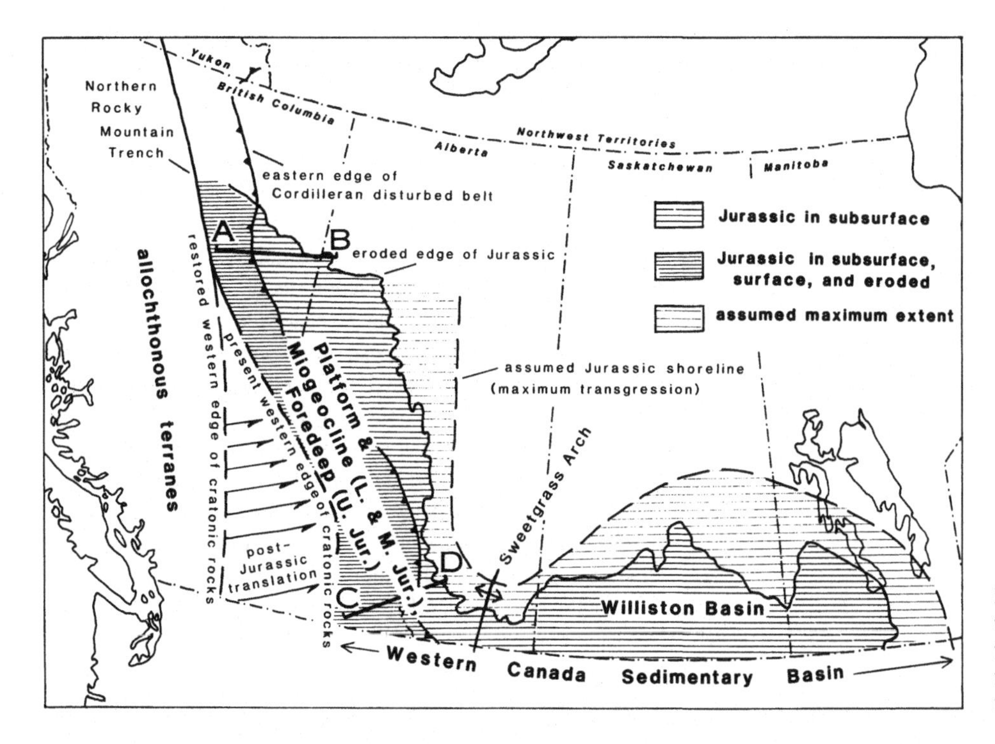


**Figure 4.4. Index map showing distributions of Jurassic rocks, major tectonic elements during Jurassic time, and locations of cross sections A–B and C–D.**

orogenic foredeep in the higher sequence (Figures 4.6 and 4.7). Erosion at the sub-Cretaceous unconformity and at unconformities within the Jurassic results in the eastern shoreline facies being almost entirely unknown.

Western source areas for the Upper Jurassic clastics are clearly indicated by the conglomerates in the westernmost facies of the latest Jurassic (Poulton 1984; Stott 1984; Gibson 1985). Radiometric studies of batholiths in southeastern British Columbia (Archibald et al. 1983), some of which cut both craton and allochthonous terranes, give evidence of Middle to Late Jurassic and Early Cretaceous uplifts of source terranes there. Plate collisions in that area may have created highlands as early as Middle Jurassic time, and Bathonian fine clastics in southeastern British Columbia may have been derived from them (Poulton 1984; Stronach 1984). Whatever lay to the west of the epicratonic sedimentary prism in Early and Middle Jurassic (preorogenic) times (Figure 4.5A) can no longer be seen at most latitudes because of its subsequent removal by large-scale northward transcurrent movements and/or erosion related to uplift of the Columbian Orogeny.

During preorogenic phases the western seas extended well into the interior of North America, forming the Williston Basin, the northern part of which is present in the subsurface of southern Canada. This basin existed from Ordovician until about Late Jurassic time (i.e., until about the same time as western accretionary tectonics became a dominant feature in the western Canada former miogeocline). In both the Williston Basin and Alberta Trough foredeep areas, expansion of the basins and accelerated basin subsidence near the beginning of the Late Jurassic preceded filling of the basins to sea level in later Jurassic time.

Regional variations on the platform include the pre–Late Jurassic uplift of a broad area across northeastern British Columbia and northwestern Alberta, straddling what earlier had been the Peace River Arch, and the Middle Jurassic inundation of the Sweetgrass Arch in southeastern Alberta. The Williston Basin (Figure 4.4) exhibits an increased rate of subsidence in the late Middle Jurassic, and it ceased to exist as a subsiding tectonic element prior to the Cretaceous.

Between Carboniferous and Middle Jurassic times, the western and northwestern margins of the Williston Basin were defined by the Sweetgrass Arch, separating it from the more westerly parts of the platform in Alberta. Whereas the Early and early Middle Jurassic platform deposits to the west consist of carbonates, clastics, cherts, and phosphorites, probable equivalents in the Williston Basin are evaporites and redbeds. The Sweetgrass Arch was transgressed in Middle Jurassic time and became the site of deposition of marine clastics and carbonates. The arch ceased to be of importance in the Early Cretaceous, when it was blanketed by Mannville sandstones like those to its east and west (Poulton 1984).

According to current interpretations, the Jurassic of western Canada extended from about 45°N to 70°N latitude, on a globe whose climatic belts were less strongly differentiated than those of today. Jurassic climates of Western Interior Canada were cooler than those of the southern United States or southern Europe, as indicated by the scarcity of limestone deposits and rarity of reef-forming organisms.

The Jurassic record of autochthonous western Canada is one of progressive, if intermittent, flooding of the cratonic margins, as it was worldwide (Hallam 1978, 1981; Vail and Todd 1981). Widespread Late Jurassic regression followed an influx of sediment eroded from the uplifted Columbian Orogeny to the west. The general transgressive trend characteristic throughout the earlier part of the Jurassic in western Canada was accomplished in at least eight major transgressive events, at about earliest Hettangian,

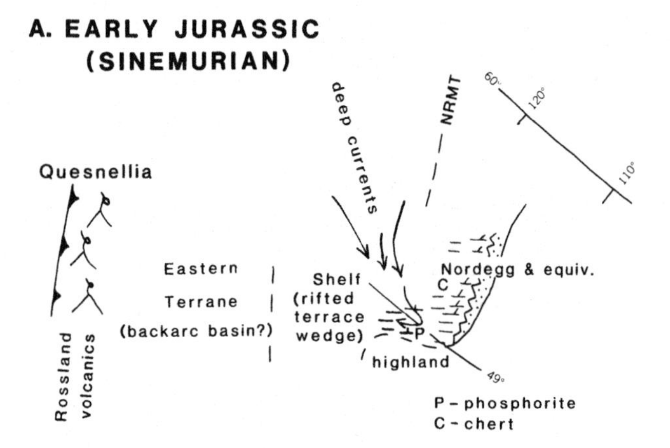


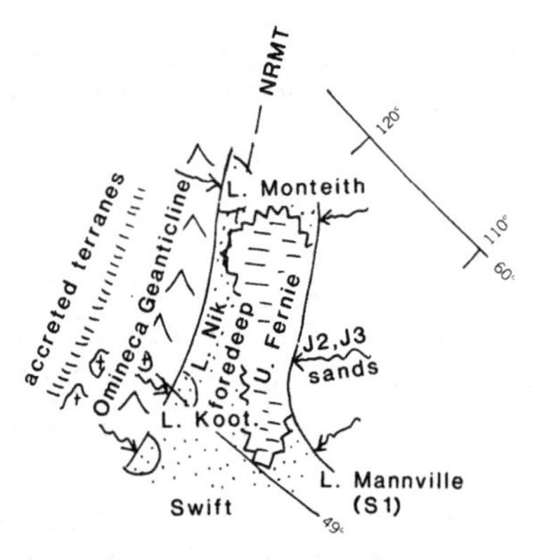


**Figure 4.5. Schematic diagrams showing (A) the general position of the allochthonous Quesnellia Terrane relative to western North America during Sinemurian (pre-accretion) time and the position of the back-arc basin between the two and (B) the disposition of accreted terranes, the Omineca Geanticline, and the distribution of formations in the foredeep to the east of the orogenic belt in Kimmeridgian (post-collision) time. For reference to modern geography, the present location of the Northern Rocky Mountain Trench (NRMT) and two lines of latitude and longitude are shown.**

earliest Sinemurian, latest Sinemurian, late Pliensbachian, early Toarcian, late early Bajocian, late Bathonian, and earliest Oxfordian times. A few of these events correspond to some of the "worldwide eustatic" events postulated by Hallam (1978) and Vail and Todd (1981), but there are also some pronounced discrepancies that reflect local tectonic events (Poulton 1988).

### Lower Jurassic

The base of the Jurassic in the platform/miogeocline is disconformable, overlying Paleozoic rocks in more easterly sections and Triassic rocks farther west. The Lower Jurassic rocks comprise the lower Fernie Formation (Frebold 1957b; Hall 1984; Poulton 1984).

The oldest dated Jurassic rocks are early Hettangian, characterized by *Psiloceras calliphyllum* (Neumayr) in northeastern British Columbia (Tozer 1982b), and overlying beds containing schlotheimiids(?). Over most of western Alberta and eastern British Columbia the basal Jurassic unit is a heterogeneous, variably pebbly, cherty or phosphatic limestone-sandstone-shale unit. It varies in thickness (0.5 m to more than 30 m) and is primarily Sinemurian in age. A banded chert and limestone in west-central Alberta, the Nordegg Member, contains sedimentary structures in its western outcrops suggesting deposition on a slope, perhaps facing the western continental margin. Its eastern facies is a massive cherty limestone, varying to nearshore sandstone near its eastern limit. This unit disappears westward into a basal Jurassic submarine unconformity and is replaced northward by basinal organic-rich limestones (Poulton, Tittemore, and Dolby 1990). Farther west and south, the basal Jurassic unit contains *Asteroceras* cf. *stellare* (Sowerby) and *Epophioceras* cf. *breoni* (Reynes) in a thin conglomerate dated as Middle Sinemurian (Hall 1987). Basal Jurassic units to the east, north, and south are undated, but all are generally assigned to the Nordegg Member.

In southwesternmost Alberta and southeastern British Columbia, phosphorite deposits with early Sinemurian *Arnioceras* (Warren 1931) occur at the base of the Jurassic. The phosphorites imply access to upwelling oceanic currents, which places constraints on the plate-tectonic models relating to accretion of allochthonous terranes from the west during the Jurassic (Figure 4.5A) (Poulton and Aitken 1989). Younger beds in these southern areas contain Upper Sinemurian *Eoderoceras, Gleviceras,* and *Crucilobiceras,* and lowest Pliensbachian *Phricodoceras* cf. *taylori* (Sowerby) (Frebold 1969).

A mixed Tethyan/Boreal ammonite fauna, including *Amaltheus* cf. *stokesi* (Sowerby), ?*Tiltoniceras propinquum* (Whiteaves), ?*Aveyroniceras,* and ?*Protogrammoceras* sp., occurs in the Red Deer Member of the Fernie Formation, representing the Upper Pliensbachian (Frebold 1966, 1969; Hall 1987).

A regional euxinic black shale, the Poker Chip Shale, overlies older beds paraconformably. Its ammonite faunas indicate that it spans most of the Toarcian, although Lower Toarcian species predominate. The lower fauna, including *Harpoceras* cf. *falciferum* (Sowerby), *Hildaites* cf. *serpentiniformis* Buckman, *Dactylioceras* cf. *athleticum* (Simpson), *Harpoceratoides* sp., and *Polyplectus* cf. *subplanatus* (Oppel), has been correlated with the Falciferum Zone (Frebold 1976; Hall 1987). In the upper parts of the unit are poorly preserved forms that have been compared with Upper Toarcian *Pleydellia, Haugia, Phlyseogrammoceras,* and *Denckmannia* (Frebold 1969, 1976; Hall 1987).

Lower Jurassic rocks are not known in the Williston Basin, and the lowest probable Jurassic beds there have not been dated. The base of the Jurassic apparently is located within a redbed/evaporite sequence that probably contains rocks of Triassic and Jurassic age (Watrous and Amaranth Formations) (Poulton, 1984, 1988).

The Lower Jurassic ammonites in Western Interior Canada mainly represent taxa that can be readily compared to northwestern European forms. Bivalves, brachiopods, crinoids, and other fossils are present in the shelly facies. Autochthonous Lower Jurassic sequences in western Canada have not yielded any diagnos-

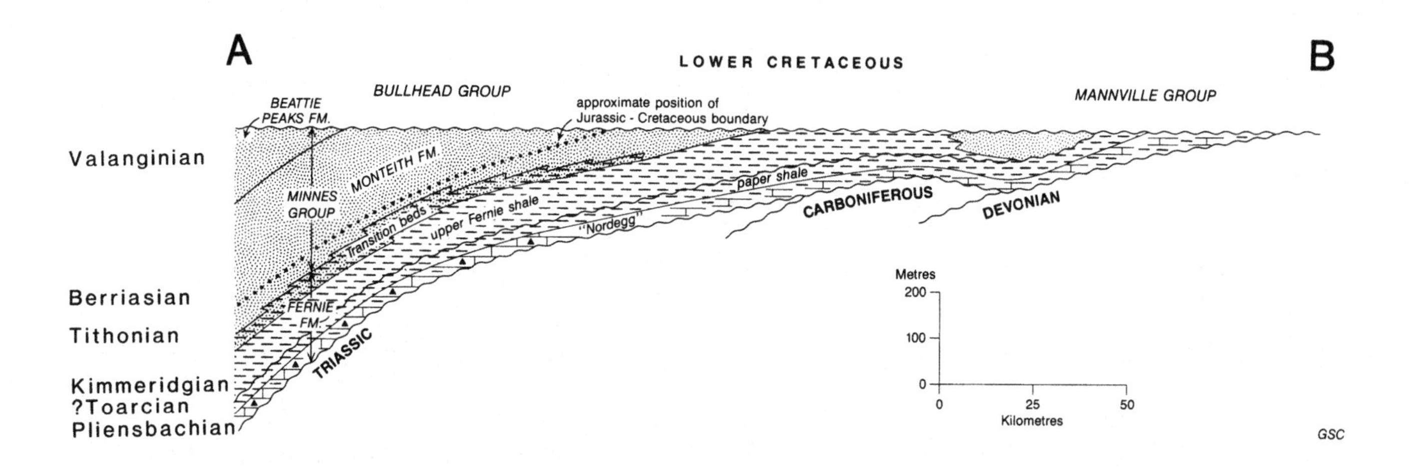


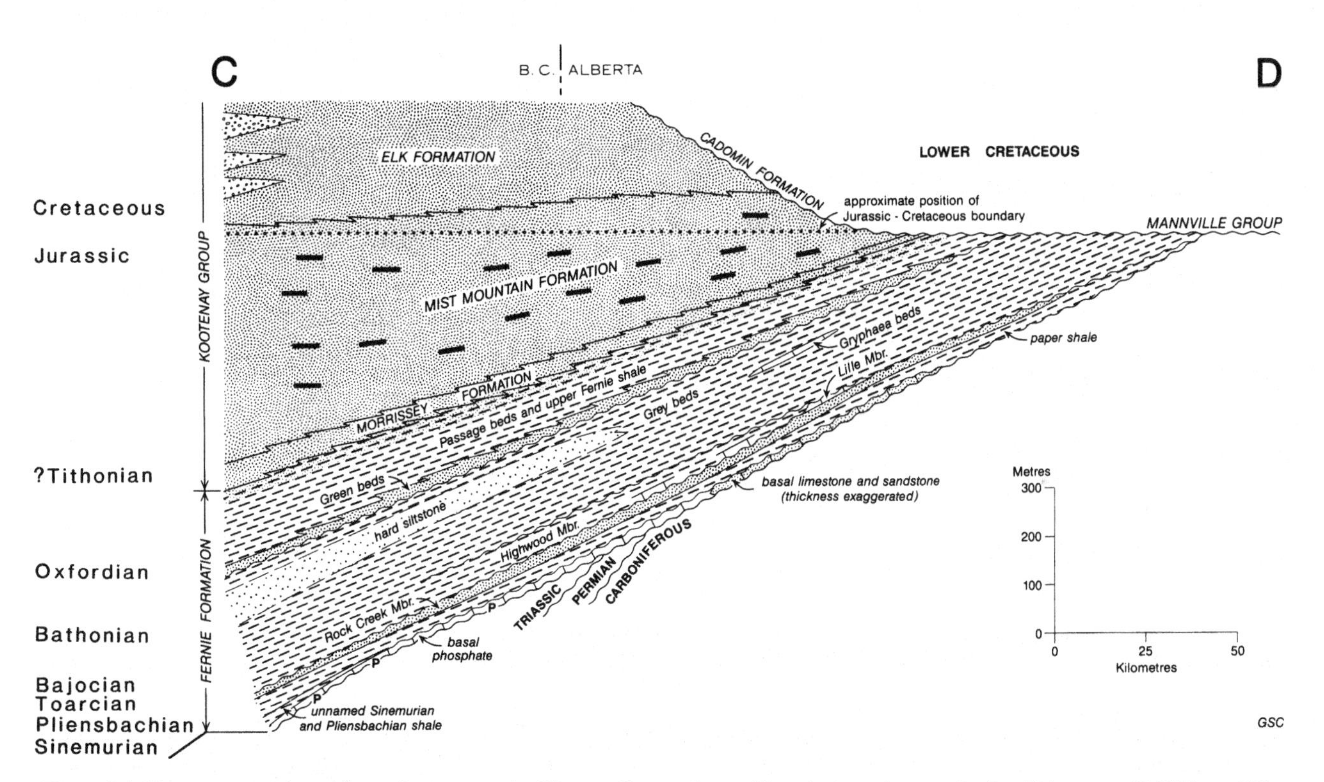


**Figure 4.6. East–west stratigraphic sections across the Western Canada Platform (Lower and Middle Jurassic) and Foredeep (Upper Jurassic): A–B, southeastern British Columbia and southwestern Alberta. C–D, northeastern British Columbia and northwestern Alberta. Note the large increase in the thicknesses of Middle and Upper Jurassic strata compared with that of the Lower Jurassic. Locations of sections shown in Figure 4.4.**

tic microfauna thus far. Palynomorphs have been described by Pocock (1970, 1972).

### Middle Jurassic

The Middle Jurassic units of the platform/miogeocline are members of the middle Fernie Formation. Middle Jurassic beds in Alberta and eastern British Columbia are known with certainty only south of about 54°N latitude, the area to the north presumably having undergone Middle or post–Middle Jurassic uplift (Davies and Poulton 1986).

The Aalenian is known in Alberta from palynomorphs from the Poker Chip Shale, and perhaps also the Rock Creek Member (Poulton et al. 1990).

Early Bajocian quartzose sandstones and siltstones of the Rock Creek Member overlie the Poker Chip Shale in southwestern

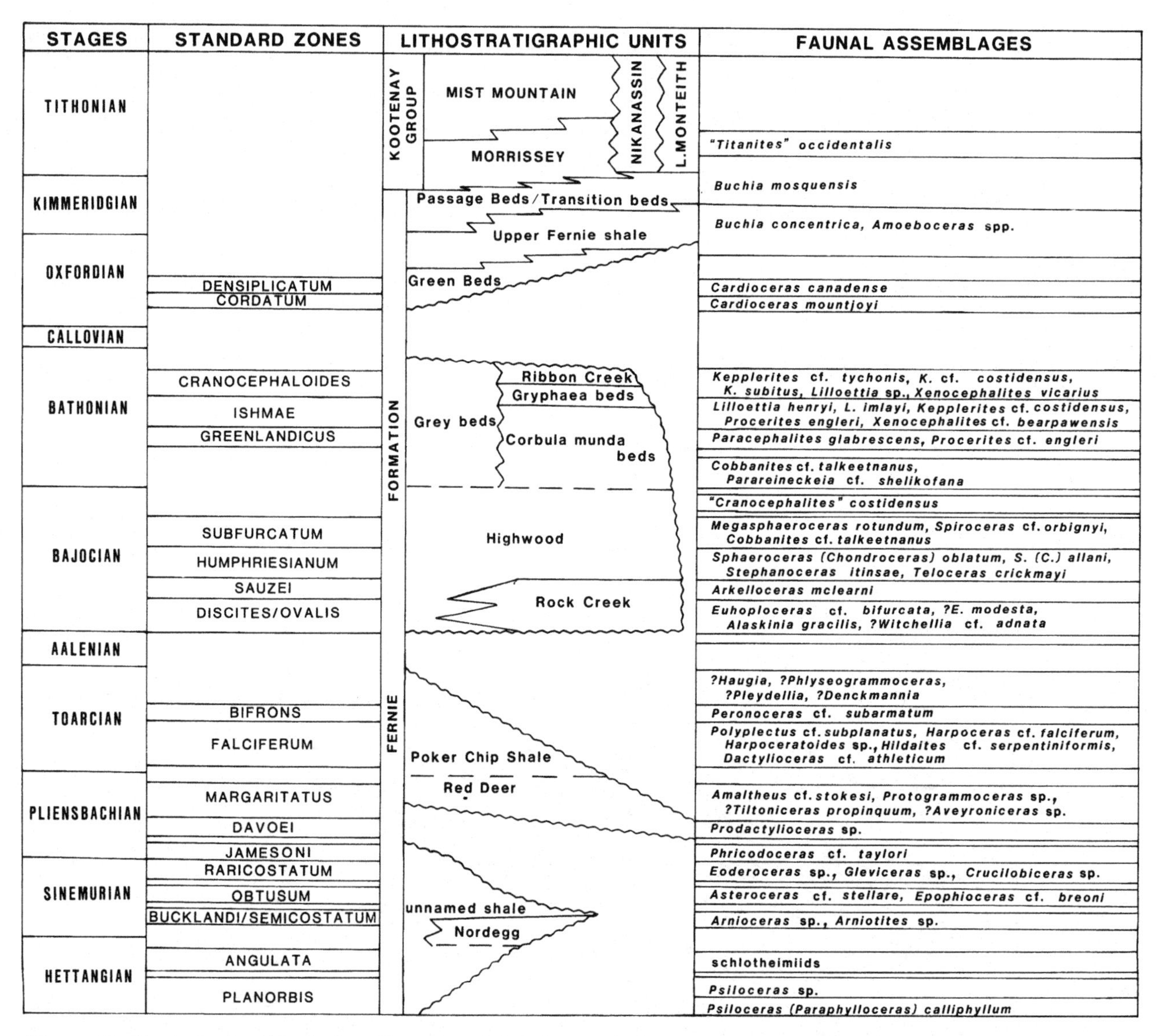


**Figure 4.7. Chart showing lithostratigraphic units and faunal assemblages in the Western Canada Sedimentary Basin. Only standard zones that are represented in the basin are shown. Stages not drawn to scale.**

Alberta and southeastern British Columbia and grade toward the west into somewhat finer, bioturbated sediments (Frebold 1957a; Stronach 1984). The Rock Creek Member has yielded a few poorly preserved ammonites at several localities, identified as *Euhoploceras* cf. *bifurcata* (Westermann), *E.*(?) *modesta* (Buckman), *Alaskinia gracilis* (Whiteaves), and *Witchellia*(?) cf. *adnata* Imlay (Frebold 1976), indicating the earliest Bajocian and, in part, the *Docidoceras widebayense* Zone of southern Alaska erected by Westermann (1969). The *Parabigotites crassicostatus* Zone, coeval with the Sauzei Zone, is indicated by an isolated find of *Arkelloceras* (Westermann 1964a). Overlying shales (Highwood Member) and black, phosphatic and oolitic limestones contain an abundant shelly fauna, including the ammonites *Sphaeroceras* (*Chondroceras*) *oblatum* (Whiteaves), *S.* (*C.*) *allani* (McLearn), *Stephanoceras itinsae* (McLearn), and *Teloceras crickmayi* (Frebold) (Hall and Westermann 1980; Hall 1984). This assemblage represents the *Sphaeroceras* (*Chondroceras*) *oblatum* Zone, which has been correlated with the middle and upper parts of the Humphriesianum Zone (Hall and Westermann 1980). (For the status of North American Bajocian zones, see Chapter 12.)

Higher in the Highwood Member, the Rotundum Zone, with *Megasphaeroceras rotundum* Imlay, *Cobbanites* (= *Leptosphinctes*) cf. *talkeetnanus* Imlay, *Spiroceras* cf. *orbignyi* (Baugier & Sauze), and stephanoceratids is coeval with the Upper Bajocian Subfurcatum Zone (Hall and Stronach 1981).

The Bathonian stage is well represented, although the chronological succession of these faunas is not yet fully understood. Whereas most of the Bajocian ammonite genera of Western Interior Canada are cosmopolitan (but not Boreal), the Bathonian species are a mixture of endemic (e.g., *Paracephalites*), East Pacific (e.g., *Iniskinites*, *Lilloettia*), and cosmopolitan (e.g., *Leptosphinctes*) forms (Callomon 1984; Taylor et al. 1984; Hall

1988). Additionally, some widespread Subboreal elements, such as *Kepplerites*, appear and are of singular importance for interbasinal correlations.

The earliest of these faunas with East Pacific and/or Boreal aspects is in the upper parts of the Highwood Member, where *Cranocephalites costidensus* Imlay occurs; elsewhere this taxon has been placed in the Upper Bajocian (Westermann 1981; Callomon 1984) or Lower Bathonian (Imlay 1980b). From a nearby section a single horizon in the overlying Grey Beds contains nuclei of *Cobbanites* cf. *talkeetnanus* Imlay and *Parareineckeia* cf. *shelikofana* (Imlay), indicating a probable early Bathonian age (Hall 1988). Elsewhere in southwestern Alberta this part of the sequence is occupied by the *Corbula munda* Beds, yielding the *Paracephalites glabrescens* (Buckman) fauna, and the superjacent *Gryphaea* Bed, with *Warrenoceras henryi* (Meek & Hayden), *Warrenoceras imlayi* (Frebold), *Procerites engleri* Frebold, *Xenocephalites* cf. *bearpawensis* Imlay, *Kepplerites* cf. *costidensus* (Imlay), and small, indeterminate perisphinctids (Frebold 1963; Hall 1988). Both faunas were placed in the Lower to Middle Bathonian by Callomon (1984). Overlying dark gray shales with large gray concretions and orange-reddish clay bands [Ribbon Creek Member of Stronach (1984)] contain, near their base, *Kepplerites* cf. *tychonis* Ravn, suggesting the lowermost Upper Bathonian (Callomon 1984; Hall 1988). At higher levels, other occurrences of kepplleritids with East Pacific genera further suggest correlation with the Upper Bathonian: *Kepplerites* cf. *costidensus* (Imlay), and an association of *Kepplerites* cf. *subitus* (Imlay), *Lilloettia* sp. indet., and *Xenocephalites vicarius* Imlay (Hall 1989).

The Bajocian to Lower or Middle Bathonian rocks in the Williston Basin are a series of shallow-marine limestones, sandstones, and shales separated by many disconformities (Gravelbourg and Shaunavon Formations). These rocks thicken and vary to more argillaceous facies southeastward in Saskatchewan toward the basin center. The younger Jurassic rocks (Vanguard Group) are mainly shale, displaying two major coarsening-upward cycles, culminating in late Middle Jurassic (Roseray Formation) and Upper Jurassic (lower Success Formation) sandstones, respectively, each overlain by a disconformable surface. The thicknesses of these cycles indicate increased subsidence rates for the basin from about late Bathonian time until the disappearance of the basin as a depositional element in the Late Jurassic (Poulton 1984, 1988).

The Callovian is perhaps indicated in the Williston Basin by *Kosmoceras knechteli* Imlay and *Kepplerites* cf. *rockymontanus* Imlay (Frebold 1963; Paterson 1968), although these specimens could possibly still be late Bathonian in age. The Callovian is not known with certainty from the western platform. In both areas, the upper parts of the Middle Jurassic are truncated to varying extents below a sub–Upper Jurassic unconformity.

Brooke and Braun (1972) erected a sequence of microfaunal zones (I to VII), based on foraminifera, ostracods, and charophytes, for subsurface units in southern Saskatchewan (the Gravelbourg, Shaunavon, and Vanguard Formations), ranging from late Bajocian to Kimmeridgian. They made a tentative comparison of their faunal associations with the zonal scheme erected by Pocock (1962, 1970, 1972) for these strata based on palynomorphs.

Weihmann (1964) described microfaunas from parts of the Fernie Formation (Poker Chip Shale, Grey Beds, Green Beds, and Passage Beds) at two localities in southeastern British Columbia. These included radiolarians, ostracods, and foraminifera; faunal associations were listed, but no zonation was proposed.

**Upper Jurassic**

A thin berthierine("chamosite")-rich unit in southwestern Alberta and adjacent areas contains *Cardioceras alphacordatum* Spath, *C. mountjoyi* Frebold, *C. canadense* (Whiteaves), *Goliathiceras* cf. *crassum* (Reeside), a giant peltoceratine, and *Buchia concentrica* (Sowerby) (Frebold 1957b; Frebold, Mountjoy, and Reed 1959; Poulton 1989c). This unit, the "Green Beds," is a condensed, shallow-marine shelf deposit, arguably the first deposit of the Alberta Trough. It overlies a regional disconformity that has been interpreted to record the passage of a tectonic forebulge (Davies and Poulton 1986). The ammonites indicate ages spanning the Oxfordian and suggest that the Green Beds are a widespread, diachronous complex of several discontinuous units.

The overlying Upper Jurassic beds in southwestern Alberta and southeastern British Columbia comprise a foredeep fill, involving a shallowing- and coarsening-upward succession of marine to nonmarine clastics. This sequence, initiated about Kimmeridgian time, consists of marine shale in the lower part, succeeded upward by shales and siltstones, with increasing amounts of sandstone, exhibiting progressively shallower depositional characteristics. *Buchia concentrica* (Sowerby) in the "Passage Beds," a series of thin-bedded strata low in the sequence of west-central Alberta (Frebold 1957b; Frebold et al. 1959), indicates a late Oxfordian or early Kimmeridgian age. In northeastern British Columbia, foraminifera and palynomorphs indicate that the Upper Jurassic shales disconformably overlie Lower Jurassic strata (Brooke and Braun 1981; Davies and Poulton 1986). A thin sheet of Upper Jurassic platformal shales and sandstones similarly lies upon Lower Jurassic (and Aalenian) strata in northern and north-central Alberta.

The Upper Jurassic shales are overlain by marginal-marine and then nonmarine coaly sandstones of the Kootenay Group in southwestern Alberta and southeastern British Columbia. The youngest marine unit is the basal Kootenay Sandstone, which contains the enigmatic large ammonite "*Titanites*" *occidentalis* Frebold (Frebold 1957b; Westermann 1966; Callomon 1984) whose approximately latest Jurassic age is supported by palynology (Poulton 1984). Approximate equivalents farther north are the marine "Transition Beds" of the upper Fernie Formation and the basal Monteith Formation in northeastern British Columbia (Stott 1967), and the lower Nikanassin Formation in central western Alberta (Poulton 1984). *Notostephanus?*, *Proniceras*, *Buchia piochii* (Gabb), *Buchia* aff. *lahuseni* (Pavlow), *B.* aff. *terebratuloides* (Lahusen), and *B.* aff. *russiensis* (Pavlow) indicate the Tithonian in northeastern British Columbia (Jeletzky 1984).

The basal Kootenay Sandstone is a coastal beach-dune-barrier and/or delta-front sheet-sand complex (Gibson 1985), indicating that the foredeep had been filled by that time. The overlying unit, straddling the Jurassic–Cretaceous boundary, is a diverse, mainly nonmarine, fluviodeltaic sandstone-siltstone-shale complex with important coal deposits, and with chert-pebble conglomeratic beds in the western localities indicating proximity to the orogenic source terrane.

The Upper Jurassic marine macrofaunas are mainly Boreal, a product of the southward expansion of the marine Boreal Realm in the late Middle Jurassic. Upper Jurassic microfossils from northeastern British Columbia, also largely Boreal, were described and zoned by Brooke and Braun (1981). Spores and pollen are the primary tools for dating and correlating the nonmarine beds of the lower Kootenay Group in southwestern Alberta and southeastern British Columbia. The Upper Jurassic Boreal palynomorphs from marine beds of the upper Fernie Formation in northeastern British Columbia have been illustrated and discussed by Davies and Poulton (1986). The change to entirely Boreal affinities for the fossils in approximately early Oxfordian time conforms with filling or elevation of the sedimentary basins of Western Interior Canada, so that southerly connections were lost, and subsidence of the northerly connected Alberta Trough foredeep. Plant macrofossils have also been used for biostratigraphy in the Upper Jurassic of the Alberta Trough (Berry 1929; Bell 1944, 1946, 1956; Brown 1946).

Only a thin sequence of Upper Jurassic strata is preserved in Williston Basin (upper Vanguard Group and Success Formation) (Christopher 1974), mainly derived from the low-lying craton to the north, east, and southeast. Lower parts are apparently conformable with older Jurassic beds, and upper parts are truncated by the sub-Cretaceous unconformity that terminated the basin (Poulton 1984).

### Ammonite zones and assemblages

*Hettangian*

1. Planorbis (Standard) Zone – *Psiloceras (Paraphylloceras) calliphyllum* (Neumayr)

*Sinemurian*

1. Bucklandi/Semicostatum Zones – *Arnioceras* sp., *Arniotites* sp.
2. Obtusum Zone – *Asteroceras* cf. *stellare* (Sowerby), *Epophioceras* cf. *breoni* (Reynes)
3. Raricostatum Zone – *Eoderoceras, Gleviceras, Crucilobiceras*

*Pliensbachian*

1. Jamesoni Zone – *Phricodoceras* cf. *taylori* (Sowerby)
2. Margaritatus Zone – *Amaltheus* cf. *stokesi* (Sowerby), ?*Tiltoniceras propinquum* (Whiteaves), ?*Protogrammoceras* sp., ?*Aveyroniceras* sp.

*Toarcian*

1. Falciferum/?Bifrons Zones – *Polyplectus* cf. *subplanatus* (Oppel), *Harpoceras* cf. *falciferum* (Sowerby), *Harpoceratoides* sp., *Hildaites* cf. *serpentiniformis* Buckman, *Dactylioceras* cf. *athleticum* (Simpson)
2. Bifrons Zone – *Peronoceras* cf. *subarmatum* (Young & Bird)
3. Upper Toarcian – ?*Haugia*, ?*Phlyseogrammoceras*, ?*Pleydellia*, ?*Denckmannia*

*Bajocian*[4]

1. Discites/Ovalis Zones – *Euhoploceras* cf. *bifurcata* (Westermann), ?*E. modesta* (Buckman), *Alaskinia gracilis* (Whiteaves), *Witchellia?* cf. *adnata* Imlay
2. *Parabigotites crassicostatus Zone* – *Arkelloceras* cf. *mclearni* Frebold (= Sauzei Zone)
3. *Sphaeroceras* Zone – *Sphaeroceras (Chondroceras) oblatum* (Whiteaves), *S. (C.) allani* (McLearn), *Stephanoceras itinsae* (McLearn), *Teloceras crickmayi* (Frebold)
4. *Megasphaeroceras rotundum* Zone – *Megasphaeroceras rotundum* Imlay, *Cobbanites* cf. *talkeetnanus* Imlay, *Spiroceras* cf. *orbignyi* (Baugier & Sauze)
5. *Cranocephalites costidensus* Imlay

*Bathonian*

1. ?Lower Bathonian – *Cobbanites* cf. *talkeetnanus* Imlay, *Parareineckeia* cf. *shelikofana* (Imlay)
2. ?Greenlandicus Zone – *Paracephalites glabrescens* Buckman, *Procerites* cf. *engleri* Frebold
3. ?Ishmae Zone – *Warrenoceras henryi* (Meek & Hayden), *Procerites engleri* Frebold, *Warrenoceras imlayi* Frebold, *Xenocephalites* cf. *bearpawensis* Imlay, *Kepplerites* cf. *costidensus* (Imlay)
4. ?Cranocephaloides Zone – *Kepplerites* cf. *tychonis* Ravn, *K.* cf. *costidensus* (Imlay), *Kepplerites* cf. *subitus* (Imlay), *Lilloettia* sp., *Xenocephalites vicarius* Imlay
5. ?Calyx Zone – *Kepplerites* cf. *rockymontanus* Imlay

*Oxfordian*

1. Cordatum Zone – *Cardioceras mountjoyi* Frebold
2. Densiplicatum Zone – *Cardioceras canadense* (Whiteaves)

*?Tithonian*

1. *"Titanites" occidentalis* Frebold

## WESTERN INTERIOR UNITED STATES[5]

### Introduction

The principal regional tectonic features (Figure 4.8) that were active and influenced sedimentation during Jurassic time include the following:

[4] For status of zones, see Chapter 12.

[5] By J. A. Peterson and R. L. Hall.

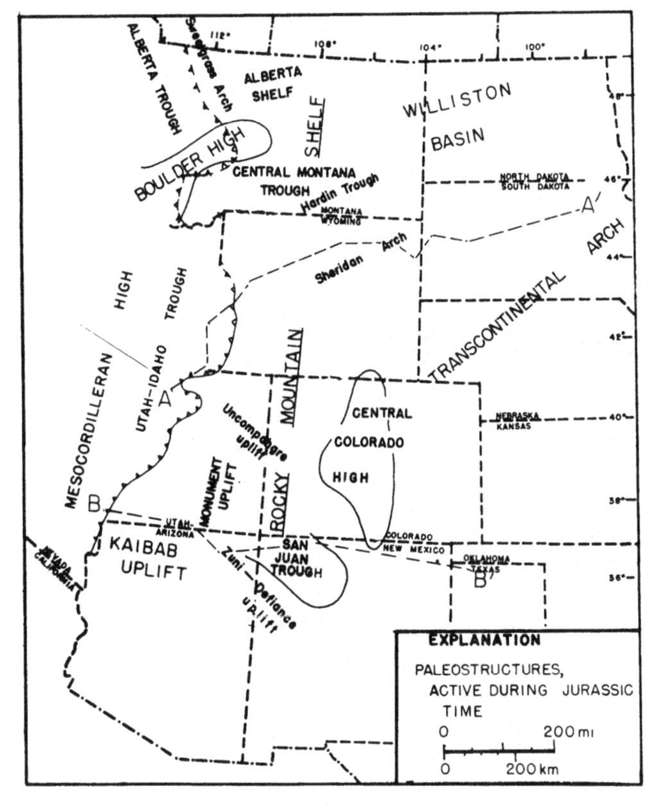


**Figure 4.8. Map showing main paleotectonic features active in Jurassic time. Cross-section lines for Figures 4.10 and 4.11 are shown.**

1. Williston Basin, an area of gentle subsidence throughout most of Paleozoic and pre-Cretaceous time.
2. The Rocky Mountain Shelf was the Paleozoic and Mesozoic epicratonic shelf.
3. The Mesocordilleran High, a north–south emergent, but low-lying, region extending from eastern Nevada to central Idaho, east of the Paleozoic Antler orogenic belt.
4. Alberta Trough, the orogenic foredeep extending southward from northeastern British Columbia into northwestern Montana.
5. The Utah–Idaho Trough, a strongly subsiding north–south trough active during Jurassic time, located just east of the Mesocordilleran High, interpreted as a foredeep associated with the early stages of the Sevier Orogenic Belt; perhaps a southern equivalent of the Alberta Trough. Local tectonic features include the Boulder High (Belt Island trend) in southwestern Montana, which separated the Alberta Trough from the Utah–Idaho Trough, the Sweetgrass Arch in northwestern Montana, the Uncomphagre Uplift in west-central Colorado, and Central Colorado Uplift, the Monument Uplift in southeastern Utah, the Zuni-Defiance Uplift in northeastern Arizona and northwestern New Mexico, the San Juan Trough in northwestern New Mexico, the Hardin Trough in south-central Montana, and the Sheridan Arch in north-central Wyoming (Figure 4.8).

The Jurassic constitutes a sequence of marine and continental clastic, carbonate, and evaporite strata ranging in thickness from more than 1,500 m in northern Utah and southeastern Idaho to less than 300 m over most of the Rocky Mountain Shelf (Figures 4.9–4.11). In parts of the northern Rocky Mountain region the Jurassic is underlain by rocks ranging from Early to Late Paleozoic. South of about central Montana the Jurassic rests on Triassic almost everywhere except central Colorado.

Regional studies of Jurassic rocks include those by Imlay (1948, 1952, 1957, 1967, 1980a,b), Stokes (1944, 1963), Nordquist (1955), Craig et al. (1955), McKee et al. (1956), J. Peterson (1957, 1972, 1986), Harshbarger, Repenning, and Irwin (1957), Pipiringos (1968), and Pipiringos, Hail, and Izett (1969).

The Jurassic sediments of the Western Interior United States were formed in a succession of six major depositional episodes, with contrasting sedimentary environments and facies distributions. The depositional episodes are separated by unconformities, some of which represent significant hiatuses and are regional in extent (Pipiringos and O'Sullivan 1978; Imlay 1980a,b): infra-Jurassic, basal Jurassic rests on rocks ranging from Precambrian to Triassic, with the hiatus greatest in the northern regions; Toarcian, probably absent throughout the Western Interior U.S. region; Aalenian–Early Bajocian, the end of the first marine cycle, affecting all areas of the Western Interior United States; middle Callovian, regression at the close of the third Jurassic marine cycle, particularly in the northern Rocky Mountain area and locally in Utah; late Callovian, in most of the Western Interior United States; late early to early middle Oxfordian, locally in the central Rocky Mountain area; late Oxfordian, the final retreat of Jurassic seas; late Tithonian, with regional unconformity at Jurassic–Cretaceous boundary.

Widespread continental sedimentation marks the lower and upper parts of the sequence, and marine sediments are dominant in four main cyclic sequences comprising the intervening section. Marine invasion of the Rocky Mountain Shelf began in early Bajocian time and continued in a series of four steadily advancing pulses through most of Oxfordian time.

### Lower Jurassic

The Early Jurassic continental phase, spanning approximately 22 m.y., is represented by the Nugget, Navajo, and Glen Canyon Sandstone Formations and their equivalents. Eolian sand and fluviatile redbed deposits are present over the central and southern Rocky Mountain region and extend westward into the Basin and Range Province of southern Nevada (Figures 4.9–4.11). These beds are as much as 600 m or more thick in southeastern Idaho, south-central Utah, and southeastern Nevada and may have been much more extensive before partial removal by Late Jurassic and Early Cretaceous erosion.

### Middle Jurassic

The first marine cycle, Aalenian and early Bajocian in age, and approximately 7 m.y. in duration, comprises the Nesson, Gypsum Spring, Temple Cap, and lower part of the Arapien Shale Formations. It represents a major invasion of marine waters from the Pacific coast region. The shelf seas were separated from the

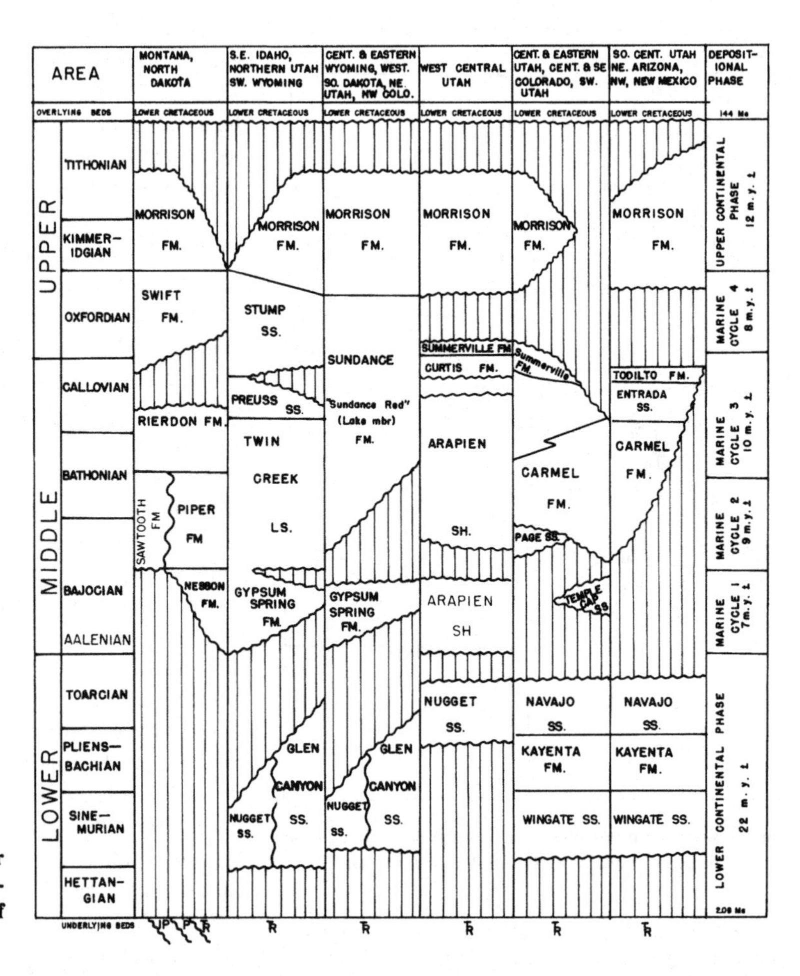


**Figure 4.9. Correlation chart for the Jurassic of the Western Interior United States. Positions of depositional phases are shown.**

Pacific by a low-lying landmass, the Mesocordilleran High (Figure 4.8). This highland furnished sand and pebbles intermittently throughout most of the Jurassic until Oxfordian time, when it became a persistent source area for clastics. The sediments are restricted-marine and fluviatile redbeds, limestone, and evaporites, including a basal salt unit in the center of the Williston Basin in North Dakota. *Chondroceras oblatum* occurring with *Stemmatoceras* sp. in the lower Sawtooth and Gypsum Spring Formations indicate a late early Bajocian (Humphriesianum Zone) age. Deposition of these beds was followed by withdrawal of marine waters from all of the Western Interior U.S. area, with subsequent erosion of older Jurassic beds in places.

The second marine cycle comprises the late early Bajocian–early Bathonian (9 m.y.). The widespread strata include normal-marine limestone and shale in the Utah–Idaho Trough of southeastern Idaho, western Wyoming, and central Utah. Marine red silt and shale, limestone, and some gypsum were deposited on the shelf area to the east. Orange and red cross-bedded sands intertongue with these beds along the southern fringes of the seaway in southeastern Utah and northern Arizona. During maximum transgression, the basin became connected with the seas to the north in Alberta through an opening in northwestern Montana north of the Boulder High (Belt Island complex). At that time, widespread normal-marine limestone was deposited across much of the Western Interior U.S. region, in the middle part of the Twin Creek Limestone and Arapien Shale of southeastern Idaho, western Wyoming, and central Utah, the Sundance Formation in Wyoming, the Piper and Sawtooth Formations of Montana and the Williston Basin, and the Carmel Formation of eastern Utah and western Colorado. Redbeds and some gypsum are interbedded with these carbonates in Utah, Montana, and Williston Basin. A minor shallowing of the sea occurred in the early Bathonian when red silt and shale were deposited throughout the Western Interior United States: sandstone, red shale, and gypsum in southern Utah, and normal-marine sandstone, silt, and shale in western Montana. The oldest fauna in this cycle consists of *Eocephalites primus,* which is associated with *Megasphaeroceras* cf. *rotundum, Spiroceras,* and stephanoceratids, allowing correlation with the late Bajocian Subfurcatum Zone. This fauna occurs in the Sliderock Member of the Twin Creek Limestone (Imlay 1967); in the overlying Rich Member, the associated endemic genera, *Sohlites* and *Parachondroceras,* may still be of late Bajocian age, but no independent evidence for this correlation exists (Callomon 1984); elements of this fauna also occur in the Piper and Carmel Formations. In the upper parts of the Sawtooth Formation of Montana occurs a sequence of faunas dominated by further endemics, *Paracephalites* and its microconch form *Xenocephalites,* associated with forms such as *Procerites* and *Kepplerites* that allow approximate corre-

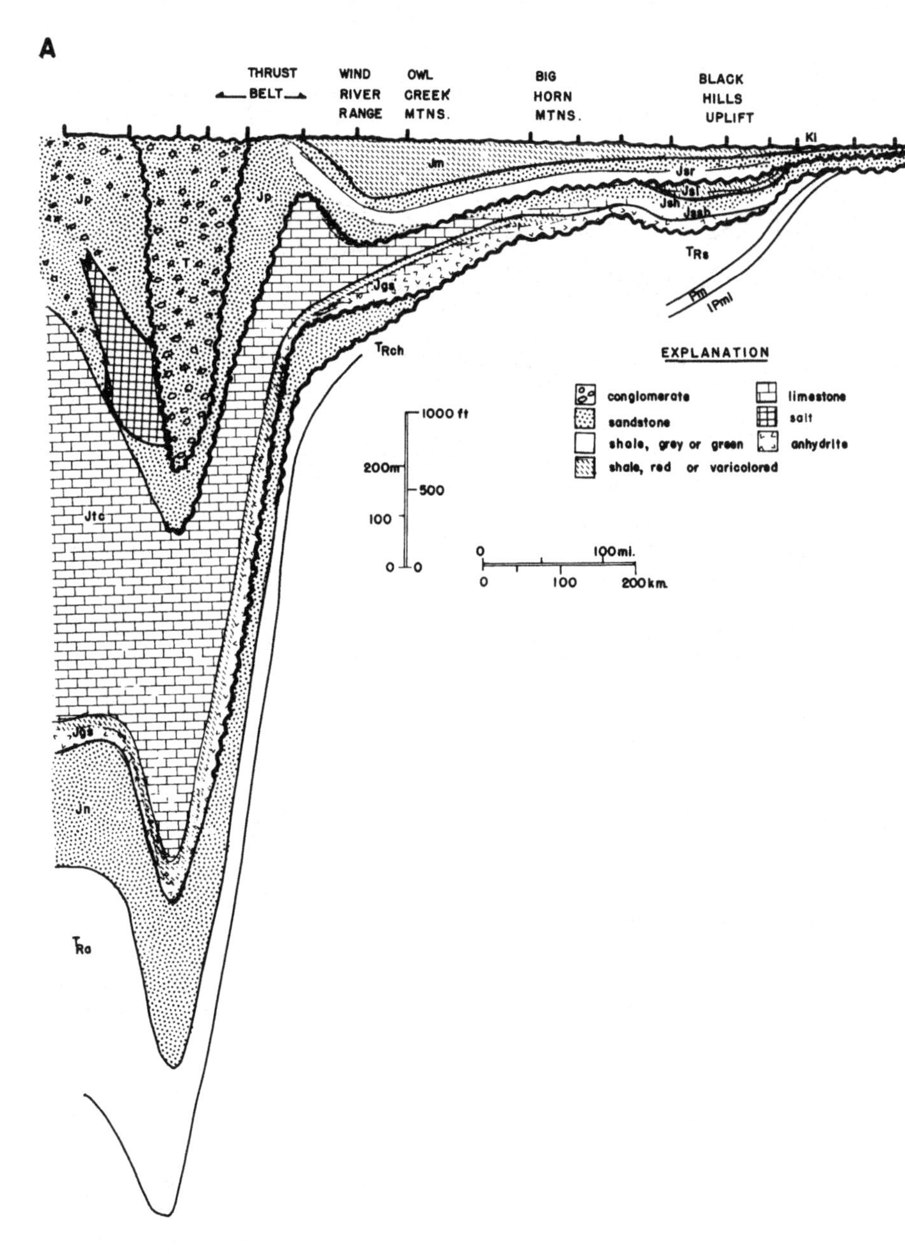


**Figure 4.10. West–southwest–northeast stratigraphic cross section A–A′, northern Utah to central South Dakota. Line of cross section shown in Figure 4.8. T, Tertiary; Jc, Carmel Formation; Je, Entrada Sandstone; Jk, Kayenta Formation; Jgs, Gypsum Spring Formation; Jm, Morrison Formation; Jn, Navajo Sandstone; Jp, Preuss Sandstone; Jpa, Page Sandstone; Jt, Todilto Formation; Jtc, Twin Creek Formation; Jw, Wingate Formation; Kl, Lower Cretaceous; TRa, Ankareh Shale (Triassic); TRch, Chinle Formation (Triassic); TRs, Spearfish Formation (Triassic); Pm, Minnekahta Limestone (Permian); lPml, Minnelusa Formation (Pennsylvanian). Sundance Formation: Jsr, Redwater Shale Member; Jsl, Lak Member; Jsh, Hulett Member; Jssb, Stockade Beaver Member.**

lations with either Boreal or western European zonal sequences. The age of the earliest of these related faunas is presumed to be early Bathonian (Callomon 1984; Hall 1988, 1989), but precise correlation with zones elsewhere is not yet possible. Callomon (1984) proposed the following sequence (oldest first): (1) *Paracephalites glabrescens* (= *sawtoothensis*) and *Xenocephalites saypoensis;* (2) *Paracephalites codyensis, Xenocephalites crassicostatum,* and *Procerites engleri;* (3) *Paracephalites muelleri, P. tetonensis, P. piperensis, Xenocephalites shoshonensis,* and *Procerites* (*Siemiradzkia*) *warmdonensis* with *Kepplerites costidensus;* (4) *Kepplerites subitus;* (5) *Kepplerites* aff. *tychonis;* (6) *Kepplerites mclearni, K. rockymontanus, K. planiventralis, Lilloettia,* and *Xenocephalites phillipsi*. The small, finely ribbed species of *Kepplerites* present in faunas (3) and (5) may be compared to similar species from eastern Greenland that first appear at the beginning of the Boreal Upper Bathonian (Callomon 1984). The giant keppleritids with long body chambers represented in fauna (6), which occur only in the Rierdon Formation of Montana and its equivalents in Jasper Park, Alberta, and the subsurface of southern Saskatchewan (Imlay 1948, 1953), have been compared to a similar assemblage in eastern Greenland from the Calyx Zone, the highest zone of the Boreal Upper Bathonian (Callomon 1984).

The third marine cycle, middle Bathonian through early middle Callovian (ca. 10 m.y.), is represented by the Rierdon Formation of the northern Rocky Mountains and Williston Basin, by the upper parts of the Twin Creek and Arapien Formations and the Preuss Sandstone of the southeastern Idaho, western Wyoming, and north-central Utah, by the lower part of the Sundance Formation of Wyoming, South Dakota, and north-central Colorado, by the Entrada Formation and the upper part of the Carmel Formation in central, southern, and eastern Utah and southwestern

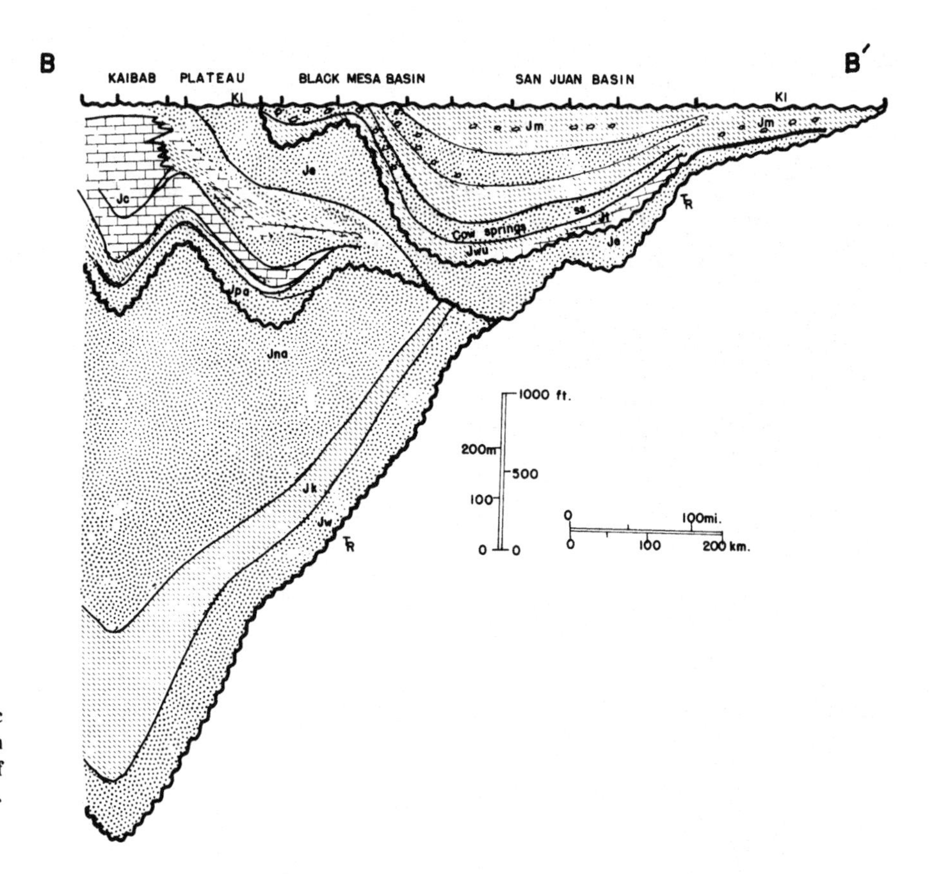


**Figure 4.11. West–east stratigraphic cross section B–B′, southwestern Utah to northwestern Texas. Line of cross section shown in Figure 4.8. Abbreviations as in Figure 4.10.**

Colorado, and by the Todilto, Entrada, and Carmel Formations in southern Utah, northeastern Arizona, and northwestern New Mexico (Figures 4.9–4.11). These formations consist primarily of shallow to moderately deep marine limestone and gray shale in the Utah-Idaho Trough. Red silt and shale with interbedded marine limestone were deposited across most of the shelf region in the early phases of the cycle. These beds grade upward into normal-marine gray shale, siltstone, and minor limestone in the central shelf region and the marine marl, shale, and limestone to the north. Marine sandstone intertongues with these beds along the eastern and southeastern borders of the seaway. Along the southern borders of the seaway, a facies of eolian, tidal flat, and near-shore marine sandstone and silt was deposited, much of it red in color. Partial withdrawal and shallowing of the sea during the closing phases of the cycle resulted in erosion and reworking of older Jurassic beds in parts of the northern Rocky Mountain Shelf region. Red sandstone, silt, and shale were deposited to the south of the Sheridan Arch trend extending from southern North Dakota across northeastern, central, and northwestern Wyoming (Figure 4.8). Salt and gypsum are interbedded with similar deposits in the Utah–Idaho Trough. Faunas of approximately middle and late Bathonian ages were discussed earlier as part of the second marine cycle.

### Upper Jurassic

The most recent marine Jurassic cycle began in the latest Callovian and extended through Oxfordian time (ca. 8 m.y.). It is represented by the Swift Formation of Montana and the Williston Basin, by the Stump Sandstone of southeastern Idaho, northern Utah, and southwestern Wyoming, and by the upper part of the Sundance Formation of Wyoming, South Dakota, and northwestern Colorado (Figures 4.9–4.11).

The cycle represents a Boreal Jurassic marine seaway that extended southward from the Alberta Trough into northern Montana and the Williston Basin. Marine shale, glauconitic sandstone, silt, and some limestone were deposited throughout the northern part of the Rocky Mountain area of the United States. During that time the western source area in Nevada, western Utah, and central Idaho (Mesocordilleran High) became more active, perhaps related to inception of the Sevier Thrust Belt of the latest Jurassic to early Tertiary.

The Upper Callovian is represented by *Quenstedtoceras collieri*, which occurs in the basal parts of the Swift Formation of Montana with *Peltoceras* and *Perisphinctes* (*Prososphinctes*) (Callomon 1984). In the Redwater Shales of the Sundance Formation and correlatives from nothern Montana to northern Utah there occurs a sequence of Boreal Cardioceratidae faunas that have been described by Reeside (1919) and Imlay (1982). Callomon (1984) has arranged these faunas in probable chronological sequence based on comparisons with European faunas (in ascending order; only one species is named to represent each fauna): (1) *Cardioceras cordiforme*, (2) *Cardioceras martini*, (3) *Cardioceras distans*, (4) *Cardioceras* (*Cawtoniceras*) *canadense*, and (5) *Cardioceras sundancense*. This sequence probably ranges throughout the Lower and Middle Oxfordian.

Retreat of the so-called Sundance Sea in the late Oxfordian resulted in deposition of continental sandstone beds of the lower Morrison Formation in the southern Rocky Mountain region. This regression, which includes the remainder of the Morrison Formation, represents the upper continental phase of late Oxfordian through middle Tithonian age (ca. 12 m.y.). Following retreat of the Oxfordian sea, a widespread blanket of fluviatile and lacustrine beds was deposited on the emergent sea floor throughout most of the Western Interior U.S. region. During the latest phases of that deposition, coal and highly carbonaceous shale and silt, probably of lacustrine origin, were deposited in central and northwestern Montana in the approximate position of the central Montana Trough.

In almost all areas of the Western Interior United States, Morrison beds are overlain unconformably by late Neocomian conglomerate and sandstone. The hiatus may represent as much as 10–15 m.y. in parts of the region.

### Ammonite assemblages

*Bajocian*

1. Humphriesianum (Standard) Zone – *Sphaeroceras (Chondroceras) allani* (McLearn), *Stephanoceras* cf. *skidegatense* (Whiteaves), *Stemmatoceras arcicostum* Imlay
2. Subfurcatum Zone – *Megasphaeroceras* cf. *rotundum* Imlay, *Eocephalites primus* Imlay, *Spiroceras* cf. *orbignyi* (Baugier & Sauzé) (see also Rotundum Zone, Chapter 12)
3. ?Upper Bajocian – *Sohlites spinosus* Imlay, *Parachondroceras andrewsi* Imlay, *P. filicostatum* Imlay

*Bathonian*

1. ?Lower Bathonian – *Paracephalites glabrescens* Buckman (= *sawtoothensis* Imlay), *Xenocephalites saypoensis* (Imlay), *Procerites engleri* Frebold
2. ?Lower Bathonian – *Paracephalites henryi* Meek & Hayden, *Paracephalites codyensis* Imlay, *Xenocephalites crassicostatum* Imlay, *Procerites engleri* Frebold

3a. ?Cranocephaloides Zone – *Paracephalites muelleri* (Imlay) (including *P. tetonensis, P. piperensis*), *Xenocephalites shoshonensis* (Imlay), *Procerites (Siemiradzkia) warmdonensis* (Imlay), *Kepplerites costidensus* (Imlay), *Kepplerites subitus* (Imlay)

3b. Cranocephaloides Zone – *Kepplerites* cf. *tychonis* Ravn

4. Calyx Zone – *Kepplerites mclearni* Imlay, *K. rockymontanus* (Imlay), *K. planiventralis* (Imlay), *K. (Torricellites) knechteli* (Imlay), *Lilloettia* sp., *Xenocephalites phillipsi* Imlay

*Callovian*

1. Lamberti Zone – *Quenstedtoceras collieri* (Reeside), *Peltoceras* sp., *Perisphinctes (Prososphinctes)* sp.

*Oxfordian*

1. Mariae Zone – *Cardioceras cordiforme* (Meek & Hayden) (including *Quenstedtoceras hoveyi, subtimidum, suspectum, tumidum* Reeside spp., *Cardioceras albaniense, auroaensis, bellefourchense, crassum, incertum, russelli* Reeside spp.), *Cardioceras latum* Reeside

2a. Lower Cordatum Zone – *Cardioceras martini* Reeside, *Cardioceras reesidei* Maire, *Cardioceras mountjoyi* Frebold, *Cardioceras mathiaspeakense* Imlay

2b. Upper Cordatum Zone – *Cardioceras distans* (Whitfield) (including *Cardioceras stantoni, hyatti, haresi* Reeside spp., and ?*Cardioceras wyomingense, crookense* Reeside spp.), *Cardioceras schucherti* Reeside, *Cardioceras minnekahtense* Imlay, *Grossouvria* cf. *trina* (Buckman)

3. Densiplicatum Zone – *Cardioceras (Cawtoniceras) canadense* Whiteaves, *Cardioceras (Vertebriceras) whiteavesi* Reeside, *Cardioceras (Maltoniceras) plattense* Reeside, *C. (M.) boghornense* Imlay, *C. (M.) reddomense* Imlay
4. Tenuiserratum Zone – *Cardioceras sundancense* Reeside (including *Cardioceras stillwelli, americanum* Reeside)

### Acknowledgements

This section benefited from critical review by T. P. Poulton, Geological Survey of Canada, and G. D. Stanley, University of Montana.

## SOUTHERN ALASKA[6]

### Introduction and terranes

It is currently accepted that southern Alaska is a collage of about 17 allochthonous tectonostratigraphic terranes (Figure 4.12) (Jones et al. 1981; Csejtey et al. 1982). The Peninsular and Wrangellia Terranes are of special interest in the reconstruction of southern Alaska; both have been studied in considerable detail (Grantz 1960a,b, 1961; MacKevett 1971, 1978; Detterman et al. 1981a,b, 1987), and both have fairly complete Mesozoic sequences, with many mutual stratal and faunal similarities (Figure 4.13) (see Chapter 24). Other terranes in southern Alaska, though not studied in great detail, add important information to the accretionary history.

The geology and paleomagnetics of the southern Alaska terranes indicate that some coalesced before final docking. Jones, Silberling, and Hillhouse (1977), and Stone, Panuska, and Packer (1982) consider that the amalgamation of Wrangellia and Peninsular Terranes was accomplished by Middle Jurassic time. An earlier amalgamation was suggested by Detterman and Reed (1980), based on similarities in the Triassic sequence. Biostratigraphic analysis adds further evidence that the two terranes were part of a single unit before docking on the craton during Middle Jurassic time.

[6] By R. L. Detterman and G. E. G. Westermann.

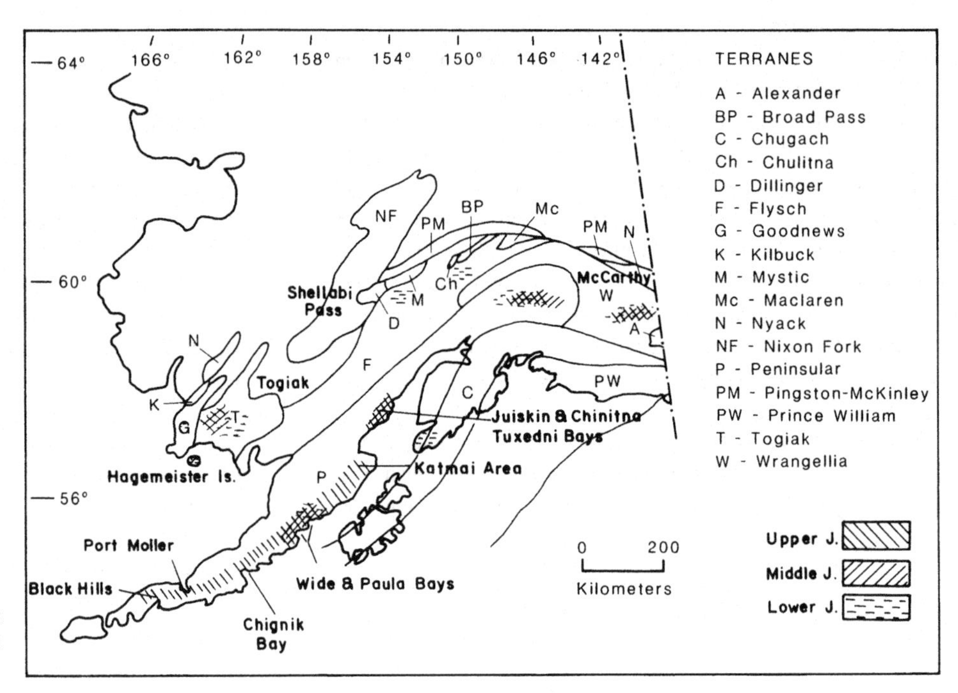


**Figure 4.12. Tectonostratigraphic terranes of southern Alaska and Jurassic outcrops.**

The complex geological histories of the numerous juxtaposed terranes in southern Alaska have only recently been investigated (Jones et al. 1977, 1978; Csejtey et al. 1982). Lithostratigraphic correlations between terranes, generally possible only for the strata deposited after the terranes became sutured together, offer little help in establishing terrane displacement. Attempts to establish the depositional sites of the southern Alaska terranes have relied mainly on paleomagnetics (Packer and Stone 1974; Hillhouse 1977; Jones et al. 1977, 1978; Stone and Packer 1979; Hillhouse and Gromme 1982; Stone et al. 1982; Plumley, Coe, and Byrne, 1983; Hillhouse, Gromme, and Csejtey 1984). The paleomagnetic data indicate that the terranes originated far to the south and migrated northward by right-lateral strike-slip motion accompanied by a clockwise rotation (Packer and Stone 1974; Hillhouse, 1977; Jones et al. 1977; Stone & Packer 1979). A considerable body of paleomagnetic data from the Nikolai Greenstone (Triassic) of the Wrangellia Terrane indicates deposition at about 15° paleolatitude (Hillhouse 1977). Ambiguities concerning the magnetic polarity at the time of deposition make it impossible to determine whether the latitude was north or south of the equator. Paleomagnetic data from the Peninsular Terrane gathered by Packer and Stone (1974) indicated to them a source possibly south of the equator for the Jurassic and Cretaceous strata in that terrane, which did not follow a simple northward movement, but at times moved southward, only to turn and move northward again. These extremely southward positions are contradicted by ammonite biogeography (Taylor et al. 1984).

### *Togiak Terrane*

The Togiak Terrane (Hoare and Coonrad 1978; Box 1983) is composed of volcanic flows, coarse volcaniclastic breccias, tuffs, and epiclastic turbidite deposits of graywacke, siltstone, and conglomerate of Late Triassic to Early Cretaceous age (Hoare and Coonrad 1978). The only Jurassic fauna from this terrane are a few Liassic bivalves (*Weyla* sp.) from Hagemeister Island. The contact between the Togiak and Goodnews Terranes is a thrust fault formed in the late Early Jurassic, when the Togiak Terrane was thrust northward (Box 1983). Turbidite deposits in the southeastern Togiak Terrane were thrust southeastward, but contact with the Peninsular Terrane is obscured by surficial deposits and Bristol Bay. Lower Middle Jurassic granitic plutons intruded the central part of the Togiak Terrane, and hornblende-gabbro ultramafic plutons intruded the northwestern Togiak and adjoining Goodnews Terrane (Shew and Wilson 1981). Middle and Upper Jurassic rocks record andesitic volcanism that was a source for the volcaniclastic sediments in the southeastern part of the Togiak Terrane.

### *Flysch Terrane*

The deformed upper Mesozoic flysch reported by Jones et al. (1981) is poorly known because of highly deformed rocks and poor exposures. Reed and Nelson (1977) have confirmed a sequence of Late Triassic to Early Cretaceous age for these highly deformed rocks. Some late Mesozoic flysch may be present, but that has not been confirmed.

Pillow basalts are interbedded with tuffaceous graywacke and siltstone that contain an abundance of Late Triassic *Monotis* species. The Triassic rocks are overturned to the northwest. Graywacke and siltstone overlying the Triassic beds contain (Sinemurian) ammonites, including *Paracaloceras* sp., *Arnioceras* sp., *Coroniceras* sp., *Arctoasteroceras* sp., and *Badouxia* sp. (Reed and Nelson 1977). In one area a *Buchia*-bearing limestone of Valanginian age was found in the flysch deposits. The stratigraphic relations of the rocks of various ages are poorly known, as are the contacts with other terranes. Plutonic rocks forming stocks and

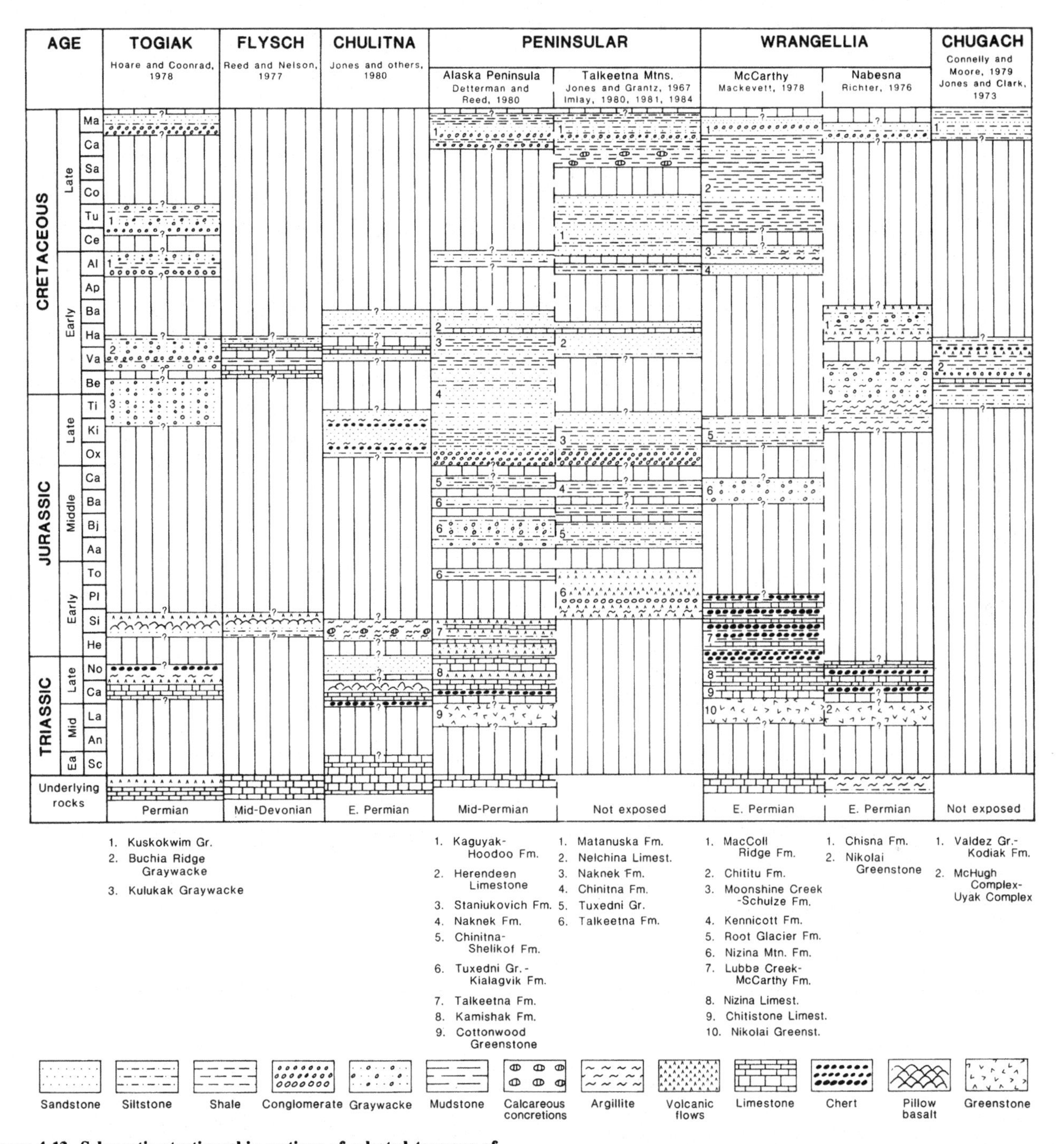


**Figure 4.13. Schematic stratigraphic sections of selected terranes of southern Alaska. Vertical lines indicate missing strata. No scale.**

batholiths are mainly of Late Cretaceous to Early Tertiary in age (Reed and Lanphere 1972).

### *Chulitna Terrane*

The Chulitna Terrane contains a sequence of Paleozoic to Lower Mesozoic rocks found nowhere else in Alaska (Jones et al. 1978, 1980; Silberling et al. 1978). Paleozoic rocks include a dismembered ophiolite of Late Devonian age (Silberling et al. 1978) and Mississippian chert and Permian volcanic conglomerate, flysch, chert, and limestone (Jones et al. 1980). Overlying this sequence are Triassic redbeds and conglomerate that contain clasts of the underlying sequence, and also rock not found elsewhere in adjoining areas, suggesting a major Late Triassic tectonic event. Marine sandstone, limestone, argillite, and chert of Late Triassic and Early Jurassic age overlie the redbed sequence, but because of numerous faults and isoclinal folding the exact relationship is unknown. The ammonites *Paracaloceras* sp. and

*Badouxia* sp. establish a Sinemurian age for part of the sequence (Jones et al. 1980).

### *Peninsular Terrane*

The Peninsular Terrane is a long (1,200 km) arcuate terrane that extends from near the tip of the Alaska Peninsula northeastward along the west side of Cook Inlet, before curving eastward to the east end of the Talkeetna Mountains. In recent years the terrane has been studied in considerable detail (Westermann 1964a, 1969; Detterman and Hartsock 1966; Detterman and Reed 1980; Detterman et al. 1981a,b; Detterman, Miller, and Case 1985). The terrane is divided into two subterranes, Iliamna and Chignik, by the Bruin Bay fault, a major fault system (Wilson, Detterman, and Case 1985). The Chignik Subterrane, east of the fault, is mainly a platform clastic sequence of Late Triassic to Late Cretaceous age. The Chignik Subterrane contains most of the Jurassic clastic sequence.

The late Early Jurassic to early Late Jurassic volcanic-plutonic arc west of the Bruin Bay fault (Reed and Lanphere 1972), with associated Late Triassic and some Early Jurassic sedimentary rocks (Iliamna Subterrane), has been considered part of the Peninsular Terrane. The possibility should be considered that the Iliamna Subterrane is actually a distinct, separate terrane, for the following reasons: The volcanic-plutonic arc is present only in the Iliamna Subterrane; Lower Jurassic rocks are unlike similar-age rocks in the Chignik Subterrane; and Mesozoic clastic rocks are nowhere present west of the Bruin Bay fault in the Iliamna Subterrane. Lithically, the Chignik Subterrane is more akin to Wrangellia than to the Iliamna Subterrane. However, if the two subterranes were separate entities, they were joined by Middle Jurassic time, as the Iliamna Subterrane provided much of the clastic debris forming the rocks of the Chignik Subterrane. The suture between the Peninsular Terrane and the Togiak Terrane is nowhere seen, as it is covered by surficial materials and Bristol Bay. The contact with the Chugach Terrane is along the Border Range fault.

### *Wrangellia*

Studies in the Wrangellia Mountains of east-central Alaska by MacKevett and Richter (1974), MacKevett (1976), Armstrong and MacKevett (1982), Richter (1976), Jones et al. (1977), and Hillhouse (1977) led to the conclusion that the Wrangellia Mountains area was a displaced terrane in southern Alaska, and subsequently to the concept that most of southern Alaska was composed of discrete tectonostratigraphic terranes.

Thin-bedded chert, siltstone, and shale of Middle Triassic age are overlain by a large basaltic lava field (Nikolai Greenstone) as much as 3,500 m thick. This is disconformably overlain by 1,400 m of Late Triassic shallow-marine sediments. A nearly complete sequence of marine Jurassic rocks consisting of Early Jurassic impure limestone, chert, and spiculite is overlain by Middle to Late Jurassic mudstone, siltstone, graywacke, and conglomerate. After a period of folding and faulting, the older rocks were overlain by a nearly complete clastic Cretaceous sequence.

## Composite sections[7] (Figure 4.14)

*West side of Cook Inlet*

Upper Jurassic
- Naknek Formation
  - Pomeroy Arkose Member – medium- to thick-bedded arkose, locally conglomeratic, siltstone interbeds (250–730 m); *Amoeboceras–Buchia concentrica* Fauna in upper 40 m
  - Snug Harbor Siltstone Member – thin to massive brownish gray siltstone with abundant limestone concretions; minor sandstone interbeds (200–250m):
    - Upper 70 m: *Amoeboceras–Bucia concentrica* Fauna
    - 200 m above base: *Cardioceras whiteavesi* Fauna
    - 60–150 m above base: *Cardioceras spiniferum* Association
    - 0–130 m above base: *Cardioceras martini* Association
  - Lower Sandstone Member – thick-bedded to massive arkosic sandstone with siltstone interbeds (150–250 m):
    - Near top: *Amoeboceras–Buchia concentrica* Fauna
    - 30 m above base to top: *Cardioceras spiniferum* Association
    - Base to top: *Cardioceras martini* Association
  - Chisik Conglomerate Member – massive cobble-boulder conglomerate, mainly granitic and metamorphic rocks, few volcanic rocks (100–170 m); no fossils.

Middle Jurassic
- Chinitna Formation
  - Paveloff Siltstone Member – massive brown siltstone with large limestone concretions; sandstone at base (260–415 m):
    - 60–260 m below top: *Longaeviceras pomeroyense* Association
    - 20–250 m below top: *Cadoceras stenoloboide* Association
    - 150 m below top: *Kepplerites loganianus* Fauna
    - 75–150 m below top: *Cadoceras comma* Fauna
  - Tonnie Siltstone Member – brown-weathering gray siltstone with abundant small limestone concretions (250–400 m):
    - 200–250 m above base: *Longaeviceras pomeroyense* Association
    - 220–250 m above base: *Cadoceras stenoloboide* Association
    - 200–225 m above base: *Cadoceras wosnessenskii* Fauna
    - 30–175 m above base: *Lilloettia stantoni* Fauna
    - 30–130 m above base: *Cadoceras tonniense* Fauna
    - Basal 30 m: *Kepplerites* cf. *abruptus* Fauna
    - Base: *Lilloettia lilloetensis* Fauna
    - Base to top: *Cadoceras comma* Fauna

[7] By R. L. Detterman.

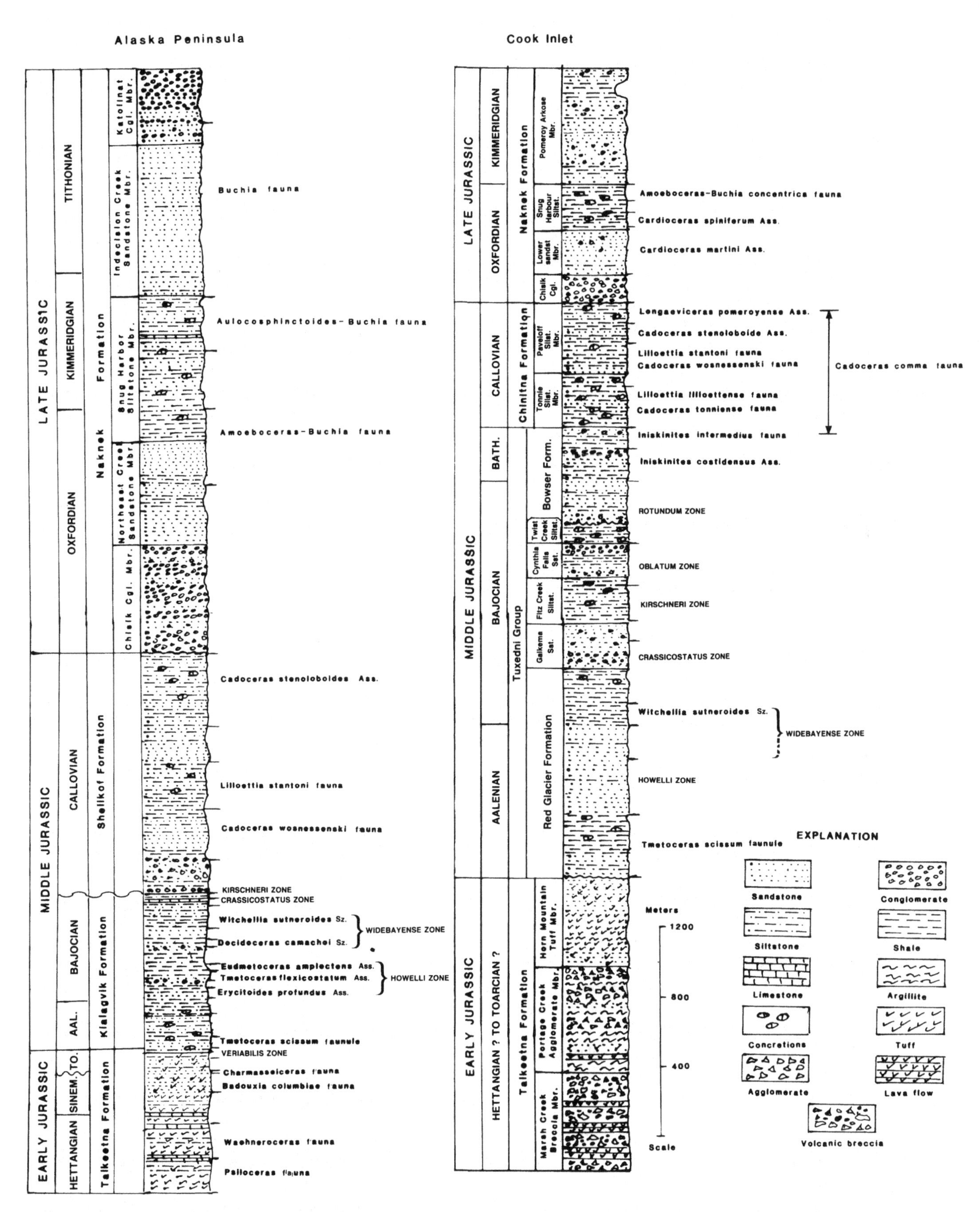


**Figure 4.14. Composite columns of the Jurassic on the Alaska Peninsula and at Cook Inlet. Ammonite zones and faunas indicated.**

Tuxedni Group

Bowser Formation – greenish gray graywacke, conglomerate, pebble-cobble, granitic and volcanic clasts, brown to gray siltstone with limestone and sandstone concretions (380–550 m); Lower Bathonian–Bajocian:

Top to 115 m below top: *Iniskinites intermedius* Fauna

125–300 m below top: *Iniskinites costidensus* Association

250–300 m below top: *Megasphaeroceras rotundum* Zone

Twist Creek Siltstone – brownish gray siltstone with limestone concretions (0–100 m):

Throughout: Rotundum Zone

Cynthia Fall Sandstone – medium- to thick-bedded greenish gray sandstone; few conglomerate beds (ca. 200 m), very few fossils:

Throughout: *Sphaeroceras oblatum* Zone

Fitz Creek Siltstone – thin-bedded dark gray siltstone with limestone concretions; minor, thin sandstone (150–390 m):

100–160 m above base: *Sphaeroceras oblatum* Zone

Mostly 120–2,240 m above base: *Stephanoceras kirschneri* Zone

Gaikema Sandstone – thin- to medium-bedded, fine-grained, greenish brown sandstone and siltstone, minor conglomerate (180–330 m):

0–150 m above base: *Parabigotites crassicostatus* Zone

Red Glacier Formation – thin- to medium-bedded brown sandstone and siltstone, minor conglomerate (550–600 m):

50–70 m below top: *St. kirschneri* Zone

200–300 m below top: *Parabigotites crassicostatus* Zone

500 m below top: *Erycitodes howelli* Zone

500 m below top to base: *Tmetoceras scissum*

Lower Jurassic

Talkeetna Formation (no marine fossils)

Horn Mountain Tuff Member – red and green tuff, tuffaceous sandstone and siltstone (650 m), no marine fossils

Portage Creek Agglomerate Member – coarse green and red agglomerate, tuff and andesite flows (850–875 m)

Marsh Creek Breccia Member – volcanic breccia and andesite flows (900–1,000 m)

*Alaska Peninsula, Wide Bay–Puale Bay area*

Upper Jurassic

Naknek Formation

Indecision Creek Member – thin-bedded, fine-grained, yellowish gray sandstone with few siltstone interbeds; siltstone increases downward; inner neritic to shore face (ca. 760 m):

Upper 260 m: *Buchia blanfordiana* (Stol.)

Base: *B. mosquensis*

Snug Harbor Siltstone Member – medium to dark gray, thin- to shaly-bedded siltstone with limestone concretions; few thick sandstone beds in upper part (ca. 800 m):

Upper 300 m: *Aulacosphinctoides–Buchia* Fauna

Basal part: *Amoeboceras–Buchia concentrica* Fauna

Northeast Creek Sandstone Member – medium- to thick-bedded olive-gray sandstone with siltstone interbeds in upper part and conglomerate in lower part (500–600 m):

Upper part: *Amoeboceras–Buchia concentrica* Fauna

Chisik Conglomerate Member – thick-bedded to massive pebble to boulder conglomerate; clasts mainly granitic and metamorphic rocks; cast size decreases upward; few sandstone interbeds (400–600 m); no fossils; unconformably overlies Shelikof Formation

Middle Jurassic

Shelikof Formation – dark gray, siliceous, thin-bedded siltstone in upper part, with greenish gray volcanogenic sandstone and conglomerate below; rapid facies change; few limestone concretions in siltstone (1,000–1,450 m):

Basal 500 m: *Cadoceras comma* Fauna

Upper several hundred meters: *Cadoceras stenoloboide* Association

ca. 500 m above base: *Lilloettia stantoni* Fauna

200–500 m above base: *Cadoceras wosnessenskii* Fauna

Kialagvik Formation – light brown, thin-bedded, fine-grained sandstone and siltstone; siltstone more abundant in upper part, sandstone in lower part (625–ca. 900 m):

Upper 30 m: *Stephanoceras kirschneri* Zone

30–70 m below top: *Par. crassicostatus* Zone

In upper 500 m: *Witchellia sutneroides* Subzone

275–300 m below top: *Docidoceras camachoi* Subzone

200–265 m below top: *Doc. widebayense* Zone

ca. 400 m below top: *Eudmetoceras*(?) *amplectens* Horizon

500–530 m below top: *Tmetoceras flexicostatum* Association

500–550 m below top: *Erycitoides profundus* Association

350–390 m above base: *Erycitoides howelli* Zone

Basal 100 m: *Tmetoceras scissum* Faunule

Lower Jurassic

Talkeetna Formation – thick- to thin-bedded, greenish gray tuffaceous sandstone, siltstone, limestone, and tuff (400–1,400 m):

Top of Talkeetna plus base of Kialagvik Formation at Puale Bay: Variabilis Zone

Upper 175 m: *Badouxia columbiae* Fauna
Upper 10 m: *Charmasseiceras* Fauna
0–200 m above base: *Waehneroceras* Fauna
Basal 200 m: *Psiloceras* Fauna

**Ammonite faunas, associations, and zones of the Peninsular Terrane[8]**

The best sections are developed at the north side of Cook Inlet, Tuxedni Bay, and Wide Bay on the Peninsula. The Lower Jurassic succession has been worked out mainly by Imlay (1981b), and the Pliensbachian sequence has been revised by Smith et al. (1988); the Middle Jurassic sequence has been described mainly by Westermann (1964a, 1969) and Hall and Westermann (1980), who also recognized the Aalenian-Bajocian assemblage zones; the post-Bajocian sequence has been described by Imlay (1981c) and Callomon (1984). Here we are using "fauna" for stratigraphically poorly known assemblages, and "association" for probable communities *sensu lato*. A sequence of Aalenian-Bajocian standard zones for North America is here described, based on former assemblage zones and with tentative stratotypes in the Cook Inlet area.

*Hettangian*

1. *Psiloceras* (*Franziceras*) cf. *ruidum* (Buckman), also *Laqueoceras* sp.
2. *Waehneroceras* cf. *tenerum* (Weumayr) and *portlocki* (Wright), *Discamphiceras* cf. *toxophorum* (Waehner), *Laqueoceras* cf. *sublaqeus* (Waehner)
3. *Schlotheimia* (*Schlotheimia*) sp.

*Sinemurian*

1. *Charmasseiceras* cf. *marmoreum* (Oppel), *Arnioceras* cf. *densicosta* (Quenstedt)
2. *Badouxia columbiae* (Frebold), *Arietites* cf. *bucklandi* (Sowerby), *Coroniceras* sp., *Paracaloceras rursicostatum* Frebold
3. *Coroniceras* (*Paracoroniceras*) sp.

*Pliensbachian*

1. Whiteavesi Zone – index *Acanthopleuroceras whiteavesi* Smith & Tipper; also *Acanthopleuroceras* sp. [*"Paltechioceras"* part of Imlay (1981b)], *Tropidoceras acteon* (d'Orbigny), *Metaderoceras* spp., *Dubariceras silvies* (Hertlein)
2. Freboldi Zone – index *Dubariceras freboldi* Dommergues, Mouterde, & Rivas; also *D. freboldi*, *Aveyroniceras italicum* (Fucini)
3. Kunae Zone – index *Fanninoceras kunae* McLearn; also *F.* cf. *fontanellense* (Gemmellaro), *Leptaleoceras* cf. *pseudoradians* (Reynes), *Arieticeras* sp., *Amaltheus stokesi* (Sowerby), *Protogrammoceras* spp.

*Toarcian*

1. *Dactylioceras* Fauna – *Dactylioceras* cf. *commune* (Sowerby), *D.* (*Orthodactylites*) *kanense* (McLearn)
2. *Haugia* Fauna – *Haugia* cf. *variabilis* (d'Orbigny), *H. compressa* and *H. grandis* Buckman spp. *Brodieia* cf. *tenuicostata* Jaw; *Pseudolioceras* sp.
3. *Grammoceras* Fauna – *Grammoceras* sp.; Thouarsense-Levesquei Zones

*Aalenian*

1. *Tmetoceras scissum* Faunule – poorly fossiliferous shales at Wide Bay, with *T. scissum* (Benecke) [in eastern USSR and Arctic Canada also with *Pseudolioceras macklintocki* = *P. macklintocki* Assemblage Zone *part.* of Scy ct al. (1986)]; Lower Aalenian
2. Howelli Zone – new standard zone for *Erycitoides howelli* Assemblage Zone of Westermann (1964b); stratotype at Wide Bay, Short Creek section, base at sample 48A-108; characteristic are *E. howelli* (White) [includes *E.* (*"Kialagvikes"*) *kialagvikensis* (White) ♂] and *Pseudolioceras* (*Tugurites*) *whiteavesi* (White); rare *P.* (*T.*) *tugurensis* Kalacheva & Sey and *Tmetoceras* spp.; Upper Aalenian (from below):
   (a) Low-diversity *Erycitoides-Pseudolioceras* fauna, with rare *Erycites imlayi* Westermann, *Abbasites platystomus* Westermann, and *Eudmetoceras?* (*Euaptetoceras?*) *amplectens* (Buckman); upper Murchisonae or Concavum Zone(?)
   (b) *Erycitoides profundus* Association – similar to (a), but also *E. profundus* ♀, *E. teres* ♀, *E. paucispinosum* ♀, *E. spinatus* ♂, and *E. levis* ♂, Westermann spp.; Concavum Zone
   (c) Diverse, mixed Boreal-Tethyan assemblage of *Erycitoides* and *Pseudolioceras,* together with *Eudmetoceras* cf. *eudmetum jaworskii* Westermann, *E.* (*Euaptetoceras*) *nucleospinosum* Westermann, *E.?* (*Euapt.?*) *amplectens, Tmetoceras kirki* Westermann; also *Partschiceras* and *Holcophylloceras;* Concavum Zone
   (d) *Tmetoceras flexicostatum* Association – *T. flexicostatum* Westermann, *T. tenue* Westermann; Concavum Zone

*Aalenian–Bajocian boundary*

1. *Eudmetoceras? amplectens* Horizon/Association – locally with *E.?* (*Euaptetoceras?*) *amplectens;* sparse *Docidoceras* (*Pseudocidoceras*) *paucinodosum* Westermann, *Praeoppelia oppeliiformis* Westermann, and *Hebetoxyites* sp.

[8] By G. E. G. Westermann.

*Bajocian*

1. Widebayense Zone – new standard zone for *Pseudocidoceras* zonule of Westermann (1969). Stratotype at Wide Bay (southeast side), with base to be defined; characterized by *D. (Pseudocidoceras) widebayense* Westermann, late Hammatoceratidae and early Sonniniidae; also rare *Pseudotoites, Holcophylloceras, Lytoceras*
   (a) *Docidoceras camachoi* Assemblage Subzone – *D. (Pseudocidoceras) camachoi* Westermann, *Planammatoceras (Pseudaptetoceras) klimakomphalum discoidale* Westermann, *Euhoploceras bifurcatum* Westermann, *Alaskinia alaskensis* Westermann, *Asthenoceras* sp., *Praeoppelia oppeliiformis* Westermann, *Pseudolioceras (Tugurites) fastigatum* Westermann; ca. Discites Zone
   (b) *Witchellia sutneroides* Assemblage Subzone – also with *W. sutneroides* Westermann, and *Pseudolioceras (Tugurites) costistriatum* Westermann; Ovalis (= lower Laeviuscula) Zone
2. Crassicostatus Zone – new standard zone for *Parabigotites crassicostatus* Assemblage Zone of Hall and Westermann (1980); widely distributed in southern Alaska, Alberta, and Oregon; stratotype to be defined at Cook Inlet; *P. crassicostatus* Imlay, *Sonninia tuxedniensis* Imlay, *Dorsetensia adnata* (Imlay), *Emileia constricta* Imlay, *Sonninia (Papilliceras)* sp.; ca. Sauzei Zone
3. Kirschneri Zone – new standard zone for *Stephanoceras kirschneri* Assemblage Zone of Hall and Westermann (1980); widely distributed in southern Alaska and Oregon; stratotype in southern Alaska to be defined in Red Glacier Formation at Cook Inlet; *S. (Skirroceras) kirschneri* Imlay, *Holcophylloceras costisparsum* Imlay
   (a) Above: *Zemistephanus richardsoni* Assemblage Subzone – *Z. richardsoni* and *Z. carlottensis* (Whiteaves), *Z. alaskensis* (Imlay), *Sphaeroceras (Defonticeras) defontii* (McLearn), *Stephanoceras (Stemmatoceras?)* cf. *palliseri* (McLearn); Romani Subzone
4. Oblatum Zone – new standard zone for *Chondroceras oblatum* Assemblage Zone of Hall and Westermann (1980); typically developed in southwest Alberta, Queen Charlotte Islands, and, especially, southern Alaska: at Tuxedno Bay (stratotype?), with *S. (Defonticeras) oblatum* (Whiteaves), *S. (D.) allani* McLearn, *Stephanoceras* sp.; upper Humphriesianum Zone
5. Rotundum Zone – new standard zone for *Megasphaeroceras rotundum* Assemblage Zone of Hall and Westermann (1980); widely distributed in southern Alaska, southern Alberta, Western Interior United States, Oregon; by extension also in southern Andes; stratotype to be defined in Twist Creek Siltstone at Cook Inlet; *M. rotundum* Imlay, *Strigoceras (Liroxyites) kellumi* Imlay, *Sphaeroceras (S.) talkeetnanum* Imlay, *Stephanoceras vigorosum* (Imlay), *Leptosphinctes (L.) cliffensis* Imlay, *L. (Prorsisphinctes?) delicatus* Imlay, *Macrophylloceras* cf. *grossimontanum* Imlay; rare *Spiroceras, Calliphylloceras, Lytoceras;* Subfurcatum Zone (note that the index species or allied forms appear to range higher up, but without *Stephanoceras* and *Spiroceras*)

*Top Bajocian–base Bathonian*

1. *Iniskinites? costidensus* Association – *I.? costidensus* ♀, *globosus* ♀, and *alaskanus* ♀, *"Tuxednites" alticostatus* ♀, *Xenocephalites cadiformis* ♂, *Parareineckeia hickersonensis, P. nelchinensis, Leptosphinctes (Cobbanites) talkeetnanus,* all Imlay spp.

*Bathonian*

1. *Iniskinites intermedius* Fauna – also *I. magniformis* ♀ and *martini, Chinitnites chinitnaensis* ♂ and *parviformis* ♂, *Kepplerites chisikensis* ♀ and *alaskanus* ♂, *Cadoceras moffiti, Choffatia irregularis,* all Imlay spp.; upper Variabile–lower Calyx Zones of Greenland
2. *Iniskinites abruptus* Association – *I. abruptus* ♀, *Kepplerites chisikensis* ♀, *alticostatus* ♀, and cf. *alaskanus* ♂, Imlay spp.; Calyx Zone

*Callovian*

1. *Cadoceras comma* Fauna – for "*Cadoceras catostoma* Zone" of Imlay (1975) (based on a *nomen dubium*); abundant *Cadoceras, Kepplerites,* and *Lilloettia* spp., all with poorly known ranges; tentatively divided by comparison with British Columbia and Greenland (Callomon 1984); Herveyi Zone ("Macrocephalus Zone")
   (a) *Kepplerites loganianus* Faunule – *K. loganianus* ♀ and *newcombii* ♂, Whiteaves spp., *K. penderi* (McLearn); basal "Macrocephalus Zone"; Keppleri Subzone
   (b) *Cadoceras tonniense* Faunule – *C. (Paracadoceras) tonniense* and *glabrum* (*part.*), Imlay spp.; Nordenskjöldi Zone of Greenland
   (c) *Lilloettia liloettensis* Faunule – *L. liloettensis* Crickmay and *Xenocephalites vicarius* Imlay
   (d) *Cadoceras wosnessenskii* Faunule – *C. wosnessenskii* (Grewingk) *sensu* Imlay ♀, *catastoma* Pompeckji *sensu* Imlay ♀, *glabrum* ♀, and *comma* ♀ (*part.*), Imlay spp.
   (e) [*Kepplerites* cf. *abruptus* Faunule – *K.* cf. *abruptus* (McLearn) and *Cadoceras* spp.]
   (f) *Lilloettia stantoni* Faunule – *L. stantoni* Imlay ♀ and *buckmani* (Crickmay) ♀, *Cadoceras* spp.
2. *Cadoceras stenoloboide* Association – *C. (Stenocadoceras) stenoloboide* ♀, *C. (Pseudocadoceras) petelini* ♂, Pompeckji spp.; Middle Callovian(?).
3. *Longaeviceras pomeroyense* Association – also *Cadoceras (Pseudocadoceras) crassicostatum* ♂ (*part.*) and *chinitnense* ♂, Imlay spp.; lower Athleta Zone

*Oxfordian*

1. *Cardioceras martini* Association – *C. martini* Reeside ♀ and ♂; Cordatum Zone, Bukowskii Subzone

2. *Cardioceras spiniferum* Association – *C. spiniferum* ♀ and *alaskense* ♀, Reeside spp., *C. distans* (Whitfield) ♂; Cordatum Zone, upper Bukowskii–lower Costicardia Subzone
3. *Cardioceras whiteavesi* Fauna – *C. whiteavesi* Reeside and *Perisphinctes* cf. *muehlbachi* Hyatt; Densiplicatum Zone
4. *Amoeboceras–Buchia concentrica* Fauna – *A.* aff. *transitorium* Spath, *Phylloceras iniskinense* Imlay, *Partschiceras, Holcophylloceras, B. concentrica* (Sowerby); Upper Oxfordian

*Volgian*

1. *Aulacosphinctoides–Buchia* Fauna – single *Aulacosphinctoides* [*"Torquatisphinctes"*] sp. and large *Subplanites?* sp. are associated with abundant *Phylloceras alaskanum* Imlay, *Buchia rugosa* (Fischer), and *B. mosquensis* (Buch); Lower Volgian

## BRITISH COLUMBIA AND ADJACENT AREAS IN CANADA[9]

### Introduction

The Jurassic rocks lying west of the Rocky Mountain Trench in British Columbia and west of the Tintina Trench in the Yukon (Figure 4.15) are thought to be mainly, if not entirely, allochthonous relative to cratonal North America. They were formed elsewhere and reached their present position as a result of translation along northwesterly trending transcurrent faults or were accreted to the margins of the continent as the Pacific plates were overridden by the continent. The assembling of the various allochthonous terranes to form the mosaic we see at present was largely complete by middle Cretaceous time according to some writers (Monger, Price, and Tempelman-Kluit 1982), or as early as Oxfordian time according to others (van der Heyden 1989), or even as early as Bajocian time by my estimate (Tipper 1984a). There is much conflicting evidence.

The Jurassic assemblages include "overlap terranes" and successor basins that contain evidence that they overlap the sutures that first welded together many discrete Triassic and Paleozoic terranes distributed along the margin of the craton. Jurassic and post-Jurassic dextral transcurrent faulting developed along major zones that broke these Jurassic overlap terranes into several fragments. At any given time these fragments had much in common, but as they continued to move and continued to receive sediments in ever-changing basins, the various resulting terranes each developed a distinct character. The ultimate product was the present mosaic of juxtaposed fragments of terranes cemented together by each succeeding overlap terrane or successor basin. The main terranes that are applicable to a discussion of the Jurassic are the Wrangellia on the west, the Stikinia and Atlin Terranes in the center, the Quesnellia lying against the craton, and the Cache Creek Terrane lying on the east side of Quesnellia (Figure 4.15).

[9] By H. W. Tipper.

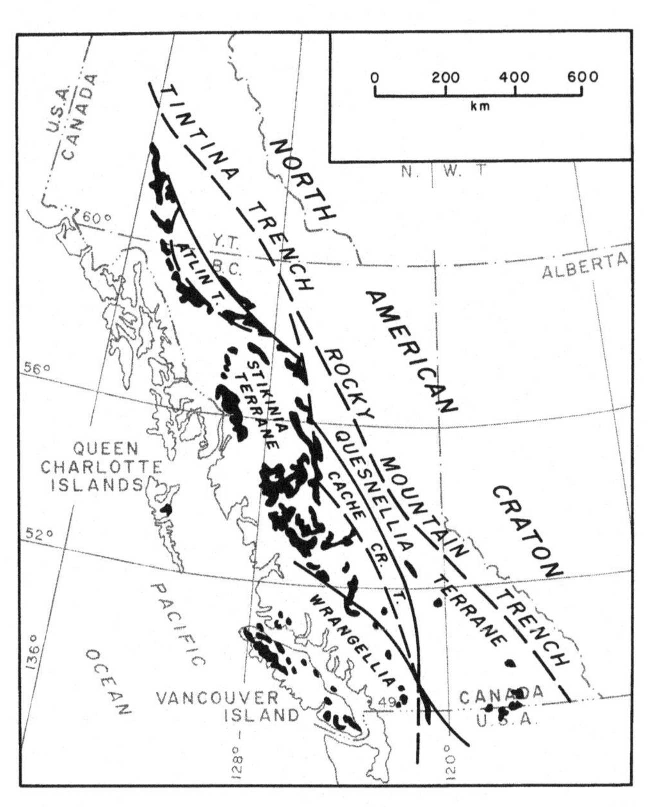


**Figure 4.15. Jurassic allochthonous terranes of the Canadian Cordillera and areas of Jurassic outcrops (shown in black).**

In general, the allochthonous terranes, except the Cache Creek Terrane, have an excellent ammonite record (Frebold and Tipper 1970). Bivalves (Poulton 1981) are, in places, abundant and well preserved. Belemnites (Jeletzky 1980) are common in the Middle Jurassic and parts of the Toarcian, and brachiopods occur abundantly in some beds. Corals occur in the Lower Jurassic, particularly the Sinemurian, and one reef has been described (Poulton 1989b). Other phyla are known from rare occurrences, such as echinoderms and fragmentary vertebrate remains (Dennison, Smith, and Tipper 1990). Plant macrofossils (Bell 1956) are not uncommon, particularly in the Upper Jurassic beds. Radiolaria (Carter, Cameron, and Smith 1988), foraminifera (Tipper and Cameron 1980), and palynomorphs (Sutherland Brown 1968, p. 76) are known from some beds, particularly in the Queen Charlotte Islands.

### Quesnellia

Of all the Jurassic terranes, Quesnellia lies closest to the craton (Figure 4.15). It is not yet clear to what extent Quesnellia is allochthonous with respect to cratonic North America, but parts, at least, appear to have been accreted by Middle Jurassic time (Tipper 1984a, p. 118). The Jurassic succession on Quesnellia is composed of two sequences – one volcanic and one sedimentary. The volcanic sequence is the oldest Jurassic in a succession made up primarily of augite-porphyry basaltic breccias and tuffs and volcaniclastic sedimentary rocks (Figure 4.16). This type of volcanism was characteristic of the underlying Upper Triassic

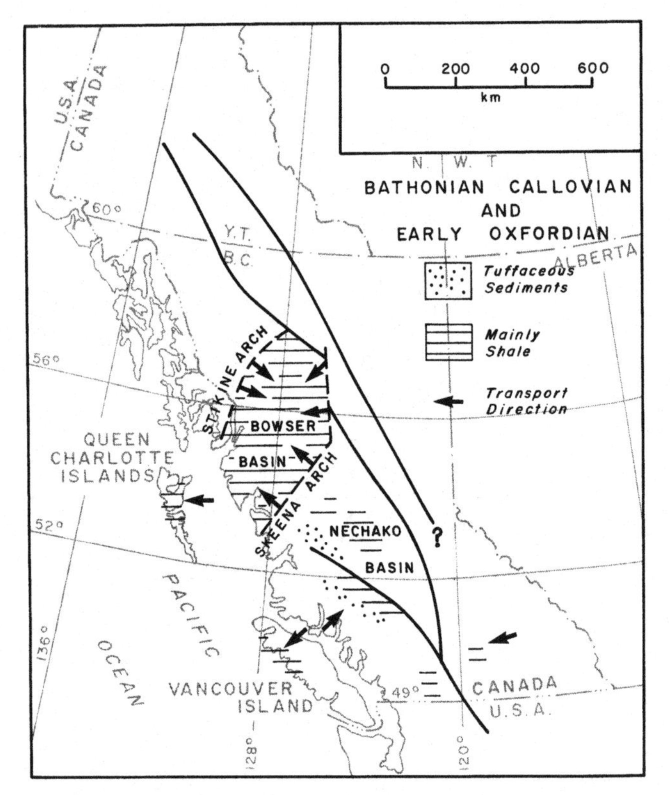


**Figure 4.16. Distribution of Sinemurian strata in the Canadian Cordillera.**

volcanics as well (Tipper 1984a, p. 117). Nowhere is there an uninterrupted succession, however, and the Sinemurian volcanics rest unconformably on the Triassic, in places marked by coarse conglomerates (Tipper and Richards 1976, pp. 9–18). These Jurassic volcanics are called the Archibald and Elise Formations of the Rossland Group in the southern part of the terrane (Little 1982), and the Quesnel River Group (Campbell 1978) in the central part; they are unnamed farther north, where they extend almost to the Yukon (H. Tipper, unpublished data).

The sedimentary sequence of the Quesnellia Terrane overlies the Sinemurian volcanics, with a marked erosional unconformity and a pronounced hiatus. The oldest sedimentary rocks are of late early Pliensbachian age (Taylor et al. 1984; Tipper 1984b) in the southern part of the terrane and of late Pliensbachian age in the central part; unnamed sedimentary rocks have been found in the north at about latitude 56°. The youngest sediments are of early Bajocian age, and all stages between Bajocian and Pliensbachian have been recognized within the belt, suggesting the presence of an uninterrupted depositional sequence. The strata are composed of marine siltstone, sandstone, and shale, with coarser sandstones and conglomerates toward the terrane margins in a few places. Detritus was derived in some places from the Cache Creek Terrane to the west and from the craton to the east. The strata are referred to the Hall Formation (Little 1982; Tipper 1984b) of the Rossland Group in the southern part of the terrane, and the Quesnel River Group in the Quesnel Lake area (Campbell 1978), but elsewhere they are unnamed.

This period of marine Jurassic sedimentation in Quesnellia ended with orogenic uplift of the Omineca Crystalline Belt. Intrusion, metamorphism, and deformation of southern parts of this belt, at least, are the results of suturing of Quesnellia to the craton. Some of the intrusions have Middle and Late Jurassic radiometric ages. Some plutons may be related to the Sinemurian volcanism, particularly in the southern half of the terrane. Deep erosion in Middle and Late Jurassic time, and in Cretaceous-Tertiary time, has almost obliterated the critical records of Jurassic history in Quesnellia.

In general, the ammonite faunas of Quesnellia are sparse, have a low diversity, and are poorly preserved. Hettangian forms are unknown. *Badouxia, Arnioceras,* and *Paltechioceras* are the most common genera of the Sinemurian. *Amaltheus, Protogrammoceras, Acanthopleuroceras,* and *Dubariceras* are known from the Pliensbachian, but are rare. *Dactylioceras* and *Harpoceras* are the only known Toarcian forms. *Erycitoides* and possibly *Tmetoceras,* known from two localities, represent the Aalenian, and Bajocian sonniniids and stephanoceratids occur at several localities. Ammonites are known only south of latitude 56°. North of this latitude, poorly preserved bivalves are the only evidence of Jurassic rocks in Quesnellia.

### Stikinia and the Atlin Terrane

The largest Jurassic terrane of the Canadian Cordillera extends from southern British Columbia into southern Yukon Territory. In the opinion of some, this is two terranes – Stikinia and the Atlin Terrane – but because the Jurassic rocks throughout their extent are intimately related and relatively inseparable, they are here considered as one terrane for Jurassic time. In southern British Columbia, south of latitude 53°, a part of the area designated as Stikinia is of low relief, heavily drift-covered, and largely obscured by younger Cretaceous and Tertiary rocks; this is only tentatively considered part of Stikinia pending further study.

Two tectonic elements dominate this terrane, namely, the Stikine Arch in northern British Columbia (Douglas et al. 1970) and the Skeena Arch (Tipper and Richards 1976) in central British Columbia (Figure 4.17). These arches are oriented generally east–west and are the sites of granitic plutons, some as old as Triassic and some as young as Middle Jurassic and younger. The plutons are genetically related to volcanic rocks of several ages. This composite terrane has Triassic and Paleozoic rocks as its basement, an amalgamation of several fragments of older terranes. Everywhere, except possibly in the southern Yukon, the Jurassic strata rest unconformably on this basement (Tipper 1984a, p. 116). Only in the southern Yukon are Hettangian beds possibly present (Lees 1934); elsewhere Sinemurian or Pliensbachian sedimentary or volcanic rocks form the base of the Jurassic succession.

Northwest of the Stikine Arch the Laberge Group overlaps the differing Triassic successions that characterize Stikinia and the Atlin Terrane. On Stikinia the base of the group consists of late Pliensbachian proximal conglomerate with interbedded sandstone and shale. On the Atlin Terrane, the base of the group is mainly Sinemurian black to gray shale and siltstone, which is succeeded by distal sandstone and shale of early and late Pliensbachian age. Clearly, the Laberge Group is evidence that these two terranes were linked by late Pliensbachian time; this linkage is apparent until early Bajocian time. The Laberge Group is characteristically coarse-clastic sedimentary rocks (Wheeler 1961), but minor volcanics, mainly tuff and breccia, occur within it.

South of the Stikine Arch, the Hazelton Group (Tipper and Richards 1976), with many formations, is widespread. In the lower parts, they are composed of calc-alkaline volcanics, mainly fragmental, ranging in age from early Sinemurian to early Bajocian. Interbedded with these volcanics, or forming sequences that represent small distinct basins associated with the volcanics, are volcanogenic sediments – shales, siltstones, and sandstones (Thomson, Smith, and Tipper 1986). The greatest thicknesses of volcanics usually occur at the base of the Hazelton Group, whether the age is Sinemurian or Pliensbachian. The volcanics are more or less centered on or flank the two arches, between which they thin and grade to tuffaceous sedimentary rocks. The source of the sediment was mainly the volcanics on the arches, but detritus was also received from the Cache Creek Terrane to the east.

South of the Skeena Arch, the Hazelton Group has been recognized over wide areas (Woodsworth 1980; Diakow and Koyanagi 1988). The Jurassic rocks disappear beneath Tertiary cover in the southern part of the terrane, so that the character of the southern part of the terrane during the Jurassic is in doubt.

After the end of volcanism in early Bajocian time, two successor basins developed (Figures 4.17 and 4.18), one south of Skeena Arch, called the Nechako Basin, which is poorly known, and one between the Skeena Arch and the Stikine Arch called the Bowser Basin (Douglas et al. 1970, p. 443). Both basins are entirely sedimentary, except for a small area in southern Bowser Basin underlain by the "Netalzul volcanics" (Tipper and Richards 1976, pp. 34–6). The strata vary from coarse sandstone, chert, pebble conglomerate to shale and siltstone. North of Stikine Arch there is no record of marine Jurassic strata younger than early Bajocian.

The Bowser Basin (Figure 4.18) received detritus from the north, east, and south and apparently opened westward to the Pacific Ocean. The basal beds are late Bajocian in age in the south and late Bathonian in the north. Marine sedimentation continued through the Callovian and Oxfordian and into the early Kimmeridgian, after which nonmarine conditions persisted, for the most part, with deltaic and fluviatile deposits. Plant fossils are abundant, and coal is locally significant. It is uncertain whether these nonmarine sediments are Jurassic or partly Jurassic and Cretaceous (Moffat, Bustin, and Rouse 1988).

Faunas of the Stikinia-Atlin Terrane are poorly to moderately well preserved and in places are relatively abundant. Diversity is fairly high. Hettangian forms are unknown except for a *Psiloceras* reported from the southern Yukon (Lees 1934). Sinemurian forms are most commonly *Arnioceras* and *Paltechioceras* and include a few unidentified arietitids and rare *Badouxia* (Souther 1972); the number of specimens is usually small. Pliensbachian ammonite faunas (Thomson 1985) are abundant and commonly are characterized by *Dubariceras, Tropidoceras, Acanthopleuroceras, Amaltheus, Protogrammoceras* and other hildoceratids, *Fanninoceras* and *Tiltoniceras.* Toarcian *Dactylioceras,* harpoceratids, *Phymatoceras,* and *Pseudogrammoceras* are common but not abundant (Frebold 1964b; Frebold and Tipper 1970). *Tmetoceras, Pseudolioceras,* and *Erycitoides* are characteristic for the Aalenian, and sparse sonniniids and stephanoceratids represent the lower Bajocian (Poulton and Tipper 1990). Within the Bowser Basin, *Iniskinites, Lilloettia, Kepplerites,* cadoceratids, perisphinctids, *Cardioceras,* and *Amoeboceras* typify the Bathonian to Kimmeridgian strata (Frebold and Tipper 1970, 1973). Bivalves (including *Buchia*) and belemnites are common and in places are predominant.

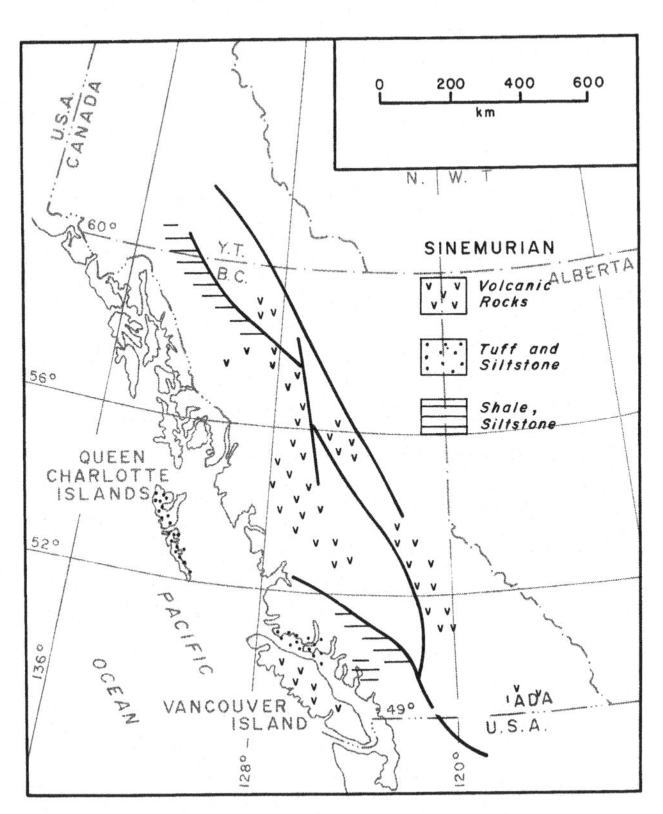


**Figure 4.17. Bathonian, Callovian, and early Oxfordian of the Cordillera overlapping Wrangellian, Stikinian, and Quesnellian Lower Jurassic strata.**

### Wrangellia

The Jurassic rocks of the Wrangellia Terrane include those of the Queen Charlotte Islands and Vancouver Island and those extending eastward to the Yalakom Fault. It is an overlap terrane deposited on the Triassic rocks of Wrangellia on Vancouver Island, on the Triassic Cadwallader Terrane near Fraser River fault, and on a Triassic terrane, possibly a part of Stikinia, between the Cadwallader Terrane and the Triassic Wrangellian Terrane.

The Jurassic rocks of Wrangellia are volcanic in the west and sedimentary in the east. The distribution is best illustrated by the Sinemurian rocks (Figure 4.16). On Vancouver Island there are extensive accumulations of varicolored calc-alkaline volcanics of the Bonanza Group (Muller, Northcote, and Carlisle 1974),

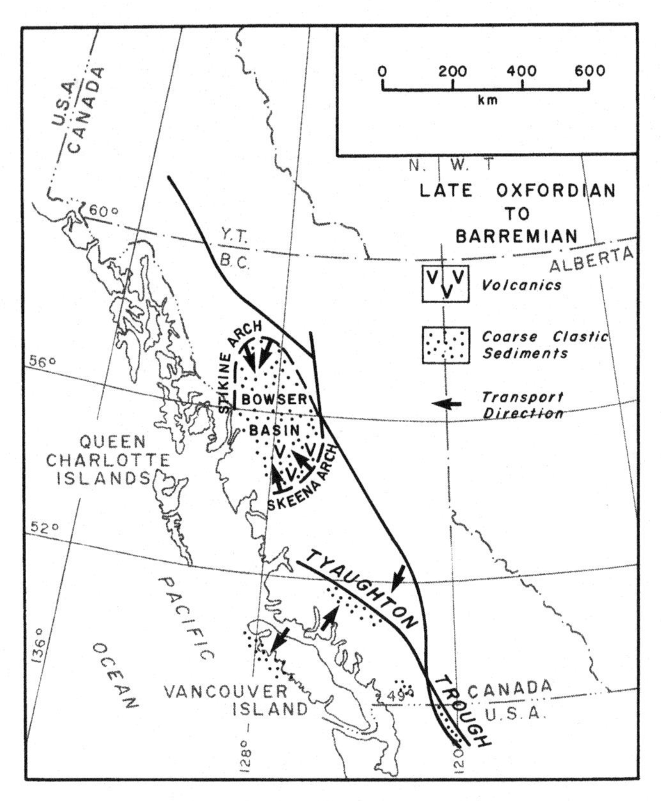


**Figure 4.18. Late Oxfordian to late Tithonian successor basins.**

with minor interbedded volcanogenic sediments. To the east of the Bonanza volcanics is the Harbledown Formation, a sequence of interbedded, finely banded tuff, siltstone, and shale that extends northwestward, where its equivalent in the Queen Charlotte Islands is called the Sandilands Formation of the Kunga Group (Cameron and Tipper 1985). East of the Harbledown Formation, within the Coast Plutonic Belt are several pendants with correlative remnants of a Sinemurian shale basin (H. Tipper, unpublished data), and on the east side of the Coast Mountains against the Yalakom Fault are extensive sections (Cameron and Tipper 1985) of shale, sandstone, and pebble conglomerate in the eastern extremity.

Wrangellia has one of the best Jurassic marine records in Canada, with, in places, a complete Lower Jurassic to Lower Bajocian succession characterized by abundant and well-preserved ammonite faunas. In the Queen Charlotte Islands, the only very good Hettangian record known in Canada (Tipper 1989) is in this terrane. Widespread Sinemurian and Pliensbachian sequences have been recognized (Muller et al. 1974; Cameron and Tipper 1985), and although not present on Vancouver Island, Toarcian to Bajocian strata are widely known elsewhere in Wrangellia. Lower Bajocian Yakoun volcanics, a mostly fragmental andesitic to basaltic sequence with marine interbeds, are widespread in the Queen Charlotte Islands (Sutherland Brown 1968) and may be correlative with the Harrison Lake volcanics east of Vancouver. After deposition of the Lower Bajocian rocks, an erosional interval, possibly accompanying deformation, changed the depositional basin to one with a more restricted character. Along and near the Yalakom Fault, a depositional basin called Tyaughton Trough was formed in which coarse clastics accumulated, derived both from the southwest and from the northeast (Figure 4.18). This trough therefore received detritus from the southern part of Stikinia and was an overlap terrane amalgamating Stikinia and Wrangellia as early as Callovian time (Jeletzky and Tipper 1968). The deposits of this trough, the Relay Mountain Group, are mainly sandstone, conglomerate, and siltstone, generally coarse and abundantly fossiliferous. Deposition in this trough continued uninterrupted into the late Lower Cretaceous. Similar sediments of Bathonian, Callovian, and Oxfordian age are known in the Queen Charlotte Islands (Cameron and Tipper 1985, p. 39) and on Vancouver Island.

Ammonite and bivalve faunas are well preserved, abundant, and highly diverse in the Queen Charlotte Islands and in the Tyaughton Creek area. Hettangian forms include *Kammerkarites, Fergusonites, Pleuroacanthites, Schlotheimia, Pseudaetamoceras, Alsatites, Franziceras* (H. Tipper, unpublished data), phylloceratids, and many other genera and species not yet studied. The possibility of the entire Hettangian stage being represented is most probable, but that has not been proved; many forms similar to those of Nevada and Oregon are present (Guex 1980). Sinemurian faunas include such genera as *Badouxia, Metophioceras,* a host of unidentified arietitids, *Arnioceras, Asteroceras, Caenisites,* oxynoticeratids, *Paltechioceras,* and *Crucilobiceras;* preservation at many localities is excellent, and specimens are numerous (Palfy et al. 1990). In the Queen Charlotte Islands, Pliensbachian faunas (Smith et al. 1988) are abundant, diverse, and well to excellently preserved; the Pliensbachian fauna here, on the whole, are the best in North America. *Pseudoskirroceras, Gemmellaroceras, Metaderoceras, Tropidoceras, Acanthopleuroceras, Dubariceras, Fanninoceras, Protogrammoceras, Arieticeras,* and *Tiltoniceras* are only a few of the multitude of genera currently being studied. The Toarcian fauna of the Queen Charlotte Islands is similarly an outstanding fauna. *Dactylioceras, Phymatoceras, Peronoceras, Haugia, Pseudogrammoceras, Hammatoceras, Grammoceras,* and *Dumortieria* are a few of the genera that have been recognized (Tipper et al. 1990). *Tmetoceras, Erycitoides,* and *Bredyia* are typical faunas in Aalenian time (Poulton and Tipper 1990), and sonniniids, *Stephanoceras, Zemistephanus,* and *Chondroceras* (Hall and Westermann 1980) are not uncommon in the Lower Bajocian.

In Tyaughton Trough, the upper part of the Middle Jurassic through to the end of the Jurassic has been identified by faunas such as *Lilloettia, Kepplerites, Cadoceras, Cardioceras,* and various perisphinctids not yet studied. The late Oxfordian to latest Jurassic beds contain an excellent record of the bivalve *Buchia* (Jeletzky and Tipper 1968), without which there would be a very incomplete Late Jurassic fossil record. Callovian *Cadoceras* and Oxfordian *Cardioceras* are present on Vancouver Island, and Bathonian to Callovian *Iniskinites, Kepplerites,* and *Cadoceras* are present in the Queen Charlotte Islands.

### Cache Creek and Bridge River Terranes

In recent years the oceanic rocks of the Cache Creek and Bridge River terranes have yielded Middle Jurassic (Bajocian)

GEOGRAPHIC LOCATION
OF STRATIGRAPHIC UNITS

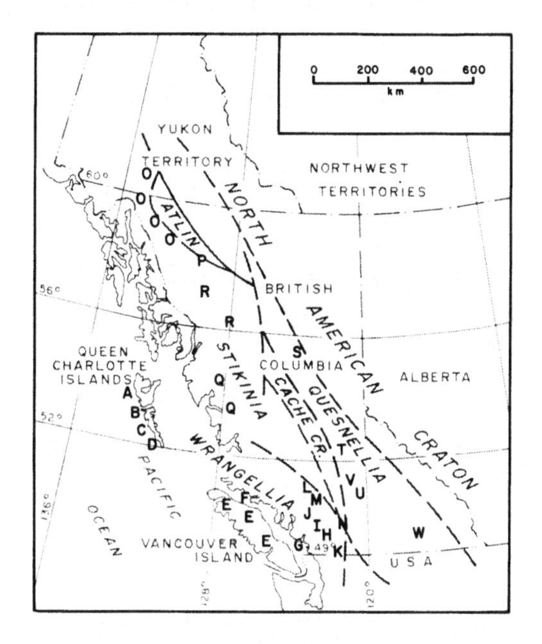


DISTRIBUTION OF UNITS
THROUGH TIME

| Period | Epoch | Stages | Ages | | | | | | | | | | | | | |
|---|---|---|---|---|---|---|---|---|---|---|---|---|---|---|---|---|
| JURASSIC | LATE | UPPER TITHONIAN | 135 | | | | | M | | | | | | | | |
| | | | | | | | | M | | | | | | | | |
| | | LOWER TITHONIAN | | | | | | M | | | | | | | | |
| | | | | | | | | M | | | | | | | | |
| | | KIMMERIDGIAN | 141 | | | | | M | | | | | | | | |
| | | | | | | | | M | | | | | | | | |
| | | OXFORDIAN | 146 | | | | | M | | | | | | | | |
| | | | | | | | J | M | | | | R | | | | |
| | MIDDLE | CALLOVIAN | 154 | | | | | | | | | | | | | |
| | | | | D | | | I | M | | | | R | | | | |
| | | BATHONIAN | 160 | D | | | | | | | | R | | | | |
| | | | | | | | | | | | | R | | | | |
| | | BAJOCIAN | 167 | | | | | | | | | R | | | | |
| | | | | C | | G | H | L | N | O | P | Q | S | T | V | |
| | | AALENIAN | 176 | | | | | | | | | | | | | |
| | | | | B | | | | L | N | O | P | Q | | T | | |
| | EARLY | TOARCIAN | 180 | B | | | | L | N | O | P | Q | | | | |
| | | | | B | | | K | L | | O | P | Q | | | | W |
| | | PLIENSBACHIAN | 187 | B | E | | K | L | | O | P | Q | S | T | | |
| | | | | B | E | | | L | | O | P | Q | | T | | W |
| | | SINEMURIAN | 194 | A | E | F | | L | | O | | Q | S | | | W |
| | | | | A | E | | | L | | O | | Q | | | U | W |
| | | HETTANGIAN | 201 | A | | | | L | | O | | | | | | W |
| | | | 205 | A | | | | | | | | | | | | |

STRATIGRAPHIC UNITS

WRANGELLIAN TERRANE

| |  |  |
|---|---|---|
| A | KUNGA GROUP | |
| | Sandilands Fm. | — siltstone, tuff. |
| B | MAUDE GROUP | |
| | Ghost Creek Fm. | — shale, siltstone. |
| | Fannin Fm. | — siltstone, sandstone, tuff. |
| | Whiteaves Fm. | — shale, siltstone. |
| | Phantom Creek Fm. | — sandstone, siltstone. |
| C | YAKOUN GROUP | — volcanic breccia, tuff, siltstone. |
| D | MORESBY GROUP | — sandstone, siltstone. |
| E | BONANZA GROUP | — volcanic breccia, tuff. flows, siltstone. |
| F | Harbledown Fm. | — siltstone, tuff. |
| G | BOWEN ISLAND GROUP | — volcanic breccia, tuff, volcanogenic sediment. |
| H | Harrison Lake Fm. | — volcanic breccia tuff. |
| I | Mysterious Creek Fm. | — shale, siltstone. |
| J | Billhook Creek Fm. | — volcanogenic sediments, breccia. |
| K | Cultus Lake Fm. | — shale. |
| L | Last Creek Fm. | — shale, siltstone, sandstone, conglomerate. |
| M | RELAY MOUNTAIN GROUP | — sandstone, siltstone conglomerate. |
| N | LADNER GROUP | — siltstone, volcanogenic sediments, conglomerate. |

STIKINIA TERRANE

| |  |  |
|---|---|---|
| O | LABERGE GROUP | |
| | Takwahoni Fm. | — conglomerate, siltstone, tuff. |
| | Inklin Fm. | — sandstone, siltstone, conglomerate. |
| P | SPATSIZI GROUP | |
| | Quock Fm. | — siltstone, tuff. |
| | Abou Fm. | — shale, tuff. |
| | Melisson Fm. | — siliceous siltstone, tuff. |
| | Joan Fm. | — siltstone, limestone, conglomerate. |
| Q | HAZELTON GROUP | |
| | Smithers Fm. | — greywacke, siltstone, volcanic breccia, tuff. |
| | Nilkitkwa Fm. | — shale, greywacke, tuff, breccia. |
| | Telkwa Fm. | — volcanic flows, breccias, tuffs. |
| R | BOWSER LAKE GROUP | |
| | Ashman Fm. | — shale, conglomerate, siltstone. |

QUESNELLIA TERRANE

| |  |  |
|---|---|---|
| S | TAKLA GROUP (in part) | — volcanic breccia, siltstone. |
| T | QUESNEL RIVER GROUP (in part) | — shale, siltstone, conglomerate. |
| U | NICOLA GROUP (in part) | — volcanic breccia, tuff. |
| V | Ashcroft Fm. | — siltstone, shale, sandstone conglomerate. |
| W | ROSSLAND GROUP | |
| | Hall Fm. | — shale, siltstone. |
| | Elise Fm. | — volcanic breccia, tuff. |
| | Archibald Fm. | — siltstone, tuff. |

**Figure 4.19. Generalized lithostratigraphy of the allochthonous terranes (excluding the Cache Creek Terrane).**

radiolaria (Cordey et al. 1987). The character and development of these terranes are still highly speculative, but it is an inescapable conclusion that at least part of these belts represent oceanic sediments of a deep-ocean basin. Their relationships to each other and to the other Jurassic terranes are still open questions, and their extents are unknown. To date, the only evidence regarding them has been derived from southern British Columbia.

### Tectonic summary

Jurassic intrusive rocks are widespread, but they may not be as extensive or prominent as Triassic or Cretaceous plutons. There appears to be a close relation in time and space with volcanic rocks. Volcanic rocks are for the most part restricted to Lower Jurassic and early Middle Jurassic time. Basic augite porphyries are (Tipper 1984a, pp. 116–17) almost entirely restricted to Quesnellia, whereas calc-alkaline volcanics typify accumulations in Wrangellia and Stikinia (Figure 4.19).

In general, the Lower Jurassic to lower Bajocian sedimentary rocks are fine-grained, except in the southern Yukon; conglomerates usually occur only in proximal settings (Wheeler 1961; Souther 1962). Middle to Upper Jurassic sedimentary rocks are mainly sandstone, conglomerate, and siltstone. Only small lenses or thin beds of limestone are known, and these commonly are in Wrangellia.

Hiatuses and unconformities (Figure 4.19) occur within the various sequences. Together with abrupt lithologic changes and cessation of volcanism, they mark major or minor events that occurred simultaneously in the various terranes, suggesting a close relationship, spatially or tectonically, between the terranes (Tipper 1984a). At the end of early Bajocian time and at the end of the early Oxfordian there were events common to all terranes, suggesting that these events were not local. Lesser changes or events are noted at the end of Sinemurian time. The base of the Jurassic is everywhere marked by an unconformity with the Triassic, except in the Queen Charlotte Islands, where sedimentation may have been continuous across the Triassic–Jurassic boundary (H. Tipper, unpublished data).

Unlike the Paleozoic (Monger and Ross 1971) and Triassic (Tozer 1982a) terranes, which have been interpreted as having been transported across the Pacific or at least for several thousand kilometers, the Jurassic terranes of the Canadian Cordillera may not have traveled far (Smith and Tipper 1986). Clearly, the Lower Jurassic rocks have been disrupted and latitudinally dislocated by dextral strike-slip faults, possibly as early as early Bajocian time, and certainly by Cretaceous and early Tertiary time. The faunas cf these Jurassic terranes have strong ties to North American cratonal raunas, suggesting they developed in the eastern Pacific; many such forms are endemic to the eastern Pacific.

### Ammonite Zones

Although there are relatively few ammonite zones established for the Jurassic of Canada and correlated with the standard zones of northwest Europe, there are nevertheless large numbers of genera and species known but not reported in the literature, and formal ammonite zones have not yet been established for most stages. Late Jurassic ammonites have not been found in abundance, and subdivision of this time has been achieved by reliance on *Buchia* faunas (Jeletzky and Tipper 1968; Jeletzky 1984). If a fauna is known to the writer to represent a particular span of time but has not been reported, its presence has been indicated in the lists that follow. The faunas are here listed by terrane.

#### *Hettangian*

The only apparently complete Hettangian section in Canada with diverse fauna was discovered recently (1988) in Wrangellia (Queen Charlotte Islands).

| Wrangellia | Stikinia | Quesnellia |
|---|---|---|
| | **Early** | |
| Present | *Psiloceras erguatum* (Bean) | No record |
| | **Middle & late** | |
| Present | No record | No record |

#### *Sinemurian*

Sinemurian rocks are widespread, but only in Wrangellia is there a strong diversity and rich fauna.

| Wrangellia | Stikinia | Quesnellia |
|---|---|---|
| | **Early** | |
| | 1. Canadensis Zone | |
| *Badouxia canadensis* (Frebold)<br>*B. columbiae* (Frebold)<br>*B. occidentalis* (Frebold)<br>*Eolytoceras tasekoi* (Frebold)<br>*Metophioceras multicostatum* (Frebold)<br>*Charmasseiceras marmoreum* (Oppel) | *Badouxia canadensis* | *Badouxia columbiae* (Frebold) |
| | 2. | |
| *Arnioceras kwakiutlanus* (Crickmay)<br>*Arnioceras sp.*<br>*Arietites sp.* | *Arnioceras* sp. | *Arnioceras kwakiutlanus* |

*Sinemurian (cont.)*

| Wrangellia | Stikinia | Quesnellia |
|---|---|---|
| | 3. | |
| Present | ? | ? |
| | **Late** | |
| *Paltechioceras harbledownensis* (Crickmay)<br>*Crucilobiceras* sp.<br>*Oxynoticeras?* sp.<br>*Asteroceras stellare* (Sowerby) | *Paltechioceras* sp. | *Paltechioceras harbledownensis* |

*Pliensbachian*

In Wrangellia (Queen Charlotte Islands) an excellent record of Pliensbachian ammonite faunas has been found and used to develop in part a North American ammonite zonation (Smith et al. 1988).

| Wrangellia | Stikinia | Quesnellia |
|---|---|---|
| | **Early** | |
| | 1. Imlayi Zone | |
| *Gemmellaroceras* spp.<br>*Polymorphites confusus* (Quenstedt)<br>*Miltoceras* aff. *sellae* (Gemmellaro)<br>*Tropidoceras* aff. *erythraeum* (Gemmellaro)<br>*T. flandrini* (Dumortier)<br>*T. actaeon* (d'Orbigny)<br>*Pseudoskirroceras imlayi* Smith & Tipper<br>*Metaderoceras evolutum* (Fucini)<br>*Phricodoceras* cf. *taylori* (Sowerby) | *Gemmellaroceras* sp.<br>*Miltoceras* sp.<br>*Polymorphites* sp. | No record |
| | 2. Whiteavesi Zone | |
| *Gemmellaroceras* spp.<br>*Polymorphites confusus* (Quenstedt)<br>*Metaderoceras evolutum* (Fucini)<br>*M.* aff. *muticum* (d'Orbigny)<br>*M. mouterdei* (Frebold)<br>*Tropidoceras flandrini* (Dumortier)<br>*T. actaeon* (d'Orbigny)<br>*T. masseanum* (d'Orbigny)<br>*Phricodoceras* cf. *taylori* (Sowerby)<br>*Acanthopleuroceras whiteavesi* Smith & Tipper<br>*A.* aff. *stahli* (Oppel)<br>*Dubariceras silviesi* (Hertlein)<br>*Liparoceras (Becheiceras) bechei* (Sowerby)<br>*Reynesocoeloceras* spp. | *Tropidocerus* sp.<br>*Metaderoceras* aff. *multicum* | *Metaderoceras* sp. |
| | 3. Freboldi Zone | |
| *Phricodoceras* cf. *taylori* (Sowerby)<br>*Dubariceras silviesi* (Hertlein)<br>*D. freboldi* Dommergues, Mouterde, & Rivas<br>*Metaderoceras* aff. *muticum* (d'Orbigny)<br>*M. mouterdei* (Frebold)<br>*Liparoceras (Becheiceras) bechei* (Sowerby)<br>*Reynesocoeloceras* spp.<br>*Aveyroniceras colubriforme* (Bettoni)<br>*Aveyroniceras* spp.<br>*Prodactylioceras* aff. *davoei* (Sowerby) | *Dubariceras freboldi* Dommergues, Mouterde, & Rivas<br>*D. silviesi* (Hertlein)<br>*Metadoceras* aff. *muticum*<br>*M. cf. mouterdei*<br>*Uptonia?* sp.<br>*Dayiceras* sp. | Present |

*Pliensbachian (cont.)*

| Wrangellia | Stikinia | Quesnellia |
|---|---|---|
| | **Late** | |
| | 4. Kunae Zone | |
| *Aveyroniceras colubriforme* (Bettoni)<br>A cf. *inaequi-oratum* (Bettoni)<br>*Fanninoceras fannini* McLearn<br>*F. latum* McLearn<br>*F. crassum* McLearn<br>*Fanninoceras* n.sp.<br>*F. kunae* McLearn<br>*Fontanelliceras* sp.<br>*Fuciniceras* aff. *intumescens* (Fucini)<br>*Fieldingiceras pseudofieldingi* (Fucini)<br>*F.* aff. *sygma* (Monestier)<br>*Arieticeras* aff. *algovianum* (Oppel)<br>*Leptaleoceras* aff. *accura tum* (Fucini)<br>*Reynesocoeloceras* cf. *indunese* (Megeghini)<br>*Protogrammo ceras* cf. *lusitanicum* (Choffat)<br>*P. varicostatum* (Behmel & Geyer)<br>*P. paltum* Guex<br>*Amaltheus stokesi* (Sowerby)<br>*Reynesoceras ragazzonii* (Hauer) | *Tiltoniceras propinquum* (Whiteaves)<br>*Liparoceras (Becheiceras) bechei* (Sowerby)<br>*Arieticeras* aff. *algovianum* (Oppel)<br>*A.* cf. *ruthenense* (Reynès)<br>*Fuciniceras* sp.<br>*F.* aff. *intumescens* (Fucini)<br>*Protogrammoceras pectinatum* (Meneghini)<br>*Protogrammoceras* sp.<br>*Amaltheus margaritatus (de Montfort)*<br>*A. stokesi* (Sowerby)<br>*Fanninoceras latum* McLearn<br>*Leptaleoceras* aff. *speciosum* (Fucini) | Present |
| | 5. Carlottense Zone | |
| *Protogrammoceras* spp.<br>*Fanninoceras latum* McLearn | *Arieticeras* cf. *algovianum*<br>*Amaltheus margaritatus* | ? |

| Wrangellia | Stikinia | Quesnellia |
|---|---|---|
| | 5. Carlottense Zone (*cont.*) | |
| *F. carlottense* McLearn<br>*Arieticeras* aff. *algovianum* (Oppel)<br>*Amaltheus margaritatus* (de Montfort)<br>*A. viligaensis* (Tuchov)<br>*Protogrammoceras paltum* Guex<br>*P. pectinatum* (Meneghini)<br>*P. allifordensis* (McLearn)<br>*Tiltoniceras propinquum* (Whiteaves)<br>*Lioceratoides* spp. | *Tiltoniceras propinquum* (Whiteaves)<br>*Lioceratoides* sp.<br>*Protogrammoceras paltum* Guex<br>*Fieldingiceras* aff. *sygma* (Monestier) | |

*Toarcian*

Toarcian faunas are widespread and relatively diverse. The best sections are in the Wrangellia.

| Wrangellia | Stikinia | Quesnellia |
|---|---|---|
| | **Early** | |
| *Tiltoniceras propinquum* (Whiteaves)<br>*Dactylioceras kanense* McLearn | *Harpoceras exaratum* Young & Bird | *Harpoceras exaratum*<br>*Dactylioceras* sp. |
| | **Middle** | |
| Present | *Peronoceras* sp. | ? |
| | **Late** | |
| *Dumortieria* sp.<br>*Phlyseogram moceras* sp.<br>*Hammatoceras* sp. | *Grammoceras* sp.<br>*Catulloceras?* sp. | ? |

*Aalenian*

The Aalenian stage is under study, and a moderately diverse fauna is being organized into ammonite zones. Prior to this study, the only record of note is of *Tmetoceras scissum* (Benecke), occurring in all three terranes at the base of the Aalenian. Higher in the Aalenian there is a diverse fauna that is comparable to those in the published records in Alaska and the Canadian Arctic. Only the latest Aalenian has not been recorded, and this may reflect collection failure.

*Bajocian*

Early Bajocian ammonoid faunas have been recorded in all three terranes (Hall and Westermann 1980), but only in Stikinia is a Late Bajocian fauna recognized.

| Wrangellia | Stikinia | Quesnellia |
|---|---|---|
| | **Early** | |
| | 1. Widebayense Zone | |
| *Fontannesia* (?) sp.<br>sonniniids | *Docidoceras* (?) sp.<br>sonniniids? | ? |
| | 2. Crassicostatus Zone | |
| *Parabigotites*(?) sp.<br>oppeliids(?)<br>sonniniids<br>*Kumatostephanus* sp. | Present | Present |
| | 3. Kirschneri Zone | |
| *Zemistephanus richardsoni* (Whiteaves)<br>*Z. carlottensis* (Whiteaves)<br>*Z. alaskensis* Hall<br>*Z. crickmayi* (McLearn)<br>*Stephanoceras* aff. *acuticostatum* Weisert<br>*Sphaeroceras (Chondroceras)* sp. | *Stephanoceras (Stemmatoceras)* sp.<br>*Chondroceras allani* (McLearn) | Present |
| | 4. Oblatum Zone | |
| *Sphaeroceras (Chondroceras) oblatum* (Whiteaves)<br>*Stephanoceras itinsae* (McLearn)<br>*Teloceras crickmayi* (Frebold) | Present | ? |
| | **Late** | |
| No record | *Megasphaeroceras?* aff. *rotundum* Imlay<br>*Epizigzagiceras crassum* Frebold | ? |

*Bathonian*

Bathonian (Frebold 1978) and younger Jurassic faunas are largely restricted to basins within Wrangellia and Stikinia.

| Wrangellia | Stikinia | Quesnellia |
|---|---|---|
| | **Early** | |
| No record | *Arctocephalites (Cranocephalites) costidensus* Imlay<br>*A. (C.)* aff. *C. pompeckji* (Madsen)<br>*Parareineckeia* cf. *P. shelikofana* (Imlay)<br>*Cobbanites talkeetnanus* Imlay<br>*Epizigzagiceras evolutum* Frebold<br>*Morrisiceras? dubium* Frebold | No record |
| | **Late** | |
| *Iniskinites mclearni* (Frebold)<br>*I. cepoides* (Whiteleaves) | *Iniskinites robustus*<br>*I. intermedius*<br>*I. tenasensis* Frebold<br>*Kepplerites (Seymourites)* sp.<br>*Xenocephalites* sp.<br>*Lilloettia lilloetensis* Crickmay | No record |

*Bathonian or Callovian*

| Wrangellia | Stikinia | Quesnellia |
|---|---|---|
| *Kepplerites (Seymourites) loganianus* Whiteaves<br>*K. (S.) ingrahami* McLearn<br>*K. (S.) newcombi* McLearn<br>*Cadoceras* cf. *C. catostoma* Pompeckji<br>*Partschiceras grantzi* Imlay | *Lilloettia* spp.<br>*Kepplerites* spp. | *Kepplerites (Seymourites)* sp. |

*Callovian*

| Wrangellia | Stikinia | Quesnellia |
|---|---|---|
| | **Early or Middle** | |
| *Cadoceras (Stenocadoceras) stenoloboide* Imlay<br>*Cadoceras* sp.<br>*C. (Stenocadoceras)* sp.<br>*Pseudocadoceras* sp. | *Cadoceras (Stenocadoceras)* sp. | *Pseudocadoceras* sp.<br>*Kepplerites* sp. |
| | **Late** | |
| No record | *Quenstedtoceras (Lamberticeras) henrici* Douvillé<br>*Q. (L.) intermissum* Buckman<br>*Phylloceras (Partschiceras) pacificum* Frebold & Tipper | No record |

*Oxfordian*

| Wrangellia | Stikinia | Quesnellia |
|---|---|---|
| | **Early** | |
| *Cardioceras (C.) lilloetense* Reeside<br>*Cardioceras (C.)* sp. | *Cardioceras (Cardioceras) cordiforme* Meek & Hayden<br>*C. (C.) lilloetense canadense* (Whiteaves)<br>*C. (Scarburgiceras) martini* Reeside<br>*Phylloceras (Partschiceras) pacificum* Frebold & Tipper | No record |
| | **Late** | |
| No record | *Amoeboceras* sp. | No record |

### Oxfordian to Tithonian *Buchia* zones of the Canadian cordillera

The Late Oxfordian to Tithonian *Buchia* fauna is the most useful for a zonation of this time period; ammonites generally are not abundant and are poorly preserved. The best record of this fauna is in Tyaughton Trough in southwest British Columbia, an overlap terrane at the approximate Wrangellia–Stikinia boundary. The fauna is unknown in Quesnellia.

| | | |
|---|---|---|
| Early/Late Oxfordian | *Buchia concentrica* Zone | *Buchia concentrica* (Sowerby)<br>*Rasenia* sp., *Amoeboceras* sp.<br>*Cardioceras* sp. |
| Kimmeridgian | *Buchia mosquensis* Zone | *Buchia mosquensis* (Buch)<br>*B. tschernyschewi* Sokolov<br>*B. volongensis* Sokolov<br>*B. lindstroemi* Sokolov<br>*B. concentrica* (Sowerby) |
| Early Tithonian | *Buchia* cf. *blanfordiana* Zone | *Buchia* cf. *blanfordiana* Stoliczka<br>*B.* aff. *russiensis* (Pavlow)<br>*B. piochii* (Gabb)<br>*?Paraberriasella* sp. |
| Late Tithonian | *Buchia piochii* Zone | *Buchia piochii* (Gabb)<br>*B. lahuseni* (Pavlow)<br>*B. fischeriana* (d'Orbigny) |
| | *Buchia fischeriana* Zone | *Buchia fischeriana* (d'Orbigny)<br>*B. lahuseni* (Pavlow), *B. terebratuloides s.l.* (Lahusen)<br>*B. piochii* (Gabb) |
| | *Buchia terebratuloides s.l.* Zone | *Buchia terebratuloides s.l.*<br>*B.* aff. *okensis* (Pavlow)<br>*B. unschensis* (Pavlow), *B. lahuseni* (Pavlow)<br>*Substeueroceras* sp.<br>*Paradontoceras reedi* (Anderson) |

## EASTERN OREGON AND ADJACENT AREAS[10]

### Introduction

The Blue Mountains island arc of eastern Oregon and adjacent areas (Blome et al. 1986; Vallier and Brooks 1986) consists of five terranes that either include Jurassic sediments in their stratigraphy or have important structural or sedimentological relationships with the Jurassic rocks of adjacent terranes (Figures 4.20 and 4.21). The terranes, each typified for the most part by its distinctive upper Paleozoic and Triassic stratigraphy, are exposed as a series of inliers trending northeastward from the center of the state toward the Idaho Batholith at the western margin of the craton. The Izee and Olds Ferry Terranes, which include thick sequences of Jurassic volcaniclastic rocks, are separated from the Wallowa Terrane to the north by a central mélange, the Baker Terrane (Dickinson 1979). The Wallowa Terrane consists of Permian to Triassic oceanic lavas, volcaniclastics, and carbonates capped unconformably by Jurassic sediments. Jones et al. (1977) have suggested that the Wallowa Terrane is a southern extension of Wrangellia.

The disrupted oceanic crust forming the Baker Terrane consists of tectonic blocks of mafic igneous rocks, radiolarian cherts, schists, marble, and fusulinid limestones (of either Tethyan or North American affinity) set chaotically in a serpentine matrix (Silberling et al. 1984). Outcrops of blueschist near Mitchell have yielded Triassic $^{40}Ar$-$^{39}Ar$ ages of 223 Ma (Hotz, Lanphere, and Swanson 1977). The sedimentary rocks normally yield late Paleozoic and Triassic faunas, but Blome et al. (1986) and Pessagno and Blome (1986) reported possible Hettangian radiolarians from cherts cropping out near John Day and Baker.

The Grindstone mélange contains chaotically arranged Devonian through Triassic rocks. Apart from the greater age of its tectonic blocks, the Grindstone Terrane differs from the Baker Terrane by its lack of serpentine (Silberling et al. 1984). The significance of the Grindstone as far as the Jurassic is concerned is that it formed the western margin of the Suplee-Izee sedimentary basin. Bajocian rocks of the Snowshoe Formation overlapped westward across the Camp Creek Fault to give the Grindstone and Izee Terranes a common stratigraphy by that time (Figures 4.20 and 4.22).

### Izee Terrane

The Jurassic sequence of the Izee Terrane reaches an aggregate thickness in excess of 8,000 m, with the thickest and most nearly continuous succession occurring east and southeast of Izee (Figures 4.21 and 4.22). Transgressive–regressive pulses across the Suplee shelf to the west resulted in accumulation of several thin, unconformity-bound sequences laid down in shallower water. Diachronism and lateral variations in facies are common. Thirty kilometers southeast of Izee the distinctive sandstones of the

[10] By P. Smith and D. G. Taylor.

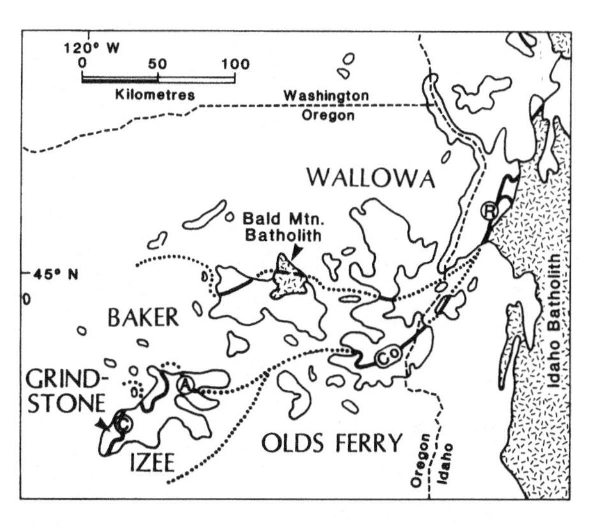


**Figure 4.20. The terranes of the Blue Mountains Island Arc as defined by Silberling et al. (1984) and Vallier and Brooks (1986). Areas of outcrop are outlined. The Camp Creek (C), Aldrich Mountain (A), Connor Creek (Co), and Rapid River (R) faults form terrane boundaries.**

Donovan Formation crop out as a small inlier in the valley of the Silvies River (Figure 4.22). The Suplee, Izee-Seneca, and Silvies River areas therefore constitute three sites of contrasting Jurassic stratigraphy (Figure 4.21), as recognized from accounts by Lupher (1941), Dickinson and Vigrass (1965), Dickinson and Thayer (1978), Smith (1980, 1981), Taylor (1981) and Imlay (1986).

The Graylock Formation, named by Dickinson and Vigrass (1965), rests conformably on the Triassic (and ?Hettangian) Rail Cabin Argillite and is overlain unconformably by the Mowich Group. It consists of dark siltstones and black limestones yielding *Discamphiceras, Schlotheimia, Sulciferites, Paracaloceras*(?), *Badouxia,* and *Sunrisites,* indicating a latest Hettangian and earliest Sinemurian age (Taylor 1986, 1988a).

The oldest Jurassic unit of the Aldrich Mountains Group is the Murderers Creek Formation, which consists of 500 m of calcarenitic turbidites that have yielded Hettangian ammonites (Dickinson and Thayer 1978). At its type locality, the Keller Creek Formation rests conformably on the Murderers Creek Formation and is overlain unconformably by the Mowich Group. The Keller Creek is poorly exposed and sparsely fossiliferous. Brown and Thayer (1966) recorded a thickness of 1,525 m, with the lower 610–760 m consisting of fine- to coarse-grained tuffaceous graywacke, and the upper part consisting of gray to black shale interbedded with some graywacke; graywacke predominates again near the top of the formation. Isolated fossil localities have yielded the bivalve *Lupherella boechiformis* (Hyatt), the trace fossil *Chondrites,* ichthyosaur and other vertebrate debris, and ammonites. The ammonite genera *Coroniceras, Arnioceras, Metaderoceras, Tropidoceras,* and *Pseudoskirroceras* indicate that the Keller Creek Formation may be allocated to the Sinemurian and basal Pliensbachian.

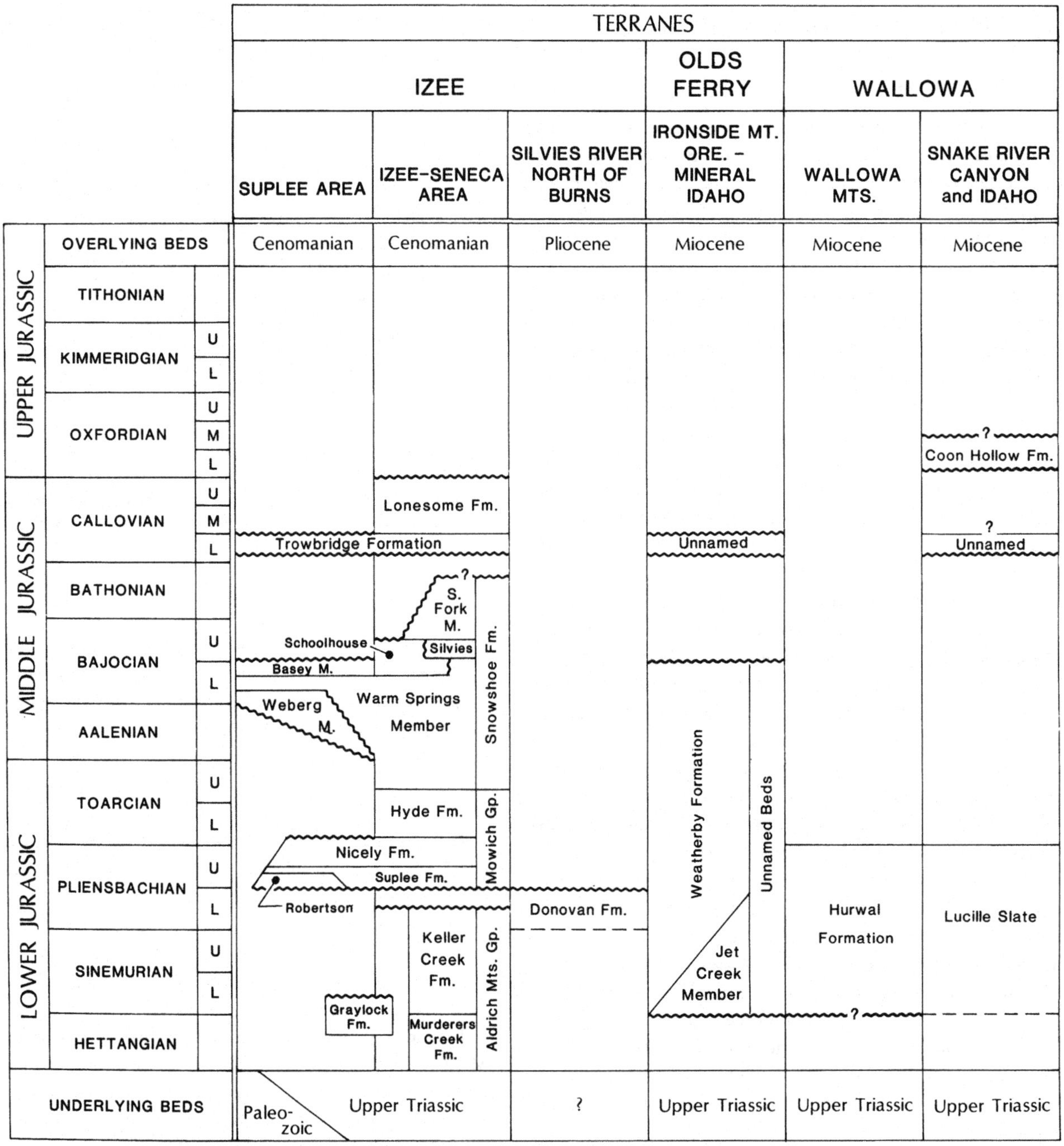

TERRANES

IZEE: SUPLEE AREA | IZEE-SENECA AREA | SILVIES RIVER NORTH OF BURNS
OLDS FERRY: IRONSIDE MT. ORE. - MINERAL IDAHO
WALLOWA: WALLOWA MTS. | SNAKE RIVER CANYON and IDAHO

OVERLYING BEDS: Cenomanian | Cenomanian | Pliocene | Miocene | Miocene | Miocene

UPPER JURASSIC: TITHONIAN; KIMMERIDGIAN (U, L); OXFORDIAN (U, M, L)
MIDDLE JURASSIC: CALLOVIAN (U, M, L); BATHONIAN; BAJOCIAN (U, L); AALENIAN
LOWER JURASSIC: TOARCIAN (U, L); PLIENSBACHIAN (U, L); SINEMURIAN (U, L); HETTANGIAN

Suplee Area / Izee-Seneca Area: Lonesome Fm.; Trowbridge Formation; ? S. Fork M.; Schoolhouse; Silvies; Basey M.; Weberg M.; Warm Springs Member; Snowshoe Fm.; Hyde Fm.; Nicely Fm.; Suplee Fm.; Mowich Gp.; Robertson; Keller Creek Fm.; Graylock Fm.; Murderers Creek Fm.; Aldrich Mts. Gp.

Silvies River North of Burns: Donovan Fm.

Ironside Mt. Ore. - Mineral Idaho: Unnamed; Weatherby Formation; Unnamed Beds; Jet Creek Member

Wallowa Mts.: Hurwal Formation; ?

Snake River Canyon and Idaho: ? Coon Hollow Fm.; ? Unnamed; Lucille Slate

UNDERLYING BEDS: Paleo-zoic / Upper Triassic | ? | Upper Triassic | Upper Triassic | Upper Triassic

**Figure 4.21. The Jurassic stratigraphy of the Izee, Olds Ferry, and Wallowa Terranes. (Adapted from Imlay 1986; Smith 1980, 1981.)**

The Donovan Formation (Lupher 1941) is 800 m thick at its type locality in the Silvies River valley (Figure 4.22), but there is evidence of faulting. Red, brown, and green sandstones are dominant through much of the section, with shales common near the base. *Tropidoceras* has been found low in the unit, and species of *Dubariceras* and *Metaderoceras* at the top, indicating that the formation was deposited during the early Pliensbachian (Smith 1983).

The Mowich Group forms a thin, unconformity-bound shelf assemblage in the Suplee area, where it is dominated by bioturbated sandstones and limestones. Locally, bioherms and biostromes are composed of the aberrant bivalve *Lithiotis* (Nauss and Smith 1988, 1989). The upper part of the Mowich Group includes black shales and mudstones of the Nicely Formation and volcaniclastic sandstones of the Hyde Formation, the latter being limited to the eastern part of the Suplee-Izee area, where it grades up into the overlying Snowshoe Formation. A diverse suite of ammonites, including species of *Arieticeras, Aveyroniceras, Protogrammoceras,*

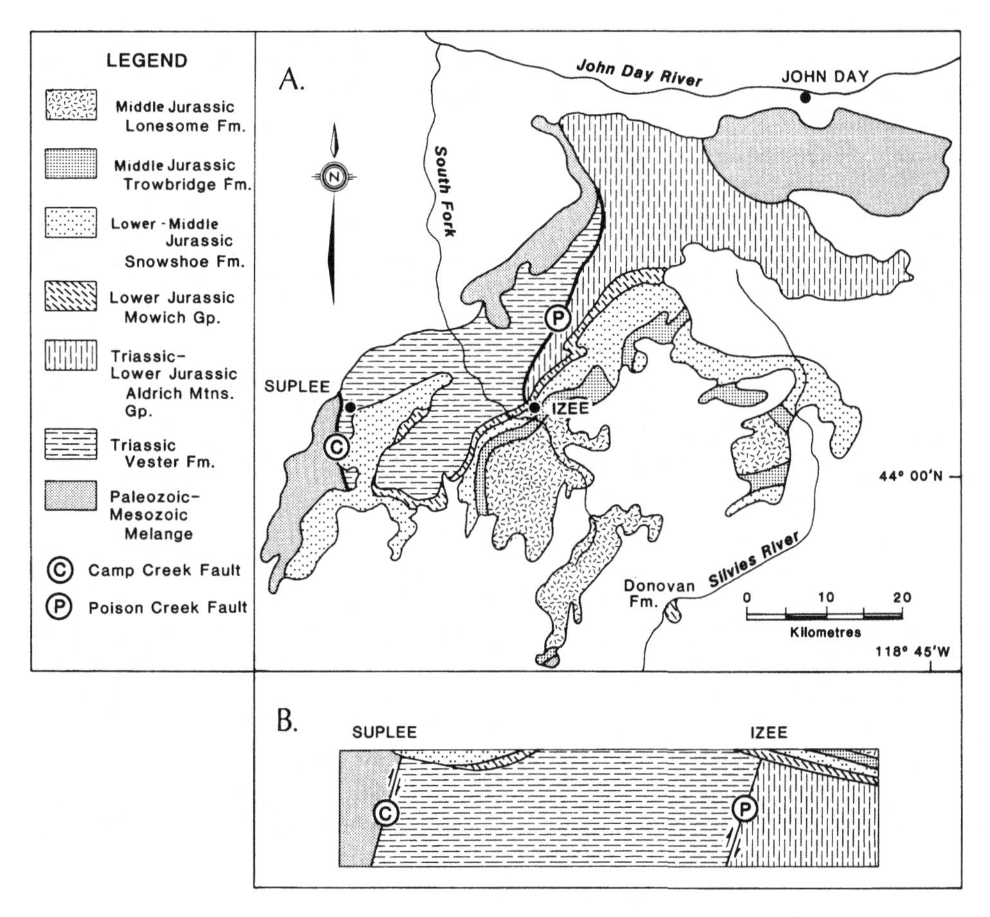


**Figure 4.22. (A) Geologic map of the Izee Terrane, with the Grindstone Terrane to the west and part of the Baker Terrane to the north (see Figure 4.20). (Modified from Dickinson and Thayer 1978 and based on maps by Brown and Thayer 1966 and Greene et al. 1972.) (B) Schematic east–west section through the Suplee-Izee area showing structural and stratigraphic relationships within the John Day Inlier (Dickinson 1979).**

Leptaleoceras, and *Fanninoceras* (Imlay 1968; Smith et al. 1988), indicates a Pliensbachian age for most of the Mowich Group, but the Hyde Formation may be as young as middle Toarcian. Radiolaria offer the best possibility of constraining the age of the Hyde; to date, species of the nassellarian genera *Farcus, Rolumbus, Jacus*(?), and *Napora* have been retrieved (Pessagno, Whalen, and Yeh 1986).

The late Toarcian to Bathonian Snowshoe Formation has been divided into seven members (Dickinson and Vigrass 1965; Smith 1980). The unconformity-bound sequence on the Suplee shelf to the west (Figures 4.21 and 4.22) consists of Aalenian and Bajocian limestones, siltstones, and sandstones that grade and interfinger eastward into finer-grained volcaniclastic rocks of the Izee Basin. Coarser and somewhat younger volcaniclastic rocks were also deposited in the Izee Basin, derived, in the case of the Silvies Member, from the South (Dickinson and Thayer 1978). Ammonites from the formation have been described by Imlay (1968, 1973, 1981a) and Taylor (1981, 1988b), and nassellarian Radiolaria have been described by Pessagno et al. (1986).

The Trowbridge Formation, which unconformably overlies the Snowshoe Formation, consists of shale and mudstone, thickening in a northeasterly direction to 1,100 m (Lupher 1941; Dickinson and Vigrass 1965; Imlay 1986). The upper part of the formation contains felsite tuffs. Rare ammonites of the genera *Lilloettia, Xenocephalites,* and *Kepplerites* indicate a Callovian age (Imlay 1964, 1981a).

Gradationally overlying the Trowbridge is the Lonesome Formation (Lupher 1941), made up of more than 3,000 m of monotonously alternating beds of graded sandstone and organic-rich mudstone. The sandstones were deposited by turbidity currents flowing northwestward into a subsiding basin from a volcanically active area where subvolcanic rocks had already been unroofed (Dickinson and Vigrass 1965; Dickinson and Thayer 1978). The unit has yielded Callovian ammonites of the genera *Xenocephalites, Lilloettia, Pseudocadoceras,* and *Kepplerites* (Imlay 1964, 1981a).

### Olds Ferry Terrane

The Jurassic rocks of the Olds Ferry Terrane rest with angular unconformity on the Upper Triassic rocks of the Huntington Formation, which consists of submarine mafic and intermediate volcanics and volcaniclastics (Brooks 1967; Brooks and Vallier 1978). The Jurassic rocks in the eastern part of the terrane are unnamed and may be as thick as 7,000 m. Those in the western inliers have been called the Weatherby Formation (Brooks 1979a; Brooks et al. 1979; Brooks and Ferns 1979).

The Weatherby Formation is a thick, monotonous flysch sequence that has been intensely deformed. Ammonite occurrences are sparse, but indicate Sinemurian to Bajocian ages (Imlay

1986). Unnamed phyllites and slates of Callovian age might also be included in the Weatherby Formation (Figure 4.21) (Brooks and Ferns 1979). The Jet Creek Member at the base of the formation consists of 250 m of shallow-marine limestone, conglomerate, and gypsum dated by ammonites as Sinemurian to Pliensbachian in age.

In the Riggins area between the Idaho Batholith and the Rapid River Fault (Figure 4.20), the Squaw Creek schists may represent a metamorphosed correlative of the Weatherby Formation (Brooks and Vallier 1978).

### Wallowa terrane

The Hurwal Formation was established by Smith and Allen (1941) for Triassic argillaceous beds conformably overlying the Martin Bridge Limestone in the northern Wallowa Mountains. Nolf (1966) later included four lithologically identical sequences yielding Lower Jurassic ammonites. These sequences are geographically isolated from Triassic rocks, however, and stratigraphic continuity cannot be demonstrated. Some of these beds, such as those exposed at Ed Smith Creek, Frances Lake, and Traverse Ridge, have yielded Lower Sinemurian fossils, including *Coroniceras, Schlotheimia,* and *Vermiceras*(?) (Smith 1981). Others have yielded a sequence of Pliensbachian ammonites (Imlay 1968; Smith 1981, 1983; Smith et al. 1988). Metamorphosed rocks in Idaho named the Lucille Slate (Hamilton 1963; Brooks and Vallier 1978) are believed to be correlative with the Hurwal Formation.

The Coon Hollow Formation, which has furnished the Boreal Oxfordian ammonite *Cardioceras* (Imlay 1964), consists of more than 1,200 m of black shale resting with angular unconformity on Triassic rocks (Morrison 1961, 1964). The only exposures are in the Snake River canyon near the Washington–Oregon border. A lithologically similar but unnamed unit cropping out in the canyon near Pittsburg Landing has yielded Callovian ammonites (Imlay 1986). Brooks (1979b) and Brooks and Vallier (1978) have suggested that these small outcrops, protected by block faulting, are the remnants of a once more extensive cover.

### Tectonic summary

Evidence that bears on the plate-tectonic history of this area includes the intrusion of plutonic rocks, breaks in sedimentation, sediment provenance, and paleomagnetic and paleobiogeographic evidence constraining terrane dispositions.

Plutons intruded into the Blue Mountains island arc show $^{87}Sr/^{86}Sr$ ratios lower than 0.704, whereas those farther east, which were intruded into continental crust, show ratios in excess of 0.704 (Armstrong, Taubeneck, and Hales 1977). The zone of transition from continent to arc is remarkably narrow and runs immediately west of the Idaho Batholith. Jurassic plutons are not known in the Izee or Grindstone Terranes, and those of the Olds Ferry, Baker, and Wallowa Terranes can be grouped into two suites on the basis of radiometric age (Armstrong et al. 1977; Brooks and Vallier 1978; Dickinson 1979):

1. small, quartz diorite plutons with late Triassic and early Jurassic ages (200–220 Ma) intruded into Upper Triassic volcanic-arc rocks of the Olds Ferry Terrane in western Idaho;
2. larger, late Jurassic and early Cretaceous granitic plutons that tend to be composite and yield maximum absolute ages of 155–160 Ma; they intrude rocks of the Baker and Wallowa Terranes, with the Bald Mountain Batholith (Figure 4.20) straddling the boundary between the two.

There are several breaks in sedimentation within the Jurassic of eastern Oregon–some firmly delineated only within one terrane, others that appear to be more regional. One of them, indicated in the Izee Basin by slight discordance to paraconformity between the Aldrich Mountains Group and the Mowich Group (Figure 4.21), represents the Ochoco Orogeny (Dickinson and Vigrass 1965; Dickinson and Thayer 1978). It is now known that the interval missing here is quite narrow, but both the missing interval and the angularity of the discordance between the groups increase markedly westward toward Suplee (Figure 4.22). The coarse deposits of the Donovan Formation to the southeast are equivalent to the missing time interval in the Izee area (Whiteavesi and Freboldi Zones of the Lower Pliensbachian) (Smith et al. 1988). Perhaps these relationships indicate uplift associated with subduction, as discussed later. The succeeding transgression may have been at least partly eustatic in origin, because a late Pliensbachian transgression is indicated elsewhere in North America and other parts of the world (Hallam 1981; Vail and Todd 1981; Smith and Tipper 1988). The stratigraphy for this interval on the Olds Ferry Terrane has not been worked out in detail; the Pliensbachian sequence of the Wallowa Terrane seems fairly complete.

In contrast to the rest of the world, where there was progressive expansion in the area of deposition during the Jurassic, the Upper Jurassic is virtually absent from eastern Oregon because of tectonic uplift and erosion. The Late Jurassic sedimentological hiatus correlates with a major phase of batholithic intrusion and pervasive structural deformation (Oldow, Lallement, and Schmidt 1984; Vallier and Brooks 1986). However, there is no sedimentological evidence of the unroofing of granites in any of the terranes or granites from a cratonic source until the middle Cretaceous. Jurassic sediments characteristically are volcaniclastic, although there is evidence that the Baker mélange provided detritus to the Weatherby Formation of the Olds Ferry Terrane during the Early Jurassic (Brooks and Vallier 1978; Blome et al. 1986). Blome et al. (1986) have also indicated the Baker Terrane as provenance for Triassic Izee Terrane clastics.

A model favored by Dickinson (1979) considers the central mélange as, in part, forming a structural high marking the seaward margin of a fore-arc basin in which Jurassic sediments of the Izee and Olds Ferry Terranes accumulated. This structural high (including the Grindstone Terrane) was formed by local backthrusting (Figure 4.22) associated with subduction in a trench to the northwest. As the subduction complex widened, Jurassic strata filled the fore-arc basin and overlapped the crest of the structural high. Volcaniclastic sediments built out from numerous centers, and the magmatic arc itself occasionally encroached into the sedimentary basin to produce local volcanics.

There is some question whether or not the Olds Ferry and Izee Terranes are distinguishable for Jurassic time. Dickinson (1979) considered their Jurassic sedimentary sequences to be of the same origin (his Mesozoic clastic terrane), but Brooks and Vallier (1978) have noted that the Olds Ferry Terrane sequence is more sedimentologically homogeneous and more tectonically deformed than is the sequence of the Izee Terrane. They have postulated that the Izee Terrane was some distance from the Olds Ferry Terrane during the Jurassic and was subsequently moved against and over it, riding on dismembered oceanic crust.

The evolution of the fore-arc basin, according to Dickinson (1979), probably was terminated during the Late Jurassic by the accretion of the Wallowa Terrane. The suture between the mélange and the Wallowa Terrane is cut by the Late Jurassic Bald Mountain Batholith (Figure 4.20) (Silberling et al. 1984), and there are plutons of that and later age in both terranes. There is doubt, on stratigraphic (Sarewitz 1983) and petrologic (Scheffler 1983) grounds, that the Wallowa Terrane is a fragment of Wrangellia, as suggested by Jones et al. (1977). Mortimer (1986) has suggested a relationship to Stikinia, rather than Wrangellia, and Pessagno and Blome (1986) suggest that all the terranes of the Blue Mountains formed a single island-arc complex.

The only paleomagnetic data that can be brought to bear on this problem indicate that the Olds Ferry Terrane had a Triassic paleolatitude of 18±4° north or south of the equator, with a large counterclockwise rotation, in post–early Cretaceous time (Hillhouse, Gromme, and Vallier 1982). Data for Wrangellia proper indicate similar paleolatitudes (Hillhouse 1977), but no data have been obtained directly from the Wallowa Terrane. Ammonite faunas show Tethyan and East Pacific affinities for the diverse Pliensbachian ammonites of the Izee and Wallowa Terranes (Imlay 1968; Smith and Tipper 1986), with communities becoming Boreal in the Izee Terrane by the Callovian as the Boreal Realm extended farther south on the craton (Taylor et al. 1984). On the basis of radiolarian endemism and diversity, Pessagno and Blome (1986) and Pessagno et al. (1986) have suggested a progressive northward displacement of the Blue Mountains island arc during the Jurassic, with a position of approximately 40°N paleolatitude being reached by the Late Jurassic.

### Ammonite zones[11]

*Hettangian*

1. Zone A of Taylor (1986) – *Sulciferites* aff. *collegnoi* (Cocchi), *Vermiceras* cf. *supraspiratus* (Waehner), *Discamphiceras silberlingi* Guex, *Eolytoceras* n.sp., *Gonioptychoceras* n.sp., *Sulciferites* n.sp. aff. *trapezoidalis* (Sowerby), *Schlotheimia* aff. *phobetica* Lange, *Paradiscamphiceras dickinsoni* Taylor, *Paradasyiceras* aff. *uermosense* (Herbich), *Nevadaphyllites compressus* Guex (Taylor 1986, 1988a); upper Angulata Zone

*Sinemurian*

1. Zone B of Taylor (1986) – *Sulciferites* n.sp., *Discamphiceras* n.sp., *Nevadaphyllites compressus* Guex, *Sunrisites sunrisensis* Guex, *Badouxia* sp., *Vermiceras* n.sp., *Paradiscamphiceras athabascanense* Taylor, *Paradasyiceras* aff. *uermosense* (Herbich) (Taylor 1986, 1988a); basal Bucklandi Zone.

*Pliensbachian*

1. Imlayi Zone – *Pseudoskirroceras imlayi* Smith & Tipper (Smith et al. 1988)
2. Whiteavesi Zone – *Metaderoceras evolutum* Fucini (Smith et al. 1988)
3. Freboldi Zone – *Dubariceras freboldi* Dommergues et al., *D. silviesi* (Hertlein 1925), *Reynesocoeloceras* cf. *baconicum* Geczy, *Metaderoceras beirense* Mouterde & Ruget, *M.* aff. *muticum* (d'Orbigny) (Smith et al. 1988)
4. Kunae Zone – The Upper Pliensbachian localities catalogued by Imlay (1968) commonly are imprecisely located and stratigraphically isolated; in addition, often it is not clear whether a collection came from one level or from an interval of unknown stratigraphic thickness: *Fanninoceras kunae*, *F.* cf. *bodegae*, *F.* cf. *carlottense*, McLearn spp., *Arieticeras* cf. *domarense* (Meneghini), *A.* cf. *algovianum* (Oppel) [?*Oregonites*] (Wiedenmayer, 1980), *A. lupheri* Imlay, *Aveyroniceras* cf. *italicum* (Meneghini), *Av.* cf. *mortilleti* (Meneghini), *Av.* cf. *meneghini* (Fucini), *Fontanelliceras* cf. *fontanellense* (Gemmellaro), *Fuciniceras* cf. *acutidorsatum* Kovacs, *F.*? cf. *intumescens* (Fucini), *Holcophylloceras* sp., *Leptaleoceras* cf. *leptum* Buckman, *L. dickinsoni* Imlay, *Liparoceras (Becheiceras)* cf. *bechei* (Sowerby), *Oregonites imlayi* Wiedenmayer, *Protogrammoceras* cf. *nipponicum* (Matsumoto & Ono), *P. meneghinii* (Bonarelli), *P.* cf. *argutum* (Buckman), *P.* cf. *mariani* (Fucini), *P.* cf. *bonarelli* (Fucini), *P.* cf. *isseli* (Fucini), *Reynesoceras ragazzonii* (Hauer) (Imlay 1968)
5. Carlottense Zone – probably represented in *ex situ* material and isolated localities in the Suplee area (Imlay 1968; Smith et al. 1988): *Fanninoceras carlottense* McLearn, *Protogrammoceras* cf. *argutum* (Buckman), *P.* (?) cf. *pseudofieldingi* (Fucini), *Arieticeras* cf. *domarense* (Meneghini), *A. lupheri* Imlay

*Toarcian*

1. An Upper Toarcian fauna from the Snowshoe Formation was described by Imlay (1968): *Dumortieria*(?) cf. *pusilla* Jaworski, *Catulloceras* cf. *dumortieria* (Thollière), *Haugia* spp., *Polyplectus* cf. *subplanatus* (Oppel), *Grammoceras*(?) spp.

[11] For comments on the standard zones, see Chapter 12.

*Aalenian*

1. *Abbasites sparsicostatus* Zone – *A. sparsicostatus* (Imlay), *Tmetoceras scissum* (Benecke) *Strigoceras harrisense* Taylor, *Eudmetoceras moerickei*(?) (Jaworski), *E.* aff. *amaltheiforme* (Vacek) (Imlay 1973; Taylor 1988a)
2. *Eudmetoceras mowichense* Zone – *E. mowichense* Taylor, *E.* aff. *amaltheiforme* (Vacek) (Imlay 1973; Taylor 1988a)
3. *Fontannesia packardi* Zone – *F. packardi, F. involuta, Eudmetoceras robertsoni, E. vigrassi, Hebetoxyites* aff. *snowshoensis, Docidoceras schnabelei, Euhoploceras crescenticostatum,* Taylor spp., *E. modestum* (Buckman) (Imlay 1973; Taylor 1988a)

*Bajocian*

1. *Euhoploceras tuberculatum* Zone – *E. tuberculatum* Taylor, *E. modestum* (Buckman), *E. acanthodes* (Buckman), *E. transiens, E. westi, E. ochocoensis, E. grantense, E. rursicostatum,* Taylor spp., *E.* cf. *crassinudum* (Buckman), *Docidoceras amundsoni* Taylor, *D. striatum* Taylor, *D. warmspringsense* Imlay, *D. lupheri* Imlay, *Asthenoceras costulum* (Imlay), *A. boreale* (Whiteaves), *Latiwitchellia evoluta* (Imlay), *Ptychopylloceras compressum, Hebetoxyites snowshoensis, Sonninia washburnensis,* Taylor spp., *Witchellia* cf. *connata* (Buckman), *Strigoceras lenticulare* Taylor, *Fissilobiceras* n.sp. (Imlay 1973; Taylor 1988a)
2. *Sonninia burkei* Zone – *S. burkei* Taylor, *Asthenoceras costulum* (Imlay), *A. boreale* (Whiteaves), *Latiwitchellia evoluta* (Imlay), *Docidoceras warmspringsense* Imlay, *D.* aff. *paucinodosum* Westermann, *Ptychophylloceras compressum* Taylor, *Holcopylloceras supleense* Taylor, *Euhoploceras donnellyense* Taylor, *E. acantherum* Buckman, *Freboldites bifurcatus* Taylor (Imlay 1973; Taylor 1988a)
3. *Parabigotites crassicostatus* Zone – *P. crassicostatus* Imlay, *Ptychophylloceras compressum* Taylor, *Euhoploceras donnellyense* Taylor, *E. acantherum* Buckman, *Sonninia burkei* Taylor, *S.* aff. *diversistriata* (Imlay), *S. grindstonensis* Taylor, *Kumatostephanus* aff. *perjucundus* Buckman, *Lissoceras hydei* Imlay, *Emileia* cf. *contracta* (Sowerby), *E. buddenhageni* Imlay, *Stephanoceras flexicostatum* Taylor, *Phaulostephanus oregonense* Imlay, *S.* cf. *dolichoecum* Buckman (Imlay 1973; Taylor 1988a)
4. *Stephanoceras kirschneri* Zone – *S. kirschneri* Imlay, *S. juhlei* Imlay, *Emileia* sp., *Dorsetensia oregonensis* Imlay, *D. romani* (Oppel), *D.* cf. *subtecta* Buckman, *Sonninia* aff. *alsatica* (Haug), *Poecilomorphus varius* Imlay, *Strigoceras* cf. *languidum* (Buckman), *Sphaeroceras allani* (McLearn), *Zemistephanus* spp. (Imlay 1973; Hall and Westermann 1980; Taylor 1988a)
5. *Sphaeroceras oblatum* Zone – possibly represented by *Sphaeroceras (Defonticeras) allani* (McLearn), *Stemmatoceras* aff. *albertense* McLearn, *Teloceras*(?) sp. (Imlay 1973; Hall and Westermann 1980; Smith 1980)
6. *Megasphaeroceras rotundum* Zone – *M. rotundum, Lupherites senecaensis, Leptosphinctes* cf. *evolutum, L.* cf. *leptum,* Imlay spp., *L. (Prorsisphinctes)* spp., *Stephanoceras (Normannites) vigorosus* (Imlay), *S. (N.) orbignyi* Buckman, *Spiroceras annulatum* (Deshayes), *S. bifurcatum* (Quenstedt), *Sphaeroceras* cf. *brogniarti* (Sowerby) (Imlay 1973; Hall and Westermann 1980; Smith 1980)

*Bathonian*

1. *Iniskinites acuticostatus, I.* cf. *intermedius, I. martini,* Imlay spp., *Bullatimorphites* sp., *Xenocephalites vicarius* Imlay (Imlay 1964, 1981a; Callomon 1984)

*Callovian (Lower)*

1. *Phylloceras* sp., *Xenocephalites vicarius* Imlay, *Kepplerites* cf. *torrensi* (McLearn), *Pseudocadoceras* cf. *grewingki* (Pompeckji), *Lilloettia buckmani* (Crickmay), *L. stantoni* Imlay, *L.* cf. *mertonyarwoodi* (Crickmay), *Parapatoceras* sp., *Grossouvria* sp. (Imlay 1964, 1981a)

*Oxfordian (Lower)*

1. *Cardioceras (Scarburgiceras) martini* Reeside (Imlay 1964)

CALIFORNIA[12]

**Introduction**

The Jurassic geology of California is complex and still poorly understood, despite more than 100 years of detailed investigations. Most Jurassic strata are structurally complex, metamorphosed, and poorly fossiliferous. In addition, those in the eastern, northern, and southern parts of the state have been intruded by upper Mesozoic plutonic rocks, thus obscuring their original relationships. Major fault zones separate the Jurassic rocks into packages with mutually distinct sequences. These packages have been designated "lithotectonic terranes," following the original analysis of Irwin (1972) in the Klamath Mountains, and as later modified by Blake et al. (1982) and Silberling et al. (1987) for the entire state. More than 15 terranes contain significant Jurassic strata, grouped here into those of the Sierra Nevada and eastern Klamath Mountains, those of the Sierra foothills and western Klamath Mountains, those of the Coast Ranges, and those of southern California (Figures 4.23 and 4.24).

**Terranes of the Sierra Nevada**

Jurassic rocks in the east-central Sierra Nevada comprise metamorphosed intermediate to silicic tuff and breccia, with thin beds of limestone, that have been assigned by Nokleberg (1983) to the Goddard Terrane. This sequence is faulted at its base against

[12] By D. L. Jones.

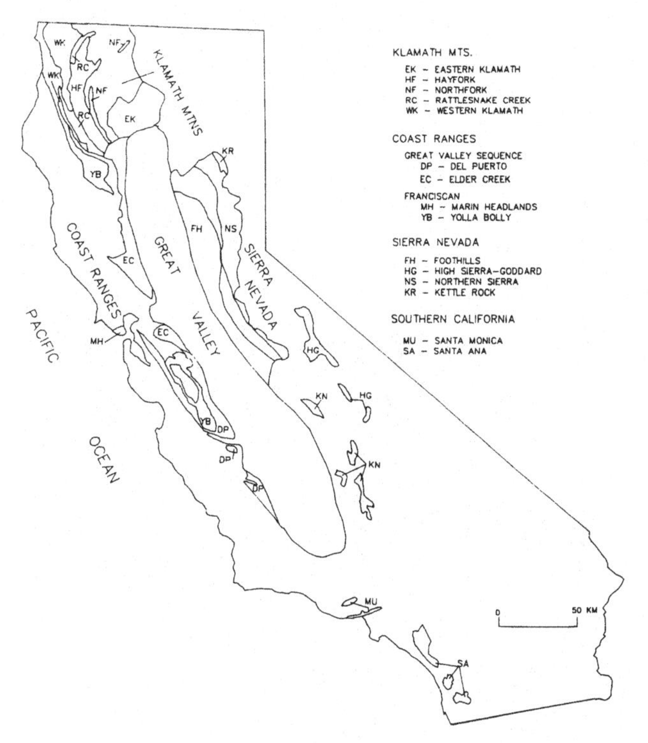


**Figure 4.23. Map showing distribution of major Jurassic terranes in California.**

older volcanic rocks and is overlain unconformably by Cretaceous volcanic breccias. The fossils from the limestone beds are mainly bivalves, including *Weyla* cf. *alata* (Buch) of probable Sinemurian or Pliensbachian age and solitary corals. Middle to Upper Jurassic volcanogenic strata may also be present in the Goddard Terrane, as suggested by radiometric ages of 153–186 Ma from zircons extracted from tuff beds (Fiske and Tobisch 1978), but no fossils of these ages have been found.

In the southern Sierra Nevada, thick sequences of quartzite, arkose, limestone, and minor silicic and andesitic tuffs and flows of the Kings Terrane have yielded sparse, poorly preserved fossils of Late Triassic and Early Jurassic age. The Sinemurian or Pliensbachian ammonite *Paracoroniceras*(?) or *Metophioceras*(?) was identified from the Boyden Cave roof pendant (Jones and Moore 1973), and an early Toarcian ammonite fauna was identified from the Mineral King roof pendant. Poorly preserved bivalves from the Lake Isabella pendant may be *Weyla*.

Jurassic strata in the northern Sierra Nevada are assigned to the Northern Sierra and Kettle Rock Terranes (Silberling et al. 1984, 1987). Those of the former consist of the Sailor Canyon Formation, a thick assemblage of poorly fossiliferous volcanic graywacke, slate, and volcanic rocks that overlies Upper Triassic limestone (Clark et al. 1962). The late Sinemurian is documented by the ammonites *Crucilobiceras* and *Arieticeras*, the Pliensbachian by *Reynesoceras*, and the Aalenian by *Tmetoceras* (Imlay 1980a). The upper, volcanic-rich part of the formation has not been dated by fossils. It is intruded by granitic rocks of middle Cretaceous age (90–94 Ma, K/Ar) (Wagner et al. 1981), and Late Jurassic granitic rocks (143 Ma) occur nearby, so a pre-Cretaceous upper limit seems reasonable.

The Kettle Rock Terrane occurs only in the northernmost Sierra Nevada, where it is best known in the Mount Jura area near Taylorsville (Crickmay 1933). More than 4,000 m of marine strata are exposed, the lower part dominantly volcanic to volcaniclastic, the upper part mixed volcaniclastic and quartzose sandstone, shale, and conglomerate. The base of the sequence is nowhere exposed, and the upper contact is a major thrust fault that separates the Kettle Rock Terrane from overlying Paleozoic rocks of the Northern Sierra Terrane. The 14 formations (McMath 1966) have yielded Sinemurian to Callovian marine fossils. The upper Sinemurian is documented by *Echioceras* sp., together with *Weyla alata* in the Lilac Argillite; the Pliensbachian is suggested by the warm-water bivalve *Plicatostylus* in the Thompson Limestone; the basal and upper Lower Bajocian are documented by *Euhoploceras*, *Stephanoceras*, and *Teloceras* in the Mormon Sandstone; and the early Upper Bajocian is suggested by *Spiroceras*(?) in the Moonshine Conglomerate. The next-higher beds dated by fossils probably are late Bathonian–early Callovian in age (Imlay 1980a, p. 57). Middle Callovian *Reineckeia*(?) *dilleri* Crickmay and supposed early Tithonian *Aulacosphinctoides* sp. are the most northern representatives of the Tethyan Realm (Crickmay 1933; Westermann 1981).

### Terranes of the eastern Klamath Mountains

A similar assemblage of Jurassic rocks occurs 120 km to the northwest along the Pit River. This sequence overlies Upper Triassic strata assigned to the eastern Klamath Terrane. The lowest unit, the Arvison Formation (Sanborn 1960), consists of more than 1,800 m of andesitic volcanic flows, agglomerate, breccia, conglomerate, tufaceous sandstone, and minor limestone containing the Sinemurian ammonites *Asteroceras* and *Arnioceras*. Imlay (1980a) correlated the Arvison with the finer-grained Lilac Argillite and Hardgrave Sandstone of the Mount Jura region and noted that some faunal elements are common to both areas. The Arvison is overlain by the undated Bagley Andesite and is intercalated with argillite and fine-grained tuffaceous sandstone of the Potem Formation, which contains the Liassic *Weyla* aff. *alata*. Sanborn (1960) suggested that the upper part of the Potem may be as young as early Bathonian, but Imlay (1980a) does not list any fossils younger than Early Jurassic.

Much of the Mount Jura sequence thus is younger than the Pit River assemblage, so only the Lower Jurassic portions can be compared. No stratigraphic unit is common to both areas, although both contain volcanigenic strata. The Pit River section records subsidence and a fining-upward sequence, with basal coarse breccias containing abundant shallow-water bivalves, indicative of high-energy environments, changing upward to deeper-water argillite containing abundant pelagic bivalves *(Bositra)*. In contrast, the Mount Jura section may record a shallowing sequence that changes upward from argillite to sandstone, with abundant shelly faunas. The significance of the differences between these occurrences remains to be established. Both the Mount Jura and Pit River sequences differ from the coeval Sailors

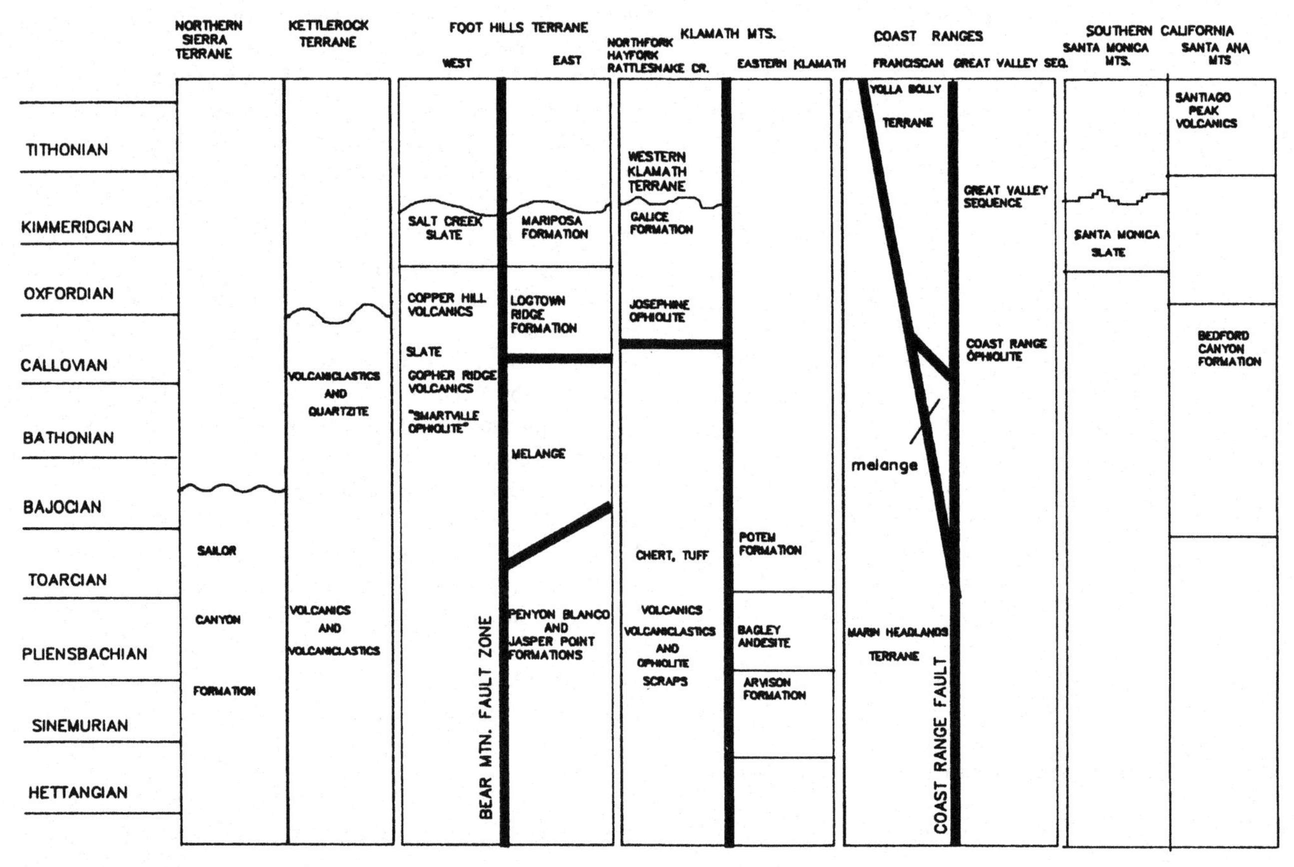


**Figure 4.24. Jurassic sequences of California.**

Canyon Formation in their shallower environments and more abundant fossils.

### Terranes of the Sierra foothills

The western foothills of the Sierra Nevada and the Western Klamath Mountains contain similar assemblages of volcanic, volcaniclastic, and clastic rocks that are mainly Late Jurassic in age. These rocks are strongly folded and metamorphosed, and their regional structural relations remain controversial. The deformation is the basis for the Nevadan Orogeny, which has been extended widely throughout the Western Cordillera (Tobisch et al. 1987). Three distinct packages, from east to west – the Motherlode, Central, and Western Belts – are separated by major fault zones with complex histories (Paterson, Tobisch, and Radloff 1987). The Motherlode Belt is the best known, most fossiliferous, and least metamorphosed.

The most fossiliferous, youngest, and best-dated unit of the Motherlode Belt is the Mariposa Formation, consisting of several thousand meters of folded slate, graywacke, and intercalated volcanic rocks. Its age ranges from middle Oxfordian to early Kimmeridgian, based on the ammonites *Perisphinctes (Discosphinctes)* sp., *P. (Dichotomosphinctes)* sp., and *Amoeboceras (Amoebites)* sp. and the bivalve *Buchia concentrica* (Sowerby). Belemnites and various bivalves are also locally abundant (Imlay 1961a, p. D–14). Comparable beds in the Klamath Mountains are much less fossiliferous and are dated solely on the presence of *B. concentrica.* The northern part of the Motherlode Belt terminates in the central foothills near Colfax, where slate and conglomerate mapped as Mariposa Formation (Chandra 1961) or "Colfax sequences" occur. Fossils include the late Bathonian to early Callovian ammonites *Grossouvria, Kepplerites,* and *K. (Gowericeras)* (Imlay 1961a). Oxfordian and Kimmeridgian fossils are unknown here, and the stratal position of the older fossils has not been determined, but may be in the lowest part of the formation (Chandra 1961).

Throughout much of the Motherlode Belt the Mariposa Formation overlies, depositionally, the Logtown Ridge Formation, several thousand meters of andesitic breccia, tuff, and volcaniclastic graywacke. Ammonites from near the base of the formation include *Pseudocadoceras,* and from the top there is *Idoceras,* indicating early Callovian to latest Oxfordian (Imlay 1980a, p. 60). The Logtown Ridge and Mariposa Formations are in structural contact with a presumed basement complex assigned to the Central Belt. In the southern foothills, this presumed basement consists of pillow basalt and chert of the Jasper Point Formation, overlain by andesitic tuff and breccia of the Penon Blanco Formation (Bogen 1985). The latter is similar to the Logtown Ridge Formation, but is older, as it is intruded by granitic rocks as old as 200 Ma (Stern et al. 1981). To the north, a mélangelike assemblage of sedimentary, volcanic, and metamorphosed rocks of uncertain age and origin occurs beneath the Logtown Ridge (Duffield and Sharp 1975). Some of the rock types present in the mélange are similar to the Jurassic strata described earlier and may be of the same age. Other lithologies, such as blocks of Permian limestone that contain Tethyan fusulinids (Douglass 1967), are not found in any nearby stratigraphic sequence and thus are exotic. Many of the other rocks present in the mélange are undated, such as disrupted quartzose sandstone and slate that locally compose the bulk of the mélange matrix (Behrman 1978). Blocks of Jurassic radiolarian cherts (D. Jones, unpublished data) in a few localities, together with slabs similar in composition to the Logtown Ridge and Mariposa Formations, suggest that the mélange formed after the Oxfordian. This age is permissible because a nearly vertical fault always separates these presumed basement rocks from the Logtown Ridge and Mariposa Formations. The amount, timing, and sense of displacement on this fault remain to be determined. Plutonic rocks dated at 143 Ma (J. Saleeby personal communication 1990) that stitch across this fault locally, intruding both the Central and Motherlode Belts, place an upper age limit on the time of mélange formation.

Another volcanic and sedimentary assemblage, the Western Belt, occurs farther west in the Sierra foothills. The boundary between the Western and Central Belts is the important Bear Mountains fault, along which significant ductile displacements have occurred. In the south, the Western Belt includes the lower Gopher Ridge Volcanics, the middle Salt Spring Slate, and the upper Copper Hill Volcanics. According to Clark (1964), these units are in depositional continuity. *Buchia concentrica* from the Salt Spring Slate indicate early Late Jurassic. This unit is thus equivalent in age and lithology to the Mariposa Formation of the Motherlode Belt, although it is more strongly deformed (Paterson et al. 1987). The underlying Gopher Ridge Formation contains more silicic volcanic rocks than the Logtown Ridge Formation (Clark 1964). The Copper Hill Volcanics are mainly tuffaceous and compositionally similar to the Logtown Ridge, a correlation made by Taliaferro (1943). Clark (1964) suggested that the Copper Hill Volcanics may range from late Oxfordian to Kimmeridgian or younger and thus may constitute the youngest bedrock unit within the Central Foothills Terrane. To the north, the Western Belt includes large bodies of mafic and minor ultramafic plutonic and extrusive rocks, plus extensive sheeted dikes sometimes referred to as the Smartville Ophiolite or the Smartville Complex (Beard and Day 1987). The volcanic portion of this complex consists of a lower pillowed basalt unit and an upper andesitic volcaniclastic unit. The latter has received a number of names, including the Oregon City Formation and the Bloomer Hill Formation. Beard and Day (1987) interpret the Smartville Complex to be the rifted edifice of a Late Jurassic island arc. They point out that the assumed ophiolitic pseudostratigraphy consisting of ultramafic rock, gabbro, sheeted dikes, and pillow basalt is not real, in that the sheeted dikes and mafic plutonic rocks intruded the pillow-basalt unit. Therefore, no oceanic crust appears to be present within the Smartville Complex, even though the arc itself originally may have formed in an oceanic setting.

The Smartville Complex is overlain by the Monte de Oro Formation, a unit coeval with both the Salt Spring Slate and the Mariposa Formation, but differing in lithology by its much greater proportion of conglomerate (Creely 1965). The Monte de Oro contains abundant plants as well as *Perisphinctes (Dichotomosphinctes)* sp. and *Buchia concentrica* of middle Oxfordian to early Kimmeridgian age (Imlay 1961a).

In summary, the Jurassic rocks of the Sierra foothills are complexes of arc-related volcanics capped by slate, sandstone, and minor conglomerate, with intercalated lenses of volcanic breccia and tuff. The regional deformation generally is ascribed to the Late Jurassic Nevadan Orogeny (140–155 Ma) (Tobisch et al. 1987). Several differing interpretations have been offered to explain the arc development and its subsequent destruction, but no consensus has emerged:

1. Oceanic-arc–continent collision along a west-dipping subduction zone (Moores 1972; Schweickert and Cowan 1975). Abundant quartz and other terrigenous clastics within the volcanic and epiclastic sequences argue against this interpretation (Behrman and Parkinson 1978). A variation involves easterly directed, low-angle thrust faults along which the arc assemblage with its plutonic components (Smartville Complex) was emplaced by thrusting onto the structurally underlying mélange. This model has been most clearly developed and documented in the northern Sierra foothills.

2. Arc development along the continental margin above an older accreted basement, with only minor structural rearrangement during the Nevadan Orogeny (Behrman 1978; Behrman & Parkinson 1978; Saleeby 1981). Depositional relations connecting the various arc and epiclastic units to underlying volcanic and mélange assemblages have not been rigorously documented. The source of the epiclastic material was in part from a nearby continent. Supporting data have been derived mainly from the southern part of the Foothills Terrane.

3. Multiple-arc collision model. The most recent interpretation, put forth by Paterson et al. (1987) for the southern part of the Foothills Terrane, argues that the Western Belt differs so significantly from the Motherlode Belt that the two segments should be considered as slivers of different arcs juxtaposed during and after the Nevadan Orogeny. They interpreted the Bear Mountains fault to be a ductile shear zone, with relative upward and northward movement of the eastern block. However, similar deposits of Mariposa-like slate and sandstone in all the belts tie them together by Oxfordian to Kimmeridgian time.

### Terranes of the western Klamath Mountains

Jurassic rocks occur here in four terranes, from east to west: the Northfork, Hayfork, Rattlesnake Creek, and Western Klamath Terranes (Smith River Subterrane). Liassic rocks are found in the Northfork and Rattlesnake Creek Terranes, Dogger is found in the Hayfork, and Middle to Late Jurassic rocks occur in the Smith River Terrane. Except for the latter, these terranes are mélangelike and lack coherent stratigraphic sequences. Fossiliferous Jurassic rocks are best known from the Northfork Terrane, where tufaceous chert, cherty tuff, and siliceous tuff-breccia have yielded Pliensbachian or older Jurassic radiolarians (Irwin 1977; Irwin, Jones, and Kaplan 1978; Irwin, Jones, and Blome 1982; Blome and Irwin 1983). Characteristic genera include *Canutus, Pantanellium, Zartus, Bagatum, Praeconocaryomma, Droltus,* and *Trillus.* No shallow-water deposits or macrofossils appear to be present. At the north end of the Northfork Terrane, Lower Jurassic argillaceous chert interbedded with argillite and graded beds of siltstone and quartzose sandstone indicates proximity to a nearby continental detrital source.

Poorly known chert in the Rattlesnake Creek Terrane contains a Jurassic radiolarian fauna similar to that of the Northfork Terrane (Irwin et al. 1982), but that equivalence remains uncertain.

Jurassic rocks in the Hayfork Terrane consist of a lower volcaniclastic unit, with minor pillow lava and pillow breccia, overlain by an epiclastic and hemipelagic sedimentary unit (Harper and Wright 1984). K/Ar ages from hornblende in volcanic rocks of the lower unit are 168–177 Ma, Middle Jurassic (Harper and Wright 1984). The upper unit contains volcaniclastic rocks as well as siliceous argillite, radiolarian chert, and polymictic pebble conglomerate derived from a mixed terrigenous source. A pre-Middle Jurassic age is probable, because metamorphic ages derived from Hayfork rocks are approximately 165–170 Ma (Harper and Wright 1984). That metamorphic event affected both the Northfork and Hayfork Terranes, indicating that they were sutured together by that time (Mortimer 1985), and it appears to have predated the intrusion of granitic plutons at about 163 Ma (Irwin 1985). The timing of events seems well constrained, but disagreements remain concerning the spatial relations of the terranes. For example, Harper and Wright (1984) believe that Jurassic volcanic rocks throughout the Klamath Mountains formed within a single west-facing magmatic arc built across a complex basement of older rocks. This is essentially the same interpretation that some authors have suggested for the Sierra Nevada foothills. Irwin (1985), on the other hand, regarded the Hayfork Terrane as a separate arc that amalgamated with other terranes during a late Middle Jurassic collisional event.

The Smith River Subterrane of the Western Klamath Terrane in California is best developed along the Smith River in the northern part of the outcrop belt. To the south, the subterrane is attenuated and separated into thin slivers by major faults. Most of the rocks are sedimentary, similar to those of the Mariposa Formation; they contain the late Oxfordian–early Kimmeridgian *Buchia concentrica* and overlie mafic and ultramafic basement rocks termed the Josephine ophiolite (Harper 1984). The latter is a nearly complete ophiolitic assemblage with a radiometric age of 157 Ma (Harper et al. 1986) and is interpreted to have formed in a back-arc basin produced by splitting of the Hayfork Terrane arc. Sedimentary rocks above the ophiolite commence with 50 m or more of interbedded argillite, minor radiolarian chert, manganiferous and iron-rich chert, and volcanic graywacke (Pinto-Auso and Harper 1985). Locally, a thick olistostrome (Lems Ridge) containing abundant ophiolitic detritus forms the base of the sequence and suggests formation of the Josephine ophiolite within an east–west-trending fracture zone (Harper, Saleeby, and Norman 1985). Graywacke and slate overlie both the Lems Ridge olistostrome and the cherty argillite unit; detrital grains in the sandstone indicate volcanic and continental sources. All rocks of the Smith River Subterrane were deformed and metamorphosed during the Nevadan Orogeny.

#### Terranes of the California Coast Ranges

Jurassic rocks occur here at the base of and within the lower part of the structurally simple Great Valley Sequence and within the structurally complex and disrupted Franciscan Complex (Bailey, Irwin, and Jones 1964). These differing assemblages are generally linked in a plate-tectonic model that equates the Franciscan with a subduction-related off-scraped sedimentary prism and the Great Valley Sequence with a fore-arc basin. Blueschist metamorphic-facies minerals within the Franciscan indicate that these rocks have been subjected to the deep burial believed to be characteristic of subduction-zone conditions. Franciscan rocks are always in tectonic contact with the adjoining Great Valley Sequence, which does not contain high-pressure minerals.

The basement of the Great Valley Sequence consists of the Coast Range Ophiolite, a nearly continuous belt of Middle to Upper Jurassic mafic and ultramafic rocks that occur between the stratified rocks of the Great Valley Sequence and the Franciscan rocks. Depositional contacts between the ophiolite and overlying sedimentary rocks have been observed in a few places (Bailey, Blake, and Jones 1970; Hopson et al. 1981), but in most places the contact is a fault or fault zone. The Coast Range Ophiolite is a composite in that different parts exhibit major differences in terms of composition, structure, and geological history (Blake and Jones 1981). For example, in the southern Coast Ranges, ophiolite within the Del Puerto Subterrane includes arc-related plagiogranite, keratophyre, and tufaceous cherty rocks with abundant radiolarians (Pessagno 1977), whereas in the northern Coast Ranges the Elder Creek Subterrane contains blocks of basalt and chert incorporated within serpentinite mélange. Near Paskenta in northern California, these rocks are faulted against another ophiolitic assemblage containing basaltic and andesitic breccia that is overlain by Kimmeridgian strata of the Great Valley Sequence (Bailey et al. 1970; Jones 1975; Shervais and Kimbrough 1985; Blake et al. 1987).

The Upper Jurassic portion of the Great Valley Sequence, often assigned to the "Knoxville," is thickest, most fossiliferous, and best studied along the west side of the Sacramento Valley. *Buchia* are locally abundant, along with belemnites, ammonites, and rarer gastropods and other forms (Jones, Bailey, and Imlay 1969; Imlay and Jones 1970), permitting zonation of the monotonous turbiditic sandstone, mudstone, and conglomerate succession, which locally reaches 10,000 m in thickness. Radiolarians and other microfossils are useful in dating parts of the section (Pessagno 1977).

The approximate Jurassic–Cretaceous boundary can be drawn consistently within the Great Valley Sequence between the *Buchia terebratuloides* aff. *okensis* Zone below and the *B. uncitoides* Zone. The former contains Late Jurassic ammonites. This local systemic boundary (Jones et al. 1969), however, does not correspond precisely with boundaries established elsewhere (Jeletzky 1984; Bralower et al. 1987).

Jurassic rocks are widespread within Franciscan terranes, but only in a few places can stratal relations be established, such as in the Marin Headlands Terrane north of San Francisco. Wahrhaftig (1984) and Murchey (1984) have documented a sequence ranging from Liassic to Cenomanian within a complexly faulted and folded chert, pillow basalt, and graywacke assemblage. The pillow basalt at the base probably represents the uppermost layer of oceanic crust. It is overlain by 80 m of fossiliferous, red manganiferous ribbon chert, with radiolarians of Pliensbachian through Albian ages. This chert sequence constitutes one of the most remarkable records of pelagic deposition known from the eastern Pacific region, recording nearly 100 m.y. of deposition below the carbonate-compensation depth in an oceanic basin removed from continentally derived sediments. It is overlain depositionally by Cretaceous coarse-grained turbiditic graywacke.

Innumerable smaller blocks of chert throughout Franciscan mélanges in the western part of the northern Coast Ranges, and possibly also in the southern Coast Ranges, are similar to the Marin Headlands Terrane (Murchey and Jones 1984). This implies that the original oceanic plate that constituted that terrane disintegrated into small blocks in proximity to the North American continental margin. Franciscan mélanges in the eastern part of the northern Coast Ranges contain Upper Jurassic cherts associated with basaltic seamounts, differing significantly from chert sequences to the west.

Upper Jurassic rocks in other parts of the Franciscan complex, as in the Yolla Bolly Terrane, include meta-graywacke and radiolarian chert. It remains to be proved that the clastic and pelagic rocks are actually interbedded (Hein & Koski 1987). Poorly preserved radiolarians from the chert are entirely of Jurassic age, perhaps later Jurassic age. If so, this sequence records the progradation of a continentally derived deep-sea turbidite-fan assemblage over an oceanic pelagic sequence during the Malm. This stratified sequence thus differs from nearby coeval chert-bearing assemblages. Apparently stabilistic models sedimentologically linking the Great Valley Sequence and the Franciscan Complex (Seiders and Blome 1987) are overly simplistic.

### Terranes of southern California

Jurassic sedimentary rocks in southern California occur in the Santa Monica Mountains (a displaced fragment of the Foothills Terrane) and to the south in the Santa Ana Mountains (Santa Ana Terrane).

The Santa Monica Slate forms the basement of the Santa Monica Mountains, part of the east–west-trending Transverse Ranges, and consists of folded slate, graywacke, and minor volcaniclastic rocks. The only fossils known are fragments of *Buchia* cf. *concentrica,* suggesting late Oxfordian to early Kimmeridgian (Imlay 1968). This unit clearly correlates with the Mariposa Formation of the Sierran foothills and differs only in its structural trends. Jones and Irwin (1975) and Jones, Blake, and Rangin (1976) therefore proposed that the Santa Monica Slate was originally continuous with the foothills belt, but that the Santa Monica Slate had been disrupted and rotated in post-Kimmeridgian time. Rotation has since been amply documented by paleomagnetics (Luyendyk, Kamerling, and Terres 1980) for the entire western part of the Transverse Ranges.

Jurassic rocks of the Santa Ana Terrane occur in two formations: the lower Bedford Canyon Formation and the overlying Santiago Peak Volcanics. The Bedford Canyon consists of more than 5,000 m of folded, thin-bedded quartzose graywacke and shale with intercalated lenses of olistostromes. These rocks are overturned and form the inverted limb of a nappelike structure (Buckley, Condra, and Cooper 1975). Fossils have been recovered only from displaced blocks within the olistostromes. Large limestone blocks (and possibly concretions) have yielded the early Bajocian *Dorsetensia* cf. *liostraca* (Buckman) and *Stephanoceras* sp. of the Humphriesianum Standard Zone and, at another locality, immature *Lilloettia* and *Hecticoceras s.l.*, indicating Early Callovian (Imlay 1964; G. E. G. Westermann and A. C. Riccardi personal communication 1989).

Triassic radiolarians occur in olistostromal chert blocks. Nearby sources for the blocks are not known, and it is possible that the Bedford Canyon Formation originated far to the south of its present position. Based on high Rb/Sr ratio in shaly beds, Criscione et al. (1978) postulated that the quartzose detritus has a probable Precambrian provenance. In contrast, the nearby Santa Monica Slate has low Rb/Sr ratios characteristic of an island-arc terrane.

Unconformably overlying the Bedford Canyon Formation is a thick andesitic volcanic and volcaniclastic assemblage termed the Santiago Peak Volcanics. This unit is poorly dated; belemnites and the Tithonian *Buchia piochii* (Fife, Minch, and Crampton 1967) have been found in black argillite only near San Diego. In the Santa Ana Mountains, the formation is unconformably overlain by shallow-water Upper Cretaceous marine deposits.

Both the Bedford Canyon Formation and the Santiago Peak Volcanics represent tectonic assemblages whose relations to other Jurassic-bearing terranes of California remain enigmatic. For example, no other *Buchia piochii*-bearing strata in California are volcanogenic, and post-Kimmeridgian marine strata are not known to occur in the volcanic-rich Mesozoic sequence of the Sierra Nevada. Likewise, the Jurassic limestone blocks of the Bedford Canyon Formation are lithically and faunally distinct from any other Jurassic strata occurring in nearby terranes or on the old continental margin of North America. These differences suggest that the entire Santa Ana Terrane was separated from the other terranes of California and perhaps did not achieve its present position until after Mesozoic time.

## NEVADA[13]

### Introduction

Strata of Early Jurassic to early Middle Jurassic age crop out extensively in western Nevada. They were deposited in an

[13] By D. G. Taylor and P. Smith.

Early Mesozoic marginal basin in western Nevada that may be delineated roughly (southern and eastern margins of basin) by the Sr. 0.706 contour (Figure 4.25). As a successor basin, it was established upon what may be interpreted either as a Paleozoic marginal basin associated with the continent (Burchfiel and Davis 1972, 1975) or as an allochthonous terrane accreted during the Permo-Triassic (Speed 1979). Marine depositional history began in the Triassic and terminated during the Middle Jurassic. In that the Mesozoic marine province is underlain by simatic crust, its history records the reestablishment of the western margin of North America, from its Late Paleozoic position in central Nevada to a Middle Jurassic location west of the Great Basin (Speed 1978a).

Significant advances have been made in recent years in interpreting the Jurassic in the back-arc basin in Nevada. Nevertheless, most stratal sequences are as yet inadequately described, and a significant body of information is based on unpublished work. Moreover, the region has been subjected to extensive Middle or Late Jurassic and Early Cretaceous thrust faulting, obscuring basin morphology. As a result of the thrust faulting, the basin has undergone considerable telescoping, with up to several hundred kilometers of northwest–southeast contraction (Oldow 1984).

The Mesozoic rocks in Nevada were divided by Oldow (1984) into eight lithotectonic assemblages (Figure 4.26), which are employed in this report. Oldow (1984, pp. 248–9) used the term "lithotectonic assemblage" to "segregate coeval and partly coeval packages of rocks which accumulated in separate though commonly related settings and/or which had different structural histories." Five of these (the Black Rock, Lovelock, Sand Springs, Pamlico, and Luning assemblages) are dismembered and were subjected to extensive thrusting. Two, the Humboldt and Gold Range assemblages, are autochthonous or para-autochthonous, having been subjected to comparatively minor thrusting, whereas one, the Pine Nut assemblage, is allochthonous and internally coherent. Of these assemblages, only the Humboldt has no record of Jurassic sedimentation.

Speed (1978a) described Early Mesozoic basinal history with reference to four "terranes," three for the Triassic through Early Jurassic phase of deposition, and one for the latest Early Jurassic through Middle Jurassic phase of basinal development. The earlier three include the "shelf terrane" for a zone of predominantly shallow-water carbonate sedimentation along the eastern margin of the basin, the "basinal terrane" for an area of largely deeper-water pelitic sedimentation in the northwestern part of the province, and the "volcanic terrane" for an area of volcanic accumulation and marine sedimentation in the southwestern part of the basin. (The extensive accumulation of volcanics in the latter is more characteristic of the Triassic phase than of the Jurassic phase of basin history.) The "orogenic terrane" of Speed (1978a) encompasses latest Early Jurassic and younger rocks deposited over much of the basin.

Approximate correspondences between the assemblages of Oldow (1984) and the three Triassic and Early Jurassic terranes of Speed (1978a) are as follows: (1) The shelf terrane is equivalent to the Humboldt assemblage plus most of the Luning assemblage; (2) the volcanic terrane corresponds essentially to the Pine Nut, Sand Springs, Pamlico, Luning (southwestern part), and Gold Range assemblages; (3) the basinal terrane corresponds to the Lovelock and Black Rock assemblages.

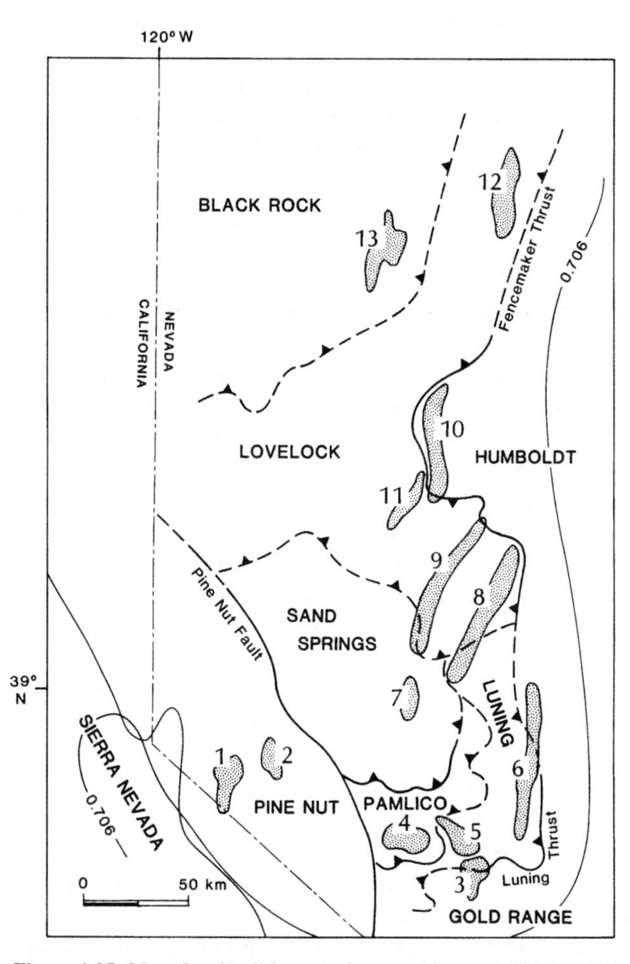


**Figure 4.25. Map showing lithotectonic assemblages of Oldow (1984). Numbers denote mountains/ranges, as follows: (1) Pine Nut Range, (2) Singatse Range, (3) Pilot Mountains, (4) Garfield Hills, (5) Gabbs Valley Range, (6) Shoshone Mountains, (7) Sand Springs Range, (8) Clan Alpine Mountains, (9) Stillwater Range, (10) Humboldt Range, (11) West Humboldt Range, (12) Santa Rosa Range, (13) Jackson Mountains.**

In delineating the lithotectonic terranes of the western conterminous United States, Silberling et al. (1984) recognized several terranes that compose the back-arc basin. The relevant terranes of Silberling et al. (1984) and the assemblages of Oldow (1984) are related as follows: The Walker Lake Terrane encompasses the Pine Nut assemblage in its western part, the Pamlico, Luning, and northern part of the Gold Range assemblages in its eastern part, and the Sand Springs assemblage in its northern part. The Jungo Terrane corresponds to the Lovelock assemblage, the Golconda Terrane corresponds to the Humboldt assemblage and part of the Gold Range assemblage, and the Jackson Terrane corresponds to the Black Rock assemblage.

The terminology of Oldow (1984) is employed, rather than that of Speed (1978a), because the former is descriptive (instead of genetic) and because Speed's threefold basinal division for the ear-

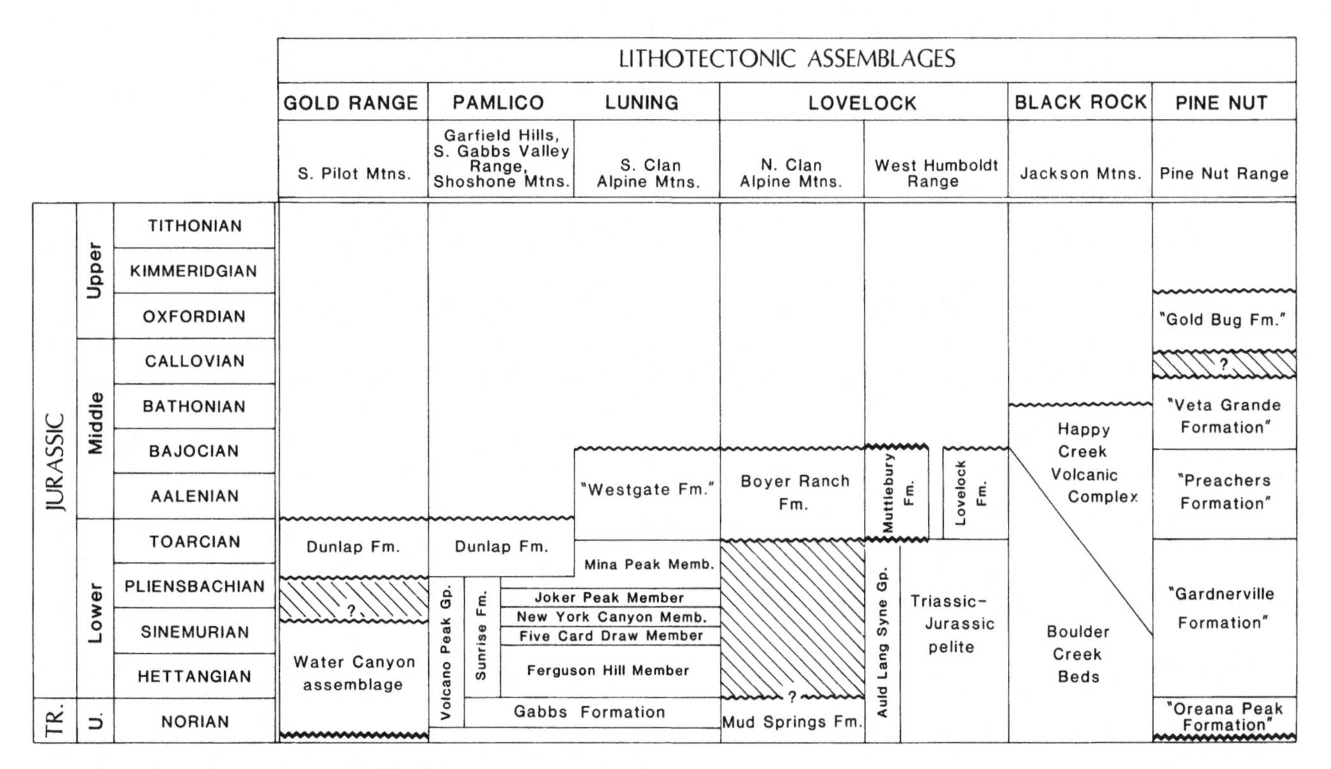


**Figure 4.26. Correlation of representative Jurassic sections in Nevada.**

lier phase of development of the marine province relates more to its Triassic history than to its Jurassic history. Although the terranes of Silberling et al. (1984) are most adequate, Oldow's terminology is employed here because it provides greater detail helpful for the description of the back-arc basin.

### Tectonic assemblages

#### *Pine Nut assemblage*

Jurassic metasedimentary and metavolcanic rocks are widespread in this assemblage, as summarized by Moore (1969). The best-known Jurassic section, from the southern Pine Nut Range (Figure 4.25), has been divided into four formations, as described in the dissertation of Noble (1962) (Moore 1969, pp. 6–7). That sequence has an aggregate thickness greater than 3,100 m.

The lowest Jurassic unit, the Gardnerville Formation (1,000 m) is composed in its lower part (900 m) of thin-bedded siltstone, with minor sandstone, carbonates, and associated volcanics, and in its upper part (100 m) of volcanic conglomerate (Noble 1962). The formation conformably overlies the Triassic Oreana Peak Formation. The Gardnerville Formation was deposited in a distal shelf setting (Oldow 1984) and encompasses most of the Early Jurassic (Noble 1962; Silberling et al. 1984). Gardnerville strata have also been recognized in the Singatse Range (Speed 1978a).

The 245-m-thick Preachers Formation of Noble (1962), which is reported to overlie the Gardnerville Formation disconformably (Oldow 1984), is a shallow-water unfossiliferous sandstone unit.

The superjacent Veta Grande (1,525 m) and Gold Bug (365 m) Formations of Noble (1962) are composed of intermediate and siliceous volcanics and associated volcanogenic sedimentary lithologies. These unfossiliferous units are Middle and/or Late Jurassic in age.

#### *Gold Range assemblage*

The most southerly rocks of Jurassic age representing the back-arc basin occur in the Gold Range assemblage. The two units of interest from that assemblage are the Water Canyon assemblage (informal) of Oldow (1981) and the Dunlap Formation.

The Water Canyon assemblage includes rocks that prior to Oldow (1981) were allocated to the Gold Range Formation (Nielsen 1963; Stanley 1971; Speed 1977; Oldow 1978). It consists of siliceous volcanic and volcanogenic rocks and includes less abundant terrigenous clastic detritus deposited in a shallow marine to largely subaerial setting. There are few age constraints for the unit, although it is considered to be largely Triassic in age. The uppermost beds of the unit in the Pilot Mountains preserve shallow-marine red mudstone and fossiliferous carbonate rocks yielding a bivalve tentatively identified by N. Silberling as *Weyla* (Nielsen 1963). *Weyla* is a post-Hettangian, Early Jurassic pelecypod.

The Dunlap Formation in the Gold Range assemblage (Muller and Ferguson 1936, 1939; Ferguson and Muller 1949; Nielsen 1963; Stanley 1971) is a lithologically heterogeneous unit of conglomerate, breccia, sandstone, mudstone, and volcanics, largely andesitic in composition. Quartz sandstone is a distinctive lithology occurring locally in the formation. The formation attains a stratigraphic thickness of approximately 1,525 m in its type area in the Pilot Mountains (Muller and Ferguson 1936, 1939). It is widespread in the Gold Range assemblage, where it overlies subjacent Mesozoic and Paleozoic rocks with angular unconformity. The formation is allocated to the latest Early Jurassic.

### *Pamlico/Luning assemblages*

The Pamlico and Luning assemblages are treated together here because they jointly preserve the Sunrise Formation, a richly fossiliferous unit (Muller and Ferguson 1936, 1939; Silberling 1959, 1984; Corvalán 1962; Hallam 1965; Guex and Taylor 1976; Guex 1980, 1981; Smith 1981; Taylor et al. 1983; Taylor 1986; Smith et al. 1988) deposited in a platformal setting. The Sunrise Formation and subjacent Gabbs Formation occur at numerous locations, and together they form the Volcano Peak Group. The Sunrise is divided into five members (Taylor et al. 1983), and the formation in the type area in the southern Gabbs Valley Range (Figure 4.25) is approximately 400 m thick. There the members delineate an alternating succession of calcareous siltstone and bioclastic limestone. The Sunrise Formation conformably overlies the late Norian (Triassic) Gabbs Formation and spans most of the Lower Jurassic. Its upper age limit ranges from Pliensbachian to Toarcian.

The Dunlap Formation is also widely distributed (Muller and Ferguson 1936, 1939; Ferguson and Muller 1949; Silberling 1959). In the Pamlico and Luning assemblages, depositional contacts with subjacent units vary from conformable in places where the formation overlies the Sunrise Formation to unconformable where it is in direct contact with Triassic and Permian units. Few fossils have been recovered from the formation that might provide an age assignment. Strata in the Garfield Hills referable either to the Dunlap or Sunrise Formation (Silberling 1984) yield the aberrant pelecypod *Lithiotis* (= *Plicatostylus*), which in North America is restricted to the Pliensbachian stage, and furnish ammonite material tentatively allocated to *Protogrammoceras* (Silberling 1984). The latter suggests a Pliensbachian or possibly earliest Toarcian allocation. The Dunlap Formation likely was deposited rapidly during the Pliensbachian and Toarcian stages of the Lower Jurassic. [A much longer age range extending perhaps into the Cretaceous is indicated by Oldow (1981).] The upper contact of the formation is unconformable with superjacent units.

In the southern Clan Alpine Mountains the Sunrise Formation is overlain conformably by the Westgate Formation of Corvalán (1962), a name adopted for USGS usage (Imlay 1980a). That unit consists of 64 m of calcareous quartz arenite overlain by 107 m of thick-bedded limestone. Although the lower beds may be as old as Toarcian, poorly preserved fossils suggest that the limestone may be as young as Bajocian (Corvalán, 1962). The sequence described by Corvalán (1962) is also discussed by Hallam (1965).

### *Sand Springs assemblage*

Although not well understood, Triassic and Jurassic sedimentary and metamorphic rocks from this assemblage are exposed in the Sand Springs Range and the southern Fairview Valley (east of the Sand Springs Range) (Willden and Speed 1974). These are overlain unconformably by Jurassic mafic rocks.

### *Lovelock assemblage*

The Lovelock assemblage preserves a thick Late Triassic (Norian)–Early Jurassic accumulation of pelitic and sandy rocks referable to the Auld Lang Syne Group (Burke and Silberling 1973; Lupe and Silberling 1985). Superjacent latest Early Jurassic to Middle Jurassic units from the assemblage include the Boyer Ranch, Muttlebury, and Lovelock Formations.

The type area for the Auld Lang Syne Group is within the Humboldt assemblage, where it overlies the Spathian through Carnian Star Peak Group (platformal). Auld Lang Syne strata on that assemblage, however, apparently do not range as high as the Jurassic.

The Triassic strata referable to the group west of the Fencemaker thrust (in the Lovelock assemblage) may be 6 km or more in thickness and are pelitic sediments deposited in a basinal setting (Compton 1960; Speed and Jones 1969; Speed 1978a,b; Heck and Speed 1987). No depositional base of the Auld Lang Syne Group is exposed in the Lovelock assemblage. The superjacent Lower Jurassic is also predominantly pelitic. The exposed stratigraphic thickness is up to 1,000 m in places, although the total thickness may once have been greater.

At the southern end of the Humboldt Range, just south of the Fencemaker thrust (Figure 4.25), are lithologically heterogeneous Lower Jurassic rocks that are referable to the group and that yield Sinemurian and possibly Toarcian ammonites (Silberling and Wallace 1969; Wallace et al. 1969) at two closely spaced but probably structurally isolated exposures.

Lower Jurassic (Auld Lang Syne) pelitic strata are widely exposed farther west, in the West Humboldt Range (Speed 1974, 1975; Willden and Speed 1974). These Jurassic strata, which have been examined in the West Humboldt Range east of Lovelock (Sulima 1970; Speed 1974, 1975), are locally metamorphosed to slate and hornfels, have an exposed thickness of at least 1,000 m in places, and are conformable with subjacent Triassic pelites.

To the north, at the south end of the Santa Rosa Range (Figure 4.25), is a series of six formations referable to the Auld Lang Syne Group (Compton 1960; Burke and Silberling 1973). The group might range into the Jurassic at that locality. There, the highest formation, the Mullinix, composed of unfossiliferous phyllites, is a thick unit (1,500 m) overlying the Andorno Formation, dated as Late Norian by *Monotis* cf. *subcircularis* and *Halorella* sp. (Burke and Silberling 1973).

A thick section of essentially Late Triassic pelitic rocks in the northern Clan Alpine Mountains herein allocated to the Auld Lang Syne Group [= Clan Alpine sequence of Speed (1978b)] may have Jurassic rocks at its top. The Mud Springs Canyon Formation (thickness 450+ m) represents the top of a Mesozoic section 5.8+ km thick having no exposed base. The formation is an accumulation of massive carbonate rocks and is reported by Speed

(1978b) to yield the late Norian bivalve *Monotis subcircularis* in its lower part. The upper part is undated, and as discussed by Speed and Jones (1969) and Speed (1978b), it could range into the Lower Jurassic. To the west, in the Stillwater Range, Late Triassic shale and sandstone are locally overlain by Jurassic strata, in places conformably (Speed 1978b).

The Boyer Ranch Formation, a latest Lower Jurassic to Middle Jurassic unit that ranges up to 150 m thick, is composed predominantly of quartz sandstone and carbonate rocks, commonly conglomeratic. That unit, restricted to isolated exposures, is widespread in the northern parts of the Clan Alpine and Stillwater Ranges. In the Clan Alpine Range the formation overlies the Mud Springs Canyon Formation, with angular unconformity.

In western Pershing County, the Late Triassic–Lower Jurassic pelitic rocks are conformably overlain by the Lovelock Formation, a 25–200-m-thick unit composed principally of limestone in its lower part (3–35 m), limeclast conglomerate and microbreccia in the middle (3–20 m), and gypsum and quartz arenite (~100 m) in the upper part (Speed 1974). Although no fossils are sufficiently preserved to provide age constraints, the formation presumably is latest Early Jurassic to Middle Jurassic in age. A probably coeval unit is the Muttlebury Formation, a carbonate nappe breccia ranging up to 70 m in thickness. It consists predominantly of carbonate breccia (rauhwacke) and includes limestone, marble, and gypsum. Marble occurs in the "nonbreccia" and was proposed to have originated through calcitization of gypsum (Speed and Clayton 1974). The breccia most probably originated through solution of gypsum, resulting in a gravity accumulation of largely carbonate fragmental residue (Speed 1975).

Basaltic rocks (locally up to 520 m or more in thickness) and a gabbroic intrusion (Humboldt lopolith) believed to be comagmatic are preserved in the Clan Alpine Mountains, northern Stillwater Range, and West Humboldt Range (Speed and Jones 1969; Willden and Speed 1974). The gabbro has yielded K/Ar dates of 150 ± 5, 154 ± 5, and 165 ± 5 Ma, the latter two from coexisting biotite and hornblende (Willden and Speed 1974), indicating a largely Late Jurassic age for the volcanics. Although it is reported that the basaltic rocks conformably overlie and locally are interbedded with quartz arenite (Willden and Speed 1974), the implications of those contact relationships for the upper age limit of the Boyer Ranch Formation, for example, are still enigmatic.

#### *Black Rock assemblage*

In the northwestern part of Nevada in the Jackson Mountains (Figure 4.25) there occur thick Upper Triassic to Jurassic accumulations termed the Boulder Creek beds and Happy Creek Complex (Russell 1984). The former consists of volcanogenic sediments, and the latter is composed of volcanic rocks and intrusives of intermediate composition, as well as minor sediments. A Rb/Sr age determination of approximately 160 Ma is thought to suggest an age assignment for "late eruptions" of the complex (Russell 1984). The Boulder Creek beds represent sedimentation in a basinal setting, and the Happy Creek Complex represents development of a magmatic arc.

### Back-arc history

Sedimentation in the back-arc region began in the Early Triassic and continued throughout the remainder of the period. In the southeastern part of the basin, Triassic marine rocks are dominated by carbonate lithologies, whereas in the southwest ("volcanic terrane") the Triassic consists in large parts of volcanics and volcaniclastic strata (Speed 1978a). The northern part of the basin on the platform (east of Fencemaker thrust) first saw Early Triassic accumulation of volcanics and volcanogenic sediments referable to the Kiopato Group, followed by deposition of Spathian through Carnian carbonate-rich lithologies of the Star Peak Group. Beginning abruptly in the Norian, the thick pile of pelitic and sandy nonvolcanic Norian sediments of the Auld Lang Syne Group accumulated. That group in the Humboldt assemblage represents a prograding delta, with sediment structures indicating northwesterly trending currents (Silberling and Wallace 1969; Lupe and Silberling 1985). The exceptionally thick (6+ km) section referable to the Auld Lang Syne Group in the Lovelock assemblage consists largely of turbiditic facies deposited in a basinal setting. Widespread quartzitic sand and silt lithologies in the group indicate a source for siliciclastic materials via a drainage system connected to the craton to the east (Lupe and Silberling 1985). Finally, the Triassic in the Jackson assemblage records the transition from a largely sediment-starved basin to the development of an oceanic magmatic arc on the western margin of the North American continent (Russell 1984).

Marine sediments accumulated throughout the basin during most of the Early Jurassic, with continued platformal sedimentation in the east and progressive "shoaling" over much of the offshore area to the west. In the southeastern part of the basin (Luning and Pamlico assemblages), continuous sedimentation from the Triassic (Gabbs Formation) to the Jurassic (Sunrise Formation) is recorded. There, the Early Jurassic Sunrise Formation is widespread, and the greatest geographic variation in its lithofacies is in the lower part of the unit. As described by Taylor et al. (1983), the lithofacies trends in the lower part of the formation indicate the shallowest-water setting in the Gabbs Valley Range, an intermediate setting in the Shoshone Mountains, and comparatively offshore (but still platformal) settings in the Garfield Hills and southern Clan Alpine Mountains (Figure 4.25). The upper members appear to be more uniformly developed, implying quite subdued paleotopography in the latter part of the Early Jurassic (through early Pliensbachian). The latest phase in Sunrise deposition records a marked regressive trend (Smith and Tipper 1988). To the south, in the Gold Range assemblage, few Lower Jurassic marine strata were preserved, and that area is perhaps near the southern limit of the marine basin.

To the west, the Gardnerville Formation, with its greater thickness, lower carbonate content, and greater abundance of volcanogenic detritus, was deposited in a setting of somewhat deeper water than was the Sunrise. In fact, the Gardnerville Formation apparently is transitional between the platform to the east (location of Sunrise Formation) and the Sierra Nevada arc to the west (Noble 1962; Speed 1978a).

No Jurassic strata are preserved in the northern part of the platform, east of the Fencemaker thrust. Jurassic strata are sparse close to that thrust fault, but are more widespread west and southwest of it (Speed 1978b). The Jurassic part of the Auld Lang Syne Group and associated rocks are not as thick and commonly are more calcareous than the Triassic part of the section, indicating waning of siliciclastic supply and shoaling.

In the Black Rock assemblage to the north, the oceanic arc apparently was well established in the Early Jurassic.

The final phase of sedimentation in the marine basin began late in the Early Jurassic and persisted into the Middle Jurassic. The phase relates to the orogenic terrane of Speed (1978a), which includes the Dunlap Formation and younger Jurassic units (Figure 4.26). Mature quartz sand is widespread at least in the lower parts of the "orogenic" sections (Speed 1978a) and represents the last (and only Jurassic) pulse of such sediments derived from a cratonal source to the east (Speed and Jones 1969; Stanley, Jordan, and Dott 1971). Jurassic quartz arenite, being widespread, occurs in all assemblages except the Sand Springs and Black Rock, occupying the western parts of the basin (Oldow 1984). The quartz sand was concentrated in lows and ranges in stratigraphic thickness from a few meters to nearly 2,000 m (Speed 1978a). Perhaps the sand reflects the westward migration of the strand line from a position close to the craton in response to regional tectonism.

In the southwestern part of the basin, the Dunlap Formation, though possessing quartz sand and various facies reflecting subtidal and intertidal environments, largely has sediments representing alluvial-fan deposits formed as a result of intensive deformation. In the Pine Nut Range, arenaceous sand (Preachers Formation) is succeeded by extensive volcanogenic accumulations. To the north, quartz sand is succeeded by shallow-marine limestone in the southern Clan Alpine Mountains, whereas sand composing the Boyer Ranch Formation was deposited farther north in the northern Clan Alpine Mountains and Stillwater Range. To the west, carbonates and gypsum were forming in the basin, which at least locally was experiencing restricted circulation. Marine sedimentation ceased in the Middle Jurassic, perhaps during Bajocian time.

The development of the Luning-Fencemaker fold and thrust belt in the back-arc basin (resulting in up to several hundred kilometers of NW–SE contraction) occurred in Middle or Late Jurassic and Early Cretaceous time (Oldow 1984). The precise timing for onset of that regional folding and thrusting phase is difficult to assess, although Oldow (1981) considered that Dunlap Formation sedimentation was related to an earlier orogenic pulse. According to him (Oldow 1984) the thrusting resulted from transpressional motion and partial coupling as the Sierra arc and back arc were being displaced relative to one another by left-lateral strike-slip movement along the Pine Nut fault.

### Ammonite zones and assemblages

#### *Hettangian*

1. Planorbis Zone – *Psiloceras pacificum* Guex, *P. polymorphum* Guex, *Transipsiloceras transiens* Guex, *Waehneroceras tenerum* (Neumayr), *Nevadaphyllites compressus* Guex, *Paradasyceras* aff. *uermoesense* (Herbich), *Fergusonites striatus* Guex (Guex 1980, 1981)
2. Liasicus Zone – *Fergusonites striatus* Guex, *Kammerkarites rectiradiatus* Guex, *K. haplotychus* (Waehner), *K. praecoronoides* Guex, *K. diplotychoides* Guex, *Discamphiceras antiquum* Guex, *D. silberlingi* Guex, *D. kammerkaroides* Guex, *Euphyllites occidentalis* Guex, *E.* cf. *struckmanni* (Neumayr), *Pleuroacanthites mulleri* Guex, *Franziceras coronoides* (Guex)
3. Angulata Zone – *Franziceras coronoides* (Guex), *Discamphiceras silberlingi, D.* aff. *mesogenus* (Waehner), *Mullerites pleuroacanthitoides* Guex, *Pleuroacanthites biformis* (Sowerby), *Alsatites latecarinatus* (Waehner), *A. nigromontanus* (Guembel), *A.* aff. *proaries* (Neumayr), *Schlotheimia angulata* (Schlotheim), *Pseudaetomoceras doetzkirchneri* (Guembel), *P. castagnolai* (Cocchi), *Sunrisites sunrisensis* Guex, *S.* aff. *hadroptychus* (Waehner), *Sulciferites* n.sp. (Guex 1980, 1981)
4. Zone B – *Sulciferites* n.sp., *Metophioceras* cf./aff. *rursicostatum* (Frebold), *Badouxia* n.sp., *Vermiceras* aff. *haueri* (Guembel), *Pseudaetomoceras doetzkirchneri* (Guembel) (Taylor 1986)

#### *Sinemurian*

1. Canadense Zone – *Badouxia canadensis* (Frebold), *Eolytoceras* n.sp., *Metophioceras rursicostatum* (Frebold), *Paracaloceras subsalinarium* (Waehner), *Pseudaetomoceras* n.sp., *Nevadaphyllites* n.sp.
2. Zone C – *Metophioceras* n.sp., *M.* n.sp. aff. *rursicostatum* (Frebold), *Vermiceras* n.sp., and "*Primarietites*" (Taylor 1986)
3. *Coroniceras* spp., *Tmaegoceras* spp., *Arnioceras* spp.

#### *Upper Sinemurian*

1. *Oxynoticeras* sp., *Eoderoceras* sp. (Hallam 1965)

#### *Pliensbachian*

1. Whiteavesi Zone – *Acanthopleuroceras* aff. *stahli* (Oppel)
2. Freboldi Zone – *Dubariceras freboldi* Dommergues, Mouterde, & Rivas, *Metaderoceras* aff. *muticum* (d'Orbigny), *Aveyroniceras* spp., *A. colubriforme* (Bettoni), *Prodactylioceras* aff. *davoei* (Sowerby) (Smith et al. 1988)
3. Kunae Zone – *Aveyroniceras* spp., *Arieticeras* aff. *algovianum* (Oppel)

4. Carlottense Zone – *Fanninoceras carlottense* McLearn, *Tiltoniceras propinquum* (Whiteaves), *Lioceratoides* sp.

*Toarcian*

1. *Tiltoniceras propinquum* (Whiteaves), *Lioceratoides* sp., *Protogrammoceras allifordensis*, dactylioceratids, *Phymatoceras*(?), *Hildaites* sp., *Grammoceras* sp., *Harpoceras* (Muller and Ferguson 1939; Silberling 1959; Corvalán 1962; Noble 1962; Hallam 1965)

### Acknowledgment

We wish to thank N. Silberling for critical review of the manuscript.

### References

Archibald, D., Glover, J. K., Price, R. A., Farrar, E., & Carmichael, D. M. (1983). Geochronology and tectonic implications of magmatism and metamorphism, southern Kootenay Arc and neighbouring regions, southeastern British Columbia. Part I, Jurassic to mid-Cretaceous. *Can. J. Earth Sci., 20,* 1891–913.

Armstrong, R. L. (1985). The Kuskokwim a-2 transect. In R. von Huene et al. (eds.), *Continent-oceans.* Trans. Geol. Soc. America, Transect Ser. Map A-2.

Armstrong, R. L., & MacKevett, E. M., Jr. (1982). *Stratigraphy and Diagenetic History of the Lower Part of the Triassic Chitistone Limestone, Alaska.* U.S. Geological Survey, Professional Paper 1212A.

Armstrong, R. L., Taubeneck, W. H., & Hales, P. O. (1977). Rb-Sr and K-Ar geochronometry of Mesozoic granitic rocks and Sr isotopic composition, Oregon, Washington and Idaho. *Geol. Soc. Am. Bull., 88,* 397–411.

Atwater, T. (1970). Implications of plate tectonics for the Cenozoic tectonic evolution of western North America. *Geol. Soc. Am. Bull., 81,* 3513–36.

Bailey, E. H., Blake, M. C., Jr., & Jones, D. L. (1970). *On-Land Mesozoic Oceanic Crust in California Coast Ranges.* U.S. Geological Survey, Professional Paper 700-C70-81.

Bailey, E. H., Irwin, W. P., & Jones, D. L. (1964). *Franciscan and Related Rocks and Their Significance in the Geology of Western California.* California Division of Mines and Geology, Bulletin 183.

Balkwill, H. R., Cook, D. G., Detterman, R. L., Embry, A. F., Håkansson, E., Miall, A. D., Poulton, T. P., & Young, F. G. (1983). Arctic North America and northern Greenland. In A. E. M. Nairn & A. Moullade (eds.), *The Phanerozoic Geology of the World. II: The Mesozoic* (pp. 1–31). Amsterdam: Elsevier.

Beard, J. S. & Day, H. W. (1987). The Smartville intrusive complex, Sierra Nevada, California: the core of a rifted volcanic arc. *Geol. Soc. Am. Bull., 99,* 779–91.

Behrman, P. S. (1978). Pre-Callovian rocks, west of the Melones fault zone, central Sierra Nevada foothills. In D. G. Howell & A. K. McDougall (eds.), *Mesozoic Paleogeography of the Western United States* (pp. 337–48). Pacific Coast Paleogeography Symposium 2. Los Angeles: Pacific Section SEPM.

Behrman, P. S., & Parkinson, G. A. (1978). Paleogeographic significance of the Callovian to Kimmeridgian strata, central Sierra Nevada foothills, California. In D. G. Howell & A. K. McDougall (eds.), *Mesozoic Paleogeography of the Western United States* (pp. 349–60). Pacific Coast Paleogeography Symposium 2. Los Angeles: Pacific Section SEPM.

Bell, W. A. (1944). Use of some fossil floras in Canadian stratigraphy. *Roy. Soc. Can. Trans.* (34d ser.), *38*(4), 1–13.

(1946). Age of the Canadian Kootenay formation. *Am. J. Sci., 244,* 513–26.

(1956). *Lower Cretaceous Floras of Western Canada.* Geological Survey of Canada, Memoir 285.

Ben-Avraham, Z. (1981). The movement of continents. *American Scientist, 69,* 291–9.

Ben-Avraham, Z., & Nur, A. (1983). An introductory overview to the concept of displaced terranes. *Can. J. Earth Sci., 20,* 994–9.

Ben-Avraham, Z., Nur, A., Jones, D., & Cox, A. (1981). Continental accretion: from oceanic plateaus to allochthonous terranes. *Science, 213,* 47–54.

Berry, E. W. (1929). *The Kootenay and Lower Blairmore Floras.* National Museum of Canada, Bulletin 58, Geology Series 50, pp. 28–54.

Blake, M. C., Jr., Howell, D. G., & Jones, D. L. (1982). *Preliminary Tectonostratigraphic Terrane Map of California.* U.S. Geol. Survey Open File Rept. 82–0593.

Blake, M. C., Jr., Jayko, A. S., Jones, D. L., & Rogers, B. W. (1987). Unconformity between Coast Range ophiolite and lower part of the Great Valley sequence, South Fork Elder Creek, Tehama County, California. In: *Geological Society of America Centennial Field Guide,* vol. 1 (pp. 279–82).

Blake, M. C., Jr., & Jones, D. L. (1981). The Franciscan assemblage and related rocks in northern California: a reinterpretation. In W. G. Ernst (ed.), *The Geotectonic Development of California* (pp. 306–28). Englewood Cliffs, N.J.: Prentice-Hall.

Blome, C. D., & Irwin, W. P. (1983). Tectonic significance of late Paleozoic to Jurassic radiolarians from the North Fork terrane, Klamath Mountains, California. In Stevens (ed.), *Pre-Jurassic Rocks in Western North American Suspect Terranes* (pp. 77–90). Los Angeles: Pacific Section SEPM.

Blome, C. D., Jones, D. L., Murchey, B. L., & Liniecki, M. (1986). Geologic implications of radiolarian-bearing Paleozoic and Mesozoic rocks from the Blue Mountains province, eastern Oregon. In T. L. Vallier & H. C. Brooks (eds.), *Geology of the Blue Mountains Region of Oregon, Idaho and Washington* (pp. 79–93). U.S. Geological Survey, Professional Paper 1435.

Bogen, N. L. (1985). Stratigraphic and sedimentologic evidence of a submarine island-arc volcano in the lower Mesozoic Penon Blanco and Jasper Point formations, Mariposa County, California. *Geol. Soc. Am. Bull., 96,* 1322–31.

Boles, J. R. (1978). Basin analysis of the Eugenia Formation (Late Jurassic), Punta Eugenia area, Baja California. In D. G. Howell & K. A. McDougall (eds.), *Mesozoic Paleogeography of the Western United States* (pp. 993–8). Pacific Coast Paleogeography Symposium 2. Los Angeles: Pacific Coast SEPM.

Boles, J. R., & Landis, C. A. (1984). Jurassic sedimentary mélange and associated facies, Baja California, Mexico. *Geol. Soc. Am. Bull., 95,* 513–21.

Box, S. E. (1983). Tectonic synthesis of Mesozoic histories of Togiak and Goodnews terranes, southwest Alaska. *Geol. Soc. Am. Ann. Meeting, Abstr. Progr., Cordill. Sect.*

Bralower, T. J., Ludwig, K. R., Obradovich, J. D., & Jones, D. L. (1987). Berriasian (Early Cretaceous) radiometric ages from the Great Valley sequence, California, and their biostratigraphic and magnetostratigraphic correlation. *Geol. Soc. Am. Abstr. Progr., 19*(7), 598.

Braman, D. R. (1985). *The Sedimentology and Stratigraphy of the Husky Formation in the Subsurface District of Mackenzie, Northwest Territories.* Geological Survey of Canada, Paper 83–14.

Brooke, M. M., & Braun, W. K. (1972). *Biostratigraphy and Microfauna of the Jurassic System of Saskatchewan.* Saskatchewan Department of Mineral Resources, Report 161.

(1981). *Jurassic Microfaunas and Biostratigraphy of Northeastern British Columbia and Adjacent Alberta.* Geological Survey of Canada, Bulletin 283.

Brooks, H. C. (1967). Distinctive conglomerate layer near Lime, Baker County, Oregon. *Ore Bin, 29,* 113–19.

(1979a). *Geological Map of the Huntington and Part of the Olds Ferry Quadrangles, Baker and Malheur Counties, Oregon.* Oregon Department of Geological and Mineral Industries Geology, Map GMS-13.

(1979b). Plate tectonics and the geologic history of the Blue Mountains. *Oregon Geol., 41,* 71–80.

Brooks, H. C., & Ferns, M. L. (1979). *Geologic Map of the Bull Run Rock Quadrangle, Oregon.* Oregon Department of Geological and Mineral Industries, Map 0-79-6.

Brooks, H. C., Ferns, M. L., Nusbaum, R. W., & Kovich, P. M. (1979). *Geologic Map of the Rastus Mountain Quadrangle, Oregon.* Oregon Department of Geological and Mineral Industries, Map 0-79-7.

Brooks, H. C., & Vallier, T. L. (1978). Mesozoic rocks and tectonic evolution of eastern Oregon and western Idaho. In D. G. Howell & K. A. McDougall (eds.), *Mesozoic Paleogeography of the Western United States* (pp. 113–46). Pacific Coast Paleogeography Symposium 2. Los Angeles: Pacific Coast SEPM.

Brown, C. E., & Thayer, T. P. (1966). *Geologic Map of the Canyon City Quadrangle, Northeastern Oregon.* U.S. Geological Survey, Miscellaneous Geological Investigation Map I-1021.

Brown, R. W. (1946). Fossil plants and Jurassic–Cretaceous boundary in Montana and Alberta. *Am. Assoc. Petrol. Geol. Bull., 30,* 238–48.

Buckley, C. P., Condra, D. A., & Cooper, J. P. (1975). The Jurassic flysch of the Santa Ana Mountains: an example of obduction? *Geol. Soc. Am. Abstr. Progr., 7*(3), 299–300.

Burchfiel, B. C., & Davis, G. A. (1972). Structural framework and evolution of the southern part of the Cordilleran orogen, western United States. *Am. J. Sci., 272,* 97–118.

(1975). Nature and controls of Cordilleran orogenesis, western United States: extensions of an early synthesis. *Am. J. Sci., 275-A,* 363–96.

Burckhardt, C. (1927). *Cefalopodos del Jurasico Medio de Oaxaca y Guerrero.* Inst. Geol. Mexico, Bol. Num. 47.

(1930). Etude synthetique sur le Mesozoique Mexicain, Premiere partie. *Mem. Soc. Paleont. Suisse, 49,* 1–123.

Burke, D. B., & Silberling, N. J. (1973). *The Auld Lang Syne Group of Late Triassic and Jurassic(?) age, North-Central Nevada.* U.S. Geological Survey, Bulletin 1394-E.

Burke, K., Cooper, C., Dewey, J. F., Mann, P., & Pindell, J. L. (1984). Caribbean tectonics and relative plate motions. *Geol. Soc. Am. Mem., 162,* 31.

Callomon, J. H. (1959). The ammonite zones of the Middle Jurassic beds of East Greenland. *Geol. Mag., 96,* 505–13.

(1984). A review of the biostratigraphy of the post-Lower Bajocian Jurassic ammonites of western and northern North America. In G. E. G. Westermann (ed.), *Jurassic-Cretaceous Biochronology and Paleogeography of North America* (pp. 143–74). Geological Society of Canada, Special Paper 27.

Cameron, B. E. B., & Tipper, H. W. (1985). *Jurassic Stratigraphy of the Queen Charlotte Islands, British Columbia.* Geological Survey of Canada, Bulletin 365.

Campbell, R. B. (1978). *Quesnel Lake Map-Area.* Geological Survey of Canada, Open File 574.

Campbell, R. H. (1967). *Areal Geology in the Vicinity of the Chariot Site, Lisburne Peninsula, Northwestern Alaska.* U.S. Geological Survey, Professional Paper 395.

Carter, E. S., Cameron, B. E. B., & Smith, P. L. (1988). *Lower and Middle Jurassic Radiolarian Biostratigraphy and Systematic Paleontology, Queen Charlotte Islands, British Columbia.* Geological Survey of Canada, Bulletin 386.

Chandra, D. K. (1961). *Geology and Mineral Deposits of the Colfax and Foresthill Quadrangles, California.* California Division of Mines, Special Report 67.

Christopher, J. E. (1974). *The Upper Jurassic Vanguard and Lower Cretaceous Mannville Groups of Southwestern Saskatchewan.* Saskatchewan Department of Mineral Resources, Report 151.

Clark, L. D. (1964). *Stratigraphy and Structure of Part of the Western Sierra Nevada Metamorphic Belt.* U.S. Geological Survey, Professional Paper 410.

Clark, L. D., Imlay, R. W., McMath, V. E., & Silberling, N. J. (1962). *Angular Unconformity between Mesozoic and Paleozoic Rocks in the Northern Sierra Nevada, California.* U.S. Geological Survey, Professional Paper 450-B.

Compton, R. R. (1960). Contact metamorphism in the Santa Rose Range, Nevada. *Geol. Soc. Am. Bull., 71,* 1383–416.

Coney, P. J. (1989). The North American Cordillera. In Z. Ben-Avraham (ed.), *The Evolution of the Pacific Ocean Margins* (pp. 43–52). London: Oxford University Press.

Cordey, F., Mortimer, N., DeWever, P., & Monger, J. W. H. (1987). Significance of Jurassic radiolarians from the Cache Creek Terrane, British Columbia. *Geology, 15*(12), 1087–222.

Corvalán, J. I. (1962). *Early Mesozoic Biostratigraphy of the Westgate Area, Churchill County, Nevada.* Ph.D. thesis, Stanford University.

Cox, A., Debiche, M. G., & Engebretson, D. C. (1989). Terrane trajectories and plate interactions along continental margins in the North Pacific basin. In Z. Ben-Avraham (ed.), *The Evolution of the Pacific Ocean Margins* (pp. 20–35). London: Oxford University Press.

Craig, L. C., Holmes, C. N., Cadigan, R. A., Freeman, V. L., Mullens, T. E., & Weir, G. W. (1955). *Stratigraphy of the Morrison and Related Formations, Colorado Plateau Region, a Preliminary Report.* U.S. Geological Survey, Bulletin 1009-E.

Creely, R. S. (1965). *Geology of the Oroville Quadrangle, California.* California Division of Mines and Geology, Bulletin 184.

Crickmay, C. H. (1933). Mount Jura investigation. *Geol. Soc. Am. Bull., 44,* 895–926.

Criscione, J. J., David, T. E., & Ehlig, P. (1978). The age and sedimentation/diagenesis for the Bedford Canyon Formation and the Santa Monica Formation in southern California: a Rb/Sr evaluation. In D. G. Howell & K. A. McDougall (eds.), *Mesozoic Paleogeography of the Western United States* (pp. 385–96). Pacific Coast Paleogeography Symposium 2. Los Angeles: Pacific Section SEPM.

Csejtey, B., Jr., Cox, D. P., Evarts, R. C., Stricker, G. D., & Foster, H. (1982). The Cenozoic Denali fault system and the Cretaceous accretionary development of southern Alaska. *J. Geophys. Res., 87,* 3741–54.

Csejtey, B., Jr., Nelson, W. H., Jones, D. L., Silberling, N. J., et al. (1978). *Reconnaissance Geologic Map and Geochronology, Talkeetna Mountains Quadrangle, Northern Part of Anchorage Quadrangle, and Southwest Corner of Healy Quadrangle, Alaska.* U.S. Geological Survey, Open-File Report 78-558A.

Davies, E. H., & Poulton, T. P. (1986). Upper Jurassic dinoflagellate cysts from strata of northeastern British Columbia. *Geol. Surv. Can. Pap., 86-1B,* 519–37.

Dennison, S. S., Smith, P. L., & Tipper, H. W. (1990). An early Jurassic ichthyosaur from the Sandilands Formation, Queen Charlotte Islands, British Columbia. *J. Paleontol., 64,* 850–3.

Detterman, R. L. (1970). Sedimentary history of Sadlerochit and Shublik Formations in northeastern Alaska. In W. L. Adkinson & M. M. Brosgé (eds.), *Proceedings of the Geologic Seminar on the North Slope of Alaska* (pp. 0-1–0-13). Pacific Section, American Association of Petroleum Geologists, Los Angeles.

(1973). Mesozoic sequence in Arctic Alaska. In M. G. Pitcher (ed.), *Arctic Geology* (pp. 376–87). Tulsa: American Association of Petroleum Geologists, Memoir 19.

Detterman, R. L., Case, J. E., Wilson, F. H., & Yount, M. E. (1987). *Geologic Map of Ugashik, Bristol Bay, and Western Part of Karluk Quadrangles, Alaska.* U.S. Geological Survey, Miscellaneous Field Studies Map I-1685.

Detterman, R. L., & Hartsock, J. K. (1966). *Geology of the Iniskin-Tuxedni Region, Alaska.* U.S. Geological Survey, Professional Paper 512.

Detterman, R. L., Miller, J. W., & Case, J. E. (1985). *Megafossil Locality Map, Checklists, and Stratigraphic Sections of Ugashik, Bristol Bay, and Part of Karluk Quadrangles, Alaska.* U.S. Geological Survey, Miscellaneous Field Studies Map MF-1359B.

Detterman, R. L., Miller, T. P., Yount, M. E., & Wilson, F. H. (1981a). *Geologic Map of the Chignik and Sutwik Island Quadrangles, Alaska.* U.S. Geological Survey, Miscellaneous Investigation Series I-1229.

Detterman, R. L., & Reed, B. L. (1980). *Stratigraphy, Structure, and Economic Geology of the Iliamna Quadrangle, Alaska.* U.S. Geological Survey, Bulletin 1368B.

Detterman, R. L., Reiser, H. N., Brosgé, W. P., & Dutro, J. T., Jr. (1975). *Post-Carboniferous Stratigraphy, Northeastern Alaska.* U.S. Geological Survey, Professional Paper 886.

Detterman, R. L., Yount, M. E., & Case, J. E. (1981b). *Megafossil Localities, Checklists, and Stratigraphic Sections, Chignik and Sutwik Island Quadrangles, Alaska.* U.S. Geological Survey, Miscellaneous Field Studies Map MF-1053N.

Diakow, L. J., & Koyanagi, V. (1988). Stratigraphy and mineral occurrence of Chikamin Mountain and Whitesail Reach map areas. In *Geological Field Work, 1987.* British Columbia Ministry of Energy, Mines and Petroleum Resources, paper 1988-1.

Dickinson, W. R. (1979). Mesozoic forearc basin in central Oregon. *Geology, 7,* 166–70.

Dickinson, W. R., & Thayer, T. P. (1978). Paleogeographic and paleotectonic implications of Mesozoic stratigraphy and structure in the John Day inlier of Central Oregon. In D. G. Howell & K. A. McDougall (eds.), *Mesozoic Paleogeography of the Western United States* (pp. 147–62). Pacific Coast Paleogeography Symposium 2. Los Angeles: Pacific Section SEPM.

Dickinson, W. R., & Vigrass, L. W. (1965). *Geology of the Suplee-Izee Area, Crook, Grant and Harney Counties, Oregon.* Oregon Department of Geological and Mineral Industries, Bulletin 58.

Dixon, J. (1982). *Jurassic and Lower Cretaceous Subsurface Stratigraphy of the Mackenzie Delta–Tuktoyaktuk Peninsula, Northwest Territories.* Geological Survey of Canada, Bulletin 349.

Douglas, R. J. W., Gabrielse, H., Wheeler, J. O., Stott, D. F., & Belyea, H. R. (1970). Geology of Western Canada. In R. J. W. Douglas (ed.), *Geology and Economic Minerals of Canada* (5th ed.). Geological Survey of Canada, Economic Geology Report 1.

Douglass, R. C. (1967). *Permian Tethyan Fusulinids from California.* U.S. Geological Survey, Professional Paper 593-A.

Duffield, W. A., & Sharp, R. V. (1975). *Geology of the Sierra Foothills Mélange and Adjacent Areas, Amador County, California.* U.S. Geological Survey, Professional Paper 827.

Duncan, R. A., & Hargraves, R. B. (1984). Plate tectonic evolution of the Caribbean region in the mantle reference frame. *Geol. Soc. Am. Mem., 162,* 81–93.

Engebretson, D. C., Cox, A., & Gordon, R. G. (1985). *Relative Motions between Oceanic and Continental Plates in the Pacific Basin.* Geological Society of America, Special Paper 206.

Fensome, R. A. (1987). Taxonomy and biostratigraphy of schizealean spores from the Jurassic–Cretaceous boundary beds of the Aklavik Range, District of Mackenzie. *Palaeontogr. Canad., 4.*

Ferguson, H. G., & Muller, S. W. (1949). *Structural Geology of the Hawthorne and Tonopah Quadrangles, Nevada.* U.S. Geological Survey, Professional Paper 216.

Fife, D. L., Minch, J. A., & Crampton, P. J. (1967). Late Jurassic age of the Santiago Peak Volcanics. *Geol. Soc. Am. Bull., 78,* 299–304.

Fiske, R. S., & Tobisch, O. T. (1978). Paleogeographic significance of volcanic rocks of the Ritter Range Pendant, central Sierra Nevada, California. In D. G. Howell & K. A. McDougall (eds.), *Mesozoic Paleogeography of the Western United States* (pp. 209–22). Pacific Coast Paleogeography Symposium 2. Los Angeles: Pacific Section SEPM.

Frebold, H. (1957a). *Fauna, Age and Correlation of the Jurassic Rocks of Prince Patrick Island.* Geological Survey of Canada, Bulletin 41.

(1957b). *The Jurassic Fernie Group in the Canadian Rocky Mountains and Foothills.* Geological Survey of Canada, Memoir 287.

(1960). *The Jurassic Faunas of the Canadian Arctic, Lower Jurassic and Lowermost Middle Jurassic Ammonites.* Geological Survey of Canada, Bulletin 59.

(1961). *The Jurassic Faunas of the Canadian Arctic, Middle and Upper Jurassic Ammonites.* Geological Survey of Canada, Bulletin 74.

(1963). *Ammonite Faunas of the Upper Middle Jurassic Beds of the Fernie Group in Western Canada.* Geological Survey of Canada, Bulletin 93.

(1964a). *The Jurassic Faunas of the Canadian Arctic, Cadoceratinae.* Geological Survey of Canada, Bulletin 119.

(1964b). *Lower Jurassic and Bajocian Ammonoid Faunas of Northwestern British Columbia and Southern Yukon.* Geological Survey of Canada, Bulletin 116.

(1966). *Upper Pliensbachian Beds in the Fernie Group of Alberta.* Geological Survey of Canada, Paper 66-27.

(1969). Subdivision and facies of Lower Jurassic rocks in the southern Canadian Rocky Mountains and foothills. *Proc. Geol. Assoc. Can., 20,* 76–89.

(1975). *The Jurassic Faunas of the Canadian Arctic, Lower Jurassic Ammonites, Biostratigraphy and Correlations.* Geological Survey of Canada, Bulletin 243.

(1976). *The Toarcian and Lower Middle Bajocian Beds and Ammonites in the Fernie Group of Southeastern British Columbia.* Geological Survey of Canada, Paper 75-39.

(1978). *Ammonites from the Late Bathonian Iniskinites Fauna of Central British Columbia.* Geological Survey of Canada, Bulletin 307.

Frebold, H., Mountjoy, E. W., & Reed, R. (1959). *The Oxfordian Beds of the Jurassic Fernie Group, Alberta and British Columbia.* Geological Survey of Canada, Bulletin 53.

Frebold, H., Mountjoy, E. W., & Tempelman-Kluit, D. J. (1967). *New Occurrences of Jurassic Rocks and Fossils in Central and Northern Yukon Territory.* Geological Survey of Canada, Paper 67-12.

Frebold, H., & Poulton, T. P. (1977). Hettangian (Lower Jurassic) rocks and faunas, northern Yukon Territory. *Can. J. Earth Sci., 14,* 89–101.

Frebold, H., & Tipper, H. W. (1970). Status of the Jurassic in the Canadian Cordillera of British Columbia, Alberta, and Southern Yukon. *Can. J. Earth Sci., 7,* 1–21.

(1973). Upper Bajocian–Lower Bathonian ammonite fauna and stratigraphy of Smithers area, British Columbia. *Can. J. Earth Sci., 10*(7), 1109–31.

Ghosh, N., Hall, S. A., & Casey, J. F. (1984). Seafloor spreading magnetic anomalies in the Venezuelan Basin. *Geol. Soc. Am. Mem., 162,* 65–79.

Gibson, D. W. (1985). *Stratigraphy, Sedimentology and Depositional Environments of the Coal-bearing Jurassic-Cretaceous Kootenay Group, Alberta and British Columbia.* Geological Survey of Canada, Bulletin 357.

Grantz, A., (1960a). *Geologic Map of Talkeetna A-1 Quadrangle, Alaska and Contiguous Area to North and Northwest.* U.S. Geological Survey, Miscellaneous Geological Investigation Map I-313.

(1960b). *Geologic Map of Talkeetna A-1 Quadrangle, and South Third of Talkeetna B-1 Quadrangle, Alaska.* U.S. Geological Survey, Miscellaneous Geological Investigation Map I-314.

(1961). *Geologic Map of the Anchorage (D-2) Triangle, Alaska. Geologic Map of North Two-thirds of the Anchorage (D-1)*

*Quadrangle, Alaska.* U.S. Geological Survey, Miscellaneous Geological Investigation Map I-343.

Greene, R. C., Walker, G. W., & Corcoran, R. E. (1972). *Geologic Map of the Burns Quadrangle, Oregon.* U.S. Geological Survey, Miscellaneous Investigation Map I-680.

Guex, J. (1980). Remarques préliminaries sur la distribution stratigraphique des ammonies hettangiennes du New York Canyon (Gabbs Valley Range, Nevada). *Bull. Lab. Geol. Lausanne, 250,* 127–40.

(1981). Quelques cas de dimorphisme chez les ammonoides du Lias inferieur. *Bull. Lab. Geol. Lausanne, 258,* 239–48.

Guex, J., & Taylor, D. G. (1976). La limite Hettangien-Sinemurien, des Prealpes romandes au Nevada. *Eclogae Geol. Helv., 69,* 521–6.

Hall, R. L. (1984). Lithostratigraphy and biostratigraphy of the Fernie Formation (Jurassic) in the southern Canadian Rocky Mountains. In D. F. Stott & D. Glass (eds.), *The Mesozoic of Middle North America* (pp. 233–47). Canadian Society of Petroleum Geologists, Memoir 9.

(1987). New Lower Jurassic ammonite faunas from the Fernie Formation, southern Canadian Rocky Mountains. *Can. J. Earth Sci., 24,* 1688–704.

(1988). Late Bajocian and Bathonian (Middle Jurassic) ammonites from the Fernie Formation, Canadian Rocky Mountains. *J. Paleontol., 62,* 575–86.

(1989). New Bathonian (Middle Jurassic) ammonite faunas from the Fernie Formation, southern Alberta. *Can. J. Earth Sci., 26,* 16–22.

Hall, R. L., & Stronach, N. J. (1981). First record of Late Bajocian (Jurassic) ammonites in the Fernie Formation, Alberta. *Can. J. Earth Sci., 18,* 919–25.

Hall, R. L., & Westermann, G. E. G. (1980). Lower Bajocian (Jurassic) cephalopod faunas from western Canada and proposed assemblage zones for the Lower Bajocian of North America. *Palaeontogr. Am., 9*(52), 1–93.

Hallam, A. (1965). Observations on marine Lower Jurassic stratigraphy of North America, with special reference to United States. *Am. Assoc. Petrol. Geol. Bull., 49,* 1485–501.

(1978). Eustatic cycles in the Jurassic. *Palaeogeogr., Palaeoclimatol., Palaeoecol., 23,* 1–32.

(1981). A revised sea-level curve for the Early Jurassic. *Quart. J. Geol. Soc. London, 138,* 735–43.

Hamilton, W. (1963). *Metamorphism in the Riggins Region, Western Idaho.* U.S. Geological Survey, Professional Paper 436.

Harper, G. D. (1984). The Josephine ophiolite, northwestern California. *Geol. Soc. Am. Bull., 95*(9), 1009–26.

Harper, G. D., Saleeby, J. B., Cashman, S., & Norman, E. (1986). Isotopic age of the Nevadan orogney in the western Klamath Mountains, California-Oregon. *Geol. Soc. Am. Abstr. Progr., 18*(2), 114.

Harper, G. D., Saleeby, J., & Norman, E. (1985). Geometry and tectonic setting of sea-floor spreading for the Josephine ophiolite and implications for Jurassic accretionary events along the California margin. In D. G. Howell (ed.), *Tectonostratigraphic Terranes of the Circum-Pacific Region* (vol. 1, pp. 239–58). Houston: Circum-Pacific Council for Energy and Mineral Resources.

Harper, G. D., & Wright, J. E. (1984). Middle to Late Jurassic tectonic evolution of the Klamath Mountains, California-Oregon. *Tectonics, 3*(7), 759–72.

Harshbarger, J. W., Repenning, C. A., & Irwin, J. H. (1957). Stratigraphy of the uppermost Triassic and the Jurassic rocks of the Navajo country. Unpublished manuscript.

Heck, F. R., & Speed, R. C. (1987). Triassic olistostrome and shelf-basin transition in the western Great Basin: paleogeographic implications. *Geol. Soc. Am. Bull., 99,* 539–51.

Hedinger, A. S. (in press). *Late Jurassic (Oxfordian-Portlandian) Foraminifera of the Aklavik Range, District of Mackenzie, Northwest Territories, Canada.* Geological Survey of Canada, Bulletin.

Hein, J. R., & Koski, R. A. (1987). Bacterially mediated diagenetic origin for chert-hosted manganese deposits in the Franciscan complex, California Coast Ranges. *Geology, 15,* 722–6.

Hertlein, L. G. (1925). New species of fossil mollusca from western North America. *Bull. Calif. Acad. Sci., 24,* 39–46.

Hillhouse, J. W. (1977). Paleomagnetism of the Triassic Nikolai Greenstone, south-central Alaska. *Can. J. Earth Sci., 14,* 2578–92.

Hillhouse, J. W., & Gromme, C. S. (1982). Limits to the northward drift of the Paleocene Cantwell Formation, central Alaska. *Geology, 10,* 552.

Hillhouse, J. W., Gromme, C. S., & Csejtey, B. (1984). Paleomagnetism of Early Tertiary volcanic rocks in the northern Talkeetna Mountains. In K. M. Reed & S. Bartsch-Winkler (eds.), *U.S. Geological Survey in Alaska: Accomplishments during 1982* (pp. 50–2). U.S. Geological Survey, Circular 939.

Hillhouse, J. W., Gromme, C. S., & Vallier, T. L. (1982). Paleomagnetism and Mesozoic tectonics of the Seven Devils volcanic arc, northeastern Oregon. *J. Geophys. Res., 87,* 3777–94.

Hoare, J. M., & Coonrad, W. L. (1978). *Geologic Map of the Goodnews and Hagemeister Island Region, Southwestern Alaska.* U.S. Geological Survey, Open-File Report OF78-9-B.

Hopson, C. A., Mattinson, J. M., & Pessagno, E. A., Jr. (1981). Coast Range ophiolite, western California. In W. G. Ernst (ed.), *The Geotectonic Development of California* (pp. 418–510). Englewood Cliffs, N.J.: Prentice-Hall.

Hotz, P. E., Lanphere, M. A., & Swanson, D. A. (1977). Triassic bluechist from northern California and north-central Oregon. *Geology, 5,* 659–63.

Howell, D. G., & Jones, D. L. (1989). Terrane analysis: a circum-Pacific overview. In Z. Ben-Avraham (ed.), *The Evolution of the Pacific Ocean Margins* (pp. 36–40). London: Oxford University Press.

Howell, D. G., Schermer, E. R., Jones, D. L., Ben-Avraham, Z., & Scheiber, E. (1983). *Tectonostratigraphic Terrane Map of the Circum-Pacific Region.* U.S. Geological Survey, Open-File Report 87-716.

Imlay, R. W. (1948). *Characteristic Marine Jurassic Fossils from the Western Interior of the United States.* U.S. Geological Survey, Professional Paper 214-B.

(1952). Correlation of Jurassic formations of North America exclusive of Canada. *Geol. Soc. Am. Bull., 63,* 953–92.

(1953). *Callovian (Jurassic) Ammonites from the United States and Alaska.* U.S. Geological Survey, Professional Paper 249B.

(1955). *Characteristic Jurassic Mollusks from Northern Alaska.* U.S. Geological Survey, Professional Paper 274-D.

(1957). Paleoecology of Jurassic seas in the western interior of the United States. In H. S. Ladd (ed.), *Treatise on Marine Paleoecology* (Chapter 17). Geological Society of America, Memoir 67, 469–504.

(1959). *Succession and Speculation of the Pelecypod Aucella.* U.S. Geological Survey, Professional Paper 324G.

(1961a). *Late Jurassic Ammonites from the Western Sierra Nevada, California.* U.S. Geological Survey, Professional Paper 374-D.

(1961b). New genera and subgenera of Jurassic (Bajocian) ammonites from Alaska. *J. Paleontol., 35*(3), 467–74.

(1964). Upper Jurassic Mollusks from Eastern Oregon and Western Idaho. U.S. Geological Survey, Professional Paper 483-D.

(1967). *Twin Creek Limestone (Jurassic) in the Western Interior of the United States.* U.S. Geological Survey, Professional Paper 540.

(1968). *Lower Jurassic (Pliensbachian and Toarcian) Ammonites from Eastern Oregon and California.* U.S. Geological Survey, Professional Paper 593-C.

(1973). *Middle Jurassic (Bajocian and Bathonian) Ammonites from Eastern Oregon.* U.S. Geological Survey, Professional Paper 756.

(1975). *Stratigraphic Distribution and Zonation of Jurassic (Callovian) Ammonites in Southern Alaska.* U.S. Geological Survey, Professional Paper 836.

(1976). *Middle Jurassic (Bajocian and Bathonian) Ammonites from Northern Alaska.* U.S. Geological Survey, Professional Paper 854.
(1980a). *Jurassic Paleobiogeography of the Conterminous United States in Its Continental Setting.* U.S. Geological Survey, Professional Paper 1060.
(1980b). *Middle Jurassic (Bathonian) Ammonites from Southern Alaska.* U.S. Geological Survey, Professional Paper 1091.
(1981a). *Jurassic (Bathonian and Callovian) Ammonites in Eastern Oregon and Western Idaho.* U.S. Geological Survey, Professional Paper 1142.
(1981b). *Early Jurassic Ammonites from Alaska.* U.S. Geological Survey, Professional Paper 1148.
(1981c). *Late Jurassic Ammonites from Alaska.* U.S. Geological Survey, Professional Paper 1190.
(1982). *Jurassic (Oxfordian and Late Callovian) Ammonites from the Western Interior Region of the United States.* U.S. Geological Survey, Professional Paper 1232.
(1983). Jurassic fossils from southern California. *J. Paleontol.*, *37*, 97–107.
(1984a). *Early and Middle Bajocian (Middle Jurassic) Ammonites from Southern Alaska.* U.S. Geological Survey, Professional Paper 1322.
(1984b). Jurassic ammonite succession in North America and biographic implications. In G. E. G. Westermann (ed.), *Jurassic-Cretaceous Biogeography and Paleogeography of North America* (pp. 1–12). Geological Society of Canada, Special Paper 27.
(1986). Jurassic ammonites and biostratigraphy of eastern Oregon and western Idaho. In T. L. Vallier & H. C. Brooks (eds.), *Geology of the Blue Mountains Region of Oregon, Idaho and Washington* (pp. 53–7). U.S. Geological Survey, Professional Paper 1435.
Imlay, R. W., & Detterman, R. L. (1973). *Jurassic Paleogeography of Alaska.* U.S. Geological Survey, Professional Paper 801.
Imlay, R. W., & Jones, D. L. (1970). *Ammonites from the Buchia Zones off Northwestern California and Southwestern Oregon.* U.S. Geological Survey, Professional Paper 6476-B.
Irwin, W. P. (1972). *Terranes of the Western Paleozoic and Triassic Belt in the Southern Klamath Mountains, California.* U.S. Geological Survey, Professional Paper 800-C.
(1977). Review of Paleozoic rocks of the Klamath Mountains. In Stewart et al. (eds.), *Paleozoic Paleogeography of the Western United States* (pp. 441–54). Pacific Coast Paleogeography Symposium 1. Los Angeles: Pacific Section SEPM.
(1985). Age and tectonics of plutonic belts in accreted terranes of the Klamath Mountains, California and Oregon. In D. G. Howell (ed.), *Tectonostratigraphic Terranes of the Circum-Pacific Region* (pp. 187–200). Houston: Circum-Pacific Council for Energy and Mineral Resources.
Irwin, W. P., Jones, D. L., & Blome, C. D. (1982). *Map Showing Sampled Radiolarian Localities in the Western Paleozoic and Triassic Belt, Klamath Mountains, California.* U.S. Geological Survey, Map MF-1399.
Irwin, W. P., Jones, D. L., & Kaplan, T. A. (1978). Radiolarians from pre-Nevadan rocks of the Klamath Mountains, California and Oregon. In D. G. Howell & A. K. McDougall (eds.), *Mesozoic Paleogeography of the Western United States* (pp. 303–10). Pacific Coast Paleogeography Symposium 2. Los Angeles: Pacific Section SEPM.
Jayko, A. S., Blake, M. C., Jr., & Harms, T. (1987). Attenuation of the Coast Range ophiolite by extensional faulting, and nature of the Coast Range "thrust," California. *Tectonics*, *6*(4), 475–88.
Jeletzky, J. A. (1965). *Late Upper Jurassic and Early Lower Cretaceous Fossil Zones of the Canadian Western Cordillera, British Columbia.* Geological Survey of Canada, Bulletin 103.
(1966). *Upper Volgian (Late Jurassic) Ammonites and Buchias of Arctic Canada.* Geological Survey of Canada, Bulletin 128.
(1967). *Jurassic and (?) Triassic Rocks of the Eastern Slope of Richardson Mountains, Northwest District of Mackenzie, 106M and 107B (part of).* Geological Survey of Canada, Paper 66–50.
(1973). Biochronology of the marine boreal latest Jurassic, Berriasian and Valanginian in Canada. In R. Casey & P. F. Rawson (eds.), *The Boreal Lower Cretaceous* (pp. 41–80). Liverpool: Seal House Press.
(1977). *Porcupine River Formation; a New Upper Jurassic Sandstone Unit, Northwest Yukon Territory.* Geological Survey of Canada, Paper 76-27.
(1980). *Dicoelitid Belemnites from the Toarcian–Middle Bajocian of Western and Arctic Canada.* Geological Survey of Canada, Bulletin 338.
(1984). Jurassic–Cretaceous boundary beds of western and Arctic Canada and the problem of the Tithonian-Berriasian stages in the Boreal realm. In G. E. G. Westermann (ed.), *Jurassic–Cretaceous Biochronology and Paleogeography of North America* (pp. 175–255). Geological Association of Canada, Special Paper 27.
Jeletzky, J. A., & Tipper, H. W. (1968). *Upper Jurassic and Cretaceous Rocks of Taseko Lakes Map-Area and Their Bearing on the Geological History of Southwestern British Columbia.* Geological Survey of Canada, Paper 67-54.
Johnson, M. G. (1977). *Geology and Mineral Deposits of Pershing County, Nevada.* Nevada Bureau of Mines and Geology, Bulletin 89.
Jones, D. L. (1975). Discovery of *Buchia rugosa* of Kimmeridgian age from the base of the Great Valley sequence. *Geol. Soc. Am. Abstr. Progr.*, *7*, 330.
Jones, D. L., Bailey, E. H., & Imlay, R. W. (1969). *Structural and stratigraphic significance of the Buchia zones in the Colyear Springs–Paskenta area, California.* U.S. Geological Survey, Professional Paper 647-A.
Jones, D. L., Blake, M. C., Jr., & Rangin, C. (1976). The four Jurassic belts of northern California and their significance to the geology of the southern California borderland. In D. G. Howell (ed.), *Aspects of the Geologic History of the California Continental Borderland* (pp. 343–62). Tulsa: American Association of Petroleum Geologists.
Jones, D. L., & Irwin, W. P. (1975). Rotated Jurassic rocks in the Transverse Ranges, California. *Geol. Soc. Am. Abstr. Progr.*, *7*(3), 330.
Jones, D. L., & Moore, J. G. (1973). Lower Jurassic ammonite from the south-central Sierra Nevada, California. *U.S. Geol. Surv. Res.*, *1*(4), 453–8.
Jones, D. L., Silberling, N. J., Berg, H. C., & Plafker, G. (1981). *Map Showing Tectonostratigraphic Terranes of Alaska, Columnar Sections, and Summary Descriptions of Terranes.* U.S. Geological Survey, Open-File Report 81-792.
Jones, D. L., Silberling, N. J., Csejtey, B., Jr., Nelson, W. H., & Blume, C. D. (1980). *Age and Structural Significance of Ophiolite and Adjoining Rocks in the Upper Chulitna District, South Central Alaska.* U.S. Geological Survey, Professional Paper 1121-A.
Jones, D. L., Silberling, N. J., & Hillhouse, J. (1977). Wrangellia – a displaced terrane in northwestern North America. *Can. J. Earth Sci.*, *14*, 2565–77.
(1978). Microplate tectonics of Alaska, significance for the Mesozoic history of the Pacific coast of North America. In D. G. Howell & K. A. McDougall (eds.), *Mesozoic Paleogeography of the Western United States* (pp. 71–4). Pacific Coast Paleogeography Symposium 2. Los Angeles: Pacific Section SEPM.
Lees, E. J. (1934). Geology of the Laberge area, Yukon. *Trans. Roy. Can. Inst.*, *1*(43), 1–48.
Leffingwell, E. de K. (1919). *The Canning River Region, Northern Alaska.* U.S. Geological Survey, Professional Paper 109.
Little, H. W. (1982). *Geology of the Rossland Trail Map-Area, British Columbia.* Geological Survey of Canada, Paper 79-26.

Lupe, R., & Silberling, N. J. (1985). Genetic relationship between Lower Mesozoic continental strata of the Colorado Plateau and marine strata of the Western Great Basin: significance for accretionary history of cordilleran lithotectonic terranes. In D. G. Howell (ed.), *Tectonostratigraphic Terranes of the Circum-Pacific Region* (pp. 263–71). Houston: Circum-Pacific Council for Energy and Mineral Resources.

Lupher, R. L. (1941). Jurassic stratigraphy of central Oregon. *Geol. Soc. Am. Bull., 52,* 219–70.

Luyendyk, B. P., Kamerling, M., & Terres, J. (1980). Geometric model for Neogene crustal rotation in southern California. *Geol. Soc. Am. Bull., 91,* 211–17.

McKee, E. D., et al. (1956). *Paleotectonic Maps of the Jurassic System.* U.S. Geological Survey, Miscellaneous Geological Investigation Map I-175.

MacKevett, E. M., Jr. (1971). *Stratigraphy and General Geology of the McCarthy C-5 Quadrangle, Alaska.* U.S. Geological Survey, Bulletin 1323.

(1976). *Geologic Map of the McCarthy Quadrangle, Alaska.* U.S. Geological Survey, Miscellaneous Field Studies Map MF-773A.

(1978). *Geological Map of the McCarthy Quadrangle, Alaska.* U.S. Geological Survey, Miscellaneous Investigation Map I-1032.

MacKevett, E. M., Jr., & Richter, D. H. (1974). The Nikolai Greenstone in the Wrangle Mountains, Alaska, and nearby areas. *Geol. Assoc. Canada, Cordill. Sect., Progr. Abstr.* pp. 13–14.

McLearn, F. H. (1932). Contributions to the stratigraphy and paleontology of Skidegate Inlet, Queen Charlotte Islands, British Columbia. *Trans. Roy. Soc. Can., Sect. IV, Geol. Sci. Mineral., Ser. 3, 26,* 51–80.

McMath, V. E. (1966). Geology of the Taylorsville area, northern Sierra Nevada. In E. H. Bailey (ed.), *Geology of Northern California* (pp. 173–83). California Department of Mines and Geology, Publication 190.

Mattson, P. H. (1984). Caribbean structural breaks and plate movements. *Geol. Soc. Am. Mem., 612,* 131.

Miller, J. W., & Detterman, R. L. (1985). The *Buchia* zones in Upper Jurassic rocks on the Alaska Peninsula. In S. Bartsch-Winkler & K. M. Reed (eds.), *The United States Geological Survey in Alaska.* U.S. Geological Survey, Circular 945.

Minch, J. C., Castil, G., Fink, W., Robinson, J., & James, A. H. (1976). Geology of the Vizcaino Peninsula. In D. G. Howell (ed.), *Aspects of the Geologic History of the California Borderland.* American Association of Petroleum Geologists, Pacific Section.

Moffat, I. W., Bustin, R. M., & Rouse, G. E. (1988). Biochronology of selected Bowser Basin strata: tectonic significance. *Can. J. Earth Sci., 25,* 1571–8.

Monger, J. W. H., & Price, R. A. (1979). Geodynamic evolution of the Canadian Cordillera – progress and problems. *Can. J. Earth Sci., 16,* 770–91.

Monger, J. W. H., Price, R. A., & Tempelman-Kluit, D. J. (1982). Tectonic accretion and the origin of the two major metamorphic and plutonic welts in the Canadian Cordillera. *Geology, 10,* 70–5.

Monger, J. W. H., & Ross, C. A. (1971). Distribution of fusilinaceans in the western Canadian Cordillera. *Can. J. Earth Sci., 8,* 259–78.

Moore, J. G. (1969). *Geology and Mineral Deposits of Lyon, Douglas, and Ormsby Counties, Nevada.* Nevada Bureau of Mines and Geology, Bulletin 75.

Moores, E. M. (1972). Model for Jurassic island-arc continental margin collision in California. *Geol. Soc. Am. Abstr. Progr., 4*(3), 202.

Morrison, R. F. (1961). Angular unconformity marks Triassic–Jurassic boundary in Snake River area of northeastern Oregon. *Ore Bin, 23,* 105–11.

(1964). Upper Jurassic mudstone unit named in Snake River Canyon, Oregon–Idaho boundary. *Northwest Sci., 38,* 83–7.

Mortimer, N. (1985). Structural and metamorphic aspects of Middle Jurassic terrane juxtaposition, northeastern Klamath Mountains, California. In D. G. Howell (ed.), *Tectonostratigraphic Terranes of the Circum-Pacific Region* (pp. 210–14). Houston: Circum-Pacific Council for Energy and Mineral Resources.

(1986). Late Triassic, arc-related, potassic igneous rocks in the North American Cordillera. *Geology, 14,* 1035–9.

Muller, J. E., Northcote, K. E., & Carlisle, D. (1974). *Geology and Mineral Deposits of Alert–Cape Scott Map-Area, Vancouver Island, British Columbia.* Geological Survey of Canada, Paper 74-8.

Muller, S. W., & Ferguson, H. G. (1936). Triassic and Lower Jurassic formations of west central Nevada. *Geol. Soc. Am. Bull., 47,* 241–52.

(1939). Mesozoic stratigraphy of the Hawthorne and Tonopah quadrangles, Nevada. *Geol. Soc. Am. Bull., 50,* 1573–624.

Murchey, B. L. (1984). Biostratigraphy and lithostratigraphy of chert in the Franciscan complex, Marin Headlands, California. In M. C. Blake, Jr. (ed.), *Franciscan Geology of Northern California* (pp. 51–70). SEPM Book 43.

Murchey, B. L., & Jones, D. L. (1984). Age and significance of chert in the Franciscan complex in the San Francisco Bay region. In M. C. Blake, Jr. (ed.), *Franciscan Geology of Northern California* (pp. 23–30). SEPM Book 43.

Nauss, A., & Smith, P. L. (1988). *Lithiotis* (Bivalvia) bioherms in the Lower Jurassic of east-central Oregon, U.S.A. *Palaeogeogr. Palaeoclimatol. Palaeoecol., 65,* 253–68.

(1989). *Lithiotis* (Bivalvia) bioherms, Lower Jurassic, Oregon. In H. H. J. Geldsetzer, N. P. James & G. E. Tebbutt (eds.), *Reefs, Canada and Adjacent Areas* (pp. 738–40). Canadian Society of Petroleum Geologists, Memoir 13.

Nielsen, R. L. (1963). *Geology of the Pilot Mountains and Vicinity, Mineral County, Nevada.* PhD thesis, University of California, Berkeley.

Noble, D. C. (1962). *Mesozoic Geology of the Southern Pine Nut Range, Douglas County, Nevada.* PhD thesis, Stanford University.

Nokleberg, W. J. (1983). *Wallrocks of the Central Sierra Nevada Batholith: A Collage of Accreted Tectono-stratigraphic Terranes.* U.S. Geological Survey, Professional Paper 1255.

Nolf, B. (1966). *Geology and Stratigraphy of Part of the Northern Wallowa Mountains, Oregon.* PhD thesis, Princeton University.

Nordquist, J. W. (1955). Pre-Rierdon Jurassic stratigraphy in northern Montana and Williston Basin. In P. J. Lewis (ed.), *Sweetgrass Arch-Disturbed Belt, Montana* (pp. 96–106). Billings Geological Society Guidebook, Sixth Annual Field Conference.

Nur, A., & Ben-Avraham, Z. (1983). Break-up and accretion tectonics. In M. Hashimoto & S. Uyeda (eds.), *Accretion Tectonics in the Circum-Pacific Regions* (pp. 3–18). Tokyo: Terra Scientific Publications.

Oldow, J. S. (1978). Triassic Pamlico Formation: an allochthonous sequence of volcanogenic-carbonate rocks in west-central Nevada. In D. G. Howell & K. A. McDougall (eds.), *Mesozoic Paleogeography of the Western United States* (pp. 223–35). Pacific Coast Paleogeography Symposium 2. Los Angeles: Pacific Section SEPM.

(1981). Structure and stratigraphy of the Luning allochthon and the kinematics of allochthon emplacement, Pilot Mountains, west-central Nevada. *Geol. Soc. Am. Bull., 92,* 888–911.

(1984). Evolution of a Late Mesozoic back-arc fold and thrust belt, northwestern Great Basin, U.S.A. *Tectonophysics, 102,* 245–74.

Oldow, J. S., Lallement, H. G., & Schmidt, W. J. (1984). Kinematics of plate convergence deduced from Mesozoic structures in the western Cordillera. *Tectonics, 3,* 210–27.

Packer, D. R., & Stone, D. B. (1974). Paleomagnetism of Jurassic rocks from southern Alaska and the tectonic implications. *Can. J. Earth Sci., 11,* 967–97.

Palfy, J., McFarlane, R. B., Smith, P. L., & Tipper, H. W. (1990). *Potential for Ammonite Biostratigraphy of the Sinemurian Part of the Sandilands Formation, Queen Charlotte Islands, British Columbia.* Geological Survey of Canada, Paper 90-1H.

Paterson, D. F. (1968). *Jurassic Megafossils of Saskatchewan with a Note on Charophytes.* Saskatchewan Department of Mineral Resources, Geological Science Branch, Sedimentary Geology Division, Report 120.

Paterson, S. R., Tobisch, O. T., & Radloff, J. K. (1987). Post-Nevadan deformation along the Bear Mountains fault zone: implications for the Foothills terrane, central Sierra Nevada, California. *Geology, 15,* 513–16.

Pessagno, E. A., Jr. (1977). Upper Jurassic Radiolaria and radiolarian biostratigraphy of the California Coast Ranges. *Micropaleontology, 23* (1), 56–113.

Pessagno, E. A., Jr., & Blome, C. D. (1986). Faunal affinities and tectonogenesis of Mesozoic rocks in the Blue Mountains province of eastern Oregon and western Idaho. In T. L. Vallier & H. C. Brooks (eds.), *Geology of the Blue Mountains Region of Oregon, Idaho and Washington* (pp. 65–78). U.S. Geological Survey, Professional Paper 1435.

Pessagno, E. A., Jr., Whalen, P. A., & Yeh, K. Y. (1986). Jurassic *Nassellariina* (Radiolaria) from North American geologic terranes. *Bull. Am. Paleontol., 91.*

Peterson, J. A. (1957). Marine Jurassic of northern Rocky Mountains and Williston Basin. In A. J. Goodman (ed.), *Jurassic and Carboniferous of Western Canada. Am. Assoc. Petrol. Geol. Bull., 41*(3), 399–440.

(1972). Jurassic system. In *Geologic Atlas of the Rocky Mountain Region, United States of America* (pp. 177–89). Denver: Rocky Mountain Association of Geologists.

(1986). Jurassic paleotectonics in the west-central part of the Colorado Plateau, Utah and Arizona. In J. A. Peterson (ed.), *Paleotectonics and Sedimentation in the Rocky Mountain Region, United States* (pp. 563–96). Tulsa: American Association of Petroleum Geologists, Memoir 41.

Peterson, J. A., & Pipiringos, G. N. (1979). *Stratigraphic Relations of the Navajo Sandstone to Middle Jurassic Formations, Southern Utah and Northern Arizona.* U.S. Geological Survey, Professional Paper 1035-B.

Peterson, J. A., & Smith, D. L. (1986). Rocky Mountain paleogeography through geologic time. In J. A. Peterson (ed.), *Paleotectonics and Sedimentation in the Rocky Mountain Region, United States.* Tulsa: American Association of Petroleum Geologists, Memoir 41.

Pinto-Auso, M., & Harper, G. D. (1985). Sedimentation, metallogenesis, and tectonic origin of the basal Galice Formation overlying the Josephine ophiolite, northwestern California. *J. Geology, 93,* 713–25.

Pipiringos, G. N. (1968). *Correlation and Nomenclature of Some Triassic and Jurassic Rocks in South-Central Wyoming.* U.S. Geological Survey, Professional Paper 594-D.

Pipiringos, G. N., Hail, W. J., Jr., & Izett, G. A. (1969). *The Chinle (Upper Triassic) and Sundance (Upper Jurassic) Formations in North-Central Colorado.* U.S. Geological Survey, Bulletin 1274-N.

Pipiringos, G. N., & O'Sullivan, R. B. (1978). *Principal Unconformities in Triassic and Jurassic Rocks, Western Interior United States – A Preliminary Survey.* U.S. Geological Survey, Professional Paper 1035-A.

Plumley, P. W., Coe, R. S., & Byrne, T. (1983). Paleomagnetism of the Paleocene Ghost Rocks Formation, Prince William Terrane, Alaska. *Tectonics, 2,* 295–314.

Pocock, S. A. J. (1962). Microfloral analysis and age determination of strata at the Jurassic-Cretaceous boundary in the western Canada plains. *Palaeontographica, B, 111,* 1–95.

(1970). Palynology of the Jurassic sediments of western Canada. Part 1. Terrestrial species. *Palaeontographica, B, 130,* 12–72.

(1972). Palynology of the Jurassic sediments of western Canada. Part 2. Marine species. *Palaeontographica, B, 137,* 85–153.

Poulton, T. P. (1978). Pre-Late Oxfordian Jurassic biostratigraphy of northern Yukon and adjacent Northwest Territories. In C. R. Stelck & B. D. E. Chatterton (eds.), *Western and Arctic Canadian Biostratigraphy* (pp. 445–71). Geological Association of Canada, Special Paper 18.

(1981). *Stratigraphic Distribution and Taxonomic Notes and Bivalves of the Bathonian and Callovian (Middle Jurassic) Upper Yakoun Formation, Queen Charlotte Islands, British Columbia. Current Research, Part B.* Geological Survey of Canada, Paper 81-1B.

(1982). Paleogeographic and tectonic implications of Lower and Middle Jurassic facies patterns in northern Yukon Territory and adjacent Northwest Territories. In A. F. Embry & H. R. Balkwill (eds.), *Arctic Geology and Geophysics* (pp. 13–27). Canadian Society of Petroleum Geologists, Memoir 8.

(1984). Jurassic of the Canadian Western Interior, from 49°N latitude to Beaufort Sea. In D. F. Stott & D. Glass (eds.), *The Mesozoic of Middle North America* (pp. 15–41). Canadian Society of Petroleum Geologists, Memoir 9.

(1987). *Boreal Middle Bathonian to Lower Callovian (Jurassic) Ammonites, Zonation and Correlation, Salmon Cache Canyon, Porcupine River, Northern Yukon.* Geological Survey of Canada, Bulletin 358.

(1988). Major interregionally correlatable events in the Jurassic of western interior, Arctic and eastern offshore Canada. In D. James & D. Leckie (eds.), *Sequences and Stratigraphy* (pp. 195–206). Canadian Society of Petroleum Geologists, Memoir 15.

(1989a). Current status of Jurassic biostratigraphy and stratigraphy, northern Yukon and adjacent Mackenzie Delta. Current research. *Geol. Surv. Can. Pap., 89-1G,* 25–30.

(1989b). A Lower Jurassic coral reef, Telkwa Range, British Columbia. In H. H. J. Geldsetzer, N. P. James & G. E. Tebbutt (eds.), *Reefs, Canada and Adjacent Area* (pp. 754–7). Canadian Society of Petroleum Geologists, Memoir 13.

(1989c). *Jurassic (Oxfordian) Ammonites from the Fernie Formation of Western Canada: A Giant Peltoceratinid, and* Cardioceras canadense *Whiteaves.* Geological Survey of Canada, Bulletin 396.

(1991). *Hettangian through Aalenian (Jurassic) Guide Fossils and Biostratigraphy, Northern Yukon and Adjacent Northwest Territories.* Geological Survey of Canada, Bulletin 410.

Poulton, T. P., & Aitken, J. D. (1989). The Lower Jurassic phosphorites of southeastern British Columbia and terrane accretion to western North America. *Can. J. Earth Sci., 26,* 1612–16.

Poulton, T. P., Leskiw, K., & Audretsch, A. P. (1982). *Stratigraphy and Microfossils of the Jurassic Bug Creek Group of Northern Richardson Mountains, Northern Yukon and Adjacent Northwest Territories.* Geological Survey of Canada, Bulletin 325.

Poulton, T. P., & Tempelman-Kluit, D. J. (1982). *Recent Discoveries of Jurassic Fossils in the Lower Schist Division of Central Yukon.* Geological Survey of Canada, Paper 82-1C.

Poulton, T. P., & Tipper, H. W. (1990). Aalenian ammonites, biostratigraphy and paleobiogeography of Canada. In *Proceedings of the 2nd International Symposium on Jurassic Stratigraphy* (pp. 181–91). Lisbon: Instituto Nacional de Investigacão Cientifica.

Poulton, T. P., Tittemore, J., & Dolby, G. (1990). Jurassic strata of northwestern (and west-central) Alberta and northeastern British Columbia. *Bulletin of Canadian Petroleum Geology, 38A,* 159–75.

Rea, D. K., & Duncan, R. A. (1986). North Pacific plate convergence; a quantitative record of the past 140 m.y. *Geology, 14,* 373–6.

Reed, B. L., & Lanphere, M. A. (1972). *Generalized Geologic Map of the Alaska–Aleutian Range Batholith Showing Potassium-Argon Ages of the Plutonic Rocks.* U.S. Geological Survey, Miscellaneous Field Studies Map MF-372.

Reed, B. L., & Nelson, S. W. (1977). *Geologic Map of the Talkeetna Quadrangle, Alaska.* U.S. Geological Survey, Miscellaneous Field Studies Map MF-870-A.

Reeside, J. B. (1919). *Some American Jurassic Ammonites of the Genera* Quenstedtoceras, Cardioceras, *and* Amoeboceras, *Family Cardioceratidae.* U.S. Geological Survey, Professional Paper 118.

Reiser, H. H., Lanphere, M. A., & Brosgé, W. P. (1965). *Jurassic Age of Mafic Igneous Complex, Christian Quadrangle, Alaska. Geological Survey Research 1965*. U.S. Geological Survey, Professional Paper 525-C.

Richter, D. H. (1976). *Geologic Map of the Nabesna Quadrangle, Alaska*. U.S. Geological Survey, Miscellaneous Investigation Series Map I-932.

Rickwood, F. K. (1970). The Prudhoe Field. In *Proceedings of the Geological Seminar on the North Slope of Alaska* (pp. L1–L11). Pacific Section, American Association of Petroleum Geologists.

Riddihough, R. P. (1982). One hundred million years of plate tectonics in western Canada. *Geosci. Can., 9*(1), 28–34.

Russell, B. J. (1984). Mesozoic geology of the Jackson Mountains, northwestern Nevada. *Geol. Soc. Am. Bull., 95*, 313–23.

Saleeby, J. (1981). Ocean floor accretion and volcanoplutonic arc evolution of the Mesozoic Sierra Nevada. In G. W. Ernst (ed.), *The Geotectonic Development of California* (pp. 132–81). Englewood Cliffs, N.J.: Prentice-Hall.

Saleeby, J., Goodin, S. E., Sharp, W. D., & Busby, C. J. (1978). Early Mesozoic paleotectonic-paleogeographic reconstruction of the southern Sierra Nevada region. In D. G. Howell & K. A. McDougall (eds.), *Mesozoic Paleogeography of the Western United States* (pp. 311–36). Pacific Coast Paleogeography Symposium 2. Los Angeles: Pacific Section SEPM.

Sanborn, A. F. (1960). *Geology and Paleontology of the Southwest Quarter of the Big Bend Quadrangle, Shasta County, California*. California Division of Mines, Special Report 63.

Sarewitz, D. (1983). Seven Devils terrane: Is it really a piece of Wrangellia? *Geology, 11*, 634–7.

Scheffler, J. M. (1983). *A Petrologic and Tectonic Comparison of the Hells Canyon Area, Oregon–Idaho, and Vancouver Island, British Columbia*. MSc thesis, Washington State University.

Schweickert, R. A., & Cowan, D. S. (1975). Early Mesozoic tectonic evolution of the western Sierra Nevada, California. *Geol. Soc. Am. Bull., 86*, 1329–36.

Sears, J. W., & Price, R. A. (1978). The Siberian connection, a case for Precambrian separation of the North American and Siberian cratons. *Geology, 6*, 267–70.

Seiders, V. M., & Blome, C. D. (1987). Stratigraphy and sedimentology of Upper Cretaceous rocks in coastal southwest Oregon: evidence for wrench-fault tectonics in postulated accretionary terrane: alternate interpretation and reply. *Geol. Soc. Am. Bull., 98*, 739–42.

Sey, I. I., Kalacheva, E. D., & Westermann, G. E. G. (1986). The Jurassic ammonite *Pseudolioceras (Tugurites)* of the Bering Province. *Can. J. Earth Sci., 23*, 1042–5.

Shervais, J. W., & Kimbrough, D. L. (1985). Geochemical evidence for the tectonic setting of the Coast Range ophiolite: a composite island arc–oceanic crust terrane in western California. *Geology, 13*, 35–8.

Shew, N., & Wilson, F. H. (1981). *Map and Tables Showing Radiometric Ages of Rocks in Southwestern Alaska*. U.S. Geological Survey, Open-File Report 81-886.

Silberling, N. J. (1959). *Pre-Tertiary Stratigraphy and Upper Triassic Paleontology of the Union District, Shoshone Mountains, Nevada*. U.S. Geological Survey, Professional Paper 322.

(1979). Stratigraphic relations of the Auld Lang Syne Group Lower Mesozoic in northwestern Nevada. *Geol. Soc. Am. Abstr. Progr., 11*, 127–8.

(1984). *Map showing localities and correlation of age-diagnostic Lower Mesozoic megafossils, Walker Lake 1° × 2° quadrangle, Nevada and California*. U.S. Geological Survey, Miscellaneous Field Studies Map MF-1382-0.

Silberling, N. J., Jones, D. L., Blake, M. C., Jr., & Howell, D. G. (1984). *Lithotectonic Terrane Maps of the North American Cordillera. Part C. Lithotectonic Terrane Maps of the Western Conterminous United States*. U.S. Geological Survey, Open-File Report 84-523.

(1987). *Lithotectonic Terrane Map of the Western Conterminous United States*. U.S. Geological Survey, Map MF-1874-C.

Silberling, N. J., Jones, D. L., Csejtey, B., Jr., & Nelson, W. H. (1978). *Interpretive Bed-Rock Geologic Map of the Upper Chulitna District*. U.S. Geological Survey, Open-File Report 78-454.

Silberling, N. J., & Wallace, R. E. (1969). *Stratigraphy of the Star Peak Group (Triassic) and Overlying Lower Mesozoic Rocks, Humboldt Range, Nevada*. U.S. Geological Survey, Professional Paper 592.

Sloss, L. L. (1982). The midcontinent province: United States. In A. R. Palmer (ed.), *Perspectives in Regional Geological Synthesis* (pp. 27–40). Geological Society of America, D-Nag Spec. Publ. 1.

Smith, P. L. (1980). Correlation of the members of the Jurassic Snowshoe Formation in the Izee basin of east-central Oregon. *Can. J. Earth Sci., 17*, 1603–8.

(1981). *Biostratigraphy and Ammonoid Fauna of the Lower Jurassic (Sinemurian, Pliensbachian and the Lowest Toarcian) of Eastern Oregon and Western Nevada*. PhD thesis, McMaster University.

(1983). The Pliensbachian ammonite *Dayiceras dayiceroides* and Early Jurassic paleogeography. *Can. J. Earth Sci., 20*, 86–91.

(1983). Suspect terranes. *Nature, 305*, 475–6.

Smith, P. L., & Tipper, H. W. (1986). Plate tectonics and paleobiogeography: Early Jurassic (Pliensbachian) endemism and diversity. *Palaios, 1*, 399–412.

(1988). Biochronology, stratigraphy and tectonic setting of the Pliensbachian of Canada and the United States. In *Proceedings of the 2nd International Symposium on Jurassic Stratigraphy* (pp. 119–38). Lisbon: Instituto Nacional de Investigacão Cientifica.

Smith, P. L., Tipper, H. W., Taylor, D. G., & Guex, J. (1988). An ammonite zonation for the Lower Jurassic of Canada and the United States: the Pliensbachian. *Can. J. Earth Sci., 25* (9).

Smith, W. D., & Allen, J. E. (1941). *Geology and Physiography of the Northern Wallowa Mountains, Oregon*. Oregon Department of Geology and Mineral Industries, Bulletin 12.

Souther, J. G. (1962). *Geology and Mineral Deposits of Tulsequah Map-Area, British Columbia*. Geological Survey of Canada, Memoir 362.

(1972). *Telegraph Creek Map-Area*. Geological Survey of Canada, Paper 71–44.

Speed, R. C. (1974). Evaporite-carbonate rocks of the Jurassic Lovelock Formation, West Humboldt Range, Nevada. *Geol. Soc. Am. Bull., 85*, 105–18.

(1975). Carbonate breccia (rauhwacke) nappes of the Carson Sink region, Nevada. *Geol. Soc. Am. Bull., 86*, 473–86.

(1977). Excelsior Formation, west-central Nevada: stratigraphic appraisal, new divisions and paleogeographic interpretations. In J. H. Stewart, C. J. Stevens & A. E. Fritche (eds.), *Paleozoic Paleogeography of the Western United States* (pp. 325–36). Pacific Coast Paleogeography Symposium 1. Los Angeles: Pacific Section SEPM.

(1978a). Paleogeographic and plate tectonic evolution of the Early Mesozoic marine province of the western Great Basin. In D. G. Howell & K. A. McDougall (eds.), *Mesozoic Paleogeography of the Western United States* (pp. 237–52). Pacific Coast Paleogeography Symposium 2. Los Angeles: Pacific Section SEPM.

(1978b). Basinal terrane of the early Mesozoic marine province of the Western Great Basin. In D. G. Howell & K. A. McDougall (eds.), *Mesozoic Paleogeography of the Western United States* (pp. 237–52). Pacific Coast Paleogeography Symposium 2. Los Angeles: Pacific Section SEPM.

(1979). Collided Paleozoic microplate in the western United States. *J. Geology, 87*, 279–92.

Speed, R. C., & Clayton, R. N. (1974). Origin of marble by replacement of gypsum in carbonate breccia nappes, Carson Sink region, Nevada. *J. Geology, 83*.

Speed, R. C., & Jones, T. A. (1969). Synorogenic quartz sandstone in the Jurassic mobile belt of Western Nevada: Boyer Ranch Formation. *Geol. Soc. Am. Bull., 89,* 2551–84.

Springer, G. D., MacDonald, W. D., & Crockford, M. B. B. (1964). Jurassic. In R. G. McCrossan & R. P. Glaister (eds.), *Geological History of Western Canada* (pp. 137–55). Calgary: Alberta Society of Petroleum Geologists.

Stanley, K. O. (1971). Tectonic and sedimentologic history of Lower Jurassic Sunrise and Dunlap formations, west-central Nevada. Unpublished manuscript, Exxon Intl., Houston, Tex.

Stanley, K. O., Jordan, W. M., & Dott, R. H., Jr. (1971). New hypothesis of Early Jurassic paleogeography and sediment dispersal for western United States. *Am. Assoc. Petrol. Geol. Bull., 55,* 10–19.

Stern, T. W., Bateman, P. C., Morgan, B. A., Newell, M. F., & Peck, D. L. (1981). *Isotopic U-Pb Ages of Zircons from the Granitoids of Central Sierra Nevada, California.* U.S. Geological Survey, Professional Paper 1185.

Stokes, W. L. (1944). Morrison Formation and related deposits in and adjacent to the Colorado Plateau. *Geol. Soc. Am. Bull., 55,* 951–92.

(1963). Triassic and Jurassic formations of southwestern Utah. In *Inter-Mountain Association of Petroleum Geologists, Guidebook, 12th Annual Field Conference, Salt Lake City* (pp. 60–4). Utah Geological and Mineralogical Survey.

Stone, D. B., & McWilliams, M. O. (1989). Paleomagnetic evidence for relative terrane motion in western North America. In Z. Ben-Avraham (ed.), *The Evolution of the Pacific Ocean Margins* (pp. 53–72). London: Oxford University Press.

Stone, D. B, & Packer, D. R. (1979). Paleomagnetic data from the Alaska Peninsula. *Geol. Soc. Am. Bull., 90,* 545–60.

Stone, D. B., & Panuska, B. C., & Packer, D. R. (1982). Paleolatitudes versus time for southern Alaska. *J. Geophys. Res., 87,* 3697–707.

Stott, D. F. (1967). *Fernie and Minnes Strata North of Peace River, Foothills of Northeastern British Columbia.* Geological Survey of Canada, Paper 67-19 (Part A).

(1970). Mesozoic stratigraphy of the Interior Platform and Eastern Cordilleran Orogen. In R. J. W. Douglas (ed.), *Geology and Economic Minerals of Canada* (pp. 438–45). Geological Survey of Canada, Economic Geology Report 1.

(1984). Cretaceous sequences of the foothills of the Canadian Rocky Mountains. In D. F. Stott & D. Glass (eds.), *The Mesozoic of Middle North America* (pp. 85–107). Canadian Society of Petroleum Geologists, Memoir 9.

Stott, D. F. & Aitken, J. D. (1982). Continental platforms and basins of Canada. In A. R. Palmer (ed.), *Perspectives in Regional Geological Synthesis* (pp. 15–26). Geological Society of America, D-Nag Special Publication 1.

Stronach, N. J. (1984). Depositional environments and cycles in the Jurassic Fernie Formation, southern Canadian Rocky Mountains. In D. F. Stott & D. J. Glass (eds.), *The Mesozoic of Middle North America* (pp. 43–67). Canadian Society of Petroleum Geologists, Memoir 9.

Sulima, J. H. (1970). *Lower Jurassic Stratigraphy in Coal Canyon, West Humboldt Range, Nevada.* MSc thesis, Northwestern University.

Sutherland Brown, A. (1968). *Geology of the Queen Charlotte Islands, British Columbia.* British Columbia Department of Mines and Petroleum Research, Bulletin 54.

Taliaferro, N. L. (1943). *Manganese Deposits of the Sierra Nevada, Their Genesis and Metamorphism.* California Division of Mines, Bulletin 125.

Taylor, D. G. (1981). *Jurassic (Bajocian) Ammonite Biostratigraphy and Macroinvertebrate Paleoecology of the Snowshoe Formation, East-Central Oregon.* PhD thesis, University of California, Berkeley.

(1986). The Hettangian–Sinemurian Boundary (Early Jurassic): reply to Bloos 1983. *Newslett. Strat., 16,* 57–67.

(1988a). *Paradiscamphiceras:* un nouveau genre d'ammonites du Lias inferieur. *Bull. Lab. Geol. Lausanne, 298,* 117–22.

(1988b). Middle Jurassic (late Aalenian and early Bajocian) ammonite biochronology of the Snowshoe Formation, Oregon. *Oregon Geology, 50,* 123–38.

Taylor, D. G., Callomon, J., Hall, R., Smith, P. L., Tipper, H. W., & Westermann, G. E. G. (1984). Jurassic ammonite biogeography of western North America: the tectonic implications. In G. E. G. Westermann (ed.), *Jurassic-Cretaceous Biochronology and Paleogeography of North America* (pp. 121–42). Geological Survey of Canada, Special Paper 27.

Taylor, D. G., Smith, P. L., Laws, R. A., & Guex, J. (1983). The stratigraphy and biofacies trends of the Lower Mesozoic Gabbs and Sunrise formations, west-central Nevada. *Can. J. Earth Sci., 20,* 1598–608.

Thomson, R. C. (1985). *Lower to Middle Jurassic (Pliensbachian to Bajocian) Stratigraphy and Pliensbachian Ammonite Fauna of the Northern Spatsizi Area, North-Central British Columbia.* MSc thesis, University of British Columbia.

Thomson, R. C., Smith, P. L., & Tipper, H. W. (1986). Lower to Middle Jurassic (Pliensbachian to Bajocian) stratigraphy of the northern Spatsizi area, north-central British Columbia. *Can. J. Earth Sci., 23*(12).

Tipper, H. W. (1984a). The allochthonous Jurassic–Lower Cretaceous terranes of the Canadian Cordillera and their relation to correlative strata of the North American craton. In G. E. G. Westermann (ed.), *Jurassic-Cretaceous Biochronology and Paleogeography of North America* (pp. 113–20). Geological Survey of Canada, Special Paper 27.

(1984b). *The Age of the Jurassic Rossland Group of Southeastern British Columbia. Current Research, Part A.* Geological Survey of Canada, Paper 84-1A.

(1989). *Lower Jurassic (Hettangian and Sinemurian) Biostratigraphy, Queen Charlotte Islands, British Columbia. Current Research, Part H.* Geological Survey of Canada, Paper 89-1H.

Tipper, H. W., & Cameron, B. E. B. (1980). *Stratigraphy and Paleontology of the Upper Yakoun Formation (Jurassic) in Alliford Bay Syncline, Queen Charlotte Islands, British Columbia. Current Research, Part C.* Geological Survey of Canada, Paper 80-1C.

Tipper, H. W., & Richards, T. A. (1976). *Jurassic Stratigraphy and History of North-Central British Columbia.* Geological Survey of Canada, Bulletin 270.

Tipper, H. W., Smith, P. L., Cameron, B. E. B., Carter, E. S., Jakobs, G. K., & Johns, M. J. (1990). *Biostratigraphy of the Lower Jurassic Formations of the Queen Charlotte Islands, British Columbia.* Geological Survey of Canada, Frontier Geoscience Report.

Tobisch, T. T., Paterson, S. R., Longiaru, S., & Bhattacharyya, T. (1987). Extent of the Nevadan orogeny, central Sierra Nevada, California. *Geology, 15*(2), 132–5.

Tozer, E. T. (1970). Marine Triassic faunas. In R. J. Douglas (ed.), *Geology and Economic Minerals of Canada* (pp. 633–40). Geological Survey of Canada, Economic Geology Report 1.

(1982a). Marine Triassic faunas of North America: their significance for assessing plate and terrane movements. *Geol. Rund., 71*(3), 1077–104.

(1982b). *Late Triassic (Upper Norian) and Earliest Jurassic (Hettangian) Rocks and Ammonoid Faunas, Halfway River and Pine Pass Map-Area, British Columbia.* Geological Survey of Canada, Paper 82-1A.

Vail, P. R., & Todd, R. G. (1981). Northern North Sea Jurassic unconformities, chronostratigraphy and sea-level changes from seismic stratigraphy. In L. V. Illing & G. D. Hobson (eds.), *Petroleum Geology of the Continental Shelf of North-West Europe* (pp. 216–35). London: Heyden & Son.

Vallier, T. L., & Brooks, H. C. (1986). Paleozoic and Mesozoic faunas of the Blue Mountains province: a review of their geologic implications and comments on papers in the volume. In T. L. Vallier & H. C. Brooks (eds.), *Geology of the Blue Mountains Region of Oregon, Idaho and Washington* (pp. 1–6). U.S. Geological Survey, Professional Paper 1435.

van der Heyden, P. (1989). *U-Pb and K-Ar Geochronometry of the Coast Plutonic Complex, 53°N to 54°N, British Columbia, and Implications for Insular-Intermontane Superterrane Boundary.* PhD thesis, University of British Columbia.

Wagner, D. L., Jennings, C. W., Bedrossian, T. L., & Bortugno, E. J. (1981). *Geologic Map of the Sacramento Quadrangle.* California Division of Mines and Geology, Regional Geologic Map Series 1A.

Wahrhaftig, C. (1984). Structure of the Marin Headland block, California: a progress report. In M. C. Blake, Jr. (ed.), *Franciscan Geology of Northern California* (pp. 31–50). SEPM Book 43.

Wallace, R. E., Silberling, N. J., Irwin, W. P., & Tatlock, D. B. (1969). *Geologic Map of the Buffalo Mountain Quadrangle, Pershing and Churchill Counties, Nevada. U.S. Geological Survey, Geological Quadrangle Map GQ-821.*

Warren, P. S. (1931). A Lower Jurassic fauna from Fernie, British Columbia. *Roy. Soc. Can. Trans., Ser. 3, Sec. IV, 25,* 105–11.

Weihmann, I. (1964). Stratigraphy and Microfauna of the Jurassic Fernie Group, Fernie Basin, Southeast British Columbia. *Bull. Can. Petrol. Geol., 12,* Special Guidebook Issue, Flathead Valley, 587–99.

Westermann, G. E. G. (1964a). Occurrence and significance of the Arctic *Arkelloceras* in the Middle Bajocian of the Alberta foothills (Ammonitina, Jurassic). *J. Paleontol., 38,* 405–9.

(1964b). The ammonite fauna of the Kialagvik Formation at Wide Bay, Alaska Peninsula. Part I. Lower Bajocian (Aalenian). *Bull. Am. Paleontol., 47,* 325–503.

(1966). The holotype (plastotype) of *Titanites occidentalis* Frebold from the Kootenay sandstone (Upper Jurassic) of southern British Columbia. *Can. J. Earth Sci., 3,* 623–5.

(1969). The ammonoid fauna of the Kialagvik Formation at Wide Bay, Alaska Peninsula, Part II. *Sonninia sowerbyi* zone (Bajocian). *Bull. Am. Paleontol., 57,* 226.

(1981) Ammonite biochronology and biogeography of the circum-Pacific Middle Jurassic. In M. R. House & J. R. Senior (eds.), *The Ammonoidea* (pp. 459–98). Systematics Association Special Volume 18.

Wheeler, J. O. (1961). *Whitehorse Map-Area, Yukon Territory.* Geological Survey of Canada, Memoir 312.

Wiedenmayer, F. (1980). Die Ammoniten der Mediterranean Provinz im Pliensbachian und unteren Toarcian aufgrund neuer Untersuchungen im Generoso-Becken (Lombardische Alpen). *Denkschriften der Schweizerischen naturforschenden Gesellschaft, 93,* 1–261.

Willden, R., & Speed, R. C. (1974). *Geology and Mineral Deposits of Churchill County, Nevada.* Nevada Bureau of Mines and Geology, Bulletin 83.

Wilson, F. H., Detterman, R. L., & Case, J. E. (1985). *The Alaska Peninsula Terrane: A Definition.* U.S. Geological Survey, Open-File Report 85-450.

Wilson, J. T. (1968). Static or mobile Earth? The current scientific revolution. *Proc. Am. Phil. Soc., 112,* 309–20.

Woodsworth, G. J. (1980). *Geology of Whitesail Lake (93E) Map-Area.* Geological Survey of Canada, Open-File Map 708.

Young, F. G., Myhr, D. G., & Yorath, C. J. (1976). *Geology of the Beaufort-Mackenzie Basin.* Geological Survey of Canada, Paper 76-11.

Zonenshayn, L. P., Savostin, L. A., & Sedov, A. P. (1984). Global paleogeodynamic reconstructions for the last 160 million years. *Geotectonics, 18*(3), 181–94.

# 5 Meso-America

A. SALVADOR, G. E. G. WESTERMANN, F. OLÓRIZ, M. B. GORDON, and H.-J. GURSKY

## WESTERN MEXICO[1]

### Introduction

Rocks of Jurassic age are known from three main areas along the Pacific margin of Western Mexico: the Baja California Peninsula; the state of Sonora, west of the Neogene volcanics of the Sierra Madre Occidental; and the states of Guerrero and Oaxaca (Figure 5.1). These Jurassic sequences were formed, in one way or another, west of the North American plate, within the Pacific domain. Other Jurassic sequences of Mexico, from the state of Chihuahua in northern Mexico to the southernmost part of the country, northeast of the Chiapas Massif, are genetically related to the early development of the Gulf of Mexico Basin (Salvador 1987).

The sites and manners of formation of the Jurassic rock sequences of western Mexico are subjects of some controversy. The Pacific margin of Mexico is believed by some authors (e.g., Coney, Jones, and Monger 1980; Campa and Coney 1983; Howell et al. 1983; Gastil 1985; Kimbrough 1985; Moore 1985) to be composed of a collage of what have been called "suspect terranes" or "tectonostratigraphic terranes" – allochthonous crustal blocks that are assumed to have collided and accreted to the North American craton mostly during the Mesozoic and early Cenozoic, after having, in some cases, traveled long distances.

The far-distant origin of the Jurassic sequences of western Mexico is not accepted by all, but there is general agreement that eastward subduction of oceanic lithosphere under the western margin of the North American plate probably was continuous in the region during the Late Triassic and Jurassic and that many of the Upper Triassic and Jurassic rock sequences of western Mexico are not now located where they were originally formed, having suffered substantial displacement as a result of either thrusting or strike-slip faulting. Evidence has been presented, for instance, to indicate that Baja California has been displaced northward at least 500 km along an important transcurrent fault system that extends from the Gulf of California to northern California.

This history of subduction and transcurrent displacement is reflected at least in Baja California and western Sonora as belts of complex structural deformation and volcanic-plutonic activity. The corresponding stratigraphic record is characterized by thick, laterally variable volcaniclastic sequences, turbidites, and debris-flow deposits shed from the nearby active volcanic arcs and by associated volcanic and plutonic rocks, generally of andesitic composition (Rangin 1979; Sedlock 1988).

If many of the Jurassic rocks of western Mexico are indeed allochthonous and have been displaced large but unknown distances from their areas of formation to their present positions, the interpretation of the geological history and paleogeography of this region during the Jurassic becomes a nearly impossible task. But even if they had not traveled long and unknown distances, the reconstruction of the Jurassic history of the region is made difficult by the limited distribution of rocks of this age, by the complexity of the structural deformation of some of the areas where the Jurassic is better exposed, and by the limited occurrences of age-diagnostic fossils.

It is possible to state, however, that the Pacific margin of Mexico was characterized from the Late Triassic to at least the end of the Jurassic by intense compressional tectonic deformation, with associated magmatic activity, the result of crustal collision and eastward subduction of oceanic lithosphere below the North American plate. During that time, the Pacific Ocean bordered Mexico to the west. The ocean covered, at various times, the western margin of the North American plate.

In Sonora, a shallow shelf bordered to the west a low continental region during the Late Triassic and the Early Jurassic. Close to the coast of this lowland, a nonmarine sedimentary section, in part coal-bearing, was deposited. It graded westward into shallow-water marine sediments. Water depth increased rapidly beyond the margin of the shelf; an oceanic trench probably was present west of continental North America during the Late Triassic and Jurassic.

[1] By A. Salvador and G. E. G. Westermann.

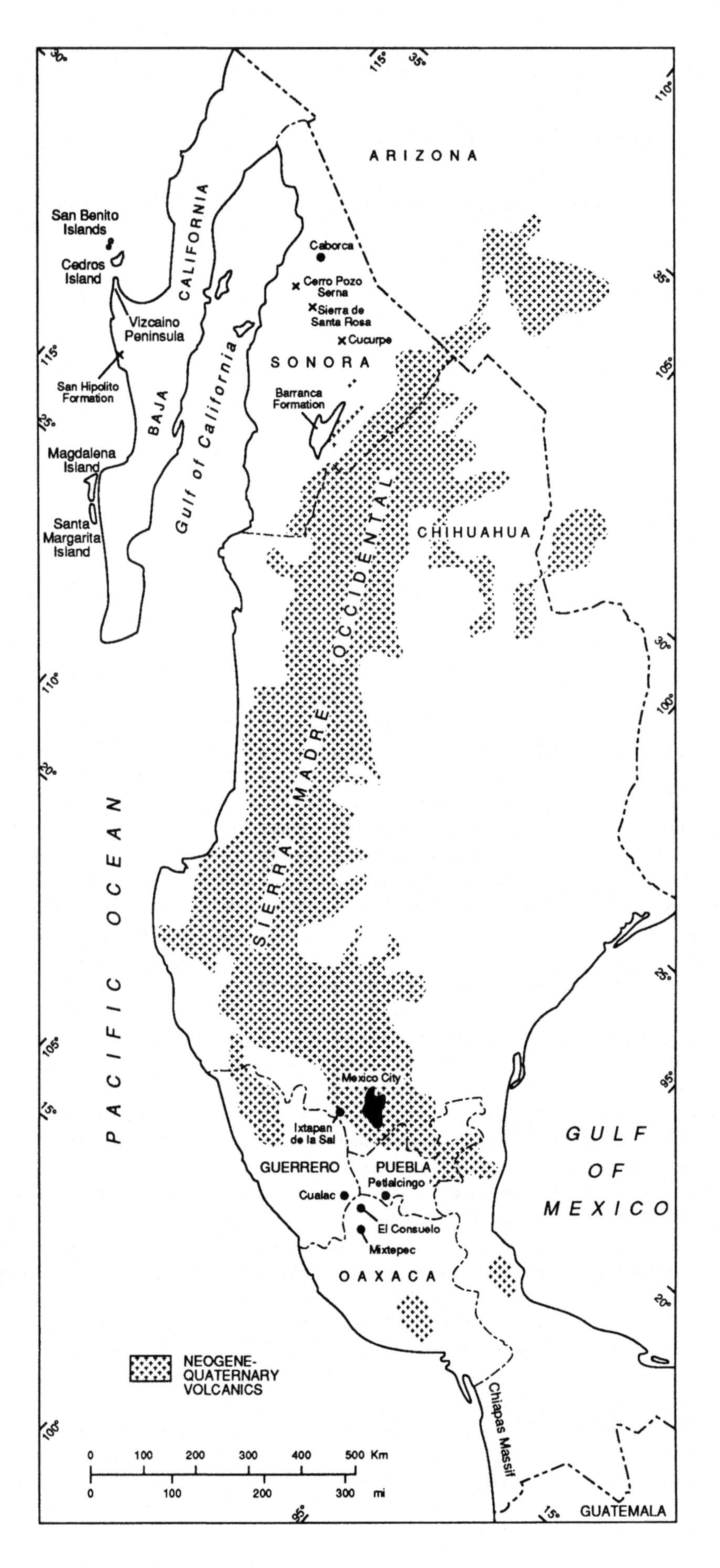


**Figure 5.1. Localities and areas where Jurassic sequences crop out in western Mexico.**

No Middle Jurassic sediments have been reported from Sonora, indicating, perhaps, that the region was uplifted or that the sea level was lower at that time, reducing the area of marine deposition in northwestern Mexico to mainly the area west of the continental border. The sea appears to have transgressed over Sonora again during the Late Jurassic, but outcrops of rocks of that age are limited and widely separated, making it difficult to estimate the magnitude and distribution of the transgression.

West of the Sonora shelf, the Late Triassic–Jurassic is represented by a complex sequence of volcaniclastic and debris-flow deposits, turbidites, and volcanic rocks, apparently deposited in deep water. They are found today in Baja California, but their original area of formation is not known.

Farther south, Salvador (1987) believes that an embayment of the Pacific Ocean occupied part of southwestern Mexico during the Late Triassic and extended into the east-central part of the country during the Early Jurassic. The Pacific embayment was restricted to southwestern Mexico (northeast Guerrero and northwest Oaxaca) during the early part of the Middle Jurassic, a further indication of uplift of western Mexico or of a lowering of the sea level at that time. Finally, an arm of the Pacific extended to the Gulf of Mexico area in the late Middle Jurassic (late Bathonian or Callovian), and during the Late Jurassic the Pacific Ocean was in continuous communication with the large ancestral Gulf of Mexico through southwestern and east-central Mexico.

Westermann's interpretation (Westermann 1984a) would have the marine upper Middle Jurassic in southwestern Mexico (Oaxaca and Guerrero) deposited in the proximity of the open Pacific Ocean, and the coeval Gulf of Mexico connected to the spreading Central Atlantic (see chapter 24).

Volcanic and plutonic activity was prevalent almost continuously along the Pacific margin of Mexico from the Late Triassic to at least the end of the Jurassic. Gastil, Morgan, and Krummenacher (1978, 1981) have postulated the presence during the Late Triassic and Early Jurassic of a volcanic island arc some distance offshore, west of the North American continent. It is represented by andesites and tonalites in Baja California. These authors also postulate the appearance during the Early Jurassic of a second volcanic arc within continental North America, manifested by rhyolites, andesites, and associated plutonic rocks in central Sonora. According to these authors, the two volcanic arcs persisted through the Middle Jurassic and into the early part of the Late Jurassic (Oxfordian). Andersen and Silver (1969, 1978, 1979) report the occurrence in southern Arizona and northern Sonora of volcanic and plutonic rocks ranging in age from 180 to 142 Ma – Middle and Late Jurassic. Volcanic and plutonic activity in late Late Jurassic time has been recognized in Cedros Island and in the Vizcaino Peninsula of Baja California, and evidence of latest Jurassic volcanic activity is also known in southwestern and southeastern Mexico.

Geographic localities mentioned in connection with the description of the Jurassic of western Mexico and some of the Jurassic outcrop areas are shown in Figure 5.1. A simplified stratigraphic correlation chart of the Jurassic sequences of western Mexico is shown in Figure 5.2.

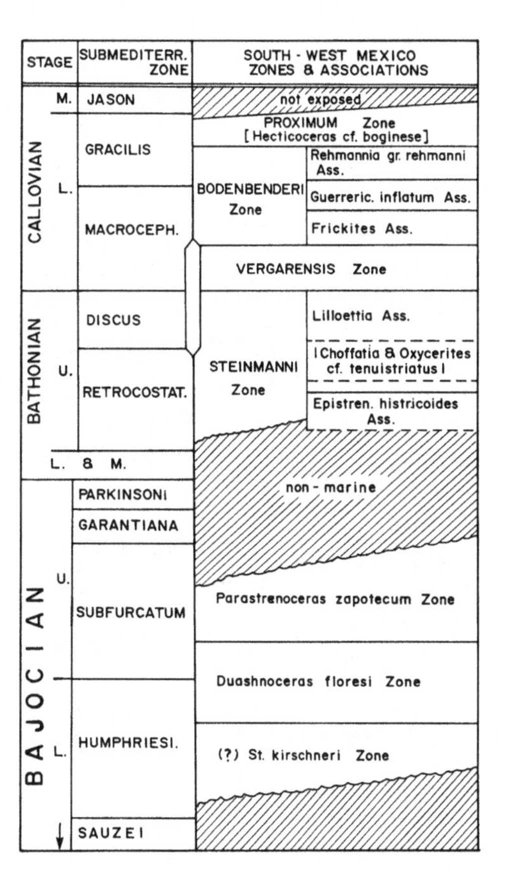


**Figure 5.2. The Middle Jurassic ammonite zones of southwest Mexico, based on the Mixtepec (Oaxaca) and Cualac (Guerrero) sections.**

## Baja California

### *Lower Jurassic*

In the Vizcaino Peninsula of Baja California the Early Jurassic is represented in the upper part of the San Hipólito Formation (Figure 5.3). This formation rests on pillow basalts, generally believed to represent oceanic crust, and is composed of four distinct units, from bottom to top (Mina 1957; Minch et al. 1976; Finch and Abbott 1977; Finch, Pessagno, and Abbott 1979; Kimbrough 1985): up to 245 m of green and red radiolarian-bearing chert; 95–210 m of interbedded chert, limestone, sandy limestone, and minor amounts of sandstone; up to 205 m of breccia formed by fragments of limestone in a matrix of volcaniclastic sandstones; and 1,800–1,900 m of volcaniclastic sandstones, shales, and tuffs. The San Hipólito Formation has been dated as Late Triassic to Early Jurassic on the basis of the occurrence in the lower chert and chert-and-limestone units of the Late Triassic pelecypods *Halobia* and *Monotis* and abundant Radiolaria (Finch et al. 1979; Pessagno, Finch and Abbott 1979; Davila-Alcocer and Pessagno 1986); and the presence of Early Jurassic Radiolaria in the lower part of the upper sandstone unit (Whalen and Pessagno 1984). The Lower Jurassic sandstone, shale, and tuff upper unit of the San Hipólito Formation has been interpreted as having been deposited by gravity flows of sediments or as turbidites shed from

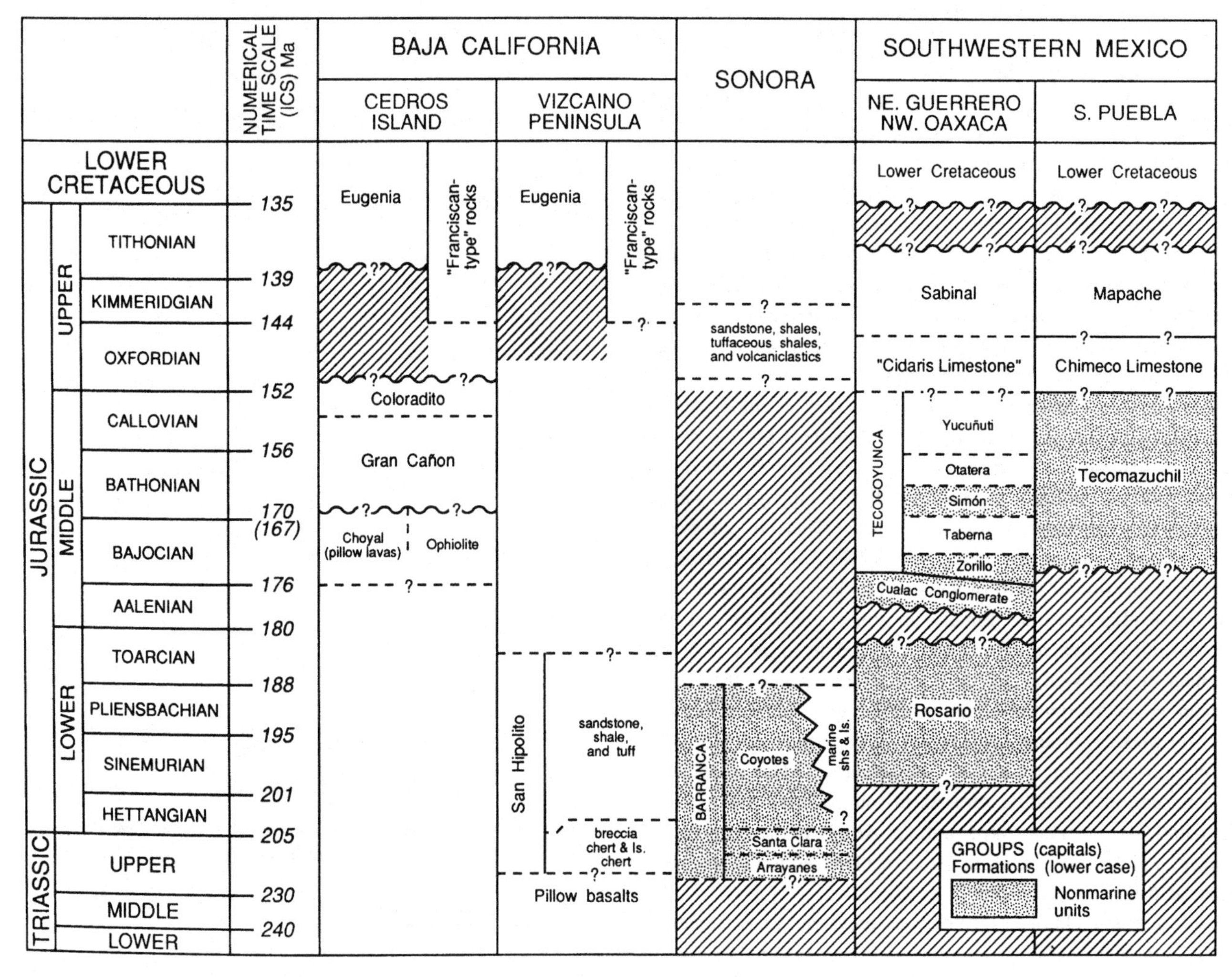


**Figure 5.3. Stratigraphic correlation chart of Jurassic sequences in western Mexico. Numerical time scale taken from Cowie and Bassett (1989).**

a nearby active volcanic arc. The relationships between the San Hipólito Formation and other stratigraphic units in the area are not yet understood, nor is the manner in which it reached its present position.

Andesitic volcanic and plutonic rocks of probable Early Jurassic age have been reported from the west coast of Baja California (Anderson et al. 1972; Gastil et al. 1978, 1981; Rangin 1978).

### *Middle Jurassic*

From Cedros Island, off the western coast of the Baja California Peninsula, Kilmer (1977, 1979) described a sequence, 2,300 m thick, of interbedded radiolarian chert, siltstone, sandstone, conglomerate, tuffs, and pillow lavas (the Gran Cañon Formation) to which he assigned a tentative Bajocian to Callovian age on the basis of fossils collected in the middle part of the formation. More recently, Kimbrough (1984, 1985) and Busby-Spera (1988) have described the Gran Cañon Formation as composed of up to 1,300 m of tuffs, tuff breccias, and volcanogenic sandstones and siltstones, with interbedded pillow lavas and conglomerate/breccia beds in its upper part, all accumulated on a deep, gently dipping sea floor mostly as gravity-flow sedimentary deposits and turbidites derived from an adjacent active volcanic arc located to the northeast. The Gran Cañon Formation overlies the tholeiitic and calc-alkaline pillow lavas of the Choyal Formation in the northern part of Cedros Island, and ophiolitic rocks in the southern part of the island. The Choyal Formation is intruded by tonalite and granodiorite stocks that have been dated as late Middle Jurassic (166–160 Ma); the ophiolite has yielded a Bajocian age (173 Ma).

Conformably above the Gran Cañon Formation, Kilmer (1979) described a sequence of thin-bedded sandstones, shales, conglomerates, tuffs, and massive megabreccias, 300–400 m thick, which he named the Coloradito Formation. He assigned to it a Late Jurassic age. Boles and Landis (1984) described the Coloradito Formation as composed of a basal thin-bedded sandstone "flysch" unit overlain by black silty argillites, tuffs, pebbly sandstones, and conglomerates. Interbedded with the argillites are olistostromal horizons containing exotic blocks of Paleozoic limestone, orthoquartzite, Triassic ribbon-banded chert, and tonalite. The blocks range in size from less than a meter to slab-shaped blocks 50–100 m long and 8–10 m thick. Tuff beds indicate contemporaneous volcanism, and the influx of continental detritus suggests proximity to an uplifted continental land mass. Gastil et al. (1978) have reported an isotopic age of 159 Ma (late Middle Jurassic) for an andesitic tuff in the upper third of the formation. Geyssant and Rangin (1979) described Callovian ammonites [two specimens of *Reineckeia* and one of *Choffatia* (*Subgrossouvria*)] from the blocks and matrix of a boulder-breccia, presumably part of the Coloradito Formation, involved in a complex thrust zone on Cedros Island. The *Reineckeia* belong to *R*. gr. *latesellata* Burckhardt (1927), otherwise known only from the middle or upper Callovian of northwestern Oaxaca. It is possible, however, that the fossils were collected from far-traveled exotic blocks. Jones, Blake, and Rangin (1976) assigned a tentative Callovian age to the upper part of a sequence of thin-bedded tuffs directly overlying pillow basalts on Cedros Island on the basis of pelecypods collected by Kilmer and identified by R. W. Imlay. These tuffs probably are part of the Gran Cañon or Coloradito Formations of Bathonian to earliest Oxfordian age (Kilmer 1977, 1979; Boles and Landis 1984; Kimbrough 1984; Busby-Spera 1988).

The ophiolitic, metamorphic, and volcanic-arc sequences of the Magdalena and Santa Margarita islands, 500 km southeast of Cedros Island (Rangin 1978; Blake et al. 1984), may be of Middle Jurassic age.

### *Upper Jurassic*

No fossiliferous rocks of Oxfordian age have been reported from Baja California. The upper part of the Coloradito Formation and parts of other volcaniclastic and andesitic volcanic sequences of this area may, however, include Oxfordian rocks.

On Cedros Island, on the San Benito Islands, and in the Vizcaino Peninsula, a metamorphic sequence with a marked lithologic resemblance to the Franciscan Complex of California occurs in intricate structural relationship with other units cropping out in these areas (Jones et al. 1976; Kilmer 1977, 1979; Rangin 1978; Kimbrough 1985). Rangin (1978) reported late Kimmeridgian–Tithonian Radiolaria, identified by E. A. Pessagno, Jr., from bedded cherts within this highly deformed sequence, but other units of this sequence could be older – early Kimmeridgian or even Oxfordian (Kilmer 1979).

The latest Jurassic is represented in Baja California by the lower part of the Eugenia Formation of the Vizcaino Peninsula and Cedros Island, which has been dated as ranging in age from Tithonian to Early Cretaceous (Mina 1957; Jones et al. 1976; Minch et al. 1976; Kilmer 1977, 1979; Boles 1978; Barnes 1984; Boles and Landis 1984). It is composed of as much as 2,700 m of interbedded graywackes, volcanogenic conglomerates, tuffs, and shales containing calcareous nodules. Pillow basalts are found interbedded with conglomerates and shales in the basal 30 m. The Eugenia Formation is intruded by numerous andesite-prophyry dikes of Early Cretaceous age. The Jurassic lower part of the formation represents deposition in submarine slopes of volcaniclastic material shed from an active volcanic island arc situated a short distance to the east.

From Magdalena Island, Blake et al. (1984) described a sequence of volcaniclastic sediments – turbidites, tuffs, conglomerates, and breccias. Limestone blocks and lenses in this sequence contain an abundant fauna of corals, bryozoa, equinoids, and brachiopods of Jurassic age. From an interval of siliceous argillites that apparently underlies the volcaniclastic sequence, R. W. Imlay identified a Jurassic ammonite of probable Kimmeridgian or younger age (Blake et al. 1984).

## **Sonora**

### *Lower Jurassic*

In the east-central part of the state of Sonora, in northwestern Mexico, a thick, predominantly clastic sequence, the Barranca Group, has been dated as Late Triassic to Early Jurassic. The Barranca Group attains thicknesses of up to 2,000 m and has been subdivided in eastern Sonora into three formations (Alencaster de Cserna 1961a): a lower Arrayanes Formation composed predominantly of quartzose and arkosic sandstones; a middle Santa Clara Formation of thin-bedded sandstones, siltstones, and plant-bearing carbonaceous shales containing locally important coal beds; and an upper arkosic and quartzose sandstone and conglomerate unit, the Coyotes Formation. This sequence has been interpreted to have been deposited in fluvial and deltaic environments. The Santa Clara Formation has been dated as Late Triassic (Carnian) on the basis of fossil plants (King 1939; Silva-Pineda 1961) and pelecypods and brachiopods (Alencaster de Cserna 1961b). The upper part of the Barranca Group in eastern and central Sonora may be of Early Jurassic age (King 1939; Alencaster de Cserna 1961a).

Westward, in northwestern Sonora, the Upper Triassic–Lower Jurassic section becomes increasingly more marine, and fossiliferous shales and limestones increase progressively in abundance. The upper part of the section has been assigned an Early Jurassic age based on a fauna of ammonites and mollusks (Jaworski 1929; Burckhardt 1930; King 1939; Imlay 1952, 1980; Gonzalez 1979). Hardy (1973) reported that an allochthonous Lower Jurassic section at Sierra Santa Rosa, 75 km southeast of Caborca, is composed of at least 2,360 m of volcaniclastic and volcanic rocks (lithic wackes and agglomerates, siltstones, shales, tuffs, and lavas), with interbedded thinner calcareous intervals.

Andesitic volcanic and plutonic rocks of probable Early Jurassic age have been reported from southwestern Sonora (Anderson et al. 1972; Gastil et al. 1978, 1981; Rangin 1978).

### *Middle Jurassic*

No rocks of Middle Jurassic age have been reported from Sonora.

### *Upper Jurassic*

Upper Jurassic rocks have been reported from only a few isolated localities in Sonora. Rangin (1977) described a thick section near Cucurpe, north-central Sonora, composed of interbedded volcaniclastic sediments and andesitic volcanics overlain by a shale interval, 100–150 m thick; containing a rich ammonite fauna identified by R. W. Imlay. The presence of *Perisphinctes (Dichotomosphinctes)* and "*Discosphinctes*" indicates a late Oxfordian age (Imlay 1980).

From western Sonora, about 60 km south of Caborca, Beauvais and Stump (1976) reported marine fossiliferous beds of late Oxfordian to early Kimmeridgian age at Cerro Pozo Serna. The section is 1,800–1,900 m thick and is composed of interbedded arkosic to lithic sandstones, shales, and calcareous shales, often tuffaceous. It contains a rich fauna of mollusks and corals and the ammonites *Amoeboceras* and *Idoceras* (the rare "*Pseudocadoceras*," a Boreal Callovian genus, is also listed).

No late Kimmeridgian or Tithonian sedimentary sequences are known from Sonora.

## Southwestern Mexico

Neither the Lower Jurassic nor the Upper Jurassic is well represented in southwestern Mexico. On the other hand, a well-exposed and fairly complete section of mostly marine Middle Jurassic rocks is known from northwestern Oaxaca and northeastern Guerrero. No similar section is known from anywhere else in Mexico. Campa and Coney (1983) included the area where the Middle Jurassic crops out in their "Mixteca tectonostratigraphic terrane," which they considered allochthonous in nature. Plate-tectonic implications of ammonoid biogeography are discussed in Chapter 24.

### *Lower Jurassic*

In the states of Guerrero and Oaxaca, the Lower Jurassic is represented by a nonmarine, plant-bearing sequence that varies in thickness from 50 to 500 m, the Rosario Formation. It is predominantly composed of interbedded fine-grained sandstones, siltstones, and dark, often carbonaceous shales that frequently contain reddish brown concretions rich in plant remains. Irregularly distributed throughout the section are conglomerate beds predominantly composed of quartz and schists derived from the underlying Acatlán Complex of Paleozoic age. Coal seams are common in the section (Burckhardt 1930; Erben 1956a, 1957). The abundant and well-preserved plant remains were studied by Wieland (1913, 1914), who assigned them an Early Jurassic age. Silva-Pineda (1970) favored a Middle Jurassic age for this fossil flora. More recently, López Ticha (1985/1988) has argued in favor of a Paleozoic age for the Rosario Formation.

From the collections of the Instituto Geológico de México, Burckhardt (1930) reported an Early Jurassic ammonite collected in the state of Guerrero that indicates the presence of marine sediments of that age in this region of Mexico. Erben (1954, 1956a) revised the identification of the ammonite, but confirmed its Early Jurassic (late Pliensbachian) age. The locality from which the ammonite apparently was collected has not been found again, and this important occurrence of a Lower Jurassic marine section in Guerrero has not been confirmed.

### *Middle Jurassic*

In northeastern Guerrero and northwestern Oaxaca, the Middle Jurassic is represented by a locally well exposed, mostly marine section 1,500–1,700 m thick (Burckhardt 1927, 1930; Guzman 1950; Erben 1956a,b, 1957; Alencaster de Cserna 1963; Westermann 1983, 1984a; Sandoval and Westermann 1986; Sandoval, Westermann and Marshall 1990). The basal part of the section is composed of a quartz conglomerate (the Cualac Conglomerate) that varies irregularly in thickness from a few meters to 400 m. It is nonfossiliferous and may be, at least in part, of Aalenian age. The Cualac Conglomerate lies unconformable on Permian sediments or on the Rosario Formation and is overlain by the Tecocoyunca Group, generally about 350–450 m thick and composed of five formations.

The lowest, the Zorillo Formation, is composed of 50–80 m of interbedded sandstones, siltstones, plant-bearing carbonaceous shales, and coals. Silva-Pineda (1970) has dated the fossil plants as Middle Jurassic. More recently, Person and Delevoryas (1982) described the flora of the Zorillo Formation, which they interpreted to have grown in a subtropical lowland near temperate highlands. They also dated it as Middle Jurassic.

The Zorillo is overlain by marine dark siltstones and calcareous shales containing abundant maroon calcareous and hematitic concretions and occasional limestone beds – the Taberna Formation. This marine unit is best developed and exposed in northwestern Oaxaca, where it is 50–60 m thick and has yielded a rich ammonite fauna of middle Bajocian age. Near Mixtepec, Sandoval and Westermann (1986) have distinguished three ammonite zones, of which the upper two are local, but may in the future be applicable to the Andean Bioprovince. The zones are as follows in ascending order:

#### *Bajocian*

1. ?*Stephanoceras kirshneri:* Zone – Hall and Westermann (1980; Chapter 12 this volume); this North American zone is indicated by *Stephanoceras* (*S.*) sp. and *S.* (*S.*) cf. *skidegatense* (Whiteaves); Humphriesianum Zone, ?Romani and Humphriesianum Subzones
2. *Duashnoceras floresi* Range Zone – with *Duashnoceras* Association (Westermann 1984a): *D. floresi* (Burckhardt), *D. undulatum* (Burckhardt), *D. paucicostatum* (Felix), *D. andinense* (Hillebrandt), and *Stephanoceras* (*S.*) *boulderense* (Imlay); also *Strigoceras (Liroxyites)* cf. *kellumi* Imlay, *Phaulostephnaus burckhardti* Sandoval & Westermann, *Stephanosphinctes buitroni* Sandoval & Westermann, *Lissoceras, Oppelia, Phylloceras,* and *Calliphylloceras;* upper Humphriesianum and lower Subfurcatums Zones

3. *Parastrenoceras zapotecum* Assemblage Zone – with *Parastrenoceras* assemblage (Westermann 1984a); *Parastrenoceras* spp., especially *P. zapotecum* (Ochoterena), *Leptosphinctes tabernai* Westermann, and *Oppelia subradiata erbeni* Westermann; basal beds also with *Duashnoceras andinense, D. paucicostatum, Leptosphinctes* cf. *davidsoni* (Buckman), and rare *Strigoceras;* abundant bivalves, especially trigoniids (Alencaster de Cserna 1963); middle and ?upper Subfurcatum Zone

The Taberna Formation is overlain by a unit of fine- to coarse-grained sandstones, siltstones, and plant-bearing dark shales, the Simon Formation, about 80–100 m thick.

The Simon Formation, in turn, is overlain by another marine unit of probable middle to late Bathonian age, the Otatera Formation, composed of interbedded fine-grained sandstones, siltstones, limestones, and distinctive pelecypod coquinas. The unit is generally 50–70 m thick, but reaches 250 m at Cualac in northeastern Guerrero.

The Otatera Formation, finally, is overlain by the Yucuñuti Formation. This unit generally ranges in thickness from 50 to 200 m, but attains 450 m at Cualac, where it is composed of calcareous shales and shaly limestones containing calcareous-hematitic concretions, occasional siltstones, and distinctive oyster and small pelecypod coquinas. The lithology and contained fauna of the Otatera and Yucuñuti Formations indicate that deposition of these units began in a shallow, brackish lagoon, continued in progressively deeper-water environments, which reached deep offshore, epicontinental-basin conditions, before returning to shallow-shelf conditions during the deposition of the uppermost part of the section (Sandoval et al. 1990).

The combined Otatera and Yucuñuti section exposed at Cualac is 700 m thick and is faulted at the top. The ammonite succession of this section (in ascending order) has recently been described by Sandoval et al. (1990), by reference to the standard zonation of the Argentine-Chilean Andes recently proposed by Riccardi, Westermann, and Elmi (1989):

*Bathonian*

1. Steinmanni (Standard) Zone – Retrocostatum and Discus Zones
   (a) Below: *Epistrenoceras histricoides* Association, with abundant *E. histricoides* Rollier; also *Choffatia constricta* (Burckhardt), *Bullatimorphites (Kheraiceras) bullatum* (D'Orbigny), *Ptychophylloceras plasticum* (Burckhardt); a few *Oxycerites (O.)* sp., *O. (Alcidellus) cualacensis* Sandoval & Westermann, *Prohecticoceras blanazense* Elmi, and *Lilloettia* sp.; abundant bivalves and brachiopods; Retrocostatum Zone
   (b) Above: *Lilloettia* Association with dominant Perisphinctidae and Eurycephalitinae, including *Choffatia suborion* (Burckhardt), *C. burckhardti* Sandoval & Westermann, *C. boehmi* (Steinmann), *C. gottschei* (Steinmann), *C.* cf. *jupiter* (Steinmann), *C. furcula* (Neumayr), *Lilloettia steinmanni* (Spath), *Xenocephalites nikitini* (Burckhardt), *X. stipanicici* Riccardi et al., *Neuqueniceras (N.) plicatum* (Burckhardt), *N. (N.) weitzi* (Burckhardt), *Oxycerites (Alcidellus) obsoletoides* Riccardi et al., *O. (A.) cualacensis* Sandovol & Westermann, *Phlycticeras mexicanum* Sandoval & Westermann, *Phylloceras,* and *Ptychophylloceras;* Discus Zone

*Callovian*

1. Vergarensis Zone – *Eurycephalites* cf. *vergarensis* (Burckhardt), *Choffatia* sp. indet. and ?*Neuqueniceras (N.) gleimi* (Stehn), *Ptychophylloceras,* abundant *Bositra* and *Cryptaulax;* ?lower Herveyi ("Macrocephalus") Zone
2. Bodenbenderi Zone – abundant *Neuqueniceras (Frickites)* cf. *bodenbenderi* (Tornquist); also varied other Reineckeidae, including *Rehmannia grossouvrei* (Petitclerc), Eurycephalitinae, Oppeliidae, and Phylloceratidae; with three assemblages (ascending order):
   (a) *Frickites* Association: dominant *Neuqueniceras (Frickites)* cf. *bodenbenderi* (Tornquist) and abundant *Eurycephalites rotundus* (Tornquist), and *Xenocephalites* gr. *nikitini* (Burckhardt); also *X. neuquensis* (Stehn), *Oxycerites (Alcidellus) cualacensis* Sandoval & Westermann, *O. (A.) obsoletoides* Riccardi et al., *O. (Paroecotraustes)* cf. *bronni* (Zeiss) *sensu* Elmi, *Phlycticeras mexicanum* Sandovol & Westermann, *Holcophylloceras, Ptychophylloceras,* and *Phylloceras;* also *Cryptaulax* and *Bositra;* upper Herveyi Zone
   (b) *Guerrericeras inflatum* Association: characterized by *G.* ["*Clydoniceras*"] *inflatum* (Westermann), the last *Frickites* and abundant eurycephalitines, especially *Eurycephalites* cf. *rotundus;* also *Parapatoceras distans* (Baugier & Sauze), *Paracuariceras, Jeanneticeras* cf. *malbosei* (Elmi), *Oxycerites* spp., *Rehmannia grossouvrei;* Phylloceratina, bivalves, and gastropods as below; approximately boundary of Herveyi and Gracilis Zones
   (c) *Rehmannia rehmanni* Association: characterized by *R.* gr. *rehmanni* (Oppel) and *Reineckeia* cf. *franconica* (Oppel); scarce *Oxycerites,* including *O. (Paroecotraustes) bronni,* and Phylloceratina; Rehmanni Subzone of Gracilis Zone
3. Proximum Zone – Andean index is *Hecticoceras proximum* Elmi; at Cualac, indicated by *H.* cf. *boginense* (Petitclerc) at the top of the Jurassic section; upper Gracilis Zone

In northwestern Oaxaca, a seemingly rich *Reineckeia* assemblage with close affinity to *R. anceps* (Reinecke) was described from an isolated locality (a mine shaft at El Consuelo) under numerous new species names (Burckhardt 1927). In fact, almost all of these probably are mere forms and variants of a single species for which the best name is *Reineckeia latesellata* Burckhardt (G. E. G. Westermann has studied the original material and the poorly fossiliferous outcrop beside the underground locality). From the same mine came "*Stephanoceras*" *mixtecorum* Burckhardt, formerly placed in *Erymnoceras*, but now also in *Reineckeia*, several *Rehmannia*, and the probably late Callovian peltoceratid "*Peltoceras*" *cricotum* Burckhardt. This assemblage, collected without stratigraphic control, probably is of middle and late Callovian age.

A similar fauna occurs near Mixtepec, where *Reineckeia* and "*Peltoceras*" are said to occur in the same beds, strongly indicating the late Callovian.

#### *Upper Jurassic*

In northeastern Guerrero and northwestern Oaxaca the Oxfordian is represented only by a sequence of interbedded shallow-water limestones, argillaceous limestones, and calcareous shales 60–100 m thick. Pelecypods and gastropods are generally common, but particularly distinctive are equinoid remains. The unit, for this reason, has received the informal name of "Cidaris Limestone" (Burckhardt 1930; Erben 1956b, 1957; Buitrón 1970; López-Ramos 1983). It is probably equivalent to the Teposcolula Limestone of Salas (1949).

From west of Petlalcingo, southern Puebla, Pérez-Ibarguengoitia, Hokuto-Castillo, and de Cserna (1965) described a sequence of limestones, shaly limestones, and calcarenites 100 m thick, which they named the Chimeco Limestone and correlated with the "Cidaris Limestone." The lower part of this unit is unfossiliferous, but the upper part contains a fauna of equinoids and pelecypods that according to Alencaster de Cserna and Buitrón (1965) indicates an Oxfordian age. The Chimeco Limestone overlies transitionally a nonmarine, unfossiliferous "red bed" sequence that Pérez-Ibarguengoitia et al. (1965) named the Tecomazuchil Formation. This unit is composed of a basal conglomerate 135 m thick and predominantly composed of quartz and metamorphic rock fragments, overlain by about 600 m of interbedded tan to red conglomerates, sandstones, and siltstones. The Tecomazuchil Formation overlies unconformably the Acatlán Complex and has been assigned a Middle Jurassic age, though it could represent at least part of the Oxfordian.

Rocks of Kimmeridgian and Tithonian age have been reported from only a very few localities in southwestern Mexico. Burckhardt (1930) referred to a specimen of the ammonite *Idoceras* of presumable early Kimmeridgian age from northwestern Oaxaca. López-Ramos (1983) reported the presence in northwestern Oaxaca and northeastern Guerrero of limestones, calcerous shales, and shales of Oxfordian age overlain by dark shales and limestones containing Kimmeridgian ammonites. He also mentioned that Tithonian beds were known from western Oaxaca, but gave neither paleontological evidence nor literature references to support those claims.

In the Petlalcingo area of southern Puebla, Pérez-Ibarguengoitia et al. (1965) mapped, conformably overlying the Chimeco Limestone, a unit of shaly limestones and interbedded calcereous siltstones and shales that they named the Mapache Formation. It reaches 400 m in thickness and is unconformably covered by Cretaceous units. The Mapache Formation contains a poorly preserved but abundant fauna predominantly composed of pelecypods and brachiopods, with scarcer gastropods and ammonites, that according to Alencaster de Cserna and Buitrón (1965) ranges from the Oxfordian to the Tithonian. The Mapache Formation grades laterally to a deeper-water sequence of dark shales and limestones, rich in organic matter, that López Ticha (1985/1988) has called the Sabinal Formation.

It is probable that additional occurrences of Kimmeridgian and Tithonian rocks may be uncovered when the southwestern Mexico area is studied and reported in more detail.

Near the town of Ixtapan de la Sal, about 90 km south–southwest of Mexico City, Campa et al. (1974) have reported the occurrence of Tithonian ammonites and tintinnids from shales in a sequence of alternating, slightly metamorphosed andesitic-dacitic pillow lavas, tuffs, and agglomerates, with lesser amounts of sandstone and limestone. Evidence of important volcanic activity during the Tithonian is also recorded in the Pimienta Formation of east-central and southeastern Mexico, which contains numerous bentonite beds (Salvador 1987).

## NORTH-CENTRAL AND EASTERN MEXICO[2]

### Introduction

A review of the geological and paleontological investigations of the Upper Jurassic in Mexico (Figure 5.4) encompasses a large amount of published information. The syntheses by Burckhardt (1930) and Imlay (1939, 1953, 1980) remain useful, as do the recent, regionally more restricted contributions, with biostratigraphic emphasis, by Cantu Chapa (1967, 1968, 1970, 1971, 1976a,b, 1977, 1979, 1984), Verma and Westermann (1973), Contreras, Martinez, and Gomez (1988), Villaseñor and Gonzalez (1988), and Olóriz and Tavera (1989). Recent works concerned with the reconstruction of sedimentary and ecological environments are by Ortuno-Arzate (1985), Araujo-Mendieta and Estavillo-Gonzalez (1987), Olóriz (1987), and Olóriz et al. (1988a), and works on ammonite biogeography include those by Jeletzky (1984), Olóriz (1988, 1990), Olóriz et al. (1990), and Olóriz and Tavera (1989).

This overview of the Upper Jurassic in north-central and eastern Mexico relates to ongoing ecostratigraphic reconstructions (Figure 5.5); geological setting and environmental reconstructions are considered; a brief appraisal is made of the significance of Mexican ammonites during the Upper Jurassic; and biostratigraphic aspects are briefly considered here and are dealt with separately in this volume by Callomon (Chapter 12).

### Geological setting

The present consensus is that the Jurassic geological histories of those parts of Mexico not bordering on the Pacific were

[2] By F. Olóriz.

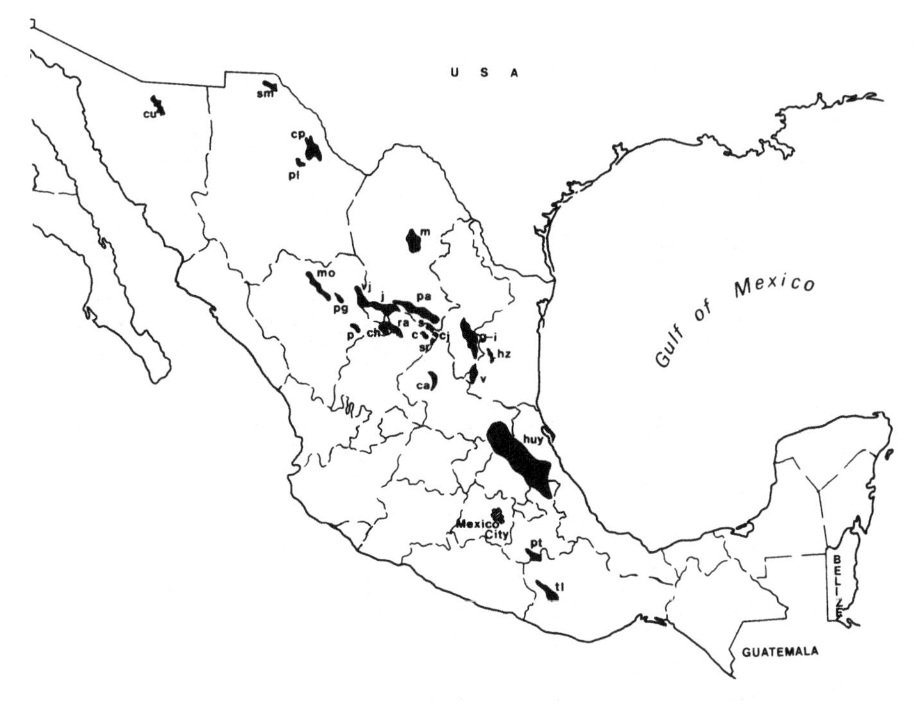


**Figure 5.4. The major Upper Jurassic outcrops in Mexico: c, Sierra Cadnelaria (Zacatecas); ca, Sierra de Catorce (San Luis Potosí); ch, Sierra del Chivo (Durango); cj, Sierra de La Caja (Zacatecas); cp, Sierra El Cuchillo Parado (Chihuahua); cu, Sierra de Cucurpe (Sonora); g-i, Sierra Galeana-Iturbide (Nuevo Leon); huy, Huayacocotla Anticlinorium [Tamazunchale (San Luis Potosí), Zacualtipan (Veracruz), Huachinango-Zacatepec (Puebla)]; hz, Huizachal Anticlinorium (Tamaulipas); j, Sierra de Kimulco (Coahuila); m, Sierra Menchaca (Coahuila); mo, Santa Maria del Oro–Sierra de La Zarca (Durango); p, Sierra de Palotes (Durango); pa, Sierra de Parres (Coahuila); pg, San Pedro del Gallo (Durango); pl, Sierra Plomosas–Placer de Guadalupe (Chihuahua); pt, Pletlalcingo–Sierra Madre del Sur (Puebla); ra, Sierra de Ramirez (Zacatecas, Durango); s, Sierra Sombreretillo (Zacatecas); sm, Sierra de Samalayuca (Chihuahua); sr, Sierra de Santa Rosa (Zacatecas); ti, Tlaxiaco–Sierra Madre del Sur (Oaxaca); v, Sierra Vieja–Arroyo Doctor (Tamaulipas).**

closely linked to the opening of the Central Atlantic and probably include a mosaic pattern of terranes (e.g., Salvador and Green 1980; Scott 1984; Westermann 1984a; Burke 1988; Michaud and Fourcade 1989). Early Jurassic sedimentation began as a result of a major eustatic rise. That event is evident in the North Atlantic Basin (Jansa 1986) and in the Huayacocotla Basin of northeast Mexico (López-Ramos 1983), with abundant Sinemurian ammonites. The overall transgressive trend of the Jurassic was punctuated by tectonoeustatic phases that affected the region of the Gulf of Mexico and the Caribbean. Proposals coherent with this hypothesis are those of Imlay (1980), for the Pliensbachian-Toarcian in the Huasteca region (eastern Mexico); Scott (1984), for the salt deposition in the Gulf of Mexico, which, in his opinion, was related to Middle Jurassic rifting; and Longoria (1984a,b), who recognized tectonic events at different time intervals during the Jurassic. New observations of abrupt changes in thickness and slumping in the La Casita/La Caja Formations in north-central Mexico, San Luis Potosí, and Zacatecas correspond well to episodically active bottom topography.

The Callovian–early Oxfordian event is of particular interest. It was recognized as a "structural movement" by Imlay (1980) to explain the entrance of seawater into the Gulf area during salt deposition. That event reflects the first important effect of the separation of the North and South American plates, including the breakup and separation of the Yucatán Block (Salvador and Green 1980; Salvador 1987). Irregularities of the sea floor were related to differential rates of subsidence, which produced differences in facies and thicknesses of the Upper Jurassic sediments, as demonstrated by Michaud and Fourcade (1989) in southeastern Mexico.

During the Upper Jurassic, the epicontinental seas developed considerably on the southern margin of the North American plate (Figure 5.5). Enos (1983) has recognized that during the Oxfordian there developed extensive, low-topography carbonate platforms with shallow, locally evaporitic basins and archipelagos. Scott (1984) has characterized the tectonosedimentary regime in the Mexican-Caribbean area and recognized a "terrigenous-clastic shelf system" well developed during the Kimmeridgian and Tithonian, with "trailing plate margins" affecting Mexico.

This suggests a major change in depositional systems that was preceded by evaporites over much of northern and northeastern Mexico (Olvido Formation, and shallow equivalents of La Gloria Formation), extending into southeastern Mexico (Campeche salt), as well as restricted shallow-marine sediments in the north (Taman Formation) and frequent hiatuses in all of northern and central Mexico. All of them predated deposition of the La Casita/La Caja Formations. Because the Kimmeridgian was generally transgressive in Mexico [for southeastern Mexico see Michaud and Fourcade (1989)], the end of mainly carbonate sedimentation during the Oxfordian was related to uplifts and block tilting. The dominance of terrigenous-clastic sediments until the end of the Jurassic, local slumping, and changes in facies and thicknesses document tectonic instability. This post-Oxfordian phase of instability may be related to pulses of opening of the Gulf of Mexico. Its "oceanization" would have taken place around the late Kimmeridgian (Salvador 1987) or late Oxfordian (Michaud and Fourcade 1989).

The precise dating of the opening of the Gulf has not yet been established, but this major event certainly affected Mexico,

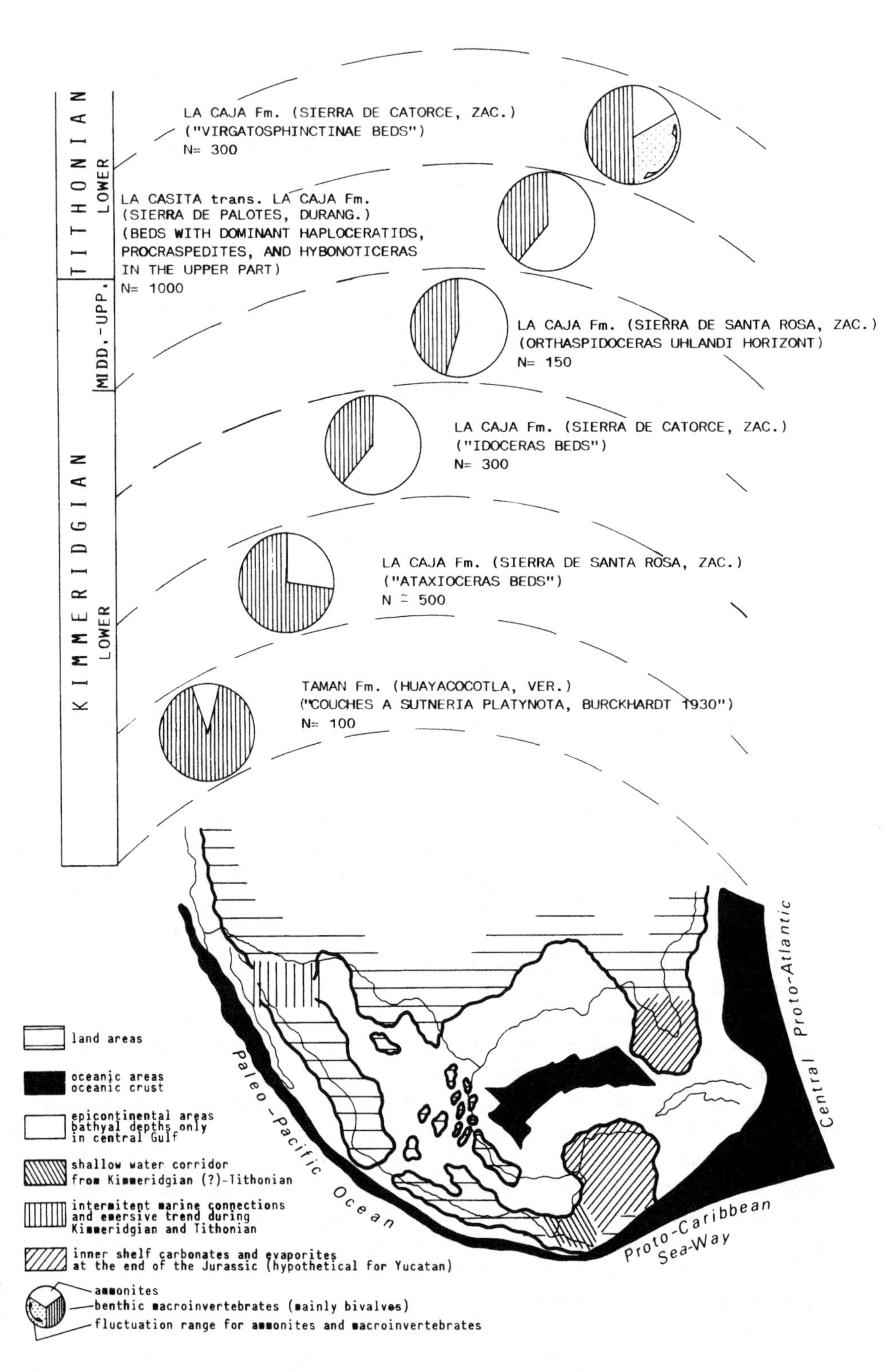


**Figure 5.5. Late Jurassic and earliest Cretaceous paleogeography of Mexico and selected invertebrate faunal spectra.**

causing important paleogeographic changes. It is therefore interesting to relate the Gulf event to the opening of intracontinental troughs, such as the Chihuahua Basin, which in turn are clearly related to the transtensive movements between the megashears of Texas and Caltam (Ortuno-Arzate 1985; Tardy, Blanchet, and Zimmermann 1989) during the Upper Jurassic and their filling with shallow-marine sediments since the lower Kimmeridgian.

The southern margin of the North American plate was tectonically unstable during the Upper Jurassic and was especially subject to vertical movements in the Gulf area after its opening (Salvador 1987); in Mexico the situation was rather complex, with transform movements in the north and south (Salvador and Green 1980; Longoria 1984a; Salvador 1987; Michaud and Fourcade 1989; Tardy et al. 1989), and tilting and/or vertical movements affecting basins in different ways (Olóriz 1987). The block tectonics proposed by Enos (1983), Ortuno-Arzate (1985), and Araujo-Mendieta and Estavillo-Gonzalez (1987), and the halokinesis recognized by Viniegra (1981), caused, moreover, more or less persistent archipelagos during the Upper Jurassic (e.g., Burckhardt 1930; Imlay 1939, 1980; Enos 1983; López-Ramos 1983; Ortuno-Arzate 1985; Salvador 1987).

## Ecological environment

### *Climate*

The Mexican Upper Jurassic faunas and lithofacies suggest a warm climate, at least subtropical, with fluctuations in rainfall. Concordant data include the overall increase in fine clastics and the reefal development on the periphery of the Gulf and in Mexico itself (Cantu Chapa 1976a; Cregg and Ahr 1984; Crevello and Harris 1984; Finneran et al. 1984), the benthic faunas with Mediterranean affinities (Alencaster de Cserna 1984; Buitrón 1984), and the climatic evolution in the North Atlantic Basin (Jansa 1986).

The analyses of the clay minerals in the Chihuahua Basin (La Casita Formation) by Ortuno-Arzate (1985) also indicate a warm climate that was relatively wet in areas of moderate, well-drained topography. Olóriz et al. (1988a) have provided isotope and mineralogical data from the middle and upper Kimmeridgian in Durango. They indicate a proximal, emerged area with relatively high hydrolysis and dominant physical erosion, as well as high rainfall and episodic coastline fluctuations in marine environments with carbonate precipitation, most probably at temperatures higher than 20°C. According to Olóriz (1987) and Doyle (1987), the presence of mollusks of Boreal affinity (Buitrón 1984) should in this case not be related to decreases in temperature nor to ingressions of cold water, as supposed by Jeletzky (1984); see Olóriz and Tavera (1989).

### *Water depth*

Whereas reconstruction of a picture of the climate does not appear to pose serious problems, the diversity of hypotheses on water depth is greater. Longoria (1984a,b), López-Ramos (1985), and Ortuno-Arzate (1985) all suggest the existence of deep seas (Longoria even suggests bathyal depths) during the Kimmeridgian and/or Tithonian. But Burckhardt (1930), Imlay (1980), Salvador (1987), Araujo-Mendieta and Estavillo-Gonzalez (1987), Olóriz (1987), Olóriz et al. (1988a), and Contreras et al. (1988) believe that only shallow to moderately shallow or medium-depth waters existed.

Olóriz (1987) and Olóriz et al. (1988a,b) analyzed macroinvertebrate assemblages, but could not exclude postmortem drifting. In the Kimmeridgian and lower Tithonian associations of Mexico, the proportion of ammonites reaches 75%, the remainder being benthic invertebrates. Such a composition would seem to indicate a depth of 80–200 m according to Ziegler (1967), 70–120 m according to Gygi (1986), or 40–150 m according to the autecological and synecological analyses of the ammonites by Westermann (1990). As shown later, the depth may have fluctuated around the lower values and may frequently have been even less locally. Under the sedimentary conditions of the Upper Jurassic terrigenous-clastic shelf system, the living conditions for benthic faunas could at times have been limited (turbidity, sporadic remobilization of soft bottoms, etc.), resulting in reduction of benthic faunas and the consequent depth overestimate based on the dominant ammonite record. Furthermore, attention should be paid to the nonlinear relationship between depth and distance from shoreline, especially when there is evidence for irregular sea floors.

Based on fieldwork during 1985, 1987, and 1989, especially in the Kimmeridgian and Tithonian, 4,500 fossils from five sections in the states of Zacatecas, Durango, San Luis Potosí, and Veracruz were evaluated statistically. Although the samples were standardized, vertical fluctuations in faunal composition were slight, despite varying sample sizes. The average spectra for different ages (Figure 5.5) show a trend toward a deterioration in life conditions for benthic faunas during the Kimmeridgian–lower Tithonian. This trend may perhaps be linked with progressive deepening and distance from shore, but bottom conditions played a significant role. On the whole, the seas were not deep in north-central and northeastern Mexico during the interval examined.

The marine environment of the Mexican Late Jurassic ammonites was that of extensive platforms, not very deep and often shallow, with irregular, unstable sea floors. The persistence of fine terrigenous-clastic inflows during the Kimmeridgian and Tithonian reveals the proximity of areas under erosion, in part due to moderate to high rainfall. Seawaters would have been warm, and salinity presumably variable, in proximal areas, according to parameters defining water types *sensu* Valentine (1973). A marine environment of rather small volume relative to its area is proposed. It was turbid, at least near the bottom, and at times partially restricted, at least for ammonites, as discussed later. In such an environment a strong but variable "platform effect" can be expected (Olóriz 1985, 1988, 1990; Olóriz et al. 1990).

## Remarks on mollusk assemblages

According to Olóriz (1987), some general features of Mexican Upper Jurassic macroinvertebrate assemblages with significant representations of ammonites are as follows: (1) faunal assemblages are formed by ammonites and bivalves, whereas other organisms are in the minority, except on reefal enclaves; (2) ammonite spectra are unbalanced, with clear predominance of one or two components; (3) high morphological diversity in ammonites; (4) extreme phenotypes occur among ammonites; (5) endemism; (6) fluctuations in ammonite assemblages apparently are related to sea-level changes, without linear or simple relationships to facies.

### *Assemblages dominated by ammonites and bivalves*

As discussed earlier, this faunal composition is related to a shallow shelf environment. Therefore, fluctuations in ammonites and bivalves reveal, in turn, changes in the proximal/distal character and in the depth. We agree with Geyer (1971) that depth is not the main ecological factor in the distribution of ammonites, as is evident from comparisons of basically different environments, such as *platform ambitus* and *basin ambitus* (Olóriz 1985, 1988, 1990). During the Kimmeridgian–Tithonian interval of shallowing in southern Germany, Gwiner (1976) noted that the *Oppelia*-type ammonites became more abundant, and perisphinctoids diminished. This contrasts clearly with the models of Ziegler (1967) and Gygi (1986). Verma and Westermann (1984) also commented on the care that must be taken in applying Ziegler's model.

Analysis of faunal spectra from the Kimmeridgian (and possibly the basal Tithonian) obtained in the section of the Cañon del Toboso (Sierra de Symon, Durango) (from *Idoceras* beds upward) shows an average composition of ammonites of 75%, with other invertebrates (mainly bivalves) at 25%. There is some fluctuation throughout this interval, and no linear correlation with data at the ammonite-zone level. Although this proportion corresponds to the 80–200-m interval of Ziegler (1967), 68% of the ammonites are Haplocerataceae ("*Oppelia* type"), and only 22.9% are Ataxioceratidae (including *Idoceras* ["*Perisphinctid* type"], with *Sutneria* at 3.97%, Aspidoceratidae (including *Hybonoticeras*) at 2.59%, and the rest (1.94%) divided between *Nebrodites-Mesosimoceras* and *Protancyloceras*. These ammonite data correspond to Ziegler's 150–200-m interval, but 25% bivalves would be excessive at those depths, according to his model. In the Gygi model (1986), our percentages of *Oppelia*-type ammonites and other invertebrates (mainly bivalves) would also fit, with some difficulty, at depths greater than 120 m. Both models imply that *Oppelia*-type ammonites are found in large numbers (20%) only at depths greater than 80 m and are restricted to shelves.

The Kimmeridgian ammonite spectra closest to those just cited are found in Swabia. Ziegler (1967, Figs. 2, 9, and 10) recognized similar ammonite compositions, but at the biozone level, and without data on ammonites versus benthic invertebrates. Invertebrates from one small area here are recorded as "quasi-*Mazapilites*" in the upper part of the Kimmeridgian: *Taramelliceras kiderleni*, according to Berckhemer and Hölder (1959). A depth of 50–100 m has been deduced for the sediments from ecological and sedimentological considerations (Gwiner 1976).

To consider the depth of the faunal assemblage found in the Cañon del Toboso to be around 60–80 m, we have only to resolve the "problem" of the significance of the *Oppelia*-type ammonites in the traditional models. Ward and Westermann (1985) and Westermann (1990) accept the existence of many oxycones in shallow waters, in clear opposition to other hypotheses that would restrict them to greater depths on the basis of the complexity of the septal suture in some genera. Moreover, the spectra from the Sierra de Symon (Durango) and from other sites in the Mexican Altiplano contradict the limitations on the development of these morphs (*Haploceras, Glochiceras*) as a consequence of the sedimentation rate of fine clastics, as interpreted by Verma and Westermann (1984) for their Kenyan case. The cause may better be sought in the limitations on ammonite entry to the area at certain times (i.e., fluctuating barrier effects would explain the discontinuous records of ammonites that occur in separate horizons). In Mexico, *Glochiceras* is frequent in the "*Idoceras* Beds," and *Haploceras* is recorded at varying frequencies at different intervals during the Kimmeridgian and Tithonian. Thus, the *Oppelia* type is present from the lower Kimmeridgian onward, regardless of the rate of clastic sedimentation.

It appears reasonable, therefore, to interpret the macroinvertebrate spectrum obtained in Sierra de Symon as indicating a depth corresponding approximately to the lower values of the ranges provided by the modified models of Ziegler (1967), Gygi (1986), and Westermann (1990). Depths of 30–80 m appear to be representative of the areas occupied by the Upper Jurassic in north-central and northeastern Mexico (Olóriz et al. 1988a).

### *Unbalanced faunal spectra*

Faunal spectra become unbalanced when the number of components (out of a minimum 2–3%) is reduced and one or two clearly dominate the association. This is caused by restricted connections with open sea. High selective predominance among ammonites reveals a strong *platform effect*. Mexican examples are the ammonite assemblages recorded at levels with "*Ataxioceras*" in the area of the Sierra de Santa Rosa (Zacatecas) and in the "Capas de Idoceras" (Cantu Chapa 1969, 1971, 1979, 1984), as well as those containing *Mazapilites* in the Altiplano and eastern Sierra Madre, and the levels with *Kossmatia* in the El Verde profile in Sierra de Catorce, San Luis Potosí.

### *Endemism*

The taxonomy of the Mexican Upper Jurassic ammonites has been excessively typological ("splitting"), and therefore endemism is recognized at relatively low levels (e.g., "species" of the genera *Idoceras, Kossmatia, Mazapilites,* etc.). But traces of endemism are also clearly recognizable when "lumping" is used, based on stratigraphically controlled populations. This results in fewer species, with greater intraspecies variation. Populations and species can easily be distinguished from those of other areas, such as the European and peri-Gondwanian platforms and the Mediterranean regions *sensu stricto*. The level of endemism in this case is greater also at the genus level.

The Mexican Upper Jurassic ammonites present traces of endemism (Olóriz et al. 1990): (1) at the subspecies level, and perhaps also the species level, in *Praeataxioceras* (upper Oxfordian) and *Hybonoticeras* (upper Kimmeridgian–basal Tithonian); (2) at the species level in *Idoceras* (lower Kimmeridgian), *Corongoceras,* and *Kossmatia* (upper Tithonian), and, at least at the species level, in *Pararasenia* (lower Kimmeridgian); and (3) at the genus level in *Epicephalites* (lower Kimmeridgian), *Procraspedites* (middle Kimmeridgian), *Mazapilites* (top Kimmeridgian–basal Tithonian), *Mazatepites, Suarites, Acevedites,* and *Salinites* (at different Tithonian levels) (cf. Cantu Chapa 1967, 1968).

Endemism at the genus and species levels holds special interest because there are some records of the same taxa from distant areas with no apparent connection between their epicontinental seas. This is the case for *Epicephalites*, which was recognized by Arkell (1956) in New Zealand in a Kimmeridgian association with similarities to the Mexican associations. The case of *Kossmatia* is also interesting, this being a genus with a wide peri-Gondwanian record in the Himalayas, Indonesia, New Zealand, and the Antarctic; the records in Argentina (Imlay 1942) and Chile, however, are doubtful, as the genus is not recognized by Leanza (1981) and Biro-Bagoczky (1984), although the Tithonian–Berriasian is well known in both regions. On the other hand, *Kossmatia* also has not been recognized in areas close to Mexico, such as Cuba, in which other taxa of the association have been found. Although an unequivocal interpretation cannot be offered at present, traditional interpretations with speculative migration routes may have to be abandoned. Possible alternatives, however, can be proposed only after understanding the regional evolving clades, and in the context of parallel-evolution phenomena as an eco-evolutionary response by genetically integrated faunas subjected to similar habitats (Olóriz 1985, 1988, 1990).

### *Faunal fluctuations, facies changes, and sea-level variations*

In the Upper Jurassic of Mexico there are two clear examples of turnovers in ammonite assemblages that can be related to facies changes. The "*Idoceras* Beds" (Burckhardt 1906, 1930; Imlay 1943, 1980; Cantu Chapa 1969, 1971, 1976a, 1984) record the lower Kimmeridgian transgression in the areas where they rest on the Zuloaga Formation or lateral equivalents. There the Ataxioceratidae faunas, which are stratigraphically separated from *Idoceras* occurrences, develop some morphotypes with a more primitive ribbing type than that of older faunas. On the other hand, the less widely recognized "unit zone with *Ataxioceras*" (Cantu Chapa 1969, 1971, 1976a, 1984) in Veracruz seems to be a case of faunal change unrelated to facies evolution. There is only a slight difference between the "clayey Taman" and the "calcareous Taman" in the Huasteca Series, which shows no marked facies changes from the upper Oxfordian to the lower Kimmeridgian ("zones" with *Discosphinctes, Ataxioceras,* and *Idoceras*). On the other hand, fieldwork in 1987 and 1989 showed that in the Sierra de Santa Rosa (Zacatecas) the levels with "*Ataxioceras*" replaced the Zuloaga Formation. Considering the mentioned morphological changes in the Ataxioceratidae without great facies changes during Taman deposition, and also the record of Mexican *Idoceras* in different sediments (clayey limestones, siltstones, and fine sandstones), it appears that Mexican "*Ataxioceras*" were also little dependent on depositional conditions. The ecological factors favoring their abundant occurrences, laterally and vertically discontinuous and often together with Glochiceratidae, must be related to barrier effects (bottom topography, marine currents, water types).

In the regions of Mazapil and Sierra de Santa Rosa (Zacatecas) and Cuecame and Symon (Durango), faunal changes sometimes are related to the deposition of a condensed level of dark to black limestone. This horizon, less than 1 m thick, contains *Procraspedites* and some perisphinctoids with prominent ornamentation, similar to the European *Crussoliceras* (together with an undescribed genus). *Haploceras* and *Glochiceras* are associated as in lower levels, but *Nebrodites, Mesosimoceras,* and *Idoceras* are absent. This condensed level is believed to document a transgressive maximum, with consequent reduction in detrital sedimentation, in areas that became more distal because of the relative rise in sea level (Olóriz et al. 1988a). The decrease in siliciclastics favored a low rate of carbonate deposition, resulting in condensation. The increase in marine ecospace had some effect on Haplocerataceae, and new extreme morphs arose among the perisphinctoids (e.g., *Procraspedites*). At higher levels, the return of predominantly clayey sedimentation was due to the recurrence of depositional and environmental conditions, presumably accompanied by a reduction of ecospace. The situation continued without significant change during *Hybonoticeras* and *Mazapilites* times. Once again, the ammonite assemblages were clearly dominated by one or two taxa (*Hybonoticeras* or *Glochiceras* and *Mazapilites*), which, moreover, developed marked endemic morphs. Thus, the associations of the upper Kimmeridgian–basal Tithonian, like those of the lower Kimmeridgian ("*Idoceras* levels" and levels with "*Ataxioceras*"), show similar ecological responses to probably equivalent environmental conditions.

A new transgressive pulse may also be represented by the "Virgatosphinctinae Beds" and lateral equivalents in north-central Mexico. This is indicated by their increased ammonite diversity, with the appearance of the semicosmopolitan *Schaireria, Aspidoceras, Simoceras,* and *Phylloceras.* The varied groups of *Torquatisphinctes,* so-called *Virgatosphinctes* and *Aulacosphinctes,* among others, requires careful analysis, because these are forms with confused taxonomy and hence biogeography. Environment reduction with or without accompanying climatic fluctuations (e.g., rainfall) and a return to fine-detrital sedimentation can be inferred from higher levels, from which monotonous assemblages (e.g., mainly *Kossmatia*) are recorded.

The faunal turnovers in the ammonite assemblages of the Mexican Upper Jurassic thus correlate with changes in the environment. Those changes were related to fluctuations in relative sea level, which presumably activated/deactivated topographical barriers to the ammonite faunas and influenced ammonite evolution.

## Remarks on biogeography and correlations

With regard to ammonite biostratigraphy and correlations (see Chapter 12), some brief observations are in order concerning their paleogeographic and paleoecological context.

According to the foregoing geological and paleontological considerations, the Upper Jurassic Mexican platforms fit a model of shallow epicontinental seas with irregular bottoms, which caused variable restriction or subdivision of basins (Olóriz et al. 1990). Limited connections with open seas or oceanic waters resulted in variable degrees of endemism imposed on populations, structured according to *r*-type ecological strategies. The considerable sizes

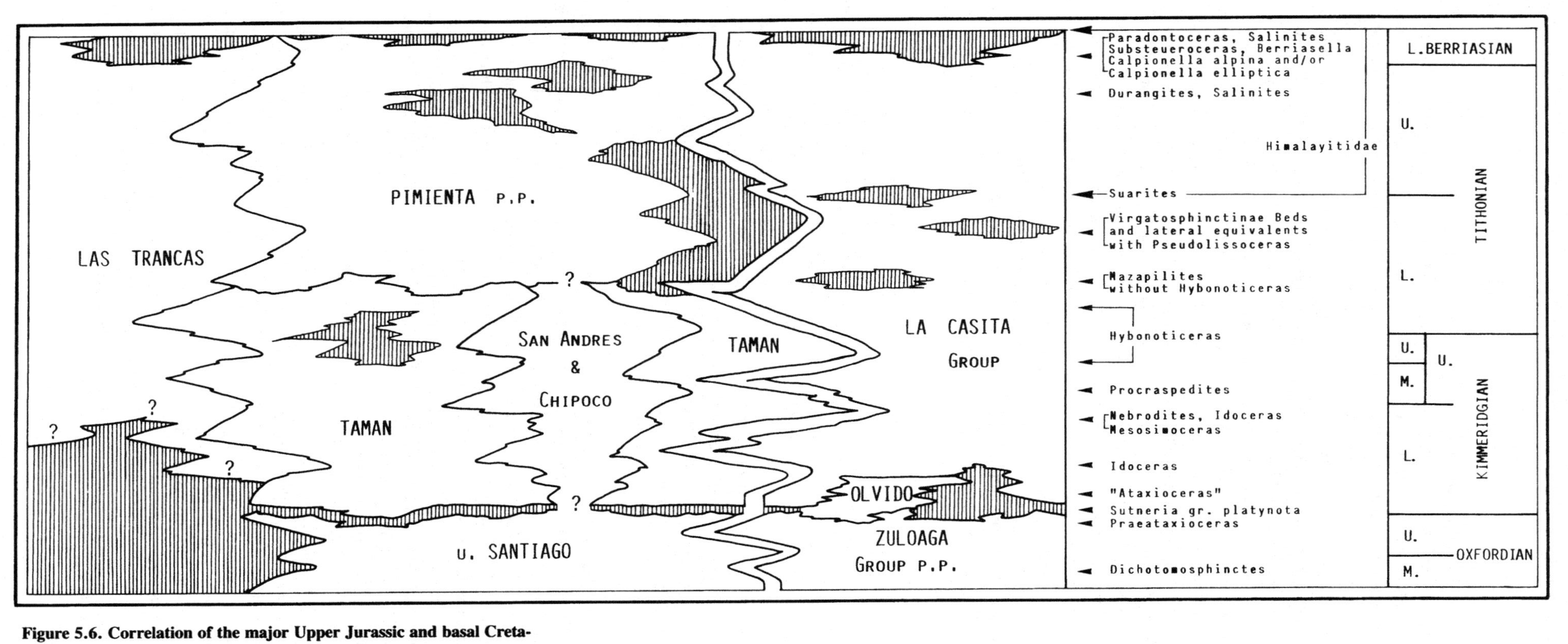


**Figure 5.6.** Correlation of the major Upper Jurassic and basal Cretaceous deposits of north-central and northeast Mexico based on ammonites and calpionellids. Upper Kimmeridgian *s. gall.* = middle and upper Kimmeridgian (Beckeri Zone) *s. mediterr.* Formations in capitals, members in lower case; vertical ruling for known hiatuses and discontinuities in the record of ammonite associations (very poor for Las Trancas Formation); only ammonite-bearing rocks are represented in the Oxfordian.

of these populations and their high intraspecific variability are characteristically shown by Ataxioceratidae, Hybonoticeratinae, Haploceratidae, and Himalayitidae during the Kimmeridgian and/or Tithonian.

In this context, the interpretation of the Mexican epicontinental seas as basically receptors during the Upper Jurassic is coherent with information from other areas of the southern margin of the North American plate.

It is proposed that the shallow epicontinental basins occupied by the ammonites often formed closely interrelated regional environments. These were frequently affected by tectonics and fluctuations in relative sea level, resulting in the recurrence of facies and episodically unfavorable conditions for ammonites. The discontinuity in the record of these faunas undoubtedly resulted from the tectonic-eustatic dynamics that also determined facies distribution. The causes of the discontinuities in the record of the often scarcely diversified ammonite assemblages must have been interrelated with the irregular sea-floor topography and, perhaps, the current patterns. That would have caused lateral variations in the successions of ammonite assemblages, and at times their compositions. Correlations are thus difficult to achieve, limiting the reliability of both biostratigraphic analysis and subsequent chronostratigraphic interpretations. The detailed correlations of the Mexican faunas with the Andean and Mediterranean faunas are still tenuous because of the existence of endemic faunas in Mexico and our limited knowledge of some key genera (Olóriz and Tavera 1989). Until new data from profiles studied bed by bed become available for evaluation of the homotaxial sequences, *sensu* Scott (1985), proposals on biostratigraphy, biostratigraphic correlations, and chronostratigraphy of the Upper Jurassic of Mexico will remain uncertain.

A tentative correlation chart for significant formations in north-central and northeastern Mexico (Figure 5.6) is based mainly on ammonites, with special attention to the known hiatuses and/or discontinuities in their record.

**Acknowledgments**

I am indebted to my collaborators in Mexico, Dra. C. Gonzales and Dra. Villaseñor, and to biologist L. Lara for fieldwork and information.

## NORTHERN CENTRAL AMERICA: THE CHORTÍS BLOCK[3]

### Introduction

Dengo (1969) and Dengo and Bohnenberger (1969) defined the Chortís block as the part of Central America south of the Motagua fault of Guatemala and north of latitude 12°30′N (in Nicaragua). Honduras, El Salvador, northern Nicaragua, and southeastern Guatemala are included in the Chortís block. Donnelly (1989, p. 302) called the Chortís block the one certain "exotic terrane" of the Caribbean region. Since at least the Eocene the Chortís block has been sliding past the North American plate along a major system of strike-slip faults. Gose (1985) proposed that the Chortís block experienced a complex rotation history during the Cretaceous. Its position in the Jurassic is uncertain, although Gose (1985) suggests that its paleolatitude in the latest Jurassic may have been close to southern Mexico. Because of the unique geology of the Chortís block relative to the regions that are currently its neighbors, this chapter is limited to just the Chortís block, instead of all of northern Central America. By studying the Jurassic rocks of the Chortís block, we can better constrain tectonic reconstructions and the paleogeography.

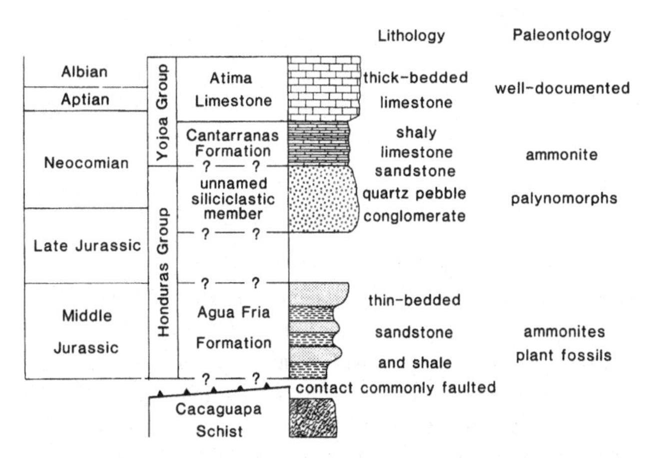


**Figure 5.7. Finch (1985) and Ritchie and Finch (1985) defined the Honduras Group as siliciclastic units that underlie the Aptian-Albian Atima Limestone and overlie the metamorphic basement. The Honduras Group is divided into two principal units. The lower unit is the Agua Fría Formation, which has been dated as Middle Jurassic. Older strata may also exist, but are not represented on the stratigraphic column. The other unit is the unnamed siliciclastic member, which has not been dated as well as the Agua Fría Formation. Palynological data indicate that some of the strata are Early Cretaceous (Emmet 1983; Gose and Finch 1987). The Cantarranas Formation is the lowest part of the Yojoa Group, in which the Atima Limestone is also included. It is commonly relatively thin. Thus, it is not shown on many geological maps of Honduras, even if it occurs in the area (e.g., King 1973). However, it is a distinct unit and has been included in the recent stratigraphic columns for Honduras (Mills et al. 1967; Finch 1981). The Atima Limestone is a well-known, prominent formation in Honduras. Mills et al. (1967) called it the "fundamental datum." Detailed studies of the Atima Limestone and overlying formations have been reported by Mills et al. (1967) and Finch (1981).**

Clastic formations underlie the Atima Limestone of Aptian–Albian age (Figure 5.7) that has been mapped throughout the Chortís block. In central Honduras, El Salvador, and southeastern Guatemala, the clastic rocks are dominantly thin fluvial deposits. However, in east-central Honduras and possibly in northern Nicaragua, marine strata with a terrigenous component crop out. Some of these clastic rocks have been dated as Jurassic on the basis of their marine invertebrates and plant fossils.

The purpose of this section is to summarize what is known about the Jurassic rocks of the Chortís block and to clarify the ex-

[3] By M. B. Gordon.

isting literature on the subject. Much of the information regarding the Chortís block is in unpublished theses and open-file reports with government agencies, and some material is not readily available. I have used as much of this information as is available to me.

*Evolution of stratigraphic nomenclature*

Although Leggett (1889) briefly described the geology of the San Juancito area (central Honduras, Figure 5.8), he did not give names to any of the geological units. Fritzgärtner (1891) introduced the name Tegucigalpa Formation for all of the sedimentary rocks exposed in Honduras. Lacking paleontological data, he considered the bulk of the strata to be Permian, based on general lithologic similarities to the Permian of North American and Europe. His Tegucigalpa Formation is clearly all of the Mesozoic rock units that have subsequently been dated paleontologically in numerous localities and assigned new formation names (Carpenter 1954; Mills et al. 1967; Finch 1981). Having used erroneous criteria to date the rocks, Fritzgärtner included the strata with the plant fossils in the upper part of the Tegucigalpa Formation instead of the lower part. Sapper (1937) also included these beds in the Tegucigalpa Formation, but he recognized that they are stratigraphically low in the section. Carpenter (1954) showed that the name Tegucigalpa Formation is inappropriate because the beds with Triassic(?)–Jurassic plant fossils are dominantly black shales, whereas the rocks in the vicinity of Tegucigalpa, Honduras, are redbeds that overlie Lower Cretaceous limestone. Carpenter named the plant-bearing strata the El Plan Formation, after a village near the fossil site. Carpenter, among others, noted that the El Plan Formation is highly deformed and that its thickness could not be measured in the San Juancito region because of this structural complexity. In spite of the problems with the San Juancito-site, Mills et al. (1967), Gallo and Van Wagoner (1978), and Finch (1981) decided to continue the use of the name "El Plan Formation." Mills et al. (1967) and Gallo and Van Wagoner (1978) extended the formation beyond the San Juancito area. After much fieldwork in eastern and southern Honduras, Ritchie and Finch (1985) introduced the name "Agua Fría Formation" for strata exposed at the Agua Fría gold mine near Danlí, Honduras (Figure 5.8). My field observations, at Agua Fría in 1987 and at San Juancito in 1989, show that deformation is less intense in the vicinity of Agua Fría than in the San Juancito area. In his field area south of the Catacamas Valley, Kozuch (1989a) replaced the name "El Plan Formation" with "Agua Fría Formation" because the strata at Agua Fría are thicker and more readily correlated with other localities in Honduras. Inasmuch as the El Plan Formation is poorly defined, and no measured section for it exists, current workers in Honduras have adopted the name "Agua Fría Formation" for similar beds found elsewhere in Honduras (Kozuch 1989a,b; Gordon, in press). Finch (1985) and Ritchie and Finch (1985) introduced the name "Honduras Group" to include all clastic units that underlie the Lower Cretaceous Atima Limestone. Thus, strata formerly referred to as the El Plan Formation are now referred to as the Agua Fría Formation of the Honduras Group (Figure 5.7). The Honduras Group also includes a sandstone-and-conglomerate unit that is commonly found below the Atima Limestone and is apparently above the Agua Fría (or El Plan) Formation. Honduran geologists (e.g., Jose María Gutierrez, Dirección General de Minas e Hidrocarburos) call this unit "the unnamed siliciclastic member of the Honduras Group" (Figure 5.7). This unit was formerly called the Todos Santos Formation, a name that should be abandoned (Gose and Finch 1987). A good type locality for this unit has not yet been designated, and no formal name has been proposed. Until a type locality is suggested, "unnamed siliciclastic member" is a temporarily appropriate term for the sandstones and conglomerates commonly found below the Atima limestone in central Honduras and in the Departmento de Olancho. These beds are lithologically quite distinct from the Agua Fría Formation and probably are younger. Finally, the Honduras Group also includes local formations that are found below the Atima Limestone, but that cannot be correlated with the above units (e.g., the "Undefined Jurassic" of San Juancito) (Carpenter 1954).

*Historical note regarding age*

As a result of silver- and gold-mining activity in the San Juancito region, the strata that we now call the Honduras Group were among the first beds to be studied paleontologically in Honduras. Nonetheless, the precise age of the beds has been a source of considerable revision and controversy over the past 100 years. New data collected during the past 10 years indicate that most of the strata exposed near San Juancito and equivalent beds elsewhere in Honduras are Middle Jurassic. However, the evolution of thought leading to this conclusion is of some interest. Furthermore, in the existing literature there is considerable confusion as to how the Jurassic age has become accepted. In the following discussion I shall attempt to make the record clear by using the original references.

Newberry (1888a,b) studied plant fossils from the San Juancito vicinity. Newberry (1888a) identified several species of cycads and gave their age as "Rhaetic or Upper Trias." In a later, more detailed report, Newberry (1888b) firmly decided on a Rhaetic age for these fossils and described them systematically. Another specimen from the same collection studied by Newberry was described by Humphreys (1916). He reiterated that the age of all the fossils was Rhaetic and did not indicate that there was any doubt about the age at that time. The samples were brought to Newberry by C. M. Rolker and T. H. Leggett. Newberry never visited the locality himself. Information regarding the location where the plant fossils were collected was given by Leggett (1889), along with a description of the geology of the site. Fritzgärtner (1891) also described the geology of the site. From the reports of the Rosario Resources Corp., R. C. Finch inferred that the plant fossils came "not from El Plan but from weakly metamorphosed clastic strata that directly overlie the basement schists in patches" (Finch 1981, 1322–3).

Knowlton (1918), in a review of the floras of the Americas, published information that questioned the age of these fossils. I could not find any evidence in his paper that Knowlton had ever

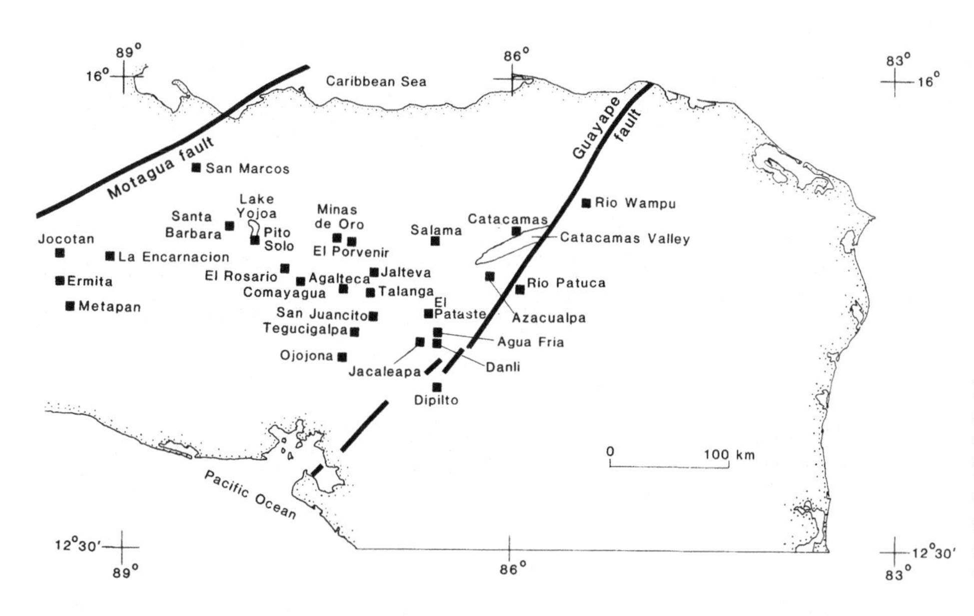


**Figure 5.8. Map showing locations cited in this report. Locations where Honduras Group strata have been described or mapped are also shown. Locations are geological sites that are not necessarily in the exact locations of the towns for which they are named.**

visited San Juancito or had seen the fossils. Instead, he cited the work of Sapper: "In a number of subsequent publications Carl [Karl] Sapper [Sapper 1896] has intimated that Newberry's age determinations should be accepted only with doubt; they are possibly Jurassic" (Knowlton 1918, p. 608). Sapper (1896) did not make a statement of that nature. However, Sapper apparently did doubt that the age of the floras was Rhaetic. Sapper (1937, p. 27) said that "Newberry bestimmte diese als rhätisch, während sie nach Böse (briefl. Mitt., 1926) nach neueren in Oaxaca, Veracruz und Sonora gewonnenen Erkenntnissen eher dem Lias angehören dürften." Thus, Sapper and Böse believed that San Juancito beds were likely Jurassic. Sapper (1937, p. 28) also reported the discovery by Fritzgärtner of a possible Jurassic ammonite, *Amaltheus margaritatus*, at Ojojona. Unfortunately, the fossil was lost during the revolution of 1894, so the identification could not be confirmed. As with Knowlton, I could not find any evidence that Sapper or Böse had visited San Juancito or that the plant fossils had been reexamined. Böse (1905) noted that "las plantas son del Triásico Superior y la fauna [flora?] tiene gran semejanza á la de las capas rheticas de Sonora y á las de las capas limítrofes entre el Keuper y Liásico de Franconia, en Baviera" (Böse 1905, p. 24). (Except for the comment in brackets, this quotation is precise.) Because the beds in Germany lie between the Keuper and the Liassic, the plant fossils could be either latest Triassic or earliest Jurassic. It should be noted that papers that have recently cited Knowlton (1918) misquote him. For example, Mills et al. (1967, p. 1718) stated that "Knowlton (1918), was able to show that the black shale sequence near Rosario [San Juancito] was of Triassic or Jurassic age," and Finch (1981, p. 1322) stated that "Knowlton (1918) suggested that a Jurassic age is equally likely." Considering the foregoing discussion, it is best to cite Sapper and Böse for recognizing the Jurassic age. No original, unambiguous data on the age appear to have been published between 1888 and 1952.

Müllerried (1942) reviewed the early work on the plant fossils. He definitely visited numerous sites in Central America and collected new specimens. From this work and his survey of literature, Müllerried (1942, p. 474) considered the plant fossils to be of "Lower and Middle Jurassic age, or Middle Lias to Lower Dogger." Some of Müllerried's discussion of new Jurassic finds is based on questionable logic, which I discuss in more detail toward the end of this section.

Maldonado-Koerdell (1952) visited San Juancito and collected a previously unknown form in the flora, *Yuccites* cf. *schimperianus* Zigno. He considered the age of the strata to be Rhaetic–Liassic. He also reported that Triassic cephalopods had been discovered at San Juancito, and he cited a letter from Lang (1917) to Dr. Luis Landa. I have not yet obtained a copy of this letter to confirm this source. According to Maldonado-Koerdell (1952, p. 294), the cephalopod is a Triassic *Tropites*. However, no photograph of it has been published in a readily available source, nor has the fossil been studied readily.

From the early literature it is not clear whether or not the plant-bearing strata are entirely restricted to the Jurassic. However, there is an indication that several early workers (Sapper, Böse, and Müllerried) did not believe the original Triassic age assigned to the fossils by Newberry. I shall argue later that most of these strata probably are restricted to the Jurassic, based on the recent literature regarding plant fossils at a new locality, and based on new faunal discoveries in several other localities. However, the foregoing survey of the literature shows that the plant fossils studied by Newberry and Humphreys are more likely Lower Jurassic than Middle Jurassic. Furthermore, Newberry's study and the possible occurrence of the *Tropites* favor a Rhaetic age. If Newberry's original date proves to be valid, it applies only to the limited, perhaps anomalous, strata sampled by Rolker and Leggett. Thus, it is possible that the specific beds sampled at San Juancito by Rolker and Leggett are significantly older than what has been sampled in the past 20 years. Recent data collected from other localities in the vicinity of San Juancito and localities throughout east-central Honduras indicate that most of the Agua Fría Formation is Middle Jurassic.

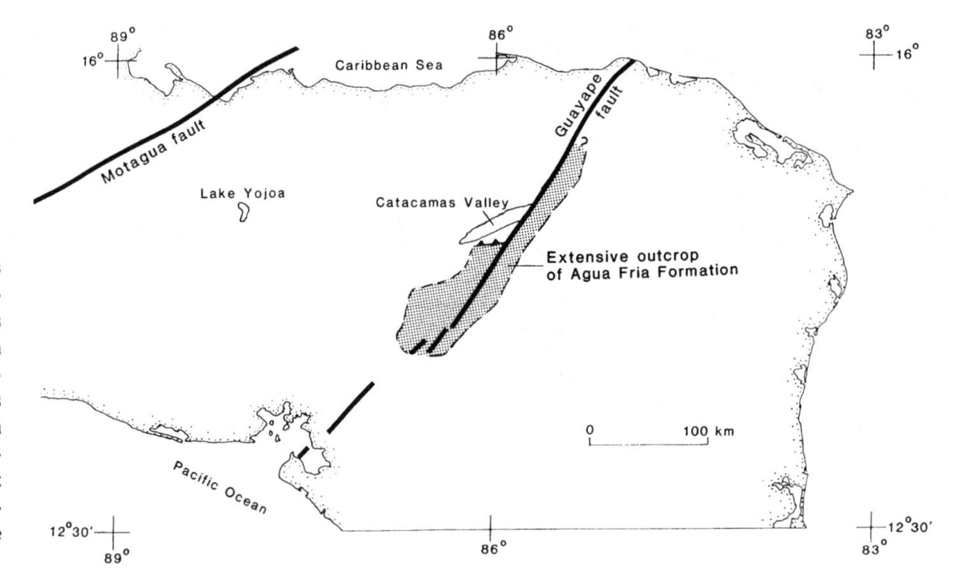


**Figure 5.9. Region of Honduras dominated by the Agua Fría Formation. The stippled pattern shows the region where the Agua Fría Formation is virtually the only Mesozoic or older formation that is found. The Agua Fría Formation has a broader distribution (see Figure 5.8), but it is not as thick at San Juancito, Minas de Oro, Jalteva, and other places as within the stippled area.**

## Agua Fría Formation

### *Distribution*

The Agua Fría Formation was first described in the Danli region (Ritchie and Finch 1985). It has been correlated with the El Plan Formation in the San Juancito district. Very similar beds have been mapped in a reconnaissance fashion between Danlí and Catacamas. The region where the Agua Fría Formation is the dominant exposed rock unit is shown in Figure 5.9. The absence of other rock units, the dips of the beds, and the topographic relief suggest that the Agua Fría Formation is very thick throughout this region. If the formation were thin, metamorphic basement would crop out extensively. The Agua Fría Formation can be mapped along the east side of Ríos Tinto, Guayape, and Guayambre. These rivers follow the Guayape fault, a major strike-slip fault in eastern Honduras (Figure 5.9). I have examined a thick section of the Agua Fría Formation along the Río Patuca on the east side of the Guayape fault. Mills et al. (1967) recognized similar strata, which they called "El Plan Formation," along the course of Río Wampú northeast of Catacamas. West of the Guayape fault, the distribution of the Agua Fría Formation is more limited. Its northward extent is abruptly terminated by a major thrust fault south of the Catacamas Valley (Kozuch 1989b). South of this thrust fault, which places basement rocks on top of the Agua Fría Formation, the formation crops out continuously on both sides of the Guayape fault until it is covered by Tertiary volcanics south of Danli. The extent of the Agua Fría Formation is much better known now than even 10 years ago because of quadrangle mapping in the Catacamas (Kozuch 1989b; Gordon, in press) and Danlí (Finch and Ritchie 1990) regions. Kozuch and I have also done reconnaissance studies along the Guayape fault confirming the regional extent of the Agua Fría Formation.

Similar strata have been reported in the literature from other locations. Strata at Jalteva (Figure 5.8) mapped by King (1973) contained plant fossils that were studied by Delevoryas and Srivastava (1981). These strata probably are equivalent to the Agua Fría Formation. King (1972) called the strata in which plant fossils were found the Todos Santos Formation (which we now refer to as the unnamed siliciclastic member of the Honduras Group). However, he described the strata as dark brown and gray-black siltstone and shale. That description is more reminiscent of the Agua Fría Formation than of the unnamed siliciclastic member. King (1972) also recognized marine fossils at one locality establishing a marine origin for at least part of the beds in the Jalteva region. The presence of marine strata, King's description of the unit, and the Middle Jurassic age strongly suggest that the plant-bearing strata at Jalteva are in the Agua Fría Formation.

Atwood (1972) described plant fossils in similar beds in the Minas de Oro region (Figure 5.8). Unfortunately, these fossils have not been studied. Atwood (1972, p. 25) called these beds "Todos Santos black shales" and said that they were similar to the El Plan Formation. They are mapped as "Todos Santos Formation" (Atwood et al. 1976). These beds could also be called Agua Fría Formation, but the correlation is uncertain until the formation is dated.

The El Plan Formation was reported by Mills et al. (1967) for strata exposed south of Salamá (Figure 5.8). The sub–Atima Limestone strata are complexly deformed in this area. Furthermore, no biostratigraphic data are available from this locality. Thus, it seems most appropriate to label these beds as Honduras Group until more data become available.

Low-grade metamorphic rocks were reported in western Honduras at La Encarnación (Figure 5.8) (Anonymous 1972). Those authors tentatively correlated these beds with the "El Plan Formation." However, they did not find any fossils with which to make the correlation more certain.

The Agua Fría Formation and equivalent Jurassic formations seem to be more common in east-central Honduras than in other areas of the Chortís block. Similar strata have not been reported in the Comayagua region, most of west-central Honduras, or southeastern Guatemala. Apparently, the known extent of the main

Middle Jurassic basin was limited to the Olancho, Francisco Morazán, and El Paraíso Departments. Future studies are likely to expand our knowledge of the basin's extent, because large areas of eastern Honduras and northern Nicaragua have not even been studied in a reconnaissance fashion.

*Lithology*

Carpenter (1954) described the El Plan Formation as composed mostly of thin-bedded dark gray shale. It has alternating beds at least 100 ft (30 m) thick that consist of (1) banded dark gray shale and siltstone beds 0.25–2 in. (0.5–5 cm) thick, with interbedded sandstone, and (2) gray, poorly bedded sandstone 1–3 ft (0.3–1 m) thick. Gallo and Van Wagoner (1978) described the El Plan Formation from two localities near San Juancito and divided it into upper and lower El Plan Formation. Their lower El Plan Formation consists of 200 m of interbedded sandstone and shale that they believe directly overlies basement schists. Their upper El Plan Formation consists of sandstone and interbedded shale. It is more quartzose, has fossiliferous layers, and includes oyster mounds.

Roberts and Irving (1957) briefly described the geology of the Agua Fría gold mine. They called the rocks alternating thin-bedded shales and gray quartzites. They estimated a thickness of at least 4,000 ft (1,200 m). Finch and Ritchie (1990) show a thick measured section of the Honduras Group along Agua Fría Creek. They also show six fossil localities within the Honduras Group in the Danlí region. Kozuch (1989a) suggests that the Agua Fría Formation could be as thick as 3,000 m in the Catacamas Valley, but does not present a measured section because the geology of his area is complicated.

Ritchie and Finch (1985) described the rocks of the Honduras Group as dark-colored shale, siltstone, immature sandstone, and quartz-pebble conglomerate. In the course of my mapping, I have found that the Agua Fría Formation consists of thin-bedded, dominantly gray shale and sandstone (turbidites?), whereas the unnamed siliciclastic member is dominantly orange, coarse-grained sandstone and conglomerate. The Agua Fría has a much higher percentage of shale, whereas the unnamed siliciclastic member has a much higher percentage of quartz-pebble conglomerate.

Bohnenberger and Dengo (1978) concluded that Honduran coal beds were part of the "El Plan Formation," based on gross lithologic character. Coal beds have been found at El Pataste and in the Azacualpa region (Figure 5.8) within the Agua Fría Formation, thus confirming the work of Bohnenberger and Dengo (1978). Kozuch (1989a) found plant fossils in the vicinity of some coal beds.

Much of the Agua Friá Formation displays a metamorphic sheen, as does the El Plan Formation at San Juancito (Carpenter 1954). Metamorphosed Agua Fría is very common in the Catacamas Valley region (Kozuch 1989a,b; Gordon in press). Carpenter (1954) suggested that the basement rocks could be metamorphosed El Plan Formation, but said that it was not likely, based on the field relationships. Recent workers (e.g., Gordon and Kozuch) have noted that metamorphosed Agua Fría Formation is commonly difficult to distinguish from the low-grade metamorphic rocks of the basement. This similarity could be due to equivalent bulk composition and does not necessarily imply that they are of the same age or metamorphic grade.

*Paleontology and age*

In the past 10 years, new data have become available on the age of Honduras Group sedimentary rocks, including the Agua Fría Formation. These data indicate that the formation is Middle Jurassic, not Early Jurassic as Sapper, Böse, and Müllerried suggested for the original plant fossils studied by Newberry. Other strata that are included in the Honduras Group vary in age from Middle Jurassic to Early Cretaceous.

Delevoryas and Srivastava (1981) studied plant fossils at a new locality near Jalteva, Honduras (Figure 5.8). They gave a systematic description of four divisions of plants collected at that locality. They relied on plant impressions, inasmuch as they were unable to recover any plant fragments or find any palynomorphs. No other paleontological data are available from that locality. They assigned the strata tentatively to the Middle Jurassic.

Ritchie and Finch (1985) reported the discovery of a Bajocian ammonite, *Stephanoceras*, at San Juancito. The ammonite is from strata in the San Juancito region that are correlated with Carpenter's El Plan Formation, but not from the exact locality where plant fossils were collected by Rolker and Leggett. Ritchie and Finch (1985) also stated that new plant fossils from San Juancito and El Pataste (Figure 5.8) have been positively identified as Jurassic and that fragmentary plant remains from Pozo Bendito (just east of Danlí) are thought to be Jurassic. The strata that yielded the ammonite are the oldest, positively dated strata within the Agua Fría (or El Plan) Formation, whereas the age of the plant fossils is less precise.

In 1987, ammonites were collected from the Agua Fría Formation in the Danlí region (location: UTM coordinates 34.6/50.5, Danlí quadrangle, hoja 2858-II, near Jacaleapa; see Figure 5.8) and were sent to Keith Young (The University of Texas at Austin) for analysis. Young identified the following fauna: a Middle Jurassic macrocephalitid, undetermined, and a Middle Jurassic stephanoceratid or pseudoperisphinctid, undetermined. Young will publish specific information on these fauna elsewhere.

Although a broader range of ages may be discovered in the course of future work on the Agua Fría Formation, the data presently available suggest that most, if not all, of the dated Agua Fría strata are of Middle Jurassic age.

*Environment of deposition*

A successful depositional model for the Agua Fría Formation needs to take into account all of the observations, which in this case are diverse. Specifically, the formation contains plant fossils, coal beds, marl beds, and marine invertebrates. Clearly, part of the formation represents nonmarine deposits. The formation probably is dominantly marine. Carpenter (1954) proposed that the El Plan Formation is a floodplain to shallow-marine deposit. Gallo and Van Wagoner (1978) called the lower El Plan

Formation a deep-marine, outer-shelf deposit, and the upper El Plan Formation progradational shelfal sands. I believe that some of the beds are turbidites, as was proposed by Gallo and Van Wagoner. However, the part of the section with coal beds must be nonmarine. Detailed sedimentological facies mapping has not yet been done in Honduras.

### Other Jurassic strata on the Chortís block?

A few dates have been obtained on Honduras Group strata that are not included in the Agua Fría (El Plan) Formation. Carpenter (1954) mapped some beds that he called "Undefined Jurassic." These beds contain numerous *Trigonia* cf. *quadrangularis, Gervillia,* and *Meretrix*(?) fossils of Jurassic age. Carpenter did not believe that he could correlate these beds with the El Plan Formation. By definition (Finch 1985; Ritchie and Finch 1985), these beds should just be included in the Honduras Group. Palynomorphs from the unnamed siliciclastic member are of Early Cretaceous (pre-Albian) age at Pito Solo, near Lake Yojoa (Gose and Finch 1987). Likewise, Emmet (1983) reported Early Cretaceous palynomorphs from the unnamed siliciclastic member. Though part of the unnamed siliciclastic member may, in fact, be Jurassic, it should not be called Jurassic until it can be dated.

Three other sites with presumed Jurassic plant fossils have been reported in the literature. However, those Jurassic ages have not been confirmed. Müllerried (1939, 1942) reported Jurassic plant remains at Jocotán, Guatemala, and Catacamas, Honduras (Figure 5.8). At Jocotán, Müllerried (1939, p. 42) reported coal, "vegetales fósiles," and cycads. However, Crane's mapping (1965) showed that the Miocene Padre Miguel Group crops out in this area. Thus, concerning Müllerried's report, Crane (1965, p. 44) said that "in view of the structural and stratigraphic relationships of the area and the poor preservation of plant remains in the lignite, Mullerried's fossils were probably not Jurassic cycads." In fact, the nearest exposure of what we now call the Honduras Group is at Ermita (Figure 5.8), which is 30 km south of Jocotán (Burkart, Clemons, and Crane 1973). At Catacamas, Müllerried (1939, p. 47) did not give any specific detail on the "madera fósil," and thus the age of the strata with fossil wood remains unclear. I have not encountered fossil wood in sub-Atima strata while mapping the Catacamas region (Gordon, in press). From Müllerried's (1939) description, I believe that the strata to which he refers belong to the unnamed siliciclastic member of the Honduras Group. If so, these strata could be Jurassic or Lower Cretaceous, but they remain undated. C. H. Wegemann, quoted by Schuchert (1935, p. 362), found "cycads" in chert 10–15 miles (16–24 km) east of Santo Tomás in Nicaragua. This town is described as being on the edge of the "great jungle," and thus I suspect, but do not know, that Wegemann was referring to the Santo Tomás at 12.04°N, 85.02°W, which is south of the Chortís block, *sensu stricto*. Those "cycads" were lost. Wegemann assumed that they were of Jurassic age, but that age was never confirmed. He described shale, sandstones, and volcanic ash at the same locality. McBirney and Williams (1965) mapped Tertiary volcanics in this region, but did not report any Mesozoic strata. Thus, this location could also be Tertiary, as the Jocotán site has proved to be. The Jurassic age at Santo Tomás was never proved and has not been confirmed.

Simonson (1977) reported carbonized plant remains in the basal siliciclastic beds of the El Porvenir region (Figure 5.8). He mapped these (Simonson 1981) as Jurassic, but the beds have not been dated. This unit could be equivalent to the Agua Fría Formation, but his description is very unlike that of the Agua Fría Formation. Leaving the unit unnamed, as Simonson did, is certainly wise. Simonson was the first worker to abandon the usage of "Todos Santos Formation" in Honduras since its adoption by Mills et al. (1967). At present, the only age constraint for the basal siliciclastic unit at El Porvenir is its stratigraphic position, which Simonson claimed to be beneath the Atima Limestone. In other words, its age is pre-Aptian, not necessarily Jurassic. From his description (Simonson 1977), some of his basal siliciclastic unit should be mapped separately, because part of it contains limestone cobbles, implying that at least part of the mapped unit overlies the Atima Limestone.

Clastic rocks have been mapped below the Atima Limestone in many localities on the Chortís block (Mills et al. 1967; Atwood et al. 1976; Fakundiny and Everett 1976; Simonson 1981). These rocks are commonly called Jurassic, because the Atima Limestone is of Early Cretaceous age. However, the latter is mostly restricted to the Aptian and the Albian. Thus, there was sufficient time to deposit the relatively thin siliciclastic member during the Early Cretaceous prior to the Aptian. These strata are appropriately dated as Jurassic to Aptian or Jurassic to Albian by some workers (Burkart 1965; Weber 1979; Curran 1981; Emmet and Logan 1983; Finch 1985).

### Jurassic igneous rocks

In most places where it has been studied, the Agua Fría Formation has a very low (negligible) volcanogenic component. However, east of the Guayape fault, a thick volcanic accumulation has been found above the sandstones and shales of the Agua Fría Formation. Mills et al (1967) were the first to report these volcanics. Along the course of Río Wampú (Figure 5.8) they measured 500 ft (150 m) of dark red, andesitic volcanic flows above the dark, very fine grained sandstones and shales of the Agua Fría Formation. Along the course of Río Patuca, just east of the Guayape fault, I have found a similar thickness of volcanics of intermediate composition above sandstones and shales of the Agua Fría Formation. The volcanics are tuffs that rest conformably on top of the sedimentary rocks. Exposure along Rio Patuca is excellent, and the relationship is unambiguous. I have also mapped (Gordon, in press) volcanics within the Agua Fría Formation in the Catacamas Valley region (Figure 5.8). These volcanics also crop out only on the east side of the Guayape fault. These volcanics have experienced low-grade metamorphism. I have found both pyroclastic rocks and lava flows, but all of the volcanics are either andesites or dacites. The Catacamas region is more complicated than the Patuca region. However, the volcanics are certainly intercalated with the sedimentary rocks. These Agua Fría volcanics apparently are limited to the east side of the Guayape fault. Much future work needs to be done on their distribution, composition, and age.

Although only a few intrusions have been well dated isotopically in Honduras, some of these are of Jurassic age. A small (5 $km^2$) pluton is exposed near San Marcos, Honduras, and intrudes the metamorphic basement (Horne, Clark, and Pushkar 1976). Its composition varies from quartz monzonite to granodiorite. With the Rb-Sr data of Horne et al. (1976) and the new IUGS decay constant ($1.42 \times 10^{-11}$ $y^{-1}$), L. E. Long (personal communication) obtained an age of $149 \pm 7$ Ma for the San Marcos pluton. The Dipilto Batholith crops out over several thousand square kilometers near the Honduras–Nicaragua border (Horne and Clark 1978). It also intrudes the basement rocks. Samples that Horne and Clark used for dating range in composition from quartz monzonite to biotite-rich tonalite. L. E. Long (personal communication) recalculated the Rb-Sr age of this pluton as $138 \pm 7$ Ma (approximately at the Jurassic–Cretaceous boundary) using the new decay constant and the data of Horne and Clark. As more samples are isotopically dated, the number of known Jurassic intrusions probably will increase.

### Conclusions

Jurassic strata crop out in a broad region of the Chortís block. Some are well dated biostratigraphically, but much of this information is the result of reconnaissance work. The Agua Fría Formation of the Honduras Group is the most widespread unit. Recent dates for the Agua Fría Formation are restricted to the Middle Jurassic. Older dates have been reported in the literature. These Late Triassic dates have not been reproduced during recent, better-documented work. Other Jurassic strata are included in the Honduras Group. These are local units that have not been correlated with the Agua Fría Formation. The Honduras Group extends into the Lower Cretaceous. The unnamed siliciclastic member of the Honduras Group may not be much older than the Aptian–Albian Atima Limestone.

Bringing the rich flora of Honduras to the attention of geologists was the object of Newberry's paper (1888b). Unfortunately, a 93-year gap followed before the next systematic study of plant fossils in Honduras (Delevoryas and Srivastava 1981). Let us hope that much more data will shortly become available for these strata and that the time gap between major publications will be much less in the future than it has been in the past.

### Acknowledgments

K. Young of The University of Texas at Austin described the new ammonites discussed in this section. His comments on the paleontology of Honduras were very helpful. Dietmar Müller helped me by translating text from German. K. Young, A. Salvador, W. R. Muehlberger, and G. E. G. Westermann provided useful reviews that helped me clarify the text. Acknowledgment is made to the Donors of The Petroleum Research Fund, administered by the American Chemical Society, for the support of this research through grant 19990-AC2 to W. R. Muehlberger. The Instituto Geográfico Nacional and Dirección General de Minas e Hidrocarburos, of Honduras, and the Geology Foundation of the University of Texas at Austin have also provided funding for my research.

## SOUTHERN CENTRAL AMERICA[4]

### Introduction

Southern Central America includes southern Nicaragua, Costa Rica, and Panama. It forms the morphotectonic connection between the North and South American continents. The adjacent continental areas of northern Central America and South America have a typically "continental" basement, whereas southern Central America represents a relatively young continentalized island arc with a Jurassic to Paleogene ophiolitic basement. The basement crops out discontinuously mainly along the Pacific coast, from northwestern Costa Rica to eastern Panama (Figure 5.10). Rock units of proven or inferred Jurassic age are restricted to a few ophiolite areas in northwestern Costa Rica: the Nicoya and Santa Elena peninsulas (Figure 5.11). Jurassic rocks may be present in the Basic Igneous Complex (R. Schmidt-Effing personal communication 1989) that forms the continuation of the southern Central American ophiolites in northwestern Colombia and western Ecuador.

### Stratigraphy of northwestern Costa Rica

The ophiolite basement is subdivided into several main units that can be briefly characterized as follows. The petrologically and geotectonically lowermost ophiolite level is exposed in the Santa Elena peninsula (Figure 5.11). It is an ultramafic complex composed mainly of serpentinized peridotites and harzburgites, with some mafic intrusions. The Santa Elena Complex is interpreted as a fragment of the upper mantle that was strongly uplifted and locally thrusted upon other ophiolite units (Tournon and Azema 1980; Bourgois et al. 1984; Wildberg 1984; Meschede, Frisch, and Sick 1988). Its age and stratigraphic relations to the other ophiolite units are unknown.

The mafic igneous units of the ophiolite basement, including associated sedimentary rocks, were defined and summarized under the term "Nicoya Complex" by Dengo (1962). The stratigraphically lowermost unit is the Lower Nicoya Complex, (*sensu* Gursky et al. (1982), hitherto identified only in northwestern Costa Rica. It is overlain by radiolarites, followed by the Upper Nicoya Complex.

1. The Lower Nicoya Complex consists of complicatedly associated basalts, gabbros, plagiogranites, basalt breccias, and local fine-grained volcaniclastic lenses. It probably represents oceanic crust formed in a spreading ridge. The exact age has not been established, but Jurassic or older is very probable. This is supported by the overlying Jurassic radiolarites in several places. However, sedimentary contacts have not yet been observed.

[4] By H.-J. Gursky.

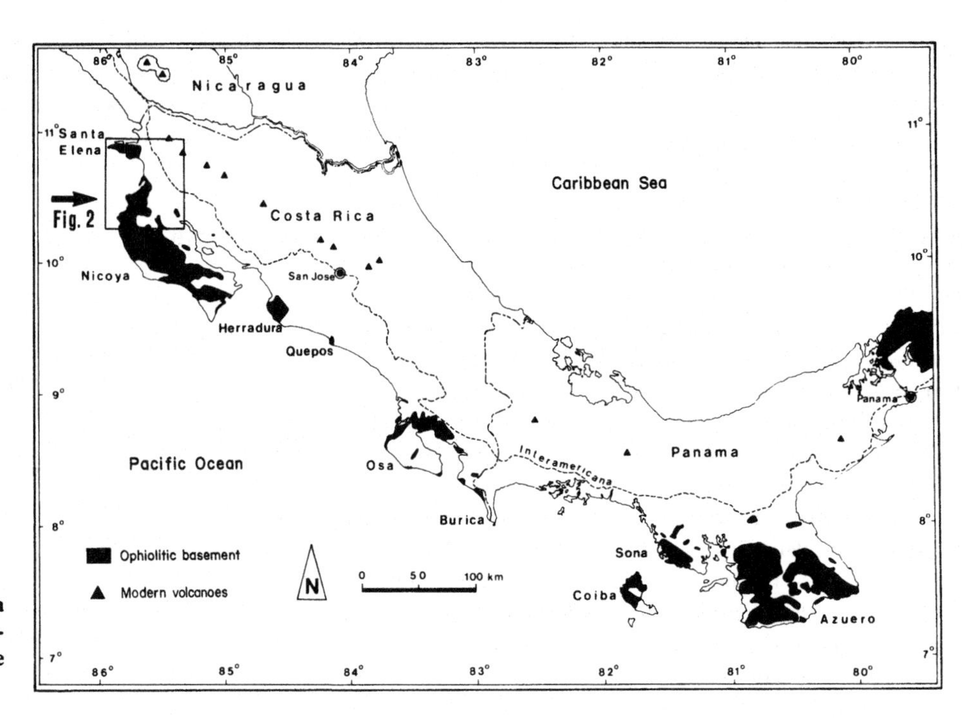


**Figure 5.10. Sketch map of Costa Rica and western Panama indicating outcrop areas of the ophiolite basement.**

2. Radiolarite sequences overlying the Lower Nicoya Complex range from Lower(?) Jurassic to Santonian (Schmidt-Effing 1979; Baumgartner 1984, 1987) and were named "Puncha Conchal Formation" by Gursky and Schmidt-Effing (1983). The radiolarite sequences are overlain and partly intruded by basalts of the Upper Nicoya Complex. The radiolarites include local horizons of sedimentary deep-sea manganese nodules (Gursky 1988).

Jurassic ages based on radiolarian faunas were determined in some of the radiolarite sequences in the peninsulas of Santa Elena and Nicoya (Pessagno, cited by Galli 1977; Schmidt-Effing 1979; Baumgartner 1984, 1987; De Wever et al. 1985). The first evidence of a pre-Campanian age in southern Central America was given by Pessagno (Galli 1977), based on a radiolarian fauna from a chert from the Brasilito coastal area in northwestern Nicoya: *Parvicingula* sp., *Pantanellium riedeli* Pessagno, *Pantanellium* sp. indet, *Archaeodictyomitra* sp., "*Pseudodictyomitra* sp." Pessagno dated this fauna as middle Tithonian–Valanginian, probably late Tithonian–Berriasian. Schmidt-Effing (1979) analyzed radiolarites from Puncha Conchal (Brasilito area) and Playa Real in northwestern Nicoya and reported, among others, *Parvicingula* cf. *hsui* Pessagno, *Praeconocaryomma* cf. *magnimamma* (Rüst), *Lithocampe* cf. *mediodilatata* Rüst, *Sphaerostylus lanceola* (Parona), *Eucyrtidium ptyctum* Riedel and Sanfilippo. He placed this fauna in the *Sphaerostylus lanceola* Zone, ?Kimmeridgian/ Tithonian–Valanginian (?Lower Tithonian or Kimmeridgian). Baumgartner (1984, 1987) indicated Berriasian–Barremian/ Aptian for radiolarites from Puncha Conchal and Playa Real, mainly based on *Parvicingula cosmoconica, Alievum helenae, Cecrops septemporatus, Sethocapsa uterculus, Thanarla pulchra, Pseudodictyomitra leptoconica, Xitus alievi*. For radiolarian cherts from the Huacas/Cartagena area (interior of northwestern Nicoya), Baumgartner (1984) reported Bathonian/Callovian–Oxfordian, mainly based on *Unuma echinatus, Stichocapsa japonica, Tricolocapsa plicarum, Podobursa helvetica, Theocapsomma cordis, Mirifusus guadalupensis, Triactoma blakei*. Baumgartner (1987) later gave ages of Kimmeridgian / Tithonian for the Brasilito area and Bajocian/Bathonian for the Huacas / Cartagena area, without presenting further paleontological data.

De Wever et al. (1985) found Jurassic radiolarites in the central Santa Elena peninsula (Río Potrero Grande valley), with *Bernoullius dicera* Baumgartner and *Bernoullius* sp. *A* indicating Callovian/Oxfordian (?upper Callovian). From Playa Santa Rosa (from a radiolarite breccia), De Wever et al. (1985) identified ?*Gigi fustis* De Wever, *Gorgansium* sp., *Jacus anatiformis* De Wever, *Protopsium libidonosum* De Wever, *Saitoum keki* De Wever, which he tentatively dated as Liassic–lower Dogger. This remains the oldest sample dated in southern Central America.

3. In the Upper Cretaceous (Schmidt-Effing 1979), extensive basaltic magmatism resulted in the formation of the Upper Nicoya Complex (*sensu* Gursky et al. 1982), which in northwestern Costa Rica forms the youngest unit of the ophiolite basement. This complex is composed of at least two genetically different series: the Oceanic Series and the Primitive Island-Arc Series, mainly distinguished chemically (Wildberg 1984). Both series consist mainly of tholeiitic basalts and volcaniclastic breccias; fine-grained sedimentary rocks were only locally deposited.

### Igneous rocks

As mentioned earlier, a Jurassic age can be assumed only for the Lower Nicoya Complex. It is a complicated association of massive and pillow basalts, dolerites, mafic plutonites partly with cumulate structures, plagiogranites, local volcaniclastic breccias and lenses of fine-grained sedimentary rocks, and massive sulfide

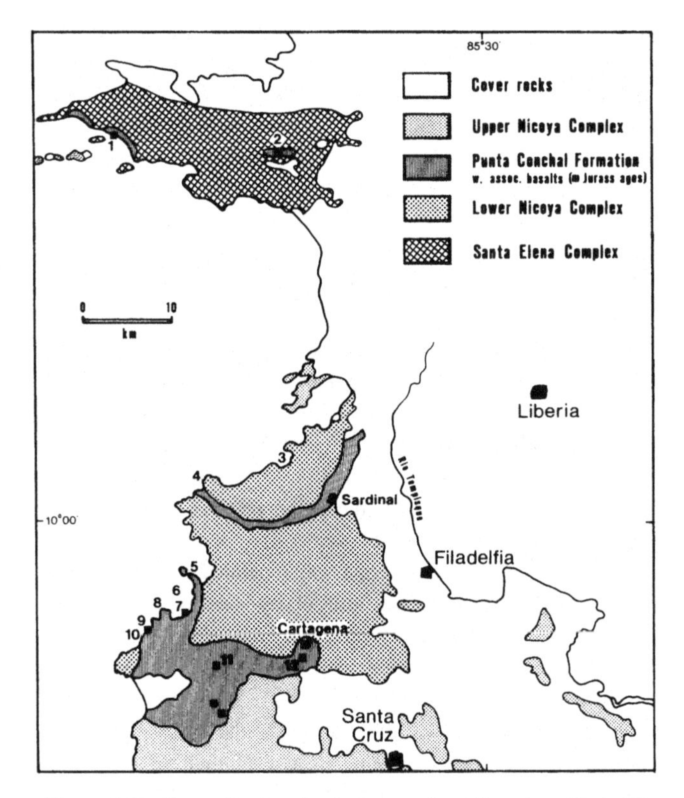


**Figure 5.11. Generalized geological map of northwestern Costa Rica (based on maps in Gursky et al. 1984). Localities with Jurassic ages are indicated according to the paleontologic references cited in the text: 1, Playa Santa Rosa; 2, Río Potrero Grande valley; 3, Playas del Coco; 4, Punta Gorda; 5, Punta Salinas; 6, Bahia de Brasilito; 7, Punta Conchal; 8, Punta Sabana; 9, Playa Real area; 10, Playa Pedregosa; 11, Haucas area; 12, Cartagena.**

mineralizations. The Lower Nicoya Complex represents the upper levels of typical oceanic crust (Wildberg 1984, 1987; Wildberg and Baumann 1987).

The *tholeiitic basalts* occur mostly as thick submarine lava flows and sills and are mainly composed of plagioclase (more than 50%; An. 55–60%) and clinopyroxenes; orthopyroxenes, olivine, and opaque phases are subordinate. Intergranular and intersertal textures are typical, and in coarse-grained varieties also porphyric, glomeroporphyric, ophitic, and subophitic textures. Variolitic textures are present in places. Most basalts show strong hydrothermal influence, with mineral alterations and the formation of zeolites, smectite, chlorite, and rare hornblende (Wildberg 1984; Gursky 1986).

The *gabbros and quartz-free diorites* occur as homogeneous, massive or stratified bodies with hundreds of meters of extension. Most contacts with adjacent rock units are tectonic; intrusive or gradational contacts are rare. The mafic plutonites are composed of 45–55% plagioclase (An. 25–80%), clinopyroxene, and subordinate orthopyroxenes, olivine, ilmenite, and magnetite. Textures are mostly nonoriented; layered textures with stratified enrichments of parallelly oriented plagioclase crystals occur in places.

The *plagiogranites* are relatively small bodies with outcrops of generally a few hundreds of square meters. The rocks are characteristically light-colored and coarse-grained. They consist of plagioclase (generally more than 50%; An. approximately 20%), quartz, augite, or hornblende. Apatite, magnetite, and ilmenite are subordinated. Hypidiomorphic granular textures are typical.

The plagiogranites are interpreted as acid residual melts formed by relatively continuous differentiation. Thus they form an integral part of the ophiolite suite (Wildberg 1987).

*Geochemical analyses* suggest that the Lower Nicoya Complex is an association of weakly to strongly fractionated magmatites. The rocks are genetically related by simple crystal fractionation or filter pressing (plagiogranites) (Wildberg 1987). Primitive magmas (i.e., direct mantle melting) do not occur.

The basalts are attributed to the MORB group (mid-ocean-ridge basalt) (Wildberg 1984). The gabbros are characterized by a clear differentiation trend from Mg-rich, primitive types to Fe-rich, developed magmas and finally to the plagiogranites. The generally K-poor gabbros are closely related to the also K-poor oceanic basalts. The plagiogranites are divided in two groups: high-level types that are closely associated to Fe-rich dolerites and occur in connection with gabbroic intrusions, and low-level types that occur at the petrologic base of the Lower Nicoya Complex within cumulate gabbros together with amphibolites, but without Fe-dolerites (Wildberg 1987). The cogenetic nature of basalts, gabbros, and high-level plagiogranites is, among others, shown by conformable $^{87}Sr/^{86}Sr$ ratios (Wildberg and Baumann 1987).

### Sedimentary rocks (radiolarites)

Fine-grained sedimentary rocks make up 1–2% by volume of the southern Central American ophiolite complexes. Among them, radiolarites are dominant and occur as bedded sequences, local lenses, xenoliths, tectonic blocks, and components of volcaniclastic units. Jurassic radiolarites are less abundant than Cretaceous ones. Because of polyphase tectonic deformations in the Upper Cretaceous and Tertiary (M. Gursky 1986, 1988), the radiolarites were repeatedly folded and faulted.

The radiolarite sequences are mostly underlain by postdepositional basalt sills intruded during the formation of the Upper Nicoya Complex (Gursky et al. 1982, 1984). Basalt flows or sills cover the radiolarites, wherever the top contacts are not faulted or eroded. Many radiolarite sections show additional intrusions of basalt dikes, some with primitive island-arc chemistry (Wildberg 1984). The sequences are up to 50 m thick. In most outcrops, however, much less is preserved because of postdepositional igneous, tectonic, and/or erosional processes. Monotonous, carbonate-free, bedded radiolarian cherts dominate, with average bed thicknesses between 1 and 10 cm. They rhythmically alternate with millimeter-fine layers of siliceous mudstone. Lithoclasts and intercalated volcaniclastic beds are very rare. Most radiolarite sections are made up of relatively pure, siliceous, biogenetic material and rare nonsiliceous components. The colors are mostly brownish red. Sections with strong thermal alteration (Gursky and Gursky 1988) are characterized by pale colors. In general, radiolarite sequences develop stratigraphically from volcaniclastically contaminated and hydrothermally influenced portions to the typical dominant radiolarian-rich cherts of the main portions of the sections (M. Gursky 1988).

Lithofacies variations are visibly documented by variable bed thicknesses, variable chert-mudstone ratios, and/or variable volcaniclastic or other admixtures or intercalations (e.g., Mn laminations, Mn nodules). Homogeneous, nonlaminated bedding and millimeter-thin parallel laminations are typical. Several types of micrograded bedding are observed. Bedding characteristics can be interpreted in terms of transport and redeposition of radiolarians and fine-grained detritus by extended low-energy currents. These currents probably were low-velocity, low-density turbidity currents or weak bottom currents (Gursky and Schmidt-Effing 1983). In the Jurassic radiolarites, radiolarians and rare sponge spicules are practically the only faunal elements. Slump folds, lenticular slide bodies, chaotic slump masses, and breccias occur in places.

Quartz/chalcedony is strongly dominant. Hematite, magnetite, plagioclase, goethite, chlorite, smectite, illite/smectite mixed-layer minerals, zeolites, and fine-grained igneous particles are accessories. The silica (88–95% in the chert beds) is nearly completely of organic origin (radiolarians, sponge spicules, and products of their diagenetic dissolution and reprecipitation). The hematite pigment (0.5–3% in chert) probably precipitated during diagenesis from the pore waters.

The iron possibly derived from unknown "hydrogenetic/hydrothermal" sources. The volcaniclastic fragments represent a selection of minerals from mafic igneous rocks (mostly plagioclase, pyroxene, ore minerals). This detritus is probably derived from the Nicoya Complex itself. Most radiolarite sections, independently of age, were formed under similar conditions, but regional and local factors influenced the thickness, kind, and distribution of the sediments. The nearly total absence of calcareous particles indicates deposition below the CCD (cf. Schmidt-Effing 1979; M. Gursky 1988). Conditions probably were oxic at the depositional surface of the radiolarites. This is shown by the high hematite contents and the intercalated manganese nodules. The sedimentation rate was very low, in relatively continuous facies development, for millions of years. This reflects a very stable, practically undisturbed, long-term, generally uniform environment of thousands of square kilometers. A prominent relief existed locally and was continuously blanketed by radiolarite sediments. However, there is no evidence, for example, of intensive volcanism during most of the time of formation of the radiolarite sequences. Present data suggest that the Nicoya radiolarites were formed relatively close to the equator at a moderate to large distance from land, in the eastern paleo-Pacific Ocean.

## Geological development

The history of the northwestern Costa Rican rock units started in the Early Jurassic in a paleo-Pacific spreading ridge that produced oceanic crust (MORB; the Lower Nicoya Complex). This crust was transported as a part of the Farallon plate to the east or northeast. The igneous rocks were hydrothermally altered by "ocean-ridge metamorphism" and "ocean-floor metamorphism." During the sea-floor spreading phase, the radiolarites were accumulated on the igneous rocks of the Lower Nicoya Complex. Vertical displacements and igneous relief probably were the causes of the syndiagenetic slumping structures in the radiolarites.

In the Upper Cretaceous, a part of this Pacific crust (= the Lower Nicoya Complex) had been transported so far between North and South America that it was tectonically influenced by the two plates: A convergence between North and South America caused the compression of the intercalated crust, resulting in a first folding of the Lower Nicoya Complex (deformation D1, post-Albian/pre-Campanian) (M. Gursky 1986, 1988). At the same time or shortly thereafter, extensions of the Caribbean "Great Flood-Basalt Event" (Donnelly 1975) reached the region of northwestern Costa Rica, and its basalts probably formed the Oceanic Series of the Upper Nicoya Complex. This resulted in contact metamorphism of parts of the radiolarite sequences under medium- or high-grade conditions.

The thickened Caribbean crust was separated from the paleo-Pacific (Farallon) plate by the onset of a subduction zone west of the Nicoya Complex in Late Cretaceous time. The Farallon plate was subducted, and its northeastern-bound movement caused the second tectonic folding (deformation D2). Approximately at the same time, an island arc started to form at the Caribbean side of this intra-oceanic plate margin because of melting processes. Primitive extensions of this island arc formed the Primitive Island-Arc Series of the Upper Nicoya Complex. In northwestern Costa Rica, the post-ophiolitic history started in the late Upper Cretaceous with the deposition of thick fore-arc sequences. Regional uplifts moved the ophiolite units into their actual morphotectonic positions.

## Acknowledgments

This research was supported by grants from the Deutsche Forschungsgemeinschaft and the Deutscher Akademischer Austauschdienst. I sincerely thank R. Schmidt-Effing, H. G. H. Wildberg, and M. M. Gursky for discussions, field trips, and technical help. The Geological Institutes at Münster, Marburg, and San Jose provided logistic and laboratory help. R. Schmidt-Effing made helpful comments on an early version of the manuscript.

## References

Alencaster de Cserna, G. (1961a). Paleontología del Triásico Superior de Sonora. Part 1: Estratigrafía del Triasico Superior de la parte central del Estado de Sonora. *Univ. Nac. Aut. México, Inst. Geol., Paleont. Mexic.*, no. 11.

(1961b). Paleontología del Triásico Superior de Sonora. Part 3: Fauna fósil de la Formación Santa Clara (Carnico) del Estado de Sonora. *Univ. Nac. Aut. México, Inst. Geol., Paleont. Mexic.*, no. 11.

(1963). Pelecipodos del Jurásico Medio del noroeste de Oaxaca y noreste de Guerrero. *Univ. Nac. Aut. México, Inst. Geol., Paleont. Mexic.*, no. 15.

(1984). Late Jurassic–Cretaceous molluscan paleogeography of the southern half of México. In G. E. G. Westermann (ed.), *Jurassic-*

*Cretaceous Biochronology and Biogeography of North America* (pp. 77–88). Geological Association of Canada, Special Paper 27.

Alencaster de Cserna, G., & Buitrón, B. E. (1965). Estratigrafía y paleontología del Jurásico Superior de la parte centroseptentrional de Estado de Puebla. Part II: Fauna del Jurásico Superior de la región de Petlalcingo, Estado de Puebla. *Paleontol. Mexic., 21,* 1–53.

Anderson, T. H., & Silver, L. T. (1969). Mesozoic magmatic events of the northern Sonora coastal region, Mexico. *Geol. Soc. Am. Abstr. Progr., 1969 Annual Meeting,* pp. 3–4.

(1978). Jurassic magmatism in Sonora, Mexico. *Geol. Soc. Am. Abstr. Progr., 10,* 359.

(1979). The role of the Mojave-Sonora megashear in the tectonic evolution of northern Sonora. In T. H. Anderson & J. Roldan-Quintana (eds.), *Geology of Northern Sonora* (pp. 59–68). Geological Society of America guidebook, Field Trip 27, 1979 Annual Meeting.

Anderson, T. H., Silver, L. T., Pearson, M., Baenteli, G., & Córdoba, D. A. (1972). Observationes geocronológicas sobre los complejos cristalinos de Sonora y Oaxaca, México. *Soc. Geol. Mexic., Mem. 2 Conv. Nat.,* 115–22.

Anonymous (1972). *Honduras Investigation of Mineral Resources in Selected Areas. The Regional Geology of N.W. Honduras.* New York: United Nations.

Araujo-Mendieta, J., & Estavillo-Gonzalez, C. F. (1987). Evolución tectónica sedimentaria del Jurásico superior y Cretácico inferior en el NE de Sonora, México. *Rev. Inst. Mexic. Petrol., 20*(3), 4–37.

Arkell, W. J. (1956). *Jurassic Geology of the World* (vol. 15, pp. 1–806). London: Oliver & Boyd.

Atwood, G., Cullen, J. L., Smith, C. H., & Simonson, B. M. (1976). *Mapa Geológico de Honduras, Minas de Oro Sheet.* Tegucigalpa, Honduras: Inst. Geogr. Nac.

Atwood, M. G. (1972). *Geology of the Minas de Oro Quadrangle, Honduras, Central America.* MA thesis, Wesleyan University, Middletown, Conn.

Barnes, D. A. (1984). Volcanic arc derived, Mesozoic sedimentary rocks, Vizcaino Peninsula, Baja California Sur, México. In V. A. Frizzell, Jr. (ed.), *Geology of the Baja California Peninsula* (vol. 39, pp. 119–80). Los Angeles: Pacific Section SEPM.

Baumgartner, P. O. (1984). El complejo ofiolítico de Nicoya (Costa Rica): modelos estructurales analizados en función de las edades de los radiolarios (Calloviense a Santoniense). In P. Sprechmann (ed.), *Manual de geología de Costa Rica. I. Estratigrafía* (pp. 115–23). San José: Edit. Univ. Costa Rica.

(1987). Tectónica y sedimentación del Cretácico superior en la zona pacífica de Costa Rica (América Central). *Actas Fac. Cienc. Tierra U.A.N.L. Linares, 2,* 251–60.

Beauvais, L., & Stump, T. E. (1976). Corals, mollusks and paleogeography of Late Jurassic strata at the Pozo Serna region, Sonora, Mexico. *Paleoclimat., Paleoecol., 19,* 275–301.

Berckhemer, F., & Hölder, H. (1959). Ammoniten aus dem Oberen Weissen Jura Suddeutschlands. *Beih. Geol. Jb. (Hannover), 35,* 1–35.

Biro-Bagoczky, L. (1984). *New contributions to the paleontology and stratigraphy of some Tithonian-Neocomian outcrops in the Chilean part of the Andean Range between 33°45' and 35° lat. S.* I.G.C.P. Project 171, Circum-Pacific Jurassic, Report 2, Special Paper 3.

Blake, M. C., Jr., Jayko, A. S., Moore, T. E., Chavez, V., Saleeby, J. B., & Seel, K. (1984). Tectonostratigraphic terranes of Magdalena Island, Baja California Sur. In V. A. Frizzel, Jr. (ed.), *Geology of the Baja California Peninsula* (vol. 39, pp. 183–91). Los Angeles: Pacific Section SEPM.

Bohnenberger, O. H., & Dengo, G. (1978). Coal resources in Central America. In F. E. Kottlowski, A. T. Cross & A. A. Meyerhoff (eds.), *Coal Resources of the Americas. Selected Papers* (pp. 65–72). Geological Society of America, Special Paper 179.

Boles, J. R. (1978). Basin analysis of the Eugenia Formation (Late Jurassic), Punta Eugenia area, Baja California. In D. G. Howell & K. A. McDougall (eds.), *Mesozoic Paleogeography of the Western United States* (pp. 493–8). Pacific Coast Paleogeography Symposium 2.

Boles, J. R., & Landis, C. A. (1984). Jurassic sedimentary mélange and associated facies, Baja California, Mexico. *Geol. Soc. Am. Bull., 95,* 513–21.

Böse, E. (1905). *Resena acerca de la geología de Chiapas y Tabasco.* Inst. Geol. Mex. Bol. 20.

Bourgois, J., Azema, J., Baumgartner, P. O., Tournon, J., Desmet, A., & Aubouin, J. (1984). The geologic history of the Caribbean–Cocos plate boundary with special reference to the Nicoya Ophiolite Complex (Costa Rica) and D.S.D.P. results (Legs 67 and 84 of Guatemala): a synthesis. *Tectonophysics, 108,* 1–32.

Buitrón, B. E. (1970). Equinoides del Jurásico Superior y del Cretácico Inferior de Tlaxiaco, Oaxaca. In L. R. Segura & R. Rodriguez-Torres (eds.), *Soc. Geol. Mexicana, Libro Guía Excursión México-Oaxaca* (pp. 154–63).

(1984). Late Jurassic bivalves and gastropods from northern Zacatecas, Mexico, and their biogeographic significance. In G. E. G. Westermann (ed.), *Jurassic-Cretaceous Biochronology and Biogeography of North America* (pp 89–98). Geological Association of Canada, Paper 27.

Burckhardt, C. (1906). *Le faune jurassique de Mazapil.* Bol. Inst. Geol. Mexico 23.

(1927). *Cefalopodos del Jurásico Medio de Oaxaca y Guerrero.* Bol. Inst. Geol. Mexico 47.

(1930). Etude synthetique sur le Mexooique Mexicain, Premiere partie. *Mem. Soc. Paleont. Suisse, 49,* 1–123.

Burkart, B. (1965) *Geology of the Esquipulas Chanmagua and Cerro Montecristo Quadrangles, Southeastern Guatemala.* Ph.D. dissertation, Rice University.

Burkart, B., Clemons, R. E., & Crane, D. C. (1973). Mesozoic and Cenozoic stratigraphy of southeastern Guatemala. *Am. Assoc. Petrol. Geol. Bull., 57,* 63–73.

Burke, K. (1988). Tectonic evolution of the Caribbean. *Ann. Rev. Earth Planet. Sci., 16,* 201–30.

Busby-Spera, C. J. (1988). Evolution of a Middle Jurassic back-arc basin, Cedros Island, Baja California, evidence from a marine volcaniclastic apron. *Geol. Soc. Am. Bull., 100,* 208–33.

Campa, M. F., Campos, F., Flores, R., & Ovidado, R. (1974). La secuencia mesozoica volcánico-sedimentaria metamorfizada de Ixtapán del la Sal, Mex. – Teloloapán, Gro. *Bol. Soc. Geol. Mexic., 35,* 7–28.

Campa, M. F., & Coney, P. J. (1983). Tectono-stratigraphic terranes and mineral resource distributions in Mexico. *Can. J. Earth Sci., 20,* 1040–51.

Cantu Chapa, A. (1967). El Limite Jurásico-Cretácico en Mazatepec, Puebla (Mexico). Instituto Mexicano del Petroleo, Sección Geología, Monografía, *1,* 3–24.

(1968). Sobre una asociación *Proniceras Durangites–"Hildoglochiceras"* del Noreste de México. *Rev. Inst. Mexic. Petrol., Sec. Geologia, 2,* 19–26.

(1969). Estratigrafía del Jurásico Medio-Superior del Subsuelo de Poza Rica, Ver. (Area de Soledad-Miquetla). *Rev. Inst. Mexic. Petrol., 1*(1), 3–9.

(1970). El Kimeridgiano Inferior de Samalayuca, Chih. *Rev. Inst. Mexic. Petrol., 2*(3), 40–4.

(1971). La Serie Huasteca (Jurásico Medio-Superior) del Centro Este de México. *Rev. Inst. Mexic. Petrol., 3*(2), 17–49.

(1976a). Nuevas localidades del Kimmeridgiano y Tithoniano en Chihuahua (Norte de México). *Rev. Inst. Mexic. Petrol., 8,* 38–45.

(1976b). El contacto Jurásico-Cretácico, la estratigrafía del Neocomiano, el hiato hauteriviano superior-Eoceno inferior y las amonitas del Pozo Bejco 6 (centro-este de México). *Bol. Soc. Geol. Mexic., 37,* 60–83.

(1977). Las Amonitas del Jurásico Superior del Pozo Chac 1, Norte de Campeche (Golfo de México). *Rev. Inst. Mexic. Petrol.*, *9*(2), 38–9.

(1979). Bioestratigrafía del la Serie Huasteca (Jurásico Medio y Superior) en el Subsuelo de Poza Rica, Veracruz. *Rev. Inst. Mexic. Petrol.*, *11*(2), 14–24.

(1984). El Jurásico Superior de Taman, San Luis Potosí. Este de México. In *Memoirs, III Congress of Latin American Paleontology* (pp. 207–15).

Carpenter, R. H. (1954). Geology and ore deposits of the Rosario Mining District and the San Juancito Mountains, Honduras, Central America. *Geol. Soc. Am. Bull.*, *65*, 23–38.

Coney, P. J., Jones, D. L., & Monger, H. W. H. (1980). Cordilleran suspect terranes. *Nature*, *288*, 329–33.

Contreras, M. B., Martinez, C. A., & Goṁez, L. M. E. (1988). Bioestratigrafía y Sedimentología del Jurásico superior en San Pedro del Gallo, Durango, México. *Rev. Inst. Mexic. Petrol.*, *20*(3), 5–49.

Cowie, J. W., and Bassett, M. G. (1989). IUGS 1989 Global Stratigraphic Chart. *Episodes*, *12*(2).

Crane, D. C. (1965). *Geology of the Jocotan and Timushan Quadrangles, Southeastern Guatemala.* PhD dissertation, Rice University.

Cregg, A. K., & Ahr, W. M. (1984). Paleoenvironment of an Upper Cotton Valley (Knowles Limestones) Patch Reef, Milam County, Texas. In W. P. S. Ventress, D. G. Bebout, B. F. Perkins & C. H. Moore (eds.), *The Jurassic of the Gulf Rim* (pp. 41–56). GCSSEPM Foundation, 3rd Ann. Res. Conf. Proc.

Crevello, P. D., & Harris, P. M. (1984). Depositional models for Jurassic reefal buildups. In W. P. S. Ventress, D. G. Bebout, B. F. Perkins & C. H. Moore (eds.), *The Jurassic of the Gulf Rim* (pp. 57–102). GCSSEPM Foundation, 3rd Ann. Res. Conf. Proc.

Curran, D. (1981). *Mapa Geológico de Honduras, Taulabe Sheet.* Inst. Geogr. Nac., Tegucigalpa, Honduras.

Davila-Alcocer, V. M., & Pessagno, E. A. (1986). Bioestratigrafía basada en radiolarios del Triásico en el noreste de la península de Vizcaino, Baja California Sur. *Univ. Nac. Aut. México, Inst. Geol. Rev.*, *6*, 136–44.

Delevoryas, T., & Srivastava, S. C. (1981). Jurassic plants from the Department of Francisco Morazan, Central Honduras. *Rev. Paleobot. Palynol.*, *34*, 345–57.

Dengo, G. (1962). *Estudio geológico de la región de Guanacaste, Costa Rica.* Inst. geogr. Costa Rica, San José.

(1969). Problems of tectonic relations between Central America and the Carribbean. *Trans. Gulf Coast Assoc. Geol. Soc.*, *19*, 311–20.

Dengo, G., & Bohnenberger, O. (1969). Structural development of northern Central America. In A. R. McBirney (ed.), *Tectonic Relations of Northern Central America and Western Caribbean – The Bonacca Expedition* (pp. 203–20). American Association of Petroleum Geologists, Memoir 11.

De Wever, P., Azema, J., Touron, J., & Desmet, A. (1985). Decouverte de materiel océanique du Lias-Dogger inferieur dans la péninsule de Santa Elena (Costa Rica, Amérique Centrale). *C. R. Acad. Sci. (Paris)*, *300*, Ser. II (15), 759–64.

Donnelly, T. W. (1975). The geological evolution of the Caribbean and Gulf of Mexico. Some critical problems and areas. In A. E. M. Nairn & F. G. Stehli (eds.), *The Gulf of Mexico and the Caribbean: The Ocean Basins and Margins* (vol. 3, pp. 663–89). New York: Plenum Press.

(1989). Geologic history of the Caribbean and Central America. In A. W. Bally & A. R. Palmer (eds.), *The Geology of North America – An Overview* (pp. 299–321). Geological Society of America.

Doyle, P. (1987). Lower Jurassic–Lower Cretaceous belemnite biogeography and the development of the Mesozoic Boreal Realm. *Palaeogeogr. Palaeoclimat. Palaeoecol.*, *61*, 237–54.

Emmet, P. A. (1983). *Geology of the Agalteca Quadrangle, Honduras, Central America.* MA thesis, University of Texas, Austin.

Emmet, P. A., & Logan, W. S. (1983). *Mapa Geológico de Honduras, Agalteca Sheet.* Inst. Geogr. Nac., Tegucigalpa, Honduras.

Enos, P. (1983). Late Mesozoic paleogeography of Mexico. In M. W. Reynolds & E. D. Dolly (eds.), *Mesozoic Paleogeography of West-Central United States* (pp. 133–57). Denver: Rocky Mountains SEPM.

Erben, H. K. (1954). Dos amonitas nuevas y su importancia para la estratigrafía del Jurásico Inferior de México. *Paleontol. Mexicana*, *1*.

(1956a). *El Jurásico Inferior de México y sus amonitas.* Presented at the 20th International Geological Congress (México).

(1956b). *El Jurásico Medio y el Calloviano de México.* Presented at the 20th International Geological Congress (México). pp. 1–140.

(1957). New biostratigraphic correlations in the Jurassic of eastern and south-central Mexico. In *El Mesozoico del hemisferio accidental y sus correlationes mundiales* (sect. 2, pp. 43–52). Transactions of the 20th International Geological Congress (Mexico).

Fakundiny, R. H., & Everett, J. R. (1976). Re-examination of Mesozoic stratigraphy of the El Rosario and Comayagua Quadrangle, Central Honduras. *Publ. Geol. ICAITI*, *5*, 5–17.

Finch, J. W., & Abbott, P. L. (1977). Petrology of a Triassic marine section, Vizcaino Peninsula, Baja California Sur, Mexico. *Sedim. Geol.*, *19*, 253–73.

Finch, J. W., Pessagno, E. A., Jr., & Abbott, P. L. (1979). San Hipolito Formation, Triassic marine rocks of the Vizcaino Peninsula. In P. L. Abbott & R. G. Gastil (eds.), *Baja California Geology* (pp. 117–20). Proceedings of the Geological Society of America 1979 Annual Meeting. San Diego: San Diego State University.

Finch, R. C. (1981). Mesozoic stratigraphy of central Honduras. *Am. Assoc. Petrol. Geol. Bull.*, *65*, 1320–33.

(1985). *Mapa Geológico de Honduras, Santa Bárbara Sheet.* Inst. Geogr. Nac., Tegucigalpa, Honduras.

Finch, R. C., & Ritchie, Q. W. (1990). *Mapa Geológico de Honduras, Danlí Sheet.* Inst. Geogr. Nac., Tegucigalpa, Honduras.

Finneran, J. M., Scott, R. W., Taylor, G. A., & Anderson, G. H. (1984). Lowermost Cretaceous ramp reefs: Knowles Limestone, Southwest Flank of the East Texas Basin. In W. P. S. Ventres, D. G. Bebout, B. F. Perkins, & C. H. Moore (eds.), *The Jurassic of the Gulf Rim* (pp. 125–34). GCSSEPM Foundation, 3rd Ann. Res. Conf. Proc.

Fritzgärtner, R. (1891). Kaleidoscopic views of Honduras. *Honduras Mining Journal (Tegucigalpa)*, *6–8*, *11*, 73–6, 89–91, 105–7, 153–4.

Galli, C. (1977). Edad de emplazamiento y período de acumulación de la ofiolita de Costa Rica. *Cienc. Tec.*, *1*, 81–6.

Gallo, J., & Van Wagoner, J. C. (1978). *Stratigraphy and Facies Analysis of Honduras*, Exxon Production Research Company, special report (in open file at Direccion General de Minas e Hidrocarburos, Tegucigalpa).

Gastil, G. (1985). Terranes of peninsular California and adjacent Sonora. In D. G. Howell (ed.), *Tectonostratigraphic Terranes of the Circum-Pacific Region* (pp. 273–83). Houston: Circum-Pacific Council for Energy and Mineral Resources.

Gastil, G., Morgan, G. J., & Krummenacher, D. (1978). Mesozoic history of peninsular California and related areas east of the Gulf of California. In D. G. Howell & K. A. McDougall (eds.), *Mesozoic Paleogeography of the Western United States* (pp. 107–16). Pacific Coast Paleogeography Symposium 2. Los Angeles: Pacific Section SEPM.

(1981). The tectonic history of peninsular California and adjacent Mexico. In W. G. Ernst (ed.), *The Geotectonic Development of California* (pp. 284–306). Englewood Cliffs, N.J.: Prentice-Hall.

Geyer, O. F. (1971). Zur palaobatymetrischen Zuverlässigkeit von Ammonoidean-Faunen Spektren. *Palaeogeogr. Palaeoclimat. Palaeoecol.*, *10*, 265–72.

Geyssant, J. R., & Rangin, C. (1979). Découverte d'ammonites calloviennes dans le complexe a blocs de l'île de Cedros (Basse California, Mexique) et implication pour l'age de la serie ophiolitique sousjacente. *C. R. Acad. Sci. [D] (Paris)*, *289*, 521–4.

Gonzalez, C. (1979). Geology of the Sierra del Alamo. In T. H. Anderson & J. Roldan-Quintana (eds.), *Geology of Northern Sonora* (pp. 23–31). Geological Society of America, guidebook, Field Trip 27.

Gordon, M. B. (in press). *Mapa Geológico de Honduras, Santa María del Real Sheet*. Inst. Geogr. Nac., Tegucigalpa, Honduras.

Gose, W. A. (1985). Paleomagnetic results from Honduras and their bearing on Caribbean tectonics. *Tectonics*, *4*, 565–85.

Gose, W. A., & Finch, R. C. (1987). Magnetostratigraphic studies of Cretaceous rocks in Central America. *Actas Fac. Cienc. Tierra U.A.N.L. Linares*, *2*, 233–41.

Gursky, H.-J. (1988). Gefüge, Zusammensetzung und Genese der Radiolarite im ophiolithischen Nicoya-Komplex (Costa Rica). *Münster. Forsch. Geol. Paläont.*, *68*, 1–189.

Gursky, H.-J., & Gursky, M. M. (1988). Thermal alteration of chert in the ophiolite basement of southern Central America. In J. R. Hein & J. Obradovic (eds.), *Siliceous Deposits of the Tethys and Pacific Regions* (pp. 217–33). Berlin: Springer-Verlag.

Gursky, H.-J., Gursky, M. M., Schmidt-Effing, R., & Wildberg, H. G. H. (1984). Karten zur Geologie von Nordwest-Costa Rica (Mittelamerika) mit Erläuterungen. *Geol. et Palaeont.*, *18*, 173–82.

Gursky, H.-J., & Schmidt-Effing, R. (1983). Sedimentology of radiolarities within the Nicoya Ophiolite Complex, Costa Rica, Central America. *Dev. Sedimentol.*, *36*, 127–42.

Gursky, H.-J., Schmidt-Effing, R., Strebin, M. M., & Wildberg, H. G. H. (1982). The ophiolite sequence in northwestern Costa Rica (Nicoya Complex): outlines of stratigraphical, geochemical, sedimentological, and tectonical data. In *Actas 5° Congr. latinoamer. Geol. Buenos Aires*, (vol. *3*, pp. 607–19).

Gursky, M. M. (1986). *Tektonische und thermische Deformationen im opiolithischen Nicoya-Komplex und seinem sedimentaren Auflager (Nicoya-Halbinsel, Costa Rica) und ihre Bedeutung für die geodynamische Entwicklung im sudlichen Zentralamerika*. PhD thesis, Philipps-Univ., Marburg.

Gursky, M. M. (1988). Análisis tectoñico de la península de Nicoya (Costa Rica) y su significado para el desarrollo estructural-geodinámico de América Central meridional. *Rev. Geol. Amer. Central*, *8*, 19–75.

Guzman, E. J. (1950). Geología del noreste de Guerrero. *Bol. Assoc. Mexic. Geol. Petrol*, *2*, 95–156.

Gwiner, M. P. (1976). Origin of the Upper Jurassic limestones of the Swabian Alb (Southwest-Germany). In H. Füchtbauer, A. P. Lisitsyn, J. D. Milliman, & E. Seibold (eds.), *Contributions to Sedimentology*, *5*, 1–75.

Gygi, R. (1986). Eustatic sea level changes of the Oxfordian (Late Jurassic) and their effect documented in sediments and fossil assemblages of an epicontinental sea. *Eclogae Geol. Helv.*, *79*(2), 455–91.

Hall, R. L., & Westermann, G. E. G. (1980). Lower Bajocian (Jurassic) cephalopod faunas from western Canada and possible assemblage zones for the Lower Bajocian of North America. *Palaeontogr. Amer.*, *9*(52).

Hardy, L. R. (1973). *The Geology of an Allochthonous Jurassic Sequence in the Sierra de Santa Rosa, Northwest Sonora, Mexico*. Unpublished Master's thesis, California State University, San Diego.

Horne, G. S., & Clark, G. S. (1978). *Mid-Mesozoic Plutonism on the Honduran Platelet*. Unpublished manuscript, Wesleyan Univ.

Horne, G. S., Clark, G. S., & Pushkar, P. (1976). Pre-Cretaceous rocks of northwestern Honduras: basement terrane in Sierra de Omoa. *Am. Assoc. Petrol. Geol. Bull.*, *60*, 566–83.

Howell, D. G., Schermer, E. R., Jones, D. L., Ben-Avraham, Z., & Scheiber, E. (1983). *Tectonostratigraphic Terrane Map of the Circum-Pacific Region*. U.S. Geological Survey, Open-File Report 87–716.

Humphreys, E. W. (1916). *Sphenozamites regersianus* Fontaine; an addition to the Rhaetic flora of San Juancito, Honduras. *J. New York Bot. Gard.*, *17*, 56–8.

Imlay, R. W. (1939). Upper Jurassic ammonites from Mexico. *Geol. Soc. Bull.*, *50*, 1–78.

(1942). Late Jurassic fossils from Cuba and their economic significance. *Geol. Soc. Am. Bull.*, *53*, 1417–78.

(1943). Jurassic formations of Gulf Region. *Am. Assoc. Petrol. Geol. Bull.*, *27*(11), 1407–533.

(1952). Correlation of Jurassic formations of North America exclusive of Canada. *Geol. Soc. Am. Bull.*, *63*, 953–92.

(1953). Las formaciones Jurásicas de México. *Bol. Soc. Geol. Mexic.*, *16*(1), 1–65.

(1980). *Jurassic Paleobiogeography of the Conterminous United States in Its Continental Setting*. U.S. Geological Survey, Professional Paper 1060.

Jansa, L. (1986). Paleoceanography and evolution of the North Atlantic Ocean Basin during the Jurassic. In P. R. Vogt & B. E. Tucholke (eds.), *The Geology of North America. Vol. M: The Western North Atlantic Region* (pp. 603–16). Geological Society of America.

Jaworski, E. (1929). Eine Liasfauna aus Nordwest-Mexico. *Schweizerische Palaeont. Gesellschaft*, *48*, 12.

Jeletzky, J. A. (1984). Jurassic–Cretaceous boundary beds of western and Arctic Canada and the problem of the Tithonian-Berriasian stages in the Boreal Realm. In G. E. G. Westermann (ed.), *Jurassic-Cretaceous Biochronology and Paleogeography of North America* (pp. 175–255). Geological Association of Canada, Special Paper 27.

Jones, D. L., Blake, M. C., Jr., & Rangin, C. (1976). The four Jurassic belts of northern California and their significance to the geology of the southern California borderland. In D. G. Howell (ed.), *Aspects of the Geologic History of the California Continental Borderland* (pp. 343–62). American Association of Petroleum Geologists, Pacific Section, Miscellaneous Publication 24.

Kilmer, F. H. (1977). Reconnaissance geology of Cedros Island, Baja California, Mexico. *S. Cal. Acad. Sci. Bull.*, *76*, 91–8.

(1979). A geological sketch of Cedros Island, Baja California, Mexico. In P. L. Abbott & R. G. Gastil (eds.), *Baja California Geology* (pp. 11–28). Proceedings of the Geological Society of America 1979 annual meeting. San Diego: San Diego State University.

Kimbrough, D. L. (1984). Paleogeographic significance of the Middle Jurassic Gran Cañon Formation. Cedros Island, Baja California Sur. In V. A. Frizzell, Jr. (ed.), *Geology of the Baja California Peninsula* (vol. 39, pp. 107–17). Los Angeles: Pacific Station SEPM.

(1985). Tectonostratigraphic terranes of the Vizcaino Peninsula and Cedros and San Benito Islands, Baja California, Mexico, In D. G. Howell (ed.), *Tectonostratigraphic Terranes of the Circum-Pacific Region* (pp. 285–98). Houston: Circum-Pacific Council for Energy and Mineral Resources.

King, A. P. (1972). *Geology of the Talanga and Cedros Quandrangles, Honduras, Central America*. Open-file report, Inst. Geogr. Nac., Tegucigalpa, Honduras.

(1973). *Mapa Geológico de Honduras, Cedros Sheet*. Inst. Geogr. Nac., Tegucigalpa, Honduras.

King, R. R. (1939). Geological reconnaissance in Northern Sierra Madre occidental of Mexico. *Geol. Soc. Am. Bull.*, *50*, 1625–1722.

Knowlton, F. W. (1918). Relations between the Mesozoic floras of North and South America. *Geol. Soc. Am. Bull.*, *29*, 607–14.

Kozuch, M. J. (1989a). *Geology of the San Francisco de Becerra Quadrangle, Honduras, Central America*. Open-file report, Inst. Geogr. Nac., Tegucigalpa, Honduras.

(1989b). *Mapa Geológico de Honduras, San Francisco de Becerra Sheet*. Inst. Geogr. Nac., Tegucigalpa, Honduras.

Lang, J. (1917). Correspondencia científica (carta al Dr. Luis Landa). *Bol. Secr. Fom. Obras Publ. (Honduras)*, *VII*(5–7), 191–2 (citation according to Maldonado-Koerdell, 1952).

Leanza, H. (1981). Faunas de ammonites del Jurásico superior y del Cretácico inferior de América del Sur, con especial consideración de la Argentina. *Comité Sudamericano del Jurásico y Cretácico: Cuencas sedimentarias del Jurásico y Cretácico de América del Sur, 2,* 559–97.

Leggett, T. H. (1889). Notes on the Rosario Mine at San Juancito, Honduras, C.A. *Trans. Am. Inst. Min. Engin., 17,* 432–49.

Longoria, J. F. (1984a). Stratigraphic studies in the Jurassic of northeastern Mexico: evidence for the origin of the Sabinas Basin. In W. P. S. Ventress, D. G. Bebout, B. F. Perkins, & C. H. Moore (eds.), *The Jurassic of the Gulf Rim* (pp. 1–93). GCSSEPM Foundation, 3rd Ann. Res. Conf. Proc.

(1984b). Mesozoic tectostratigraphy domains in east-central Mexico. In G. E. G. Westermann, (ed.), *Jurassic-Cretaceous Biochronology and Paleogeography of North America* (pp. 55–76). Geological Association of Canada, Special Paper 27.

López-Ramos, E. (1983). *Geologia de México. Part 3* (3rd ed.). México, D.F.: Edición Escolar.

(1985). *Geología de México. Part 2* (3rd ed.) Mexico, D.F.: Edición Escolar.

López Ticha, D. (1988). Revisión de la estratigrafía y potencial petrolero de la Cuenca de Tlaxiaco. *Bol. Asoc. Mexic. Geol. Petrol., 37,* 49–92. (Original work completed 1985).

McBirney, A. R., & Williams, H. (1965). Colcanic history of Nicaragua. *Univ. Calif. Publ. Geol. Sci., 55.*

Maldonado-Koerdell, M. (1952). Plantas del Rético-Liásico y otros fósiles Triásicos de Honduras. *Ciencia, 12,* 294–6.

Meschede, M., Frisch, W., & Sick, M. (1988). Interpretación geodinámica de los complejos ofiolíticos de Costa Rica. *Rev. Geol. Amer. Central, 8,* 1–17.

Michaud, F., & Fourcade, E. (1989). Stratigraphie et Paléogéographie du Jurassique et du Crétace du Chiapas (Sud-Est du Mexique). *Bull. Soc. Geol. France, 8*(5), 639–50.

Mills, R. A., Hugh, K. E., Feray, D. E., & Swolfs, H. C. (1967). Mesozoic stratigraphy of Honduras. *Am. Assoc. Petrol. Geol. Bull., 51,* 1711–86.

Mina, U. F. (1957). Bosquejo geológico del Territorio Sur de la Baja California. *Bol. Assoc. Mexic. Geol. Petrol., 9,* 139–269.

Minch, J. C., Castil, G., Fink, W., Robinson, J., & James, A. H. (1976). Geology of the Vizcaino Peninsula. In D. G. Howell (ed.), *Aspects of the Geologic History of the California Borderland* (pp. 136–95). American Association of Petroleum Geologists, Pacific Section, Miscellaneous Publication 24.

Moore, T. E. (1985). Stratigraphy and tectonic significance of the Mesozoic tectonostratigraphic terranes of the Vizcaino Peninsula, Baja California Sur, Mexico. In D. G. Howell (ed.), *Tectonostratigraphic Terranes of the Circum-Pacific Region* (pp. 315–29). Houston: Circum-Pacific Council for Energy and Mineral Resources.

Müllerried, F. K. G. (1939). Investigaciones y Exploraciones Geográfico-Geológicas en la Porción Nor-Oeste de la América Central. *Inst. Panamer. Geogr. Hist., 38.*

(1942). Contribution to the geology of northwestern Central America. In *Proceedings of the Eighth American Scientific Congress, Vol. 4: Geological Sciences* (pp. 469–82).

Newberry, J. S., (1888a). Triassic plants from Honduras. *Trans. New York Acad. Sci., 7,* 113–15.

(1888b). Rhaetic plants from Honduras. *Am. J. Sci., 32,* ser. 3, 342–51.

Olóriz, F. (1985). Paleogeography and ammonites in the Upper Jurassic, outlines for a pattern. In C. Pallini (ed.), *Proc. Symp. Don Raff. Piccinini, Pergola 1984* (pp. 1–9). Roma: Com. Cont. R. Piccinini.

(1987). El significado biogeográfico de las plataformas mexicans en el Jurásico superior. Consideraciones sobre un modelo eco-evolutivo. *Rev. Soc. Mexic. Paleontol., 1*(1), 219–47.

(1988). "Ammonites and dispersal biogeography." Is that all? In *Proceedings of the 2nd International Symposium on Jurassic Stratigraphy* (pp. 563–80). Lisbon: Instituto Nacional de Investigacão Cientifica.

(1990). Ammonite phenotypes and ammonite distributions. Notes and comments. In C. Pallini et al. (eds.), *Fossili Evoluzione Ambiente* (pp. 417–26). Atti 2. Conv. Intl. Pergola 1987. Rome: Com. Cent. R. Piccinini.

Olóriz, F., López-Galindo, A., Villaseñor, A. B., & González, C. (1988a). Analisis isotópicos y consideraciones paleoecológicas en el Jurásico superior de Mexico (Fm. La Casita, Cuéncame, Durango). Datos preliminares. In *II Congr. Geol. España, Granada* (vol. 1, pp. 144–8).

Olóriz, F., Marques, B., & Moliner, L. (1988b). The platform effect: an example from Iberian shelf areas in the lowermost Kimmeridgian. In *2nd International Symposium on Jurassic Stratigraphy, Lisboa, Portugal, Sept. 12–21, 1987* (pp. 543–62).

Olóriz, F., & Tavera, J. M. (1989). The significance of Mediterranean ammonites with regard to the traditional Jurassic–Cretaceous boundary. *Cretac. Res., 10,* 221–37.

Olóriz, F., Villaseñor, A. B., González, C., & Westermann, G. E. G. (1990). Las plataformas mexicanas durante el Jurásico superior. Un ejemplo de áreas de recepción en biogeografía. *Actas Paleontol.* (Salamanca), *4,* 277–87.

Ortuno-Arzate, F. (1985). *Evolution sedimentaire mesozoique du Basin Rift de Chichuahua le long d'une transversale Aldama-Ojinaga (Mexique). Implications geodynamiques.* These, Université Pau, Pays de L'Adour 206.

Pérez-Ibarguengoitia, J. M., Hokuto-Castillo, A., & de Cserna, Z. (1965). Estratigrafía y paleontología del Juraśico Superior de la parte centroseptentrional del Estado de Puebla. Part I: Reconocimiento geológico del area de Petlalcingo-Santa Cruz, Municipio de Acatlán, Estado de Puebla. *Paleontól. Mexicana, 21.* 1–22.

Person, C. P., & Delevoryas, T. (1982). The Middle Jurassic flora of Oaxaca, Mexico. *Palaeontographica B, 180,* 82–119.

Pessagno, E. A., Jr., Finch, W., & Abbott, P. L. (1979). Upper Triassic Radiolaria from the San Hipolito Formation, Baja California. *Micropaleontology, 25,* 160–97.

Rangin, C. (1977). Sobre la presencia del Jurásico Superior con amonitas en Sonora septentrional. *Univ. Nac. Aut. México, Rev. Inst. Geol., 1,* 1–4.

(1978). Speculative model of Mesozoic geodynamics, central Baja California to northeast Sonora (Mexico). In D. G. Howell & K. A. McDougall (eds.), *Mesozoic Paleography of the Western United States* (pp. 85–106). Pacific Coast Paleogeography Symposium 2. Los Angeles: Pacific Section SEPM.

(1979). Evidence for superimposed subduction and collision processes during Jurassic-Cretaceous time along Baja California continental borderland. In P. L. Abbott & R. G. Gastil (eds.), *Baja California Geology* (pp. 37–51). Proceedings of the Geological Society of America 1979 annual meeting. San Diego: San Diego State University.

Riccardi, A. C., Westermann, G. E. G., & Elmi, S. (1989). The Middle Jurassic ammonite zones of the Argentine–Chilean Andes. *Geobios, 22,* 557–97.

Ritchie, A. W., & Finch, R. C. (1985). Widespread Jurassic strata on the Chortis block of the Caribbean plate. *Geol. Soc. Am. Abstr. Prog., 17,* 700–1.

Roberts, R. J., & Irving, E. M. (1957). *Mineral Deposits of Central America.* U.S. Geological Survey, Bulletin 1034.

Salas, G. P. (1949). Bosquejo geológico de la cuenca sedimentaria de Oaxaca. *Bol. Asoc. Mexic. Geol. Petrol., 1,* 79–156.

Salvador, A. (1987). Late Triassic-Jurassic paleogeography and origin of Gulf of Mexico Basin. *Am. Assoc. Petrol. Geol. Bull., 71,* 419–51.

Salvador, A., & Green, A. R. (1980). Opening of the Caribbean Tethys (origin and development of the Caribbean and the Gulf of Mexico). In J. Auboin, J. Debelmas, & M. Lattrelle (eds.), *Geology of the Alpine Chains Born of the Tethys. Bur. Res. Geol. Min., 115,* 224–9.

Sandoval, J., & Westermann, G. E. G. (1986). The Bajocian (Jurassic) ammonite fauna of Oaxaca, Mexico. *J. Paleontol., 60,* 1220–71.

Sandoval, J., Westermann, G. E. G., & Marshall, M. C. (1990). Ammonite fauna, stratigraphy and ecology of Bathonian–Callovian (Jurassic) Tococoyunca Group, South Mexico. *Palaeontographica, A, 210,* 93–149.

Sapper, Carlos [Karl] (1896). *Sobre la geografía física y la geología de la Península de Yucatán.* Inst. Geol. Mex., Bol. 3.

(1937). *Mittelamerika: Hanbuch der regionalen Geologie, 4a Abteilung.* Heidelberg: Enke.

Schmidt-Effing, R. (1979). Alter und Genese des Nicoya-Komplexes, einer ozeanishen Paläokruste (Oberjura bis Eozan) im südlichen Zentralamerika. *Geol. Rdsch., 68,* 457–94.

(1980). The Huayacocotla aulacogen in Mexico (lower Jurassic) and the origin of the Gulf of Mexico. In: R. H. Pilger, Jr. (ed.), *The Origin of the Gulf of Mexico and the Early Opening of the Central North Atlantic Ocean* (pp. 79–86). Baton Rouge: Louisiana State University.

Schuchert, C. (1935). *Historical Geology of the Antillean Caribbean Region.* New York: Wiley.

Scott, G. H. (1985). Homotaxy and biostratigraphical theory. *Paleontology, 28*(4), 777–82.

Scott, R. W. (1984). Mesozoic biota and depositional systems of the Gulf of Mexico–Caribbean region. In G. E. G. Westermann (ed.), *Jurassic-Cretaceous Biochronology and Paleogeography of North America* (pp. 49–69). Geological Association of Canada, Special Paper 27.

Sedlock, R. L. (1988). Tectonic setting of blueschist and island arc terranes of west-central Baja California, Mexico. *Geology, 16,* 623–6.

Silva-Pineda, A. (1961). Paleontología del Triaśsico Superior de Sonora. Part 2: Flora fósil de la Formación Santa Clara (Carnico) del Estado de Sonora. *Univ. Nac. Aut. México, Inst. Geol., Paleont. Mexic.,* no. 11.

(1970). Plantas fósiles del Jurásico medio de la región de Tezoatlán, Oaxaca. In L. R. Segura & R. Rodriguez-Torres (eds.), *Libro Guiá de la Excrusión Mexíco-Oaxaca* (pp. 129–53). Mexico, D.F.: Soc. Geol. Mex.

Simonson, B. M. (1977). *Geology of the El Porvenir Quadrangle, Honduras, Central America.* Open-file report, Inst. Geogr. Nac., Tegucigalpa, Honduras.

(1981). *Mapa Geologico de Honduras, El Porvenir Sheet.* Inst. Geogr. Nac., Tegucigalpa, Honduras.

Tardy, M., Blanchet, R., & Zimmerman, M. (1989). The Texas and Caltam lineaments from the American Cordillera to the Mexican Sierras Madres: Nature, origin and structural evolution. *Bull. Centres Rech. Explor.-Prod. Elf-Aquitaine, 13*(2), 219–27.

Tournon, J., & Azema, J. (1980). Sobre la estructura y la petrología del macizo ultrabásico de Santa Elena (provincia de Guanacaste, Costa Rica). *Informe semestral Inst. geogr. Costa Rica* (enero a junio), 17–54.

Valentine, J. W. (1973). *Evolutionary Paleoecology of the Marine Biosphere.* Englewood Cliffs, N.J.: Prentice-Hall.

Verma, V., & Westermann, G. E. G. (1973). The Tithonian (Jurassic) ammonite fauna and stratigraphy of Sierra Catorce, San Luis Potosi, Mexico. *Bull. Am. Paleont., 63* (277), 107–320.

(1984). *The Ammonoid Fauna of the Kimmeridgian–Tithonian Boundary Beds of Mombasa, Kenya.* Life Sciences Contributions, Royal Ontario Museum, *135,* 1–123.

Villaseñor, A. B., & González, C. (1988). Fauna de amonitas y presencia de *Lamellaptychus murocostatus* Trauth del Jurásico superior de la Sierra de Palotes, Durango. *Univ. Nat. Aut. México, Inst. Geol., Rev., 7* (1), 71–7.

Viniegra, O. F. (1981). Great carbonate bank of Yucatan, Southern Mexico. *J. Petrol. Geol., 3*(3), 247–78.

Ward, P. D., & Westermann, G. E. G. (1985). Cephalopod paleoecology. In D. S. Bottjer, C. S. Hickman, & P. D. Ward (eds.), *Molluska, Note for Short Course. University of Tennessee, Department of Geological Science, Studies in Geology, 13,* 1–18.

Weber, H. -S. (1979). On the lithology and stratigraphy of the "Estratos de Metapan" in the Republic of El Salvador, Central America. *Geol. Jb., B, 37,* 31–54.

Westermann, G. E. G. (1983). The Upper Bajocian and Lower Bathonian (Jurassic) ammonite faunas of Oaxaca, Mexico, and West-Tethyan affinities. *Paleontol. Mexicana, 46,* 1–63.

(1984a). The Late Bajocian Duashnoceras Association (Jurassic Ammonitina) of Mixtepec in Oaxaca, Mexico. In *Memoirs, III Cong. Latinamer. Paleontol.* (pp. 193–8).

(1984b). Summary of symposium papers on the Jurassic–Cretaceous biochronology and paleogeography of North America. In G. E. G. Westermann (ed.), *Jurassic-Cretaceous Biochronology and Paleogeography of North America* (pp. 307–15). Geological Association of Canada, Special Paper 27.

(1990). New developments in Jurassic-Cretaceous ammonite ecology. In G. Pallini et al. (eds.), *Fossili Evoluzione Ambiente* (pp. 459–78). Atti 2. Conv. Intl. Pergola 1987. Roma: Com. Cent. R. Piccinini.

Westermann, G. E. G., Corona, R., & Carrasco, R. (1984). The Andean Mid-Jurassic Neuqueniceras ammonite assemblage of Cualac, Mexico. In G. E. G. Westermann (ed.), *Jurassic-Cretaceous Biochronology and Paleogeography of North America* (pp. 99–112). Geological Association of Canada, Special Paper 27.

Whalen, P. A., & Pessagno, E. A., Jr. (1984). Lower Jurassic Radiolaria, San Hipolito Formation, Vizcaino Peninsual, Baja California Sur. In V. A. Frizzell, Jr. (ed.), *Geology of the Baja California Peninsula* (pp. 53–65). Los Angeles: Pacific Station SEPM.

Wieland, G. R. (1913). The Liassic flora of the Mixteca Alta of Mexico – its composition, age, and source. *Am. J. Sci.,* ser. 4, *36,* 251–81.

(1914). *La Flora Liasica de la Mexteca Alt.* Inst. Geol. Mex., Bol. 31.

Wildberg, H. G. H. (1984). Der Nicoya-Komplex, Costa Rica, Zentralamerika: Magmatismus und Genese eines polygenetischen Ophiolith-Komplexes. *Münster. Forsch. Geol. Paläont., 62,* 1–123.

(1987). High level and low level plagiogranites from the Nicoya Ophiolite Complex, Costa Rica, Central America. *Geol. Rundsch., 76,* 285–301.

Wildberg, H. G. H., & Baumann, A. (1987). The ophiolitic Nicoya Complex, Costa Rica, Central America: genetic implications by the Sr isotopic composition of igneous rocks. In *Transactions of the 11th Caribbean Geological Conference, Barbados* (pp. 23.1–23.7).

Ziegler, B. 1967. Ammoniten-Okologie am Beispiel des Oberjura. *Geol. Rundsch., 60,* 439–64.

Zonenshayn, L. P., Savostin, L. A., & Sedov, A. P. (1984). Global paleogeodynamic reconstructions for the last 160 million years. *Geotectonics, 18*(3), 181–94.

# 6 Western South America and Antarctica

A. C. RICCARDI, C. A. GULISANO, J. MOJICA, O. PALACIOS, C. SCHUBERT, and M. R. A. THOMSON

## REGIONAL SETTING[1]

### Morphostructural units

Most of eastern South America, between the Orinoco, the Atlantic Ocean, and the La Plata River, is formed by the Precambrian Brazilian and Guyanas Shields, consisting of several crustal domains (Cordani et al. 1988). To the west the Brazilian shield is flanked, from north to south, by the Llanos, Beni-Chaco, and Pampas plains (Figure 6.1).

In the west and south, smaller Precambrian to Lower Paleozoic tectonic units are present in southern Peru, central Argentina, and eastern Patagonia. The latter two areas form the nucleus of the Pampean, North Patagonian, and Deseado massifs, which behaved as positive stable blocks throughout the Phanerozoic.

Along the western margin of South America, from its southern tip to the Caribbean sea, the Andes are formed by several mountain ranges. These ranges are tectonically divided along strike into several segments (Gansser 1973; Sillitoe 1974; Barazangi and Isacks 1976; Hervé et al. 1988; Megard 1988). This division is probably related to variations in plate geometry along the subduction zone of the Nazca Plate underneath the South American Plate, caused by consumption of the Carnegie, Grijalba, Nazca, and Juan Fernandez aseismic ridges and the active Chile Rise (Nur and Ben-Abraham 1981; Ben-Abraham and Nur 1988). Although the Andean ranges are best developed in the Central Andes, we will proceed from the north to the south.

#### *Northern Andes*

The northern Andes extend from 5°S to the Caribbean coast. The southern limit is marked by the Huancabamba deflection (Peru), a boundary roughly aligned with the Grijalba and Carnegie aseismic ridges. North of the Huancabamba deflection (5°S) are, from west to east, the Western Cordillera, the Royal or Central Cordillera, and, in Colombia, the Eastern Cordillera. Structural units are heterogeneous and complex (Megard 1988; Restrepo and Toussaint 1988).

Near the Ecuadorian–Peruvian border, at the Huancabamba deflection, the strike of the Western and Central cordilleras changes from NW to NE, and along the Caribbean coast the Western and Central Cordilleras give way to a complex of subdued branches, basins, and uplifts. The Eastern Cordillera, also with a northeast (NE) strike, splits into two branches northeast of Bogotá: the Sierra de Perijá to the northwest and the Mérida Andes to the southeast. The Sierra de Perijá is cut off by a major fault, and the Mérida or Venezuelan Andes pass over (at 69°W) to the Caribbean Mountain System. This structurally complex system, which may be divided in a Coast Range and an Interior Range, extends along the Caribbean coast as far as Trinidad.

The Royal Cordillera of Colombia and Ecuador is composed of several belts of Precambrian–Paleozoic metamorphics and Mesozoic intrusives, which have largely behaved as a stable positive block. The Central Cordillera may therefore be the continuation of the Coastal Cordillera that from 15°S to 5°S is interpreted as a submarine outer-shelf high (Shepherd and Moberly 1981; Thornburg and Kulm 1981).

Precambrian and Paleozoic rocks are also present in the Eastern Cordillera as well as in the Sierra de Perijá and Mérida Andes. But these ranges are mostly characterized by Mesozoic and Cenozoic rocks. The Jurassic consists of continental sequences possibly partly derived from the eastern Guyana Shield and emerged areas in western Venezuela, locally including marine Early and Late Jurassic and volcanic intercalations. Cretaceous and Cenozoic rocks contain marine and continental facies.

The Western Cordillera, north of the Huancabamba deflection, and some parts of the Central Cordillera of Colombia, as well as all the region west of the Dolores-Guayaquil Megashear Zone, consist of terranes of oceanic origin (ophiolitic complexes, basalts, and Cretaceous and Tertiary flysch sequences), which were

[1] By A. C. Riccardi.

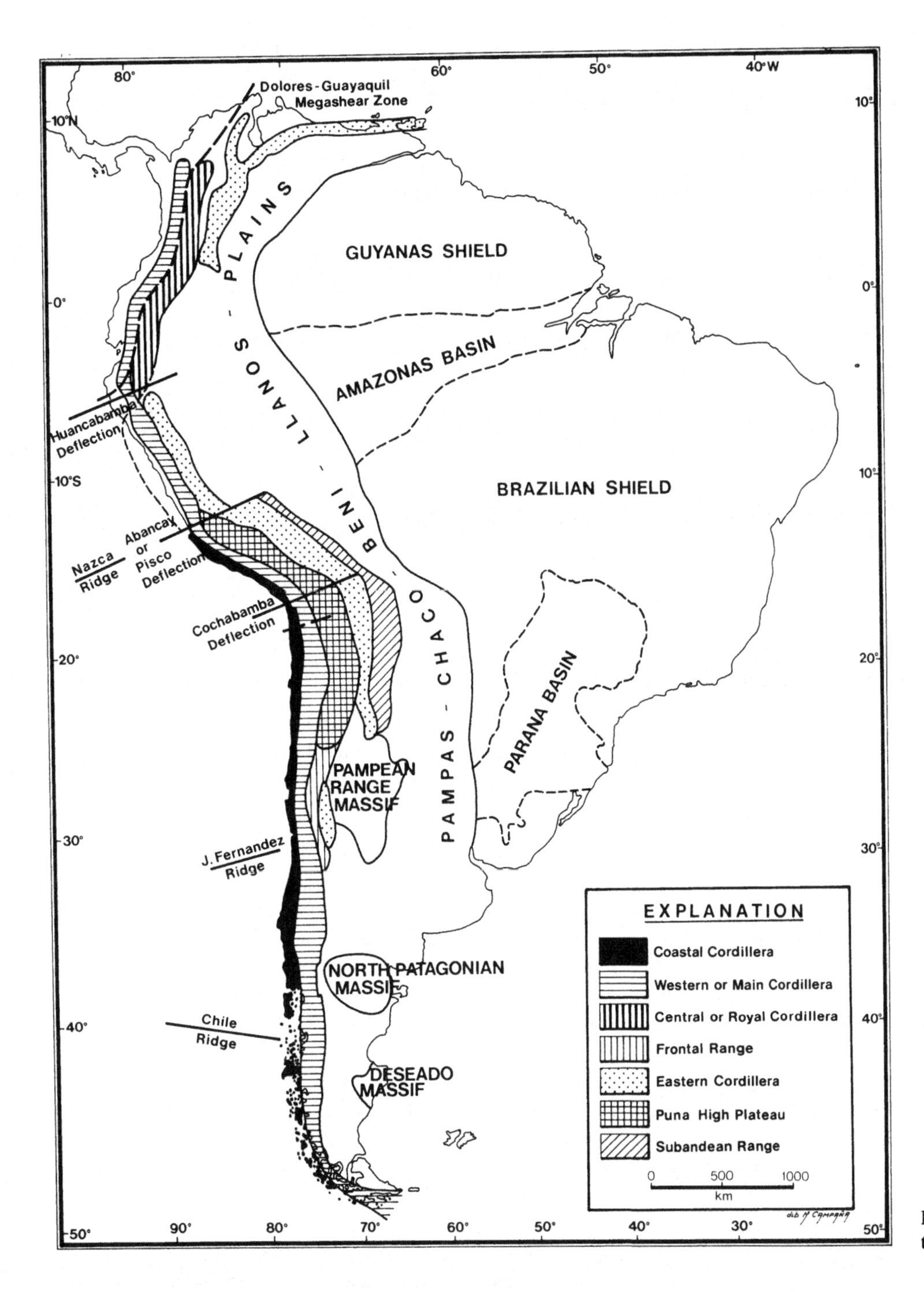


**Figure 6.1. Main morphostructural units of South America.**

accreted in two Cretaceous and one Early Tertiary episodes (Shepard and Moerly 1981; Megard 1988).

*Central Andes*

The Andean ranges are best developed between 5°S and 39°S, where they are named Chilean Peruvian or Central Andes (Cobbing and Pitcher 1983; Hervé et al. 1988; Megard 1988).

Between 5°S and 28°S they form two main mountain belts, in the east the Subandean Range and Eastern Cordillera, and in the west the Western and, marginal to the continent, the Coastal Cordilleras. These eastern and western belts are separated, along the boundary between Peru, Bolivia, northern Chile, and Argentina, by the Puna High Plateau. A subdivision occurs at Nazca (15°S), where the Coastal Cordillera continues north as a submarine outer-shelf high. There the Abancay or Pisco deflection coincides with the eastern Pacific Nazca aseismic ridge. South of the Subandean Range, at 28°S, are the Pampean Ranges, which extend to 36°S and are flanked to the west by the Precordillera and the Frontal, Principal, and Coastal

Cordilleras. South of 36°S only the Coastal and Principal Cordilleras are present.

The Coastal Cordillera is an old Precambrian–Paleozoic range that has remained a positive block since the Late Paleozoic–Early Mesozoic. The Main or Western Cordillera is the result of Mesozoic–Cenozoic uplift and is formed mostly by Mesozoic and Cenozoic rocks that in Peru overlie Precambrian–Paleozoic basement.

North of 28°S the Puna High Plateau is an old block characterized by widespread Cenozoic volcanics. It was uplifted by the Ocloyic (Ordovician–Silurian and Late Devonian) tectonic phases, but attained its present altitude (4,000 m) only in the Neogene. The Eastern Cordillera and Subandean Range are foldbelts composed of Paleozoic, Mesozoic, and Cenozoic sequences. The Eastern Cordillera has been uplifted and thrust eastward along high-angle reverse faults.

At 28–36°S the Frontal Range and the Precordillera are old foldbelts consisting mainly of Paleozoic rocks. They have behaved as stable units since the Late Paleozoic.

The Central Andes form an orogen related to the Early Jurassic to Present subduction, with variations in the development of volcanism along strike. Thus, Late Cenozoic volcanics are well developed between 15°S and 24°S, but absent between 28°S and 33°S (Hervé et al. 1988).

### *Southern Andes*

At 40–42°S the Coastal Cordillera becomes subdued in the Chilean archipelago, and the Andes consist of a single range, the Patagonian (to 52°S) or Fueguian (south of 52°S) Andes.

Between 39°S and 47°S there are widespread Upper Jurassic to Cenozoic volcanics, with Neogene to Recent high-angle or transcurrent faults (Haller and Lapido 1982; Ramos et al. 1982; Hervé et al. 1988). South of the triple junction (47°S) between the Nazca, Atlantic, and South American plates, where the active Chile Rise collides with the western margin of South America, the Patagonian or Southern Andes are formed by widespread Middle-Upper Jurassic acidic volcanics, Upper Jurassic–Cretaceous marine strata and intrusives, and Cenozoic sequences affected by low-angle faulting (Riccardi and Rolleri 1980; Gust et al. 1985; Hervé et al. 1988; Riccardi 1988).

### *Extra-Andean region*

Several fault-bounded epicratonic basins are present on the Brazilian shield. The largest is the Amazonas Basin at the boundary of Brazil, Paraguay, Uruguay, and Argentina. They are the result of reactivated Precambrian and Late Jurassic–Cretaceous fractures (Almeida 1972) and are characterized by Paleozoic and Mesozoic continental volcanics and sediments.

Other smaller basins are present between the Brazilian Shield and the Patagonian and Deseado Massifs, as well as on the eastern and western sides of the cordilleran units mentioned earlier. Most of these basins were formed during the Mesozoic, beginning with extensional tectonics related to the fragmentation of Gondwana, and followed by repeated compressive phases related to subduction of the Farallones-Nazca Plate under South America, leading to the rising of the Andean belt.

## Distribution and boundaries of the Jurassic

### *Distribution*

Except for continental and volcanic rocks of the Paraná Basin and of central and eastern Patagonia, the Jurassic in South America is mostly restricted to the Andean belt.

The Lower Jurassic south of 39°S consists mostly of continental rocks, but Pliensbachian-Toarcian marine strata are developed in central Patagonia. North of 39°S, the marine Lias forms an almost continuous belt along the Pacific margin of South America (Coastal and Principal Cordilleras of Argentina and Chile, Coastal, Western, and Eastern Cordilleras of Perú, and Eastern Cordillera of Ecuador and Colombia). The Hettangian is mostly restricted to the westernmost areas, where it forms a continuous succession with marine Upper Triassic. Middle and Upper Jurassic have similar distributions. South of 39°S and north of 12°S the Dogger is exclusively continental and/or volcanic and/or missing, except for a local occurrence in Venezuela. Complete marine Malm appears to be present exclusively in Chile (and Argentina). Marine Tithonian is widespread, and Jurassic–Cretaceous boundary beds are present at the coast of Perú and in Colombia.

### *Boundaries*

The oldest Jurassic beds documented by Hettangian (Planorbis Standard Zone) ammonites occur in relatively restricted areas of Chile, Peru and west-central Argentina. In Peru and Chile they overlie marine Triassic with Norian (include "Rhaetian") ammonites (Cecioni and Westermann 1968; Hillebrandt 1987). In Argentina, Hettangian overlies 300 m of turbidites (base not exposed) that could belong to the Triassic (Riccardi et al. 1988a). In other areas the marine Jurassic begins with younger strata and rests unconformably on continental or volcanic Triassic or older rocks.

In west-central Argentina and central Chile most of the Jurassic is characterized by a major marine sedimentary cycle extending from the Hettangian to the Oxfordian, followed by Kimmeridgian continental volcanics and sediments. A second major marine cycle began in the Tithonian and extended well into the Early Cretaceous. The Tithonian of Argentina and Chile is therefore more closely related to the Lower Cretaceous than to the Jurassic succession, so that Harrington (1962) commented: "there is little doubt that if geology had been born in South America instead of Europe, the Jurassic–Cretaceous boundary would have been drawn between the Kimmeridgian and Tithonian." This situation is also clearly evident in southern Patagonia, Peru, and Colombia, where marine Upper Jurassic is poorly represented at the base of a thick Cretaceous fossiliferous sequence.

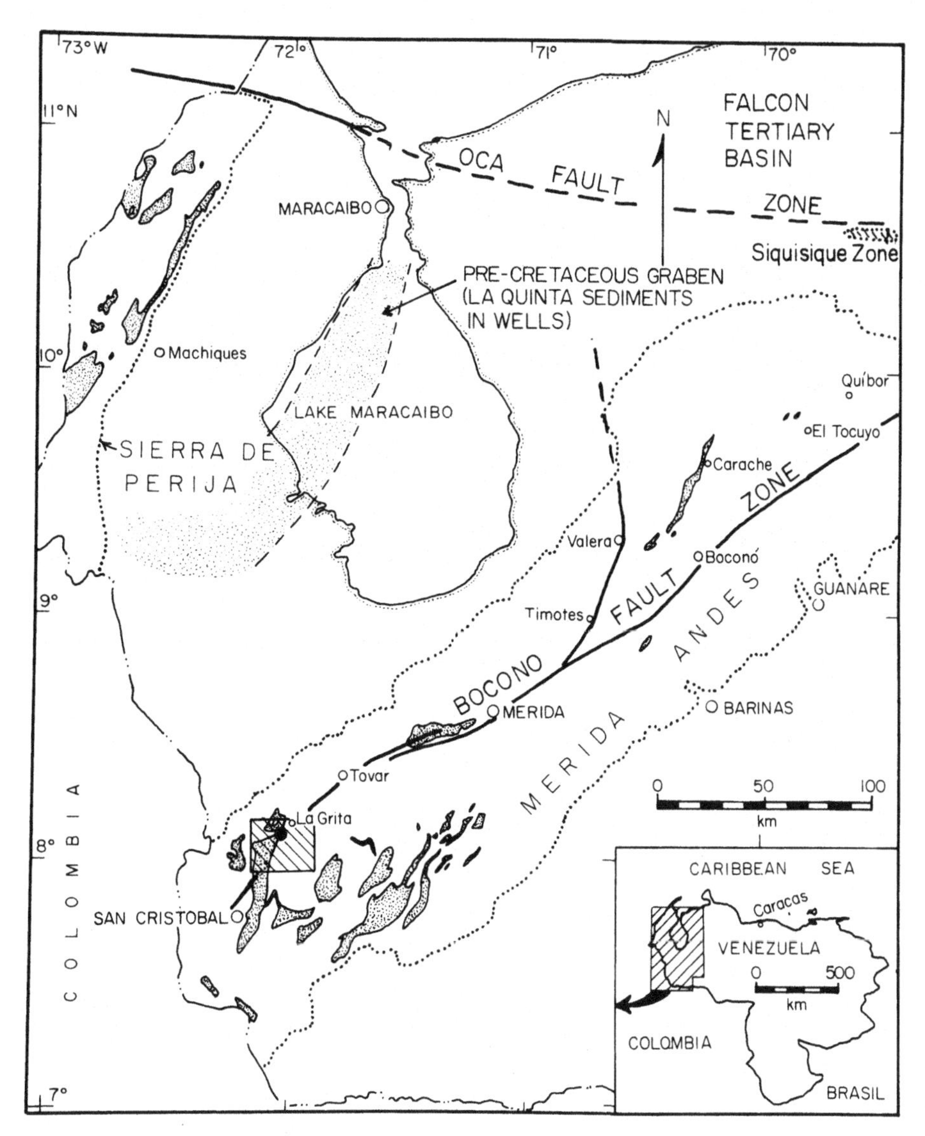


**Figure 6.2. Jurassic outcrops and subcrops in Venezuela.**

The Jurassic–Cretaceous (Tithonian–Berriasian) boundary has been clearly defined on the basis of ammonites in west-central Argentina by Leanza (1947a), that is between the *Substeueroceras koeneni* Zone and the *Argentiniceras noduliferum* Zone. The zonal sequence was recently discussed by Leanza and Hugo (1978), Wiedmann (1980a,b), Zeiss (1983, 1986), Jeletzky (1984), Huber and Wiedmann (1987), and Riccardi (1984a, 1988).

Wiedmann (1980a) correlated the *koeneni* Zone with the Jacobi Standard Zone, which he placed in the topmost "Ardescian," below the Berriasian. Furthermore, Wiedmann even placed the Berriasian within the Tithonian, and the Jurassic–Cretaceous boundary between the Tithonian and the Valanginian. Zeiss (1983, 1986), on the other hand, includes both the Jacobi Zone and the *koeneni* Zone in the Berriasian basal Cretaceous. Uppermost Tithonian for the *koeneni* Zone has been upheld in a detailed discussion by Jeletzky (1984), who correlated it with the Transitorius Standard Zone (including part of the "Jacobi Zone") of the Submediterranean Province.

Material close to *Substeueroceras koeneni* has been figured from Peru (Rivera 1951; Geyer 1983) and Colombia (Bürgl 1960; Haas 1960), documenting the presence of the uppermost Tithonian in those countries.

## VENEZUELA[2]

Nonmetamorphic Jurassic crops out mainly in the Mérida Andes, the Sierra de Perijá (Figure 6.2), and the Siquisique region, all in north-central Venezuela. Jurassic rocks have been drilled in the Lake Maracaibo Basin. Metamorphic Jurassic occurs in the Paraguaná Peninsula in northwestern Venezuela, and a possible Jurassic igneous–metamorphic complex crops out in the Paria Peninsula in northeastern Venezuela (Pumpin 1978; González de Juana, Iturralde de Arozena, and Picard 1980,

[2] By C. Schubert.

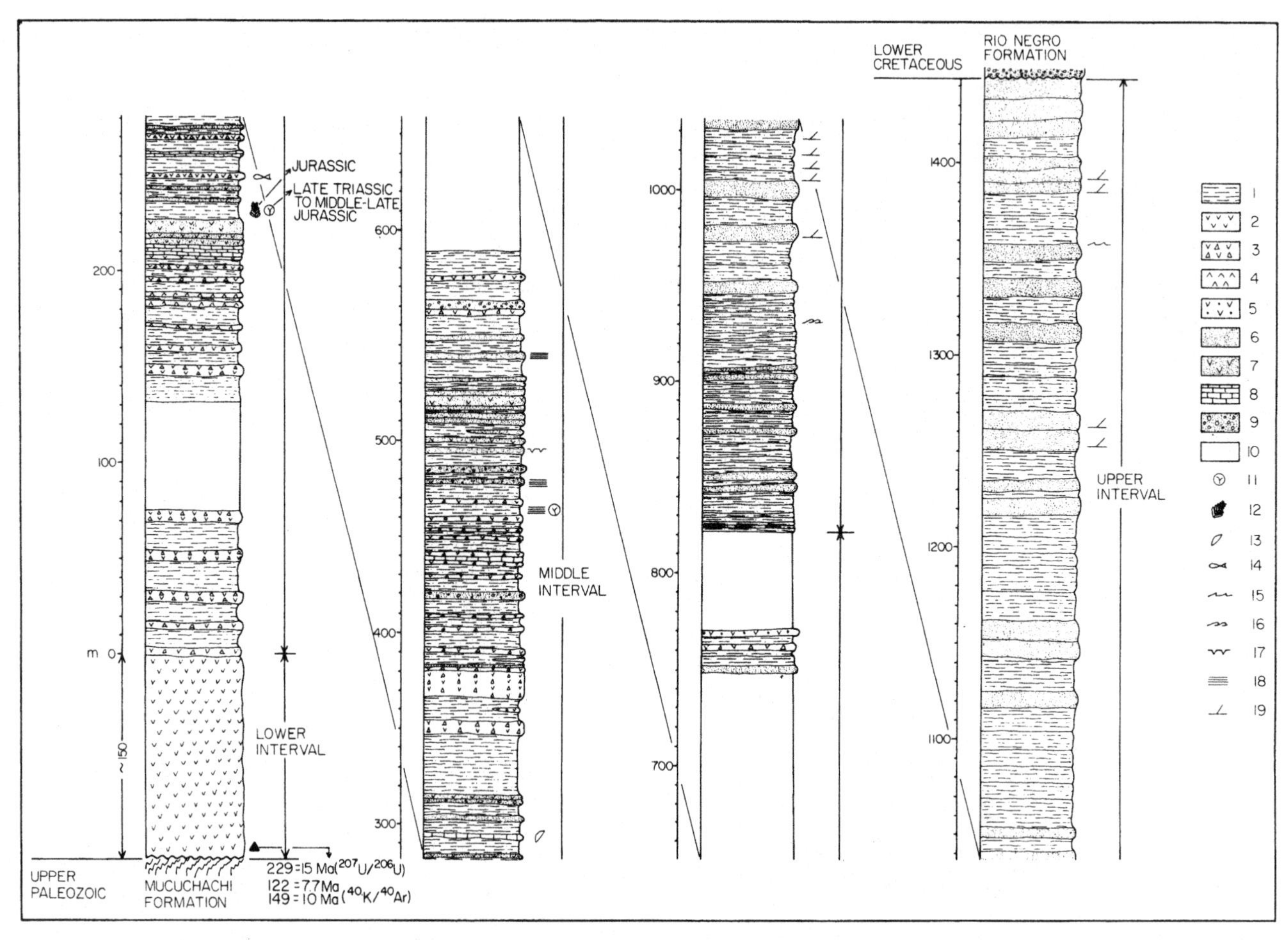


**Figure 6.3. Jurassic lithostratigraphy of Venezuela.**

pp. 158–61; Bartok, Renz, and Westermann 1985). I shall deal mainly with the little-altered rocks because they are the best known and dated.

### *Mérida Andes*

Jurassic rocks belong to the La Quinta Formation (Kundig 1938; Schubert 1986) and crop out mainly along steep cliffs with characteristic brick-red to chocolate-brown color in Táchira state (Figures 6.2 and 6.3). The La Quinta Formation occurs in graben-like structures, is extensively folded, and varies greatly in thickness from about 1,300 m to 3,400 m, but only 300 m in Barinas Piedmont. There are three members: the lower consists of violet vitric tuff, about 150 m thick; the middle member of about 840 m of interbedded green, whitish, gray, and violet tuff, conglomerate, siltstone, and thin beds of limestone; and the upper of about 620 m of interbedded brick-red or chocolate-brown siltstone and sandstone with some tuff. The fraction of pyroclastics decreases from southwest to northeast, and near Valera-Carache and southeast of La Grita there are clastics only (Arnold 1961; Schubert 1986). The La Quinta Formation lies unconformably on greenish gray phyllite of the Upper Paleozoic Mucuchachí Formation and is usually disconformably overlain, but locally transitional to, the Lower Cretaceous Rio Negro Formation. Faulted contacts with Paleozoic and Mesozoic rocks are frequent.

Fossils have been reported by Kundig (1938), Sutton (1946), Bock (1953), Useche and Fierro (1972), Benedetto and Odreman (1977), González de Juana et al. (1980, p. 171), and Schubert (1986). Fish coprolites, scales, plates, teeth, and earbones belong to *Lepidotus;* conchostracans include *Isaura olssoni* Bock, *Cyoris valdensis* Sowerby, and *Howellites colombianus* Bock; among palynomorphs are *Circulina meyeriana* Klaus, *Classopolis papillatus,* and *Caytopollenites pallidus;* and among plants the genera *Dictyophyllum, Nilssonia, Ptilophyllum, Zamites,* and *Otozamites.* All indicate a Jurassic age. Radiochronological data for the basal tuff differ greatly: $^{207}U/^{206}U$ ages average 229 ± 15 Ma; $^{40}K/^{40}Ar$ ages average 122 ± 7.7 and 149 ± 10 Ma.

### *Sierra de Perijá*

Jurassic rocks here were grouped into the La Gé Group (Hea and Whitman 1960), which consists of (ascending order) the Tinacoa, Macoita, and La Quinta Formations. Plutonic and volcanic rocks are also present (González de Juana et al. 1980, pp. 162–8).

The Tinacoa Formation consists of a lower member of tuffaceous and carbonaceous shale and tuffaceous aphanitic limestone, and an upper member of graywacke with intercalations of pyroclastic rocks and tuff, alternating with calcareous and carbonaceous shale. The thickness is poorly known and may vary between 685 m and 1,680 m (Hea and Whitman 1960; Bowen 1972). The fish *Lepidotus*, the plant genera *Ptilophyllum* and *Otozamites*, and numerous conchostracans belonging to *Cyzicus (Eustheria)* spp. and *C. (Lioestheria) colombianus* (Bock) have been identified (Bowen 1977; Odreman and Benedetto 1977), indicating Early to Middle Jurassic ages (*Ptilophyllum-Otozamites* association). The basal contact is unknown. The upper boundary with the Macoita Formation is transitional or partly discordant. The upper part grades laterally into the La Gé Volcanics.

The Macoita Formation, up to several thousand meters thick, consists of olive-gray calcareous shale and siltstone, interbedded with graywacke, tuffaceous sandstone, and lithic tuff, frequently cross-bedded. The plants *Ptilophyllum* sp. and *Phlebopteris*(?) sp. suggest a Middle Jurassic age (Odreman and Benedetto 1977). The Macoita intergrades laterally into the Tinacoa Formation and the La Gé Volcanics and is equivalent to the La Quinta Formation.

The La Gé Volcanics consist of gray to greenish gray massive tuff and agglomerate, with a few layers of black siliceous limestone in the lower part.

The La Quinta Formation consists here of a thick sequence of red silty shale, siltstone, arkose, and conglomerate. It is locally replaced, to varying degrees, by andesitic tuff, breccia, and intermediate to basic dikes of the El Totumo–Inciarte Volcanic-Plutonic Complex (Moticska 1975). At other places the upper conglomeratic part is free of volcanic rocks and has been called the Seco Conglomerate. *Posidonomya* and *Estheria* have been recorded from the Sierra de Perijá (Francken 1956; Bowen 1972). The La Quinta Formation at Perijá is equivalent to the only upper part of the La Quinta Formation in the Andes, and the Tinacoa and Macoita Formations equate to the lower and middle parts of the La Quinta of the Andes, respectively.

Radiochronology ($^{40}$K/$^{40}$Ar) for granitic rocks (biotite, chlorite, amphibole) of the El Palmar Granite, which unconformably underlies the La Quinta Formation, indicates an average age of 185 ± 7.5 Ma (Espejo et al. 1980); $^{87}$Rb/$^{86}$Sr whole-rock ages of 156–174 Ma and $^{40}$K/$^{40}$Ar whole-rock ages of 155 ± 5 and 146 ± 7 Ma were determined for the La Quinta Volcanics (El Totumo–Inciarte Complex) (Maze 1984).

## COLOMBIA[3]

### Outcrop distribution and subsurface occurrence

Figure 6.4 illustrates the Jurassic outcrops in a SSW–NNE belt. The belt comprises large areas of the Eastern Cordillera (including its bifurcation in the Sierra de Perijá and the Sierra de Mérida), the inter-Andean Middle and Upper Magdalena Valley, the eastern flank of the Central Cordillera, the Sierra Nevada de Santa Marta, and the northern half of the Guajira Peninsula. This distribution coincides quite well with the so-called Eastern Andes [Oriente Andino of Hubach and Alvarado (1934)], the region between the western border of the Guayana shield and the Romeral Masterfault (Dolores–Guayaquil Megashear) that supposedly separates continental (E) from oceanic (W) crust; the oceanic crust corresponds to the "Western Andes" (Occidente Andino). Thus the sedimentary units in the Eastern Andes region have epicontinental/miogeoclinal affinities with shallow-marine to terrestrial environments.

In the pericratonic Llanos Orientales (Subandean Basins) the Jurassic has been found in a few oil wells near the foothills of the Eastern Cordillera. These occurrences are in the Girón Formation in the northern Arauca–Casanare area and the Motema Formation in the Western Putumayo Basin.

Little is known about the Jurassic of the Western Andes, and its presence remains doubtful. Several paleogeographic reconstructions (e.g. Weeks 1947; Bürgl 1961; Harrington 1962; Cediel 1968; Geyer 1973; Barrero 1979; Mojica 1982) have postulated a paleo-Pacific oceanic domain, but that was based on the assumption, mainly by Hubach and Alvarado (1934) and Nelson (1957), that the Dagua Group is Jurassic in age. Recently, Etayo-Serna, Parra, and Rodriguez (1982) found some fauna that, to them, indicated Upper Cretaceous.

### Stratigraphy

The pre-Cretaceous in Colombia (Figure 6.5) is mainly terrestrial and volcanogenic, including redbeds, pyroclastics, and effusives, and has yielded only scarce fossils. Triassic is thus difficult to separate from undoubted Jurassic, and "dating" is commonly stratigraphic only, above and below fossiliferous Paleozoic or Cretaceous. These rocks have therefore often been referred to as "Triassic-Jurassic," "Jura-Cretaceous," or "Infra-Cretaceous."

Undoubted Triassic is present in the Payandé Formation, which occurs in a narrow belt near the Ecuadorian border (Bürgl 1961; Mojica and Dorado 1987). The approximately 600 m of limestones contain a Norian fauna (Trümpy 1943; Geyer 1973). The Montebel and Los Indios Formations, consisting of gray to black silty and sandy sediments, have yielded fossil plants, ostracods, and brachiopods, which suggest a late Norian–Liassic age, and brackish to paralic environments (Trümpy 1943; Geyer 1973).

Marine Jurassic deposits with fauna occur locally in the Liassic Batá and Morrocoyal Formations and the Malmic Cosinas Group. The bulk of the Jurassic, however, is terrestrial and consists of acidic to intermediate volcanics of the Motema, Saldaña, Jordán, Guatapurí, Corual, and Ranchogrande Formations, redbeds of the Girón and La Quinta Formations, and dark gray sedimentary rocks of the Bocas Formation. Although the fossil content of these units is normally very poor, some discoveries made during recent years have permitted dating. The Bocas Formation, which previously was considered as Paleozoic, based on the flora first studied by Langenheim (1961), has now yielded the Liassic plants

[3] By J. Mojica.

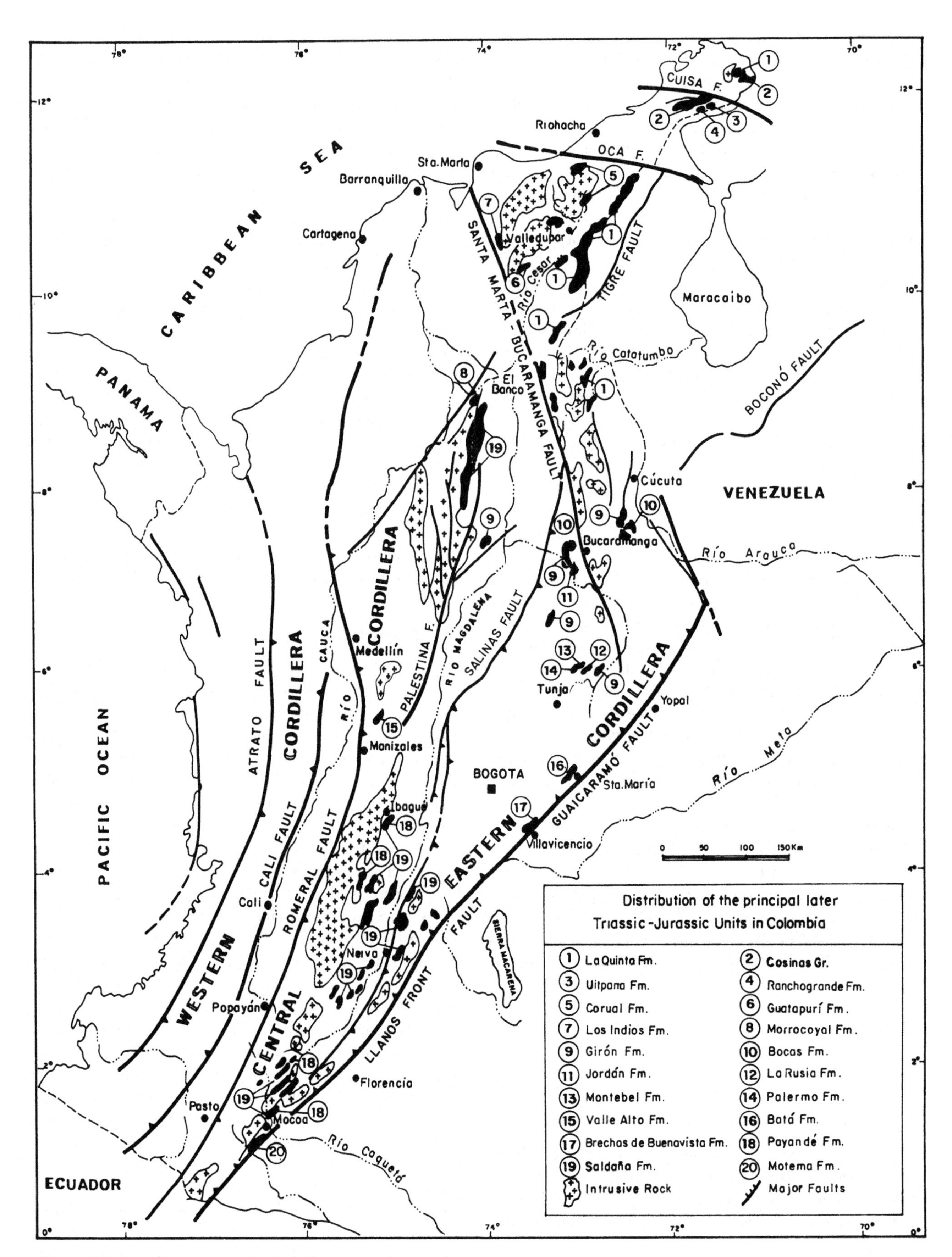


**Figure 6.4. Jurassic outcrops and principal structures in Colombia.**

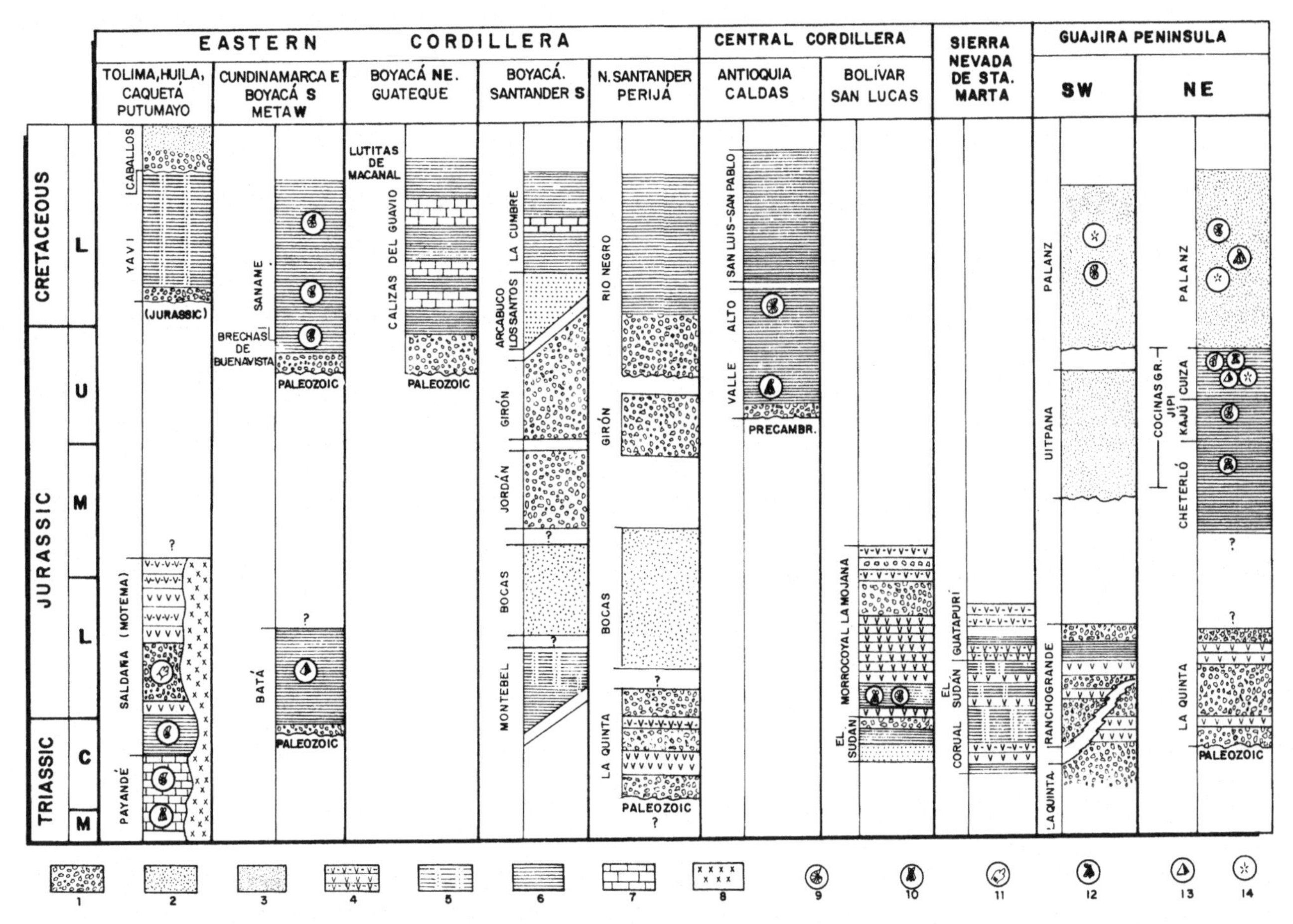


**Figure 6.5. Lithostratigraphy of the pre-Cretaceous Mesozoic in Colombia: 1, continental beds; 2, terrestrial gray facies; 3, quartzitic sandstone, continental/marine; 4, volcanics, mainly continental; 5, paralic-fluviatile facies; 6, shallow marine; 7, marine limestones; 8, granodioritic intrusions; 9, ammonites; 10, bivalves; 11, footprints; 12, oysters; 13, trigoniids; 14, corals.**

*Phleboteris branneri, Gleicheniidites* sp., and *Classopolis* sp. (Remy et al. 1975). The overlying redbeds of the Jordán Formation may consequently be Middle Jurassic, and the Girón has yielded latest Jurassic ostracods (Rabe 1974) – not the generally postulated "Jura-Triassic" range assumed years ago (Cediel 1968).

Because the lower part of the Saldaña Formation [Chicalá member of Mojica and Llinás (1984)] is late Norian (Wiedmann and Mojica 1980) and the Morrocoyal Formation is Sinemurian in age (Geyer 1967, 1973), the "Saldaña Vulcanism" appears to have included a latest Triassic–early Liassic marine phase and a Middle Jurassic (and Early Jurassic?) terrestrial volcaniclastic phase.

The Valle Alto Formation (González 1980; Huber 1982), Buenavista Breccias (Renzoni 1968; Dorado 1984), and Arcabuco Sandstones (Renzoni, Rosas, and Etayo 1983) are believed to be Tithonian to lowermost Cretaceous, based on fossils, including plants, ammonites, and bivalves, in the first two formations, and the stratic position of the third. Poorly preserved, probably latest Tithonian ammonites occur west of Villavicencio in the Sáname Formation (Bürgl 1960, 1961; Geyer 1973; Wiedmann 1980a,b).

PERU[4]

The Jurassic System of Peru is best exposed in the south (Figure 6.6), but also is known from the central and northern regions. The correlation chart (Figure 6.7) summarizes the formations and their ages.

### Lower Jurassic

The Lias is well represented in the Andean belt by marine calcareous rocks, the Pucará Group, which to the northwest (6–8°S) passes into sandy limestones and lithic sandstones. In the southwest (15–18°S) is a marine volcanic succession, the Chocolate volcanics.

The Pucará Group extends from Ecuador, where it has a N–S strike and is known as the Santiago Formation, along the central Andes with a NW–SE strike, to the Abancay deflection near Cuzco (13–14°S). The carbonatic sequence spans Upper Triassic (Norian) to Toarcian, and in parts the Aalenian. In central Peru the basin becomes deeper toward the Subandean Belt, across a fault axis passing through Cerro de Pasco-Junin.

[4] By O. Palacios.

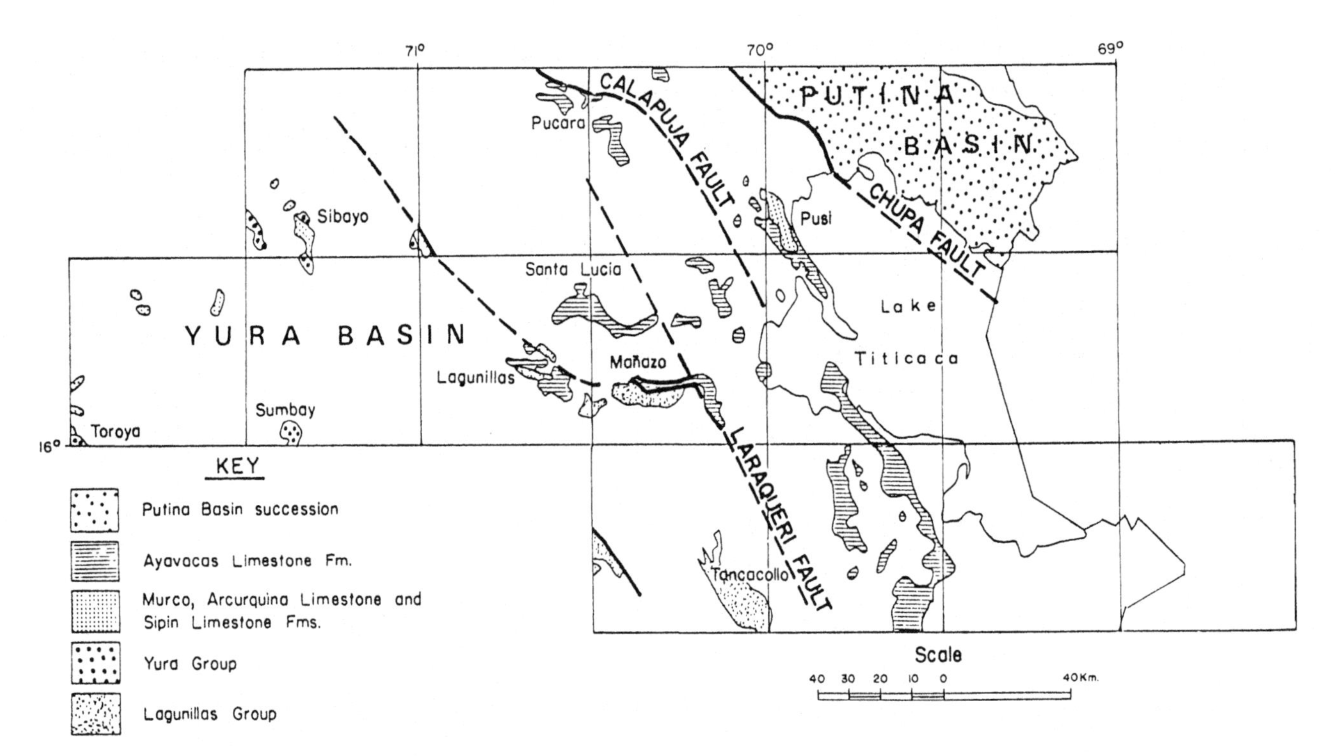


**Figure 6.6. Jurassic–Cretaceous outcrop in southern Peru.**

The Pucará outcrops in north Peru occur in long NW–SE-striking belts along the Eastern Cordillera and the Altiplano: from the Marañon Valley to the headwaters of the Huallaga River, Cerro de Pasco, Junín Lake, Mantaro River valley, Pampas River (south of Ayacucho), to Abancay and Chuquibambilla. In the Subandean Belt a thick succession extends from San Ramón, Oxapampa, Tingo María, and Chachapayas. In the Utcubamba River valley, Weaver (1942) recognized Hettangian and Sinemurian in two calcareous series, the Chillingote and Suta Formations. They lie conformably on the Upper Triassic Utcubamba Formation and are unconformably overlain by Upper Jurassic continental redbeds.

In central Peru, Harrison (1940–3) and Megard (1968) divided the Pucará Group into the Chambara (Norian–Rhaetian), Aramachay, and Condorsinga (Liassic) Formations. The Aramachay Formation consists of some 300–400 m of brown-weathering dark limestones, shales, and sandstones with high organic content. The Hettangian ammonites *Psiloceras, Caloceras,* and *Schlotheimia* and the Sinemurian ammonites *Eparietites* and *Paltechioceras* have been recorded. The Condorsinga Formation consists almost entirely of banked oolitic or bioclastic limestones and is said to contain the bivalve *Weyla alata* (v. Buch), echinoids, corals, sponges, and the Sinemurian to Toarcian ammonite genera *Oxynoticeras, Coeloceras, Androgynoceras, Uptonia, Harpoceras, Esericeras,* and *Phymatoceras* (Megard 1968). The record of *Stephanoceras* dates the top of this formation at Huancayo (Harrison 1956) as late Lower Bajocian.

In southern Peru there are the Early Jurassic Chocolate Volcanics consisting of marine andesites interbedded with reef limestones, and to the east the Sinemurian limestones and clastics of the Lagunillas Group (Newell 1949; Portugal 1974).

### Middle Jurassic

The Dogger succession has been dated by marine fauna in the Central Andean region of central Peru and in southern Peru. In central Peru, the Bajocian outcrops are restricted to a NW–SE belt extending between Huancayo and Abancay. The calcareous succession of the Chunumayo region, known as the Chunumayo Formation (Weaver 1942; Megard 1968), has yielded a small ammonite assemblage described by Jaworski (1926) and revised by Westermann et al. (1980): *Sonninia* (*Papilliceras*) cf. *mesacantha* Waagen, *S.* (*P.*) aff. *arenata* (Quenst.), *S.* (*P.?*) *peruana* Jaw., *Emileia multiformis* (Gott.), and *E.* cf. *giebeli submicrostoma* (Gott.). This formation extends to the Bajocian–Bathonian boundary in the Pacuyacu River, south of Ayacucho. In that area a shaley succession conformably overlies limestones with *Spiroceras orbigny* (Baug. & Sauze) and *Leptosphinctes* (*Cobbanites*) cf. *talkeetnanus* Imlay [= ?*Bigotites martinsi* (d'Orb.)] (Westermann et al. 1980). The Bathonian could be represented in continental facies that are overlain by Callovian strata.

At the southern coast, near Nazca, the limestones and sandstones of the Río Grande Formation have yielded supposedly Aalenian, Bajocian, and, perhaps, Bathonian fauna (Rüegg 1968). Farther south, in Arequipa, the marine Lower Bajocian, with *Sonninia* sp., consists of the limestones of the Socosani Formation, and in Tacna of carbonatic clastics of the San Francisco Formation. East of Arequipa, in Lagunillas, the Lagunillas Group includes Sinemurian–Callovian fauna (Newell 1949; Portugal 1974; Westermann et al. 1980).

In northeastern Peru the epicontinental succession of the Boquerón (lower Sarayaquillo) Formation (?Bajocian) lies conformably on the Pucará Group.

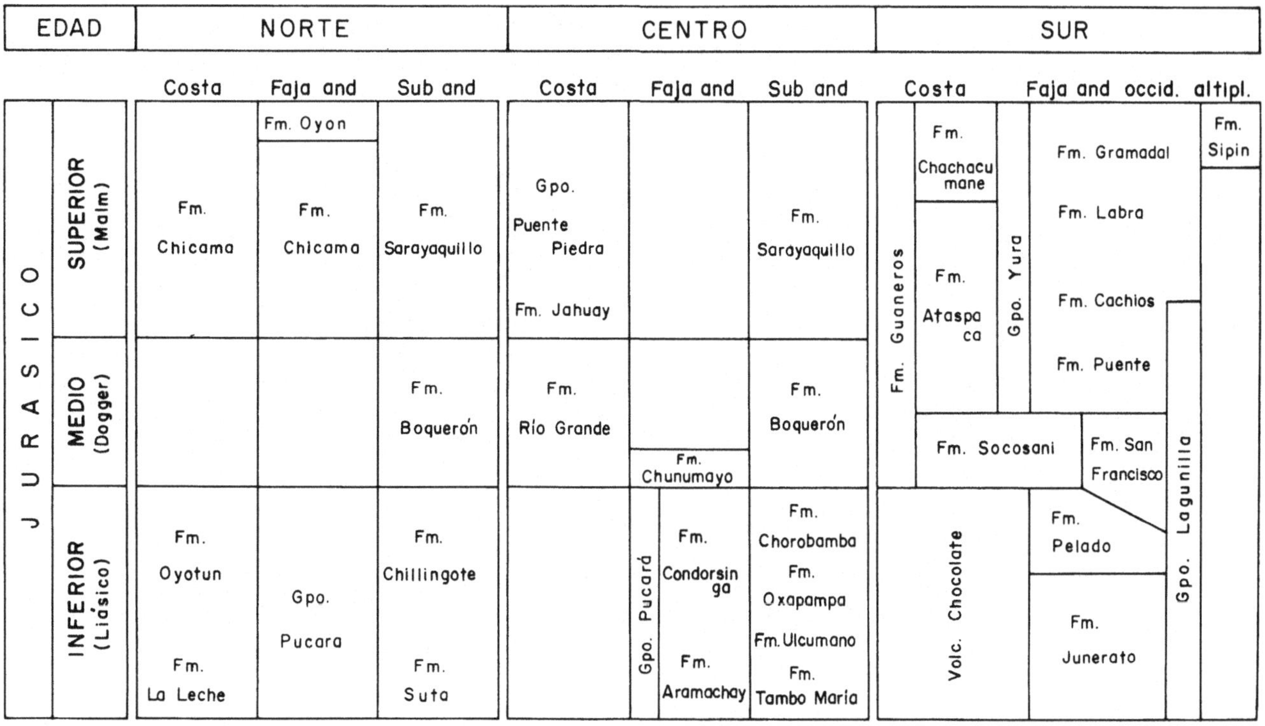

| EDAD | | NORTE | | | CENTRO | | | SUR | |
|---|---|---|---|---|---|---|---|---|---|
| | | Costa | Faja and | Sub and | Costa | Faja and | Sub and | Costa | Faja and occid. altipl. |
| JURASICO | SUPERIOR (Malm) | Fm. Chicama | Fm. Oyon<br>Fm. Chicama | Fm. Sarayaquillo | Gpo. Puente Piedra<br>Fm. Jahuay | | Fm. Sarayaquillo | Fm. Guaneros<br>Fm. Chachacumane<br>Fm. Ataspaca | Gpo. Yura: Fm. Gramadal, Fm. Labra, Fm. Cachios, Fm. Puente<br>Fm. Sipin |
| | MEDIO (Dogger) | | | Fm. Boquerón | Fm. Río Grande | Fm. Chunumayo | Fm. Boquerón | Fm. Guaneros<br>Fm. Socosani | Fm. Socosani<br>Fm. San Francisco<br>Gpo. Lagunilla |
| | INFERIOR (Liásico) | Fm. Oyotun<br>Fm. La Leche | Gpo. Pucara | Fm. Chillingote<br>Fm. Suta | | Gpo. Pucará: Fm. Condorsinga, Fm. Aramachay | Fm. Chorobamba<br>Fm. Oxapampa<br>Fm. Ulcumano<br>Fm. Tambo María | Volc. Chocolate | Fm. Pelado<br>Fm. Junerato<br>Gpo. Lagunilla |

**Figure 6.7. Correlation chart for the Jurassic of Peru.**

Whereas a Bathonian hiatus may be present in the northern and central Andes of Peru, the poorly exposed lower sandy shales of the Yura Group in southern Peru have yielded *Eurycephalites* and *Reineckeia* above *Epistrenoceras*, indicating a continuous upper Bathonian–Callovian succession. The Yura Group intergrades westward with andesitic volcanics of the Guaneros Formation. To the east, the Callovian consists of the calcareous shales of the upper part of the Lagunillas Group, with a fauna similar to that of the Yura Group.

The Callovian succession extends north of Arequipa to Huancavelica and to the Ayacucho area (Pumani River–Huancapi), again with predominantly sandy and calcareous shales. It yielded *Gryphaea* cf. *tricarinata* Philippi and *Eurycephalites* cf. *bosei* (Burckhardt) in the Querabamba Quadrangle (E. Guevara, personal communication).

### Upper Jurassic

Malm is present between 40′S and 18°00′S and is quite thick in southern Peru, where the Yura Group bears faunas indicating Callovian to Tithonian. The group has been divided in the Puentes, Cachios, Labra, Gramadal, and Huallhuani Formations (the last probably Early Cretaceous) (Jenks 1948; Benavides 1962). The clastic succession consists of shales, quartzites, and sandstones, with limestones in the upper part.

In northeastern Peru, and Pucará Group is unconformably overlain by carbonaceous black shales and gray sandstones with tuffaceous levels, the Chicama Formation. It has yielded the Tithonian ammonite genera *Berriasella* and *Aspidoceras*.

Northwest of Lima, Wilson (1963) placed in the Upper Jurassic a succession of shales and carbonaceous siltstones interbedded with sandstones and quartzites, the Oyón Formation. It has yielded only plant fragments, including *Wechselia* and *Otozamites*. In the Oyón River valley, this formation forms the base of the Goyllar Group, which along the northwestern Andean Belt is Early Cretaceous.

Along the central coast near Lima, the Tithonian consists of marine andesitic volcanics interbedded with clayish and sandy strata, the Puente Piedra Formation, bearing fauna and flora. A similar lithologic succession, exposed between Ica and Nazca, is included in the Río Grande Formation. Southeast of Lima, in the Huaytara River valley, whitish quartzites and black shales with *Mytilus* and *Otozamites* were placed in the Malm (Bellido 1956).

In the Late Jurassic, the Northeastern Basin, extending from Ecuador to about 12°00′S, was separated from the marine Northwestern Basin by the Marañon Geoanticline. The Northeastern Basin has red continental molasse, the Sarayaquillo Formation.

### Geological evolution

The evolution of the Jurassic basins of northern and southern Perú is graphically summarized in Figure 6.8.

## ARGENTINA AND CHILE[5]

Jurassic rocks in Argentina and Chile are present over extensive areas and include a large variety of marine and continental

[5] By A. C. Riccardi and C. A. Gulisano.

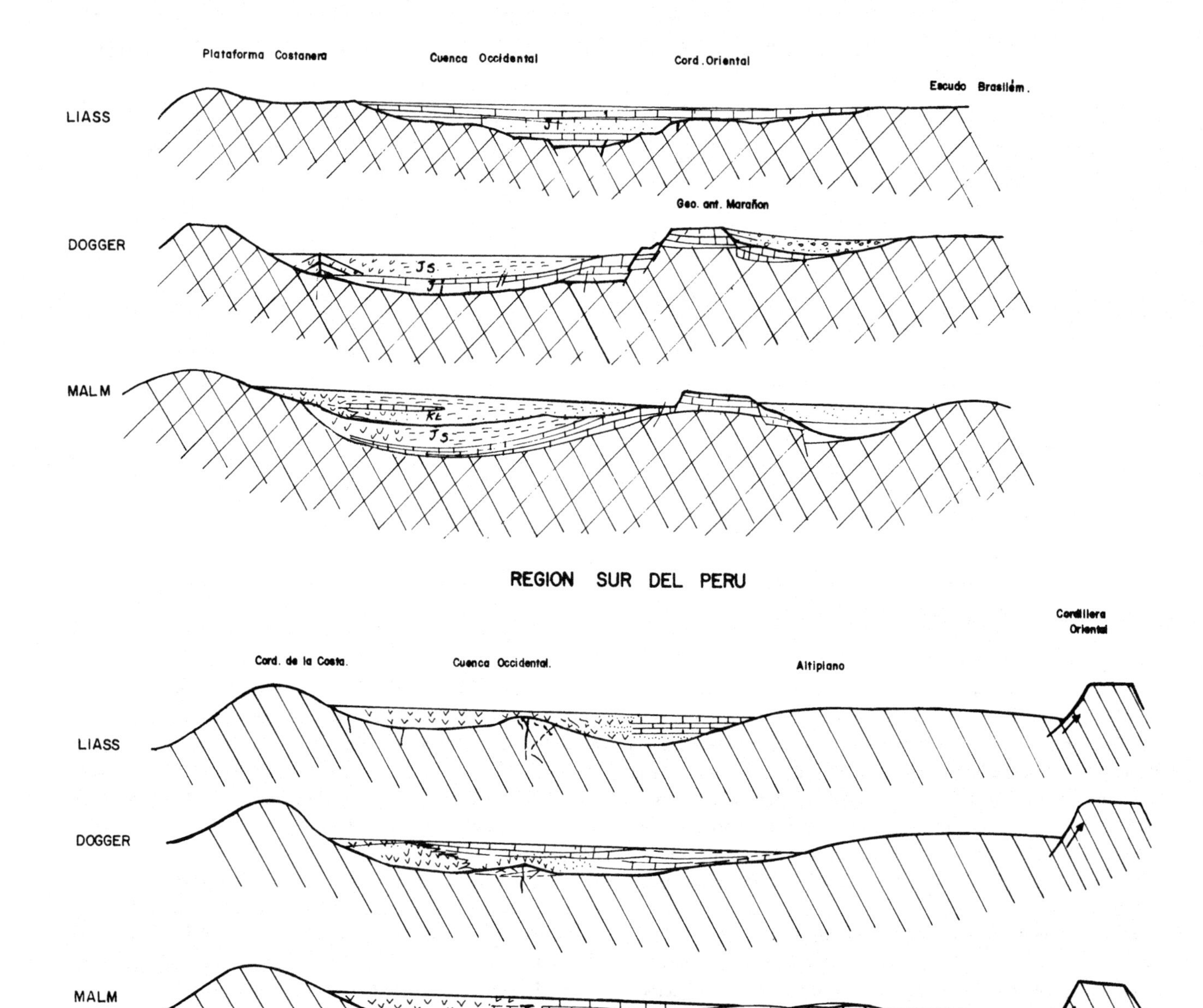


**Figure 6.8. Geological evolution of Jurassic basins in northern and southern Peru.**

facies (Riccardi 1983). South of 39°S the Jurassic is widespread in Patagonia, where it is represented by continental and marine volcanics interbedded with sedimentary rocks in some areas. Between 39°S and 32°S, Jurassic outcrops are mostly in Argentina along a north–south belt roughly coincident with the boundary between Argentina and Chile, whereas north of 31°S the Jurassic is unknown in Argentina. To the east of this belt, the Jurassic consists mainly of marine strata, and to the west it is a thick volcanic sequence with marine intercalations. Jurassic rocks are also known from northeastern Argentina, Uruguay, Paraguay, and southern Brazil, where the continental sandstones of the Botucatú Formation are overlain by Late Jurassic–Early Cretaceous extensive basaltic flows, the Serra Geral Formation (Riccardi 1988).

### Southern Patagonia[6]

The older rocks in this area are Paleozoic sediments and metamorphics and Jurassic volcanics. They are overlain by Upper Jurassic–Cenozoic strata.

The Jurassic volcanics are up to 2,000 m thick and consist of porphyrites, rhyodacites, andesites, and their breccias and tuffs. They originated in an extensive volcanism that covered most of Patagonia and were variously named in different areas (Riccardi and Rolleri 1980), as discussed later. In southern Patagonia they are included in the El Quemado Complex (Andes) or Tobífera Formation (subsurface). These volcanics are mostly continental, but in some areas along the boundary between Chile and Argentina their upper levels interfinger with and are overlain by marine strata, the Zapata Formation. It has yielded different species of *Aulacosphinctes* and *Berriasella* (Feruglio 1936–7; Leanza 1968; Olivero 1987), which indicate several Tithonian-Berriasian ammonite zones recognized in west-central Argentina. This is probably also the age of the youngest volcanics. The Kimmeridgian *Retroceramus haasti* (Hochstetter), *Glochiceras* sp., *Uhligites*(?) sp., and *Taramelliceras* cf. *trachinotum* (Oppel) have been reported from southern Chile overlying and underlying the volcanics (Fuenzalida and Covacevich 1988). At Lake Fontana, volcanic rocks of the El Quemado Complex overlie marine Toarcian (Malumian and Ploszkiewicz 1977). Radiochronology (Rb-Sr and K-Ar) indicates values ranging between 158 ± 10 Ma and 136 Ma (Halpern 1973; Charrier et al. 1979; Nullo, Proserpio, and Ramos 1979), which is a mostly Late Jurassic age.

A belt of basaltic rocks, the "green rocks" or Sarmiento Complex, exposed to the west in Chile, has the same stratic position and age as the El Quemado Complex (Fuenzalida and Covacevich 1988). Toward the south, similar rocks are known as the Tortugas Complex. Farther southwest and south there is a magmatic-arc suite of Upper Jurassic–Lower Cretaceous rocks of rhyolitic to basaltic composition, the Hardy Formation (Suárez and Pettigrew 1976).

The volcanic rocks of the El Quemado Complex are overlain by Late Jurassic–Early Cretaceous shales and sandstones.

Continental to marine sandstones of the Springhill Formation are well developed on the platform. They are 30–40 m (but locally 130–150 m) thick and usually restricted to the topographic depressions of the El Quemado Complex. Regionally the Springhill consists of three or more backstepping (but individually prograding) sandstone sequences, which may or may not be continuous (Biddle et al. 1986). Although the ammonites *Jabronella* aff. *michaelis* (Uhlig), *Neocosmoceras* spp., and *Delphinella* sp. (Riccardi 1976, 1977; Olivero 1983) are Berriasian in age, the reported presence of *Aspidoceras* cf. *haupti* Krantz, *Virgatosphinctes* sp., and *Aulacosphinctoides* sp. (Blasco, Nullo, and Proserpio 1979b) indicates Tithonian. On the basis of microfossils (Sigal et al. 1970; Natland et al. 1974; Kielbowicz, Ronchi, and Stach 1984) this formation has been dated as Oxfordian–early Valanginian (Riccardi 1988), the wide range being attributed to diachronism. Plants and palynomorphs, probably of Tithonian age, have been described by different authors (Riccardi 1988).

Intrusives of adamellitic to tonalitic composition are present in the southern Andes. Isotopic analysis indicated that the oldest intrusives are Late Jurassic (Halpern 1973; Hervé et al. 1981).

### Deseado Massif[7]

This structurally positive block near the Atlantic coast (47–49°S) behaved as a craton during the Mesozoic-Cenozoic Andean orogeny. A major part of the massif is covered by Middle to Late Jurassic volcanic and pyroclastic rocks.

The older Jurassic rocks consist of gray and green tuffs, siltstones, and coarse sandstones of continental origin, the Roca Blanca Formation. This unit is 990–1,200 m thick and has yielded the anuran *Vieraella herbsti* (Reig 1961; Casamiquela 1965) and fossil plants of the genera *Equisetites, Coniopteris, Thaumatopteris, Otozamites, Cladophlebis,* and *Sagenopteris* (Herbst 1965). The Roca Blanca Formation has been dated as Toarcian-Aalenian. It is probably conformably overlain by 200–600 m of basalts and volcanic agglomerates, which in some areas are interbedded with sandstones, tuffs, and conglomerates of the Bajo Pobre Formation.

Unconformably overlying these two continental units is the volcanic and pyroclastic Bahía Laura Group or Complex, the extra-Andean equivalent of the El Quemado Complex. The Bahía Laura Group has been divided in the Chon Aike and La Matilde Formations, chiefly volcanic and mainly pyroclastic, respectively. The La Matilde has yielded vertebrate tracks, anurans (Stipanicic and Reig 1957; Casamiquela 1964), and plants, including the genera *Phellinites, Osmundacaulis, Ruffordia, Gleichenites, Hausmannia, Sphenopteris, Cladophlebis, Otozamites, Ptilophyllum, Podozamites, Paraaraucaria, Araucarites, Podocarpus,* and *Masculostrobus* (Stipanicic and Bonetti 1970). Radiochronology (K-Ar) gave 161 Ma (Cazeneuve 1965) and 166 ± 5 Ma (Creer, Mitchell, and Abou Deeb 1972), indicating Middle to Late Jurassic for the Bahía Laura Group.

### Central and northern Patagonia[8]

In west-central Patagonia the Jurassic succession is mostly continental (Figure 6.9). It has been interpreted as the infilling of NNW-striking grabens formed on the Paleozoic "basement" during Late Triassic–Early Jurassic time (Gust et al. 1985; Uliana et al. 1985). However, there appears to be a facies change in a NE–SW direction, from coarse to fine-grained sediments (Cortiñas 1984).

In the northeast the Jurassic consists of up to 640 m of coarse-grained, gray-greenish andesitic conglomerates, sandstones, and siltstones, the Cajón de Ginebra Formation. To the southwest, similar conglomerates are 250 m thick, the El Córdoba Formation, and are overlain by, and in part interfingered with, up to 300 m of continental to marine gray siltstones and yellowish

[6] By A. C. Riccardi.

[7] By A. C. Riccardi. [8] By A. C. Riccardi.

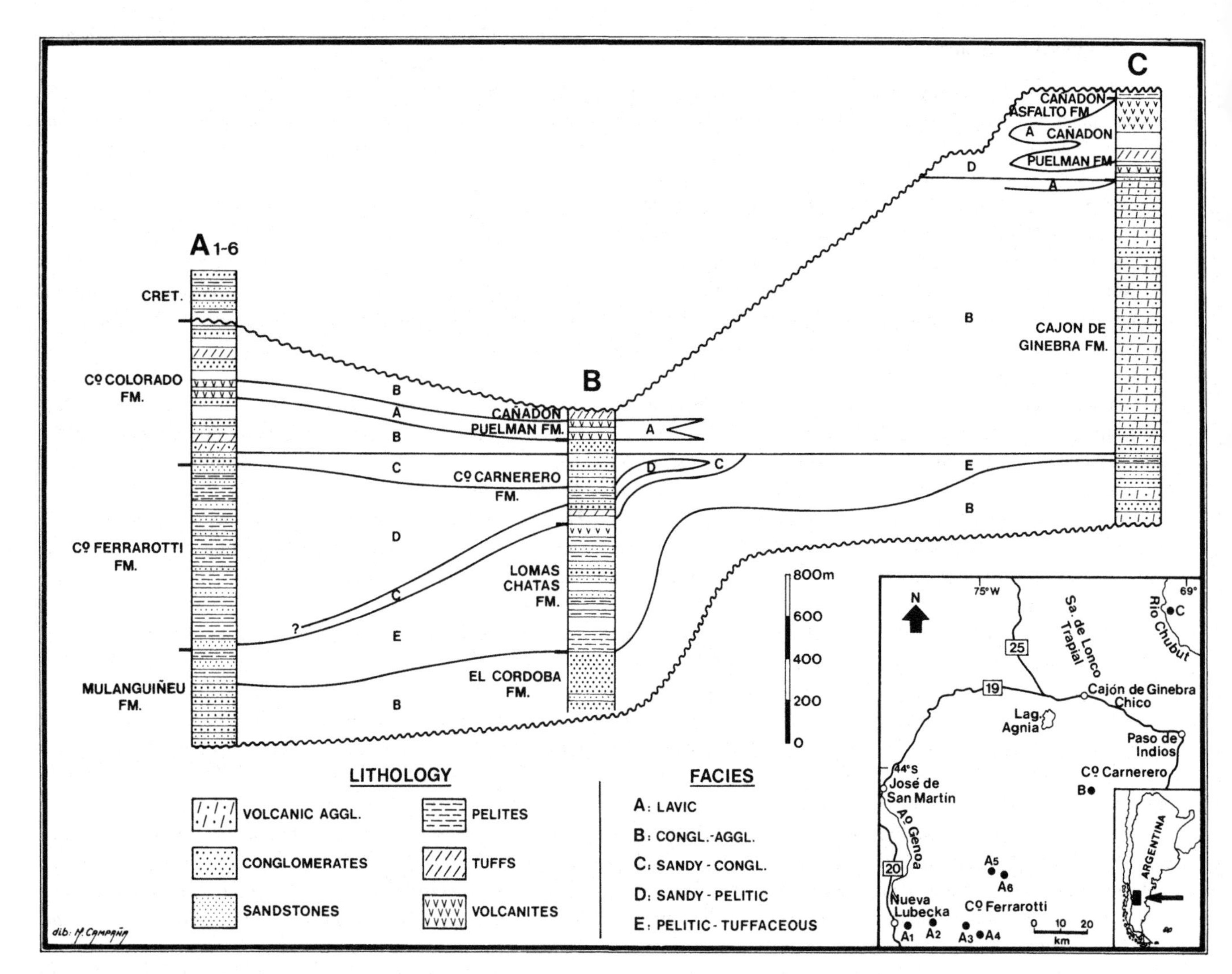


**Figure 6.9. Geological columns with facies changes for the Jurassic of central Patagonia, from northeast to southeast. (Adapted from Cortiñas 1984.)**

sandstones, tuffs, tuffites, and conglomerates. The marine facies, called the Osta Arena or Lomas Chatas and the Mulanguiñeu Formations, yielded late Pliensbachian–early Toarcian ammonites (Musacchio and Riccardi 1971; Blasco et al. 1979a; Hillebrandt 1987): *Fanninoceras oxyconum* Hill., of the *disciforme* Zone; *Dactylioceras* (*Eodactylites*) *simplex* Fucini, of the *tenuicostatum* Zone; *D.* (*Orthodactylites*) *helianthoides* Yok., *Nodicoeloceras* cf. *crassoides* (Simp.), of the *hoelderi* Zone; *Peronoceras spinatum* (Frebold), *P.* cf. *vorticellum* (Simpson), *P. tenellissimum* (Monestier), *Catacoeloceras* cf. *crassum* (Young & Bird), of the *chilensis* Zone. The Puntudo Alto Formation, with marine and continental facies, bears plants of the genus *Coniopteris, Goeppertella, Sclerôpteris, Otozamites, Sagenopteris, Ptylophyllum,* and *Elatocladus.*

Marine facies have been recorded as far south as Lake Fontana, with ammonites of the *pacificum* Zone (Blasco, Levy, and Ploszkiewicz 1980). To the west-northwest, marine strata are interbedded with shales, limestones, sandstones, and conglomerates, alternating with dacites, rhyodacites, and andesites, named the Piltriquitrón or Epuyen-Cholila Formation, the Millaqueo-(?) and Montes de Oca Formations (González Bonorino 1974; González Diaz and Nullo 1980; Lizuain 1980), and the Futaleufu Group (Thiele et al. 1979).

The Osta Arena and Mulanguiñeu Formations probably are unconformably overlain by 350 m of red and white tuffs and reddish and greenish siltstones with intercalated conglomerates of the Carnerero or Cerro Ferrarotti Formation, which yielded the sauropod *Amygdalodon patagonicus* Cabrera, followed by 100–175 m of andesites and andesitic tuffs, conglomerates, and agglomerates of the Pampa de Agnia or Cañadon Puelman or Cerro Colorado Formation. The Cerro Carnerero, Cajón de Ginebra, and Cañadon Puelman Formations constitute the Lonco Trapial Group, which is coeval to part of the Bahía Laura Group of the Deseado Massif, and to the El Quemado Complex of southern Patagonia. Stratigraphic relationships suggest a Toarcian to Middle Jurassic age for the Lonco Trapial Group; isotope analysis of its volcanics gave

158 ± 6 Ma and 147 Ma, that is, Late Jurassic (Stipanicic and Rodrigo 1970; Franchi and Page 1980).

To the east-northeast, close to the Atlantic coast (41–44°S), the Lonco Trapial Group interfingers with a thick sequence of rhyolitic, rhyodacitic, and dacitic volcanics, the Marifil Complex (Cortés 1981; Lapido, Lizuain, and Nuñez 1984). It has been divided into (ascending order) the Puesto Piris Formation (conglomerates, tuffs, tuffites, volcanic agglomerates, and ignimbrites), the Aguada del Bagual Formation (rhyolitic intrusives), and the La Porfia Formation (tuffites, sandstones, volcanic agglomerates, and rhyodacitic ignimbrites). Radiochronology (K-Ar) gave values ranging between 211 ± 10 Ma and 161 ± 10 Ma, that is, Early to Middle Jurassic (Cortés 1981).

Probably conformably above the Lonco Trapial Group follows a 300-m-thick lacustrine sequence of tuffs, conglomerates, sandstones, siltstones, limestones, and bituminous shales, the Cañadon Asfalto Formation. It bears plants, freshwater invertebrates (Tasch and Volkheimer 1970), and fish and has been dated as Callovian–Oxfordian.

In northern Chubut (ca. 42–43°S and 69–70°W), a succession of volcanic and pyroclastic rocks, the Taquetren Formation, has furnished a flora similar to that of the Cañadon Asfalto Formation. But it is probably younger and of latest Jurassic–earliest Cretaceous age. These formations have yielded the plant genera *Equisetites, Gleichenites, Clathropteris, Thaumatopteris, Sphenopteris, Cladophlebis, Scleropteris, Sagenopteris, Williamsonia, Otozamites, Zamites, Araucarites, Elatocladus, Athrotaxis, Brachyphyllum,* and *Pagiophyllum* (Herbst and Anzoátegui 1968; Stipanicic and Bonetti 1970). Isotope analysis gave 136 ± 6 Ma (Nullo and Proserpio 1975; Franchi and Page 1980; Lapido et al. 1984), indicating Tithonian–Berriasian.

## West-central Argentina[9]

In the Neuquén, Mendoza, Río Negro, and La Pampa provinces of west-central Argentina (Figure 6.10) a 6,000-m-thick Upper Triassic–Lower Tertiary sedimentary succession is present within a back-arc basin, the Neuquén Basin. The axis of the Neuquén Basin (Figures 6.10 and 6.11) is roughly coincident with that of the Andes, with a southeastward expansion. It was bounded in the west by an active volcanic arc and in the east by the land formed by Paleozoic and Triassic volcanics, plutonics, and metamorphics.

The Jurassic sedimentary succession of the Neuquén Basin has been included in the Araucanian Synthem (Riccardi and Gulisano 1992) and the lower part of the Andean Cycle (Groeber 1946). The lower part of the Araucanian Synthem includes uppermost Triassic, and within the lower part of the Andean Cycle is the Upper Jurassic–Lower Cretaceous transition. The paleogeographic evolution of the area between 135 and 210 Ma is depicted in several maps.

The Jurassic of the Neuquén Basin has been dealt with in numerous papers. Among the most recent syntheses are those by Digregorio (1972, 1978), Yrigoyen (1979), Digregorio and Uliana (1980), Riccardi (1983), Orchuela and Ploszkiewicz (1984), and Legarreta and Gulisano (1989). The evolution of the Jurassic succession has been analyzed using a seismostratigraphic or sequential stratigraphic approach (Gulisano 1981; Mitchum and Uliana 1982; Gulisano 1984; Gulisano, Gutierrez Pleimling, and Digregorio 1984; Legarreta and Kozlowsky 1984; Mitchum 1985; Hinterwimmer, Bravo de Laguna, and Orchuela 1986; Orchuela, Cibanal, and Palade 1987).

Riccardi and Gulisano (1992) used the Jurassic of this area as an example of discontinuity-bounded units, *sensu* Chang (1975).

### Araucanian Synthem

The Araucanian Synthem consists of more than 5,000 m of Upper Triassic (Rhaetian) to Upper Oxfordian marine and continental strata, with subordinate tuffs and volcanics. It lies unconformably (Supra-Triassic Discontinuity) on Upper Triassic (Carnian–Norian) continental deposits, Permian volcanics (Choiyoi Group), and granites and metamorphites of probable Devonian age.

As a result of a late Oxfordian–early Kimmeridgian fall in sea level, coupled with expansion of the volcanic arc, marginal areas of the Neuquén Basin were subjected to erosion, and continental deposition occurred in central parts of the basin. In southern Neuquén, faulting and uplift resulted in a regional unconformity (Intra-Malmic Discontinuity), and in other areas of the basin the Andean Cycle was deposited paraconformably on different parts of the Araucanian Synthem.

*Sañico Subsynthem.* This is a thick epiclastic and volcanic succession bounded by the Supra-Triassic and Intra-Malmic Discontinuities. The Sañico Subsynthem is equivalent to the Precuyan Cycle (Gulisano et al. 1984) and includes the Piedra del Aguila, Sañico, Chacaico, and Lapa Formations of western Neuquén province, the Planicie Morada Formation of eastern Neuquén, Río Negro, and La Pampa provinces, and the Remoredo Formation of southern Mendoza province.

As a result of the Supra-Triassic and previous discontinuities, this subsynthem rests on units of different ages. Its upper limit is defined by the Intra-Liassic Discontinuity, above which are the marine and continental deposits of the Cuyo Subsynthem.

Taphrogenic depressions produced by extensive basement faulting were filled by the related volcanics and the sedimentary rocks of the Sañico Subsynthem. Irregularly distributed wedge deposits, up to 1,000 m thick, are the result of this rifting event.

The Sañico Subsynthem is dated as Late Triassic–early Hettangian, as it rests unconformably on Carnian-Norian deposits and is covered by strata of the Cuyo Subsynthem, whose oldest paleontologic record indicates late Hettangian.

*Cuyo Subsynthem.* This includes the first Mesozoic marine strata recorded in the Neuquén Basin. It is bounded by the Intra-Liassic (Gulisano et al. 1984) and Intra-Callovian (Dellape et al. 1979) regional discontinuities. The subsynthem includes the

[9] By C. A. Gulisano.

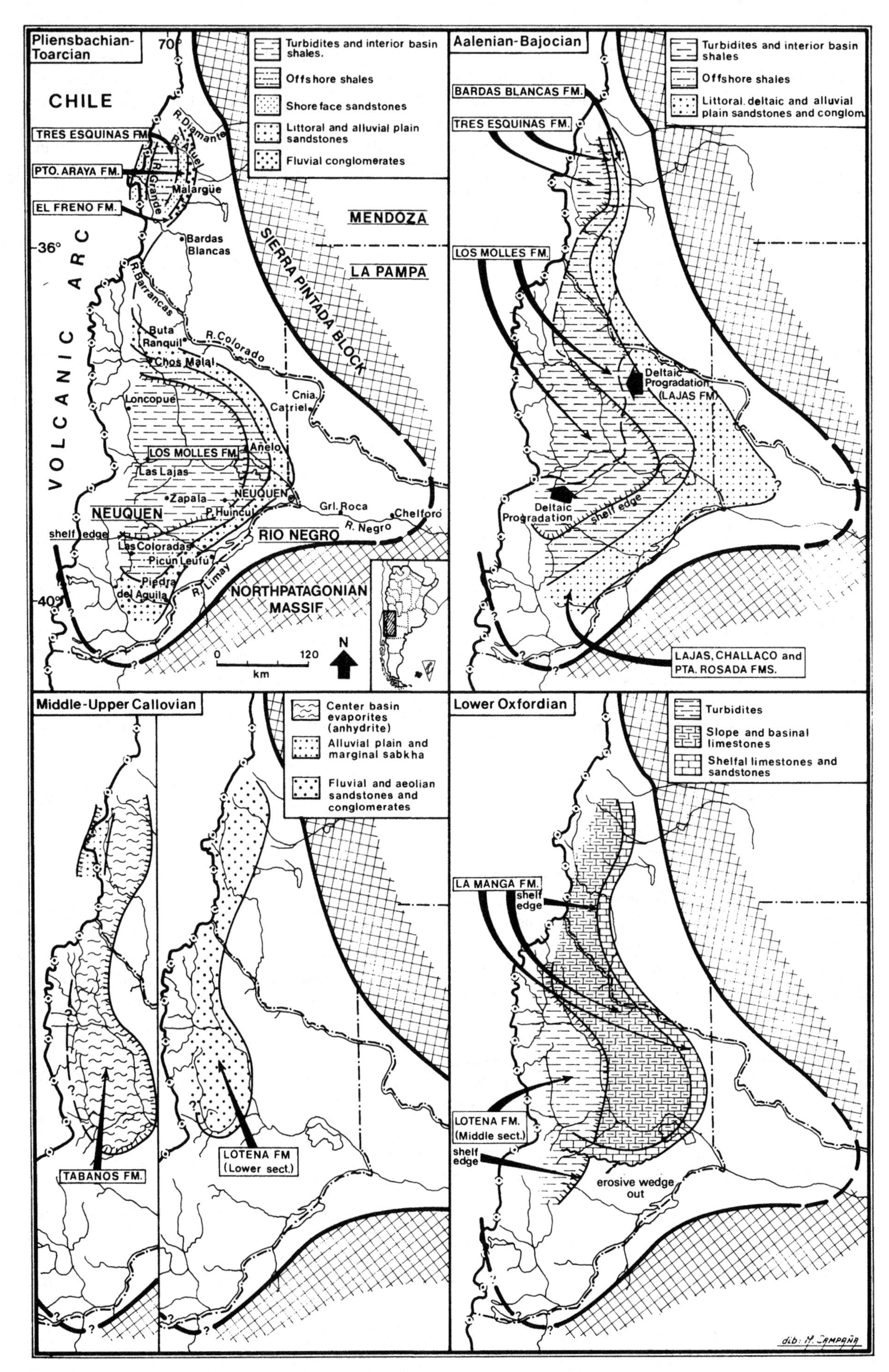


**Figure 6.10. Paleogeographic maps of west-central Argentina: Pliens-**

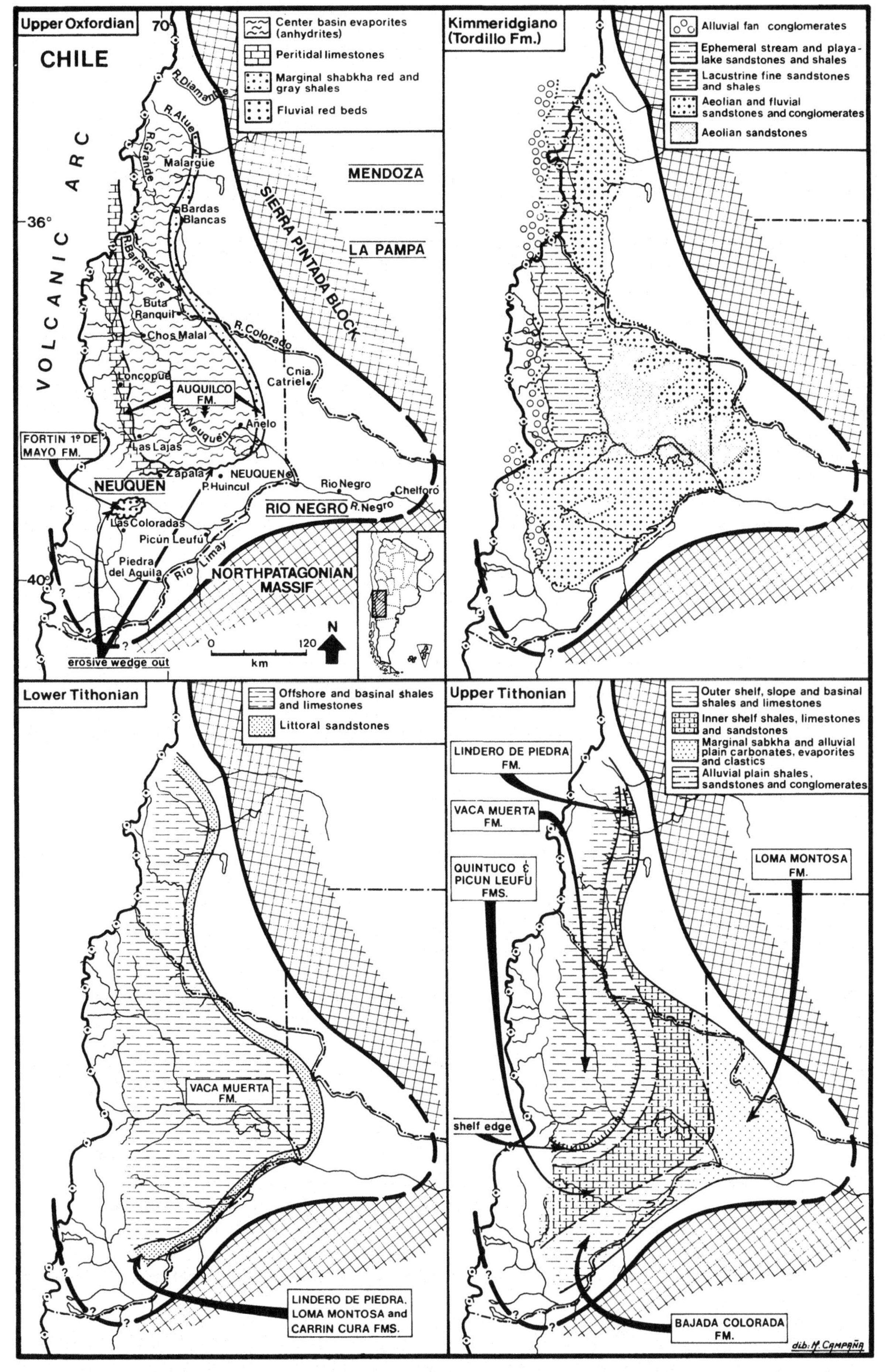


Figure 6.11. Paleogeographic maps of west-central Argentina: late Ox-

following lithostratigraphic units: Cuyo Group and Los Molles, Piedra Pintada, Tres Esquinas, El Freno, Puesto Araya, Bardas Blancas, Lajas, Loma Negra, Challaco, Punta Rosada, Calabozo, and Tabanos Formations. These rocks rest unconformably on different units of the pre-Jurassic basement and the Sañico Subsynthem. Upward it is separated from the Lotena Subsynthem by the Intra-Callovian Discontinuity. In some places the Lotena Subsynthem is missing because of erosion or nondeposition, and the Cuyo Subsynthem is covered by the Mendoza Group. The maximum recorded thickness is 3,500 m.

Tensional tectonism of the Neuquén Basin was followed by a marked sea-level rise, and the resulting transgressive episode caused the deposition of the Cuyo Subsynthem.

The lower part of the Cuyo Subsynthem consists of coarse continental deposits (El Freno Formation in southern Mendoza, and the basal conglomerate of the Los Molles Formation in central and southern Neuquén), which are overlain by littoral sandstones (Puesto Araya Formation of southern Mendoza) and/or outer-shelf to basinal black pelites (Tres Esquinas Formation in Mendoza and Los Molles and Piedra Pintada Formations from southern Mendoza to southern Neuquén). In southern Mendoza this subsynthem was deposited in increasingly deep water. Regressive sandstones occur only in the upper part of some sections. In central and southern Neuquén, the succession is clearly regressive, as evident by progradation on dark pelitic facies (Los Molles Formation) of marginal marine (Lajas Formation) and fluvial (Challaco and Punta Rosada Formations) facies.

The uppermost part of this subsynthem is represented by central-basin evaporites, consisting of 20–100 m of gypsum and stromatolitic limestones (Tabanos Formation).

Based on the method developed by Vail et al. (1977), the succession has been divided into at least seven depositional sequences or miosynthems (Gulisano et al. 1984; Legarreta and Gulisano 1989). Each depositional sequence is bounded by second-order discontinuities.

The oldest Hettangian levels have been found in the Atuel River area of southern Mendoza (Riccardi et al. 1988a). The Tabanos Formation, at the top of the Cuyo Subsynthem, is Middle Callovian, as it rests on strata with Lower Callovian ammonites and is overlain by strata of the Lotena Subsynthem with Middle–Upper Callovian ammonites (Dellape et al. 1979).

*Lotena Subsynthem.* This Subsynthem includes deposits belonging to the final stage of marine sedimentation within the Araucanian Synthem. It is bounded by the Intra-Callovian Discontinuity below and the Intra-Malmic Discontinuity above. Included lithostratigraphic units are the Lotena, La Manga, Barda Negra, Auquilco, and Fortin 1 de Mayo Formations. The maximum thickness is 800 m.

After deposition of the Tabanos evaporites, the terminal miosynthem (C7) of the Cuyo Subsynthem, the basin became continental, and fluvial and aeolian conglomerates and sandstones were deposited in deep areas. This continental succession is transgressive over basin margins and in some places has caused significant erosion of the evaporites. Toward the basin margins the Lotena Subsynthem rests on older units of the Cuyo Subsynthem.

The sedimentary succession included in the Lotena Subsynthem has been divided into five miosynthems (Legarreta and Gulisano 1989): L1 consists of coarse continental clastics overlain by marine sandstones, pelites, and carbonates; L2 consists of deep-marine deposits (turbidites), on top of which are basinal and outer-shelf dark pelites and shelf carbonate facies; L3 contains pelitic and sandy shelf facies on which prograde basinal to shelf carbonates; L4 and L5 are gypsum and anhydrite in the central part of the basin, associated with stromatolitic calcareous facies toward marginal areas.

The clastic facies belong to the Lotena Formation, the calcareous facies of the middle part of the La Manga Formation (or Barda Negra Formation of Neuquén subsurface), and the evaporites with marginal calcareous facies to the Auquilco Formation.

Paleontologic data and stratigraphy indicate that this subsynthem spans the Middle–Upper Callovian to Upper Oxfordian.

### Andean Cycle

This sedimentary cycle, originally confined by Groeber (1946) to Tithonian–Coniacian deposits and excluding the Kimmeridgian continental rocks of the Tordillo Formation, was redefined to include the Kimmeridgian (Stipanicic 1969) and, at the upper boundary, the Albian (Digregorio and Uliana 1980). The lower limit is defined by the Intra-Malmic Discontinuity. Thus, the basal deposits of the Mendoza Group (Tordillo and Quebrada del Sapo Formations) rest on different parts of the Lotena and Cuyo Subsynthems. The lower part of this cycle is bounded above by the Intra-Valanginian Unconformity (Gulisano et al. 1984), so that it spans the Kimmeridgian–Lower Valanginian and includes the Tordillo, Sierra Blanca, Catriel, and Quebrada del Sapo Formations of the Mendoza Group. Upward, and laterally intergrading, are the Vaca Muerta, Carrin Cura, Lindero de Piedra, Quintuco, Picún Leufu, Loma Montosa, and Bajada Colorada Formations. They consists of marine and continental clastic, carbonate, and subordinate evaporites.

Within the Andean Cycle, nine sequences have been recognized (Legarreta and Gulisano 1989). The first five range from the Upper Tithonian to the Lower Berriasian, and the fifth includes the Jurassic–Cretaceous transition.

The lower part of the Andean Cycle is characterized by a 1,500-m-thick clastic continental sequence, the Tordillo Formation and equivalents, which filled the entire basin. Continental sedimentation was interrupted by an early Tithonian sea-level rise, resulting in a condensed (*sensu* Vail, Hardenbol, and Todd 1984) marine interval with pelitic and bituminous carbonates of the starved-basin type, the lower Vaca Muerta Formation.

The following eight depositional sequences of Tithonian–Valanginian age include basinal to outer-shelf dark pelites of the Vaca Muerta Formation and inner-shelf limestones, pelites, and sandstones of the Quintuco, Picún Leufu and Lindero de Piedra Formations. These facies are the marginal sabkha carbonates, evaporites, and clastics of the Loma Montosa Formation and coarse continental clastics of the Bajada Colorada Formation. In Mendoza province these sequences are 500 m thick and follow gently diverging axes toward the basin center. In Neuquén, Río Negro, and La Pampa provinces they are about 2,000 m thick, with a rapidly prograding, sigmoidal outline (Mitchum and Uliana 1985).

### Biostratigraphy of west-central Argentina[10]

The Cuyo Subsynthem (Hettangian–Callovian) has yielded a rich ammonite fauna (Jaworski 1925, 1926; Westermann 1967, 1985; Westermann and Riccardi 1972, 1973, 1975, 1979, 1982; Riccardi 1983, 1984a,b, 1985; Hillebrandt 1987; Riccardi, Westermann, and Elmi 1988b; Riccardi et al. 1988a; Ricardi, Westermann, and Damborenea 1989b; Riccardi, Damborenea, and Mancenido 1990; Riccardi and Westermann 1991a, b).

#### *Hettangian*

The first Hettangian and early Sinemurian ammonites of Argentina now have been reported (Riccardi et al. 1988a) from the Puesto Araya Formation:

1. *Psiloceras* Assemblage Zone – *Psiloceras* sp., *Caloceras* cf. *johnstoni* (Sow.), *C.* cf. *peruvianum* (Lange), *P.* cf. *plicatulum* (Pomp.), *Alsatites* cf. *liasicus* (d'Orb.).
2. *Schlotheimia* Assemblage Zone – *Waehneroceras longipontinum* (Fraas), *Schlotheimia* cf. *angulata* (Schl.) & *polyeides* Lange, *S.* cf. *polyptycha* Lange & *complanata* Koenen, *Sulciferites* cf. *stenorhynchus*, *Vermiceras* (*Paracaloceras*) cf. *coregonense* (Sow.).

#### *Sinemurian*

1. *Badouxia canadensis* Assemblage Zone – *B. canadensis* (Frebold), *Vermiceras* (*Paracaloceras*) cf. *rursicostatum* Frebold, *Sulciferites* cf. *marmoreum* (Oppel).
2. *Vermiceras* Assemblage Zone – *V.* cf. *gracile* (Spath), *V.* spp.
3. *Agassiceras* Assemblage Zone – *A.* cf. *scipionianum* (Qu.), *Coroniceras* (*Paracoroniceras*) cf. *charlesi* Don., *Euagassiceras* sp.
4. *Epophioceras* Faunule – *E.* cf. *cognitum* Guer.-Fran., ?*Microderoceras* cf. *birchi* (Sow.).

#### *Pliensbachian*

Pliensbachian ammonites are widespread and well known in west-central Argentina (Jaworski 1913, 1926; Groeber, Stipanicic, and Mingramm 1953; Stipanicic 1969; Hillebrandt 1981, 1987; Riccardi 1983, 1984a,b; Riccardi et al. 1989b):

1. *Miltoceras* Faunule – *M.* cf. *sellae* (Gemm.), *Apoderoceras* sp., *Eoderoceras* sp.
2. *Dubariceras* Faunule – *Tropidoceras* cf. *stahli* (Oppel), *T.* ex gr. *flandrini* (Dum.), *Eoamaltheus meridianus* Hill., *Dubariceras* sp.
3. *Fanninoceras* Assemblage Zone – *F. behrendseni* (Jaw.), *Protogrammoceras* cf. *normanianum* (d'Orb.), *Arieticeras* sp., *Fuciniceras* sp.

#### *Toarcian*

1. Tenuicostatum Standard Zone – *Dactylioceras* (*Eodactylites*) sp., *D.* (*Orthodactylites*) sp.
2. *Dactylioceras hoelderi* Assemblage Zone – *D.* (*Orthodactylites*) *hoelderi* Hill. & Schm.-Eff., *Nodicoeloceras* sp., *Harpoceratoides* cf. *alternatus* (Simps.), *Hildaites* cf. *levinsoni* (Simps.).

3–5. *Peronoceras largaense*, *P. pacificum*, and *Collina chilensis* Assemblage Zones – *Peronoceras* cf. *subarmatum* (Y. & B.), *P.* cf. *vortex* (Simps.), *Harpoceras* cf. *subexaratum* Bon., *Frechiella* cf. *helvetica* Renz, *Phymatoceras* ex gr. *erbaense* (Hauer), *Maconiceras* sp., *Polyplectus* sp., *Collina* sp.

6. *Phymatoceras* Faunule – *P. copiapense* (Moricke), *P.* ex gr. *lilli* (Hauer).
7. *Phlyseogrammoceras tenuicostatum* Assemblage Zone – *P. tenuicostatum* (Jaw.), "*Witchellia*" *obscurecostata* Jaw., *Hammatoceras insigne* (Schubler), *Sphaerocoeloceras* cf. *brochiiforme* Jaw.
8. *Dumortieria* Faunule – *D. pusilla* Jaw., *Pleydellia* spp., *Sphaerocoeloceras brochiiforme* Jaw., *Hammatoceras* sp.

#### *Aalenian*

1. *Bredyia manflasensis* Assemblage Zone – *B. manflasensis* West., *B. delicata* West., *Ancolioceras*(?) *chilense* West., *Sphaerocoeloceras brochiiforme* Jaw.
2. "*Zurcheria*" *groeberi* Assemblage Zone – "*Z.*" *groeberi* West. & Ricc., *Planammatoceras* cf. *planiforme* Buck., *Podagrosiceras athleticum* Maub. & Lamb., *Tmetoceras* sp.
3. Malarguensis Standard Zone (new) – for *Puchenquia malarguensis* Assemblage Zone of Westermann and Riccardi (1979). Stratotype at Cerro Puchenque, Mendoza province. Base: bed 16 (6). With *Puchenquia* (*P.*) *malarguensis* (Burck.), *P.* (*Gerthiceras*) *compressa* West. & Ricc., *P.* (*G.*) *mendozana* West. & Ricc., *Planammatoceras* (*Pseudaptetoceras*) *klimakomphalum* (Vacek), *P.* (*P.*) *moerickei* (Jaw.), *P.* (*P.*) *tricolore* West. & Ricc., *P. gerthi* (Jaw.), *Eudmetoceras amplectens* (Buckman), *Euhoploceras amosi* West. & Ricc., *Praeleptosphinctes jaworskii* West.

#### *Bajocian*

1. Singularis Standard Zone (new) – for *Pseudotoites singularis* Assemblage Zone of Westermann and Riccardi (1979). Stratotype at Cerro Puchenque, Mendoza province. Base: Bed 12 (342). With *Pseudotoites singularis* (Gottsche), *P.*

[10] By A. C. Riccardi.

*sphaeroceroides* (Tornq.), *P. crassus* West. & Ricc., *P. transatlanticus* (Tornq.), *P. argentinus* Arkell, *Sonninia (Euhoploceras) amosi* West. & Ricc., *S. (Fissilobiceras) zitteli* (Gott.), *S. (Papilliceras) espinazitensis altecostata* Tornq.

2. Giebeli Standard Zone (new) – for *Emileia (E.) multiformis* Assemblage Zone of Westermann and Riccardi (1979). Stratotype at Carro Quebrado, Neuquén province. Base: Bed 149. With *E. (E.) multiformis* (Gott.), *E. (E.)* aff. *brochii* (Sow.), *E. (E.) vagabunda* Buck., *E. (Chondromileia) giebeli* (Gott.), *Chondroceras recticostatum* West. & Ricc., *Stephanoceras (Skirroceras)* cf. *macrum* (Quenst.), *Sonninia (Papilliceras) espinazitensis* Tornq., *Witchellia* sp., *Dorsetensia mendozai* West. & Ricc., *D. blancoensis* West. & Ricc.
3. Humphriesianum Standard Chronozone – *Stephanoceras (S.) pyritosum* (Qu.), *S. (Stemmatoceras)* aff. *frechi* (Renz), *Duashnoceras paucicostatum chilense* (Hill.), *Teloceras crikmayi chacayi* West. & Ricc., *Dorsetensia romani* (Oppel), *D. liostraca* (Buckman).
4. Rotundum Standard Chronozone – *Megaspaheroceras magnum* Ricc. & West., *Cadomites* cf. *daubenyi* (Gemm.), *C.* aff. *deslongchampsi* (d'Orb.), *Leptosphinctes* sp., *Lobosphinctes intersertus* Buckman.

*Bathonian*

1. *Cadomites*-Tulitidae Mixed Assemblage – *Cadomites* ex gr. *orbignyi-bremeri, Tulites*(?) (*Rugiferites?*) cf. *davaiacensis* (Liss.), *Bullatimorphites (Kheraiceras)* cf. *bullatus* (Orb.), *Choffatia (Homeoplanulites)* ex gr. *aequalis* (Roem.).
2. Steinmanni Standard Zone – *Lilloettia steinmanni* (Spath), *L. australis, Iniskinites crassus, I. gulisanoi,* Ricc. & West. spp., *Xenocephalites* cf. *araucanus* (Burck.), *Choffatia jupiter* (Stein.), *Neuqueniceras (N.) steinmanni* Stehn, *N. (N.) biscissum* (Stehn), *Oxycerites obsoletoides* Ricc. et al., *Stehnocephalites gerthi* Ricc. et al.

*Lower Callovian*

1. Vergarensis Standard Zone – *Eurycephalites vergarensis* Burck., *Neuqueniceras (N.) steinmanni* Stehn, *Xenocephalites gottschei* (Tornq.).
2. Bodenbenderi Standard Zone – *Neuqueniceras (Frickites) bodenbenderi* (Tornq.), *Eurycephalites rotundus* (Tornq.), *E. extremus* (Tornq.), *Xenocephalites stipanicici* Ricc. et al., *X. involutus* Ricc. & West.
3. Proximum Standard Zone – *Hecticoceras (H.) proximum* Elmi, *H. (H.)* cf. *hecticum* (Rein.), *H. (H.) boginense* (Pet.), *H. (Chanasia) navense* Roman, *H. (Ch.) ardescicum* Elmi, *Xenocephalites stipanicici* Ricc. et al., *X. involutus* Ricc. & West., *Neuqueniceras (Frickites)* cf. *antipodum* (Gott.), *Rehmannia (R.)* cf. *paucicostata* (Tornq.), *R. (R.) brancoi* (Stein.), *R. (R.) stehni* (Zeiss), *Oxycerites (Paroxycerites) oxynotus* (Leanza).

The Lotena Subsynthem has yielded a relatively rare and poorly preserved ammonite fauna (Weaver 1931; Leanza 1947a; Stipanicic 1951; Stipanicic, Westermann, and Riccardi, 1976; Riccardi, Leanza, and Volkheimer 1989a):

*Middle Callovian*

1. *Rehmannia patagoniensis* Horizon – abundant *R. (Loczyceras) patagoniensis* (Weaver).

*Oxfordian*

1. *Perisphinctes-Araucanites* Assemblage Zone – *P. (Kranaosphinctes)* spp., *P. (Arisphinctes)* spp., *Peltoceratoides* cf. *constantii* (Orb.), *Araucanites mulai, A. reyesi, A. stipanicici,* West. & Ricc. spp., *Euaspidoceras* aff. *waageni* Spath.

The base of the *Mendoza Group* in west-central Argentina has yielded a rich ammonite fauna (Behrendsen 1891–2; Steuer 1897; Haupt 1907; Douville 1910; Krantz 1928; A. Leanza 1945, 1947b, 1949; Indans 1954; H. Leanza 1975, 1980, 1981a,b; Leanza and Hugo 1978; Riccardi 1983; Riccardi et al. 1989a):

*Tithonian*

1. *Virgatosphinctes mendozanus* Assemblage Zone – *V. mendozanus* (Burck.), *V. andesensis* (Douv.), *V. denseplicatus rotundus* Spath, *V. evolutus* Leanza, *Pseudinvoluticeras douvillei* Spath, *P. windhauseni* (Weav.), *Choicensisphinctes choicensis* (Burck.), *Ch. erinoides* (Burck.).
2. *Pseudolissoceras zitteli* Assemblage Zone – *P. zitteli* (Burck.), *P. pseudoolithicum* (Haupt), *Glochiceras steueri* Leanza, *Hildoglochiceras wiedmanni* Leanza, *Parastreblites comahuensis* Leanza, *Aspidoceras cieneguitensis* (Steuer).
3. *Aulacosphinctes proximus* Assemblage Zone – *A. proximus* (Steuer), *Subdichotomoceras* sp., *Pseudhimalayites steinmanni* (Steuer), *Aspidoceras andinum* (Steuer), *A. neuquensis* Weaver.
4. *Windhauseniceras internispinosum* Assemblage Zone – *W. internispinosum* (Krantz), *Wichmanniceras mirum* Leanza, *Pachysphinctes americanensis* Leanza, *Hemispiticeras* aff. *steinmanni* (Steuer), *Subdichotomoceras araucanense* Leanza, *S. windhauseni* (Weaver), *Parapallasiceras* aff. *pseudocolubrinoides* Oloriz, *P.* aff. *recticosta* Oloriz, *Aulacosphinctoides* aff. *hundesianus* (Uhlig), *Aspidoceras euomphalum* Steuer.
5. *Corongoceras alternans* Assemblage Zone – *C. alternans* (Gerth), *C. mendozanum* (Behr.), *C. rigali* Leanza, *Aulacosphinctes mangaensis* (Steuer), *Micracanthoceras tapiai* Leanza, *M. lamberti* Leanza, *Berriasella australis* Leanza, *B. pastorei* Leanza, *B. krantzi, B. bardensis* Krantz.
6. *Substeueroceras koeneni* Assemblage Zone – *S. koeneni* (Steuer), *S. exstans* Leanza, *Aulacosphinctes azulensis*

*Leanza, Berriasella inaequicostata* Gerth, *Parodontoceras calistoides* Behr., *Blanfordiceras vetustum* (Steuer), *Himalayites andinus* Leanza, *Spiticeras acutum* Gerth.

## Central and northern Chile[11]

### *Central Chile*

A Middle to Upper Jurassic succession similar to that in west-central Argentina is exposed in the Chilean part of the Principal Cordillera. Between 37°S and 34°S the Jurassic begins with a comprehensive unit, the Nacientes del Teno Formation, which can be divided into numerous members. The lowest (unnamed) beds consist of 1,000 m of conglomerates, sandstones, shales, tuff, and tuffites. Above follow 450 m of sandstones, limestones, and shales of the Rinconada Member, in the upper levels with Bathonian-Oxfordian fauna. The uppermost 100 m of the formation are the gypsum of the Santa Elena Member (Klohn 1960; González and Vergara 1962; Muñoz 1984; Davidson 1988). At 36–37°S the Nacientes del Teno has yielded (Muñoz and Niemeyer 1984) the ammonites *Paltarpites* sp. of the late Pliensbachian–early Toarcian, *E. vergarensis* (Burck.) of the late Bathonian–early Callovian, and *Euaspidoceras* sp., *Peltoceratoides* sp., *Parawedekindia* sp., and *Perisphinctes* (*Dichotomosphinctes*) sp. of the Oxfordian.

Between 32°S and 34°S the oldest Jurassic rocks are equivalent to the upper part of the Nacientes del Teno Formation. The gypsum of the Santa Elena Member is here interbedded with, or forms diapirs in, calcareous shales, limestones, and fine-grained conglomerates of the Río Colina and Lagunilla Formations. They bear Oxfordian ammonites (Thiele 1980; Moscoso, Nasi, and Salinas 1982a; Moscoso, Padilla, and Rivano 1982b).

The Nacientes del Teno and Río Colina Formations are conformably overlain by up to 5,000 m of continental red sandstones, conglomerates, breccias, and shales placed in the Río Damas Formation. It is Kimmeridgian in age and equivalent to the Tordillo Formation of west-central Argentina. A sequence up to 1,200 m thick of probably lacustrine turbidites consists of breccias, sandstones, and siltstones. It is in part coeval to the Rio Damas Formation and has been included in the Leñas Espinoza Formation (Charrier 1982).

Above follow, to the south, approximately 400 m of Tithonian calcareous sandstones, limestones, and shales of the Baños del Flaco Formation, and to the north 1,300 m of Tithonian-Hauterivian limestones, shales, and sandstones, with andesitic lava flows, the Los Valdes Formation (Biro-Bagoczky 1980; Hallam, Biro-Bagoczky, and Perez 1986). Tithonian levels bear ammonites representing the zones of *Virgatosphinctes andesensis, Windhauseniceras internispinosum, Corongoceras alternans,* and *Substeueroceras koeneni.*

To the west, in the Coastal Cordillera, north of 35°20′S, Jurassic rocks form a very thick volcaniclastic sequence, with minor marine intercalations. The southernmost outcrops in the Vichuquen-Tilicura and Hualane-Curepto areas expose 1,300–1,800 m of sandstone and shales, the Laguna Tilicura and Rincón de Nuñez Formations, which conformably overlie marine Triassic. From this succession the following Hettangian-Sinemurian species were reported: *Caloceras* ex gr. *johnstoni* (Sow.), *Schlotheimia* cf. *angulata* (Schl.), "*Charmasseiceras*" sp., *Coroniceras*(?) sp., *Arnioceras semicostatum* (Young & Bird), and *Pararnioceras* cf. *gallicum* Guerin-Franiatte (Corvalan 1976, 1982; Minato 1977; Escobar 1980; Nasi 1984).

The Rincón de Nuñez Formation is overlain by approximately 2,000 m of andesitic flows and breccias, interbedded with sandstones called the Altos de Hualmapu Formation, which is assigned to the Middle and Upper Jurassic.

A similar succession is present between 32°S and 33°S, in the Los Vilos–Los Molles region (Piraces 1976), where up to 800 m of sandstones and shales of the Los Molles and Quebrada El Pobre Formations yielded Hettangian ammonites: *Psiloceras planorbis tilmanni* Lange, *P. planorbis erugatum* (Phill.), *Caloceras peruvianum* (Lange), *Schlotheimia* ex gr. *angulata* (Schloth.) (Cecioni and Westermann 1968). These formations are conformably overlain by thousands of meters of volcanics interbedded with and overlain by marine sandstones, shales, and limestones bearing Early Jurassic to Bajocian invertebrates. These sediments are called the Ajial (post–Sinemurian-Aalenian) and Cerro Calera Formations. Ammonites present in the Cerro Calera include the Aalenian *Tmetoceras* sp. and *Eudmetoceras* sp. and the middle Bajocian *Sphaeroceras* (*Chondroceras*) sp., *Leptosphinctes* sp., *Duashnoceras paucicostatum* (Felix), *Teloceras* cf. *blagdeniformis* (Roche), *Stephanoceras* (*S.*) sp., *Skirroceras* sp., and *Megasphaeroceras* ex gr. *rotundum* Imlay (Covacevich and Piraces 1976; Hillebrandt 1977; Nasi and Thiele 1982). North of 32°S, the Los Molles and Ajial Formations are missing.

The Cerro Calera Formation is overlain by continental volcanic and sedimentary rocks, the Horqueta Formation, of post-Bajocian to pre-Neocomian age (Nasi 1984; Nasi and Thiele 1982; Rivano 1984).

Radiochronology of the Jurassic intrusives along the Pacific coast on the western side of the Coastal Cordillera gave Pb-a and K-Ar ages of 175–140 Ma (Nasi 1984). Two age clusters can be recognized: 230–170 Ma (Late Triassic–Early Jurassic) and 170–140 Ma (Middle–Late Jurassic) (Drake et al. 1982), or 190–170 Ma (Early Jurassic) and 160–140 Ma (Middle–Late Jurassic) (Aguirre 1983).

### *Copiapo–Ovalle area*

Between approximately 27°S and 31°S most of the Jurassic outcrops are in the High Cordillera. Marine facies are restricted to the Lower and Middle Jurassic, and the Upper Jurassic is represented by a continental sequence (Segerstrom 1959, 1968; Reutter 1974; Jensen and Vicente 1977; Rivano 1980; Moscoso et al. 1982a,b; Sepulveda and Naranjo 1982; Cornejo, Nasi, and Mpodozis 1984).

The Jurassic rests unconformably on Upper Paleozoic or Triassic rocks. To the west, the succession begins with approximately

[11] By A. C. Riccardi.

390–650 m of fine-grained conglomerates, calcareous sandstones, and andesitic volcanic flows, called the Tres Cruces Formation or the Los Pingos and Punilla Formations. Both bear Sinemurian-Bajocian ammonites. This sequence changes toward the east. At first, the volcanic flows disappear, the facies become more limy and marly, the Lautaro Formation. The easternmost outcrops consist of 300 m of gypsum and calcareous limestones, with poorly preserved invertebrates, the El Tapado Beds. The sea transgressed from west to east from Sinemurian to Toarcian time, and then regressed during the late Toarcian–late Bajocian interval. Marine Callovian beds are known only north of about 28°S (Hillebrandt 1973b; Hillebrandt and Westermann 1985).

In this region is the long-known locality of Manflas. The fossils have been described by Burmeister and Giebel (1861), Steinmann (1881), Möricke (1894), Philippi (1899), Westermann and Riccardi (1972, 1979), and Hillebrandt (1977). The rich ammonite fauna includes the following ammonites from regional assemblage zones (Westermann and Riccardi 1972, 1979; Hillebrandt 1977, 1981, 1987; Hillebrandt and Schmidt-Effing 1981; Hillebrandt and Westermann 1985):

*Sinemurian*

1. "*Arietites bucklandi*" Zone – *Arietites* ex gr. *bucklandi.*
2. "*Asteroceras obtusum*" Zone – *Asteroceras* cf. *obtusum* (Sow.).
3. "*Oxynoticeras oxynotum*" Zone – *Oxynoticeras* cf. *lymense* (Wright), *Cheltonia* cf. *retentum* (Simpson).
4. "*Echioceras raricostatum*" Zone – *Pseudoskirroceras wiedenmayeri* Hill.

*Pliensbachian*

1. "*Apoderoceras-Eoderoceras*" Zone – *Eoderoceras pinguecostatum* (Bremer).
2. "*Tropidoceras*" Zone – *Tropidoceras* sp., *Eoderoceras* cf. *unimacula* (Quenst.), *Apoderoceras* (*Miltoceras*) sp.
3. *Eoamaltheus meridianus* Zone – *E. meridianus* Hill., *Uptonia* cf. *obsoleta* (Simpson), *U.* cf. *angusta* (Quenst.).
4. *Fanninoceras behrendseni* Zone – *F. behrendseni* (Jaw.).
5. *Fanninoceras disciforme* Zone – *F. disciforme* Hill., *Paltarpites* cf. *argutus* (Buckman), *Protogrammoceras* cf. *normannianum* (d'Orb.), *Leptaleoceras* (*L.*) *timai* (Gemm.), *L.* (*L.*) *pulcherrimum* (Fucini), *Bouleiceras* sp.

*Toarcian*

1. Tenuicostatum Standard Zone – *Dactylioceras* (?*Eodactylites*) *simplex* Fucini, *Nodicoeloceras* cf. *eikenbergi* (Hoff.), *N.* cf. *pseudosemicelatum* (Maub.), *Bouleiceras* cf. *chakdallaense* Fatmi, *Dactylioceras* (*Orthodactylites*) *tenuicostatum chilense* Hill. & Schm.-Eff.; divided into *Dactylioceras simplex* and *D. tenuicostatum* Subzones.
2. *Dactylioceras hoelderi* Zone – *D.* (*Orthodactylites*) *hoelderi* Hill. & Schm.-Eff. *D.* (*O.*) cf. *directum* (Buckman), *D.* (*O.*) *anguinum* (Reinecke), *Nodicoeloceras* cf. *crassoides* (Simp.), *Harpoceratoides* cf. *alternatus* (Simp.), *Hildaites* cf. *levinsoni* (Simp.).
3. *Peronoceras largaense* Zone – *P. largaense* Hill. & Schm.-Eff., *P.* cf. *desplacei* (d'Orb.), *P.* cf. *choffati* (Renz), *P.* cf. *subarmatum* (Young & Bird), *P.* cf. *renzi* (Pinna & L.-S.), *Harpoceras* cf. *chrysanthemum* (Yok.).
4. *Peronoceras pacificum* Zone – *Peronoceras pacificum* Hill. & Schm.-Eff., *P.* cf. *verticosum* (Buckman), *Polyplectus* cf. *discoides* (Ziet.), *Frechiella* cf. *helvetica* Renz, *Harpoceras* sp., *Maconniceras* cf. *connectens* (Haug).
5. *Collina chilensis* Zone – *C. chilensis*, *Peronoceras bolitoense*, *P. moericke*, Hill. & Schm.-Eff. spp., *P.* cf. *planiventer* (Guex), *P.* cf. *crassicostatum* (Guex), *P.* cf. *vortex* (Simp.), *Phymatoceras* cf. *erbaense* (Hauer), *P.* cf. *robustum* Hyatt, *P.* cf. *iserense* (Oppel), *Harpoceras* cf. *subexaratum* (Bonarelli), *Atacamiceras glabrum*, *A. parvicostatum*, *Hildaitoides retrocostatus*, *H. atacamensis* Hill. spp.
6. *Phymatoceras toroense* Zone – *P. toroense* Hill.; divided into *P. bolitoense* and *P. moerickei* Subzones.
7. *Phymatoceras copiapense* Zone – *P. copiapense* Mor., *P.* cf. *pseudoerbaense* Gab.
8. *Phlyseogrammoceras tenuicostatum* Zone – *P.* cf. *tenuicostatum* Jaw.; divided into "*P. dispansum*" and *P. tenuicostatum* Subzones.
9. "*Pleydellia lotharingica*" Zone – *P.* cf. *lotharingica* (Branco).
10. "*Pleydellia fluitans*" Zone – *P.* cf. *fluitans* (Dum.), *Dumortieria pusilla*, *Sphaerocoeloceras brochiiforme*, Jaw. spp.

*Aalenian*

1. *Bredyia manflasensis* Zone – *Bredyia manflasensis* West., *B. delicata* West., *Ancolioceras*(?) *chilense* West., *Sphaerocoeloceras brochiiforme* Jaw.
2. "*Zurcheria*" *groeberi* Zone – *Parammatoceras jenseni* West.

*Aalenian–Bajocian boundary*

1. Malarguensis Standard Zone – *Puchenquia* (*P.*) *malarguensis* (Burck.), *P.* (*Gerthiceras*) *compressa* West. & Ricc., *P.* (*G.*) *mendozana* West. & Ricc., *Planammatoceras* (*Pseudammatoceras*) *klimakomphalum* (Vacek), *P.* (*P.*) *tricolore* West. & Ricc., *Eudmetoceras endmetum jaworskii* West., *Tmetoceras* spp., *Podagrosiceras* sp.

*Bajocian*

1. Singularis Standard Zone – *Sonninia* (*Fissilobiceras*) *zitteli* (Gott.), *Pseudotoites singularis* (Gott.), *P.* cf. *argentinus* (Arkell), *S.* (*Papilliceras*) *espinazitensis altecostata* Tornq.
2. Giebeli Standard Zone – *Emileia giebeli* (Gott.), *E. multiformis* (Gott.), *E.* cf. *vagabunda* Buckman, *Pseudotoites*

*sphaeroceroides* (Tornq.), *Sonninia (Papilliceras) espinazitensis* Tornq., *Sphaeroceras (Chondroceras)* cf. *defontii* (McLearn).
3. Humphriesianum Standard Zone – *Stephanoceras* cf. *humphriesianum* (Sow.), *S. pyritosum* (Quenst.), *Duashnoceras paucicostatum* (Felix), *Teloceras* spp., *Dorsetensia* aff. *deltafalcata* (Quenst.), *Lupherites dehmi* (Hill.).
4. Rotundum Standard Zone – *Megasphaeroceras magnum* Ricc. & West.

The Tres Cruces and Lautaro Formations are conformably overlain by up to 2,000 m of andesitic rocks and reddish shales, sandstones, and conglomerates of continental origin, the Mostazal, Algarrobal, and Picudo Formations, which are placed in the Upper Jurassic (Thiele 1964; Dedios 1967; Moscoso et al. 1982a,b; Cornejo et al. 1984).

In the Coastal Cordillera, at about 28–29°S, a 2,100-m-thick succession of sandstones, conglomerates, shales, and limestones of the Canto del Agua Formation lies unconformably on Paleozoic. It includes levels with Middle Triassic and Early Jurassic marine fauna. Hettangian and Sinemurian are dated by *Psiloceras (Caloceras)* sp. and *Arnioceras* sp. This formation is overlain by Neocomian andesites of the Bandurrias Group (Figure 6.12) (Moscoso and Covacevich 1982; Moscoso et al. 1982a,b).

On the Pacific coast, along the Coastal Cordillera, Jurassic tonalites, granodiorites, granites, and diorites intrude Paleozoic basement and the Canto del Agua Formation. Radiochronology (Pb-α) ages average 158 ± 20 Ma, or Middle to Late Jurassic (Moscoso et al. 1982a,b).

*Northern Atacama and Antofagasta regions*

Marine and continental Jurassic deposits are quite well represented here. Along the Pacific coast between Chañaral and Taltal the Jurassic succession begins with up to 690 m of gray-yellowish, calcareous shales interbedded with gray-greenish fine to coarse sandstones, the Pan de Azúcar Formation. *Psiloceras* sp., *Angulaticeras* sp., *Schlotheimia* cf. *angulata* (Schl.), *Agassiceras* sp., and *Arnioceras* sp. indicate Hettangian-Sinemurian ages (Cecioni 1960; Mercado 1980; Naranjo and Puig 1984).

The Pan de Azúcar Formation lies unconformably on Triassic or Paleozoic rocks. It grades laterally and upward into approximately 500 m of marine andesitic flows and breccias, and sandstones of the Posada de los Hidalgo Formation, that include bivalves and the Sinemurian ammonites *Angulaticeras* cf. *ventricosum* (Sow.) and *Arnioceras semicostatum* (Young & Bird) (Hillebrandt 1981, 1987; Naranjo and Puig 1984).

Similar rocks found northeast of Antofagasta have been named the Cerros de Cuevitas Formation and the Rencoret Beds. The former yielded *Caloceras* ex gr. *peruvianum* (Lange), *Schlotheimia* sp., *Paracaloceras* sp., *Agassiceras* ex gr. *scipionianum*, *Arnioceras* sp., and *Spiriferina* aff. *walcotti* (Sow.); the latter include *Psiloceras (Caloceras)* sp., *Schlotheimia*(?) sp., *Coroniceras* sp., and *Arnioceras* sp. (Hillebrandt 1981; Di Biase 1985; Muñoz, Venegas, and Tellez 1988). This fauna dates both units as Hettangian to early Sinemurian.

Above follow, unconformably in some areas and conformably or interfingering in others, 5,000–7,000 m of andesitic, dacitic, and basaltic flows and breccias, the La Negra Formation. It has interbedded marine sediments with the ammonites *Fanninoceras* sp., *Arieticeras* sp.(?) of the Pliensbachian, and *Emileia giebeli* (Gott.), *E. multiformis* (Gott.), *Stephanoceras (Skirroceras)* sp., and bivalves of the Lower Bajocian (Davidson, Godoy, and Covacevich 1976; Naranjo and Puig 1984). The La Negra extends along the Coastal Cordillera for more than 550 km, mostly between Taltal and Inquique (Naranjo, Puig, and Suárez 1982; Suárez, Naranjo, and Puig 1982, 1985). It possibly spans the entire Jurassic (Boric 1981; Skarmeta and Marinovic 1981), because the ages given to the intruded granodiorites and diorites vary between 167 and 147 Ma (Maksaev 1984) or 200 and 150 Ma (Davidson 1984).

Intrusions of monzonitic, granodioritic, tonalitic, and granitic composition are well exposed along the Coastal Cordillera between 23°S and 29°S, and even farther north (Boric 1981; Skarmeta and Marinovic 1981). Radiochronology (K-Ar and Rb-Sr) gave values between 198 ± 3.2 and 187 ± 6 Ma (earliest Jurassic) and between 159.7 ± 1.6 and 137 ± 4.4 Ma (Late Jurassic) (Zentilli 1974, cited by Moscoso et al. 1982a,b; Mercado 1980; Damm and Pichowiak 1981; Berg and Breitkreuz 1983; Naranjo and Puig 1984; Berg and Baumann 1985).

Toward the east, a Jurassic sedimentary sequence that partly interfingers with the La Negra Formation is exposed in a belt west of about 69°W. In the south of this belt is a well-exposed section in the Pedernales–Quebrada Asientos area (Harrington 1961; Pérez 1982). The succession begins with a basal conglomerate and approximately 500 m of grayish black limestones and shales, the Montandon Formation, which has yielded a rich ammonite fauna (Westermann and Riccardi 1972, 1979; Hillebrandt 1977; Hillebrandt and Schmidt-Effing 1981; Pérez 1982; Hillebrandt and Westermann 1985):

*Pliensbachian*

1. *Eoamaltheus meridianus* Zone – *Uptonia* cf. *ignota* (Simp.).
2. *Fanninoceras behrendseni* Zone – *F. behrendseni* (Jaw.).
3. *Fanninoceras fannini* Zone – *F. fannini* McLearn, *F.* cf. *carlottense* (McLearn), *F.* cf. *kunae latum* McLearn, *Protogrammoceras* cf. *normanianum* (d'Orb.).
4. *Fanninoceras disciforme* Zone – *F. oxyconum* Hill., *F.* cf. *lowrii* McLearn, *Arieticeras* cf. *fucinii* (Del Campana), *A.* cf. *simplex* (Fucini), *Paltarpites argutus* (Buckman), *Reynesocoeloceras (Bettoniceras)* cf. *colubriforme* (Bettoni), *R. (B.)* cf. *mortilleti* (Meneghini).

*Toarcian*

1. Tenuicostatum Standard Zone – *Dactylioceras (Orthodactylites) directum* (Buckman).
2. *Dactylioceras hoelderi* Zone – *D. (O.) anguinum* (Rein.), *D. (O.)* cf. *helianthoides* Yok., *Harpoceratoides* cf. *alternatus*

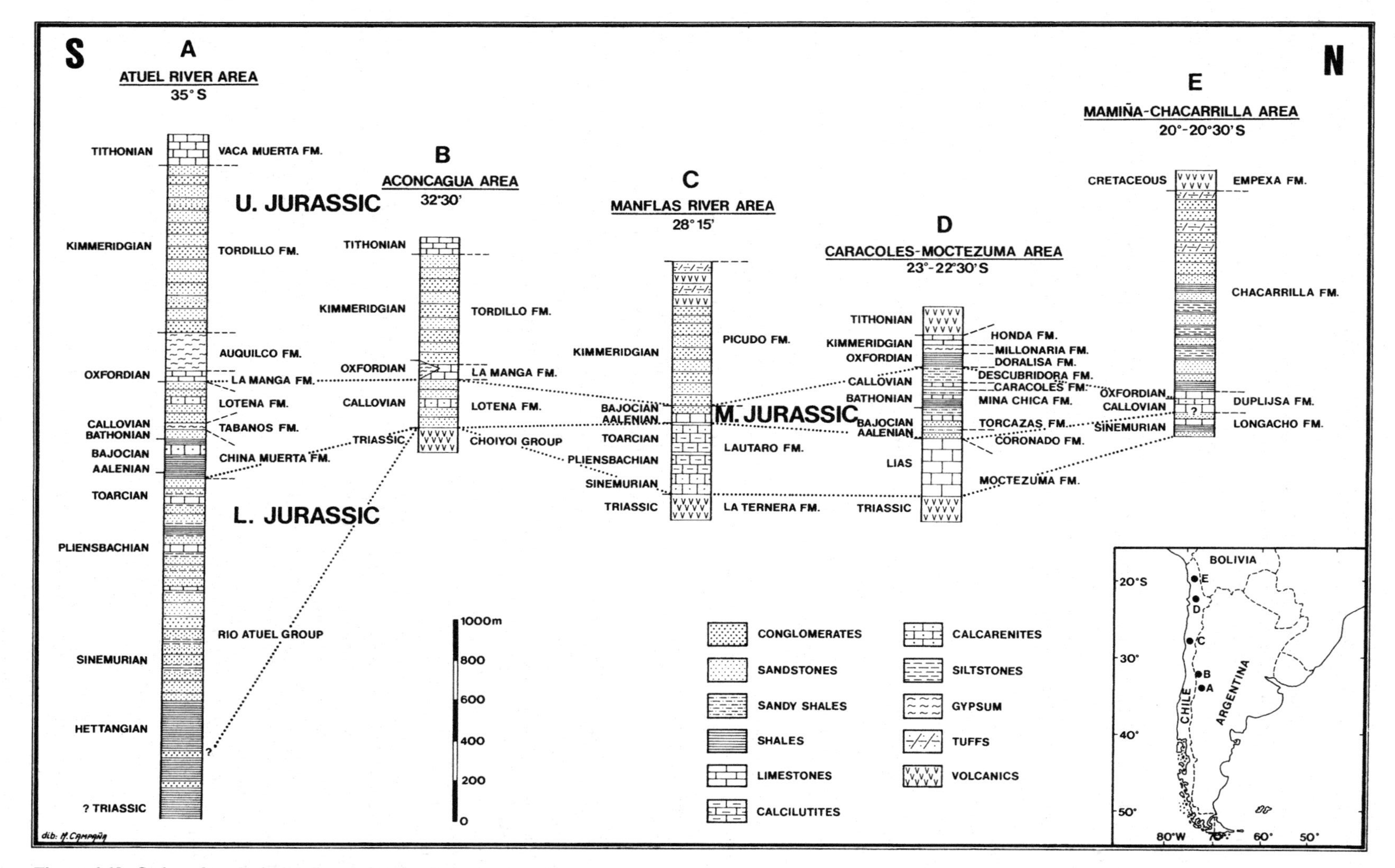


Figure 6.12. Series of geological columns for the Jurassic of west-central Argentina to northern Chile, south to north. (Adapted from Riccardi 1983.)

(Simp.), *Hildaites* cf. *serpentiniformis* (Buckman), *Harpoceras* sp.

3. *Peronoceras largaense* Zone – *P.* cf. *subarmatum* (Young & Bird), *Harpoceras* cf. *falcifer* (Sow.).
4. *Peronoceras pacificum* Zone – *P.* cf. *verticosum* (Buckman).
5. *Collina chilensis* Zone – *Peronoceras* cf. *vorticellum* (Simp.), *P. moerickei* Hill. & Schm.-Eff., *Harpoceras* cf. *subexaratum, Atacamiceras parvicostatum* Hill.
6. "*Pleydellia fluitans*" Zone – *P.* cf. *fluitans* (Dum.).

*Aalenian*

1. *Bredyia manflasensis* Zone – *B. manflasensis* West., *B. delicata* West.
2. "*Zurcheria*" *groeberi* Zone – *Zurcheria groeberi* West. & Ricc., *Parammatoceras jenseni* West., *Podagrosiceras athleticum* Maub. & Lamb.

*Aalenian–Bajocian boundary*

1. Malarguensis Standard Zone – *Puchenquia* (*Gerthiceras*) *compressa, P.* (*G.*) *mendozana, Planammatoceras* (*Pseudaptetoceras*) *tricolore,* West. & Ricc. spp., *P.* (*P.*) *klimakomphalum* (Vac.), *Eudmetoceras eudmetum jaworskii* West., *Puchenquia* (*P.*) *malarguensis* (Burck.), *Tmetoceras* spp., *Podagrosiceras maubeugei* West. & Ricc.

*Bajocian*

1. Giebeli Standard Zone – *Pseudotoites* cf. *sphaeroceroides* (Tornq.).
2. Humphriesianum Standard Zone – *Stephanoceras* cf. *humphriesianum* (Sow.), *S. pyritosum* (Quenst.), *S.* (*Stemmatoceras*?) aff. *frechi* (Renz), *Dorsetensia* aff. *deltafalcata* (Quenst.), *D. romani* (Oppel), *D. liostraca* Buckman.

A late Bajocian–middle Bathonian hiatus seems to be present (Pérez 1982), and the Montandon Formation is overlain by 280–560 m of massive blue limestones and brown, sandy limestones, with intercalations of shales, the Asientos Formation. It has yielded the late Bathonian and Callovian ammonites *Epistrenoceras* sp., "*Macrocephalites*" sp., *Neuqueniceras* cf. *bodenbenderi* (Tornq.), *Reineckeia* spp., and *Parawedekindia*? sp. (Martincorena and Tapia 1982). The Asientos is unconformably followed by the Tithonian–Lower Cretaceous marine sediments of the Pedernales Formation. Tithonian levels include *Virgatosphinctes lenaensis* (Corvalan), *Windhauseniceras internispinosum* (Krantz), *Himalayites* sp., *Aulacosphinctes*(?) sp., and *Hemispiticeras* sp. (Pérez 1982).

A more complete succession is exposed farther north in the Domeyko Cordillera (22°20′–25°30′S) (Chong 1977). The Jurassic begins with Hettangian marine facies, which extend as far north as approximately 21°30′S (Niemeyer et al. 1985). Marine Sinemurian to Oxfordian strata are also widespread along this belt. This succession reached a thickness of about 1,200 m in the southern part of the Domeyko Range (24°30′S and 25°30′S), where it has been named the Profeta Formation (Bogdanic and Chong 1985). In the Caracoles-Moctezuma region, in the north, it is 820 m thick and has been divided (Harrington 1961; Garcia 1967) into several formations, in ascending order: the Moctezuma, Coronado, Torcazas, Mina Chica, Caracoles, Descubridora, and Doralisa. In many sections, limestones with Oxfordian ammonites of the Doralisa Formation are overlain by gypsum beds and limestones of the Millonaria and Honda Formations. Part or all of these formations have been included in the Caracoles Group (Garcia 1967; Ramírez and Gardeweg 1982; Marinovic and Lahsen 1984). In the north, this succession is overlain by Kimmeridgian red continental sandstones, named the Cerritos Bayos, Quehuita, and Quinchamaele Formations and the Sierra San Lorenzo beds (Skarmeta and Marinovic 1981; Maksaev 1984). Marine Kimmeridgian has been found only in a few southern localities (Chong 1977; Forster and Hillebrandt 1984; Groschke et al. 1988). Formational names have also been used for the interfingering sequence between the western volcanics and the eastern sedimentary facies (i.e., Candeleros and Cholita Formations (Maksaev 1984).

The Jurassic ammonite succession of the Domeyko Cordillera is the most complete in Chile (Westermann et al. 1980; Bogdanic and Chong 1985; Bogdanic, Hillebrandt, and Quinzio 1985; Riccardi et al. 1989a–c):

*Hettangian*

1. "*Psiloceras planorbis*" Zone – *P. plicatulum* (Pomp.), *P.* cf. *reissi* (Tilmann), *Caloceras johnstoni* (Sow.), *C. peruvianum* (Lange).
2. "*Alsatites liasicus*" Zone – *A.* cf. *platysoma* (Lange), *Ectocentrites* cf. *petersi* (Hauer).
3. "*Schlotheimia angulata*" Zone – *Schlotheimia* sp.

*Sinemurian*

1. "*Arietites bucklandi*" Zone – *Arietites* sp., *Megarietites* sp.
2. "*Arnioceras semicostatum*" Zone – *A.* cf. *ceratitoides* (Quenst.), *Coroniceras* sp., *Agassiceras* cf. *scipionianum* (Quenst.).
3. "*Asteroceras obtusum*" Zone – *A. obtusum, Eparietites* cf. *undaries* (Quenst.), *Epophioceras* cf. *cognitum* Guerin-Franiatte.
4. "*Oxynoticeras oxynotum*" Zone – *O.* cf. *lymense* (Wright).
5. "*Echioceras raricostatum*" Zone – *Microderoceras* cf. *oosteri* (Hug), *Plesechioceras arcticum* Freb., *Paltechioceras* sp.

*Pliensbachian*

1. "*Apoderoceras-Eoderoceras*" Zone – *Atractites* sp.
2. *Fanninoceras fannini* Zone – *F.* cf. *fannini* McLearn.
3. *Fanninoceras disciforme* Zone – *F.* cf. *oxyconum* Hill., *Protogrammoceras* sp., *Arieticeras* sp.

*Toarcian*

1. Tenuicostatum to *Collina chilensis* Zones – Dactylioceratidae indet.
2. *Collina chilensis* Zone – *Atacamiceras glabrum* Hill.
3. *Phlyseogrammoceras*(?) *tenuicostatum* Zone – *P.*(?) *tenuicostatum* (Jaw.).
4. "*Pleydellia lotharingica–fluitans*" Zones – *Pleydellia* sp., *Dumortieria pusilla* Jaw.

*Aalenian*

1. *Bredyia manflasensis* Zone – *B. manflasensis* West., *Sphaerocoeloceras* sp.

*Aalenian–Bajocian boundary*

1. Malarguensis Standard Zone – *Puchenquia* (*P.*) *malarguensis* (Burck.), *Euhoploceras amosi* West. & Ricc.

*Bajocian*

1. Singularis Standard Zone – *Pseudotoites transatlanticus* (Tornq.).
2. Giebeli Standard Zone – *Emileia multiformis* (Gott.), *Sonninia* (*Papilliceras*) *espinazitensis* Tornq., *S.* (*S.*) *alsatica* Haug.
3. Humphriesianum Standard Zone – *Duashnoceras andinense* (Hill.), *D. paucicostatum chilense* (Hill.), *Stephanoceras* (*S.*) *arcicostum caracolense* West. & Ricc., *S.* (*Stemmatoceras*) cf. *dowlingi* (McLearn), *S.* (*St.*) *allani* (Warren), *Teloceras* cf. *crickmayi chacayi* West. & Ricc., *Cadomites* n. sp., *Lissoceras* cf. *oolithicum* (d'Orb.), *Lupherites dehmi* (Hill.), *L. chongi* (Hill.).
4. Rotundum Standard Zone – *Megasphaeroceras magnum* Ricc. & West., *Spiroceras orbignyi* (Baug. & Sauz.), *Strenoceras* cf. *latesulcatum* (Quenst.).

*Bathonian*

1. Steinmanni Standard Zone – *Lilloettia steinmanni* (Spath), *Oxycerites* (*Paroxycerites*) *exoticus* (Stein.), *Hecticoceras* (*Prohecticoceras*) *blanazense* Elmi, *Eohecticoceras* sp., *Choffatia subbakeriae* (d'Orb.), *Ch. jupiter* (Stein.), *Epistrenoceras* sp., *Hecticoceras* (*Prohecticoceras*) retrocostatum (De Gross.).

*Callovian*

1. Bodenbenderi Standard Zone – *Neuqueniceras* (*Frickites*) *bodenbenderi* (Tornq.), *Oxycerites* (*Alcidellus*) *obsoletoides* Ricc. et al., *Xenocephalites* spp., *Eurycephalites* spp.
2. *Reineckeia* Assemblage – *Reineckeia* sp.

*Oxfordian*

1. *Peltoceratoides-Parawedekindia* Zone – *Pachyceras* (*Tornquistes*) sp., *Parawedekindia* cf. *arduennense*, *Peltoceratoides* cf. *athletoides* (Lahusen).
2. "Plicatilis-Transversarium Zones" – *Gregoryceras* sp., *Otosphinctes* sp., *Perisphinctes* (*Dichotomosphinctes*) sp., *Ochetoceras* cf. *canaliculatum* Buckman, *Euaspidoceras* sp.
3. "Bifurcatum Zone" – *Cubaspidoceras* sp.

*Kimmeridgian*

1. "Mutabilis-Eudoxus Zones" – *Orthaspidoceras* sp.

*Tarapacá region*

East of the Coastal Cordillera, in south-central Tarapacá (ca. 21°S), 150 m of shales, mudstones, fine-grained sandstones, and limestones belong to the Longacho Formation. *Arietites* sp., *Coroniceras* sp., and *Arnioceras* sp. (Garcia 1967) date the beds as Sinemurian. This unit is overlain (unknown relationship) by some 1,100–3,600 m of mudstones and sandstones interbedded with few trachyte flows, the Chacarilla Formation, which grades upward from near-shore to continental facies. The Chacarilla bears poorly preserved Oxfordian ammonites, bivalves, plants (Galli and Menéndez 1968), and dinosaur footprints.

Farther north the formation possibly intergrades (unknown relationship) with approximately 90 m of limestones with intercalated sandstones, the Duplijsa Formation, which has yielded Callovian-Oxfordian ammonites (Galli 1957, 1968; Galli and Dingman 1962; Dingman and Galli 1965; Thomas 1967).

Other small outcrops of Jurassic rocks are farther north, east, and southeast of Arica (Garcia 1967).

Jurassic rocks are well exposed along the Coastal Cordillera between Iquique and Arica (Salas et al. 1966; Thomas 1970; Silva 1976; Vila 1976; Vogel and Vila 1980). In the south the succession begins with approximately 2,000 m of andesites, the Oficina Viz Formation, which is coeval to the La Negra Formation farther south. Toward the north, the Oficina Viz interfingers with calcareous sandstones and limestones, named the Camaraca Formation. *Duashnoceras paucicostatum chilense* (Hill.), *Teloceras bladgeni* (Sow.), and Eurycephalitinae (Möricke 1894; Hillebrandt 1977; Vogel and Vila 1980) have been recorded, indicating the latest Lower and Upper Bajocian (and ?Callovian). Above follow 1,500–2,100 m of sedimentary rocks, which have been divided into two partially interfingering units, the Caleta Ligate (below) and Guantajaya (above) Formations. They bear middle Bajocian, Callovian, and Oxfordian invertebrates.

In the Arica region the Camaraca Formation grades upward into 600 m of dark shales, interbedded with limestones, quartzites, and andesites, the Los Tarros Formation, which includes Oxfordian ammonites. Other small Jurassic outcrops also occur east of the Coastal Cordillera (Garcia 1967).

## PALEOGEOGRAPHY

### Venezuela[12]

Jurassic redbed sequences along the Atlantic Ocean margins have been associated with early rifting during the breakup of Pangea in the Jurassic (McConnell 1969; Burke 1976; Turner 1980, pp. 35–59).

The postulated Maracaibo Graben contains the Triassic-Jurassic redbeds of western Venezuela and central-eastern Colombia (Cediel 1969; Burke 1976; Bartok, Reijers, and Juhasz 1981). Seismic data indicate a north-northeast-trending graben (or basin) in the Lake Maracaibo subsurface, in which the Jurassic sediments and volcanics were deposited (Pumpin 1978). This graben was associated with tensional tectonics and widespread volcanism, in particular in the western part, and probably was part of the western end of a Middle Jurassic seaway between the Tethys and the Pacific (Bartok et al. 1985). Possible mechanisms for the breakup of the Maracaibo Basin, as indicated by differences in the Jurassic rock sequences, include transcurrent motion between the Mérida Andes and the Sierra de Perijá (Maze 1984).

### Colombia[13]

Tensional tectonics began during the Triassic in the intercordilleran, ancestral Central and Eastern Cordilleras of Colombia (Macia and Mojica 1981), producing continental graben or rift structures, which were transversely divided into depressions and horsts. The graben accumulated and preserved, in chronological order, continental redbeds of the Luisa Formation, shallow-marine limestones of the Norian Payandé Formation, and marine to terrestrial, mainly Liassic volcanics of the Saldaña/Motema, Morrocoyal, Corual, and Guatapurí Formations. When the volcanic activity ended, possibly in the Middle Jurassic, mainly the northern portion of the graben was filled with Girón-type redbeds. During the Tithonian a new taphrogenic phase began, preceding the Cretaceous transgression (Figure 6.13).

The Jurassic intrusives, generally granodioritic to dioritic, locally affected the previously deposited volcaniclastics. Radiometric (K/Ar) ages of the intrusions vary mostly between 187 and 146 Ma (late Liassic–Dogger). Jurassic batholiths and stocks occur mainly in the Central Cordillera, the Santander Massif and the Sierra Nevada de Santa Marta (Figure 6.4).

The poorly known ages of the predominantly terrestrial sediments and volcanics in the Colombian Jurassic, and the consequent difficulty in understanding spatial relationships, hinder paleogeographic reconstructions.

The shallow-marine (in some cases brackish to paralic) environments in the Early and Late Jurassic (Figure 6.13) indicate paleo-Pacific transgressions. But the marine connections remain difficult to locate and explain, because the Central Cordillera seems to have been uplifted (Interandean Uplift or Barrier) during most of the Triassic and Jurassic (Nelson 1957; Bürgl 1961, 1964; Macia et al. 1985).

With due consideration given to earlier and recent reconstructions (Weeks 1947; Bürgl 1961, 1964; Harrington 1962; Irving 1971; Geyer 1973, 1979), I would tentatively propose the paleogeographic schemes (present globe) illustrated in Figure 6.13.

### Peru[14]

The paleogeographic limits of the Early Jurassic marine basin were defined to the east by emergent land close to the Brazilian shield, and to the west by the forerunners of the Coastal Cordillera. Liassic is unknown in the central part of the coastal belt (Figure 6.14).

Between 8°00′S and 14°00′S the western emerged area was therefore wider, and in the south a basin reached northern Chile, which presumably was connected with the Central Peruvian Basin. In this basin the volcanism, restricted in the Triassic to the Arica region, migrated northeast and reached the Nazca area in the Middle Jurassic. The sea was then restricted to southwestern and, probably, central Peru, becoming epineritic in the northeast.

During the Dogger and Malm, epeirogenic movements (Nevadian) produced an uplift in central and northern Perú, the Marañon-Mantaro Geoanticline, where today is the Eastern Cordillera. This uplift separated the western marine basin, limited to the west by the Coastal Cordillera, and the eastern continental basin, bounded to the east by the Brazilian Shield. The western basin extended northward to Lima, where an island arc developed at the end of the Jurassic. The Early and Middle Jurassic eastern basin began its maximum expansion in the Callovian and reached the Lima area (Huaytara–Río Pisco, 13°30′S).

### Argentina and Chile

#### *Regional overview*[15]

During the Late Triassic and Early Jurassic, extensional tectonics affected the western margin of Gondwana, and a series of fault-bounded, north-northeast-striking grabens were formed in a belt wider than the present continental region of Patagonia. In these grabens volcanic and volcaniclastic rocks were accumulated, together with continental and, locally, marine sediments. Among the several sedimentary basins were those of San Jorge and Neuquén. During the Middle and Late Jurassic the sea also transgressed on the Coastal Cordillera between 26°S and 37°S (all present latitudes) (Stipanicic 1983).

During the Early Jurassic the Pacific margin in southern South America developed differently north and south of 40–45°S. In the north, a back-arc basin was formed, the Neuquén Basin; in the south, extension probably began affecting the pre-Mesozoic basement. Thus, the Liassic is unknown from the Magallanes

[12] By C. Schubert. [13] By J. Mojica.

[14] By O. Palacios. [15] By A. C. Riccardi.

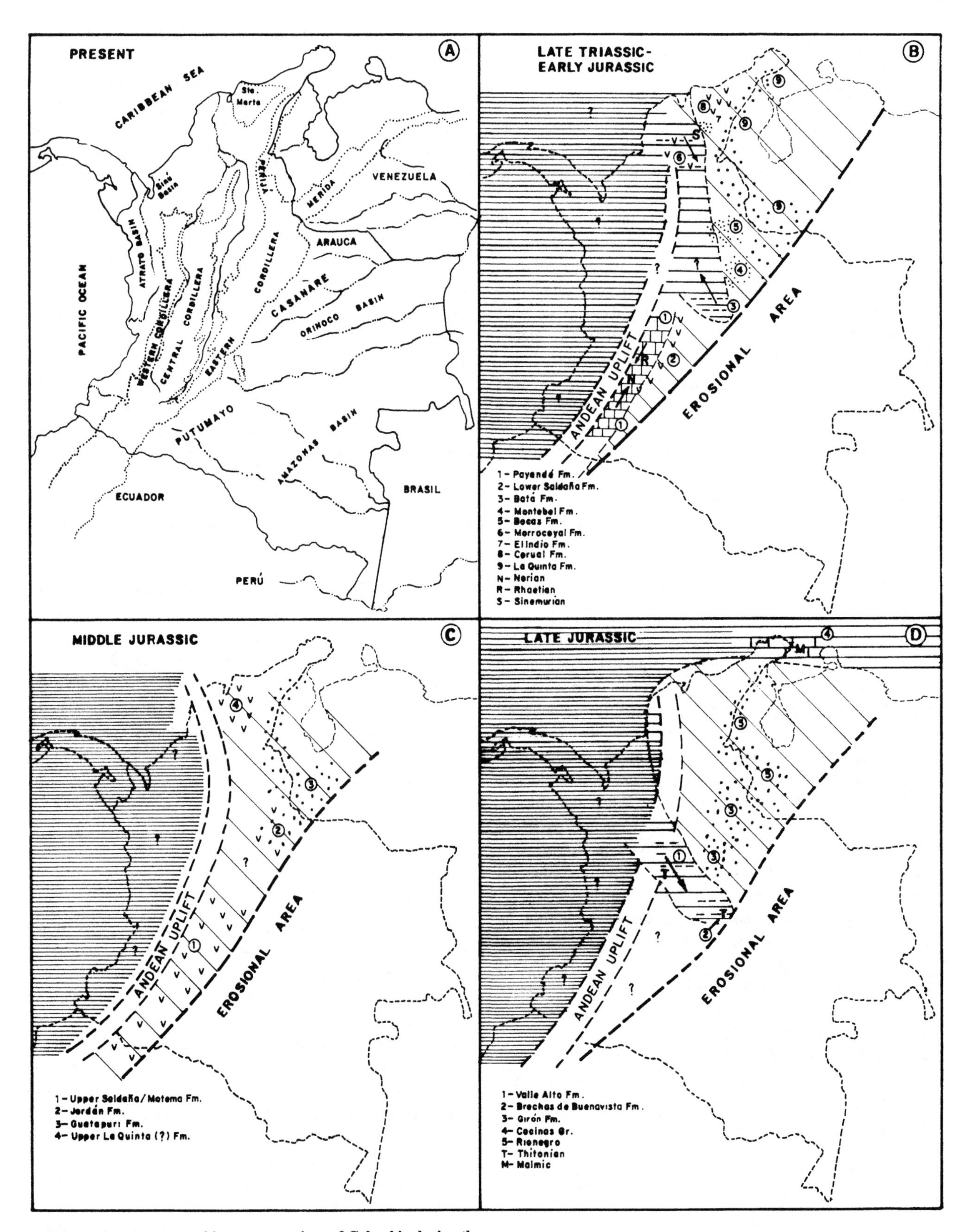


**Figure 6.13. Paleogeographic reconstructions of Colombia during the Jurassic.**

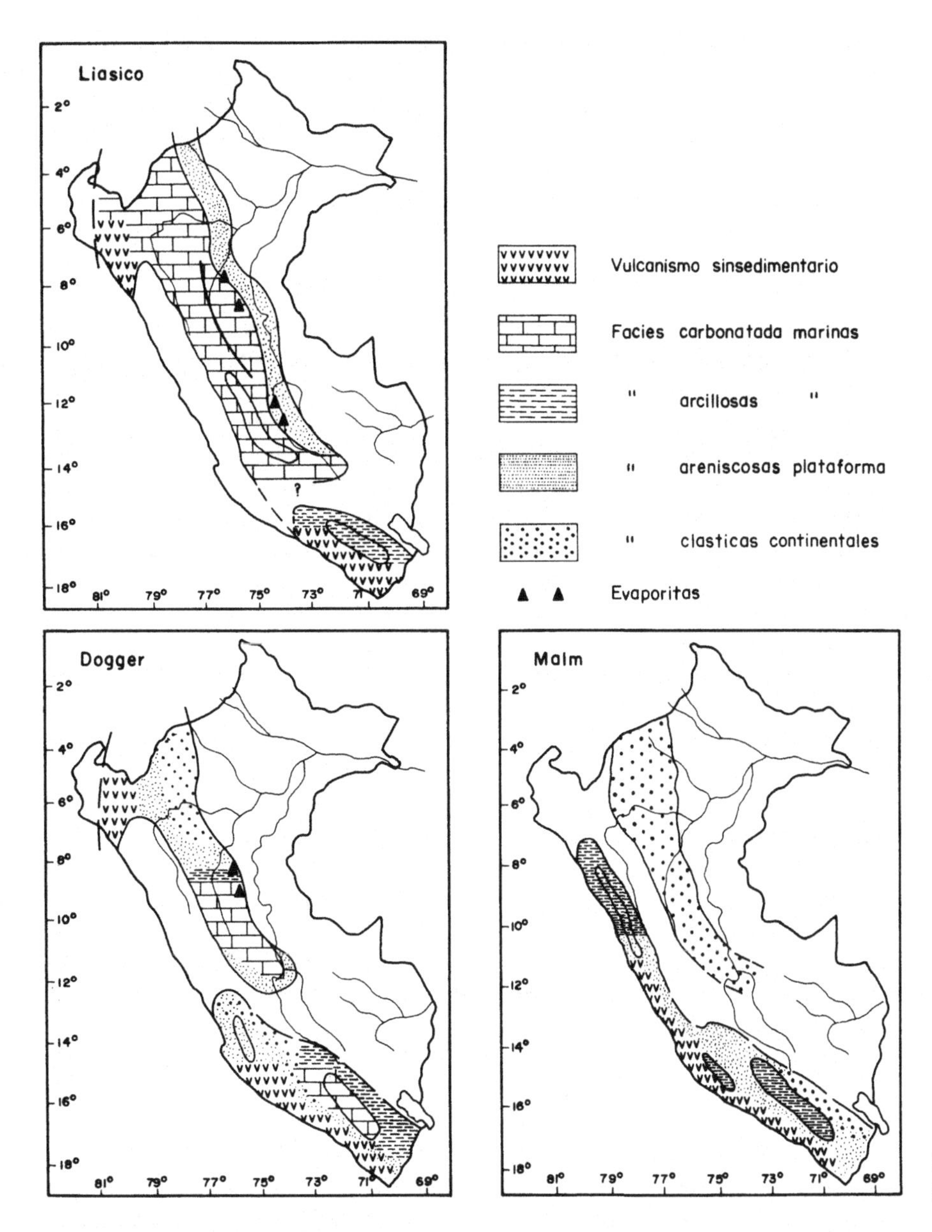


**Figure 6.14. Jurassic paleogeography of Perú.**

Basin, whereas Liassic volcanics and continental deposits are well developed in the taphrogenic depression of the San Jorge and Neuquén basins.

The first Jurassic marine encroachment occurred north of 40°S. The Hettangian-Sinemurian sea covered coastal areas of Chile at the present latitudes of 35–35°15′S and 28–29°15′S, and especially 21–26°S, reaching as far east as the Río Atuel area at 34°50′S (Riccardi et al. 1988a). From the Sinemurian onward thick volcaniclastic sequences and related intrusives developed north of 40°S, along the volcanic arc of the Coastal Cordillera of Chile. Eastward of this magmatic arc, marine sedimentation was generally restricted to an ensialic back-arc basin. At times during the Jurassic this basin would become an intra-arc basin because of intermittent development of two volcanic arcs (Ramos 1988). Smaller marine intra-arc and fore-arc basins were also present, at least during the Hettangian-Sinemurian (Suárez et al. 1982; Davidson 1984; Nasi 1984). The maximum eastward expansion of the Liassic sea was attained in the late Pliensbachian, when most of the Neuquén embayment in west-central Argentina was flooded. Farther south, the Pliensbachian–early Toarcian sea only reached west-central Patagonia.

During the Middle Jurassic, most of central and southern Patagonia underwent extensional tectonics. Volcanism, with minor continental sedimentation, became widespread later in the Jurassic, such as in the El Quemado Complex, Bahía Laura and Lonco Trapial Groups, Taquetrén and Marifil Formations (Uliana et al. 1985).

Along the Pacific margin, north of 40°S, parts of the volcanic arc became continental, with subaerial volcanism (La Negra Formation and equivalents), although small intra-arc basins were also present (Suárez et al. 1985). To the east, marine sedimentation was continuous in the back-arc basin. However, the coastline changed repeatedly in transgressive–regressive cycles as a result of global sea-level variation and local tectonics. A transgressive maximum occurred during the Bajocian north of 28°S and south of 31°S, and regression affected the intermediate region since the Toarcian (Jensen et al. 1976).

A regional regression occurred during most of the Bathonian in west-central Argentina and northern Chile, so that in marginal areas and between 26°S and 37°S the Callovian rests directly on Bajocian or older strata. Thus, the Bathonian record is incomplete and is restricted to northern Chile (Domeyko Range) and the Chacay Melehue area in west-central Argentina. The Lower Bathonian appears to be missing entirely. This regression resulted in the creation of two basins separated by an emergent land, the Tarapaquean Basin in the north, and the Aconcagua-Neuquénian Basin in the south. The Bathonian regressive phase extended into the early Callovian and was followed by a renewed transgression of the middle to late Callovian sea, which, however, did not reach the extent of the earlier Bajocian sea. Marine Callovian has therefore not been recorded between 28°S and 30–31°S.

In central and southern Patagonia, intensive volcanism continued throughout most of the Late Jurassic, producing thick deposits of volcanics and pyroclastics. In central Patagonia the Lower Jurassic marine embayment was obliterated by these rocks, which spread over most of Patagonia. In the west, andesites and basalts dominated, and in the east, rhyolitic ignimbrites. Continental sediments, partly lacustrine (Cañadon Asfalto Formation), were also deposited in this region.

In southern Patagonia, large-scale extensional tectonics, related to the opening of the South Atlantic, produced silicic volcanics (El Quemado Complex or Tobífera Formation) by anatexis of continental crust within a broad rift zone. Opening of a marginal basin resulted in pillow lavas, dolerites, and gabbros. On the oceanic side of this basin, a volcanic arc, related to subduction of Pacific lithosphere under South America, produced calc-alkalic rocks (the Hardy Formation). Subsidence, followed by marine transgression, resulted in continued sedimentation throughout the latest Jurassic and Cretaceous.

At the Pacific margin, north of 40°S, a major regressive phase culminated during the latest Oxfordian–early Kimmeridgian, whereas thick evaporite sequences were deposited in the central parts of the basin, and terrestrial sediments in most marginal areas.

The progressive shallowing of basins of west-central Argentina and northern Chile during the Callovian-Oxfordian was due in part to the continuous uplift of the Coastal Cordillera, which acted as a source of clastics and volcanics. The resulting deposition from the west and displacement of the basin axis to the east climaxed during the Kimmeridgian, when the volcanism reached the Principal Cordillera. This area was again flooded during the Tithonian–Early Cretaceous by a new major transgression.

### *Paleogeographic evolution of west-central Argentina*[16]

Eighteen depositional sequences have been recognized within the Jurassic of the Neuquén Basin (Legarreta and Gulisano 1989). They have been divided into synthems bounded by major discontinuities that are correlated with important paleogeographic changes. These discontinuities are the result of marked changes in sea level that in turn depend on eustasy subsidence, and sediment influx. It should be noted that, excepting the Intra-Malmic Discontinuity in one area of central Neuquén province, these discontinuities are marked by paraconformities (i.e., bedding parallel below and above the discontinuity surfaces). This is especially true along the Andean axis of Neuquén and Mendoza provinces, where angular relationships should be expected if compressive tectonics existed.

[16] By C. A. Gulisano.

The paleogeographic maps (see Figures 6.10 and 6.11) show the principal characters for each group of the depositional sequences. Knowledge of the western border of the basin, however, is limited to isolated outcrops, because of erosion and extensive Cenozoic volcanism. Thus, the reconstructions are incomplete and tentative for that area. The Sañico Subsynthem (lower Araucanian Synthem) developed during the Late Triassic–earliest Jurassic has not been included because its continental deposits are poorly known.

#### *Cuyo Subsynthem*

The oldest deposits, dated as Hettangian-Sinemurian (Riccardi et al. 1988a), are known only from a narrow strip along the upper valley of the Atuel River, in southern Mendoza province (Figure 6.10). They consist of basinal to shelf pelites and sandstones, which are absent toward the south, where the Cuyo Subsynthem begins with the Pliensbachian.

During the Pliensbachian–early Toarcian (Figure 6.10, top left) the sea covered extensive areas, with marked eastward transgression. Shelf and basinal areas were clearly defined by an articulation axis. When the sea level was low in the late Pliensbachian, important turbiditic deposits (Los Molles Formation) were formed in the basinal area, and when the sea level was high, outer-shelf pelites (Los Molles Formation), littoral sandstones (Puesto Araya Formation), and continental facies (El Freno Formation) were developed.

During the Aalenian-Bajocian (Figure 6.10, top right) the shelf margin moved basinward, and the area of turbidite deposition was reduced to a brief, low-sea-level stage in the early Aalenian. The following sea-level rise expanded the shelf area, and high clastic input from the south and east resulted in deltaic progradation (Lajas Formation, shown by dashed line). A similar sedimentary pattern was present during the Bathonian–early Callovian.

During the middle Callovian (Figure 6.10, part 3) the basin became strongly restricted, and evaporites were deposited in its center. At the western, Andean margin, fluvial and marginal sabkha deposits were formed, which toward the east intergraded with the central basin evaporites. The paleogeographic evolution of the Cuyo Subsynthem shows a progressive expansion of the sedimentation area, followed by restriction to central areas of the basin.

#### *Lotena Subsynthem*

During the middle and late Callovian the basin dried up as a result of an important sea-level fall. Thus, during the early late Callovian (Figure 6.10, bottom left) the shelf areas were partially eroded, and in the deepest areas fluvial and aeolian continental sandstones and conglomerates were deposited (lower Lotena Formation). A strong sea-level rise followed, with a marked expansion of the area of marine sedimentation. The Lotena Formation thus transgressed over evaporites, deltaic, and even continental facies of different ages of the Cuyo Subsynthem.

During the early Oxfordian (Figure 6.10, bottom right), as a result of a sea-level rise, a turbiditic succession developed in the Andean area (middle Lotena Formation), followed by pelites that transgressed the shelf area (upper Lotena Formation). Toward the end of this stage, prograding carbonates developed (La Manga Formation, shelf margin shown by dashed line).

The Lotena Subsynthem ends in the late Oxfordian (Figure 6.11, top left), with evaporitic sedimentation in the central basin (Auquilco Formation). To the west the evaporites intergraded with inter-tidal carbonates, and to the east with marginal–sabkha fine clastics. Toward the south, fluvial redbeds developed (Fortin 1 de Mayo Formation).

*Andean Cycle (lower part)*

During the Kimmeridgian (Figure 6.11, top right), as a result of a strong sea-level fall related to marked growth of the volcanic arc, the marginal areas were sites of strong erosion, and deposition occurred in the deepest parts of the basin. Thus, a complex setting of continental deposits was developed, with deposition of clastic sediments of alluvial fans, braided and ephemeral streams, "beach lakes," permanent lakes, and aeolian facies (Tordillo, Quebrada del Sapo, Sierra Blanca, and Catriel Formations).

In the middle Lower Tithonian (Figure 6.11, bottom left), a sea-level rise caused fast submergence of the basin and strong expansion of the marine area, with deposition of marls and bituminous limestones (lower Vaca Muerta Formation). Deposition of littoral sandstones (Carrin Cura and Lindero de Piedra Formations) was restricted to some marginal areas.

During the late Tithonian (Figure 6.11, bottom right), as a result of strong progradation of clastic carbonates from the south and east (Neuquén, Río Negro, and La Pampa provinces), the area of sedimentation was restricted (Vaca Muerta Formation), the inner-shelf and continental facies became widespread (Quintuco, Picún Leufu, Loma Montosa, and Bajada Colorada Formations).

## ANTARCTICA[17]

Marine Jurassic rocks in Antarctica are confined to the Antarctic Peninsula region and the neighboring South Shetland Islands (Figure 6.15). For much of Mesozoic and Cenozoic time the peninsula was an active magmatic arc that formed part of the Pacific margin of Gondwana. Exposed successions relate mainly to fore- and back-arc basins (Fossil Bluff Group and Latady Formation, respectively), but there are fragments of an Upper Jurassic anoxic basin sequence in northern Graham Land (Nordenskjöld Formation), and Lower Jurassic volcaniclastic rocks form part of a major accretionary complex that occupies most of central and western Alexander Island. All of these rocks contain marine fossils that locally may be relatively abundant. The presence of fossilized leaves and fronds within these marine deposits suggests proximity to land, and plant-bearing fluvial sequences, associated with terrestrial volcanic successions, form part of the arc assemblages. By contrast, East Antarctica was a relatively stable cratonic area, and the only exposed Jurassic sedimentary rocks are localized lacustrine interbeds, containing plant beds and freshwater crustaceans, within the Kirkpatrick Basalts of the Transantarctic Mountains, as reviewed by Thomson (1983a). This review will concentrate on the marine sequences of the peninsula region and their stratigraphically diagnostic invertebrate faunas.

### Lower Jurassic

Marine faunas of the Lower Jurassic are known from only two localities in Antarctica (Thomson and Tranter 1986). They occur in highly volcaniclastic rocks in the Lully Foothills, which seem to represent part of an exotic seamount sequence within an extensive accretionary complex (LeMay Group) that covers the greater part of Alexander Island. Although only one locality contains age-diagnostic fossils, in the absence of better evidence the proximity and lithological similarity of the two are taken to suggest that they are of approximately similar (Sinemurian) age.

Locality 1: *Epophioceras*(?) sp., *Protremaster uniserialis* Smith.
Locality 2: *Cardinia* sp. nov.

### Middle Jurassic

Middle Jurassic marine faunas are almost as rare as those of the Lower Hettangian; Bajocian to early Callovian ammonites and bivalves have been described from only a few isolated nunataks in the southern Behrendt Mountains of eastern Ellsworth Land (Quilty 1970, 1977, 1983). The faunas occur in a thick, widely exposed sequence of back-arc mudstones, sandstones, and conglomerates (Latady Formation) of mainly Late Jurassic age, as discussed later. However, the Latady Formation is strongly folded into large, east-facing chevrons; it has been studied at a reconnaissance level only, and knowledge of its stratigraphy is rudimentary. Thus, although the Middle Jurassic faunas appear to be poorly diverse and geographically restricted, further discoveries within the Latady Formation would not be surprising.

Mid-Bajocian: *Megasphaeroceras, Teloceras, Stephanoceras.*
Bathonian–Lower Callovian: eurycephalitine ammonites (G. E. G. Westermann personal communication 1991), *Meleagrinella.*

### Upper Jurassic

Fossiliferous marine rocks of Late Jurassic age occur in three main areas of the Antarctic Peninsula region: northeastern Graham Land, eastern Alexander Island, and the Lassiter-Orville coast area of southeasternmost Antarctic Peninsula. However,

[17] By M. R. A. Thomson.

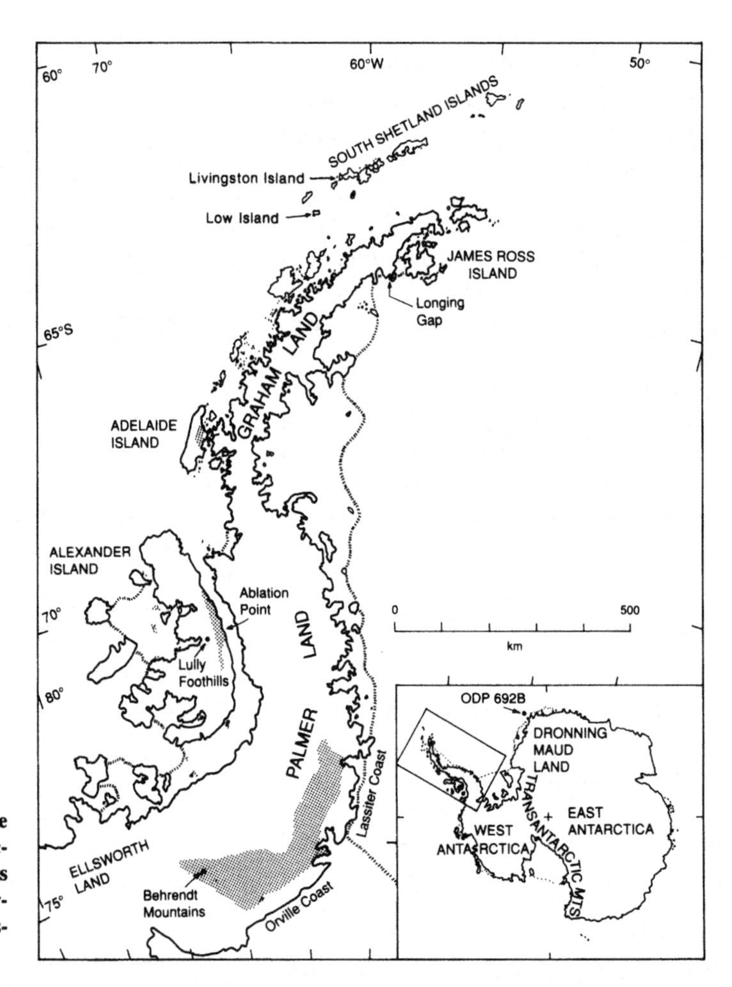


**Figure 6.15. Sketch map of the Antarctic Peninsula with Antarctica (inset) to show main localities mentioned in the text. Stippled areas denote principal areas of Jurassic marine strata.**

continuing discoveries of localized Late Jurassic marine faunas in volcaniclastic rocks of the fore arc [e.g., South Shetland Islands (Smellie, Davies, and Thomson 1980; Thomson 1982) and Adelaide Island (Thomson 1972)] suggest that rocks of this age are widespread.

*South Shetland Islands*

Late Jurassic marine rocks have been described from two small localities at the western end of the South Shetland Islands: northwestern Low Island (Thomson 1982) and Byers Peninsula, and westernmost Livingston Island (Smellie et al. 1980). Indurated black mudstones on Cape Wallace, intruded by an Early Cretaceous granodiorite, contain localized occurrences of ammonites and bivalves. The presence of the ammonite *Epimayites* indicates an Oxfordian age, whereas associated retroceramid bivalves, related to *Retroceramus haasti* Hochstetter, would favor a slightly younger age.

On Livingston Island, Late Jurassic mudstones and interbedded tuffaceous sandstones, with lithological similarities to parts of the Nordenskjöld Formation of northeast Graham Land, are now known to be more localized than was first thought (Smellie et al. 1980). They are faulted against poorly indurated Early Cretaceous mudstones, with "*Spiticeras*" and *Bochianites*, and are not transitional with them, as was suspected. Apart from ichnofossils, notably *Chondrites*, other fossils are scarce. Some squashed ammonites have ribbing reminiscent of some Kimmeridgian-Tithonian perisphinctids. The belemnites are better preserved and include guards of *Hibolithes marwicki marwicki* Stevens and *Belemnopsis stoleyi* Stevens, suggestive of an early Tithonian age.

*Northeast Graham Land*

Patchily distributed exposures of well-laminated mudstones, with interbedded air-fall tuffs, in northeast Graham Land have been independently described under the names Ameghino Formation (Medina and Ramos 1981) and Nordenskjöld Formation (Farquharson 1983). The principal outcrop of the Nordenskjöld Formation is at Longing Gap (450 m), where it can be divided into the Longing and Ameghino Members, but it has been reported from isolated localities between Joinville Island and Cape Fairweather, a distance of nearly 350 km; see Whitham and Doyle (1989) for a review and preference for use of the name Nordenskjöld Formation. Derived clasts and gigantic blocks of it are important constituents of Albian conglomerates on western James Ross Island (Ineson 1985). At certain levels there are varied molluscan faunas (mainly ammonites and bivalves); they are in the process of detailed study and indicate an age range of

Kimmeridgian to Berriasian for the succession at Longing Gap. However, faunas at other localities suggest a slightly greater age range for the formation as a whole (?Oxfordian–Berriasian) (Whitham and Doyle 1989, Fig. 5). Key Jurassic ammonite and bivalve genera from the Longing Member (Kimmeridgian-Tithonian) are as follows:

Ammonites: *Taramelliceras* (*Metahaploceras*), *Torquatisphinctes, Hybonoticeras, Neochetoceras, Virgatosphinctes, Lithacoceras, Kossmatia*(?) and *Anavirgatites*(?).

Bivalves: *Retroceramus, Anopaea, Buchia,* and *Aulacomyella.*

The succeeding Ameghino member contains an ammonite fauna dominated by berriasellids and *Spiticeras* and was considered to be mainly Berriasian in age. However, the possible presence of *Blanfordiceras* could signal the presence of the latest Tithonian.

The Nordenskjöld Formation is believed to represent part of a widespread Late Jurassic–Early Cretaceous anoxic event that affected much of the South Atlantic region (Farquharson 1983). The most recent discovery of rocks pertaining to this event was in hole 692B of ODP Leg 113, which penetrated Nordenskjöld Formation–like Berriasian strata, less than 100 km offshore of northwestern Dronning Maud Land (Doyle, Crame, and Thomson 1990).

### *Alexander Island*

The best-studied marine Jurassic strata in Antarctica belong to the Ablation Point and Himalia Ridge Formations of eastern Alexander Island (Thomson 1979; Crame and Howlett 1988; Howlett 1989). These two formations from part of a >4,000-m-thick fore-arc basin sequence (Fossil Bluff Group) that crops out over a distance of 260 km along the southeastern margin of Alexander Island. The known age range of the Fossil Bluff Group is Kimmeridgian–Albian, but its base has not been precisely dated, and the unit is not seen to be overlain by any other; for the most part, to the west it lies in faulted contact with an extensive sequence of mainly Mesozoic accretionary-complex rocks (LeMay Group) but an unconformity between the two is exposed at one locality (Nell and Storey 1991, p. 245).

The Ablation Point Formation lies entirely within a very thick (350 m) slump unit (Butterworth et al. 1988). Apart from belemnites, fossils are rare, but include the ammonite *Pachysphinctes* (Thomson 1979) and the bivalve *Retroceramus* (Crame 1982). Howlett (1989, text Figs. 8 and 9) grouped these faunas into one belemnite biozone, *Belemnopsis* cf. *aucklandica,* of Kimmeridgian age.

The succeeding Himalia Ridge Formation of coarse-grained channeled complexes and mudstone interbeds is more fossiliferous. It spans the upper and lower Tithonian and the lower part of the Berriasian. It can be divided into three ammonite zones; and it lies entirely within a single belemnite biozone of *Hibolithes belligerundi.* Typical ammonites reported by Howlett (1989) are as follows:

1. *Virgatosphinctes* biozone (Lower Tithonian): *V.* cf. *kagbenensis* Helmstaedt, *V.* cf. *mexicanus* (Burckhardt), *V.* cf. *andesensis* (Douvillé), *V.* cf. *rotundidoma* Uhlig, *V.* cf. *frequens* Uhlig, *V.* cf. *haydeni* Uhlig, *V. denseplicatus* (Waagen), *V. falloti* Collignon, *Pterolytoceras* cf. *exoticum* (Oppel), and *Aulacosphinctoides* spp.

Barren interval

2. *Blanfordiceras* biozone (lower Upper Tithonian): *B. acuticosta* (Uhlig), *B. weaveri* Howlett, and *Lytohoplites* cf. *burckhardti* (Mayer-Eymar).
3. *Haplophylloceras* interval biozone [uppermost Tithonian–Lower Berriasian (poorly constrained)]: *Haplophylloceras.*

### *Orville and Lassiter coasts*

Because of its remoteness and structural complexity (Kellogg and Rowley 1989), the biostratigraphy of the mainly Upper Jurassic Latady Formation of the southeastern Antarctic Peninsula is poorly known. However, the rocks are widespread and comprise an important geological unit. For the most part, fossils are not well preserved, and the dispersal of outcrops in numerous isolated nunataks and mountains hampers interpretation of the few collections made during reconnaissance surveys (cf. Laudon et al. 1983). Bivalves (Quilty 1977; Crame 1982, 1983), belemnites (Mutterlose 1986), and ammonites (Thomson 1983b) suggest the occurrence of Kimmeridgian and Tithonian strata, although the presence of the Oxfordian (Quilty 1970) still requires confirmation.

A general Kimmeridgian–Tithonian age is suggested by the following:

Bivalves: *Retroceramus galoi* (Boehm), *R. haasti* (Hochstetter), *R. subhaasti* (Wandel), *Malayomaorica malayomaorica* (Krumbeck).

Belemnites: *Conodicoelites* sp., *Hibolithes* aff. *marwicki* Stevens, *H.* aff. *verbeeki* Kruizinga, *H.* aff. *arkelli* Stevens, *Belemnopsis* aff. *keari* Stevens.

Ammonites: *Katroliceras, Pachysphinctes, Torquatisphinctes*(?), *Subdichotomoceras, Virgatosphinctes, Kossmatia.*

The youngest strata, which occur in the extreme southeastern part of the area, contain the ammonites *Berriasella*(?), and *Blanfordiceras* and the belemnite *Produvalia* aff. *neyrivensis* (Favre) and probably are latest Tithonian in age.

## References

Aguirre, L. (1983). Granitoids in Chile. *Geol. Soc. Am. Mem., 159,* 293–316.

Almeida, F. F. M. (1972). Tectono-magmatic activation of the South American platform and associated mineralization. In *Proceedings of the International Geological Congress, 24th Session, Section 3, Tectonics* (pp. 339–46).

Archangelsky, S. (1977). Vegetales fośiles de la Formación Springhill, Cretácico, en el subsuelo de la Cuenca Magallánica, Chile. *Ameghiniana, 13*(2), 141–58.

Arnold, H. C. (1961). *The Pre-Cretaceous of the Venezuelan Andes.* Unpublished report, Exploration Department, Compañia Shell de Venezuela, Caracas.

Baranzangi, M., & Isacks, B. L. (1976). Spatial distribution of earthquakes and subduction of the Nazca plate beneath South America. *Geology, 4,* 686–92.

Barrero, D. (1979). Geology of the Western Cordillera, west of Buga and Roldanillo, Colombia. *Publ. Geol. Eso. Ingeominas (Bogotá), 4,* 1–75.

Bartok, P., Reijers, T. J. A., & Juhasz, I. (1981). Lower Cretaceous Cogollo Group, Maracaibo Basin, Venezuela: sedimentology, diagenesis, and petrophysics. *Am. Assoc. Petrol. Geol. Bull., 65,* 1110–34.

Bartok, P., Renz, O., & Westermann, G. E. G. (1985). The Siquisique ophiolites, northern Lara State, Venezuela: a discussion of their Middle Jurassic ammonites and tectonic implications. *Geol. Soc. Am. Bull., 96,* 1050–5.

Behrendsen, O. (1891–2). Zur Geologie des Ostabhanges der Argentinischen Kordillere. *Z. Deutsch. Geol. Ges., 43,* 396–420; *44,* 1–42.

Bellido, E. (1956). Geología del curso medio del Río Huaytará, Huancavelica. *Bol. Soc. Geol. Perú, 30,* 33–47.

Bellizzia, A. (coord.). (1976). *Mapa Geológico-Estructural de Venezuela.* Caracas: Dirección de Geología, Ministerio de Minas e Hidrocarburos.

Ben-Abraham, Z., & Nur, A. (1988). Effects of collisions at trenches on oceanic ridges and passive margins. *Am. Geophysical Union, Geodynamics Ser., 18,* 9–18.

Benavides, C. V. (1962). Estratigrafía pre-Tertiaria de la región de Arequipa. *Bol. Soc. Geol. Perú, 30,* 49–79.

Benedetto, G., & Odreman, O. (1977). Nuevas evidencias paleontológicas en la Formación La Quinta, sus edad y correlación con las unidades aflorantes en la Sierra de Perija y Cordillera Oriental de Colombia. *Mem. V Congr. Geol. Venezolano, 1,* 87–106.

Berg, K., & Baumann, A. (1985). Plutonic and metasedimentary rocks from the Coastal Range of northern Chile: Rb-Sr and U-Pb isotopic systematics. *Earth Planet. Sci. Lett., 75,* 101–15.

Berg. K., & Breitkreuz, C. (1983). Mesozoische Plutone in der nordchilenischen Küsten-Kordillere: Petrogenese, Geochronologie, Geochimie und Geodynamik mantelbetonter Magmatite. *Geotekt. Forsch., 66,* 1–107.

Biddle, K. T., Uliana, M. A., Mitchum, R. M., Fitzgerald, M. G., & Wright, R. C. (1986). The stratigraphic and structural evolution of the Magallaes Basin, southern South America. *Special Publications Internat. Assoc. Sediment., 8,* 41–61.

Biro-Bagoczky, L. (1980). Estudio sobre el límite entre el Titoniano y el neocomiano en la Formación Lo Valdés, Provincia de Santiago (33 50′ lat. sur), Chile: principalmente sobre la base de ammonoideos. *II Congr. Argent. Paleont. Bioestr. & I Congr. Latinoam. Paleont., Actas, 1,* 1237–52.

Blasco, G., Levy, R., & Nullo, F. (1979a). Los amonites de la formación Osta Arena (Liásico) y su posición estratigráfica, Pampa de Agnia (Provincia de Chubut). *VII Congr. Geol. Argent., Actas, 2,* 407–29.

Blasco, G., Levy, R., & Ploszkiewicz, V. (1980). Las calizas toarcianas de Loncopán, Depto. Tehuelches, Provincia del Chubut, República Argentina. *II Congr. Argent. Paleont. Bioestr. & I Congr. Latinoam. Paleont., Actas, 1,* 191–200.

Blasco, G., Nullo, F., & Proserpio, C. (1976b). *Aspidoceras* en Cuenca Austral, Lago Argentino, Prov. de Santa Cruz. *Asoc. Geol. Argent. Rev., 34*(4), 282–93.

Bock, W. (1953). American Triassic estherids. *J. Paleontol., 39,* 62–76.

Bogdanic, T., & Chong, G. (1985). Bioestratigrafía del Jurásico de la zona preandina chilena entre los 24 30′–25 30′ Lat. Sur. *IV Congr. Geol. Chileno, 1,* 38–57.

Bogdanic, T., Hillebrandt, A. von, & Quinzio, L. A. (1985). El Aaleniano de Sierra de Varas, Cordillera de Domeyko, Antofagasta, Chile. *IV Congr. Geol. Chileno, 1,* 58–75.

Boric, R. (1981). *Cuadrangulos Estación Colúpito y Toco, Región de Antofagasta.* Inst. Invest. Geol., Chile, Carta Geológica, No. 49–50, 1–52.

Bowen, J. M. (1972). Estratigrafía del Pre-Cretácico en la parte norte de la Sierra de Perija. *Bol. Geol. (Venezuela), Pub. Esp., 5*(2), 729–61.

Bürgl, H. (1960). El Jurásico e Infracretaceo del Río Bata. Boyaca. *Bol. Géol. (Bogotá), 6*(1–3), 169–211.

(1961). Historia Geológica de Colombia. *Rev. Acad. Col. Cienc. Exac. Fis. Nat. (Bogotá), 9*(43), 137–91.

(1964). El Jura-Triásico de Colombia. *Bol. Géol. (Bogotá), 12,* 5–31.

Burke, K. (1976). Development of graben associated with the initial ruptures of the Atlantic Ocean. *Tectonophysics, 36,* 93–112.

Burmeister, H., & Giebel, C. (1861). Die Versteinerungen von Juntas um Thal des Rio Copiapo. *Naturforsch. Ges. Halle, Abh., 6,* 122–32.

Butterworth, P. J., Crame, J. A., Howlett, P. J., & MacDonald, D. I. M. (1988). Lithostratigraphy of Upper Jurassic–Lower Cretaceous strata of eastern Alexander Island, Antarctica. *Cretaceous Research, 9,* 249–64.

Casamiquela, R. (1964). *Estudios Icnológicos. Problemas y Métodos de la Icnología con Aplicación al Estudio de Pisadas Mesozoicas (Reptilia, Mammalia) de la Patagonia.* Buenos Aires.

(1965). Nuevo material de *Vieraella herbstii* Reig. Reinterpretación de la ranita liásica de la Patagonia y consideraciones sobre filogenia y sistemática de los anuros. *Mus. La Plata Rev. Paleont., 4*(27), 265–317.

Cazeneuve, H. (1965). Datación de una toba de la Formación Chon Aike (Jurásico de Santa Cruz, Patagonia) por el método de Potasio-Argón. *Ameghiniana, 4*(5), 156–8.

Cecioni, G. (1960). La Zona de Psiloceras planorbis en Chile. *Univ. Chile, Fac. Cienc. Fisc. Mat., Inst. Geol. Comunic., 1* (1), 1–19.

Cecioni, G., & Westermann, G. E. G. (1968). The Triassic/Jurassic marine transition of Coastal Central Chile. *Pac. Geol., 1,* 41–75.

Cediel, F. (1968). El Grupo Giroń, una Molasa mesozoica de la Cordillera Oriental. *Bol. Géol. (Bogotá), 16,* 5–96.

(1969). Die Giron-Gruppe. Eine fruh-mesozoische Molasse der Ostkordillere Kolumbiens. *N. Jb. Geol. Palaont, Abh., 133,* 111–62.

Cediel, F., Mojica, J., & Macia, C. (1980). Definición Estratigráfica del Triaśico de Colombia, Sudamerica. Formaciones Luisa, Payande y Saldana. *Newsl. Stratigr., 9*(2), 73–104.

Chang, K. H. (1975). Unconformity-bounded stratigraphic units. *Geol. Soc. Am. Bull., 86,* 1544–52.

Charrier, R. (1982). La Formación Leñas Espinoza: Redefinición, Petrografía y Ambiente de Sedimentación. *Rev. Geol. Chile, 17,* 71–82.

Charrier, R., Linares, E., Niemeyer, H., & Skarmeta, J. (1979). Edades Potasio-Argón de vulcanitas mesozoicas y cenozoicas del sector chileno de la meseta Buenos Aires, Aysen, Chile y su significado geológico. *VII Congr. Geol. Argent., Actas, 2,* 23–41.

Chong, G. (1977). Contribution to the knowledge of the Domeyko Range in the Andes of northern Chile. *Geol. Rund., 66*(2), 374–404.

Cobbing, E. J., & Pitcher, W. S. (1983). Andean plutonism in Peru and its relationship to volcanism and metallogenesis at a segmented plate edge. *Geol. Soc. Am. Mem., 159,* 277–91.

Cordani, U. G., Teixeira, W., Tassinari, C. C. G., Kawashita, K., & Sato, K. (1988). The growth of the Brazilian Shield. *Episodes, 11*(13), 163–7.

Cornejo, P., Nasi, C., & Mpodozis, C. (1984). La Alta Cordillera entre Copiapó y Ovalle. *Serv. Nac. Géol. Mineria, Chile, Misc., 4,* H1–45.

Cortés, J. M. (1981). El Sustrato Precretácico del extremo Noreste de la Provincia del Chubut. *Asoc. Geol. Argent. Rev., 36*(3), 217–35.

Cortiñas, J. S. (1984). Estratigrafía y facies del Jurásico entre Nueva Lubecka, Ferrarotti y Cerro Colorado. Su relación con los depósitos coetáneos del Chubut Central. *IX Congr. Geol. Argent.*, *2*, 283–99.

Corvalan, J. (1976). El Triásico y Jurásico de Vichuquen-Tilcura y de Hualane, Prov. de Curicó. Implicaciones Paleogeográficas. *I Congr. Geol. Chileno*, *1*, A137–54.

(1982). El límite Triásico-Jurásico en la cordillera de la Costa de las Provincias de Curicó y Talca. *III Congr. Geol. Chileno*, *3*, F63–85.

Covacevich, V., & Piraces, R. (1976). Hallazgo de ammonites del Bajociano superior en la Cordillera de la Costa de Chile central entre la Cuesta de Melón y Limache. *I Congr. Geol. Chileno*, *1*, C67–85.

Crame, J. A. (1982). Late Jurassic inoceramid bivalves from the Antarctic Peninsula and their stratigraphical significance. *Palaeontology*, *25*(3), 555–603.

(1983). The occurrence of the Upper Jurassic bivalve *Malayomaorica malayomaorica* (Krumbeck) on the Orville Coast, Antarctica. *J. Molluscan Studies*, *49*, 61–76.

Crame, J. A., & Howlett, P. J. (1988). *Late Jurassic and Early Cretaceous Biostratigraphy of the Fossil Bluff Formation, Alexander Island*. British Antarctic Survey Bulletin No. 78.

Creer, K. M., Mitchell, J. G., & Abou Deeb, J. (1972). Palaeomagnetism and radiometric age of the Jurassic Chon-Aike Formation from Santa Cruz province. *Earth Planet. Sci. Lett.*, *14*(1), 131–8.

Dalmayrac, B., Laubacher, C., & Marocco, R. (1977). *Etude Géologique des Andes Peruviennes a partir de Trois Transversales*. Tesis, Montpellier.

Damm, K.-W., & Pichowiak, S. (1981). Geodynamik und Magmengenese in der Küstenkordillere Nordchiles zwischen Taltal und Chañaral. *Geotekt. Forsch.*, *61*, 1–166.

Davidson, J. (1984). El Paleozoico y Mesozoico inferior de Atacama. *Serv. Nac. Geol. Minería Chile, Misc.*, *4*, F1–15.

(1988). El Jurásico y Cretácico inferior en las Nacientes del Teno (Chile): una revisión. *V Congr. Geol. Chile*, *1*, A453–8.

Davidson, J., Godoy, E., & Covacevich, V. (1976). El Bajociano marino de Sierra Minillas (70°30′ L.S.) y Sierra Fraga (69°50′ L.O.–27 L.S.), Provincia de Atacama, Chile: Edad y Marco Geotectónico de la Formación La Negra en esta latitud. *I Congr. Geol. Chileno*, *1*, A255–72.

Dedios, P. (1967). Cuadrangulo Vicuña, Provincia de Coquimbo. *Inst. Invest. Geol., Carta Geol., Chile*, *16*, 5–65.

Dellape, D. A., Mombru, C., Pando, G. A., Riccardi, A. C., Uliana, M. A., & Westermann, G. E. G. (1979). Edad y Correlación de la Formación Tabanos en Chacay Melehue y otras localidades de Neuquén y Mendoza. Con consideraciones sobre la distribución y el significado de las sedimentitas Lotenianas. *Mus. La Plata Obra Cent.*, *5*, 81–105.

Di Biase, F. (1985). Noticia Preliminar sobre el hallazgo del Liásico marino en los Cerros de Cuevitas, Provincia de Antofagasta. *IV Congr. Geol. Chileno*, *1*, 249–61.

Digregorio, J. H. (1972). Neuquén. In A. F. Leanza (ed.), *Geología Regional Argentina* (pp. 439–506). Córdoba: Acad. Nac. Cienc.

(1978). Estratigrafía de las Acumulaciones Mesozoicas. In E. O. Rolleri (ed.), *Geologia y Recursos Naturales del Neuquén. VIII Congr. Geol. Argent. (Relatorio)*.

Digregorio, J. H., & Uliana, M. A. (1980). Cuenca Neuquína. In *Seg. Simp. Geologia Regional Argentina* (vol. 2, pp. 985–1032). Córdoba: Acad. Nac. Cienc.

Dingman, R. J., & Galli, C. (1965). *Geology and Groundwater Resources of the Pica Area, Tarapaca Province, Chile*. U.S. Geological Survey Bulletin 1189, 1–113.

Dorado, J. (1984). *Contribución al Conocimiento de la Estratigrafía de la Formación Brechas de Buenavista (Limite Jurásico–Cretácico), Oeste de Villavicencio, Meta*. Tésis de Grado, Univ. Nal., Depto. Geociencias, Bogotá.

Douville, R. (1910). Cephalópodes Argentins. *Soc. Geol. Fr. Mém., Pal.*, *43*, 1–22.

Doyle, P., Crame, J. A., & Thomson, M. R. A. (1990). Late Jurassic–Early Cretaceous macrofossils from Leg 113, Hole 692B, eastern Weddell Sea. In P. F. Barker, J. P. Kennett, et al. (eds.), *Proceedings of the Ocean Drilling Program, Scientific Results, Vol. 113, Weddell Sea, Antarctica* (pp. 443–8). College Station, Tex.: Ocean Drilling Program.

Drake, R., Vergara, M., Munizaga, F., & Vincente J.-C. (1982). Geochronology of Mesozoic-Cenozoic magmatism in Central Chile, Lat. 31–36 S. *Earth Sci. Rev.*, *18*, 353–63.

Escobar, F. (1980). *Paleontología y Bioestratigrafía del Triásico Superior y Jurásico Inferior en el Area de Curepto, Provincia de Talca*. Inst. Inv. Geol. Chile, Bol. 351, 1–78.

Espejo, A., Etchart, H., Cordani, U., & Kawashita, K. (1980). Geocronología de intrusivas ácidas en la Sierra de Perija. *Bol. Geol. (Venezuela)*, *14*(26), 245–54.

Etayo-Serna, F. (1985). El límite Jurásico–Cretácico en Colombia. In *Proyecto Cretácico. Publ. Esp. Ingeomin.* (Bogotá), *16*, XXIII-1–4.

Etayo-Serna, F., Parra, E., & Rodriguez, G. (1982). Análisis Facial del "Grupo Dagua" con base en secciones aflorantes al oeste de Toro (Valle del Cauca). *Geol. Nordandina (Bogotá)*, *5*, 3–12.

Farquharson, G. W. (1983). *The Nordensjuöld Formation of the Northern Antarctic Peninsula: An Upper Jurassic Radiolarian Mudstone Sequence*. British Antarctic Survey Bulletin No. 60.

Feruglio, E. (1936–7). *Palaeontographía Patagónica*. Ist. Geol., Univ. Padova, Mem. 11, 1–384.

Forster, R., & Hillebrandt, A. von. (1984). Das Kimmeridge des Profeta-Jura in Nordchile mit einer *Mecochirus favreina*-Vergesellschaftung (Crustacea, Decapoda – Ichnogenus). *Mitt. Bayer. Staatsslg. Paläont. hist. Geol.*, *24*, 67–84.

Franchi, M. R., & Page, R. F. N. (1980). Los Basaltos Cretácicos y la Evolución Magmática del Chubut Occidental. *Asoc. Geol. Argent. Rev.*, *35*(2), 208–29.

Francken, C. (1956). Triásico y Jurásico en Venezuela. In *Léxico Estratigráfico de Venezuela* (pp. 648–50). Caracas: Ministerio de Minas e Hidrocarburos.

Fuenzalida, R., & Covacevich, V. (1988). Volcanismo y Bioestratigrafía del Jurásico superior y Cretácico inferior en la cordillera Patagónica, Región de Magallanes, Chile. *V Congr. Geol. Chileno*, *3*, H159–83.

Galli, C. (1957). Las formaciones geológicas en el borde occidental de la Puna de Atacama, sector de Pica, Tarapacá. *Minerales*, *56*, 3–15.

(1968). Cuadrangulo Juan de Morales, Provincia de Tarapacá. *Inst. Inv. Geol., Carta Geol., Chile*, *18*, 5–53.

Galli, C., & Dingman, R. J. (1962). Cuadrangulos Pica, Alca, Matilla y Chacarilla, Provincia de Tarapacá. *Inst. Inv. Geol., Carta Geol., Chile*, *3*(2–5), 7–125.

Galli, C., & Menéndez, C. A. (1968). Geología de la Quebrada Juan de Morales, Tarapacá, Chile y su flora Jurásica. *Terc. J. Geol. Argent., Actas*, *1*, 163–71.

Gansser, A. (1973). Facts and theories on the Andes. *J. Geol. Soc. London*, *129*(2), 93–131.

Garcia, F. (1967). Geología del Norte Grande de Chile. In *Simp. Gesinclinal Andino, 1962*. Santiago: Edic. ENAP.

Geyer, O. (1967). Das Typus-Profil der Morrocoyal-Formation (Unterlias; Depto. Bolivar, Kolumbien). *Mitt. Inst. Colombo-Aleman Invest. Cient. (Santa Marta)*, *1*, 53–63.

(1973). Das praekretazische Mesozoikum von Kolumbien. *Geol. Jahrb. (Hannover)*, *B5*, 1–156.

(1979). Zur Paläogeographie mesozoischer Ingressionen und Transgressionen in Kolumbien. *N. Jb. Geol. Paläont., Mh.*, *6*, 349–68.

(1983). Obertithonische Ammoniten-Faunen von Peru. *Zbl. Geol. Paläont.*, J.H., *3/4*, 335–50.

González, H. (1980). Geología de las Planchas 167 (Sonson) y 187 (Salamina). *Bol. Geol. Ingeominas, 23*(1), 1–174.
González, O., & Vergara, M. (1962). *Reconocimiento Geológico de la Cordillera de los Andes entre los Paralelos 35 y 38°S Latitud Sur.* Univ. Chile, Fac. Cienc. Fis. Mat., Inst. Geol., Publ. 24.
González Bonorino, F. (1974). La Formación Millaqueo y la "Serie Porfiritica" de la Cordillera Nordpatagónica: Nota Preliminar. *Asoc. Geol. Argent. Rev., 29*(2), 145–53.
González de Juana, C., Iturralde de Arozena, J. M., & Picard, X. (1980). *Geología de Venezuela y de sus Cuencas Petrolíferas.* Caracas: Ediciones Foninves.
González Diaz, E. F., & Nullo, F. E. (1980). Cordillera Neuquina. In *Segundo Simp. Geol. Reg. Argent.* (vol. 2, pp. 1099–147). Cordoba: Acad. Nac. Cienc.
Groeber, P. (1946). Observaciones geológicas a lo largo del Meridiano 70, 1. Hoja Chos Malal. *Asoc. Geol. Argent. Rev., 1*(3), 177–208.
Groeber, P., Stipanicic, P. N., & Mingramm, A. (1953). Jurásico. In *Geografía de la República Argentina. T. II: Mesozoico* (pp. 143–347). Buenos Aires: GAEA.
Groschke, M., & Hillebrandt, A. von. (1985). Trias und Jura in der mittleren Cordillera Domeyko von Chile (23°30'–24°30'). *N. Jb. Geol. Palaont., 170*(2), 129–66.
Groschke, M., Hillebrandt, A. von, Prinz, P., & Quinzio, L. A. (1988). Marine Mesozoic in northern Chile between 21–26°S. *Lect. Notes Earth Sci., 17,* 105–17.
Grose, L. T., & Szekely, T. S. (1968). Lower Jurassic Pucara Group of Central Peru. *Geol. Soc. Am. An. Mtz. Mexico, Progr. Abstr.,* 199.
Gulisano, C. A. (1981). Síntesis Estratigráfica de la Cuenca neuquina: Jurásico. In E. Mutti, C. A. Gulisano, L. Legarreta, Peroni, G., & Cazau, L. (eds.), *Guía de Campo de la 1ra Escuela de Análisis de Facies Clásticas.* Buenos Aires: YPF.
(1984). Esquema estratigráfico de la secuencia jurásica del oeste de la Provincia del Neuquén. *IX Congr. Geol. Argent., Actas, 1,* 236–59.
Gulisano, C. A., Gutierrez Pleimling, A. R., & Digregorio, R. E. (1984). Análisis Estratigráfico del Intervalo Tithoniano-Valanginiano (Formaciones Vaca Muerta, Quintuco y Mulichinco) en el suroeste de la Provincia del Neuquén. *IX Congr. Geol. Argent., Actas, 1,* 221–35.
Gust, D. A., Biddle, K. T., Phelps, D. W., & Uliana, M. A. (1985). Associated Middle to Late Jurassic volcanism and extension in southern South America. *Tectonophysics, 116,* 223–53.
Haas, O. (1960). Lower Cretaceous ammonites from Colombia, South America. *Am. Mus. Nov., 2005,* 1–62.
Hallam, A., Biro-Bagoczky, L., & Perez, E. (1986). Facies analysis of the Lo Valdes Formation (Tithonian-Hauterivian) of the High Cordillera of central Chile, and the paleogeographic evolution of the Andean Basin. *Geol. Mag., 123*(4), 425–35.
Haller, M. J., & Lapido, O. R. (1982). The Jurassic-Cretaceous volcanism in the Septentrional Patagonian Andes. *Earth Sci. Rev., 18,* 395–410.
Halpern, M. (1973). Regional geochronology of Chile south of 50 latitude. *Geol. Soc. Am. Bull., 84,* 2407–22.
Haq, G. U., Hardenbol, J., & Vail, P. R. (1987). Chronology of fluctuating sea levels since the Triassic. *Science, 235,* 1156–67.
Harrington, H. J. (1961). Geology of parts of Antofagasta Provinces, Northern Chile. *Am. Assoc. Petrol. Geol. Bull., 45*(2), 169–97.
(1962). Paleogeographic development of South America. *Am. Assoc. Petrol. Geol. Bull., 46*(10), 1773–814.
Harrison, J. V. (1940–3). The geology of the Central Andes in part of the Province of Junin. *Geol. Soc. London, 99.*
Harrison, J. V. (1956). Geología de la carretera Huancayo–Santa Beatriz en el Perú Central. *Bol. Soc. Geol. Perú, 28,* 5–52.
Haupt, O. (1907). Beiträge zur Fauna des oberen Malm und der unteren Kreide in der argentinischen Cordillere. *N. Jb. Min. Geol. Paläont., 23,* 187–236.
Hea, J. P., & Whitman, A. B. (1960). Estratigrafía y petrología de los sedimentos precretácicos de la parte norte-central de la Sierra de Perija, Estado Zulia, Venezuela. *Bol. Geol. (Venezuela), Pub. Esp., 3*(1), 351–76.
Herbst, R. (1965). La flora fósil de la Formación Roca Blanca (Prov. de Santa Cruz, Patagónia), con consideraciones geológicas y estratigráficas. *Op. Lilloana, 12,* 3–11.
(1966). La flora liásica del Grupo Pampa de Agnia, Chubut, Patagónia. *Ameghiniana, 4*(9), 337–47.
Herbst, R., & Anzoátegui, L. M. (1968). Nuevas plantas de la flora del Jurásico medio (Matildense) de Taquetren, Prov. de Chubut. *Ameghiniana, 5*(6), 183–90.
Hervé, F. (1984). Geología de la Región al sur de los Canales Beagle y Ballenero. *Serv. Geol. Miner. Chile, Misc., 4,* P1–13.
Hervé, F., Godoy, E., Parada, M. A., Ramos, V., Rapela, C., Mpodozis, C., & Davidson, J. (1988). A general view on the Chilean-Argentine Andes, with emphasis on their early history. *Am. Geophysical Union, Geodyn. Ser., 18,* 97–113.
Hervé, F., Nelson, E., Kawashita, K., & Suarez, M. (1981). New isotopic ages and the timing of orogenic events in the Cordillera Darwin, southernmost Chilean Andes. *Earth Planet. Sci. Lett. 55,* 257–65.
Hillebrandt, A. von. (1970). Zur Biostratigraphie und Ammoniten-Fauna des südamerikanischen Jura (insbes. Chile). *N. Jb. Geol. Paläont. Abh., 136*(2), 166–211.
(1973a). Die Ammoniten-Gattungen *Bouleiceras* und *Frechiella* im Jura von Chile und Argentinien. *Ecl. Geol. Helv., 66*(2), 351–63.
(1973b). Neue Ergebnisse über den Jura in Chile und Argentinien. *Münster. Forsch. Geol. Paläont., 31/32,* 167–99.
(1977). Ammoniten aus dem Bajocien (Jura) von Chile (Südamerika). Neue Arten der Gattungen *Stephanoceras* und *Domeykoceras* n. gen. (Stephanoceratidae). *Mitt. Bayer. Staatsslg. Paläont. hist. Geol., 17,* 35–69.
(1981). Faunas de amonites del liásico inferior y medio (Hettangiano hasta Pliensbachiano) de América del Sur (excluyendo Argentina). In *Cuencas Sedimentarias del Jurásico y Cretácico de América del Sur* (vol. 2, pp. 499–538). Buenos Aires:
(1987). Liassic ammonite zones of South America and correlations with other provinces. In W. Volkheimer (ed.), *Bioestratigrafía de los Sistemas Regionales del Jruásico y Cretácico de América del Sur* (vol. 1, pp. 111–29). Mendoza: Com. Sudam. Juras. Cretac.
Hillebrandt, A. von, & Schmidt-Effing, R. (1981). Ammonites aus dem Toarcium (Jura) von Chile (Sudamerika). *Zitteliana, 6,* 1–74.
Hillebrandt, A. von, & Westermann, G. E. G. (1985). Aalenien (Jurassic) ammonite faunas and zones of the southern Andes. *Zitteliana, 12,* 3–55.
Hinterwimmer, G. A., Bravo de Laguna, M. A., & Orchuela, I. A. (1986). Análisis de facies y sismoestratigrafía de una progradación clástica en sedimentitas cuyanas en la zona de Borde Montuoso, Provincia del Neuquén. *Prim. Reun. Argent. Sedm., Resum. Expand. (La Plata),* 181–4.
Howlett, P. J. (1989). *Late Jurassic–early Cretaceous cephalopods of eastern Alexander Island, Antarctica. Special Papers in Palaeontology* No. 41.
Hubach, E. (1957). Contribución a las unidades estratigráficas de Colombia. *Serv. Geol. Nal. Informe (Bogotá), 1212,* 1–166.
Hubach, E., & Alvarado, B. (1934). Geología de los Departamentos del Valle y Cauca, en especial del Carbón. *Serv. Géol. Nal. Informe (Bogotá), 224,* 467.
Huber, K. (1982). Geologie der jurassischen Valle-Alto Formation in der Zentralkordillere Kolumbiens. *Asrb. Inst. Geol. Paläont. Univ. Stuttgart, NF 77,* 1–74.
Huber, K., & Wiedmann, J. (1987). Sobre el limite Jurásico-Cretácico en los alrededores de Villa de Leiva, Depto. de Boyacá, Colombia. *Geol. Colombiana, 15,* 81–92.

Huff, K. F. (1949). Sedimentos del Jurásico Superior y Cretáceo inferior en el este del Perú. *Soc. Geol. Perú, Vol. Jubilar 25 Aniv.*, Pt. 2, fasc. 15.

Indans, J. (1954). Eine Ammonitenfauna aus dem Untertithon der argentinischen Kordillere in Sud Mendoza. *Palaeontographica, A105*, 96–132.

Ineson, J. R. (1985). Submarine glide locks from the Lower Cretaceous of the Antarctic Peninsula. *Sedimentology, 32*, 659–70.

Ingeominas (1983). *Mapa de Terrenos Geológicos de Colombia*. Publ. Geol. Esp. 14 (Edic. preliminar), Bogotá.

Irving, E. (1971). Evolución estructural de los Andes más Septentrionales de Colombia. *Bol. Geol. (Bogotá), 19*, 1–90.

Jaworski, E. (1913). Beiträge zur Kenntnis des Jura in Süd-Amerika, Teil I. *N. Jb. Min. Geol. Paläont., 37*, 285–342.

(1925). Contribución a la Paleontología del Jurásico Sudamericáno. *Dir. Gen. Miner. Geol. Hidrol., Sect. Geol., 4*, 1–160.

(1926). La Fauna del Lias y Dogger de la Cordillera Argentina en la parte meridional de la provincia de Mendoza. *Acad. Nac. Cienc. Córdoba, Actas, 9*, 135–317.

Jeletzky, J. A. (1984). Jurassic–Cretaceous boundary beds of western and Arctic Canada and the problem of the Tithonian-Berriasian stages in the Boreal Realm. In G. E. G. Westermann (ed.), *Jurassic–Cretaceous Biochronology and Paleogeography of North America* (pp. 175–255). Geological Association of Canada, Special Paper 27.

Jenks, W. F. (1948). Geología de la hoja de Arequipa al 200,000. *Bol. Inst. Geol. Peru, 9*, 204.

Jensen, O., & Vicente, J. C. (1977). Estudio geológico del área de "Las Juntas" del Río Copiapó. *Asoc. Geol. Argent. Rev., 31*(3), 145–73.

Jensen, O., Vicente, J. C., Davidson, J., & Godoy, E. (1976). Etapas de la evolución marina jurásica de la Cuenca Andina externa (mioliminar) entre los paralelos 26 y 29, 30 Sur. *I Congr. Geol. Chileno, 1*, A273–95.

Julivert, M. (1968). Colombie (premiere partie). In *Lexique Stratigr. Internat.* (vol. 4a, pp. 1–650). Paris: Centre Nat. Rech. Sci.

Kellogg, K. S., & Rowley, P. D. (1989). *Structural Geology and Tectonics of the Orville Coast Region, Southern Antarctic Peninsula, Antarctica.* U.S. Geological Survey Professional Paper No. 1498.

Kielbowicz, A. A., Ronchi, D. I., & Stach, N. H. (1984). Foraminiferos y ostrácodos Valanginianos de la Formación Springhill. *Asoc. Geol. Argent. Rev., 38*(3–4), 313–39.

Klohn, C. (1960). *Geología de la Cordillera de los Andes de Chile Central, Provincias de Santiago, O'Higgins, Colchagua y Curicó.* Inst. Invest. Geol. Chile, Bol. 8.

Krantz, F. (1928). La Fauna del Tithono superior y medio de la parte meridional de la provincia de Mendoza. *Acad. Nac. Cienc. Córdoba, Actas, 10*(4), 1–57.

Kummel, B. (1948). Estratificación de la región de Santa Clara, Estudio Preliminar. *Bol. Soc. Geol. Am., 59*, 1217–64.

Kundig, E. (1938). Las rocas pre-Cretácicas de los Andes centrales de Venezuela, con algunas observaciones sobre su tectónica. *Bol. Geol. Miner. (Venezuela), 2*(2–4), 21–43.

Langenheim, J. H. (1961). Late Paleozoic and early Mesozoic plant fossils from the Cordillera Oriental of Colombia and correlation of the Girón Formation. *Bol. Géol. (Bogotá), 8*, 95–132.

Lapido, O. R., Lizuain, A., & Nuñez, E. (1984). La Cobertura Sedimentaria Mesozoica. In *IX Congr. Geol. Argent. Relatorio* (pp. 139–62.

Laudon, T. S., Thomson, M. R. A., Williams, P. L., Milliken, K. L., Rowley, P. D., & Boyles, J. M. (1983). The Jurassic Latady Formation, southern Antarctic Peninsula. In R. L. Oliver, P. R. James, & J. B. Jago (eds.), *Antarctic Earth Science* (pp. 308–14). Cambridge University Press.

Leanza, A. F. (1945). Ammonites del Jurásico superior y del Cretácico inferior de la Sierra Azul, en la parte meridional de la Provincia de Mendoza. *Mus. La Plata, An., N.S., 1*, 1–99.

(1947a). Upper Limit of the Jurassic System. *Geol. Soc. Am. Bull., 58*, 833–42.

(1947b). Descripción de la Faunula Kimmeridgiana de Neuquén. *Dir. Nac. Geol. Minas, Inf. Prelim. Comunic., 1*, 3–15.

(1949). Sobre *Windhauseniceras humphreyi* n. sp. del Tithoniano de Neuquén. *Asoc. Geol. Argent. Rev., 4*(3), 239–42.

(1968). Anotaciones sobre los fósiles Jurásico–Cretácicos de Patagónia Austral (Coleccion Feruglio) conservados en la Universidad de Bologna. *Acta Geol. Lilloana, 9*, 121–86.

Leanza, H. A. (1975). *Himalayites andinus* n. sp. (Ammonitina) del Tithoniano superior del Neuquén, Argentina. *I Congr. Argent. Paleont. Bioestr., Actas, 1*, 581–8.

(1980). The Lower and Middle Tithonian ammonite fauna from Cerro Lotena, province of Neuquén, Argentina. *Zitteliana, 5*, 1–49.

(1981a). Faunas de Ammonites del Jurásico superior y del Cretácico inferior de América del Sur, con especial consideración de la Argentina. *Cuencas Sediment. Jurásico Cret. América del Sur, 2*, 559–97.

(1981b). The Jurassic–Cretaceous boundary beds in west-central Argentina and their ammonite zones. *N. Jb. Geol. Paläontol., 161*(1), 62–92.

Leanza, H. A., & Hugo, C. A. (1978). Sucesión de ammonites y edad de la Formación Vaca Muerta y sincrónicas entre los paralelos 35 y 40 l.s., Cuenca Neuquina-Mendozina. *Asoc. Geol. Argent. Rev., 32*(4), 248–64.

Legarreta, L., & Gulisano, C. A. (1989). Análisis estratigráfico secuencial de la Cuenca Neuquina (Triásico superior–Terciário inferior), Argentina.

Legarreta, L., & Kozlowsky, E. (1984). Secciones Condensadas del Jurásico-Cretácico de los Andes del Sur de Mendoza: estratigrafía y significado tectosedimentario. *IX Congr. Geol. Argent., Actas, 1*, 286–97.

Lizuain, A. (1980). Las Formaciones Suprapaleozóicas y Jurásicas de la Cordillera Patagónica, Provincias de Río Negro y Chubut. *Asoc. Geol. Argent. Rev., 35*(2), 174–82.

McConnell, R. B. (1969). Fundamental fault zones in the Guiana and West African shields in relation to presumed axes of Atlantic spreading. *Geol. Soc. Am. Bull., 80*, 1775–82.

Macia, C., & Mojica, C. (1981). Nuévos puntos de vista sobre el magmatismo Triásico Superior (Fm. Saldana) en el Valle Superior del Magdalena, Colombia. *Zbl. Geol. Paläont., 1*(3/4), 243–51.

Macia, C., Mojica, J., & Colmenares, F. (1985). Consideraciones sobre la importancia de la paleogeografía y las áreas de aporte pre-Cretácicas en la prospección de hidrocarburos en el Valle Superior del Magdalena. *Geol. Colombiana, 14*, 49–70.

MacLaughlin, D. (1924). Geology and physiography of the Peruvian Cordillera, departments of Junin and Lima. *Geol. Soc. Am. Bull., 35*, 595–632.

Maksaev, V. (1984). Mesozoico a Paleógéno de la región de Antofagasta. *Serv. Nac. Geol. Minería Chile, Misc., 4*, C1–20.

Malumian, N., & Ploszkiewicz, J. V. (1977). El Liásico fosilífero de loncopán, Departamento Tehuelches (Provincia de Chubut, República Argentina). *Asoc. Geol. Argent. Rev., 31*(4), 279–80.

Marinovic, N., & Lahsen, A. (1984). *Hoja Calàma, Región de Antofagasta.* Serv. Nac. Geol. Minería Chile, Carta Geológica Chile, No. 58, 1–140.

Martincorena, L., & Tapia, I. (1982). Situación Estratigráfica del Jurásico del Área de Pedernales – Quebrada Asientos. *III Congr. Geol. Chileno*, A159–75.

Maze, W. B. (1984). Jurassic La Quinta Formation in the Sierra de Perija, northwestern Venezuela: geology and tectonic environment of red beds and volcanic rocks. *Geol. Soc. Am. Mem., 162*, 263–82.

Medina, F. A., & Ramos, A. M. (1981). Geología de las inmediaciones del Refugio Ameghino (64°26′/58°59′), Tierra de San Martín, Península Antártica. *Actas VIII Congreso Geológico Argentino, 2*, 871–82.

Megard, F. (1968). *Geología del Cuadrangulo de Huancayo*. Serv. Geol. Min. Bol. 18, Serie A.

(1988). Cordilleran Andes and marginal Andes: a review of Andean geology north of the Arica elbow (18°S). *Am. Geophysical Union, Geodyn. Ser., 18,* 97–113.

Mercado, M. (1980). *Area Pan de Azúcar, Región de Atacama*. Inst. Invest. Geol. Chile, Carta Geol. Chile, No. 37, 1–30.

(1982). *Hoja Laguna del negro Francisco*. Serv. Nac. Geol. Minería Chile, Carta Geol. Chile, No. 56, 1–73.

Minato, M. (1977). Brief note on the Lower Jurassic ammonites from the Vichuquen region, central Chile. In T. Ishikawa & L. Aguirre (eds.), *Comparative Studies on the Geology of the Circum-Pacific Orogenic Belt in Japan and Chile* (pp. 119–23). Tokyo: 1st. Rep. Japan Soc. Promotion Sci.

Mitchum, R. M., & Uliana, M. A. (1982). Estratigrafía sísmica de las Formaciones Loma Montosa, Quintuco y Vaca Muerta, Jurásico superior y Cretácico inferior de la Cuenca Neuquina, República Argentina. In *I Congr. Nac. Hidroc. Conf.* (pp. 439–84). Buenos Aires.

(1985). Seismic stratigraphy of carbonate depositional sequences, Upper Jurassic–Lower Cretaceous, Neuquén Basin, Argentina. *Am. Assoc. Petrol. Geol., Mem., 39,* 255–74.

Mojica, J. (1982). Observaciones acerca del estado actual del conocimiento de la Formación Payande (Triásico Superior), Valle Superior del Río Magdalena, Colombia. *Géol. Colombiána, 11,* 67–91.

Mojica, J., & Dorado, J. (1987). El Jurásico anterior a los movimientos intermálmicos en los Ándes Colombianos, Parte A; Estratigrafía. In W. Volkheimer (ed.), *Bioestratigrafía de los Sistemas Regionales del Jurásico y Cretácico de America del Sur. 1: El Jurásico Anterior a los Movimientos Intermálmicos* (pp. 49–110). Mendoza.

Mojica, J., & Llinás, R. (1984). Nuevos datos sobre la petrografía y la estratigrafía de la parte baja (Miembro Chicala) de la Formación Saldana en los alrededores de Payande, Tolima, Colombia, *Geol. Colombiana, 13,* 81–128.

Möricke, W. (1894). Versteinerungen des Lias und Unteroolith von Chile. *N. Jb. Geol. Paläont. Abh., 9,* 1–100.

Moscoso, R., & Covacevich, V. (1982). Las sedimentitas Triásico-Jurásicas al sur de Canto del Agua, Cordillera de la Costa, Región de Atacama, Chile: Descripción de la Formación Canto del Agua. *III Congr. Geol. Chileno,* F179–96.

Moscoso, R., Nasi, C., & Salinas, P. (1982a). *Hoja Vallenar y parte norte de La Serena, regiones de Atacama y Coquimbo*. Serv. Geol. Chile, Carta Geol. Chile, No. 55, 1–100.

Moscoso, R., Padilla, H., & Rivano, S. (1982b). *Hoja Los Andes, Región de Valparaíso*. Serv. Nac. Geol. Minería Chile, Carta Geol. Chile, No. 52, 1–67.

Moticska, P. (1975). Sierra de Perija. Excursión No. 2: Complejo volcánico-plutónico de El Totumo-Inciarte. *Bol. Geol. (Venezuela), Pub. Esp., 7*(1), 306–11.

Muñoz, G. N., Venegas, R., & Tellez, C. (1988). La Formación La Negra: Nuevos antecedentes estratigráficos en la Cordillera de la Costa de Antofagasta. *V Congr. Geol. Chileno, 1,* A283–311.

Muñoz, J. (1984). Cordillera Principal de los Andes de Chile entre Talca (35 L.S.) y Temuco (39 L.S.). *Serv. Geol. Miner. Chile, Misc., 4,* N1–15.

Muñoz, J., & Niemeyer, H. (1984). *Hoja Laguna del Maule, Regiones del Maule y del BioBio*. Serv. Nac. Geol. Minería Chile, Carta Geol. Chile, No. 64, 1–98.

Musacchio, E. A., & Riccardi, A. C. (1971). Estratigrafía, principalmente del Jurásico, en la Sierra de Agnia, Chubut, República Argentina. *Asoc. Geol. Argent. Rev., 26*(2), 272–3.

Mutterlose, J. (1986). *Upper Jurassic Belemnites from the Orville Coast, Western Antarctica, and Their Palaeobiological Significance*. British Antarctic Survey Bulletin No. 70.

Naranjo, J. A., & Puig, A. (1984). *Hojas Taltal y Chañaral, Regiones de Antofagasta y Atacama*. Serv. Nac. Geol. Minería Chile, Carta Geol. Chile, No. 62–3, 1–140.

Naranjo, J. A., Puig, A., & Suárez, M. (1982). Nuevos antecedentes estratigráficos del Triásico superior–Jurásico de la Cordillera de la Costa, sector Meridional de la región de Antofagasta, Chile. *III Congr. Geol. Chileno, 1,* A189–206.

Nasi, C. (1984). Geología de la Cordillera de la Costa de Chile Central. *Serv. Nac. Geol. Miner. Chile, Misc., 4,* L1–17.

Nasi, C., & Thiele, R. (1982). Estratigrafía del Jurásico y Cretácico de la Cordillera de la Costa, al sur del Río Maipo, entre Melipilla y Laguna de Aculeo (Chile Central). *Rev. Geol. Chile, 16,* 81–99.

Natland, M. L., Gonzalez, E., Cañon, A., & Ernst, M. (1974). A system of stages for correlation of Magallanes Basin sediments. *Geol. Soc. Am. Mem., 139,* 126.

Nell, P. A. R., & Storey, B. C. (1991). Strike-slip tectonics within the Antarctic Peninsula fore-arc. In M. R. A. Thomson, J. A. Crame, & J. W. Thomson (eds.), *Geological Evolution of Antarctica* (pp. 443–8). Cambridge University Press.

Nelson, H. B. (1957). Contribution to the geology of the Central and Western Cordillera of Colombia in the sector between Ibague and Cali. *Leidsche Geol. Meded. (Leiden), 22,* 1–75.

Newell, N. D. (1945). Investigaciones geológicas en las zonas circunvecinas del Lago Titicaca. *Bol. Soc. Geol. Perú, 18.*

Newell, N. D. (1949). Geology of the Lake Titicaca region, Peru and Bolivia. *Geol. Soc. Am. Mem., 36,* 1–111.

Niemeyer, H., Venegas, R., Baeza, C., & Soto, H. (1985). Reconocimiento geológico del sector Sud-Occidental del Cuadrangulo Cerro Yocas, ubicado en la Zona de Falla Quebrada Blanca-Chuquicamata, Región de Antofagasta. *IV Congr. Geol. Chileno, 1,* 629–53.

Nullo, F., & Proserpio, C. (1975). La Formación Taquetren en Cañadon del Zaino (Chubut) y sus relaciones estratigráficas en el ámbito de la Patagónia, de acuerdo a la flora, República Argentina. *Asoc. Geol. Argent. Rev., 30*(2), 133–50.

Nullo, F., Proserpio, C., & Ramos, V. A. (1979). Estratigrafía y Tectónica de la vertiente Este del Hielo Patagónico, Argentina-Chile. *VII Congr. Geol. Argent., Actas, 1,* 455–70.

Nur, A., & Ben-Abraham, Z. (1981). Volcanic gaps and the consumption of aseismic ridges in South America. *Geol. Soc. Am. Mem., 154,* 729–40.

Odreman, O., & Benedetto, G. (1977). Paleontología y edad de la Formación Tinacoa, Sierra de Perija, Estado Zulia, Venezuela. *Mem. V Congr. Geol. Venez.,* pp. 15–32.

Olivero, E. B. (1983). Ammonoideos y Bivalvos Berriasianos de la cantera Tres Lagunas, Chubut. *Ameghiniana, 20*(1–2), 11–20.

(1987). Cefalopodos y Bivalvos Titonianos y Hauterivianos de la Formación Lago La Plata, Chubut. *Ameghiniana, 24*(3–4), 181–202.

Orchuela, I. A., Cibanal, C., & Palade, S. H. (1987). Sismoestratigrafía del Ciclo Cuyano en el sector central del Engolfamiento Neuquino, Provincia del Neuquén, Argentina. *X Congr. Geol. Argent., Actas, 1,* 157–67.

Orchuela, I. A., & Ploszkiewicz, J. V. (1984). La Cuenca Neuquina. In V. A. Ramos (ed.), *Geología y Recursos Naturales de la Provincia de Río Negro*. Buenos Aires: IX Congr. Geol. Argent., Relatorio.

Orchuela, I. A., Ploszkiewicz, J. V., & Vines, R. (1981). Reinterpretación estructural de la denominada "Dorsal Neuquina." *VIII Congr. Geol. Argent., Actas, 3,* 281–93.

Palacios, O., & Castillo, M. (1983). Compendio estratigráfico del Jurásico-Cretácico en el Perú – Faja Costañera y Andina. *V Congr. Geol. Peruano, Anal., Soc. Geol. Perú, 71.*

Pérez, E. (1982). *Bioestratigrafía del Jurásico de Quebrada Asientos, Norte de Potrerillos, Región de Atacama*. Serv. Nac. Geol. Miner. Chile, Bol. 37.

Philippi, R. A. (1899). *Los Fósiles Secundarios de Chile*. Santiago.

Piraces, R. (1976). Estratigrafía de la Cordillere de la Costa entre la Cuesta El Melón y Limache, Provincia de Valparaíso, Chile. *I Congr. Geol. Chileno, Actas, 1*, A65–82.

Portugal, J. A. (1974). Mesozoic and Cenozoic stratigraphy and Tectonic events of Puno-Sta. Lucía area; Dep. Puno, Peru. *Am. Assoc. Petrol. Geol. Bull., 58.*

Portugal, J. A., & Gordon, L. (1976). Geologic history of southern Peru. *II Congr. Latinoam. Geol. Mem. (Caracas), 2*, 789–819.

Pumpin, V. F. (1978). *The Structural Setting of Northwestern Venezuela.* Unpublished report No. EPC-6094. Maraven, S.A., Caracas.

Quilty, P. G. (1970). Jurassic ammonites from Ellsworth Land, Antarctica. *J. Paleontol., 44*(1), 110–16.

(1977). Late Jurassic bivalves from Ellsworth Land, Antarctica: their systematics and paleogeographic implications. *N.Z. J. Geol. Geophys., 20*(6), 1033–80.

(1983). Bajocian bivalves from Ellsworth Land, Antarctica. *N.Z. J. Geol. Geophys., 26*(4), 395–418.

Rabe, E. (1974). *Zur Stratigraphie des ostandinen Raumes von Kolumbien. I: Die Abfolge Devon bis Perm der Ost-Kordillere nördlich von Bucaramanga. II: Die Präkretazische mesozoische Abfolge der Ost-Kordillere nördlich von Bucaramanga. III: Conodonten des jüngeren Paläozoikums der Ost-Kordillere, Sierra Nevada de Santa Marta und der Sierra de Perija, Kolumbien.* I, 1–46; II, 1–37. Dissertation, Justus Liebig Universität, Giessen.

Ramírez, C. F., & Gardeweg, M. (1982). *Hoja Toconao, Región de Antofagasta.* Serv. Nac. Geol. Minería Chile, Carta Geol. Chile, No. 54, 1–117.

Ramos, V. A. (1988). The tectonics of the Central Andes, 30 to 33° S latitude. *Geol. Soc. Am. Spec. Pap., 218*, 31–54.

Ramos, V. A., Niemeyer, H., Skarmeta, J., & Muñoz, J. (1982). Magmatic evolution of the Austral Patagonian Andes. *Earth Sci. Rev., 18*, 411–43.

Reig, O. (1961). Noticias sobre un nuevo anuro fósil del Jurásico de Santa Cruz (Patagónia). *Ameghiniana, 2*(5), 73–8.

Remy, W., Remy, R., Pfefferkorn, H. W., Volkheimer, W., & Rabe, E. (1975). Neueinstufung der Bocas-Folge (Bucaramanga, Kolumbien) in den Unteren Jura anhand einer Phlebopterisbranneri und Classopollis-Flora. *Argum. Palaobot. (Münster), 4*, 55–77.

Renz, O. (1956). Cretaceous in western Venezuela and the Guajira (Colombia). *XX Sess. Internatl. Geol. Congr., Mexico City,* pp. 1–13.

(1960). Geología de la parte sureste de la Península de La Guajira (República de Colombia). *Bol. Geol. Publ. Espec. 3, Mem. III Congr. Geol. Venez. (Caracas), 1*, 317–439.

Renzoni, G. (1968). Geología del Macizo de Quetame. *Geol. Colombiana, 5*, 75–127.

Renzoni, G., Rosas, H., & Etayo, F. (1983). *Mapa Geológico de la Plancha 171, Duitama.* Bogotá: Ingeominas.

Restrepo, J. J., & Toussaint, J. F. (1988). Terranes and continental accretion in the Colombian Andes. *Episodes, 11*(3), 189–93.

Reutter, K. J. (1974). Entwicklung und Bauplan der chilenischen Hochkordillere im Bereich 29 südlicher Breite. *N. Jb. Geol. Paläont. Abh., 146*(2), 153–78.

Riccardi, A. C. (1976). Paleontología y Edad de la Formación Springhill. *X Congr. Geol. Chileno, 1*, C42–56.

(1977). Berriasian invertebrate fauna from the Springhill Formation of southern Patagonia. *N. Jb. Geol. Paläont. Abh., 155*(2), 216–52.

(1983). The Jurassic of Argentina and Chile. In M. Moullade & A. E. M. Nairn (eds.), *The Phanerozoic Geology of the World. II: The Mesozoic, B* (pp. 201–63). Amsterdam: Elsevier.

(1984a). Las Asociaciones de Amonitas del Jurásico y Cretácico de la Argentina. *IX Congr. Geol. Argent., Actas, 4*, 569–95.

(1984b). Los Eurycephalitinae en America del Sur. *III Congr. Argent. Paleont. Bioestr. (Corrientes, 1982), Actas,* pp. 151–61.

(1985). Los Eurycephalitinae Andinos (Ammonitina, Jurásico medio): Modelos Evolutivos y Resolucion Paleontológica. *Bol. Gent. Inst. Fitol., 13*, 1–27.

(1988). The Cretaceous System in southern South America. *Geol. Soc. Am. Mem., 168*, 1–161.

Riccardi, A. C., Damborenea, S. E., & Manceñido, M. O. (1990). Lower Jurassic of South America and Antarctic Peninsula. In G. E. G. Westermann & A. C. Riccardi (eds.), *Jurassic Taxa Ranges and Correlation Charts for the Circum-Pacific. Newsl. Stratigr., 21*, 75–104.

Riccardi, A. C., Damborenea, S. E., Manceñido, M. O., & Ballent, S. C. (1988a). Hettangiano y Sinemuriano marinos en Argentina. *V Congr. Geol. Chileno, 2*, C359–73.

Riccardi, A. C., & Gulisano, C. A. (1992). Unidades Limitadas por Discontinuidades. Su aplicación al Jurásico Andino. *Asoc. Geol. Argent. Rev., 45*, 346–64.

Riccardi, A. C., Leanza, H. A., & Volkheimer, W. (1989a). Upper Jurassic of South America and Antarctic Peninsula. In G. E. G. Westermann & A. C. Riccardi (eds.), *Jurassic Taxa Ranges and Correlation Charts for the Circum-Pacific. Newsl. Stratigr., 21*, 129–47.

Riccardi, A. C., & Rolleri, E. O. (1980). Cordillera Patagónica Austral. In *Segundo Simposio de Geología Regional Argentina. Acad. Nac. Cienc., Cordoba, 2*, 1173–306.

Riccardi, A. C., Westermann, G. E. G., & Damborenea, S. E. (1989b). Middle Jurassic of South America and Antarctic Peninsula. In G. E. G. Westermann & A. C. Riccardi (eds.), *Jurassic Taxa Ranges and Correlation Charts for the Circum-Pacific. Newsl. Stratigr.*

Riccardi, A. C., Westermann, G. E. G., & Elmi, S. (1988b). Las Zonas de Amonites del Bathoniano-Caloviano inferior de los Andes Argentino-Chilenos. *V Congr. Geol. Chileno, 2*, C415–26.

(1989c). The Bathonian-Callovian ammonite zones of the Argentine-Chilean Andes. *Geobios, 22*, 553–97.

Riccardi, A. C., & Westermann, G. E. G. (1991a). Middle Jurassic ammonoid fauna and biochronology of the Argentine-Chilean Andes. III: Bajocian–Callovian Eurycephalitinae, Stephanocerataceae. *Palaeontographica, 216*, 1–110.

Riccardi, A. C., & Westermann, G. E. G. (1991b). Midde Jurassic ammonoid fauna and biochronology of the Argentine-Chilean Andes. Part IV: Bathonian-Callovian Reineckeiidae. *Palaeontographica, 216*, 111–45.

Rivano, S. (1980). *Cuadrangulos D86, Las Ramadas, Carrizal y Paso Río Negro, Región de Coquimbo.* Inst. Invest. Geol. Chile, Carta Geol. Chile, No. 41–4, 1–68.

(1984). Geología del Meso-Cenozoico entre los 31 y 33 Lat. Sur. *Serv. Nac. Geol. Miner. Chile, Misc.,4*, K1–17.

Rivera, R. (1951). La Fauna de los estratos Puente Inga, Lima. *Bol. Soc. Geol. Perú, 22*, 5–23.

Rivera, R., Petersen, G., & Rivera, M. (1975). Estratigrafía de la Costa de Lima. *Bol. Soc. Geol. Perú, 45.*

Rüegg, W. (1961). Hallazgos y posición estratigráfica-tectónica del Titoniano de la Costa sur del Perú. *Bol. Soc. Geol. Perú, 36*, 203–8.

(1968). Mil Kilometros de geología en la Faja Pacífica del Perú. *Actas Terceras J. Geol. Argentina,* I.

Salas, R., Kast, R. F., Montecinos, F., & Salas, I. (1966). *Geología y recursos minerales del Departamento de Arica, Provincia de tarapaca.* Inst. Invest. Geol. Chile, Bol. 21, 7–114.

Schubert, C. (1986). Stratigraphy of the Jurassic La Quinta Formation, Merida Andes, Venezuela: type section. *Z. Deutsch. Geol. Ges, 137*, 391–411.

Schubert, C., Sifontes, R., Padron, V., Velez, J., & Loaiza, P. (1979). Formacion La Quinta (Juraśico) Andes Meridenos: Geología de la Sección tipo. *Acta Cient. Venezolana, 30.*

Segerstrom, K. (1959). *Cuadrangulo Los Loros, Provincia de Atacama.* Inst. Invest. Geol. Chile, Carta Geol. Chile, No. 1 (1), 5–33.

(1968). *Geología de las Hojas Copiapó y Ojos del Salado, Provincia de Atacama.* Inst. Invest. Geol. Chile, Bol. 24, 5–58.

Sepulveda, P., & Naranjo, J. A. (1982). *Hoja Carrera Pinto, Región de Atacama*. Serv. Nac. Geol. Minería Chile, Carta Geol. Chile, No. 53, 1–60.

Shepherd, G. L., & Moberly, R. (1981). Coastal structure of the continental margin, northwest Peru and southwest Ecuador. *Geol. Soc. Am. Bull., 154*, 351–91.

Sigal, J., Grekoff, N., Singh, N. P., Cañon, A., & Ernst, M. (1970). Sur l'age et les affinites "gondwaniennes" de microfaunes (Foraminiferes et Ostracodes) malgaches, indiennes, et chiliennes au sommet du Jurassique et a la base du Cretace. *C. R. Acad. Sci. [D] (Paris), 271*, 24–7.

Sillitoe, R. H. (1974). Tectonic segmentation in the Andes: implications for magmatism and metallogeny. *Nature, 250*, 542–5.

Silva, I. (1976). Antecedentes estratigráficos del Jurásico y estructurales de la Cordillera de la Costa en el Norte Grande de Chile. *I Congr. Geol. Chileno, 1*, A83–95.

Skarmeta, J., & Marinovic, N. (1981). *Hoja Quillagua, Región de Antofagasta*. Inst. Invest. Geol. Chile, Carta Geol. Chile, No. 51, 1–63.

Smellie, J. L., Davies, R. E. S., & Thomson, M. R. A. (1980). *Geology of a Mesozoic Intra-Arc Sequence on Byers Peninsula, Livingston Island, South Shetland Islands*. British Antarctic Survey Bulletin No. 50, pp. 55–76.

Steinmann, G. (1881). Zur Kenntniss der Jura und Kreideformation von Caracoles (Bolivia). *N. Jb. Geol. Paläont. Abh., 1*, 239–301.

Steinmann, G., & Lisson, C. (1924). *Mapa Geológico de la Cordillera del Perú*. Heidelberg: Inst. Geology of Perú.

Steuer, A. (1897). Argentinische Jura-ablagerungen. *Palaont. Abh., N.F., 7*,(3), 127–222.

Stipanicic, P. N. (1951). Sobre la presencia del Oxfordense en el Arroyo de la Manga (Provincia de Mendoza). *Asoc. Geol. Argent. Rev., 6*(4), 213–39.

(1969). El avance de los conocimientos del Jurásico Argentino a partir del esquema de Groeber. *Asoc. Geol. Argent. Rev., 24*, 367–88.

(1983). The Triassic of Argentina and Chile. In M. Moullade & A. E. M. Nairn (eds.), *The Phanerozoic Geology of the World. II. The Mesozoic B* (pp. 181–99). Amsterdam: Elsevier.

Stipanicic, P. N., & Bonetti, M. I. R. (1970a). Posiciones estratigráficas y edades de las principales floras Jurásicas Argentinas. I: Floras Liásicas. *Ameghiniana, 7*(1), 57–78.

(1970b). Posiciónes estratigráficas de las principales floras Jurásicas Argentinas. II: Floras Doggerianas y Málmicas. *Ameghiniana, 7*(2), 101–18.

Stipanicic, P. N., & Reig, A. O. (1957). El Complejo Porfírico de la Patagónia Extraandina y su fauna de anuros. *Acta Geol. Lilloana, 1*, 185–297.

Stipanicic, P. N., & Rodrigo, F. (1970). El diastrofismo Eo- y Mesocretácico en Argentina y Chile, con referencia a los movimientos Jurásicos de la Patagónia. *Cuart. J. Geol. Argent. Actas, 2*, 337–52.

Stipanicic, P. N., Westermann, G. E. G., & Riccardi, A. C. (1976). The Indo-Pacific ammonite *Mayaites* in the Oxfordian of the Southern Andes. *Ameghiniana, 7*, 57–78.

Suárez, M. (1979). A Late Mesozoic island-arc in the southern Andes, Chile. *Geol. Mag., 116*(3), 181–90.

(1985). Estratigrafía de la Cordillera de la Costa, al Sur de Taltal, Chile: Etapas iniciales de la evolución andina. *Rev. Geol. Chile, 24*, 19–28.

Suárez, M., Naranjo, J. A., & Puig, A. (1982). Volcanismo Liásico inferior en la región costera de Antofagasta meridional: Piroclástitas en la Formación Pan de Azúcar e implicancias paleogeográficas. *Rev. Geol. Chile, 17*, 83–90.

(1985). Estratigrafía de la Cordillera de la Costa, al Sur de Taltal, Chile: Etapas iniciales de la evolución andina. *Rev. Geol. Chile, 24*, 19–28.

Suárez, M., & Pettigrew, T. H. (1976). An Upper Mesozoic island-arc–back-arc system in the southern Andes and South Georgia. *Geol. Mag., 113*(4), 305–28.

Sutton, F. A. (1946). Geology of Maracaibo Basin, Venezuela. *Am. Assoc. Petrol. Geol. Bull., 30*, 1621–741.

Tasch, P., & Volkheimer, W. (1970). Jurassic conchostracans from Patagonia. *Kans. Univ., Paleontol. Contrib. Pap., 50*, 1–23.

Thiele, R. (1964). Reconocimiento geológico de la Alta Cordillera de Elqui. *Univ. Chile, Fac. Cienc. Fis. Mat., Inst. Geol., Publ., 27*, 135–97.

(1980). *Hoja Santiago, Región Metropolitana*. Inst. Invest. Geol. Chile, Carta Geol. Chile, No. 39, 1–51.

Thiele, R., Castillo, J. C., Heim, R., Romero, G., & Ulloa, M. (1979). Geología del sector fronterizo de Chile Continental entre los 43°00′–43°50′ latitud sur (Comúnas de Futaleufu y de Paléna). *VII Congr. Geol. Argent., Actas, 1*, 577–91.

Thomas, A. (1967). *Cuadrángulo Mamina, Provincia de Tarapacá*. Inst. Invest. Geol. Chile, Carta Geol. Chile, No. *17*, 5–49.

(1970). *Cuadrangulos Iquique y Caleta Molle, Provincia de Tarapacá*. Inst. Invest. Geol. Chile, Carta Geol. Chile, No. 21–2, 5–52.

Thomson, M. R. A. (1972). *New Discoveries of Fossils in the Upper Jurassic Volcanic Group of Adelaide Island*. British Antarctic Survey Bulletin No. 30, pp. 95–101.

(1979). *Upper Jurassic and Lower Cretaceous Ammonite Faunas of the Ablation Point Area, Alexander Island*. British Antarctic Survey Scientific Reports No. 97.

(1982). *Late Jurassic Fossils from Low Island, South Shetland Islands*. British Antarctic Survey Bulletin No. 56, pp. 25–35.

(1983a). Antarctica. In M. Moullade, & A. E. M. Nairn, (eds.), *The Phanerozoic Geology of the World. II: The Mesozoic, B* (pp. 391–422). Amsterdam: Elsevier.

(1983b). Late Jurassic ammonites from the Orville Coast, Antarctica. In R. L. Oliver, P. R. James, & J. B. Jago (eds.), *Antarctic Earth Science* (pp. 315–19). Cambridge University Press.

Thomson, M. R. A., & Tranter, T. H. (1986). *Early Jurassic Fossils from Central Alexander Island and Their Geological Setting*. British Antarctic Survey Bulletin No. 70, pp. 23–39.

Thornburg, T., & Kulm, L. D. (1981). Sedimentary basins of the Peru continental margin: structure, stratigraphy and Cenozoic tectonics from 6°S to 16°S latitude. *Geol. Soc. Am. Mem., 154*, 393–422.

Trümpy, D. (1943). Pre-Cretaceous of Colombia. *Geol. Soc. Am. Bull., 54*, 1281–304.

Turner, P. (1980). *Continental Red Beds*. Amsterdam: Elsevier.

Uliana, M. A., Biddle, K. T., Phelps, D. W., & Gust, D. A. (1985). Significado del vulcanismo y extensión mesojurásicos en el extremo meridional de Sudamérica. *Asoc. Geol. Argent. Rev., 40*(3–4), 231–53.

Useche, A., & Fierro, I. (1972). Geología de la región de Pregonero, Estados Tachira y Mérida. *Bol. Geol. (Venezuela), Pub. Esp., 5*(2), 963–98.

Vail, P. R., Hardenbol, J., & Todd, R. G. (1984). Jurassic unconformities, chronostratigraphy and sea-level changes from seismic stratigraphy and biostratigraphy. In J. S. Schlee (ed.), *Interregional Unconformities and Hydrocarbon Accumulation. Am. Assoc. Petrol. Geol. Mem., 36*, 129–44.

Vail, P. R., Mitchum, R. M., Todd, R. G., Widmier, J. M., Thompson, S., Sangree, J. B., Bubb, J. N., & Hatlelid, W. G. T. (1977). Seismic stratigraphy and global changes of sea level. *Am. Assoc. Petrol. Geol. Mem., 26*, 49–212.

Vergara, H. (1984). *Hoja Collacagua, Región de Tarapacá*. Serv. Nac. Geol. Min., Carta Geol. Chile, No. 59, 1–79.

Vicente, J. C. (1981). Eleméntos de la estratigrafía mesozoica surperuana. In W. Volkheimer (ed.), *Cuencas Sediment. Jurásico y*

*Cretácico de América del Sur. 1,* 319–51. Buenos Aires: Com. Sudam. Jura. Cretac.

Vila, T. (1976). Secuencia estratigráfica del Morro de Arica, Provincia de Tarapacá. *I Congr. Geol. Chileno, 1,* A1–10.

Vogel, S., & Vila, T. (1980). *Cuadrángulos Arica y Poconchile, Región de Tarapacá.* Inst. Invest. Geol. Chile, Carta Geol. Chile, No. 35, 1–24.

Weaver, C. H. (1931). Paleontology of the Jurassic and Cretaceous from West Central Argentina. *Mem. Univ. Washington, 1,* 1–496.

(1942). A general summary of the Mesozoic of Southern America and Central America. *Proc. 8th Amer. Sci. Cong. (1940) Washington, Geol., 4,* 149–53.

Weeks, L. G. (1947). Paleogeography of South America. *Am. Assoc. Petrol. Geol. Bull., 31*(7), 1194–241.

Westermann, G. E. G. (1967). Sucesión de amonites del jurásico medio en Antofagasta, Atacáma, Mendoza y Neuquén. *Asoc. Geol. Argent. Rev., 22*(1), 65–73.

(1985). Taxonomy. In A. von Hillebrandt & G. E. G. Westermann (eds.), *Aalenian (Jurassic) Ammonite Faunas and Zones of the Southern Andes. Zitteliana, 12,* 3–55.

Westermann, G. E. G., & Riccardi, A. C. (1972). Middle Jurassic ammonoid fauna and biochronology of the Argentine-Chilean Andes. I: Hildocerataceae. *Palaeontographica, A, 140,* 1–116.

(1973). Amonitas y estratigrafía del Aaleniano-Bayociano en los Andes argentino-chilenos. *Ameghiniana, 9,* 357–89.

(1975). Edad y Taxonomía del género *Podagrosiceras* Maubeuge et Lambert (Ammonitina, Jurásico medio). *Ameghiniana, 12*(3), 242–52.

(1979). Middle Jurassic ammonoid fauna and biochronology of the Argentine-Chilean Andes. Part II: Bajocian Stephanocerataceae. *Palaeontographica, A164,* 85–118.

(1980). The Upper Bajocian ammonite *Strenoceras* in Chile: first circum-Pacific record of the Subfurcatum Zone. *Newsl. Stratigr., 9*(1), 19–29.

(1982). Ammonoid fauna from the early Middle Jurassic of Mendoza province, Argentina. *J. Paleontol., 56,* 11–41.

Westermann, G. E. G., Riccardi, A. C., Palacios, O., & Rangel, C. (1980). *Jurásico medio en el Perú.* Inst. Geol. Min. Metal., Bol. 9, Ser. D, 1–47.

Whitham, G., & Doyle, P. (1989). Stratigraphy of the Upper Jurassic–Lower Cretaceous Nordenskjöld Formation of eastern Graham Land, Antarctica. *J. South American Earth Sciences, 2*(4), 371–84.

Wiedmann, J. (1980a). El límite Jurásico-Cretácico: Problemas y soluciones. *II Congr. Argent. Paleontol. Bioestr., Actas, 5,* 103–20.

(1980b). Palaogeographie und Stratigraphie im Grenzbereich Jura/Kreide Sudamerikas. *Münst. Forsch. Geol. Palaont., 51,* 27–61.

Wiedmann, J., & Mojica, J. (1980). Obertrias-Ammoniten der Saldana-Formation, Tolima-Kolumbien. In *7. Geowiss. Lateinam. Kolloquium, Tagungshefte.* Heidelberg.

Wilson, J. (1963). Cretaceous stratigraphy of Central Andes of Peru. *Bull. Am. Assoc. Petrol. Geol., 47,* 1–34.

Yrigoyen, M. R. (1979). Cordillera Principal. In *Segundo Simposio Geología Regional Argentina, Vol. 1* (pp. 651–94). Córdoba: Academia Nacional de Ciencias.

Zeiss, A. (1983). Zur Frage der Aequivalenz der Stufen Tithon/Berrias/Volga/Portland in Eurasien und Amerika. Ein Beitrag zur Klärung der weltweitn Korrelation der Jura/Kreide-Grenzschichten in marinen Bereich. *Zitteliana, 10,* 427–38.

(1986). Comments on a tentative correlation chart for the most important marine provinces at the Jurassic–Cretaceous boundary. *Acta Geol. Hungar., 29*(1–2), 27–30.

# 7 Australasia

M. BRADSHAW and A. B. CHALLINOR

## AUSTRALIAN REGION[1]

As part of the joint Bureau of Mineral Resources (BMR) and Australian Petroleum Industry Research Association (APIRA) Palaeogeographic Maps Project (Cook 1988), environmental reconstructions have been prepared for Australia for ten Jurassic time slices (Bradshaw and Yeung 1990). Some of the results of this work have been reported by Bradshaw et al. (1988) and by Struckmeyer, Yeung, and Bradshaw (1990), and the following relies heavily on those compilations.

The BMR-APIRA Jurassic paleogeographic maps were compiled from well and outcrop information for both onshore and offshore Australia. The "compromise time scale" (Chapter 1) was used, and correlation was based on the palynological zonation of Helby, Morgan, and Partridge (1987). Figure 7.1 shows the distribution of Jurassic rocks in Australia. There was a thin but extensive platform cover spread across the continent through the Eromanga and adjacent basins, and the thickest depocenters were along the western margin (Barrow-Dampier Subbasin and Perth Basin) and to the east (Clarence-Moreton Basin) and north (Papuan Basin). Marine to fluviodeltaic deposition occurred along the western and northern margins of the continent that bordered Tethys, and nonmarine sequences were deposited in eastern Australia. Deposition was initiated in a terrestrial rift-valley system along the southern margin between Australia and Antarctica. The widespread fluviolacustrine facies attest to a temperate and humid climatic regime. Redbeds are restricted to the Early Jurassic along the western margin, and marine sequences are dominated by clastics, with only rare carbonates.

Australia's Jurassic tectonic regime was dominated by the breakup of Gondwana. At the beginning of the Jurassic, Greater India and possibly other parts of Asia lay to the west, and Antarctica filled the curve of the southern margin. Tensional rift environments encircled the continent from the northwest to the southeast. In contrast to this divergent stress regime, a Chilean-type magmatic arc (Veveers 1984), present off the eastern margin, was indicative of a convergent plate boundary in the northeast quadrant of the continent. In response to rifting along the southern margin, large volumes of dolerite were intruded into the Permo-Triassic sediments of Tasmania, and there are many small basic to intermediate intrusives of Jurassic age throughout southeastern Australia (Figure 7.1).

The shifting pattern of environments across Australia during the Jurassic was controlled by the global climate and sea level, the tectonic regime, and the continent's paleogeographic position in high southerly latitudes, on the eastern margin of Gondwana, facing north and west into Tethys. Eastern Australia was partially enclosed from the "palaeo-Pacific" by the Lord Howe Rise, the Bunker Ridge, Queensland Plateau, New Caledonia, and other possible land areas to the east of the present continent.

Figure 7.2 shows sketch maps of the distribution of environments for six of the ten time slices used by Bradshaw and Yeung (in press). Through the Jurassic there was a gradual expansion of the area of deposition in onshore Australia and an increasing marine influence on sedimentation along the northwest margin.

*Time slice 1* (Figure 7.2A): Hettangian–Sinemurian (205–192 Ma). the Jurassic–Triassic boundary occurs within the *A. reducta* and *P. crenulatus* spore-pollen zones and close to the base of the *D. priscum* dinoflagellate zone (Helby et al. 1987). The real disjunction in the biostratigraphy and break in the stratigraphy occurred within the Hettangian, with the change from the *Falcisporities* to the *C. dampieri* Superzone.

Isolated patches of continental deposition occur in eastern Australia and a broad fringe of paralic environments on the northwestern margin, facing onto Tethys. In eastern Australia, fluvial networks developed over and around the irregular topography of the Palaeozoic to Triassic Tasmanides foldbelt. Arkosic sandstones and conglomerates of the basal Balimbu Formation were shed from the granitic Oriomo Ridge into the Papuan Basin. To the south, fluvial sands were deposited in eastern Queensland, the Precipice Sandstone and its equivalents. Isolated outcrops have been interpreted as erosional remnants of a once-continuous cover.

[1] By M. Bradshaw.

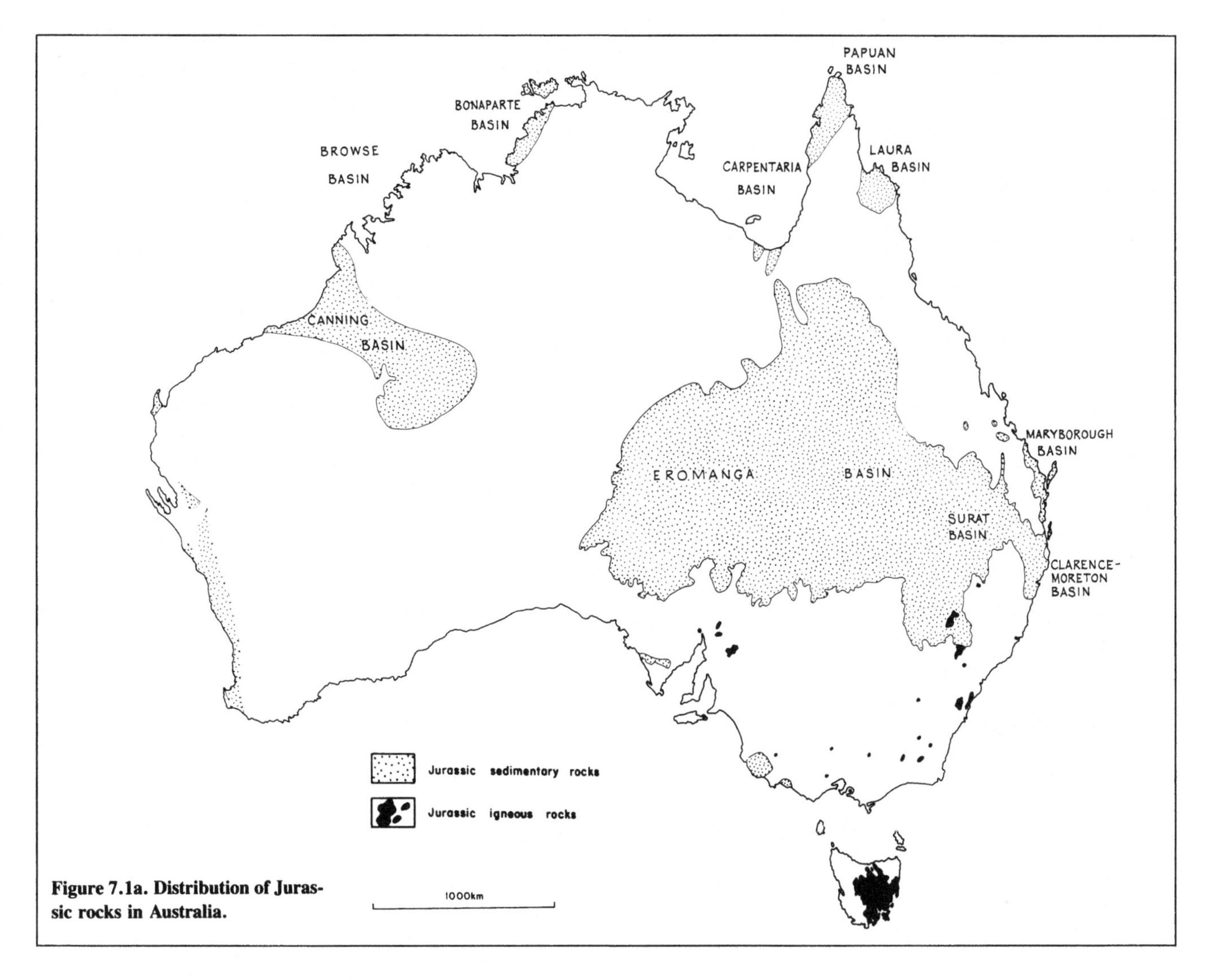


**Figure 7.1a. Distribution of Jurassic rocks in Australia.**

Paleocurrent measurements and variations in sediment thicknesses and facies indicate that the general paleoslope was to the east. East-flowing streams may have deposited earliest Jurassic sediments on the Queensland Plateau, where a thick sequence of pre-Eocene sediments is interpreted from seismic records (D. Choy personal communication). An isolated pocket of fluviolacustrine sediments was deposited in the Poolowanna Trough in central Australia.

Along the northwest margin there was a broad arc of paralic and alluvial facies trending northeast–southwest. The alluvial facies were redbeds, indicating a monsoonal climate with seasonal aridity and lowered water tables.

*Time slice 2* (Figure 7.2B). The base corresponds to the Sinemurian–Pliensbachian boundary at 194 Ma. In eastern Australia there was a change from sand deposition to more shale-prone facies, which is best exemplified by the Precipice Sandstone/Evergreen Formations succession in the Surat Basin. On the northwest margin there was a facies change from marginal marine clastic sediments to shallow-water limestone, and in central Australia deposition commenced in the Eromanga Basin.

In comparison to time slice 1, this was a time of relatively higher sea level. There was a marine transgression across part of the northwest margin, and in the continental regime of eastern Australia there was a corresponding rise in base level that produced a change to low-energy fluviolacustrine deposition. A distinctive Jurassic cast to the continent developed. There was expansion and coalescence of fluvial systems, the underlying foldbelt basement topography had less control on depositional patterns, and facies became increasingly marine on the northwestern margin. Limestone was deposited over the marginal-marine clastics of time slice 1.

*Time slice 3* (Figure 7.2C). This is a narrow interval of 2 million years within the Toarcian (ca. 185 Ma). Its base is placed slightly above the *C. torosa*/*C. turbatus* boundary of Helby et al. (1987). The cycle chart of Hag, Hardenbol, and Vail (1987) places thepeak high sea level of the Early Jurassic in the middle Toarcian, corresponding to the interval chosen for this time. There is a distinct change in lithology and depositional environment in Surat and other eastern basins that corresponds to the incoming of *Expollenites* palynomorphs and the deposition of ironstone oolites (Evans 1966).

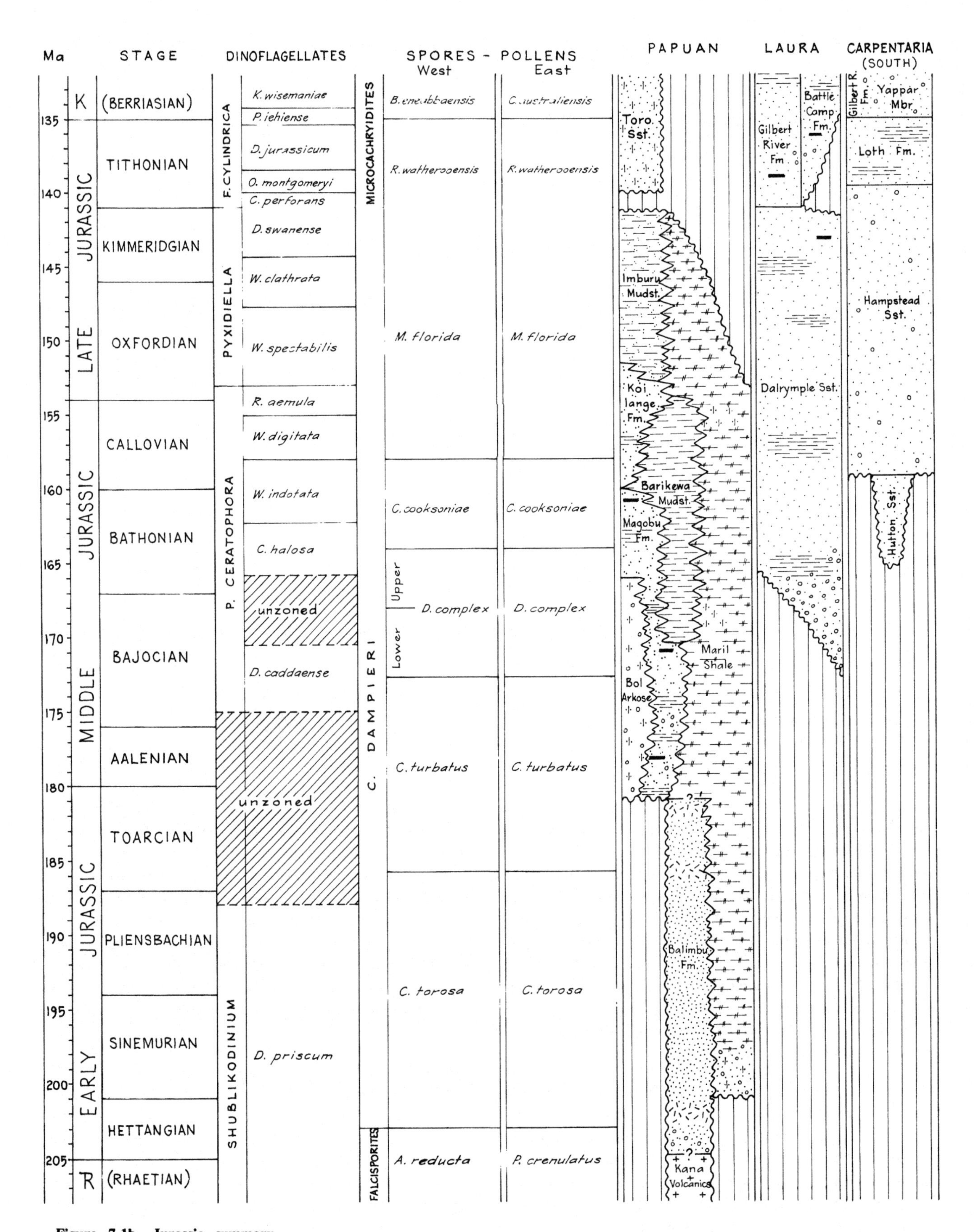


**Figure 7.1b. Jurassic summary columns.**

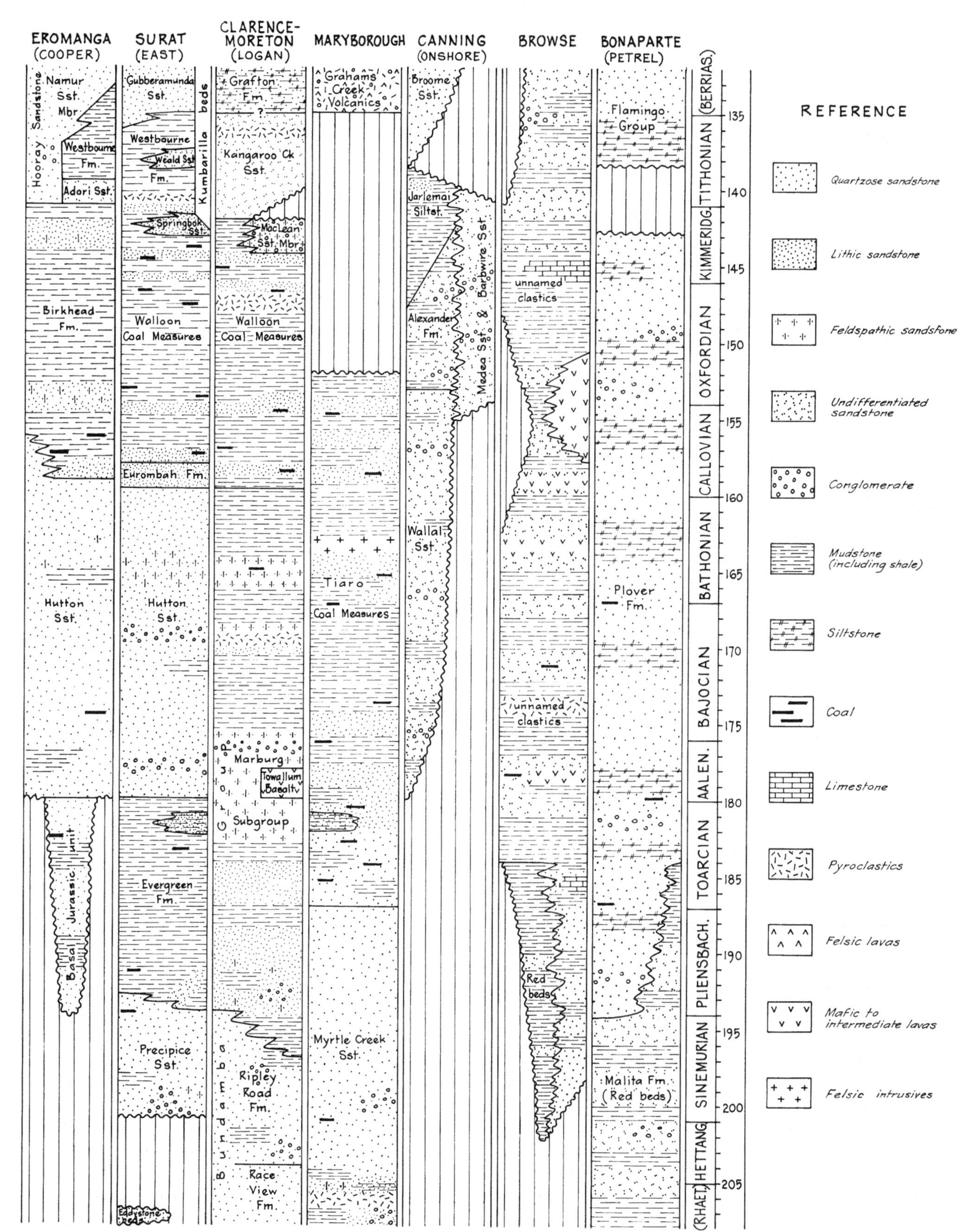


**Figure 7.1b (cont.)**

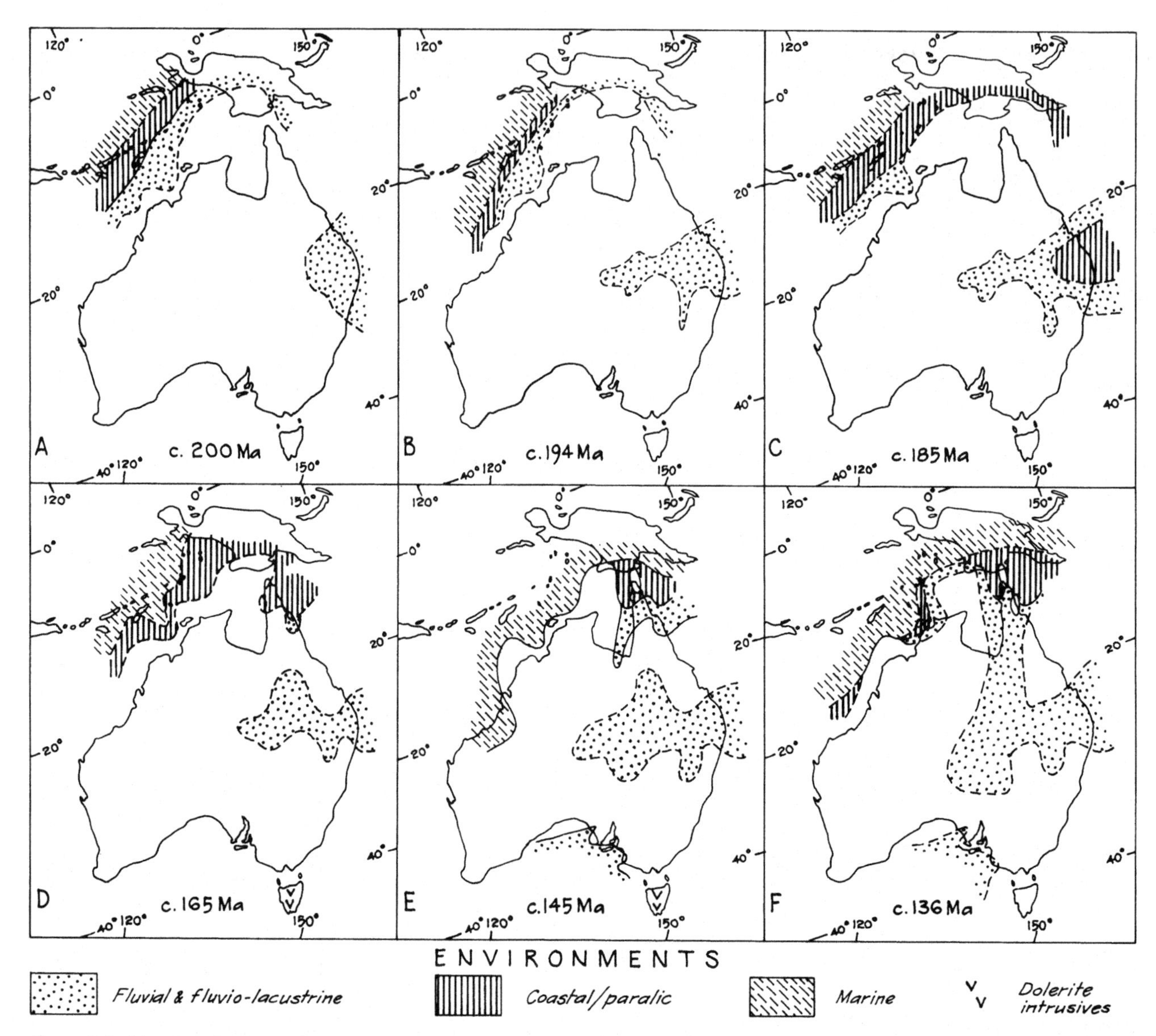


**Figure 7.2. Jurassic paleogeography.**

The area of oolite deposition, which extends through the Maryborough, Nambour, Clarence Moreton, and Surat Basins into the Eromanga Basin, is depicted as a paralic environment. Rising sea level had caused the encroachment of brackish water into the lake systems of eastern Australia. In the shallow, mildly salty waters of this large lagoon, chamosoitic ooids formed. The inflow of fresh water from the surrounding hinterland of fluvial channels, lakes, and coal swamps was sufficient to prevent fully marine conditions from developing.

This time also corresponds to the end of redbed deposition on the northwestern margin and the further advance of paralic conditions into the Bonaparte Basin. A broad fringe of paralic environments – sand bars, lagoons, barriers – stretched along the margin. Areas of fluvial to deltaic deposition occurred where supplies of sediment were greatest, such as along fault scraps and at the river mouths.

*Time slice 6* (Figure 7.2D). This slice (170–158 Ma) encompasses the later Bajocian, all of the Bathonian, and the early Callovian. The base is well controlled biostratigraphically. It lies between the *C. turbatus/D. complex* spore-pollen zone boundary (Helby et al. 1987) and is coeval with the top of the *D. caddaense* dinoflagellate zone. This time is notable in eastern Australia by the commencement of deposition in the Laura and Carpentaria Basins

Paralic environments expanded southward into the Carpentaria Basin and perhaps into the offshore extension of the Laura Basin. Tuff beds and tuffaceous sandstones deposited in the Cape York area indicate the proximity of a volcanic source somewhere to the east at the convergent plate boundary. Fluviolacustrine deposition occurred in the Eromanga and Surat Basins, with some expansion of the depositional area northward. Fluviolacustrine conditions persisted in the most easterly parts of the drainage network, in the Maryborough Basin. In contrast to the blanket of quartz sand spread across the interior basins, thicker pockets of sediment, including lithic and feldspatic sandstones and tuffs, accumulated in the Clarence Moreton Basin, reflecting the proximity of granitic sources in the New England Highland and the input of volcanic material from the east.

In Tasmania there was an episode of dolerite intrusion related to the developing rift between Australia and Antarctica, where there are dolerites of similar age. Along the northwestern margin the pattern of environments established in earlier time slices persisted. There was a band of paralic and deltaic environments trending northwesterly along the entire length of the margin and passing westward into marine facies.

*Time slice 8* (Figure 7.2E). This slice (150–141 Ma) covers most of the Oxfordian and all of the Kimmeridgian. The base is the base of the *W. spectablis* dinoflagellate zone, and the top corresponds to major zonation boundaries in both the dinoflagellate and spore-pollen schemes of Helby et al. (1987) (see Chapter 13).

This was the time of maximum transgression in the Jurassic. The top boundary coincides with an unconformity on the North West Shelf and in the Papuan and Laura Basins, and with a facies change in many other basins, notably the Birkhead-to-Adori transition in the Eromanga Basin.

Fluviolacustrine and fluvial environments held sway over most of eastern and central Australia. Many sediments show the input of volcanic material derived from an arc of probably andesitic composition located somewhere to the east of the continent. Dolerite intrusions continued in Tasmania. A significant feature is the recognition of widespread deposition along the southern margin. Sediments of this age have been intersected in half a dozen wells from the western Otway Basin to the Eyre Subbasin. Deposition occurred in a series of fault-bounded valleys along the general Australian/Antarctic rift zone, within the interior of Gondwana.

This time corresponds to a phase of eustatic high sea level, with a peak in the Kimmeridgian (Haq et al. 1987) that is evident on the northwestern margin. The sea transgressed significantly across the present-day coastline into the Canning Basin, and two seaward deep-marine environments extended into fault-controlled troughs.

*Time slice 10* (Figure 7.2F). This is a short time interval in the later Tithonian (138–135 Ma). The base corresponds to the base of the *D. jurassicum* dinoflagellate zone (Helby et al. 1987). The top is the Jurassic–Cretaceous boundary, which lies within the narrow *P. iehensis* dinoflagellate zone and coincides with the *R. watherooensis*/*C. australiensis* boundary in the Helby et al. (1987) spore-pollen zonation.

This time corresponds to a transgressive phase, following the regression of time slice 9. There was an expansion of the fluvial network in eastern Australia and marine transgression in northern Australia.

Time slice 10 was a prelude to the Cretaceous. The sea was lapping at the edges of the Carpentaria Basin prior to its Aptian invasion into the heart of the continent. Australia had commenced its separation from Gondwana: fully marine conditions had been established on the northwest margin, to be followed by sea-floor spreading along the western, southern, and southeastern margins in the Cretaceous. Australia was on its way to becoming the island continent.

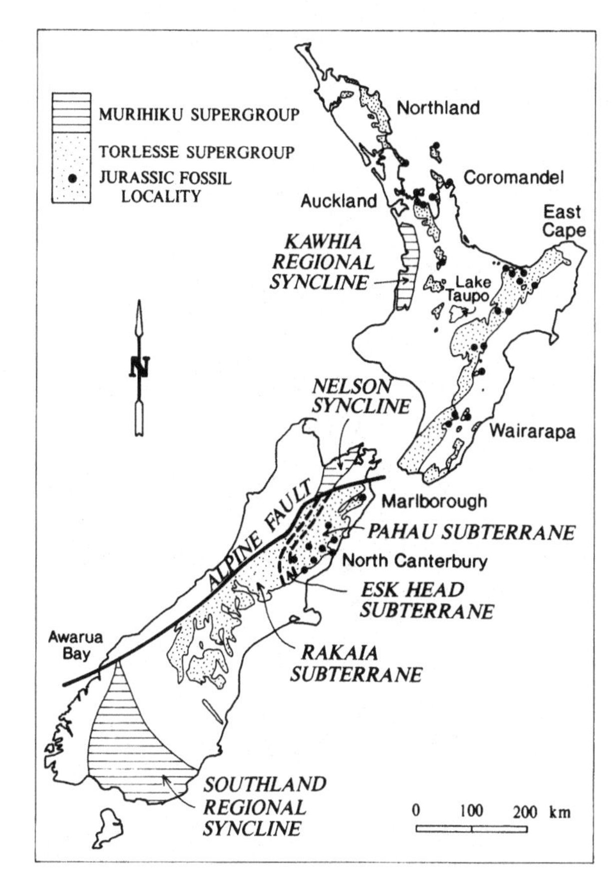


**Figure 7.3. Distributions of Murihiku and Torlesse Supergroup rocks.**

## NEW ZEALAND[2]

Present-day New Zealand is the emergent part of an extensive tract of Late Paleozoic and Mesozoic continental sediments (partially veneered with later sediments) extending to the north as far as New Caledonia and to the east and southeast across the Chatham Rise and Campbell Plateau. New Zealand Jurassic rocks occur in two distinct assemblages of contrasting lithologies, structure, and provenance. Those of the western belt [part of the Murihiku Supergroup of Campbell and Coombs (1966)] occur in the axes of the Kawhia, Nelson, and Southland Regional Synclines, all part of a once-continuous structure dislocated by some 400 km of lateral displacement along the Alpine Fault. Those of the eastern belt [part of the Torlesse Group of Suggate (1961), Supergroup of subsequent authors] occur in parts of the axial ranges of North Island and South Island (Figure 7.3). Murihiku Jurassic rocks are abundantly fossiliferous, whereas the Torlesse is sparsely fossiliferous, with Jurassic confined to the east and north.

[2] By A. B. Challinor.

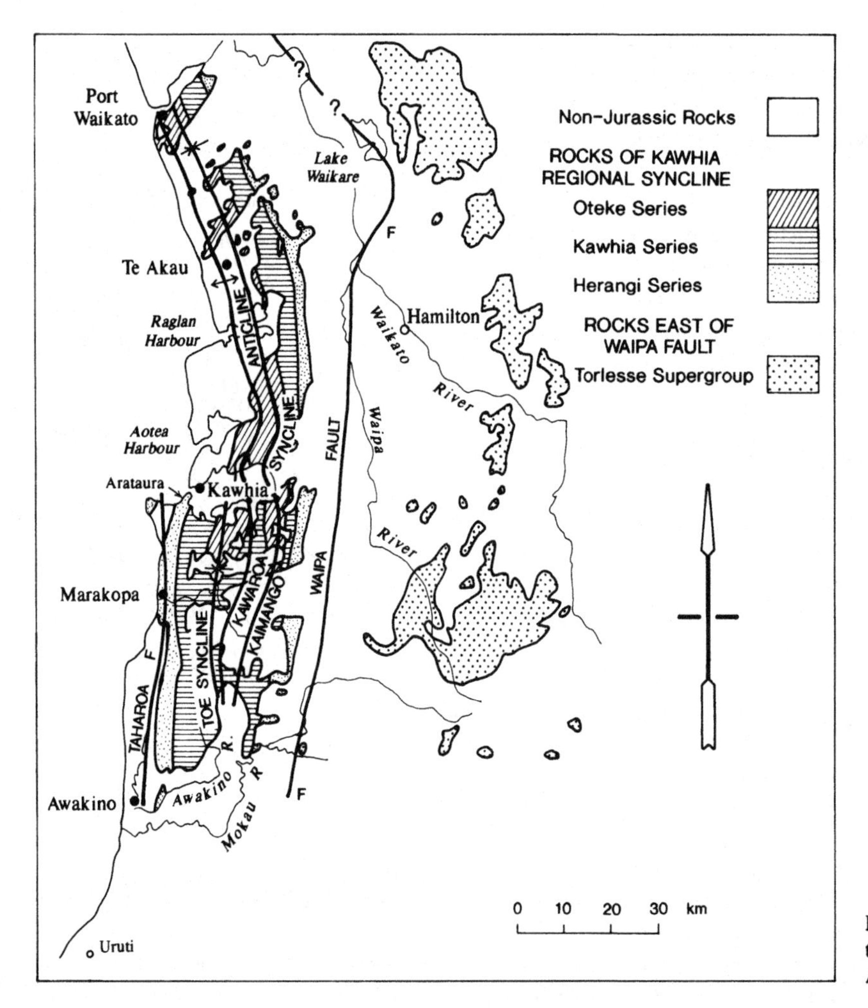


**Figure 7.4. Structure and distribution of Jurassic rocks in southwest Auckland province.**

## Lithostratigraphy

### *Murihiku Supergroup*

Jurassic rocks of the Murihiku Supergroup are present in the core of the Kawhia Regional Syncline in southwest Auckland, flanked to east and west by Triassic strata. The syncline abuts the Waikato Fault in the north and extends some 140 km southward. It is partly hidden by, and eventually disappears beneath, Tertiary rocks in the Awakino region (Figure 7.4). It continues at depth; Jurassic foraminifera have been recovered from a depth of about 350 m near Uruti, some 30 km south of the southernmost surface outcrop (Hornibrook 1953).

The syncline (properly a synclinorium) has minor folds to east and west of its axis (Figure 7.4), and Jurassic beds are well exposed in the Port Waikato, Te Akau, Kawhia, and Awakino regions. Dips are generally steeper on west-dipping limbs (Fleming and Kear 1960), and the structure plunges gently northwards at about 2° from Marakopa to Te Akau (Kear 1966), and to the south at about 2° near Port Waikato (Purser 1961).

The sediments are predominantly mudstones or siltstones (some 70% of the total), often massive, with less common sandstone/siltstone sequences, sandstones, and conglomerates. The latter contain volcanic, sedimentary, granitic, and metamorphic pebbles (in decreasing frequency), and tuffaceous beds are minor but widespread (Bartrum 1935; MacDonald 1954; Fleming and Kear 1960; Purser 1961). Late Jurassic formations generally thicken to the west, and Triassic and Jurassic conglomerates are coarser and thicker in the same direction, suggesting derivation from a western landmass. Early Jurassic beds are more indurated than Middle or Late Jurassic beds.

The Jurassic is in sequence with the Triassic southwest of Kawhia Harbour, but is nowhere known to be in sequence with the Cretaceous, as discussed later. The sediments are mostly marine, but Middle Jurassic nonmarine and terrestrial beds some 800 m thick are present in the Kawhia region, and thick nonmarine sequences occur in the latest Jurassic at Port Waikato and Te Akau (Purser 1961; Kear 1966). Detailed sections through most of the Jurassic have been described from Port Waikato (Purser 1961) and Te Akau (Kear 1966), and less complete sections from Kawhia (Fleming and Kear 1960; Martin 1967; Meesook 1989) and regions to the north, south, and east (Player 1958; Clarke 1959; Grant-Mackie 1959; MacFarlan 1975; Hasibuan 1980; Hudson 1983). Thicknesses range up to almost 8,000 m. Although different in detail, sections from throughout the syncline are broadly similar, except for those near Kawhia, with their

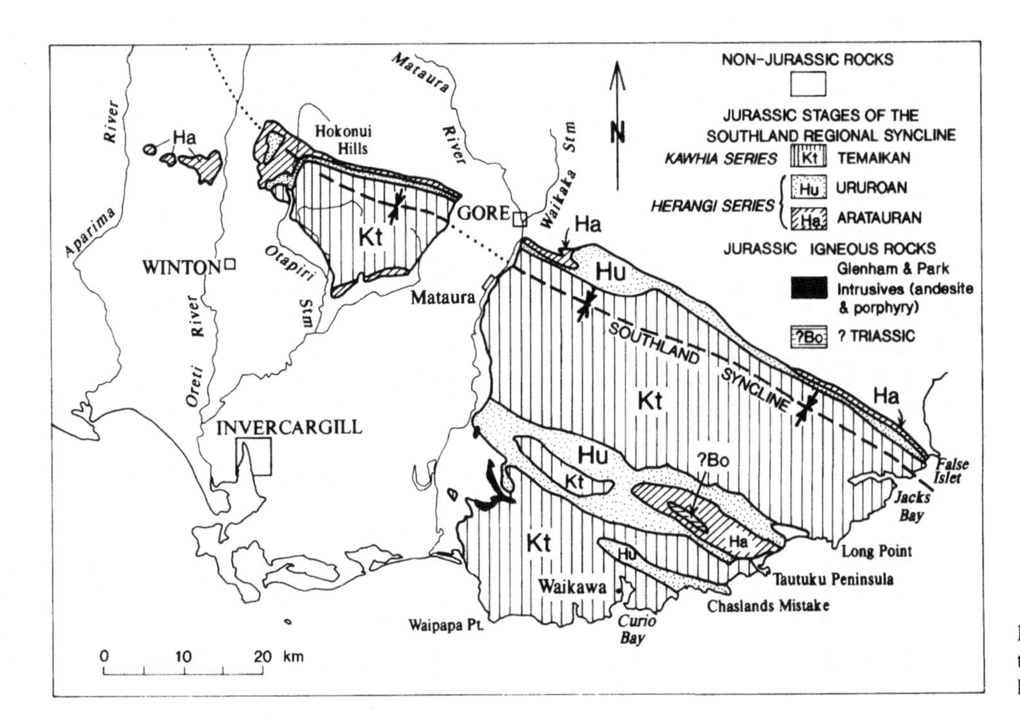


**Figure 7.5. Structure and distribution of Jurassic rocks in the Southland Regional Syncline.**

atypical nonmarine Middle Jurassic sequences. The Te Akau section is summarized in Table 7.1. Ages are in terms of New Zealand stages, the fossil assemblages and correlations of which are discussed later.

Triassic Murihiku rocks appear in the core of the Nelson Syncline west of the Alpine Fault (Figure 7.3), and Jurassic strata of probable nonmarine origin have recently been identified (Johnston, Raine, and Watters 1987). They consist of a fault-bounded sequence of fine to coarse volcanic sediments, the finer of which contain carbonaceous debris and have yielded a macroflora of undifferentiated Jurassic age, including *Cladophlebis* cf. *indica* (Oldh. & Morris) Sahni & Rao, *Maitaia podocarpoides* Ettingsh. ex Townrow, *Rissikia talbragarensis* White, *Podozamites* sp., *Taeniopteris thomsoniana* Arber, and *T.* cf. *crassinervis* (Feistm.) Arber. The Southland Regional Syncline appears east of the Alpine Fault in northwest Otago and strikes southeast to the coast in eastern Southland (Figure 7.3). Similarities in lithology and structure between the Triassic of the Nelson and Southland Synclines support the hypothesis of an originally continuous structure.

Jurassic rocks occupy the axial zone of the Southland Syncline from just west of the Hokonui Hills for about 120 km southeast to the coast at False Islet (Figure 7.5). They form a narrow strip of steeply dipping strata (75–90° to slightly overturned) some 3–5 km wide on the north limb. Dips are less on the south limb (<50°), and several secondary folds result in extensive outcrops (Speden 1971). Sandstones and siltstones are the dominant lithologies, with less common thinly bedded alternating sandstone/siltstone sequences. Carbonaceous beds occur throughout the Temaikan, and conglomerates, usually laterally discontinuous, occur at a number of horizons (Speden 1961, 1971). Pebbles from a Southland conglomerate include dolerite, basalt, andesite, granite, rhyolite, and sedimentary and possibly metamorphic lithologies (Wood 1956). Volcanic clasts (92%) dominate a Southland conglomerate, with minor granitic and other pebbles present (Speden 1971).

The Jurassic conformably overlies the Triassic on the north limb, and on the south limb both conformable (probably) and unconformable contacts are known. The sequence ranges in age from Aratauran to Temaikan (Hettangian–Bathonian); it is 4,000–5,000 m thick and mostly marine, except for an upper nonmarine section 600–700 m thick at the top of the Temaikan. Local unconformities are present in the Temaikan, and the formations coarsen to the southwest. Mudstones thicken, and sandstones thin, to the northeast, indicating a probable shoreline to the southwest (Speden 1961, 1971). Table 7.2 summarizes the stratigraphy of the north limb.

*Possible extrasynclinal Murihiku rocks.* Small areas of possible Murihiku rocks are present outside the regional synclines in northwest Otago and in the eastern foothills of Canterbury. Those of Canterbury [Clent Hills Formation of Haast (1872)] consist of carbonaceous conglomerate, sandstone, and shale and unconformably overlie the Torlesse Supergroup. They contain a flora of probable Middle to Late Jurassic age and were included in discussions on Murihiku rocks by Warren (1978), but Andrews, Speden, and Bradshaw (1976) and MacKinnon (1983) included them and similar small outcrops in the Torlesse Supergroup.

An isolated, fault-bounded strip of Late Jurassic strata [Barrier Formation of Mutch (1964)] surrounded by Permian volcanic and sedimentary rocks occurs in northwest Otago east of Awarua Bay and just east of the Alpine Fault (McKellar 1978). It contains *Belemnopsis* similar to those of the *B. aucklandica* group, of Late Jurassic (Ohauan–Puaroan) age, and suggests that Murihiku rocks were possibly more extensive than those preserved today.

Table 7.1. *Jurassic stratigraphy of the Te Akau district, southwest Auckland*

| Regional series | Regional stages | Stratigraphic unit, lithology, and paleontology | Average thickness (m) |
|---|---|---|---|
| Oteke | Puaroan | Huriwai Group | |
| | | Matira Siltstone: interbedded siltstone and sandstone with plant impressions (*Taeniopteris daintreei, Caldophlebis australis*) | 925+ |
| | | Mangatara Measures: carbonaceous sandstone and conglomerate with plant impressions (*T. daintreei, C. australis*) | 370 |
| | | Apotu Group | |
| | | Waikorea Siltstone: siltstone, with some sandstone beds and carbonaceous material; rare *Buchia plicata* | 400 |
| | | Coleman Conglomerate: conglomerate and sandstone; *B. plicata* and *Belemnopsis a. aucklandica* | 310 |
| | | Owhiro Subgroup | |
| | | Puti Siltstone: siltstone, with thin tuff beds and minor sandstone beds, including Ruakiwi Sandstone Member at base; *Buchia plicata, Buchia hochstetteri, Belemnopsis a. aucklandica,* and *Hibolithes* cf. *arkelli* | 250 |
| | | Siltstone, with sandstone beds, especially near base, and same belemnites as Puti Siltstone | 430 |
| Kawhia | Ohauan | Waiharakeke Conglomerate: conglomerate and sandstone, rare siltstone; *Belemnopsis spathi* | 185 |
| | | Ahuahu Subgroup | |
| | | Kinohaku Siltstone: siltstone, with graded sandstone beds near base; *Hibolithes minor* | 770 |
| | | Takatahi Formation: sandstone, conglomerate, and siltstone; rare *Belemnopsis a. trechmanni* | 125 |
| | | Kirikiri Group | |
| | | Siltstone, with some thin sandstone, tuffaceous, and concretionary beds; *Malayomaorica malayomaorica* throughout; *Retroceramus haasti* in bottom 525 m | 670 |
| | Heterian | Bedded siltstone, sandstone, conglomerate | 15 |
| | | Massive or bedded siltstone, with thin tuffaceous and concretionary beds; *M. malayomaorica, Retroceramus galoi* | 690 |
| | | Rengarenga Group | |
| | | Wilson Sandstone: sandstone, siltstone, rare conglomerate, vitric tuff, and coal; *R.* cf. *galoi* (small), *Conodicoelites* n. sp. | 250 |
| | Temaikan | Putau Siltstone: siltsone, with sandstone and rare conglomerate; *Hibolithes catlinensis* | 460 |
| | | Ohautira Conglomerate: conglomerate, sandstone, siltstone, with wood, coal, and *Meleagrinella* cf. *echinata, Cryptorhynchia kawhiana,* and *Belemnopsis mackayi* | 215 |
| Herangi | Ururoan | Newcastle Group | |
| | | Pongawhakatiki Siltstone: indurated siltsone and rare sandstone, with tuffaceous horizons, basal sandstone and conglomerate, and three higher thin coarse members (Kouratahi Sandstone, Hetherington Conglomerate, Ruawaro Sandstone); *Pseudaucella marshalli* throughout | 1,700 |
| | Aratauran | Te Pake Sandstone: massive fine to coarse sandstone, rare siltstone beds; unfossiliferous except for plant impressions at top (?lower part Triassic) | 310 |

*Source:* Based on Kear (1966).

Table 7.2. *Jurassic stratigraphy of north limb of Southland Syncline*

| Regional series | Regional stages | Formation | Lithology and paleontology |
|---|---|---|---|
| Kawhia | Upper Temaikan | False Islet | 185 m; a basal conglomerate of variable thickness, thinner on the north limb, overlain by coarse, frequently rudaceous sediments, interbedded carbonaceous bed sequences, and fine- to medium-grained sandstones; plant fragments and pebbles common; no invertebrate fossils |
| | | Pounawea | 1000 m of dominantly sandy mudstones and fine-grained graywackes, with interbedded sandstones, which are more numerous in coarser phases in the lower 400 m and upper 150 m; fossils, including *Haastina haastiana,* occur sporadically; the Cannibal Bay Beds near the base contain Callovian ammonites; *Nucula cuneiformis, Entolium, Ostrea,* and *Haastina haastiana* common on south limb |
| | Lower Temaikan | Sweetwater | 725 m of sandstones, with thin interbedded finer-grained sediments and rare carbonaceous-bed sequences; plant fragments and pebbles present; no invertebrate fossils |
| | | Tucks Bay | 120 m of thin-bedded silty mudstones, with thin beds of laumontitized siltstone and fine-grained sandstone; pebbles and plant remains common, sparse faunules containing *M.* cf. *echinata, Astarte* n. sp. A, and *Pleuromya* spp. |
| | | Ironwood | 540 m of dominantly sandstones, with minor fine-grained beds, thin conglomerates, and one carbonaceous-bed sequence; fossils, including *P. milleformis* and *Retroceramus inconditus,* common at base and top of formation |
| | | Boatlanding Bay | 125 m of thin-bedded sandy mudstones and very fine grained graywackes, often concretionary; fossils occur throughout the formation and include *Grammatodon (Indogrammatodon)* n. sp. A, *Pleuromya* spp., and *Cylindroteuthis* sp. |
| | | Omaru | 190 m of sandstones, with interbedded sandy mudstoncs and rare rudaceous beds; Temaikan fossils occur in the upper 115 m and include *P. milleformis, Astarte* n. sp. A, and *Meleangrinella* cf. *echinata* (Smith) |
| Herangi | Ururoan | Otekura | 680 m of thin-bedded dominantly sandy mud-mudstones and very fine grained graywackes, often concretionary and with rare sandstones; the sequence is markedly finer-grained than those on the south limb; a graded-bed sequence forms the upper 75 m; fossils are rare except for *Pseudaucella marshalli,* which extends through a zone of 65 m commencing at 150 m above the base |
| | Aratauran | Pre-Otekura beds | 455 m of interbedded thin, fine- to coarse-grained sandstones, with minor fine-grained graywackes, siltstones, and thin discontinuous conglomerates; *Otapiria marshalli* (Trechmann) and other Aratauran fossils are known from one locality; psiloceratid ammonites |
| | Triassic | | |

*Source:* Based on Speden (1971).

*Igneous rocks.* Small areas of igneous rocks [feldsparphyric basalt and quartz porphyry, the Park Intrusives of Mutch (1964)] occur 15–25 km north of Waipapa Point on the southern limb of the Southland Syncline. They have been K-Ar dated (K feldspar) at 159 Ma (Callovian).

### *Torlesse Supergroup*

Rocks of the Torlesse Supergroup form most of the axial ranges of North and South Islands (Figure 7.3). They grade into their metamorphic equivalents, the Haast Schist, in the south and west of South Island; in the northwest they are bounded by the Alpine Fault. They are exposed on North Island from Wellington to the East Cape Peninsula, and less extensively from west of Lake Taupo north-northwestward through South Auckland (where they are separated from Murihiku rocks by the Waipa Fault) into eastern Northland.

Torlesse rocks form a structurally complex, variably indurated, weakly metamorphosed (zeolite and prehnite-pumpellyite grade), quartzofeldspathic suite of continental aspect, locally variable in lithology and of unknown (but very great) thickness. They consist in the main of blue-gray "graywacke" and dark "argillite" (of New Zealand usage), alternating sandstone/siltstone graded sequences, minor but locally important conglomerates (Andrews et al. 1976), spilitic lavas and tuffs, chert, limestone, and manganese-bearing rocks. Detrital conglomerates include boulders up to 1.3 m in diameter (Maxwell 1964), which include microgranite and granophyre, rhyolite, trachyte, andesite, basalt, gneiss, quartz schist, chert, Torlesse-type sandstones, calcareous sandstones, calcareous mudstones, and impure limestone (Andrews et al. 1976). The Torlesse is very sparsely fossiliferous, with large tracts apparently barren, but fossils ranging in age from Carboniferous to Cretaceous are known, although within this range there are numerous large gaps in the record. Although fossil occurrences are isolated, they are sufficiently consistent over large areas to indicate that Permo-Triassic (with Carboniferous locally) (Jenkins and Jenkins 1971) occurs to the west and south, Jurassic to the north and east, and Cretaceous to the east.

Although most authors have agreed on a western origin for the Murihiku Supergroup, there has been no consensus on the source of the Torlesse. Some proposed a western origin, with deposition in an axial (deep-water) setting, as opposed to a shelf (marginal, shallow-water) setting, for Murihiku. The complex structure of Torlesse was explained in terms of redeposition by gravity flow at the foot of the continental slope. No convincing method for transporting a quartzofeldspathic suite of sediments across a shelf on which a volcaniclastic suite was being deposited was advanced. The faunally, geographically, and temporally restricted nature of the fossil record remained unexplained.

Andrews et al. (1976), from a study of depositional facies, detrital conglomerate, plant fossils, and other data, favored derivation from the east, with accumulation of a clastic wedge or wedges of continental sediment over the site of a west-dipping subduction zone, with deposition in different subbasins at different times (hence the nonrandom distribution of faunas). No particular continental source was proposed, and an eastern origin for the sediment has not been generally accepted.

MacKinnon (1983) suggested that both the Murihiku and Torlesse accumulated in a subduction setting near the coast of Gondwana, but in laterally distinct environments. Murihiku rocks were deposited in a marginal sea/volcanic arc/trench setting to the west, and the Torlesse in a trench, slope, or borderline basin setting fronting a continental volcano-plutonic arc to the east. The Torlesse is thought to have originated as an accretionary complex, becoming progressively younger seaward from a convergent margin. The fossil record suggests that deposition was episodic, with most activity in the Permian, Middle and Late Triassic, and Late Jurassic–Early Cretaceous. The Murihiku and Torlesse were later juxtaposed by strike-slip movement subparallel to the Gondwana coast, with the Torlesse emplaced oceanward of the Murihiku. Recurving of the parallel strip of the conjugated Murihiku/Torlesse and later dislocation by movement along the Alpine Fault have resulted in a general younging of the Torlesse to the north and east, with Jurassic beds restricted to those regions. Explanations that have both the Murihiku and Torlesse accumulating adjacent to Gondwana are gaining acceptance (Sporli 1987).

The Esk Head Subterrane of the Torlesse Terrane (Bradshaw, Adams, and Andrews 1981) of North Canterbury separates fossil-bearing Jurassic outcrops to the northeast from older outcrops (Figure 7.3). It is a tectonic mélange containing blocks of various lithologies and is interpreted as a slice of upper oceanic plate and associated sediments. It contains Triassic conodonts and Radiolaria of Late Triassic and Early, Middle, and Late Jurassic ages. Torlesse rocks to the northeast [Pahau Subterrane of Bradshaw et al. (1981)] contain Late Jurassic and Early Cretaceous macrofossils, those to the southwest (Rakaia Subterrane) mostly Permian and Triassic.

North Island Jurassic Torlesse is more extensive but less fossiliferous. Scattered late Jurassic fossils occur from the Wairarapa to East Cape and in the Waikato, western Bay of Plenty, and North Auckland regions. Two Early Jurassic localities containing *Pseudaucella marshalli* (Trechmann) are known. One is in the North Island (near Taneatua, eastern Bay of Plenty) (Fleming 1953); the other is a float boulder from northern Canterbury (Kaiwara River) (Andrews et al. 1976). Study of radiolarian-bearing Torlesse rocks (e.g., Sporli and Aita 1988) is providing additional evidence of more widespread Jurassic. Table 7.3 summarizes the paleontology of the Jurassic Torlesse.

## Chronostratigraphy

### *New Zealand Jurassic stages and their correlation*

Strong endemism in New Zealand faunas led Marwick (1951, 1953) to propose a series of regional Jurassic stages [later revised by Fleming (1958, 1960)] based on relatively few, geographically widespread but stratigraphically restricted fossils, mostly bivalves. Faunal definition of stage boundaries has tended to resolve itself into definition in terms of first occurrence of single species believed to have dispersed rapidly (Fleming and Kear 1960). Regional stages, though criticized by some, facilitate discussion and provide a stable base for stratigraphic units and intraregional correlation. New Zealand Jurassic faunas are dominated by bivalves; ammonites and other taxa useful for overseas correlation are sparsely distributed, correlation being possible only at scattered horizons, particularly in the Early and Middle Jurassic. More precise correlations are possible in the Late Jurassic using ammonites and bivalves, but even these are not fully accepted by dinoflagellate workers.

The New Zealand Jurassic system is divided into three series: Herangi, Kawhia, and Oteke (Table 7.1). Six stages are recognized, all originally based on the sequence at Kawhia Harbour, but the Aratauran, Ururoan, and Temaikan are better represented in Southland and Otago. The base of the Jurassic is well exposed in both regions and is marked by the first appearance of psiloceratid ammonites and/or *Otapiria marshalli* (Trechmann). The Cretaceous is nowhere known to be in sequence with the Jurassic. The youngest Murihiku beds [the Huriwai Formation of Purser (1961)] are nonmarine, and although their plant fossils were originally assigned to the Neocomian (Arber 1917), the Late Jurassic (McQueen 1955; Purser 1961; Norris 1964, 1968) or Tithonian–Neocomian (Harris 1962) is preferred by later workers.

Table 7.3. *Jurassic fossils of Torlesse Supergroup*

| New Zealand stages | Northland, east & south Auckland, Coromandel | Eastern & southern North Island | Marlborough, north Canterbury |
|---|---|---|---|
| Puaroan | *Hibolithes* cf. *marwicki marwicki, Hibolithes* cf. *arkelli* | *Buchia* aff. *hochstetteri, Stomiosphaera,* microplankton, spores | *Retroceramus* aff. *everesti, Buchia* aff. *subpallasi, Belemnopsis* ex. gr. *aucklandica, Hibolithes arkelli, Buchia plicata, Anopaea* aff. *verbeeki, Grammatodon* sp., *Retroceramus* aff. *gracilis, Psilotrigonia zelandica, Burmirhynchia warreni, Holcothyris*(?) *kaiwaraensis, Stomiosphaera moluccana,* rhynchonellid, terebratulid, decapod crustacean, spores, microplankton |
| | [radiolaria] | | |
| Ohauan | *Belemnopsis* ex. gr. *aucklandica, R. haasti* | *Belemnopsis* ex. gr. *aucklandica* | |
| | *Malayomaorica malayomaorica* | *Malayomaorica malayomaorica* | |
| Heterian | *Retroceramus galoi* | | *Idoceras speighti, Retroceramus galoi, Belemnopsis* cf. *kerari* |
| | [radiolaria] | | |
| Temaikan | | | |
| Ururoan | *Dentalina,* ?*Cylindroteuthis* sp. | *Pseudaucella marshalli* | *Pseudaucella marshalli* |
| | [radiolaria] | | |
| Aratauran | | | |

*Source:* Based on Suggate (1978, Table 4.14) and Sporli and Aita (1988).

*Herangi Series*

*Aratauran stage (Hettangian–Sinemurian).* The base of the stage is marked by the first appearance of *Otapiria marshalli* (Trechmann) and psiloceratid ammonites and is well illustrated in the coastal section of southeast Otago and in the Hokonui Hills, Southland (Campbell and McKellar 1956; Speden 1971; Stevens 1978). *Otapiria marshalli* appears a few metres above *O. dissimilis* (Cox) (indicator for the Otapirian – correlated with Rhaetian) near Awakino (Grant-Mackie 1959) and is known from near Marakopa (Stevens 1978). Neither *O. marshalli* nor psiloceratid ammonites were known from the type section near Kawhia (Campbell 1956; Martin 1967), and Martin (1967) used the appearance of *Pseudolimea fida* Marwick to mark the boundary.

The Aratauran was correlated with Hettangian–Sinemurian by Marwick (1953) and confirmed by Arkell (1953, 1956) and by Stevens (1978), who recognized Hettangian correlatives and the Sinemurian Bucklandi Zone in Southland. Martin (1967) collected *Primarietites* cf. *rotiforme* (Sowerby) of the Bucklandi Zone near Arataura, and Hettangian correlatives have been recognized near Marakopa (Stevens 1978). An association of a possibly psiloceratid ammonite, *O. marshalli,* and *Atractites* has recently been found near Arataura (J. A. Grant-Mackie personal communication).

Representative taxa of the Aratauran Stage include *Otapiria marshalli* (Trech.), *Pseudolimea fida, Kalentera mackayi, Entolium fossatum, Chlamys (Camptochlamys) wunschae* Marwick spp., *Pleuromya, Schlotheimia, Waehneroceras, Primarietites, Saxoceras, Discamphiceras, Charmacisseras,* and *Atractites.*

*Uroroan stage (Pliensbachian–Toarchian, ?Aalenian).* In the Otago (Speden and McKellar 1958), Awakino (Grant-Mackie 1959), southwest Kawhia (Martin 1967), Te Akau (Kear 1966), and Port Waikato (Purser 1961) regions the first occurrence of *Pseudaucella marshalli* (Trech.) is considered to mark the base of the Ururoan Stage. At Kawhia, *P. marshalli* ranges through 140 m, early Toarcian *Dactylioceras* occurs 190 m above the topmost *P. marshalli* (Fleming and Kear 1960), and undescribed *Lytoceras* occurs just above (Stevens 1984).

Grant-Mackie (1959) considered the lower Ururoan to be characterized by *P. marshalli* and rare *Chlamys,* and the upper Ururoan by *Dactylioceras* and some characteristic but undescribed rhynchonellid brachiopods. *Dactylioceras* is not known from Awakino, and here Grant-Mackie (1959) used the first appearance of the rhynchonellids to indicate basal upper Ururoan. Martin (1967) revised the Ururoan of southwest Kawhia and defined the lower Ururoan as strata between the first *Pseudaucella marshalli* and beds with a fauna including rhynchonellid brachiopods. He included in the upper Ururoan some beds with rhynchonellids [including "*Rhynchonella*" *bartrumi* (Marwick)] and strata extending up to the lowermost Temaikan fauna.

MacFarlan (1975) pointed out that the most common rhynchonellid from the type Ururoan (Tetrarhynchiinae n. gen. et sp.) occurs close to the base of the stage at Marakopa. Hudson (1983) defined the lower Ururoan as between the appearances of *P. marshalli* and an upper Ururoan fauna [redefined as including any of *Mentzelia radiata* (Hector), *M.* cf. *radiata, Lobothyris* sp., and *Paleoneilo* n. sp.]. He pointed out that the *Chlamys* included by Grant-Mackie (1959) in the lower Ururoan appears to be confined to the upper part of the Ururoan sequence at Kawhia. He also found that all the common rhynchonellids from Martin's upper Ururoan (except "*Rhynchonella*" *bartrumi*) also occur with *Pseudaucella marshalli,* making Martin's upper Ururoan invalid.

The upper Ururoan [*Dactylioceras* assemblage of Grant-Mackie (1959)] is correlated with the Toarcian by Spath (1923), Marwick (1953), Grant-Mackie (1959), and Hudson (1983). The (lower) Toarcian ammonites *Harpoceras* cf. *falcifer* (Sowerby) and *Alocolytoceras*(?) have been identified by Stevens (1978) from the upper Ururoan. The lower Ururoan cannot be directly correlated. It is inferred as Pliensbachian from its position above Sinemurian and below Toarcian, but neither the Sinemurian–Pliensbachian nor the Pliensbachian–Toarcian boundary can be accurately located in New Zealand. Similarly, it is unknown how much of the Ururoan, if any, is of Aalenian age.

Characteristic Ururoan taxa include *Spiriferina radiata* (Hector), "*Rhynchonella*" *bartrumi* Marwick and some other Rhynchonellidae, *Pseudaucella marshalli* (Trech.), *Pleuromya urnula* Marwick, *Pseudolimea, Camptonectes (Camptochlamys) wunschae* Marwick, "*Inoceramus*" *ururoaensis* Speden, *Parainoceramus martini* Speden, and the early Toarcian *Dactylioceras* cf. *anguinum* (Rein.), *D.* aff. *timorense* (Boehm), *D.* aff. *commune* (Sow.), and *Harpoceras* cf. *falcifer* (Sow.).

### *Kawhia Series*

*Temaikan stage (?Aalenian, Bajocian–Bathonian).* The Temaikan type section at Kawhia Harbour consists of 23 m of marine Opapaka Sandstone between 520 m of nonmarine Urawitiki Measures below and 290 m of nonmarine Wharetanu Measures above. Marwick (1953) listed a small fauna and suggested *Meleagrinella* [= *M.* cf. *echinata* (Smith)] as a guide fossil. Fleming and Kear (1960) used the same indicator to recognize basal Temaikan at Kawhia, as did Purser (1961) at Port Waikato. *Meleagrinella* was later shown to extend down into the Aratauran (Speden 1971). Grant-Mackie (1959) placed the Ururoan–Temaikan boundary at the appearance of "*Inoceramus*" [the Ururoan-Puaroan inoceramid bivalves previously placed in *Inoceramus* belong mostly in *Retroceramus,* according to Crame (1982) and Crampton (1988)]. Martin (1967) placed the boundary at the first "*Inoceramus*" ex. gr. *R. inconditus* (Marwick), later described as "*I.*" *ururoaensis* Speden and shown to extend well down into the Ururoan (Hudson 1983).

Speden (1971) suggested that the appearance of *Meleagrinella* cf. *echinata* and *Retroceramus inconditus* above *Pseudaucella,* preferably together with other Temaikan species, is the best indicator of the stage. He recognized a twofold division in Otago, a lower Temaikan with a *Pleuromya milleformis* Assemblage Zone (with *M.* cf. *echinata* and *R. inconditus*) and an upper Temaikan with a *Haastina haastiana* Assemblage Zone [with *Retroceramus marwicki* (Speden) and *Tancredia allani* Marwick].

Hudson (1983) suggested a tripartite division of the southwest Auckland Temaikan (ages revised by G. E. G. Westermann, based on new ammonoid finds):

1. Lower Temaikan (Bajocian) – base at the appearance of *Belemnopsis mackavi* Stevens and/or *B. deborahae* Challinor (synonyms?, see below).
2. Middle Temaikan (Bajocian–Bathonian?) – base at appearance of *Retroceramus inconditus, R. brownei* Marwick or *Meleagrinella* sp.
3. Upper Temaikan (Bajocian–Bathonian–Callovian) – with "*Macrocephalites (Kamptocephalites)* cf. *beta-gamma* Boehm" of Stevens (1978), according to G. E. G. Westermann (personal communication), an undescribed endemic sphaeroceratid genus that is associated with *Sphaeroceras (Sph.)* sp. of mid-Bajocian age; *Stephanoceras* gr. *humphriesianum* (Sowerby) has also been identified (Figure 7.6); slightly higher is *Macrocephalites parki* (Wilkens) (Speden 1960, 1971), which is closest to the middle Bathonian *M. bifurcatum* Boehm from the Sula Islands and New Guinea (G. E. G. Westermann personal communication); and, near the top, with *Xenocephalites grantmackiei* Westermann & Hudson, which is also known from the *M. apertus* Zone of Papua New Guinea (Westermann and Callomon 1988) and *Lilloettia* aff. *boesei* (Burckhardt), both indicating late or latest Bathonian (Westermann and Hudson 1990). In the Awakinio River section, this assemblage is separated from the Heterian by roughly 100 m of poorly fossiliferous shales, which could be Callovian (G. E. G. Westermann and N. Hudson personal communication).

McKellar (1969) had earlier used *Belemnopsis mackayi* to indicate the basal Temaikan in the Southland, and this taxon [together with the very similar and possibly conspecific *B. deborahae;* see Hudson (1983)] seems to be the most useful indicator of the basal Temaikan, although it is still unknown from southeast Otago (Speden 1971).

Hudson (1983) and Hudson, Grant-Mackie, and Helby (1987) have recently recorded *Trigonia signicollina* Flemming from the base of the marine Oraka Sandstone, immediately above the nonmarine Wharetanu Measures at Kawhia Harbour, establishing a Temaikan (approximately Bajocian–Bathonian) age at that level (Fleming 1987). The Bajocian ages of the Temaikan were based on the presence of the belemnite *Brachybelus zieteni* (Werner)

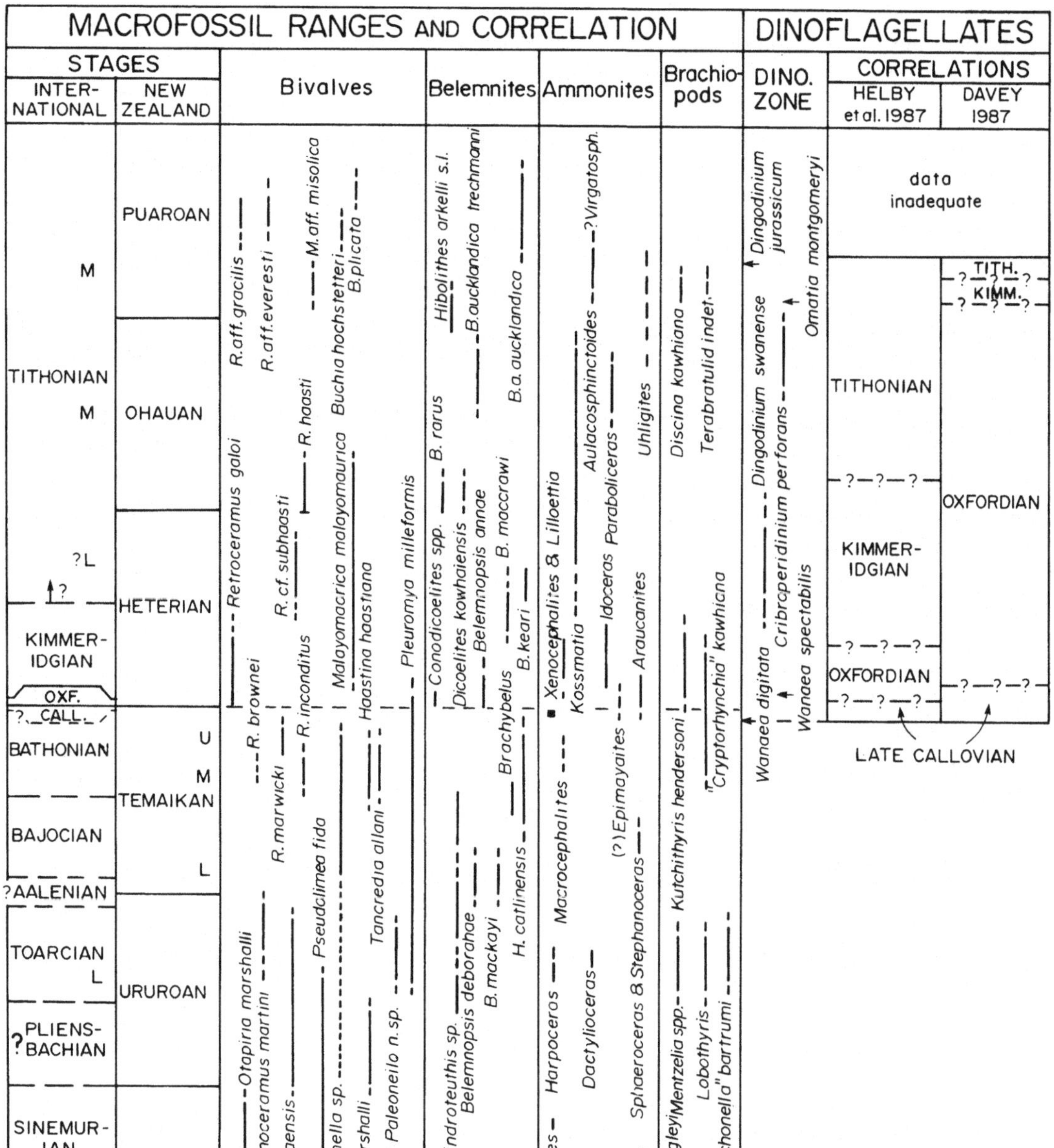

**Figure 7.6. New Zealand stages, with correlations and fossil ranges. Bajocian-Oxfordian ammonites according to G. E. G. Westermann (personal communication).**

in southeast Auckland (middle Liassic to Bajocian in Europe) (Stevens 1965) and the general resemblance of *Belemnopsis mackayi* to Bajocian *Belemnopsis* of Europe (Stevens 1965). *Nannolytoceras* in the middle Temaikan (Type) at Kawhia and the upper Temaikan at Port Waikato (both southwest Auckland) (N. Hudson personal communication), as well as in the lower Temaikan of South Otago, suggests Bajocian–Bathonian, according to its range in Europe and North America (Stevens 1978). Earlier references to Callovian *Macrocephalites* from New Zealand, such as "*M. etheridgei* Spath" (Stevens 1978), however, are now referred to *M.* gr. *parki* (Wilkins) and because of their resemblance to *M. bifurcatum* Boehm, are tentatively dated as middle to late Bathonian. Aalenian and Callovian have thus not been identified in New Zealand, and a Bajocian–Bathonian correlation of the Temaikan seems likely.

Characteristic Temaikan taxa include *Palaeonucula cuneiformis* Sow., *Retroceramus (Fractoceramus) inconditus* (Marwick), *Retroceramus marwicki* (Speden), *Retroceramus brownei* (Marwick), *Meleagrinella, Astarte, Pleuromya milleformis* Marwick, *Haastina haastiana* Wilkens, *Tancredia allani* Marwick, *Trigonia signicollina* Fleming, *Vaugonia spedeni* Fleming, *V. (Hijitrigonia) kahuika* Fleming, *Orthotrigonia waipahiensis* Fleming, *Myo-*

*phorella (Scaphogonia) hokonuiensis* Fleming, *Laevitrigonia (Malagasitrigonia) macfarlani* Fleming, *Leptomaria* sp.(?), *Nannolytoceras* sp., *Macrocephalites parki* (Wilkens), n. gen. et sp. [''*M. beta-gamma*'' Stevens, non-Boehm], *Xenocephalites grantmackiei* West. & Hudson, *Lilloettia* sp., *Stephanoceras* sp., and *Sphaeroceras* sp. (G. E. G. Westermann personal communication), *Belemnopsis deborahae* Challinor, *Belemnopsis mackayi* Stevens, *Cylindroteuthis*(?) sp., *Hibolithes catlinensis* (Hector), and *Brachybelus zieteni* (Werner).

*Heterian stage [?late Callovian, Oxfordian to middle(?) Tithonian].* *Retroceramus galoi* (Boehm) was regarded by Marwick (1951, 1953) as the best indicator of Heterian age, and the base of the stage was later defined or recognized (Fleming and Kear 1960; Purser 1961; Hudson 1983) by its first appearance.

MacFarlan (1975) recognized three zones within the Heterian: a lower *Retroceramus galoi* Zone, a middle *Malayomaorica malayomaorica* Zone, and an upper *Retroceramus* cf. *subhaasti* Zone. *R. galoi* has been dated as middle Oxfordian to late Tithonian (Crame 1982), and *M. malayomaorica* as Kimmeridgian–early Tithonian (Crame 1983), but the taxonomic status and correlative value of *R.* cf. *subhaasti* are unclear (Crame 1982).

Hudson et al. (1987) recorded an association of *Retroceramus galoi* and *Epimayaites* with a dinoflagellate assemblage, apparently predating the Australian *Rigaudella aemula* and *Wanaea spectabilis* Zones (Callovian–Oxfordian) (Helby et al. 1987) and probably equivalent to the upper *Wanaea digitata* Zone (late Callovian) (Helby et al. 1987), about 20 m above the base of the Oraka Sandstone at Kawhia Harbour. The Kimmeridgian *Epicephalites* was said to be found at about the same level (Fleming and Kear 1960) (but see footnote 3). *Conodicoelites* was found at and above this horizon, and *R. galoi* was subsequently found about 3 m below the other fossils (Helby, Wilson, and Grant-Mackie 1988). *Epimayaites* has been dated as early to middle Oxfordian (Sato et al. 1978; Thomson 1982), and *Conodicoelites* appears in eastern Europe in the Bajocian and is known from middle(?) Bathonian to early Oxfordian in Indonesia and Papua New Guinea (Challinor and Skwarko 1982; Challinor 1990, 1991). Thus, a condensed sequence apparently of late(?) Callovian to Kimmeridgian age occurs in the lower Oraka Sandstone, and beds between this horizon and almost to the base of the formation may be Callovian.

*R. galoi* is associated at Kawhia with *Idoceras* (Kimmeridgian) (Fleming and Kear 1960; Verma and Westermann 1973; Stevens and Speden 1978), and the cosmopolitan *Kossmatia,* dated as middle to late Tithonian,[3] occurs at about the midpoint of the range of *R.* cf. *subhaasti.* Thus the Heterian appears to range from the (?late Callovian) Oxfordian to middle Tithonian. *Retroceramus, Malayomaorica,* and belemnites are valuable for correlation within New Zealand, but the Heterian ammonites (except for *Idoceras speighti,* the holotype of which was found in a float boulder in Canterbury, and phylloceratids at Port Waikato) are not known outside Kawhia.

Characteristic taxa of the Heterian stage include *Retroceramus galoi* (Boehm), *Retroceramus* cf. *subhaasti* (Wandel), *Malayomaorica malayomaorica* (Krumbeck), *Pinna kawhiana* Marwick, *Indogrammatodon taylori* Marwick, *Astarte morgani* Trech., *Neocrassina* cf. *spitiensis* (Stol.), *Pleuromya milleformis* Marwick, *Vaugonia kawhiana* (Trech.), *Kutchithyris hendersoni* Marwick, ''*Cryptorhynchia*'' *kawhiana* (Trech.), *Partschiceras* sp., *Phylloceras polyolcum* (Ben.), *Lytoceras taharoaense* Stevens, *Epimayaites*(?) sp., *Epicephalites* cf. *epigonus* (Burckhardt) [most recently placed in the Andean genus *Araucanites* (G. E. G. Westermann personal communication)], *Idoceras*(?) *speighti* (Marshall), *Kossmatia* sp., *Belemnopsis keari* Stevens, *Belemnopsis annae* Challinor, *Belemnopsis maccrawi* Challinor, *Conodicoelites flemingi* Stevens, and *Conodicoelites orakaensis* Stevens.

*Ohauan stage [middle(?) Tithonian].* The Ohauan stage contains scattered cephalopods and a bivalve assemblage of low diversity in its lower part and abundant cephalopods in its upper part. The base is recognized by the appearance of *Retroceramus haasti* (Hochst.), but this horizon is difficult to locate because a plexus of forms believed to be transient between *R. galoi* and *R. haasti* (and including *R.* cf. *subhaasti*) appears in the upper part of the Heterian. In addition, the upper Heterian to basal Ohauan section is poorly exposed at Kawhia. *Malayomaorica malayomaorica* (Krumbeck) continues from Heterian into the lower Ohauan. *R. haasti* is considered Tithonian by Crame (1982).

The lower Ohauan is characterized by *Kossmatia* (continued from upper Heterian), and the upper by *Paraboliceras.* Spath (1923) assigned the Ohauan to the late Tithonian on the presence of ''*Berriasella*'' *novoseelandica* (Hochst.), but this specimen was regarded by Arkell (cited by Fleming and Kear 1960) as a crushed *Kossmatia* or *Paraboliceras.* Arkell correlated the Ohauan with early Tithonian (*sensu* Harland et al. 1982), but Enay (1973) regarded *Kossmatia* and *Paraboliceras,* as well as ammonites from the overlying Puaroan stage, as late Tithonian (bifold division) and suggested that the early Tithonian is poorly represented or missing in the western Pacific. However, *Uhligites* is present in the upper Ohauan (Fleming and Kear 1960), and Enay (973) suggested that this genus occurs in the lower Tithonian of Spiti and in coeval beds in Anatolia and the Malagasy Republic. Recent work in the Spiti Shales (Krishna, Kumar, and Singh 1982) suggests, however, a middle Tithonian age (but see footnote 3).

Characteristic taxa of the Ohauan stage include *Retroceramus haasti* (Hochst.), *Malayomaorica malayomaorica* (Krumbeck), *Calliphylloceras empedoclis* Gemm., *Phylloceras salina* Krumbeck, *Holcophylloceras polyolcum* (Ben.), *Kossmatia novoseelandica* (Hochst.), *Paraboliceras* n. spp., *Partschiceras* n. sp., *Hibolithes minor* (Hauer), *Hibolithes* sp., *Dicoelites kowhaiensis* Chall., *Belemnopsis rarus* Chall., and *Belemnopsis aucklandica trechmanni* Stevens.

[3] Editor's comment: Possibly a late Bathonian–early Callovian *Lilloettia*, Stevens (pers. comm.) and recent palynology would suggest, however, that *Kossmatia* appears already in the Kimmeridgian, in New Zealand, and elsewhere in Australasia. The New Zealand *Idoceras* may be an endemic homeomorph.

*Oteke Series*

*Puaroan stage [middle(?) Tithonian].* Marwick (1953) listed three taxa as the type fauna, but of these only *Buchia plicata* (Zittel) can be used to define the stage. The original specimens have been lost, and similar specimens have not been recollected from the same level; *Buchia plicata* s.s. is now known to characterize beds much higher in the Puaroan stage (Player 1958; Purser 1961). Challinor (1977a) redefined the stage by the appearance of *Hibolithes arkelli grantmackiei,* occurring in several sections close to the horizon at which Marwick's *Buchia "plicata"* was recorded and signifying a change from *Belemnopsis* below to *Hibolithes* above. He recognized a Mangaoran substage (early Puaroan) dominated by *Hibolithes* and a Waikatoan substage (late Puaroan, not present at Kawhia, but recognized in beds to the north) marked by a return to *Belemnopsis.* There is now a valid argument for placing the Ohauan–Puaroan boundary in beds presently classed as upper Ohauan, because new species of *Hibolithes* (with one abundant) occur toward the top of the Ohauan stage as currently defined (A. B. Challinor unpublished data).

*Malayomaorica* aff. *misolica* (Krumbeck) occurs in both Mangaoran and Waikatoan substages, *Buchia hochstetteri* Fleming and *B. plicata* (Zittel) occur in the Waikatoan substage, and *Aulacosphinctoides*? *brownei* (Marshall) (a ?*Torquatisphinctes*) (G. E. G. Westermann personal communication) appears not far above the Ohauan–Puaroan boundary, as defined by Challinor (1977a). The lower Puaroan ammonites include also *Kossmatia, Uhligites,* and relatively abundant *Aulacosphinctoides*? [or *Subplanites*] cf. *marshalli* Spath (determined by G. E. G. Westermann in University of Auckland collections assembled mainly by J. Grant-Mackie). *A.*? cf. *marshalli* resembles the septate whorls of the *Virgatosphinctes densiplicatus* (Waagen) group occurring mainly in the middle Tithonian *Hildoglochiceras-Virgatosphinctes* Assemblage (Krishna et al. 1982) of the Himalayan Spiti Shales (G. E. G. Westermann, personal communication). Li and Grant-Mackie (1988) have similarly dated the *Malayomaorica* aff. *misolica, Buchia hochstetteri,* and *B. plicata* zones with the "early upper Tithonian" of a bifold division. A middle Tithonian age is therefore indicated.

Some 700 m of beds above marine Puaroan at Te Akau and Port Waikato contain abundant plant fossils [*Taeniopteris daintreei, Cladophlebis australis* (Morris), *Coniopteris, Nilsonnia*]. Norris (1964, 1968) recognized two microfloral assemblages older than Berriasian but post–early Tithonian in the nonmarine Puaroan.

Characteristic taxa of the Puaroan stage include *Palaeonucula putiensis* (Marwick), *Buchia plicata* (Zittel), *Buchia hochstetteri* Fleming, *Malayomaorica* aff. *misolica* (Krumbeck), *Retroceramus* aff. *everesti* Oppel, *Retroceramus* aff. *gracilis* Holdhaus, *Placunopsis striatula* Zittel, *Otapiria masoni* Marwick, *Discina kawhiana* Boehm, *Myophorella (Scaphogonia) purseri* Fleming, *Holcophylloceras polyolcum* (Benecke), *Calliphylloceras empedoclis* (Gemm.), *Aulacosphinctoides*? *brownei* (Marshall), *A.*? [or ?*Subplanites*] cf. *marshalli* Spath, *Kossmatia* sp., *Uhligites motutaraensis* (Boehm), *Torquatisphinctes?, Hibolithes arkelli arkelli* Stevens, *Hibolithes arkelli grantmackiei* Chall., *Hibolithes marwicki marwicki* Stevens, *Belemnopsis aucklandica aucklandica* (Hochst.), *Taeniopteris daintreei* McCoy, *Carnoconites cranwelli* Harris, *Pentoxylon, Cladophlebis "australis"* (Morris) [= *hochstetteri* (Unger)], *Coniopteris hymenophylloides* (Brong.), *Mataia podocarpoides* Ettingsh. ex Townrow.

The approximate range zones of characteristic taxa of the New Zealand Jurassic are illustrated in Figure 7.6. Most of this information is from Stevens (1965, 1984), Speden (1971), Challinor (1977a,b, 1979, 1980), Hudson (1983), Hudson et al. (1987), and Helby et al. (1988), with some information from other sources mentioned in the text.

The relative precision of the bivalve and belemnite data reflects the nature of the fossil record and the interests of New Zealand paleontologists. The bivalve ranges depicted are thought to be reasonably accurate, except for the upper Temaikan, where the ranges of *Retroceramus inconditus, R. brownei,* and *R. marwicki* and the amounts of overlap between them are uncertain. The status of the Ururoan *Meleagrinella* and its relationship to *M.* cf. *echinata* Smith are likewise uncertain. Except for the upper Ohauan, most of the Jurassic is reasonably well zoned by bivalves. Belemnite ranges are also reasonably well established, particularly in the Late Jurassic, and most taxa have been recorded in several regions of southwest Auckland. The Middle Jurassic *Belemnopsis mackayi* and *Hibolithes catlinensis* are known from both the Kawhia and Southland Synclines, but *Brachybelus* has so far been recorded only from the Awakino region. The systematic status of New Zealand *Cylindroteuthis*? is uncertain, and its long range (middle Ururoan to middle Temaikan) is unusual for a New Zealand belemnite. It may prove to be a composite taxon, possibly not including *Cylindroteuthis* s.s. The Aratauran to middle Ururoan record is poor, with only the belemnoid *Atractites* known from the basal Aratauran.

The ammonite record is less comprehensive, and except for the work of Stevens (1984), Hudson et al. (1987), Helby et al. (1988), and Westermann & Hudson (1990), little has been published. Little information on ammonite ranges in the Aratauran–Temaikan is available, most records being of single or very localized occurrences; the taxa listed are mostly the unrevised original determinations. More is known of ranges in the Heterian–Puaroan, but even here the record is mostly of occurrences or distributions of genera, and some generic placings are suspect (Enay 1973). Furthermore, the Heterian–Puaroan interval is almost entirely contained within the Late Jurassic, and its ammonite record is therefore only a small part of the whole. Data on the distribution of brachiopods are limited, and they have not been extensively used as zonal fossils. Research continues, mainly by G. E. G. Westermann and G. Stevens.

The correlations between New Zealand and international stages adopted here (Figure 7.6) are based mostly on macrofossils, but it must be admitted that the time ranges of the correlating taxa [see Helby et al. (1988) for discussion of Middle–Late Jurassic correlations] are imperfectly known. Alternative correlations based on a preliminary study of dinoflagellate assemblages of the Heterian–Puroan of Kawhia Harbour have been proposed. The dinoflagellate date in Figure 7.6 are from Helby et al. (1988), and

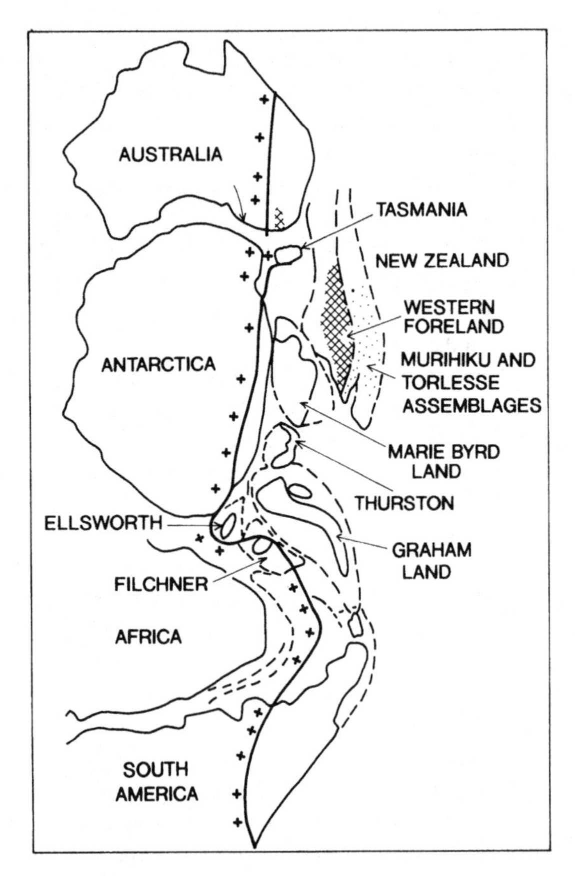


**Figure 7.7. Middle Cretaceous reconstruction of the Gondwana margin. (Adapted from De Wit 1977, 1981; Sporli 1987.) Names of western Antarctic microplates are from De Wit (1981). The Jurassic positions of continents and microplates were essentially the same.**

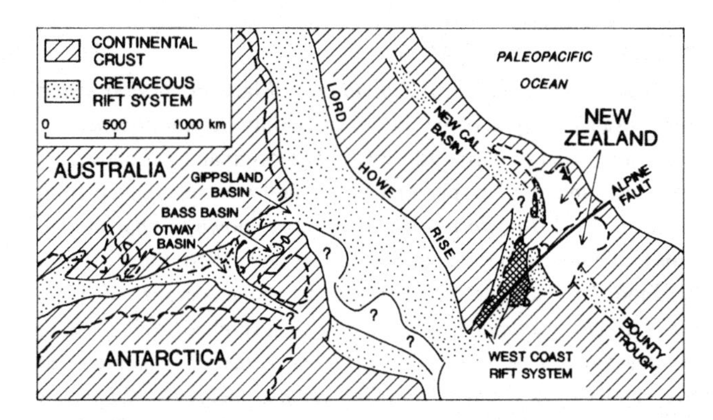


**Figure 7.8. Cretaceous rift systems in the New Zealand sector of Gondwana. (Adapted from Laird 1981; Sporli 1987.) For structural detail of New Zealand, see Figure 7.3.**

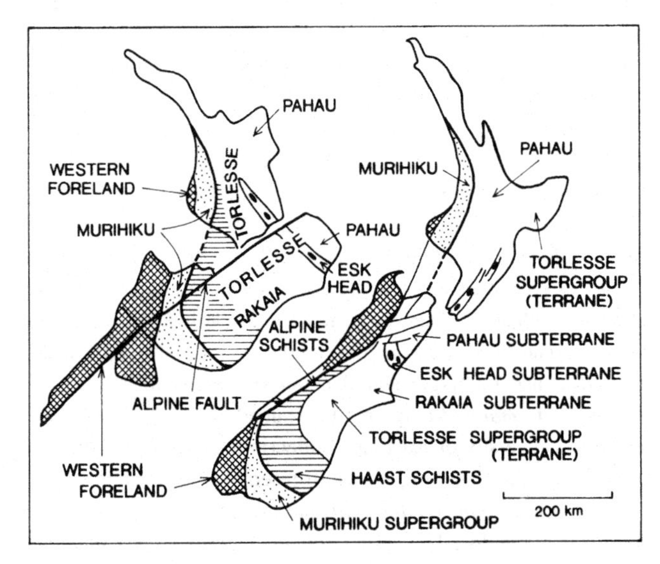


**Figure 7.9. Present-day New Zealand before and after dislocation on Alpine fault. (Adapted from Bradshaw et al. 1981; Sporli 1987.)**

suggested age correlations [note that stage boundaries, based on Helby et al. (1988, Figs. 2 and 23), are highly provisional] are based on the dinoflagellate zonations of Helby et al. (1987) and Davey (1987). The former was developed from a study of Australian assemblages (mostly from the west and north), with some data from Papua New Guinea, the latter from the Strickland River gorge of Papua New Guinea. The three schemes produce significantly different correlations for the Heterian–Puaroan interval.

## Paleogeography

The oldest rocks of the New Zealand microcontinent (Sporli 1987) are today confined mostly to Fiordland and Nelson. They are continental clastics, limestones, basaltic and andesitic assemblages, and volcaniclastic sediments of Late Precambrian to Early Devonian age (Adams 1975; Cooper 1975, 1979; Hume 1977). The plate processes that resulted in their accumulation and consolidation are obscure, but generally are referred to as the Tuhua orogeny, a multiphase event with episodes of deformation in the (?)Late Precambrian, Late Cambrian, Silurian–Devonian, and latest Devonian–Carboniferous. During the later episodes, these assemblages were intruded by granitic plutons (Sporli 1987). They form the western foreland, part of Gondwana (Figure 7.7) against which the younger basement rocks of New Zealand (Murihiku and Torlesse Supergroups) accumulated.

Murihiku and Torlesse rocks (locally Carboniferous, mostly Permian to Early Cretaceous) were consolidated, deformed, and metamorphosed during the Rangitata orogeny, with episodes of deformation in the Late Triassic producing the Haast Schist (Bradshaw et al. 1981), and in the Early Cretaceous producing a New Zealand–wide unconformity (Sporli 1987). An extentional regime commenced in the Early Cretaceous (Figure 7.8), with rifting between the New Zealand microcontinent and Australia and Antarctica. With increasing separation between Australia and Antarctica, the presently active Pacific/Indian convergent transcurrent plate boundary, of which the Alpine Fault is the surface expression in New Zealand (Figures 7.8 and 7.9), was established. Current thinking favors a recent time frame for much of the movement on the Alpine Fault and development of New Zealand's present configuration (Figure 7.9). Present-day strike-slip and dip-slip movements on the Alpine Fault are 40 mm/year and 20 mm/year, respectively (Sporli 1987). Cutten (1979) suggests a 420-km offset on the fault in the past 11 m.y. and strong uplift on its eastern side over the past 1.8 m.y. The total displacement may be as great as 1,300 km (Sporli 1987).

### References

Adams, C. J. (1975). Discovery of Precambrian rocks in New Zealand, age relations of the Greenland Group and Constant Gneiss, West Coast, South Island. *Earth Planet. Sci. Lett., 281,* 98–104.

Andrews, P. B. Speden, I. G., & Bradshaw, J. D. (1976). Lithological and paleontological content of the Carboniferous-Jurassic Canterbury Suite, South Island, New Zealand. *N.Z. J. Geol. Geophys., 19,* 179–89.

Arber, A. E. N. (1917). *The Earlier Mesozoic Floras of New Zealand.* N.Z. Geol. Surv. Paleont. Bull. No. 6.

Arkell, W. J. (1953). Two Jurassic ammonites from South Island, New Zealand, and a note on the Pacific Ocean in the Jurassic. *N.Z. J. Sci. Technol. B, 35,* 259–64.

(1956). *Jurassic Geology of the World.* London: Oliver & Boyd.

Bartrum, J. A. (1935). Metamorphic rocks and albite-rich igneous rocks from Jurassic conglomerates at Kawhia. *Trans. R. Soc. N.Z., 65,* 95–107.

Bradshaw, J. D., Adams, C. J., & Andrews, P. B. (1981). Carboniferous to Cretaceous on the Pacific margin of Gondwana: the Rangitata phase of New Zealand. In M. N. Cresserll & P. Vella (eds.), *Gondwana Five.* Rotterdam: Balkema.

Bradshaw, M. T., Yeates, A. N., Beynon, R. M., Brakel, A. T., Langford, R. P., Toterdell, J. M., & Yeung, M. (1988). Palaeogeographic evolution of the North West Shelf Region. In P. G. Purcell (ed.), *The North West Shelf of Australia* (pp. 29–54). Proc. Petrol. Explor. Soc. Australia Symp., Perth, 1988.

Bradshaw, M. T., & Yeung, M. (in press). *Palaeogeographic Atlas of Australia. Volume 8: Jurassic.* Canberra: Bur. Mine. Res.

Campbell, J. D. (1956). The Otapirian Stage of the Triassic System of New Zealand, Part 2. *Trans. R. Soc. N.Z., 84,* 45–50.

Campbell, J. D., & Coombs, D. S. (1966). Murihiku Supergroup (Triassic–Jurassic) of Southland and south Otago. *N.Z. J. Geol. Geophys., 9,* 393–8.

Campbell, J. D., & McKellar, I. C. (1956). The Otapirian Stage of the Triassic System of New Zealand, Part 1. *Trans. R. Soc. N.Z., 83,* 695–704.

Challilnor, A. B. (1977a). Proposal to redefine the Puaroan Stage of the New Zealand Jurassic System. *N.Z. J. Geol. Geophys., 20,* 17–46.

(1977b). New Lower or Middle Jurassic belemnite from southwest Auckland. *N.Z. J. Geol. Geophys., 20,* 249–62.

(1979). The succession of *Belemnopsis* in the Heterian Stratotype, Kawhia Harbour, New Zealand. *N.Z. J. Geol. Geophys., 22,* 105–23.

(1980). Two new belemnites from the lower Ohauan (?middle Kimmeridgian) Stage, Kawhia Harbour, New Zealand. *N.Z. J. Geol. Geophys., 23,* 257–65.

(1990). A belemnite biozonation for the Jurassic–Cretaceous of Papua New Guinea and a faunal compairson with eastern Indonesia. *BMR J. Aust. Geol. Geophys., 11,* 429–47.

(1991). Revision of the belemnites of Misool and a review of the belemnites of Indonesia. *Palaeontographica, A218,* 87–164.

Challinor, A. B., & Skwarko, S. K. (1982). Jurassic belemnites from Sula Islands, Moluccana, Indonesia. *Geol. Res. Devel. Centre, Pal. Ser., 3,* 1–89.

Clarke, L. N. (1959). *The Stratigraphy of the Mesozoic Rocks of the Hauturu Area, Southwest Auckland.* Unpublished MSc thesis, University of Auckland.

Cook. P. J. (1988). *Palaeogeographic Atlas of Australia. 1: Cambrian.* Canberra: Bur. Mine. Res.

Cooper, R. A. (1975). New Zealand and South East Australia in the early Paleozoic. *N.Z. J. Geol. Geophys., 18*(7), 1–20.

(1979). Lower Paleozoic rocks of New Zealand. *J. Roy. Soc. N.Z., 1,* 29–84.

Crame, J.A. (1982). Late Jurassic inoceramid bivalves from the Antarctic Peninsula and their stratigraphic use. *Palaeontology, 25,* 555–603.

(1983). The occurrence of the Upper Jurassic bivalve *Malayomaorica malayomaorica* (Krumbeck) on the Orville Coast, Antarctic. *J. Moll. Stud., 49,* 61–76.

Crampton, J. S. (1988). Comparative taxonomy of the bivalve families Isognomonidae, Inoceramidae and Retroceramidae. *Palaeontology, 31,* 965–96.

Cutten, H. N. C. (1979). Rappahannok Group: Late Cenozoic sedimentation and tectonics contemporaneous with Alpine Fault movement. *N.Z. J. Geol. Geophys., 22*(5), 535–53.

Davey, R. J. (1987). *Palynological Zonation of the Lower Cretaceous, Upper and Uppermost Middle Jurassic in the Northwestern Papuan Basin of Papua Guinea.* Papua New Guinea Mem. 13.

De Wit, M. J. (1977). The evolution of the Scotia Arc as a key to the reconstruction of southwestern Gondwanaland. *Tectonophysics, 37,* 53–81.

(1981). Gondwana reassembly. In M. M. Cresswell & P. Vella (eds.), *Gondwana Five.* Rotterdam: Balkema.

Enay, R. (1973). Upper Jurassic (Tithonian) ammonites. In A. Hallam, *Atlas of Paleobiography* (pp. 297–307). Amsterdam: Elsevier.

Evans, P. R. (1966). Mesozoic stratigraphic palynology in Australia. *Australas. Oil Gas J., 12*(6), 58–63.

Fleming, C. A. (1953). Lower Jurassic fossils from Taneatua, Bay of Plenty, New Zealand. *N.Z. J. Sci. Technol., B, 35,* 129–33.

(1958). Upper Jurassic fossils and hydrocarbon traces from the Cheviot Hills, north Canterbury. *N.Z. J. Geol. Geophys., 1,* 375–94.

(1960). The Upper Jurassic sequence at Kawhia, New Zealand, with reference to the ages of some Tethyan guide fossils. *Rep. 21st Intl. Geol. Congr., 21,* 264–9.

(1987). *New Zealand Mesozoic Bivalves of the Superfamily Trigoniacea.* N.Z. Geol. Surv. Paleont. Bull. 53.

Fleming, C. A., & Kear, D. (1960). The Jurassic sequence at Kawhia Harbour, New Zealand. N.Z. Geol. Surv. Bull. 67.

Grant-Mackie, J. A. (1959). Hokonui stratigraphy of the Awakino-Mahoenui area, southwest Auckland. *N.Z. J. Geol. Geophys., 2,* 755–87.

Haast, J. (1872). *Report on the Geology of the Malvern Hills, Canterbury.* N.Z. Geol. Surv. Rep. Geol. Explor. 1871–2, No. 7.

Haq, B. U., Hardenbol, J., & Vail, P. R. (1987). Chronology of fluctuating sea levels since the Triassic. *Science, 235,* 1156–67.

Harland, W. B., Cox, A. V., Llewellyn, P. G. Pickton, C. A. G., Smith, A. G., & Walters, R. (1982). *A Geologic Time Scale.* Cambridge University Press.

Harris, T. M. (1962). The occurrence of the fructification *Carnoconites* in New Zealand. *Trans. R. Soc. N.Z., Geol., 1,* 17–27.

Hasibuan, F. (1980). *The Mesozoic Sequence in the Waikawau Area, Southwest Auckland, New Zealand.* Unpublished MSc thesis, University of Auckland Library.

Helby, R., Morgan, R., & Partridge, A. D. (1987). A palynological zonation of the Australian Mesozoic. In P. A. Jell (ed.), *Studies in Australian Mesozoic Palynology.* Association of Australasian Palaeontologists, Sydney, Memoir 4.

Helby, R., Wilson, G. J., & Grant-Mackie, J. A. (1988). A preliminary biostratigraphic study of Middle to Late Jurassic dinoflagellate assemblages from Kawhia, New Zealand. *Mem. Assoc. Australas. Palaeontol., 5,* 125–66.

Hornibrook, N. de B. (1953). Jurassic foraminifera from New Zealand. *Trans. R. Soc. N.Z., 81,* 375–8.

Hudson, N. (1983). *Stratigraphy of the Ururoan, Temaikan and Heterian Stages; Kawhia Harbour to Awakino Gorge, Southwest Auckland.* Unpublished MSc thesis, University of Auckland.

Hudson, N., Grant-Mackie, J. A., & Helby, R. (1987). Closure of the New Zealand "Middle Jurassic Hiatus"? *Search, 18,* 146–8.

Hume, B. J. (1977). The relation between Charleston Metamorphic Group and the Greenland Group in the Central Paparoa Range, South Island, New Zealand. *J. Roy. Soc. N.Z.*, *7*(3), 379–92.

Jenkins, D. G., & Jenkins, T. B. H. (1971). First diagnostic Carboniferous fossils from New Zealand. *Nature, 233(5315)*, 117–18.

Johnston, M. R., Raine, J. I., & Watters, W. A. (1987). Drumduan Group of East Nelson, New Zealand: plant-bearing Jurassic arc rocks metamorphosed during terrane interaction. *J. R. Soc. N.Z.*, *17*(3), 275–301.

Kear, D. (1966). *Sheet N55 – Te Akau (1st ed.), "Geological Map of New Zealand 1:63, 360."* Wellington, N.Z.: D.S.I.R.

Krishna, J. Kumar, S., & Singh, I. B. (1982). Ammonoid stratigraphy of the Spiti shale (Upper Jurassic), Thethya Himalaya, India. *N. Jb. Geol. Palaont. Mh.*, *10*, 580–92.

Laird, M. G. (1981). The late Mesozoic fragmentation of the New Zealand fragment of Gondwana. In M. M. Cresswell & P. Vella (eds.), *Gondwana Five*. Rotterdam: Balkema.

Li, X. & Grant-Mackie, J. A. (1988). Upper Jurassic and Lower Cretaceous *Buchia* (Bivalvia) from Southern Tibet, and some wider considerations. *Alcheringa*, *12*, 249–68.

MacDonald, H. A. H. (1954). The petrography of some Jurassic conglomerates at Kawhia, New Zealand. *Trans. R. Soc. N.Z.*, *82*, 223–30.

MacFarlan, D. A. B. (1975). *Mesozoic Stratigraphy of the Marakopa Area*. Unpublished MSc thesis, University of Auckland Library.

McKellar, I. C. (1969). *Sheet S169 – Winton (1st ed.). "Geological Map of New Zealand, 1:63,360."* Wellington, N.Z.: D.S.I.R.

(1978). "Jurassic, NW Otago." In R. P. Suggate (ed.), *The Geology of New Zealand*. N.Z. Geol. Surv., D.S.I.R. Publ., N.Z. Govt. Printer.

MacKinnon, T. C. (1983). Origin of the Torlesse terrane and coeval rocks, South Island, New Zealand. *Geol. Soc. Am. Bull.*, *94*, 967–85.

McQueen, D. R. (1955). Revision of supposed Jurassic angiosperms from New Zealand. *Nature*, *175*, 177.

Martin, K. R. (1967). *The Mesozoic Sequence of Southwest Kawhia, New Zealand*. Unpublished MSc thesis, University of Auckland.

Marwick, J. (1951). Series and stage divisions of New Zealand Triassic and Jurassic rocks. *N.Z. J. Sci. Technol.*, *B*, *32*, 8–10.

(1953). *Divisions and Faunas of the Hokonui System (Triassic and Jurassic)*. N.Z. Geol. Surv. Paleont. Bull. 21.

Maxwell, P. A. (1964). *Structural Geology and Pre-Quaternary Stratigraphy of the Kaiwara District, North Canterbury, New Zealand*. Unpublished MSc thesis, University of Canterbury, Christchurch, N.Z.

Meesook, A. (1989). *Upper Jurassic Sequence of the Kawhia Harbour to Te Anga Area, Southwest Auckland*. Unpublished Msc thesis, University of Auckland Library.

Mutch, A. R. (1964). *Sheet S159 – Morley (1st ed.), "Geological Map of New Zealand 1:63, 360."* Wellington, N.Z.: D.S.I.R.

Norris, G. (1964). *Report on Spore and Pollen Assemblages from the Puaroan B and Huriwai Formation of Port Waikato, Southwest Auckland*. N.Z. Geol. Surv. Rep. File N51.

(1968). Plant fossils from the Hawk's Crag breccia, S.W. Nelson, New Zealand. *N.Z. J. Geol. Geophys.*, *11*, 312–44.

Player, R. A. (1958). *The Geology of North Kawhia*. Unpublished MSc thesis, University of Auckland.

Purser, B. H. (1961). *Geology of the Port Waikato Region (Onewhero Sheet N51)*. N.Z. Geol. Surv. Bull. 69.

Sato, T., Westermann, G. E. G., Skwarko, S. K., & Hasibuan, F. (1978). Jurassic biostratigraphy of the Sula Islands, Indonesia. *Bull. Geol. Surv. Indonesia*, *4*, 1–28.

Spath, L. F. (1923). On ammonites from New Zealand. *Quart. J. Geol. Soc. London*, *79*, 286–308.

Speden, I. G. (1960). The Jurassic age of some supposedly Triassic mollusca described by Wilkens (1927) from Mt. St. Mary. *N.Z. J. Geol. Geophys.*, *3*, 510–23.

(1961). *Sheet S184 – Papatowai (1st ed.), "Geological Map of New Zealand, 1:63,360."* Wellington, N.Z.: D.S.I.R.

(1971). *Geology of Papatowai Subdivision, Southeast Otago*. N.Z. Geol. Surv. Bull. 81.

Speden, I. G., & McKellar, I. C. (1958). The occurrence of Aratauran beds south of Nugget Point, south Otago, New Zealand. *N.Z. J. Geol. Geophys.*, *1*, 647–52.

Sporli, K. B. (1987). Development of the New Zealand microcontinent. In J. W. H. Monger & J. Fracheteau (eds.), *Circum-Pacific Orogenic Belts and Evolution of the Pacific Ocean Basin* (pp. 115–32). Am. Geophysical Union, Geodynam. Ser. 18.

Sporli, K. B., & Aita, Y. (1988). *Field Guide to Waipapa Basement Rocks, Kawakawa Bay*. Geol. Soc. N.Z. Misc. Publ. 39.

Stevens, G. R. (1965). *The Jurassic and Cretaceous Belemnites of New Zealand and a Review of the Jurassic and Cretaceous Belemnites of the Indo-Pacific Region*. N.Z. Geol. Surv. Pal. Bull. 36.

(1978). Jurassic paleontology. In R. P. Suggate (ed.), *The Geology of New Zealand* (pp. 215–28). N.Z. Geol. Surv., D.S.I.R. Publ., N.Z. Govt. Printer.

(1984). A revision of the Lytoceratinae (subclass Ammonoidea) including *Lytoceras taharoaense* n. sp., Upper Jurassic, New Zealand. *N.Z. J. Geol. Geophys.*, *28*, 153–85.

Stevens, G. R., & Speden, I. G. (1978). New Zealand. In M. Moullade & A. E. M. Nairn (eds.), *The Phanerozoic Geology of the World. II: The Mesozoic, A* (pp. 263–8). Amsterdam: Elsevier.

Struckmeyer, H. I. M., Yeung, M., & Bradshaw, M. T. (1990). Mesozoic palaeogeography of the northern margin of the Australian Plate and its implications for hydrocarbon exploration. In G. J. & Z. Carman (eds.), *Proceedings of the 1st PNG Petroleum Convention* (pp. 137–52). Port Morseby, 1990.

Suggate, R. P. (1961). Rock stratigraphic names for the South Island schists and undifferentiated sediments of the New Zealand geosyncline. *N.Z. J. Geol. Geophys.*, *4*, 392–9.

(ed.). (1978). *The Geology of New Zealand*. N.Z. Geol. Surv., D.S.I.R. Publ., N.Z. Govt. Printer.

Thomson, M. R. A. (1982). Late Jurassic fossils from Low Island, South Shetland Islands. *Brit. Ant. Surv. Bull.*, *56*, 25–35.

Veevers, J. J. (ed.) (1984). *Phanerozoic Earth History of Australia*. Oxford: Clarendon Press.

Verma, H. M., & Westermann, G. E. G. (1973). The Tithonian (Jurassic) ammonite fauna and stratigraphy of Sierra Catorce, San Luis Potosí, Mexico. *Bull. Amer. Paleont.*, *63*, 103–320.

Warren, G. (1978). Stratigraphy, mid-Canterbury. In R. P. Suggate (ed.), *The Geology of New Zealand* (pp. 240–1). N.Z. Geol. Surv., D.S.I.R. Publ., N.Z. Govt. Printer.

Westermann, G. E. G., & Callomon, J. H. (1988). The Macrocephalitinae and associated Bathonian and Early Callovian (Jurassic) ammonoids of the Sula Islands and New Guinea. *Palaeontographica*, *A*, *203*, 1–90.

Westermann, G. E. G., & Hudson, N. (1990). First find of Eurycephalitinae (Jurassic Ammonitina) from New Zealand and biogeographic implications. *J. Paleontol. 65*, 689–93.

Wood, B. L. (1956). *The Geology of the Gore Subdivision (S170)*. N.Z. Geol. Surv. Bull. 53.

# 8 Indonesia and Papua New Guinea

R. SUKAMTO and G. E. G. WESTERMANN

## INDONESIA[1]

### Introduction: tectonics and Jurassic outcrops

The Indonesian Archipelago represents a convergence between three megaplates: Eurasian, Indoaustralian, and Pacific. The three megaplates have moved against each other since Late Triassic time and are at present still moving, so that the Indonesian region belongs to a highly mobile zone. Consequently, this region has one of the most complex geomorphologic and geologic development patterns in the world. The archipelago at present has two stable plates, Sunda and Sahul, which are intertwined by island arcs, magmatic arcs, and subduction trenches, whereby lateral faults are interrupting the regular form of the arcs (Figure 8.1).

This complex region is the result of combined development of arc-trench systems, accretions, and collisions initiated by movement of the megaplates. The three tectonic regimes are represented by accreted terranes, arc–microplate collisions, arc–continent collisions, and arc–arc collisions. Northwestern Indonesia, including west Sulawesi, is part of the Eurasian plate, whereas southern and eastern Indonesia and all of Irianjaya belong to the Indoaustralian plate.

The basement of the sedimentary basins consists of continental and oceanic terranes of incompletely known ages. Radiometric ages for metamorphic and mafic basement rocks are limited and become unreliable for older epochs because of alteration and repeated metamorphism by tectonic events. The most acceptable dating for those rocks is therefore based on their stratic positions relative to fossil-bearing beds. Biostratigraphy is therefore very important in terrane analysis.

Figure 8.2 shows the Jurassic outcrops of Indonesia, numbered as in the descriptions and in Figure 8.3. The last figure depicts the distinctive Jurassic features from each area, often tentative, in rock composition, environment of deposition, and stratic position.

[1] By R. Sukamto and G. E. G. Westermann.

A comprehensive bibliography on Indonesian invertebrate fossils has recently been published by Skwarko (1982).

#### *Irianjaya (1 and 2)*

Nonmarine sediments are overlain by marine sequences totaling more than 2,500 m in thickness. The sediments constitute the Tipuma Formation at the bottom and the Kembelangan Group at the top.

The Tipuma Formation (1, ca. 500 m), nonmarine (terrestrial and fluviatile), consists of fine and coarse-grained clastics deposited in an oxidizing environment. It is unfossiliferous and, to some extent, overlies the Permo-Carboniferous shallow-marine Aifam Group (Pigram and Panggabean 1983; Pieters, Hakim, and Atmawinata 1985; Dow et al. 1986; Robinson et al. 1987).

The Kembelangan Group (>2,000 m), marine, follows disconformably and consists of varied shales and sandstones (van Bemmelen, 1949; Pieters, Hartono, and Amri 1982; Pigram and Sukanta 1982; Pigram and Panggabean 1983; Dow, Trail, and Hamonangan 1984; Pieters et al. 1985; Robinson et al. 1987). Sedimentation was sublittoral, on open shelf, extending from the Bajocian to the Early Cretaceous. The following Jurassic formations are distinguished, in descending order:

1. *Kopai Formation* (2a): quartz sandstones, siltstones, and mudstones, with intercalations of conglomerate and carbonate
2. *Woniwogi sandstone* (2b): orthoquartzites, with minor siltstones and mudstones
3. *Tamrau Formation* (2c): shales, siltstones, and slates, with intercalations of quartzwacke and carbonate; concordant under Cretaceous sediments (Piniya mudstone, 2d; Jass Formation, 2e; Ekmai sandstone, 2f)

The Jurassic sediments occur throughout the Central Mountains range (extending to western Papua New Guinea, as described later), the Neck, and the Bird's Head. Their early Bajocian to early Callovian, Oxfordian, and Tithonian ammonites came

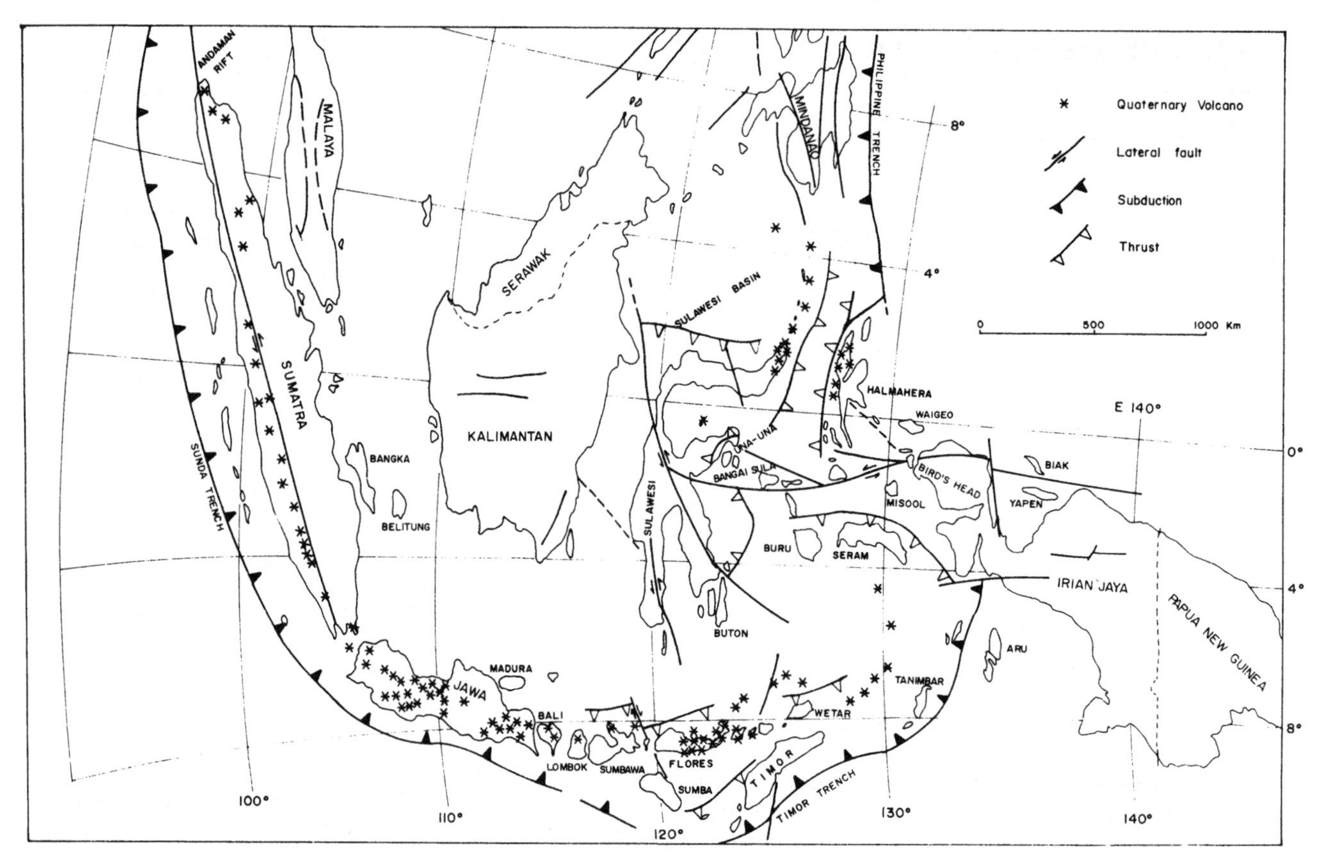


Figure 8.1. Structural patterns of Indonesia and surroundings.

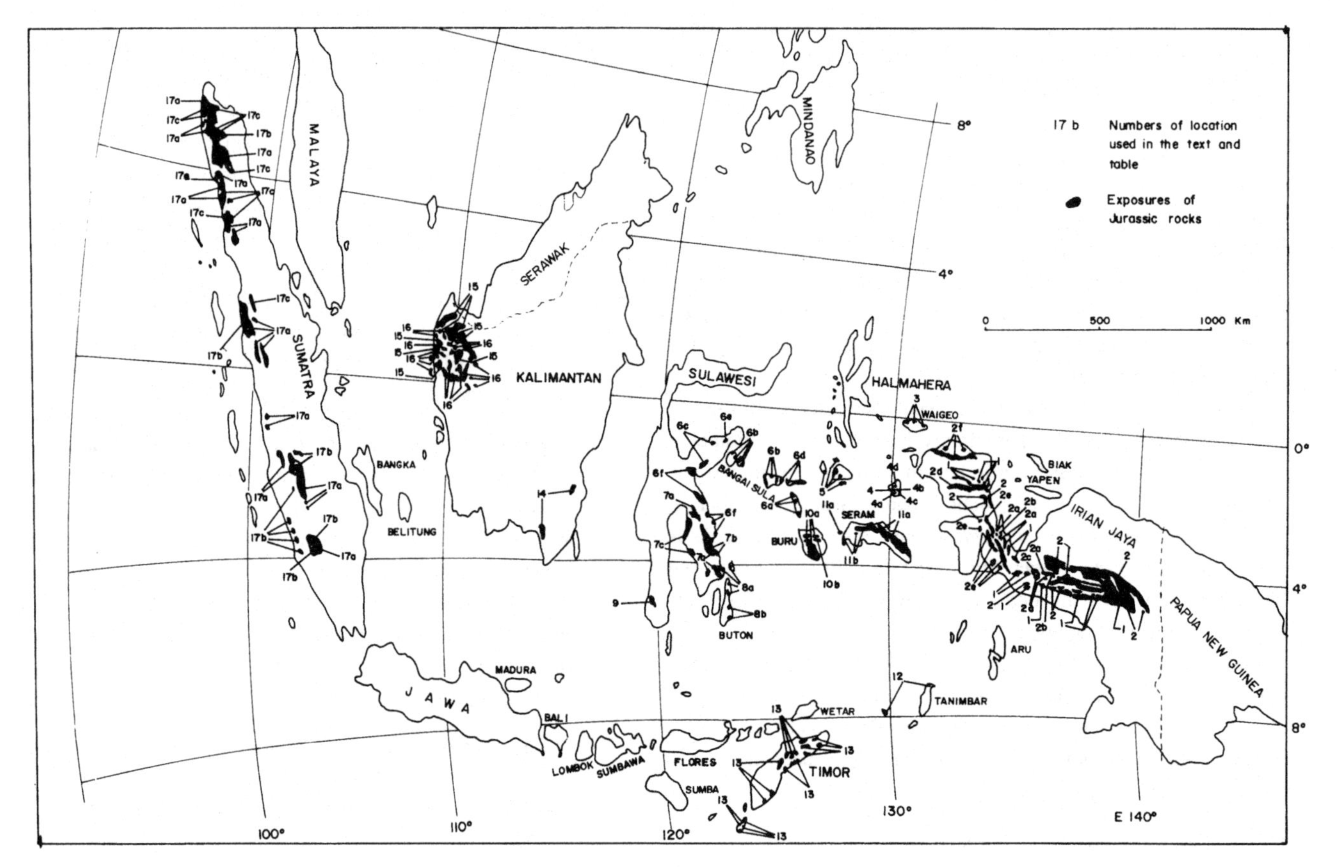


Figure 8.2. Distribution of Jurassic outcrops in Indonesia.

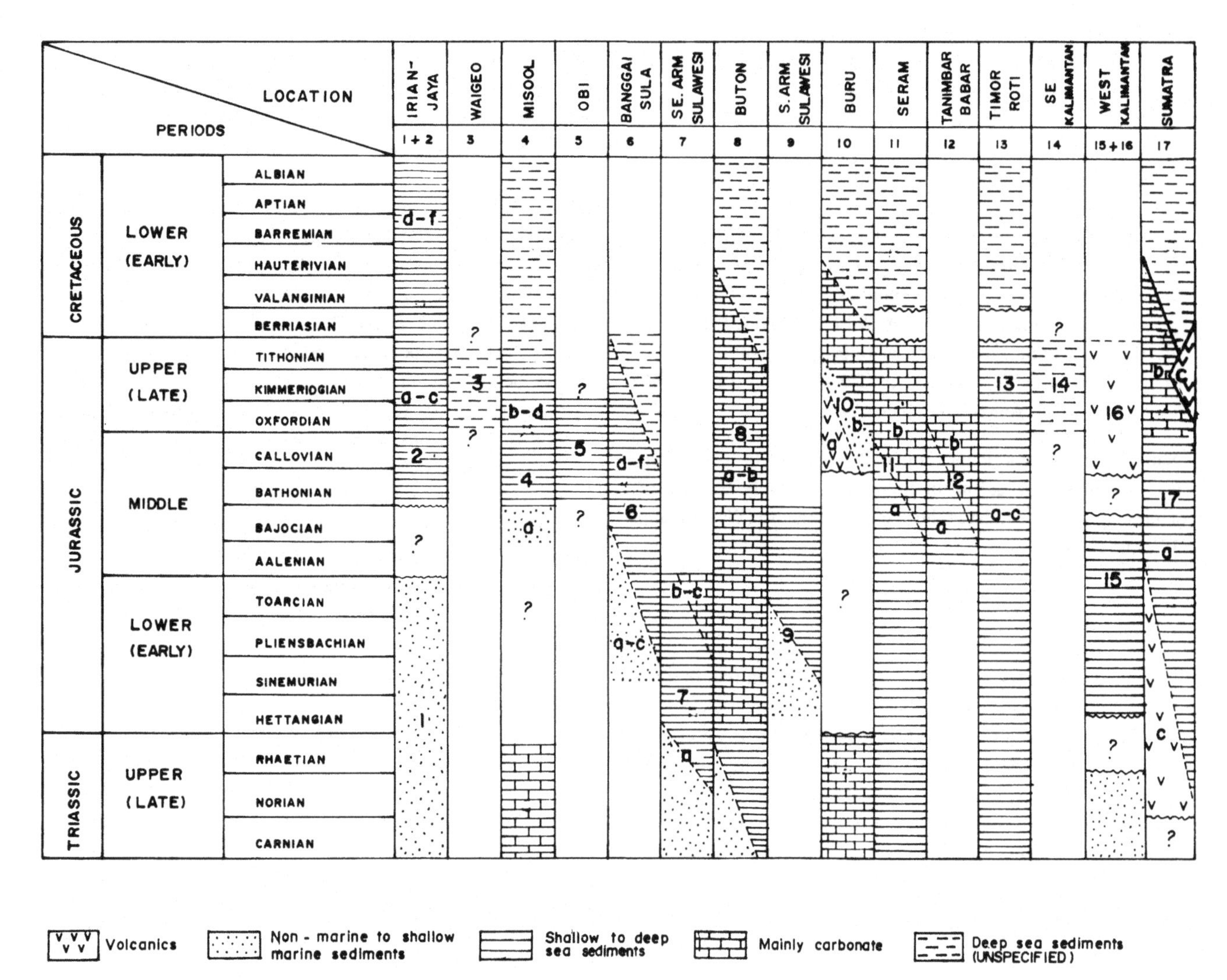


**Figure 8.3. Stratigraphic columns of Indonesian Jurassic sequences.**

almost exclusively from surface collections (often provided by natives). They were described by Boehm (1913), Etheridge (1890), Gerth (1927, 1965), Visser and Hermes (1962), and Westermann and Getty (1970), reviewed by Sato (1975), most recently recorded by Helmcke, Barthel, and Hillebrandt (1978), and partly revised by Westermann and Callomon (1988) (see also the descriptions for Papua New Guinea). The belemnite succession was described by Challinor (1989a, 1991).

Outcrop conditions are extremely poor in the mountainous, heavily forested, and steeply eroded Central Mountains, where more recent field investigations by S. Susuki (Kyoto University) and G. E. G. Westermann have again yielded only transported material in the famous Kemaboe Valley (Westermann and Getty 1970; G. E. G. Westermann personal communication). Across the border, in Papua New Guinea, however, small stratigraphic sequences have recently been investigated and collected, as discussed later. The classical ammonite faunas of the Neck and Bird's Head (Vogelkop) in western Irianjaya (Boehm 1913) have not yet been reinvestigated in the field. The following ammonite assemblages, mostly known from the Sula Islands, appear to be present in descending sequences:

1. *Haplophylloceras strigile* (Boehm), *Blanfordiceras wallichi* Gerth; Tithonian–Berriasian boundary.
2. *Kossmatia desmidoptycha* Uhlig, *Paraboliceras* cf. *polysphinctum* Uhlig; Lower–Middle Tithonian.

3a. *"Pseudoparaboliceras"* [nom. nud.] gr. *sularum* (Boehm), *"P." "aramaraii"* Gerth [nom. nud.], *"P." novaguinense* (Gerth) [nom. nud.] [= ?*"P." boehmi* (Kruizinga)]; Upper Oxfordian–Kimmeridgian(?).

3b. *Mayaites (Epimayaites)* gr. *palmarum* (Boehm), *Perisphinctes (Kranaosphinctes)* gr. *burui* and *moluccanum* Boehm; Middle Oxfordian.

4. ?*Mayaites (Epimayaites)* sp., *Peltoceratoides* sp.; Lower Oxfordian.
5. *Macrocephalites keeuwensis* Boehm, *M. etheridgei* (Spath), *M. wichmanni* (Boehm MS) Spath; Upper Bathonian–Lower Callovian.

6. *Satoceras satoi* West. & Call., *S. boehmi* West. & Getty, *Tulites godohensis* (Boehm), *Cadomites* cf. *rectelobatus* (Hauer); Lower–Middle Bathonian.
7. *Praetulites kruizingai* West.; Upper Bajocian–Lower Bathonian.
8. *Leptosphinctes (Cobbanites?)* aff. *engleri* Freb.; Upper Bajocian.
9. *Stephanoceras etheridgei* (Gerth), *S.* gr. *humphriesianum* (Sow.), *Teloceras indicum* (Kruiz.), *Sphaeroceras (Chondroceras) boehmi* West., *Irianites moermanni* (Kruiz.), *Labyrinthoceras? costidensum* (West. & Getty), *Pseudotoites* sp., middle Lower to basal Upper Bajocian.
10. *Fontannesia kiliani* (Kruiz.), *''Docidoceras''* cf. *longalvum* (Vacek); Aalenian–Bajocian boundary.

Note that no Kimmeridgian ammonites have so far been identified from Irianjaya and all of Indonesia, probably because of biofacies conditions similar to those of the Himalaya (Westermann and Wang 1988).

*Waigeo (3)*

The deep-sea sediments of Tanjung Formation (ca. 400 m) in eastern and western Waigeo Island consist of graywackes, siltstones, red shales, and cherts, but their Jurassic age is uncertain. This sequence is in fault-contact with ophiolites and is unconformably overlain by Paleocene sediments. *Microglobigerina* and *Calpionella* suggest a Late Jurassic age and a deep-sea environment (P. T. Shell 1977, cited by Supriatna, Apandi, and Simandjuntak 1984a).

*Misool (4)*

Misool Island contains a rather complete and fossiliferous Mesozoic sequence. The Fageo Group (ca. 3,250 m) comprises four formations:

1. *Yefbie Formation* (4a): fine clastics, with coarse clastics at the base; belemnites range from the Bathonian–early Callovian to the Neocomian (Challinor 1988, 1991), and a Liassic hiatus is indicated (Skwarko 1981); the invertebrate fauna is currently being described by F. Hasibuan, as discussed later.
2. *Demu Formation* (4b): mainly sandy carbonates and shales.
3. *Lelinta Formation* (4c): a series of clastics and carbonates, intergrading into the Facet Formation.
4. *Facet Formation* (4d): mainly carbonates, with some cherts and shales, and tuffs with Tithonian to Cenomanian fossils.

The Fageo Group lies unconformably on Triassic carbonates of the Bogal Formation and, in turn, is partially overlain by Cretaceous sediments. The sequence resembles the Kemberlangan Group of Irianjaya (Skwarko 1981; Pigram et al. 1982; Rusmana, Hartono, and Pigram 1983; Challinor 1988). The Fageo depositional environment was at first fluviatile, followed by a shallow-marine environment for most of the Jurassic, and bathyal at the Jurassic–Cretaceous boundary.

During a recent visit, one of us (G.W.) examined the new ammonite collections assembled by Hasibuan in the Misool Archipelago. Above a basal conglomerate on the Upper Triassic (dated by Hasibuan) occur what appear to be Early Aalenian *Bredyia* sp. and Early Bajocian *Fontannesia kiliani* (Kruiz.); other sections at similar levels have yielded poorly preserved Sphaeroceratidae, together with a small *Witchellia* (or ?*Fontannesia*) and, elsewhere, *Pseudotoites* gr. *transatlanticus* (Torn.) & *argentinus* Arkell, mainly indicating the Early Bajocian Ovalis and Laeviuscula Standard Zones. Hasibuan also pointed out that Soergel (1913) had described a Toarcian harpoceratid fauna from the archipelago. It consists of Late Toarcian or latest Toarcian grammoceratines and *Haugia* sp. The only other Jurassic ammonite faunas known from the archipelago consist of Tithonian (and ?Kimmeridgian) *Kossmatia* gr. *tenuistriata* (Gray) Uhlig, *Paraboliceras* sp., and *Aulacosphinctoides* sp., collected and recognized by Hasibuan. The marine upper Oxfordian and Kimmeridgian are indicated by the byssate bivalves *Retroceramus subhaasti* (Wandel) and *Malayomaorica malayomaorica* Krumbeck, respectively (Wandel 1936).

*Obi (5)*

Jurassic sediments are exposed in the northwestern part of the island and on the islands of Obilatu and Gommu: the Leleobasso Formation (ca. 500 m), consisting of slightly metamorphic sandstones, claystones, and shales (Sudana and Yassin 1983). No older rocks are known, and Oligocene-Miocene sediments lie unconformably on the Jurassic. Brouwer (1924a) reported *''Phylloceras, Stephanoceras* and *Macrocephalites''* indicating Bajocian to Late Bathonian–Early Callovian and shallow-marine conditions.

*Bangai-Sula platform (6)*

A thick Jurassic sedimentary sequence (1,000–2,500 m) is extensively exposed in the area from eastern Sulawesi to Mangole in the eastern Sula Islands and contains the richest, best-preserved, and best-known Jurassic molluscan fauna of Indonesia. The lower part of the sequence comprises the terrestrial to shallow-marine Kabauw (6a, ca. 200 m), Bobong (6b, ca. 200 m) and Nanaka (6c, 800–2,000 m) Formations, consisting mainly of coarse clastics with some coal and of fine-grained clastics. These formations extend from Sanana Island (south Sula Archipelago), in the east, and to the east arm of Sulawesi, in the west. They are overlain by the Buya (6d), Nambo (6e), and Tetambahu (6f) Formations, deposited in shallow to deep-sea environments. The Buya Formation is highly fossiliferous, as in the classic ammonite faunas on the islands of Mangole, Taliabu, and Peleng; the Nambo (ca. 300 m) and Tetambahu (ca. 500 m) Formations, on eastern Sulawesi, consist mainly of carbonates, partly with chert intercalations (Sukamto 1975; Simandjuntak, Surono, and Supandjono 1982b; Surono, Simandjuntak, and Situmorang 1983; Rusmana,

Koswara, and Simandjuntak 1984; Surono and Sukarna 1985; Supandjono and Haryono 1986).

The Buya Formation (6d; 1,500–2,500 m) consists of shales, with intercalations of sandstone, limestone, marl, and conglomerate in the lower part; the fossil discoveries began as early as 1705, were described especially by Boehm (1904–12), Jaworsky (1933), and Kruizinga (1926), and more recently have been reviewed by Westermann and Getty (1970), Sato (1975), and Westermann (1981). G. E. G. Westermann, T. Sato, S. Skwarko, and F. Hasibuan made 14 stream traverses in the Sula Islands in 1976 and visited most of the classical Jurassic localities (Sato et al. 1978). They concluded that the Buya Formation ranges from the late Toarcian to the late Tithonian, with the Aalenian and Bathonian missing or poorly represented. Westermann and Callomon (1988), in a recent monographic description of the Middle Jurassic ammonite faunas collected in situ during the 1976 expedition, however, redated most of the alleged early Callovian ammonite faunas as Bathonian. A Middle Bathonian–Early Callovian sequence of five or six ammonite associations was worked out for the Sula Islands. Several *Macrocephalites* associations were also observed in westernmost Papua New Guinea. They are therefore here distinguished as the first Jurassic ammonite assemblage zones in the region.

*Upper Jurassic.* The latest Tithonian is represented by the *Blanfordiceras*-Himalayitidae assemblage (Boehm 1904–12), but the localities have not been reinvestigated. Locally occur also *Uhligites* and *Bochianites* spp., resembling the fauna of the Upper Spiti Shales of the Himalayas (Uhlig 1903–10; Arkell 1956; Krishna 1983). Lower and Middle Tithonian outcrops are unknown, and the *Kossmatia* described by Kruizinga (1926) probably does not belong to that Tithonian genus (Arkell 1956).

Kimmeridgian mudstones are present only in *Retroceramus-Belemnopsis* biofacies, again resembling the Spiti Shale (Westermann and Wang 1988), and the only supposedly Kimmeridgian ammonite recorded, *''Idoceras'' mihanum* Boehm, is a late Bajocian leptosphinctid, probably *Caumontisphinctes* (Sato et al. 1978, loc. 6B; Westermann and Callomon 1988, p. 5).

Oxfordian ammonites have been richly found in situ at Boehm's Wai Galo locality on Taliabu (Sato et al. 1978), and two associations are present. The upper one is Middle Oxfordian and yields *Mayaites (Epimayaites)* spp., together with *Persphinctes (Kranaosphinctes)* spp. and *Retroceramus galoi* (Boehm); the lower one is Early Oxfordian and contains similar *Epimayaites* spp., together with the dimorphic pair of *Peltoceratoides (Peltomorphites) tjapalului* (Boehm) ♀ and *P. (Parawedekindia)* aff. *arduennensis* (d'Orb.) ♂. *Calliphylloceras malayanum* (Boehm) occurs throughout.

*Middle Jurassic.* Evidence for Late and Middle Callovian is missing. The Early Callovian and middle part of the Bajocian are well recorded on Taliabu Island in the Sula Archipelago, and the type sections of the three new macrocephalitid assemblage zones and subzones (formerly associations) and one association are in the Keeuw area along the upper Wai Miha, with tributaries Kalepu and Betino, and at Tikong (Sato et al. 1978, Fig. 4; Westermann and Callomon 1988, Fig. 1). They are in descending order:

1. *Macrocephalites keeuwensis* Association: typically developed at loc. 4, with *M. keeuwensis* Boehm, *Oxycerites sulaensis* West. & Call. and *Choffatia* cf. aff. *furcula* Neum.; probably late Early Callovian.
2. *Macrocephalites apertus* Assemblage Zone (for *apertus-mantataranus* Association): locality 3b, with *M. apertus* (Spath) and *M. mantataranus* Boehm; also rare *Oxycerites* cf. *sulaensis* West. & Call. and *O.* cf. *tenuistriatus* (De Gross.) and a single *Xenocephalites grantmackiei* West. & Hudson (1991); Late Bathonian.
3. *Macrocephalites bifurcatus* Assemblage Zone: type locality 3f, with *M. bifurcatus* Boehm, *M.* cf. *etheridgei* (Spath); also *Bullatimorphites ymir* (Oppel) and *Cadomites* cf. *rectelobatus* (Hauer); rare *Oxycerites* ex gr. *biflexuosus* (d'Orb.)–*costatus* (Roem.); (late) Middle Bathonian:
   (a) *Cadomites ymir* Assemblage Subzone: with *C. ymir* (Oppel), and as for zone.
   (b) *Macrocephalites bifurcatus intermedius* Assemblage Subzone: with *M. b. bifurcatus* Boehm, and as for zone.
4. The Middle Bathonian is also represented by the holotypes of *Tulites (Rugiferites) godohensis* (Boehm) and *T.*? *(R.*?) *sofanus* (Boehm) from Keeuw, but the outcrop has probably disappeared (Sato et al. 1978).
5. *Satoceras* Association: with *Satoceras satoi* West. & Call., *S.*? *subkamptum* (Boehm), (?)*S. boehmi* (West.), and rare *Cadomites* ex gr. *rectelobatus* (Hauer); Early to Middle Bathonian.
6. *Praetulites kruizingai* level: with *P. kruizingai* West.; latest Bajocian–earliest Bathonian.
7. The Late Bajocian has been established at Wai Mbono by a small assemblage of *Caumontisphinctes* sp., *Cadomites* aff. *deslongchampsii* (d'Orb.), and *Holcophylloceras* cf. *mediterraneum* (Neum.) (Sato et al. 1978).
8. The late Early Bajocian or earliest Late Bajocian is indicated by the holotype of *Teloceras*(?) *indicum* (Kruiz.), the middle Early Bajocian by *Pseudoites* cf. *robiginosus* (Crick.) (Kruizinga 1926, pl. 6, Figs. 1a,b) and possibly by *Irianites moermanni* (Kruiz.) (see the section on Irianjaya), and a similar age by *Fontannesia klarkei* (Kruiz.) (Sato et al. 1978; Westermann and Wang 1988). All were found loose.

The Aalenian, if present, has yielded no dateable fauna.

*Lower Jurassic.* The Upper Toarcian occurs in outcrops at the Wai Menanga on Taliabu (Sato et al. 1978), with abundant shallow-water invertebrate fauna and *Hammatoceras molukkanum* Cloos. The oldest identified fossil from the Sula Islands is probably *Fuciniceras*(?) *arietiforme* (Kruiz.), found loose in north Taliabu (Arkell 1956).

The major study of belemnites collected during the expedition in 1976 by Westermann and others was made by Challinor and Skwarko (1982), supplemented by Challinor (1989b).

*Southeast arm of Sulawesi (7)*

The Meluhu Formation (7a, 750–2,000 m), consisting of partly metamorphosed shales, sandstones, and limestones, is considered Late Triassic–Early Jurassic, based on belemnites and bivalves (*Halobia, Daonella*). The sediments are terrestrial and shallow-marine, and they are associated with carbonates of the Laonti (7b, ca. 750 m) and Tamborasi (7c, ca. 1,500 m) Formations (Simandjuntak et al. 1981; Rusmana, Sukido, and Sukarna 1987). This sequence lies unconformably on Late Paleozoic metamorphics.

*Buton (8)*

The Jurassic of Buton Island is represented by the carbonates of the Ogena (8a, ca. 960 m) and Rumu (8b, ca. 150 m) Formations, deposited in shallow- to deep-sea environments. They lie conformably on Triassic flysch of the Winto Formation and are in turn conformably overlain by the Cretaceous to Paleocene carbonates of the Tobelo Formation. From the Early Jurassic Ogena Formation, *Phylloceras* sp., *Psiloceras* sp. (Hettangian), and *Arietites* sp. (Sinemurian) have tentatively been identified; from the Late Jurassic Rumu Formation, *Belemnopsis gerardi* Oppel, *B. alfurica* Boehm, and "*Aucella*" cf. *malayomaorica* (Kimmeridgian–early Tithonian), *Stomiosphaera moluccana*, and the benthic foraminifera *Trocolina* sp., *Spirillina (Involutina) liassica* Jones, and *Epistomina* sp. have also been identified (Bothe 1927; Hetzel 1936; Indonesian Gulf Oil 1972, cited by Sikumbang and Sanyoto 1981). This fauna is also in urgent need of thorough taxonomy, for stratigraphic and biogeographic reasons.

*South Sulawesi (9)*

The terrestrial to shallow-marine sandstone sequence of the Paremba Sandstone contains ammonites (e.g., reportedly the middle Liassic *Fuciniceras*), gastropods, and brachiopods of the Early and Middle Jurassic. The rocks are strongly tectonized and imbricated with Triassic and Lower Cretaceous rocks, which formed a mélange complex. The fossils have been identified as Turonian–Santonian (J. Grant-Mackie personal communication 1979; Sukamto 1986).

*Buru (10)*

At the base of the Jurassic sequence of Buru Island are the volcaniclastics of the Mefa Formation (10a, ca. 300m). It interfingers with the lower part of the Kuma Formation (10b, ca. 2,000 m) consisting of calcilutite with intercalations of chert and clastics. Belemnites, foraminifers, and the Late Jurassic ammonites *Perisphinctes (Kranaosphinctes) burui* Boehm, *P. (Dichotomosphinctes)* cf. *rotoides* (Boehm), *Calliphylloceras malayanum* (Boehm), and *Taramelliceras* cf. *flexuosum* (Münster) (Arkell 1956) occur in the carbonates of both formations, documenting the Middle Oxfordian (?Plicatilis Zone) and indicating shelf to slope environments. The fauna has not been revised. The sequence lies unconformably on Triassic carbonates that resemble the Fageo Group of Misool (Wanner 1922; Hummel 1923; Siregar 1977, cited by Tjokrosapoetro, Budhitrisna, and Rusmana 1981).

*Seram (11)*

The flysch-type sediments of the Kanikeh Formation (11a, 1,500–2,000 m) and the carbonates with chert and oolites of the Manusela Formation (11b, ca. 1,000 m) have yielded Jurassic ammonites and the Late Triassic bivalves *Monotis* and *Halobia* and the brachiopod *Halorella*. The corals *Lovcenipora, Pseudocyclammina, Chaffatella,* and *Pachypora* and the alga *Clypeina* in limestones of the Manusela are Late Jurassic. The sequence is underlain by Late Paleozoic metamorphics and unconformably overlain by Cretaceous carbonates (Van Bemmelen 1949; Gafoer, Suwitodirdjo, and Suharsono 1984; Tjokrosapoetro, Achdan, and Suwitodirdjo 1987a; Tjokrosapoetro, Rusmana, and Turkandi 1987b).

*Tanimbar and Barbar (12)*

Clastics and carbonates in mélange complexes of these islands have yielded Jurassic ammonites (12b, Middle Jurassic on Barbar) and belemnites. The islands belong to the nonvolcanic Outer Banda Arc. The rocks are chaotically mixed and unconformably overlain by Miocene carbonates (Tanimbar) or by Pleistocene carbonates (Barbar) (Wiryosujono 1976; Sukardi and Sutrisno 1985; Suparman, Agustyanto, and Achdan 1987).

*Timor and Rotti (13)*

The Mesozoic of Timor Island was studied mainly by Brouwer (1924b) and summarized by Van Bemmelen (1949). The Jurassic of West Timor belongs to the Sonnebait Series (13c), the Fatu Complex, and the Palelo Series. Shallow- to deep-sea limestones with marls and cherts dominate and, on West Timor, are divided into several tectonic units. On East Timor, approximately 1,000-m-thick Jurassic flysch-type sediments are placed in the Wailuli Formation (13a). These shallow- to deep-sea deposits lie conformably on Late Triassic sediments of the Aitutu Formation (13b) and are unconformably overlain by the Cretaceous pelagic sediments of the Nakfunu Formation (Audley-Charles 1968; Rosidi et al. 1979).

The following ammonites (especially Liassic) have been reported (Rutten 1927; Arkell 1956; Sato 1975):

*Timor.* *Paracaloceras* cf. *coregonense* (Sow.) Waehner and *Ectocentrites* aff. *italicus* (Canav.) Waehner (Hettangian–?Early Sinemurian); *Arnioceras subgeometricum* Jaworski (Early Sinemurian); *Echioceras* cf. *radiatum* Trueman & Will. (Late Sinemurian); *Uptonia* sp. (Early Pliensbachian); *Esericeras timorense* (Krumb.) (Late Toarcian); *Perisphinctes s.l.* (Late Oxfordian).

*Rotti.* *Arnioceras* cf. *fortunatum* Buck., *A. mendix* Fucini, *A. subgeometricum* Jaw., *A. ceratitoides* (Qu.), *Arietites* aff. *lyra*

Hyatt, *A.* cf. *rotiformis* (Sow.), and *Microderoceras landaui* (Boehm) (Early Sinemurian); *Uptonia* sp., *Phricodoceras subtaylori* (Krum.), *Liparoceras rotticum* Krum., *L.* aff. *kilsbiense* Spath, together with some lytoceratids and phylloceratids (Early Plienbachian); *Dactylioceras s.l.* sp. (Early Toarcian); *Esericeras timorense* (Krum.) (late Toarcian); *Macrocephalites ("Dolicephalites")* sp. (Middle Bathonian–Early Callovian); *Taramelliceras* sp. and *Perisphinctes s.l.* (Late Oxfordian).

### *Kalimantan (Indonesian Borneo) (14–16)*

Van Bemmelen (1949) and Marks (1957) dated the Alino Formation and Paniungan Beds of south Kalimantan (14) as Jurassic–Cretaceous, based on radiolarians and gastropods (*Cylindrites*). But the cherts, siliceous shales, and limestones of these formations are now known to belong to a mélange imbricated with metamorphics and ultramafics. The mélange rocks are overlain by Early Cretaceous flysch of the Pitap Formation (Rustandi, Nila, and Sanyoto 1981; Sikumbang and Heryanto 1981; Supriatna 1983; Supriatna, Rustandi, and Heryanto 1984b).

The Jurassic of west Kalimantan consists of sediments (15) and volcanics (16) that continue across the border into neighboring Sarawak (Sato 1975). The sediments and volcanics belong to the Bengkayang Group, placed in the Banan (ca. 1,500 m), Kalung (ca. 150 m), Riampelaya (ca. 300 m), and Sungaibetung 2,000–5,000 m) Formations. They consists of shallow-marine shales and sandstones, with intercalations of carbonate and tuff, and range from Late Triassic to Middle Jurassic. The volcanics are basaltic-andesitic (Serian and Jirak Volcanics) and dacitic-basaltic (Belango Volcanics), and their probable age is Middle to Late Jurassic (Rusmana and Pardede 1987; Supriatna and Margono 1987; Suwarna and de Keyser 1987).

The Jurassic of west and south Kalimantan was reviewed by Sato (1975), Hirano et al. (1981), and Fontaine et al. (1983). The Sinemurian was documented by the ammonite *Xipheroceras* sp., and the Early and Late Toarcian, respectively, by *Harpoceras s.s.* and *Dactylioceras (Orthodactylites)* sp. and by *Dumortieria* sp. The Upper Jurassic is developed in shallow-water bivalve facies, yielding rare Kimmeridgian (*Lithacoceras* or *Subplanites*) and late Tithonian ammonites only in neighboring southwest Sarawak.

### *Sumatra (17)*

Thick Jurassic sequences formerly held to be piles of bathyal to neritic foredeep accumulations are now considered to be the product of volcanic arcs. Mesozoic sequences are widely distributed in Sumatra, especially Upper Jurassic–Lower Cretaceous, but the Lower Jurassic appears to be missing (Fontaine et al. 1983).

The Jurassic Woyla Group of north (and south?) Sumatra is divided into three subgroups based on the predominance of clastics, carbonate with chert, or volcanics. The Clastic Subgroup (17a, ca. 2,000 m) is marine and has associated volcanics; the Carbonatic Subgroup (17b, ca. 300 m) is partly in fringing-reef facies and partly contains chert in pelagic facies; the Volcanic Subgroup (17c, ca. 450 m) includes marine clastics, carbonates, and cherts. Most rocks are slightly metamorphosed. They reflect deep- to shallow-sea, even subaerial environments along a submarine volcanic arc. The basement consists of late Paleozoic sediments, metasediments, and volcanics: the Tapanuli and Peusangan Groups in the north and the Palepat and Mengkarang Formations in the south. The Mesozoic is unconformably overlain by Tertiary sediments and volcanics (Rosidi, Tjokrosapoetro, and Pendowo 1976; Bennet et al. 1981a,b; Aspden et al. 1982; Cameron et al. 1982a,b, 1983; Simandjuntak et al. 1982b; Rock et al. 1983; Pardede and Brata 1984; Suwarna and Suharsono 1984; Gafoer, Burhan, and Purnomo 1986; Pardede, Cobrie, and Gafoer 1986; Pardede and Gafoer 1986).

From north Sumatra, Jurassic radiolarians, foraminifers (*Pseudocyclammina*), algae (*Clypeina, Permocalculus, Lithocodium*), and corals have been reported, and from south Sumatra, Jurassic pelecypods (*Astarte, Modiolus, Leda, Oxytoma*), gastropods (*Glauconia*), crinoids (*Pentacrinus*), echinoids (*Pygurus*), algae (*Pseudolithocodium, Stromatolites, Bacinella*), and foraminifers (*Orbitolina*), as well as Cretaceous ammonites (*Holcostephanus, Desmoceras, Mortoniceras, Neocomites, Thurmannia*) (Van Bemmelen, 1949). They indicate Early Jurassic to Early Cretaceous ages and fringe-reef to deep-sea environments. Fontaine et al. (1983) suggest that these earlier records imply Late Jurassic to Early Cretaceous ages for north Sumatra, and Middle Jurassic–Early Cretaceous ages for south Sumatra (Asai and Rawas Formations). Recent work on hexacoral assemblages by Fontaine and Beauvais (1985) yielded Middle Oxfordian to Aptian ages for limestones of north Sumatra (Raba, Lamno, Sise, Tapaktuan), and Oxfordian to Tithonian for south Sumatra (Padang, Asai, Mengkadai).

## PAPUA NEW GUINEA[2]

### Geology and Jurassic stratigraphy

Outcropping Jurassic rocks are known only from the Central Ranges of the mainland, where they occur in a broad, structurally delimited belt stretching from the Bird's Head (Vogelkop) in western Irianjaya (discussed earlier) to the Bismarck and Kubor Ranges in the Central Highlands. Petroleum-exploration wells have recently revealed extensive Jurassic subcrops south of the Central Ranges, including submarine outcrops in the Coral Sea (Figure 8.4). The Jurassic is part of the thick Middle Triassic–Late Cretaceous clastics of the Papuan Basin on the northeastern margin of the Australo-Indian plate. They lie unconformably on Paleozoic basement with Triassic intrusions, part of the Australian continent.

Distal and proximal lithofacies are readily distinguished in the Jurassic successions (Figure 8.5) (Francis and Diebert 1988; Francis et al. 1990; Denison and Anthony 1991). Along the southern margin of the outcrop belt and in the subcrop, the nearshore

[2] By G. E. G. Westermann.

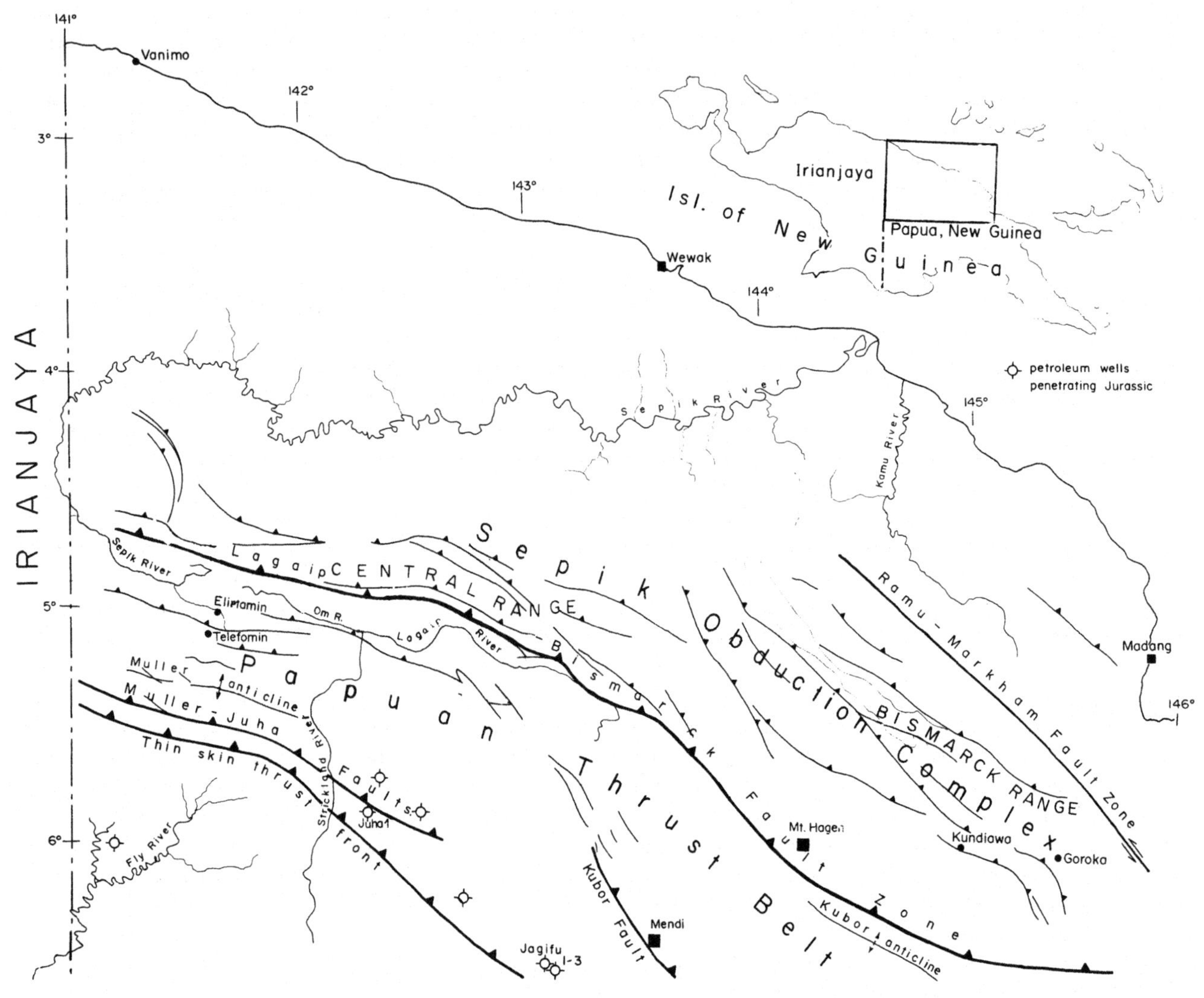


**Figure 8.4. The principal Jurassic outcrops in Papua New Guinea.**

clastic Kuagben Group includes several arenaceous formations and members ranging in environment of deposition from fluvial and deltaic to mid-neritic. Ammonites are excedingly rare, and the dating, from Early Jurassic to Early Cretaceous, has been achieved mainly by palynology. Interdigitating toward the north is the Wahgi Group, mainly mudstones deposited in mid-neritic to bathyal environments. The upper parts of the "Jimi and Balimbu Greywackes," in the lower part of the sequence, and the Maril Formation, above, have yielded diverse microfaunas and macrofaunas and floras, including sparse ammonoids of Early Jurassic to Early Cretaceous ages, as discussed later. Detailed stratigraphy, however, has remained elusive owing to poor outcrop conditions and extreme structural complications.

For most of Jurassic and Cretaceous time, the stratotectonic setting was a passive continental margin (Francis et al. 1990). During the middle and late Cretaceous there was alkaline magmatism associated with rifting, which preceded the Paleocene opening of the Coral Sea Basin and the proto–Solomon Sea. In the latest Cretaceous and Paleocene, local uplift was associated with rifting and marginal-basin spreading. During the Oligocene there occurred a change from a passive margin to an active margin with a subduction zone (New Guinea Trobriand Trench) and associated volcanic arc (Maramuni Volcanic Arc).

Structural interpretations have greatly changed in recent years (Francis and Deibert 1988; Hilyard, Rogerson, and Francis 1988; Francis et al. 1990). The geologists mapping the area in the fifties and sixties concluded that the Mesozoics formed synclines and anticlines dissected by near-vertical faults, with some smaller, high-angle thrusts. Recent work has revealed that low-angle, southwestward thrusting is an essential feature, including the large-scale obduction of bathyal and abyssal Jurassic-Cenozoic thrust sheets. In the West Sepik area (Sandaun province), klippen of ?Paleogene ophiolite or early arc rocks occur up to 200 km south of the most hindward exposure of continental crust, superjacent to metamorphosed Jurassic clastics. At 20–25 km of depth, thrust planes interpreted from seizmicity are believed to be high-angle, suggesting that deep crustal thrusting is occurring at present. At a smaller scale, isoclinal folds and various types of faults can be abundantly observed in the Central Highlands (e.g., Francis et al. 1990, Fig. 12), as well as in the West Sepik area.

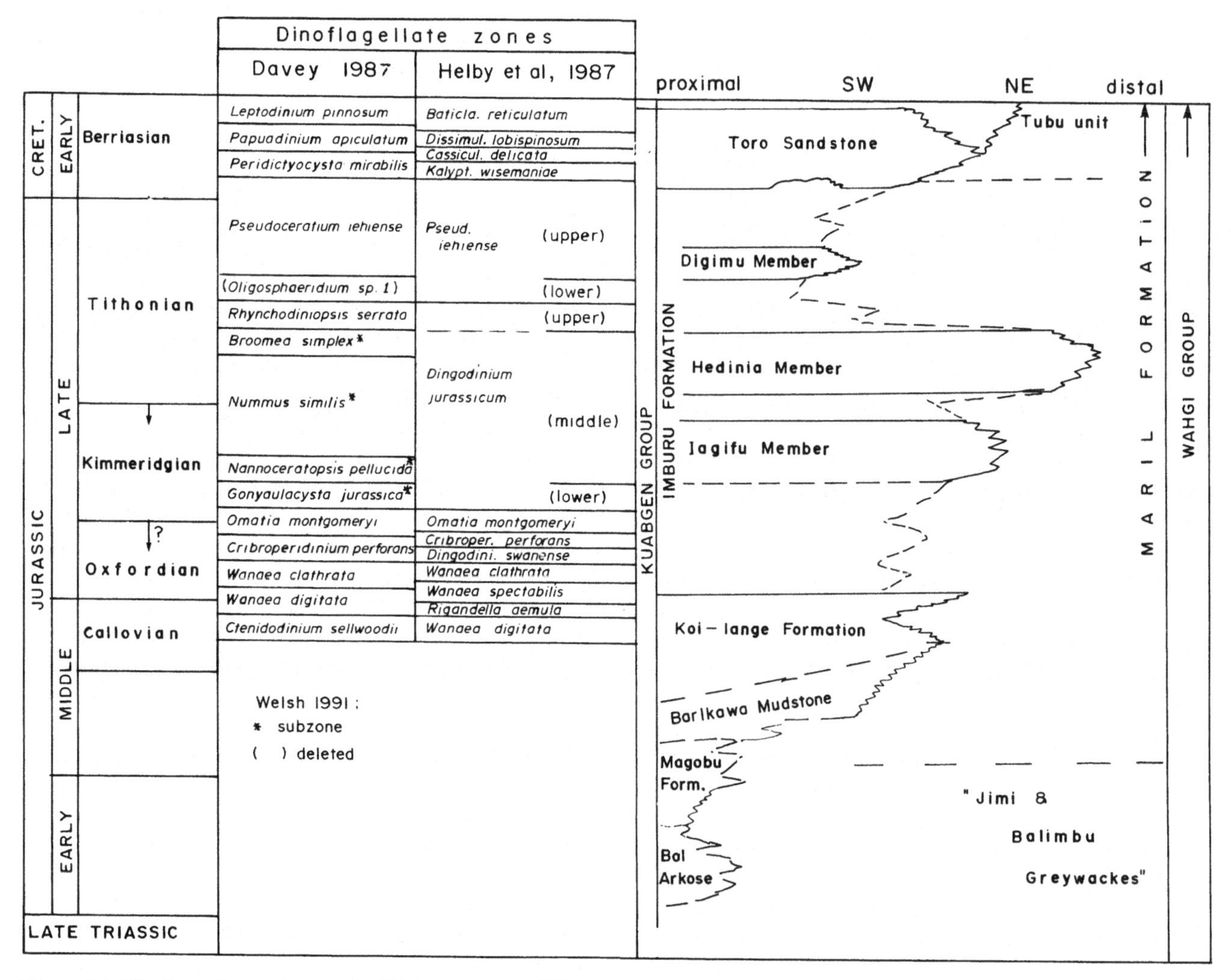


**Figure 8.5. The Jurassic nearshore and offshore sequences of Papua New Guinea, according to dinoflagellate zonations; stage correlations tentative.**

The main ammonite-bearing outcrops, in the Maril Formation, consist of homogeneous mudstones with small concretions and are particularly prone to tectonic deformation as well as to erosion and sliding. The resistant units consist of arenaceous beds that form the exposed cliffs and rapids, but have yielded no ammonites. Furthermore, rain forest covers essentially all slopes, and regrowth on the slope faces exposed by slides occurs within a few years.

### Biostratigraphy and ammonite fauna

Surprisingly few data have been published on Jurassic macrofossils of Papua New Guinea, and relative dating of the thick Jurassic sequences has been achieved only very recently, mainly by palynostratigraphy. Two parallel sets of dinoflagellate zones for the Upper Jurassic and Lower Cretaceous have been developed simultaneously and applied sucessfully over much of Australasia and Indonesia by Helby, Morgan, and Partridge (1987) and by Davey (1987) (see Chapters 13, 16, and 17). The fundamental differences in the presumably Kimmeridgian–early Tithonian parts of these zonations have most recently been reconciled by Welsh (1991): (1) He restricted the *Dingodinium jurassicum* Zone of Helby et al. (1987) to the former "lower" and "middle" parts of that zone, the "upper" part becoming the *Rhynchodiniopsis serrata* Zone of Davy. (2) He reduced the rank of the *Gonyaulacysta jurassica–Broomea simplex* Zones of Davey to subzones of the *D. jurassicum* Zone (Figure 8.5).

The extensive Upper Jurassic–Lower Cretaceous subcrop has recently been explored using dinoflagellate biostratigraphy, especially the proximal facies that include thick arenaceous packages (Figure 8.5) yielding important hydrocarbon resources (Denison and Anthony 1991; Welsh 1991).

The different correlations of the Late Jurassic dinoflagellate zones with the global stages, however, remain highly controversial. The regional ammonite biostratigraphy and biochronology required to solve this problem have thus far mostly eluded us; the published works on ammonoids of New Guinea have been based on *ex situ* collections. The exception is a small Bathonian section formed by rapids of Strickland River headwaters, where the contiguous *Macrocephalites bifurcatus* and *M. apertus* Zones were found by S. Skwarko and identified by Westermann and Callomon (1988). The closest useful ammonite-bearing sequences are known from the Moluccas, but there the Tithonian, represented by most Jurassic ammonite faunas of Papua New Guinea, is poorly known.

The only Early Jurassic ammonoids described from Papua New Guinea were collected loose from the upper part of the "Balimbu Greywacke" in the Central Highlands. The few poorly preserved specimens were tentatively placed in the Sinemurian–Pliensbachian genera *Tropidoceras* (or *Asteroceras*?), *Paltechioceras*, and *Arieticeras* (or *Arnioceras?*) by Skwarko (1973). The only early Middle Jurassic ammonites known from Papua New Guinea are two specimens of the Late Bajocian *Praetulites* that were collected by Australian geologists in the Central Highlands and identified by G. E. G. Westermann (unpublished data). Late Jurassic ammonites have not been described, although they are locally relatively common in the Sepik headwaters region, as discussed later.

The Jurassic-Cretaceous belemnites of Papua New Guinea have recently been described and zoned by Challinor (1990).

By far the largest Jurassic mollusk collection of Papua New Guinea was assembled in the seventies by Ms. Betty Crouch, then a senior nurse at the Baptist Mission Hospital in the village of Telefomin (or Telefolmin, with airstrip), West Sepik area (Sandaun province) (Figure 8.4). The fossils were all acquired from native people who found them loose in riverbeds and on large slumps in an area 10–15 km north of Telefomin, around the small settlements of Konduvip and Eliptamin (with airstrip). The original collection comprised roughly a thousand specimens, dominantly Late Jurassic ammonoids, together with belemnoids (ca. 6%, mainly phragmocones of *Belemnopsis*) and bivalves (ca. 20%, infaunal as well as the epifaunal *Retroceramus*, *Bositra*, and ?*Buchia*). The collection was inspected by Westermann during a 1972 geologic field trip to the area with S. Skwarko. At that time B. Crouch kindly gave a representative set of the ammonoids to Westermann, and the host of the collection was later acquired by the Royal Ontario Museum, Toronto; only about 40 representative specimens comprising mostly large ammonoids (mainly *Paraboliceras*), as well as several *Belemnopsis* and *Retroceramus*, remain in the Telefomin Baptist Mission.

The Late Jurassic ammonoid collections from Telefomin presently at the Royal Ontario Museum (ROM) and McMaster University (to be combined at the ROM) contain about 460 specimens, which belong to the following taxa:

- 3% Phylloceratina (almost exclusively *Ptychophylloceras*)
- 97% Ammonitina:
- 40% Mayaitinae (almost exclusively *Epimayaites*)
- 23% Perisphinctaceae [mostly "*Pseudoparaboliceras*" (nom. dub.), with the possible species and synonyms "*P. aramaraii*" Gerth (nom. dub.), "*Perisphinctes*" *moluccanus* and *sularus* Boehm, "*Per.*" *boehmi* and "*Kossmatia*" *kiliani* Kruizinga, and ?"*Per. novaguinensis*" Gerth (nom. nud.)]
- 23% Berriasellinae(?) (*Paraboliceras* and *Kossmatia* in similar numbers)
- 9% Streblitinae (*Gymnodiscoceras* and a few *Uhligites*)
- 2% Virgatosphinctinae (*Aulacosphinctoides* and *Virgatosphinctes*)

The Mayaitinae and some of the Telefomin forms tentatively placed in "*Pseudoparaboliceras*" can be closely matched with Oxfordian species of the Sula Islands (Boehm 1904–12); the latter seem to be endemic to the New Guinea–Moluccas area and could possibly range upward into the Kimmeridgian. Together they comprise two-thirds of the Telefomin fauna in the collection. The remaining one-third of the ammonites (*Kossmatia, Paraboliceras, Aulacosphinctoides, Virgatosphinctoides, Gymnodiscoceras*, and *Uhligites*) can be closely matched with the many morphospecies described by Uhlig (1903–10) from the upper Spiti Shales of the western Himalayas (J. Krishna and G. E. G. Westermann unpublished data), which are dated as Early and Middle Tithonian by Krishna (1983). Ammonites elsewhere known to be diagnostic for the Kimmeridgian remain unknown from Papua New Guinea, just as from the Himalayas and Indonesia, where an ammonitophobic biofacies has been invoked (Westermann and Wang 1988; Gradstein et al. 1989). Significantly, the lithofacies of the Maril Formation (also known as Om Beds) and equivalents in both Irianjaya and Papua New Guinea (i.e., dominant mudstone with concretions) are close to those of the Himalayan Spiti Shales and equivalents (Gradstein et al. 1989).

Recent progress in Australasian dinoflagellate stratigraphy (Davey 1987; Helby et al. 1987) and New Zealand ammonite stratigraphy (G. Stevens personal communication), however, suggests that in Australasia some elements of the Tithonian fauna, especially the cosmopolitan *Kossmatia*, may have made their appearance already in the Kimmeridgian (or even Oxfordian, G. Francis personal communication). A Kimmeridgian appearance should not come as too much of a surprise, considering the mentioned overall absence of ammonites in the southeast Tethyan Kimmeridgian (even though *Kossmatia* is said to have appeared in the Himalayas only in the Middle Tithonian) (Krishna 1983).

In 1991, G. Francis and G. E. G. Westermann attempted to solve the stratigraphic problems of the Late Jurassic ammonoid faunas found loose in the Maril Formation north of Telefomin. They revisited the most promising areas, but they essentially failed because of complete overgrowth of the major rock faces exposed by the large slides of the late sixties near Konduvip, the complex tectonics of the sequences exposed in riverbeds and canyons, and the differential erosion of the recessive mudstones with fossiliferous concretions. Furthermore, it has become evident that the size of the Betty Crouch collection resulted from the diligence of the numerous native people who collected over several years immediately after a time of major fresh slides. The collection does not reflect the true conditions as seen in the present outcrops and secondary deposits: Ammonites are generally scarce, concretions are restricted to a few horizons, and only a few of them yield single ammonites (even fewer have inocerami or belemnite phragmocones); only belemnite rostra and bivalve shells are locally abundant in arenite beds. A single locality yielded to us a couple of crushed macrocephalitids in the mudstone matrix.

We hope to salvage data on the faunal-floral associations and perhaps limited biostratigraphy from the 1991 field trip, based on larger samples of ammonite-bearing concretions with matrix, which we hope will permit zonation by dinoflagellates and other microfossils and nannofossils. Geochemical "fingerprinting" of concretions for possible horizon identification will also be attempted.

Finally, there is an apparent contradiction concerning the environment of deposition. The ammonoid faunas consist overwhelmingly of "trachyostraca," indicating mid-shelf (offshore neritic, 100–150 m depth) conditions (Westermann 1990), whereas the benthic foraminifers and lithofacies suggest outermost shelf to upper continental slope (150–500 m) (G. Francis personal communication). A similar depth of 200–300 m was estimated for parts of the Ammonitina-bearing Spiti Shales equivalents of the Nepal Himalayas, based also on benthic foraminifers and belemnite phragmocones very similar to those found near Telefomin (Gradstein et al. 1989). G. Francis (personal communication) suggests that the scattered ammonites of the Maril Formation might have been transported downslope from mid-shelf and redeposited on the shelf margin and upper slope. The scarcity of bathyal Phylloceratina, and especially Lytoceratina ("leiostraca," deposited near their habitats), could then be due to the consequent intensive muddying of the bottom waters on the continental slope.

### References


Aldiss, D. T., Whandoyo, R., Sjaefudin, A. G., & Kusjono, A. (1983). *The Geology of the Sidikalang Quadrangle, Sumatra*. Bandung: GRDC.

Arkell, W. J. (1956). *Jurassic Geology of the World*. London: Oliver & Boyd.

Aspden, J. A., Kartawa, W., Aldiss, D. T., Djunuddin, A., Whandoyo, R., Diatma, D., Clarke, M. C. G., & Harahap, H. (1982). *The Geology of the Padangsidempuan and Sibolga Quadrangle, Sumatra*. Bandung: GRDC.

Audley-Charles, M. G. (1968). The geology of Portuguese Timor. *Geol. Soc. London, Mem., 4*, 76.

Bennet, J. D., Bridge, D. M., Cameron, N. R., Djunuddin, A., Ghazali, S. A., Jeffery, D. H., Kartawa, W., Keats, W., Rock, N. M. S., Thompson, S. J., & Whandoyo, R. (1981a). *Geologic Map of the Banda Aceh Quadrangle, Sumatra*. Bandung: GRDC.

Bennet, J. D., Bridge, D. M., Cameron, N. R., Djunuddin, A., Ghazali, S. A., Jeffery, D. H., Kartawa, W., Keats, W., Rock, N. M. S., & Thompson, S. J. (1981b). *The Geology of the Calang Quadrangle, Sumatra*. Bandung: GRDC.

Boehm, G. (1904–12). Die Südküsten der Sula-Inseln Taliabu and Mangoli. 1. Grenzschichten zwischen Jura und Kreide (1904); 2. Der Fundpunkt am oberen Lagoi auf Taliabu (1907); 3. Oxford des Wai Galo (1907); 4. Unteres Callovien (1912). *Palaeontogr. Suppl., 4*, 1–179.

(1913). Unteres Callovium und Coronaten-Schichten zwischen MacCluer Golf und Geelvink-Bai. Nova Guinea (6). *Geologie (Amsterdam), 1*, 1–20.

Bothe, A. C. D. (1927). Voorloopige mededeeling betreffende de geologie van Zuidoost Celebes. *Mijningenieur, 8*(6), 97–103.

Brouwer, H. A. (1924a). Bijdrage tot de geologie der Obi-eilanden. *Jaarb. Mijnw. N.I. 1923, Verh., 5*, 62.

(1924b). Summary of the geological results of the expeditions. In *Geol. Exped. Lesser Sunda Islands* (vol. 4, pp. 345–402). Amsterdam.

Cameron, N. R., Aspden, J. A., Bridge, D. M., Djunuddin, A., Ghazali, S. A., Harahap, H., Hariwidjaja, Johari, S., Kartawa, W., Keats, W., Ngabito, H., Rock, N. M. S., & Whandoyo, R. (1982a). *The Geology of the Medan Quadrangle, Sumatra*. Bandung: GRDC.

Cameron, N. R., Bennet, J. D., Bridge, D. M., Clarke, M. C. G., Djunuddin, A., Ghazali, S. A., Harahap, H., Jeffery, D. H., Keats, W., Ngabito, H., Rock, N. M. S., & Thompson, S. J. (1983). *The Geology of the Takengon Quadrangle, Sumatra*. Bandung: GRDC.

Cameron, N. R., Bennett, J. D., Bridge, D. M., Djunuddin, A., Ghazali, S. A., Harahap, H., Jeffery, D. H., Kartawa, W., Keats, W., Rock, N. M. S., & Whandoyo, R. (1982b). *Geologic Map of the Tapaktuan Quadrangle, Sumatra*. Bandung: GRDC.

Challinor, A. B. (1989a). Early Cretaceous belemnites from the central Bird's Head, Irian Jaya, Indonesia. *Geol. Res. Devel. Centre Indonesia (Bandung), Paleont. Ser., 5*, 1–25.

(1989b). The succession of *Belemnopsis* in the Late Jurassic of eastern Indonesia. *Palaeontology, 32*, 571–96.

(1990). A belemnite zonation for the Jurassic-Cretaceous of Papua New Guinea and a faunal comparison with eastern Indonesia. *BMR J. Aust. Geol. Geophys., 11*, 429–47.

(1991). Recision of the belemnites of Misool and a review of the belemnites of Indonesia. *Palaeontographica, A218*, 87–164.

Challinor, A. B., & Skwarko, S. K. (1982). *Jurassic Belemnite from Sula Islands, Moluccas, Indonesia*. Geol. Res. Dev. Centre, Pal. Series, No. 3.

Davey, R. J. (1987). *Palynological Zonation of the Lower Cretaceous, Upper and Uppermost Middle Jurassic in the Northwestern Papuan Basin of Papua Guinea*. Papua New Guinea Mem. 13.

Denison, C. N., & Anthony, J. S. (1991). New Late Jurassic subsurface lithostratigraphic units, PPL-100, Papua New Guinea. In G. J. Carman & Z. Carman (eds.), *Petroleum exploraton in Papua New Guinea* (pp. 153–8). Proc. 1 PNG Petrol Conv., Port Moresby, Feb. 1990.

Dow, D. B., Robinson, G. P., Hartono, U., & Ratman, N. (1986). *Geological Map of Irian Jaya, Indonesia*. Bandung: GRDC.

Dow, D. B., Trail, D. S., & Hamonangan, B. H. (1984). *Geological Data Record, Enarotali, 1:250,000. Sheet Area, Irian Jaya*. Open-file report. Bandung: GRDC.

Etheridge, R. (1890). Our present state of knowledge of palaeontology of New Guinea. *Rec. Geol. Surv. New South Wales, 1*, 172–9.

Fontaine, H., & Beauvais, L. (1985). Stratigraphic units, fossil localities, oil wells, radiometric dating, paleogeography. In *The Pre-Tertiary Fossils of Sumatra and Their Environments, 22nd CCOP session, Guangzhou*.

Fontaine, H., David, P., Pardede, R., & Suwarna, N. (1983). *Marine Jurassic in Southest Asia*. UN-ESCAP-CCOP Tech. Bull. 16.

Francis, G., & Deibert, D. H. (1988). *Petroleum Potential of the North New Guinea Basin and Associated Infra-basins*. Papua New Guinea Geol. Sur. Rep. 88/77.

Francis, G., Rogerson, R., Hilyard, D., & Haig, D. W. (1990). Excursion guide to the Wahgi and Chimbu Gorges. In *Proceedings Papua New Guinea Petroleum Convention, Port Moresby, 1990* (pp. 1–55). Port Moresby: Geol. Sur. Papua New Guinea.

Gafoer, S., Burhan, G., & Purnomo, J. (1986). *The Geology of the Palembang Quadrangle, Sumatra*. Bandung: GRDC.

Gafoer, S., Suwitodirdjo, K., & Suharsono (1984). *Laporan Geologi Lembar Bula dan Kep. Watubela*, Open-file report. Bandung: GRDC.

Gerth, H. (1927). ?Ein neues Vorkommen der bathyalen Cephalopoden-Fazies des mittleren Jura in Niederländisch Neu-Guinea. *Leidsche Geol. Mededeel., 2*, 225–8.

Gerth, H. (1965). Ammoniten des mittleren und oberen Jura und der älteren Kreide vom Nordabhang des Schneegebirges in Neu Guinea. *N. Jb. Geol. Paläontol. Abh., 121*, 209–18.

Gradstein, F., Gibling, M. R., Jansa, L. F., Kaminski, M. A., Ogg, J. G., Sarti, M., Thurow, J. W., Rad, U. von, & Westermann, G. E. G. (1989). *Mesozoic Stratigraphy of Thakkhola, Central Ne-*

pal. Centre Marine Geol., Dalhousie University Spec. Rep. 1.
Hamilton, W. (1979). *Tectonics of the Indonesian Region.* U.S. Geological Survey Professional Paper 1078.
Hasibuan, F., & Janvier, P. (1985). Lepidates *sp. (Achinopteygii, Halecustami), a Fish from the Lower Jurassic of Misool Island.* Geol. Res. Dev. Centre, Pal. Series, No. 5.
Helby, R., Morgan, R., & Partridge, A. D. (1987). A palynological zonation of the Australian Mesozoic. In P. A. Jell (ed.), *Studies in Australian Mesozoic Palynology. Mem. Assoc. Australas. Palyn., 4,* 1–94.
Helmcke, D. von, Barthel, K. W., & Hillebrandt, A. von (1987). Notes on the Jurassic and Cretaceous of the Central Mountains chain of Irian Jaya (Indonesia). *N. Jb. Geol. Palaont. Mh., 11,* 674–84.
Hetzel, W. H. (1936). *Verslag van het onderzoek naar het voorkomen van asfaltgesteenten op het eiland Boeton.* Versl. Ned. Ind. Delfst. N. 21, dienst Mijnb. Ned. Ind., Batavia.
Hilyard, D., Rogerson, R., & Francis, G. (1988). *Accretionary Terranes and Evolution on the New Guinea Orogen, Papua New Guinea.* Papua New Guinea Geol. Sur. Report 88/9.
Hirano, H., Ichihara, S., Sunarya, Y., Nakajima, N., Obata, I., & Futakami, M. (1981). Lower Jurassic ammonites from Bengkayang, West Kalimantan Province, Republic of Indonesia. *Bull. Geol. Res. Dev. Centre, 4,* 21–6.
Hummel, K. (1923). Die Oxford-Tuffite der Insel Buru and ihre Fauna. *Palaeontographica,* Suppl. IV.
Jaworsky, J. A. (1933). Revision der Arieten, Echioceraten und Dactylioceraten des Lias von Niederländisch-Indien. *N. Jb. Miner. Paläont.Beil.-Bd., 70,* 251–333.
Kastowo & Leo, G. W. (1973). *Geologic Map of the Padang Quadrangle, Sumatra.* Bandung: GSI.
Krishna, J. (1983). Callovian-Albian ammonoid stratigraphy and paleobiogeography in the Indian sub-continent, with special reference to the Tethys Himalaya. *Himalayan Geol., 11,* 43–72.
Kruizinga, P. (1926). Ammonieten en eenige andere Fossielen uit de Jurassische Afzettingen de Soela Eilanden. *Jb. Mijn. Ned. Oostind., 4,* 11–85.
Marks, P. (1957). *Stratigraphic Lexicon of Indonesia.* Pub. Keilmuan N. 31, Ser. Geol. Bandung: GSI.
Pardede, R., & Brata, K. (1984). *Laporan Geologi Lembar Sungaipenuh dan Ketaun, Sumatra.* Open-file report. Bandung: GRDC.
Pardede, R., Cobrie, T., & Gafoer, S. (1986). *Laporan Geologi Lembar Bengkulu, Sumatra.* Open-file report. Bandung: GRDC.
Pardede, R., & Gafoer, S. (1986). *Laporan Geologi Lembar Baturadja, Sumatra.* Bandung: GRDC.
Pieters, P. E., Hakim, A. S., & Atmawinata, S. (1985). *Geological Data Record, Ransiki.* Open-file report. Bandung: GRDC.
Pieters, P. E., Hartono, U., & Amri, C. H. (1982). *Geological Data Record, Mar, 1:250,000 Sheet Area; Irian Jaya.* Open-file report. Bandung: GRDC.
Pieters, P.E., Pigram, C. J., Trail, D. S., Dow, D. B., Ratman, N., & Sukamto, R.(1983). The stratigraphy of Western Irian Jaya. In *Proc. 12th Ann. Conv. IPA,* Jakarta.
Pigram, C. J., Challinor, A. B., Hasibuan, F., Rusmana, E., & Hartono, U. (1982). *Geological Results of the 1981 Expedition to the Misool Archipelago, Irian Jaya.* Bull. No. 6. Bandung: GRDC.
Pigram, C. J., & Panggabean, H. (1983). *Geological Data Record, Waghete (Yapenkopra), 1:250,000, Sheet Area, Irian Jaya.* Open-file report. Bandung: GRDC.
Pigram, C. J., & Sukanta, U. (1982). *Geological Data Record, Taminabuan, 1:250,000 Sheet Area, Irian Jaya.* Open-file report. Bandung: GRDC.
Robinson, G. P., Ryburn, R. J., Tobing, S. L., & Achdan, A. (1987). *Geological Data Record, Steenkool (Wasier) – Kaimana, 1:250,000 Sheet Area, Irian Jaya.* Open-file report. Bandung: GRDC.
Rock, N. M. S., Aldiss, D. T., Aspden, J. A., Clarke, M. C. G., Djunuddin, A., Kartawa, W., Miswar, Thompson, S. J., & Whandoyo, R. (1983). *The Geology of the Lubuksikaping Quadrangle, Sumatra.* Bandung: GRDC.
Rosidi, H. M. D., Tjokrosapoetro, S., Gafoer, S., & Suwitodirdjo, K. (1979). *Geological Map of the Kupang-Atambua, Timor.* Bandung: GRDC.
Rosidi, H. M. D., Tjokrosapoetro, S., & Pendowo, B. (1976). *Geologic Map of the Painan and Northeastern Part of the Muarasiberut Quadrangles, Sumatra.* Bandung: GSI.
Rusmana, E., Hartono, U., & Pigram, C. J. (1983). *The Geology of Misool Quadrangle, Irian Jaya.* Open-file report. Bandung: GRDC.
Rusmana, E., Koswara, A., & Simandjuntak, T. O. (1984). *La poran Geologi Lembar Luwuk, Sulawesi.* Open-file report. Bandung: GRDC.
Rusmana, E., & Pardede, R. (1987). *Geologic Map of the Sambas and Siluas Quadrangle.* Bandung: GRDC.
Rusmana, E., Sukido, & Sukarna, D. (1987). *Laporan Geologi Lembar Kendari, Sulawesi.* Bandung: GRDC.
Rustandi, E., Nila, E. S., & Sanyoto, P. (1981). *Laporan Geologi Lembar Kotabaru, Kalimantan.* Open-file report. Bandung: GRDC.
Rutten, L. M. R. (1927). *Soemba, Roendjawa, Savoe en Rottie, Voordrachten over de Geologie van Nederlansch Oost-Indie.* Grooningen: Den Haag, Wolters J.B.
Sato, T. (1975). Marine Jurassic formations and faunas in Southeast Asia and New Guinea. *Geol. Palaeont. Southeast Asia, 15,* 151–89.
Sato, T., Westermann, G. E. G., Skwarko, S. K., & Hasibuan, F. (1978). Jurassic biostratigraphy of the Sula Islands, Indonesia. *Bull. Geol. Surv. Indonesia, 1*(4).
Sikumbang, N., & Heryanto, R. (1981). *Laporan Geologi Lembar Banjarmasin, Kalimantan.* Open-file report. Bandung: GRDC.
Sikumbang, N., & Sanyoto, P. (1981). *Laporan Geologi Lembar Buton dan Muna, Sulawesi.* Open-file report. Bandung: GRDC.
Silitonga, P. H., & Kastowo (1975). *Geologic Map of the Solok Quadrangle, Sumatra.* Bandung: GSI.
Simandjuntak, T. O., Rusmana, E., Surono, & Supandjono, J. B. (1981). *Laporan Geologi Lembar Malili, Sulawesi.* Open-file report. Bandung: GRDC.
Simandjuntak, T. O., Sukardi, H. Budhitrisna, T., & Surono, (1982a). *Laporan Geologi Lembar Muarabungo, Sumatra,* Open-file report. Bandung: GRDC.
Simandjuntak, T. O., & Sukido (1984). *Laporan Geologi Lembar Kolaka, Sulawesi.* Open-file report. Bandung: GRDC.
Simandjuntak, T. O., Surono, & Supandjono, J. B. (1982b). *Laporan Geologi Lembar Poso, Sulawesi.* Open-file report. Bandung: GRDC.
Skwarko, S. K. (1973). First report of Domerian (Lower Jurassic) marine Mollusca from New Guinea. *Bur. Min. Res. Geol. Geophys. Canberra Bull., 140,* 105–12.
Skwarko, S. K. (1981). History of investigation of the Misool Archipelago. *Geol. Res. Dev. Centre, Indonesia, Pal. Series, 2,* 53–66.
(1982). *Bibliography of Invertebrate Macrofossils of Indonesia (with Cross-references).* Geol. Res. Dev. Centre (Bandung), Spec. Publ. 3. 1–66.
Soergel, W. (1913). Lias und Dogger von Jefbie und Filialpopo (Misolarchipelago). *N. Jb. Miner. Geol., Beil. Bd., 36B,* 586–612.
Sudana, D., & Yassin, A. (1983). *Laporan Geologi Lembar Obi, Maluku.* Open-file report. Bandung: GRDC.
Sukamto, R. (1975). Geologi daerah Kepulauan Banggai dan Sula. *Geologi Indonesia, 2,* 23–8.
(1986). *Tektonik Sulawesi Selatan dengan acuan khusus ciri-ciri himpunan batuan daerah Bantimala.* Dissertation, ITB, Bandung.
Sukardi & Sutrisno (1985). *Laporan Geologi Lembar Kepulauan Tanimbar, Maluku.* Open-file report. Bandung: GRDC.
Supandjono, J. B., & Haryono, E. (1986). *Laporan Geologi Lembar Banggai, Sulawesi.* Open-file report. Bandung: GRDC.
Suparman, M., Agustyanto, D., & Achdan, A. (1987). *Laporan Geologi Lembar Babar, Maluku.* Bandung: GRDC.

Supriatna, S. (1983). *Stratigrafi regional daerah Kalimantan Tenggara.* Lap. Tah. 1982–1983, Proyek PGIF, P3G, 34–6.

Supriatna, S., Apandi, T., & Simandjuntak, W. (1984a). *Laporan Geologi Lembar Waigeo, Irian Jaya.* Open-file report. Bandung: GRDC.

Supriatna, S., & Margono, U. (1987). *Geologic Map of the Sanggau Quadrangle, Kalimantan.* Bandung: GRDC.

Supriatna, S., Rustandi, E., & Heryanto, R. (1984b). *The Geology of Sampanahan Quadrangle, Kalimantan.* Open-file report. Bandung: GRDC.

Surono, Simandjuntak, T. O., & Situmorang, R. L. (1983). *Laporan Geologi Lembar Batui, Sulawesi.* Open-file report. Bandung: GRDC.

Surono & Sukarna, D. (1985). *Laporan Geologi Lembar Sanana, Sulawesi.* Open-file report. Bandung: GRDC.

Suwarna, N., & de Keyser, F. (1987). *Geologic Map of the Singkawang Quadrangle, Kalimantan.* Bandung: GRDC.

Suwarna, N., & Suharsono (1984). *Laporan Geologi Lembar Bangko (Sarolangun), Sumatra.* Open-file report. Bandung: GRDC.

Tjokrosapoetro, S., Achdan, A., & Suwitodirdjo, K. (1987a). *Laporan Geologi Lembar Mahsoi, Maluku.* Bandung: GRDC.

Tjokrosapoetro, S., Budhitrisna, T., & Rusmana, E. (1981). *Laporan Geologi Lembar P. Buru, Maluku.* Open-file report. Bandung: GRDC.

Tjokrosapoetro, S., Rusmana, E., & Turkandi, T. (1987b). *Laporan Geologi Lembar Ambon, Maluku.* Bandung: GRDC.

Uhlig, V. (1903–10). The fauna of the Spiti Shale. *Palaeont. Indica,* part 4, fasc. 1–3, 1–511.

Van Bemmelen, R. W. (1949). *The Geology of Indonesia.* The Hague: Government Printing Office.

Visser, W. A., & Hermes, J. J. (1962). *Geological Results of the Exploration for Oil in Netherlands New Guinea Koninklijk.* Neder. Geol. Mijn. Genootsch, Verhand. Geol. Ser. No. 20.

Wandel, G. (1936). Beiträge zur Kenntnis der Jurassischen Molluskenfauna von Misol, Ost-Celebes, Butan, Seran und Jamdena. In J. Wanner (ed.), *Beiträge zur Palaeontologie des Ostindischen Archipelago, XIII. N. Jb. Miner. Geol. Palaeont., Beil. Bd., 75B,* 447–526.

Wanner, J. (1922). Beiträge zur Geologie der Insel Beoreo, Geol. Ergebnisse der Reisen K. Denningers. *Palaeontographica,* Suppl. IV, Abt. III, Lief. 3.

Welsh, A. (1991). Applied Mesozoic biostratigraphy in the western Papuan Basin. In G. J. Carman & Z. Carman (eds.), *Petroleum Exploration in Papua New Guinea* (pp. 369–79). *Proc. 1 PNG Petrol. Conv.*, Port Moresby, Feb. 1990.

Westermann, G. E. G. (1981). Ammonite biochronology and biogeography of the circum-Pacific Middle Jurassic. In M. R. House & J. R. Senior (eds.), *The Ammonoidea* (pp. 433–58). Syst. Assoc. Spec. Vol. 18.

(1990). New developments in Jurassic-Cretaceous ammonite ecology. In G. Pallini et al. (eds.), *Fossili Evoluzione Ambiente* (pp. 459–78). Atti. 2. Conv. Int. Pergola 1987. Roma: Com. Cent. R. Piccinini.

Westermann, G. E. G., & Callomon, J. H. (1988). The Macrocephalitinae and associated Bathonian and Early Callovian (Jurassic) ammonoids of the Sula Islands and New Guinea. *Palaeontographica, A, 203,* 1–90.

Westermann, G. E. G., & Getty, T. A. (1970). New Middle Jurassic Ammonitina from New Guinea. *Bull. Am. Palaeontol., 57*(256), 227–321.

Westermann, G. E. G., & Wang, Y. (1988). Middle Jurassic ammonites of Tibet and the age of the lower Spiti Shales. *Palaeontology, 31,* 20–5.

Wiryosujono, S. (1976). Mélange asemblage in Babar Islands. *Berita Dit. Geol., 9*(6), 71–5.

# 9 Southeast Asia and Japan

T. SATO

The main part of the Eurasian continent was above sea level during the Jurassic. Continental deposits are widely distributed on the continental nucleus of China and surrounding areas, whereas marine Jurassic occurs mostly at the continental margins. There are narrow Jurassic exposures in northeastern China, Hongkong, and the Indochinese Peninsula that represent ingressions onto the continental areas. The Jurassic deposits of Japan, the Philippines, Sulawesi, and Kalimantan, as well as the Thai–Burmese border region, Yunnan, and Tibet, were largely under open-sea conditions, consisting mostly of orogenic sediments.

## JAPAN

### Tectonostratigraphic terranes

The areas of the Jurassic exposures in Japan can be classified into six blocks or terranes, each with more or less characteristic lithology, sequence, and faunas. This is a reflection of the complicated tectonic setting of the Japanese Jurassic (Figure 9.1).

*Hida Terrane.* The Hida Terrane includes the Hida and Hida Marginal Belts in the northern part of central Japan and their westward extension in Chugoku. The Jurassic is exposed in isolated local basins. The Toyora, Tetori, and Kuruma areas are major basins located along the boundary of the Hida and Hida Marginal Belts, and the smaller Higuchi and Yamaoku areas are in the Paleozoic belt of Chugoku and the Sangun Belt, respectively.

*Tamba Terrane.* The Tamba Terrane (Tamba, Mino, Kiso, Ashio, and Yamizo provinces) is a narrow belt stretching on the northern side of the Median Tectonic Line, occupying the axial part of Southwest Japan and its northeastward extension in Northeast Japan. The Jurassic consists mostly of olistostrome or mélange-type sediments that accumulated in an essentially continuous sedimentary basin.

*Chichibu Terrane.* This corresponds to the Sambagawa–Chichibu Belt in the outer belt of Southwest Japan, a narrow tectonic zone parallel to the island axis. The high-pressure-type Sambagawa Metamorphic Belt occupies the northern part and is lithologically continuous with the Chichibu Belt (*s.s.*). The Soma area at the eastern margin of the Abukuma Mountains in Northeast Japan is tentatively included in this province.

*Southern Kitakami Terrane.* This is the southern part of the Kitakami Mountains in Northeast Japan. The Jurassic is exposed in two rows of more or less isolated areas, each of which represents the axial part of a synclinorium running roughly north–south. The Shizukawa, Hashiura, and Mizunuma areas (also the small Chonomori area) are disposed along the axis of the western synclinorium, whereas the Karakuwa and Ojika areas form the eastern synclinorium.

*Northern Kitakami Terrane.* This comprises the Iwaizumi and Taro Belts at the northeastern margin of the Kitakami Mountains. The terrane is oriented NNW–SSE and extends northward to the southwestern part of Hokkaido. The Jurassic is exposed in isolated small areas under the Tertiary cover, except in the Iwaizumi and Taro Belts.

*Axial and Eastern Hokkaido Terranes.* In Hokkaido, the Jurassic is known only in a few localities in the Sorachi–Kamuikotan Belt of the backbone range and in the Tokoro Belt of Eastern Hokkaido, where the areal extent is unknown. These seem to be fragments of oceanic crust, representing the basal part of the subduction complex.

### Representative Jurassic sequences

The Japanese Jurassic occurs in localized basins, and no single continuous section of the entire Jurassic is known. The

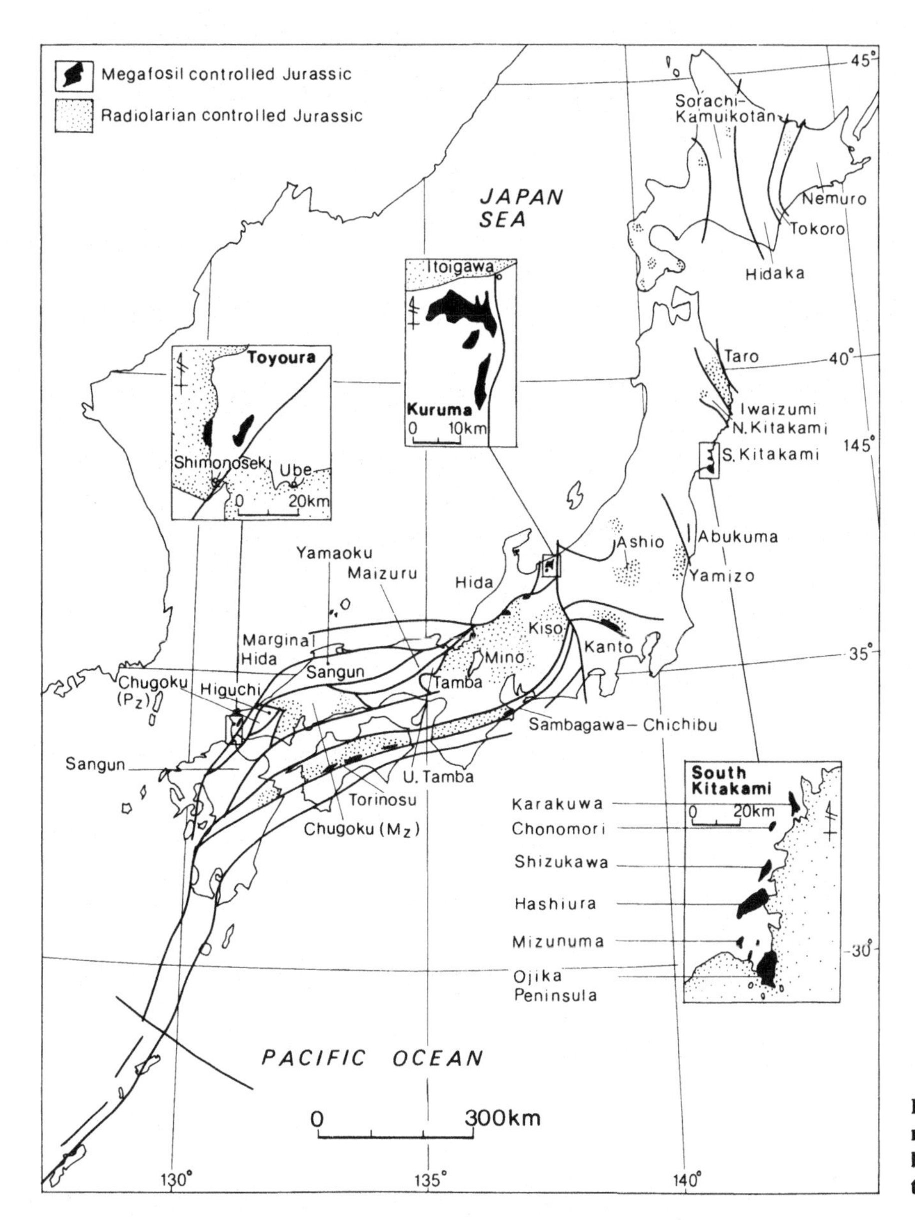


**Figure 9.1. Distribution of the Jurassic exposures in Japan. Solid lines show the boundaries of tectonic belts.**

stratigraphic sequences established in the individual basins are shown in Figure 9.2.

### *Hida Terrane*

Postorogenic sediments accumulated in local basins mostly along the boundary between the Hida and Hida Marginal Belts. The basins are limited in size and have the outlines of embayments(?) or elongated depressions, and some were created as pull-apart or oblique-slip basins. The sediments are generally marine, with nonmarine or brackish beds intercalated locally.

*Toyora Group.* The Toyora Group is a series of marine, relatively fine grained sediments deposited mostly in basin-floor environments. The group lies unconformably on the metamorphic rocks of the Hida Marginal Belt. The stratigraphy and its ammonite zonation were established by Hirano (1971, 1973a,b). The age of the group ranges from Hettangian to Bathonian (Figure 9.3).

The group is rich in ammonites, and three assemblage zones ranging from Upper Pliensbachian to Lower Toarcian have been established in the Nishinakayama Formation. Several other horizons have been dated by the occurrences of isolated ammonites (Figure 9.3). The faunas have generally Tethyan affinities, except for the Boreal *Amaltheus* in the Higashinagano Formation (Sato 1960). In the Upper Pliensbachian, this site was located in the overlap of Boreal and Tethyan realms.

Bivalves are not uncommon, and seven faunules were identified by Hayami (1961b). *Inoceramus* is abundant in the pelitic parts, suggesting an offshore environment.

The Nishinakayama Formation yields a land flora characterized by *Gleichenites*, *Phlebopteris*, and *Sphenopteris* and unique for eastern Asia, including the nearby Kuruma area. Kimura (1988)

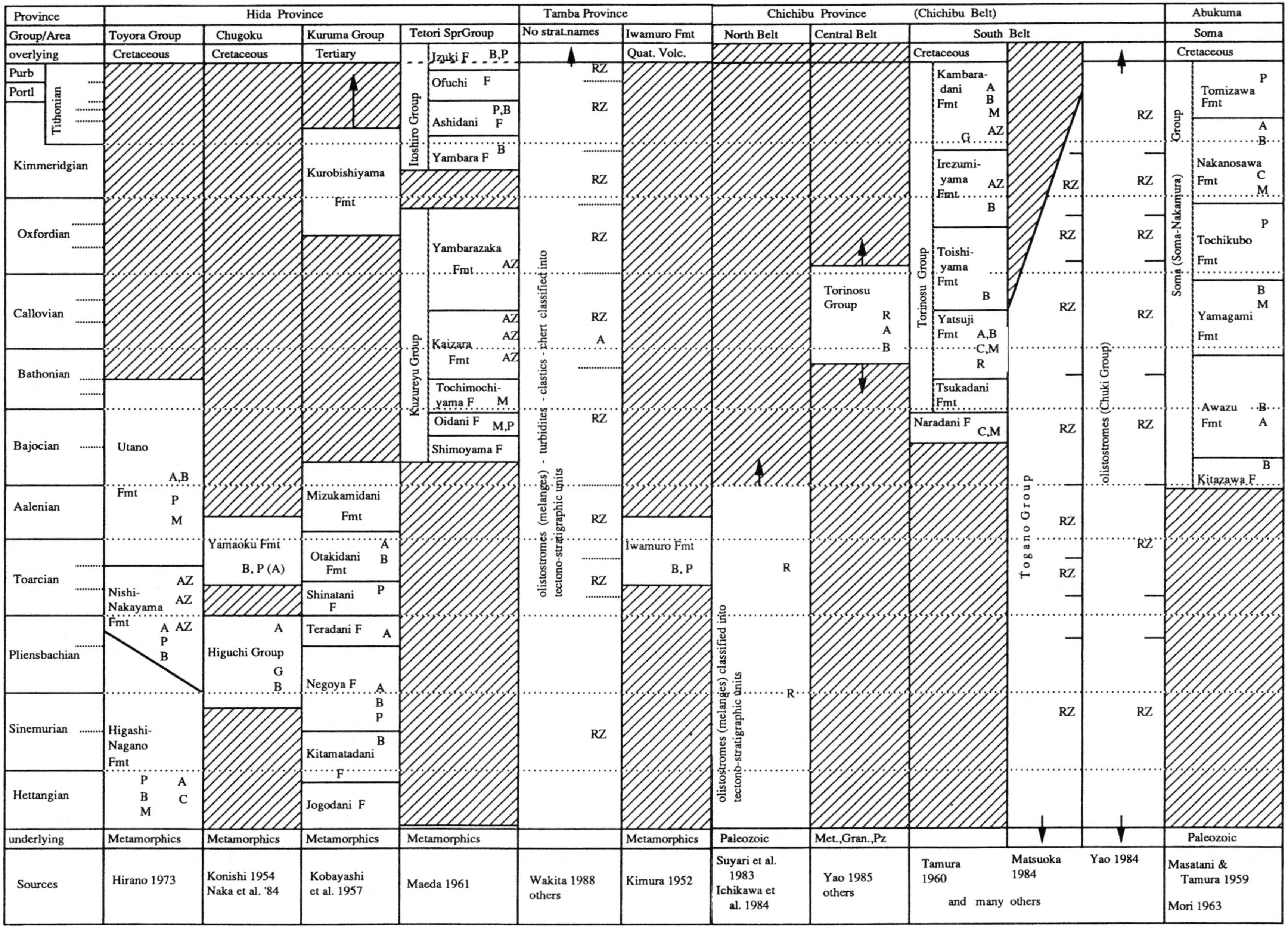

| Province | Hida Province | | | | Tamba Province | | Chichibu Province (Chichibu Belt) | | | | | Abukuma |
|---|---|---|---|---|---|---|---|---|---|---|---|---|
| Group/Area | Toyora Group | Chugoku | Kuruma Group | Tetori SprGroup | No strat.names | Iwamuro Fmt | North Belt | Central Belt | South Belt | | | Soma |
| overlying | Cretaceous | Cretaceous | Tertiary | | | Quat. Volc. | | | Cretaceous | | | Cretaceous |
| Tithonian (Purb, Portl) | | | | Itoshiro Group: Izuki F B,P; Ofuchi F; Ashidani F P,B | RZ; RZ | | | | Torinosu Group: Kambara-dani Fmt A B M AZ G | | RZ | Soma (Soma-Nakamura) Group: Tomizawa Fmt P A B |
| Kimmeridgian | | | Kurobishiyama Fmt | Yambara F B | RZ | | | | Irezumi-yama Fmt AZ B | RZ | RZ | Nakanosawa Fmt C M |
| Oxfordian | | | | Kuzureyu Group: Yambarazaka Fmt AZ | RZ | | | | Toishi-yama Fmt B | RZ | RZ | Tochikubo Fmt P |
| Callovian | | | | Kaizara Fmt AZ AZ AZ | RZ A | | | Torinosu Group R A B | Yatsuji Fmt A,B C,M R | RZ | RZ | Yamagami Fmt B M |
| Bathonian | | | | Tochimochi-yama F M | | | | | Tsukadani Fmt | | | Awazu Fmt |
| Bajocian | Utano Fmt A,B P M | | | Oidani F M,P; Shimoyama F | RZ | | | | Naradani F C,M | RZ | RZ | Awazu Fmt B A; Kitazawa F B |
| Aalenian | | | Mizukamidani Fmt | | RZ | | | | | Togano Group RZ | olistostromes (Chuki Group) RZ | |
| Toarcian | Nishi-Nakayama Fmt AZ AZ AZ | Yamaoku Fmt B, P (A) | Otakidani Fmt A B; Shinatani F P | | olistostromes (melanges) - turbidites - clastics - chert classified into tectono-stratigraphic units RZ | Iwamuro Fmt B, P | olistostromes (melanges) classified into tectono-stratigraphic units R | | | RZ | | |
| Pliensbachian | A P B | Higuchi Group A G B | Teradani F A; Negoya F A B P | | | | | | | | | |
| Sinemurian | Higashi-Nagano Fmt | | Kitamatadani F B | | RZ | | R | | | RZ | RZ | |
| Hettangian | P A B C M | | Jogodani F | | | | | | | | | |
| underlying | Metamorphics | Metamorphics | Metamorphics | Metamorphics | | Metamorphics | Paleozoic | Met.,Gran.,Pz | | | | Paleozoic |
| Sources | Hirano 1973 | Konishi 1954 Naka et al. '84 | Kobayashi et al. 1957 | Maeda 1961 | Wakita 1988 others | Kimura 1952 | Suyari et al. 1983 Ichikawa et al. 1984 | Yao 1985 others | Tamura 1960 and many others | Matsuoka 1984 | Yao 1984 | Masatani & Tamura 1959 Mori 1963 |

**Figure 9.2. Correlation chart of the Jurassic formations in Japan. Ammonite (AZ) and radiolarian (RZ) assemblage zones are shown. Letters in the columns denote the occurrences of fossils. Classification**

| Province | South Kitakami | | | North Kitakami | | Hokkaido | | Ammonite assemblage zones (ammonite horizons) | Representative faunules of Bivalves | Radiolarian assemblage zones (Radiolarian interval zones) |
|---|---|---|---|---|---|---|---|---|---|---|
| Group/Area | Western Belt | Karakuwa | Ojika | Iwaizumi Belt | Taro Belt | Sorachi Belt | Tokoro Belt | | | |
| overlying | Cretaceous | Cretaceous | | Tertiary | Cretaceous | | | | | ↑ |
| Purb / Portl (Tithonian) | [Hashiura Group] Sodenohama Fmt B A | [Karakuwa Group] Kogoshio A P B | [Ojika (Oshika) Group] ↑ Ayukawa F F | R C | R C | [Sorachi Group] Upper part R | R | Substeueroceras<br>Corongoceras<br>Aulacosphinctoides cf.steigeri Association | | Pseudodictyomitra cf.carpatica AZ<br>Pseudodictyomitra primitiva - P.sp.A. AZ |
| Kimmeridgian | A | Fmt A | A | M Iwaizumi Group | M Rikuchu | | | Virgataxioceras<br>Ataxioceras kurisakense AZ<br>Aspidoceras | Myophorella crenulata f. | (Pseudodictyomitra primitiva Z.)<br>Tricolocapsa yaoi AZ (Cingulotturis carpatica Z.) |
| Oxfordian | Arato AZ | Mone A | Oginohama P AZ | [Shimokita]<br>[Tsugaru]<br>[Oshima ] | Group | Lower part | Nikoro Group | Perisphinctes s.s.<br>Kranaosphinctes matsushimai AZ | Nipponitrigonia sagawai f.<br>Latitrigonia tetoriensis f. | Gongylothorax sakawaensis - Stichocapsa naadaniensis AZ (Stilocapsa (?) spiralis Z.) |
| Callovian | Fmt B A | P Fmt | Formation | | | | | Oxycerites cf.sulaensis AZ<br>Kepplerites japonicus AZ | Inoceramus maedae faunule | Guexella nudata AZ (Tricolocapsa conexa Z.) |
| Bathonian | | Ishiwari-toge Fmt | | | | | | Neuqueniceras yokoyamai AZ<br>Sphaeroceratids | Latitrigonia unicarinata f.<br>Retroceramus utanoensis f. | |
| Bajocian | AZ B | B Tsunakizaka Fmt AZ | Tsukinoura Fmt A AZ | | | | | Leptosphinctes association<br>Sonninia Ass.Zone | Latitrigonia pyramidalis f.<br>Kobayashites hemicylindricus faunule<br>Trigonia sumiyagura faunule | Unuma echinatus AZ (Tricolocapsa plicarum Z.) |
| Aalenian | Aratozaki F<br>[Shizukawa Group] AZ AZ | Kosaba F B | B | | | | | Planammatoceras planinsigne AZ<br>Hosoureites ikianus AZ | Inoceramus ? kudoi faunule | Hsuum hisuikyoensis (Laxtorum (?) jurassicum Z.) |
| Toarcian | B M P | | | ↓ | ↓ | | | Dactylioceras helianthoides AZ<br>Protogrammoceras nipponicum AZ | Bakevellia magnissima faunule | Parahsuum sp. D AZ (Archicapsa pachyderma Z.) |
| Pliensbachian | Hosoura Formation | | | | | | | Fontanelliceras fontanellense Ass.Zone<br>Eoderoceras ? | Radulonectites japonicus f. | |
| Sinemurian | A | | | | | | | Arnioceras | Prosogyrotrigonia inouyei f.<br>Cardinia toriyamai faunule | Parahsuum simplum AZ (Parahsuum sp. C Z.) |
| Hettangian | Niranohama F P A B | | | | | | | Alsatites | Trigonia senex faunule<br>Geratrigonia hosourensis f.<br>Burmesia japonica faunule | |
| underlying | Triassic | Triassic | Triassic | | | | | | | |
| Sources | Hayami 1961<br>Sato 1962<br>Takahashi 1969<br>others | Hayami 1961<br>Takizawa 1985<br>others | Takizawa 1985<br>others | Minoura 1983<br>Sugimoto 1974 | Sugimoto '74<br>Minoura & Taketani '82 | Komatsu 1985 | Iwata et al. 1983 | Sato 1962<br>Hirano 1973<br>Sato & Westermann 1985 | Hayami 1961a, 1985 | Matsuoka & Yao 1986 (see also Pessagno & Mizutani, this volume) |

A: Ammonites, B: Bivalves, C: Corals, M: Miscellaneous, P: Plants, R: Radiolarians, AZ: ammonite assemblage zones, RZ: radiolarian assemblage zones

**Figure 9.2** (*cont.*)

| Stage | Formation | m | Lithology | Representative fossils | | Biostrat. unit |
|---|---|---|---|---|---|---|
| Bathonian<br>Bajocian<br>Aalenian<br>Toarcian (part) | Utano Formation | 510 - 1000 | silty shale,<br>sandy shale,<br>sandstone<br>in alternation | Ammonites<br>Belemnites<br>Bivalves<br>Plants | Bositra<br>Inoceramus<br>Harpophylloceras<br>Planammatoceras<br>Pseudolioceras<br>Phymatoceras<br>Grammoceras | Inoceramus utanoensis faunule<br>Bositra gr.ornati faunule<br>Utano flora |
| Toarcian<br>Pliensbach | Nishi-Nakayama Formation | 220 - 250 | Black shale,<br>silty shale &<br>sandstone<br>in upper part | Ammonites<br>Bivalves<br>Plants | Dactylioceras<br>Protogrammoceras<br>Fontanelliceras | Dactylioceras helianthoides AZ<br>Protogrammoceras nipponicum AZ<br>Fontanelliceras fontanellense AZ<br>Parainoceramus lunaris faunule<br>Parainoceramus matsumotoi f.<br>Nishinakayama flora |
| Sinemurian<br>Hettangian | Higashi-Nagano Formation | 350 | sandy shale<br>fine sand-stone<br>basal conglo-merate | Ammonites<br>Bivalves<br>Gastropods<br>Scaphopods<br>Brachiopods<br>Corals<br>Crinoids | Amaltheus<br>Arieticeras<br>Harpophylloceras<br>Vermiceras | Oxytoma kobayashii faunule<br>Prosogyrotrigonia inouyei f.<br>Cardinia toriyamai faunule |
| Sources | stratigraphy: Hirano 1973<br>ammonites: Hirano 1971,73,73a<br>Sato & Westermann 1985<br>bivalves: Hayami 1961a, 75 | | | plants: Kimura 1988 | | |

**Figure 9.3. Succession of the Toyora Group in the Toyora area, western Chugoku.**

suggested that this flora may be of a southern type that flourished at the southern margin of Tethys.

The Higuchi Group, in the Paleozoic belt of the Chugoku, is similar in lithology and fauna to the Toyora Group. The shallow-marine sediments lie unconformably on deformed Paleozoic. Upper Pliensbachian *Fontanelliceras* occurs along with *Amaltheus* (Naka et al. 1985).

*Kuruma Group*. The thick, coarse-grained clastics (Kobayashi et al. 1957) constitute superposed fan-delta systems and lie unconformably on metamorphic rocks. The sedimentary facies range from fluvial-littoral to offshore environments, separated by two pelitic horizons representing a basin-floor environment (i.e., the Teradani and Otakidani Formations). Tectonically the group represents the postorogenic sediments after the pre-Jurassic collision of Inner Southwest Japan with the Hida landmass. The basin is rounded, with the original outline apparently rather well preserved (Figure 9.4).

Mollusks, especially bivalves, and plants are abundant almost throughout the group. Ammonites are known only from two offshore mudstone sequences. The total range of the group is probably Hettangian–Toarcian. The Upper Pliensbachian ammonite fauna of the Teradani Formation is characterized by the coexistence of Boreal *Amaltheus* and Tethyan *Canavaria*. This suggests that the Kuruma area was in the biogeographic overlap (Sato 1960).

The bivalve faunas are rather endemic, except for a few elements common between the coeval Yamaoku and Iwamuro faunas with similar tectonic setting (Hayami 1961b). The land flora of the Kuruma Group (Kuruma flora) is characteristic for the Early Jurassic in Japan and South China, but elements in common with the coeval Nishinakayama flora are missing (Kimura 1988).

The Yamaoku Formation is exposed in a small area within the Sangun metamorphic belt in central Chugoku. The lithology is comparable to that of the Kuruma Group, and the bivalve faunule indicates Jurassic (Konishi 1954).

*Tetori Supergroup*. This is a Jurassic-Cretaceous clastic sequence exposed in a series of localized basins along the boundary of the Hida and Hida Marginal Belts. The sequence differs somewhat in the basins, but is always postorogenic, lying unconformably on more or less metamorphosed basement. In the type

| Stage | Formation | m | Lithology | Representative fossils | | characteristic faunas/floras |
|---|---|---|---|---|---|---|
| Bajocian<br>Aalenian | Mizukami-dani Formation | 2000 | conglomerate coarse sand-stone, shale | barren | | |
| Toarcian | Otakidani Formation | 850 | sandy shale congl., s.s. | Ammonites<br>Belemnites<br>Bivalves | Pseudogrammoceras<br>Hammatoceras | Geratrigonia kurumensis f. |
| Toarcian | Shinatani Formation | 700 | sandstone, shale | Bivalves<br>Plants | | Kuruma flora<br>Bakevellia magnissima f.<br>Camptonectes sp. faunule |
| Pliensbach | Teradani Formation | 600 | sandy shale | Ammonites<br>Bivalves | Amaltheus<br>Canavaria | Pleuromya hashidatensis f. |
| Pliensbach / Sinemurian | Negoya Formation | 1600 | sandstone-shale in alternation | Ammonites<br>Bivalves<br>Plants | Eoderoceras | Bakevellia negoyensis f. |
| Sinemurian | Kitamata-dani Formation | 2000 | sandstone/ shale in alternation | Bivalves<br>Plants | | Bakevellia ohishiensis f. |
| Hettangian | Jogodani Formation | 700 | conglomerate shale | barren | | |
| Sources | stratigraphy: Kobayashi et al. 1957<br>Chihara et al. 1979 | | | ammonites: Sato 1962<br>bivalves: Hayami 1961a<br>plants: Kimura 1988 | | |

**Figure 9.4. Succession of the Kuruma Group in the Kuruma area eastern Hida Mountains.**

area at the Kuzuryu River, a complete sequence was established (Figure 9.5). The supergroup is generally divided into three groups: The Kuzuryu Group, below, is marine; the Akaiwa Group, above, is lacustrine (Cretaceous); the Itoshiro Group, middle, is transitional between the other two and mostly brackish in facies (Maeda 1961).

Fossils are common in all the groups. Ammonites are abundant in the Kuzuryu Group, especially in the Kaizara and Yambarazaka Formations, where four Upper Bathonian to Lower Oxfordian ammonite assemblage zones have been established (Sato and Westermann 1985) (Figure 9.5). These zones represent a standard for the Middle Jurassic biochronology of Japan and its surrounding areas.

The bivalves of the basins vary in composition, indicating different biofacies. Trigonia-bearing sandstone is developed in the Kuzuryu Group, and the "Corbiculid"-bearing shale is dominant in the Itoshiro Group (Hayami 1961b, 1975). The middle Itoshiro Group yields a rich macroflora, named Tetori flora, a representative Upper Jurassic–Lower Cretaceous flora in eastern Asia (Kimura 1988).

### *Tamba Terrane*

The Tamba Terrane, as here designated, includes the Tamba, Mino, and Kiso areas in Southwest Japan and equivalent areas of Ashio and Yamizo in Northeast Japan.

The Jurassic to earliest Cretaceous ages of the sediments are based on radiolarians, mostly from the pelitic matrix, but until the 1980s a late Paleozoic age had been accepted based on fusulines and corals from the exotic limestone bodies embedded in the matrix. The beds are of mélange or olistostrome type, comprising exotic blocks of limestone, greenstone, chert, and other clastics of late Paleozoic, Triassic, and even Jurassic ages. As the ordinary stratigraphic procedure cannot be used here, tectonostratigraphic classification is introduced throughout the province. Each tectonostratigraphic unit forms a tectonic slice bounded on both sides by thrusts and is characterized by a particular association of exotic blocks of different lithologies and ages.

The sequence varies regionally, and no composite section can yet be presented. Wakita (1988) established six tectonostratigraphic units in the Mino area. Brief descriptions of the units of this area are as follows (Figure 9.6):

*Sakamototoge unit.* This is a mélange, including clasts of limestone, greenstone, chert, and siliceous shale of Carboniferous to Jurassic age, intercalated with disrupted turbidite. The age of the matrix is Pliensbachian–Toarcian; *Parahsuum* sp. *D* to *Unuma echinatus* Radiolarian Zones.

*Samondake unit.* This is a massive sandstone intercalated with turbidite beds, and in the lower part with exotic blocks of

| Stage | Group/Formation | | m | Lithology | Representative fossils | Biostrat. unit |
|---|---|---|---|---|---|---|
| Cretaceous | Itoshiro G. | Izuki Fmt | 50 | non marine shale | non-marine bivalves | Myrene tetoriensis f. |
| Tithonian | | Ofuchi Fmt | 150 | non marine conglomerate | non-marine bivalves Plants | Tetori flora |
| | | Ashidani Fmt | 180 | sandstone/shale | Bivalves Ostrea | |
| Kimmeridge | | Yambara Fmt | 50 | conglomerate | Bivalves | Vaugonia yambaraensis f. |
| | hiatus | | | | | |
| Oxfordian | Kuzuryu Group | Yambarazaka Formation | 120 | conglomerate, sandstone, shale in alternation | Ammonites Belemnites Bivalves Plants | Inoceramus cf.nitescense f. Latitrigonia tetoriensis f. Kranaosphinctes matsushimai Ass.Zone |
| Callovian | | Kaizara Formation | 200 230 | sandy shale | Ammonites Bivalves | Oxycerites cf.sulaensis AZ Kepplerites japonicus AZ Neuqueniceras yokoyamai AZ Inoceramus hamadae faunule |
| Bathonian | | Tochimochi-yama Fmt | 200 | coarse/fine sandstone | Belemnites | |
| Bajocian | | Oidani Fmt | 300 | sandstone/shale | Belemnites, Plants | |
| | | Shimoyama F. | 300 | conglomerate, sandstone | barren | |
| Aalenian | | | | | | |
| Sources | stratigraphy: Maeda 1961<br>ammonite zones: Sato & Westermann 1985 | | | | ammonites: Sato 1962<br>bivalves: Hayami 1961a<br>plants: Kimura 1988 | |

**Figure 9.5. Succession of the Tetori Supergroup in the type section of the Kuzuryu area, Hida Mountains.**

Permian–Triassic chert. The shale layers are dated Bathonian–Callovian; *Unuma echinatus* to *Guexella nudata* Radiolarian Zones. A solitary occurrence of lower Callovian *Kepplerites* sp. (Sato, Kasahara, and Wakita 1985) is consistent with this age.

*Funafuseyama unit.* This is stacked slices of mélanges, including disrupted turbidites, huge masses of limestone, and exotic blocks of chert and greenstone; Permian–Triassic; shaly matrix, Middle Jurassic (Bathonian?); *Hsuum hisuikyoense* to *Unuma echinatus* Radiolarian Zones.

*Nabi unit.* This is a chaotic mixture of slices of sandstone and shale, chert, limestone, disrupted turbidite, and other things; shale matrix, Oxfordian; *Guexella nudata* to *Gongylothoras sakawaensis–Stichocapsa naradaniensis* Radiolarian Zones.

*Kanayama unit.* This is composed of mélanges containing clasts of sandstone, siliceous shale, greenstone, and chert; Triassic to Jurassic age; shale matrix, Oxfordian–Berriasian; *Gongylothorax sakawaensis–Stichocapsa naradaniensis* to *Pseudodictyomitra* cf. *carpatica* Radiolarian Zones.

*Kamiaso unit.* This is sandstone, turbidite, siliceous shale, and chert, locally conglomeratic. The conglomerate contains pebbles of Precambrian gneiss. Sediments are at least partly Upper Bathonian–Callovian, judged by *Choffatia* (*Subgrossouvria*) sp. from shale bed near Inuyama (Sato 1974); radiolarian ages are consistent. Mesozoic petrified woods are also known (Nishida, Adachi, and Kondo 1974).

Though this stratigraphy may not have general validity, similar chaotic sediments are widely distributed elsewhere: the Tamba (Ishiga 1983) and Kiso areas (Otsuka 1988), the Ashio and Yamizo mountains in Northeast Japan (Aono 1985), and the westward extension of Tamba in western Chugoku (Hayasaka, Isozaki, and Hara 1983). The ages based on rare ammonites are consistent with those based on radiolarians. Besides the mentioned *Kepplerites* and *Choffatia* from the Mino Belt, middle Oxfordian ammonites [i.e., *Perisphinctes* (*s.s.*)] are known from the mélange matrix of the Kuga Group in western Chugoku (Sato, Hashimoto, and Suyama 1986), and *P.* (*Kranaosphinctes*) from the Yamizo Mountains in the North Kanto region (Suzuki and Sato 1972).

### *Chichibu Terrane*

This terrane corresponds, tectonically, roughly to the Sambagawa–Chichibu Belt. This is a narrow tectonic belt that extends the whole of Outer Southwest Japan (Figure 9.7). The Chichibu Belt (*s.s.*) is divided tectonically into three subbelts. Jurassic is known from all the subbelts, but facies and structures differ. The rocks are roughly classified into two types: chaotic deposits of olistostrome type and ordinary clastic sediments (Ichikawa, Hada, and Yao 1985; Yao 1985). Equivalent rocks extend from southern Shikoku westward to central Kyushu and eastward to the Kanto Mountains, passing through the Kii Peninsula and the central

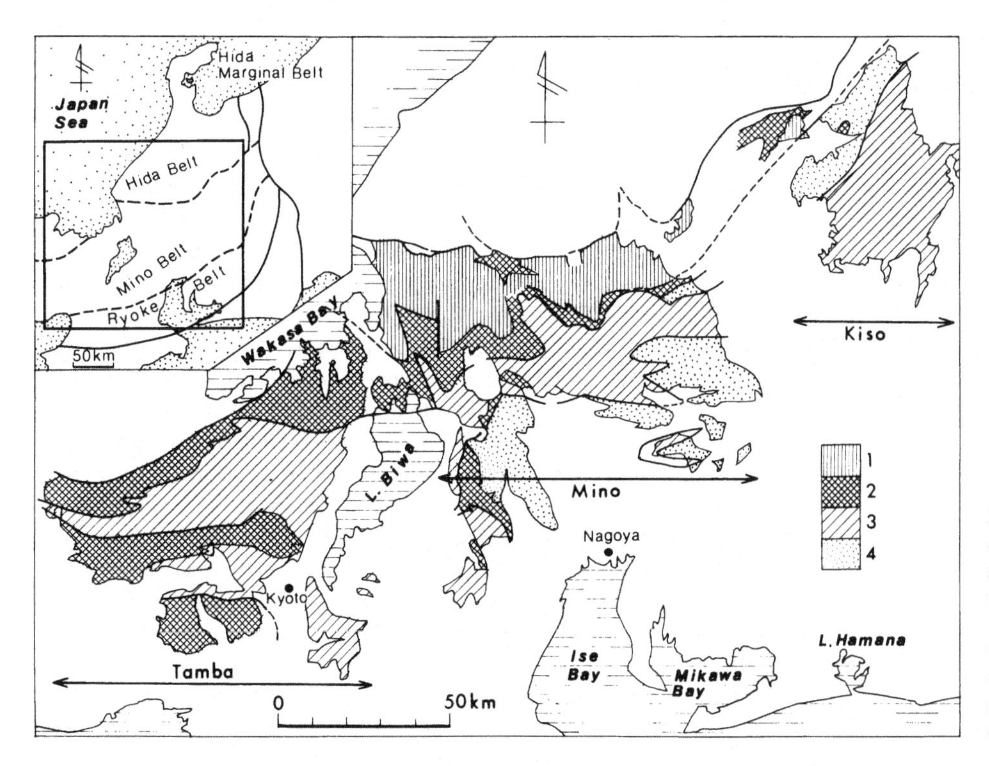


**Figure 9.6. Geological sketch map of the Tamba-Mino-Kiso areas showing the tectonostratigraphic units. (Adapted from Wakita 1988.) 1, Sakamototoge and Samondake units; 2, Funafuseyama and Nabi units; 3, Kanayama unit; 4, Kamiaso unit. See text for explanation of the tectonostratigraphic units.**

Chubu Region. This belt lies on the southern side of the Sambagawa Metamorphic Belt.

*Northern subbelt.* The Jurassic rocks here are classified into two facies groups: One is characterized by the association of sandstone and shale, accompanied by chaotic deposits, including exotic blocks and sheets of greenstone and chert; the other is characterized by chaotic beds of olistostrome type comprising exotic blocks of Paleozoic limestone, greenstone, and chert, and Triassic chert. The fine-grained clastics contain Lower and Middle Jurassic radiolarian assemblages.

The Jurassic beds are discontinuous, and radiolarian assemblages procured sporadically from the fine matrix range from Early to Middle Jurassic. This interval should include the emplacement of the olistostromes. A representative sequence is in south-central Shikoku (Suyari, Kuwano, and Ishida 1983). Similar rocks are exposed elsewhere in Chichibu province, such as the western part of Kii Peninsula (Ichikawa et al. 1985) and the northern belt of the Kanto Mountains (Sashida et al. 1982; Hisada 1983).

*Central subbelt.* Jurassic rocks occur here as small lenticular bodies sandwiched between older tectonic slices. The Jurassic is generally sandstone and shale, subject to ordinary stratigraphy; rare ammonites and radiolarians indicate the Middle Jurassic. This unit is classified as a member of the classical Torinosu Group, developed largely in the southern subbelt.

*Southern subbelt.* This subbelt has an overall imbricate structure; narrow tectonical slices are disposed roughly parallel to the general trend of the Chichibu Belt, and each slice consists of a somewhat distinct lithologic sequence. The Jurassic differs lithologically from one sheet to the other and is classified into three lithofacies groups in southern Shikoku (Yao 1985).

The first group is characterized by clastics and chert, partly with siliceous shale, collectively called the Togano Group. A compound sequence of the tectonic slices indicates that the group ranges from Middle Triassic to Late Jurassic. Eight radiolarian zones and interval zones were established for almost the entire Jurassic (Matsuoka 1984; Matsuoka and Yao 1986) (Figure 9.2).

The second facies group is the olistostrome sequence developed in the corresponding southern subbelt in western Kii (Yao 1984). The exotic blocks are Middle Triassic to Early Jurassic chert, Paleozoic–Triassic limestone and greenstone, and so on. The subbelt is formed by stacked tectonic units, each separated by thrusts. The total sequence is not observable in any single sheet and has been composed. Seven radiolarian assemblage zones are recognized in the Jurassic (Matsuoka 1983) (Figure 9.2).

The third facies group consists of sandstone, shale, and reef-limestone lenses. This is the classical Torinosu Group, whose stratigraphy was established in central Shikoku (Tamura 1960) (summary in Figure 9.8). Different Callovian to Oxfordian ages are suggested by ammonites; the lower Tithonian is indicated by the *Aulacosphinctoides* Association, which is distributed widely in Japan, including southern Kitakami. The *Ataxioceras kurisakense* Association, found in an isolated outcrop lithologically similar to the type Torinosu Group in the Kurisaka area of Shikoku, is the only Kimmeridgian ammonite assemblage established in Japan. It is characterized by *Ataxioceras* and *Euaspidoceras.* The uppermost Jurassic seems to be continuous with the Lower Cretaceous of western Kyushu (Sato 1961). The Lower Cretaceous is suggested for the Torinosu Group by Neocomian radiolarians

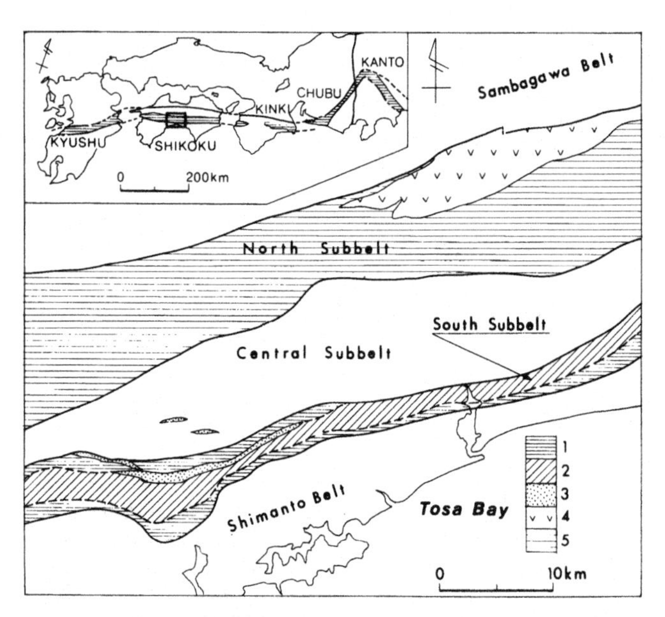


**Figure 9.7. Sketch map of the southern part of Shikoku, showing the subbelts and the distribution of Jurassic. (Adapted from Yao 1985.) 1, olistostrome-type Jurassic–Lower Cretaceous in Southern Subbelt; 2, chert-clastics of Middle Triassic to Jurassic (Togano Group) in Southern Subbelt; 3, Torinosu-type Jurassic; 4, metavolcanics; 5, olistostrome-type Triassic–Jurassic in Northern Subbelt.**

(Suyari and Ishida 1985). The Torinosu Group is generally rich in bivalves and other mollusks, brachiopods, corals, stromatopores, and calcareous algae known mostly from the limestone lenses of the Yatsuji Formation (Tamura 1961) and widely distributed along the Chichibu Belt, from western central Kyushu to the Kanto Mountains.

The Soma Group is exposed in a narrow N–S belt in the eastern Abukuma Mountains. Lithologically similar to the Torinosu Group in Shikoku, but somewhat thicker, the group consists of sandstone and shales, with bituminous limestone lenses (Masatani and Tamura 1959; Mori 1963) (Figure 9.9). It appears to be bounded by faults. The depositional environment is believed to have been a fan-delta system developed upon the eroded Upper Paleozoic (Takizawa 1985). Ammonites are rather common, but generally poorly preserved. Some were recorded from the limestone lenses and given partly different ages. The lower Tithonian is documented by the *Aulacosphinctoides* Association, as in southern Kitakami. The Upper Bajocian is suggested by a solitary occurrence of a supposed *Bigotites* (Sato 1962). The calcareus beds in the middle of the sequence have yielded ammonites, bivalves, gastropods, brachiopods, corals, stromatopores, and calcareous algae, indicating Upper Jurassic age (Tamura 1961; Mori 1963).

### *Southern Kitakami Terrane*

The Jurassic consists of shallow-marine and partly fluvial sediments deposited in a fan-delta environment. Unlike other provinces, the Jurassic is part of an essentially continuous Upper Paleozoic to Lower Cretaceous sequence, with a minor disconformity between Jurassic and Upper Triassic. No drastic crustal movements took place at the systems' boundaries.

*Shizukawa and Hashiura Groups.* Although the Jurassic is now exposed in four isolated basins, lithology and succession vary little, and the sediments are judged to have been deposited in an essentially continuous basin. A two-fold division is generally accepted (Hayami 1961a; Sato 1962; Takahashi 1969): a lower Shizukawa Group and an upper Hashiura Group (Figure 9.10). Except for the Sodenohama Formation at the top, the same stratigraphic units are distinguished in the three areas.

The Shizukawa Group represents one transgressive cycle, beginning with shore-face to breaker-zone sandstone and conglomerate (Niranohama Formation) and grading into offshore mud-facies shale (Hosoura Formation). Fossils are abundant, especially ammonites and bivalves, and two ammonite assemblage zones are distinguished in the upper part of the Hosoura Formation (Sato 1962). Upper Hettangian and lower Sinemurian are indicated by rare ammonites. At least six bivalve faunules can be distinguished (Hayami 1961b). The group ranges from Hettangian to Aalenian.

The overlying Hashiura Group represents the second transgressive cycle, again beginning with shore-face sandstone beds (Aratozaki Formation) grading to offshore mud (Arato Formation). Thick sandstone (Sodenohama Formation) overlying this group represents another incomplete cycle. Fossils are less common than in the Shizukawa Group. The lower Upper Bajocian *Leptosphinctes* gr. *davidsoni* Assemblage Zone was distinguished in the lower part of the Arato Formation (Sato and Westermann 1985). The equivalents of the middle Oxfordian *Perisphinctes matsushimai* Assemblage Zone and the early Tithonian *Aulacosphinctoides* Association are recognized by ammonites. Some other Upper Jurassic horizons are indicated by a few ammonites. Four bivalve faunules are known (Hayami 1961b). The group is dated as Bajocian–Tithonian.

*Karakuwa Group.* This group represents two superposed fan-delta systems that extended from fluvial to offshore environments (Takizawa 1985). It lies with paraconformity on the Upper Triassic Inai Group and is continuous into the Lower Cretaceous. The age of the group is late Aalenian to Tithonian (Figure 9.11).

The lower two formations form a transgressive cycle, beginning with nearshore sandstone (Kosaba Formation) and ending with basin-mud facies (Tsunakizka Formation). Fossils are common. The upper Lower Bajocian *Sonninia* Assemblage Zone is established in the Tsunakizaka Formation (Sato and Westermann 1985). Bivalves are abundant in the Kosaba Formation, forming a nearshore *Trigonia* faunule (Hayami 1961b). The Ishiwaritoge Formation is an alluvial-fan deposit at the base of the second transgressive cycle. The Mone Formation is in basin-mud facies, yielding some rare ammonites. The uppermost Kogoshio Formation is again in alluvial-fan facies, containing rather rich ammonite and bivalve faunas. Fossils are rather common in these formations. Three bivalve faunules can be distinguished (Hayami

| Stage | Formation | m | Lithology | Representative fossils | | Biostrat. unit |
|---|---|---|---|---|---|---|
| Tithonian | Kamabaradani Formation | 70 | shale arkose sandstone | Ammonites Belemnites Bivalves Brachiopods | Aulacosphinctes Corongoceras | Aulacosphinctoides Assoc. |
| Kimmeridge | Irezumizawa F | 130 | shale, sandstone | Bivalves Plants | | |
| | Toishiyama F. | 15 | sandstone | Bivalves | | |
| Oxfordian | Yatusji Formation | 100 - 350 | shale, sandstone limestone lenses | Ammonites Bivalves Gastropods Brachiopods Corals Stromatopores Algae | Lithacoceras Virgataxioceras Euaspidoceras | |
| Callovian | | | | | Horioceras | |
| Bathonian | Tsukadani Formation | 40 - 70 | sandstone conglo-merate | | | |
| Bajocian | | | | | | |
| Sources | stratigraphy: Tamura 1960; ammonites: Sato 1962; bivalves: Tamura 1961; Hayami 1975 | | | fossils in general: Tamura 1961 | | |

**Figure 9.8. Succession of the Torinosu Group at its type section in southern Shikoku.**

1961b) in the Mone and Kogoshio Formations. The Mone flora, with typically Japanese Upper Jurassic–Lower Cretaceous development, is also known from both formations (Kimura 1988).

This group is overlain by Lower Cretaceous pyroclastics, including Berriasian ammonite beds. The topmost horizon of the Japanese Jurassic is dated in this area by the late Tithonian *Substeueroceras* (Sato 1962).

The group forms a rather open N–S syncline, with well-developed axial, slaty cleavage (Iwamatsu 1969); the fossils are all strongly crushed and elliptically deformed.

*Ojika Group.* This group is extensively developed in the Ojika Peninsula and has been dated to the Middle–Upper Jurassic; Lower Jurassic is again lacking here (Takizawa 1985). The group consists mostly of sandstone and shale representing fan-delta systems. The general succession has been established (Figure 9.12). The Tsukinoura Formation represents littoral sand overlain by quiet-basin mud. The Oginohama Formation is another fan-delta system, representing alluvial-fan to basin-floor mud facies. Fluvial-fan deposits are intercalated, as well as lake or marsh deposits accumulated behind the delta. The group is strongly folded into a train of tight NE–SW folds.

Fossils are common only in some horizons; ammonites are rare (Sato 1962). The Lower Bajocian is indicated by rare *Stephanoceras*, and the middle Oxfordian by occasional *Perisphinctes s.s.* and *P.* (*Kranaosphinctes*), all found in the offshore mud facies (Oginohama Formation); rare occurrences of *Satoceras* indicate lower Bathonian (Westermann and Callomon 1988); *Choffatia* gr. *recuperoi* suggests Lower Callovian; the Tithonian is identified by *Aulacosphinctoides*. Bivalves occur in at least five faunules (Hayami 1961b).

### *North Kitakami Terrane*

The rocks of the Iwaizumi and Taro Belts are mostly clastics of various grain sizes, intercalated with lenses of limestone, chert, and pyroclastics, including pillow lavas. The overall lithology and facies are similar to those in the Chichibu Terrane. The sediments are considered to form normal sequences and have been dated at least partly as Jurassic by poorly preserved corals, stromatopores, and calcareous algae found in the limestone lenses (Sugimoto 1974).

Recent discoveries of Permian–Triassic conodonts from cherts and of Middle–Upper Jurassic radiolarians from the siliceous shales (Taketani and Minoura 1984) indicate that, like the chaotic beds of Chichibu province, these sediments are of olistostrome type (Minoura 1983). The radiolarian-bearing matrix, dated Late Jurassic, should represent the age of the emplacement of the formations.

Early Jurassic radiolarians are also known from the northern Kitakami Terrane, on the southwestern side of the Iwizumi Belt (Matsuoka 1988).

| Stage | Formation | m | Lithology | Representative fossils | biostrat. unit |
|---|---|---|---|---|---|
| Cretraceous | Koyamada F | 160 | shale/sandst. | Ammonites Thurmanniceras | |
| Tithonian | Tomizawa Fmt | 400 | coarse sand-stone & shale | | |
| Kimmeridge | Nakanosawa Fmt | 140 | arkose and/or calcareous sand-stone, sapropeli-tic limestone lenses | Stromatopores, Hydrozoans, Corals, Ammonites, Bivalves, Brachiopods, Gastropods, Echinoids,Crinoids, Calcareous Algae | Aulacosphinctoides Ass.<br>Lima faunule<br>Trigonia faunule |
| Oxfordian | Tochikuba Fmt | 350 | arkose sand-stone, shale, coal seams | Plants | Ryoseki-type flora |
| Callovian | Yamagami Fmt | 200 | massive sand-stone, alter-nation of sandstone & shale | Bivalves massive sand-stone, alternation of sandstone and shale | Trigonia faunules |
| Bathonian | Awazu Fmt | 160 | shale, basal conglomerate | Ammonites Bigotites<br>Bivalves | |
| Bajocian | Kitazawa F | ? | coarse - medium sandstone | | |
| Aalenian | | | | | |
| Sources | stratigraphy. Masatani & Tamura 1959; Mori 196?, Takizawa 1985<br>ammonites: Sato 1962<br>bivalves: Tamura 1961 | | | corals: Mori 1963<br>plants:Kimura 1988<br>other fossils: Tamura 1961 | |

**Figure 9.9. Succession of the Soma Group in the Soma area, eastern Abukuma Mountains.**

Similar rocks of olistostrome type are exposed on northeastern Honshu and southwestern Hokkaido and have been dated by radiolarians as Jurassic.

Noteworthy is the presence of Late Jurassic–Early Cretaceous chert pebbles derived from basaltic-andesitic pyroclastics that form the basement of the Kabato Mountains in western Hokkaido (Kido, Nagata, and Araida 1985). This suggests the presence of an ocean to the west of the Hokkaido axial range, which began to close in the Late Jurassic.

### *Axial and Eastern Hokkaido Terranes*

No rocks of definitely Jurassic age are known. Some earlier reports of Jurassic stromatopores and ammonites are considered unreliable because the fossils probably were collected from the exotic blocks embedded in younger deposits. In the N–S-running axial range of Hokkaido, the oldest, ophiolitic rocks are overlain by terrigenous sediments of the Sorachi Group, forming the main constituent of the Sorachi-Kamuikotan belt. The basal part of this group is dated Upper Jurassic–lowest Cretaceous by radiolarians and by the Torinosu-type bryozoa-coral-gastropod fauna (Jolivet, Nakagawa, and Kito 1983; Kiminami and Kontani 1983). This might represent a subduction complex in a fore-arc basin developed since the Early Cretaceous. In eastern Hokkaido, the Tokoro Belt comprises an Upper Jurassic (Kimmeridgian to Tithonian?) sequence as part of the basement of the belt (Iwata et al. 1983), interpreted as the cover rocks of the oceanic crust of the Okhotsk plate (Komatsu 1985).

## Ammonite zones

Thirteen ammonite assemblage zones have been established in the Japanese Jurassic (Sato and Westermann 1985, 1991) (Figures 12.1–3), assembled from the local sequences shown in the correlation table (Figure 9.2) and ranging from Upper Pliensbachian to Lower Tithonian. Besides these zones, there are some very local associations or horizons dated by rare ammonites, which also are cited in Figures 9.3–4, 8, 10–12 and briefly described in the section of representative sequences. The ammonite assemblage zones and associations are as follows, in ascending order:

### *Lower Jurassic*

1. *Fontanelliceras fontannellense* Assemblage Zone – Dominant are *Fontanelliceras fontanellense, Paltarpites toyoranus, P. paltus, Fuciniceras primordium, Dactylioceras*

| Stage | Group | Formation | m | Lithology | Representative fossils | Biostrat. unit |
|---|---|---|---|---|---|---|
| Tithonian | Hashiura Group | Sodenohama Formation | 180 | sandstone, sandy shale | Ammonites Bivalves | Aulacosphinctoides Assoc. |
| Kimmeridge | | Arato Formation | 350 | sandy shale, with sheet sandstone intercalations | Taramelliceras | |
| Oxfordian | | | | | | Kranaosphinctes matsushimai Ass.Zone |
| Callovian | | | | | Ammonites Bivalves | |
| Bathonian | | | | | Kepplerites /?Cadomites Satoceras | Inoceramus hashiurensis faunule |
| Bajocian | | | | | Leptosphinctes Holcophylloceras | Leptosphinctes gr.davidsoni AZ |
| | | Aratozaki Formation | 100 | conglomerate sandstone | Bivalves, brachiopods belemnites, corals | Kobayashites hemicylindricus faun. Trigonia sumiyagura faunule Vaugonia kodaijimensis faunule |
| Aalenian | Shizukawa Group | Hosoura Formation | 130 | sandy shale with sheet sandstone intercalations | Planammatoceras Hosoureites | Planammatoceras planinsigne AZ Hosoureites ikianus AZ |
| Toarcian | | | | | Ammonites Belemnites Bivalves Vertebrates Plants; Ichthyosaur Dadoxylon | Inoceramus kudoi faunule |
| Pliensbach. | | | | | | Variamussium sp. faunule |
| Sinemurian | | | | | Arnioceras | Meleagrinella sp. faunule Trigonia senex faunule |
| Hettangian | | Niranohama Formation | | | Ammonites Bivalves Alsatites | Geratrigonia hosourensis faunule Burmesia japonica faunule |
| Sources | stratigraphy: Hayami 1961, others; ammonites: Sato 1962, Takahashi 1969, Sato & Westermann 1985; bivalves: Hayami 1961a, plants: Kimura 1988 | | | | | |

**Figure 9.10. Succession of the Shizukawa and Hashiura Groups in the Shizukawa area, southern Kitakami.**

(*Prodactylioceras*) aff. *italicum*, *Amaltheus* cf. *stokesi*, *Canavaria japonica*, and *Arieticeras* sp.; Upper Pliensbachian; Spinatus and Margaritatus (Standard) Zones.

2. *Protogrammoceras nipponicum* Assemblage Zone – Characteristic are *Fuciniceras nakayamense*, *Protogrammoceras nipponicum*, *P. onoi*, *P. yabei*, *Lioceratoides yokoyamai*, *L. matsumotoi*, *Polypelectus okadai*, *Harpocras chrysanthemus*, *H. inouyei*, *H.* cf. *exaratum*, etc.; Tenuicostatum and Falciferum Zones. (Coexistence of *Fuciniceras* and *Protogrammoceras* suggests that the zone extends down into the Upper Pliensbachian.)
3. *Dactylioceras helianthoides* Assemblage Zone – Dominant are *Dactylioceras helianthoides*, *Peronoceras subfibulatum*, *Harpoceras chrysanthemus*, *H.* cf. *exaratum*, *Hildoceras* aff. *bifrons*, *Lioceratoides yokoyamai*, *L. matsumotoi*, and *Polyplectus okadai;* correlation with the standard zones is not easy, because genera generally regarded as Lower Toarcian (*Harpoceras*, *Dactylioceras* and *Peronoceras*) coexist with the Middle Toarcian *Hildoceras;* ?upper Lower Toarcian to lower Upper Toarcian; Falciferum, Bifrons, and Variabilis Zones.

*Middle Jurassic (southern Kitakami and Fukui prefecture)*

4. *Hosoureites ikianus* Assemblage Zone – Dominant are *Hosoureites ikianus*, *Tmetoceras scissum*, *Leioceras* cf. *opalinum*, *L.* cf. *striatum*, *Pseudolioceras* cf. *maclintocki*, and ?*Hyperlioceras* sp.; latest Toarcian(?) to lower Aalenian; Levesquei(?) and Opalinum Zones.
5. *Planammatoceras planinsigne* Assemblage Zone – Dominant are *Tmetoceras scissum*, *Pseudolioceras* (*Tugurites*) cf. *tugurense*, *Hosoureites ikianus*, and *Planammatoceras planinsigne;* Aalenian; Murchisonae–?Concavum Zones.
6. *Sonninia* Assemblage Zone – with *Sonninia* gr. *sowerbyi*, *S.* (*Pelekodites*?) cf. *spatians*, *S.* (*P.*) cf. *pelekus*, *Strigoceras* cf. *longuidum*, rare *Emileia* (*Otoites*) sp., and *Stephanoceras* cf. *plicatissimum;* Bajocian; Sauzei–Humphriesianum Zones (and late Laeviuscula?).
7. *Leptosphinctes* Assemblage Zone – with *Leptosphinctes* gr. *davidsoni*, *L.* gr. *coronarius*, *Oppelia* cf. *genicularis*, and *Cadomites bandoi;* Bajocian; Subfurcatum Zone.
8. *Pseudoneuqueniceras yokoyamai* Assemblage Zone – characterized by *P. yokoyami;* ?latest Bathonian.
9. *Kepplerites japonicus* Assemblage Zone – Dominant are *Kepplerites japonicus*, *K. acuticostum*, *Choffatia laeviradiata*, and *Ch.* cf. *subtilis;* also *Macrophylloceras grossimontanum* from isolated locality; Callovian; lower Macrocephalus [Herveyi] Zone.
10. *Oxycerites* Assemblage Zone – with *Oxycerites* cf. *sulaensis*, *O.* gr. *subcostarius*, *Bullatimorphites* cf. *microstoma*, and perisphinctids; Lower Callovian.

| Stage | Formation | m | Lithology | Representative fossils | | Biostrat. unit |
|---|---|---|---|---|---|---|
| Tithonian<br>Kimmeridge | Kogoshio Formation | 420 - 570 | Coarse sandstone with shale intercalations | Ammonites<br>Bivalves<br>Gastropods<br>Echinoids<br>Corals, sponges<br>Forams, Plants | Substeueroceras<br>Aspidoceras | Myophorella obsoleta f.<br>Parallelodon kesennumensis faunule<br>Mone flora |
| Oxfordian<br>Callovian | Mone Formation | 150 - 360 | Shale with sheet sandstone intercalations | Ammonites<br>Bivalves<br>Plants | Perisphinctes s.s. | Myophorella crenulata f.<br>Mone flora |
| Bathonian | Ishiwaritoge Fmt | 200 | conglomerate sandstone | Plants | | |
| Bajocian | Tsunakizaka Formation | 390 | sandy shale sandsonte | Ammonites,<br>Bivalves | Sonninia<br>Stephanoceras | Sonninia AZ<br>Inoceramus sumiyagura faun. |
| Aalenian | Kosaba Formation | 200 | conglomerate coarse sandstone | Bivalves | | Trigonia sumiyagura faun. |
| Toarcian | | | | | | |
| Sources | stratigraphy: Takizawa 1985<br>ammonites: Sato & Westermann 1985<br>Takahashi 1969 | | | bivalves: Hayami 1961a<br>plants: Kimura 1988 | | |

**Figure 9.11. Succession of the Karakuwa Group in the Karakuwa area, southern Kitakami.**

*Upper Jurassic (Fukui prefecture and Shikoku)*

11. *Perisphinctes matsushimai* Assemblage Zone – Characteristic are *P.* (*Kranaosphinctes*) *matsushimai*, *Peltoceratoides* sp., and phylloceratids; Lower–Middle Oxfordian.
12. *Ataxioceras kuriskakense* Assemblage Zone – with *Ataxioceras kurisakense, Euaspidoceras* sp., and *?Lithacoceras* sp.; Middle Kimmeridgian.
13. *Aulacosphinctoides* cf. *steigeri* Association – with *Aulacosphinctoides* cf. *steigeri;* Lower Tithonian.

### Radiolarian zones

Until the discovery of Jurassic radiolarians (from matrix), most of these mélange-type rocks had formerly been dated as Upper Paleozoic. Megafossils such as ammonites or bivalves are too sporadic, and radiolarians are therefore used for this type of sediments.

The radiolarian assemblage zones established to date by different authors differ more or less from one Jurassic bioprovince to another. One of these zonal schemes, established in the Mino area of central Japan (Yao 1986), is listed here:

*Upper Jurassic*

1. *Pseudodictyomitra primitiva–P.* sp. *A* Assemblage Zone
2. *Tricolocapsa yaoi* Assemblage Zone
3. *Gongylothorax sakawaensis–Stichocapsa naradaniensis* Assemblage Zone

*Middle Jurassic*

4. *Guexella nudata* Assemblage Zone
5. *Unuma echinatus* Assemblage Zone

*Lower Jurassic*

6. *Hsuum hisuikyoensis* Assemblage Zone
7. *Parahsuum* sp. *D* Assemblage Zone
8. *Parahsuum simplum* Assemblage Zone

Jurassic radiolarian biohorizons have also been introduced for all of Japan (Matsuoka and Yao 1986) (Figure 9.2).

### PHILIPPINES

In the Philippine archipelago, Jurassic rocks are exposed in very limited areas in the southwest, where sediments dated by fossils are known (Figure 9.14). The Mansalay Formation, controlled by ammonites, occurs in southern Mindoro, along lower reaches of the Bongabong River and in the Mansalay area. Fossiliferous Jurassic, without ammonites, is developed in North Palawan and its neighboring island groups, such as the Linapacan, Culion, and Busuanga islands (Sato 1975; Fontaine et al. 1983).

The igneous and metamorphic basement rocks known from various places in the Philippines have been provisionally dated as Jurassic, but this age is based only on stratigraphic position below beds of known age. Igneous rocks of probably ophiolite origin, mostly ultramafics and mafics, are preliminarily placed in the Jurassic, but their age also has not been firmly established.

| Stage | Formation | m | Lithology | Representative fossils | Biostrat. unit |
|---|---|---|---|---|---|
| Cretaceous | Ayukawa Fmt | 640 | sandstone/shale | Ammonites (Berriasian) | (mostly Lower Cretaceous) |
| Tithonian | Oginohama Formation | 1500 | coarse sand-stone - shale in alternation | Aulacosphinctoides | |
| Kimmeridge | | | | | Myophorella orientalis f. |
| Oxfordian | | | | Ammonites Bivalves Plants; Perisphinctes Kranaosphinctes | Ryoseki type flora |
| Callovian | | | | Choffatia | |
| Bathonian | | | | Satoceras | |
| Bajocian | Tsukinoura Formation | 500 | shale in upper part sandstone/con-glomerate in lower part | Ammonites Bivalves; Stephanoceras | Chlamys kobayashii faunule; Kobayashites hemicylindricus f.; Trigonia sumiyagura f.; Vaugonia kodaijimensis f. |
| Aalenian | | | | | |
| Toarcian | | | | | |
| Sources | stratigraphy: Takizawa 1985; ammonites: Sato & Westermann 1985, Takahashi 1969 | | | bivalves: Hayami 1961a; plants: Kimura 1988 | sedimentary facies: Takizawa 1985 |

**Figure 9.12. Succession of the Ojika Group on the Ojika Peninsula, southern Kitakami.**

## Representative sequences

### *North Palawan region*

The weakly metamorphic rocks distributed in the northern part of Palawan Island and its dependent island groups are the oldest sediments of the Philippines. They were named Malampaya Sound Group and were dated Permian–Triassic (Hashimoto and Sato 1973). Subsequently, Triassic and Jurassic fossils were discovered from the neighboring islands of Calamian (Fontaine 1979). The whole sequence is more or less metamorphosed and folded into chevron folds, with the NNW–SSE axes plunging slightly south. The following (descending) succession of the Malampaya Sound Group is now accepted.

*Kapoas Granite:* Late Jurassic (field relations)
*Guinlo Formation:* barren sandstone and conglomerate; age unknown
*Limestone, shale, and sandstone beds:* sandstone-and-shale sequence, with pollen of Jurassic affinity, on Busuanga Island; limestone beds unconformably on Triassic conodont-bearing chert (Liminangcong Formation and equivalent); Upper Jurassic calcareous algae, smaller foraminifers, Cnidaria (Fontaine 1979; Bassoullet 1983; Beauvais 1983; Fontaine et al. 1983)
Unconformity
*Liminangcong Formation:* more or less metamorphosed, bedded chert; Middle–Upper Triassic (conodonts)
*Minilog Formation:* thick limestone bed; Upper Permian (fusulines)
*Bacuit Formation:* slump beds, including chert sharpstones; middle Permian (conodonts)
*Barton Metamorphics:* age unknown

### *Southern Mindoro*

In the southeastern part of Mindoro, the ammonite-bearing Mansalay Formation (Andal et al. 1968) consists of terrigenous clastics, mostly medium-grained sandstone intercalated with shale; both upper and lower limits are unknown, but below an unconformity appears to be metamorphic basement. The cumulative thickness is about 4,000 m, but the sequence has been folded at least twice in different directions. The depositional environment of the Mansalay Formation probably was postorogenic on older metamorphic rocks, and deformation occurred before the Middle Jurassic. Three fossil horizons are distinguished (ascending order):

1. *Caromata Hill Horizon* – alternation of black shale and carbonaceous medium-grained sandstone ($\geq$30 m), relatively rich in ammonites: *Hecticoceras* (*Zieteniceras*) sp., *H.* (*Lunuloceras*) cf. *lunula*, *H.* (*Sublunuloceras*) cf. *guthei*, etc.; also belemnites and *Cladophlebis*-like plant impressions; Lower–Middle Callovian.
2. *Parucpoc Hill Horizon* – sandy shale with calcareous concretions ($\sim$200 m), rich in ammonites and bivalves; the *Parawedekindia arduennensis* Assemblage Zone includes *P. arduennensis*, *Peltoceratoides* sp., *Perisphinctes* (*Kranaosphinctes*) cf. *bullingdonensis*, etc., *Myophorella* sp., and other bivalves; Lower–Middle Oxfordian (Cordatum–Transversarium Zones).

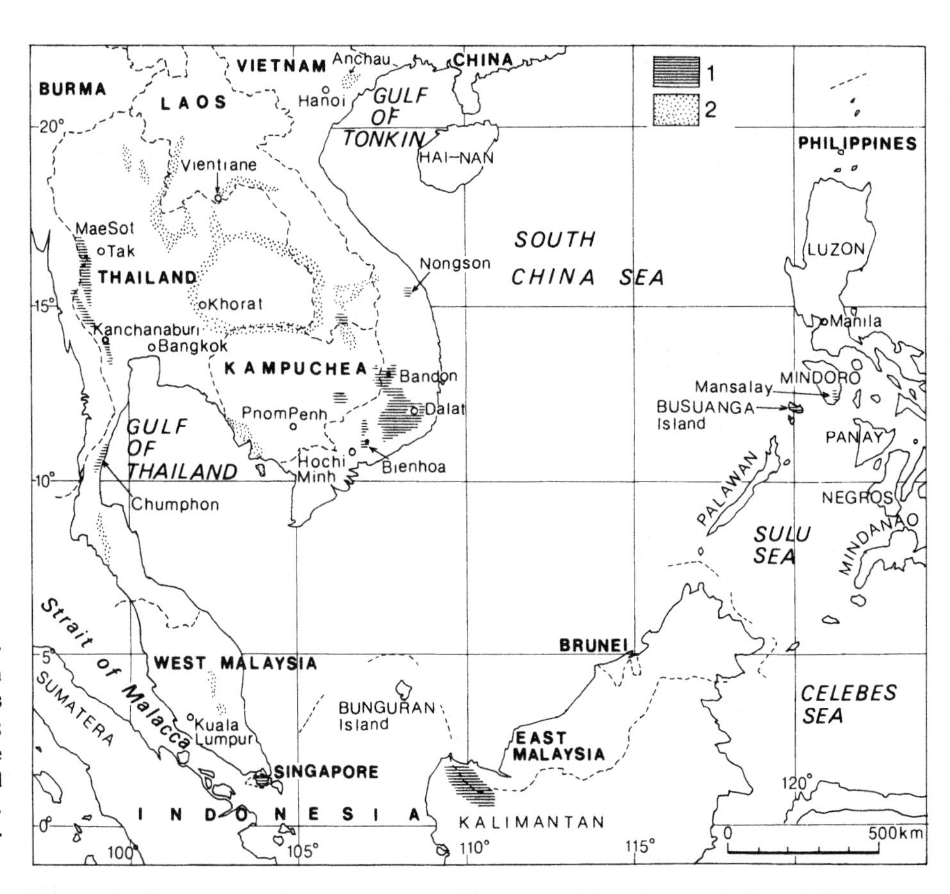


**Figure 9.13. Map showing the distribution of Jurassic exposures in Southeast Asia: 1, hachured areas denote marine Jurassic exposures; 2, stippled areas show nonmarine Jurassic exposures, but the areal extents are not clearly demarcated. Marine horizons are locally intercalated in nonmarine sequences.**

3. *Amaga River Horizon* – relatively homogeneous sandy shale and argillaceous siltstone (~350 m); ammonites are common; the *Euaspidoceras* cf. *hypselum* Assemblage Zone includes *Perisphinctes* (*Kranaosphinctes*) cf. *bullingdonensis, Euaspidoceras* cf. *hypselum, Paraspidoceras* cf. *knechti, Taramelliceras* cf. *trachinotrum, Myophorella orientalis,* and other bivalves; Middle–Upper Oxfordian (lower limit defined by disappearance of *Parawedekindia*).

## THAILAND AND MALAY PENINSULA

Jurassic rocks are widely distributed in Thailand and adjoining areas. Nonmarine Jurassic presumably covers the entire Khorat Plateau, but is exposed only along its margin. Similar rocks occur in the northern part of the Yunnan–Malay Mobile Belt (Hahn, Koch, and Wittekindt 1986) and in peninsular Thailand.

Nonmarine Jurassic rocks of clearly postorogenic nature are exposed in the central range of the Malay Peninsula (Burton 1973). The marine Jurassic is relatively restricted in a narrow zone along the Thai–Burmese border. The area around Mae Sot, Tak province, exposes marine Jurassic with ammonites. In the Si Sawat area north of Kanchanaburi there is a marine sequence controlled by small foraminifers.

In Singapore, marine bivalves were described by Newton (1906) from Mount Guthrie. Burton (1973) placed the beds in the Lower Jurassic.

### Nonmarine sequence

The standard sequence of nonmarine Mesozoic, the Khorat Group, is present in the Khorat Plateau, northeastern Thailand, and ranges from Upper Triassic to Cretaceous. The Jurassic parts belong to the Phu Kradung, Phra Wihan, and Sao Khua Formations, based on sporadic occurrences of bivalves, vertebrates, and pollen (Chonglakmani, Meesok, and Suteetorn 1985). They are, in ascending order:

*Phu Kradung Formation.* Red to purple siltstone and sandstone (~1,000 m); rare nonmarine bivalves; *Unio thailandica, Neomiodon*(?) *khoratensis* (Kobayashi, Takai, and Hayami 1963; Hayami 1968), plesiosaur teeth (Kobayashi et al. 1963), mesosuchian crocodile *Sunosuchus thailandicus* (Hahn 1982; Buffetaut and Ingavat 1984), and some pollen *Classopollis* sp.; Lower Jurassic.

*Phra Wihan Formation.* Cross-bedded sandstone and siltstone (60–200 m); rare fossils: *Brachypyllum* sp. and *Sphenopteris* sp. (Kon'no and Asama 1973), and pollen similar to that in the Phu Kradung Formation (Hahn 1982); age uncertain.

*Sao Khua Formation.* Siltstone, sandstone partly conglomeratic (400–700 m); rare marine Jurassic bivalves include *Cardinioides magnus* and *Mytilus rectangularis* (Kobayashi et al. 1963); plants include *Sphenopteris geopperti* (Kon'no and Asama 1973) and

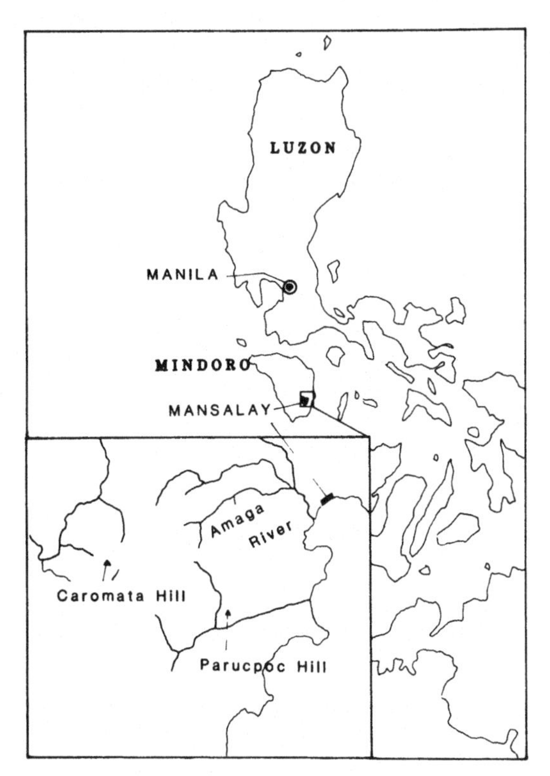


**Figure 9.14. Map showing the Jurassic outcrops in the Philippines.**

pollen (Hahn 1982); abundant reptiles, including turtles, *Acrodus* sp., *Composognathus* sp., and *Goniophiolis phuwiangensis* (Ingavat and Taquet 1978); Upper Jurassic.

These mostly nonmarine formations cover the stable Khorat-Kontum microcontinent. Equivalent formations are known also from central and peninsular Thailand. The existence of marine Jurassic at the Thai–Burmese border region suggests that they represent Jurassic mountain ranges along the western margin of the continent.

The nonmarine sediments in the Central Belt of the Malay Peninsula are placed in the Tembeling Formation (Koopmans 1988). It consists of coarse red clastics, intercalated with thin layers of shale, and is locally up to 3,000 m thick, with plants (Smiley 1970a,b) and palynomorphs (Koopmans 1968). The formation lies unconformably on weakly metamorphosed Upper Triassic.

### Marine sequence

There are two types of marine Jurassic sequences in western Thailand. One, consisting of shale and marl, is developed around Mae Sot and the Kamawakale Gorge; the other, essentially limestone, is along the Thai–Burmese border (Chonglakmani et al. 1985).

#### *Mae Sot and Kamawakale area*

The Jurassic is the upper part of a subcontinuous Triassic–Jurassic sequence, the Mae Moei Group (Braun and Jordan 1976). The Jurassic sequence consists of approximately 1,200 m of silty shale, sandy shale, and marly limestone beds, known mostly from isolated, discontinuous outcrops, and it contains ammonites. Four faunas are distinguished (Sato and Westermann 1991), in ascending order:

1. *Pseudolioceras* Fauna – *Pseudolioceras* sp. together with rare *Lytoceras* (*Alocolytoceras*) *ophioneum*, *Pseudolioceras* sp., *Onychoceras* sp., and ?*Haugia* sp.; Upper Toarcian (Braun and Jordan 1976).
2. *Tmetoceras* Fauna – *Tmetoceras scissum* [''*T. dhanarajatai*''], *T.* cf. *kirki* [''*regleyi*''], *Ancolioceras*(?) [''*Graphoceras concavum*''] (J. Sandoval and A. Linares personal communication 1989), and ?*Erycites* [or ?Bajocian *Pseudotoites*]; middle or upper Aalenian (Komalarjun and Sato 1964, revised).
3. *Skirroceras* Fauna – *Stephanoceras* (*Skirroceras*) ex. gr. *macrum* [''*Docidoceras* cf. *longalvum,*'' but suture strongly retracted] and *Sonninia* (*Papilliceras*?) sp. [''*Eudmetoceras* (*Planammatoceras*) sp.'']; Bajocian; Sauzei Zone (Braun and Jordan 1976, revised).
4. *Epimayaites*? Fauna – ?*Epimayaites* cf. *falcoides*, *Phylloceras* sp., ?*Glochiceras* sp., in addition to *Bositra alpina*, echinids, crinoids, and foraminifers; ?middle Oxfordian (Zeiss, cited by Braun and Jordan 1976). Note that this ?*Epimayaites* occurrence would be the only one known from Jurassic Asia [possibly suggesting that this area of Thailand may also have originated in northern Gondwana (G. E. G. Westermann personal communication)].

#### *Kanchanaburi area*

In the Si Sawat area in Kanchanaburi province, Jurassic foraminifers occur in bedded limestone (200–300 m) associated with red clastics and conglomerates (Hagen and Kemper 1976). The three assemblages are *Orbitopsells dubari* and associated species, *Lucasella kaempferi* and associated species, and *Kurnubia* etc., claimed to be Early, Middle and Late Jurassic in age, respectively (Kemper 1976).

#### *Chumphon area*

The Jurassic bivalve *Eomiodon chumphonensis* was reported from argillaceous sandstone beds at the mouth of the Chumphon River in peninsular Thailand (Hayami 1960). The fossiliferous bed occurred within a nonmarine redbed sequence of Khorat affinity, and similar rocks with Jurassic ammonites and bivalves were recently discovered at Khao Lak, north of Chumphon (Chonglakmani et al. 1985). The marine Jurassic is generally in heteropic facies within the continental Khorat Group, which lies conformably on Triassic. The Triassic–Jurassic Mae Moei Group, in turn, lies on the eroded surface of deformed Paleozoic basement (Hahn et al. 1986).

INDOCHINA

The Jurassic was a period of episodic marine ingressions onto the Indochinese microcontinent. The rocks and faunas were described by Sato (1975) and Vu Khuc (1985a).

Following the Triassic, regression was predominant on the Indochinese Peninsula, and marine Jurassic was restricted to the Kontum Massif and surrounding areas. The major occurrences in the "Dalat Depression" of the southern Kontum Massif (formerly dated as Paleozoic) extend from the Trian-Loduc area northward to the Darlac-Bandon area and westward to southern Laos and northeast Kampuchea. Isolated from this area, the marine Jurassic is also exposed in the "Nongson depression," including the famous Huunien area at the northern margin of the Kontum Massif.

Continental Jurassic is known from the Hacoi-Auchau and Nampo areas in North Vietnam, where it includes coal. The *Cardinia*-bearing marine beds reported by Mansuy (1919), however, could not be found by recent investigators (Vu Khuc 1985a).

**Nongson Depression**

At the northern margin of the Kontum Massif, the locally outcropping marine Jurassic (Counillon 1908; Bourret 1925) resembles that in the Huunien area in central Vietnam, in the Nongson Depression. Two formations are distinguished:

*Banco Formation.* Conglomerate and coarse-grained sandstone (800–1,200 m); base mostly conformable, on coal-bearing Upper Triassic; spores and pollen occur in basal sandstone; *Coniopteris* sp., *Classopollis* sp., *Mattonia* cf. *triassica*, etc.; Lower Jurassic.

*Kheren Formation.* Sandstone, siltstone, and mudstone with intercalations of marl and limestone (~130–150 m). Two ammonite zones are distinguished (Sato and Westermann 1991) (from below):

1. *Psiloceras* Assemblage Zone – with *Psiloceras* sp. and *Saxoceras* sp., bivalves, including *Cultriopsis, Goniomya, Protocardia, Cardinia, Astarte* (*Nicaniella*), etc.; upper Hettangian (Vu Khuc and Nguyen 1981).

An isolate locality in the Mondulkiri area of Kampuchea has yielded *Psiloceras* gr. *planorbis;* lower Hettangian (Saurin 1965).

2. *Xipheroceras*? Assemblage Zone – with ?*Xipheroceras dudressieri* and rich bivalve fauna, including *Goniomya, Modiolus, Myophorella, Thracia, Astarte* (*Nicaniella*), etc.; ?Lower Sinemurian (Hayami 1972).

A corresponding horizon extends into southern Laos.

The upper part of the sequence is characterized by red siltstone, sandstone, and mudstone (400 m), yielding silicified wood: *Protophyllocladoxylon* and *Brachioxylon;* ?Middle Jurassic (Serra 1968).

**Dalat Depression**

Marine Jurassic with ammonites and bivalves is widely exposed in the Dalat Depression. The composite succession established from several sections is as follows (Nguyen 1982; Vu Khuc 1983, 1985a):

*Draylinh Formation.* Calcareous sandstone, clayey shale with basal conglomerate, some intercalated marls, calcareous siltstone and sandstone (~700 m); unconformable on Precambrian, Carboniferous or Middle Triassic; four Lower Jurassic Zones and assemblages are distinguished (Sato and Westermann 1991) (ascending order):

1. *Coroniceras rotiforme* Assemblage Zone – with *Arietites* sp., *Coroniceras rotiforme, Pseudasteroceras* sp., *Arnioceras* sp., *Asteroceras stellare*, and *Oxynoticeras* cf. *oxynotum;* also *Cardinia, Otapiris,* and others, mostly Upper Sinemurian (Ta Tran Tan 1968).

A corresponding horizon is known from Dakder Valley in eastern Kampuchea (Saurin 1965), yielding *Vermiceras spiratissimum.*

2. *Paltarpites* Assemblage – with *Paltarpites* sp., *Arieticeras* sp., and bivalves *Lima, Entolium,* etc.; Upper Pliensbachian (Nguyen 1982; Vu Khuc 1984a).
3. *Ovaticeras* Assemblage – with *Ovaticeras* sp. and *Peronoceras* sp.; Lower Toarcian (Vu Khuc 1984a).
4. *Dumortieria lantenoisi* Assemblage Zone – *Pseudogrammoceras*? *loducensis, Dumortieria lantenoisi, D.* cf. *metita, Hammatoceras* sp., *Pseudammatoceras molukkanum,* and rich bivalve fauna of *Parvamussium, Myophorella, Myopholas, Pholadomya, Thracia, Modiolus,* etc.; Upper Toarcian (Mansuy 1914; Sato 1972a).

*Langa Formation.* Clayey shale, siltstone, and some sandstone intercalations (700 m); conformable on Draylinh Formation (Nguyen 1982); with *Planammatoceras* and abundant *Bositra bronni* in lower part; *Tmetoceras* sp. is recorded, but not described (Vu Khuc 1984a, 1985a); Aalenian.

Marine Jurassic in Indochina is therefore confined to the Lower Jurassic and Aalenian of the Kontum Massif area. The sequence lies on the older rocks, including nonmarine Triassic in central Vietnam. The upper part of this section again grades into nonmarine beds.

North Vietnam became land in the Late Triassic. Nonmarine Jurassic is known from the Auchau and Nampo Basins, in the east and west, respectively. The sequence of the Auchau Basin consists of conglomerate, sandstone, red siltstone, and shale, with marls and carbonaceous(?) shale intercalations. It is generally 1,000–2,000 m thick and yields megaflora, including *Coniopteris, Equisetites,* and *Podozamites,* and rare fish and estherians. The similar rocks of the Nampo Basin are unfossiliferous and lie on coal-bearing Triassic.

**References**

Andal, D. R., Esguerra, J. S., Hashimoto, W., Reyes, B. P., & Sato, T. (1968). The Jurassic Mansalay Formation, southern Mindoro, Philippines. *Geol. Palaeont. SE Asia, 4,* 179–97.

Aono, H. (1985). Geologic structure of the Ashio and Yamizo Mountains with special reference to its tectonic evolution. *Sci. Rept., Inst. Geosci., Univ. Tsukuba, Sect. B, 6,* 21–57.

Bando, Y., Sato, T., & Matsumoto, T. (1987). Paleobiogeography of the Mesozoic Ammonoides, with special reference to Asia and the Pacific. In A. Taira & M. Tashiro (eds.), *Historical Biogeography and Plate Tectonic Evolution of Japan and Eastern Asia* (pp. 65–95). Tokyo: Terrapub.

Bassoullet, J. P. (1983). Jurassic microfossils from the Philippines. *CCOP Tech. Bull., 16,* 31–8.

Beauvais, L. (1983). Jurassic Cnidaria from the Philippines and Sumatra. *CCOP Tech. Bull., 16,* 39–67.

Bourret, R. (1925). La chain annamitique et les plateaux du Bas Loas, a l'Ouest de Hue. *Bull. Serv. geol. Indoch., 14,* 50–6.

Braun, E. von, & Jordan, R. (1976). The stratigraphy and paleontology of the Mesozoic sequence in the Mae Sot area in western Thailand. *Geol. Jb., B21,* 5–51.

Buffetaut, E., & Ingavat, R. (1984). The lower jaw of *Sunosuchus thailandicus,* a mesosuchian crocodile from the Jurassic of Thailand. *Palaeontology, 27,* 199–206.

Bureau of Mines and Geo-Sciences (1981). *Geology and Mineral Resources of the Philippines, Vol. 1.* Manila: Bureau of Mines and Geo-Sciences.

Burton, C. K. (1973). Mesozic. In D. J. Gobbett & C. S. Hutchison (eds.), *Geology of the Malay Peninsula* (pp. 97–141). New York: Wiley.

Chihara, K., Komatsu, M., Uemura, T., Hasegawa, Y., Shiraishi, S., Yoshimura, T., & Nakamizu, M. (1979). *Geology and Tectonics of the Omi-Range and Joetsu Tectonic Belts.* (5). Niigata Univ., Ser. E., Geol. & Mineral., Sci. Rept. 5.

Chonglakmani, C., Meesok, A., & Suteetorn, V. (1985). *Jurassic Stratigraphy of Thailand.* IGCP Project #171, Circum-Pacific Jurassic, Special Paper No. 14.

Counillon, R. (1908). Sur le gisement liasique de Huu Nien, province de Quangnam. *Bull. Soc. géol. France, 4e Sér., 8,* fasc. 7–8, 524–32.

Dovzhikov, A. E., et al. (1965). *Geology of North Viet Nam.* Hanoi: Science and Technics Publ. House.

Fontaine, H. (1966). Nouveau gisement d'ammonites sinémuriennes au Darlac (Viétnam meridional). *Arch. Géol. Viétnam, 8,* 5–8.

(1979). Note on the geology of the Calamian Islands. *CCOP Newsletter, 6,* 40–72.

Fontaine, H., David, P., Pardede, R., & Suwarna, N. (1983). Marine Jurassic in Southeast Asia. *CCOP Tech. Bull., 16,* 3–30.

Hagen, D., & Kemper, E. (1976). Geology of the Thong Pha Phum area (Kanchanaburi Province, Western Thailand). *Geol. Jb., B21,* 53–91.

Hahn, L. (1982). Stratigraphy and marine ingressions of the Mesozoic Khorat Group in northeastern Thailand. *Geol. Jb., B43,* 7–35.

Hahn, L., Koch, E. E., & Wittekindt, H. (1986). Outline of the geology and the mineral potential of Thailand. *Geol. Jb., B59,* 3–49.

Hashimoto, W., & Sato, T. (1973). Geologic structure of North Palawan, and its bearing on the geologic history of the Philippines. *Geol. Palaeont. SE Asia, 15,* 151–89.

Hayami, I. (1960). Two Jurassic pelecypods from west Thailand. *Trans. Proc. Palaeont. Soc. Japan, N.S., 38,* 284.

(1961a). Succession of the Kitakami Jurassic. *Japan. J. Geol. Geogr., 32,* 159–77.

(1961b). On the Jurassic pelecypod faunas in Japan. *J. Fac. Sci., Univ. Tokyo, Sec. II, 32,* 243–343.

(1968). Some non-marine bivalves from the Mesozoic Khorat Group of Thailand. *Geol. Palaeont. SE Asia, 4,* 100–8.

(1972). Lower Jurassic Bivalvia from the environ of Saigon. *Geol. Palaeont. SE Asia, 10,* 179–230.

(1975). *A Systematic Survey of the Mesozoic Bivalvia from Japan.* Univ. Mus., Univ. Tokyo, Bull. 10.

Hayasaka, Y., Isozaki, Y., & Hara, I. (1983). Discovery of Jurassic radiolarians from the Kuga and Kanoashi Groups in the western Chugoku district, Southwest Japan. *J. Geol. Soc. Japan, 89,* 527–30.

Hirano, H. (1971). Biostratigraphic study of the Jurassic Toyora Group, Part I. *Mem. Fac. Sci., Kyushu Univ., Ser. D, 21,* 93–128.

(1973a). Biostratigraphic study of the Jurassic Toyora Group, Part III. *Mem. Fac. Sci., Kyushu Univ., Ser. D, 90,* 45–71.

(1973b). Biostratigraphic study of the Jurassic Toyora Group, Part II. *Trans. Proc. Palaeont. Soc. Japan, N.S., 89,* 1–14.

Hisada, K. (1983). Jurassic olistostrome in the southern Kanto Mountains, central Japan. *Sci. Rep., Inst. Geosci., Univ. Tsukuba, Sect. B, 4,* 99–119.

Ichikawa, K., Hada, S., & Yao, A. (1985). Recent problems of Paleozoic-Mesozoic microbiostratigraphy and Mesozoic geohistory of southwest Japan. *Mem. Geol. Soc. Japan, 25,* 1–18.

Ingavat, R., & Taquet, P. (1978). First discovery of dinosaur remains in Thailand. *J. Geol. Soc. Thailand, 3,* 1–6.

Ishiga, H. (1983). Two suites of stratigraphic succession within the Tamba Group in the western part of the Tamba Belt, southwest Japan. *J. Geol. Soc. Japan, 89,* 443–54.

Iwamatsu, A. (1969). Structural analysis of the Tsunakizaka syncline, in southern Kitakami mountainous land, northeast Japan. *Earth Sci. (Chikyu Kagaku), 23,* 227–35.

Iwata, K., Watabe, M., Nakamura, K., & Uozumi, S. (1983). Occurrence of Jurassic and Cretaceous radiolarians from the pre-Tertiary systems around Lake Saroma, northeast Hokkaido (preliminary report). *Earth Sci. (Chikyu Kagaku), 37,* 225–8.

Jolivet, L., Nakagawa, M., & Kito, N. (1983). Uppermost Jurassic unconformity in Hokkaido. Evidence for an early tectonic stage. *Proc. Japan. Acad., 59*(B), 153–7.

Kemper, E. (1976). The foraminifera in the Jurassic limestone in west Thailand. *Geol. Jb., B21,* 129–53.

Kemper, E., Maronde, H. D., & Stoppel, D. (1976). Triassic and Jurassic limestone in the region northwest and west of Si Sawat (Kanchanaburi Province, northern Thailand). *Geol. Jb., B21,* 93–127.

Kido, N., Nagata, M., & Araida, K. (1985). Stratigraphy and geologic age of the Kumanejiri Group. In *61st Ann. Meeting, Geol. Soc. Japan, Abstr.,* 175.

Kiminami, K., & Kontani, Y. (1983). Mesozoic arc-trench systems in Hokkaido, Japan. In M. Hashimoto & S. Uyeda, (eds.), *Accretion Tectonics in the Circum-Pacific Regions* (pp. 107–22). Tokyo: Terra Pub.

Kimura, T. (1952). On the geological study of the Iwamuro Formation, Tone Gun, Gunma Prefecture. Ser. 1. *J. Geol. Soc. Japan, 58,* 457–68.

(1988). Jurassic macrofloras in Japan and palaeophytogeography in East Asia. *Bull. Tokyo Gakugei Univ., Sect. IV, 40,* 147–64.

Kobayashi, T. (1947). On the occurrence of Seymourites in Nippon and its bearing on the Jurassic palaeogeography. *Japan. J. Geol. Geogr., 20,* 19–31.

Kobayashi, T., Konishi, K., Sato, T., Hayami, I., & Tokuyama, A. (1957). On the Lower Jurassic Kuruma Group. *J. Geol. Soc. Japan, 63,* 182–94.

Kobayashi, T., Takai, F., & Hayami, I. (1963). On some Mesozoic fossils of east Thailand and a note on the Khorat Series. *Japan. J. Geol. Geogr., 34,* 187–91.

Komalarjun, P., & Sato, T. (1964). Aalenian (Jurassic) ammonites from Mae Sot, northwestern Thailand. *Japan J. Geol. Geogr., 35,* 149–61.

Komatsu, M. (1985). Structural framework of the axial zone in Hokkaido – its composition, characters and tectonics. *Mem. Geol. Soc. Japan, 25,* 137–55.

Konishi, K. (1954). Yamaoku Formation (a Jurassic deposit recently discovered in Okayama Prefecture). *J. Geol. Soc. Japan, 60,* 325–32.

Kon'no, E., & Asama, K. (1973). Mesozoic plants from Khorat, Thailand. *Geol. Palaeont. SE Asia, 12,* 149–71.

Koopmans, B. N. (1968). The Tembeling Formation – a lithostratigraphic description (west Malaysia). *Bull. Geol. Soc. Malaysia, 1,* 23–43.

Luong, T. D., & Nguyen, X. B. (1979). On the geological map of Viet Nam, scale 1:500,000. *Geol. Min. Res. Viet Nam (Hanoi), 1,* 5–8.

Maeda, S. (1961). On the geological history of the Mesozoic Tetori Group in Japan. *Japan. J. Geol. Geogr., 32,* 375–96.

Mansuy, H. (1914). Gisement liasique des schistes de Trian (Cochinchine). *Mém. Serv. géol. Indoch. III, 2,* 37–9.

(1919). Faunes triasiques et liasiques de Na Cham. *Mém. Serv. géol. Indoch., VI, 1.*

Masatani, K., & Tamura, M. (1959). A stratigraphic study of the Jurassic Soma Group on the eastern foot of the Abukuma Mountains, Northeast Japan. *Mém. Serv. géol. Indoch., 30,* 245–57.

Matsuoka, A. (1983). Middle and late Jurassic radiolarian biostratigraphy in the Sakawa and adjacent areas, Shikoku, Southwest Japan. *J. Geosci., Osaka City Univ., 26,* 1–48.

(1984). Togano Group of the southern Chichibu terrane in the western part of Kochi Prefecture, southwest Japan. *J. Geol. Soc. Japan, 90,* 455–77.

(1988). Discovery of Early Jurassic radiolarians from the North Kitakami belt (s.s.), northeast Japan. *Earth Sci. (Chikyu Kagaku), 42,* 104–6.

Matsuoka, A., & Yao, A. (1986). A newly proposed radiolarian zonation for the Jurassic of Japan. *Marine Micropaleont., 11,* 91–105.

Minoura, K. (1983). Geology of North Belt of Kitakami. *Monthly Chikyu, 5,* 480–7.

Mori, K. (1963). Geology and paleontology of the Jurassic Somanakamura Group, Fuskushima Prefecture, Japan. *Sci. Rept., Tohoku Univ., 2nd Ser., 35,* 33–65.

Naka, T., Tokuoka, T., Sano, S., Watase, H., Nishimura, K., Kuinose, M., & Hashimoto, K. (1985). Stratigraphy and structure of the Higuchi Group (Lower Jurassic), Muikaichicho, Shimane Prefecture. *Geol. Rept., Shimane Univ., 4,* 91–104.

Newton, R. B. (1906). Notice on some fossils from Singapore discovered by John B. Scrivenor, F. G. S., geologist to the Federated Malay States. *Geol. Mag., 43,* 487–96.

Nguyen, D. H.(1982). Some Early Jurassic ammonites recently discovered in South Vietnam. *Contr. Palaeont. Vietnam, 1.*

Nishida, M., Adachi, M., & Kondo, N. (1974). Fossil fragments of petrified wood from pre-Tertiary formations in the northern area of Inuyama City, Aichi Prefecture, and their bearing on geology. *J. Japan. Bot., 49,* 265–72.

Otsuka, T. (1988). Paleozoic-Mesozoic sedimentary complex in the eastern Mino terrane, central Japan and its Jurassic tectonism. *J. Geosci., Osaka City Univ., 31,* 63–122.

Sashida, K. (1988). Lower Jurassic multisegmented *Nassellaria* from the Itsukaichi area, western part of Tokyo Prefecture, central Japan. *Sci. Rep., Inst. Geosci., Univ. Tsukuba, Sect. B, 9,* 1–27.

Sashida, K., Igo, H., Takizawa, S., Hisada, K., Shibata, T., Tsukada, K., & Nishimura, H. (1982). On the Jurassic radiolarian assemblages in the Kanto district. *News Osaka Micropal., Spec. Vol., 5,* 51–66.

Sato, T. (1960). A propos des courants océaniques froids prouves par l'existence des ammonites d'origine arctique dans le Jurassic japonais. *Rept. I.G.C., XXI Session, Part XII,* 165–9.

(1961). La limite jurassico-cretacée dans la stratigraphie japonaise. *Japan. J. Geol. Geogr., 32,* 533–51.

(1962). Etudes biostratigraphiques des Ammonites du Jurassique du Japon. *Mém. Soc. géol. France, N.S., 41,* 1–122.

(1972a). Some Bajocian ammonites from Kitakami, northeast Japan. *Trans. Palaeont. Soc. Japan, N.S., 85,* 280–92.

(1972b), Ammonites du Toarcien au Nord de Saigon. *Geol. Palaeont. SE Asia, 10,* 231–42.

(1974). A Jurassic ammonite from near Inuyama, north of Nagoya. *Geol. Palaeont. SE Asia, N.S., 96,* 427–32.

(1975). Marine Jurassic formations and faunas in Southeast Asia and New Guinea. *Geol. Paleont. SE Asia, 15,* 151–89.

Sato, T., Hashimoto, K., & Suyama, Y. (1986). Discovery of a Late Jurassic ammonite *Perisphinctes* from the Kuga Group in Yamaguchi Prefecture. *Bull. Yamaguchi Pref., Yamaguchi Mus., 12,* 1–5.

Sato, T., Kasahara, Y., & Wakita, K. (1985). Discovery of a Middle Jurassic ammonite *Kepplerites* from the Mino Belt, central Japan. *Trans. Proc. Palaeont. Soc. Japan, N.S., 139,* 218–21.

Sato, T., & Westermann, G. E. G. (1985). *Range Chart and Zonations in Japan.* Rept. IGCP Project #171, Circum-Pacific Jurassic, III Field Conference, Tsukuba, 73–97.

Sato, T., & Westermann, G. E. G. (1991). Japan and South-East Asia. In G. E. G. Westermann & R. C. Riccardi (eds.), *Jurassic Taxa Ranges and Correlation Charts for the Circum-Pacific Newsl. Strat., 24,* 81–108.

Saurin, E. (1965). Deux ammonites du Lias inferieur du Cambodge oriental. *Archiv. Géol. Vietnam, 7,* 52–8.

Serra, C. (1968). Sur quelques vegetaux mésozoiquest de las région de Vungrua (province de Quang-nam). *Arch. Géol. Vietnam, 11,* 1–9.

Smiley, C. J. (1970a). Later Mesozoic flora from Maran, Pahang, West Malaysia. Pt. 1: Geologic considerations. *Bull. Geol. Soc. Malaysia, 3,* 77–88.

(1970b). Later Mesozoic flora from Maran, Pahang, West Malaysia. Pt. 2: Taxonomic considerations. *Bull. Geol. Soc. Malaysia,* 89–113.

Sugimoto, M. (1974). The northern Kitakami Massif and its bearing on the geologic strucutre of the Japanese islands. *Mem. Geol. Soc. Japan, 10,* 29–40.

Suyari, K., & Ishida, K. (1985). Radiolarian age of the Torinosu Group, Shikoku, Japan. *Mem. Geol. Soc. Japan, 18,* 83–101.

Suyari, K., Kuwano, Y., & Ishida, K. (1983). Biostratigraphic study of the north subbelt of the Chichibu belt in central Shikoku. *J. Sci., Univ. Tokushima, 16,* 143–67.

Suzuki, A., & Sato, T. (1972). Discovery of Jurassic ammonite from Toriashi Mountain. *J. Geol. Soc. Japan, 78,* 213–15.

Ta Tran Tan (1968). Ammonites sinemuriennes à Chau-thoi (Bien hoa). *Arch. Géol. Vietnam, 11,* 103–12.

Takahashi, H. (1969). Stratigraphy and ammonite fauna of the Jurassic system of the southern Kitakami massif, northeast Honshu, Japan. *Sci. Rep., Tohoku Univ., II Ser. (Geol.), 41,* 93.

Taketani, Y., & Minoura, K. (1984). *Radiolarian Fossils Discovered from the Pre-Miyako Group in the Eastern Margin of Kitakami Mountain.* Preprint, 91st Annual Meeting, Geol. Soc. Japan, Abstr. 205.

Takizawa, F. (1985). Jurassic sedimentation in the south Kitakami belt, northeast Japan. *Bull. Geol. Surv. Japan, 36,* 203–320.

Tamura, M. (1960). A stratigraphic study of the Torinosu Group and its relatives. *Mem. Fac. Educ., Kumamoto Univ., Suppl., 8,* 1–40.

(1961). The Torinosu series and fossils therein. *Japan. J. Geol. Geogr., 32,* 219–51.

Vu Khuc (1983). Stratigraphy of Mesozoic deposits in the south of Vietnam. *J. Earth Sci., 1.*

(1984a). *Characteristic Fossils in the South of Viet Nam.* Hanoi: Science and Technics Publ. House.

(1984b). *Jurassic of Vietnam.* IGCP Project #171, Circum-Pacific Jurassic, Rept. 2.

(1985a). *Stratigraphy of Marine Jurassic in Vietnam and Neighbouring Countries.* Geol. Conf. Vietnam, Hanoi, Rept. 2.

(1985b). *Stratigraphy of Marine Jurassic in Vietnam in the Light of New Data.* IGCP Project #171, Circum-Pacific Jurassic, III Field Conference.

Vu Khuc & Nguyen, D. H. (1981). On the presence of Hettangian faunas in Tholam–Songbung area (Quangnam-Danang). *J. Earth Sci., 4*(3).

Vu Khuc, Nguyen, X. B., & Le, V. C. (1986). Stratigraphy and sedimentary basins of Viet Nam. *Escap Atlas of Stratigraphy, 6,* 1–11.
Wakita, K. (1988). Origin of chaotically mixed rock bodies in the early Jurassic to early Cretaceous sedimentary complex of the Mino Terrane, central Japan. *Bull. Geol. Surv. Japan, 39,* 675–757.
Westermann, G. E. G., & Callomon, J. (1988). The Macrocephalitinae and associated Bathonian and Early Callovian (Jurassic) ammonoids of the Sula Islands and New Guinea. *Palaeontographica, A203,* 1–90.
Yao, A. (1984). Subdivision of the Mesozoic complex in Kii-Yura area, southwest Japan, and its bearing on the Mesozoic basin development in the southern Chichibu terrane. *J. Geosci., Osaka City Univ., 27,* 41–103.
(1985). Recent advances of geologic study of the Paleozoic-Mesozoic complexes in the Chichibu belt. *Earth Sci. (Chikyu Kagaku), 39,* 44–56.
(1986). Geological age of Jurassic radiolarian zones in Japan and their international correlations. *News Osaka Micropal., Spec. Vol., 7,* 63–74.

# 10 Eastern China

Y. G. WANG, S. E. WANG, B. P. LIU, and J. S. YÜ

## GEOLOGY AND STRATIGRAPHY[1]

East China is approximately the vast area east of longitude 105°E, north of latitude 35°N, and east of longitude 97°E south of 35°N (Figure 10.1).

Later Triassic (Indo–Sinian) tectonism caused large-scale marine regression in South China and greatly influenced all of China. During the Jurassic, South China was uplifted and joined with North China to form a landmass. Terrestrial Jurassic rocks therefore predominate in East China. Marine transgressions occurred regionally in the areas of Guangdong, south Hunan, south Fujian, northeast Heilongjiang, and west Yunnan. Evidence for Jurassic in Taiwan is missing.

During the Palaeozoic and Triassic, three latitudinal tectonic belts, from north to south the Yin-shan, Qinling, and Nanling Belts, controlled geological events of East China. Since the Indo-Sinian tectonism along the western active margin of the Pacific plate, East China has been differentiated from West China. Easternmost China was under shear stress, resulting in two sets of orthogonal faults, northeast and northwest, and in many small sedimentary block basins, with active volcanism along the faults. Volcanic activity gradually increased while the depositional basins moved from west to east. Consequently, the vast area west of longitude 105°E is the Stable Giant Basin region.

Two major facies are distinguished: terrestrial and alternating marine and terrestrial, each with stable and mobile shelf environments. Many facies assemblages have also been distinguished (Table 10.1). Based on these facies, the Jurassic of East China is divided into two regions and eight districts (Figure 10.1). The Jurassic succession is as follows (Table 10.2) (numbered as in Figure 10.1).

[1] By Y. G. Wang and S. E. Wang.

### Active coastal volcanic region of eastern East China (I)

#### *Wandashan–Jixi district (I.1)*

*Wandashan subdistrict.* The Lungzhagou Group, with marine intercalations, occurs mainly in the east at Yunshan, Lungzhagou, and along the Oihulin River in Hulin county, at Zhushan, Peide, Neizhishan, and Xincun in Mishan county, at Dahushu Baomi bridge in Baoqing county, and in Raohe county. The sequence is as follows (ascending order):

- *Peide Formation* (1,000 m, unconformable on Paleozoic granites): conglomerate, coarse sandstone, with intermediate volcaniclastics in the middle part and thin coal beds in the upper part, yielding the Middle Jurassic plants *Cladophlebis argutula, Neoclamites nathorsti,* etc.
- *Qihulin Formation* (300–400 m, conformable or disconformable): marine black mudstone and siltstone, with sandstone and tuff intercalations, gastropods, bivalves, and the rare, poorly preserved ammonites ?*Morphoceras* sp. [G. E. G. Westermann personal communication 1985; formerly "*Arctocephalites* (*Cranocephalites*) *hulinensis*"], *Holcophylloceras* sp., and *Lytoceras* sp.; Bathonian–Callovian.
- *Yunshan Formation* (1,500–2,000 m, conformable): alternating marine and terrestrial, mainly clastics, intercalated with a few coal beds in the upper part and tuffaceous breccia in the lower part, yielding brachiopods, gastropods, the bivalves *Buchia orbicularis* and *B.* cf. *tenuistriata,* ostracods, and plants; Oxfordian–Lower Kimmeridgian.
- *Zhushan Formation* (conformable): coal-bearing clastics intercalated with tuff and tuffaceous breccia, with plants

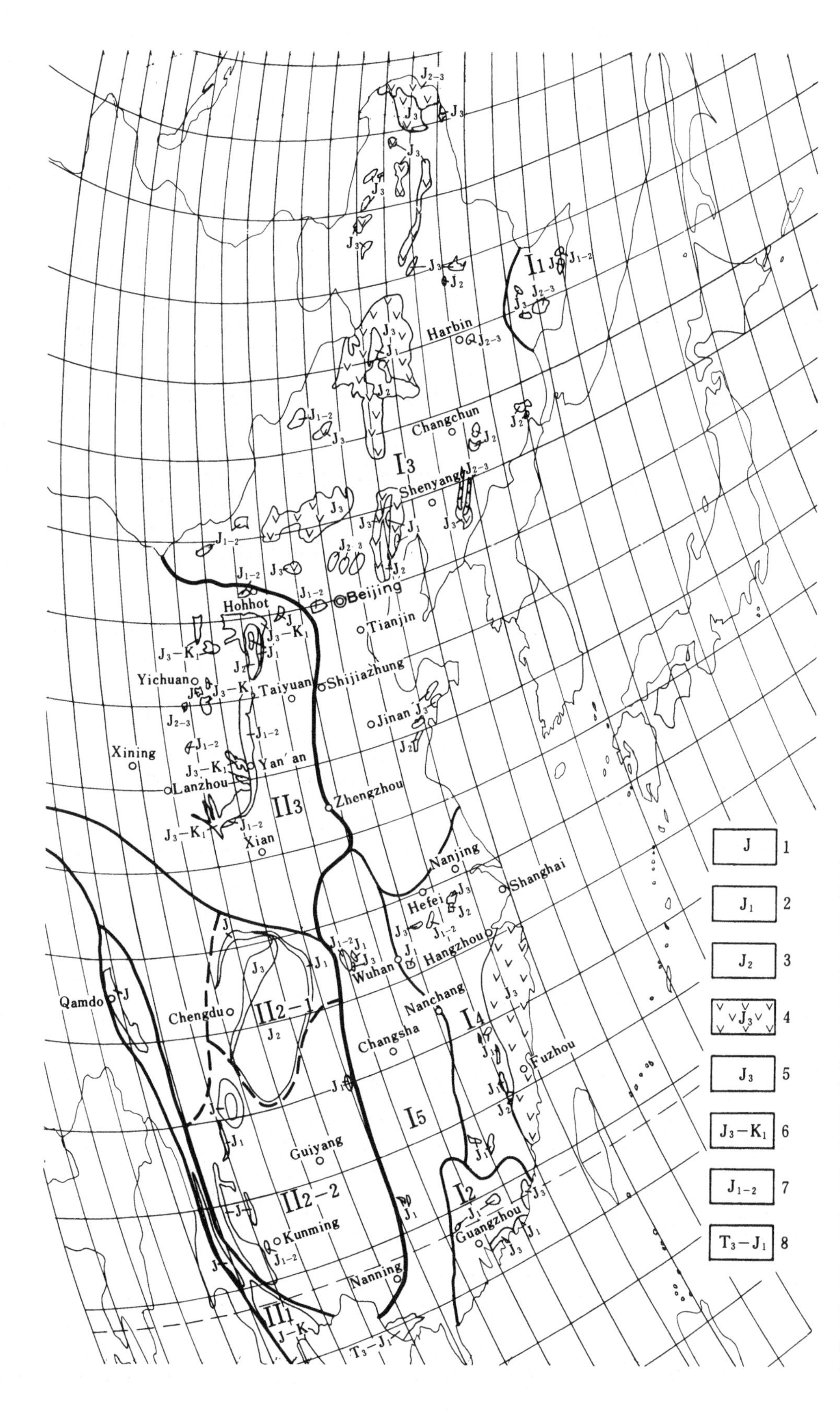


Figure 10.1. Map showing geologic regions of East China: I, active coastal volcanic region of eastern East China; $I_1$, Wanda Shan–Jixi district of northeast Heilonjiang; $I_2$, Guangdong–south Hunan–south Fujian district; $I_3$, northeast and North China district; $I_4$, Anhui-Zhejiang-Fujiang-Jiangxi district; $I_5$, Hubei-Hunan-Gangxi district; II. Stable Giant basin region of western East China; $II_1$, west Yunnan district; $II_2$, Sichuan-Yunnan district; $II_{2-1}$, Sichuan subdistrict; $II_{2-2}$, central and east Yunnan subdistrict; $II_3$, Ordos (Shaanxi-Gansu-Ningxia) district. Stratigraphy: 1, Jurassic; 2, Lower Jurassic; 3, Middle Jurassic; 4, Upper Jurassic volcanics; 5, Upper Jurassic; 6, Upper Jurassic–Lower Cretaceous; 7, Lower–Middle Jurassic; 8, Upper Triassic–Lower Jurassic.

Table 10.1. *The lithologic megafacies of eastern China*

| | | Facies | region | district | subdist. | |
|---|---|---|---|---|---|---|
| Alternating marine and terretrstrial | Active | Coal–bearing clastics and lake lutites with intercalated marine beds( Bathonian–Callovian) | Coast active volcanoes region of east part of Eastern China (I) | Wanda Shan –Jixi of N. E. Heilongjiang( $I_1$ ) | Wanda Shan ( $I_{1-1}$ ) | Pacific |
| | | | | | Jixi ( $I_{1-2}$ ) | |
| | | Great thick clastics with volcanics | | Guangdong–S. Hunan–S. Fujian ( $I_2$ ) | N. & C. part ( $I_{2-1}$ ) | |
| | Stable | Coal–bearing clastics with intercalatered marine beds ( Hett.–Sinem.) | | | East part ( $I_{2-2}$ ) | |
| | Stable | Mainly neritic and littoral pelitic carbonates with continent clastic deposits in the lower & upper parts | Stable giant basin region (II) | Western Yunnan ( $II_1$ ) | | Mediterr. |
| Terrestrial | Active | Coal–bearing clastics and red beds of giant basins, mostly lake deposits | Coast active volcanoes region of east part of Eastern China (I) | N. E. & N. China ( $I_3$ ) | | |
| | | Coal–bearing beds, red beds and vocanics | | Anhui–Zhejiang–Fujian–Jiangxi ( $I_4$ ) | | |
| | | Coal–bearing clastics | | Hubei–Hunan–Guangxi ( $I_5$ ) | | |
| | Stable | Coal–bearing clastics and red beds of giant basins, mostly lake deposits | Stable giant basin region of west part of Eastern China (II) | Ordos( Shaanxi–Gansu–Ningxia) ( $II_3$ ) | | |
| | | Mostly terrestrial red beds of giant basins | | Sichuan–Yunnan ( $II_2$ ) | Sichuan ( $II_{2-1}$ ) | |
| | | | | | C. & E. Yunnan ( $II_{2-2}$ ) | |

by Wang Yi–gang

*Coniopteris saportana, Onychiopsis elongata, Cladoohlebis contracta, Nilssonia* cf. *sinensis,* etc.; ?latest Jurassic.

*Jixi subdistrict.* Only entirely continental Upper Jurassic is present, that is, the Jixi Group of Jixi, Muling, and Boli counties. The sequence is (ascending order) clastics, conglomerate, sandstone, and siltstone, coal-bearing in the middle part and with volcanics in the upper part, yielding the plants *Onychiopsis elongata, Coniopteris* cf. *suessi, C.* cf. *burejensis, Nilssonia sinensis,* etc.

*Chengzihe Formation* (500–2,000 m, disconformable): mainly continental, coal measures, with submarine and tuffaceous intercalation, yielding plants, the bivalve *Buchia tenuistrata,* insects, and fishes; Kimmeridgian.

*Muleng Formation* (300–1,000 m, conformable): continental fine-grained sandstone and mudstone, intercalated with coal beds and tuff and tuffaceous clastics, yielding plants, the bivalve *Sphaerium subplanum,* and the gastropod *Bellamya* cf. *clavilithiformis.*

Recently, some Late Triassic–?Early Jurassic radiolarians and Late Jurassic ammonites and bivalves were found in the northeastern part of this district.

*Guangdong–south Hunan–south Fujian district (I.2)*

*Lower Jurassic.* Marine clastics intercalated with coal measures are widespread, conformable or disconformable on Upper Triassic continental deposits, in South Hunan with unconformity.

*Central and north Guangdong–south Hunan (including Hongkong) subdistrict.* Mostly neritic fine clastics (200–400 m) with coal measures; typically developed at Jinji, Kaiping county, central Guangdong (Figure 10.2).

*East Guangdon–south Fujian subdistrict.* Thick clastic deposits (2,000–5,000 m, conformable on Upper Triassic), including gray, grayish black, and light yellow argillaceous siltstone, mudstone, and fine to medium sandstone, intercalated with volcaniclastics in the east. The Huizhai section, Jiexi county, has yielded the ammonite *Sulciferites hongkongensis* and also arietitids. *Juraphyllites* sp. and *Eparietites* sp. occur at Shanlongshui in Huidong county, and *Arnioceras ceratitoides* occurs at Xeilin in Xinfong county and at Lantang in Zijin county; Sinemurian.

Middle and Upper Jurassic continental deposits without marine intercalation occur in the south Hunan–north Guangdong and Guangdong–south Fujian subdistricts. The sequence of the first subdistrict is usually much thinner than in the second and is composed of intermontane-lake clastic, unconformable on Lower Jurassic. The second subdistrict has thick (2,000–5,000 m) volcaniclastics, with limnic intercalations in the lower part, which lies unconformably on Lower Jurassic or pre-Jurassic.

*Northeast and North China district (I.3)*

During the Jurassic, this district belonged to the Circum-Pacific Tectonic Belt, with intense tectonism, magmatism, and volcanism. Jurassic rocks occur in many separate fault basins and consist of interlayeral volcanics and sediments. The Lower and Middle Jurassic sediments are characterized by coal measures, with the *Coniopteris-Phoenicopsis* flora, and the Upper Jurassic sediments are characterized by red clastics and dark rocks, intercalated with coal beds yielding the Rehe (Jehol) fauna (e.g., the *Lycoptera-Eoestheria-Ephemeropsis* assemblage).

The Jurassic of the Yanliao area (Hebei and Liaoning provinces), which is representative of this district, has been divided as follows. The Lower Jurassic includes two formations. The Xinlongou Formation (500 m) consists of basalt, andesite, andesitic breccia lava, volcanic agglomerate, and tuff, and it lies disconformably on the Upper Triassic Kuntoubolo Formation. The Beipiao Formation (over 1,000 m) is mainly sandstone, shale, and conglomerate, intercalated with coal measures, and it yields the plants *Neocalamites carrerei, Thallites pingsiangensis, Thaumatopteris pusilla, Cladophlebis asiatica, Todites williamsoni,* etc.

The Middle Jurassic includes two formations: (1) The Haifanggou Formation (268 m) consists of yellowish green to yellowish gray conglomerate, sandstone, and shale, intercalated with tuffaceous sandstone, yielding the conchostracan *Euestheria haifanggouensis,* the insect *Sunoplecia liaoningensis,* the bivalve

Table 10.2. *Correlation chart*

| | | Wanda Shan–Jixi district ( I1 ) | | Guangdong–S. Hunan–S. Fujian( I2 ) | | N. E. N. China(I3) | ( I4 ) |
|---|---|---|---|---|---|---|---|
| | | Dongshan Fm. ( K1 ? ) | Jixi( I1–2 ) | N. & C. part ( I2–1 ) | East part ( I2–2 ) | Hebei–Liaoning area | N. & C. Anhui |
| overlying beds | | Dongshan Fm. ( $K_1$ ? ) | Dongshan Fm. ( $K_1$ ) | Q | Q | Banlashan Fm. ( K1. ) | Red beds ( K ?) |
| U. Jurassic | Tith. | Lungzhagou Group: Zhushan Formation ( 930m. ) | Jixi Group: Muleng Formation ( 1000m. ) | Baizushan Group ( 170–1450 m. ) | Kaojiping Formation (>4000m. ) | Fuxin Formation ( 1000m. ); Jiufotang Fm. ( 1000–2000m. ) | Heishidu Formation (>1300m. ) |
| | Kimm. | Yushan Formation ( 1500–2000m. ) | Chengzihe Fm. ( 500–2000 m. ) | | | Yixian Fm. ( 2000m); Dabeigou Fm. (150m) | |
| | Oxf. | | Didao Fm. ( 0–400m. ) | | | Zhangjiakou Fm. ( 2300–2700m. ); Tuchengzi Fm. ( 800–2900m. ) | Maotangchang Formation ( 967m. ) |
| M. Jurassic | Call. | Qihulin Formation ( 300–400m. ) | | | Zhangping Formation ( >1000m. ) | Lanqi Formation ( 500m. ) | Yuantangshan Formation ( 1339m. ) |
| | Bath. | Peide Fm. ( 1000m. ) | ? Effusive Formation | | | | |
| | Baj. | ? | ? | | | Jaifanggou Formation ( 268m. ) | |
| | Aal. | | | | | | |
| L. Jurassic | Toar. | | | | ? | Beipiao Formatinon ( 1000m. ) | Fanghushan Formation ( 400–500m. ) |
| | Plein. | | | | | | |
| | Sinem | | | Jinji Formation ( 200–400 m. ) | Jinji Formation ( >2000m. ) | Xinlonggou Formation ( 500m. ) | |
| | Hett. | | | | | | |
| undering beds | | Granites or Permian | Granites or Mashan Gr. ( Pt. ) | Shiaoping Fm. $T_3$ | Shiaoping Fm. $T_3$ | Kuntoubuluo Fm. $T_3$ | Xiaurenchung Fm. ( Sinian) |

| | | Anhui–Zhejiang–Fujian–Jiangxi district ( I4 ) | | | Hubei–Hunan–Guangxi district ( I5 ) | | | |
|---|---|---|---|---|---|---|---|---|
| | | S. Jiangsu, S. Anhui | Zhejiang | Fujian | W. Hubei | E. Hunan | S. W. Hunan | S. E. Guangxi |
| overlying beds | | Gecun Fm. | Heshan Fm. ( K) | Shaxian Fm. ( K) | Donghu Fm. ( K) | | | Napa Fm. ( $K_1$ ) |
| U. Jurassic | Tith. | Dawanshan Fm.; Yunheshan Fm. (3000m. ) | Shouchang Fm. ( 1160m. ); Huangjian Fm. ( 175–700m. ) | Bantou Fm. ( 275–1300 ); Nanyuan Fm. ( 1790m. ) | | | | Kuali Formation ( 1182m. ) |
| | Kimm. | Longwangshan Fm.; Xihengshan Fm | Laocun Fm. ( 1600–200m. ) | Changlin Fm. ( 1955m. ) | Penglaizhen Formation | | | |
| | Oxf. | | | | Suining Fm. | | | |
| M. Jurassic | Call. | Xiangshan Group ( 1500m. ) | Yushanjian Formation ( 3039m. ) | Zhangping Formation ( 1043m. ) | Shaximiao Formation | | Yanglukou Formation ( 184m. ) | Natang Formation ( 995m. ) |
| | Bath. | | | | | | | |
| | Baj. | | | | | | Shixi–jiang Formation ( 146m. ) | |
| | Aal. | | | | | | | |
| L. Jurassic | Toar. | | Majian Formation ( 345m. ) | Lishan Formtion (>579m. ) | Ziliujing Fm. or Xiangxi Fm. | Menkou–shan Fm. ( 378m. ) | Guangyintan Formation ( 500m. ) | Baixing Formation ( 1300m. ) |
| | Plein. | | | | | | | |
| | Sinem. | | | | | Zhaoshang Formation ( 200m. ) | | Wangmen Formation ( 520m. ) |
| | Hett. | | | | | | | |
| unhdering beds | | Permian | Wozao Fm. $T_3$ | Wenbinshan Fm. $T_3$ | Shazhengxi Fm. $T_3$ | Sanqiutian Fm. $T_3$ | | Fulongao Fm. $T_3$ |

*Ferganoconcha sibirica*, and the plants *Coniopteris hemenophylloides* and *Todites williamson*. (2) The Lanqi Formation (500 m) consists of andesitic breccia, volcanic agglomerate, and andesitic tuff, intercalated with tuffaceous siltstone, yielding the plants *Coniopteris* cf. *hymenophylloides*, *Hausmannia* sp., *Podozamites landceoalatus*, etc., from the sediments.

The Upper Jurassic includes six formations (from below): (1) Tuchengzi Formation (800–2,900 m), red conglomerate and sand-

Table 10.2 (*cont.*)

| | | W. Yunnan district ($II_1$) | Sichuan–Yunnan district ($II_2$) | | | | Ordos district ($II_3$) |
|---|---|---|---|---|---|---|---|
| | | | Sichuan ($II_{2-1}$) | | E. & C. Yunnan ($II_{2-2}$) | | (Shaanxi–Gansu–Ningxia) |
| | | | centeral | northern | Lufeng basin (E.) | C. Yunnan | |
| overlying beds | | ? | Tienmashan Fm. ($K_1$) | Jiemenguan Fm. ($K_1$) | | Gaofongshi Fm. ($K_1$) | Zhidan Group (2100m.) |
| U. Jurassic | Tith. | Longkan Formation (280–1390m.) | Penglaizhen Formation (200–1120m.) | Lianhuakou Formation (1720m.) | | Tuodian Formation (374m.) | |
| | Kimm. | | | | | | |
| | Oxf. | | Suining Formation (300–600m.) | | | Shedian Fm. (1086m.) | Fenfanghe Fm. (1000m.) |
| M. Jurassic | Call. | Liuwan Formation (>700m.) | Shaximiao Formation | u. part (450–1700m.) | Upper Lufeng Formation (270m.) | Zhanghe Formation (80–1000m.) | Anding Fm. (50–91m.) |
| | Bath. | | | l. part (100–500m.) | | | Zhilo Fm. (250m.) |
| | Baj. | | Xiniangou Fm. (100–200m.) | Qiantoyan Fm. (100–350m.) | | | Yanan Fm. (147–250 m.) |
| | Aal. | | | | | | |
| L. Jurassic | Toar. | Mengga Fm. (>500m.) ? | Ziliujing Formation (267–415m.) | Baitianba Formation (60–470m.) | Lower Lufeng Formation (731m.) | Fengjiahe Formation (166m.) | Fuxian Formation (100m.) |
| | Plein. | | | | | | |
| | Sinem. | | | | | | |
| | Hett. | Fault | | | | | |
| undering beds | | Permian | Xujiahe Formation ($T_3$) | | Yipinglang Formation ($T_3$) | | Yanchang Fm. ($T_3$) |

stone, intercalated with grayish green mudstone and shale, yielding the conchostracans *Mesolimnadia jinlingensis* and *Nestoria orbita*, the ostracods *Wolburgia polyphema* and *Stenestroemia subcalaris*, and the reptile *Chaoyoungosaurus liaosiensis;* (2) Zhangjiakou Formation (2,300–2,700 m), mainly quartz-porphyry, rhyolite, trachyte tuff, and tuffaceous breccia, intercalated with andesite and sandstone, yielding the insect *Coptoclava* sp.; (3) Dabeigou Formation (150 m), yellowish green to grayish green coarse sandstone, fine-grained sandstone, and shales, intercalated with calcareous and siliceous shales; in some basins developed as thick (over 1,000 m) volcanics, intercalated with sandstone and shale, yielding the conchostracans *Neostoria pissovi* and *Keratestheria gigantia*, the fish *Peipiaotus pani*, the insect *Ephemeropsis trisetalis*, the *Eoparacypris-Luanpingella-Pseudoparacypris* ostracod assemblage, and some bivalves; (4) Yixian Formation (over 2,000 m), andesite and basalt, intercalated with sediments, yielding the typical Rehe fauna (*Lycoptera-Eoestheria-Ephemeropsis* assemblage); (5) Jiufotang Formation (1,000–2,000 m), alternating sandstone, siltstone, and shale, intercalated with conglomerate, yielding bivalves, ostracods, insects, conchostracans, and fish; (6) Fuxin Formation (1,000 m), sandstone, shale, and conglomerate, intercalated with coal beds, yielding the conchostracans *Pseudestheria qinghemenensis* and *Diestheria yangliutunensis*, the bivalves *Ferganoconcha curta* and *Sohaerium jeholense*, the gastropods *Probaicalia gerassimori* and *Bellamya clavilithiformis*, and the early *Ruffordia-Onychiopsis* flora.

*Anhui-Zhejiang-Fujian-Jiangxi district (I.4)*

The Jurassic occurs in very small basins and lies disconformably on older rocks. The Lias is characterized by coal-bearing deposits, the Dogger by yellowish green to purplish red rocks, and the Malm by volcanics and sediments.

In northern Anhui, chiefly in the Hefei Depression and along the southern and northern piedmonts of the Dabie Mountains, occur the Lower Jurassic Fanghushan Formation, the Middle Jurassic Yuantongshan Formation, and the Upper Jurassic Maotanchang and Heishidu Formations. The Fanghushan (400–500 m) consists of yellow to grayish white sandstone, siltstone, and conglomerate, intercalated with carbonaceous shale and coal seams, yielding the plants *Neocalamites carrerei* and *Pityophyllum nordenskioldi* and the bivalve *Sibireconcha* cf. *anodontoides*. The Yuantongshan is mainly unfossiliferous, purplish red to yellowish green sandstones, intercalated with silty mudstone. The Maotanchang (976 m) consists of andesite, tuff, and tuffaceous sandstone intercalated with dark gray siltstone and shale, yielding the conchostracans *Eoestheria* cf. *middendorfii* and *Diestheria suboblonga*, the bivalve *Ferganoconcha sibirica*, and gastropods. The Heishidu has two members: The lower (838 m) consists of dark purple conglomerate, tuffaceous conglomerate, and sandstone; the upper (over 480 m) consists of yellowish green to gray sandstone and conglomerate, intercalated with grayish black shale, yielding plants, bivalves, gastropods, and conchostracans.

In the lower Yangtzi area, including south Jiangsu and south

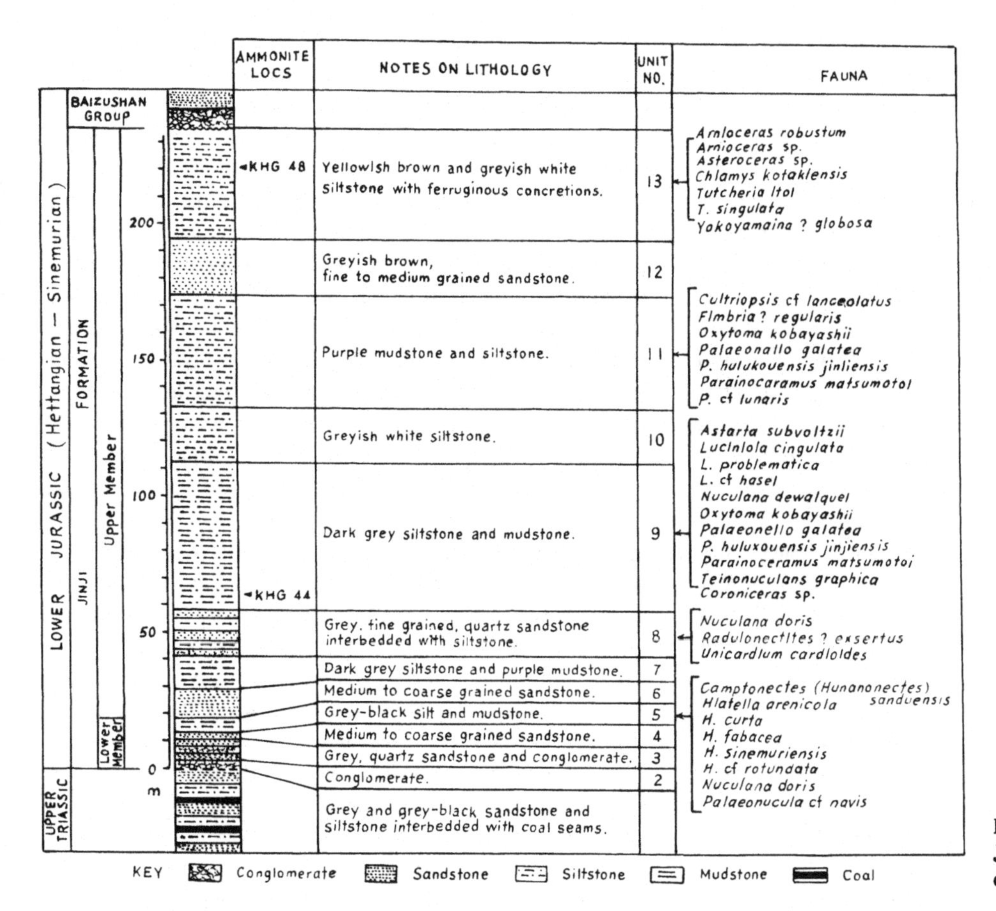


**Figure 10.2. The Jinji Formation at Jinji Hill, Kaiping county, Guangdong province.**

Anhui, Lower and Middle Jurassic are represented by the Xianshan Group, with (1) a lower member (400 m) of mainly brownish yellow to grayish white sandstone, silty mudstone, and conglomerate, intercalated with black shale and coal measures, yielding the plants *Equisetites sarrani*, *Clathrooteris meniscioides*, and *Cladophlebis denticulata* and the bivalves *Undulatula perlonga* and *Unio* cf. *pictoriformis*, indicating Early Jurassic, and (2) a Middle Jurassic upper member of purplish red to grayish white sandstone. The Upper Jurassic is represented by the Xihengshan, Longwangshan, Yuanheshan, and Dawangshan Formations (3,000 m), which are chiefly conglomerate, sandstone, and volcanics and yield the plants *Cladophlebis* cf. *browniana*, *Brachyphyllum obesum*, and *Sphenopteris* cf. *nitidula*, spores, pollen, the gastropod *Probaicalia*, the insect *Ephemeropsis trisetalis*, and the conchostracans *Dictyestheria*, *Orthestheria*, and *Yanjiestheria*.

The Lias of west Zhejiang is called the Majian Formation (345 m). It consists of grayish yellow to yellowish green conglomerate, sandstone, and silty mudstone, intercalated with coal measures, yielding the plants *Clathropteris meniscoides* and *Coniopteris hymenophylloides* and the bivalve *Pseudocardinia sibirensis*. The Middle Jurassic Yushanjian Formation (3,000 m) is mainly an alternation of yellowish green conglomerate, sandstone, and mudstone, in the lower part, grading into purple siltstone, sandstone, and conglomerate, and yields the plants *Cladophlebis raciborskii* and *Coniopteris hymenophylloides* and the bivalves *Pseudocardinia* sp. and *Talluella* sp. The Malm has three formations (ascending order): (1) Laocun Formation (1,600–2,000 m), mainly purplish red sandstone, mudstone, and conglomerate, intercalated with rhyolite and tuff, containing the insects *Mesopanorpa yaojiashanensis* and *Linicorixa odota*, the bivalves *Ferganoconcha curta* and *F. liaoxiensis*, the insect *Ephemeropsis triselalis*, the conchostracans *Yanjiestheria chekiangensis* and *Eosestheria qingtanensis*, ostracods, and the fish *Mesoclupea showchangensis*, (2) a middle member (360 m) of rhyolite-porphyry and tuff, and (3) an upper member (776 m) of chiefly yellowish green fine-grained sandstone, siltsone, and mudstone, yielding the bivalves *Sphaerium jeholense* and *S. selenginense*, the gastropods *Probaicalia vitimensis* and *P. gerassimovi*, the conchostracans *Eosestheria shouchangensis* and *Shouchangestheria zhexiensis*, the ostracods *Darwinula oblonga*, etc., and the fishes *Mesocluoea showchangensis*, *Sinamia huananensis*, and *Fuchunkiangia chinensis*. The Lower and Middle Jurassic of east Zhejiang are dominated by volcanic rocks.

In Fujian, the Lower Jurassic Lishan Formation consists of coal measures, the Middle Jurassic Zhangping Formation of mainly redbeds, and the Upper Jurassic (Changlin Nanyuan and Bantou Formations) of alternating volcanic and sedimentary rocks, with biotas very similar to those of west Zhejiang.

In east Jiangxi, only the Malm is developed, and it consists of volcanic and sedimentary rocks, with flora and fauna similar to those of west Zhejiang.

### *Hubei–Hunan–Guangxi district (I.5)*

Except for the Zigui Basin of west Hubei and the Shiwandashan area of southeast Guangxi where the entire Jurassic is present, the Middle and Upper Jurassic are poorly developed in this district. In the Zigui Basin occur the Lower Jurassic Xiangxi or Ziliujing Formation, the Middle Jurassic Shaximiao Formation, and the Upper Jurassic Suining and Penglaizhen Formations. The Xiangxi consists of grayish green sandstone and shale, intercalated with coal beds, yielding the plants *Marattiposis asiatica* and *Clathropteris obovata.* The Ziliujing, Shaximiao, Suining, and Penglaizhen Formations are similar to those of the Sichuan Basin (see Sichuan-Yunnan district).

The Lias of southeast Hubei is represented by the Wuchang Formation (168 m), which consists of grayish yellow to brownish yellow sandstone, mudstone, and conglomerate at the base, intercalated with shale and coal beds, yielding the plant *Coniopteris hymenophylloides.* The Dogger is mainly brownish yellow mudstone and fine-grained sandstone (115 m) and yields the bivalves *Pseudocardinia khadjakalanensis* and *P. kweichowensis.*

In east Hunan, only the Lias is developed, and it is subdivided into the Zhaoshan and Menkoushan Formations. The Zhaoshan (2,000 m) consists of gray to grayish white sandstone, mudstone, and conglomerate, yielding the bivalves *Waagenoperna mytiloides* and *Lilingella simplex,* the ostracod *Gomphocythere(?) liuyangensis,* and the plant *Nilssonia polymorpha.* The Menkoushan (378 m) consists of grayish green sandstone and mudstone, intercalated with coal beds, yielding the bivalve *Lilingella robusta* and the plant *Nilssonia polymorpha.*

The Jurassic rocks of southwest Hunan can be subdivided into the Lower Jurassic Guangyintan Formation and the Middle Jurassic Shixijiang and Yanglukou Formations. The Guangyintan (500 m) is sandstone and mudstone, with a coal measure in the middle, yielding the bivalves *Waagenoperna mytiloides, Lilingella simplex,* and *Qiyangia loxos,* the ostracod *Gomphocythere*(?) *liuyangensis,* and the *Marattiopsis-Otozamites* plant assemblage. Both the Shixijiang and Yanglukou are mainly purplish red to light yellow sandstone and mudstone, yielding bivalves and plants. The Jurassic rocks of the Shiwandashan area have been subdivided into four formations: (1) the Liassic Wangmen Formation (520 m), brownish yellow to grayish white sandstone and purplish red mudstone with basal conglomerate, intercalated with coal stringers, containing spores and pollen; (2) the Baixing Formation (1,300 m), purplish red sandstone and mudstone intercalated with grayish green mudstone, yielding plants *Equisetites* cf. *sarrani, Coniopteris* cf. *hymenophylloides,* and probably late Liassic; (3) the Middle Jurassic Natang Formation (955 m), almost entirely redbeds, unfossiliferous; and (4) the Upper Jurassic Kuali Formation (1,182 m), alternating purplish red to brownish yellow sandstone and silty mudstone, yielding few plants. The Jurassic rocks and fossils of northeast Guangxi are similar to those of southwest Hunan.

## Stable Giant Basin region of western East China (II)

### *Western Yunnan district (II.1)*

The Jurassic of western Yunnan was formerly subdivided into four formations dated as follows: (1) the Mengga (Aalenian–Bajocian?), (2) Liuwan (Bathonian?), (3) Longhai (Callovian?), and (4) Longkan (Malm). The first three yield abundant fossils and were commonly considered to be Middle Jurassic, mainly Bathonian. The main stratigraphic problem of this district involves the age and facies change of the Liuwan Formation. Recently, Sun et al. (1983) pointed out that its lower part, together with the upper part of the Mengga Formation, could be Lias and that the Longhai correlates with the middle Liuwan Formation. In typical sections, the Liuwan is underlain by the Mengga Formation (over 800 m), which consists of purple to grayish yellow mudstone, marl siltstone, and fine sandstone and yields the bivalve *Luciniola* cf. *problematica.* The Liuwan (709 m, conformable) consists of gray to grayish black calcareous mudstone, limestone, yellowish green or purple sandstone, and sandy shale. It has yielded the foraminifers *Textularia* and *Glomospira* and conchostracan *Quenstedia* sp. in the upper part, the brachiopods *Kutchithyris oliveformis, Burmirhynchia prastans, B. shanensis, Holcothyris tangulaica,* and *Avonothyris distarta,* the bivalves *Pseudolimea duplicta, Parvamussium pumillum, Camptonectes lens,* and *Liostrea birmanica,* and gastropods in the middle part, and the foraminifer *Labyrinthina* and the bivalve *Parvamussium* in the lower part. The occurrence in the middle part of the *Burmirhynchia-Holcothyris* brachiopod and *Camptonectes lens–Liostrea birmanica* bivalve assemblages, which are widely distributed also in the supposed Bathonian of south Qinghai, north and east Tibet, and south Yunnan, suggests a Bathonian age. The upper Liuwan is presumably Callovian. The overlying Longkan Formation (357 m) consists of purple mudstone and sandstone, yielding the bivalve *Modiolus* sp.

### *Sichuan–Yunnan district (II.2)*

The Jurassic is well developed in the Sichuan and central Yunnan Basins, but the Malm is missing in some intermontane basins of east Yunnan. Redbeds dominate, except along the northern margin of the Sichuan Basin, where the Lias has coal measures.

The Lias includes the Ziliujing Formation of central Sichuan and the Baitianba Formation of north Sichuan. The Ziliujing (267–415 m) consists of purplish red mudstone and silty mudstone, intercalated with red marl and dark shale, yielding the reptiles *Lufengosaurus* sp., *Sinopliosaurus weiyuanensis,* and *Sanpasaurus yaoi* and the bivalve *Acuneopsis luochangensis.* The Baitianba (60–470 m) consists of yellowish green fine sandstone and shale, intercalated with black shale and coal beds, yielding the plant *Cladophlebis deticulata,* the conchostracans *Palaeolimnadia batianbaensis* and *P. sichuanensis,* and bivalves. The Dogger includes the Xintiangou Formation of central Sichuan (or the Qianfoyan Formation of north Sichuan) and the Shaximiao Formation. The Xintiangou (100–2,000 m) is gray to yellowish green shale and sandy shale, intercalated with siltstone, purplish red mudstone, and shelly limestone. The Qianfoyan, coeval with the Xintiangou Formation, is mainly sandstone and shale and yields the bivalve *Lamprotula* (*Eolamprotula*) *cremeri* and plants. The lower part of the Shaximiao Formation (100–500 m) consists of purplish red mudstone, intercalated with gray sandstone, which at the top has a shale bed that yields the *Shunosaurus* fauna (Dong, Zhou, and Zhang 1983) and the *Euestheria ziliujingensis* conchostracan fauna. The upper part of the Shaximiao (450–1,700 m) is also purplish red mudstone and sandstone and yields the *Mamenchisaurus* fauna, the fish *Yuchowlepis szechuanensis*, bivalves, and the ostracod *Darwinula impudica*. The Malm includes two formations: (1) the Suining (300–600 m), which consists of brownish red to light red mudstone and sandstone, yielding the bivalve *Unio martynovae*, the ostracod *Darwinula impudica*, and the conchostracan *Eosestheria dianzhongensis;* (2) the Penglaizhen (200–1,120 m), of central Sichuan, which consists of light red to brownish red sandstone, siltstone, and mudstone, with the bivalve *Danlengiconcha elongata*, the conchostracan *Eosestheria dianzhongensis*, and the ostracod *Darwinula legunminella*. In north Sichuan the Lianhuakou Formation (1,720 m) is correlated with the Penglaizhen Formation and consists of light red conglomerate, sandstone, and mudstone.

The Lias in the central Yunnan Basin and the Lufeng Basin of east Yunnan is represented by the Fengjiahe and lower Lufeng Formations, respectively. The lower Lufeng consists of redbeds, yielding the rich *Lufengosaurus* fauna. The Dogger includes the Zhanghe Formation of the central Yunnan Basin and the upper Lufeng Formation of the Lufeng Basin, both consisting of redbeds. The Malm consists of (ascending order) (1) the Shedian Formation (1,086 m), purplish red coarse sandstone and medium sandstone, intercalated with conglomerate and mudstone, and (2) the Tuodian Formation (374 m), an alternation of purplish red sandstone and mudstone, yielding the conchostracan *Eosestheria dianzhongensis* and the ostracod *Darwinula sarytirmenensis*.

### *Ordos Basin district (II.3)*

The Ordos (Shaanxi-Gausu-Ningxia) Basin is a giant cratonic basin on the North China Platform, where Jurassic rocks are well developed. They have been subdivided into the Liassic Fuxian Formation, the Middle Jurassic Yanan, Zhilo, and Anding Formations, and the Upper Jurassic Fengfanghe Formation and Zhidan Group. The typical Fuxian (100 m), lying disconformably on the Upper Triassic Yanchang Formation, consists of purplish red mudstone and grayish green sandstone. It changes northeastward (Fugu area) into grayish green sandstone, black shale, and variegated mudstone, with conchostracans, bivalves, plants, spores, and pollen. The Yanan Formation (147–250 m) consists of grayish green to grayish yellow sandstone and dark shale in the basin center, but toward the basin margins it becomes coal-bearing, with the plants *Ginkgoites sibiricus, Phoenicopsis speciosa*, and *Sphenobasera longifolia* and the bivalves *Ferganoconcha sibirica, Tutuella forma, Margaritifera isfarensis*. The Zhilo Formation (250 m) consists of yellowish gray to variegated sandstone, siltstone, and mudstone and yields the bivalves *Sibiriconcha anodontoides*, spores, pollen, and plants. The Anding Formation (50–91 m) can be divided into two members in the center of the basin, the lower consisting of gray to black sandstone, shale, and oil-shale, intercalated with calcareous shale, and the upper consisting mainly of light red marl, yielding the ostracod *Darwinula sarytirmenensis*, the bivalve *Psiluniosuni*, the Pisces *Baleiichthys antingensis*, gastropods, spores, and pollen. Toward the southern basin margin the facies changes to unfossiliferous purple mudstone and sandstone. The Fenfanghe Formation (over 1,000 m) occurs only in the Qianyang area in the southern part of the basin and consists of variegated conglomerate, lying unconformably on the Anding Formation. The Zhidian Group (2,100 m) outcrops in the northern and western parts of the basin and can be divided into six members or formations, consisting chiefly of redbeds with grayish green shale and mudstone and yielding Reptilia, as well as ostracods, conchostracans, bivalves, gastropods, etc. In the upper part occur fishes, as well as ostracods, conchostracans, bivalves, gastropods, etc.

## PALEOGEOGRAPHY[2]

In the Nadan Hada area of the Wanda Mountains of eastern Heilongjiang province, northeastern China, some argillaceous and siliceous sediments contain Late Triassic to ?Early Jurassic radiolarians and basic volcanics, resembling the Sichote-Alin belt of Russia. Both suggest bathyal marginal-sea environments. The Middle Jurassic intercalated marine and nonmarine coal-bearing sediments differ markedly from the underlying sequence. Quite likely, the Nadan Hada area represents an oceanic exotic terrane that was accreted to the eastern margin of the Asian continent prior to the Middle Jurassic, when it turned into an epicontinental basin undergoing inundations from the Arctic (Figure 10.3).

The Late Triassic Indo–Sinian orogeny, which affected North China, Nei Mongol and northeast China to the north of the Qinling-Dabie Mountains, resulted in large-scale uplift and erosion. Several inland fluvial-lacustrine coal basins, with rich flora and some bivalves, developed here in the late Early to Middle Jurassic. In the west of this area was the large stable basin of Ordos, without volcanism, and in the east, a series of small basins, sometimes with basic to median volcanics at the base, indicative of inland rift depressions. Both types of basins have industrial coal measures, indicating a temperate, wet climate. Paleomagnetic data for 34.4°N for the Middle Jurassic have been obtained recently in the Shanxi area of North China, consistent with the stratigraphic and paleontologic analysis.

In the coastal areas of southeast China, south of the Qinling-Dabie Mountains, the early Liassic transgression formed the Jiangxi-Hunan-Guangdong embayment. Shallow-marine sediments containing *Arnioceras, Sulciferites*, and some bivalves and brachiopods cover the areas of Hongkong, north Guangdong, and south Hunan. The marine fauna includes cosmopolitan elements, reflecting the linking between the Tethyan and the circum-Pacific areas. Farther north, in Hunan, Jiangxi, and northeast Guangxi, brackish to freshwater bivalve faunas (e.g., *Qiyangia* and *Lilingella*) occur in argillaceous sediments with coal seams and plants, indicating a estuarine environment. Alternating marine–nonmarine environments also occurred in east Guangdong and

[2] By B. P. Liu and J. S. Yü.

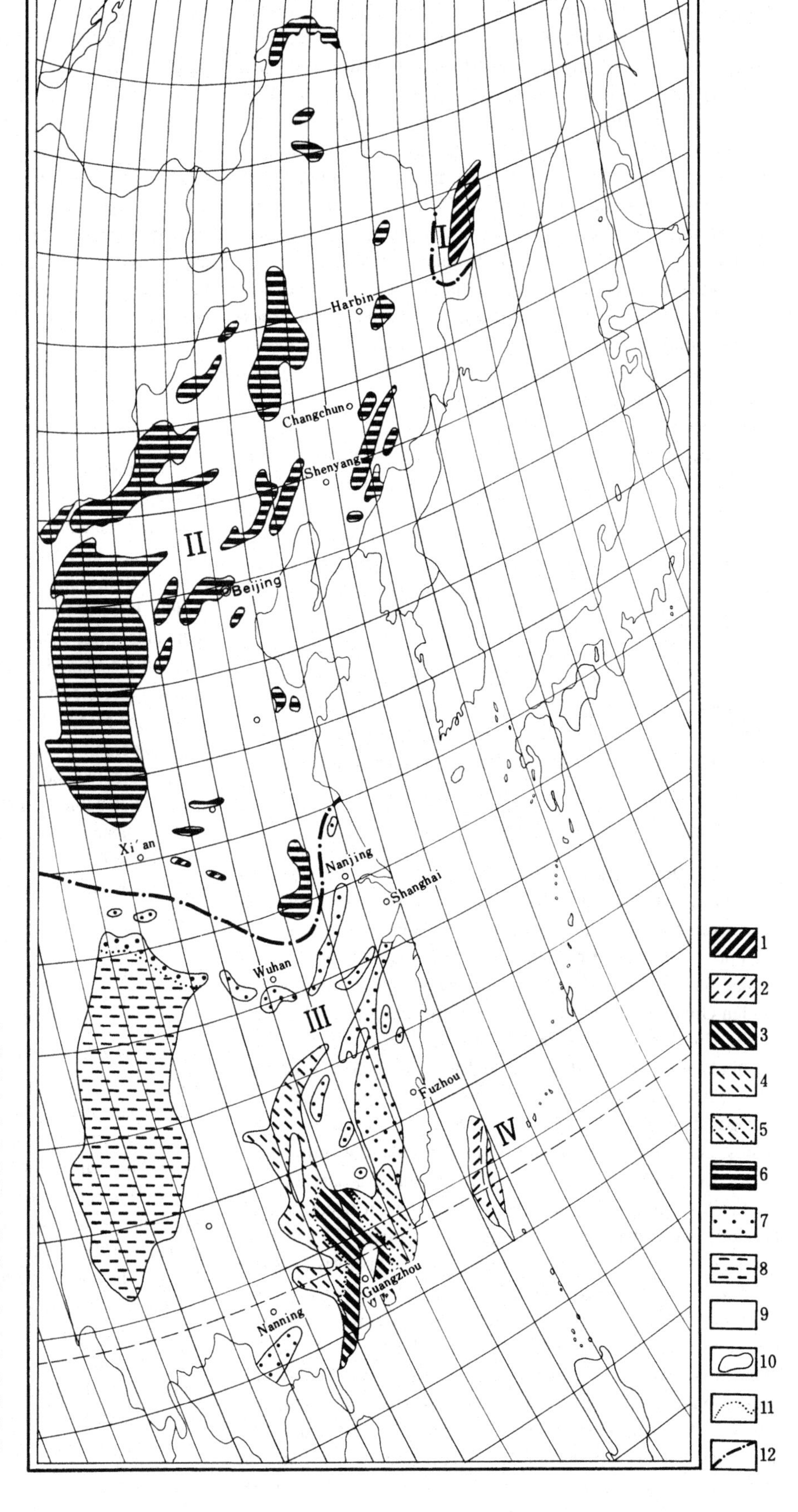


**Figure 10.3. Early and Middle Jurassic paleogeographic map of East China: 1, bathyal siliceous and argillaceous sediments; 2, presumed bathyal; 3, shallow-marine argillaceous and clastic sediments; 4, brackish clastic and argillaceous sediments; 5, paralic clastic and argillaceous sediments; 6, fluvio-lacustrine clastics and argillites (coal-bearing); 7, inland-basin clastics (locally coal-bearing); 8, lacustrine argillites and marls; 9, uplifted erosional regions; 10, shorelines and sedimentary-basin boundaries; 11, facies boundaries; 12, boundaries of paleogeographic provinces (I, Wandashan region; II, northern region; III, southern region; IV, Taiwan region).**

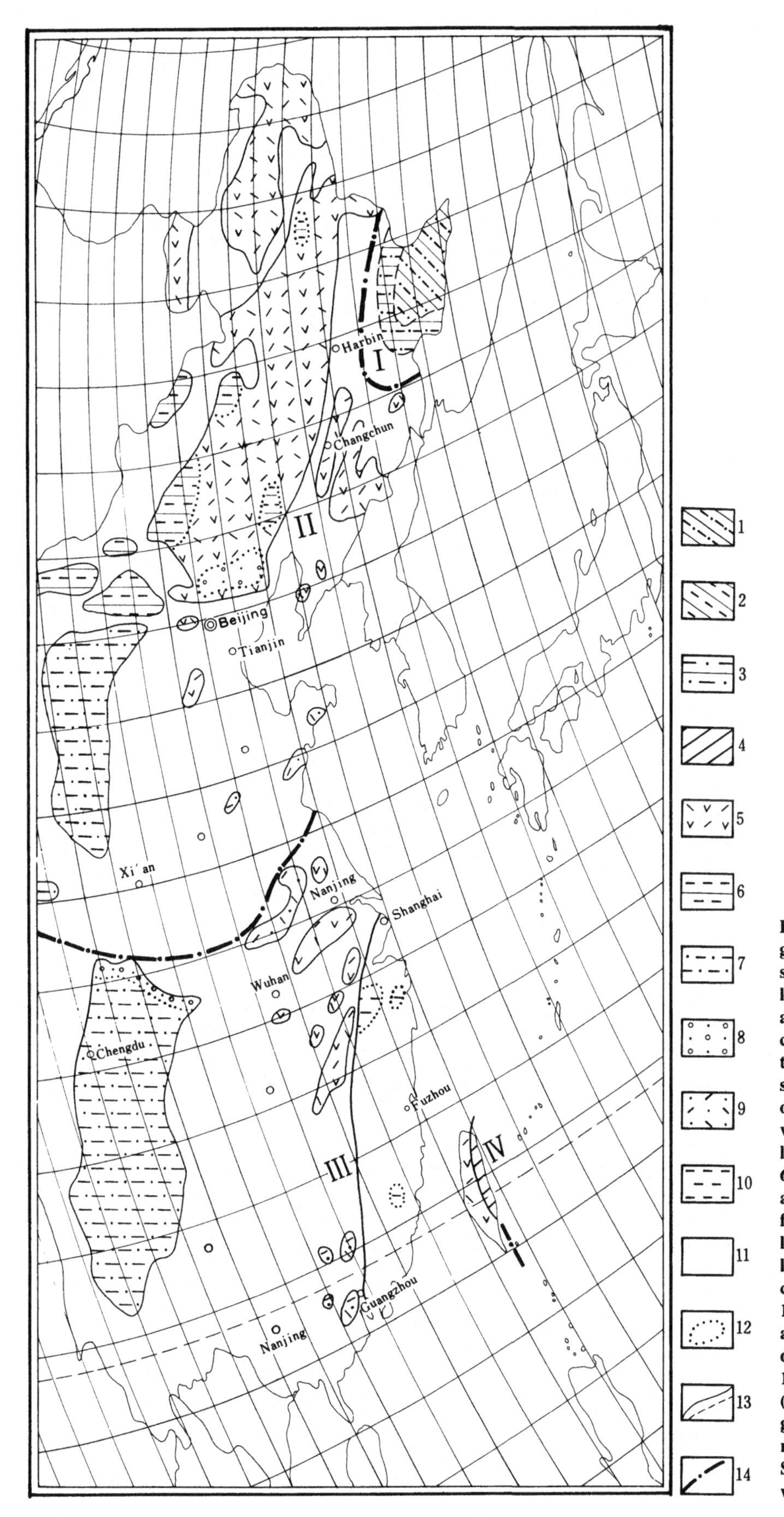


**Figure 10.4.** Late Jurassic paleogeographic map of East China: 1, shallow ocean (arenites and argillites); 2, paralic (arenites and argillites with coal and marine clastic intercalation); 3, continental margin (arenites and argillites, similar to 2); 4, shallow to deep ocean; 5, terrestrial volcanics and volcaniclastics, with coal and fluvio-lacustrine clastic intercalations; 6, inland lacustrine (arenites and argillites, with coal); 7, inland fluvio-lacustrine (clastics); 8, inland fluvial (coarse clastics); 9, inland fluvio-lacustrine (clastics with coal and volcanic intercalations); 10, inland lacustrine (arenites and argillites); 11, inland regions under erosion; 12, facies boundaries; 13, sedimentary-basin boundaries (observed and assumed); 14, paleographic boundaries [I, Wandashan region; II, Northern region; III, Southern region; IV, Yuli belt (Taiwan) region].

southwest Fujian, with ammonites and plants, regionally bearing thin coal seams and volcanics. No indisputable Jurassic fossils are known from Taiwan. It is presumed that the island was a deep-sea terrane subsequently accreted to the Asian continent.

The northern limits of the Early Jurassic sea of southeast China remain unknown. It is commonly held that the coastal area of Fujian and Zhejiang and eastern Hubei belonged to inland basins that regionally developed coal-bearing clastics. But some brackish bivalves (e.g., *Qiyangia*) have been found in east Hubei and central Fujian, suggesting that brief Early Jura§sic inundations may have reached these areas. By the Middle Jurassic, marine conditions had totally disappeared from the southeastern area; relief differentiation increased, basins shrank, and the earliest Jurassic redbeds appeared; some lenticular thin coal seams occur in the coastal areas of Zhejiang. The Early Jurassic flora and Middle Jurassic thick-shelled unionids in this area indicate that the areas to the south of the Qinling-Dabie Mountains had a warm-temperate to subtropical, semihumid climate, whereas the biogeographic area of the Early Jurassic flora and freshwater bivalves coincides with the Qinling-Dabie Belt. The Middle Jurassic large, thick-shelled unionids adapted to a temperate climate are limited to the Ordos Basin in northern Shaanxi and the Jiyuan Basin to the north of Zhengzhou, indicating a northward extension of the warm climate. It is also apparent that the paleolatitudes indicated by the paleoclimatic belts are not parallel to present-day latitudes.

Drilling data indicate that the Late Jurassic sea extended to the Suibin area east of Wandashan (Figure 10.4). Here the sediments consist of three types (1) neritic, arenaceous pelites, yielding the bivalve *Buchia* and ammonites – Raohe area between the Dahezhen fault and the Wusuli River; (2) littoral and paralic, alternating clastics and pyroclastics, yielding the *Morphoceras*(?)–*Entolium* assemblage in the lower part and the *Liostrea-Sinepseammolia* assemblage in the upper part – between Mishan Youth Reservoir and Dahe (Longzhuagou Group); (3) continental margin, mainly terrestrial coal measures – area west of the Mishan Youth Reservoir.

The northern region includes the areas Dahingganling, Yanliao, east Jilin, east Liaoning, and west Wandashan. North of Dabie Mountain and west of Mount Wanda, macroclastic rocks predominate in the early Malm (Tuchengzi and Houchengzi Formations). In the Yanliao Fault Basin, for example, the sediments are greatly thickened, and their gravels are well rounded. In the middle Malm, extensive andesites and basalts rest on different strata, change rapidly in lithofacies and thickness, and are sporadically intercalated with lacustrine sediments. The latest Jurassic is characteristically lacustrine, arenaceous-pelitic, with coal measures (Fuxin and Hegang coalfields), and the well-known Rehe biota (*Lycoptera-Eoestheria-Ephemeropsis* assemblage) occurs in freshwater lakes.

The southern region, west of Dabie Mountain, includes several fault depressions with acidic volcanics (rhyolite and dacite) and pyroclastics, intercalated with fluvio-lacustrine sediments. These sequences are thinner and less widely distributed than in the northern region, for sedimentation was more unstable; significant coal measures are unknown, and the biota differ.

## References


Cai Shaoying & Liu Xiezhang (1978). Nonmarine bivalvia (in Chinese). In *Palaeontological Atlas of Southwest China, Sichuan, Vol. 2* (pp. 365–402). Beijing: Geol. Publ. House.

Chen Jin-hau (1982). Liassic bivalve fossils from Mt. Jinji of Guangdong (in Chinese, with English abstract). *Acta Palaeont. Sin., 21*(4), 404–15.

Chen Pei-ji & Shen Yan-bin (1982). Late Mesozoic conchostracans from Zhejiang, Anhui and Jiangsu Provinces (in Chinese and English). *Palaeont. Sin., N.S., B, 161,* 17.

Dong Zhi-ming, Zhou Shi-wu, & Zhang Yi-hong (1983). The Dinosauria remains from Sichuang Basin, China (in Chinese and English). *Palaeont. Sin., N.S., D, 162,* 23.

Gu Zhi-wei (1980). Correlation chart of the Jurassic in China, with explantory text (in Chinese). In *Stratigraphic Correlation Chart in China, with Explanatory Text* (pp. 223–40). Beijing: Sci. Press.

Gu Zhi-wei & Chen Dao-kuo (1983). A brief note on stratigraphy (in Chinese, with English abstract). In *Fossils from the Middle–Upper Jurassic and Lower Cretaceous in Eastern Heilongjiang Province, China* (vol. 1, pp. 4–9). Harbin: Heilongjiang Sci. & Techn. Publ. House.

Li Zi-shun, Wang Si-en, Yu Jingshan, Huang Huai-zen, Zheng Shao-lin, & Yu Xi-han (1982). On the classification of the Upper Jurassic in North China and its bearing on the Juro-Cretaceous boundary (in Chinese, with English abstract). *Acta Geol., 56*(4).

Lin, C. C. (1961). On the occurrence of Jurassic ammonite newly found in Taiwan. *Acta Geol. Taiwan., 9,* 79–81.

Lin, C. C., & Huang, T. Y. (1974). A new name for the ammonite fossil from Peikang, Yunlin. Taiwan. *Acta Geol. Taiwan., 17,* 37.

Matsumoto, T. (1979). Restudy a phylloceratid ammonites from Peikang, Taiwan. *Petrol. Geol. Taiwan, 16,* 15–17.

Pang Qiqing, Zhang Lixian, & Wang Qiang (1984). Ostracoda (in Chinese). In *Palaeontological Atlas of North China, Vol. 3*. Beijing: Geol. Publ. House.

Sun Ding-li, Chen Pei-ji, Cao Mei-zhen, & Pan Hua-zhong (1983). Restudy on the marine Jurassic of western Yunnan (in Chinese). In *A Report of Expedition in the Mts. Hengduan* (vol. 1, pp 66–73). People's Press of Yunnan.

Wang Si-en (ed.). (1985). *The Jurassic System of China, Stratigraphy of China* (in Chinese) (pp. 11, 43–50, 116–25, 204–41, 252–4). Beijing: Geol. Publ. House.

Wang Si-en et al. (1980) *The Mesozoic Stratigraphy and Palaeontology of the Shaan.-Gan.-Ning. (Ordos) Basin, China* (in Chinese) (vol. 1, pp. 31–2; vol. 2, p. 87). Beijing: Geol. Publ. House.

(1981). On Upper Jurassic phyllopods (Conchostraca) from northern Hebei and Daxinganling and their significance (in Chinese). *Bull. Inst. Geol. Chin. Acad. Geol. Sci., 3,* 97–114. Beijing: Geol. Publ. House.

Wang Yigang & P. L. Smith (1986). Sinemurian (Early Jurassic) ammonite fauna from the Guangdong region of Southern China. *J. Paleontol., 60*(5), 1075–85.

Wang Yigang & Sun Dongli (1983). A survey of the Jurassic of China. *Can J. Earth Sci., 20*(11), 1646–56.

(1985). The Triassic and Jurassic paleogeography and evolution of the Qinghai-Xizang (Tibet) Plateau. *Can J. Earth Sci., 22*(2), 195–204.

Ye Meina & Li Baoxian (1982). Subdivision and correlation of the Jurassic plant-bearing strata in China (in Chinese). In *Stratigraphic Correlation Chart in China, with Explanatory Text* (pp. 242–3). Beijing: Sci. Press.

# 11 Eastern Russia

I. I. SEY, Y. S. REPIN, E. D. KALACHEVA, T. M. OKUNEVA, K. V. PARAKETSOV, and I. V. POLUBOTKO

## INTRODUCTION[1]

### Diastrophism

The Jurassic of eastern Russia is part of the Mesozoic tectonosedimentary megacycle, from Late Triassic to Neocomian, which formed the Mesozoids of Northeast Asia and laid the foundations of its present structural plan. The Jurassic changed the geodynamic setting from active geoclinal to continentalization, except for the Pacific margin, where new geoclines were initiated at the close of the Jurassic.

Lower and Middle Jurassic deposition occurred during the major cycle of geoclinal evolution of the Mesozoids (North-East Russia: Yukagir cycle), and it comprises a single structural complex together with the Upper Triassic (commonly Norian) deposits. The lower boundary of this complex in the North-East is characterized by hiatuses of differing lengths and minor structural changes (Figure 11.1). In areas of stable down-warping (Oldzhoi and Inyali–Debin Troughs), a weak compressional phase at the beginning of the Yukagir cycle resulted in small thrusts and olistostromes (Parfenov 1984). In Far East Russia, the emplacement of this complex was preceded by a major restructuring accompanied by folding (Figure 11.1).

The upper boundary of the Yukagir cycle is almost ubiquitously marked by angular and stratigraphic uncomformities. This regressive phase was of long duration. In the North-East, uncomformities are most distinct in the Bathonian, and less pronounced in the Oxfordian. In the Far East, the first folding occurred in the middle Bajocian and the main phase in the Callovian; the latter is associated with a total inversion of most of the Amur-Okhotsk Geocline. These uncomformities are regarded as the lower boundary of the late geoclinal cycle (North-East: Kolyma) of Mesozoid evolution that ended at the close of the Neocomian.

[1] By I. I. Sey and Y. S. Repin.

### Volcanism

Volcanism was most pronounced in the Early and Middle Jurassic of the North-East, particularly along its eastern margin. Three types of volcanic activity are distinguished (Figures 11.1 and 11.2):

1. Viliga suboceanic type of the Early Jurassic: At the upper Indigirka River, the Kobyuma volcanic structure consists of layers of Hettangian to late Pliensbachian subalkaline hyalobasalts (800 m); at the middle Viliga River, the late Toarcian hyalobasalts are up to 250 m thick.
2. Trachybasalts and similar rocks are the prevailing constituents of Early Jurassic rifts (late Sinemurian to early Pliensbachian). This type occurs only in the Omolon Massif, where it is subaerial and subaqueous.
3. The island-arc type, represented by andesitic and andesite-basaltic associations, with associated pyroclastic, occurs throughout the Jurassic of the North-East and is characteristic of the eastern peripheral part of the region (Figure 11.2).

In the Far East, the Early and Middle Jurassic are generally amagmatic, with the exception of eugeoclinal oceanic areas (Figures 11.1 and 11.2). Intense submarine volcanism in the axial zone of the Sikhote-Alin Geocline resulted in spilite-diabase associations, 200–300 m thick, within the Lower to Middle Jurassic siliceous-terrigenous sequences. These formations are now interpreted as suboceanic ophiolite association (Parfenov 1984).

In the Late Jurassic, rift-volcanism occurred in the North-East, with an association of continental olivine basalts and andesite-rhyolite complexes (Iliin–Tas). The South Anyui zone is characterized by spherulitic and ropy lavas of subalkaline and tholelitic basalts and is included in the ophiolite complex (Parfenov 1984) or assigned to the basaltic association of continental rift zones (Gusev et al. 1985).

A rift zone presumably was present in the Late Jurassic of central Sikhote-Alin, east of the Khanko Massif, where Late Jurassic

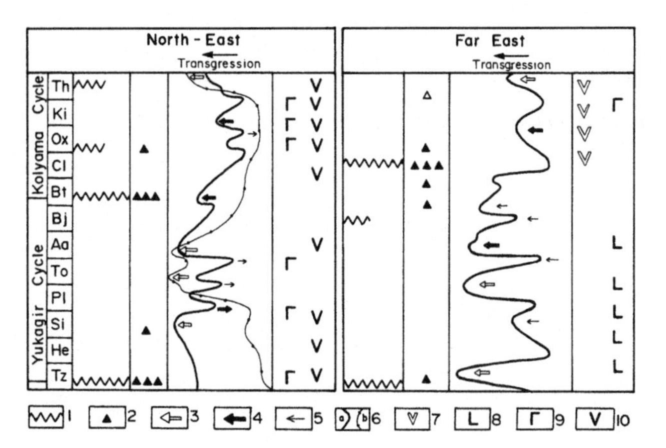


**Figure 11.1. Principal geological events in the Jurassic of northeastern Asia: 1, diastrophic phases; 2, levels on which olistostromes are formed. Transgressive–regressive activity: 3, eustatic (global); 4, mixed or of uncertain origin; 5, structural origin (regional); 6, curve of transgression–regression (a, for Far East and North-East; b, for the Siberian Platform). Volcanism: 7, subaerial; 8, eugeoclinal; 9, rift; 10, island arc.**

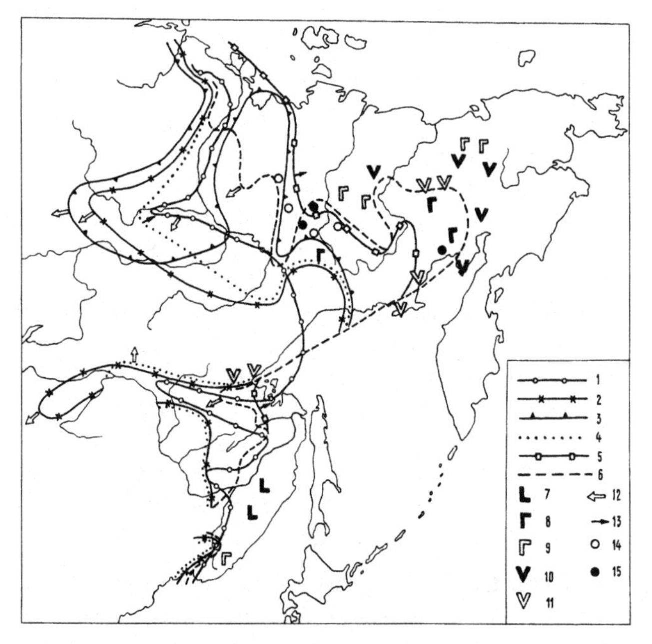

**Figure 11.2. Boundaries of major Jurassic transgressions and regressions and areas with volcanics in northeastern Asia. Position of coastline: 1, Hettangian–Sinemurian; 2, late Pliensbachian–Toarcian of Far East and late Pliensbachian of North-East; 3, Toarcian (North-East); 4, Aalenian–Bajocian; 5, Kimmeridgian; 6, Volgian. Volcanism: 7, eugeoclinal; 8, 9, rift zones (8, Early Jurassic; 9, Late Jurassic); 10, island arcs. Coastline position: 11, land; 12, transgressions; 13, regressions. Olistostromes: 14, Norian; 15, Bathonian.**

flows, dikes, and extrusive alkaline basaltoid bodies are known (Parfenov 1984; Mazarovich 1985). The Late Jurassic in the Far East is noted for its subaerial volcanic activity in the west and north of the region, dominated by andesites. In the North-East, continental-type Late Jurassic volcanism occurred along the northern margin of the Omolon Massif.

### Transgressions and regressions

In the North-East, the Norian to early Sinemurian transgression resulted mainly in deepening of the basin and slight extension (Figure 11.1). It was predominantly eustatic, as evident from the extensive occurrence of cosmopolitan ammonite genera (*Psiloceras, Waehneroceras, Schlotheimia,* and *Arietites*) and some Pacific bivalves (i.e., *Otapiria*). In the middle Sinemurian begins a progressive shallowing trend in the North-East, a consequence of the local inversion of some structures.

In the Far East, however, the earliest Jurassic was regressive, with the sea remaining only in the deepest depressions. The transgressive cycle began in the Sinemurian with a small-scale ingression (Figure 11.1). The Early Jurassic transgression in eastern Russia reached its maximum in the late Pliensbachian–early Toarcian, when the margins of the sea basin extended westward to the western Transbaikal area and the eastern Siberian Platform (Figure 11.2). This eustatic transgression provided broad connections between basins and resulted in the dispersion of amaltheids, pandemic dactylioceratids, and hildoceratids. The subsequent regression at the end of the Toarcian was associated with the general uplift of Northeast Asia.

The Middle Jurassic of eastern Russia is characterized by a thalassocratic regime resulting from local geodynamic processes. The extent of the Aalenian-Bajocian transgression was similar to that of the Early Jurassic. Against the background of a generally high sea level, there were repeated coast-line fluctuations, resulting from local uplifting mainly in the west of the area. Whereas the transgressive phase is early Bathonian in the North-East, it is late Bathonian in the Far East.

The close of the Middle Jurassic was marked by a slow regression. It was associated with diachronous upliftings and folding in the North-East and Far East during Bathonian to Oxfordian times.

The Late Jurassic transgression, which in the Far East began in the latest Callovian, reached its first peak in the Oxfordian–early Kimmeridgian, but the marine basin was much smaller than in the preceding epochs. The transgressive peak was in the middle to late Volgian, eustatic in origin, and extensive in northern Eurasia.

Eastern Russia comprises two major regions, the Far East and the North-East, which differ notably in geological structure and history.

## FAR EAST RUSSIA[2]

The Russian Far East, south of the Stanovoi Range, reaches from Sakhalin Island, in the east, to the Eastern Transbaikal area, in the west, where the Jurassic structures are closely similar. Within this area of complex structures, the Jurassic deposits are divided into different structural-facies units that are based on specific features of the rock sequences and also their biota. The Jurassic is most widespread within geoclinal-fold systems, that is,

[2] By I. I. Sey, E. D. Kalacheva, and T. M. Okuneva.

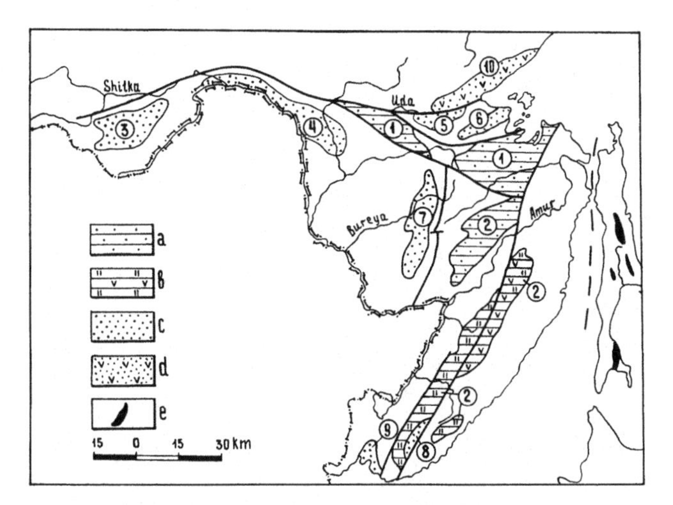


**Figure 11.3. Distribution of main types of Jurassic deposits in Far East Russia. Geoclinal-fold systems: 1, Amur-Okhotsk; 2, Sikhote-Alin. Subplatform troughs: 3, East Transbaikal; 4, upper Amur; 5, Uda; 6, Torom; 7, Bureya; 8, Okrainsky; 9, Southern Primorye. Mesozoic structures: 10, Uda-Okhotsk Volcanic Molasse Trough. Rock types: a, miogeoclinal deposits of the Amur-Okhotsk and western Sikhote-Alin geoclinal systems; b, eugeoclinal terrigenous-siliceous-volcanic formations of the central and eastern Sikhote-Alin Geocline; c, terrigenous deposits of subplatform troughs; d, volcanic and volcanosedimentary rocks in zones of Mesozoic activity; e, jasper complex of Sakhalin Island.**

the sublatitudinal Amur-Okhotsk and submeridional Sikhote-Alin systems (Figure 11.3), and is up to 10 km thick. The dominant rock types are miogeoclinal sandy to shaley flysch and flyschoid deposits, particularly in the Amur-Okhotsk Geoclinal System. The axial part of the Sikhote-Alin Geocline contains abundant silicovolcanics that mark zones of eugeoclinal sedimentation (Figure 11.4) All these strata are intensely folded, with extensively imbricated overthrust structures and mictite. These probably comprise large structural plates.

The Jurassic geoclinal deposits of relatively deep-sea basins yield very few macrofossils, and the study of microfauna, mainly radiolarians, has just begun. The biostratigraphy of Jurassic geoclinal strata is thus known inadequately and only in general terms.

The second type of deposit corresponds to the so-called subplatform troughs, which, in the Jurassic, formed on the margins of rigid, unconsolidated structures around geoclinal systems. Near the Amur-Okhotsk Geoclinal System are the Eastern Transbaikal, Upper Amur, Uda, and Torom Troughs (from west to east). The Bureya and Southern Primorye Troughs are conjugated with the Sikhote-Alin Geocline (Figure 11.3). The Okrainsky Block, which is presently assumed to be an intrageoclinal uplift (Mazarovich, 1985), has a similar sequence.

These structures are noted for incomplete sections, with unconformities, the presence of coastal-marine and continental facies, and more diverse rock types: sandy to silty sediments dominate, together with coarse-grained and tufogenic rocks. Their thicknesses (to 8,000 m) are comparable to those of the coeval adjacent geoclinal systems (Figures 11.5 and 11.6).

Subplatform troughs have relatively simple geological structures and are, as a rule, large synclinoriums complicated by minor folds and faults with local subhorizontal attitudes.

Paleogeographically, subplatform troughs represent shallow-sea basins, with rich benthic and nektonic fauna that resulted in the relative abundance of fossils. Sections of this type, therefore, predominate in the construction of the detailed stratigraphic scale for the Far East.

The third type of deposit comprises the Jurassic in Sakhalin Island (Figure 11.4). It is part of the Mesozoic Volcanic Jasper Complex, regarded as accumulation of an open-oceanic basin, that is, the Paleopacific (Rikhter 1986).

In addition to these major types, there are also the sedimentary-volcanic formations in volcanic depressions in the north of the region (Dzhugdzhur Ridge) and coarse-clastic molassoid sediments of orogenic intermontane areas of the Bureya Massif and surrounding fold systems.

### Lower Jurassic

The Lower Jurassic sections in the Far East all have incomplete fossil records. The outcrops are isolated and the successions separated by uncomformities, so that only a few biostratigraphic units can be distinguished.

Hettangian macrofossils are unknown from the Far East. Sinemurian deposits occur in isolated exposures only in the east of the region. Lower Sinemurian sandstone and siltstone (300 m) with *Otapiria omolonica* and *O. pseudooriginalis* crop out in a tectonic block in the western Sikhote-Alin Geoclin area (Kur River Basin). Upper Sinemurian sandstone, siltstone, and clay (to 260 m) form small exposures in the southern Bureya and South Primorye Troughs and yield *Otapiria limaeformis* and *Pseudomytiloides rassochaensis;* in Primorye occurs also *Angulaticeras* cf. *ochoticum.* This assemblage resembles coeval faunas of the North-East, and the same ammonite zones are recognized: the *Otapiria omolonica* Zone (Lower Sinemurian) and the *Angulaticeras kolymicum* and *Otapiria limaeformis* Zones (Upper Sinemurian).

The so-called Kiselevka Assemblage (lower Amur River), with *Juraphyllites amurensis, Chlamys textoria,* abundant *Cardinia, Plagiostoma,* and other bivalves, gastropods, and corals (Kiparisova 1952) holds a peculiar position among the lower Liassic faunas of the region. (E. V. Krasnov also identified the scleractinians *Anabacia* and *Montlivalcia.*) This fauna is confined to a large limestone lens within a thick (to 2,000 m) sequence of jasper, siliceous-clayey rocks, and volcanics (Kiselevka Formation) that forms a tectonic block among the Upper Jurassic–Lower Cretaceous deposits. The Kiselevka Assemblage may therefore be allochthonous in origin.

Lower Liassic deposits in geoclinal facies appear to be widely distributed in the continental parts of the Far East. They comprise undivided geoclinal complexes that have been dated as Late Triassic–Early Jurassic. Here also belong the Kurnal Formation (sandstone, siltstone, tuff sandstone, gritstone, breccia, basic effusive lenses, 3,500 m), in the central Amur-Okhotsk fold system, and the Sorukan Formation (sandstone, siltstone, 2,000) in the

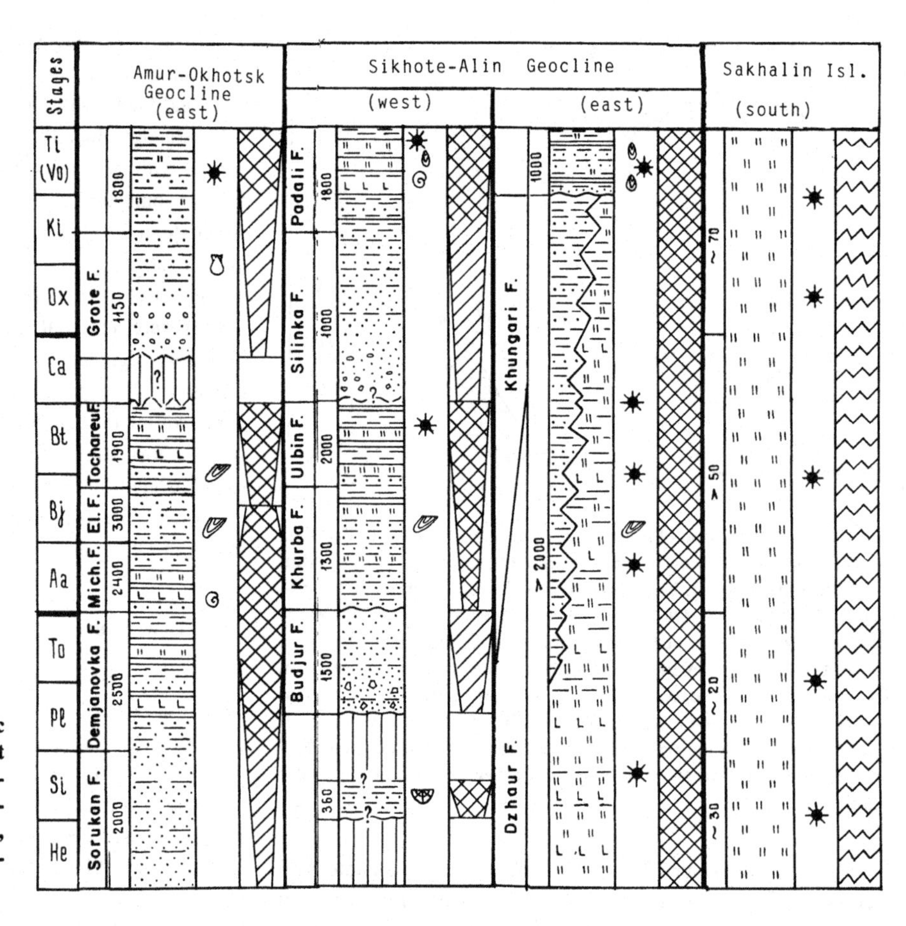


**Figure 11.4. Important Jurassic geoclinal sequences in Far East Russia and the oceanic (pregeoclinal) Jurassic sequence of Sakhalin Island. Formations: Mich., Michalitsynskaja; El., Elgon. Legend same as in Figure 11.5.**

east of this structure, as well as siliceous-volcanic rocks in the lower part of the Dzhaur Formation in the northern Sikhote-Alin folded area. Their Jurassic age is confirmed in some places by radiolarians (Tikhomirova 1986).

Lower Liassic oceanic facies probably occur in Sakhalin Island (Rikhter 1986), where, according to N. Y. Bragin, jaspers (30 m) in the vicinity of the Yunona Mount (South Sakhalin) contain the Hettangian–Sinemurian radiolarians *Parahsuum simplum, Parahsuum, Dictyomitrella,* and *Lithostrobus.*

Lower Pliensbachian cannot be documented by fossils. The beginning of late Pliensbachian is marked by the peak Jurassic transgression, reaching the Eastern Transbaikal area and forming the base of the sedimentary sequences in all subplatform troughs. Coarse-grained layers (to 400 m) occur ubiquitously at the bases of the sections, yielding abundant *Oxytoma (Palmoxytoma) cygnipes* and *Harpax spinosus.* These deposits are traceable from the Eastern Transbaikal area to Southern Primorye and are distinguished as the *O. cygnipes* and *H. spinosus* Beds.

The most complete Upper Pliensbachian section is recorded in the Bureya Trough from the Desh Formation. Earlier these deposits were placed into the Umalta Formation (Krymholts, Meseshnikov, and Westermann 1988). The latter is at present subdivided into two formations: the Desh (Upper Pliensbachian–Lower Toarcian) and the Sinkaltu (Aalenian–Lower Bajocian) (Sey et al. 1984). It includes (ascending order):

1. *Amaltheus stokesi* Zone – siltstone, with sandstone interbeds, coarse-grained sandstone, breccia, and conglomerate at the base (to 340 m)
2. *A. margaritatus* Zone – siltstone, sometimes sandstone; also with *A.* cf. *complanatus* (160 m)
3. *A. viligaensis* Beds – siltstone, less frequently sandstone (80 m)

Most of the section is characterized by the bivalves *Ochotochlamys bureiensis, Chlamys (Ch.) torulosa, Kolymonectes* ex gr. *staeschei, Amuropecten solonensis,* i.e., the *Ochotochlamys bureiensis* Beds. These units are traceable in the Eastern Transbaikal Trough and Okrainsky Block (Ontagaja and Okrainka Formations, respectively).

In the Okrainsky Block, the upper Pliensbachian, in addition to *Amaltheus,* yields the predominant Tethyan genera *Arieticeras, Fontanelliceras, "Dactylioceras,"* and *Protogrammoceras.* The upper part of the section without amaltheids is distinguished as *"Paltarpites"* Beds.

Toarcian deposits in subplatform troughs are fragmented due to regional pre–Middle Jurassic uplift. In the Bureya (Desh Formation), Uda, and Torom Troughs, sandstone and siltstone (300 m) rest conformably or transitionally on Upper Pliensbachian. In the Eastern Transbaikal area, the Toarcian has a coarse clastic basal horizon (Sivachi Formation, to 600 m), and up the section the

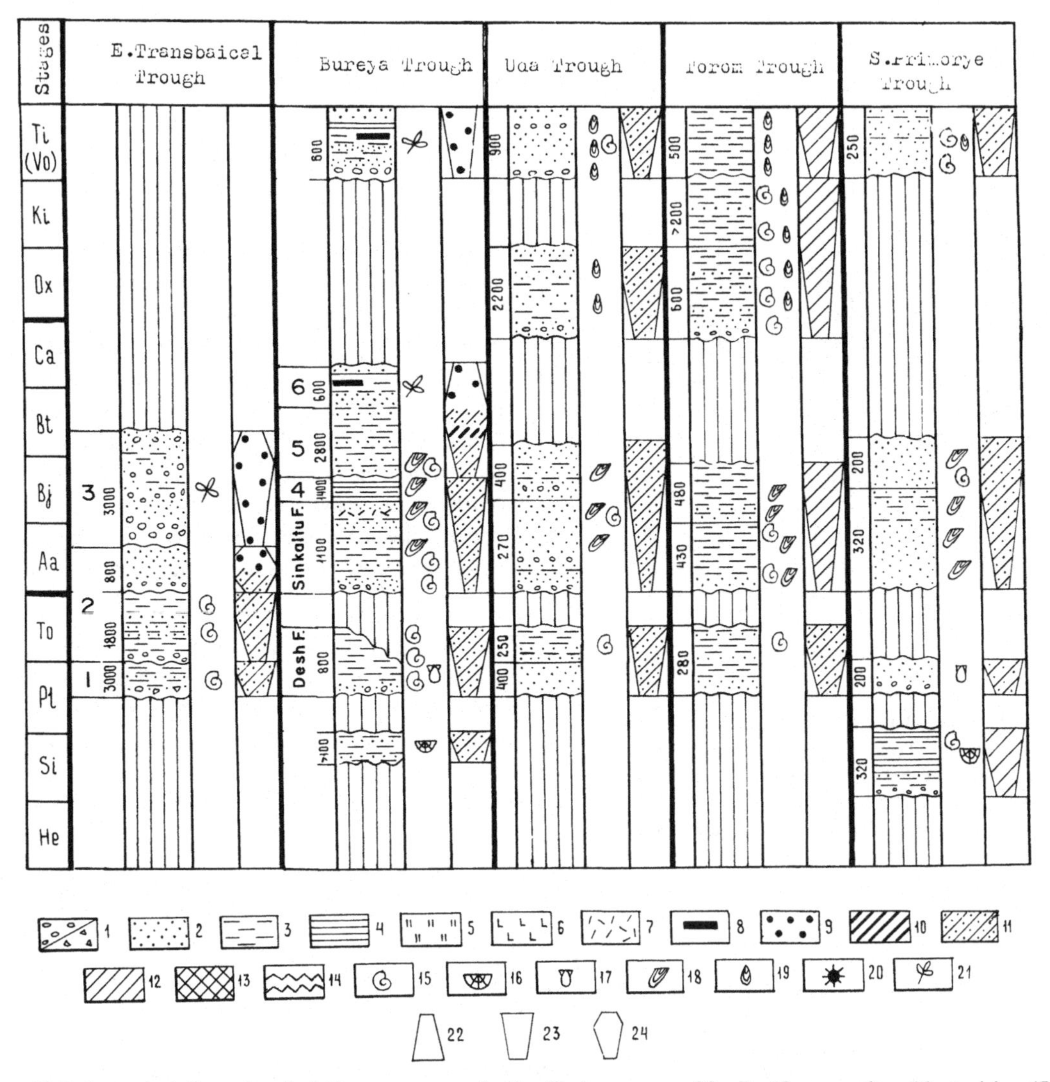

**Figure 11.5. Important Jurassic subplatform sequences in Far East Russia. Rock types: 1, conglomerates and breccias; 2, sandstone; 3, siltstone; 4, mudstone and shale; 5, siliceous rocks; 6, basic effusives; 7, acidic effusives; 8, coals. Facies: 9, lacustrine alluvial; 10, coastal marine; 11, upper and middle sublittoral; 12, lower sublittoral; 13, deep sea (pseudoabyssal); 14, oceanic (pelagic). Main groups of fossils: 15, ammonites; 16, otapirias; 17, pectenids; 18, mytilocerams; 19, buchias; 20, radiolarians; 21, flora. Cycles: 22, regressive; 23, transgressive; 24, more or less symmetric. Formations (with thickness in m): 1, Ontagaja; 2, Sivachi and Onon-Borzja; 3, upper Gazimur; 4, Epikan; 5, Elga and Chagany; 6, Talyndzhan; 7, Dublikan; 8, Komarovka.**

sediments are more diverse, from conglomerate to siltstone (Onon-Borzja Formation). The total thickness in this area reaches 2,500 m (Okuneva 1973).

The generalized Toarcian section is divisible into four units of zonal rank: *Harpoceras falciferum* (Eastern Transbaikal area), *Dactylioceras athleticum* (Uda and Bureya Troughs), *Zugodactylites monestieri* (Eastern Transbaikal area, Uda and Torom Troughs), and *Poroceras spinatum* (Bureya Trough). This succession is similar to that of North-East Russia and is correlated with the standard scale. The Upper Toarcian of the Eastern Transbaikal area also contains the ammonite *Pseudolioceras* cf. *rosenkrantzi*, indicating the *P. rosenkrantzi* Zone.

Pliensbachian and Toarcian in geoclinal facies can be distinguished only approximately, proceeding from the continuous character of the strata. In the Amur-Okhotsk Geoclinal Fold System, sandy to silty deposits (2,500 m) (Amkan, Nimelen, and Demyanovka Formations) may be of this age; in the Sikhote-Alin Geoclinal Fold System, both terrigenous and volcanosiliceous rocks are possibly of the same age. The latter occur mainly in northern Sikhote-Alin (Dzhaur Formation). In the south (central Sikhote-Alin), part of the Late Triassic–Middle Jurassic black shale may be Lower Jurassic. This is a structurally complex, sandy to silty sequence with mictite and olistostromes (Mazarovich 1985). In Sakhalin Island, N. Y. Bragin identified the Pliensbachian–Toarcian radiolarian association in the volcanic-jasper oceanic complex (20 m) (Rikhter 1986): *Dultus* aff. *hecattaenus*, *Praeconocaryomina immodica*, *Trillus*, *Canutus*, and *Lupherium*.

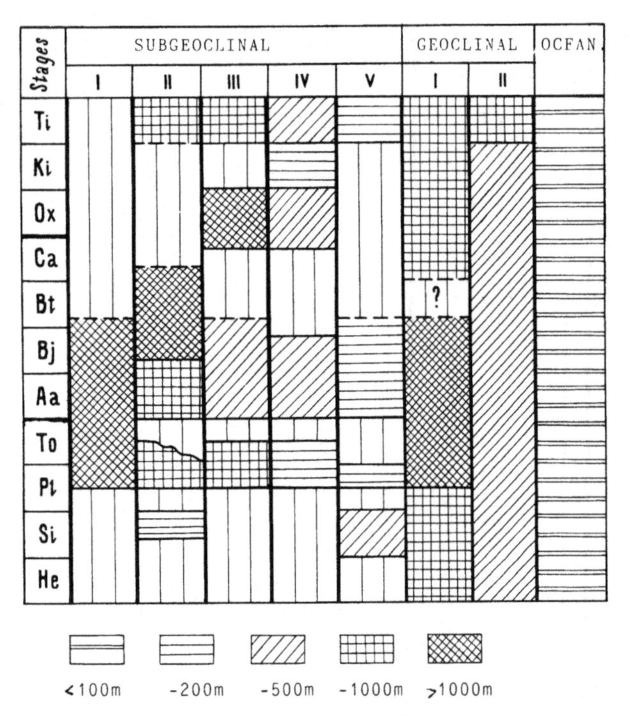


**Figure 11.6. Thickness of Jurassic deposits in Far East Russia. Subplatform: I, Eastern Transbaikal Trough; II, Bureya Trough; III, Uda Trough; IV, Torom Trough; V, Southern Primorye Trough. Geoclinal: I, Amur-Okhotsk Geocline (eastern part); II, Sikhote-Alin Geocline (northern part). Oceanic: pregeoclinal, Sakhalin Island. Thickness of deposits per stage: 1, less than 100 m; 2, to 200 m; 3, to 500 m; 4, to 1,000 m; 5, to over 1,000 m.**

## Middle Jurassic

The beginning of Middle Jurassic in the Far East is marked by a major transgression similar in extent to that in the late Pliensbachian. But it hardly affected the Eastern Transbaikal area, reflecting a gradual area reduction of marine sedimentation from west to east (Figure 11.7). All subplatform troughs have a basal Aalenian unconformity that is also traceable in the marginal parts of geoclinal systems.

The Aalenian transgression favored an extensive distribution of mytiloceram, which, as dominant Middle Jurassic macrofossils, are used along with ammonites for detailed biostratigraphy of the Middle Jurassic.

The reference Aalenian section in this region is on the western coast of the Tugur Bay in the Okhotsk Sea (Torom Trough), where the zonal stratotypes for eastern Russia are located. This section is as follows (ascending order):

1. *Pseudolioceras beyrichi* Zone – fine-grained sandstone, with some pebbles (64 m); also with *P. (P.)* aff. *beyrichi, Mytiloceramus priscus,* and *M. subtilis.*
2. *P. maclintocki* Zone – black, thin-bedded, platy siltstone (100 m); *Mytiloceramus priscus* and *M. quenstedti;* in upper part with accumulations of peculiar *P. (Tugurites)* [*''Grammoceras''*]
3. *P. tugurense* Zone – black, platy, massive siltstone (260 m); *P. (Tugurites) whiteavesi;* in the upper part, *Erycitoides (E.) howelli* and *E. (Kialagvikes) spinatus;* mytiloceram dominated by *Mytiloceramus obliquus, M. polyplocus, M. tugurensis, M. anilis,* and, rarely, *M. jurensis* and *M. morii*

Correlation with the standard scale is accomplished by the ammonites via southern Alaska and Canada (Frebold 1957, 1960; Westermann 1964). The two lower zones (1–2) approximately correspond to the Lower Aalenian, and the *tugurense* Zone (3) to Upper Aalenian. At Tugur Bay are also the stratotypes of Aalenian mytiloceram zones (i.e., *M. priscus* and *M. obliquus* Zones), which approximately correspond to the Aalenian substages. Aalenian ammonite and mytiloceram zonal assemblages are known from almost all subplatform troughs, where Aalenian sandstones and siltstones are up to 700 m thick (lower Sinkaltu Member in Bureya Trough, lower Bonivur Formation in southern Primorye).

The Bajocian of subplatform troughs contains abundant fossils for detailed biostratigraphy. Most stratotypes are in the Bureya Trough (upper Sinkaltu Member), where the Lower Bajocian section along the Bureya River is as follows (ascending order):

1. *Pseudolioceras fastigatum* Zone/*Mytiloceramus jurensis* Zone – equigranular sandstone and siltstone (225 m); in upper part, with rare *P. (Tugurites) costistriatum;* throughout abundant mytiloceram
2. *Arkelloceras tozeri* Zone/*Mytiloceramus lucifer* Zone – siltstone, with sandstone interbeds (200 m); rare *A. tozeri* and *A. elegans,* abundant *M. lucifer* and *M.* ex gr. *lucifer*

The ages of both ammonite zones are also determined via southern Alaska and western Canada (Westermann 1969; Hall 1984). The *fastigatum* Zone is approximately coeval with the Discites and Laeviuscula Standard Zones (= Widebayense Zone of North America), and the *tozeri* Zone with the Sauzei Zone (= Crassicostatus Zone). The lower boundary of the Bajocian in eastern Russia is drawn at the base of the *fastigatum* Zone.

The stratotype of the superjacent *Mytiloceramus clinatus* Zone is at the Tugur Bay coast (Torom Trough), where deposits with *M. lucifer* are overlain by 230-m-thick siltstone with large *M. clinatus.*

In the Bureya Trough, coeval siltstones along the Soloni River grade into the *Mytiloceramus kystatymenis* Beds. Here, both units are placed in the Epikan Formation (1,400 m). Along the Soloni River, the Bajocian section ends with the *Umaltites era* Beds (horizon), fine-grained sandstone and siltstone (300 m), which overlie siltstone with rare *Liroxyites* cf. *kellumi* (lower Elga Formation).

Dating of the *Umaltites era* Beds presented great difficulties (Sey and Kalacheva 1980). At present, by analogy with similar faunas of western Canada (Frebold and Tipper 1973), they are considered as Late Bajocian, approximately Garantiana and Parkinsoni Standard Zones (Callomon 1984; Sey and Kalacheva 1987). Westermann (personal communication; Sey and Kalacheva 1988) believes that this assemblage is closely akin to the early Late Bajocian *Megasphaeroceras-Liroxyites* assemblage of the Rotundum Zone widely developed in western North America.

The thickness of the Bajocian in the Bureya Trough reaches 2,000 m, probably a maximun for Bajocian subplatform deposits.

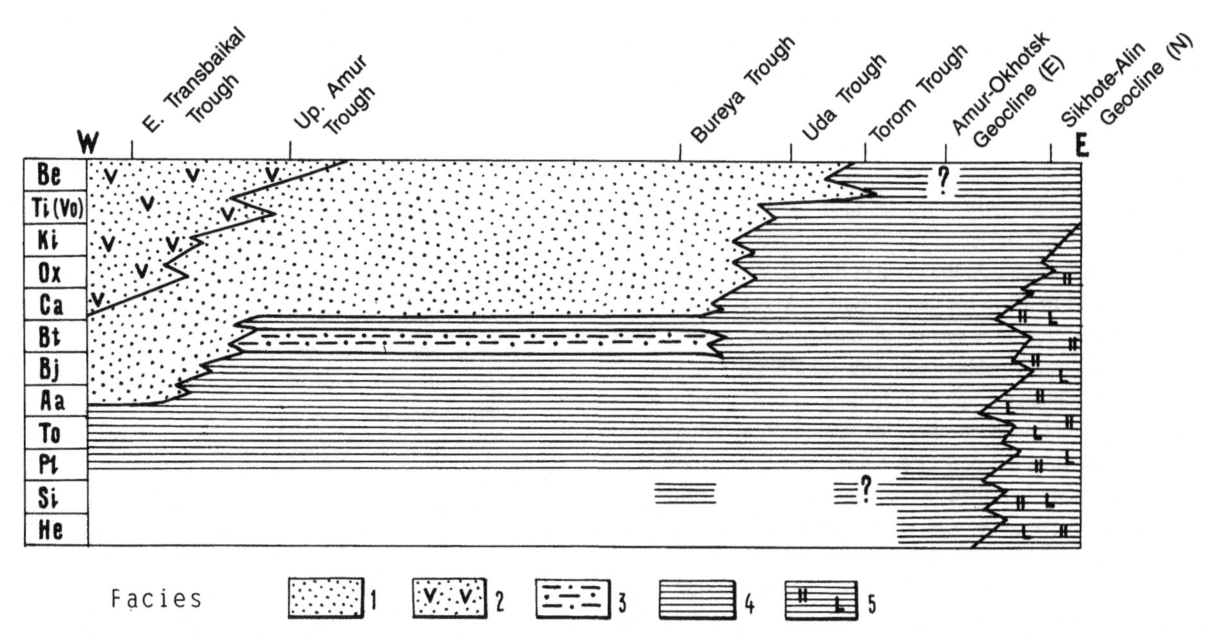


**Figure 11.7. Jurassic lithofacies in northern Far East Russia. Facies: 1, continental terrigenous; 2, continental volcanic-terrigenous; 3, coastal marine; 4, marine terrigenous; 5, marine siliceous-volcanic-terrigenous.**

The *Mytiloceramus lucifer* Zone is widespread. It is recorded in all subplatform sections (with the exception of the Eastern Transbaikal area, where marine sedimentation apparently ended at the beginning of the Aalenian) and is traceable in geoclinal troughs. It is thus important for intraregional and interregional correlation.

At present, no macrofossils of undisputed Bathonian age are known from the Far East. In the Bureya Trough, the *Umaltites era* Beds are overlain by siltstones with sandstone interbeds (ca. 1,600 m), yielding rare, poorly preserved ammonites that could be Bathonian (upper Elga and Chagany Formations). Marine, coastal-marine, and continental sediments with similar stratigraphic positions are also found in other subplatform troughs. Callovian deposits are known only from the Torom Trough, Tugur Bay coast, where Lower Bajocian is overlain with a distinct angular unconformity by Upper Callovian sandstone and conglomerate (90 m) with *Longaeviceras* cf. *keyserlingi*.

The analysis of the different structural-facies zones of the Far East indicates significant tectonic activity, that is, uplifting and folding, most likely in the Callovian in both the subplatform and geoclinal troughs.

The Middle Jurassic of geoclinal-fold systems, similar to the Lower Jurassic, yields extremely scarce fossils, making biostratigraphy impossible. Local ammonite and mytiloceram levels are known, and the most common among the latter are the Lower Bajocian *Mytiloceramus lucifer* Beds.

The miogeoclinal deposits of the Amur-Okhotsk and western Sikhote-Alin Systems (left bank of Amur River) are subdivided into several formations (i.e., Mikhalitsynskaja, Elgon, and Tochareu in the first area, Churba and Ulbin in the second). They consist of sandy-silty-clayey, flyschoid sediments with siliceous-clayey rock and basic effusive inclusions (over 5,000 m, Figure 11.4). Their Middle Jurassic age is approximately determined by rare finds of ammonites, mytiloceram, and radiolarians (*Resolutions . . .* 1982; Tikhomirova 1986).

In the eugeoclinal zone of the northern Sikhote-Alin Fold System the Middle Jurassic section is still dominated by silicovolcanic rocks of the Dzhaur Formation, which laterally grade into sandy shale accumulations (Khungari Formation). The Middle Jurassic age of both is confirmed by mytiloceram and radiolarians.

In central Sikhote-Alin, clayey rocks of black shale yield the radiolarians *Gorgansium silviensense, Zortus* cf. *jonesi, Acanthocircus* aff. *dicvanocanthus, Parvicingula,* and *Hsuum,* which, according to N. Y. Bragin, indicate Middle Jurassic (Bajocian–Bathonian?) (Mazarovich 1985). Radiolarian assemblages of different (Bajocian–Callovian) ages have been identified from isolated exposures of siliceous, terrigenous rocks in east-central and south Sikhote–Alin (Tikhomirova 1986).

Pre-late Callovian tectonic movements were most distinct in the Amur-Okhotsk Geoclinal System. They are associated with the inversion of its western and central parts and intense folding, complicated by large thrusts, which resulted in an appreciable narrowing of the geoclinal zone and the imbricated thrust structure of the Jurassic (Kirillova and Turbin 1979; Parfenov 1984).

The Middle Jurassic oceanic facies, similar to those of the Lower Jurassic, are part of the volcanic jasper section in Sakhalin Island and yield Middle Jurassic radiolarians. The thickness appears to vary greatly (minimum ca. 50 m) (Rikhter 1986).

### Upper Jurassic

Marine Upper Jurassic occurs only in the eastern Far East, indicating continued migration of marine sedimentary basins toward the Paleopacific (Figure 11.7).

In the western Far East, thick continental coal measures (to 5,000 m) follow marine deposits in the upper Amur Trough (Ajak, Dep, and other formations) and Bureya Basin (Dublikan Formation). The narrow fault-bounded depressions surrounding the Amur-Okhotsk Fold System have coarse, continental molasses (over 3,000 m). In the north of the region (Dzhugdzhur Ridge), accumulation of sedimetary volcanics of the Dzhelon Formation (ca. 3,000 m), which started in the Middle Jurassic, continued in the volcanic depressions. At the close of the Jurassic, the destruction of an arched uplift in the central Bureya Massif (lower Zeya River Basin) resulted in a system of intermontane basins filled with molassoid, sandy-pebbly sediments (to 500 m).

Following a hiatus subsequent to folding, Late Jurassic marine sedimentation continued in the Uda and Torom Subplatform Troughs, which have the best Upper Jurassic sections in the Far East. They are supplemented by sections in southern Primorye, where the marine regime was renewed only in the Tithonian (Volgian).

A specific feature of the Upper Jurassic in this region is the scarcity of ammonoids, except in the Tithonian of southern Primorye. The predominant fauna are bivalves, mainly buchiids, which served to define *Buchia* zones commonly with the range of substages (Sey and Kalacheva 1985). The correlation with the ammonite zones is, in most cases, only approximate.

On the Tugur Bay coast (Torom Trough) the following zonal succession is observed (total 1,000 m; ascending order):

1. *Praebuchia impressae* Beds – fine-grained sandstone and siltstone (257 m); *Cardioceras (Scarburgiceras) praecordatum, C. (S.)* cf. *gloriosum,* and *Partschiceras pacificum* (= *Scarburgiceras* spp. beds); Lower Oxfordian
2. *Praebuchia lata–Buchia concentrica* Zone – siltstone (233 m); below, *Perisphinctes (Dichotomosphinctes)* cf. *müehlbachi* and *Maltoniceras* aff. *schellwiene;* above, *P. (Dichotomosphinctes)* sp.; Middle-Upper Oxfordian
3. *B. concentrica–B. tenuistriata* Zone – mudstone (220 m); also *B. ochotica, B. lindstroemi,* and *Amoeboceras (Amoebites)* cf. *dubium* (= *Amoeboceras* ex. gr. *kitchini* beds); Lower Kimmeridgian
4. *B. tenuistriata–B. rugosa* Zone – siltstone (5 m), possibly with *Ochetoceras elgense;* Upper Kimmeridgian

The Volgian is better represented in the Uda Trough (Urmi and Gerbikan Rivers) (total 3,300 m; ascending order):

5. *B. rugosa–B. mosquensis* Zone – conglomerate, sandstone (over 300 m); Lower–Middle Volgian
6. *B. mosquensis–B. russiensis* Zone – sandstone, siltstone, and coquina interbeds (180 m); rare *B. rugosa* and *B. trigonoides;* Middle Volgian
7. *B. russiensis–B. fischeriana* Zone – sandstone, siltstone (230 m), and coquina; *B. trigonoides,* rare *B. mosquensis* and *B. piochii;* rare ammonites *Durangites* sp. ind. and *Partschiceras schetuchaense;* Middle Volgian
8. *B. terebratuloides–B. piochii* Zone – sandstone, siltstone; *B. fischeriana, B. tirgonoides, B.* ex gr. *unschensis* and *B. lahuseni* (170 m); Upper Volgian

In Southern Primorye, Upper Jurassic deposits yield Tithonian ammonites and Volgian buchias. Their succession is as follows (ascending order):

1. Upper Lower Tithonian (Promyslovka Village) – calcareous sandstone (30 m), with *Virgatosphinctes* cf. *mexicanus.*

Middle Tithonian (Putyatin Island); calcareous shelly sandstone (80 m):

2. *Pseudolissoceras zitteli* Zone – with abundant Haploceratidae, Oppeliidae, and rare *Subplanitoides, Parapallasiceras,* and *Torquatisphinctes*
3. *Aulacosphinctes proximus* Zone – with *Aulacosphinctes, Sublithaccoceras, Lemencia, Subplanitoides, Parapallasiceras,* and rare Haploceratidae
4. Upper Tithonian–lower Berriasian (Ussuri Bay coast) – sandstone, rare siltstone, with conglomerate at base (800 m); throughout with *Buchia piochii, B. terebratuloides, B.* ex gr. *unschenssis,* and *B. fischeriana,* rare *B. volgensis;* middle and upper parts also with early Berriasian ammonites

The Upper Jurassic sections of geoclinal systems resemble the Lower and Middle Jurassic sections. In the strongly reduced Amur-Okhotsk Geocline and western Sikhote-Alin Geocline these terrigenous strata (Grote, Silinka, and Padali Formations, to 3,500 m) yield rare Middle Volgian buchias and Upper Jurassic radiolarians. The axial part of the Sikhote-Alin Geocline is also characterized by silicovolcanics. In northern Sikhote-Alin, silicovolcanic and terrigenous strata are unconformably overlain by sandy-silty-clayey deposits (to 1,500 m) with Middle and Upper Volgian buchias. In central Sikhote-Alin, mainly at its eastern flank, siliceous rocks of the Erdagou Formation (to 2,000 m) and its equivalents yield rich Oxfordian to Valanginian radiolarians (Tikhomirova 1986).

In the oceanic zone, Upper Jurassic (including Kimmeridgian–Tithonian) radiolarian assemblages are known from the upper part of the volcanic jasper section (ca. 70 m) of southern Sakhalin Island (Rikhter 1986). In more northern parts of the island the presumably much thicker volcanic-siliceous-terrigenous strata with limestone lenses (Ostraya Formation and lower Khoe Formation), according to E. V. Krasnov, yield a fauna of reef-building and reefophilic corals. Among these are the Kimmeridgian–Tithonian *Calamophyllia flabellum, Cryptocoenia sexradiata,* and *Convexastraed funuzawensis.* According to new radiolarian data, the silicovolcanics surrounding the limestones are of Albian-Cenomanian age (Kazintsova 1987).

The Jurassic–Cretaceous boundary is indistinct in most of the Far East, and in almost all sedimentary basins, from continental to oceanic, the boundary is drawn within continuous, often uniform strata. In marine facies, the boundary is mainly determined by a change in the *Buchia* assemblages, that is, conditionally between the *B. terebratuloides-B. piochii* Zone and the Berriasian *B. okensis* Zone. The find of a lower Berriasian ammonite assemblage in Southern Primorye may permit a more precise definition of the system boundary, in both the *Buchia* and regional ammonite scales.

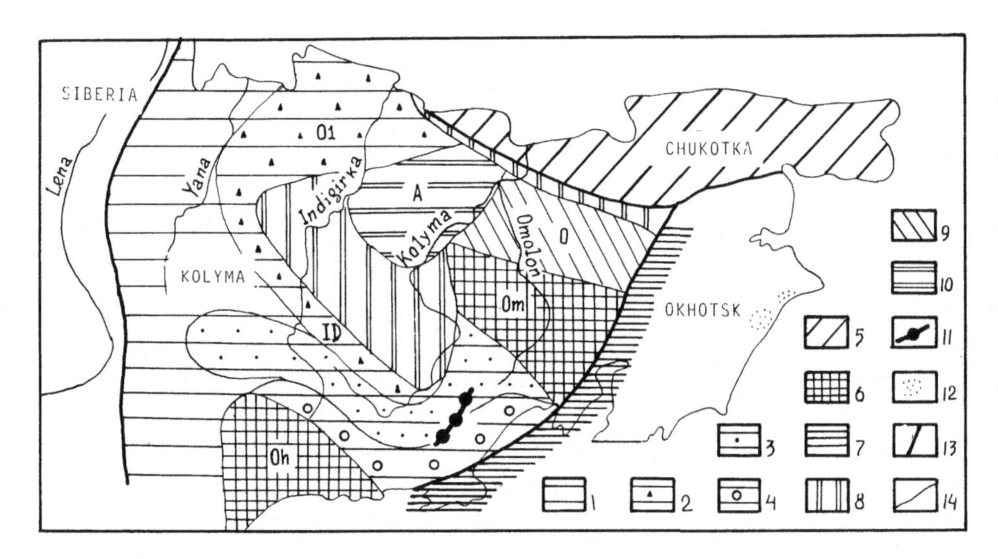


**Figure 11.8. Distribution of Jurassic depositional types in North-East Russia. 1–5, terrigenous (miogeoclinal) [1, Verkhoyanye (shelf); 2, Debin (complete); 3, Delinya-Sugoi (discontinuous); 4, Viliga (complete, slightly volcanogenic); 5, Chukotka]; 6, subplatform (Omolon) type; 7–10, volcanoterrigenous (eugeoclinal) [7, Taigonos (island arc); 8, Ilin-Tas and South Anyui (rift zones); 9, Oloi; 10, Alazeya (discontinuous)]; 11, Olyn Island Arc; 12, Jurassic exposures in Koryakia; 13, boundaries of structural blocks; 14, boundaries between depositional types.**

## NORTH-EAST RUSSIA[3]

North-East Russia is east of the Lena River and north of latitude 59°N. This area is a structurally complex region with heterogeneous geological setting, corresponding in general to the Mesozoic folding area. Distinguished within its limits are the geoclinal folded systems of Verkhoyanye–Kolyma, Chukotka, and Western Kamchatka–Koryak, the Alazeya Fold-Block System, the narrow geoclinal folded zones of Ilin–Tas and South Anyui, the Omolon and Okhotsk Massifs, the Taigonos and Eastern Kamchatka Island-Arc Systems, and a number of other structures (*Geological Structure* . . . 1984). Jurassic deposits are, to varying extents, recorded from all of these major structural areas.

Several types of Jurassic sequences (Figure 11.8) are distinguished on the basis of rock composition, completeness, thickness, biota, and geological evolution.

### Terrigenous miogeoclinal sequences

The main occurrence is confined to the Verkhoyanye–Kolyma Geoclinal-Fold System. Lower and Middle Jurassic deposits comprise a geoclinal complex that, together with the underlying Norian rocks, belongs to the Yukagir Complex, formed at the final stage of geoclinal development. Upper Jurassic deposits (sometimes from Volgian) form late geoclinal complexes, together with Lower Cretaceous rocks.

The thick, most nearly complete and continuous, Jurassic sequence of miogeoclinal type is in the Inyali–Debin Megasynclinorium (ID) and Oldzhoj Trough (OL) (Figures 11.9 and 11.10). During the Triassic, the maximum down-warping was farther west (Yana and upper Indigirka Basins), and structures were displaced in the Jurassic (ID and OL). The Lower Jurassic of ID (Kadykchan Formation and lower part of Aren Formation) is represented by fine clastics (mudstone, siltstone), often clayey-siliceous rocks with an admixture of fine volcanic material of intermediate–basic composition (1,000–1,500 m). Most of the Middle Jurassic (Aalenian–Bathonian) consists of 1,800–4,000 m terrigenous three-component flysch (upper part of Aren and Mereduj Formations). Callovian, Oxfordian, and Kimmeridgian sediments (Koster Formation) gradually become less rhythmical and more diverse (to 2,300 m). The total thickness of the Jurassic in ID reaches 7 km. In OL, the Lower Jurassic (2,000 m) consists of sandstone, the Middle Jurassic (1,000–2,500 m) of claystone, and the Upper Jurassic (3,000–5,000 m, Oxfordian–Volgian) of alternating mudstone, siltstone, and sandstone. The total thickness of the section exceeds 8 km.

The terrigenous miogeoclinal type represents sediments of an abyssal trough (pseudoabyssal zone) that during the Jurassic underwent a progressive shallowing. Due to the relatively poor and irregular Jurassic fossil record in ID and OL, lithostratons play the leading role in the subdivision (Figure 11.9). Several formations are distinguished that are traceable over considerable distances. Rare and poorly preserved ammonites make it possible to distinguish only the zones of *Alsatites liasicus, Coroniceras siverti, Angulaticeras kolymicum, Amaltheus viligaensis, Cranocephalites vulgaris, Arctocephalites elegans, Amoeboceras alternans*, and *Cardioceras cordatum*.

To the west, between the Siberian Platform and the ID and OL structures, the Jurassic fills a number of minor synclinal structures in the Verkhoyanye Megaanticlinorium. The total thickness of the Jurassic does not exceed 3.5 km. Presumably, there is a Toarcian hiatus. No Upper Jurassic deposits are recorded. Early (Sakkyryr Group) and Middle (Ulaga Group) Jurassic sediments were accumulated on a broad shelf and are lower to middle sublittoral facies, whereas the Late Jurassic sediments are middle to upper sublittoral.

Rough biostratigraphy is based on rare ammonite finds: Hettangian and Sinemurian on *Psiloceras* sp., *Waehneroceras frigga*, and *Schlotheimia angulata;* Aalenian to Lower Bathonian on *Pseudolioceras beyrichi, Boreiocephalites borealis* (also placed in *Cranocephalites*), *Stephanoceras (Itinsaites)* sp., *Cranocephalites pompeckji, Arctocephalites elegans, Oxycerites jugatus*, and *Longaeviceras keyserlingi.*

Southwestward, at the right banks of the upper Kolyma and Indigirka Rivers, Jurassic exposures again become more extensive,

[3] By Y. S. Repin, K. V. Paraketsov, and I. V. Polubotko.

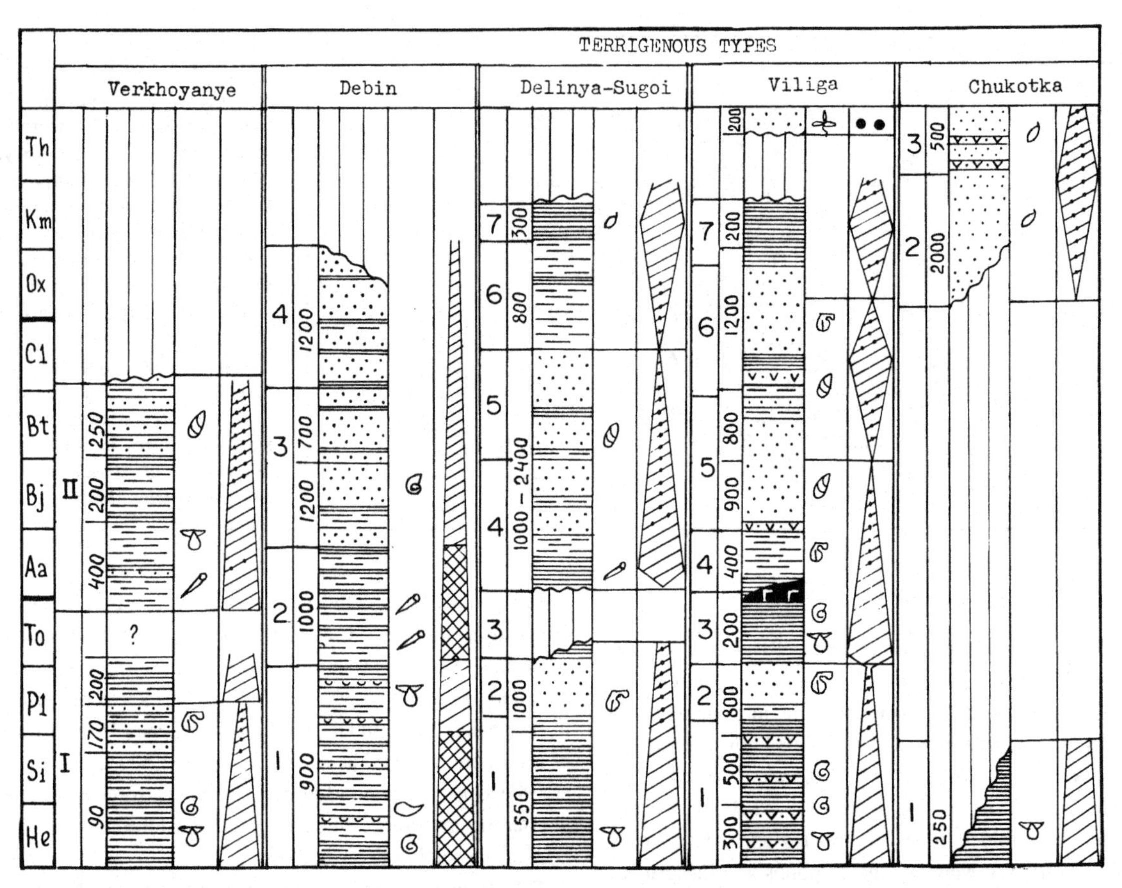


**Figure 11.9. Important terrigenous Jurassic sequences in North-East Russia. Legend as in Figure 11.11. Groups and formations: Verkhoyanye: I, Sakkyryr Group; II, Ulaga Groups. Debin: 1, Kadykchan Formation; 2, Aren Formation; 3, Mereduj Formation; 4, Koster Formation. Delinya-Sugoi: 1, Penalti Formation; Tabornyi Formation; 3, Marat Formation; 4, Memechen Formation; 5, Oktyabrina Formation; 6, Kuchukan Formation; 7, Baza Formation. Viliga: 1, Mokhovoj Formation; 2, Makarych Formation; 3, Zazor Formation; 4, Yaschan Formation; 5, Mongke Formation; 6, Viliga Formation; 7, Kalkutty Formation. Chukotka: 1, Kytepveem Formation; 2, Rauchua Formation; 3, Netepneiveem Formation.**

with the Delinya–Sugoi depositional type (Figures 11.9 and 11.10). These areas are characterized by continental-crust blocks that subsided during certain Jurassic stages, resulting in thick terrigenous-geoclinal, often flyschoid, sediments. At other times (pre-Aalenian and post-Callovian), down-warping ceased, and troughs were filled by sediments and usually elevated above sea level. Hiatuses at the Lower–Middle and, locally, the Middle–Upper Jurassic boundaries are traceable extensively and sometimes show slight unconformities. These haituses are associated with periods of uplifting or cessation of down-warping.

The Lower Jurassic characterizes the Delinya–Sugoi depositional type: considerable thickness (1,000–2,500 m), flyschoid siltstone–mudstone alternation, extensive subaqueous slumping, and poor fossil record (Penalti and Taborny Formations). Toarcian and part of the Upper Pliensbachian are usually missing over most of the record. The Middle Jurassic is locally reduced, but better developed in other areas, with, below, mudstone (Marat Formation) and above, sandstone (Memechen and Oktyabrina Formations). Upper Jurassic consists of Oxfordian–Kimmeridgian mudstone, siltstone, and terrigenous volcanic sandstone (400–800 m; Kuchukan and Baza Formations).

South of the Delinya–Sugoi-type deposits, in (river) basins of the Sea of Okhotsk, the similar but more complete Viliga depositional type is developed. It has more abundant fossils, particularly Middle Jurassic mytilocerams. Only the Upper Jurassic is incomplete (i.e., Volgian is lacking). In completeness and thickness, this section is similar to the Debin type. The sediments have admixed volcanics, due to the Early Jurassic Kobjume Rift in the northwest and the Olyn Volcanic Arc in the southeast (Okhotsk–Kolyma divide, boundary between sedimentary types) (Repin 1975).

The Lower Jurassic sandstone and siltstone sequences of the Viliga Basin (1,800 m) have, below, Hettangian–Lower Pliensbachian tufogenic flysch and, at the top, late Toarcian basalt (250 m). The Middle Jurassic (2,800 m) is siltstone and sandstone, flyschoid and tufogenic in the lower part. The Upper Jurassic (ca. 2,000 m) in its lower part consists of sandy-silty rocks, with abundant Lower Oxfordian *Meleagrinella ovalis*, and, in the upper part, of predominantly clayey rocks, with rare Oxfordian–Kimmeridgian buchias. The Viliga sequence is exceptionally complete and fossiliferous and is a reference section for the geoclinal type, with stratotypes and parastratotypes of some zones yielding ammonites and bivalves.

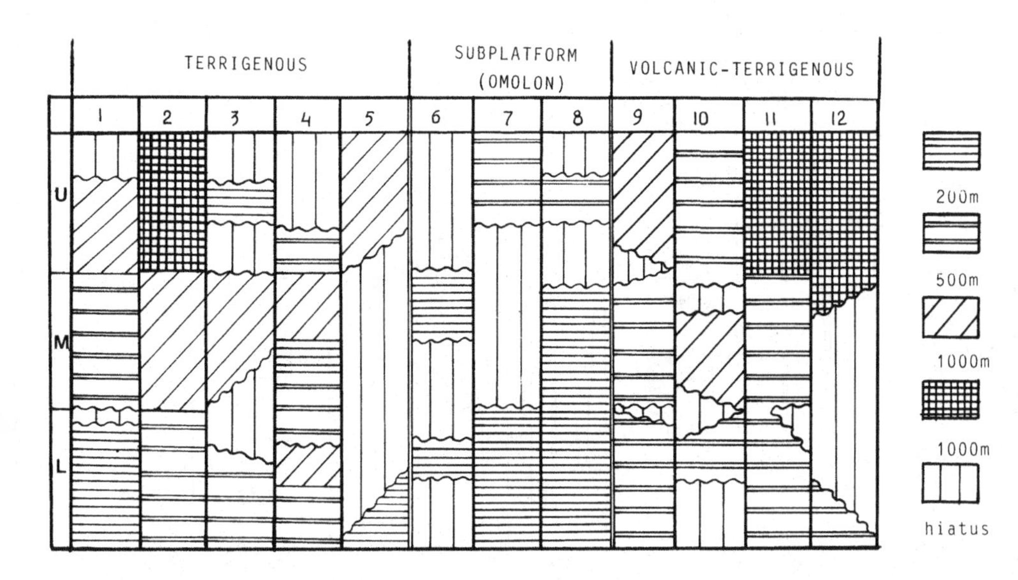


**Figure 11.10. Sedimentation rates for the main types of Jurassic deposits in North-East Russia; thicknesses of stages in meters: 1, Verkhoyanye; 2, Debin; 3, Delinya-Sugoi; 4, Viliga; 5, Chukotka; 6, Russian River; 7, Kedon River; 8, Paren River; 9, Oloi; 10, Alazeya; 11, Taigonos; 12, Ilin-Tas.**

The Jurassic section of the Chukotka Geoclinal-Fold System (Chukotka type) is terrigenous. Lower Jurassic (Hettangian–Sinemurian) mudstone and clayey siltstone (200–300 m), with rare *Otapiria*, are known only from the small Kytepveem Basin in the west (Kytepveem Formation). Chukotka-type deposition is characterized by Upper Jurassic, filling several late geoclinal depressions that are mainly superposed on an eroded Triassic surface. On the Kolyma–Rauchua Divide, the Myrgovaam Depression is filled with thick arkoses, with single mudstone and siltstone interbeds (to 2,000 m), yielding rare Oxfordian–Kimmeridgian buchias (Rauchua Formation). The large Rauchia Depression along the southwestern coast of Chaun Bay and a number of small depressions farther east were emplaced during the middle Volgian. They are filled by molassoid sediments: polymictic and volcanoterrigenous sandstones, siltstone, and mudstone, with intermediate and acidic tuffite and tuff, tuffaceous gritstone, and conglomerate (600–700 m; Netpneiveem Formation). The deposits yield middle and late Volgian buchias.

### Volcanic terrigenous (eugeoclinal) sequences

The Jurassic sequences are volcanic-terrigenous in the Alazeya Fold-Block System, the Oloi Marginal Trough, the Ilin-Tas and South Anyui Suture geoclinal-fold zones, and the Taigonos Island-Arc System, in separate small rift structures, and in the late geoclinal and orogenic Late Jurassic depressions (Figure 11.11). The most complete and thickest Jurassic sequence of this type (Taigonos) is known from the Taigonos Peninsula, which belongs to the Taigonos Island-Arc System (Nekrasov 1976; *Geological Structure . . .* 1984).

The Lower Jurassic (to 1,500 m) consists of andesitic and basaltic tuffs, with sedimentary interbeds (Tikass Group); the Middle Jurassic (3,400 m), of basalts, andesites, tuffs, sandstone, and siltstone; the Upper Jurassic (to 5,400 m), of alternating andesite, basalt, dacite, liparite, tuff, siltstone, and sandstone (Vnutrennyaya Group). The fossil record consists only of levels with *Zugodactylites* sp., *Porpoceras polare*, *Pseudolioceras beyrichi*, *Arctocephalites* sp., and *Costacadoceras bluethgeni*.

Southwest of the Taigonos Peninsula, on the Koni and Pyagin Peninsulas and at the Babushkin Bay coast, the Jurassic is generally similar. The upper Middle Jurassic yields the Callovian ammonites *Iniskinites* cf. *magniformis*, *Costacadoceras* cf. *bluethgeni*, *Cadoceras*, and, slightly higher, *Longaeviceras* aff. *nikitini*. The Upper Jurassic (ca. 3,000 m) has abundant volcanics and is crowned by flora-bearing Upper Volgian continental beds (Severo-Taigonos Group). Northeast of the Taigonos Peninsula, the structurally complex Western Kamchatka–Koryak Geoclinal-Fold System has only small blocks of the Lower and Middle Jurassic. Silicoterrigenous beds yield a relatively poor fauna resembling the fauna in the remainder of the North-East: *Angulaticeras* ex gr. *kolymicum* (Sinemurian), *Amaltheus* cf. *stokesi* (Upper Pliensbachian), and *Mytiloceramus*, *Pseudolioceras (Tugurites)* ex gr. *maclintocki*, and *Arkelloceras* (Aalenian–Bajocian). In the Kojverelan Basin (right tributary of the Velikaya River) volcanosiliceous, terrigenous strata (ca. 100 m) have yielded Tethyan Callovian ammonites, unusual for the North-East: *Putealiceras (Zieteniceras?)* cf. *zieteni*, *Lunuloceras*, and *Choffatia*. Some workers regard this block as allochthonous.

The Upper Jurassic in the Western Kamchatka–Koryak System is represented only by Middle–Upper Volgian. The deposits of some areas (Pekulnei Ridge, Velikaya Basin, Lake Mainits area) are typically eugeoclinal, with spilite, andesite, tuffs with interbeds of jasper, siliceous rocks, sandstone, siltstone, and conglomerate (500–1,000 m). The sediments contain Volgian buchias, whereas the siliceous deposits yield radiolarians. In other areas occur terrigenous or volcanoterrigenous rocks with Volgian buchias.

The Alazeya–Oloi depositional type occurs in the area between the Indigirka River in the west to the Kolyma–Anadyr Divide in the east. Compared with the Taigonos type, it is characterized by hiatuses and reduced thicknesses, and diversity complexity of deposition, sometimes with intergrading marine and continental facies. Lower and Upper Jurassic have the greatest proportion of

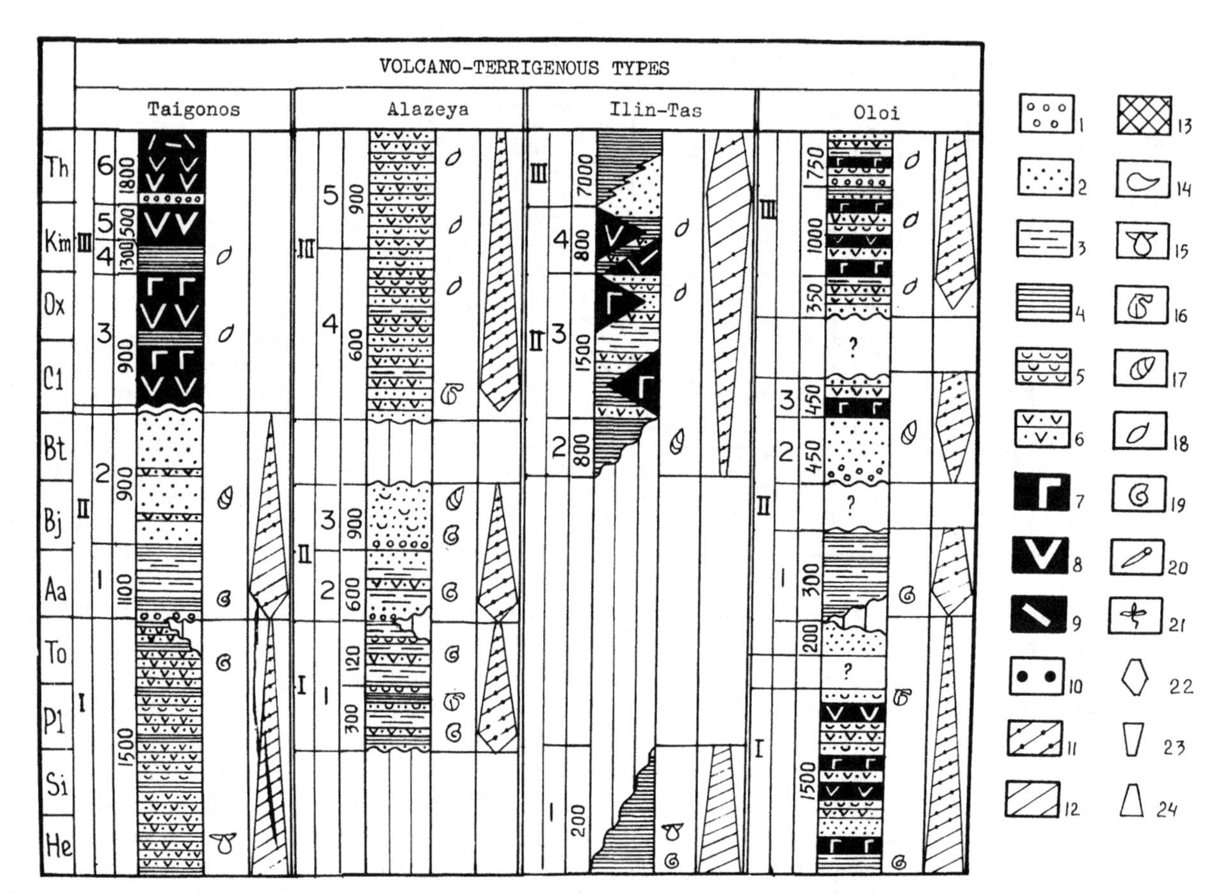


**Figure 11.11. Important volcanoterrigenous Jurassic sequences in North-East Russia. Rock types: 1, conglomerate; 2, sandstone; 3, siltstone; 4, mudstone; 5, tuffite and ash tuffs; 6, tuffs and lavas (7, basic; 8, intermediate; 9, acid). Facies: 10, lacustrine alluvial; 11, middle to upper sublittoral; 12, lower sublittoral; 13, deep sea (pseudoabyssal). Benthic bivalves: 14, pseudoabyssal; 15, lower sublittoral [Hettangian-Sinemurian *Otapiria*, *"Pseudomytiloides," Kolymonectes;* Aalenian *Oxytoma, Propeamussium*]; 16, upper sublittoral [late Pliensbachian *"Velata," Radulonectites;* Callovian *Maclearnia*]; 17, mytiloceramus; 18, buchias. Nectobenthos: 19, ammonites; 20, belemnites; 21, flora cycles [22, subsymmetric; 23, transgressive; 24, regressive]. Groups and formations: Taigonos: I, Tikass Group; II, Vnutrennaya Group [1, Kuchuveen Formation; 2, Khalpil Formation]; III, Severo-Taigonos Group [3, Talnaveem Formation; 4, Srednij Formation; 5, Gyryangin Formation; 6, Vavachun Formation]. Alazeya: I, Egelyakh Group, upper part [1, Pologij Formation]; II, Fedotikha Group [2, Kurunyur Formation; 3, Syustinnyakh Formation]; III, Khoska Group [4, Abagylakh Formation; 5, Ikki-Kyunnyakh Formation]. Ilin-Tas: 1, Olguya Formation; II, Ilin-Tas Group [2, Taskan Formation; 3, Talbygyr Formation; 4, Kyureter Formation]; III, Bastakh Group. Oloi: I, Krichal Group; II, Privalnaya Group [1, Kojguveem Formation; 2, Losikha Formation; 3, Karkasnaya Formation]; III, Tantyn Group.**

volcanics. The Dogger is composed of terrigenous siltstone and sandstone and, locally (Alazeya Plateau), of sandstone and conglomerate. Maximal thickness is in the Oloi Marginal Trough (5.5 km), and somewhat less in the Alazeya System (3.5 km).

The Ilin-Tas depositional type occurs in the middle Indigirka and Kolyma Basins, where it corresponds to the Momo-Zyryan Trough and adjacent Late Jurassic grabens. Lias (Olguya Formation) is represented locally only by Hettangian–Sinemurian fine-clastic, terrigenous deposits (200–1,000 m) that are closely associated with the subjacent Upper Triassic. Above are fine- and coarse-clastic Bathonian–Callovian (600–1,500 m), resting with a considerable hiatus and unconformity on the Hettangian–Sinemurian and older rocks. Malm follows conformably or disconformably. Oxfordian–Kimmeridgian is terrigenous-volcanic, with lateral facies changes (1,000–2,500 m) and with volcanics: basalt, andesite, dacite, liparite, and their tuffs. Volgian is terrigenous-marine in the lower part and lagoonal-continental in the upper, with maximum thickness in the Momo-Zyryan Trough (7–8 km).

In the South Anyui Zone, only the Upper Jurassic can be established: volcanosiliceous strata with spilitic lavas, tuffs, siliceous rocks, and jasperoids. Volgian buchias and radiolarians occur in fine-clastic terrigenous interbeds.

### Subplatform (Omolon-type) sequences

The Omolon (subplatform) type of deposition has much more limited distribution than the other types, but is important for Jurassic biozonation in North-East Russia. The principal occurrence is in the Omolon Median Massif (Figure 11.12). This sequence is characterized by relatively low thickness (entire Jurassic, 1,300–2,800 m), hiatuses, and lagoonal-continental

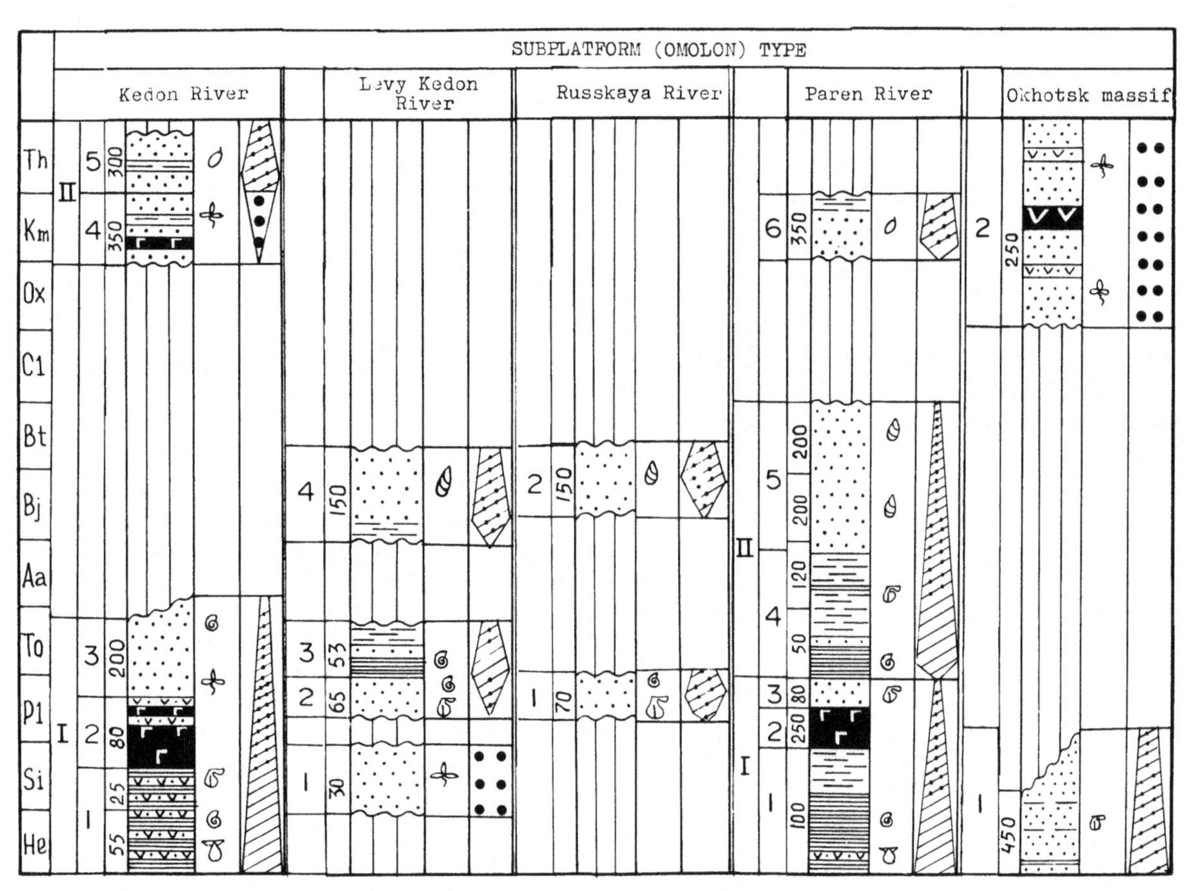


**Figure 11.12. Important subplatform-type Jurassic sequences in North-East Russia. Legend as in Figure 11.11. Groups and formations: Kedon River: I, Ust-Finish Group [1, Kinas Formation; 2, Omkuchan Formation; 3, Chirok Formation]; II, Ust-Kedon Group [4, Namyndykan Formation; 5, Ajnene Formation]. Levy Kedon River: 1, Brodnaya Formation; 2, Nalednyi Formation; 3, Start Formation; 4, Emlyndzha Formation. Paren River: I, Uglovaya Group [1, Rybnaya Formation; 2, Tumma Formation; 3, Tokchikin Formation]; II, Chaivaveem Group [4, Eksin Formation; 5, Emlyndzha Formation]; 6, Provorotnyi Formation. Okhotsk Massif: 1, Nyadbaka Formation; 2, Khavakchan Formation.**

sediments, as well as marine, predominantly terrigenous rock dominated by sandstone (mainly Upper Pliensbachian and Middle Jurassic), together with basic and intermediate lavas, tuffs, and tephrogenic coarse-clastic rocks (Upper Sinemurian–Lower Pliensbachian, Upper Volgian), in some areas adjacent to fault zones and in late geoclinal Late Jurassic depressions. Abundant fossils include bivalves, ammonites, belemnites, and brachipods in the Lias, mytiloceramss in the Dogger, and buchiids and other bivalves in the Malm. However, as in most regions in the North-East, other taxa (e.g., microfauna) are unknown. Here are the stratotypes of the majority of the Liassic regional ammonite zones.

The thickest (to 2,800 m) subcontinuous section, except for a Callovian–Oxfordian hiatus, is at the northwestern margin of the Omolon Massif, in the Rassokha and Tokur–Yuryakh Basins. Farther northwestward this section becomes geoclinal, with increased thicknesses and impoverished fossil record. Jurassic occurs at the eastern and southeastern margins of the Paren River Massif, where the Lias and Dogger are well developed (900–1,200 m) and the Malm is absent, except for local Kimmeridgian sandstones (to 300 m). Sequences within the massif are more discontinuous and thinner, varying in the different blocks. The most reduced section is in the Russkay River valley (Omolon Basin), in the center of the massif, with only Upper Pliensbachian (Nalednyi Formation, 75 m) resting on Upper Norian and Bathonian (Emlyndzha Formation, ca. 200 m). The Jurassic deposits of the Omolon type are open-shoal facies around an archipelago, connected freely with epicontinental seas and the ocean.

Unlike the Omolon Massif, the Okhotsk Massif has only small incomplete Lower Jurassic (Nyadbaka Formation, Hettangian–Sinemurian and Pliensbachian), with sandstone and siltstone (to 600 m), overlain with a considerable hiatus by the Upper Jurassic (Khavakchan Formation) continental, sedimentary volcanic, flora-bearing strata (ca. 250 m).

#### Ammonite and bivalve zones

The zonation of Lower Jurassic in North-East Russia is based on the subplatform sections of the Omolon Massif and is supplemented by geocline-type sections (Viliga River, etc.). Most stratotypes and reference sections of zonal units, established on ammonites and bivalves, are in the Omolon Massif.

### *Hettangian*

The best section is on the left bank of the Kedon River, just below the mouth of Omkuchan Creek, where the siliceous, slightly tufogenic Upper Norian (including Rhaetian) mudstones, with *Oxytoma mojsisovicsi* and *Tosapecten efimovae*, are conformably overlain by finely alternating mudstone and tufogenic and siliceous mudstone with lenticular limestone (55 m). The thicknesses of the standard zones are Primulum Zone 9 m, Planorbis Zone 7 m, Liasicus Zone ca. 32 m, and Angulata Zone 5 m (Polubotko and Repin 1972, 1981). The bivalve assemblage consists of *Otapiria originalis, Meleagrinella subolifex, Oxytoma orientalis, Ochotochlamys kiparisovae, "Pseudomytiloides" sinuosus,* and *"P." latus.*

### *Sinemurian*

The zonation is based on several depressions in the central Omolon Massif. The composite sequence is as follows (ascending order):

1. Bucklandi Standard Zone – tuffaceous mudstone (12 m); *Arietites libratus, Paradasyceras* sp.
2. *Coroniceras siverti* Zone – alternation of mudstone and tuffite with carbonite lenses (15 m); also with *Primarietites* cf. *bisulcatus* and *P. reynesi;* upper part with *Eparietites* cf. *denotatus*
3. *Angulaticeras kolymicum* Zone – alternation of tuffaceous mudstone and siltstone and tuffite (35 m); *Angulaticeras (Gydanoceras) ochoticum.*

The two lower zones are characterized by the bivalves *Otapiria omolonica, O. tailleuri, Oxytoma orientalis, "Pseudomytiloides" rassochaensis,* and *Kolymonectes kedonensis* and the brachiopod *Ochotorhynchia omolonensis.* The upper zone contains *Otapiria limaeformis, O. affecta, "Monotis" inopinata, Kolymonectes staeshei, Anomia lemniscata,* and *"Pseudomytiloides" rassochaensis.*

### *Pliensbachian*

Lower Pliensbachian is documented by ammonites finds only in the Bol.-Anyui Basin (*Polymorphites*) (Afitsky 1970) and in the Viliga Basin (Polymorphitidae) (Polubotko and Repin 1974b; Milova 1980). In other areas (Viliga, Bulun, and Gizhiga rivers) the lower substage is distinguished by its bivalve assemblage, including *"Amonotis"* sp., *Kolymonectes staeschei,* and *Chlamys tapensis.*

Upper Pliensbachian is widespread in most of the North-East and is divided into three zones in sections of the Omolon Massif (Brodnaya River in the Levy Kedon Basin) and the Viliga River:

1. *Amaltheus stokesi* Zone – Viliga River; fine quartz-plagioclase sandstone (200 m); also *A. bifurcus* and *A. complanatus*
2. *Amaltheus talrosei* Zone – Brodnaya River; polymictic sandstone and siltstone (13 m); also *A. ventrocalvus* and *A. subbifurcus*
3. *Amaltheus viligaensis* Zone – volcanomictic and tufogenic sandstone (15 m); also *A. brodnaensis, A. extremus,* and *A. orientalis*

The bivalve assemblage of the *stokesi* Zone consists of *"Velata" viligaensis, Kolymonectes mongkensis, Meleagrinella ansparsicosta,* and *Lima philatovi;* that of the *talrosei* and *viligaensis* Zones, of *K. mongkensis, K. levis, Ochotochlamys grandis, Veteranella (Glyptoleda) formosa, Meleagrinella ptchelincevae, M. oxytomaeformis, Harpax laevigatus, Tancredia omolonensis,* etc.

### *Toarcian*

The reference section of the stage and stratotypes of all zones are in the Omolon Massif (upper Levy Kedon River) (Dagis and Dagis 1965; Polubotko and Repin 1966, 1974b; Dagis 1968, 1974 (ascending order):

1. *Tiltoniceras propinquum* Zone – mudstone (12–18 m); *Kedonoceras compactum, K. asperum, Tiltoniceras propinquum, T. costatum, T. capillatum, Arctomercaticeras costatum,* and *A. tenue;* Lower Toarcian
2. *Eleganticeras elegantulum* Zone – mudstone (3 m); *Eleganticeras alajaense, E. planum, E. confragosum,* and *E. connexivum*
3. *Harpoceras falciferum* Zone – mudstone and siltstone (10 m); *Harpoceras falciferum, H. exaratum,* and *Hildoceratoides levisoni*
4. *Dactylioceras athleticum* Zone – polymictic inequigranular sandstone (10 m); *Dactylioceras commune, D. athleticum, D. kanense, D. amplum, Harpohildoceras grande, Kolymoceras cognatum, K. crebrinodum,* and *Hildoceratoides chrysanthemum*
5. *Zugodactylites monestieri* Zone – mudstone and siltstone (5 m); *Zugodactylites braunianus, Z. monestieri, Z. pseudobraunianus, Z. exilis, Z. moratus, Z. latus, Catacoeloceras proprium, C. manifestum, Kolymoceras* ex gr. *viluiense, Pseudolioceras lythense,* and *P. kedonense*
6. *Porpoceras spinatum* Zone – pisolitic tuffites (1.8 m); *Porpoceras polare, P. spinatum, Collina mucronata, C. orientalis,* and *Pseudolioceras gradatum*
7. *Pseudolioceras rosenkrantzi* Zone – clayey-silty sandstone (5 m); also *Pseudolioceras compactile;* Upper Toarcian

Toarcian bivalves form two assemblages: below (*propinquum, elegantulum,* and *falciferum* Zones), *Meleagrinella substriata, Mytilus mytileformis,* and *"Pseudomytiloides"* aff. *amygdaloides;* above, *Meleagrinella faminaestriata, Oxytoma startensis, Propeamussium pumilum, Pseudomytiloides marchaensis, Vaugonia literata, Protocardia striatula,* and *Goniomya rhombifera*

### *Aalenian*

The reference section of the stage in geoclinal development is in the Viliga Basin (Polubotko and Repin 1974a):

1. *Pseudolioceras maclintocki* Zone – alternating siltstone, mudstone, and dacitic tuffs (90 m); also *P. (P.) replicatum, P. (P.) beyrichi, Mytiloceramus priscus, M. subtilis, Oxytoma jacksoni,* and *Propeamussium olenekense*
2. *Pseudolioceras tugurense* Zone – alternation of siltstone and sandstone (200 m); with *P. (Tugurites) tugurense,* also *Mytiloceramus elegans* and *M. popovi*

*Bajocian*

The biozones of the lower substage are based on sections in the Omolon Massif (Kegali River), Alazeya Upland (Sededema River), and Anadyr Basin (Repin 1972; *Resolutions* . . . 1978).

1. *Pseudolioceras fastigatum* Zone – also *P. (Tugurites) costistriatum, Mytiloceramus jurensis,* and *M. menneri;* Discites and Laeviuscula Zones (= Widebayense Zone of Southern Alaska)
2. *Arkelloceras tozeri* Zone – *A. elegans, A.* aff. *maclearni, Bradfordia alaseica, Holcophylloceras costisparsum, Zetoceras, Mytiloceramus lucifer,* and *M. omolonensis;* Sauzei Zone (= Crassicostatus Zone)
3. Beds with *Chondroceras* cf. *marshalli* and *Lissoceras* sp. – also *Mytiloceramus porrectus;* Humphriesianum Zone

The base of the Upper Bajocian does not contain ammonites and is defined by the *Mytiloceramus clinatus* Assemblage Zone:

4. *Boreiocephalites borealis* Zone – single finds of index species (ed. note: also placed in *Cranocephalites*) in Adycha River area; Upper Bajocian
5. *Cranocephalites vulgaris* Zone – also *C. nordvikensis, C. inconstans,* and *C. furcatus*

*Bathonian*

1. *Arctocephalites elegans* Zone – only a single zone is recognizable here; also *A. ellipticus, A. stepankovi,* and *"Oxycerites" jugatus*
2. Beds with *Arcticoceras–Costacadoceras*

*Callovian*

1. Beds with *Cadoceras anabarense* – lower Callovian
2. Beds with *Longaeviceras keyserlingi* – upper Callovian

Because of the extreme scarcity of ammonites, the zonation of the Malm in the North-East is based exclusively on buchiids. The first buchiid, *Praebuchia anyuensis,* occurs in the Dogger, probably directly above the last mytilocerams (*anyuensis* Beds, upper Bathonian). The *Buchia* succession is most complete in the exceptionally continuous Malm section of the Bolshoi Anyui Basin. Here is also the majority of stratotypes of regional *Buchia* zones.

*Oxfordian*

1. Beds with *Praebuchia? impressae* – tufogenic sandstone with conglomerate, tuff, siltstone, and mudstone (ca. 300 m); dated by stratigraphic position and analogy with the Far East, where *P.? impressae* is associated with *Scarburgiceras;* Lower Oxfordian
2. *Praebuchia kirghisensis* (= *P. lata*)–*Buchia concentrica* Zone – mudstone and siltstone, sometimes silicified (to 430 m); *B. linstroemi, B. mosquensis tenuistriata,* and *B. jeropolensis;* in other areas of the North-East, also *Cardioceras* sp.

Underlain by beds with *Quenstedtoceras;* overlain by sediments with *Amoeboceras* ex gr. *glosense* and *Amoebites* sp.; Middle–Upper Oxfordian.

*Kimmeridgian*

1. *Buchia concentrica–B. mosquensis tenuistriata* Zone – alternation of mudstone, siltstone, sandstone, tuffaceous sandstone, and tuffite (ca. 200 m); also *B. jeropolensis, B. lindstroemi,* and *B. vuquaamensis,* as well as *Oxytoma (O.) expansa, Meleagrinella ovalis, Isognomon embolicus, Maclearnia broenlundi, Pseudolimeaa borealis,* and *Amoeboceras (Amoebites)* ex gr. *kitchini* throughout the zone; Lower Kimmeridgian
2. *Buchia rugosa–B. mosquensis paradoxa* Zone – siltstone, sandstone, agglomerate, and tuffs (to 200 m); also *B. vuquaamensis* and *B. mosquensis tenuistriata;* remainder of bivalve assemblage similar to that found below; Upper Kimmeridgian

*Volgian*

1. *Buchia mosquensis mosquensis–B. russiensis* Zone – tufogenic rocks and tuffs (ca. 200 m); also *B. rugosa;* Lower Volgian
2. *Buchia russiensis–B. fischeriana* Zone – siltstone, mudstone, and sandstone, frequently tufogenic (to 200 m); rare *B. mosquensis mosquensis, B. rugosa;* in upper part, *B. flexuosa;* age based on *Dorsoplanites* cf. *transitorus* and *Partschiceras schetuchaense* in this zone and by correlation with adjacent regions; Middle Volgian
3. *Buchia tenuicollis–B. terebratuloides* Zone – alternation of basic and intermediate tephroids, tuffaceous sandstone, siltstone, mudstone, gritstone, and conglomerate (over 1,500 m); also *B. flexuosa, B. lahuseni, B. obliqua,* and endemic *B. circula;* age based on stratigraphic position and ?*Chetaites;* Upper Volgian

## PALEOGEOGRAPHY AND GEODYNAMICS OF NORTHEAST ASIA[4]

### Paleogeography

Mesozoic eastern Russia was part of the Pacific Mobile Belt. This implies diverse paleoenvironments, from mountainous areas in the west to ocean in the east.

[4] By Y. S. Repin and I. I. Sey.

During the Hettangian and Sinemurian the sea flooded the margin of the Siberian landmass from the east. It comprised platforms and basins with different depths and sediments. A belt of continental shelf, bounded in the northeast by the Debin Linear Suboceanic Basin, extended parallel to the Siberian landmass, with moderate relief. The Anyui Basin separated the Chukotka and Kolyma Blocks. In the central parts of the region, major island systems (Kolyma, Omolon) were surrounded by shelf seas and relatively deep basins. Along the northern and eastern margins of the Kolyma Block, volcanic archipelagoes existed throughout the Jurassic (Taigonos, Anyui).

In all marine areas, accumulation of clays and silts proceeded mainly in lower-sublittoral environments. The principal sources of sediments were the Siberian landmass and the Okhotsk Peninsula. Local provenances were also significant, providing volcanogenic constituents.

In most of Far East Russia, continental regimes began in the earliest Jurassic, subsequent to the Late Triassic transgressive cycle. Marine environments continued only in the linear Amur-Okhotsk and Sikhote-Alin deep-sea depressions, which were paleogeographically stable throughout most of the Jurassic.

In the Early and Middle Jurassic, the intracratonic Amur-Okhotsk Marine Basin, far inland, was filled with sandy-silty-clayey rhythmites, siliceous silts, and volcanites.

The perioceanic Sikhote-Alin Basin apparently corresponded to the continental slope and was separated from the ocean by a series of nonvolcanic marginal uplifts. During most of the Jurassic, volcanosiliceous strata accumulated in the basin, with the maximum amount of volcanites in the Early Jurassic. This sea was surrounded by a denudation plain, which in the west and north graded into the Bureya and Stanovoi uplands, the main sources of clastics during the Jurassic.

In late Pliensbachian–early Toarcian time, a eustatic rise resulted in transgression, especially in the Far East, where the sea covered the Eastern Transbaikal area (Figure 11.13). In the North-East, the Vilyui Lowland was flooded. Mainly fine sand and silt were deposited in the shallow-sea basins under upper and middle sublittoral conditions, generally with low relief of the surrounding land.

The early Toarcian transgression was modified by the development of several geanticlines. This resulted in the drainage of extensive areas in the Lena–Yana interfluve, forming a peninsula, and in extension of the Kolyma Archipelago. The Lena–Yana Peninsula separated the Yakutsk Sea from the markedly reduced Kolyma Basin.

The late Toarcian in eastern Russia was a time of regional uplift and regression, when seaways to the northwestern European sea and to the Paleopacific were absent or reduced.

The transgressive cycle in the Aalenian–early Bajocian greatly enlarged the sea in eastern Russia (Figures 11.13 and 11.14). In some cases the sea deepened slightly, but remained mainly upper–middle, rarely lower, sublittoral. The prevailing sandy-silty sediments indicate weak denudation, rather than a deepening of the basins. This is confirmed by the character of the benthos, the presence of plant remains, and the sedimentary textures.

Beginning with the late Bajocian, gradual shoaling and regression occurred, with repeated coastline fluctuations. The lithofacies are therefore diverse, from deep sea to littoral and continental. Suboceanic basins were filled by sediments and shoaled. The Yakutsk Sea became an area of continental sedimentation. The contrasting thicknesses in sedimentary basins, increased height and relief of the land, and new source areas resulted from increased tectonic activity.

The latest Middle Jurassic and earliest Late Jurassic are noted for major tectonic and geographic changes in the region; some structures were reversed, resulting in the drainage of extensive areas.

In the Late Jurassic, the continentalization of northeastern Asia continued, interrupted by Oxfordian–early Kimmeridgian and middle–late Volgian (Tithonian) transgressions. The area of marine sedimentation was displaced toward the northern and eastern margins of the region, with the maximum transgression during the Volgian (Figures 11.1 and 11.3). Relatively deep sea was retained in the reduced Amur-Okhotsk and Sikhote-Alin Basins, but terrigenous sediments increased in abundance and coarseness. During the earliest Volgian, the Momo-Zyryan deep-sea depression was formed and the Anyui linear oceanic basin appeared, which was open to the Paleopacific. In other basins, upper-sublittoral conditions prevailed, often extremely shallow and with sandy-silty and sandy-pebbly sediments. Subaqueous volcanic activity intensified, particularly peripheral to the Kolyma Block.

Uplifting resulted in increased elevation and relief of the land. In the south of the region, the Stanovye Mountains formed and included molassoid and volcanic depressions. Several, briefly marine, molassoid depressions also formed in North-East Russia. Coastal and lacustrine alluvial plains were widespread and included coal measures.

Continentalization of eastern Russia was completed in the Hauterivian, when almost the entire area became land. An aqueous regime was retained only at the eastern margin of the region (Sakhalin Island, Kamchatka, Koryak Upland), which throughout the Jurassic was peripheral to the Paleopacific. Accumulation of siliceous silts and terrigenous siliceous sediments proceeded on the abyssal plains and on intraoceanic uplifts of the Paleopacific.

### Geodynamics

The proposed scheme of geodynamic evolution of Eastern Russia in Jurassic time is based on the ideas and notions of several workers who have studied the problems of geodynamics of northeastern Asia in the Mesozoic: Gusev et al. (1985), Zonenschain, Kuzmin, and Natapov (1987), Karasik, Ustritsky, and Khramov (1984), Mazarovich (1985), Parfenov (1984), Rikhter (1986), Stavsky (1988), Khanchuk, Panchenko, and Kemkin (1988), Chudinov (1985). The geological and biogeographic data available to the authors make it necessary to give a slightly modified model.

The geodynamic setting in northeastern Asia by the beginning of the Jurassic had formed under the influence of major riftogenic processes, which started as far back as the Early Paleozoic, followed by subsequent restructuring. Significant restructuring

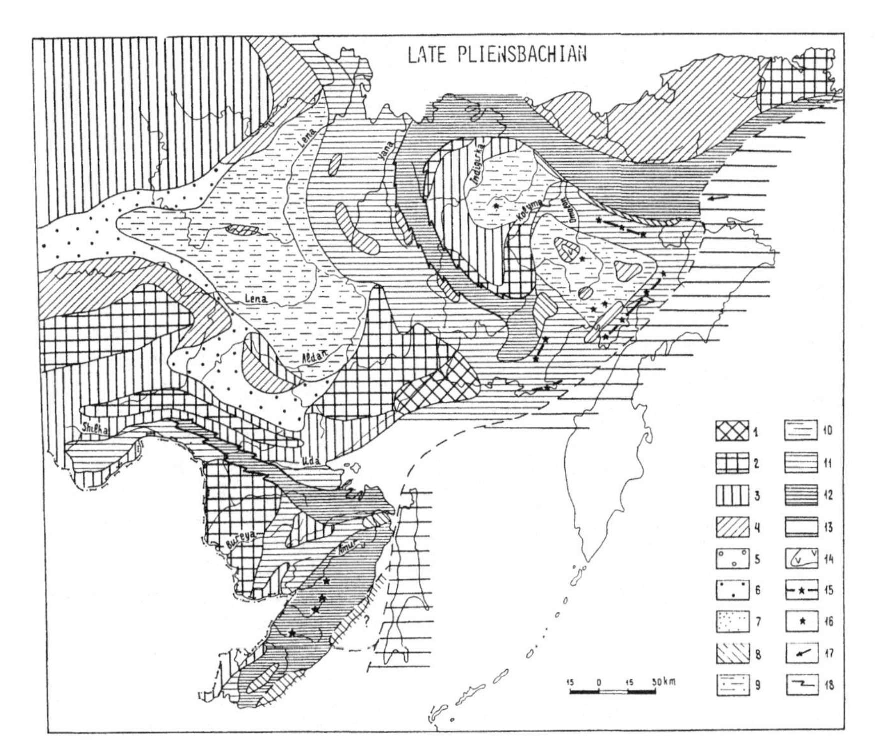


**Figure 11.13. Paleogeographic maps of northeast Asia (on the Present globe). Land: 1, high mountains; 2, low mountains; 3, denudation elevation; 4, denudation plains; 5, intermontane depressions; 6, lacustrine alluvial plains; 7, coastal marine plains. Sea: 8, submarine rises; 9, upper sublittoral; 10, lower sublittoral; 11, sublittoral without differentiation; 12, deep sea (pseudoabyssal); 13, oceanic basin (pelagic). Other symbols: 14, areas of subaerial volcanic activity; 15, volcanic island arcs; 16, centers of submarine volcanic activity; 17, direction of main currents; 18, boundaries of compression and rift zones.**

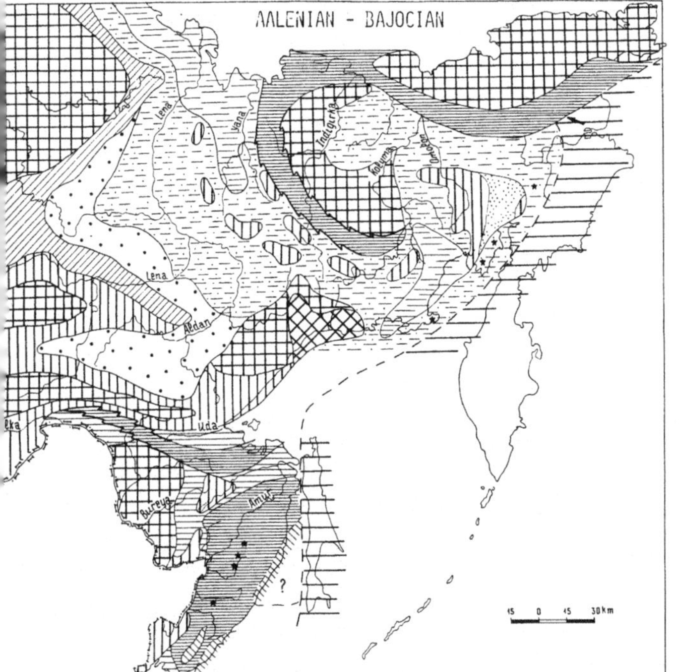


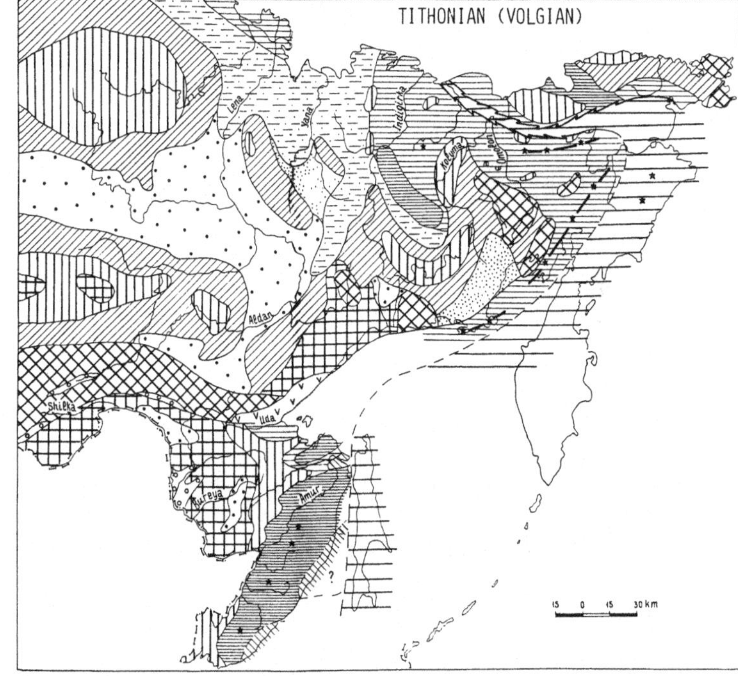


occurred in the Norian, when deep-sea troughs of suboceanic basins were formed and surrounded the eastern extremity of the Siberian Continent, and island-arc systems were emplaced.

Along the eastern margin of the Siberian Continent was the rather broad Verkhoyanye Shelf (Figure 11.14), which in turn was surrounded by the Debin Deep-Sea Trough (a pull-apart suboceanic basin). The steep flanks of the trough, which represented the continental slope, are noted for the development of olistostromes (Gusev et al. 1985), whereas the western boundary of the trough in modern plan is traced by the Adycha-Taryn Tectonic Suture. The Debin Deep-Sea Trough was emplaced in late Norian.

The Debin Basin in the east adjoined a system of smaller basins, which separated a number of blocks and microcontinents. Along the northeastern boundary, a linear chain of blocks was lo-

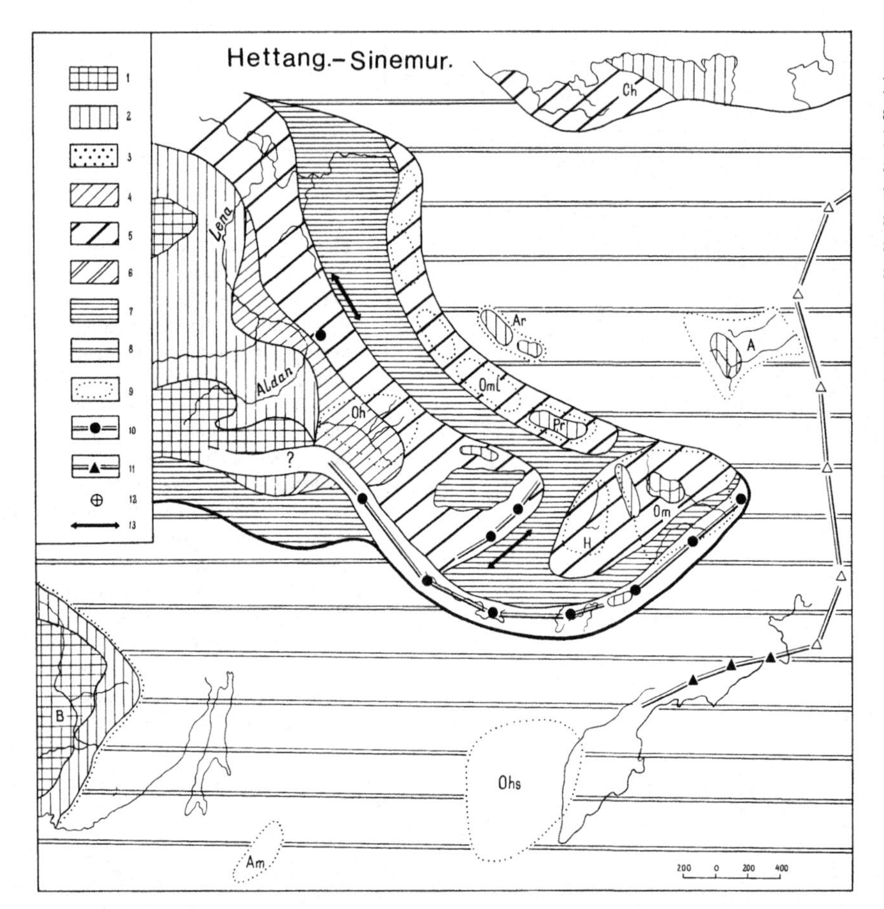


Figure 11.14. Palinspastic reconstructions of northeastern Asia. Land: 1, mountains; 2, uplifts; 3, accumulation plains and intermontane depressions. Sea: 4, shallow shelf (upper sublittoral), sandy sediments predominate; 5, deep shelf (low sublittoral), silty sediments predominate; 6, undivided shelf, sandy-silty sediments; 7, deep-sea troughs on a thinned continental and partly oceanic crust, sandy-silty-clayey flyschoid deposits; 8, deep-sea basins of oceanic type. Other symbols: 9, contours of blocks and microcontinents, later subject to displacement and collision; 10, volcanoes and volcanic arcs on continental crust (black) and their presumed extensions; 11, volcanic arcs on oceanic crust (black) and their presumed extensions; 12, volcano-plutonic belts; 13, contour currents. Microcontinents and blocks: A, Alazeya; Am, Amur; Ar, Arga-Tas; B, Bureya-Khanka; Ch, Chukotka; H, Khetagchan; Oh, Okhotsk; Ohs, Sea of Okhotsk; Om, Omolon; Oml, Omulevka; Pr, Prikolymje.

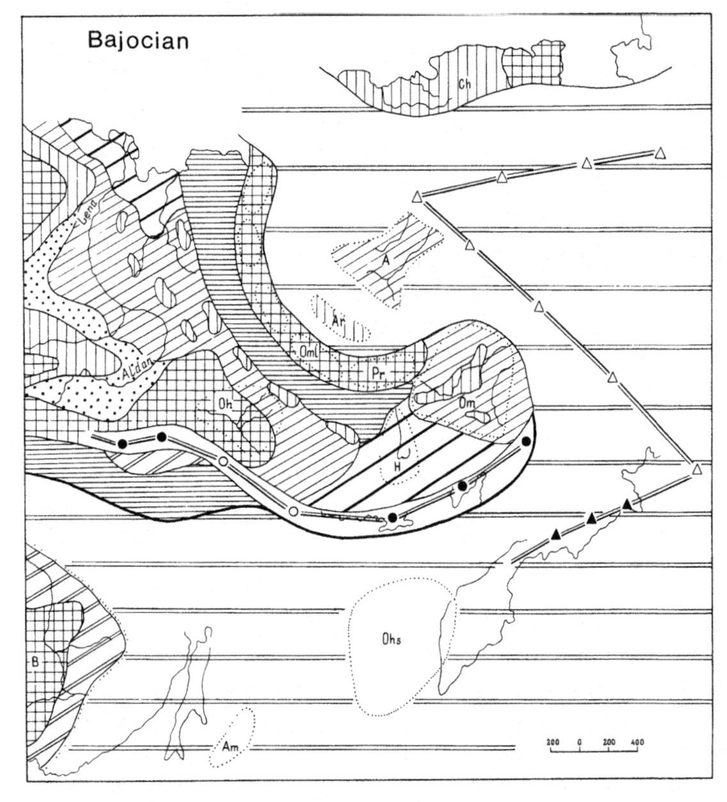


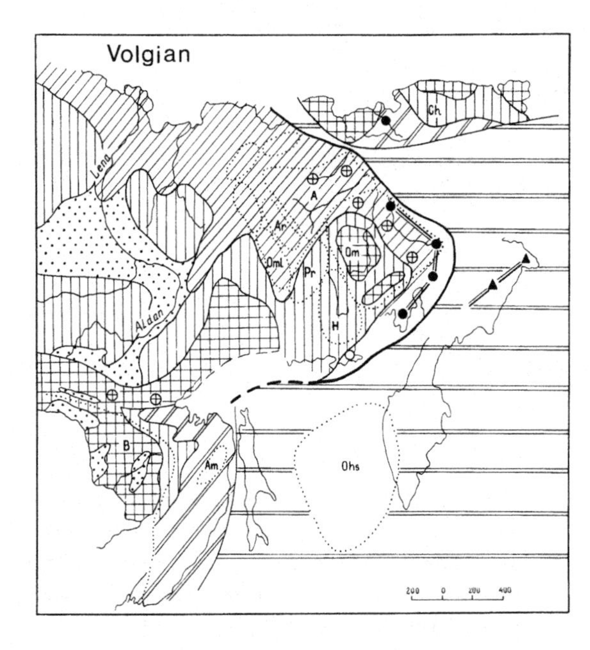

cated (Kolyma, Omulevka, Tas-Khayakhtak, etc.), with the continental crust making up a peculiar foreland of the Siberian Continent. Its marginal, southern member was the Omolon Microcontinent. The southeastern margin of the Siberian Continent was bounded by the Taigonos Volcanic Arc emplaced in the late Norian due to under-thrusting of the Pacific oceanic plate under the continent. The same process also caused the emplacement of the Olyn Volcanic Arc during its short evolution.

Chukotka was separated from the Siberian Continent by the Anyui Intercontinental Oceanic Basin, whose possible fragment is the Argatas Range. Alazeya, apparently, represented a continental block. This is consistent with the Late Triassic and Early and Middle Jurassic shallow-water facies and faunas. According to Gusev et al. (1985, p. 138, Fig. 36), the trend of the Carboniferous and Early Mesozoic basalts on the Alazeya Upland is similar to that of the Siberian Platform traps. This block was most remote from its present location. It is represented by distinct early Middle Jurassic biota, including the Subtethyan and East-Pacific element *Bradfordia* and the Bering element *Erycitoides*. Pliensbachian bivalve communities are characterized by the presence of *Weyla*, a characteristic Tethyan element, widespread along the eastern Pacific coast. *Pinna* is also found, a representative of relatively thermophilic faunas, not recorded in coeval Aalenian communities of other structures of North-East Russia.

The Siberian Continent was bounded in the south by the Amur-Okhotsk (Mongolo-Okhotsk) Intracontinental Oceanic Basin, whose emplacement was in the Early Paleozoic, and which was revived by rifting in the early Norian. The southern extremity of the continent was taken up by the Uda Volcanic Zone, which represented a continental extension of the Uda–Murgal or Taigonos Volcanic Arc. In this area, the Siberian Continent was surrounded by the Uda and Torom Troughs, which joined the continental slope at a deep-sea trench.

The southern extremity of the Amur-Okhotsk Basin was the Bureya Land (Bureya–Khanka Megablock), around whose periphery in the Early and Middle Jurassic periodically appeared a system of marginal epicontinental seas (upper Amur, Bureya, etc.).

East of the Bureya-Khanka Megablock was an oceanic basin, the margin of the Paleopacific. Parallel to the Bureya-Khanka Land, in place of the axial part of the Sikhote-Alin Fold System, extended a deep-sea trench in which Jurassic sedimentation was accompanied by emplacement of olistostrome sequences. Eastward, including Sakhalin Island, was an oceanic plate with oceanic-type sedimentation throughout the Jurassic (i.e., deposition of extremely thin siliceous and siliceous-terrigenous sediments).

Thus, the Jurassic of northeastern Asia included a large spectrum of paleogeographic settings.

The Hettangian and Sinemurian of the eastern Siberian Continent were characterized by a moderately rugged topography. Adjacent seas were mainly sublittoral (Verkhoyanye Shelf, Omolon and Kolyma Archipelagoes), with clayey-silty sediments. Similar sediments also deposited on the shelf east of the Bureya Land, which probably had slightly higher elevations.

The late Pliensbachian–early Toarcian transgression resulted from a eustatic rise in sea level on the Siberian Continent; the Vilyui Lowland was flooded. Around the periphery of the Bureya Land, including the Eastern Transbaikal area, a chain of shallow marginal seas was formed. An extensive basin of continental sedimentation appeared on the eastern Siberian Continent (Vilyui and South Yakutsk Lacustrine-Alluvial Plains). Simultaneously, the topography became more dissected in eroded areas. Generally, in the late Pliensbachian, middle and upper sublittoral, sandy-clayey sediments were deposited, sometimes with coarse-clastic rocks. In the early Toarcian, the development of transgression in the northeast was weakened by the growth of geoanticlinal structures, which resulted in the formation of a large peninsula in the Lena-Yana interfluve.

Throughout the Early Jurassic, in suboceanic basins and deep-sea trenches, thick, often flyschoid, sandy-silty-clayey and siliceous-clayey sequences are formed.

The late Toarcian in eastern Russia was a time of regional uplifts, reduction, and segregation of margin areas. At the beginning of the Middle Jurassic, a regional phase of folding occurred on the Alazeya Upland. Apparently, this event was caused by collision of the Chukotka Continent with the Alazeya Block. In adjacent areas this event is indicated by conglomerates.

A new transgressive cycle in the Aalenian–early Bajocian considerably revived the marine basins of northeastern Asia (Figure 11.14) and favored stabilization and leveling of the late Toarcian disturbances. In the late Bajocian, a progressive continentalization of the territory started; the area of shelf and epicontinental seas was reduced and shallowing proceeded; suboceanic basins became shallower and were filled by sediments.

The end of the Middle Jurassic is noted for a major geodynamic and paleogeographic restructuring. In North-East Russia, the major phase of folding was in the Bathonian. It was the result of complicated interactions between the Chukotka Continent, the Alazeya Block, the foreland of the Siberian Continent, and the Pacific plate with the Omolon Microcontinent. Thus, the Alazeya and Omolon Microcontinents formed the main structural constituents in this part of the region.

The subduction along the southern front of the Uda-Murgal Volcanic Arc and the southern boundary of the Siberian Continent in the Middle Jurassic resulted in the northward movement of the Bureya-Khanka Massif. Apparently, at the Middle–Late Jurassic boundary, the Amur-Okhotsk Oceanic Basin became closed, and most of its marine regime was eliminated, though the collisional processes doubtless occurred through the entire Late Jurassic.

In the Late Jurassic the process of continentalization of eastern Russia continued, complicated by marine transgressions in the Oxfordian–early Kimmeridgian and in the middle–late Volgian. The area of marine sedimentation was displaced toward the northern and eastern margins of the region (Figure 11.14). Uplifting was accompanied by a further rise of elevation and a more contrasting land topography; a system of molassoid depressions was emplaced; subaerial volcanic activity was intensified. In the Oxfordian–Kimmeridgian–early Volgian, volcanic belts were formed: Uyanda–Yasachnaya, in front of the Alazeya Block, and

Uda on the southern margin of the Siberian Continent within the Stanovoy Arched Uplift. Over all of northeastern Asia, coastal and lacustrine alluvial plains were developed, where coal-bearing deposits accumulated.

The shelf area in the Sikhote-Alin Oceanic Basin increased. However, the deep-sea trench, with thick terrigenous sedimentation, continued to exist there up to the end of the Neocomian, when, because of subduction, the oceanic plate was thrust over the terrigenous complexes of the paleotrench. In Sakhalin, the abyssal-plain setting was retained to the end of the Early Cretaceous. Probably at the close of the Jurassic, the Amur (Kiselev) Block, associated with the exotic Early Jurassic fauna, was accreted. The fauna included Tethyan or Subtethyan elements (*Juraphyllites, Ctenostreon,* numerous *Cardinia*) and had no equivalents among coeval communities of northeastern Asia.

The opposite direction in the movement of the Pacific plate and the Chukotka Continent, which continued in the Late Jurassic, promoted a further piling-up of blocks and microcontinents and ended in the collision of the Siberian Continent and Chukotka in the Neocomian.

Modern Koryakiya apparently is a microblock (terrane) mosaic, part of which is characterized by exotic (extra-Boreal) invertebrate communities. In the Norian of the Kenkeren Range, shallow-water faunas practically lack bivalve genera and species in common with similar fauna in the inner region of North-East Russia. *Costatoria, Cassianella, Septocardia, Pinna, and Maoritrigonia,* as well as the hermatypic coral *Astraemorpha,* were reported from there (Bychkov 1985). The Callovian sediments in the Koiverelen Basin contain representatives of the Tethyan ammonite genera *Lunuloceras, Zieteniceras,* and *Choffatia* (Sey and Kalacheva 1985). Accretion of Koryakiya probably took place in post-Jurassic time.

## References


Afitsky, A. I. (1970). *Biostratigraphy of Triassic and Jurassic Deposits of the Great Anuy River Basin.* (in Russian). Moscow: Nauka.

Byckhov, Y. M. (1985). Late Triassic mollusks of the Kenkeren Range (Koryak Upland). In *Mesozoic Bivalves and Cephalopods of Northeast USSR* (in Russian) (pp. 5–24). Magadan.

Callomon, J. H. (1984). A review of the biostratigraphy of the post–lower Bajocian Jurassic ammonites of western and northern North America. *Geol. Assoc. Canada Spec. Pap., 27,* 143–74.

Chudinov, Y. V. (1985). *Geology of Active Oceanic Margins and Global Tectonics* (in Russian). Moscow: Nedra.

Dagis, A. A. (1968). *Toarcian Ammonites (Dactylioceratidae) of North Siberia* (in Russian). Moscow: Nauka.

(1974). *Toarcian Ammonites (Hildoceratidae) of North Siberia* (in Russian). Novosibirsk: Nauka.

(1976). *Late Pliensbachian Ammonites (Amaltheidae) of North Siberia* (in Russian). Novosibirsk: Nauka.

Dagis, A. A., and Dagis, A. S. (1965). On zonal subdivision of Toarcian deposits in the North-East of USSR. In *Stratigraphy and Paleontology of Mesozoic Deposits of Siberia* (in Russian) (pp. 15–26). Moscow: Nauka.

Frebold, H. (1957). *Fauna, Age and Correlation of the Jurassic Rocks of Prince Patrik Island.* Geol. Surv. Canada Bull. 41.

(1960). *The Jurassic Faunas of the Canadian Arctic. Lower Jurassic and Lowermost Middle Jurassic Ammonites.* Geol. Surv. Canada Bull. 59.

Frebold, H., & Tipper, H. W. (1973). Upper Bajocian–Lower Bathonian ammonite fauna and stratigraphy of Smithers Area, British Columbia. *Can. J. Earth Sci., 10,* 1109–31.

*Geological Structure of the USSR and Distribution Pattern of Mineral Deposits. Vol. 8: East USSR* (1984) (in Russian). Leningrad: Nedra.

Gusev, G. S., Petrov, A. F., Fradkin, G. S., et al. (1985). *Structure and Evolution of the Earth's Crust in Yakutia.* Moscow: Nauka.

Hall, R. L. (1984). Lithostratigraphy and biostratigraphy of the Fernie Formation (Jurassic) in the Southern Canadian Rocky Mountains. *Can. Soc. Petrol. Geol. Mem., 9,* 233–47.

Karasik, A. M., Ustritsky, V. I., & Khramov, A. N. (1984). History of formation of the Arctic Ocean. In *Geology of the Arctic* (pp. 151–9).

Kazintsova, L. I. (1987). Late Cretaceous Radiolaria assemblages from siliceous rocks of the eastern Sakhalin Mountains (in Russian). *Ann. All-Union Paleontol. Soc., 30,* 67–76.

Khanchuk, A. I., Panchenko, I. V., & Kemkin, I. V. (1988). *Geodynamic Evolution of Sikhote-Alin and Sakhalin in Paleozoic and Mesozoic* (in Russian). Vladivostok: Nauka.

Kiparisova, L. D. (1952). *New Lower Jurassic Fauna of Priamurye* (in Russian). Moscow: Gosgeoisdat.

Kirillova, G. L., & Turbin, M. T. (1979). *Formations and Tectonic of Dzhagdy Part of Mongol Okhotsk Folded Area* (in Russian). Moscow: Nauka.

Krymholts, G. Y., Meseshnikov, M. S., & Westermann, G. E. G. (eds.). (1988). *The Jurassic Ammonite Zones of the Soviet Union.* Geol. Soc. Am., Special Paper 223.

Mazarovich, A. O. (1985). Tectonic development of South Primorye in Paleozoic and Early Mesozoic (in Russian). *GIN Trans., 392.*

Milova, L. V. (1980). Reference section of Pliensbachian deposits of the Viliga Basin (Pryochotie). In *Biostratigraphy and Correlation of Mesozoic Deposits of the North-East of the USSR* (in Russian) (pp. 28–46). Magadan.

Nekrasov, G. E. (1976). *Tectonic and Magmatism of the Taigonos and the North-West Kamchatka* (in Russian). Moscow: Nauka.

Okuneva, T. M. (1973). Stratigraphy of the marine Jurassic deposits of Eastern Transbaical area and its paleontological grounds (in Russian). *VSEGEI Trans., 219,* 3–117.

Parfenov, L. M. (1984). *Mesozoic Continental Margin and Island Arcs of North-East Asia* (in Russian). Novosibizsk: Nauka.

Polubotko, I. V., & Repin, Y. S. (1966). Stratigraphy and ammonites of Toarcian stage of the central part of Omolon Massif. In *Materials on Geology and Mineral Resources of the North-East USSR, Issue 19* (in Russian) (pp. 30–55). Magadan.

(1972). Ammonites and zonal subdivision of the Lower Lias of the North-East USSR. In *Materials on Geology and Mineral Resources of the North-East USSR, 20* (in Russian) (pp. 97–116). Magadan.

(1974a). Biostratigraphy of Aalenian stage of the North-East USSR. In *Biostratigraphy of the Boreal Mesozoic* (in Russian). *Trans. Inst. Geol. Geophys., USSR Acad. Sci., 136,* 91–101.

(1974b). Biostratigraphy of the Lower Jurassic deposits of the North-East USSR. In *The Main Problems of Biostratigraphy and Paleogeography of the North-East USSR* (in Russian) (pp. 68–89). Magadan.

(1981). On establishment of new ammonite zone at the base of Jurassic system (in Russian). *Rep. Acad. Sci. USSR, 261*(N6), 1394–8.

(1988). Lower and Middle Jurassic of the North-East Soviet Union. *Newsl. Stratig., 19*(½), 1–17.

Repin, Y. S. (1972). Bajocian ammonites of the North-East USSR. In *Materials on Geology and Mineral Resources of the North-East USSR, 20* (in Russian) (pp. 117–24). Magadan.

(1975). South-East Prykolymie and North Pryochotie at Triassic and Jurassic. In *Mesozoicum of the North-East USSR* (in Russian) (pp. 28–30). Magadan.

*Resolutions of the 2nd Interdepartmental Regional Stratigraphic Meeting on the Precambrian and Phanerozoic of the North-East of the USSR* (1978) (in Russian). Magadan.

*Resolutions of the 3rd Interdepartmental Regional Stratigraphic Meeting on the Precambrian and Phanerozoic of the Far East of the USSR* (1982) (in Russian). Magadan.

Rikhter, A. V. (1986). Structure and tectonic development of the Sakhalin in Mesozoic (in Russian). *GIN Trans., 411.*

Sey, I. I., Brudnitskaya, E. P., Kalacheva, E. D., et al. (1984). New Lower and Middle Jurassic subdivisions for Bureinsky Depression (in Russian). *Geol. Geophys., 6,* 28–37.

Sey, I. I., & Kalacheva, E. D. (1980). Biostratigraphy of the Lower and Middle Jurassic deposits of the Far East (in Russian). *VSEGEI Trans., N.S., 285.*

(1983). On invasions of Tethyan ammonites into Late Jurassic Boreal basins of the East USSR. In *Mesozoic of the Soviet Arctic* (in Russian). *Trans. Inst. Geol. Geophys., Sib. Branch, USSR Acad. Sci., 555,* 61–72.

(1985). Biostratigraphic chart for the Upper Jurassic marine deposits of the northern Far East of USSR (in Russian). *Geol. Geophys., 5,* 136–8.

(1987). The problem of Bajocian and Bathonian stages in the east and in the north of the USSR (in Russian). *Sov. Geol., 4,* 51–7.

(1988). Ammonites and bivalves of the Far East, Soviet Union. *Newsl. Stratig., 19*(½), 35–60.

Stavsky, A. P. (1988). Accretional tectonics of the Argatas zone (northeast USSR) (in Russian). *Geotectonics, 2,* 84–91.

Tikhomirova, L. B. (1986). Jurassic radiolarians of the Far East (in Russian). *Proc. Acad. Sci. USSR, Geol. Ser., 9,* 123–6.

Westermann, G. E. G. (1964). The ammonite fauna of the Kialagvik Formation at Wide Bay, Alaska Peninsula. I: Lower Bajocian (Aalenian). *Bull. Am. Paleontol., 47*(216), 329–503.

(1969). The ammonite fauna of the Kialagvik Formation at Wide Bay, Alaska Peninsula. II: *Sonninia sowerbyi* Zone (Bajocian). *Bull. Am. Paleontol., 57*(255).

Zonenschain, L. P., Kuzmin, M. I., & Natapov, L. M. (1987). Phanerozoic palinspastic reconstruction in the USSR territory (in Russian). *Geotectonics, 6,* 3–19.

# Part IV: Biochronology

## 12 Ammonite zones of the circum-Pacific region

A. VON HILLEBRANDT, P. SMITH, G. E. G. WESTERMANN, and J. H. CALLOMON

LOWER JURASSIC[1]

### Introduction

In recent years, many publications on Lower Jurassic ammonites from the western side of the Pacific Ocean have been published, and together with earlier publications they provide a much more complete understanding of regional biostratigraphy (Table 12.1). Various Russian authors have described Lower Jurassic ammonites from eastern and northeastern Asia, and a detailed biozonation has been developed based, for the most part, on ammonites; for summaries, see Repin (1984) and Krymholts, Mesezhnikov, and Westermann (1988). Lower Jurassic ammonites are rare in most areas in Japan, and only a few papers have been published since the synthesis by Arkell (1956). A summary, regional zonation for part of the Lower Jurassic, along with range charts, has been provided by Hirano (1973, 1985) and Sato and Westermann (1985, 1991). Lower Jurassic ammonites from the Pacific border of southern China are mainly known from the Guangdong province, but only three Sinemurian assemblages have been established (Wang and Smith 1986). In Indochina, Lower Jurassic ammonites are mainly represented in Vietnam. All stages have been proved, but a detailed biozonation is not yet possible (Vu Khuc 1984). From Thailand (Mae Sot region), only late Toarcian ammonites are known (Braun and Jordan 1976). No Lower Jurassic ammonites have been described from the Malay Peninsula, the Philippines, or Sumatra. Some ammonites of Sinemurian and Toarcian age were found in Borneo (Krause 1896, 1911; Hirano et al. 1981).

Lower Jurassic ammonites are known from various Indonesian islands, but there have been few publications since that by Arkell (1956, p. 441), which gave a summary of known stages, genera, and species. The most complete collections were described from Timor and Rotti, but without sections. Two ammonites of doubtful generic affinity were reported from New Guinea (Skwarko 1967), supposedly of late Sinemurian and early Pliensbachian age. New Caledonia and New Zealand are said to have similar faunas (Stevens 1977), but ammonites are rare and have not allowed a detailed biozonation. The Pliensbachian has not yet been proved by ammonites in either country.

Since Arkell's summary (1956), many monographs and papers dealing with Lower Jurassic ammonites of the eastern border of the Pacific Ocean have been published. In North America this progress is mainly recorded in the publications of Frebold and Imlay, completed and extended in recent years by different authors. For some regions a detailed biozonation based on ammonites typical for North America is now possible, especially the Pliensbachian of the allochthonous terranes (Smith et al. 1988). In the cratonic region the biostratigraphic record of ammonites is mostly less complete (Hall 1984). In eastern Mexico, a detailed Sinemurian ammonite succession is known; it was first recorded by Burckhardt (1930), described and figured by Erben (1956), and biostratigraphically revised by Schiatter and Schmidt-Effing (1984).

In the past 20 years, some monographs and many papers on Lower Jurassic ammonites of different countries of the Andes have been published. From Colombia and Ecuador only Sinemurian ammonites have been described (Geyer 1973, 1974, 1976). The most complete ammonite successions are found in Peru, Chile, and Argentina where a detailed biozonation, partly based on endemic East Pacific ammonites, is now possible (Hillebrandt 1987, 1988, 1990a; Riccardi et al. 1988).

From the Antarctic Peninsula, the first Lower Jurassic ammonite has recently been described (Thomson and Trantner 1986).

In various regions of the Pacific Ocean, Lower Jurassic ammonite zones often have different definitions in the sense of chronostratigraphic units (assemblage, range, etc., zones). European standard zones are used when correspondence with the Pacific area is possible, or if the fossil record or taxonomy is not yet good enough for a local biozonation. Continuous successions from one zone to another are rarely found, and thus range zones probably are often "artificial."

See Appendix, Plates 1–91.

[1] By A. von Hillebrandt and P. Smith.

Table 12.1. *Lower Jurassic ammonite zones of the circum-Pacific region.*

| Stage | Substage | NW – Europe: Zones | NW – Europe: Subzones | Mediterranean (Braga et al. 1984 (a,b); Elmi et al. 1974) | NE-Asia (Repin 1988; Kalacheva 1988) | Japan (Sato & Westermann 1985) | S-China (Wang & Smith 1986) | Canada & USA (Smith et al., 1988; Frebold & Poulton 1977; Guex 1987) | Cratonal Canada (Hall 1984, 1987) | South America (v. Hillebrandt 1987, 1988 and in press) |
|---|---|---|---|---|---|---|---|---|---|---|
| Toarcian | U | LEVESQUEI | AALENSIS<br>MOOREI<br>LEVESQUEI<br>DISPANSUM | P. aalensis<br>D. meneghinii<br>H. insigne | Pseudolioceras rosenkrantzi | H. ikianus Ass.Z. | | not yet zoned | | "P. fluitans"<br>"P. lotharingica"<br>P. tenuicostatum |
| | | THOUARSENSE | FALLACIOSUM<br>THOUARSENSE | G. thouarsense | | | | | | P. copiapense |
| | M | VARIABILIS | | B. alticarinata<br>B. gradata<br>C. gemma | | Dactylioceras helianthoides Ass.Zone | | | TO 3 | P. toroense<br>C. chil. {P. moerickei, P. bolitoens.} |
| | | BIFRONS | CRASSUM<br>FIBULATUM<br>COMMUNE | H. semipolitum<br>H. bifrons<br>H. lusitanicum<br>H. sublevisoni | P. polare<br>Z. monestieri<br>D. athleticum | | | | TO 2 | P. pacificum<br>P. largaense |
| | L | FALCIFERUM | FALCIFERUM<br>EXARATUM | H. serpentinus | H. falciferum | | | | TO 1 | D. hoelderi |
| | | TENUI-COSTATUM | SEMICELATUM<br>TENUICOSTATUM<br>CLEVELANDICUM<br>PALTUM | D. tenuicostatum | Tiltoniceras propinquum | Protogrammocer. nipponicum Ass.Zone | | | | D. tenuicostatum<br>D. simplex |
| Pliensbachian | U | SPINATUM | HAWSKERENSE<br>APYRENUM | E. emaciatum<br>? | A. viligaensis | Fontanelliceras fontanellence Ass.Zone | | F. carlottense | | F. disciforme |
| | | MARGARI-TATUS | GIBBOSUS<br>SUBNODOSUS<br>STOKESI | A. algovianum<br>F. lavinianium | A. talrosei<br>A. stokesi | | | Fanninoceras kunae | P 2 | F. fannini |
| | L | DAVOEI | FIGULINUM<br>CAPRICORNU<br>MACULATUM | P. dilectum | Poly-morphites | | | Dubariceras freboldi | | F. behrendseni |
| | | IBEX | LURIDUM<br>VALDANI | ? | | | | A. whiteavesi | | E. meridianus<br>"Tropidoceras" |
| | | JAMESONI | MASSEANUM<br>JAMESONI<br>BREVISPINA<br>POLYMORPHUS<br>TAYLORI | T. demonense<br>G. aenigmatum | | | | Pseudoskirroce-ras imlayi | P 1 | "Eoderoceras" + "Apoderoceras" |
| Sinemurian | U | RARI-COSTATUM | APLANATUM<br>MACDONELLI<br>RARICOSTATOIDES<br>DELICATUM | I | Angulaticeras kolymicum | | | not yet zoned | S 3 | "E. raricostatum" |
| | | OXYNOTUM | OXYNOTUM<br>SIMPSONI | | | | | | | "O. oxynotum" |
| | | OBTUSUM | DENOTATUS<br>STELLARE<br>OBTUSUM | H | Coroniceras siverti | | Assemblage 3 | | S 2 | "A. obtusum" |
| | L | TURNERI | BIRCHII<br>BROOKI | G | | | | | | "C. turneri" |
| | | SEMI-COSTATUM | SAUZEANUM<br>SCIPIONIANUM<br>REYNESI | F | | | Assemblage 2 | | S 1 | A. semicostatum |
| | | BUCKLANDI | BUCKLANDI<br>ROTIFORME<br>CONYBEARI | E<br>S. marmorea | A. bucklandi | | Assemblage 1 | | | "A. bucklandi" |
| Hettangian | U | ANGULATA | DEPRESSA<br>COMPLANATA<br>EXTRANODOSA | | S. angulata | | | B. canadense | | "S. angulata" |
| | M | LIASICUS | LAQUEUS<br>PORTLOCKI | K. megastoma | A. liasicus | | | not yet zoned | | D. reissi<br>"Curviceras" ssp. |
| | L | PLANORBIS | JOHNSTONI<br>PLICATULUM<br>PLANORBIS | P. calliphyllum | P. planorbis<br>P. primulum | | | | H 1 | P. rectocostatum<br>P. primocostatum<br>P. tilmanni |

## Hettangian

Hettangian ammonite successions with different biozones are known from northeastern Asia (Repin 1984), North America (Frebold 1967; Guex 1980), and South America (Riccardi et al. 1988; Hillebrandt 1988, 1990a,b). Only one Middle or Late Hettangian ammonite is known from Japan (i.e., *Yebisites onoderai*) (Matsumoto 1956). "*Psiloceras longipontinum* Oppel" from Vietnam (Counillon 1908) is said to be a Late Sinemurian *Xipheroceras dudressieri* (Orb.) (Vu Khuc 1984). Vu Khuc (1984) mentions the Middle and/or Late Hettangian ammonites *Laqueoceras* (= *Alsatites*) and *Saxoceras* from Vietnam. Some ammonites described from Timor and Babar are at least partly of Hettangian age (Arkell 1956, p. 441). Hettangian ammonites figured from New Caledonia (Avias 1953) are Middle and Late Hettangian. Similar genera and species were described from New Zealand (Stevens 1977; Suggate, Stevens, and Te Punga 1978). Reviews of Hettangian ammonites known from Canada and Alaska were given by Frebold and Poulton (1977) and Imlay (1981). Continuous sections with uppermost Triassic ("Rhaetian") *Choristoceras* and Lower Hettangian *Psiloceras* are known only from Nevada (Guex 1980), British Columbia (Frebold 1967), Peru (Prinz 1985a,b; Hillebrandt in press), and Chile (Hillebrandt 1990a). Recently, Middle and Late Hettangian ammonites have also been discovered in Argentina (Riccardi et al. 1988; Hillebrandt 1990b).

In northeastern Asia the Hettangian (see Plate 15 in the Appendix) was subdivided into four zones (Repin 1984). Polubotko and Repin (1981) established the *Primapsiloceras primulum* Zone, which is said to be older than the Planorbis Zone (Repin 1984). Unfortunately, no older ammonites are found below the *P. primulum* Zone. Guex (1987, p. 460) believes that *Primapsiloceras primulum* is younger than the oldest real *Psiloceras* and belongs to the Planorbis Subzone. For the upper three zones, names of the European standard zonation were used.

Above the *P. primulum* Zone, an ammonite assemblage follows that is characterized by "*Psiloceras viligense* and *P. suberugatum*" Chud. and Polub. spp. This assemblage is correlated with the Planorbis Subzone of Europe. Guex (1987, p. 462) states that both species are intermediate between *Psiloceras* s. str. and the Pleuroacanthidae *Pleuroacanthites* and *Euphyllites.* Those two genera are not known from the earliest Hettangian, but they exist up to the Late Hettangian and are also found in Middle and Upper Hettangian strata of Chile.

In North America (Plates 1, 5, 6, 15), a regional biozonation is in preparation by P. Smith. The first well-preserved ammonite above the last "Rhaetian" *Choristoceras* in Nevada (New York Canyon) is *Psiloceras pacificum* Guex (1980, 1982). This species is synonymous with *P. tilmanni* Lange, first described from Peru (Tilmann 1917), and is found above beds with *Choristoceras* (Prinz 1985b; Hillebrandt in press).

Hillebrandt (1988, 1990a) established in South America five Early and Middle Hettangian zones characterized by endemic species. The Late Hettangian European Angulata Zone was recently subdivided (Hillebrandt in press) in Peru and Chile, by means of species of the genus *Schlotheimia,* into two subzones. Many Hettangian genera and sometimes also species found around the Pacific are very similar to those described from Europe, especially the Mediterranean region, but the European forms frequently have much more complicated septal sutures. The significance of this phenomenon is not well understood and is a subject of some controversy (G. Bloos personal communication; Guex 1982). The definitions of some Hettangian ammonite genera and their phylogenetic relationships are still in flux, and genera are often used inconsistently.

The South American (Plate 8) *Psiloceras tilmanni* Zone (1) is coeval with the European Planorbis Subzone. An exact comparison between *P. planorbis* and *P. tilmanni* is not possible, because the first species is always crushed, so that cross section and septal suture are unknown. The *Psiloceras primocostatum* Zone (2) (Hillebrandt 1988) is characterized by the *P. psilonotum–calliphyllum* group. A specimen of this group was also figured by Tozer (1982) from Canada. This zone is coeval with the upper part of the Planorbis Subzone s. str. or the lower part of the Plicatulum Subzone in Europe. The Andean *Psiloceras rectocostatum* Zone (3) is characterized by ammonites similar to those found in Europe in the upper part of the Plicatulum Subzone or in the lower part of the Johnstoni Subzone. *Psiloceras polymorphum* Guex (1980) from Nevada probably also belongs to this species group. The next zone (4) (Hillebrandt 1988) was said to be characterized by *Caloceras peruvianum* Lange, which was described from northern Peru. Recent investigations at the type area of this species (Hillebrandt in press) have proved that *C. peruvianum* is of Late Hettangian age and belongs to the genus *Sunrisites.* In northern Chile above the *P. rectocostatum* Zone (3), an ammonite zone (4) is found which yielded specimens similar to "*C.*" *peruvianum* and which is characterized by the first species of *Curviceras* [e.g., a form similar to *C. subangulare* (Oppel), and a species nearly identical with *Curviceras*(?) *armanense* (Repin)]. The latter species was first described form northeastern Asia. This zone with *Curviceras* ssp. (Hillebrandt in press) may be correlated with the upper part of the Johnstoni Subzone or the lowest part of the Liasicus Zone in Europe. The *Discamphiceras*(?) *reissi* Zone (5) [for "*Badouxia*(?) *reissi* Zone" of Hillebrandt (1988)] is characterized by ammonites with a test morphology similar to those of some Alpine and Mediterranean species of *Discamphiceras* and *Kammerkarites,* but with simpler septal sutures. Similar species are also found in North America and northeastern Asia. This zone may be coeval with the European Liasicus Zone, as also suggested by Repin (1984), who subdivided this zone into two subzones.

The European Upper Hettangian is characterized and subdivided by species of *Schlotheimia.* This genus is also found in North America and South America. The lower part of the Angulata Zone–at least in South America–is dominated by *Schlotheimia* cf. *montana* (Wähner), and the upper part by species of the "*Schlotheimia marmorea* group" (Hillebrandt in press). In North and South America, *Schlotheimia* occurs with species of *Sunrisites* first described from North America (Guex 1980) and also known from the northern Alps. At some localities, *Sunrisites* and giant psiloceratids are more frequent than *Schlotheimia. Badouxia* occurs together with *Schlotheimia* cf. *montana* and "*Schlotheimia*" of the "*Sch. marmorea* group." *Badouxia* was first described from North America (Guex and Taylor 1976), is also found in South America, and was recorded from northeastern Asia (Repin 1984, p. 3). A fauna with *Badouxia* cf. *canadensis, Alsatites* cf. *platystoma,* and *Ectocentrites* cf. *petersi* was described by Hillebrandt (1981) and dated as Middle Hettangian. Since then, the same horizon has yielded *Schlotheimia* ex gr. *marmorea* and *Schlotheimia* cf. *montana* at its basis. Similar *Alsatites* s.l. (including *Paracaloceras* and *Alpinoceras*) together with *Badouxia* cf. *canadensis* and *Schlotheimia* ex gr. *marmorea* have recently been found in Argentina, where these beds lie above a horizon with *Schlotheimia* s. str. On that basis the North American *B. canadensis* Zone has also been recognized in Argentina (Riccardi et al. 1988; Hillebrandt 1990b).

The *Badouxia canadensis* Zone (Taylor 1986) is correlated with the upper part of the *marmorea* Zone of the northern Alps and the Conybeari Subzone, the first subzone of the Sinemurian in northwestern Europe. The probably condensed *marmorea* Zone includes the northwestern European Extranodosa, Complanata, and Depressa Subzones and possibly the Conybeari Subzone. The Depressa Subzone was considered coeval with the Conybeari Subzone by Taylor (1986), but *Schlotheimia depressa* was never found in horizons comparable to the Conybeari Subzone; the Depressa Subzone is, at least momentary, regarded as an independent subzone between the Complanata and the Conybeari Subzones by Bloos (1983), 1988a,b).

### Sinemurian

Ammonites of Sinemurian age are distributed more widely around the Pacific than are those of the Hettangian. The Arietitidae are most important for Lower and Middle Sinemurian biostratigraphy, but in the Mediterranean and northwestern Europe this family already appears in the Upper Hettangian. An exact correlation with Europe is therefore possible only at the species level. Furthermore, the generic classification of the Arietitidae is not uniform among authors. Late Sinemurian biostratigraphy is mostly based on Echioceratidae, which evolved from Arietitidae in middle Sinemurian time (Getty 1973). In most regions, Oxynoticeratidae and Eoderoceratidae are less frequent but still important for correlation. Thus far, a detailed subdivision in zones and subzones as in northwestern Europe has not been possible in any Pacific region. On the western side of the Pacific a subdivision of the stage was possible in northeastern Asia (Repin 1984) and southern China (Wang and Smith 1986). Sinemurian ammonites were recorded from Japan (Sato and Westermann 1985), Vietnam (Vu Khuc 1984), various islands in Indonesia (Arkell 1956), New Caledonia (Avias 1953), and New Zealand (Stevens 1978b). The North American Sinemurian has not yet been subdivided into zones. A subdivision into various ammonite horizons was possible in Alaska (Imlay 1981), the Canadian Arctic (Frebold 1975), British Columbia (Frebold 1964; Cameron and Tipper 1985), Alberta (Hall 1984), and Nevada (Hallam 1965). In eastern Mexico a thick series of Sinemurian sediments is exposed, and numerous ammonite horizons have been recognized (Schlatter and Schmidt-Effing 1984). In Colombia and Ecuador, equivalents of only the European Semicostatum Zone and its subzones could be proved (Geyer 1973, 1974). Biozones of northern and central Peru were published by Prinz (1983, 1985a,b). In northern and central Chile, most European zones and some subzones were found (Hillebrandt 1987). A detailed zonation of the Lower Sinemurian and lower Upper Sinemurian was given by Quinzio (1987). In Argentina, ammonites of Sinemurian age occur only in the Mendoza province (Riccardi et al. 1988; Hillebrandt 1989, 1990b).

The base of the European Sinemurian (Conybeari Subzone) is defined by species of *Metophioceras* and the very similar genus *Vermiceras*. Specimens similar to *M. conybeari* from Chile were described by Quinzio (1987) (Plate 9). *M.* cf. *conybeari* was also recorded from the lower part of the northeastern Asian Bucklandi Zone (Repin 1984). The stratigraphic position of the earliest arietitids in Canada is controversial (Plate 2) (Guex and Taylor 1976; Bloos 1983, 1988a; Taylor 1986). Species of *Coroniceras* similar to those of the European Rotiforme Subzone are found in Chile (Plate 9) (Hillebrandt 1987; Quinzio 1987), North America (Frebold 1975), and southern China (Wang and Smith 1986). *Arietites libratus* (Repin) (Plate 16) from northeastern Asia is said to belong to the species group of *A. bucklandi* (Repin 1984).

The cosmopolitan genus *Arnioceras*, described from many Pacific regions, often is the most frequent and sometimes the only Sinemurian ammonite known (Plates 2, 5, 9, 14). It is however, long-ranging, at least from the upper Lower Sinemurian to the lower Upper Sinemurian. In South America, the last *Coroniceras* s.l. or related genera – partly also gigantic – are found together with the first *Arnioceras*. This assemblage probably is also found in North America and may be coeval with the European Bucklandi or Reynesi Subzones. In northeastern Asia, *Arnioceras* is unknown, and the *Coroniceras siverti* Zone comprises three ammonite assemblages (Repin 1984) coeval with at least three European zones (Plate 16). Almost the same interval was subdivided into three assemblages in southern China (Wang and Smith 1986). Contrary to the case of northeastern Asia, *Arnioceras* is the most frequent genus in this region. The different species of *Arnioceras*, in combination with other genera, provide, at least for South America, a detailed biozonation (Geyer 1973; Quinzio 1987). Different European species are found, and the Mediterranean species *A. ceratitoides* is widely distributed on both sides of the Pacific.

In the entire Pacific area, including regions promising more detailed biozonation, direct proof of the European Turneri Zone is difficult. The Obtusum Zone is subdivided (three subzones) by means of species of *Asteroceras* and *Eparietites*. *Epophioceras* is also important (Prinz 1983) and thus far is the only Liassic genus found on the Antarctic Peninsula (Thomson and Trantner 1986). Different species of these genera resembling European forms have been figured from South America, North America, and southern China. In South America, the Obtusum Zone can be subdivided into two or three subzones (Quinzio 1987). In northeastern Asia, *Eparietites* cf. *denotatus* was reported from the upper *Coroniceras siverti* Zone (Repin 1984).

The European Oxynotum Zone is characterized by species of *Oxynoticeras*. This genus and other Oxynoticeratidae (i.e., *Gleviceras* and *Cheltonia*) occur in different Pacific regions, but in South America, for example, they are less frequent than the first Echioceratidae (Plate 10). *O. oxynotum* was described by Frebold (1960) from the Canadian Arctic together with *Arctoasteroceras* [= *?Aegasteroceras*] (Plate 5) (cf. Hallam 1965; Imlay 1981). *Arctoasteroceras* occurs in Alaska (Imlay 1981) together with "*Paltechioceras* (*Orthechioceras*?) sp.," a possible *Epophioceras*. The age of this assemblage probably is Obtusum Zone, rather than latest Sinemurian (Imlay 1981).

The last zone of the European Sinemurian is characterized by Echioceratidae. *Echioceras* (= *Plesechioceras*) *arcticum* Frebold (1975) occurs in South America (Hillebrandt 1987) and probably is typical for the lower Raricostatum Zone. *Paltechioceras* was described from Mexico (Schmidt-Effing 1980) and South America (Hillebrandt 1987). Echioceratids also occur in the Indonesian islands of Rotti, Jamdena, and Timor (Jaworski 1933) and are found in North America below the first appearance of Pliensbachian ammonites (Smith et al. 1988).

In northeastern Asia the *Angulaticeras kolymicum* Zone was correlated with the European Oxynotum and Raricostatum Zones (Plate 16). In Europe the genus *Angulaticeras* ranges from Bucklandi to Oxynotum Zones. Species very similar to *A. kolymicum* occur in South America (Prinz 1985b; Hillebrandt 1987; Quinzio 1987), mainly in beds coeval with the Bucklandi Zone of Europe.

Eoderoceratidae are sometimes important for correlations with Europe, but too rare in the Pacific area to establish biozonations.

### Pliensbachian

Pliensbachian ammonites are known from many regions around the Pacific Ocean, but correlation is often difficult because pandemic genera and species are less frequent. In the Upper Pliensbachian, a differentiation of ammonites into realms and subrealms (Boreal, Tethyan, and East Pacific) is observed for the first time (Taylor et al. 1984) (Chapter 24).

In northeastern Asia the Lower Pliensbachian is practically devoid of ammonites, except for rare *Polymorphites* (Repin 1984). The Upper Pliensbachian is known also from eastern Asia, where it is characterized and subdivided by the Boreal genus *Amaltheus*, with many endemic species (Plate 17). Tethyan *Arieticeras, Fontanelliceras, Protogrammoceras,* and "*Dactylioceras*" are associated with *Amaltheus* in Far East Russia (Sey and Kalacheva 1980).

In Japan, the Lower Pliensbachian is indicated by only one specimen of *Eoderoceras?* sp. (Sato and Westermann 1985). The Upper Pliensbachian is represented by the *Fontanelliceras fontanellense* Zone and the lower part of the *Protogrammoceras nipponicum* Zone (Plate 14) (Hirano 1973). Both zones are characterized by Tethyan ammonites and some Boreal elements that are important for the correlation between realms. The Tethyan *Paltarpites* and *Arieticeras* were recorded from Vietnam (Vu Khuc 1984). Pliensbachian ammonites were figured from the Indonesian islands of Rotti and Timor (Krumbeck 1922, 1923; cf. Arkell 1956, p. 441). Only "*Uptonia*" and *Tropidoceras* are restricted to the Lower Pliensbachian; *Phricodoceras subtaylori* and *Liparoceras* range into the Margaritatus Zone (Wiedenmayer 1980). No detailed biozonation can be recognized based on species of these genera. *Tropidoceras?* sp. (Lower Pliensbachian) occurs also in New Guinea (Skwarko 1967). Pliensbachian ammonites are unknown from New Caledonia and New Zealand.

In North and South American, Pliensbachian ammonites often are more frequent than those of the Upper Sinemurian. They occur in South America mainly in Chile and Argentina, less frequently in Peru (Hillebrandt 1987). In North America a detailed biozonation has been worked out (Smith et al. 1988), and, significantly, the associations include species typical for the East Pacific and the Mediterranean, as well as for northeastern and northwestern Europe.

Eoderoceratidae and Acanthopleuroceratinae are the most important zonal markers for the Lower Pliensbachian of North and South America (Plates 3, 11). The East Pacific genus *Fanninoceras* is important for the Upper Pliensbachian. It evolved from Eoderoceratidae in the upper part of the Lower Pliensbachian, and transitional forms occur in South America [cf. *Galaticeras*(?) of Hillebrandt (1987)]. The lower stage boundary in the Americas is drawn above the last Echioceratidae. The North American Imlayi Zone (Smith et al. 1988) is characterized by *Pseudoskirroceras imlayi,* which is closely related to species of the genus *Miltoceras.* Similar specimens were figured from South America as *Apoderoceras (Miltoceras)* cf. *sellae* and occur together with species of *Eoderoceras* in the "*Apoderoceras* and *Eoderoceras* Zone" (Hillebrandt 1987). *Acanthopleuroceras whiteavesi* is confined to the North American Whiteavesi Zone (Smith et al. 1988). This species is found together with *Tropidoceras* spp. The South American "*Tropidoceras* Zone" (Hillebrandt 1987) includes species that could belong to *Acanthopleuroceras* (e.g., *T.* cf. *stahli*). A species closely resembling *A. whiteavesi* occurs in the upper part of this zone, which therefore is coeval with the uppermost Imlayi Zone and the (lower) Whiteavesi Zone. In North America, *Dubariceras freboldi* marks the base of the Freboldi Zone (Smith et al. 1988) and is associated with the closely related *D. silviesi,* which extends from the Whiteavesi Zone into the lower Freboldi Zone. Different species of *Metaderoceras* range from the Imlayi to the Freboldi Zone. The *Eoamaltheus meridianus* Zone of South America includes ammonites resembling various species of *Uptonia* and *Dayiceras* (Hillebrandt 1987). These forms are partly transitional from *Metaderoceras* to *Dubariceras* and partly nearly identical with *Dubariceras silviesi.* Forms transitional to *D. freboldi* exist also. Specimens very similar to *Metaderoceras* aff. *muticum* (Smith et al. 1988) are also found. *Metaderoceras* gr. *evolutum-venarense-uhliqi* appear in the lower part of the zone (Hillebrandt 1981). The *meridianus* Zone is coeval with the uppermost Whiteavesi Zone and most of the Freboldi Zone of North America.

In the Americas, many genera of the Lower Pliensbachian of northwestern Europe are missing, and the succession is similar to that of the Mediterranean region (Plates 3, 11). The appearance of the East Pacific genus *Fanninoceras* is used in North America as base for the Kunae Zone. *F. fannini* is the first species appearing and overlaps with *F. kunae* and *F. latum.* Between the South American zones of *E. meridianus* and *F. fannini* is found the zone of *Fanninoceras behrendseni.* Transitional forms between *Eoamaltheus* and *F. behrendseni* [e.g., *Galaticeras*(?) of Hillebrandt (1987) and additional forms of Hillebrandt (1990b)] appear between the *meridianus* and *behrendseni* Zones. *F. behrendseni* and *F. fannini* are also connected by transitional forms.

The Carlottense Zone, the uppermost standard zone of the North American Pliensbachian, includes similarly highly involute species of *Fanninoceras* as in the South American *Fanninoceras disciforme* Zone. Direct correlation of the Upper Pliensbachian with the Boreal Realm is possible only in Canada, where species of *Amaltheus* occur together with *Fanninoceras* (Plate 5). The index species of the northeastern European Stokesi Subzone is reported from the lower part of the Kunae Zone, before the first appearance of *F. fannini* (Smith et al. 1988). *A.* (*Nordamaltheus*) *viligaense* Repin, described from the lower part of the Carlottense Zone, is the index for the *A.* (*N.*) *viligaense* Zone in northeastern Asia (Plate 17). *Tiltoniceras propinquum* ranges in North America from the Carlottense Zone to the lower part of the Lower Toarcian; in northeastern Asia it is the index of the *propinquum* Zone, which below contains the first northeastern Asian Dactylioceratidae, that is, *Kedonoceras* [= *Dactylioceras (Orthodactylites),* cf. Howarth (1973)] and is considered the first zone of the Toarcian (Plate 18).

In Nevada, dactylioceratids are similar to species of *Nodicoeloceras* from the basal Toarcian of Chile and "*Kedonoceras*" of northeastern Asia (Smith et al. 1988). *Dactylioceras simplex* and *D. polymorphum* occur in the Upper Pliensbachian of Far East Russia (Sey and Kalacheva 1980). Both species in Europe are typical for the lowermost Toarcian, and *D. simplex* was also figured from South America (Hillebrandt and Schmidt-Effing 1981). Pliensbachian ammonites in the Canadian Arctic are low in diversity and abundance (Frebold 1975; Hall 1984, 1987). They resemble species from northwestern Europe, and North American endemics appear to be absent.

Different genera and species of typical Mediterranean Hildoceratidae are found in the Upper Pliensbachian of North and South America. They are mostly less frequent than *Fanninoceras*, but useful for correlations with the Tethyan Realm. Species of this family also permit correlation with the Japanese biozonation. Pliensbachian Dactylioceratidae (*sensu* Dommergues 1986) are more frequent in North America than in South America and are likewise important for the correlation with Europe, mainly the Tethyan Realm.

The Upper Pliensbachian has been recorded from western Mexico with *Arieticeras guerrense* (Erben 1954).

### Toarcian

Provincialism is less strongly marked than in Pliensbachian time, and some genera and species are widespread. Toarcian (mainly lower) ammonites are distributed more widely on the western side of the Pacific than are those of Pliensbachian age, and a detailed ammonite zonation was recognized in northeastern Asia (Repin 1984). Most Japanese Lower Jurassic ammonites are of Toarcian age, but many species are said to have a longer range than in other regions, and the zones are coeval with several zones of more detailed biozonations. Only a few early Toarcian ammonites were figured from Borneo (Hirano et al. 1981), Rotti (Jaworski 1933), and New Zealand (Stevens 1978b). Late (in part also middle) Toarcian ammonites without detailed stratigraphy were described from Vietnam (Sato 1972; Vu Khuc 1984), Thailand (Braun and Jordan 1976), Borneo (Sato 1975), the Sula Islands (Kruizinga 1926; Sato et al. 1978), the Misol Archipelago (Soergel 1913), and Timor and Rotti (Krumbeck 1922, 1923; Arkell 1956, p. 440).

On the eastern side of the Pacific Ocean, Toarcian ammonites occur mainly in Canada, Alaska, southern Peru, Chile, and Argentina. The Arctic Toarcian is subdivided into four ammonite assemblages (Frebold 1975). Lower, Middle, and Upper Toarcian ammonites were reported from south-central Alaska and the Canadian Cordillera. A diverse and abundant fauna from the Queen Charlotte Islands was said to represent most of the Toarcian (Taylor et al. 1984). From the craton of Canada, diverse ammonite faunas were reported, and three early and middle Toarcian assemblages were recognized (Hall 1984). Earliest and latest Toarcian ammonites were reported from Nevada (Hallam 1965) and figured from eastern Oregon (Imlay 1968; Taylor et al. 1984). In Chile, the most complete biozonation of Toarcian ammonites around the Pacific has been established (Hillebrandt and Schmidt-Effing 1981; Hillebrandt 1987): 10 zones, some with subzones. Several of them were also reported from southern Peru, and ammonites of many of these zones were also described from Argentina as far south as Chubut province.

The base of the Toarcian is marked by different species of *Dactylioceras* s.l., some of them with a wide geographic distribution (Plates 4, 5, 7, 12, 14). Genera and species of the Hildoceratidae frequently are less useful because their provincialism is usually greater and their vertical range often longer. The definition of the base of the Toarcian has recently been discussed by Fischer (1984). In northeastern Asia it is the *Tiltoniceras propinquum* Zone, characterized below by different species of *Tiltoniceras* and *Kedonoceras*. The latter genus was said to be endemic for northeastern Asia, but according to Howarth (1973, p. 273), similar species occur in the lower part of the Tenuicostatum Zone (Clevelandicum Subzone) of Yorkshire, and *Kedonoceras* is synonymous with *Dactylioceras* (*Orthodactylites*). In the Queen Charlotte Islands the densely ribbed *Dactylioceras kanense* are typical for the basal Toarcian. The lowest subzone in South America is characterized by the Mediterranean species *Dactylioceras* (*Eodactylites*) *simplex*, which is also found in northwestern Europe (Schlatter 1982) at the base of the Toarcian.

In Japan, *Dactylioceras* (*Orthodactylites*) *helianthoides* (Plate 14) is said to range from Upper Pliensbachian to Middle Toarcian and is most frequent in the *D. helianthoides* Zone (Sato and Westermann 1985). But most specimens are crushed, making comparison with other species difficult. This species was also figured from western Canada (Frebold 1964, Pl. 5, Figs. 7–8) and Chile (Hillebrandt and Schmidt-Effing 1981), where it is restricted to the *Dactylioceras hoelderi* Zone. This zone can be subdivided by species of *Harpoceratoides* ["*Hildoceratoides*" of Hillebrandt (1987, Table 1)] and *Hildaites*. Species of these genera are also found in North America and northeastern Asia, together with species of *Harpoceras* similar to European forms and in similar stratigraphic positions. Northeastern Asian *Eleganticeras* spp. are described from the lower part of the European Falcifer Zone, which is coeval with the South American *hoelderi* Zone. Dactylioceratids from the Indonesian island of Rotti probably are Lower and Middle Toarcian.

The Middle Toarcian of northwestern Europe was subdivided into two zones. The Bifrons Zone is characterized by *Hildoceras*, a genus also frequent in the Mediterranean, but not yet found around the Pacific. On the other hand, the Bifrons Zone of northwestern Europe is subdivided into three subzones based on Dactylioceratidae that also occur in different Pacific regions.

*Dactylioceras* (*D.*) *commune* is the index of a subzone marking the base of the Middle Toarcian (Plates 4, 5, 18). This species appears in northeastern Asia in the *D. athleticum* Zone (Repin 1984) and occurs also in the Canadian Arctic and Alberta (Frebold 1975; Hall 1984). In northeastern Asia *D. commune* was recorded together with *Harpoceras chrysanthemum*, described originally from the Japanese *D. helianthoides* Zone, which is coeval with two or more zones of other Pacific regions. *H. chrysanthemum* was also cited from western Canada (Cameron and Tipper 1985) and is found in South America together with the first *Peronoceras*. In South America and the Mediterranean, *Peronoceras* appears earlier than in northwestern Europe. The genus *Peronoceras* (= *Porpoceras*) is important for the biostratigraphy of South America, North America, and northeastern Asia, where it occurs mainly in the middle and upper parts of the Middle Toarcian.

The South American zone of *Peronoceras pacificum* (Plate 12) is more or less coeval with the European zone of *Peronoceras fibulatum*. In Europe this zone is also characterized by *Zugodactylites braunianus*. This and similar species were found in northeastern Asia in the *Z. monestieri* Zone (Plate 19) (Repin 1984) and occur also in the Canadian Arctic (Frebold 1975). *Zugodactylites* seems to be somewhat restricted to the Boreal Realm. *Peronoceras spinatum* and *P. polare* are species typical for the Canadian Arctic (Frebold 1975) and northeastern Asia (Repin 1984). *P. spinatum* was also described from the South American *Peronoceras bolitoense* Subzone (Plate 12) (Hillebrandt 1987) and was found together with *Catacoeloceras* cf. *crassum*. *C. crassum* is the index for the northwestern European Crassum Zone. The beds with *P. spinatum* and *P. polare* are coeval with at least part of the Crassum Zone and perhaps with the lower Variabilis Zone. Above beds with dactylioceratids of the *bolitoense* Subzone in South America, large *Peronoceras* characterizing the *moerickei* Subzone are found (Plate 13). In both subzones the Mediterranean genus *Collina* (*chilensis* Zone) occurs. The endemic hildoceratid genera *Atacamiceras* and *Hildaitoides* were also reported from this zone (Hillebrandt 1987).

The northwestern European Variabilis Zone is characterized by species of the genus *Haugia*, which is more frequent in the Boreal Realm (Plates 4, 7). Species of this genus were figured from the Arctic (Imlay 1981) and the Toarcian of cratonic Canada (Frebold 1976; Hall 1984), where they appear together with *Brodieia* and *Phymatoceras*. *Haugia* was cited together with "*Denckmannia*" (= *Phymatoceras* s.l.) from western Canada (Cameron and Tipper 1985) and was also described from Oregon (Hallam 1965). Ammonite fragments figured by Imlay (1968) as *Haugia* are more similar to *Hammatoceras*. *Phymatoceras* is more frequent in the Tethyan Realm and is important for the biostratigraphy of the upper Middle and lower Upper Toarcian of South America (Plate 13). Two zones, the *Phymatoceras toroense* Zone and *P. copiapense* Zone, are distinguished, based on endemic species of this genus (Hillebrandt 1987).

The Upper Toarcian genera *Grammoceras* and *Pseudogrammoceras* are important for the northwestern European biostratigraphy. These genera were described from western Canada (Frebold and Tipper 1970), the Arctic and Alaska (Imlay 1981), Oregon (Imlay 1968) Japan (Hirano 1973), Vietnam (Plate 14) (Sato 1972; Vu Khuc 1984), and the Sula Islands (Kruizinga 1926; Arkell 1956, p. 440). Species of these genera similar to European ones occur in Japan together with endemic species of *Phymatoceras*, and above the *D. helianthoides* Zone (Sato and Westermann 1985). *Grammoceras* and *Pseudogrammoceras* are unknown from South America. The northwestern European Dispansum Zone is characterized by species of *Phlyseogrammoceras*. The South American *Phlyseogrammoceras tenuicostatum* Zone is defined by species of this genus. A species similar to *P. dispansiforme* and *P. werthi* was figured from the Upper Toarcian of southwestern British Columbia (Frebold, Tipper, and Coates 1969). Some species of *Esericeras* are very similar to *Phlyseogrammoceras* and are said to occur on Timor (Arkell 1956, p. 440).

*Dumortieria* and *Pleydellia* (Plates 4, 13, 14) are restricted to the uppermost part of the European Toarcian. *Dumortieria* was described from Vietnam (Sato 1972; Vu Khuc 1984). A species similar to *D. pusilla* was figured from Oregon (Imlay 1968). This species is typical for the uppermost Toarcian of South America and may range into the lowermost Aalenian (Bogdanic, Hillebrandt, and Quinzio 1985). Specimens similar to European species of *Pleydellia* were figured from the Misol Archipelago (Soergel 1913) and the Sula Islands (Kruizinga 1926). The uppermost Toarcian of South America is characterized by two species of *Pleydellia* that are very similar to the European *P. lotharingica* and *P. fluitans*. The Upper Toarcian genus *Hammatoceras* was described from different localities around the Pacific (Plate 4), mainly South America and Southeast Asia, but also Australia. An improved taxonomy would be important for correlations. The Mediterranean genus *Catulloceras* is of latest Toarcian age and occurs also in North America (Frebold 1964; Imlay 1968) and South America (Hillebrandt and Westermann 1985, p. 7).

The lower part of the *Hosoureites ikianus* Zone of Japan (Plate 14) probably belongs to the uppermost Toarcian (Sato and Westermann 1985, p. 75). In northeastern Asia (Repin 1984), northern Alaska (Imlay and Detterman 1973), and the Canadian Arctic (Frebold 1975) the Upper Toarcian (and Aalenian) is dominated by species of *Pseudolioceras* (Plates 5, 19). At this time the Bering Province (Taylor et al. 1984) was established.

## MIDDLE JURASSIC[2]

Middle Jurassic correlations of ammonite assemblages and zones among circum-Pacific areas (Table 12.2) and with the European–eastern Greenland chronostratigraphic standard successions are inhibited by well-developed provincialism. In the Aalenian this is particularly due to the virtual absence of Graphoceratidae, on which the standard zonation is based, whereas from the late Bajocian to the early Callovian the entire East Pacific develops an ammonite fauna largely of its own (subrealm). A number of regional ammonite zones are therefore distinguished in the different circum-Pacific regions, particularly in the eastern Pacific, where several standard zones have now been defined.

To avoid unnecessary repetition of Part III, which presents the biostratigraphy with full references, the only references given here are the most recent. For each stage, we shall proceed from the Boreal Realm to the Tethyan Realm and, usually, from north to south.

### Toarcian–Aalenian boundary

The lower boundary of the Aalenian (Lower–Middle Jurassic boundary) can be placed with some precision only in

[2] By G. E. G. Westermann.

Table 12.2. *Middle Jurassic ammonite zones of the circum-Pacific region.*

| Stage | Substage | STANDARD ZONES – EUROPE: Submedit. & NW. Europe | STANDARD ZONES – E. GREENLAND: Boreal-Subbor. | SOVIET UNION: Northeast | SOVIET UNION: Far East | ARCTIC SLOPE NORTH AMERICA | S. ALASKA & BRIT. COLUM. | E. OREGON |
|---|---|---|---|---|---|---|---|---|
| CALLOVIAN | U | LAMBERTI | LAMBERTI | | ? | | ? | |
| | | ATHLETA | ATHLETA | | *Longaeviceras d. keyserlinig* | Longaeviceras stenolobum Z. | L. pomeroy. As. | ? |
| | M | CORONATUM | CORONATUM | | | Cadoceras stenoloboide As. | Cadoceras stenoloboide As. | Cadoceras stenoloboide As. |
| | | JASON | JASON | | | | ? | ? |
| | L | GRACILIS | CALLOVIENSE | | | Cadoceras articum Z. | *Lilloettia, Cadoceras, Keppl.* | *Lilloettia, Cadoceras, Keppl.* |
| | | MACROCEPHALUS | NORDENSKOLDI | *Cad. anabarense As.* | | | Cadoceras coma As. | C. tonniense Fl. |
| | | MACROCEPHALUS | APERTUM | | | C. bodylevski Z. | | Keppl. loganianus Fl. |
| BATHONIAN | U | DISCUS | CALYX | ? | ? | Cad. barnstoni Z. | Iniskinites apruptus As. | |
| | | RETROC. / ORBIS | VARIABILE | | | *Cadomites, Iniskinites. Loucheuxia. Kepplerites* | | *Iniskinites acuticostatus* |
| | | BREMERI / HUDSONI | CRANOCEPHAL. | | | | *Iniskinites pp.* | |
| | M | MORRISI / SUBCONTRACTUS | ISHMAE | *Arcticoceras & pseudocadoc.* | | ISHMAE Z. – ISHMAE Sz.; harlandi Sz. | ? | *Cobbanites Parareinekeia* |
| | | PROGRACILIS | GREENLANDICUS | | | Arctoceph. frami Z.; A. amundseni Z. | | |
| | L | ZIGZAG | ARCTICUS | Arctoc. elegans Z. – jugatus Sz. | | A. porcupinese Z.; A. spathi Z. | | |
| BAJOCIAN | U | PARKINSONI | POMPECKJI | Cranocephal. 'vulgaris' Z. | Megasphaerocera B. | | *Iniskinites ? costidensus.* | |
| | | GARANTIANA | | | | | *Epizigzagiceras & Megasphaer.* | |
| | | SUBFURCATUM (NIORTENSE) | INDISTINCTUS | | | INDISTINCTUS Z. | ROTUNDUM Z. | ROTUNDUM Z. |
| | | | BOEALIS | BOREALIS Z. | | BOREALIS Z. | | |
| | L | HUMPHRIESIANUM | non-marine | ? | ? | | OBLATUM Z.; KIRSCHNERI Z. | OBLATUM Z.; Richardsoni Sz. |
| | | SAUZEI | non-marine | Arkelloceras B. / A. tozeri B. | Arkelloceras B. / A. tozeri B. | Arkelloceras B. / A. tozeri B. | CRASSICOTATUS Z | CRASSICOTATUS Z |
| | | LAEVIUSCULA | non-marine | Pseudolioceras fastigatum B. / Z. | Pseudolioceras fastigatum B. / Z. | ? | ?; Witch. sudner. Sz. | Sonn. burkei Z. |
| | | OVALIS | non-marine | | | | WIDEBAYENSIS Z. | Euhopl. tuberculatum Z. – E. ochoc. Sz. |
| | | DISCITES | non-marine | | | | Doc. camachoi Sz. | E. westi Sz. |
| AALENIAN | U | CONCAVUM | non-marine | | HOWELLI Z. | HOWELLI Z. | HOWELLI Z. | "Font". packardi Z.; Eud. "movichense" Z. |
| | | MURCHISONAE | non-marine | P. tugurense B. / Z. | P. tugurense B. / Z. | ? | | sparsocostatum Z. |
| | L | SCISSUM | non-marine | P. maclintocki B. / Z. | P. maclintocki B. / Z. | | *Tmetoceras scissum* | |
| | | OPALINUM | non-marine | | P. beyrichi B. | | | |

North-East Russia and in the Southern Andes. Elsewhere the boundary lies in a hiatus and/or among poorly known faunas.

In North-East Russia, within the Bering Province of the Boreal Realm, the boundary is placed between the *Pseudolioceras rosenkrantzi* Zone (Upper Toarcian, Table 12.1) and the *P. beyrichi* beds of the *P. maclintocki* Zone (Plate 73) (Kalacheva 1988). The *beyrichi* beds are also known from the Far East, where the upper Toarcian is missing (Table 12.2). *P. beyrichi* (Schloenb.) is well dated in Europe, from Germany to the northern Caucasus, where it occurs in the lower Opalinum Zone.

In northern Chile (and west-central Argentina; A. Riccardi personal communication 1990), in an apparently continuous sedimentary sequence, the upper Levesquei Zone, with *Pleydellia* (*Walkericeras*) spp., is overlain by the Manflasensis Zone (Plates 50, 51) (Hillebrandt and Westermann 1985). Besides dominant *Bredyia*, this zone includes what appear to be rare Leioceratinae; *"Leioceras comptum" chilense* West., indicating the Opalinum Zone, however, may be an *Ancolioceras*, a European (mainly Mediterranean) genus indicating the Scissum and Murchisonae Zones (J. Sandoval personal communication 1990). *Bredyia* occurs in Europe in the Opalinum and Scissum Zones. Therefore, the presence of Opalinum Zone equivalents in the Andes, although highly probable, cannot be clearly demonstrated.

Table 12.2 (*cont.*)

| WESTERN INTER. U.S.A. | S. MEXICO | S. ANDES | JAPAN | E.INDONESIA & IRIANJAYA | | |
|---|---|---|---|---|---|---|
| *Quenstedtoceras. collieri* | ? | | ? | ? | U | CALLOVIAN |
| ? | [? Peltoceratids] | | | | | |
| | *Reineckeia* s.s. *Rehmannia* gr. *densestriata* | *Rehm. patagoniensis* & *Oxyceri. oxynotus* | | | M | |
| | PROXIMUM Z. | | Oxycerites Z. | Macrocephalites keeuwensis. As. | L | |
| | rehmanni As.<br>inflatum As.<br>Frickites As. | BODENDENDERI Z. | | | | |
| | VERGARENSIS Z. | | Kepplerites japonicus Z. | ? | | |
| Kepplerites macavoi As..../ | Lilloettia As. | jupiter H. / gerthi H. | "Neuquenic" Yokoyamai Z. | Macr. apertus Z. | U | BATHONIAN |
| K. maclearni As. | STEINMANNI Z. | | | | | |
| K. subitus As. | | | | | | |
| K. costidensus As. | Histricoides As. | | | | | |
| Warrenoceras henryi & codyen. As. | non-marine | *Cadomites-Tulitidae* As. | ? | Macr. bifurcatus Z. s. str. / intermed. Sz. | M | |
| | | | | *Tulitidae* | | |
| Paracephalites glabrescens As. | | | | *Satoceras* [*Praetulites*] | L | |
| Parachondroc. andrewsi As. | | *Lobosphinctes* | | | U | BAJOCIAN |
| | | *Megasph. magnum* ss. | | *Leptosphinctinae* | | |
| | P. zapotec. Z. | ROTUNDUM Z. / dehmi Sz. | Leptosphinctes Z. | | | |
| *Chondroceras*, *Stemmatoceras* | D. floresi Z. | HUMPHR. Z. chilense Sz. / ROMANI Sz. | | *Teloceras*, *Stephanoceras*, *Chondroc.* | | |
| | *Stephanoceras* | | Sonninia Z. | | | |
| non-marine | non-marine | GIEBELI Z. blancoensis H. / MULTIFORME Sz. / SUBMICROST. Sz. | | *Pseudotoites* *Fontannesia kiliani*. | L | |
| | | SINGULARIS Z. | ? | ? | | |
| | | MALARG. Z. maubeugei Sz. / mendozana Sz. / compressa Sz. | | | U | AALENIAN |
| | | "Zurch." groeberi Z. | Planammatoceras planinsigne Z. | | | |
| | | MANFLASENSIS Z. | Hosour. ikianus Z., part. | | L | |

## Aalenian

In northeastern Asia the *Pseudolioceras maclintocki* Zone is approximately coeval with the Lower Aalenian (i.e., Opalinum and Scissum Zones (Plates 73, 74). It is followed by the *P. tugurense* Zone, which approximately comprises the Upper Aalenian, if the "*Erycitoides howelli* Beds," at the top, are included (Kalacheva 1988). The Howelli Zone of southern Alaska, however, is now known from the entire Bering Province and therefore is here elevated to a standard zone.

In Alaska and Canada, only the Howelli Zone (new standard zone, see Chapter 4) can be clearly recognized in the Aalenian, from the North Slope to central [and southern(?)] British Columbia (Plates 20, 23, 24). Best known from the Alaska Peninsula (Westermann 1964, 1969), the Subboreal zonal assemblage overlies beds yielding only the pan-Tethyan *Tmetoceras scissum* (Ben.), ranging through the Aalenian stage. The Howelli Zone includes dominant *Erycitoides* spp., *Pseudolioceras* (*Tugurites*) *whiteavesi* (White), and rare *P.* (*T.*) *tugurense* as endemic Bering elements, together with some Tethyan hammatoceratids – *Eudmetoceras* cf. *eudmetum* (Buck.) and *E.*(?) (*Euaptetoceras*?) *amplectens* (Buck,) – clearly dating this assemblage as late Aalenian, approximately Concavum Zone. *E.*(?) *amplectens* ranges upward into the clearly Bajocian Widebayense Zone, as discussed later.

The richest late Aalenian and early Bajocian faunas, with strong Tethyan affinities, occur in eastern Oregon (Plates 42, 43) (Taylor 1988). Five new late Aalenian assemblage zones are said to be coeval with about three European standard zones, based on a multitude of newly named, incompletely known taxa. Detailed publication is required for the sections and faunas, with their regional distributions. The succession is in ascending order: (1) The *Abbasites sparsicostatus* Zone, with *Tmetoceras scissum* (Ben.) and *Strigoceras*, is coeval with the upper Murchisonae Zone. (2) The "*E. movichense*" Zone is named after "*Eudmetoceras (Planammatoceras) movichense*" Taylor, based on a small, distorted holotype of uncertain affinities resembling either the pan-Tethyan *Planammatoceras (Pseudaptetoceras)* gr. *klimakomphalum* (Vacek) or the Andean *Puchenquia*, and probably is coeval with part of the Concavum Zone. (3) The "*Fontannesia*" *packardi* Zone, named after a new species placed in *Fontannesia*, but bearing ventrolateral spines, and also found with the first *Euhoploceras*, including *E. modestum* (Buck.), *Hebetoxyites*, and the unique "*Eudmetoceras* (*Planammatoceras*)" *vigrassi* Taylor (with radial septal suture), probably is correctly dated as late Concavum Zone.

Aalenian faunas are richly represented in the Andes, from Peru to Mendoza province in Argentina (Plates 51, 52). The best-known sections, in northern Chile (Hillebrandt and Westermann 1985), show the Manflasensis Zone overlain by the "*Zurcheria*" *groeberi* Zone. Originally defined in Mendoza (Westermann and Riccardi 1972–9), it coincides with the range zone of the endemic *Parammatoceras jenseni* West. in northern Chile. In Mendoza the *groeberi* Zone includes *Planammatoceras* cf. *planiforme* Buck. (Westermann and Riccardi 1982), indicating the Murchisonae Zone (s.l.); this species is also known from central Peru (Westermann et al. 1980).

From Peru to Mendoza, there follows the Malarguensis (new standard zone for the *Puchenquia malarguensis* Assemblage Zone of Westermann and Riccardi 1972–9), which appears to span late Murchisonae–early Discites Zone ages and was recently subdivided into two or three subzones (Hillebrandt and Westermann 1985). The *Puchenquia compressa* and *P. mendozana* Subzones include the Tethyan *Eudmetoceras eudmetum* Buck., *Planammatoceras (Pseudaptetoceras) klimakomphalum* (Vacek), and common *Tmetoceras* spp., including *T.* cf. *flexicostatum* West. All are found also in the Howelli Zone of southern Alaska and clearly indicate the latest Aalenian, Concavum Zone age. At the top of the northern Chilean sequence, the *Podagrosiceras maubeugei* horizon probably already represents the earliest Bajocian, as discussed later.

Farther south in Pacific Asia, the Aalenian is again well documented by ammonites only in Japan (Sato and Westermann (1991). The *Hosoureites ikianus* Zone, based on the endemic *H. ikianus* (Yokuy.), includes below the late Toarcian *Polyplectus*(?), whereas common *Tmetoceras* and rare *Leioceras* cf. *opalinum*, *L.* cf. *striatum*, and *Pseudolioceras* cf. *maclintocki* clearly document the Lower Aalenian, approximately Opalinum Zone and Scissum(?) Zones. The *Planammatoceras planinsigne* Zone includes *Planammatoceras* (*P.*) spp. closely resembling the Mediterranean-Submediterranean species group of the Murchisonae Zone (s.l.). It also includes rare *Pseudolioceras (Tugurites)* cf. *tugurense* Kal., suggesting correlation with the lower (restricted) part of the northeastern Asian zone, discussed earlier.

In western Indochina, the small assemblage from Mae Sot in western Thailand consists of *Tmetoceras scissum*, a single *Ancolioceras*(?) sp. (for "*Graphoceras concavum*"; J. Sandoval and A. Linares personal communication 1990), and *Erycites*(?) (or possibly a *Pseudotoites*), tentatively suggesting the Scissum-Murchisonae Zones. *Planammatoceras* and *Tmetoceras* have also been recorded from Vietnam (Vu Khuc 1984).

Aalenian ammonites are unknown from the whole of Australasia.

### Lower Bajocian

The Lower Bajocian south of the Boreal Realm is among the best-known and best-correlated Jurassic substages of the circum-Pacific area; its well-established zones include one widely recognized, formerly only European standard zone, the Humphriesianum Zone, and its Romani Subzone.

In Boreal northeastern Asia, the ammonite fauna, as usual, has low diversity and resembles that of Arctic North America. Tentative zonation is based not only on the rare ammonites but also on the more abundant inoceramid bivalves (Krymholts et al. 1988). In North-East and Far East Russia (Plates 73, 74), beds with *Pseudolioceras fastigatum* and *Mytiloceramus jurensis* are overlain by beds with *Arkelloceras* and *M. lucifer* (Plate 122). Both assemblages are correlated with the European standard zones via southern Alaska, where *P. (Tugurites) fastigatum* West. occurs in the (lower) Widebayense Zone, coeval with the Discites-Laeviuscula Zones, and *Arkelloceras* occurs in the Crassicostatus Zone of Sauzei Zone age, as discussed later. The highest beds (presumably still lower Bajocian) are characterized by *Mytiloceramus clinatus* (Kosh.), but have yielded no stratigraphically useful ammonites.

The Lower Bajocian of western North America is well defined by the *Euhoploceras* assemblage below and the *Defonticeras* assemblage above (Hall and Westermann 1980). In southern Alaska the Bajocian sequence begins with the Widebayense Zone (Plates 25–27) (here elevated to standard zone from assemblage zone). Its base is poorly known, but lies above beds and a turbidite with *Eudmetoceras*(?) *amplectens*, a good marker for the Aalenian–Bajocian boundary in Europe. The dominant endemic *Docidoceras (Pseudocidoceras)* spp., *Alaskinia*, and the Bering Province *Pseudolioceras (Tugurites) fastigatum* West. are accompanied by a few Pacific *Pseudotoites* spp. and the Euro-American *Asthenoceras*. The *Docidoceras camachoi* Subzone, below, includes the Tethyan *Planammatoceras (Pseudaptetoceras) klimakomphalum* (Vacek), *Eudmetoceras* (*E.*) gr. *eudmetum* (Buck.), and *Euhoploceras*, indicating the Discites Zone. The index species of the superjacent *Witchellia sutneroides* Subzone resembles early European *Witchellia* gr. *sutneri* (Branco), indicating an Ovalis–lower Laeviuscula Zone age. Isolated finds of *Euhoploceras* and *Alaskinia*(?) gracilis (White.)

in southern Alberta and British Columbia suggest the presence of coeval beds.

In eastern Oregon (Plates 43–46), a rich fauna with strong Tethyan affinities is developed, and two assemblage zones (see the earlier section "Aalenian") coeval with the Subboreal, southern Alaskan Widebayense Zone have recently been defined (Taylor 1988): (1) The *Euhoploceras tuberculatum* Zone, with *E. acanthodes* (Buck.) said to range throughout, as well as *Euhoploceras* and *Docidoceras* spp., is subdivided into two subzones: (a) the *Euhoploceras westi* Subzone, coeval with the Discites Zone, and (b) the *E. ochocoensis* Subzone, with *Sonninia* (*Fissilobiceras*), *Witchellia(?), Latiwitchellia,* and *Asthenoceras,* coeval mainly with the Ovalis Zone. (2) The *Sonninia burkei* Zone is said to include the last *Euhoploceras,* including *E. acanthodes* (Buck.), and the new genus "*Freboldites*" with the type (and only) species "*F. bifurcatus*" Taylor based on a poorly known, serpenticonic early stephanoceratid [*Stephanoceras* (*Skirroceras*) cf. *dolichoecus* (Buck.), of Imlay (1973, Pl. 45, Fig. 8)]; but the single illustration does not display the supposed diagnostic features. The *burkei* Zone supposedly is coeval with only the early Laeviuscula Zone and is followed by the Crassicostatus Zone, said to still contain *Euhoploceras acantherum* (Buck.). The new taxa and zones need confirmation through full descriptions of sections and fauna.

In southern Alaska, Oregon, and southern Alberta (see Chapter 4) the Lower Bajocian and basal Upper Bajocian are well developed, fossiliferous, and well known. Complete sections with a continuous fossil record are rare, but some seem to be present in southern Alaska (Chapter 4). Several former assemblage zones with known superregional distributions are here raised to standard zones, contributing to a consistent zonation of the Middle Jurassic along the eastern Pacific margin. The fact that these standard zones have lacked precise definitions of their bases need not concern us too much, remembering that most of the well-established European standard zones have fared no better in this respect.

The Crassicostatus Zone (Plates 28, 29, 46) [new standard zone for *Parabigotites crassicostatus* Assemblage Zone of Hall and Westermann (1980)] includes a rich assemblage, including the European *Emileia* gr. *polyshides* (Waag.), *Sonninia* (*Papilliceras*) cf. *arenata* (Qu.), and *Stephanoceras* (*Skirroceras*) of clearly Sauzei Zone age; the Bering *Arkelloceras* is also associated with this southern Alaska assemblage (Imlay 1964). The Cook Inlet in southern Alaska has the most promising stratotype. In southern Alberta, *Arkelloceras* and *Stephanoceras* (*Skirroceras*) indicate the Crassicostatus Zone. In eastern Oregon, the rich assemblage includes also the Andean subgenus *Emileia* (*Chondromileia*) (i.e., *E. buddenhageni* Imlay).

The Kirschneri Zone (Plates 29, 30) [new standard zone for *Stephanoceras kirschneri* Assemblage Zone of Hall and Westermann (1980)] was named after *S. kirschneri* Imlay, common in southern Alaska (Cook Inlet; future stratotype) and eastern Oregon. The assemblage is also well developed in Alberta (Chapter 4). At least in Oregon, the lower part is still of Sauzei Zone age. This is indicated by *S.* (*Skirroceras*) and early *Dorsetensia,* resembling the fauna of the Hebridica Subzone of Scotland and the *Dorsetensia blancoensis* horizon of the Andean Giebeli Zone, as discussed later. The upper part of the Kirschneri Zone, the *Zemistephanus richardsoni* Subzone, includes besides this endemic genus also *Sphaeroceras* (*Defonticeras*) spp., *Stephanoceras* (*Stemmatoceras*) spp., and, in British Columbia, *Poecilomorphus crickmayi* (Freb.), a genus almost restricted to the lower Humphriesianum Zone in Europe.

The Oblatum Zone (Plates 29, 36) [new standard zone for *Chondroceras oblatum* Assemblage Zone of Hall and Westermann (1980)] includes *Stephanoceras* s. str. and rare *Teloceras* [i.e., *T. crickmayi* (Frebold)], strongly indicating a late Humphriesianum Zone age. The zone is well developed in southwest Alberta, southern Alaska (especially Tuxedno Bay), and the Queen Charlotte Islands.

In the Southern Andes (Westermann and Riccardi 1972–9), from central Peru to Mendoza, the Lower Bajocian probably begins with the upper part of the Malarguensis Zone (Plates 52, 53) (new standard zone for *Puchenquia malarguensis* Assemblage Zone), that is, all or part of the *Podagrosiceras maubeugei* horizon with *Euhoploceras amosi* West. & Ricc., as discussed earlier. The Singularis Zone (Plates 53, 54, 56, 57) (new standard zone for *Pseudotoites singularis* Assemblage Zone) yields abundant *Sonninia* (*Fissilobiceras*) spp., including rare *S.* (*F.*) *ovalis* (Qu.), as from the European Ovalis Subzone. The Giebeli Zone (Plates 53, 57) (new standard zone for *Emileia giebeli* Assemblage Zone) is approximately coincident with the range zones of the genus *Emileia* and subgenus *Sonninia* (*Papilliceras*). The *E. submicrostoma* Subzone, below, with late *Pseudotoites,* is dated as Laeviuscula Zone, mainly by stratigraphic interpolation; the *E. multiformis* Subzone includes *Emileia* (*E.*) cf. *vagabunda* Buck., *E.* (*E.*) aff. *brochii* (Sow.), and *Stephanoceras* (*Skirroceras*) spp., an association clearly dated as Sauzei Zone. The top of the Giebeli Zone is marked by the *Dorsetensia blancoensis* horizon, a species closely resembling *D. hebridica* Morton from Scotland, index of a subzone in the highest Sauzei Zone.

The fauna of the superjacent beds in the Chile-Argentine Andes (Plates 55, 57) is specifically so close to that of the formerly only European Humphriesianum Zone that no new zonal name is required. Even the Romani Subzone, in the lower part of the zone, is clearly marked by the index species *Dorsetensia romani* (Opel), as well as by the characteristic *D.* gr. *complanata-tecta* Buck. and *Stephanoceras* (*S.*) *pyritosum* (Qu.) (Westermann and Riccardi 1972–9). In northern Chile, the uppermost part of this zone is distinguished as the *Duashnoceras chilense* Subzone; the index fossil has recently been placed as a subspecies in the southern Mexican species *Duashnoceras paucicostatum* (Felix) of the *D. floresi* Zone, as discussed later. The subzone includes *Stephanoceras* gr. *umbilicum* (Qu.), *S.* (*Stemmatoceras*) spp., and *Teloceras* spp., indicating the Blagdeni Subzone [and "Humphriesianum"(?) Subzone].

In Guerrero state in southern Mexico, the oldest marine Middle Jurassic has yielded *Stephanoceras* (*S.*) cf. *skidegatense* (White.) and *S.* (*S.*) gr. *mutabile* (Qu.)–*itinsae* (McL.) and has been placed in the North American Kirschneri Zone (Sandoval and Westermann 1988). Above follows the *Duashnoceras floresi* Zone (Plate

47), which, besides that Andean endemic genus, has yielded *Stephanoceras* (*S.*) *boulderense* (Imlay) in the lower part, still indicating the Humphriesianum Zone.

In Pacific Asia south of Russia, marine Lower Bajocian is absent from eastern China, but is documented by isolated Tethyan assemblages in Japan (Plate 72) (Sato and Westermann 1991). The *Sonninia* Zone includes *S.* gr. *sowerbyi*, *Stephanoceras* cf. *plicatissimum* (Qu.), *Strigoceras* cf. *longuidum* Imlay, and *Emileia* sp., indicating Sauzei and early Humphriesianum Zone ages.

The only record of Lower Bajocian from Southeast Asia is from western Thailand, where *Stephanoceras* (*Skirroceras*) cf. *macrum* (Qu.) and *Sonninia* (*Papilliceras*) sp. indicate the Sauzei Zone (for *''Docidoceras''* and *''Planammatoceras''* of Braun and Jordan 1976).

The few records of Lower Bajocian from Pacific Australasia are entirely based on old, ex situ collections from eastern Indonesia (Plate 66): *Pseudotoites* sp. and *Fontannesia kiliani* indicate the presence of the *Fontannesia-Pseudotoites* fauna of Western Australia, which is dated as Laeviuscula Zone (Westermann and Wang 1988); *Sphaeroceras* (*Chondroceras*) *boehmi* West., *Stephanoceras etheridgei* (Gerth), *S.* cf. *humphriesianum* (Sow.), and *Teloceras indicum* (Kruiz.) document the presence of the Humphriesianum Zone.

The Humphriesianum Zone is also indicated in New Zealand, where *Stephanoceras* gr. *humphriesianum, Sphaeroceras* s.l., and an endemic new genus have recently been found by N. Hudson, C. D. Campbell, F. Hasibuan and G. E. G. Westermann (unpublished data).

## Upper Bajocian

The Lower–Upper Bajocian boundary is marked throughout most of the circum-Pacific area by a strong faunistic change, that is, the sudden beginning of extreme provincialism, especially in the east, the East Pacific Subrealm. Boundary ''events'' include the birth of the Boreal Cardioceratidae and the East Pacific Eurycephalitinae.

In the Boreal northern Pacific the Upper Bajocian begins with the *Cranocephalites* (*''Boreiocephalites''*) *borealis* Zone in North-East Russia (Krymholts et al. 1988), the earliest of the eastern Greenland Cardioceratid zones, which may be as old as Subfurcatum Zone (Callomon 1985), followed by the *C. ''vulgaris''* (= *pompeckji*) Zone. Locally more abundant and useful, however, is the inoceramid zonation based on *Mytiloceramus.* In cratonic North America the eastern Greenland zones are indicated in Arctic Can-ada, where a regional zonation has been introduced (Poulton (1987), but are poorly known from the Alaskan North Slope (Callomon 1984).

In Far East Russia (Plates 74, 75), as well as in the allochthonous North American terranes and the Andes, the first Upper Bajocian faunas are characterized by the first eurycephalitid, *Megasphaeroceras* (includes *Umaltites*), of about Subfurcatum-Garantiana Zone age. In the Far East the *M. era* beds (Krymholts et al. 1988) include *Epizigzagiceras* and overlie beds with *Strigoceras* (*Liroxyites*) cf. *kellumi* Imlay; in southern Alaska, both species are associated in the Rotundum Zone, as discussed later.

The Rotumdum Zone [new standard zone for *Megasphaeroceras rotundum* Assemblage Zone of Hall and Westermann (1980)] occurs in southern Alaska, the Western Interior United States, and eastern Oregon (Plates 30, 31, 39, 46). The best stratotype probably can be found in the Twist Creek Siltstone at Cook Inlet. The zone includes an association of *Leptosphinctes, Spiroceras, Sphaeroceras* s. str., and late Stephanoceratinae, documenting Subfurcatum Zone age (Hall and Westermann 1980; Hall and Stronach 1981). As in the Andes (discussed later), *Megasphaeroceras* ranges upward, here into the *Epizigzagiceras* associations (faunas B2,3 of Callomon 1984) of central British Columbia (and possibly southern Alaska), which are therefore still considered late Bajocian.

In southern Mexico, the upper *Duashnoceras floresi* Zone of Oaxaca (Plate 47) has yielded also *Strigoceras* (*Liroxyites*) *kellumi,* unknown from the Rotundum Zone, *Stephanoceras* cf. *boulderense* (Imlay), and the Mediterranean *Subcollina lucretia* (Orb.) and is thus dated as early Subfurcatum Zone, approximately Banksi Subzone (Sandoval and Westermann 1988). There are also common elements with the *Lupherites dehmi* Subzone of the Andes. Curiously, however, the coeval *Megasphaeroceras* is missing from the Mexican association, presumably owing to differences in biofacies (perhaps the Oaxaca Sea having been too shallow). The superjacent *Parastrenoceras zapotecum* Zone includes *Oppelia subradiata* (Sow.) and endemic *Leptosphinctes* species throughout; and near the base occurs also *Leptosphinctes* cf. *davidsoni* (Buck.), known from the Polygyralis and Baculatum Subzones of Europe. *Parastrenoceras* occurs rarely in the Mediterranean Subfurcatum Zone.

In the Southern Andes, only a single fragment of an indubitable *Strenoceras* has been discovered in northern Chile, although faunas of Subfurcatum Zone age are richly developed there and in west-central Argentina (Riccardi, Westermann, and Elmi 1989). In Chile, the Rotundum Zone (Plate 58) includes large *Spiroceras* and *Teloceras,* and below also *Lupherites* [*''Domeykoceras''*], known elsewhere only from eastern Oregon (and ?*''Otoites'' filicostatus* Imlay from southern Alaska) and marking the *L. dehmi* Subzone. In Neuquén province (Plate 59), *Teloceras, Lepto- sphinctes coronoides* Buck., *Megasphaeroceras magnum spissum* Ricc. & West., and a single minute *Strenoceras*(?) (or ?*Parastrenoceras*) occur in the Rotundum Zone. *Megasphaeroceras magnum magnum* Ricc. & West. ranges somewhat higher up, without the mentioned other taxa, into beds of approximately Garantiana Zone age (Riccardi et al. 1989; Riccardi and Westermann 1991). In the absence of Parkinsoniidae, the uppermost Bajocian cannot be identified with certainty; but the Parkinsoni Zone is suggested by the occurrence of *Lobosphinctes intersertus* Buck. somewhat higher in the famous Chacay Melehue section.

Returning to Pacific Asia, the only known occurrence of marine Upper Bajocian south of Russia is in Japan (Sato and Westermann 1991). The *Leptosphinctes* Zone includes *L.* gr. *davidsoni-coronarius* Buck., together with *Cadomites* and *Oppelia,* and is dated as Subfurcatum Zone.

The only known Upper Bajocian marine record in Australasia is from the Sula Islands and New Guinea (Plate 67) (Westermann and Callomon 1988). The previously alleged Kimmeridgian ammonite *"Idoceras" mihanum* (Boehm) is a leptosphinctid. A similar *Leptosphinctes* cf. *engleri* (Frebold) has been found in situ. The endemic genera *Irianites* and *Praetulites* still cannot be dated with any certainty.

### Bathonian

The Boreal zones of North-East Russia (Plate 75) resemble those of cratonic northern Alaska. The *Arctocephalites elegans* Zone is equated with the early Bathonian by Russian authors (Meledina 1988) and includes a lower, *Oxycerites jugatus* Subzone. The higher beds, with *Arcticoceras* sp. and *Pseudocadoceras,* were placed in the *A. "kochi"* (= *ishmae*) Zone and were considered to be of middle Bathonian or early late Bathonian age (Callomon 1984). No significant Bathonian ammonite faunas are known from Far East Russia or Pacific China.

The unusual Subboreal fauna (with phylloceratids!) from cratonic northern Yukon (Plates 20–22) (Poulton 1987) is divided into six Bathonian zones, five of them new. The *Arctocephalites spathi* Zone probably is coeval with the Arcticus Zone of Boreal eastern Greenland; the *A. porcupinensis* Zone may still be late Arcticus Zone; the *A. amundsoni* and *A. frami* Zones probably are both of Greenlandica Zone age; the *Arcticoceras harlandi* "Zone" is correlated with the lower Ishmae Zone, and the "*A. ishmae* Zone" with the upper Ishmae Zone (so that the new "zones" are here considered subzones); and the *Cadoceras barnstoni* Zone is correlated with the late Callovian Variabile Zone.

In southern Alaska, probably only the late Bathonian or latest Bathonian (Plate 32) is documented by the *Iniskinites intermedius* and *interruptus* associations (faunas B7a,c of Callomon 1984), which include *Kepplerites* spp. and *Cadoceras* spp. (i.e., Subboreal and East Pacific representatives).

Several *Iniskinites* associations are present in British Columbia (Plat 35) and eastern Oregon (Plate 46) and also partly include early *Kepplerites* [faunas B5a–c of Callomon (1984)]; thus far they have been difficult to place stratigraphically and to date (middle to late Bathonian).

In the Shoshonean Province of the Western Interior United States and Alberta (Plates 37, 38, 40, 41, 78) there is a succession of *Paracephalites, Warrenoceras,* and *Kepplerites* associations [faunas A3–10 of Callomon (1984)]. The lowest, the *P. glabrescens* Association, with probable *Procerites,* probably is still Lower Bathonian; the two *Warrenoceras* assemblages, of which at least one (A4a) includes the oldest *Kepplerites* and the long-ranging *Cobbanites engleri* Freb. (Hall and Stronach 1981), is dated as Middle Bathonian. Of the three *Kepplerites* associations, *K. costidensus* Imlay (fauna A5) is compared by Callomon (1984) with the earliest *Kepplerites* from eastern Greenland, dated as earliest Upper Bathonian. The *K. maclearni* Association (A8) includes the first North American *Lilloettia,* which is also present in the next association (A9); the keppleritids of this and of the *K. macevyi* Association (A10) resemble those of the Calyx Zone of eastern Greenland, latest Bathonian.

In Mexico the marine Upper Bathonian of Guerrero (Plate 48) is clearly dated by the Tethyan taxa within an Andean fauna (Sandoval, Westermann, and Marshall 1989). The Andean Steinmanni Zone (discussed later) is identified by the index *Lilloettia steinmanni* (Spath). It includes, below, the *Epistrenoceras histricoides* Association, with *Hecticoceras* (*Prohecticoceras*) *blanazense* Elmi characteristic of the lower Retrocostatum Zone of the Mediterranean and Submediterranean Provinces, and the *Lilloettia* Association, with *Neuqueniceras* s. str., dated as upper Retrocostatum–(?)Discites Zones.

In the Southern Andes, Lower and Middle Bathonian (Plates 60, 61) are poorly documented and possibly are largely missing. The only pre–late Bathonian ammonites are known from reworked submarine mudflows of the *Cadomites*–Tulitidae mixed assemblage in the famous Chacay Melehue section of Neuquén province (Riccardi et al. 1989). Apparently reworked *Tulites?* (*Rugiferites*?) cf. *davaiacensis* (Liss.), *T.*? (*R.*?) aff. *sofanus* (Boehm), and *Cadomites* ex gr. *orbignyi-bremeri* clearly indicate Middle Bathonian, whereas *Bullatimorphites* (*Kheraiceras*) cf. *bullatus* (Orb.) in the matrix indicates Upper Bathonian. In northern Chile, similar *Cadomites* occur, partly together with *Epistrenoceras,* indicating either (1) that *Epistrenoceras* appears earlier in the Andean Province than in Europe, where the genus is cryptogenic in the upper Retrocostatum Zone, or (2) that these *Cadomites* are in fact late Bathonian, a local occurrence in Europe.

The Steinmanni Zone (Riccardi et al. 1989) (Plates 60–62) contains, besides the dominant East Pacific Eurycephalitinae and *Neuqueniceras* s. str., also cosmopolitan *Choffatia,* as well as *Hecticoceras* (*Prohecticoceras*) *blanazense* Elmi and *Eohecticoceras* sp., indicating the early Retrocostatum Zone, as in Mexico. The upper part of the zone is in northern Chile marked by the *Choffatia jupiter* Horizon, with, or just above, *Hecticoceras* (*Prohecticoceras*) *retrocostatum,* guide of the upper Retrocostatum Zone in Europe and northwest Africa. In Neuquén, the uppermost Steinmanni Zone has the *Stehnocephalites gerthi* Horizon (Riccardi et al. 1989), a strictly endemic genus accompanied by Andean species of *Choffatia* that are also known from the upper Steinmanni Zone of southern Mexico. The Steinmanni Zone is therefore dated as Retrocostatum–(?)Discites Zones.

Returning to the western Pacific: The only Bathonian ammonite faunas of Japan (Plate 72) are from the *Pseudoneuqueniceras yokuyamai* Zone of western Honshu, which, besides this close homeomorph of the Andean quasi-coeval *Neuqueniceras* (Riccardi and Westermann 1991), also yields cosmopolitan long-ranging *Choffatia;* its late Bathonian age is indicated by the superjacent *Kepplerites japonicus* Zone (discussed later).

The only indisputable Bathonian ammonite fauna from the southwestern Pacific is known from eastern Indonesia and Papua New Guinea (Plates 67–70) (Westermann and Callomon 1988). Lower Bathonian is suggested for the endemic genera *Satoceras* (also in Japan!) and possibly *Praetulites,* because they appear to underlie Middle Bathonian faunas. The Middle Bathonian with western Tethyan ammonites is the best-documented in the entire Pacific area: *Tulites* (*Rugiferites*) godohensis (Boehm) and *T.*? (*R.*?) *sofanus* (Boehm) (or ?*Bullatimorphites* s. str.) are known only from old collections from the famous locality "Keeuw," without stratigraphic control; but *Bullatimorphites* (*B.*) *ymir* (Oppel) and *B.* (*B.*) cf. *costatus* (Arkell), as well as *Cadomites* cf. *rectelobatus* (Hauer), close to *C. bremeri* Tser., have been collected from the two *Macrocephalites bifurcatus* Associations, here together named the *M. bifurcatus* Assemblage Zone. This Mediterranean assemblage is typical of the Bremeri Zone, upper Middle Bathonian (S. Medit.). The superjacent *Macrocephalites apertus* Assemblage Zone (new for *M. apertus-mantataranus* Association) includes late Bathonian *Oxycerites* (*Alcidellus*) gr. *tenuistriatus* (Gross.), a species known from the European Orbis and coeval Retrocostatum Zones, *M.* cf. *madagascariensis* Lem., a species probably occurring already in the Upper Bathonian of Madagascar and India (Krishna and Westermann 1987), and even a single *Xenocephalites grantmackiei* West. and Hudson, a species otherwise known only from New Zealand and closely resembling *X. neuquensis* (Stehn) of the late Bathonian Steinmanni Zone of the Andean Province (i.e., southern Mexico and the Southern Andes) (Westermann and Hudson 1991).

### Callovian

In Pacific Russia the stage is very incompletely developed, and only partly marine (Plate 76) (Meledina 1988). In North-East Russia, *Cadoceras* (*Paracadoceras*) cf. *anabarense* Bodyl. indicates the early Callovian *C. elatmae* Zone; in Far East Russia, *Longaeviceras?* cf. *keyserlingi* suggests the late Callovian *Cadoceras keyserlingi* Zone, of Athleta Zone age.

The Boreal North American ammonite succession appears to be similarly incomplete (Callomon 1984; Poulton 1987). Only three faunas are distinguished on the Alaskan North Slope: the early Callovian *Cadoceras arcticum* Zone (?Calloviense Zone), the ?middle Callovian *Stenocadoceras* cf./aff. *canadense* Frebold fauna, and the late Callovian *Longaeviceras stenolobum* Zone (Athleta Zone). Only the early Callovian *Cadoceras bodylewskyi* Zone is recorded from Yukon (Plate 22). The southern Alaskan ammonite associations (Plate 33) are better known and can be better dated because of their Subboreal composition, especially in equivalents of the Herveyi Zone (formerly Macrocephalus Zone). But relative stratigraphic positions of the usually isolated occurrences are tentative and partly based on comparisons with other bioprovinces (Callomon 1984). In ascending, probable sequence: The *Cadoceras comma* Association, with the *Kepplerites loganianus* faunule, is dated as Keppleri Subzone (basal Callovian) and probably is coeval with the Japanese *Kepplerites japonicus* Zone (discussed later); the *Cadoceras tonniense* faunule is coeval with the Nordenskjöldi Zone of eastern Greenland, apparently followed in British Columbia by an alternating sequence of East Pacific (*Lilloettia*) and Subboreal (*Cadoceras, Kepplerites*) faunules. All these faunas may still be of Herveyi Zone age. (Note that in the Andes, *Lilloettia* is now firmly dated as late Bathonian, followed in the early Callovian by the endemic *Eurycephalites*, as discussed later.) The *Cadoceras stenoloboide* Association reaches from Arctic Canada and Russia through southern Alaska and British Columbia, to Oregon (and possibly northern California), and is tentatively dated as middle Callovian. In southern Alaska the *Longaeviceras pomeroyense* Association, with *Cadoceras* (*Pseudocadoceras*), is late Athleta Zone in age. The highest Callovian appears to be absent. A small late Callovian fauna with *Quenstedtoceras* is known from Montana (Plate 79).

In Mexico, Lower Callovian is well documented only from Guerrero state (Plate 49) (Sandoval et al. 1989), whereas the record of the other substages is poor, as usual. The ammonite fauna clearly belongs to the Andean Province, and the same Lower Callovian standard zones (discussed later) are recognized here. The Vergarensis Zone is poorly developed, with rare *Eurycephalites* cf. *vergarensis* (Burck.), and overlies beds with the last *Neugueniceras* (*N.*) spp.

The superjacent Bodenbenderi Zone, coeval with the late Herveyi Zone and early Gracilis Zone, is dominated by Andean species, but has also yielded an exceptionally high proportion of Mediterranean species that permit close correlations: The Frickites Association, below, includes rare *Oxycerites* (*Paroecotraustes*) cf. *bronni* (Zeiss); the *Guerrericeras inflatum* Association, named after this endemic hecticoceratid genus (also in northern Chile?) (Gröschke and Zeiss 1990), contains the Mediterranean *Rehmannia* cf. *grossouvrei* (Petit.), *Parapatoceras distans* (Baug. and Sauze), *Paracuariceras* cf. *incisum* Schind., and *Jeanneticeras malbosi* Elmi, all of the top Herveyi to basal Gracilis Zones; the *Rehmannia* gr. *rehmanni* Association includes, below, the last *Neuqueniceras* (*Frickites*) and, above, *Reineckeia* cf. *franconica* (Qu.), clearly dated in Europe as early Gracilis/Calloviense Zone. The highest Callovian beds in the Cualac area belong to the Proximum Zone, having yielded *Hecticoceras* cf. *boginense* Petit. and *Reineckeia* sp., and are late Gracilis/Calloviense Zone in age.

Middle to Upper(?) Callovian is indicated by a number of *Reineckeia* s. str. "species," here placed in *R.* gr. *latesellata* Burck., and by possible peltoceratids ("*Peltoceras cricotum*") described by Burckhardt (1927) from the old mine shaft of El Consuelo in northwestern Oaxaca, but collected without stratigraphic control. The sparse record of *Rehmannia* gr. *densestriata* Burck. from the same mine also suggests the presence here of some older, Middle(?) Callovian beds. The small surface outcrops beside the collapsed shaft are poorly fossiliferous.

The Lower Callovian of the Southern Andes (Plates 63, 64), recently divided into standard zones at the famous Chacay Melehue section in Neuquén province (Riccardi et al. 1989), begins with the Vergarensis Zone. Besides the dominating East Pacific Eurycephalitinae (Riccardi and Westermann 1991), this zone has yielded only cosmopolitan and rather long-ranging perisphinctids and oppeliids with endemic species, so that its precise age cannot be established. The stratigraphic position of the Vergarensis Zone between the better-dated Steinmanni and Bodenbenderi Zones, however, indicates earliest Callovian rather than latest Bathonian. The superjacent Bodenbenderi Zone, with *Neuqueniceras* (*Frickites*) spp., *Choffatia* spp., and rare *Eurycephalites*, is dated as Middle Lower Callovian (upper Herveyi to lower Calloviense Zones) via south Mexico, where several Mediterranean guide species are associated, as discussed earlier. The Proximum Zone, with a Tethyan hecticoceratid fauna of *Hecticoceras* (*H.*) *proximum* Elmi, *H. boginense* (Pet.) and *H.* cf. *hecticum* (Rein.), and *H.* (*Chanasia*) *navense* Roman and *H.* (*C.*) *ardescicum* Elmi, can be clearly dated as upper Gracilis/Calloviense Zone, Patina and Enodatum Subzones, respectively.

*Rehmannia* (*R.*) *brancoi* and *douvillei* (Stein.) occur in northern Chile (e.g., Caracoles) within a few meters of the thick sequence, together with *Neuqueniceras* (*Frickites*) or *Jeanneticeras* cf. *meridionale* Elmi, indicating uppermost Lower Callovian. But precise stratigraphic collecting is still needed here. *R.* (*R.*) *stehni* (Zeiss) was found loose at Chacay Melehue in Neuquén and possibly came from the *Hecticoceras* beds (i.e., Proximum Zone) (Riccardi and Westermann 1991).

Middle (and upper?) Callovian (Plate 65) is incompletely developed in the Andean Basin and frequently in ammonitophobic facies (e.g., evaporites). The reineckeiids (Riccardi and Westermann 1991) of the highest marine Middle Jurassic levels in Neuquén, especially in the Chacay Melehue section, have yielded a few *Rehmannia* (*Lodzyceras*) *patagonica* (Weaver), together with *Oxycerites* (*Paroxycerites*) *oxynotus* (Leanza), probably indicating lowest Middle Callovian.

In eastern Asia south of Russia, Lower Callovian is known from western Honshu, Japan (Plate 72) (Sato and Westermann 1991). The *Kepplerites japonicus* Zone is very probably basal Callovian because of the close resemblance of the *Kepplerites* spp. to those of the *K. loganianus* Association of the Queen Charlotte Islands, British Columbia. The *Oxycerites* Zone, above, includes *O.* cf. *sulaensis* West. & Call., *Bullatimorphites* cf. *microstoma* (Orb.), and *Hecticoceras* (*Chanasia*) cf. *chanasiensis* (Per. & Bon.), strongly indicating late early Callovian.

Farther south, Callovian hecticoceratids occur in the Philippines (Plate 72) (Sato and Westermann 1991).

A probably late Lower Callovian, isolated ammonite fauna is known from eastern Indonesia (Plate 71) (Westermann and Callomon 1988); but again, there is no fossil evidence for Middle and Upper Callovian marine deposits. (Note that the previous records of Middle Callovian "*Eucycloceras, Idiocycloceras,* and *Subkossmatia*" were all concerned with Bajocian-Bathonian Sphaeroceratinae.) The *Macrocephalites keeuwensis* Association includes *M.* cf. *folliformis* Buck. and *Choffatia* cf./aff. *furcula* (Neum.), both known from the upper Lower Callovian of Europe, but the latter species being long-ranging; the endemic *Oxycerites* (*O.*) *sulaensis* West. & Call., formerly dated as early Bathonian, probably recurs in Japan (as discussed earlier) and Nepal, as indicated by a single large fragment from the Ferruginous Oolite (Mouterde 1971, Pl. 1, Fig. 1), which in the Thakkola Valley is now tentatively dated as earliest Callovian, *M. madagascariensis* Association (G. E. G. Westermann unpublished data). Another specimen of *O. sulaensis* has recently been found here together with several Lower Callovian macrocephalitids (E. Cariou personal communication). The closest European species of *O. sulaensis* appears to be *O. tilli* Loczy from the Middle Callovian of Hungary. [A single, distorted fragment from the late Bathonian Orbis Oolith of the Franconian Alb recently compared by Dietl and Callomon (1988) with *M. keeuwensis,* and their consequent chronostratigraphic remarks concerning the Indonesian assemblage, cannot be taken seriously at this time; i.e., the angular umbilical edge is common to most Indonesian *Macrocephalites* species.]

## UPPER JURASSIC, ESPECIALLY OF MEXICO[3]

Marine sediments of Upper Jurassic age are known to be widespread all round the Pacific. In marked contrast to the highly detailed successions that have been worked out in the shelf carbonates of the western Tethys and the epicontinental sediments of the adjacent Eurasian cratonic basins, our knowledge of circum-Pacific faunal successions remains highly fragmentary. The distribution of the Upper Jurassic ammonites is particularly patchy. Many of the shallow-shelf sediments that tended to be their favorite habitats have disappeared because of subduction or other megatectonic processes (western and northwestern Pacific) or have been subjected to intense synsedimentary perturbations of volcanic origin, giving rise to turbiditic sequences poor in fossils of any kind and now very hard to unravel (American Cordillera). Yet other areas seem to have been ecologically unfavorable to ammonites, whose place in the fossil record was taken by buchiids (northern Pacific, North America) or inocerami (southern Pacific). Good successions of rich and diverse ammonite faunas have so far been described from only two regions: in Mexico and in the Chilean and Argentine Andes. A standard zonation has been worked out only for the Tithonian of the Andes. Everywhere else, the best that can be done at present is to record the ammonite assemblages that have been identified and to indicate their ages against the standard European scales as far as this is possible. In many cases this is made additionally difficult by yet further problems, those of bioprovincialism. Some of the spot collections consist of forms so far unknown elsewhere that are hence not closely correlatable. The most famous example must still be the "*Titanites*" *occidentalis* from the topmost marine Upper Jurassic level of the southern Canadian Rockies (discussed later).

The ammonite successions are summarized in Table 12.3 clockwise from southern Alaska, and the discussion is in the form of notes.

The faunas of most of Canada and the United States (first three columns) were recently reviewed by Callomon (1984). They were labeled by numbers, and for convenience these numbers are included here (B12–17, A11–18). The discussions will not be repeated, and the reader is referred to that review for details and references. The geographic classification follows current assignments to tectonic terranes (e.g., Coney, Jones, and Monger 1980; Monger, Price, and Templeman-Kluit 1982).

*Wrangellia Terrane* (southern Alaskan Peninsula, Talkeetnas, Queen Charlotte Islands, and Vancouver Island). For the most recent review, see Imlay 1981 (Plate 77).

*Central British Columbia* (intermontane Stikinia and Quesnellia Terranes, Taseko-Tyaughton and Manning Park areas). For the Lower Oxfordian, there are rich but scattered collections from

[3] By J. H. Callomon.

Table 12.3. *Correlation chart for the circum-Pacific Upper Jurassic. Data for Antarctic Peninsula, New Zealand, Indonesia and Papua New Guinea, and Japan and the Philippines compiled by Westermann (ed.) from Part III and from Riccardi et al. (1990)*

| Stage | Substage | STANDARD EUROPEAN ZONES | S. ALASKA & W. BRITISH COL. (WRANGELLIA Terrane) | CENTRAL BRITISH COLUMBIA | WESTERN INTERIOR | CALIFORNIA & SW OREGON | MEXICO |
|---|---|---|---|---|---|---|---|
| BERRIASIAN | | OCCITANICA | | [Buchia okensis] | | Neocosmoceras | Subthurmannia mazatepensis M22 |
| | | EUXINUS | | | | Proniceras | Protacanthodiscus densiplicatus M21 |
| TITHONIAN | U | Durangites spp. | [Buchia fischeriana] | [B. fischeriana] | | Substeueroceras; Durangites ? | Substeuerocceras, Kossm M19-20; Corongocoras cordobai M18 |
| | | MICRACANTHUM | | | | ? | Suarites bituberculatus M17; Kossmatia victoris M16 |
| | M | PONTI | | [B. lahusemi] | MMF | [Buchia piochii] | Durangites acanthicus M15; Proniceras neohispan. M14 |
| | | FALLAUXI | "Subplanites" & Aulacosphinctoides B17 | | ? | | Torquatisphinctes ? potosinus M13 |
| | | SEMIFORME | | [B. piochii] | ? | | Virgatosphinctes aquilari b M12 |
| | L | DARWINI | [Buchia rugosa, B. mosquensis] | | | | Mazapilites symonensis a |
| | | HYBONOTUM | | [B. mosquensis] | | | Hybonot. autharis M11; Lithacoceras mexicanum M10 |
| KIMMERIDGIAN | U | BECKERI | | | "Titanites" occidentalis A18 | | Hybonitic. beckeri M9 |
| | | EUDOXUS | Phylloceras alaskanum | ? | | Amoeboc. dubium B16 | Haploceras transatlanticum M8 |
| | | ACANTHICUM | | | ? | | Epicephalites epigonum b M7; Idoceras santarosanum a |
| | L | DIVISUM | | | | | Streblites uhligi b M6 |
| | | HYPSELOCYCLUM | | Discosphinctoides ? [B. concentrica] | | [B. concentrica] | Idoceras durangense a |
| | | PLATYNOTA | | | | | Ataxioceras ? spp. M5 |
| OXFORDIAN | U | PLANULA | | | Amoeboceras [B. concentrica] A17 | Subnebrodites ? | |
| | | BIMAMMATUM | | | | ? Perisphinctes virgulatiformis | Ochetoceras mexicanum M4 |
| | | BIFURCATUM | Amoeboceras cf. transitorium B15 | | MF | | |
| | M | TRANSVERSARIUM | | | Cardioc. sundancense A16 | ? | Perisphinctes elizabethaeformis M3 |
| | | PLICATILIS | Cardioc. whiteavesi B14 | | C. canadense A15 | Perisphinctes muehlbachi | Perisphinctes durangensis M2 |
| | L | CORDATUM | C. spiniferum B13; C. martini B12 | Cardioc cf. alaskense; C. cf. mountjoyi | C. distans, C. hyatti A14; C. mountjoyi, C. reesidei A13 | ? | ? |
| | | MARIAE | | C. cf. praecoadatum, C. cf. scarburgense B11 | C. cordiforme A12 | Metapeltoceras ? | Per.(Prososphinctes ?) sp, Fehlmanites sp.? M1 |
| | | | | Quenstedtoceras cf. lamberti | Quenst. collieri A11 | | |

Table 12.3 (*cont.*)

| ANDES (*) | ANTARCTIC PENINSULA* | NEW ZEALAND * | REGIONAL STAGES | INDONESIA & PAPUA NEW GUINEA * | JAPAN & PHILIPINES * | FAR EAST of USSR * |
|---|---|---|---|---|---|---|
| Spiticeras damesi Zone | | | | | | |
| *Argentiniceras* noduliferum Z. | Spiticeras | (non-marine) | PUARUAN | Himalayitids | | |
| *Substeueroceras* koeneni Z. | Haploph. strigile | | | Haplophyll.strigile | Substeueroceras Corongoceras | |
| Corong. alternans Z. | Blanfordiceras | | | Blanford.wallichi | | |
| *Windhauseniceras* interspinosum Z. | Corong. lotenaense | | | | | |
| *Anlacosphinctes* proximus Z. | Kossmatia carsensis | | | | | Aulacosphinctes proximum Zone |
| | | | | Uhligites | | |
| *Pseudolissoceras* zitteli Z. | Uhligites Aulacosphinctes Virgatosphinctes | Uhligites Aulacosph. ? browni | | Aulacosphinctoides Kossmatia desmidoptycha Paraboliceras pseudopar. | Aulacosphinctoid. cf. steigeri Assoc. | Pseudoliss. zitteli Z. |
| "*Virgatosphinctes*" mendozanus Z. | | Virgatosph.marshalli Paraboliceras | OHAUAN | Pachyplanulites novaguin. | | Virgatosph. cf. mexicanus |
| | | Kossmatia novaseel. | | ? | | |
| ? | "Pachysphinctes" Subdichotomoc. | ? | ? | | | |
| Torquatisphinctes | | | | | | |
| Orthaspidoceras | | | | (no ammonites) | | Ochetoceras elgense |
| | | ? | HETERIAN | | Ataxioceras kurisakense Assoc. | Amoeboc. cf. dubium |
| (gypsum) | | Idoceras | | | | A. ex gr. kitcheni beds |
| | | Epicephalites | | | | |
| | ? | ? | | ? | | Perisphinctes cf. muehlbachi |
| Euaspid. hypselum | | Epimayaites & Retrocer.galoi | | | Kranaosph. matsuch. Z / Euaspid cf. hypselum Z. | Dichotomosphinctes spp. beds |
| Cubaspidoceras | Epimayaites aff. transiens | ? | | Perisphinctes burui-molucc. [Retrocer.galoi] | | |
| Perisphinctes-Araucanites Assoc. | | | ? | Epimayaites gr.palmarum | | |
| | | | | | | Scarburgic spp. beds |
| Peltoceratoides | ? | | | Peltoceratoides Parawedekindia | Parawedekindia arduennensis Z. | Cardioc praecordatum |

many places in British Columbia (Plate 78), many as yet unpublished. For the Kimmeridgian, see Poulton, Zeiss, and Jeletzky (1988).

*Western Interior United States and Canada* (Rocky Mountain foothills, Montana-Wyoming-Utah) (Plates 78, 79). Morrison Formation (MF): mostly nonmarine sediments sharply terminating the Middle Oxfordian south of the 49th parallel. Mist Mountain Formation (MMF): the main coal-bearing formation of the southern Rocky Mountains of Alberta and British Columbia. Farther north, in the Peace River area of northeastern British Columbia, marine sedimentation may have been more or less continuous into the Neocomian (Monteith Formation), but no positively identifiable ammonites have been recorded. The age of *"Titanites" occidentalis* and hence the age of the MMF continue to be highly uncertain (Callomon 1984).

Upper Jurassic appears to have been identified at only one locality in the important, possible allochthonous Jurassic window of eastern Oregon. It lies at the most northeasterly corner, near the junction of Oregon, Washington, and Idaho. The ammonites are all Lower Oxfordian *Cardioceras martini*, suggesting that this locality is tectonically still part of the Western Interior United States (Imlay 1964, p. D15, Pl. 2, Figs, 4 and 5, text Fig. 1, locality 1; Imlay 1980, pp. 66, 41, text Fig. 18, locality 13).

*California and southwestern Oregon* (Franciscan Mélange; Coast Ranges; Sonomia Terrane: Sierra Nevada). Upper Jurassic rocks occur in two separate major NW–SE-trending belts belonging to separate allochthonous terranes. Ammonites are very rare, accumulated over the years from isolated spot occurrences, often lying in enormous thicknesses of slope sediments and poorly preserved. Additional and more widespread evidence comes from buchiids. All one can do, therefore, is to give very broad outlines of the geology, with tentative identifications of the ammonites and their ages. Most of them are perisphinctids. The special difficulties in identifying scattered specimens of this group are well known, arising from the tendency to produce iterative homeomorphs of quite different ages. Claims to have identified particular genera, such as *Dichotomosphinctes, Discosphinctes,* or *Subdichotomoceras,* with connotations of their implied ages, can be illusory, especially when the correlations are between regions as distant as California and Europe. The assignments in column 4 of Table 12.3 are essentially those of Imlay (1961, 1980), who reviewed the available evidence, such as it is.

The faunas in the lower half of the column were found in the Sierra Nevada, on the Sonomia Terrane. The *"Metapeltoceras"*?, based on a single, stratigraphically unlocalized specimen, may well be a peltoceratid, although it is not closely comparable with known species. It could still be Callovian. The perisphinctids above do give the general impression of being Oxfordian, and Imlay's comparison with undoubtedly Upper Oxfordian faunas from Cuba may well be correct. *Subnebrodites* is the proper name for the "early *Idoceras*" group from the Planula Zone, with which the Californian material was compared. *Idoceras* proper (type species *I. balderum*) comes from the Divisum Zone at the top of the Lower Kimmeridgian. The only indubitably identifiable fauna from Sonomia remains the *Amoeboceras dubium* (Hyatt) from the Mariposa Formation. If the comparison with closely similar forms from Europe is correct, its age is Eudoxus Zone. Its appearance today so far south lends support to the suggestion that the Sonomia Terrane has been tectonically displaced southward. Many of the other, earlier faunas found in the region, such as Callovian *Kepplerites* and *Cadoceras,* also point to a northerly, Subboreal provenance.

In contrast, all the faunas shown in the upper part of this column, from the Coastal Ranges, strongly indicate more southerly affinities, particularly with Mexico and the Andes. The faunas are still sparse in the Jurassic, but become rich and diverse in the Lower Cretaceous (e.g., Imlay 1960; Imlay and Jones 1970). The southerly, warm-water aspect of the ammonites continues, however, to be balanced by the presence of quantities of species of *Buchia.*

*Mexico.* The Jurassic of Mexico can be divided into five regions:

1. Sonora in the northwest, including Baja California,
2. Coahuila in the northeast, extending into southern Texas and beyond,
3. north-central Mexico: Chihuahua, Durango, Zacatecas, Neuvo León,
4. southeastern borders of the Gulf of Mexico: Veracruz,
5. southern Mexico, Pacific borders: Guerrero, Oaxaca.

Of these, 1 and 5 lie on allochthonous terranes that may have traveled over considerable distances. They retain little Upper Jurassic and will not be considered further here.

Regions 2 and 4, although also traversed by tectonic discontinuities, are not greatly displaced from their relative positions in the Jurassic, which were closely contiguous. They retain what was early recognized to be one of the most complete and important Upper Jurassic–Neocomian ammonite successions in the world. Unfortunately, with few exceptions, there has been little modern revisionary work, and we are still largely dependent on the classical accounts by Burckhardt (1906, 1912, 1919–21) for our knowledge of the faunas.[4] The early work has been repeatedly summarized: by Burckhardt (1930), by Imlay (1939), who revisited and recollected many of Burckhardt's localities, by Arkell (1956), and, most recently, again by Imlay (1980). It was based largely on the area in region 3, between San Pedro del Gallo (Durango) and San Lazaro in the Sierra Madre Oriental (Nuevo León), some 500 km from east to west, via the classical localities of Symón and Mazapil. The lithostratigraphy of the area of San Pedro del Gallo has recently been redescribed by Contreras Montero, Martinez Cortes, and Gomez Luna (1988), who also described some new collections of ammonites.

The earliest Jurassic rocks in the region consist of redbeds, shales, sandstones, and shallow-water carbonates, probably of early Oxfordian age. They contain few ammonites. They are followed by thick sediments in predominantly argillaceous facies

[4] Editor's comment: The classical Upper Jurassic outcrops of Mexico have also been recollected recently by F. Olóriz in a joint project with C. Goncales and G. E. G. Westermann, but the studies have not been completed (see also Chapter 5).

characteristic of more distal shelf basins, ranging in age from Upper Oxfordian into the Cretaceous. Formations typically are tens to hundreds of meters thick and change only slowly with distance, so that lithostratigraphic correlations can be made very widely. With some exceptions, such as levels full of *Buchia*, the faunas are dominated by well-preserved ammonites that tend to occur in concretions in shales. Previous biostratigraphy has had to be based on material from rather broad stratigraphic intervals lumped together, but perusal of the associations in the old collections shows that these change significantly from one locality to another. This suggests strongly that faunal successions here are as locally discontinuous and incomplete as elsewhere and that, conversely, a synthesis of more carefully collected assemblages should make it possible to construct a more finely resolved succession that could become the basis for a new, regional Upper Jurassic standard zonation. Such a zonation would be of immense value as a fixed reference point for regional correlations in what is a pivotal region in the Hispanic Corridor, at the gateway between the East Pacific and the western Tethys. An attempt has therefore been made to identify and list distinguishable faunal horizons, as summarized in Table 12.3. For ease of reference, they have been numbered from below, with the name of one characteristic species selected in each to label them. As usual, the time duration represented in any of the faunal horizons is not known and can only be guessed at. They are therefore represented in the table by a discontinuous succession of boxes at levels, more or less uncertain, indicating the closest correlation with the European standard that is possible at this time.

*M1*. Based on a sparse fauna described by Imlay (1939 pp. 33–4) as *Indosphinctes*(?) (his Pl. 7, Fig. 7), *Subgrossouvria*, and *Pseudopeltoceras*(?) (unidentifiable fragments) from Las Cuevas, east of San Pedro del Gallo. Of these, the *"Indosphinctes"* bears some resemblance to new discoveries in the Lower Oxfordian of British Columbia (Poulton, Callomon, and Hall 1991). Another indication may be a small *Perisphinctes* illustrated by Imlay (1982, Pl. 26, Figs. 8–10) from the topmost Callovian of Montana. A further possibly early Oxfordian fauna may be present in the Santiago Formation of east-central Mexico, from which Cantu (1971,p. 23) cites, but does not figure, *Fehlmannites*. All these species could, however, still be Callovian.

*M2*. *Perisphinctes* (*Dichotomosphinctes*) *durangensis* Burckhardt (1912, Pl. 3, Fig. 2; Pl. 4, Fig. 6) and other species (Pl. 2, Figs. 13–14; Pl. 3, Figs. 3–6; Pl. 4). As Arkell (1956, p. 563) has already pointed out, these are so similar to European forms that their position in the Plicatilis Zone is in little doubt. *P. durangensis* is barely distinguishable from *P. antecedens* Salfeld, indicating the upper part of the Plicatilis Zone.

*M3*. *Perisphinctes* (*Dichotomosphinctes*) *elizabethaeformis* Burckhardt (1912, Pl. 6; Pl. 7, Figs. 1–3) strongly indicates the lowest, Parandieri Subzone of the Transversarium Zone. The *Perisphinctes* beds of San Pedro are 150 m thick, so a spread of ages and faunas is not unexpected.

*M4*. *Ochetoceras mexicanum* Burckhardt (1912) and other species. Very close to European *O. canaliiculatum*, but to avoid possible circular arguments in precise correlation, it seems safest at this stage to retain local names if such exist. The rest of the fauna is also similar to that of the European Bimammatum Zone (Arkell 1956, p. 563), with, in addition, a Cuban specialty in *Perisphinctes carribbeanus* Jaworski (1940, Pl. 3, Figs. 1–2; Burckhardt 1912, Pl. 7, Figs. 4–14; Arkell 1956, pp. 563, 573).

These three faunas are of special interest in being the oldest to show close and unambiguous relationships with those of Cuba, Portugal, the western Tethys, and northwestern Europe as a whole, rather than with adjacent regions of the eastern Pacific. They therefore date the fully marine, faunally unrestricted opening of the connection between the central East Pacific and the western Tethys along the Hispanic Corridor (early Central Atlantic) as certainly no later, and probably not much before, Middle Oxfordian. This connection continued without further break into the Cretaceous.

*M5*. "*Ataxioceras* Zone" of Cantu (1969, p. 7; 1971, p. 25). This lies in the lower part of the Taman Formation of east-central Mexico and is said to be characterized by abundant but fragmentary *Ataxioceras* aff. *subinvolutum* (Siemiradzki) and *Taramelliceras* aff. *subnereum* (Wegele) and to account for a citation of *Sutneria* cf. *platynota* by Burckhardt (1930, pp. 61, 91). The ammonites have not been figured, and their generic identifications must remain in doubt, if only because they would be the first and only representatives of *Ataxioceras* known outside the confines of the western Tethys and its adjacent epicratonic shelves. The *Sutneria*, if correctly identified, would indicate basal Kimmeridgian.

*M6*. "Beds with *Idoceras* group of *durangense*" of Imlay (1939). Faunas described almost entirely from San Pedro del Gallo (Burckhardt 1912), where good, clean sections appear to be rare. Most of the material came from unspecified levels in 400 m of "couches inférieures de San Pedro del Gallo." The area was revisited by Imlay (1939), and the assemblages he obtained from different localities indicate at least two separable faunas:

*M6(a)*. *Idoceras durangense* Burckhardt and 12 other species [M], with *I. sautieri* (Fontannes) *sensu* Burckhardt [m], *Aspidoceras neohispanicum* B. and other spp., *Nebrodites quenstedti* B. [M] and other species, *N. haizmanni* B. [m], and *Sutneria* aff. *cyclodorsata* (Moesch).

*M6(b)*. *Streblites uhligi* B. [M], with *S.* (*"Glochiceras"*) *auriculatus* B. [m] and *Idoceras* spp. as in M6(a). Fauna M6(b) is found at "locality 4" only (Imlay 1939, p. 15) and lacks the aspidoceratids of M6(a). Imlay remarks that it probably represents only the upper part of the beds with *Idoceras*.

Close correlation with Europe is not easy. The idoceratids of the group of *I. durangense* are unknown there. Arkell (1956, p. 563) regarded them as "clearly related closely to *Ataxioceras*," but this seems highly unlikely, for although compressed, involute and densely ribbed, the style of the secondary ribbing is quite distinct. The ataxioceratids are characterized above all by their unmistakable virgatotome ribbing. In the Mexican forms, instead, the division of the primary ribs into secondaries is bidichotomous. The *I. sautieri* recalls *Subnebrodites* (formerly *Idoceras*) of the *planula* group. The aspidoceratids, although not strongly diagnostic, closely resemble the forms often referred to *Aspidoceras binodum* (Quenstedt) and include some *Physodoceras*, both commonest in Swabian White Jura $\gamma$, Hypselocyclum Zone. Similarly, *Streblites uhligi* is very close to *Str. tenuilobatus* (Oppel), which is commonest in the Hypselocyclum-Divisum Zones. On balance, this seems the most probable age.

This horizon appears to be the one that has yielded the unique ''*Pictonia*'' (*Colladites*) *granadillensis* Cantu (1967).

*M7*. ''Beds with *Idoceras* group of *balderum* (Oppel)'' of Imlay (1939). The faunas were first described from the region of Mazapil by Burckhardt and were followed further eastward to the Sierra Madre Oriental by Imlay. They do not seem to occur either at San Pedro del Gallo or at Symón. Again, there appear to be at least two separable assemblages:

*M7(a). Idoceras santarosanum* Burckhardt [M] (including *I. soteloi, subdedalum, canalense, viverosi,* and *balderum sensu* Burckhardt), with *I. neogaenum* B. [m] (including *I. humboldti, submalleti, mexicanum, ?zacatecanum,* cf. *hospes,* and *laxevolutum sensu* Burckhardt), *Nebrodites aguilerae* B. [M]. *N.* cf. *doublieri* (D'Orb.) B. [m], *Aspidoceras quemadense* B., *A. contemporaneum* (Favre), *Streblites sanlazarensis* (Imlay), *Taramelliceras boesei* B., *Metahaploceras* aff. *nereus* (Fontannes), and *Sutneria* cf. *cyclodorsata* (Moesch).

*M7(b). Taramelliceras harpoceroides* B., with *T.* aff. *pugile* (Neumayr) (Burckhardt 1906, Pl. 3, Fig. 5), *Epicephalites epigonus* (Burckhardt), and *Subneumayria ordonezi* (B). The near mutual exclusiveness of these two assemblages may be seen in comparing Burckhardt's sections at Mazapil, Casa Sotelo [Burckhardt 1930, p. 50, Fig. 13a, bed H; assemblage M7(a)], and Puerto Blanco [Burckhardt 1930, p. 51, Fig. 13b, bed L; assemblage M7(b)]. Assemblage M7(a) also occurs elsewhere (Imlay 1939, Table 4), but M7(b) seems to be known only from the one locality. Perhaps even more striking than the forms that do occur in M7(b) are the ones that do not. With the possible exception of two specimens recorded from Puerto Blanco by Imlay, it is the commonest ones of M7(a) that are missing there: *Idoceras* ex gr. *balderum* and *Aspidoceras.* The two assemblages share *Sutneria* cf. *cyclodorsata* and the local curiosity, *Subneumayria ordonezi.*

Correlation with Europe is fairly clear. The close similarity between Mexican *Idoceras* and the European type species, *I. balderum* (Oppel), maintained by all previous authors, seems secure. The age of *I. balderum* is Balderum Subzone at the top of the Divisum Zone [Swabian White Jura $\gamma 6$ in Aldinger's notation of 1945 – see Ziegler (1977)]. The assemblage as a whole, however, making the fairly safe assumption that all of Burckhardt's 12 or so ''species'' (and specimens) are merely variants of a single dimorphic biospecies, is not the same as the analogous assemblage of *I. balderum.* As usual, there can be two reasons for this disparity: a slight difference in ages, or bioprovincial separation into geographic subspecies – there is no telling which. It is therefore best not to use the name of the German species to label the Mexican assemblage, as was already advised by Ziegler (1959). The names used here – *santarosanum* [M]/*neogaenum* [m] – are those of what seem to be the most typical forms. *Aspidoceras quemadense* seems difficult to distinguish from *A. acanthicum* (Oppel), and the *A. contemporaneum* (Favre) seems to be correctly identified. Both species in Europe are commonest in the Acanthicum Zone (White Jura $\delta$1–2), although they do occur earlier. The *Nebrodites* are close to *N. herbichi* (Neumayr) [M]/*hospes* (Neumayr) [m], which in Swabia range from the upper Divisum into the Acanthicum Zone. This is also the acme of *Sutneria cyclodorsata.* The *Taramelliceras harpoceroides* is close to *T. pseudoflexuosum* (Favre 1877), which occurs in Europe in the Acanthicum Zone and a little higher. For this admittedly not very strong reason, and because of the persistence of the ''pseudo-rasenids'' like *Epicephalites* and *Subneumayria* (alias *Procraspedites*) into the next horizon above, fauna M7(b) is here placed above M7(a).

There has also been uncertainty about the relative levels of assemblages M6 and M7, and direct stratigraphic proof is still lacking. Burckhardt made no attempt to separate them, referring them simply to ''Couches à *Idoceras*'' or ''Couches inférieures de San Pedro.'' The first to realize that they were distinct appears to have been Imlay. He placed them in the opposite order to that adopted here. He gave no explicit reasons, but seems to have arrived at his conclusion by attempts to correlate them with the European succession. There were two difficulties. First, the levels of many European species that most resembled Mexican forms, such as *Idoceras balderum, Physodoceras pavlowi, Aspidoceras acanthicum,* and *Streblites tenuilobatus,* were themselves not precisely known, locked either in the ''White Jura $\gamma$-$\delta$'' (Quenstedt) or comprehensive ''Zone des *Amm. tenuilobatus*'' (Oppel) duality of nomenclatures, little changed since the nineteenth century. Second, there were assumed generic affinities that probably are incorrect, locked in circular argument with their age assignment. This includes the presumption of affinity between Mexican *Idoceras* ex gr. *durangense* and *Ataxioceras* (discussed earlier), *Subneumayria* and *Involuticeras,* and *Pararasenia* and *Aulacostephanus.* Imlay was followed by Arkell (1956, p. 563). Since that time there have been comprehensive revisions of the stratigraphy of the Submediterranean Kimmeridgian, notably by Hölder, Ziegler, and Geyer, and of the systematics of some of its ammonites. The revised correlations of the Mexican faunas were indicated earlier.

*M8.* ''Beds/Zone with *Haploceras* ex gp. *fialar* (Oppel)'' of Burckhardt, Imlay, and Cantu. The identification of the microconchs so abundantly illustrated by Burckhardt (1906, Mazapil) with *Haploceras* or *Glochiceras fialar* was already questioned by Imlay and rejected by Ziegler (1958). The macroconchs, by common consent, are typical *Haploceras* s.s. close to *H. subelimatum* (Fontannes, 1879). Once again, to avoid prejudging correlations, it seems best to retain local names to label the Mexican fauna. Several are available. The ones chosen here are *Haploceras transatlanticum* Burckhardt, 1906 [M] (including *H. zacatecanum, mexicanum, cornutum, felixi,* and *costatum* B.) and *Glochiceras mazapilense* (Aguilera, 1895) [m] [including *G. carinatum* (Ag.)]. Imlay combined this fauna with the *Idoceras balderum* fauna below, M7, but gave no reasons. In this he was followed by Arkell. But both the stratigraphic evidence and the compositional evidence make it unambiguously clear that faunas M7 and M8 are distinct. This was reconfirmed by Cantu (1971). Correlation with European standards is again fairly close. *H. subelimatum* (Font.) occurs in the Beckeri Zone (White Jura $\epsilon$1–2). The microconchs perhaps closest to *G. mazapilense* are *G. modestum* Ziegler (Eudoxus Zone, δ3–4) and *G. solenoides* (Quenstedt) (Beckeri Zone and upward). Minor elements include *Metahaploceras* aff. *strombecki* (Oppel), a species of the Hypselocyclum Zone, and *Taramelliceras* cf. *trachynotum* of the Divisum Zone. A level somewhere in the Eudoxus Zone therefore seems most probable, above M7 of Acanthicum age and below M9 of undoubted Beckeri age.

The systematic positions of *"Craspedites" praecursor* Burckhardt, type species of *Procraspedites* Spath, 1930, and the similar *"C." mazapilensis*, which occur in this fauna, remain uncertain.

*M9.* "Beds with *Waagenia*" of Mazapil and Symón (Burckhardt 1906; Imlay 1939). All the *Hybonoticeras* that have been figured from these beds seem to indicate the Beckeri Zone. There are insufficient other forms associated with the *Hybonoticeras* to help much with correlation.

*M10.* "Zona con *Virgatosphinctes mexicanus* y *Aulacomyella neogeae*" of Cantu (1971). *Lithacoceras? mexicanus* (Burckhardt 1906). An isolated fauna of perisphinctids from the Cañon de San Matias near Mazapil (Burckhardt 1930, p. 50, Fig. 13, bed E; 1906, Pl. 30, Figs. 4 and 8; Pl. 31, Figs. 5–9; Pl. 32, Figs. 1–2) stands out by virtue of its prominent and abundant genuine virgatotome secondary ribbing, in contrast to the idoceratids below, and many of the "virgatosphinctids" above. The style recalls that of the European forms found around the Kimmeridgian–Tithonian boundary [e.g., *Lithacoceras albulus* (Quenstedt) (Berckhemer and Hölder 1959, Figs. 48 and 51)]. The position in Mexico is immediately above the *beckeri* fauna, which would also fit, but there are no other time-diagnostic elements in the fauna. Imlay (1939, Table 10) placed it higher, above *Mazapilites*, assigning it to *Virgatosphinctes* and *Subplanites*, but he had no new stratigraphic evidence. Cantu (1971, p. 27) confirms its position at the level inferred here. *"Virgatites" mexicanus* is often cited as occurring in the (*"Virgatosphinctes"*) Mendozanus Zone of the Southern Andes, but the forms illustrated under this name differ substantially from the Mexican ones.

*M11.* "Beds with *Waagenia* and *Mazapilites*" at Toboso, Sierra de Symón (Burckhardt 1919–21; Imlay 1939, p. 10, bed 7). The *"Waagenia"* (Burckhardt 1919–21, Pl. 4, Fig. 11) seems to be certainly identifiable as *Hybonoticeras* cf. *autharis* (Oppel), the microconch of *H. hybonotum* (Oppel). This level sees the first *Mazapilites*, which may have a plausible ancestry in *Taramelliceras kiderleni* of Berckhemer and Hölder (1959) (Enay 1972, p. 373) of the Beckeri Zone.

*M12.* "Beds with *Mazapilites.*" There appear to be two slightly different faunas:

*M12(a):* Sierra de Symón, Toboso (Imlay 1939, p. 10, bed 8): *Mazapilites symonensis* Burckhardt and three other species; *Torquatisphinctes? tobosensis* [M]/*neohispanicus* [m], *T.*? *titan* [M]/ *boesei* [m], all Burckhardt (1919–21) species; and several other species. The generic assignment of these forms has varied widely, including *Aulacosphinctes*, *Aulacosphinctoides*, *Virgatosphinctes*, *Pachysphinctes*, and *Epivirgatites*. While the coiling and general style of ribbing do certainly resemble those of some *Virgatosphinctes*, there are two major differences. The characteristic virgatotome secondaries of *Virgatosphinctes* seem to be wholly absent from the Mexican forms, discounting perhaps occasional virgatitid-like triplicates at constrictions, while the Mexican forms bear frequent simple ribs not found in *Virgatosphinctes*. These characteristics are shared by the earlier Idoceratinae and some later Himalayitidae. They appear also to be characteristic of Indo-Ethiopian *Torquatisphinctes*, and although it may be widening the scope of this genus somewhat, it seems the most appropriate one for the Mexican faunas among those currently available (cf. Enay 1972, p. 373). The more evolute variants (e.g., *T. boesei*) are certainly very typical. Ancestors may include *"Perisphinctes" cimbricus* Neumayr, 1873 (Berckhemer and Hölder 1959, Figs. 24–26) of the Beckeri Zone.

*M12(b).* Sierra de la Caja, Mazapil (Burckhardt 1930, p. 52, Fig. 14, bed G): *Mazapilites zitteli* Burckhardt (1906), with *Aspidoceras cyclotum* (Oppel) (incl. *phosphoriticum*, *cajaense* B., etc.), *Virgatosphinctes? aguilari* Burckhardt (1906, Pl. 27, Figs. 6–9), and *Torquatisphinctes nikitini* (B.). There is no clear evidence as to the order of 12(a) and 12(b). Fauna 12(a) covers an extended interval of shaly limestones, whereas 12(b) occurs in what appears to be a phosphatized, probably condensed, thin terminal member of a sequence. It contains ?*Virgatosphinctes* more closely resembling the forms found in the series M13 above than those below, and *Virgatosphinctes* was also recorded by Imlay at the top of M12(a). Hence M12(b) is here presumed to lie above M12(a).

*M13.* "Virgatosphinctinae beds of Sierra Catorce" of Verma and Westermann (1973): *Torquatisphinctes? potosinus* (Aguilera [including *T. laurei* (Ag.), *"Subdichotomoceras"* aff. *inversum* Spath–Verma and Westermann, and *Virgatosphinctes? sanchezi* V. & W.], with *Mazapilites mexicanus* (Aguilera), *Haploceras elimatum* (Oppel), *Pseudolissoceras zitteli* (Burckhardt), *Simoceras (Volanoceras)* cf. *volanense* (Oppel), and *Aspidoceras alamitoense* (Ag.). This appears also to be the horizon in northern Mexico from which Imlay (1943) described *"Virgatosphinctes" chihuahuaensis*, *adkinsi*, and *"Subplanites" fresnoensis*, as far as one can tell from the limited material.

For the distinctions between *Torquatisphinctes* and *Virgatosphinctes* the same remarks apply as under M12. The similarities are closer in M13, but it is better to follow Enay (1972, pp. 373, 377) in restricting the name *Virgatosphinctes* to the Middle/ early Upper Tithonian forms of the Spiti Shales. The age of M13 appears to be firmly Middle Tithonian. In Europe, *Simoceras* of the *volanense* group ranges through the Fallauxi-Ponti Zones; *Aspidoceras rafaeli* (Oppel), which is barely distinguishable from *A. alamitoense*, ranges from the Darwini to the Semiforme Zone; and *Pseudolissoceras bavaricum* Barthel, possibly a synonym of *Ps. zitteli*, ranges through the whole of the Middle Tithonian.

*M14.* *"Proniceras* beds" of Burckhardt (1919–21, 1930): *Proniceras neohispanicum* Burckhardt [M], with *P. victoris* B. [m], *P. idoceroides* B. [m], and other species, *Aulacosphinctes torresiani* B. [m], and *Au. wilfridi* B. [m]. The horizon is best developed north of Torres in the Sierra Ramirez, where it occupies a single thin bed of limestone. Its position in the faunal succession there cannot be closely determined, other than that it lies above *Mazapilites* [M12(a)] and below *Substeueroceras* (M19–20). What appears to be the same bed, however, occurs similarly isolated at San Pedro del Gallo (Burckhardt 1930, p. 63, Fig. 21, bed B, "loc. 22"). Here it lies below *Durangites* (ibid., bed C, "loc. 23"), horizon M15.

*Proniceras* is also found at higher levels. In Europe it ranges with little change into the Berriasian, changing imperceptibly into *Spiticeras*. Imlay's discovery (1939, p. 11, Fig. 4, bed 15) of some fragments at Toboso, near Symón, above *Kossmatia* (another barely identifiable fragment, fauna M16?) does not therefore necessarily mean that the principal *Proniceras* horizon has to lie above the main level of *Kossmatia*, as Imlay implied (1939, p. 23, followed i.a. by Verma and Westermann 1973, p. 156). In Europe, *Proniceras* appears to begin at the base of the Upper Tithonian (Enay and Geyssant 1975), but in Kurdistan it may already occur a little earlier (Enay 1972, p. 374). Either way, the Mexican fauna must be among the earliest known.

*M15*. "*Durangites* beds" of Burckhardt (1912, 1930). This appears to be a widely recognizable horizon, namely, at San Pedro del Gallo (Burckhardt 1930, p. 63, Fig. 21, bed C, "loc. 23"); Sierra de Jimulco (Imlay 1939, p. 8, Fig. 3, bed 4, collections K2, K4) and Sierra de Parras (loc. 34); Cerra de la Cruz, near San Pedro del Gallo (Imlay 1939, p. 17); and Galeana, Nuevo León (Cantu 1968). The fauna includes *Durangites acanthicus, vulgaris* Burckhardt [m], *D. astillerensis* Imlay, 1939 [m], *D. galeanensis* Cantu, 1968 [m], *Micracanthoceras* ("*Durangites*") *densestriatum* (B.) [M], *M.* ("*D.*") *rarifurcatum* Imlay [M], *M.* ("*Blandfordia*") cf. *wallichi* (Gray) B. [M], *M. acanthellum, aguajitense* Imlay, *Kossmatia interrupta, pectinata* B. [m], *Proniceras* aff. *subpronum* B. (Cantu 1968, Pl. 5, Fig. 6), *Hildoglochiceras (Salinites) grossicostatum* Imlay. An essentially similar assemblage was recorded from a slightly higher level at San Pedro (Burckhardt 1930, Fig. 21, bed D, "loc. 24").

*M16*. "*Kossmatia* beds" of Burckhardt (1906, 1930. p. 50, Fig. 13, bed C): Sierra de Santa Rosa, Mazapil; beds with *Kossmatia victoris* and *Pseudolissoceras* of Cantu (1967), Mazatepec, Pueblo. *Kossmatia victoris* (Burckhardt) and other species, *K. sanatarosana* (B.) [M]. These perisphinctids – relatively small, compressed, densely and finely ribbed, with strong ventral projection of the secondaries and only mildly variocostate macroconchs – make up one of the most striking and widespread elements of the Tithonian faunas worldwide. They have been described under a variety of generic names, and it remains to be determined to what extent taxa such as *Lemencia, Richterella, Berriasella,* and *Substeueroceras* may intergrade morphologically and stratigraphically. The oldest available name is *Kossmatia* Uhlig, 1907, and despite the fact that the type species, *Amm. tenuistriatus* Gray, 1830, came from an unknown horizon in the Spiti Shales, its use for the eastern Pacific faunas seems safe.

In Mexico, *Kossmatia* is common and seems to have some vertical range. The principal horizon of *K. victoris* occurs essentially in isolation, so that its position in the succession is to some extent inferred.

Another assemblage has been described by Imlay (1943) in an occasional collection made in Chihuahua, northern Mexico: *Kossmatia varicostata* Imlay, *K. kingi* I. and *K. rancheriasensis* I. It is said to occur together with various other perisphinctids, but as Imlay pointed out, the implied associations are so unlikely that stratigraphic failure seems a more probable explanation.

Cantu's assemblage from about this level contains some interesting forms, including *Mazatepites* [=*Virgatosimoceras*?] *arrendonensis* nov. In Europe, *Virgatosimoceras* occurs in about the Semiforme-Fallauxi Zones (i.e., somewhat lower than the age of the *Kossmatia victoris* assemblage suggested here). But the Mazatepec collections may cover some range of ages.

*M17*. "*Suarites* beds" of Cantu (1967, Mazatepec, Pueblo, southern Mexico; 1976, Sierra de Chorréras, Chihuahua): *Suraites bituberculatus* Cantu, *S. floreslopezi* C., *S. velardensis* C., *Acevedites acevedense* C., *Wichmanniceras hernandense* C., and *Corongoceras* cf. *filicostatum* Imlay. This assemblage is made up almost wholly of what appear to be local specialties whose wider affinities are not clear. The closest resemblance seems to be to the fauna of the Interspinosum Zone of the Andes, with *Wichmanniceras* and *Corongoceras lotenoense*.

*M18*. *Corongoceras cordobai* Verma & Westermann. The ranges of *Corongoceras, Kossmatia,* and *Durangites* may to some extent overlap, but there do appear to be some localities in the Sierra Catorce (Verma and Westermann 1973, sec. 1 and 2 and loc. 22) at which *Corongoceras* characterizes a separable horizon, below *Substeueroceras*.

*M19*. *Kossmatia exceptionalis* (Aguilera), Sierra Catorce, San Luis Potosí, first described by Castillo and Aguilera (1895). This is a second horizon dominated by *Kossmatia* spp. [including *K. bifurcata, flexicostata, alamitoensis* (Ag.), *purisma* Verma & Westermann], but including now more coarsely ribbed, inflated forms (variants?) usually ascribed to *Substeueroceras: S. catorcense* Verma & Westermann, *S. koeneni tabulatum* V. & W. Correlation with Koeneni Zone of the Andes seems secure.

*M20*. Principal *Substeueroceras* beds of San Pedro del Gallo (Burckhardt 1912, p. 207, loc. 25; 1930, p. 63, Fig. 21, bed E) and the Sierra Jimulco (Imlay 1939, p. 8, Fig. 3, bed 2, faunas K1, K5, K25), with *Substeueroceras kellumi* Imlay (including *S. alticostatum, subquadratum,* and "*Berriasella*" *coahuilensis* Imlay), *S. lamellicostatum* (Burckhardt) (including *S. durangense* B. and "*Berriasella*" spp.), *Micracanthoceras alamense* I., *Proniceras jimulcense* I. [M], *P. scorpionum* I. [m] (with lappet), *Hildoglochiceras alamense* I., and *H. inflatum* I. This fauna also seems to correlate with the Koeneni Zone of the Andes.

There remains the question of the position of the Jurassic–Cretaceous boundary. This has been conventionally placed at the top of the *Substeueroceras koeneni* Zone of the Andes, but the correlation with the type region in southern France must still be regarded as tentative and uncertain. The Mexican faunas M19–20 could already be in part Berriasian.

Higher faunal horizons in Mexico certainly are the following:

*M21*. *Protacanthodiscus densiplicatus* Cantu (1967), Mazatepec, with *Proniceras larense* C. and *Calpionella alpina*.

*M22*. *Subthurmannia mazatepensis* Cantu (1967) and other species, including ?*Kilianella dominguensis* C., *Groebericeras problanense* C., *Calpionella alpina*, and *C. elliptica*.

*M23. Spiticeras uhligi* Burckhardt (1912, p. 207, loc. 20) and other species, San Pedro del Gallo, with *"Acantodiscus" euthymiformis, "A." transatlanticus* B., *"Necomites" densestriatus,* and *"N." praeneocomensis* B. This fauna may lie already near the Berriasian–Valanginian boundary.

*South America: the Andes.* Upper Jurassic rocks with ammonites occur along the length of western South America and continue into the Antarctic Peninsula; see the review by Riccardi (1983). What are now the coastal ranges of Chile formed a long volcanic arc or chain in Jurassic times. Pyroclastic sediments alternate with igneous intrusives or flows, and fossils are scattered and sparse. Little of interest can be said of their ammonites. More extensive and complete faunal successions occur in the sediments filling the wider, shallow mioclinal back-arc basins to the east, lapping on the craton, in what are now largely the high Andes. Upper Jurassic ammonites occur commonly in two principal regions: Caracoles–Sierra de Domeyko in northern Chile (22–25°S); and in the Mendoza–Neuquén Basins of western Argentina (33–39°S).

Oxfordian faunas have been widely reported, but few have yet been described. To judge from published faunal lists, much of the Lower, Middle, and early Upper Oxfordian is represented in northern Chile (Förster & Hillebrandt 1984; Gröschke and Hillebrandt 1985).

The similarities are close with Submediterranean Europe and Cuba. The faunas can be dated approximately in terms of European standards, but they are as yet insufficiently known to attempt any local zonations. Of particular interest is the presence of undoubted *Gregoryceras* of the Middle Oxfordian (Gygi & Hillebrandt 1991), and of the absence of *Taramelliceras.*

Kimmeridgian ammonites are almost unknown so far. An *Orthaspidoceras* has been reported from sediments above the gypsum in the Sierra de Domeyko (Forster & Hillebrandt 1984).

A rich succession of ammonite faunas begins in the Tithonian of the Mendoza-Neuquen basins (Plates 80–82) and continues upward into the Neocomian. It is summarized elsewhere in the present volume, and forms the basis of a regional standard zonation that has been worked out by H. Leanza (1981). As in Mexico, the faunal similarities with the Tithonian and Berriasian of the western Tethys are sufficiently close to allow broad correlations to be made. There are, however, also importnt differences. Consequently, the Jurassic–Cretaceous boundary cannot as yet be closely located in South America.

Editor's note: For the Late Jurassic ammonite faunas of Australasia, Indonesia, and eastern Asia, see Chapters 7–11 and Plates 83–91. For ranges in Andes see Riccardi et al. (1990).

## References


Arkell, W. J. (1956). *Jurrassic Geology of the World.* London: Oliver & Boyd.

Avias, J. (1953). Contribution a l'étude stratigraphique et paléontologique des formations antecrétacées de la Nouvelle Caledone Centrale. *Science de la Terre, 1*(1–2), 1–276.

Berckhemer, F., & Hölder, H. (1959). Ammoniten aus dem Oberen Weissen Jura Süddeutschlands. *Beih. Geol. Jb. (Hannover), 35,* 1–135.

Bloos, G. (1983). The zone of *Schlotheimia marmorea* (Lower Lias) – Hettangian or Sinemurian? *Newsl. Stratigr., 12*(3), 123–31.

(1988a). *Ammonites marmoreus* Oppel (Schlotheimiidae) im unteren Lias (angulata-Zone, depressa-Subzone) von Württemberg (Südwestdeutschland). *Stuttgarter Beitr. Naturk., Ser. B. 141,* 1–47.

(1988b). On the stage boundary Hettangian/Sinemurian in Northwest Europe and in Eastern Alps. In R. B. Rocha & A. F. Soares (eds.), *Proceedings of the 2nd International Symposium on Jurassic Stratigraphy* (pp. 71–83). Lisboa: Centro Geocien. Univ. Coimbra.

Bogdanic, T., Hillebrandt, A. von, & Quinzio, L. A. (1985). El Aaleniano de Sierra de Varas, Cordillera de Domeyko, Antofagasta, Chile. *IV Congreso Geológico Chileno, 1,* 58–75.

Braga, J. C., Comas-Rengifo, M. J., Goy, A., & Rivas, P. (1984a). The Pliensbachian of Spain: ammonite successions, boundaries and correlations. In O. Michelson & A. Zeiss (eds.), *International Symposium on Jurassic Stratigraphy* (vol. 1, pp. 159–76). Copenhagen: Int. Subcom. Jurassic Stratigraphy.

Braga, J. C. Rivas, P. & Martin-Algarra, A. (1984b). Biostratigraphic sketch of the Lower Liassic of the Betic cordilleras. In O. Michelson & A. Zeiss (eds.), *International Symposium on Jurassic Stratigraphy* (vol. 1. pp. 177–90). Copenhagen: Int. Subcom. Jurassic Stratigraphy.

Braun, E. von, & Jordan, R. (1976). The stratigraphy and paleontology of the Mesozoic sequence in the Mae Sot Area in Western Thailand. *Geol. Jb. (Hannover), B 21,* 5–51.

Burckhardt, C. (1906). *La Fauna Jurásica de Mazapil.* Inst. Geol. Mexico, Bol. 23.

(1912). *Faunes Jurassiques et Crétaciques de San Pedro del Gallo.* Inst. Geol. Mexico, Bol. 39.

(1919–21). *Faunas Jurásicas de Symon (Zacatecas).* Inst. Geol. Mexico, Bol. 33.

(1927) *Cefalopodeos de Jurásico Medio de Oaxaca y Guerrero.* Inst. Geol. Mexico, Bol. 47.

(1930). Etude synthétique sur le Mésozoique mexicain. *Mem. Soc. paleont. Suisse, 49, 1–123.*

Callomon, J. H. (1984). A review of the biostratigraphy of the post–Lower Bajocian ammonites of western and northern North America. In G. E. G. Westermann (ed.), *Jurassic-Cretaceous Biochronology and Paleogeography of North America. Geological Association of Canada Special Papers, 27,* 143–74.

(1985). The evolution of the Jurassic ammonite family Cardioceratidae. *Spec. Pap. Palaeont., 33,* 49–90.

Cameron, B. E. B., & Tipper, H. W. (1985). *Jurassic Stratigraphy of the Queen Charlotte Islands, British Columbia.* Geol. Surv. Canada, Bull. 365.

Cantu Chapa, A. (1967). Estratigrafía del Jurásico de Mazatepec, Puebla (México). 1: El límite Jurásico-Cretácico en Mazatepec, Puebla (México). *Inst. Mexi. Petról. Tecn. Explor. Sec. Geol. Monografía, 1,* 3–24.

(1968). Sobre una asociación *Proniceras-Durangites* – "Hildoglochiceras" del noreste de México. *Inst. Mexi. Petrol. Tecn. Explor. Sec. Geol. Monografia, 2,* 19–26.

(1969). Estratigrafía del Jurásico Medio-Superior del subsuelo de Poza Rica, Ver. (area de Soledad-Miquetla). *Rev. Inst. Mexic. Petrol., 3,* 3–9.

(1971). La serie Huastrea (Jurásico Medio-Superior) del centro este de México. *Rev. Inst. Mexic. Petrol., 3,* 17–40.

(1976). Nuevas localidades del Kimeridgiano y Tithoniano en Chihuahua (norte de México). *Rev. Inst. Mexic. Petrol., 8,* 38–49.

Castillo, A. D., & Aguilera, J. G. (1895). *Fauna fósil de la Sierra de Catorce, San Luis Postosí.* Com. Geol. Mexico, Bol. 1.

Coney, P. J., Jones, D. L., & Monger, J. W. H. (1980). Cordilleran suspect terranes. *Nature, 288,* 329–33.

Contreras Montero, B., Martinez Cortés, A., & Gómez Luna, A. E. (1988). Bioestratigrafía y sedimentología del Jurásico Superior en *San Pedro del Gallo, Durango, México. Rev. Inst. Mexic. Petrol., 20,* 5–49.

Counillon, H. (1908). Sur le gisement liasique de Huu-Nien, province de Quang-Nam (Annam). *Bull. Soc. Géol. France, 8*(4), 524–32.

Dietl, G., & Callomon, J. H. (1988). On the Orbis Oolite in the upper Bathonian (Middle Jurassic) of Sengenthal Ofr., Franconian Alb, and its significance for the correlation and subdivision of the Orbis Zone. *Stuttgarter Beitr. Naturkunde, B142,* 1–31.

Dommergues, J. -L. (1986). Les Dactylioceratidae du Domérien basal, un groupe monophyletique. Les Reynesocoeloceratinae nov. subfam. *Bull. Sci. Bourg., 39*(1), 1–26.

Elmi, S., Atrops, F., & Mangold, C. (1974). Les Zones d'Ammonites du Domérien-Callovien de l'Algérie Occidentale. *Docum. Lab. Géol. Fac. Sci. Lyon, 61,* 83.

Enay, R. (1972). Paléobiogéographie des ammonites du Jurassique terminal (Tithonique/Volgien/Portlandien s.l.) et mobilité continentale. *Geobios, 5,* 355–407.

Enay, R., & Geyssant, J. (1975). Faunes tithoniques des chaînes bétiques (Espagne meridionale). In *Colloque sur la limite Jurassique-Crétacé. Mém. Bureau Rech. Géol. Min. Paris, 86,* 39–55.

Erben, H. K. (1954). Dos amonitas nuevos y su importancia para la estratigrafía del Jurásico inferior de México. *Univ. Nac. Auton. México, Instit. Geol. Paleont. Mexicana, 22.*

(1956). El Jurásico inferior de México y sus amonitas. In *XX Congr. Geol. Intern., XII.*

Favre, E. (1877). La Zone a Ammonites acanthicus dans les Alpes de la Suisse et de la Savoie. *Mem. Soc. Paleont. Suisse, 4,* 1–113.

Fischer, R. (1984). Some problems of Toarcian biostratigraphy. In O. Michelson & A. Zeiss (eds.), *International Symposium on Jurassic Stratigraphy* (vol. 1, pp. 30–44). Copenhagen: Int. Subcom. Jurassic Stratigraphy.

Förster, R., & Hillebrandt, A. von (1984). Das Kimmeridge des Profeta-Jura in Nordchile mit einer Mecochirus-Favreina-Vergesellschaftung (Decapoda, Crustacea – Ichnogenus). *Mitt. Bayer, Staatsamlung Palaont. Hist. Geol., 24,* 67–84.

Frebold, H. (1960). *The Jurassic Faunas of the Canadian Arctic. Lower and Lowermost Middle Jurassic Ammonites.* Geol. Surv. Can., Bull. 59.

(1964). *Lower Jurassic and Bajocian Ammonoid Faunas of Northwestern British Columbia and Southern Yukon.* Geol. Surv. Can., Bull. 116.

(1967). *Hettangian Ammonite Faunas of the Taseko Lakes Area.* Geol. Surv. Can., Bull. 158.

(1975). *The Jurassic Faunas of the Canadian Arctic. Lower Jurassic Ammonites, Biostratigraphy and Correlations.* Geol. Surv. Can., Bull. 243.

(1976). *The Toarcian and Lower Middle Bajocian Bed and Ammonites in the Fernie Group of Southeastern British Columbia and Parts of Alberta.* Geol. Surv. Can., Paper 75–39.

Frebold, H., & Poulton, T. P. (1977). Hettangian (Lower Jurassic) rocks and faunas, northern Yukon Territory. *Can. J. Earth Sci., 14,* 89–101.

Frebold, H., & Tipper, H. W. (1970). Status of the Jurassic in the Canadian Cordillera of British Columbia, Alberta, and southern Yukon. *Can. J. Earth Sci., 7,* 1–21.

Frebold, H., Tipper, H. W., & Coates, J. A. (1969). *Toarcian and Bajocian Rocks and Guide Ammonites from Southwestern British Columbia.* Geol. Surv. Can., Paper 67–10.

Getty, T. A. (1973). *A revision of the Generic Classification of the Family Echioceratidae (Cephalopoda, Ammonoidea) (Lower Jurassic).* Univ. Kansas Paleont. Contr., Paper 63.

Geyer, O. F. (1973). Das prakretazische Mesozoikum von Kolumbien. *Geol. Jb. (Hannover), B5.*

(1974). Der Unterjura (Santiago-Formation) von Ekuador. *N. Jb. Geol. Palaont. Mh., 9,* 525–41.

(1976). La fauna de amonites del perfil típico de la formación Morrocoyal. In *Congr. Colomb. Geol. 1969, Mem.,* 111–34.

Gröschke, M., & Hillebrandt, A. von (1985). Trias und Jura in der mittleren Cordillera Domeyko von Chile (23°30'–24°30'). *N. Jb. Geol. Palaont., 170,* 129–66.

Gröschke, M., & Zeiss, A. (1990). Die ersten Hecticoceraten und Distichoceraten (Ammonitina) aus dem Callovium der Zentral-Anden. *N. Jb. Geol. Paläont., 178,* 267–83.

Guex, J. (1980). Remarques préliminaires sur la distribution stratigraphique des ammonites hettangiennes du New York Canyon (Gabbs Valley Range, Nevada). *Bull. Geol. Lausanne, 250,* and *Bull. Soc. Vaud. Sc. Nat. 75.2* (358), 127–40.

(1982). Relation entre le genre Psiloceras et les Phylloceratida au voisingage de la limite Trias-Jurassique. *Bull. Geol. Lausanne, 260,* and *Bull, Soc. Vaud. Sc. Nat., 76.2,* 47–51.

(1987). Sur la phylogenese des ammonites de Lias inferieur. *Bull. Geol. Lausanne, 292,* and *Bull Soc. Vaud. Sc. Nat., 78.4,* 455–569.

Guex, J., & Taylor D. (1976). La limite Hettangien–Sinemurien des Prealpes romandes au Nevada. *Eclogae Geol. Helv., 69*(2), 521–6.

Gygi, R. A., & Hillebrandt, A. von (1991). Ammonites (mainly *Gregoryceras*) of the Oxfordian (Late Jurassic) in northern Chile and time-correlation with Europe. *Schweiz. paläont. Abh. 113,* 135–85.

Hall, R. L. (1984). Lithostratigraphy and biostratigraphy of the Fernie Formation (Jurassic) in the Southern Canadian Rocky Mountains. *Can. Soc. Petrol. Geol. Mem., 9,* 233–47.

(1987). New Lower Jurassic ammonite faunas from the Fernie Formation, southern Canadian Rocky Mountains. *Can. J. Earth Sci., 24,* 1688–704.

Hall, R. L., & Stronach, N. H. (1981). The first record of Late Bajocian (Jurassic) ammonites in the Fernie Formation, Alberta. *Can. J. Earth Sci., 18,* 919–25.

Hall, R. L., & Westermann, G. E. G. (1980). Lower Bajocian (Jurassic) cephalopod faunas from western Canada and proposed assemblage zones for the Lower Bajocian of North America. *Paleontogr. Amer., 9*(52), 1–93.

Hallam, A. (1965). Observation on marine Lower Jurassic stratigraphy of North America, with special reference to United States. *Am. Assoc. Petrol. Geol. Bull., 49,* 1485–501.

Harrington, H. J. (1962). Paleogeographic development of South America. *Am. Assoc. Petrol. Geol. Bull., 46,* 1773–814.

Hillebrandt, A. von (1981). Faunas de amonites del liásico inferior y medio (Hettangiano hasta Pliensbachiano) de América del Sur (excluyendo Argentina). In W. Volkheimer & E. A. Musacchio (eds.), *Cuencas sedimentarias del Jurásico y Cretácio de América del Sur* (vol. 2, pp. 499–538). Com. Sudamer. Jurásico Cretácico, Buenos Aires.

(1987). Liassic ammonite zones of South America and correlations with other provinces – description of new genera and species of ammonites. In W. Volkheimer (ed.), *Bioestratigrafia de los Sistemas Regionales del Jurásico y Cretácico en América del Sur* (pp. 111–57). Mendoza: Com. Sudam. Jurassic Cretaceous.

(1989). The Lower Jurassic of the Rio Atuel Region, Mendoza Province, Argentina. In *IV Congr. Argentino Paleont. Biostratigr., Mendoza, 1986, Actas* (vol. 4, pp. 39–43). Mendoza.

(1988). Ammonite biostratigraphy of the South American Hettangian – description of two new species of *Psiloceras.* In R. B. Rocha & A. F. Soares (eds.), *2nd International Symposium on Proceedings of the Jurassic Stratigraphy* (pp. 13–27). Lisboa: Centro Geocien. Univ. Coimbra.

(1990a). The Triassic/Jurassic boundary in northern Chile. In *Meeting on the Triassic-Jurassic Boundary* (pp. 27–53). Lyon: Inst. Cathol. Lyon, *Cahiers, ser. sci.*

(1990b). Der Untere Jura im Gebiet des Rio Atuel (Provinz Mendoza, Argentinien). *N. Jb. Geol. Paläont., 181*(1–3), 143–57.

(in press). The Triassic/Jurassic boundary and Hettangian biostratigraphy in the area of the Utcubamba Valley (northern Peru). In *3rd International Symposium on Jurassic Stratigraphy,* Poitiers, 1991 (to be published in *Geobios*).

Hillebrandt, A. von, & Schmidt-Effing. R. (1981). Ammoniten au dem Toarcium (Jura) von Chile (Sudamerika). Die Arten der Gattungen Dactylioceras, Nodicoeloceras, Peronoceras und Collina. *Zittelania, 6,* 3–71.

Hillebrandt. A. von, & Westermann, G. E. G. (1985). Aalenian (Jurassic)ammonite faunas and zones of the Southern Andes. *Zittelania, 12*, 3–55.

Hirano, H. (1973). Biostratigraphic study of the Jurassic Toyora Group. Part III. *Trans. Proc. Palaeont. Soc. Japan, N.S., 90*, 45–71.

(1985) Recent status of the Middle–Upper Liassic biostratigraphy of the Inner Belt of Southwest Japan. In *IGCP Project 171: Field Conference III, Circum-Pacific Jurassic* (pp. 95–8). Tsukuba.

Hirano, H., Ichihara, S., Sunarya, Y., Nakajima, N., Obata, I., & Futakami, M. (1981). Lower Jurassic ammonites from Bengkayang, West Kalimantan Province, Republic of Indonesia. *Bull. Geol. Res. Dev. Centre, 4*, 21–6.

Howarth, M. K. (1973). The stratigraphy and ammonite fauna of the Upper Liassic grey shales of the Yorkshire Coast. *Bull. British Mus. Nat. Hist. Geol., 24*, 235–77.

Imlay, R. W. (1939). Upper Jurassic ammonites from Mexico. *Geol. Soc. Am. Bull., 50*, 1–77.

(1943). Upper Jurassic ammonites from the Placer de Guadalupe, Chihuahua, Mexico. *J. Paleontol., 17*, 527–43.

(1960). *Ammonites of Early Cretaceous Age (Valanginian and Hauterivian) from the Pacific Coast States.* U.S. Geological Survey Professional Paper 334-F.

(1961). *Late Jurassic Ammonites from the Western Sierra Nevada, California.* U.S. Geological Survey Professional Paper 374-D.

(1964). *Upper Jurassic Mollusks from Eastern Oregon and Western Idaho.* U.S. Geological Survey Professional Paper 483-D.

(1968). *Lower Jurassic (Pliensbachian and Toarcian) Ammonites from Eastern Oregon and California.* U.S. Geological Survey Professional Paper 593-C.

(1973). *Middle Jurassic (Bajocian) Ammonites from Eastern Oregon.* U.S. Geological Survey Professional Paper 756.

(1980). *Jurassic Paleobiogeography of the Conterminous United States in Its Continental Setting.* U.S. Geological Survey Professional Paper 1062.

(1981a) *Early Jurassic Ammonites from Alaska.* U.S. Geological Survey Professional Paper 1148.

(1981b) *Late Jurassic Ammonites from Alaska.* U.S. Geological Survey Professional Paper 1190.

(1982). *Jurassic (Oxfordian and Late Callovian) Ammonites from the Western Interior Region of the United States.* U.S. Geological Survey Professional Paper 1232.

Imlay, R. W., & Detterman, R. L. (1973). *Jurassic Paleobiogeography of Alaska.* U.S. Geological Survey Professional Paper 801.

Imlay R. W., & Jones, D. L. (1970). *Ammonites from the Buchia Zones in Northwestern California and Southwestern Oregon.* U.S. Geological Survey Professional Paper 647-B.

Jaworski, E. (1933). Revision der Arieten, Echioceraten und Dactylioceraten des Lias von Niederlandisch-Indien. *N. Jb. Min. Geol., Beil. -B., 70B*, 2251–334.

(1940). Oxford-Ammoniten von Cuba. *N. Jb. Min. Geol., Beil. -B., 83B*, 87–137.

Kalacheva, E. D. (1988). Toarcian. In G. Y. Krymholts et al. (eds.), *The Jurassic Ammonite Zones of the Soviet Union* (pp. 1–116). Geol. Soc. Am. Spec. Pap. 223.

Krause, P. G. (1896). Uber den Lias von Borneo. *Jb. Mijnewezen Ned. Oost-Indie, 25*, 28; *Samml. Reichs. Mus. Leiden, Ser. I, 5/3.*

(1911). Uber unteren Lias von Borneo. *Samml. Reichs. Mus. Leiden, Ser. I, 9/1.*

Krishna, J., & Westermann, G. E. G. (1987). The faunal associations of the Middle Jurassic ammonite genus *Macrocephalites* in Kachchh, western India. *Can. J. Earth Sci., 24*, 1570–82.

Kruizinga, P. (1926). Ammonieten en einige andere Fossielen uit de Jurassische Afzettingen der Soela Eilande. *Jb. Mijnewezen Ned. Oost-Indie, 54.*

Krumbeck, L. (1922). Zur Kenntnis des Jura der Insel Rotti. *Jb. Mijnewezen Ned. Oost-Indie, 1920, Verh. 3*, 107–219.

(1923). Zur Kenntnis des Juras der Insel Timor sowie des Aucellen-Horizontes von Seran und Buru. *Palaontologie von Timor, 12.*

Krymholts, G. Y., Mesezhnikov, L. M. S., & Westermann, G. E. G. (1988). *The Jurassic Ammonite Zones of the Soviet Union.* Geol. Soc. Am. Spec. Pap. 223.

Leanza, H. A. (1981). Faunas de ammonites del Jurásico Superior y del Cretácico Inferior de América del Sur, con especial consideración de la Argentina. *Cuencas sedimentarias del Jurásico y Cretácico de América del Sur, 2*, 559–97 (Comité Sudamericano del Jurásico y Cretácico, Buenos Aries).

McLean, F. H. (1932). Contributions to the stratigraphy and paleontology of Skidegate Inlet, Queen Charlotte Islands, British Columbia. *Trans. Roy. Soc. Canada, Sect. IV, Geol. Sci. Mineral, Ser. 3, 26*, 51–80.

Matsumoto, T. (1956). *Yebisites*, a new Lower Jurassic ammonite from Japan. *Trans. Proc. Pal. Soc. Japan, N.S., 23*, 205–12.

Meledina, S. V. (1988). Bathonian, Callovian. In G. Y. Krymholts et al. (eds.), *The Jurassic Ammonite Zones of the Soviet Union* (pp. 29–38). Geol. Soc. Am. Spec. Pap. 223.

Monger, J. W. H., Price, R. A., & Templeman-Kluit, D. J. (1982). Tectonic accretion and the origin of the two major metamorphic and plutonic welts in the Canadian Cordillera. *Geology, 10*, 70–5.

Mouterde, R. (1971). Les formations Mésozoïques de la Thakkhola. In P. Bordet, M. Colchen, D. Krummenacher, P. le Fort, R. Mouterde, & M. Remy (eds.), *Recherches géologiques dans l'Himalaya du Népal, région de la Thakkhola* (pp. 119–86). Paris: Edit. Centre Natl. Rech. Scien.

Polubotko, I. V., & Repin, Y. S. (1981). On the separation of a new ammonite zone at the basis of the Jurassic system. *Dok. Akad. Nauk SSSR, 261*, 1394–8.

Poulton, T. P. (1987). Zonation and correlation of middle Boreal Bathonian to Lower Callovian (Jurassic) ammonites, Salmon Cache Canyon, Porcupine River, Norther Yukon. *Geol. Surv. Can. Bull., 358*, 1–155.

Poulton, T. P., Callomon, J. H., & Hall, R. L. (1991). *Bathonian through Oxfordian (Middle and Upper Jurassic) Marine Macrofossil Assemblages and Correlations, Bowser Lake Group, West-Central Spatsizi Map Area, Northwestern British Columbia.* Geological Survey of Canada Paper 91-1A.

Poulton, T. P., Zeiss, A., & Jeletzky, J. A. (1988). New molluscan faunas from the Late Jurassic (Kimmeridgian and Early Tithonian) of western Canada. *Geol. Surv. Can. Bull., 379*, 103–15.

Prinz, P. (1983). Uber unterjurassische Ammoniten aus Mittelperu. *Zbl. Geol. Palaont., 1*(3–4), 329–34.

(1985a). Zur Stratigraphie und Ammonitenfauna der Pucara-Gruppe bei San Vicente (Depto. Junin, Peru). *Newsl. Stratigr. 14*, 129–41.

(1985b). Stratigraphie und Ammonitenfauna der Pucara-Gruppe (Obertrias-Unterjura) von Nord-Peru. *Palaeontographica, A188*, 153–97.

Quinzio, S. L. A. (1987). Stratigraphische Untersuchungen im Unterjura des Sudteils der Provinz Antofagasta in Nord-Chile. *Berliner Geowiss. Abh., A87.*

Repin, Y. S. (1984) *Lower Jurassic Ammonite Standard Zones and Zoogeography in North-East Asia.* IGCP Project 171: Circum-Pacific Jurassic, Report 2, Special Paper 1.

Riccardi, A. C. (1983). The Jurassic of Argentina and Chile. In N. Moullade & A. E. M. Nairn (eds.), *The Phanerozoic Geology of the World. II. The Mesozoic, B*, (pp. 201–63). Amsterdam: Elsevier.

Riccardi, A. C., Damborenea, S. E., Manceñido, M. O., & Ballent, S. C. (1988). Hettangiano y Sinemuriano marinos en Argentina, In *V. Congr. Geol. Chileno (Santiago)* (vol. 2, pp. 359–73). Santiago.

Riccardi, A. C., Leanza, H. A., & Volkheimer, W.(1990). Upper Jurassic of South America and Antarctic Peninsula. In G. E. G. Westermann & A. C. Riccardi (eds.), *Jurassic Taxa Ranges and Correlation Charts for the Circum-Pacific. 3: South America and Antarctic Peninsula. Newsl. Stratig., 21*(2), 129–47.

Riccardi, A. C., & Westermann, G. E. G. (1991). The Middle Jurassic ammonoid fauna and biochronology of the Argentine–Chilean Andes. III: Eurycephalitinae, Stephanocerataceae. IV: Bathonian–Callovian Reineckeiidae. *Palaeontographica, A, 216*, 1–110.

Riccardi, A. C., Westermann, G. E. G., & Elmi, S. (1989). The Middle Jurassic Bathonian–Callovian ammonite zones of the Argentine–Chilean Andes. *Geobios, 22,* 553–97.

Sandoval, J., & Westermann, G. E. G. (1988). The Bajocian (Jurassic) ammonite fauna of Oaxaca, Mexico. *J. Paleontol., 60,* 1220–71.

Sandoval, J., Westermann, G. E. G., & Marshall, M. C. (1989). Ammonite fauna, stratigraphy and ecology of the Bathonian–Callovian (Jurassic) Tecocoyunca Group, south Mexico. *Palaeontographica, A210,* 93–149.

Sato, T. (1972). Ammonites du Toarcien au Nord de Saigon (Sud Viet Nam). *Geol. Palaeont. SE Asia, 10,* 231–42.

(1975) Marine Jurassic formations and faunas in South-east Asia and New Guinea. *Geol. Palaeont. SE Asia, 15,* 151–89.

Sato, T., & Westermann, G. E. G. (1985). Range Chart and Zonations in Japan. In G. E. G. Westermann (ed.), *IGCP Project 171: Field Conference III, Circum-Pacific Jurassic* pp. 73–94. Tsukuba.

(1991). Japan and Southeast Asia. In G. E. G. Westermann & R. C. Riccardi (eds.), *Jurassic Taxa Ranges and Correlation Charts for the Circum-Pacific. Newsl. Strat. 24,* 81–108.

Sato, T., Westermann, G. E. G., Skwarko, S. K., & Hasibuan, F. (1978). Jurassic biostratigraphy of the Sula Islands, Indonesia. *Bull. Geol. Survey Indonesia, 4,* 1–28.

Schlatter, R. (1982). Zur Grenze Pliensbachian-Toarcian im Klettgau (Kanton Schaffhausen, Schweiz). *Ecolgae Geol. Helv., 75,* 759–71.

Schlatter, R., & Schmidt-Effing, R. (1984). Bioestratigrafía y fauna de amonites del Jurásico inferior (Sinemuriano) del Area de Tenengo de Doria (Estado de Hidalgo, México). In C. Perrilliat (ed.), *III Congr. Latinamericano de Paleont.* Mexico: Inst. Geol. UNAM.

Schmidt-Effing, R. (1980). The Huayacoctla Aulacogen in Mexico (Lower Jurassic) and the origin of the Gulf of Mexico. In R. H. Pilger, Jr. (ed.), *The Origin of the Gulf of Mexico and the Early Opening of the Central North Atlantic Ocean* (pp. 79–86). Baton Rouge: Louisiana Geol. Surv.

Sey, I. I., & Kalacheva, E. D. (1980). *Biostratigraphy of the Lower and Middle Jurassic Deposits of the Far East.* Minist. Geol. SSSR, Geol. Inst., Publ. n.s. 285.

Skwarko, S. K. (1967). *Mesozoic Mollusca from Australia and New Guinea.* Ept. Natl. Dev. Bur. Min. Res. Geol. Geophys. Bull. 75.

Smith, P. L., & Tipper, H. W. (1986). Plate tectonics and paleobiogeography: Early Jurassic (Pliensbachian) endemism and diversity. *Palaios, 1,* 399–412.

Smith, P. L., Tipper, H. W., Taylor, D. G., & Guex, J. (1988). An ammonite zonation for the Lower Jurassic of Canada and the United States: the Pliensbachian. *Can. J. Earth Sci., 25,* 1503–23.

Soergel, W. (1913). Lias und Dogger von Jefbie und Fialpopo (Misolarchipel). *N. Jb. Geol. Palaont., B.B., 36,* 586–650.

Stevens, G. R. (1977). *Mesozoic Chronostratigraphy, New Zealand–New Caledonia.* IGCP Project 8, Rep. 5.

(1978a). Jurassic paleontology. In R. P. Suggate, G. R. Stevens, & M. T. Te Punga (eds.), *The Geology of New Zealand* (pp. 215–28). Wellington: Keating.

(1978b). New Zealand Jurassic. In M. Moullade & A. E. M. Nairn (eds.), *The Phanerozoic Geology of the World: The Mesozoic A* (pp. 261–71). Amsterdam: Elsevier.

Stevens, G. R., & Speden, I. G. (1978) New Zealand (Jurassic). In M. Moullade & A. E. M. Nairn (eds.), *The Phanerozoic Geology of the World. II: The Mesozoic A* (pp. 251–328). Amsterdam: Elsevier.

Suggate, R. P., Stevens, G. R., & T. Punga, M. T. (eds.), (1978). *The Geology of New Zealand.* Wellington.

Taylor, D. G. (1986). The Hettangian–Sinemurian boundary (Early Jurassic): reply to Bloos 1983. *Newsl. Stratigr., 16,* 57–67.

(1988). Middle Jurassic (late Aalenian and early Bajocian) ammonite biochronology of the Snowshoe Formation, Oregon. *Oregon Geol., 50,* 123–38.

Taylor, D. G., Callomon, J. H., Hall, R., Smith, P. L., Tipper, H. W., & Westermann, G. E. G. (1984). *Jurassic Ammonite Biogeography of Western North America: The Tectonic Implications.* Geol. Assoc. Can., Spec. Pap. 27.

Thomson, M. R. A., & Trantner, H. (1986). Early Jurassic fossils from Central Alexander Island and their geological setting. *Br. Antarct. Surv. Bull., 70,* 23–39.

Tipper, H. W., Smith, P. L., Cameron, B. E. B., Carter, E. S., Jakobs, G. K., & Johns, M. J. (1990). Biostratigraphy of the Lower Jurassic Formations of the Queen Charlotte Islands, British Columbia In *Evolution and Petroleum Potential of the Queen Charlotte Basin, British Columbia.* Geol. Surv. Can., Pap. 90-10.

Tozer, E. T. (1982). *Late Triassic (Upper Norian) and Earliest Jurassic (Hettangian) Rocks and Ammonoid Faunas, Halfway River and Pine Pass Map Areas, British Columbia.* Geol. Survey. Can., Pap. 82-1A.

Verma, H. M., & Westermann, G. E. G. (1973). The Tithonian (Jurassic ammonite fauna and stratigraphy of Sierra Catorce, San Luis Potosí, Mexico. *Bull. Am. Paleont., 63* (277), 107–320.

Vu Khuc (1984). *Jurassic of Vietnam.* IGCP Project 171: Circum-Pacific Jurassic, Spec. Pap. 7.

Wang, Yi-Gang, & Smith, P. L. (1986). Sinemurian (Early Jurassic) ammonite fauna from the Guangdong Region of Southern China. *J. Paleontol., 60,* 1075–85.

Westermann, G. E. G. (1964). The ammonite fauna of the Kialagvik Formation. 1. Lower Bajocian (Aalenian). *Bull. Am. Paleont., 48*(216), 329–503.

(1969). The ammonite fauna of the Kialagvik Formation. 2. *Sonninia Sowerbyi* Zone (Bajocian). *Bull. Am. Paleont., 57*(255), 5–226.

Westermann, G. E. G., & Callomon, J. H. (1988). The Macrocephalitinae and associated Bathonian and Early Callovian (Jurassic) ammonoids of the Sula Islands and New Guinea. *Palaeontographica, A203,* 1–90.

Westermann, G. E. G., & Hudson, N. (1991). The first find of Eurycephalitinae (Jurassic Ammonitina) in New Zealand and its biogeographic implications. *J. Paleont., 65,* 689–93.

Westermann, G. E. G., & Riccardi, A. C. (1972–9). Middle Jurassic ammonoid fauna and biochronology of the Argentine-Chilean Andes. I. (1972) Hildoceratacea. *Palaeontolographica, A140,* 1–116; II. (1979) Bajocian Stephanocerataceae. *Palaeontographica, A164,* 85–188.

(1975) Systematics. In P. N. Stipanicic, G. E. G. Westermann, & A. C. Riccardi (eds.), The Indo-Pacific Ammonite Mayaites in the Oxfordian of the Southern Andes. *Ameghiniana, 12,* 281–305.

(1982). Ammonoid fauna from the early Middle Jurassic of Mendoza province, Argentina. *J. Paleont., 56,* 11–41.

(1985). *Middle Jurassic Ammonite Evolution in the Andean Province and Emigration to Tethys. Lecture Notes Earth Sci., 1,* 6–34. Berlin: Springer-Verlag.

Westermann, G. E. G., Riccardi, A. C., Palacios, O., & Rangel, C. (1980). Jurásico Medio en el Perú. *Inst. Geol. Min. Metal. Bol. (Lima), 9,* 1–47.

Westermann, G. E. G., & Wang, Y. (1988). Middle Jurassic ammonites of Tibet and the age of the lower Spiti Shales. *Paleontology, 31,* 295–339.

Wiedenmayer, F. (1980). Die Ammoniten der mediterranen Provinz im Pliensbachian und unteren Toarcian aufgrund neuer Untersuchungen im Generoso-Becken (Lombardische Alpen). *Mem. Soc. Helv. Sci. Nat., 93.*

Ziegler, B. (1958). Monographie der Ammonitengattun *Glochiceras* im epikontinentalen Weissjura Mitteleuropas. *Palaeontographica, A110,* 91–164.

(1959). *Idoceras* und verwandte Ammoniten-Gattungen im Oberjura Schwabens. *Eclogae Geol. Helv., 52,* 19–56.

(1977). The "White" (Upper) Jurassic in southern Germany. *Stuttgarter Beitr. Naturk., B26.*

# 13 Jurassic palynomorphs of the circum-Pacific region

W. A. S. SARJEANT, W. VOLKHEIMER and W.-P. ZHANG

In the Jurassic, the palynomorphs available for study include representatives of eight groups of organisms – cysts of dinoflagellates (Dinophyceae); phycomae and zoosporangia (tasmanitids) of prasinophycean algae; representatives of two groups of colonial algae (Xanthophyceae in freshwater sediments, Chlorophyceae in marine sediments); acritarchs (mostly small polygonomorphs and acanthomorphs, loosely termed "micrhystridia," but with also some sphaeromorphs, herkomorphs and larger acanthomorphs); scolecodonts; the organic linings of the proloculus and early chambers of foraminifera; and the spores and pollen of higher plants. Of these groups, most serve as paleoenvironmental guides, both in broad terms and as indicators of water depth or degree of shoreline proximity. Only three – dinoflagellates, spores and pollen – are of demonstrated applicability in biostratigraphy; however, preliminary scanning-electron-microscope studies of micrhystridia and researches on foraminiferal linings by R. P. W. Stancliffe suggest that these groups will also soon prove utilizable.

Spores and pollen are of particular importance in determining the sequence of non-marine sediments and afford the important secondary asset of enabling firm correlations to be made between non-marine and marine sediments. However, even in the relatively equable world climates of the Jurassic, plants showed a considerable regionality, directly reflected by the distribution (by wind, water or small animals) of their reproductive bodies. Moreover, their taxonomy is presently in a confused state, with many different generic or specific names being applied to virtually identical taxa and with regional preferences for these different names causing difficulties in perceiving such inter-regional similarities as do exist. Nevertheless, spores and pollen are being used very successfully as biostratigraphic tools within regions.

Dinoflagellates, though certainly circumscribed in distribution by oceanic water circulation patterns and by water depth and nutrient availability, provide a readier means for long-distance correlation. Unfortunately, rich cyst assemblages have hitherto been recorded in detail only from the southwestern Pacific (Australia, Papua New Guinea and New Zealand). Elsewhere, they have been reported only in preliminary fashion or sought in vain. Their absence results sometimes from adverse paleoecological circumstances, but appears more often to be a consequence of low-grade metamorphism of the sediments, carbonizing and destroying all contained palynomorphs.

In the account that follows, the geographic sequence accords with that adopted in the regional accounts. Negative as well as positive records are presented, since they serve as a guide to whether further studies would be worthwhile.

See Appendix, Plates 92–98.

### New Zealand

The earliest published spore-pollen studies were by R. A. Couper (1953; 1960) and G. Norris (1968; J. B. Waterhouse and Norris 1972). Norris's observations of dinoflagellates and acritarchs in the Torlesse Group were noted by J. D. Campbell and G. Warren (1965). Further discoveries of dinoflagellates were reported by G. J. Wilson (1978, 1982, 1984a,b). When presenting a dinoflagellate zonation for the Late Jurassic to Eocene strata of New Zealand, Wilson (1984b) proposed two dinoflagellate-based zones for the Late Jurassic, the *Scriniodinium crystallinum* Zone, equivalent to the Heterian and Ohauan Stages of the New Zealand Jurassic, and the *Belodinium dysculum* Zone, considered equivalent to the Puaroan. However, a restudy by Wilson and R. Helby (1988) has shown the latter zone to be not of Jurassic but of Early Cretaceous (Neocomian) age.

Wilson and Helby (1987) described briefly a probably Oxfordian dinoflagellate assemblage from North Canterbury, South Island; and N. Hudson, J. A. Grant-Mackie and Helby (1987) gave a preliminary mention of forms from the Oraka Sandstone of the Kawhia Series of North Island. Subsequently Helby, Wilson and Grant-Mackie (1988) gave a more extensive account of Late Jurassic assemblages, comparing them with the Australian dinoflagellate zonation (see Table 13.2):

*Oraka Sandstone*. Preliminary studies revealed dinoflagellates of Late Callovian age, notably *Atopodinium prostatum, Reutlingia gochtii* and *Stephanelytron* sp. nov., indicating an age older than the Late Callovian *Rigaudella aemula* Zone.

*Ohineruru Formation.* Two suites of assemblages were recognized. The lower, considered of Oxfordian age, includes *Pyxidiella pandora, Rigaudella aemula* and *Rhynchodiniopsis cladophora,* along with undescribed proximate cysts; it correlates with the lower *Wanaea spectabilis* Zone. The upper, considered of Early Kimmeridgian age, includes *Dingodinium swanense, Scriniodinium crystallinum* and *S. anceps,* along with undescribed taxa; it correlates with the upper *Wanaea clathrata* or lower *Dingodinium swanense* Zone.

*Waikatakuta and Kowhai Point Siltstones.* These siltstones contain diverse assemblages, including *D. swanense* and *Tubotuberella missilis,* and are respectively considered to correlate with the middle and upper parts of the *Dingodinium swanense* Zone.

*Kinohaku Siltstone.* Rich microfloras were recovered, with up to 15% microplankton: *Belodinium dysculum* and accompanying dinoflagellates indicated a correlation with the *Cribroperidinium perforans* Zone of the Late Kimmeridgian (≡ Early Tithonian).

*Puti Siltstone.* The earlier assemblages included *B. dysculum, Komewuia glabra* and species of *Biorbifera,* indicating a correlation with the *Dingodinium jurassicum* Zone. Later assemblages included *Omatia montgomeryi* and *Herendeenia pisciformis,* indicating the *Omatia montgomeryi* Zone. The total time range for this unit thus appears to be from Early Kimmeridgian to as late as Early Portlandian (≡ Middle Tithonian).

For the moment, exact correlations with European assemblages and stages are equivocal, but the indications from dinoflagellates are that the beds of the Kawhia Series range in age from Late Callovian to Late Kimmeridgian (≡ Early Tithonian) or even Early Portlandian (≡ Middle Tithonian).

### Australia

The terrestrial Jurassic palynofloras of Australia have received especially intensive and careful study. I. C. Cookson (1953) inaugurated this work, with a report of a palynoflora from a South Australian borehole; later studies in that state have been by G. Playford and M. E. Dettmann (1965), P. R. Evans (1966) and W. K. Harris (1966, 1970). Work in Western Australia was begun by B. E. Balme (1957, 1961, 1963, 1966; Balme in McWhae et al. 1958) and followed up by J. Backhouse (1974, 1975, 1978, 1988) and J. Filatoff (1975). Much work has been done in Queensland, primarily by N. J. De Jersey (1959, 1960, 1963, 1965, 1971a,b, 1972, 1973, 1976; De Jersey and R. J. Paten 1963, 1964), but also by G. Playford and K. D. Cornelius (1967), R. F. Reiser and A. J. Williams (1969), J. L. McKellar (1974, 1975, 1981a–c, 1982), D. Burger and B. R. Senior (1979), J. Stevens (1981) and Burger (1989). The sequences of palynomorphs in Queensland's Great Artesian Basin were utilized by Burger (1986) in defining sedimentary cycles.

Jurassic dinoflagellates were first reported from Australia by G. Deflandre and I. C. Cookson (1955), subsequently by I. C. Cookson and A. Eisenack (1958, 1960, 1974, 1982), J. F. Wiseman (1980) and R. Helby and L. E. Stover (1987; Stover and Helby 1987a,b). A reassessment of Cookson and Eisenack's stratigraphical conclusions was published by W. A. S. Sarjeant (1968, pp. 235, 240).

A comprehensive overview of Australian Jurassic palynostratigraphy was furnished by Helby, R. Morgan and A. D. Partridge (1987); this work is especially valuable in that it incorporates much unpublished data. Their conclusions, and the zonations provided, are embodied in Tables 13.1–13.3. Three tables are necessary, since the microfloras of eastern Australia and of Western Australia indicate that the Jurassic terrestrial floras of those regions were different enough in detail to justify separate treatment, whereas the dinoflagellates of Australian marine deposits are of more constant character. These zonations have provided a basis for biostratigraphic reference for the whole southwestern Pacific. Key taxa of pollen and spores and of dinoflagellates are illustrated in the Appendix (Plates 92, 93).

### Papua New Guinea

Accounts of dinoflagellate cysts and acritarchs from Papua were given first by Deflandre and Cookson (1955) and Cookson and Eisenack (1958, 1960); their stratigraphical conclusions were briefly reassessed by Sarjeant (1968, p. 258). Helby, Morgan and Partridge (1987), Helby and Stover (1987) and Stover and Helby (1987a,b) discussed dinoflagellate taxa from Papua, while fiches by Helby and Partridge, appended in a slip-case to Jell (ed., 1987), embody palynological analyses of samples from the Papuan Basin.

R. J. Davey (1988) described the marine microflora of Jurassic sections along the Strickland River in western Papua New Guinea. His zonal scheme corresponds in broad outline with that of Helby, Morgan and Partridge (1987); however, it is more elaborate and Davey's interpretation of international stage equivalences differs from theirs. Twelve zones are recognized; these are briefly characterized below, but Davey's paper should be consulted for fuller details of his rich assemblages.

1. *Ctenidodinium sellwoodii* Zone. Base: first *Wanaea digitata.* Top: last *Dichadogonyaulax* (= *Ctenidodinium*) *sellwoodii.* Prominent species include *Adnatosphaeridium caulleryi.* Middle Callovian [≡*Wanaea digitata* Zone of Helby et al. (1987)].
2. *Wanaea digitata* Zone. Base: first *Rigaudella aemula* and *R. filamentosa.* Top: last *Wanaea digitata, Endoscrinium galeritum* and *Gonyaulacysta eisenackii.* Prominent species include *Scriniodinium luridum, S. ceratophorum* and *Stephanelytron scarburghense.* Late Callovian to Early Oxfordian [approx. ≡ *Wanaea spectabilis* and *Rigaudella aemula* Zones of Helby et al. (1987)].
3. *Wanaea clathrata* Zone. Base: common *W. clathrata,* with *W. spectabilis, W. fimbriata* and *Dingodinium jurassicum.*

Table 13.1. *Biostratigrahic zonation of stratigraphically important pollen and spores in eastern Australia*

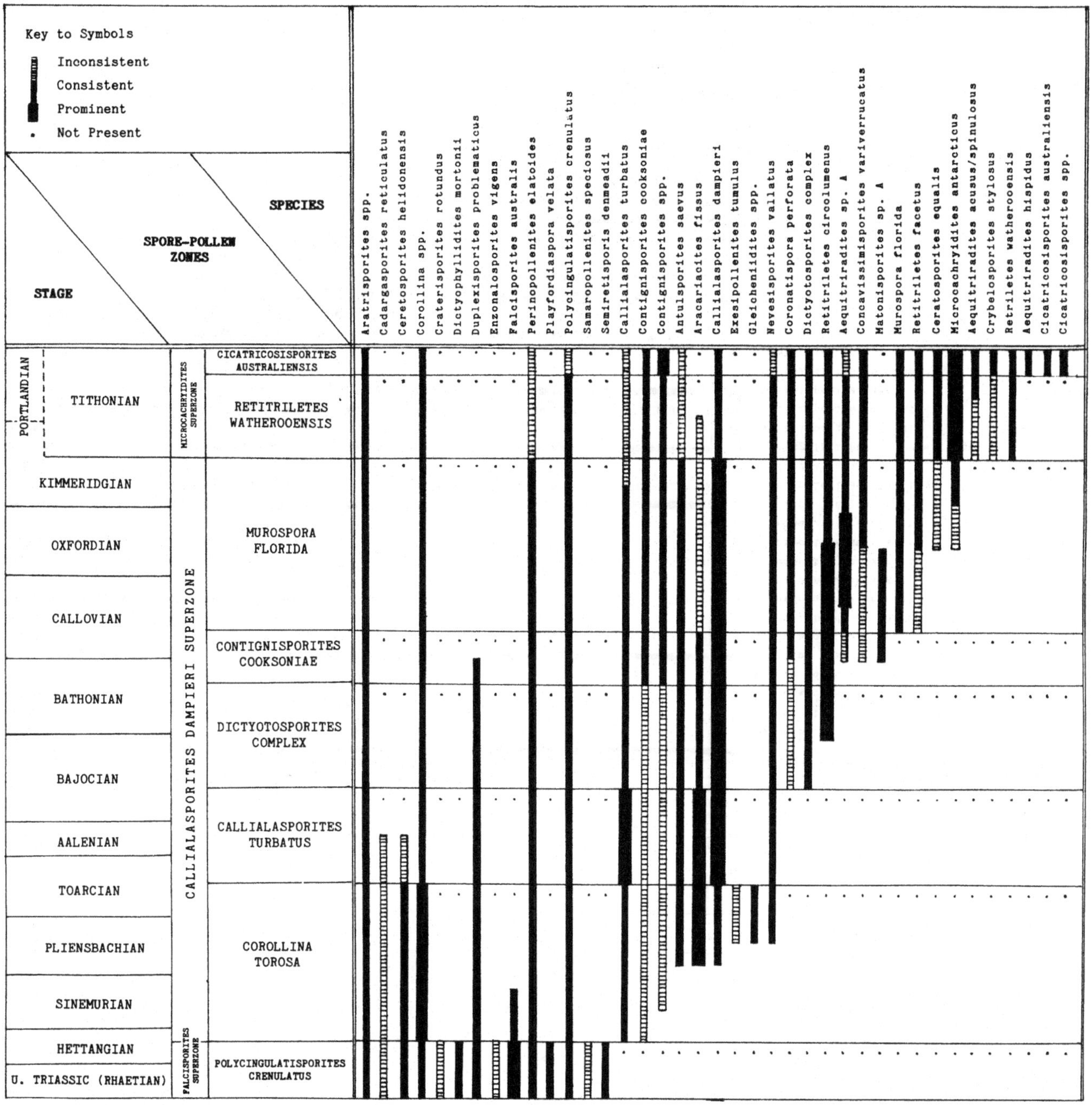

Key to Symbols

- Inconsistent
- Consistent
- Prominent
- . Not Present

| STAGE | SPORE-POLLEN ZONES (Superzone) | SPORE-POLLEN ZONES (Zone) |
|---|---|---|
| PORTLANDIAN / TITHONIAN | MICROCACHRYIDITES SUPERZONE | CICATRICOSISPORITES AUSTRALIENSIS |
| PORTLANDIAN / TITHONIAN | MICROCACHRYIDITES SUPERZONE | RETITRILETES WATHEROOENSIS |
| KIMMERIDGIAN | CALLIALASPORITES DAMPIERI SUPERZONE | MUROSPORA FLORIDA |
| OXFORDIAN | CALLIALASPORITES DAMPIERI SUPERZONE | MUROSPORA FLORIDA |
| CALLOVIAN | CALLIALASPORITES DAMPIERI SUPERZONE | MUROSPORA FLORIDA / CONTIGNISPORITES COOKSONIAE |
| BATHONIAN | CALLIALASPORITES DAMPIERI SUPERZONE | CONTIGNISPORITES COOKSONIAE / DICTYOTOSPORITES COMPLEX |
| BAJOCIAN | CALLIALASPORITES DAMPIERI SUPERZONE | DICTYOTOSPORITES COMPLEX / CALLIALASPORITES TURBATUS |
| AALENIAN | CALLIALASPORITES DAMPIERI SUPERZONE | CALLIALASPORITES TURBATUS |
| TOARCIAN | CALLIALASPORITES DAMPIERI SUPERZONE | COROLLINA TOROSA |
| PLIENSBACHIAN | CALLIALASPORITES DAMPIERI SUPERZONE | COROLLINA TOROSA |
| SINEMURIAN | CALLIALASPORITES DAMPIERI SUPERZONE | COROLLINA TOROSA |
| HETTANGIAN | FALCISPORITES SUPERZONE | POLYCINGULATISPORITES CRENULATUS |
| U. TRIASSIC (RHAETIAN) | FALCISPORITES SUPERZONE | POLYCINGULATISPORITES CRENULATUS |

SPECIES:
Aratrisporites spp.; Cadargasporites reticulatus; Ceretosporites helidonensis; Corollina spp.; Craterisporites rotundus; Dictyophyllidites mortonii; Duplexisporites problematicus; Enzonalasporites vigens; Falcisporites australis; Perinopollenites elatoides; Playfordiaspora velata; Polycingulatisporites crenulatus; Samaropollenites speciosus; Semiretisporis denmeadii; Callialasporites turbatus; Contignisporites cooksoniae; Contignisporites spp.; Antulsporites saevus; Aracariacites fissus; Callialasporites dampieri; Exesipollenites tumulus; Gleicheniidites spp.; Neveisporites vallatus; Coronatispora perforata; Dictyotosporites complex; Retitriletes circolumenus; Aequitriradites sp. A; Concavissimisporites variverrucatus; Matonisporites sp. A; Murospora florida; Retitriletes facetus; Ceratosporites equalis; Microcachryidites antarcticus; Aequitriradites acusus/spinulosus; Crybelosporites stylosus; Retriletes watherooensis; Aequitriradites hispidus; Cicatricosisporites australiensis; Cicatricosisporites spp.

Top: last *W. clathrata*. Prominent species include *Ellipsoidictyum cinctum*. Early to Middle Oxfordian [approx. ≡ *Wanaea clathrata* Zone of Helby et al. (1987)].

4. *Cribroperidinium perforans* Zone. Base: first *C. perforans*, with *Hystrichosphaeridium pachydermum* and *Peridictyocysta mirabilis*. Top: last *C. perforans*. Prominent species include *Productodinium chenii*. Late Oxfordian [approx. ≡ *Cribroperidinium perforans* and *Dingodinium swanense* Zones of Helby et al. (1987)].
5. *Omatia montgomeryi* Zone. Base: first *Fistulacysta simplex*. Top: last *O. montgomeryi*. Prominent species include *H. pachydermum*, *Gonyaulacysta jurassica* and the acritarch *Nummus similis*. Late Oxfordian to Early Kimmeridgian [approx. ≡ *Omatia montgomeryi* Zone of Helby et al. (1987)].
6. *Gonyaulacysta jurassica* Zone. Base: first *Oligosphaeridium* sp. 1 of Davey. Top: latest abundance of *G. jurassica*. Prominent species include *Nannoceratopsis pellucida*. Early Kimmeridgian [≡ lower *Dingodinium jurassicum* Zone of Helby et al. (1987)].
7. *Nannoceratopsis pellucida* Zone. Base: numerous *Peridictyocysta mirabilis* and last *Chlamydophorella wallala*. Top:

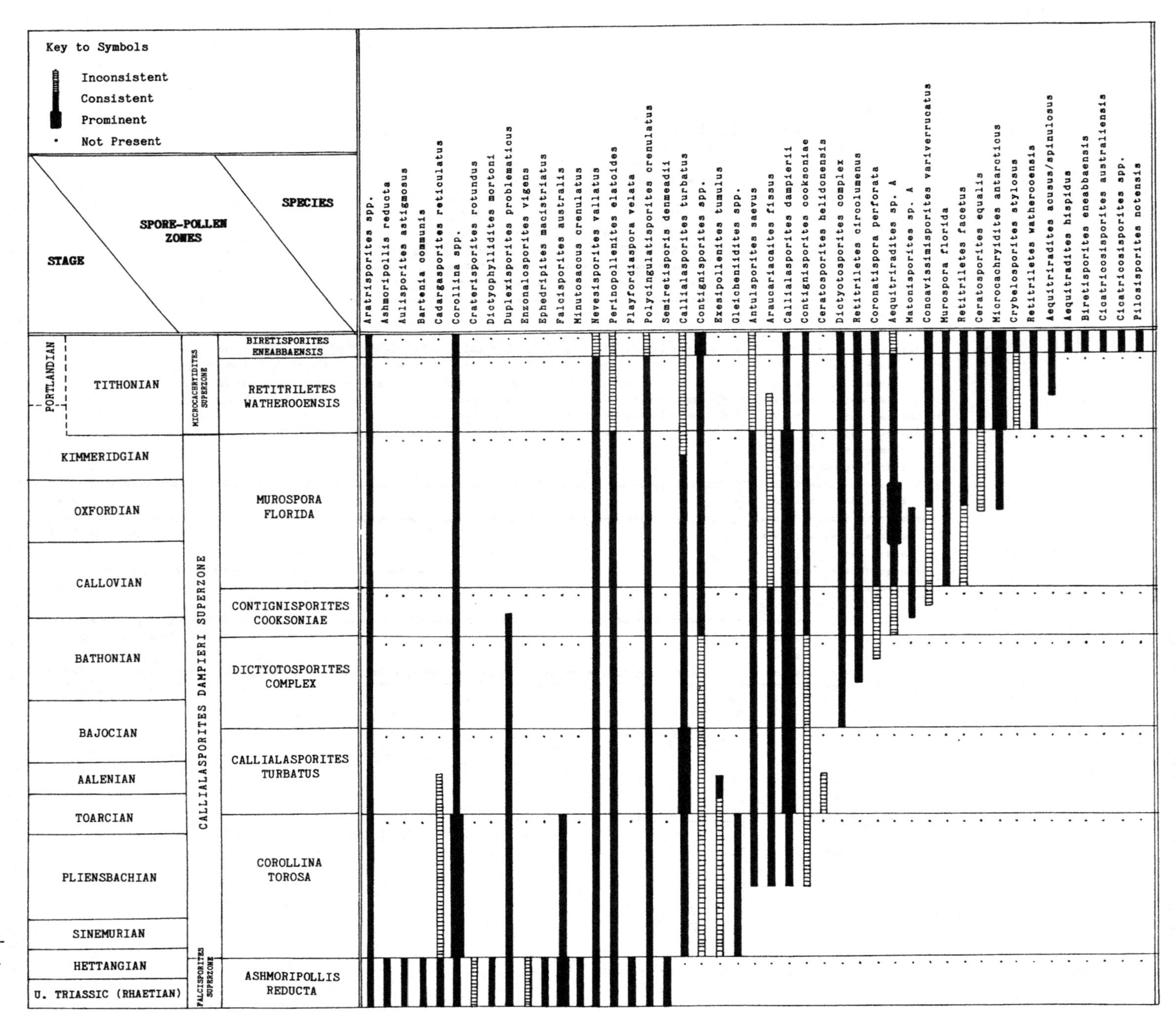


Table 13.2. *Biostratigraphic zonation of stratigraphically important pollen and spores in Western Australia*

Table 13.3. (opposite) *Biostratigraphic zonation of stratigraphically important dinoflagellates in Australia*

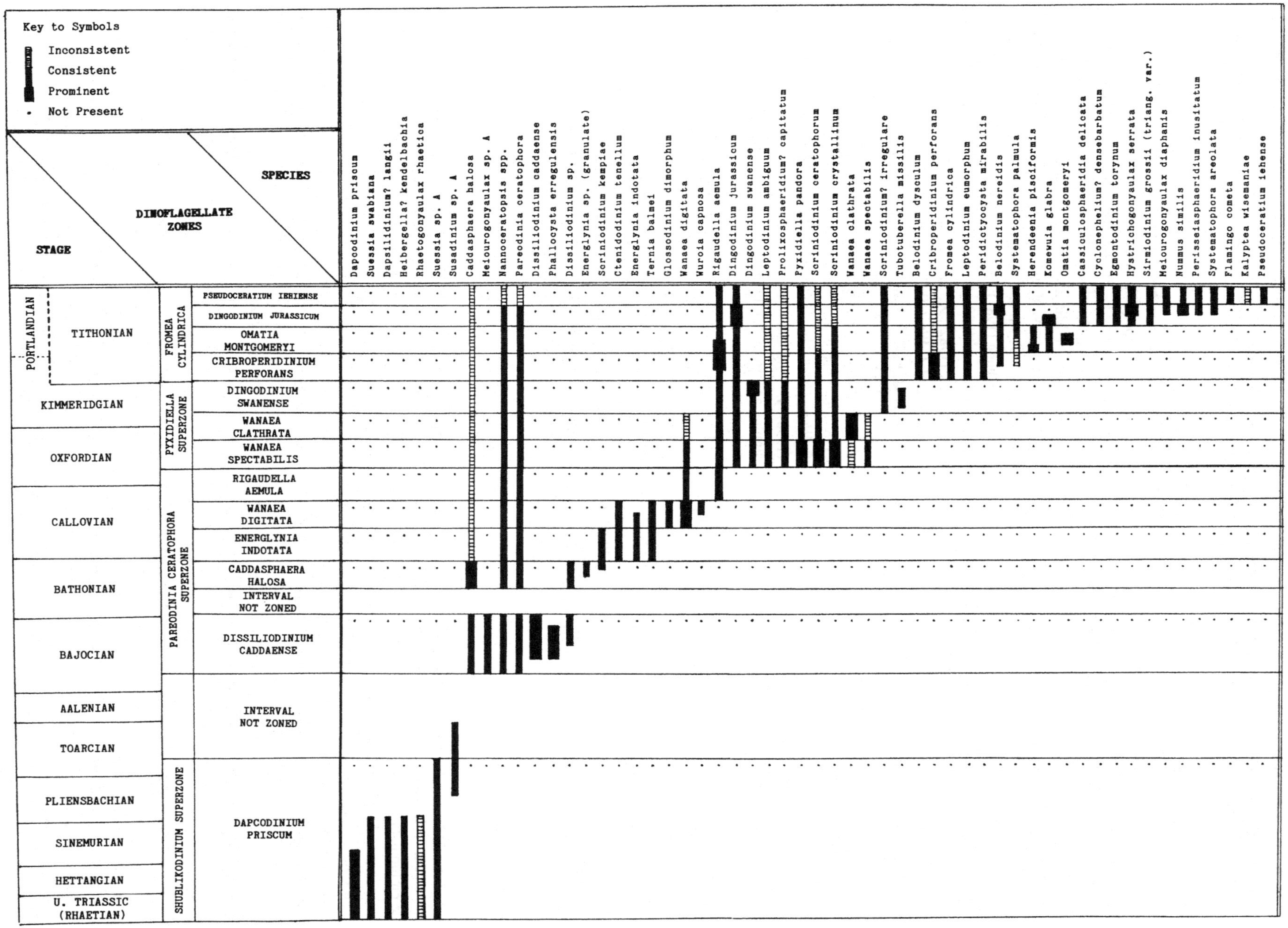

Key to Symbols

- Inconsistent
- Consistent
- Prominent
- • Not Present

| STAGE | DINOFLAGELLATE ZONES | | SPECIES |
|---|---|---|---|
| PORTLANDIAN / TITHONIAN | FROMEA CYLINDRICA | PSEUDOCERATIUM IEHIENSE | |
| | | DINGODINIUM JURASSICUM | |
| | | OMATIA MONTGOMERYI | |
| | | CRIBROPERIDINIUM PERFORANS | |
| KIMMERIDGIAN | PYXIDIELLA SUPERZONE | DINGODINIUM SWANENSE | |
| | | WANAEA CLATHRATA | |
| OXFORDIAN | | WANAEA SPECTABILIS | |
| | PAREODINIA CERATOPHORA SUPERZONE | RIGAUDELLA AEMULA | |
| CALLOVIAN | | WANAEA DIGITATA | |
| | | ENERGLYNIA INDOTATA | |
| BATHONIAN | | CADDASPHAERA HALOSA | |
| | | INTERVAL NOT ZONED | |
| BAJOCIAN | | DISSILIODINIUM CADDAENSE | |
| AALENIAN | | INTERVAL NOT ZONED | |
| TOARCIAN | | | |
| PLIENSBACHIAN | SHUBLIKODINIUM SUPERZONE | DAPCODINIUM PRISCUM | |
| SINEMURIAN | | | |
| HETTANGIAN | | | |
| U. TRIASSIC (RHAETIAN) | | | |

Species (left to right):

Dapcodinium priscum
Suessia swabiana
Dapsilidinium? langii
Heibergella? kendelbachia
Rhaetogonyaulax rhaetica
Suessia sp. A
Susadinium sp. A
Caddasphaera halosa
Meiourogonyaulax sp. A
Nannoceratopsis spp.
Pareodinia ceratophora
Dissiliodinium caddaense
Phallocysta erregulensis
Dissiliodinium sp.
Energlynia sp. (granulate)
Scriniodinium kempiae
Ctenidodinium tenellum
Energlynia indotata
Ternia balmei
Glossodinium dimorphum
Wanaea digitata
Wuroia capnosa
Rigaudella aemula
Dingodinium jurassicum
Dingodinium swanense
Leptodinium ambiguum
Prolixosphaeridium? capitatum
Pyxidiella pandora
Scriniodinium ceratophorum
Scriniodinium crystallinum
Wanaea clathrata
Wanaea spectabilis
Scriniodinium? irregulare
Tubotuberella missilis
Belodinium dysculum
Cribroperidinium perforans
Fromea cylindrica
Leptodinium eumorphum
Peridictyocysta mirabilis
Belodinium nereidis
Systematophora palmula
Herendeenia pisciformis
Komewuia glabra
Omatia montgomeryi
Cassiculosphaeridia delicata
Cyclonephelium? densebarbatum
Egmontodinium torynum
Hystrichogonyaulax serrata
Sirmiodinium grossii (triang. var.)
Meiourogonyaulax diaphanis
Nummus similis
Perisseiasphaeridium insusitatum
Systematophora areolata
Flamingo cometa
Kalyptea wisemaniae
Pseudoceratium iehiense

latest abundance of *N. pellucida*. Prominent species include *Stiphrosphaeridium dictyophorum* in upper part. Middle to Late Kimmeridgian [≡ early middle *Dingodinium jurassicum* Zone of Helby et al. (1987)].

8. *Nummus similis* Zone. Base: not well characterized, but *Levisphaera crassicingulata* and last *G. jurassica* are in lower part. Top: latest abundance of *N. similis*. Prominent species include *S. dictyophorum, Broomea simplex* and *Peridictyocysta mirabilis*. Late Kimmeridgian to Early Tithonian [≡ middle *Dingodinium jurassicum* Zone of Helby et al. (1987)].
9. *Broomea simplex* Zone. Base: *Papuadinium apiculatum* and *Omatidinium amphiacanthum* present. Top: last *B. simplex*. Prominent species include *H. pachydermum, P. mirabilis* and *Cyclonephelium densebarbatum*. Middle Tithonian [≡ late middle *Dingodinium jurassicum* Zone of Helby et al. (1987)].
10. *Rhynchodiniopsis serrata* Zone. Base: first *Cassiculosphaeridia delicata*. Top: last *R. serrata*. Prominent species include *P. mirabilis*. Middle Tithonian [approx. ≡ upper *Dingodinium jurassicum* Zone of Helby et al. (1987)].
11. *Oligosphaeridium* sp. 1 Zone. Base: *Hystrichodinium pulchrum* and *Cassiculosphaeridia magna*. Top: latest abundance of *P. mirabilis*. Prominent species include *O.* sp. 1 and *Papuadinium apiculatum*. Middle Tithonian [≡ lower *Pseudoceratium iehiense* Zone of Helby et al. (1987)].
12. *Pseudoceratium iehiense* Zone. Base: *Systematophora palmula* and the *Gonyaulacysta helicoidea/G. cretacea* group. Top: last *P. iehiense, O.* sp. 1 and *Pseudoceratium weymouthense*. Late Tithonian to Berriasian [≡ upper *Pseudoceratium iehiense* Zone of Helby et al. (1987)].

### Indonesia and Southeast Asia

Palynological studies of samples from Indonesia were inaugurated some years ago at the University of Toronto, under the supervision of G. Norris, but no results have been forthcoming. The preliminary results of an unpublished study (Helby and F. Hasibuan in press) of the Fageo Group (Toarcian to basal Cretaceous) of Misool, an island west of New Guinea, have been courteously communicated to us by Dr. Robin Helby. From comparisons with Australian dinoflagellate ranges, considerable advances in understanding have resulted. The Yefbie Shale has yielded both a characteristic Toarcian suite, including *Susadinium*, and a restricted Bajocian to Early Bathonian *Dissiliodinium* association. A suite of dinoflagellates assignable to the Late Callovian to basal Oxfordian *Rigaudella aemula* Zone is present at the base of the Demu Formation, while unequivocal *Wanaea spectabilis* Zone associations (Late Oxfordian to Early Kimmeridgian) were encountered throughout the rest of this formation. The lowest assemblages from the Lelinta Shale also fall within that zone; later assemblages suggest an extension as high as the *Kalyptea wisemaniae* Zone (Early Berriasian).

No studies have yet been published on the Jurassic palynofloras of southeast Asia.

### China

Studies in China have been principally upon spores and pollen rather than upon dinoflagellate cysts. They appear to have been inaugurated by L. Zhang (1965), who examined an Early-Middle Jurassic palynological assemblage from the Yima coal-bearing series of Mianchi county, Henan. Subsequently, much work in Jurassic palynology has been done by L. Zhang (1978); Z. Zhang (1978); Y. Xu and W. Zhang (1980); W. Li and Y. Shang (1980); Y. Shang (1981); Z. Lei (1981); Z. Liu (1982); B. Du, X. Li and W. Duan (1982); R. Pu and H. Wu (1982, 1985); C. Wang and G. Tong (1982); X. Li, W. Duan and B. Du (1982); L. Qian, C. Zhao and J. Wu (1983); D. Su et al. (1983); H. Wu and X. Zhang (1983); Y. Bai, M. Lu, L. Chen and R. Long (1983); S. Miao et al. (1984); W. Zhang and Q. Zhao (1985); B. Du (1985); Z. Gan (1986); W. Zhang and Z. Zhang (1987); L. Qian and J. Wu (1987); L. Qian et al. (1987); L. Qu, W. Zhang and J. Yu (1987); S. Wang (1988); W. Li (1989); W. Zhang (1989, 1990); and Z. Liu (1990). J. Yang and S. Sun (1982, 1987) recorded megaspores of similar age from the Junggar Basin, Xinjiang and the Yangtze Gorge, Hubei.

The work by the third author in Eastern China demonstrates that the microfloras are divisible into Northern and Southern Provinces. In the Early Jurassic, the boundary between these provinces was approximately the ancient Qinling Range (located near latitude 33°N). Its position in the Middle Jurassic is not clear, but by the Late Jurassic it had moved northward, giving an almost uniform aspect to the terrestrial palynofloras. The assemblages are indicated in Table 13.4 and selectively illustrated in Plate 94. The palynofloras of the Shangganing Basin typify those of the Northern Province. Four assemblages can be distinguished:

1. *Protoconiferus-Cycadopites-Cyathidites* assemblage, to be found in the Fuxian Formation, is characterized by a predominance of gymnospermous pollen, with some pteridophyte spores. Its principal elements are *Cyathidites minor, C. infrapunctatus, Cibotiumspora juncta* and species of the genera *Lycopodiumsporites, Osmundacidites, Crassitudisporites, Protoconiferus, Pinuspollenites, Quadraeculina* and *Cycadopites*. Most of these taxa are long-ranging, but specifically Jurassic taxa such as *Cerebropollenites carlylensis, C. papillopolus* and *Callialasporites* are present, along with a few surviving representatives of *Taeniaesporites*. This is an Early Jurassic assemblage, indicating a warm, humid, temperate to subtropical climate.
2. The *Cyathidites-Piceaepollenites* assemblage, to be found in the Yanan Formation, principally comprises a rich variety of bisaccate pollen and spores, but with no ancient striate conifer pollen. *Cyathidites* is the most abundant genus; other common elements are *Deltoidospora, Cibotium-*

Table 13.4. *Distribution of spore/pollen associations in China*

| Epoch | | | NORTH CHINA | | | | | SOUTH CHINA | |
|---|---|---|---|---|---|---|---|---|---|
| | | | **Shangganning Basin** | **Shanxi Province** | **Hebei Province** | **Liaoning Province** | **Nei Mongol Zizhiqu** | **Yangzi Gorges Region** | **Xianggan Region** |
| Jurassic | Late | Oxfordian - Tithonian | Fenfanghe Formation<br><br>Anding Formation<br>*Classopollis-*<br>*Concavissimisporites*<br>Association | Tianchihe Formation | Dabeigou Form/Lower Member<br>*Piceites- Podocarpidites-*<br>*Todisporites* Assoc.<br><br>Houcheng Formation<br>*Classopollis-*<br>*Callialasporites-*<br>*Schizaeoisporites*<br>Association | Tuchengzi Formation<br>*Classopollis-*<br>*Cicatricosisporites*<br>Association | | Penglaizhen Formation<br>*Classopollis-*<br>*Couperisporites*<br>Association<br><br>Suining Formation | |
| | Middle | Bathonian - Callovian | Zhiluo Formation<br>*Cyathidites minor-*<br>*Quadraeculina*<br>Association | Yungang Formation<br>*Cyathidites minor-*<br>*Cycadopites-*<br>*Quadraeculina*<br>Association | Diaojishan Formation<br><br>Jiulongshan Formation<br>*Cyathidites minor-*<br>*Quadraeculina-*<br>*Classopollis*<br>Association | Lanqi Formation | Changhangou and<br>Zhaogou Formations<br>*Cyathidites minor-*<br>*Classopollis-*<br>*Quadraeculina*<br>Association | Shangshaximiao Formation<br>*Cyathidites minor-*<br>*Classopollis- Neorasrickia*<br>Association<br><br>Xiashaximiao Formation | |
| | | Aalenian - Bajocian | Yanan Formation<br>*Cyathidites minor-*<br>*Piceaepollenites*<br>Association | Datong Formation<br>*Cyathidites minor-*<br>*Cycadopites*<br>Association | Xiahuayan Formation<br>*Cyathidites minor-*<br>*Cycadopites-*<br>*Perinopollenites*<br>Association | Guojiadian Formation<br>*Cyathidites minor-*<br>*Cycadopites-*<br>*Neoraistrickia*<br>Association | | Chenjiawan Formation<br><br>Xietan Formation<br>*Cyathidites minor-*<br>*Callialasporites-*<br>*Classopollis* Association | |
| | Early | Pliensbachian - Toarcian | Fuxian Formation<br>*Protoconiferus-*<br>*Cyathidites*<br>Association | Yongdinzhuang Formation<br>*Cyathidites-*<br>*Pseudopicea-*<br>*Triquitrites*<br>Assocation | Nandaling Formation | Beipiao Formation<br>*Protoconiferus-*<br>*Cycadopites-*<br>*Cyathidites*<br>Association | Wudanggou Formation<br>*Cycadopites-*<br>*Cyathidites-*<br>*Triquitrites*<br>Association | Xiangxi Formation<br>Upper Member<br>*Dictyophyllidites-*<br>*Cyathidites-*<br>*Classopollis*<br>Subassociation | Menkoushan Formation<br>*Classopollis-*<br>*Cyathidites-*<br>*Cerebropollenites*<br>Subassociation |
| | | Hettangian - Sinemurian | | | Xinshikou Formation | Xinlonggou Formation<br><br>Kuntonbuoluo Formation | | Xiangxi Formation<br>Lower Member<br>*Dictyophyllidites-*<br>*Classopollis-*<br>*Cerebropollenites*<br>Subassociation | Zishang Formation<br>*Marattisporites-*<br>*Classopollis-*<br>*Cerebropollenties*<br>Subassociation |

Table 13.5. *Biostratigraphic distribution of spores and pollen in the Jurassic of north-central Siberia*

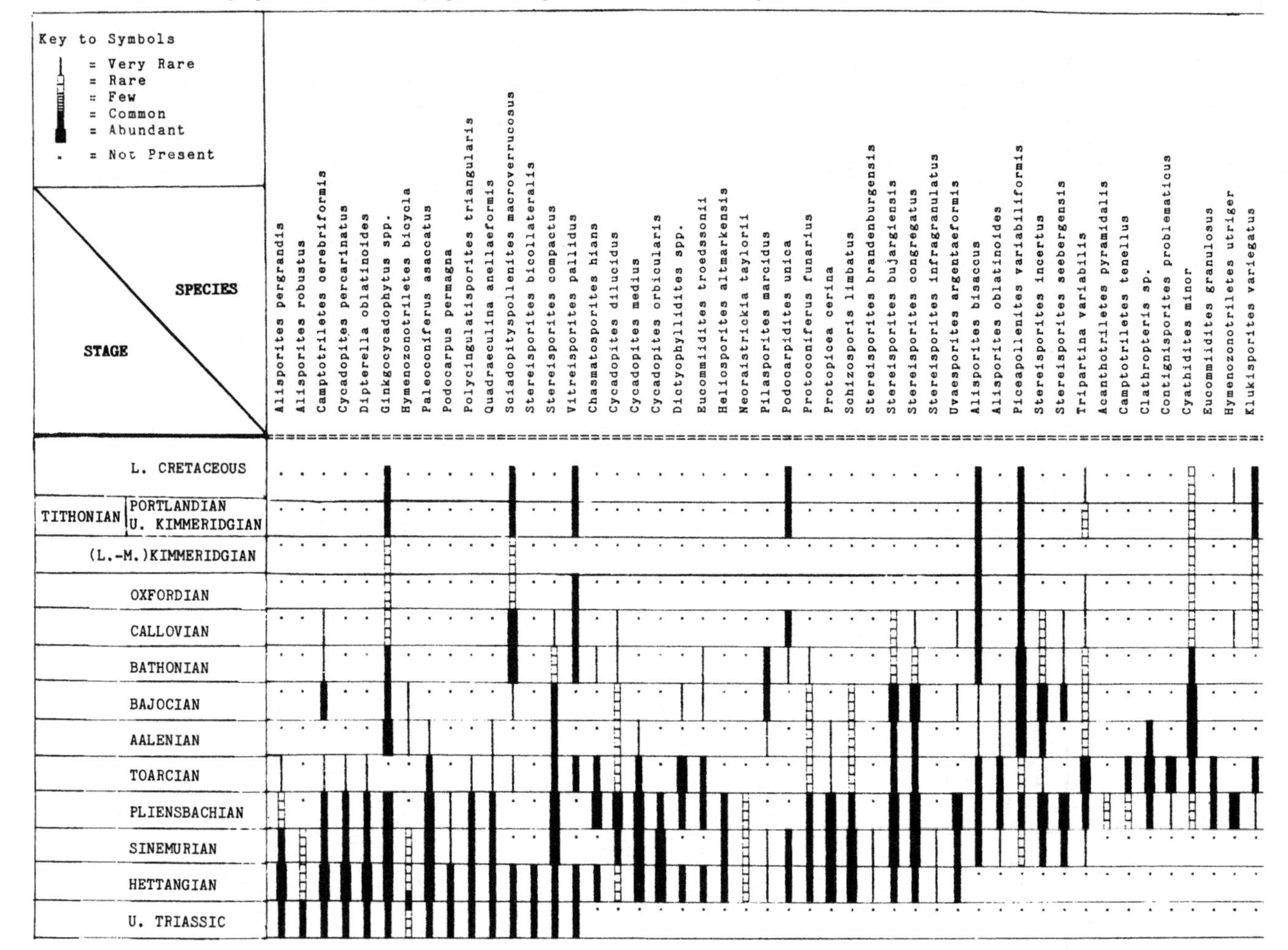


*spora, Todisporites, Lycopodiumsporites, Neoraistrickia, Osmundacidites, Leptolepidites, Klukisporites, Cycadopites, Callialasporites* and *Cerebropollenites.* This is an early Middle Jurassic assemblage, indicating again a warm, humid, temperate to subtropical climate.

3. The *Cyathidites-Quadraeculina* assemblage, to be found in the Zhiluo Formation, has gymnospermous pollen comprising 70% and the spores of pteridophytes 30%. In general, its composition is similar to that of the preceding assemblage, but less diverse. *Quadraeculina* and *Classopollis* have become much commoner; and *Concavissimisporites irroratus* (≡ *C. variverrucatus* s.l.) is present. The age is probably late Middle Jurassic, the climate now semi-arid.
4. The *Classopollis-Concavissimisporites* assemblage, to be found in the Anding Formation, has gymnospermous pollen overwhelmingly predominant, with pteridophyte spores rare. *Classopollis* is very abundant and shows a variety of form. The genera *Trilobosporites* and *Cicatricosporites* appear; and species of *Cyathidites, Deltoidospora, Klukisporites, Gleicheniidites, Concavissimisporites, Pinuspollenites, Podocarpidites, Quadraeculina, Cerebropollenites* and *Callialasporites* are well represented. This is a Late Jurassic assemblage of a hot, arid subtropical to tropical climate.

The palynofloras of the Yangzi Gorges region typify those of the Southern Province. Two subassemblages, succeeded by three assemblages, may be recognized:

1. The *Dictyophyllidites-Classopollis-Cerebropollenites* subassemblage, to be found in the Lower Member of the Xianxi Formation, has pteridophyte spores and gymnospermous pollen in almost equal numbers. The spores comprise *Dictyophyllidites, Concavisporites* and *Auritulinasporites,* with some *Cyathidites, Cibotiumspora, Crassitudisporites, Lycopodiumsporites, Osmundacidites* and *Marattisporites. Classopollis* is prominent among the pollen, with *Cycadopites, Chasmatosporites* and various conifer pollen. Specifically Jurassic elements are *Cerebropollenites, Eucommiidites* and rare surviving representatives of *Canalizonospora, Kyrtomisporites* and *Zebrosporites.* This is an early Early Jurassic assemblage of a hot and humid, subtropical to tropical climate.
2. The *Dictyophyllidites-Cyathidites-Classopollis* subassemblage, to be found in the Upper Member of the Xianxi For-

Table 13.5. (*cont.*)

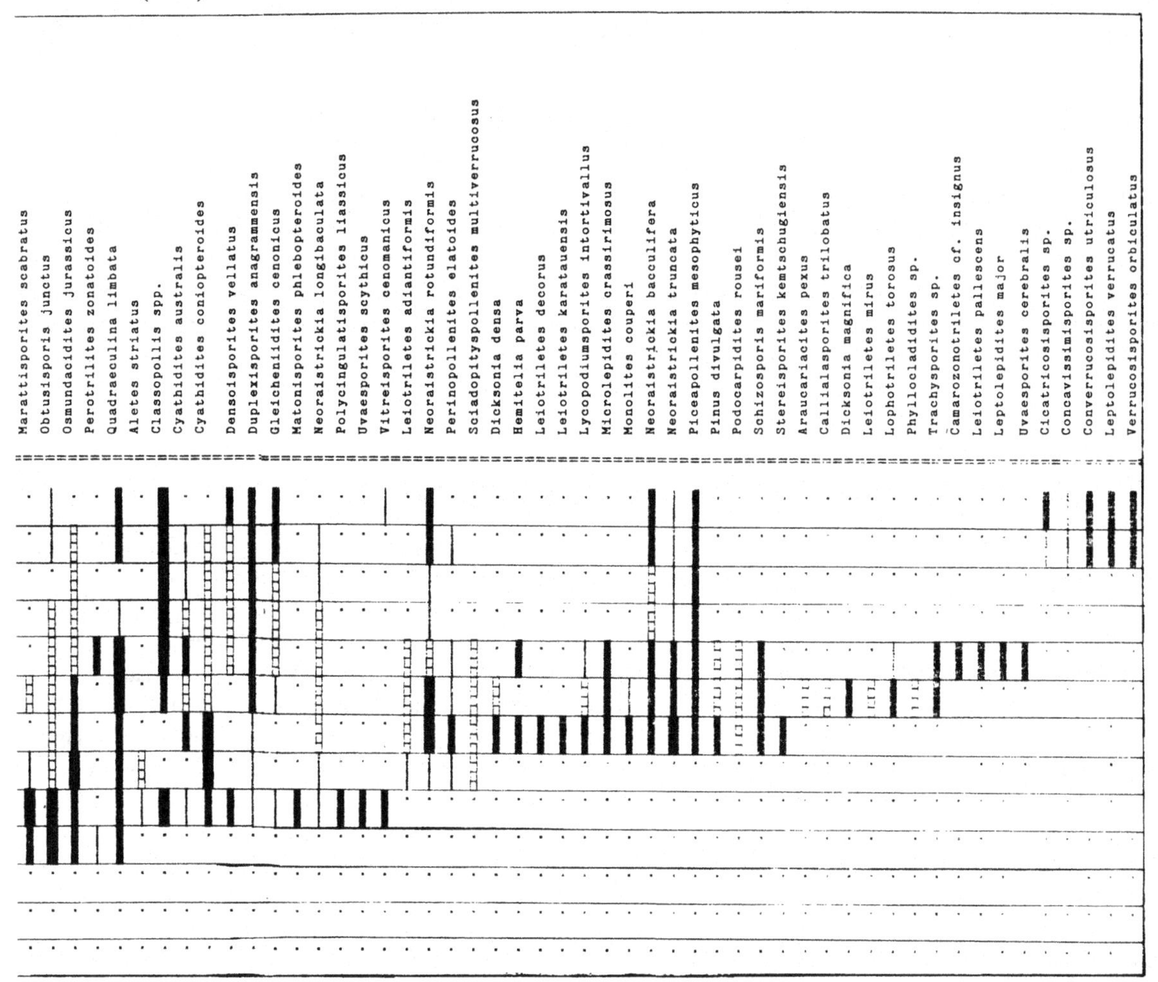


mation, is similar but includes some new forms, such as *Klukisporites variegatus*, *K. pseudoreticulatus*, *Converrucosisporites veritus* and *Cerebropollenites macroverrucosus*. The content of *Classopollis* and *Cyathidites* has increased markedly; *Canalizonospora* is still present. This is a late Early Jurassic assemblage, again of a hot and humid, subtropical to tropical climate.

3. The *Cyathidites-Callialasporites-Classopollis* assemblage, to be found in the Xietan Formation, in contrast has gymnospermous pollen more numerous than pteridophyte spores. The principal pollen are *Classopollis*, with some *Callialasporites*, *Perinopollenites*, *Psophosphaera* and bisaccate taxa. *Cyathidites minor* is the most prominent spore, with lesser numbers of *Osmundacidites*, *Cibotiumspora*, *Todisporites*, *Dictyophyllidites*, *Crassitudisporites*, *Lycopodiumsporites*, *Neoraistrickia* and *Klukisporites*. This is an early Middle Jurassic microflora of a semi-arid, subtropical to tropical climate.
4. The *Cyathidites-Classopollis-Neoraistrickia* assemblage, to be found in the Shangshaximiao Formation, differs from the last only in having a higher content of *Neoraistrickia*, *Podocarpidites* and *Piceites*, with somewhat fewer *Classopollis*. It is of late Middle Jurassic age, from similar climatic conditions.
5. The *Classopollis-Couperisporites* assemblage, to be found in the Penglaizhen Formation, consists almost wholly of gymnospermous pollen, *Classopollis* alone accounting for 90% of the assemblage, with few *Cyathidites*, *Neoraistrickia*, *Osmundacidites*, *Callialasporites*, *Quadraeculina* and *Couperisporites*. This is a Late Jurassic assemblage of a hot and dry, subtropical to tropical climate.

Biostratigraphical studies based upon marine palynomorphs were begun by Yu Jingxian (1982), in a study of the Shihebei Formation (Jurassic–Cretaceous) of eastern Heilongjiang. Two assemblages were distinguished, defining very broad stratigraphic zones. The lower of these, the *Gonyaulacysta-Pareodinia-Hystrichogonyaulax* assemblage, includes species of Late Jurassic type, such as *Apteodinium nuciformis*, *Millioudodinium longicornis*, *Pareodinia ceratophora*, *Leptodinium crassinervum*, *Hystrichogonyaulax cladophora* and *Scriniodinium galeritum;* but it includes also Early Cretaceous species such as *Rhynchodiniopsis aptiana*, *Cribroperidinium cornutum*, *Canningia brevispinosa* and *Meiourogonyaulax bulloidea*. It is assigned a Late Kimmeridgian to Berriasian age and considered to represent a shallow marine environment. Y. Xu (1982) reported a new Early to Middle Jurassic microflora from Nylamo, Tibet, in the same year.

**Siberia (eastern Russia)**

Study of the Jurassic spores and pollen of this region has been intensive. It was inaugurated by V. S. Maliavkina (1949, 1961, 1962) and N. A. Bolkhovitina (1952, 1956, 1959, 1966, 1968). Other noteworthy works were by L. N. Gutova and M. M. Odintsova (1964); Gutova and V. A. Ilyina (1969); Gutova, I. Kilesso and Odintsova (1969); G. V. Shramkova (1966); I. I. Charudo (1969); L. G. Markova (1969); O. V. Shugaevskaya (1962, 1969); and L. I. Bogdashova (1969). Seven important papers on this theme were presented at the Third International Conference on Palynology, held in Novosibirsk in 1971 and published in 1973 – by L. V. Rovnina; Markova; Markova and J. V. Teslenko; Odintsova and Gutova; Ilyina; L. D. Petrova and A. F. Fradkina; and Shugaevskaya, V. S. Markevich and P. I. Bitutskaya. A. M. Medvedeva and L. P. Klimushina (1984) reported palynomorphs from Jurassic oils and condensates; and Markova et al. (1988) reviewed Early–Middle Jurassic assemblages.

Unquestionably the most prolific contributor to Siberian studies has been V. A. Ilyina, in a long series of papers commencing in 1965 and continuing to 1988. Her fullest summation of biostratigraphical knowledge on Siberian terrestrial palynomorphs was presented three years earlier (1985b); it is epitomized in the accompanying chart (Table 13.5), which, although dealing specifically with assemblages from north-central Siberia, conveniently serves as basis for all correlations in the northwestern Pacific region.

Studies of dinoflagellates from this region were inaugurated by T. F. Vozzhennikova (1967, 1969; V. I. Teodorovich and Vozzhennikova 1971). Recently, J. K. Lentin has restudied Vozzhennikova's type material and has marketed a set of transparencies of these, through L.I.B. Consultants of Calgary. However, though a taxonomic redescription appears imminent, a biostratigraphical update appears unlikely to be forthcoming.

**Japan**

No palynological studies have yet been published. Consequently, during the Circum-Pacific Jurassic Field Conference in Japan (1986), the first author collected samples for examination which, after some had been lost in shipment, were supplemented by others collected during 1987, through the courtesy of Dr. T. Sato. The samples examined, arranged in approximate descending stratigraphical order, are shown in the accompanying Table 13.6.

As may be seen, the results are disappointing. Most samples seem to have undergone low-grade metamorphism, which has carbonized the wood and destroyed the rest of the organic residue. Pollen were abundant in only two samples (2862 and 2863), in the former accompanied by fungal spores, and sparsely present in two others (2871 and 2882). No marine palynomorphs were encountered. Such results furnish no encouragement to further palynological research in the Japanese Jurassic; reliance must instead, it seems, be placed upon radiolarians for biostratigraphical correlation.

**Alaska (United States of America)**

Rich palynological assemblages are widely present in the Alaskan Jurassic, but await adequate description. Caytonialean micropores gained mention in a paper presented by J. Gray and J. L. Browning (1959), and dinoflagellate cysts in papers by D. I. Wharton (1987) and N. Albert (1990), but these were published only in abstract. Alaskan representatives of the dinoflagellate family Pareodiniaceae were mentioned in a review by V. D. Wiggins (1975). As yet, no detailed biostratigraphical accounts have been forthcoming.

**British Columbia (Canada)**

Through the courtesy of Dr. Bruce Cameron and the Geological Survey of Canada, 17 samples from the Jurassic of the Queen Charlotte Islands were furnished to the first author for study. Samples from the upper Kunga Formation (Late Sinemurian) of East Sandilands Island and the lowest Maude Formation (Early Pliensbachian to Early Toarcian) of Maude Island yielded only carbonized wood fragments. Samples from higher levels in the Maude Formation of Graham Island proved more variable. The lowest (Middle Toarcian) yielded many *Classopollis,* with black and brown wood and pale plant fragments. Two succeeding samples (Late Toarcian) again contained only carbonized wood fragments. The two highest samples (Early Bajocian) each contained black and brown wood and plant fragments, the uppermost having also bisaccate pollen and trilete miospores that were too poorly preserved for confident identification. In five samples from the overlying Yakoun Formation (Middle Bajocian to Early Callovian), carbonized wood fragments and light-colored plant debris were present; poorly preserved pollen grains and miospores were present in the Late Bathonian sample, but microfossils were otherwise absent. In view of the rich invertebrate faunas recovered from the Queen Charlottes, these results are both surprising and disappointing.

Though receiving passing mention in some publications (e.g., W. H. Ziegler and S. A. J. Pocock 1960), Jurassic palynomorphs from the western regions of mainland British Columbia still await proper biostratigraphical description. The dinoflagellate assemblages reported by E. H. Davies and T. P. Poulton (1986) from northeastern British Columbia were derived from a basin connected with the interior seaway traversing North America during the Mesozoic; they bear no direct relation to the marine microflora of the circum-Pacific region.

**Western United States of America**

Thirteen samples from the Jurassic of Oregon were courteously furnished for palynological study to the first author by Dr. Emile Pessagno. All contained abundant carbonized wood; many

Table 13.6. *Results of palynological study of samples from northern Honshu, Japan*

| Specimen Number | Description | Locality | Age | Comments on Residues |
|---|---|---|---|---|
| 2862 | Shale about 3m down road from uppermost limit of outcrop. | Namiyanagi Pass | Late Kimmeridgian [=Early Tithonian] | Abundance of pollen: some fungal spores, plant plant fragments and much wood, carbonized or not. |
| 2863 | Radiolarian chert | Namiyanagi Pass | Jurassic? | Abundance of pollen of simple type, plant fragments and carbonized wood. |
| 2864 | Siltstone immediately above boundary with Lower Jurassic chert. | Hanakazuki Forestry Road, Locality 2. | Late Kimmeridgian [=Early Tithonian] | Very few plant fragments, with carbonized wood. |
| 2865 | Radiolarian-bearing shale between two massive sandstones. | Quarry behind sawmill, Taniida | Jurassic | Carbonized wood only. |
| 2867 | Shaly siltstone close to middle of fold structure. | Hatsuneyama Quarry | Middle Jurassic? | Carbonized wood only. |
| 2868 | Shale about 3m above last. | Hatsuneyama Quarry | Middle Jurassic? | Carbonized wood only. |
| 2869 | Sandstone, unlocalized. | Hatsuneyama Quarry | Middle Jurassic? | Carbonized wood only. |
| 2870 | Middle part of Black Oolite, Koike Limestone, Nakonasawa Formation. | North corner of Oyama Quarry | Late Kimmeridgian [=Early Tithonian] | A few plant fragments plus carbonized wood. |
| 2871 | Top of outcrop, Koike Limestone, Nakonasawa Formation. | Tomizawa Quarry | Late Kimmeridgian [=Early Tithonian] | Few pollen, with plant fragments and carbonized wood. |
| 2872 | Terrigenous sandstone of Lower Member, Niranohama Formation. | Foreshore outcrop at south side of bay, Niranohama. | Hettangian | Carbonized wood only. |
| 2875 | Black shale of *Hosoureites ikianus* Zone. | Outcrop closest to village road, Gongen. | Aalenian | Carbonized wood only. |
| 2876 | Shales of Barren Zone. | Base of cliff at southernmost end of outcrop, Gongen. | Late Hettangian | Carbonized wood only. |
| 2877 | Clays with *Tmetoceras*. | Outcrop at north end of shore, close to road to village, Gongen. | Aalenian | Carbonized wood only. |
| 2878 | *Planammatoceras planinsigne* Zone, Hosoura Formation, near top. | Inland outcrop, Hosoura. | Aalenian | Carbonized wood only. |
| 2879 | Lower part of Aratozaki Formation. | First outcrop beyond harbour, Shizuhama. | Probably Bajocian | Carbonized wood only. |
| 2880 | Tsumakizaka Formation, Karakuwa Group. | Section at foot of Tadakoshi Pass | Late Bajocian | Carbonized wood only. |
| 2881 | Shales in upper part of Mone Formation, Shishiori Group. | Tadakoshi Pass | Late Jurassic? | Carbonized wood only. |
| 2882 | Kosaba Formation about 3m above base, Karakuwa Group. | Promontory at Kosaba | Prob. Late Aalenian to Early Bajocian | A few pollen, with plant fragments and some wood. |
| 2883 | Sandstone of Kosaba Formation. | Middle of outcrop, quarry at Shuku. | Prob. Late Aalenian to Early Bajocian. | Plant fragments and carbonized wood. |

contained pale-colored plant debris also; but none yielded recognizable microfossils of any kind.

On the basis of very poorly preserved palynomorphs, A. Traverse (1972) assigned a Late Jurassic to Early Cretaceous age to part of the Franciscan mélanges of the Coast Ranges of California. D. Habib and J. S. Warren (1973) examined dinoflagellates from west of the Sacramento Valley that were of disputable date – latest Jurassic or earliest Cretaceous – and decided in favor of the latter age. Subsequently, Warren (1973) discussed variation in the dinoflagellate *Sirmiodinium grossii* from the Jurassic rocks of that region, but his full biostratigraphical study remains unpublished.

## Mexico

The palynostratigraphy of Mexico has, to date, attracted little study. Through the courtesy of Dr. Gerd E. G. Westermann, the first author was furnished with 41 samples from the Middle Jurassic of Oaxaca and Guerrero provinces. Samples from Oaxaca were collected by Dr. José Sandoval in 1984 and from Guerrero by Michael Marshall in 1986. Twenty-four samples were from the Barranco de La Boleta, La Isleta and Los Rebajes sections of Oaxaca, from the Taberna, Simon, Otatera and Yucuñuti Formations (Lower Bajocian–Callovian); nineteen samples were from the Coauilote composite section of Guerrero, from the Simon, Otatera and Yucuñuti Formations (Bathonian–Lower Callovian). Stratigraphical details may be found in Westermann, Corona and Carrasco (1984), Sandoval and Westermann (1986) and Sandoval, Westermann and Marshall (1990).

The samples contained no marine palynomorphs but were, in general, extremely rich in spores and pollen. Consequently, there is the promise that terrestrial palynomorphs will furnish a strong basis for future biostratigraphical and palaeoenvironmental studies in that region.

## Colombia

Jurassic palynomorphs from the Lower Jurassic Bocas Formation, Eastern Cordillera near Bucaramanga, were illustrated by Remy et al. (1975). The predominant genera are *Classopollis, Cycadopites, Vitreisporites, Dictyophyllidites, Gleicheniidites, Verrucosisporites* and *Concavisporites;* they are associated with a *Phlebopteris branneri* megaflora. The palynomorphs are so strongly carbonized that recognition to specific level was not possible.

Two main biotopes could be recognized: (1) drier ones, where Cheirolepidiaceae (which produced *Classopollis*) prevailed; (2) less dry biotopes with monocultures of *Phlebopteris branneri*, producing laevigate spores of *Dictyophyllidites* type.

### Peru

The only Jurassic palynomorphs known from Peru are from the Arequipa Basin, southern Peru where, out of more than 50 Lower, Middle and Upper Jurassic palynological samples processed by the second author, only two from the Middle Jurassic yielded assemblages. These consisted of trilete spores and pollen so extremely carbonized that, even at generic level, the identification of taxa proved difficult; however, *Deltoidospora, Gleicheniidites, Inaperturopollenites* and *Classopollis* could be recognized. They indicate an epineritic to near-shore environment.

### Chile

Nearly 100 samples have been processed from northern and central Chile by the second author. These were obtained from Tarapacá Province, east of Iquique; Quebrada Juan de Morales; the Antofagasta area; Quebrada del Profeta; the Alta Cordillera at 23°30′S; and Quebrada Asientos. Apart from carbonized remnants, all samples proved barren. The causes must have been the intrusive and extrusive volcanic activity during and after the Jurassic, combined with the effects of tectonic deformation.

Nevertheless, there remains a good prospect for recovery of palynomorphs in less deformed areas – in particular, the Principal Cordillera of central Chile.

### Argentina

Studies of Jurassic spores and pollen were begun by C. A. Menéndez (1956, 1968a,b) and have been continued by Volkheimer and his associates, most notably M. E. Quattrocchio, in a succession of papers which have embraced also dinoflagellate cysts (Volkheimer 1968, 1969, 1970, 1971a,b, 1972, 1974a,b; Volkheimer, M. O. Manceñido and S. E. Damborenea 1978; Volkheimer and Quattrocchio 1975a,b, 1977, 1981; Volkheimer, Quattrocchio et al. 1976; Volkheimer and A. M. Moroni 1981; O. González-Amicón and Volkheimer 1982a,b; M. Arguijo, Volkheimer and U. Rosenfeld 1982; Quattrocchio and Volkheimer 1983; Quattrocchio 1981, 1984). Middle Jurassic microfloras were also reported by L. H. Scarfati and M. A. Morbelli (1984).

The stratigraphical results of their studies of pollen and spores provide a solid basis for correlations in the southeastern part of the circum-Pacific region. Recently, in three papers, Volkheimer, Quattrocchio and A. M. Zavattieri (1992), Volkheimer, Quattrocchio and Scarfati (1992) and Volkheimer, Quattrocchio and Prámparo (1992) have brought together the results; their work is utilized in our charts. Because the palynofloras are so rich, it has been necessary to display the ranges in two tables (Tables 13.7 and 13.8), representative forms being illustrated in Plate 95.

Jurassic megaspores from Argentina were first reported by Morbelli (1990) from the Los Molles Formation of Neuquén province; their date was believed to be Aalenian to Bajocian and their relationships to be with the extant Selaginellales.

Work on dinoflagellates is proceeding apace. Assemblages from strata of Bajocian to Late Tithonian (≡ Portlandian) age were reported from the Cura Niyeu, Lotena, Lower Vaca Muerta and Picún Leufú formations from the Neuquén Basin by Quattrocchio and Volkheimer (1990). In a subsequent study by Quattrocchio and Sarjeant (1992) of the Middle and Upper Jurassic dinoflagellates and acritarchs of that basin (the Lotena, Lower Vaca Muerta, and Picún Leufú Formations), five zones were distinguished:

1. *Endoscrinium galeritum reticulatum* Zone, defined by the total range of that subspecies and by the presence of *Lithodinia jurassica reburrosa, Barbatacysta* cf. *brevispinosa* and *Dissiliodinium volkheimeri*.
2. *Acanthaulax downiei* Zone, beginning with the appearance of *Acanthaulax downiei* and ending at the appearance of *Diacanthum* cf. *hollisteri*. Also prominent are *Escharisphaeridia pocockii* and *Hystrichosphaerina neuquina*.
3. *Millioudodinium nuciforme* Zone, defined by the total range of that species and with lower boundary marked by the appearance of *Diacanthum* cf. *hollisteri*. *A. downiei* is still present; other prominent species are *Pilosidinium cactosum* and *Sentusidinium villersense*.
4. *Dichadogonyaulax culmula* var. *curtospina* Zone, beginning with the appearance of that taxon and ending at the appearance of *Aptea notialis*. *D.* cf. *hollisteri* and *S. villersense* are still present; *Hystrichosphaerina neuquina* and *Rhynchodiniopsis* cf. *setcheyensis* are prominent.
5. *Aptea notialis* Zone, defined by the total range of that species. *Dingodinium* sp. and *Paraevansia mammillata* are also noteworthy.

These Argentinian assemblages are not rich in taxa and compare poorly with others from the circum-Pacific region, suggesting a strong regionality in the dinoflagellate floras of that time interval.

### Antarctica

The earliest work to be done in this region was by G. Norris (Norris in H. S. Gair, Norris and J. Ricker 1965; Norris 1965), a presumed Early Jurassic microflora being reported from the Ferrar Group of Victoria Land. This included *Sphagnumsporites tenuis, Concavisporites mortoni, ?Osmundacidites parvus, Verrucosisporites cameroni, Ginkgocycadophytus nitidus, Inapertisporites minutus, Circulisporites parvus* and species of *Deltoidospora, Monosulcites, Araucariacites* and *Undulatisporites*, as well as fragmentary bisaccates. The strongest comparisons were with Australian microfloras.

Subsequently, P. Tasch and J. M. Lammons (1977) reported a rich Early Jurassic palynoflora from lacustrine interbeds among the basalt flows of the Transantarctic Mountains. Forty-five taxa were recognized, including *Alisporites grandis, A. lowoodensis, A. similis, Araucariacites australis, A. fissus, Calamospora mesozoica, Callialasporites dampieri, C. microvelatus, C. segmentatus, C. trilobatus, Cibotiumspora juriensis, Classopollis anasillos, C. chateaunovi, C. classoides, C. simplex, Contignisporites cooksonii, Cycadopites follicularis, Duplexisporites problematicus, Inaperturopollenites turbatus, Ischysporites crateris, I.*

*marburgensis, Klukisporites lacunus, K. neovariegatus, Laricoidites triquetrus, Lycopodiumsporites australoclavatoides, L. rosewoodensis, Pinuspollenites globosaccus, Taschia antarctica, Todisporites minor, Trilobosporites antiquus* and *Vitreisporites pallidus*, together with unidentified species of *Bharadwajapollenites, Chasmatosporites, Clavatipollenites, Gingkocycadophytus, Patinasporites, Podocarpidites, Protopinus, Pteruchipollenites* and *Undulatisporites*. This is definitely a Gondwana microflora, again showing strongest comparisons with Australia.

The first Late Jurassic (Portlandian ≡ Middle to Late Tithonian) microflora from Antarctica was reported by R. A. Askin (1981) from Livingston Island, South Shetland Islands; it included the spores *Cicatricosisporites australiensis* and *C. ludbrookii*, together with a dinoflagellate comparable to *Broomea simplex*. Subsequently J. B. Riding (1988) briefly reported Kimmeridgian and Tithonian dinoflagellate cysts from the Antarctic Peninsula. A fuller study of these materials has not yet been forthcoming; and indeed, our knowledge of Antarctic palynofloras remains meagre.

**Acknowledgments**

The authors are indebted to Dr. Robin Helby, Dr. Roger Morgan, Dr. Alan D. Partridge and Dr. Graeme J. Wilson for furnishing unpublished data and illustrative material and to Dr. Victor Zakharov for assistance concerning data on Siberian Jurassic palynostratigraphy. Dr. Helby courteously read and commented in detail on this chapter in manuscript; Mr. Patrick Cashman gave generous help with Russian translations; and Mrs. Linda Dietz, research assistant to the first author, aided in the preparation of the tables and in many other ways.

## References

*New Zealand*

Campbell, J. D., & Warren, G. (1965). Fossil localities of the Torlesse Group in the South Island. *Trans. R. Soc. N.Z., Geology, 3,* 99–137.

Couper, R. A. (1953). Upper Mesozoic and Cainozoic spores and pollen grains from New Zealand. *N.Z. Geol. Surv., Paleont. Bull., 22,* 1–77.

(1960). New Zealand Mesozoic and Cainozoic plant microfossils. *N.Z. Geol. Surv., Paleont. Bull., 32,* 1–87.

Helby, R., Wilson, G. J., & Grant-Mackie, J. A. (1988). A preliminary biostratigraphic study of Mid to Late Jurassic dinoflagellate assemblages from Kawhia, New Zealand. In P. A. Jell & G. Playford (eds.), *Palynological and Palaeobotanical Studies in Honour of Basil E. Balme. Mem. Assoc. Australas. Palaeontol., 6,* 125–66.

Hudson, N., Grant-Mackie, J. A., & Helby, R. (1987). Closure of the New Zealand 'Middle Jurassic Hiatus'? *Search, 18,* 146–8.

Norris, G. (1968). Plant microfossils from the Hawks Crag Breccia, southwest Nelson, New Zealand. *N.Z. J. Geol. Geophys, 11,* 312–44.

Waterhouse, J. B., & Norris, G. (1972). Paleobotanical solution to a granite conundrum. Hawks Crag Breccia of New Zealand and the tectonic evolution of the southwest Pacific. *Geosci. Man, 4,* 1–15.

Wilson, G. J. (1978). *Kaiwaradinium*, a new dinoflagellate genus from the Late Jurassic of North Canterbury, New Zealand. *N.Z. J. Geol. Geophys, 21,* 81–4.

(1982). Dinoflagellate assemblages from the Puaroan, Ohauan and Heterian Stages (Late Jurassic), Kaiwara Valley, North Canterbury, New Zealand. *Rep. N.Z. Geol. Surv., Palaeont., 59,* 1–20.

(1984a). Two new dinoflagellates from the Late Jurassic of North Canterbury, New Zealand. *J. R. Soc. N.Z., 14,* 215–21.

(1984b). New Zealand Late Jurassic to Eocene dinoflagellate biostratigraphy – a summary. *Newsl. Stratigr., 13,* 104–17.

Wilson, G. J., & Helby, R. (1987). A probable Oxfordian dinoflagellate assemblage from North Canterbury, New Zealand. *N.Z. Geol. Surv. Rec., 20,* 119–25.

(1988). Early Cretaceous dinoflagellate assemblages from Torlesse rocks near Ethelton, North Canterbury. *N.Z. Geol. Surv. Rec., 35,* 38–43.

*Australia and Papua New Guinea*

Backhouse, J. (1974). Stratigraphic palynology of the Watheroo Line boreholes, Perth Basin. *Ann. Rep. Geol. Surv. West. Aust., 1973,* 99–103.

(1975). Palynology of the Yarragadee Formation in the Eneabba Line boreholes. *Ann. Rep. Geol. Surv. West. Aust., 1974,* 107–9.

(1978). Palynological zonation of the Late Jurassic and Early Cretaceous sediments of the Yarragadee Formation, central Perth Basin, Western Australia. *Rep. Geol. Surv. West. Aust., 7,* 1–53.

(1988). Late Jurassic and Early Cretaceous palynology of the Perth Basin, Western Australia. *Bull. Geol. Surv. West. Aust., 135,* 1–233.

Balme, B. E. (1957). Spores and pollen grains from the Mesozoic of Western Australia. *Coal Res., Commonw. Scient. Ind. Res. Orgn., 25,* 1–48.

(1961). Palynological examination of samples from Samphire Marsh No. 1. *Publ. Bur. Miner. Resour., Geol. Geophys., Aust., 5,* 22–6.

(1963). Palynological reports on samples from BMR 4A. *Rep. Bur. Miner. Resour., Geol. Geophys., 60,* 71–4.

(1964). The palynological record of Australian pre-Tertiary floras. In L. M. Cranwell (ed.), *Ancient Pacific Floras: The Pollen Story* (pp. 49–80). Honolulu: University of Hawaii Press.

(1966). Jurassic and Triassic microflora from Shallow Borehole (UWA 10), Enanty Hill. *Bull. Commonw. Aust. Bur. Miner. Resour., 92,* 212–15.

Burger, D. (1977). Palynomorphs from Eromanga Basin formations in QDM Aramac 1 well. *Qd. Govt. Min. J., 78,* 331–6.

(1986). Palynology, cyclic sedimentation, and palaeoenvironments in the Late Mesozoic of the Eromanga Basin. *Geol. Soc. Austral. Spec. Publ., 12,* 53–70.

(1989). Stratigraphy, palynology, and palaeoenvironments of the Hooray Sandstone, eastern Eromanga Basin, Queensland and New South Wales. *Qd. Dept. Mines Rep., 3,* 1–28.

Burger, D., & Senior, B. R. (1979). A revision of the sedimentary and palynological history of the northeastern Eromanga Basin, Queensland. *J. Geol. Soc. Aust., 26,* 121–32.

Cookson, I. C. (1953). Difference in miospore composition of some samples from a bore at Comaum, South Australia. *Aust. J. Bot., 1,* 462–73.

Cookson, I. C., & Eisenack, A. (1958). Microplankton from Australian and New Guinea Upper Mesozoic sediments. *Proc. R. Soc. Vict., 70,* 19–79.

(1960). Upper Mesozoic microplankton from Australia and New Guinea. *Palaentology, 2,* 243–61.

(1974). Mikroplankton aus Australischen Mesozoichen und Tertiären Sedimenten. *Palaeontographica, B148,* 44–93.

(1982). Mikrofossilien aus Australischen Mesozoischen und Tertiären Sedimenten II. *Palaeontographica, B184,* 23–63.

Davey, R. J. (1988). Palynological zonation of the Lower Cretaceous, upper and uppermost Middle Jurassic in the northwestern Papuan

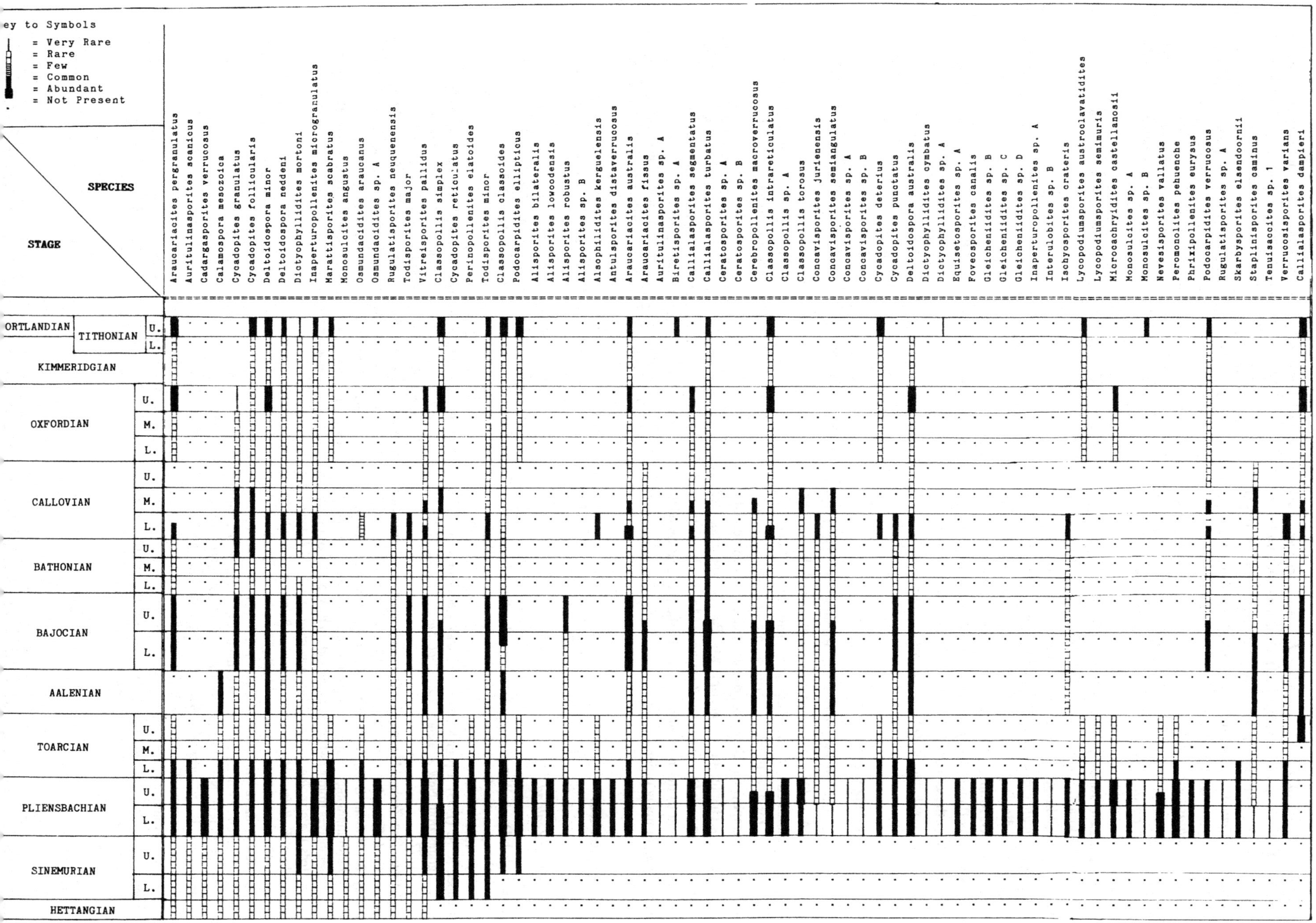

ey to Symbols

- = Very Rare
- = Rare
- = Few
- = Common
- = Abundant
- . = Not Present

STAGE

SPECIES

Araucariacites pergranulatus
Aurituliinasporites scanicus
Cadargasporites verrucosus
Calamospora mesozoica
Cycadopites granulatus
Cycadopites follicularis
Deltoidospora minor
Deltoidospora neddeni
Dictyophyllidites mortoni
Inaperturopollenites microgranulatus
Marattisporites scabratus
Monosulcites angustus
Osmundacidites araucanus
Osmundacidites sp. A
Rugulatisporites neuquenensis
Todisporites major
Vitreisporites pallidus
Classopollis simplex
Cycadopites reticulatus
Perinopollenites elatoides
Todisporites minor
Classopollis classoides
Podocarpidites ellipticus
Alisporites bilateralis
Alisporites lowoodensis
Alisporites robustus
Alisporites sp. B
Alsophilidites kerguelensis
Antulsporites distaverrucosus
Araucariacites australis
Araucariacites fissus
Aurituliinasporites sp. A
Biretisporites sp. A
Callialasporites segmentatus
Callialasporites turbatus
Ceratosporites sp. A
Ceratosporites sp. B
Cerebropollenites macroverrucosus
Classopollis intrareticulatus
Classopollis sp. A
Classopollis torosus
Concavisporites juriensis
Concavisporites semiangulatus
Concavisporites sp. A
Concavisporites sp. B
Cycadopites deterius
Cycadopites punctatus
Deltoidospora australis
Dictyophyllidites cymbatus
Dictyophyllidites sp. A
Equisetosporites sp. A
Foveosporites canalis
Gleicheniidites sp. B
Gleicheniidites sp. C
Gleicheniidites sp. D
Inaperturopollenites sp. A
Interulobites sp. B
Ischyosporites crateris
Lycopodiumsporites austroclavatidites
Lycopodiumsporites semimuris
Microcachryidites castellanosii
Monosulcites sp. A
Monosulcites sp. B
Nevesisporites vallatus
Peromonolites pehuenche
Phrixipollenites eurysus
Podocarpidites verrucosus
Rugulatisporites sp. A
Skarbysporites elsendoornii
Staplinisporites caminus
Tenuisaccites sp. 1
Verrucosisporites varians
Callialasporites dampieri

ORTLANDIAN
TITHONIAN U.
L.
KIMMERIDGIAN
OXFORDIAN U.
M.
L.
CALLOVIAN U.
M.
L.
BATHONIAN U.
M.
L.
BAJOCIAN U.
L.
AALENIAN
TOARCIAN U.
M.
L.
PLIENSBACHIAN U.
L.
SINEMURIAN U.
L.
HETTANGIAN

Table 13.7. (opposite) *Biostratigraphic distribution of Jurassic spores and pollen in Argentina: species with first appearance in Early Jurassic*

Table 13.8. *Biostratigraphic distribution of Jurassic spores and pollen in Argentina: species with first appearance in Middle to Late Jurassic*

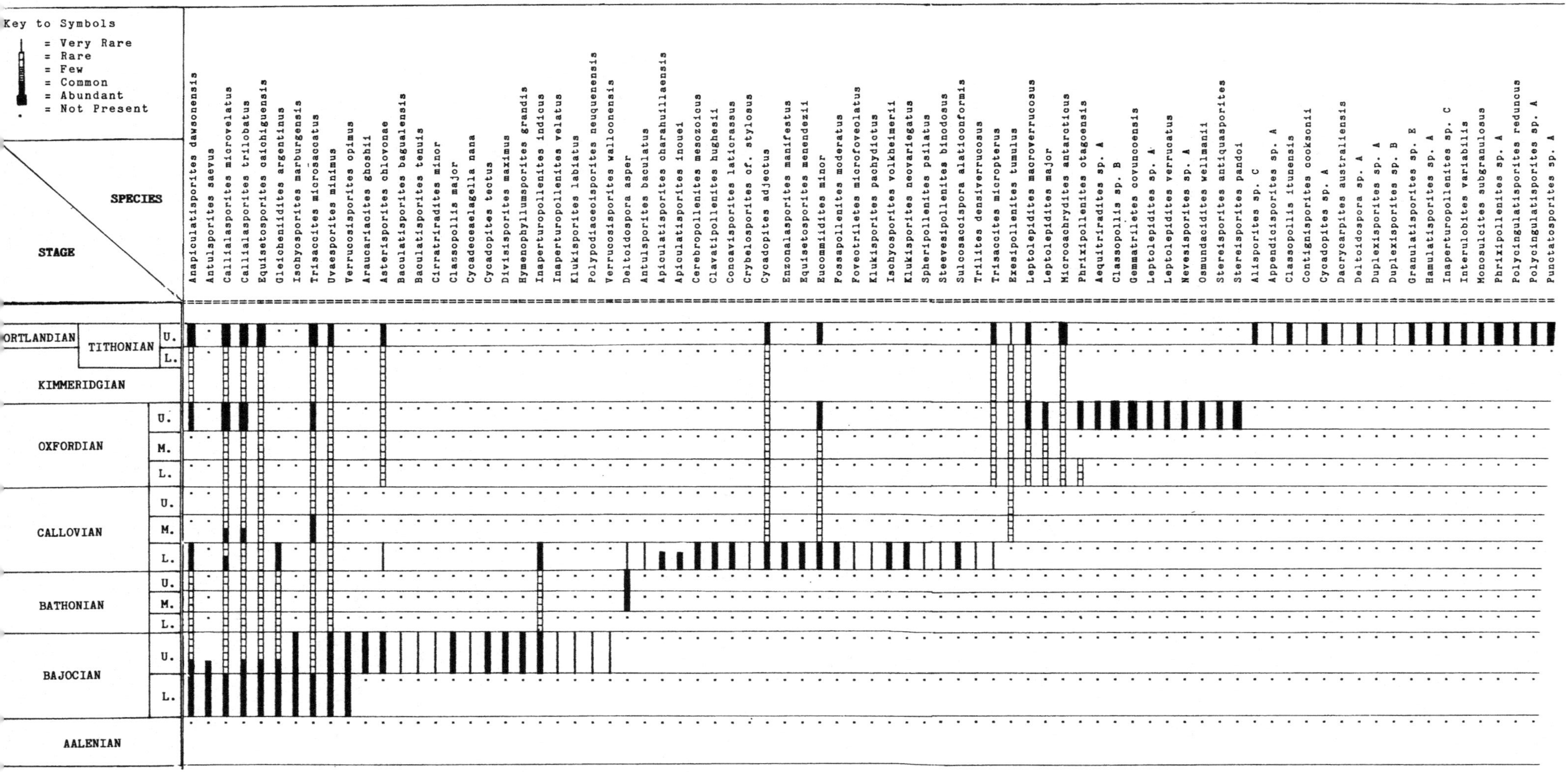

Key to Symbols

- = Very Rare
- = Rare
- = Few
- = Common
- = Abundant
- • = Not Present

**STAGE** / **SPECIES**

Stages:

- PORTLANDIAN / TITHONIAN: U.
- TITHONIAN: L.
- KIMMERIDGIAN
- OXFORDIAN: U., M., L.
- CALLOVIAN: U., M., L.
- BATHONIAN: U., M., L.
- BAJOCIAN: U., L.
- AALENIAN

Species:

Anapiculatisporites dawsonensis
Antulsporites saevus
Callialasporites microvelatus
Callialasporites trilobatus
Equisetosporites caichiguensis
Gleicheniidites argentinus
Ischyosporites marburgensis
Trisaccites microsaccatus
Uvaesporites minimus
Verrucosisporites opimus
Araucariacites ghoshii
Asterisporites ohlovonae
Baculatisporites bagualensis
Baculatisporites tenuis
Cirratriradites minor
Classopollis major
Cycadeoeaelagella nana
Cycadopites tectus
Divisisporites maximus
Hymenophyllumsporites grandis
Inaperturopollenites indicus
Inaperturopollenites velatus
Klukisporites labiatus
Polypodiaceoisporites neuquenensis
Verrucosisporites walloonensis
Deltoidospora asper
Antulsporites baculatus
Apiculatisporites oharahuillaensis
Apiculatisporites inouei
Cerebropollenites mesozoicus
Clavatipollenites hughesii
Concavisporites laticrassus
Crybelosporites cf. stylosus
Cycadopites adjectus
Enzonalasporites manifestus
Equisetosporites menendezii
Eucommiidites minor
Fossapollenites moderatus
Foveotriletes microfoveolatus
Klukisporites pachydictus
Ischyosporites volkheimerii
Klukisporites neovariegatus
Spheripollenites psilatus
Steevesipollenites binodosus
Sulcosaccispora alaticonformis
Trilites densiverrucosus
Trisaccites micropterus
Exesipollenites tumulus
Leptolepidites macroverrucosus
Leptolepidites major
Microcachrydites antarcticus
Phrixipollenites otagoensis
Aequitriradites sp. A
Classopollis sp. B
Gemmatriletes covunccensis
Leptolepidites sp. A
Leptolepidites verrucatus
Nevesisporites sp. A
Osmundacidites wellmanii
Stereisporites antiquasporites
Stereisporites pandoi
Alisporites sp. C
Appendicisporites sp. A
Classopollis itunensis
Contignisporites cooksonii
Cycadopites sp. A
Dacrycarpites australiensis
Deltoidospora sp. A
Duplexisporites sp. A
Duplexisporites sp. B
Granulatisporites sp. E
Hamulatisporites sp. A
Inaperturopollenites sp. C
Interulobites variabilis
Monosulcites subgranulosus
Phrixipollenites sp. A
Polycingulatisporites reduncus
Polycingulatisporites sp. A
Punctatosporites sp. A

Basin of Papua New Guinea. *Mem. Geol. Surv. Papua/New Guinea, 13*(for 1987), 1–86.

Deflandre, G., & Cookson, I. C. (1955). Fossil microplankton from Australasian Late Mesozoic and Tertiary sediments. *Aust. J. Mar. Freshwat. Res., 6,* 241–313.

De Jersey, N. J. (1959). Jurassic spores and pollen grains from the Rosewood coalfield. *Qd. Govt. Min. J., 60,* 346–66.

(1960). Spore distribution and correlation in the Rosewood coalfield. *Publ. Geol. Surv. Qd., 295,* 1–27.

(1963). Jurassic spores and pollen grains from the Marburg Sandstone. *Publ. Geol. Surv. Qd., 313,* 1–15.

(1965). Plant microfossils in some Queensland crude oil samples. *Publ. Geol. Surv. Qd., 329,* 1–9.

(1971a). Early Jurassic miospores from the Helidon Sandstone. *Publ. Geol. Surv. Qd., 351,* 1–27.

(1971b). Palynological evidence for a facies change in the Moreton Basin. *Qd. Govt. Min. J., 72,* 1–8.

(1972). Palynology of a sample from the Jurassic of the Mundubbera area. *Qd. Govt. Min. J., 73,* 273.

(1973). Palynology of core samples from the Helidon, Toowoomba and Kulpi areas. *Qd. Govt. Min. J., 74,* 128–44.

(1976). Palynology and time relationships in the Lower Bundamba Group (Moreton Basin). *Qd. Govt. Min. J., 77,* 461–5.

De Jersey, N. J., & Paten, R. J. (1963). Palynology of samples from Union-Kern-A.O.G. Moonie Nos. 1 and 3 wells. Appendix 2 of *Petroleum Search Sub-Acts.* publ. 45 (Moonie No. 1). Geological Survey of Queensland, pp. 29–55.

(1964). Jurassic spores and pollen grains from the Surat Basin. *Publ. Geol. Surv. Qd., 322,* 1–18.

Evans, P. R. (1966). Mesozoic stratigraphic palynology in Australia. *Australas. Oil Gas J., 12,* 58–63.

Filatoff, J. (1975). Jurassic palynology of the Perth Basin, Western Australia. *Palaeontographica, B154,* 1–113.

Harris, W. K. (1966). Delhi-Santos Gidgealpa No. 1. Palaeontology, Part A. *Publ. Bur. Miner. Resour. Aust., 73-A3,* 87–93.

(1970). An Upper Jurassic microflora from the western margin of the Great Artesian Basin, South Australia. *Q. Notes Geol. Surv. S. Aust., 35,* 3–8.

Helby, R., Morgan, R., & Partridge, A. D. (1987). A palynological zonation of the Australian Mesozoic. In P. A. Jell (ed.), *Studies in Australian Mesozoic Palynology. Mem. Assoc. Australas. Palynol., 4,* 1–94.

Helby, R., & Stover, L. E. (1987). *Ternia balmei* gen. et sp. nov., Jurassic dinoflagellate with possible dinophysealian affinity. In P. A. Jell (ed.), *Studies in Australian Mesozoic Palynology. Mem. Assoc. Australas. Palynol., 4,* 135–41.

Jell, P. A. (ed.). (1987). Studies in Australian Mesozoic Palynology. *Mem. Assoc. Australas. Palynol., 4,* 341 pp, 5 fiches.

McKellar, J. L. (1974). Jurassic miospores from the upper Evergreen formation, Hutton Sandstone and basal Injune Creek Group, northeastern Surat Basin. *Publ. Geol. Surv. Qd., 361,* 1–89.

(1975). Jurassic miospore assemblages from the lower Tiaro Coal Measures. *Qd. Govt. Min. J., 76,* 355–9.

(1981a). Palynostratigraphy of samples from the Narangba area, Nambour Basin. *Qd. Govt. Min. J., 82,* 268–73.

(1981b). Palynostratigraphy of samples from GSQ Ipswich 24 and 25. *Qd. Govt. Min. Jr., 82,* 479–87.

(1981c). Palynostratigraphy of samples from the Lockyer Valley, Moreton Basin. *Qd. Govt. Min. J., 82,* 540–4.

(1982). Late Triassic ("Rhaetian") and Jurassic palynostratigraphy of the Surat Basin. In P. S. Moore & T. J. Mount (eds.), *Eromanga Basin Symposium, Summary Papers* (pp. 172–3). Adelaide: Geological Society of Australia and Petroleum Exploration Society of Australia.

McWhae, J. R. H., Playford, P. E., Lindner, A. W., Glenister, B. F., & Balme, B. E. (1958). The stratigraphy of Western Australia. *J. Geol. Soc. Aust., 4,* 1–161.

Playford, G., & Cornelius, K. D. (1967). Palynological and lithostratigraphic features of the Razorback Beds, Mount Morgan district, Queensland. *Pap. Dept. Geol. Univ. Qd., 6,* 81–94.

Playford, G., & Dettmann, M. E. (1965). Rhaeto-Liassic plant microfossils from the Leigh Creek Coal Measures, South Australia. *Senckenberg. Leth., 46,* 127–81.

Reiser, R. F., & Williams, A. J. (1969). Palynology of the Lower Jurassic sediments of the northern Surat Basin, Queensland. *Publ. Geol. Surv. Qd., 339,* 1–24.

Sarjeant, W. A. S. (1968). Microplankton from the Upper Callovian and Lower Oxfordian of Normandy. *Rev. Micropaléont., 10,* 221–42.

Stevens, J. (1981). Palynology of the Callide Basin, east-central Queensland. *Pap. Dept. Geol. Univ. Qd., 9,* 1–35.

Stover, L. E., & Helby, R. (1987a). Some Australian Mesozoic microplankton index fossils. In P. A. Jell (ed.), *Studies in Australian Mesozoic Palynology* (pp. 101–34). Sydney: Association of Australasian Palynologists.

(1987b). The Jurassic dinoflagellate *Omatia* and allied genera. In P. A. Jell (ed.), *Studies in Australian Mesozoic Palynology* (pp. 143–58). Sydney: Association of Australasian Palynologists.

Wiseman, J. F. (1980). Palynostratigraphy near the "Jurassic-Cretaceous boundary" in the Carnarvon Basin, Western Australia. *Proc. IV Intern. Palynological Conference, Lucknow (1976–1977), 2,* 330–49.

### *Indonesia*

Helby, R., & Hasibuan, F. (in press). A preliminary palynological study of the Fageo Group (Toarcian to basal Berriasian) of Misool, Indonesia.

### *China*

The listing that follows is the most complete index to Chinese palynological works on the Jurassic yet to be published in the Latin alphabet. Most works are in Chinese, but many have titles and summaries in English appended.

Bai Y., Lu M., Chen L. & Long R. (1983). Mesozoic spores and pollen. In Chengdu Institute of Geology and Mineral Resources (ed.), *Paleontological Atlas of Southwest China, Microfossil Volume* (pp. 520–649). Beijing: Geological Publishing House.

Du B. (1985). Sporo-pollen assemblages from the Middle Jurassic in the Wagnjiashan Basin of Jingyuan, Gansu and their stratigraphic and paleogeographic significance. *Geol. Rev./Dizhi Lun-P'ing (Beijing), 31,* 131–41.

Du B., Li X. & Duan W. (1982). Sporo-pollen assemblages from Yanan and Zhiluo Formation in Chongxin county, Gansu province. *Acta Palaeont. Sin., 21,* 597–605.

Gan Z. (1986). Spores and pollen from the Xiahuayuan Formation in Hebei and their stratigraphical significance. *Acta Palaeont. Sin., 25,* 87–93.

Lei Z. (1981). The sporo-pollen assemblage of the Nandian red beds, Lancang, Yunnan. *Acta Bot. Sin., 23,* 235–42.

Li W. (1989). Palynomorphs from the Shouchang Formation and their stratigraphical significance. *Acta Palaeont. Sin., 28,* 724–9.

Li W. & Shang Y. (1980). Sporo-pollen assemblages from the Mesozoic coal series of western Hubei. *Acta Palaeont. Sin., 19,* 201–19.

Li X., Duan W. & Du B. (1982). Palynological assemblage and its age from the Fuxian Formation in Chongxin county, Gansu province. In Palynological Society of China (ed.), *I Congress of the Palynological Society of China, Tianjin (1979)* (pp. 105–9). Beijing: Science Press.

Liu Z. (1982). Early and Middle Jurassic sporo-pollen assemblages from the Shiguai coalfield of Baotou, Nei Monggol. *Acta Palaeont. Sin., 21,* 371–9.

(1990). Sporo-pollen assemblage from Middle Jurassic Xishanyao Formation of Shawan, Xinjiang, China. *Acta Palaeont. Sin., 29*, 62–83.

Liu Z., Shang Y. & Li W. (1981). Triassic and Jurassic sporo-pollen assemblages from some localities of Shaanxi and Gansu, North-west China. *Bull. Nanjing Inst. Geol. Palaeont. Acad. Sin., 3*, 131–210.

Miao S., Yu J., Qu L., Zhang W., Zhang Q. & Zhang D. (1984). Mesozoic spores and pollen. In Tianjin Institute of Geology and Mineral Resources (ed.), *Palaeontological Atlas of North China. Vol. III: Micropaleontological Volume* (pp. 440–638). Beijing: Geological Publishing House.

Pu R. & Wu H. (1982). Sporo-pollen assemblages of the Middle-Late Jurassic of western Liaoning region. *Bull. Shenyang Inst. Geol. Miner. Resour., Chin. Acad. Geol. Sci., 4*, 169–86.

(1985). Mesozoic sporo-pollen assemblages in western Liaoning and their stratigraphic significance. In L. Zhang, R. Pu & H. Wu (eds.), *Mesozoic Stratigraphy and Paleontology of Western Liaoning* (vol. 2, pp. 121–89). Beijing: Geological Publishing House.

Qian L., Bai Q., Xiong C., Wu J., Hu D., Zhang X. & Xu M. (1987). *Jurassic Coal-bearing Strata and the Characteristics of Coal Accumulation from Northern Shaanxi* (pp. 1–202). Xian: Northwestern University Press.

Qian L. & Wu J. (1987). Sporo-pollen assemblages of Mesozoic coal-bearing strata from south China. In Academy of Geological Exploration (ed.), *Mesozoic Coal-bearing Strata from South China* (pp. 91–114). Beijing: Coal Industry Publishing House.

Qian L., Zhao C. & Wu J. (1983). *Mesozoic Coal-bearing Strata and Fossils from Hunan-Jiangxi Area. Pt. III: Sporo-pollen Assemblages* (pp. 1–140). Beijing: Coal Industry Publishing House.

Qu L., Zhang W. & Yu J. (1987). Advance in Mesozoic palynological researches for thirty years. *Prof. Pap. Stratigr. Palaeont.*, Institute of the Chinese Academy of Geological Sciences (Beijing), *17*, 29–46.

Shang Y. (1981). Early Jurassic sporo-pollen assemblages in southwestern Hunan, northeastern Guangxi. *Acta Palaeont. Sin., 20*, 428–42.

Su D., Li Y., Yu J., Zhang W., Zhang L., Pu R. & Yang R. (1983). Late Mesozoic biostratigraphy of nonmarine Ostracoda and pollen and spores in China. *Acta Geol. Sin., 22*, 329–46.

Wang C. & Tong G. (1982). New materials of Middle Jurassic spores and pollen from Fengxian, Jiangsu. In Palynological Society of China (ed.), *I Congress of the Palynological Society of China, Tianjin (1979)* (vol. 1, pp. 100–4). Beijing: Science Press.

Wang S. (1988). Sporo-pollen assemblage of Jurassic-Cretaceous in Tonghua, Jilin. *Acta Palaeont. Sin., 27*, 729–36.

Wu H. & Zhang X. (1983). Palynological assemblage from the upper coal-bearing member of the Beipiao Formation of Liaoning. *Acta Palaeont. Sin. 22*, 564–72.

Xu Y. (1982). Occurrence and significance of the Early to Middle Jurassic microflora in Nylamo, Xizang (Tibet). *Earth Sci. J. Wuhan Coll. Geol., 3*, 121–30.

Xu Y. & Zhang W. (1980). Jurassic spores and pollen. In Institute of Geology, CAGS (ed.), *Mesozoic Stratigraphy and Palaeontology from Shangganning Basin, Upper Part* (pp. 143–86). Beijing: Geological Publishing House.

Yang J. & Sun S. (1982). The discovery of Early and Middle Jurassic megaspores from the Junggar Basin, Xinjiang, and their stratigraphical significance. In Geological Society of China (ed.), *Abstracts, Symposium on Mesozoic and Cenozoic Geology, 60th Anniversary of the Geological Society of China, Beidaihe, 1982* (pp. 42–3). Beijing: Geological Publishing House.

(1987). Fossil megaspores. In Yichang Institute of Geology and Mineral Resources (ed.), *Biostratigraphy of the Yangtze Gorge Area. Triassic and Jurassic* (vol. 4, pp. 310–15). Beijing: Geological Publishing House.

Yu J. (1982). Late Jurassic and Early Cretaceous dinoflagellate assemblages of eastern Heilongjiang province, China. *Bull. Shenyang Inst. Geol. Miner. Resour., Chin. Acad. Geol. Sci., 5*, 227–62.

Zhang L. (1965). Pollen assemblages and their significance in the Yima coal-bearing series from Mianchi County, Henan province. *Acta Palaeont. Sin., 13*, 160–96.

(1978). Mesozoic spores and pollen grains from the volcanic clastic sedimentary rocks in Zhejiang, with their stratigraphic significance. *Acta Palaeont. Sin., 17*, 180–92.

Zhang W. (1989). Jurassic sporo-pollen assemblages from some parts of Eastern China. In *Tectonic-Magmatic Evolution and Metallogeny of Eastern China. 2: The Paleontology and Stratigraphy of the Jurassic and Cretaceous in Eastern China* (pp. 1–20). Beijing: Geological Publishing House.

(1990). Jurassic spores and pollen assemblages in Junggar Basin of Zinjiang. In *Permian to Tertiary Strata and Palynological Assemblages in the North of Xinjiang* (pp. 57–96). Beijing: China Environmental Science Press.

Zhang W. & Zhang Z. (1990). Jurassic spores and pollen. In Yichang Institute of Geology and Mineral Resources (ed.), *Biostratigraphy of the Yangtze Gorge Area. Triassic and Jurassic* (vol. 4, pp. 282–310). Beijing: Geological Publishing House.

Zhang W. & Zhao Q. (1985). Early Jurassic sporo-pollen assemblages of the Tandonggou Formation in Yaojie district, Gansu province. *Geol. Rev./Dizhi Lun-P'ing (Beijing), 31*, 13–22.

Zhang Z. (1978). Mesozoic spores and pollen. In Hubei Institute of Geological Sciences (ed.), *Palaeontological Atlas of South-Central China. Microfossil Volume* (vol. 4, pp. 440–513). Beijing: Geological Publishing House.

### *Siberia (Russia)*

Most works listed below are in Russian, though sometimes with brief English summaries. To facilitate the identification by readers of relevant stratigraphical works, translated titles are furnished; however, for library ordering purposes, the title of volumes in which a paper is contained is given in Russian. The titles of books are given first in Russian and afterward in English translation.

Bogdashova, L. I. (1969). Megaspores of the Jurassic deposits of the central part of the Siberian Platform. In *Voprosi Biostratigrafii i Paleogeografii Sibirskoy Platformy* (pp. 65–72). Moscow: Akademiya Nauk SSSR.

Bolkhovitina, N. A. (1952). Pollen of conifers in Mesozoic deposits and its value for stratigraphy. *Izv. Akad. Nauk SSSR, Ser. Geol., 5*, 105–20.

(1956). Atlas spor i pyl'tsy iz Yurskikh i Nizhemelovikh otlozhenii Vilyiskoi Vladiny (Atlas of spores and pollen from Jurassic and Lower Cretaceous deposits of the Vilyuysk Basin). *Trudy Geol. Inst. Kazan. Fil., 2*, 1–132.

(1959). Sporovo-pyl'tseny kompleksy Mesozoiskikh otlozhenii Vilyiskoi Vladinyi ikh zhnachenye dlya stratigrafiya (Spore-pollen complexes from the Mesozoic deposits of the Vilyuy Basin and their importance for stratigraphy). *Trudy Geol. Inst. Kazan. Fil., 24*, 1–185.

(1966). The fossil spores of the ferns of the Family Gleicheniaceae: taxonomy and distribution. In *Important Palynological, Analytical, Stratigraphical and Palaeofloristic Investigations, II International Palynological Congress, Utrecht* (pp. 65–75). Moscow: Akademiya Nauk SSSR.

(1968). Spory Gleikheniyevykh palorotnikov i ikh stratigraficheskoye znacheniye. (The spores of the Family Gleicheniaceae ferns and their importance for the stratigraphy). *Trudy Akad. Nauk SSSR, 186*, 1–116.

Charudo, I. I. (1969). Mesozoic spore and pollen assemblages of Siberia and the Far East. *Trudy Inst. Geol. Geofis. Acad. Nauk SSSR, 91*, 1–105.

Gutova, L. N., & Ilyina, T. I. (1969). Vegetation of Early and Middle Jurassic epochs of the Irkutsk Coal Basin. In *Voprosy Biostratigrafii i Paleogeografii Sibirskoi Platformy* (pp. 73–88). Moscow: Akademiya Nauk SSSR.

Gutova, L. N., Kilesso, I., & Odintsova, M. M. (1969). Palynologic characteristics of the Jurassic deposits of the central part, Sayano-Vilyusk Depression. In *Voprosy Biostratigrafii i Paleogeografii Sibirskoy Platformy* (pp. 89–96). Moscow: Akademiya Nauk SSSR.

Gutova, L. N. & Odintsova, M. M. (1964). Solutions of problems of taxonomy and nomenclature obtained from the study of spores and pollen occurring in the Jurassic deposits of the Siberian Platform. In V. N. Sachs & A. F. Khlonova (eds.), *Sistematika i Metody Izucheniya Iskopaemykh Pyl'tsy i Spor* (pp. 53–60). Moscow: Nauka.

Ilyina, V. I. (1965). Age of Jurassic deposits on the left bank of the River Tom. *Geol. Geofis.*, *10*, 10–18.

(1966). Correlation of spore and pollen assemblages of Middle Jurassic deposits of Chulym-Yenisei Depression. In *Reports of Soviet Scientists. Papers, II International Palynological Conference (Utrecht)* (pp. 7–26). Moscow: Akademiya Nauk SSSR.

(1968). *Sravnitel'nyi Analyz sporovo-pyl'tzevykh kompleksov Yurskikh otlozhenyi Izhnoi Chasti Zapadnoi Sibiri*. (Comparative analysis of spore-pollen complexes of Jurassic deposits of the southern part of Western Siberia). Moscow: Instityta Geologii i Geofisika, Akademiya Nauk SSSR, 111 pp.

(1969a). Spore and pollen assemblages of Lower Jurassic deposits of middle course of the Viljul River. In *Sporovo-pyl'tzeviye Komplexii Mesozoya Sibirii i Dal'nego Vostoka* (pp. 70–88). Moscow: Akademiya Nauk SSSR.

(1969b). Spore and pollen characteristics of upper horizons of the Middle Jurassic of Chulym-Yenisei region. In *Sporovo-pyl'tzeviye Komplexii Mesozoya Sibirii i Dal'nego Vostoka* (pp. 102–6). Moscow: Akademiya Nauk SSSR.

(1971a). Palynological characterization of Jurassic deposits. In *Stratigrafiya i Palinologichevskiy Svoictvo Paleozoyskego i Mesozoyskego Kuznetsko Basyena, Papers, III International Conference on Palynology (Novosibirsk)* (pp. 47–54). Moscow: Nauka.

(1971b). Palynological characteristics of Jurassic deposits of Siberia. In V. N. Sachs (ed.), *Mikrofossilii Mesozoya Sibirii i Dal'nego Vostoka, Papers, III International Conference on Palynology (Novosibirsk)* (pp. 6–50). Moscow: Nauka.

(1973). Biostratigraphical significance of Toarcian spore-pollen assemblages of Siberia. In A. F. Khlonova (ed.), *Palinologiya Mezofita, Proceedings III International Conference on Palynology (Novosibirsk)* (pp. 75–9). Moscow: Nauka.

(1976). Comparative analysis of Jurassic palynological assemblages of marine and non-marine deposits of Siberia. In *Palynology in U.S.S.R., Reports, Soviet Palynologists, IV International Palynological Conference (Lucknow)* (pp. 76–9). Moscow: Nauka.

(1978a). Palynological substantiation for the stratigraphic division of the Middle Jurassic in Northern Siberia. *Geol. Geofis.*, *9*, 16–23.

(1978b). Concerning the possible identification of the Middle Jurassic in Northern Siberia according to palynological data. In *Noviye dannye po stratigrafii i faune Yury i Mela Sibiri* (pp. 86–96). Novosibirsk: Akademiya Nauk SSSR, Instituta Geiologii i Geofisika, Sibirskoe Otdelenie.

(1979a). The pollen genus *Eucommiidites* in the Jurassic deposits of Siberia and its stratigraphic importance. In *Stratigrafiya i palinologiya Mesozoya i Cainozoya Sibiri* (pp. 5–18). Novosibirsk: Nauka.

(1979b). The first find of the pollen genus *Chasmatosporites* in the Jurassic of Siberia. In *Stratigrafiya i palinologiya Mesozoya i Cainozoya Sibiri* (pp. 19–25). Novosibirsk: Nauka.

(1980). Palynological evidence in the continental stratigraphy of the Middle Jurassic in southern Siberia. In *Paleopalinologiya Sibiri* (pp. 28–38). Moscow: Nauka.

(1981a). Division and correlation of the Middle Jurassic deposits of eastern Siberia according to palynological data. *Geol. Geofis.*, *5*, 9–19.

(1981b). Jurassic palynostratigraphy. In *Mesozoya i Cainozoya Sibirii Dal'nego Vostoka* (pp. 45–53). Novosibirsk: Akademiya Nauk SSSR, Instituta Geologii i Geofisika, Sibirskoe Otdelenie.

(1984). Methodological bases of subdivision and correlation of Jurassic deposits according to palynological data. In A. F. Khlonova (ed.), *Problemei Sovremennoi Palinologii* (pp. 22–7). Moscow: Nauka.

(1985a). Jurassic marine and continental palynostratigraphy of Siberia. *Sov. Geol. Geophys.*, *26*, 9–17.

(1985b). *Palinologiya Jurski Sibiri* (Jurassic Palynology of Siberia). Moscow: Nauka, 237 pp.

(1988). Palynology characteristics of deposits near the Middle and Upper Jurassic boundary of southern West Siberia. In S. B. Shatskiy (ed.), *Mikrofitofossilii i stratigrafiya mezozoya i kaynozoya Sibirii. Trudy Inst. Geol. Geofiz.*, *697*, 42–51.

Maliavkina, V. S. (1949). *Key to Determination of Pollen and Spores from Jurassic and Cretaceous Deposits.* Leningrad and Moscow: Naphta-Instityta, 138 pp.

(1961). Podocarpaceae. In S. R. Samoilovich (ed.), *Pyl'tsa i Spory Zapadny-Sibirii, Yura-Paleotsen* (pp. 127–31). Leningrad: VNIGRI.

(1962). Spores and pollen of the Aalenian of the Siberian Plain. In *Reports of the Soviet Palynologists, I International Conference on Palynology (Tucson, Arizona)* (pp. 149–50). Moscow: Akademiya Nauk SSSR.

Maliavkina, V. S., & Rovina, L. V. (1961). Pinaceae. In S. R. Samoilovich (ed.), *Pyl'tsa i Spory Zapadny-Sibirii, Yura-Paleotsen* (pp. 137–49). Leningrad: VNIGRI.

Markova, L. G. (1969). *Istoriya Razvitya Yurskoi i Rannemelovoi Flory Zapadno-Sibirskoi Nizmennosti (po dannym Palynologii)* [History of the development of the Jurassic and Early Cretaceous flora of the western Siberian lowlands (from palynological data)]. Tomsk, 50 pp.

(1973). On the evolution of Jurassic and Early Cretaceous flora of West Siberia. In A. F. Khlonova (ed.), *Palinologiya Mezofita. Proceedings, III International Conference on Palynology (Novosibirsk)* (pp. 117–24). Moscow: Nauka.

Markova, L. G., Skuratenko, A. V., Tkacheva, L. G., & Chesnokova, V. S. (1988). Palynostratigraphy of Upper Jurassic Tomsk Oblast. In V. M. Podobina (ed.), *Materialy po Paleontologii i Stratigrafii Zapadnoy Sibiri* (pp. 74–82). Tomsk: Tomsk Gosudarstvennii Univ. imeni V. V. Kuybysheva.

Markova, L. G., & Teslenko, J. V. (1973). Palynological evidence as applied to stratigraphic classification of the non-marine Mesozoic and Cenozoic deposits in West Siberia. In A. F. Khlonova (ed.), *Palinologiya Mezofita. Proceedings, III International Conference on Palynology (Novosibirsk)* (pp. 67–71). Moscow: Nauka.

Medvedeva, A. M., & Klimushina, L. P. (1984). Microfossils in oils and condensates of the Jurassic deposits of West Siberia. In A. F. Khlonova (ed.), *Problemei Sovremennoi Palinologii* (pp. 27–30). Moscow: Nauka.

Odintsova, M. M., & Gutova, L. N. (1973). Early and Middle Jurassic spore-pollen assemblage of Siberian Platform and their relationship with palaeolandscapes. In. A. F. Khlonova (ed.), *Palinologiya Mezofita. Proceedings, III International Conference on Palynology (Novosibirsk)* (pp. 71–5). Moscow: Nauka.

Petrova, L. D., & Fradkina, A. F. (1973). Palynological characteristic of Pliensbachian and Toarcian deposits of the Malo-Botuobinsky region (Vilyui River basin). In A. F. Khlonova (ed.), *Palinologiya Mesofita. Proceedings, III International Conference on Palynology (Novosibirsk)* (pp. 80–3). Moscow: Nauka.

Rovina, L. V. (1973). Development of flora during the Early Mesozoic in West Siberia and its possible relationships with synchronous floras of some other regions. In A. F. Khlonova (ed.), *Palinologiya Mesofita. Proceedings, III International Conference on Palynology (Novosibirsk)* (pp. 38–41). Moscow: Nauka.

Shramkova, G. V. (1966). Spore and pollen complexes of the Lower Volgian Stage and the Neocomian Superstage on the Territory of Kursk magnetic anomaly. In *Reports of Soviet Scientists. Papers, II International Conference on Palynology (Utrecht)* (pp. 109–12). Moscow: Akademiya Nauk SSSR.

Shugaevskaya, O. V. (1962). Spore and pollen assemblages of Upper Jurassic and Lower Cretaceous deposits of certain regions of the southern Far East. In V. L. Komarova (ed.), *Reports of the Jubilee Session, 30th Anniversary, Far East Branch, Akademiya Nauk SSSR* (pp. 19–21). Vladivostok: Akademiya Nauk SSSR.

(1969). Spores *Duplexisporites* in Upper Mesozoic deposits of the River Gerbilean (Uda Depression). In *Fossil Fauna and Flora of the Far East* (vol. 1, pp. 153–60). Moscow: Akademiya Nauk SSSR.

Shugaevskaya, O. V., Markevich, V. S., & Bitutskaya, P. I. (1973). Spores and pollen from the coal sequences of the Bureyan Basin, with reference to their stratigraphic significance. In A. F. Khlonova (ed.), *Palinologiya Mesofita. Proceedings, III International Conference on Palynology (Novosibirsk)* (pp. 147–9). Moscow: Nauka.

Teodorovich, V. I., & Vozzhennikova, T. F. (1971). O morskom genezhice Sredniorsko-Nizhnekelloveiskoi Tolzhii Pritiman'ya. *Byull. Mosk. Obstich. Ispyt. Prir., Otdel. Geol., 46*, 62–8.

Vozzhennikova, T. F. (1967). *Iskopaemye Peridinei Yurskikh, Melovyikh i Paleogenovikh otlozhenii SSSR* (Fossilized peridinid algae in the Jurassic, Cretaceous and Palaeogene deposits of the U.S.S.R.). Moscow: Nauka, 347 pp.

(1969). *Dinotsisty i ikh stratigraficheskoe znachenie*. (Dinoflagellate cysts and their stratigraphic significance). Novosibirsk: Nauka, Sibirskoe Otdelenie, 135 pp.

### *Alaska (U.S.A.)*

Albert, N. R. (1990). Dinoflagellate cysts from the Upper Jurassic Naknek Formation, southern Alaska (abstract). In D. K. Goodman (ed.), *Abstracts and Proceedings of the 22nd Annual Meeting of the American Association of Stratigraphic Palynologists. Palynology, 14*, 209.

Gray, J., & Browning, J. L. (1959). Caytonialean microspores from the Jurassic and Cretaceous of Alaska (abstract). *Bull. Geol. Soc. Am., 70*, 1722.

Wharton, D. L. (1987). Dinoflagellates from Middle Jurassic sediments of Alaska (abstract). *Program and Abstracts of the 20th Annual Meeting of the American Association of Stratigraphic Palynologists* (p. 146). Also published in 1988, *Abstracts of the Proceedings. Palynology, 12*, 248.

Wiggins, V. D. (1975). The dinoflagellate family Pareodiniaceae: a discussion. *Geosci. Man, 11*, 95–115.

### *British Columbia (Canada)*

Davies, E. H., & Poulton, T. P. (1986). Upper Jurassic dinoflagellate cysts from strata of northeastern British Columbia. *Current Research, B: Geol. Surv. Can. Pap., 86-1B*, 519–37.

Ziegler, W. H., & Pocock, S. A. J. (1960). The Minnes Formations. In *Guide Book, II Annual Field Conference, Edmonton Geological Society* (pp. 43–71). Edmonton, Alberta: Edmonton Geological Society.

### *Western United States of America*

Habib, D., & Warren, J. S. (1973). Dinoflagellates near the Cretaceous–Jurassic boundary. *Nature (London), 241*, 217–18.

Traverse, A. (1972). A case of marginal palynology: a study of the Franciscan mélanges. *Geosci. Man, 4*, 87–90.

Warren, J. S. (1973). Form and variation of the dinoflagellate *Sirmiodinium grossi* Alberti, from the Upper Jurassic and Lower Cretaceous of California. *J. Paleontol., 47*, 101–14.

### *Mexico*

The three references that follow are stratigraphical and do not contain palynological data:

Sandoval, J., & Westermann, G. E. G. (1986). The Bajocian (Jurassic) ammonite fauna of Oaxaca, Mexico. *J. Paleontol., 60*, 1220–71.

Sandoval, J., Westermann, G. E. G., & Marshall, M. C. (1990). Ammonite fauna, stratigraphy, and ecology of Bathonian-Callovian (Jurassic) Tococoyunca Group, South Mexico. *Palaeontographica, A210*, 93–149.

Westermann, G. E. G., Corona, R., & Carrasco, R. (1984). The Andean mid-Jurassic *Neuqueniceras* ammonite assemblage of Cualac, Mexico. In G. E. G. Westermann (ed.), *Jurassic-Cretaceous Biochronology and Paleogeography of North America. Spec. Pap. Geol. Assoc. Canada, 27*, 99–112.

### *Colombia*

Remy, W., Remy, R., Pfefferkorn, H. W., Volkheimer, W., & Rabe, E. (1975). Neueinstufung der Bocas-Folge (Bucaramanga, Kolumbien) in den Unteren Jura anhand einer *Phlebopteris branneri* und *Classopollis* Flora. *Argumenta Palaeobot., 4*, 55–77.

### *Argentina*

Arguijo, M., & Volkheimer, W. (1985). Palinología de la Formación Piedra Pintada, Jurásico Inferior, Neuquén, República Argentina. Descripciones sistemáticas. *Revta Esp. Micropaleont., 17*, 65–92.

Arguijo, M., Volkheimer, W., & Rosenfeld, U. (1982). Estudio palinológico de la Formación Piedra Pintada, Jurásico Inferior de la Cuenca Neuquina (Argentina). In *Paleobotânica e Palinológia na América do Sul. Boln Inst. Geociênc. Univ. São Paulo, 13*, 100–4.

González-Amicón, O., & Volkheimer, W. (1982a). Datos palinológicos del Bayociano (Formación Cura Niyeu) de la Sierra de Chacai Có, Cuenca Neuquina, Argentina. In *Paleobotânica e Palinológia na América do Sul. Boln Inst. Geociênc. Univ. São Paulo, 13*, 108–15.

(1982b). Palinológia estratigráfica del Jurásico de la Sierra de Chacai Có y adyacencias (Cuenca Neuquina, República Argentino). III Descripciones sistemáticas de los palinomorfos de la Formación Cura Niyeu (Bayociano). *Ameghiniana, 19*, 165–78.

Menéndez, C. A. (1956). Jurassic flora of Bajo de Los Baguales in Plaza Huincul, Neuquén. *Acta Geol. Lilloana, 1*, 315–38.

(1968a). Palynological study of the Middle Jurassic from Picún Leufú, Neuquén. *Ameghiniana, 5*, 379–405.

(1968b). Palynological record of pre-Tertiary floras of Argentina. *Publnes Mus. Argent. Cienc. Natur. (Buenos Aires), Paleontol., 1*, 231–42.

Morbelli, M. A. (1990). Austral South American megaspores. *Rev. Palaeobot. Palyn., 65*, 209–16. Republished in E. M. Truswell & J. A. Owen (eds.), *Proceedings of the 7th International Palynological Congress, Part II*.

Quattrocchio, M. E. (1980). Contribución al conocimiento de la palinología estratigráfica del Jurásico Superior en la Cuenca Neuquina. *Op. Lilloana, 31*, 1–59.

(1981). Palinomorfos del Bayociano de Loan Mahuida (Cuenca Neuquina, Argentina). Descripciones sistemáticas. *Memorias III Congreso Latinoamericano de Paleontología*, 175–83.

(1984). Sobre el posible significado paleoclimático de los quistes de dinoflagelados en el Jurásico y Cretácico inferior de la Cuenca Neuquina. *II Congreso Argentino de Paleontológia y Bioestratigráfia, Corrientes*, 107–13.

Quattrocchio, M. E., & Sarjeant, W. A. S. (1992). Dinoflagellate cysts and acritarchs from the Middle and Upper Jurassic of the Neuquén Basin, Argentina. *Revta Españ. Micropaleont., 24*.

Quattrocchio, M. E., & Volkheimer, W. (1983). Datos palinológicos de la Formación Picún Leufú (Jurásico superior) en su localidad tipo, Provincia de Neuquén. Pt. I: Especies marinas. *Revta Asoc. Geol. Argent., 38*, 34–48.

(1990). Jurassic and Lower Cretaceous dinocysts from Argentina. Their biostratigraphic significance. *Rev. Palaeobot. Palyn., 65*, 319–30. Republished in E. M. Truswell & J. A. Owen (eds.), *Proceedings of the 7th International Palynological Congress, Part II*.

Scarfati, L. H., & Morbelli, M. A. (1984). Nuevos datos palinológicos de la Formacion Lajas, Jurásico Medio de la Cuenca Neuquina. *Actas III Congreso Argentino de Paleontológia e Bioestratigráfia, Corrientes, 3*, 73–105.

Volkheimer, W. (1968). Esporas y granos de polen del Jurásico de Neuquén (República Argentina). I. Descripciones sistemáticas. *Ameghiniana, 5*, 333–70.

(1969). Esporas y granos de polen del Jurásico de Neuquén (República Argentina). II. Associaciones microflorísticas, aspectos paleoecológicos y paleoclima. *Ameghiniana, 6*, 127–45.

(1970). Jurassic microfloras and paleoclimates in Argentina. *Proceedings & Papers, II Gondwana Symposium, South Africa*, 543–9.

(1971a). Algunos adelantos en la microbioestratigrafía del Jurásico en la Argentina y comparación con otras regiones del hemisferio Austral. *Ameghiniana 8*, 341–55.

(1971b). Zur stratigraphischen Verbreitung von Sporen und Pollen im Unter- und Mitteljura des Neuquén-Beckens (Argentinien). *Münster. Forsch. Geol. Paläontol., 20/21*, 297–321.

(1972). Estudio palinológico de un carbón caloviano de Neuquén y consideraciones sobre los paleoclimas jurásicos de la Argentina. *Revta Mus. La Plata, n.s., 6*, 101–57.

(1974a). Palinológia estratigráfica del Jurásico de la Sierra de Chacai Có y adyacencias (Cuenca Neuquina, República Argentina). I. Estratigrafía de las formacíones Sierra Chacai Có (Pliensbachiano), Los Molles (Toarciano, Aaleniano), Cura Niyeu (Bayociano) y Lajas (Caloviano inferior). *Ameghiniana, 10*, 105–29.

(1974b). Palinología estratigráfica del Jurásico de la Sierra de Chacai Có y adyacencias (Cuenca Neuquina, República Argentina). II. Descripción de los palinomorfos de Jurásico inferior y Aaleniano (Formaciónes Sierra Chacai Có y Los Molles). *Ameghiniana, 11*, 135–72.

Volkheimer, W., Manceñido, M. O., & Damborenea, S. E. (1978). Zur Biostratigraphie des Lias in der Hochkordillere von San Juan, Argentinien. *Münster. Forsch. Geol. Paläontol., 44/45*, 205–35.

Volkheimer, W., & Moroni, A. M. (1981). Datos palinológicos de la Formación Auquinco, Jurásico superior de la Cuenca Neuquina, Argentina. *Actas VIII Congreso Geológico Argentino, San Luis, 4*, 795–812.

Volkheimer, W. & Quattrocchio, M. E. (1975a). Palinología estratigráfica del Titoniano (Formación Vaca Muerta) en el área de Caichigüe (Cuenca Neuquina). Pte. A: Especies terrestres. *Ameghiniana, 12*, 193–241.

(1975b). Sobre el hallazgo de microfloras en el Jurásico superior del borde austral de la Cuenca Neuquina (República Argentina). *Actas I Congreso Argentino de Paleontológia e Bioestratigráfia, Tucumán, 1*, 589–615.

(1977). Palinología estratigráfica del Titoniano (Formación Vaca Muerta) en el área de Caichigüe (Cuenca Neuquina). Pte. B. Especies marinas. *Ameghiniana, 14*, 162–9.

(1981). Distribución estratigráfica de los palinomorfos jurásicos y cretácicos en la Faja Andina y áreas adyacentes de América del Sur austral con especial consideración de la Cuenca Neuquina. In Comité Sudaméricano del Jurásico y Cretácico (eds.), *Cuencas sedimentarias del Jurásico y Cretácico de América del Sur, 2*, 407–43.

Volkheimer, W., Quattrocchio, M. E., & Prámparo, M. (1992). Stratigraphic distribution of Upper Jurassic palynomorphs in central western Argentina. *Newsl. Stratigr.* (in press)

Volkheimer, W., Quattrocchio, M. E., Salas, A., & Sepúlveda, E. (1976). Caracterización palinológica de formacíones del Jurásico superior y Cretácico inferior de la Cuenca Neuquina (República Argentina). *Actas VI Congreso Geológico Argentino, 1*, 593–608.

Volkheimer, W., Quattracchio, M. E., & Scarfati, L. (1992). Stratigraphic distribution of Middle Jurassic palynomorphs in central western Argentina. *Newsl. Stratigr.*,

Volkheimer, W. Quattracchio, M. E., & Zavattieri, A. M. (1992). Stratigraphic distribution of Lower Jurassic palynomorphs in central western Argentina. *Newsl. Stratigr.*,

*Antarctica*

Askin, R. A. (1981). Jurassic-Cretaceous palynology of Byers Peninsula, Livingston Island, Antarctica. *Antarctic J. U.S., 16*, 11–12.

Gair, H. S., Norris, G., & Ricker, J. (1965). Early Mesozoic microfloras from Antarctica. *N.Z. J. Geol. Geophys., 8*, 231–5.

Norris, G. (1965). Triassic and Jurassic miospores and acritarchs from the Beacon and Ferrar Groups, Victoria Land, Antarctica. *N.Z. J. Geol. Geophys., 8*, 236–77.

Riding, J. B. (1988). Preliminary palynological investigation of the Mesozoic from the Antarctic Peninsula (abstract). *Abstracts of the Proceedings of the 20th Annual Meeting of the American Association of Stratigraphic Palynologists. Palynology, 12*, 246.

Tasch, P., & Lammons, J. M. (1977). Palynology of some lacustrine interbeds of the Antarctic Jurassic. *Palinologia, (Número extraordinario), 1*, 455–61.

# 14 Radiolarian biozones of North America and Japan

E. PESSAGNO, JR., and S. MIZUTANI

The correlation chart presented herein (Figure 14.1) was compiled through a critical examination of published and unpublished data for both Japan and North America. Because of the vast amount of unpublished data from both North America and Japan, it was first necessary for the authors to meet and to examine hundreds of scanning electron photomicrographs of North American and Japanese Radiolaria. Without such a meeting, any attempt at correlating North American and Japanese Jurassic radiolarian biozones would have been, at best, an exercise in futility (plates 99–102).

The radiolarian zonation for North America followed herein represents an emended version of that presented by Pessagno et al. (1987b). The North American zonation shown in Figure 14.1 is the result of over 10 years of field and laboratory investigations by Pessagno and his associates. North American Jurassic samples were analyzed from Alaska, the Queen Charlotte Islands (British Columbia), east-central Oregon, California, Baja California Sur, and east-central Mexico. In establishing a zonal scheme for the North American Jurassic, Pessagno et al. (1987b) calibrated the radiolarian biostratigraphy with those of the ammonites, calpionellids, *Buchia,* and other well-studied fossil groups. This amalgamation of data allowed a fit of the radiolarian zonation to the ammonite-based chronostratigraphic scale. Figures showing the approximate correlation of radiolarian zonal units with ammonite standard zones were presented by Pessagno et al. (1987b) and are not included here.

Emendations to the original zonal scheme are minor. They include the introduction of a new subzone, Subzone 2 gamma, and the subsequent redefinition of Subzone 2 beta (Pessagno, Six, and Yang 1989). Subzone 2 gamma is defined from new data (Smith River Subterrane, Klamath Mountains, northwestern California) (Pessagno and Blome 1988, 1990) that document an interval of concurrence between the first appearance of *Mirifusus* Pessagno and the final appearance of *Xiphostylus* Haeckel (*primary marker taxa*). The base of overlying Subzone 2 beta (Figure 14.1) is redefined to occur in the interval immediately above the final occurrence of *Xiphostylus,* whereas its top occurs immediately below the first occurrence of *Parvicingula* Pessagno *s.s.* Further discussion of the biostratigraphic, chronostratigraphic, and geochronological significance of Subzone 2 gamma is presented later.

The Japanese radiolarian zonation presented herein is that of Matsuoka and Yao (1986). This zonal scheme has become the standard for the western Pacific. Ammonites and other megafossils are very rare or totally absent from the Japanese radiolarian-bearing succession (e.g., Mino terrane). Hence, the chronostratigraphic assignment of Japanese Jurassic radiolarian biozones is based on North American data or data from elsewhere.

Correlation of base of Japanese "*Pseudodictyomitra*" *primitiva* Zone (Pp) with the upper part of Zone 3, Subzone 3 alpha: In North America, "*Pseudodictyomitra*" *primitiva* Matsuoka & Yao has been found in Upper Tithonian strata in east-central Mexico, assignable to Zone 4, Subzone 4 beta (Q. Yang, personal communication). The Mexican strata ("upper thin-bedded member" of the Taman Formation) contain a rich radiolarian assemblage that can be related to co-occurring ammonites and calpionellids (Pessagno et al. 1987a,b). In Japan, no Zone 4 (Subzone 4 beta) marker taxa have been found in Zone Pp. Nevertheless, "*Pseudodictyomitra*" *primitiva* appears to occur above the last appearance of both *Hsuum maxwelli* Pessagno and *Turanta* Pessagno & Blome, which make their final appearance within North American Subzone 3 alpha. The last taxon is regarded as a more reliable marker than *Hsuum maxwelli* simply because it is easy for all workers to identify correctly. In summary, our analysis of both the North American and Japanese data indicates that the base of Zone Pp and, for that matter, all of Zone Pp (Matsuoka and Yao 1986) correlate with the upper part of Subzone 3 alpha and occur in the lower Tithonian. Thus far, there is no indication that the equivalent of North American Zone 4 (upper Tithonian) exists in Japan.

Correlation of top of Japanese *Tricolocapsa plicarum* Zone (Tp) with top of North American Subzone 2 gamma: In North America, Subzone 2 gamma is defined by the concurrence of *Mirifusus* and *Xiphostylus.* In the Smith River terrane, Subzone 2 gamma

See Appendix, Plates 99–102.

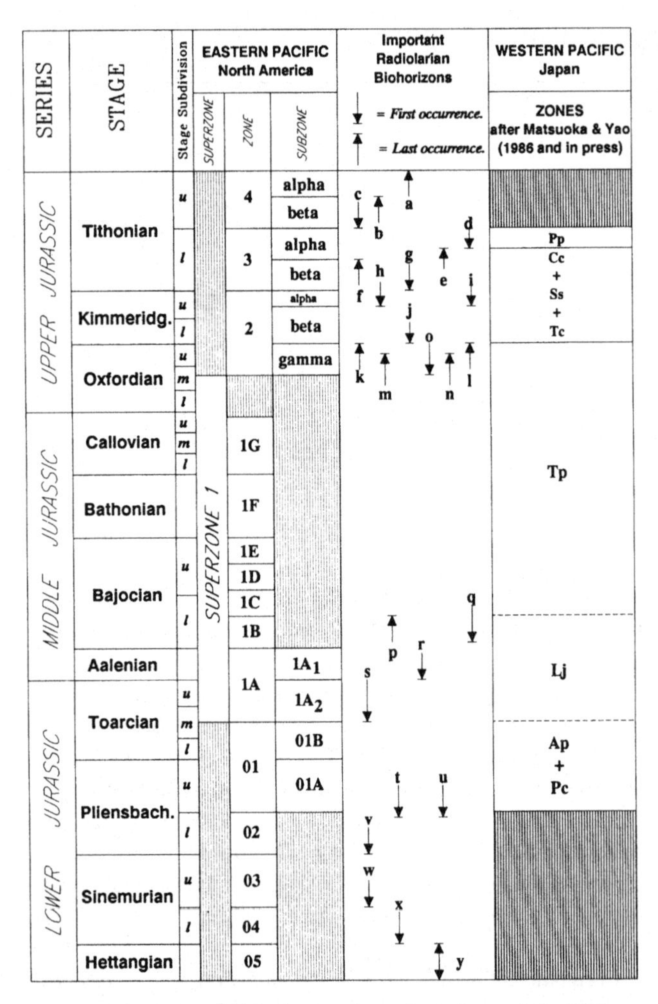

**Figure 14.1. Guide radiolarian species and zonal correlation between United States and Japan: a, *Ristola altissima* (Rüst)*; b, *Perispyridium dumitricai**; c, *Acanthocircus dicranocanthos* (Squinabol)*; d, "*Pseudodictyomitra*" *primitiva* Matsuoka & Yao*; e, *Turanta* Pessagno & Blome*; f, *Mirifusus guadaleupensis* Pessagno*; g, *Mirifusus baileyi* Pessagno*; h, *Parvicingula* s.s. (with horn and elongate terminal tube that may be closed or open on final postabdominal chamber)*; i, first appearance of pseudodictyomitrids*; j, *Eucyrtidiellum ptyctum* (Riedel & Sanfilippo)*; k, *Xiphostylus* Haeckel*; l, last appearance of *Unuma echinatus* Ichikawa & Yao*; m, *Acanthocircus suboblongus* (Yao, 1972; non-Baumgartner, 1984)*; n, *Tetraditryma praeplena* Baumgartner; o, first appearance of *Mirifusus* Pessagno, with three rows of pore frames between circumferential ridges*; p, *Parasaturnalis* Kozur & Mostler*; q, first appearance of *Unuma echinatus* Ichikawa & Yao*; r, *Hilarisirex* Takemura & Nakaseko; s, *Perispyridium dumitricai;* t, *Paracanoptum anulatum* Pessagno & Poisson*; u, *Parahsuum simplum* Yao*; v, *Noritus lillihornensis* Pessagno & Whalen*; w, *Canutus* Pessagno & Whalen s.s.*; x, *Katroma* Pessagno & Poisson*; y, *Canoptum merum* Pessagno & Whalen.* Light gray shading, interval not subdivided into superzones, zones, or subzones. Dark gray shading, interval thus far not reported from Japan. *Important marker taxa or potential important taxa in both North America and Japan.**

has been recognized in strata occurring from 17.6 m to 91.4 m above the contact with the underlying Josephine ophiolite (Pessagno, Six, and Yang 1989). Saleeby (1987) obtained a concordant U/Pb date (zircon in plagiogranite) from the Josephine ophiolite at this locality of 162 ± 1 m.y. Moreover, Harper et al. (1986) determined that the sedimentary succession overlying the Josephine ophiolite was older than 150 ± 1 m.y. Hence, it now can be established from geochronometric and geochronological studies (Smith River subterrane) that both the *Mirifusus* first-occurrence event and the *Xiphostylus* final-occurrence event are younger than 162 ± 1 m.y. (latest Callovian) and older than 150 ± 1 m.y. (earliest Tithonian) (Pessagno and Blome 1987). The placement of the Zone 2 (Subzone 2 gamma)–Superzone 1 boundary in the middle Oxfordian is somewhat arbitrary; it could just as easily be placed in the lower Oxfordian. However, in that the upper part of Subzone 2 gamma occurs in strata (Galice Formation *s.l.*) that also contain *Buchia concentrica* (Sowerby) at nearby localities in the Smith River subterrane, it is likely that at least part of the biozone is assignable to the middle Oxfordian. (Range of *Buchia concentrica* in North America = middle Oxfordian to upper Kimmeridgian) (Imlay 1980; Pessagno, Blome, and Longoria 1984). A similar zone of concurrence between *Mirifusus* and *Xiphostylus* occurs in the upper part of the *Unuma echinatus* Zone (= Tp Zone) of Japan (Mizutani and Koike 1982, Table 1, p. 122). It is significant to note that both Subzone 2 gamma and its equivalent in Japan contain *Acanthocircus suboblongus* (Yao) and *Tetraditryma praeplena* Baumgartner; both of these taxa also occur in the upper part of underlying Superzone 1 and have been recognized in upper Superzone 1 strata in Yugoslavia (Gorican 1987). *Eucyrtidiellum ptyctum* (Riedel & Sanfilippo) is absent from the interval of concurrence between *Mirifusus* and *Xiphostylus* in both North America and Japan. In North America, *E. ptyctum* is present in the lower part of Subzone 2 beta. To date, we have not observed *E. ptyctum* in association with either *A. suboblongus* or *T. praeplena* in either the North American or Japanese radiolarian assemblages.

Correlation of the top of Japanese *Laxtorum*(?) *jurassicum* Zone (Lj) with top of North American Zone 1, Subzone 1B: This correlation is less certain than those presented earlier. It is based on the occurrence of *Parasaturnalis* spp. at the top of North American Subzone 1B (lower Bajocian) and its co-occurrence in Japanese Zone Lj. Such a correlation is also supported by the presence of *Hilarisirex* in both the North American and Japanese assemblages.

Correlation of the base of Japanese *Laxtorum*(?) *jurassicum* Zone (Lj) with the base of North American Superzone 1, Zone 1A, Subzone $1A_2$: This correlation is based on the presence of *Perispyridium dumitrica* in Japanese Zone Lj and in North American Zone 1A, Subzone $1A_2$. In North America, *Perispyridium* first appears at the base of Subzone $1A_2$ (middle Toarcian) (Pessagno et al. 1987b).

Correlation of the base of Japanese *Parahsuum* sp. C Zone (Pc) with the base of North American Zone 01, Subzone 01A: The base of Japanese Zone Pc is defined by the first occurrence of *Parahsuum simplum* Yao (Matsuoka and Yao 1986, p. 95). *Parahsuum* Yao (type series = *P. simplum* Yao, 1982) is a senior synonym of *Lupherium* Pessagno & Whalen (1982; type species = *Lupherium snowshoense* Pessagno & Whalen, 1982). In the sense of its type species, *Parahsuum* Yao (1982) includes all hsuids with straight, continuous, closely spaced costae. In North America, *Parahsuum* s.s. first appears at the base of Subzone 01A (upper Pliensbachian). *Parahsuum* sp. C of Matsuoka and Yao (1986, Pl. 1, Fig. 3) is not a true *Parahsuum* and may possibly represent a broken specimen of *Bagotum* Pessagno & Whalen (1982). Moreover, *Parahsuum* sp. D of Matsuoka and Yao (1986, Pl. 1, Fig. 7) is not assignable to *Parahsuum* and is more closely related to *Hsuum* Pessagno (1977). The latter form differs from *Hsuum* s.s. only by lacking a horn. Hsuids of this type first appear in North American Zone 02 (lower Pliensbachian) and range throughout most of the North American Jurassic.

We see no evidence of Hettangian, Sinemurian, or lower Pliensbachian strata in the Japanese radiolarian-bearing successions. This interval in North America is represented by Zones 05, 04, 03, and 02.

## Acknowledgments

Pessagno's investigations were supported by grants from the National Science Foundation (GA-35094, DES-72-01528-A01, EAR-76-22029, EAR-77-22457, EAR-78-12934, EAR-8121550, EAR-8305894, EAR-8415676, EAR-8615790) and by funding from the Arco Oil and Gas Company, the Exxon Production Research Company, and the Mobil Oil Corporation.

## References

Baumgartner, P. O. (1984). A Middle Jurassic–Early Cretaceous radiolarian zonation based on unitary associations and age of Tethyan radiolarities. *Ecl. Geol. Helvet.*, *77*, 729–837.

Gorican, S. (1987). Jurassic and Cretaceous radiolarians from the Budva Zone (Montenegro, Yugoslavia). *Rev. Micropaléont.*, *30*, 177–96.

Harper, G. D., Saleeby, J. B., Cashman, S., & Norman, E. A. S. (1986). Isotopic age of the Nevadian Orogeny in the western Klamath Mountains, California–Oregon: Cordilleran Section. *Geol. Soc. Am. Abstr. Progr.*, *18*(2), 114.

Imlay, R. W. (1980). *Jurassic Paleobiogeography of Conterminous United States in Its Continental Setting*. U.S. Geological Survey Professional Paper 1062.

Matsuoka, A., & Yao, A. (1986). A newly proposed radiolarian zonation for the Jurassic of Japan. *Mar. Micropaleont.* *11*, 91–105.

Mizutani, S., & Koike, T. (1982). Radiolarians in the Jurassic siliceous shale and in the Triassic bedded chert of Unuma, Kagamigahara City, Gifu Prefecture, Central Japan. In K. Nakaseko (ed.), *Proceedings of the First Japanese Radiolarian Symposium. JRS 81 Osaka. News Osaka Micropaleont., Spec. Vol. 5*, 117–34.

Pessagno, E. A., Jr. (1977). Upper Jurassic radiolaria and radiolarian biostratigraphy of the California Coast Ranges. *Micropaleontology, 23*, 56–113.

Pessagno, E. A., Jr., & Blome, C. D. (1988). Biostratigraphic, chronostratigraphic, and U/Pb geochronometric data from the Rogue and Galice formations, Western Klamath terrane, (Oregon and California) – their bearing on the age of the Oxfordian–Kimmeridgian boundary and the *Mirifusus* first occurrence event. In *Proceedings of 2nd International Symposium on Jurassic Stratigraphy.* Lisbon: Int. Subcom. Jurassic Stratigraphy, Centro de Estratigrafia e Paleobiologia UNL (INIC).

(1990). Implications of new Jurassic stratigraphic, geochronometric, and paleolatitudinal data from the Western Klamath terrane (Smith River and Rogue Valley subterranes). *Geology, 18*, 665–8.

Pessagno, E. A., Jr., Blome, C. D., Carter, E. S., Macleod, N., Whalen, P. A., & Yeh, K. (1987b). Studies of North American Jurassic Radiolaria. Part II: Preliminary radiolarian zonation for the Jurassic of North America. *Cushman Found. Foraminif. Res. Spec. Publ., 23*, 1–18.

Pessagno, E. A., Jr., Blome, C. D., & Longoria, J. F. (1984). A revised radiolarian zonation for the Upper Jurassic of western North America. *Bull. Am. Paleontol.*, *87*(320), 1–51.

Pessagno, E. A., Jr., Longoria, J. F., Macleod, N., & Six, W. M. (1987a). Upper Jurassic (Kimmeridgian–Upper Tithonian) Pantanelliidae from the Taman Formation, east-central Mexico: tectonostratigraphic, chronostratigraphic, and phylogenetic implications. *Cushman Found. Foraminif. Res. Spec. Publ., 23*, 1–51.

Pessagno, E. A., Jr., Six, W. M., & Yang, Q. (1989). The Xiphostylidae Haeckel and Parvivaccidae, n. fam. (Radiolaria) from the North American Jurassic. *Micropaleontology, 35*(3), 193–255.

Pessagno, E. A., Jr., & Whalen, P. A. (1982). Lower and Middle Jurassic radiolaria (Multicyrtid Nassellariina) from California, east-central Oregon, and the Queen Charlotte Islands, B.C. *Micropaleontology, 28*, 111–69.

Passagno, E. A., Jr., Whalen, P. A., & Yeh, K. (1986). Jurassic Nassellariina (radiolaria) from North American geologic terranes. *Bull. Am. Paleonol.*, *91*(326), 1–75.

Saleeby, J. B. (1987). Discordance patterns in Pb/U zircon ages of the Sierra Nevada and Klamath Mountains. *Enos, 68*, 1514–15.

Yao, A. (1982). Middle Triassic to Early Jurassic radiolarians from the Inuyama area, Central Japan *J. Geosci. Osaka City Univ., 25*, 53–70.

# 15 Ostracods of western Canada

W. K. BRAUN and M. M. BROOKE

Jurassic strata are well known from many boreholes throughout the southern prairie regions of western Canada, which, in Middle to Late Jurassic time, were part of the large intercratonic Williston Basin (see Figure 19.1, Chapter 19). Its relatively undisturbed sedimentary pile yields a varied and abundant microfauna on the basis of which a detailed biostratigraphic framework was constructed by Brooke and Braun (1972). This zonal scheme serves also as a template for correlations of Jurassic microfaunas recorded from other Western Interior regions of North America.

The exposures and the subsurface occurrences of Jurassic sequences in the Rocky Mountains to the west, the former miogeocline and Western Canada Basin, are geographically disjointed and tectonically disturbed. The rocks are by far not as fossiliferous, and the faunas not as varied and well preserved, as they are in the Williston Basin. They differ furthermore in overall composition and belong in part to a different faunal province and realm (see Chapter 19), adding measurably to the difficulties in correlation.

The biozones (assemblages, faunas), identified by Roman numerals, are based on all identifiable taxa of Ostracoda and Foraminifera recovered to date, with zonal boundaries chosen along marked breaks in the faunal spectrum. Appearances and/or disappearances of key species are used to denote the boundaries, with the assemblages providing substance to the zones. They are a mixture, therefore, of taxon range zones and assemblage zones following traditional micropaleontological practice. By experience, the ostracods were found to be more distinctive, shorter-ranging, and easier to deal with paleontologically than the Foraminifera and therefore were chosen as the guide microfossils. Only in those zones or along those boundaries where they were absent, or were represented by nondiagnostic or too few specimens, were the Foraminifera used instead.

As it is impossible to do justice to all species recovered so far (120 species of ostracods and 330 foraminiferal taxa), only the ranges of the common ostracods are tabulated (Figures 15.1 and 15.2). Chapter 19 gives additional details regarding key Foraminifera and ostracods and the assumed correlations between East Pacific and Boreal microfaunas.

Although provincialism and facies-controlled faunas pose problems in interbasinal correlations, the zonal sequence for each of the regions and basins discussed is valid and therefore a most useful biostratigraphic yardstick. Zonal boundaries mirror exactly or closely lithostratigraphic boundaries – as is to be expected – and for that reason the lithostratigraphic framework is included in the two faunal charts provided. Less precise, and in some cases still tentative, is the alignment of the microfossil zones within the stage-time framework, for ammonites or other guide megafossils are notoriously rare (or too often absent) in many of the Jurassic sequences of western Canada, especially in the subsurface. The microfossils therefore play a larger role in stratigraphic studies than in central Europe with its famous ammonite sequences. Indeed, those microfossil zones, in particular biozones IV and V, VII, and X, from which at least some ammonites are known, can be readily used as chronostratigraphic substitutes in all those many areas where ammonites are absent.

In Figure 15.1, the ranges of the more common ostracods in the northern part of the Williston Basin are indicated. The reconstruction is based on cores from 30 boreholes across southern Saskatchewan, and one composite outcrop section of the Little Rocky Mountains of north-central Montana, from which a few ammonites were reported also. The area covered is bounded by longitude 102°W and 110°W and latitude 48°N and 51°N, including the southern prairie region of Canada and the northern part of the Great Plains of the United States. The tabulation is adapted from Brooke and Braun's (1972) composite charts 20 and 21, incorporating some newer findings.

So far, no ostracods have been recovered from the Lower Jurassic black shales and mudstone sequences of the Canadian Rocky Mountains, which have yielded only very few Foraminifera, collectively assigned to biozone I. Biozone II, spanning most of the Bajocian, is characterized by irregularly distributed fossil lenses with abundant freshwater to brackish-water ostracods

See Appendix, Plates 106–108.

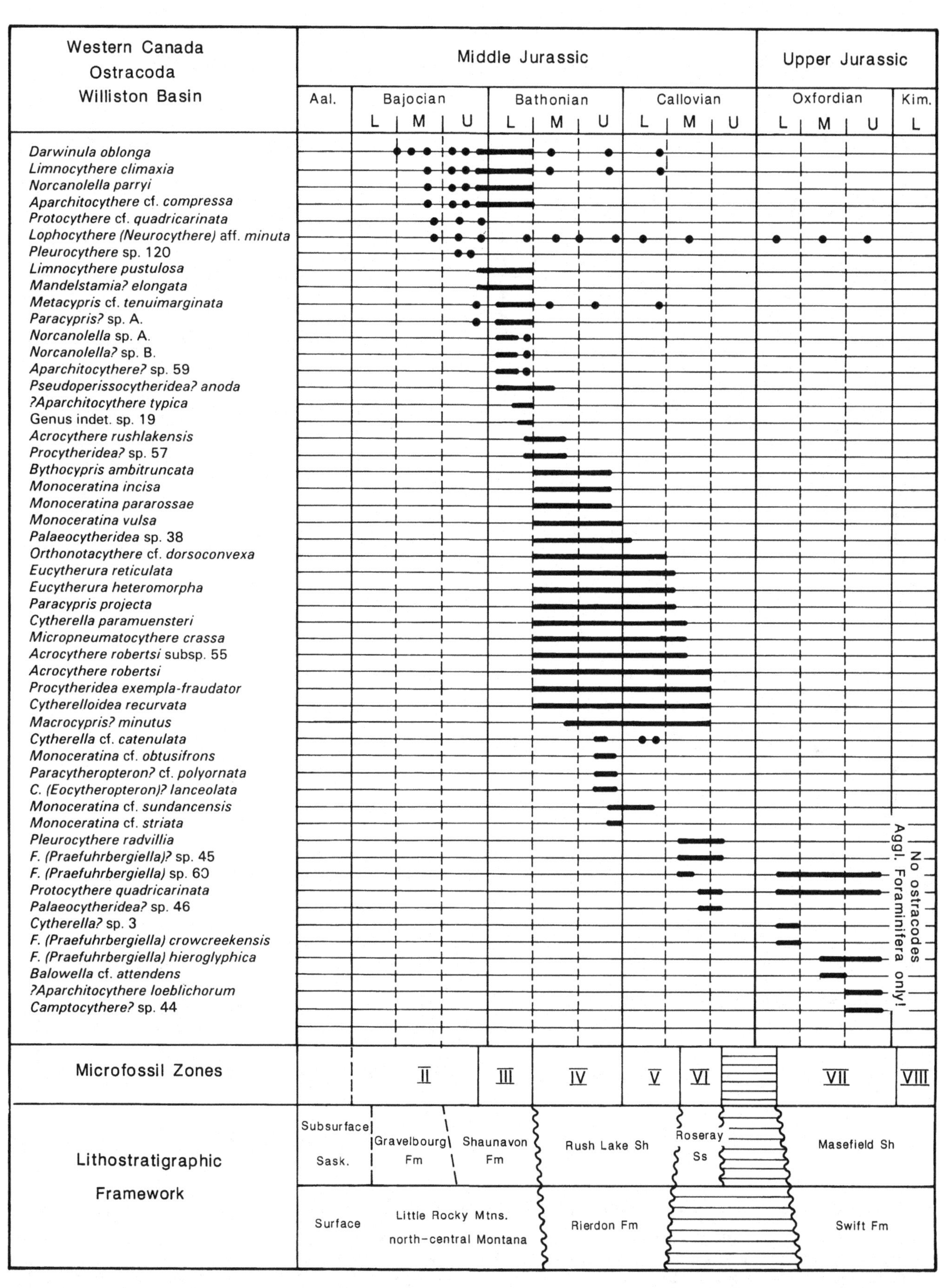


**Figure 15.1. Range chart for common Middle and Late Jurassic Ostracoda in the Williston Basin and Western Interior regions of North America. Solid lines indicate more or less continuous distribution within biozones and rock units; dots represent erratic patterns. (Adapted from Brooke and Braun 1972.)**

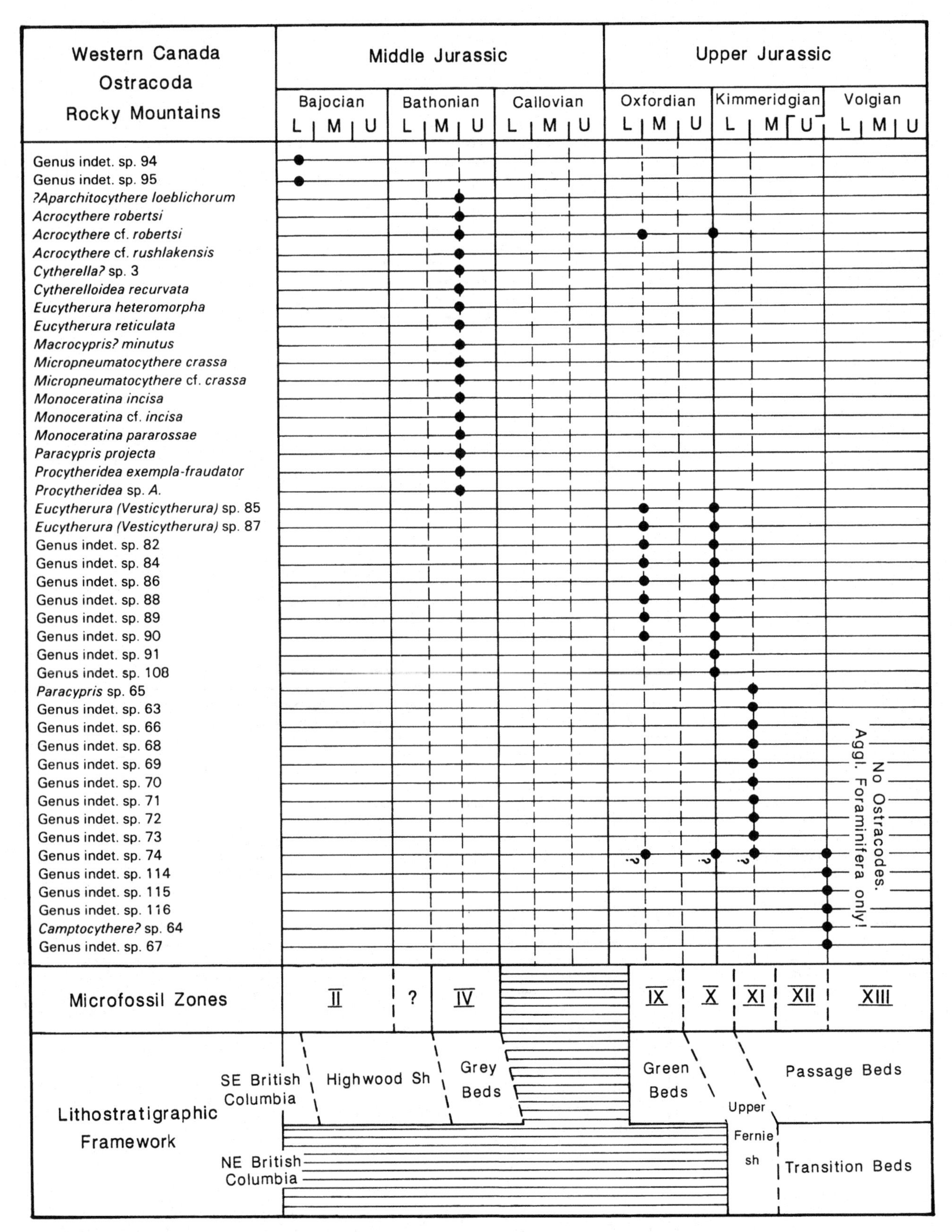


**Figure 15.2. Distribution of Middle and Late Jurassic Ostracoda in the Western Canada Basin and Canadian Rocky Mountains. Dots indicate the occurrence in the respective biozones and rock units, without exact ranges. Placements within substages or series are tentative because of lack of ammonites or chronostratigraphically recognized guide species.**

within otherwise nonfossiliferous sections, and with a few minor marine horizons and some marine ostracods intercalated. This fauna is known mainly from the northern part of the Williston Basin, where it developed into the conspicuously abundant and continuously distributed freshwater ostracod fauna of biozone III. Both biozones are summarily referred to as the Shaunavon Fauna. The stratigraphic level of biozone III is occupied by a marine shale-limestone sequence in south-central Montana, from which a moderately diverse foraminiferal fauna was described by Lalicker (1950), and which was dated by ammonites to be early Bathonian in age (Imlay 1980).

The succeeding biozones IV and V, the Rierdon Fauna, in stark contrast, are fully marine and contain a great variety and number of ostracods, Foraminifera, fragments of various echinoderm and other invertebrate groups indicating transgressive conditions. Only in a few rock samples of the easternmost regions of the basin were some of the freshwater indicators of biozone III found intercalated with the marine faunas. The Rierdon Fauna flourished from about middle Bathonian to middle Callovian time, when a major regression started. Biozone IV is represented in both the Williston and Western Canada Basins and is well known from both surface and subsurface strata. In effect, it is the most widespread of the Jurassic microfaunas of the North American continent, and it has been reported from many Canadian and United States localities. Zone V, in contrast, is largely confined to the Williston Basin.

A wedge of sandstone in the subsurface of southwestern Saskatchewan, with a few distinctive ostracod species (Roseray faunule, biozone VI), is the only evidence preserved of the regressive pile and of an age straddling the middle–late Callovian boundary. This ostracod fauna is missing everywhere else, in particular from the Rocky Mountains, where the hiatus includes, in addition, most or all of biozone V.

When the sea returned in early (but not earliest) Oxfordian time to both basins, the ostracods and Foraminifera developed in two different faunal provinces. In the Williston Basin, biozones VII to VIII (the Swift Fauna) still contain about the same mix of genera and a number of through-ranging species from the Rierdon Fauna. They belong to the Shoshonean province (formerly Western Interior province) of the East Pacific Subrealm (Tethyan Realm), as do their immediate predecessors. However, certain Boreal influences can be detected in biozone VIII, a regressive unit without ostracods in which agglutinated Foraminifera dominate.

In the Rocky Mountains and the Western Canada Basin, the Green Bed Fauna (Figure 15.2) of southeastern British Columbia and adjoining Alberta contains microfaunal elements with Shoshonean, central European, and Boreal affinities. It had to be assigned its own status as biozone IX for this reason and could not be referred to the lower to middle part of the Swift Fauna and biozone VII, which seemingly occupies the same stratigraphic level from the middle part of the Lower to the Middle Oxfordian. Because of the perfect transition of lithology and gradual changes in the faunal spectrum, the succeeding biozones X to XII of the Passage and Transition Beds cannot be precisely delineated either, and faunal boundaries are tentatively drawn. These microfaunas – essentially agglutinated Foraminifera with only few calcareous Foraminifera and ostracods – compare closely or are identical with those reported and recovered from Arctic North America and Siberia. Undoubtedly, they are Boreal Realm faunas. There are no ostracods left in biozone XIII, the last marine unit.

There are still major difficulties in correlating Upper Jurassic strata between the southern and northern Rocky Mountains, despite the fact that the faunas belong to the same faunal province. The problem arises from the highly fragmented and disturbed outcrop pattern, the paucity and erratic distribution of the microfaunas, and the lack of ammonites for anchoring at least a few of the fossiliferous horizons. Nevertheless, the local zonation worked out by Brooke and Braun (1981) for the Upper Jurassic microfaunas of northeastern British Columbia is integrated, and the former biozones I to III are incorporated as zones XI to XIII in the comprehensive scheme (see Chapter 19).

In Figure 15.2, the ranges of the ostracods are not traced out, and are summarily indicated by dots only within the respective rock sequence and biozone. This is a reflection of the basic difficulties mentioned earlier. There are solid indications, however, that the ranges of the ostracods in biozone IV, and whatever is preserved of biozone V (the Rierdon Fauna), are the same as those indicated for the Williston Basin. The main problems are with the scattered ostracods in the succeeding biozones, which had to be defined mainly on the basis of changes in foraminiferal faunas, for ostracods are erratically distributed and occur in only low numbers in these uppermost Jurassic sequences. Furthermore, they all were found as carapaces that could not be pried open to observe the interior valve features, and they were mostly pyritized and not too well preserved. Comparisons thus are most difficult to make, and precise identifications impossible to achieve on the basis of the material recovered thus far. However, photos of most of these unassigned species are shown in Plate 108, with plates 106 and 107 showing the key ostracods of the Shaunavon, Rierdon, and Roseray Faunas and biozones II to VI. The upper part of Plate 108 shows the diagnostic Swift ostracods (biozone VII), followed by representatives of biozones IX to XII of the Rocky Mountains.

## References


Brooke, M. M., & Braun, W. K. (1972). *Biostratigraphy and Microfaunas of the Jurassic System of Saskatchewan*. Saskatchewan Dept. Min. Res. Rept. 161.

(1981). *Jurassic Microfaunas and Biostratigraphy of Northeastern British Columbia and Adjacent Alberta*. Geol. Surv. Canada Bull. 283.

Imlay, R. W. (1980). *Jurassic Paleobiogeography of the Conterminous United States and Its Continental Setting*. U.S. Geological Survey Professional Paper 1062.

Lalicker, C. G. (1950). Foraminifera of the Ellis Group, Jurassic, at the type locality. *Univ. Kansas Paleontol. Contr., Protozoa, 2*, 3–20.

# 16 Bivalve zones and assemblages of the circum-Pacific region

S. E. DAMBORENEA, I. V. POLUBOTKO, I. I. SEY, and K. V. PARAKETSOV

The use of Late Jurassic bivalves (buchiids) in biostratigraphy has been successful in the northern circum-Pacific for a long time. In recent years the increase of detailed studies on other bivalve faunas has extended the potential biostratigraphic usefulness of bivalves to cover also the Early and Middle Jurassic of other circum-Pacific regions. Only some bivalve groups have proved useful, notably monotaceans and pectinaceans for the Lower Jurassic, inoceramids (s.l.) for the Middle Jurassic, and buchiids for the Upper Jurassic. As knowledge of systematics and vertical ranges of bivalve species is still very uneven, many more studies are needed to obtain a comprehensive picture. A correlation chart for several circum-Pacific regions has been compiled showing approximate equivalences (Table 16.1). This chart is mainly intended to stimulate further discussions on this subject.

As seen in Table 16.1, many of the biostratigraphic units recognized thus far are based on and named after local species belonging to mainly circum-Pacific genera, such as *Otapiria, Radulonectites, Parainoceramus, Retroceramus* (or *Mytiloceramus*), and *Buchia*. Although this is only a consequence of the fact that correlation tables are commonly based on offshore species, it makes comparison between circum-Pacific regions easier. This analysis should not overlook the progressive displacement of local vertical ranges of some of these bivalve taxa along their geographic ranges. Bivalve biostratigraphic units have only local or, at best, regional value, and up to now correlation between them has been done by independent reference of the local zonations to the European standard zonation, thus introducing a certain degree of ambiguity and inaccuracy.

A zonation for the whole Jurassic based on bivalves is available only for northern Russia, but data from South America and New Zealand are sufficiently complete to merit further comments. For other regions, see the references cited in the source note to Table 16.1.

See Appendix, Plates 115–127.

## Eastern Russia[1]

Bivalves have the most extensive distribution in the Jurassic deposits of northeastern Asia and are indispensable when dividing and correlating the enclosing deposits. The role of bivalve groups in this process varies. Of paramount significance are monotids, inocerams, and buchiids (for Lower, Middle, and Upper Jurassic, respectively), because of their wide distribution, high evolution rates, and slight dependence on facies. Pectinids and some other groups are also of great significance, particularly for Pliensbachian and Toarcian, where no representatives of the mentioned three groups are found.

Hettangian and Sinemurian are divisible into three zones on the basis of an evolutionary succession of the genus *Otapiria* (i.e., *O. originalis–omolonica–limaeformis*). This series displays the following evolutionary trend in the ornamentation of the right valves: from differentiated fine ribbing in *O. originalis* to a coarser and variable ribbing in species of the *O. omolonica* Zone, up to the appearance of *"Monotis" inopinata* with coarsely ribbed left and right valves in the upper part of this level, and further to the most specialized species of the *O. limaeformis* Zone, with ribbed left and smooth right valves.

The level with *"Monotis" inopinata* is distinguished as a marker at the base of the *O. limaeformis* Zone, coinciding with the appearance of the late Sinemurian *Angulaticeras kolymicum*. The age of the *omolonica* Zone is defined by the presence of *Arietites bucklandi* and *Coroniceras siverti*, and that of the *originalis* Zone by *Schlotheimia angulata* and *Waehneroceras* spp. In the Planorbis Zone, otapirias are extremely rare. This genus does not go beyond the Sinemurian–Pliensbachian boundary. Otapirias are accompanied by *Meleagrinella subolifex* (Hettangian) and different species of *"Pseudomytiloides,"* as well as *Kolymonectes*.

A Lower Pliensbachian bivalve fauna is poorly developed at single sections and is noted for small species.

[1] By I. V. Polubotko, I. I. Sey, and K. V. Paraketsov.

Upper Pliensbachian is marked by the appearance of numerous representatives of *Harpax,* whose range in eastern Russia may coincide with the Upper Pliensbachian boundaries. However, pectinids *Kolymonectes, "Velata"* (= *Eopecten*), *Chlamys, Ochotochlamys,* and *Radulonectites* are of major stratigraphic significance. These forms allow the distinction of two levels in North-East Russia: The *"Velata" viligaensis* Beds correspond to the larger, upper part of the *Amaltheus stokesi* Zone and lower *A. talrosei* Zone, with the marker level *Kolymonectes mongkensis* at the base of the Stokesi Zone, and the *Radulonectites hayamii* Beds correspond to the rest of the Upper Pliensbachian. In Far East Russia, the *Ochotochlamys bureiensis* Beds, comprising *Kolymonectes* and *Amuropecten,* correspond to most of the Upper Pliensbachian. In both regions, pectinids are accompanied by different species of *Meleagrinella, Oxytoma, Myophoria,* etc.

Toarcian faunal communities include few bivalves. Two levels are distinguished in the North-East: *Meleagrinella substriata* Beds, which approximately correspond to the three lower zones of the stage, and *M. faminaestriata* Beds, comprising the rest of the stage. The latter comprise two marker levels: beds with *Vaugonia literata* (*polare* Zone) and beds with *Pseudomytiloides marchaensis* (upper *rosenkrantzi* Zone). In the Far East, only one level with *Galinia borsjaensis* is recorded, approximately in the range of the *Z. monestieri* Zone.

The Toarcian–Aalenian boundary in the Far East and in the Anadyr Basin is marked by the appearance of the first mytiloceramsof the *M. priscus* group, and over the rest of the area by an association of large *Oxytoma jacksoni* and *Propeamussium olenekense,* ranging throughout the Aalenian and Lower Bajocian, and by the *Trigonia alta* Beds of a narrower age range, corresponding approximately to the *Pseudolioceras (P.) beyrichi* level.

Beginning with the Upper Aalenian, biozonation of the Middle Jurassic is based exclusively on mytiloceramsv(Polubotko and Sey 1981). The Upper Aalenian and Lower Bajocian mytiloceram assemblage, which is found together with *Pseudolioceras (Tugurites),* is composed of morphologically similar species: *M. elegans, popovi, obliquus, lungershauseni, jurensis, morii, ochoticus, anilis, karakuwensis,* etc., on the basis of which the *M. popovi* (North-East) and *M. obliquus* (Far East) Zones are distinguished in the Upper Aalenian, and a common regional *M. jurensis* Zone in the Lower Bajocian. In the upper part of the latter in the North-East and northern Siberia, beds with large elongated *M. menneri* are recorded, which are not found in the Far East. The *M. lucifer* Zone is the most distinct marker level for the entire area under discussion. In the Far East, it is easily distinguished on the basis of the predominant species *M. ussuriensis, M. formosulus,* etc., belonging to the *M. lucifer* group. In the North-East, *M. omolonensis* prevails. Typical *M. lucifer* is known from the Alazeya Plateau and Anadyr and Bureya Basins, where *Arkelloceras elegans* and *A. tozeri* were found together with *M. lucifer.*

The *M. clinatus* level is most closely associated with the marker level *M. lucifer,* because it contains direct descendants of the *M. lucifer* group. Species of this level are characterized by an elongated shell with numerous constrictions. No ammonites are reported from these beds.

In the *M. retrorsus* and *M. electus* Beds, which are conditionally correlated with the Upper Bajocian, rapid speciation is recorded: the *M. retrorsus* and *M. tongusensis* groups appear, and the *M. porrectus–M. kystatymensis–M. bulunensis* and *M. electus–M. almus* lineages evolve, characterized by gigantism at certain phylogenetic stages (some mytiloceramsare up to 60 cm along the growth axis). Some species extend into the Bathonian and dominate the *M. bulunensis* Beds, which in the North-East have yielded *Arctocephalites elegans,* thus enabling the placement of these beds at the base of the Bathonian.

Mytiloceramsin the Far East are not found higher than the *M. bulunensis* Beds, whereas in the North-East another assemblage (i.e., *M. polaris–M. vagt*) is widespread above beds with *Arctocephalites.* It is related to the underlying assemblage, but is composed of characteristic species with dense regular concentric folds and a convex anterior margin (index species, *M. ultimus, M. tantus,* etc.). Mytiloceramsin the Jurassic of eastern Russia are, in fact, not found above these beds. Overlying beds with *Cadoceras* yield no mytiloceramss.

Biozonation of the Upper Jurassic in northeastern Asia is based entirely on buchias, which make up several successive assemblages in the sections. This succession is generally similar to that observed on the Russian Platform, in northern Siberia, and in western North America. Even taking into account the considerable age range of certain buchia species and a possible diachroneity of acme zones in different regions, the buchia scale of eastern Russia uses concurrent range zones, which provide a more reliable and detailed stratification of deposits (Paraketsov 1980; Sey and Kalacheva 1985). The range and boundaries of buchia zones are in most cases approximate, because of impoverishment of ammonite faunas within the area discussed, and they are defined both by rare joint occurrences of ammonites in the region itself and by correlation with adjacent regions, where buchia scales are confirmed by ammonites (Imlay and Jones 1970; Zakharov 1981; Jeletzky 1984).

The first rare, small buchiids [i.e., *Praebuchia*? *anyuensis* (Parak.)] appear possibly as early as the Bathonian, above the last mytiloceramss. The basis of the Upper Jurassic in the buchia scale is marked by small, elongated, finely ornamented *Praebuchia*?, similar to the European species *impressae.* In the Far East, they are found together with the Lower Oxfordian *Scarburgiceras.*

The next two zones (i.e., *P. lata–B. concentrica* and *B. concentrica–B. tenuistriata*) have reliable equivalents in other regions (Zakharov 1981). Their ages in eastern Russia are confirmed by occurrences of the Middle Oxfordian *Maltoniceras* and *Dichotomosphinctes* and Lower Kimmeridgian *Amoeboceras,* respectively. A zonal assemblage of the Lower Kimmeridgian zone, in addition to the index species, also comprises *B. lindstroemi* (Sok.), *B. ochotica* (Sey) (1986), and *B. vaquaamensis* (Parak.); both zones in the North-East contain *B. jeropolensis* (Parak.).

The age range of the Upper Kimmeridgian–Volgian buchia zones is most problematic, possibly resulting in discrepancies

| Stage | Substage | NORTHEAST USSR (1) | FAR-EAST USSR (2) | ALASKA (3) | WESTERN CANADA (4) | WESTERN USA (5) | SOUTH AMERICA (6) |
|---|---|---|---|---|---|---|---|
| VOLGIAN | U | *B. tenuicollis* – *B. terebratuloides* Zone | *B. piochii* – *B. terebratuloides* Zone | *B. subokensis* | *B. unschensis* *B.* aff. *okensis*; *B. terebratuloides* *B. fischeriana* *B. lahuseni* | *B. unschensis* *B.* aff. *okensis*; *B. terebratuloides* | *R.* aff. *everesti* |
| | M | *B. russiensis* – *B. fischeriana* Z. | *B. russiensis* – *B. fischeriana* Z.; *B. mosquensis* – *B. russiensis* Z. | *B. piochii* | *B piochii*; *B.* cf. *blanfordiana* *B.* aff. *russiensis* | *B. piochii* *B. fischeriana* *B. lahuseni*; *B. russiensis* | |
| | L | *B. m. mosquensis* – *B. russiensis* Z. | *B. rugosa* – *B. mosquensis* Zone | *B. rugosa* *B. mosquensis* | *B. mosquensis* | | |
| KIMMERIDGIAN | | *B. rugosa* – *B. mosquensis paradoxa* Zone | *B. tenuistriata* – *B. rugosa* Z. | ? | ? | | *R.* aff. *haasti* |
| | | *B. tenuistriata* – *B. concentrica* Zone | *B. tenuistriata* – *B. concentrica* Z. | *B. concentrica* | *B. concentrica* | *B. concentrica* | |
| OXFORDIAN | U | *B. lata* – *B. concentrica* Zone | *B. concentrica* Zone | | | | |
| | M | | *Praebuchia lata* Zone | *B. spitiensis* | | | |
| | L | *Praebuchia ? impressae* Beds | *Praebuchia impressae* Beds | | | | |
| CALLOVIAN | U | | | | | | *R.* aff. *galoi* |
| | M | | | | | | |
| | L | *Praebuchia ? anyuensis* Beds | | | | | *R. stehni* Beds |
| BATHONIAN | U | *M. polaris* – *M. vagt* Beds | | | *R. obliquiformis* | | *R. patagonicus* |
| | M | | | | | | |
| | L | *M. bulunensis* Beds | | *R. ambiguus* | | | |
| BAJOCIAN | U | *M. retrorsus* – *M. electus* Beds | *M. bulunensis* Beds; *M. kystatymensis* | | *R. lucifer* | | *R.* cf. *marwicki* |
| | L | *M. clinatus* Z.; *M. lucifer* Z.; *M. menneri* Beds; *M. jurensis* Z | *M. clinatus* Z.; *M. lucifer* Z.; *M. jurensis* Zone | *R. lucifer* | *R. ferniensis* | | *P. ? westermanni*; *R.* cf. *inconditus* |
| AALENIAN | | *M. popovi* Z.; *M. priscus* Z.; *T. alta* | *M. oblicuus* Z.; *M. priscus* Z. | | | | |
| TOARCIAN | U | *Meleagrinella faminaestriata* Beds | *G. borsjaensis* | | | | |
| | L | *Meleagrinella substriata* Beds | | | | | *Propeamussium pumilum* Ass. Z.; *Posidonotis cancellata* Ass. Z. |
| PLIENSBACHIAN | U | *Radulonectites hayamii* Beds; *E. viligaensis* Beds | *Ochotochl. bureiensis* | | | *L. boechiformis* | *Radulonectites sosneadoensis* Ass. Zone |
| | L | *Chlamys tapensis* Beds | *Palmoxytoma cygnipes*; *H. spinosus* Beds | | *Posidonotis semiplicata* | *Posidonotis semiplicata* | *O neuquensis* Ass. Zone |
| SINEMURIAN | U | *O. limaeformis* Z.; "*Monotis*" *inopinata* | *O. limaeformis* Zone | *O. tailleuri* | | | |
| | L | *O. omolonica* Zone | *O. omolonica* Zone | | | | *O. pacifica* Ass. Zone |
| HETTANGIAN | | *O. originalis* Beds | | | | | |

| ANTARCTICA (7) | NEW ZEALAND (8) | | INDONESIA (9) | | CHINA (10) | JAPAN (11) |
|---|---|---|---|---|---|---|
| *R. everesti* | *R. aff. everesti* | | *B. cf. plicata* | | *B. piochii* | |
| *B. blanfordiana*<br>*B. spitiensis* | *B. plicata* | | *Malayomaorica ? misolica* | | *B. rugosa* | |
| *A. stoliczkai* | *B. hochstetteri*<br>*M. ? aff. misolica*<br>*B. cf. plicata* | | | | *B. concentrica* | |
| *R. haasti* | *R. haasti* | *Malayomaorica malayomaorica* | *R. haasti* | *Malayomaorica malayomaorica* | *B. spitiensis* | |
| *R. galoi* | *R. galoi* | | *R. galoi* | *B.* sp. | *Praebuchia kirghisensis* | |
| | | | *Praebuchia kirghisensis* | | | |
| | | | *Praebuchia* cf. *orientalis* | | *Mesosaccella morrisi – Entolium demissum* Ass. | |
| | *R. stehni*<br>*R.* cf. *patagonicus* | | | | | *R. maedae* |
| | | | | | | *R. utanoensis* |
| | *R. marwicki* | | | | | |
| | *R. inconditus* | | | | *C. lens – Liostrea birmanica* Ass. | *R. morii* |
| | ? | | | | | *I. ? kudoi* |
| | | | | | | *P. matsumotoi* |
| | *I. ururoaensis* | | | | *Astarte* cf. *voltzi – W. ambongoensis* | *Posidonotis* sp.<br>*Radul. japonicus* |
| | *P. martini*<br>*Pseudaucella marshalli* | | | | | |
| | ? | | | | | |
| | *O. marshalli* | | | | *Parainoceramus – Teinonuculana* Ass. | |
| | | | | | *Hiatella* Ass. | |

Table 16.1. *Correlation chart based on bivalve biostratigraphic units*

Compiled by S. Damborenea from the following literature: (1) Paraketzov (1980), Polubotko & Repin (1988), Polubotko & Sey (1981), Zakharov, Paraketzov, & Paraketzova (1988); compiled by Sey, Polubotkov, Paraketzov, and Okuneva (chapters herein). (2) Polubotko & Sey (1981), Sey (1984, 1986), Sey & Kalacheva (1980, 1985, 1988); compiled by Sey, Polubotko, Paraketzov, and Okuneva (chapters herein). (3) Imlay (1959, 1967), Imlay & Detterman (1973). (4) Frebold (1964), Jeletzky (1965, 1966, 1984), Zakharov, Paraketzov, & Paraketzova (1988). (5) Imlay (1967), Imlay & Jones (1970), Jeletzky (1984), Jones, Bailey, & Imlay (1969). (6) Damborenea (1990), Damborenea in Riccardi, Damborenea, & Manceñido (1990). (7) Crame (1982b, 1984). (8) Fleming (1958, 1959), Fleming & Kear (1960), Li & Grant-Mackie (1988), Marwick (1953), Purser (1961), Speden (1970), Stevens (1965, 1978), Stevens & Speden (1978); new data herein. (9) Li & Grant-Mackie (1988), Sato, Westermann, Skwarko, & Hasibuan (1978). (10) Li & Grant-Mackie (1988), Wang (1988), Wang & Sun (1983). (11) Hayami (1960, 1985).

between zonal scales of the North-East and Far East by approximately a substage, whereas the succession of buchia assemblages in the section remains the same. Guide-ammonite occurrences are known only in the Middle Volgian *B. russiensis–B. fischeriana* Zone: These are *Dorsoplanites* cf. *transitorius* Spath in the North-East, and *Durangites* sp. ind. in the Far East.

The zonal units crowning the Jurassic section in both regions apparently have similar ranges and identical buchia assemblages. Their different naming may be connected with problems of nomenclature. Along with the index species, they also comprise *B. fischeriana* (predominant), *B.* ex gr. *unschensis* [= *B. circula* (Parak.)], *B. lahuseni,* and *B. trigonoides.* They correlate well with the Upper Tithonian buchia assemblages of British Columbia and California (Jones, Bailey, and Imlay 1969; Imlay and Jones 1970; Jeletzky 1984). The interval of the above units possibly also includes the lower Berriasian.

The Upper Jurassic in eastern Russia, besides buchia, yields abundant representatives of other groups of bivalves, particularly in the lower (Oxfordian) part of the section. These are *Mclearnia, Camptonectes* s.s., *Isognomon, Aguilerella, Meleagrinella, Tancredia, Pseudolimea, Pleuromya,* etc. Higher, buchias dominate, and some levels in the Middle and Upper Volgian also yield abundant *Oxytoma, Astarte,* and *Modiolus,* as well as large *Mclearnia, Isognomon,* and other bivalves.

### South America[2]

A clear picture of the stratigraphic distribution of Jurassic South American bivalves is only just emerging. In most places ammonoids are present, and so it is possible to use the local zonation based on them. As a consequence, up until very recently no proper attention had been paid to the accurate vertical ranges of bivalve species. An attempt to compile the stratigraphic distribution of Chilean trigoniid species was made by Pérez and Reyes (1977), but a thorough, systematic revision is still wanting. On the other hand, an earlier chronostratigraphic scheme based on bivalves was proposed by Leanza (1942) for the Early Jurassic of a small area in southern Neuquén (Argentina), but the assemblages he recognized were later referred to different biofacies with an ecological rather than a temporal meaning (Damborenea, Manceñido, and Riccardi 1975).

Thus far, only a preliminary chronostratigraphic table for part of the Early Jurassic of South America can be provided. A systematic revision of Early Jurassic bivalves from Argentina (Damborenea 1987a,b), based on material collected from the same sections and levels that provided the ammonites identified by Riccardi (1984), is yielding promising results. Five assemblage zones for Argentina and Chile, calibrated against local ammonite zonations, were recently proposed (Damborenea in Riccardi, Damborenea, and Manceñido 1990), based on the vertical ranges of bivalves in relatively offshore facies. With more data, these assemblage zones could perhaps also be recognized in Perú and Columbia–Ecuador.

[2] By S. E. Damborenea.

The study of Hettangian–Sinemurian faunas has not been completed yet. An assemblage zone characterized by *Otapiria pacifica* Covacevich & Escobar can be recognized for the latest Hettangian to early Sinemurian of Chile and Argentina (Plate 115, Figs. 3, 4). *Otapiria pacifica* belongs to the morphological group of *O. limaeformis.* This bivalve appears with latest Hettangian (*B. canadensis* Zone) to late Lower Sinemurian (*Agassiceras* Zone) ammonites. In littoral facies, *Quadratojaworskiella* sp. and the first South American *Weyla* may represent a more or less coeval association. An ill-defined Hettangian bivalve faunule with *''Inoceramus''* sp. (Escobar 1980) and *Palmoxytoma* sp. (Riccardi et al. 1988) (Plate 115, Fig. 1) occurs below it, but it cannot yet be identified as a distinctive assemblage zone. Similarly, the study of late Sinemurian bivalve faunas will allow the recognition of one or more assemblage zones of that age. *Cardinia* cf. *listeri* (Sow.) is a conspicuous element of the bivalve faunas of this age in littoral facies (Plate 115, Fig. 10).

Two assemblage zones were recognized for the Pliensbachian. The first one spans most of the Lower Pliensbachian (upper part of *Miltoceras* Ammonite Zone to lower *Fanninoceras* Zone) and is characterized by *Otapiria neuquensis* Damborenea (Plate 115, Figs. 5, 6), *Palaeoneilo patagonidica* (Leanza), a local species of *Kalentera* (Plate 115, Figs. 7, 8), and *Parainoceramus apollo* (Leanza) (Plate 115, Fig. 9), whereas the Upper Pliensbachian zone (rest of *Fanninoceras* Ammonite Zone) is characterized by *Radulonectites sosneadoensis* (Weaver) (Plate 116, Figs. 6, 7) and *Kolymonectes coloradoensis* (Weaver) (Plate 115, Figs. 11, 12). In certain facies, the pectinid *Camptochlamys wunschae* (Marwick) (Plate 116, Figs. 9, 10), characteristic of the lower Ururoan in New Zealand, occurs also in this assemblage, together with *Plicatula* (*Harpax*) *rapa* Bayle & Coquand, of the same group of species as the coeval specimens from Far East Russia referred to *Harpax spinosus* (Sow.) (Plate 116, Figs. 13, 14). During the Pliensbachian an especially rich and varied littoral bivalve fauna flourished in the South American Andes, with several species of *Weyla, Parallelodon, Cucullaea, Jaworskiella, Frenguelliella, Cardinia,* other pectinids, mytilids, astartids, etc. The *R. sosneadoensis* Assemblage Zone is nearly equivalent to the *R. hayamii* Beds of northeastern Asia.

During the latest Pliensbachian and basal Lower Toarcian (*simplex* and *chilense* Ammonite Zones) the *Posidonotis cancellata* Assemblage Zone can be recognized, typically with *P. cancellata* (Leanza) (Plate 117, Figs. 9–11) and *Weyla alata angustecostata* (Philippi) (Plate 117, Fig. 12). In South America, *Posidonotis* consistently appears at the Pliensbachian–Toarcian boundary, but beds with *Posidonotis* spp. in western North America seem to be older (Damborenea 1989). Subsequently, another association characterized by the appearance of *Propeamussium pumilus* (Lamarck) (Plate 117, Figs. 13–15), *Bositra ornati* (Quenstedt), and *Meleagrinella* sp. extend over most of the Toarcian. A revision of South American late Toarcian bivalves may provide arguments to divide it. Latest Toarcian–earliest Aalenian bivalve faunas probably belong to a different assemblage zone.

Current knowledge of South American Middle Jurassic bivalve faunas is far from complete (except for a few species). Though less abundant than in northern circum-Pacific regions, inoceramids have proved useful. Four informal biostratigraphic units for the Middle Jurassic of Argentina have been proposed (Damborenea 1990). These faunas occur at the same sections that have provided data for the ammonite zonations (Riccardi, Westermann, and Elmi 1989) and thus are accurately dated.

The succession is highly congruent with the distribution in time of the same group in the northern Pacific. A *Retroceramus*? sp. found in early Bajocian beds belongs to the same morphological group as the Aalenian–Bajocian *R. mongkensis* (Kosh.) and *R. popovi* (Kosh.) and the early Temaikan *R. inconditus* (Marwick). A new species tentatively referred to *Parainoceramus* (Plate 118, Fig. 1) is characteristic of the Humphriesianum Ammonite Zone. Three species of *Retroceramus* occur between the Upper Bajocian and the lowest Callovian: *R.* cf. *marwicki* (Speden) (Plate 118, Fig. 8) in the Rotundum Ammonite Zone, *R. patagonicus* (Philippi) (Plate 118, Fig. 9) in the Steinmanni Ammonite Zone, and *R. stehni* Damborenea (Plate 118, Fig. 13) spanning the Vergarensis, Bodenbenderi, and perhaps the lower part of the Proximum Ammonite Zones. *Retroceramus* cf. *marwicki* is only slightly younger than the similar Boreal *R. lucifer* (Eichwald), and the Bathonian–Callovian lineage *R. patagonicus–R. stehni* is, again, somewhat younger than the Boreal group of *R. bulunensis–vagt* (Kosh.).

Knowledge of potentially useful Upper Jurassic bivalves from South America is in a very preliminary state, and no attempt to recognize associations has been made so far. *Retroceramus* aff. *galoi* (Boehm) was recently identified from Callovian beds of the Neuquén Basin (Damborenea 1990), while Fuenzalida and Covacevich (1988) mentioned the presence of *R.* aff. *haasti* (Hochst.) in Kimmeridgian beds of the Magallanes Basin. The precise ages and affinities of the inoceramid and buchiid faunas described by Sokolov (1946) and Feruglio (1937) remain ill-known. The material figured by Feruglio as *Inoceramus* cf. *steinmanni* Wilckens from the Tithonian of the Austral Basin in Argentina was later referred by Crame (1982a) to *Retroceramus everesti* (Oppel).

### New Zealand[3]

New Zealand Jurassic faunas are rich in bivalves, but ammonites are scarce and have not been revised yet.[4] It is only natural, then, that bivalves have great stratigraphic value, and they are even used to recognize the lower boundaries of local stages (Marwick 1951, 1953; Fleming 1958, 1960; Fleming and Kear 1960; Speden 1970; Stevens 1978; Stevens and Speden 1978), as follows. Aratauran: *Otapiria marshalli*. Ururoan: *Pseudaucella marshalli*. Temaikan: *Meleagrinella* cf. *echinata* and *Retroceramus inconditus* have been used, but they have not proved satisfactory, and other invertebrates, such as belemnites, are now used. Heterian: *Retroceramus galoi*. Ohauan: *Retroceramus haasti*. Puaroan: *Buchia hochstetteri*. The temporal ranges of some of these key species still are not accurately known, making correlation with the standard European zonation very difficult. Nevertheless, New Zealand stages have been used in other Indo-Pacific regions (New Caledonia, New Guinea, Indonesia) and have even been extended to Antarctica (Crame 1982b, 1984). Apart from these bivalve taxa, there are other bivalve assemblages that are potentially useful for New Zealand biostratigraphy and have thus been included in the correlation table.

*Otapiria marshalli* (Trechmann) is widely distributed within New Zealand and has also been found in New Caledonia. This species is associated with Hettangian and Sinemurian ammonites, but does not appear together with the lowermost Hettangian ammonite fauna in the Hokonui Hills. A distinctive bivalve fauna occurs with *O. marshalli*, among which *Kalentera mackayi* Marwick, *Pseudolimea fida* Marwick, and *Sphaeriola leedae* Marwick are especially abundant.

In the upper part of the Lower Jurassic two other bivalve assemblages can be recognized (Grant-Mackie 1959), the older one with *Pseudaucella marshalli* (Trechmann) and, locally, *Camptochlamys wunschae* (Marwick), *Plicatula* (*Harpax*) cf. *rapa* Bayle & Coquand, and *Parainoceramus martini* Speden. This assemblage probably includes most of the Pliensbachian. *Camptochlamys wunschae* and *Plicatula* (*Harpax*) *rapa* appear in lower Upper Pliensbachian beds in the South American Andes. The other assemblage includes *Inoceramus ururoaensis* Speden and species of *Palaeoneilo*, *Parallelodon*, and *Astarte* and is dated as Lower Toarcian on account of harpoceratid and dactylioceratid ammonoids.

It is generally agreed that the Aalenian is not well developed in New Zealand, and no diagnostic bivalve faunas of this age have been identified so far.

As in other circum-Pacific regions, during the Middle Jurassic a varied inoceramid fauna developed in New Zealand. The sequence and distribution of local inoceramid species are complex and still are being examined. The geographic and stratigraphic ranges of different species, as well as their intraspecific variation and affinities, are poorly understood. Nevertheless, this group of bivalves is a potentially useful biostratigraphic tool, and a local bivalve zonation based on inoceramids could be established for New Zealand in the near future. Four different successive faunas are here recognized for the Middle Jurassic. This is only a very sketchy scheme that needs further study, but it is highly congruent (see correlation table) with the inoceramid successions found in South America, with most taxa in common even at specific level, and also, but to a lesser degree, with those of the Northern Hemisphere (Russia, Alaska, Canada).

*Retroceramus inconditus* (Marwick), *"Inoceramus" brownei* Marwick, *Pleuromya milleformis* Marwick, and *Meleagrinella* cf. *echinata* (Smith) form the next bivalve assemblage, which has been variously assigned to the Bajocian or Bathonian. *Retroceramus inconditus* belongs to the group of *R. popovi* (Kosch.) that in northern Russia appears in the topmost Aalenian, and a very similar species occurs in Lower Bajocian deposits in Argentina. Other bivalves from this assemblage are *Propeamussium clamosseum* (Marwick), again very similar to *P. andium* (Tornquist) from the

[3] By S. E. Damborenea.

[4] Editor's note: A taxonomic revision of the ammonites is currently being carried out by G. Stevens and G. E. G. Westermann.

Andes, *Camptonectes* cf. *laminatus* (Sow.), *Solemya*? *maikaensis* (Marwick), and several trigoniid species.

This association is followed by *Retroceramus marwicki* (Speden), *Haastina haastiana* Wilckens, and *Tancredia allani* Marwick and has been dated as Callovian (Speden 1970). *Retroceramus marwicki* belongs to the group of *R. lucifer* (Eichwald), widely distributed during the early Bajocian in Arctic regions (see correlation table). This, and the fact that in the Andes *R.* cf. *marwicki* appears in lower Upper Bajocian beds, may indicate a similar age for the New Zealand fauna. Significantly, the small ammonite fauna on which the "Callovian" age had been based has just been revised to mid-Bajocian by G. E. G. Westermann (personal communication).

Above this fauna, a different inoceramid species, which was in part erroneously referred to *R. galoi* in the past, appears instead to be related to the South American *R. patagonicus* (Philippi) from the late Bathonian. The late Bathonian–early Callovian species *R. stehni* Damborenea has also been identified with certainty among material from the North Island housed in Auckland University. In New Zealand these two taxa had been mentioned previously only as indeterminate species in an unpublished thesis (Hudson 1983), but their affinities are only now understood. The age of these species in New Zealand has not yet been determined independently, such as by accompanying ammonites, but it may not greatly differ from that of the well-documented South American occurrences (Damborenea 1990). The presence of the *patagonicus–stehni* lineage in New Zealand fills, in part, the apparent time gap between the Temaikan and Heterian.

The following assemblage is characterized by *R. galoi* (Boehm), which in New Zealand has traditionally been referred to the Kimmeridgian (Speden 1970; Stevens and Speden 1978), whereas in its type locality in the Sula Islands it occurs in middle or late Oxfordian beds (Sato et al. 1978). On the basis of independent evidence, some authors now think that the first appearance of *R. galoi* may be older, even late Callovian (Grant-Mackie, Hudson, and Helby 1986; Hudson, Grant-Mackie, and Helby 1987; Helby, Wilson and Grant-Mackie 1988), but a reassessment of the whole inoceramid sequence in New Zealand and Indonesia is still needed to adjust the correlation of this part of the sequence to the international standard.

The upper part of the range of *R. galoi* is shared by *Malayomaorica malayomaorica* (Krumbeck), which extends into the next inoceramid association, characterized by *R. haasti* (Hochstetter), dated as late Kimmeridgian or early Tithonian (Speden 1970; Stevens and Speden 1978). Intermediate forms between *R. galoi* and *R. haasti* have been referred to *R. subhaasti* (Wandel).

Though less abundant than in Arctic regions, *Buchia* species are common elements of Late Jurassic New Zealand faunas. The recognized succession is as follows: *B.* cf. *plicata*, *Malayomaorica*(?) aff. *misolica* (Krumbeck), *B. hochstetteri* Fleming, and *B. plicata* (Zittel) (Fleming, 1959; Stevens 1978; Li and Grant-Mackie 1988). The ages of these faunas are not yet known, but it is generally agreed that in New Zealand *Buchia* appeared later than in other regions (Li and Grant-Mackie 1988); they are all regarded as Tithonian. Other bivalves from this part of the Jurassic are *Otapiria masoni* Marwick and *Retroceramus* aff. *everesti* (Oppel).

## References


Crame, J. A. (1982a). Late Jurassic inoceramid bivalves from the Antarctic Peninsula and their stratigraphic use. *Palaeontology, 25,* 555–603.

(1982b). Late Mesozoic bivalve biostratigraphy of the Antarctic Peninsula region. *J. Geol. Soc. London, 139,* 771–8.

(1984). Preliminary bivalve zonation of the Jurassic–Cretaceous boundary in Antarctica. In *Mem. III Cong. Latinoamer. Paleont., Mexico.* (pp. 242–54).

Damborenea, S. E. (1987a). Early Jurassic Bivalvia of Argentina. Part 1: Stratigraphical introduction and superfamilies Nuculanacea, Arcacea, Mytilacea and Pinnacea. *Palaeontographica, A199,* 23–111.

(1987b). Early Jurassic Bivalvia of Argentina. Part 2: Superfamilies Pteriacea, Buchiacea and part of Pectinacea. *Palaeontographica, A199,* 113–216.

(1989). El género *Posidonotis* Losacco (Bivalvia, Jurásico inferior): su distribución estratigráfica y paleogeográfica. *IV Cong. Argentino Paleont. Biostrat., Mendoza. 4,* 45–51.

(1990). Middle Jurassic inoceramids from Argentina *J. Paleont. 64,* 736–59.

Damborenea, S. E., Manceñido, M. O., & Riccardi, A. C. (1975). Biofacies y estratigrafía del liásico de Piedra Pintada, Neuquén, Argentina. *Actas I Congreso Argentino de Paleontología y Bioestratigrafía, 2,* 173–228.

Escobar, F. (1980). Paleontología y bioestratigrafía del Triásico superior y Jurásico inferior en el área de Curepto, Provincia de Talca. *Bol. Invest. Geol. Chile, 35,* 1–78.

Feruglio, E. (1937). Palaeontographia Patagonica. *Mem. Inst. Geol. Univ. Padova, 11,* 1–384.

Fleming, C. A. (1958). Upper Jurassic fossils and hydrocarbon traces from the Cheviot Hills, North Canterbury. *N.Z. J. Geol. Geophys. 1*(2), 375–94.

(1959). *Buchia plicata* (Zittel) and its allies, with a description of a new species, *Buchia hochstetteri. N.Z. J. Geol. Geophys., 2*(5), 889–904.

(1960). The Upper Jurassic sequence at Kawhia, New Zealand with reference to the ages of some Tethyan guide fossils. *Rep. 21st Internat. Geol. Cong., 21,* 264–9.

Fleming, C. A., & Kear, D. (1960). The Jurassic sequence at Kawhia Harbour, New Zealand (Kawhia Sheet, N73). *Bull. N.Z. Geol. Surv., 67,* 1–50.

Frebold, H. (1964). Illustrations of Canadian fossils. Jurassic of western and Arctic Canada. *Paper, Geol. Surv. Canada, 63*(4), 1–107.

Fuenzalida P. R., & Covacevich C. V. (1988). Vulcanismo y bioestratigrafía del Jurásico superior y Cretácico inferior en la Cordillera Patagónica, región de Magallanes, Chile. *Actas 5 Cong. Geol. Chileno, 3,* H159–83.

Grant-Mackie, J. A. (1959). Hokonui stratigraphy of the Awakino–Mahoenui area, southwest Auckland. *N.Z. J. Geol. Geophys., 2,* 755–87.

Grant-Mackie, J. A., Hudson, N., & Helby, R. (1986). The Murihiku Callovian–Oxfordian "hiatus" seems to be closed. *Geol. Soc. N.Z., Misc. Publ. 35A,* 55.

Hayami, I. (1960). Jurassic inoceramids in Japan. *J. Fac. Sci. Univ. Tokyo, Sect. 2, 12*(2), 277–328.

(1985). Range chart of selected bivalve species of Japanese Jurassic. In *1985 Circum-Pacific Jurassic Field Conference, III* (pp. 44–50).

Helby, R., Wilson, G. J., & Grant-Mackie, J. A. (1988). A preliminary study of Middle to Late Jurassic dinoflagellate assemblage from Kawhia, New Zealand. *Mem. Assoc. Australas. Palaeont., 5,* 125–66.

Hudson, N. (1983). *Stratigraphy of the Ururoan, Temaikan, and Heterian Stages; Kawhia Harbour to Awakino Gorge, South-West Auckland.* Unpublished masters in science thesis, University of Auckland, 1–168, pl. 1-2.

Hudson, N., Grant-Mackie, J. A., & Helby, R. (1987). Closure of the New Zealand "Middle Jurassic Hiatus"? *Search, 18,* 146–8.

Imlay, R. W. (1959). *Succession and Speciation of the Pelecypod Aucella.* U.S. Geological Survey Professional Paper, 314-G (pp. 155–69). Washington.

(1967). *The Mesozoic Pelecypods* Otapiria *Marwick and* Lupherella *Imlay, New Genus, in the United States.* U.S. Geological Survey, Professional Paper 573-B.

Imlay, R. W., & Detterman, R. L. (1973). *Jurassic Paleobiogeography of Alaska.* U.S. Geological Survey Professional Paper 801.

Imlay, R. W., & Jones, D. L. (1970). *Ammonites from the Buchia Zones in Northwestern California and Southwestern Oregon.* U.S. Geological Survey, Professional Paper 647-B.

Jeletzky, J. A. (1965). Late Upper Jurassic and early Lower Cretaceous fossil zones of Canadian Western Cordillera, British Columbia. *Bull. Geol. Surv. Canada, 103,* 1–70.

(1966). Upper Volgian (latest Jurassic) ammonites and buchias of Arctic Canada. *Bull. Geol. Surv. Canada, 128,* 1–51.

(1984). Jurassic–Cretaceous boundary beds of western and Arctic Canada and the problem of the Tithonian–Berriasian stages in the Boreal realm. In G. E. G. Westermann, (ed.), *Jurassic–Cretaceous Biochronolgy and Paleography of North America. Geol. Assoc. Canada, Spec. Paper, 27,* 175–256.

Jones, D. L., Bailey, E. H., & Imlay, R. W. (1969). Structural and stratigraphic significance of the *Buchia* Zones in the Colyear Springs–Paskenta Area, California. *U.S. Geological Survey, Professional Paper* 647-A.

Leanza, A. F. (1942). Los pelecípodos del Lias de Piedra Pintada, en el Neuquén. *Rev. Museo Plata (n.s.), Paleont.,* 2 (10), 143–206.

Li Xiaochi & Grant-Mackie, J. A. (1988). Upper Jurassic and Lower Cretaceous *Buchia* (Bivalvia) from southern Tibet, and some wider considerations. *Alcheringa, 12,* 249–68.

Marwick, J. (1951). Series and stage divisions of New Zealand Triassic and Jurassic rocks. *N.Z. J. Sci. Techn. B32*(3), 8–10.

(1953). Divisions and faunas of the Hokonui System (Triassic and Jurassic). *Palaeont. Bull. N.Z. Geol. Surv., 21,* 1–142.

Paraketzov, K. V. (1980). The problem of the Upper Jurassic and Lower Cretaceous buchia zonation of the North-East of the USSR. In *Biostratigraphy and Correlation of the Mesozoic Sediments of the North-East USSR* (in Russian) (pp. 91–106). Magadan.

Pérez, E., & Reyes, R. (1977). Las trigonias jurásicas de Chile y su valor cronoestratigráfico. *Bol. Inst. Invest. Geol. Chile, 3,* 1–58.

Polubotko, I. V., & Repin, Y. S. (1988). Lower and Middle Jurassic of the North-East. In G. E. G. Westermann & A. C. Riccardi (eds.), *Jurassic Taxa Ranges and Correlation Charts for the Circum-Pacific. 1. Soviet Union. Newsl. Strat., 19,* 1–17.

Polubotko, I. V., & Sey, I. I. (1981). Division of Middle Jurassic deposits of the USSR eastern part on mytilocerams (in Russian). *Proc. Acad. Sci. USSR, Geol. Ser., 12,* 63–70.

Purser, B. H. (1961). Geology of the Port Waikato Region (Onewhero Sheet N 51). *Bull. N.Z. Geol. Surv., 69,* 1–36.

Riccardi, A. C. (1984). Las asociaciones de amonitas del Jurásico y Cretácico de la Argentina. *Actas IX Congr. Geol. Argentino, 4,* 559–95.

Riccardi, A. C., Damborenea, S. E., & Manceñido, M. O. (1990). Lower Jurassic of South America and Antarctic Peninsula. In G. E. G. Westermann & A. C. Riccardi (eds.), *Jurassic Taxa Ranges and Correlation Charts for the Circum Pacific. 3. South America and Antarctic Peninsula. Newsl. Strat., 21*(2), 75–104.

Riccardi, A. C., Damborenea, S. E., Manceñido, M. O., & Ballent, S. C. (1988). Hettangiano y Sinemuriano marinos en Argentina. *Actas V Congr. Geol. Chileno, 2,* C359–73.

Riccardi, A. C., Westermann, G. E. G., & Elmi, S. (1989). The Middle Jurassic Bathonian–Callovian Ammonite Zones of the Argentine–Chilean Andes. *Géobios, 22*(5), 553–97.

Sato, T., Westermann, G. E. G., Skwarko, S. K., & Hasibuan, F. (1978). Jurassic biostratigraphy of the Sula Islands, Indonesia. *Bull. Geol. Surv. Indonesia, 4*(1), 1–28.

Sey, I. I. (1984). Late Pliensbachian bivalvia of the Bureya Trough. In *New Data on Detailed Biostratigraphy of Phanerozoic of the Far East* (pp. 86–97). Vladivostok.

(1986). Oxfordian and Kimmeridgian buchiids of Western Okhotsk area (Far East) (in Russian) In *Biostratigraphy of Siberia and Far East* (pp. 127–33). Novosibirsk: Nauka.

Sey, I. I., & Kalacheva, E. D. (1980). Biostratigraphy of the Lower and Middle Jurassic of the Far East. VSEGEI Transact., n.s., 285. *L. Nedra,* 1–186.

(1985). Scheme of biostratigraphy of the marine deposits of the northern part of the Far East (in Russian). *Geol. Geophys., 5,* 136–8.

(1988). Ammonites and bivalves of the Far East. In G. E. G. Westermann & A. C. Riccardi (eds.), *Jurassic Taxa Ranges and Correlation Charts for the Circum-Pacific. 1. Soviet Union. Newsl. Strat., 19*(1–2), 35–65.

Sokolov, D. N. (1946). Algunos fósiles suprajurásicos de la República Argentina. *Rev. Asoc. Geol. Argentina, 1*(1), 7–16.

Speden, I. G. (1970). Three new inoceramid species from the Jurassic of New Zealand. *N.Z. J. Geol. Geophys. 13*(3), 825–51.

Stevens, G. R. (1965). The Jurassic and Cretaceous belemnites of New Zealand and a review of the Jurassic and Cretaceous belemnites of the Indo-Pacific Region. *Palaeont. Bull. N.Z. Geol. Surv., 36,* 1–231.

(1978). Jurassic paleontology. In R. P. Suggate, G. R. Stevens, & M. T. Te Punga (eds.), *The Geology of New Zealand. N.Z. Geol. Surv., 1,* 215–28.

Stevens, G. R., & Speden, I. G. (1978). New Zealand. In M. Moullade & A. Nairn, (eds.), *The Phanerozoic Geology of the World. II. The Mesozoic, A.* (pp. 251–328). Amsterdam: Elsevier.

Wang Yi-gang (1988). China. In G. E. G. Westermann & A. C. Riccardi (eds.), *Jurassic Taxa Ranges and Correlation Charts for the Circum-Pacific. Newsl. Strat. 19*(1–2), 95–130.

Wang Yi-gang & Sun Dong-li (1983). A survey of the Jurassic System of China. *Can. J. Earth Sci., 20*(11), 1646–56.

Zakharov, V. A. (1981). Buchiids and biostratigraphy of the boreal Upper Jurassic and Neocomian (in Russian). *Nauka,* 1–272.

Zakharov, V. A., Paraketzov, K. V., & Paraketzova, G. I. (1988). Callovian and Upper Jurassic of the North-East of USSR. In G. E. G. Westermann & A. C. Riccardi (eds.), *Jurassic Taxa Ranges and Correlation Charts for the Circum-Pacific. Newsl. Strat., 19*(1–2), 19–34.

# 17 Belemnites of the southwest Pacific

A. B. CHALLINOR

Belemnitida are straigraphically useful within the southwest Pacific region both for local correlation and for testing hypotheses on interregional correlations.

### Indonesia and Papua New Guinea

Several genera and species, all members of the suborder Belemnopseina Jeletzky, have been recognized widely within eastern Indonesia and Papua New Guinea (Kruizinga 1920; Stolley 1929; Stevens 1965; Challinor and Skwarko 1982; Challinor 1991), and zones based on them have regional significance. Easily recognized, partly overlapping assemblages (Figure 17.1) characterized either by their dominant genera (assemblages 1 and 3) or by genera absent or poorly reperesented in other intervals (assemblage 2) are as follows:

1. *Dicoelites–Conodicoelites* assemblage (late Bajocian–early Oxfordian); *Belemnopsis* is common in the Callovian–Oxfordian
2. *Hibolithes* assemblage (late Callovian–latest Oxfordian); *Belemnopsis* present throughout
3. *Belemnopsis* assemblage (basal Oxfordian–latest Tithonian) equivalent to the range of the *moluccana* lineage (Challinor 1991); scattered *Hibolithes* are present in the Tithonian

Only assemblages 1 and 3 have thus far been recognized in Papua New Guinea.

Closer correlation may be achieved utilizing individual species. The Callovian *Dicoelites* is distinctive and has been recognized at other localities in eastern Indonesia (Challinor 1991). Early Oxfordian to late Tithonian *Belemnopsis* of the *moluccana* lineage are widely distributed (Challinor 1989) and form an evolving group whose species are useful for closer correlation. Interspecific changes allow more precise correlation than can be attained by using individual species (Challinor 1989).

See Appendix, Plates 128–132.

### New Zealand

Local stages are used in New Zealand, and problems in correlation with international stages exist, particularly in the Late Jurassic. Alternative correlations are indicated in Figure 17.1, where New Zealand A follows Stevens (1978), and New Zealand B follows more recent workers such as Helby, Wilson, and Grant-Mackie (1988).

The suborder Belemnitina Jeletzky is present from early Toarcian to middle Bathonian. *Cylindroteuthis*(?) occurs in the early Toarcian, early Bajocian, and early Bathonian, and *Brachybelus* in the late Bajocian–early Bathonian (Stevens 1965; Hudson 1983), but occurences are few, and distributions probably are not completely known.

The suborder Belemnopseina Jeletzky appears in the early Bajocian and extends into the late Tithonian, but no genus has a continuous distribution. The Late Jurassic, in particular, is marked by rapid replacement of both genera and species, many of which are useful for local correlation (Stevens 1965; Challinor 1975, 1977, 1979a,b, 1980).

Belemnitida of the New Zealand and Indonesian regions are considered to represent different stocks derived from different centers of distribution (see Chapter 23). This is indicated by the absence of the suborder Belemnitina from Indonesia, by regional infrageneric differences (see Chapter 23), by disjunct generic distributions no matter which correlation between New Zealand and international stages is adopted (Figure 17.1), and by a complete lack of species common to both regions.

New Caledonia cannot be assigned with confidence to either regional fauna and appears to contain elements of both the New Zealand and Indonesian assemblages (see Chapter 23), but none of its species occurs in the other regions (Challinor & Grant-Mackie 1989).

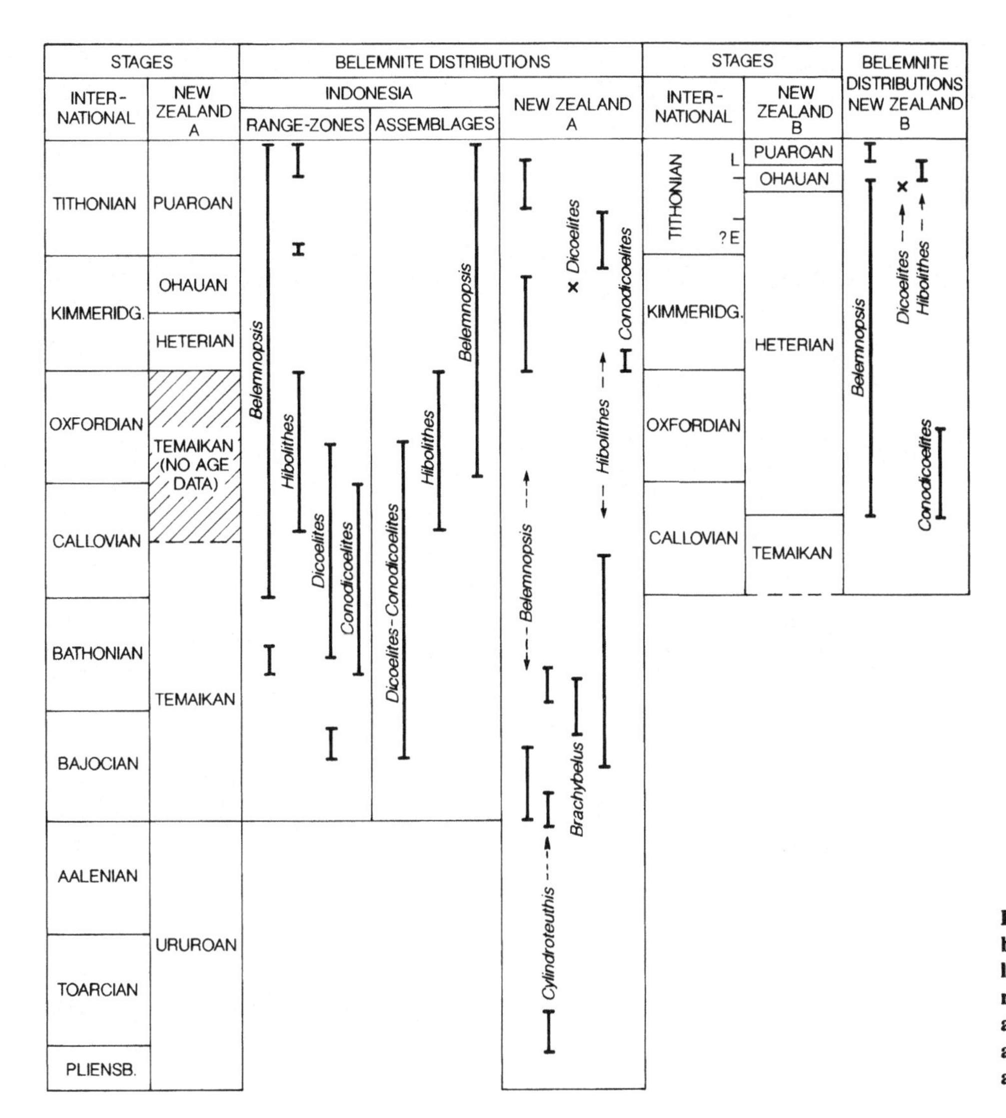


**Figure 17.1. Time distribution of belemnite genera from New Zealand and Indonesia. Alternative correlations of New Zealand stages; A, according to Stevens et al. (1980); B, according to Hudson et al. (1987) and Helby et al. (1988).**

## References

Challinor, A. B. (1975). New Upper Jurassic belemnite from southwest Auckland, New Zealand. *N.Z. J. Geol. Geophys.*, *18*, 361–71.

(1977). New Lower or Middle Jurassic belemnite from southwest Auckland, New Zealand, *N.Z. J. Geol. Geophys.*, *20*, 249–62.

(1979a). The succession of *Belemnopsis* in the Heterian stratotype, Kawhia Harbour, New Zealand. *N.Z. J. Geol. Geophys.*, *22*, 105–23.

(1979b). Recognition and distribution of Heterian *Belemnopsis* in southwest Auckland. *N.Z. J. Geol. Geophys.*, *22*, 267–75.

(1980). Two new belemnites from the lower Ohauan (?middle Kimmeridgian) Stage, Kawhia Harbour, New Zealand. *N.Z. J. Geol. Geophys.*, *23*, 257–65.

(1989). The succession of *Belemnopsis* in the Late Jurassic of eastern Indonesia. *Paleontology, 32*, 571–96.

(1991). Revision of the belemnites of Misool and a review of the belemnites of Indonesia. *Paleontographica*, *A218*, 87–164.

Challinor, A. B., & Grant-Mackie, J. A. (1989). Jurassic Coleodea of New Caledonia. *Alcheringa*, *13*, 269–304.

Challinor, A. B., & Skwarko, S. K. (1982). Jurassic belemnites from Sula Islands, Moluccas, Indonesia. *Geological Research and Development Centre, Indonesia, Paleontology Series*, *3*, 1–89.

Helby, R., Wilson, G., & Grant-Mackie, J. A. (1988) A preliminary biostratigraphic study of Middle to Late Jurassic dinoflagellate assemblages from Kawhia, New Zealand. *Mem. Assoc. Australas. Palaeont.*, *5*, 125–66.

Hudson, N., Grant-Mackie, J. A., & Helby, R. (1987). Closure of the New Zealand "Middle Jurassic hiatus"? *Search*, *18*, 146–8.

Kruizinga, P. (1920). Die Belemniten uit de Jurassische afzettingen van de Soela-Eilanden. *Jaarboek van het Mijnwezen in Nederlandsch OOst-Indie*, *49*, 161–89.

Stolley, E. (1929). Uber Ostindische Jura-Belemniten. *Palaeontology of Timor*, *16*, 91–213.

Stevens, G. R. (1965). The Jurassic and Cretaceous belemnites of New Zealand and a review of the Jurassic and Cretaceous belemnites of the Indo-Pacific region. *N.Z. Geol. Surv., Paleont. Bull.*, *36*, 1–231.

Stevens, G. R. (1978). Jurassic paleontology. In R. P. Suggate, G. R. Stevens, & M. T. Te Punga (eds.), *The Geology of New Zealand*, *1*, 215–28.

# Part V: Biogeography

# 18 Macroflora of eastern Asia and other circum-Pacific areas

T. KIMURA, E. L. LEBEDEV, E. M. MARKOVICH, and V. A. SAMYLINA

Studies of Mesozoic plants have for a long time centered on Europe, where every Mesozoic flora was regarded as a world standard, such as the Yorkshire flora. Studies of the Mesozoic plants and floras of the circum-Pacific regions have advanced rapidly during the past three decades, and many new data are now available. Representative specimens are illustrated in Plates 96–98.

In Japan, Mesozoic plants are generally abundant, though not throughout the section, and they occur mostly in marine strata or in plant beds sandwiched between marine beds (Kimura 1987, 1988). Elsewhere in the Pacific region, Mesozoic plant beds are nonmarine and thus difficult to date precisely. In most Mesozoic plant assemblages of Japan the details of reproductive organs of ferns and cuticles of gymnosperms have been lost because of tectonic and repeated igneous activities. Most Mesozoic floras of Japan, of every period and stage, may nevertheless be regarded as the most important ones in the Pacific.

In this chapter, Kimura reviews the current status of the Japanese Jurassic floras and that of coeval floras in other cirum-Pacific regions, and Lebedev, Markovich, and Samylina review the Jurassic floras in the Pacific sector of Russia.

## Early Jurassic

### *Japan*

Early Jurassic plants are known from the Higashinagano Formation and the lower part of Shizugawa Group, but they are rare and insignificant (Figure 18.1).

The upper Liassic bears two distinct floras (Kimura and Tsujii 1980–4; Kimura, Naito, and Ohana 1986; Kimura, Ohana, and Tsujii 1988): (1) The Kuruma-type flora is known from the lower Pliensbachian Negoya Formation and lower Toarcian Shinadani Formation of the Kuruma Group and their equivalents in the Inner Zone of Japan. It is characterized by an abundance of varied osmundaceous ferns, *Ptilophyllum, Nilssonia, Sphenobaiera,* and *Podozamites,* and the presence of abundant *Marattia, Phlebopteris, Dictyophyllum, Cladophlebis, Phoenicopsis,* and *Taeniopteris. Zamites* and conifers with scale-like leaves have not been found except for ''*Widdringtonia*''-like leafy shoots. The presence of varied *Ptilophyllum* species and other thermophilic plants in the Kuruma-type floras of Japan and China (Xiangxi flora) would correspond to the Toarcian amelioration suggested by Vakhrameev (1987). (2) The Nishinakayama-type flora, ranging from Pliensbachian to early Toarcian, is known from the Nishinakayama Formation, Toyora district in southwest Japan. This unique flora is characterized by doubtful ferns, *Ctenozamites, Otozamites, Zamites, Pseudoctenis, Brachyphyllum, Cupressinocladus, Elatides, Geinitzia,* and *Araucarites.* Dipteridaceous ferns, ginkgoaleans, czekanowskialeans, and *Podozamites* have not been found in spite of our intensive collecting for 30 years. Plant species common to the floras of both areas are unknown, resembling the difference in ammonites. Coal seams are present in the Kuruma Group and its equivalents, but not in the Nishinakayama Formation. Redbeds and evaporites are absent from the latter.

It appears, therefore, that the Toyora Group, consisting of the Higashinagano, Nishinakayama, and Utano Formations and containing the second flora, is allochthonous. The actual boundary between the Jurassic Toyora Group and the marine Triassic, plant-bearing Mine Group has not been recognized in the Toyora district.

### *East Asia*

The main Liassic plant sites in eastern Asia were listed by Kimura (1988) and Kimura et al. (1988). According to Vakhrameev (1964, 1987), Early Jurassic floras in eastern Siberia are characterized by an abundance of ginkgoaleans and czekanowskialeans and the sporadic presence of a few bennettitaleans and cycadaleans; conifers with scale-like leaves have not been found.

*Pacific sector of Russia.* The Hettangian flora of southern Primorye (Figure 18.2, loc. 1) (Krassilov and Schorochova 1973) bears abundant *Phoenicopsis* and *Czekanowskia,* indicating the Siberian Palaeofloristic Area. However, it includes representatives

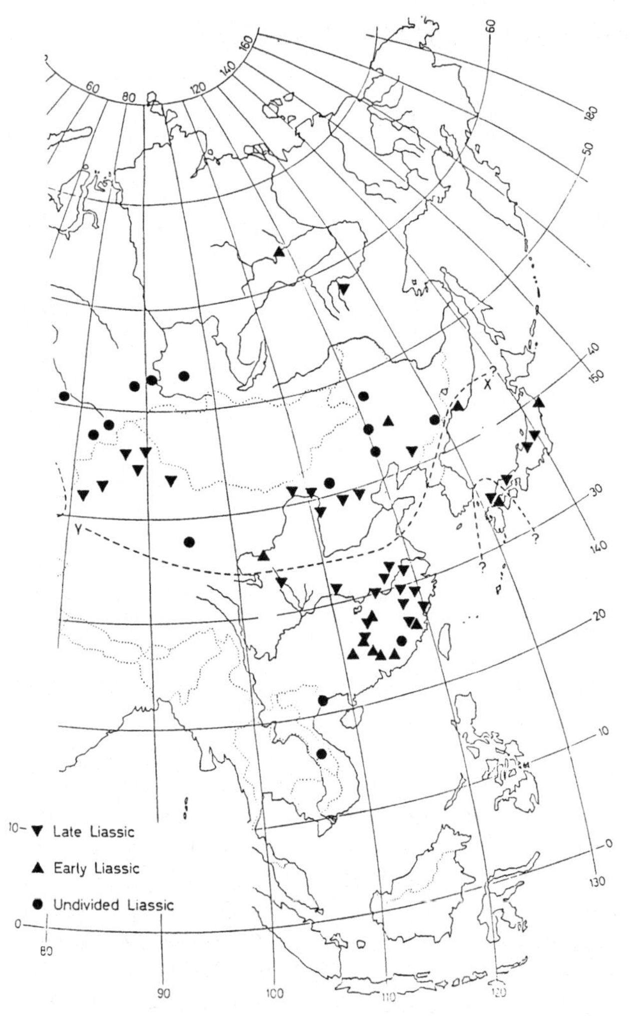


**Figure 18.1. Main Early Jurassic plant sites of eastern Asia and Southeast Asia.**

from the southernmost Euro–Sinian Area. This flora is dated by the associated ammonites, such as *Franziceras* and *Caloceras*, found in the lower part of the section. In addition, the presence of *Marattiopsis hoerensis* and *Phlebopteris angustiloba* permits correlation with the *Thaumatopteris* Zone of Greenland (Harris 1946).

A small flora in the Okhotsk area (Figure 18.2, loc. 5) includes *Cladophlebis*, *Nilssonia*, and *Czekanowskia* (Lebedev 1987). Several small Early Jurassic floras dated by faunas are known from the coastal-marine deposits of North-East Russia (Samylina and Efimova 1968). In the Korkodon Basin (loc. 9), the lower Liassic has yielded *Thaumatopteris* and *Cladophlebis*. *Taeniopteris* leaves were collected from the middle Liassic along the Russkaya River (loc. 10). On the Levy Kedon River (loc. 11), the Toarcian includes "*Dicroidium*" and *Ptilophyllum*. In the Bolshoi–Maly Anyui Interfluve (loc. 14), *Neocalamites*, *Marattiopsis*, *Cladophlebis*, and others are known from the plant beds of probably Liassic age (Samylina and Efimova 1968).

*China*. Early Jurassic floras are extensively distributed in both northern and southern China (Kimura 1984; Kimura et al. 1988). The Guanyintan flora (Zhou 1984) in the early Liassic floras of southern China is of the Kuruma type, but is characterized by varied *Otozamites* and *Brachyphyllum*. The late Liassic Xiangxi flora (Wu, Ye, and Li 1980) of southern China apparently is also of the Kuruma type. In contrast, the Early Jurassic floras known from northern China are of the Siberian type.

### *South America*

Herbst (1965) reviewed the supposedly Liassic floras from the nonmarine plant beds of Santa Cruz, Chubút, Neuquén, and Mendoza provinces, Argentina. They are characterized by an abundance of varied *Otozamites* and *Ptilophyllum* and the presence of dicksoniaceous and dipteridaceous ferns, bennettitaleans such as *Zamites* and *Dictyozamites*, and conifers such as *Araucaria*, *Podocarpus*, *Brachyphyllum*, and *Pagiophyllum*. Ginkgoaleans and czekanowskialeans apparently have not been recorded. Herbst (1965) compared these floras with those of New Zealand and Australia, although the latter are poorly known.

The list of plants presented by Herbst (1965) indicates that the Liassic floras of South America are of the Nishinakayama type, except for the presence of dipteridaceous ferns. Typical Gondwana-type elements are absent except for *Araucaria* and *Podocarpus*.

## Middle Jurassic

### *Japan*

The fossiliferous, marine Utano Formation of the Toyora district ranges from late Toarcian to Callovain/early Oxfordian, but ammonites are not associated with the plants. Generally ammonites are restricted to the northeastern half and plants to the southwestern half of the district. The Utano flora was described by Kimura and Ohana (1987a,b) from three main localities (063, 065, and 061). They belong to the lowest, middle, and uppermost parts, respectively, of the middle Utano Formation.

The flora consists of younger Mesozoic-type taxa. The boundary between the older and younger Mesozoic-type floras in Japan is about the middle Toarcian. The Utano flora is the only known Dogger flora of Japan and is fundamentally of the Ryoseki type, but a few Tetori-type elements are included: *Coniopteris* sp., *Onychiopsis elongata*, *Dictyozamites naitoi*, *Ginkgoites* ex gr. *sibiricus* and *Czekanowskia* ex gr. *rigida* (the first and last two represented by single specimens).

### *East Asia*

*Pacific sector of Russia*. A possible Dogger flora occurs in the Dzhelon Formation at the southern coast of the Sea of Okhotsk (Figure 18.2, loc. 4) (Lebedev 1973), including *Cladophlebis*, *Czekanowskia*, *Ferganiella*, and others. *Ferganiella* appeared in the Upper Triassic and was widespread in the Lower and Middle Jurassic of central Asia, as well as in the Dogger of the Tuva and Eastern Transbaikal areas.

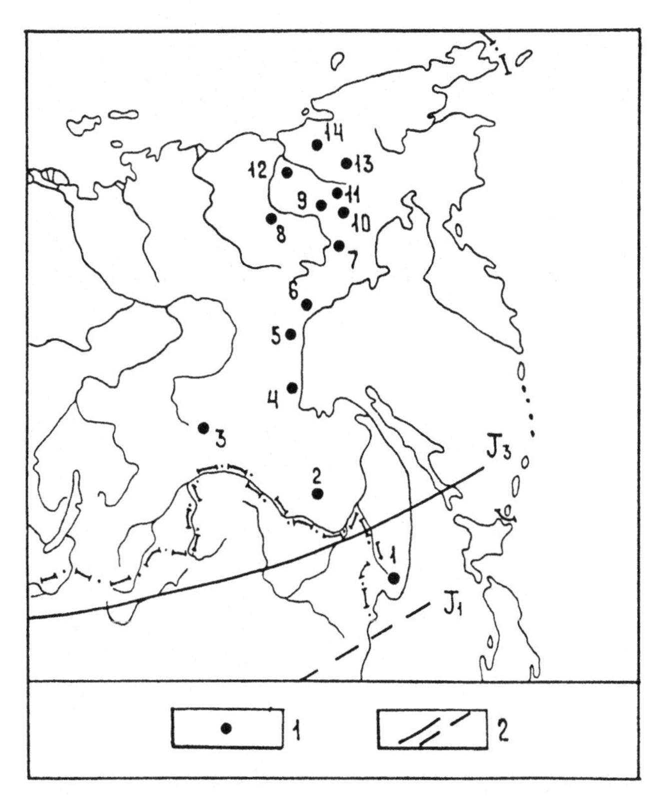

**Figure 18.2. Localities with Jurassic floras in the Pacific sector of Russia: 1, localities, as explained in the text; 2, boundaries between the Siberian and Euro-Sinian Palaeofloristic Areas in Early and Late Jurassic.**

*China.* There are many Middle Jurassic plant sites in China (Kimura and Ohana 1987a, p. 48). Most are nonmarine deposits, except for the partly marine lower part of the Longzhaogou Group in the southeastern part of northeastern China. As a whole, the floras north of the Yangtze (or Yangzi) River differ in composition from the Utano flora and resemble those of Siberia.

In southern China, the poorly known fossils are characterized by varied dicksoniaceous ferns and northern-type taxa, except for *Ptilophyllum* sp. No floras resembling the Utano flora are known from China.

### *Middle and South America*

The Middle Jurassic flora of Oaxaca described by Person and Delevoryas (1982) is rather close to the Utano flora, because it contains a few northern-type taxa such as *Coniopteris* and many southern-type taxa.

In Argentina, several floras said to be Middle Jurassic were reviewed by Herbst (1965). They contain dipteridaceous ferns, *Ptilophyllum*, and conifers such as *Podocarpus*, *Araucaria*, and *Pagiophyllum*.

### *Australia*

The Waloon Coal Measures flora (Gould 1981) is characterized by an abundance of varied *Pachypteris*, *Ptilophyllum*, and conifers belonging to Araucariaceae and Podocarpaceae, and by the absence of ginkgoaleans.

In general, Middle Jurassic floras of the Southern Hemisphere seem to be characterized by an abundance of varied podocarpaceous and araucariaceous conifers. These Jurassic floras are not as nearly unique as suggested previously, although they differ regionally in floristic composition according to climatic and geographic conditions.

## Late Jurassic

### *Japan*

There are two distinct floras, the Tetori-type and Ryoseki-type floras (Figure 18.3) The Tetori type occurs in the Inner Zone of central Japan, in the marine Kuzuryu Group and the nonmarine Lower Cretaceous Itoshiro Group. This flora consists of abundant gleicheniaceous and varied dicksoniaceous ferns and such characteristic ferns as *Onychiopsis elongata* and *Raphaelia*, as well as common and varied *Dictyozamites*, *Otozamites*, *Neozamites*, *Nilssonia* spp., including *N. lobatidentata*, czekanowskialeans, ginkgoaleans, *Podozamites*, and conifers with needle-like leaves. Apparently absent are *Aldania* and *Heilungia*, characteristic for the Siberian region, as well as matoniaceous ferns, *Ptilophyllum*, *Zamites*, and *Nilssonia*, with long and very narrow lamina (*N. schaumburgensis*-type leaves), and conifers with scale-like leaves.

The Ryoseki-type floras occur in the Outer Zone of Japan and are known from the Upper Jurassic Shishiori, Oshika, and Somanakamura Groups in northeastern Japan and the Lower Cretaceous plant beds (Kimura 1984, 1987, 1988). About 100 plant taxa have been recognized (e.g., Kimura and Hirata 1975; Kimura 1988). This flora is characterized by the following: abundant gleicheniaceous and matoniaceous ferns (*Weichselia* in Lower Cretaceous) and such characteristic ferns as *Onychiopsis yokoyamai* and *Acrostichopteris;* rare dicksoniaceous ferns with single species of *Eboracia* and *Conipteris* (Upper Jurassic–Lower Cretaceous plant beds in northeastern Japan) and ferns of uncertain affinity (e.g., *Cladophlebis*) with multipinnate leaves bearing small-sized pinnules with convex upper surface and reflexed margins; abundant and varied bennettitaleans, *Zamites* and *Ptilophyllum*, and the cycadalean *Nilssonia;* abundant conifers, *Brachyphyllum*, *Frenelopsis*, and *Cupressinocladus*, all with scale-like leaves, and locally common *Araucaria* and *Podocarpus* (*Nageia*). Apparently absent are czekanowskialeans, ginkgoaleans, and *Podozamites* [many leafy shoots recorded by Oishi (1940) as *Podozamites lanceolatus* from the Upper Jurassic Somanakamura Group almost certainly belong to *Podocarpus* (*Nageia*) *ryosekiensis*].

It is noteworthy that no species have been found in common between the Tetori-type and coeval Ryoseki-type floras (Kimura 1984, 1987, 1988).

Cuticles of *Zamites*, *Ptilophyllum*, *Nilssonia*, and *Frenelopis* occur in the marine Lower Cretaceous Choshi Group. They are generally thinner than those of coeval floras in South America,

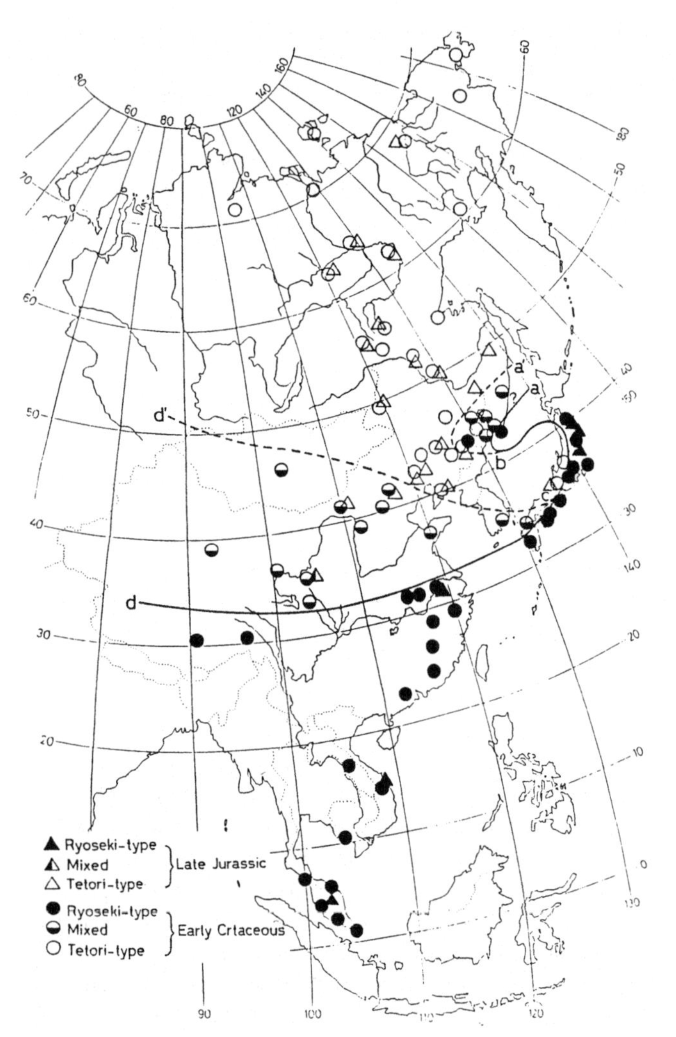


**Figure 18.3. Main Late Jurassic–Early Cretaceous plant localities and phytogeographic boundaries in eastern Asia and Southeast Asia (after Kimura 1988): line a–b–c–d, boundary between Tetori-type and Ryoseki-type floras; areas between the a–b and a′–b lines and c–d and c–d′ lines are mixed zones; line x–y, boundary between the Siberian-Canadian and Euro-Sinian Regions, according to Vakhrameev (1987, 1988).**

Australia, India, the Middle East, and Europe. In addition, the papillae of *Zamites* and *Ptilophyllum* generally are undeveloped, and their outer surface of epidermal cells is exceptionally smooth, contrasting with the heavily cutinized and papillate epidermal cells of bennettitaleans known from other regions where "redbeds" or evaporites are extensively developed. These cuticular features suggest a maritime rather than arid or continental climate. This is supported by the marine origin of most plant-bearing beds in the Outer Zone of Japan. "Red" or variegated plant beds have not been found, except for the thin and local redbed in the lowest barren horizon of the Lower Cretaceous Ryoseki Formation in southwestern Japan. Most plant-bearing beds in the Tetori Supergroup consist of black shales or medium-fine-grained sandstones and coal seams.

The boundary between the Tetori-type and Ryoseki-type floras is quite distinct and sharp in most of Japan (Figure 18.3). In the Toyora district, however, the Late Liassic Nishinakayama flora is uniquely of the Ryoseki type; the Middle Jurassic Utano flora is fundamentally of the Ryoseki type, but includes a few Tetori-type taxa; and the latest Jurassic (or earliest Cretaceous) Kiyosue flora includes both Tetori-type and Ryoseki-type taxa. Similar Early Cretaceous mixed floras are recognized in South Korea (the Nagdong or Naktong flora) and in northern and northwestern China, as well as in southern Primorye and the eastern border region between northeastern China and western Primorye (Figure 18.3). These mixed floras supposedly evolved in the Early Cretaceous, except for the Toyora district of Japan (Kimura 1987).

*Continental East Asia*

A number of Late Jurassic–Early Cretaceous plant sites are known from eastern Siberia and China. Vakhrameev (1964) introduced the concept of the Siberian and Indo-European Palaeophytogeographic Areas in Eurasia during the Jurassic and Early Cretaceous. Later, Vakhrameev (1971) divided his Siberian Palaeophytogeographic Area during Late Jurassic–Early Cretaceous into the Lena and Amur Subprovinces. The coeval plant sites in northeastern China and Japan were included in this Indo-European Palaeophytogeographic Area, respectively, as the Warm-Temperate Siberian (Late Jurassic) and Siberian–Canadian (Early Cretaceous) Regions, and as the subtropical to tropical Euro-Sinian region (Late Jurassic–Early Cretaceous). He consistently included both the Tetori-type and Ryoseki-type plant sites of Japan in the East Asian Province of the Euro-Sinian Region.

*Pacific sector of Russia.* Malm floras are most abundant and diverse. In the Amur Basin (Figure 18.2, loc. 2) along the Bureya River, the Talyndzan and Dublikan Formations yielded *Coniopteris, Eboracia, Cladophlebis, Raphaelia, Anomozamites, Nilssonia, Heilungia, Ctenis, Sphenobaiera, Phoenicopsis, Pseudotorellia, Czekanowskia,* and others (Vakhrameev and Doludenko 1961; Krassilov 1972). The Dep and Ayak Formations along the Zeya River (Lebedev 1965) include the characteristic *Cladophlebis, Raphaelia,* and *Heilungia.* To the north, along the upper Burgagylkan River (Figure 18.2, loc. 5), deposits underlying Cretaceous volcanics of the Okhotsk-Chukotka Belt yielded a Late Jurassic assemblage with *Cladophlebis, Raphaelia, Heilungia, Phoenicopsis, Czekanowskia,* and others (Lebedev 1987).

Many fossil plant sites are known in North-East Russia. In the Anyui Basin (Figure 18.2, loc. 12) the Pezhenka flora from marine Volgian (Paraketsov 1970) contains *Coniopteris, Cladophlebis, Heilungia, Czekanowskia,* and *Pheonicopsis* (Samylina 1974). Beds along the Kolyma River, resting on Volgian with *Buchia* (loc. 8), yielded *Cladophlebis* and *Heilungia.* Along the Omolon River (loc. 12), *Cladophlebis* and *Raphaelia* occur below *Buchia*-bearing beds (Vakhrameev 1964). The lower Ozhogina Formation of the Kolyma River in the Silyap Basin (loc. 8) has yielded characteristic *Equisetites, Raphaelia, Coniopteris, Czekanowskia,* and *Phoenicopsis* (Samylina 1974).

Coal-bearing deposits of southern Yakutia contain Early to Late Jurassic floras (Markovich 1985).

*China.* The Late Jurassic–Early Cretaceous floras of northeastern China are of the Tetori type, except for those in the southeastern part of northeastern China. In fact, the floristic composition of the Tetori-type floras in the Inner Zone of Japan and in most parts of northeastern China differ more or less from that of the coeval floras in Vakhrameev's Siberian or Siberian–Canadian Region. The latter floras have been less varied bennettitaleans and cycadaleans, except for *Nilssonia,* and include such characteristic genera as *Aldania* and *Heilungia* and varied czekanowskialeans and gikgoaleans; the former floras have varied and abundant gleicheniaceous ferns and *Dictyozamites.*

It is usually difficult to recognize the Jurassic–Cretaceous boundary in nonmarine sequences. Many Jurasso-Cretaceous plant sites are known from northeastern, northern, and northwestern China, but their precise ages are still under discussion.

The strong differences between the Tetori-type and Ryoseki-type floras make the inclusion of both types of floras into a single phytogeographic province inappropriate (Kimura's opinion). It appears that the Tetori-type floras of Japan and northeastern China inhabited the southern margin of Vakhrameev's Siberian Region (Figure 18.3) (Kimura 1987, 1988). The sharp boundary between the Tetori-type and Ryoseki-type plant distributions in Japan corresponds to the Median Tectonic Line in southwest Japan. The marked floristic boundary and the presence of the mixed zones (Figure 18.3) can be explained tectonically: Major parts of Japan and adjacent areas were well apart during the Jurassic and accreted in the Early Cretaceous.

### *Malaysia*

A flora of supposedly Late Jurassic age was described by Smiley (1970) from Maran, West Malaysia. It is clearly of the Ryoseki type.

### *North America*

Floras of supposedly Late Jurassic age are known from the Kootenay and Morrison Formations of western interior Canada and the United States (Bell 1956; Miller 1987). Their compositions are fundamentally of the Tetori type (Kimura's opinion), but both include several Ryoseki-type elements. The flora from Douglas County, Oregon (ward 1905), is similar.

### *Antarctic Peninsula*

The age of the flora from Hope Bay (Halle 1913) is still controversial. It appears to be Late Jurassic and, as typical for the Southern Hemisphere, of the mixed type.

### *Australia*

The Talbragar Fish-Bed Flora in New South Wales, revised by White (1981), includes such characteristic genera as *Agathis, Rissikia, Pentoxylon,* and *Allocladus,* but the precise age remains uncertain.

## References


Bell, W. A. (1956). Lower Cretaceous floras of Western Canada. *Geol. Surv. Can. Mem., 285.*

Gould, R. (1981). The coal-forming flora of the Walloon Coal Measures. *Coal Geol. (Australia), 1*(3), 83–105.

Halle, T. G. (1913). The Mesozoic flora of Graham Land. *Wissensch. Ergebnisse d. Schwedischen Südpolar Exped., 1901–1903., 3*(14), 1–123.

Harris, T. M. (1946). Liassic and Rhaetic plants collected in 1936–38 from East Greenland. *Medd. om Gronland, 114,* 1–38.

Herbst, R. (1965). La flora fosil de la formacion Roca Blanca, provincia Santa Cruz, Patagonia. Con consideraciones geológicas y estratigráficas. *Opera Lilloana, 12,* 7–101.

Kimura, T. (1984). Mesozoic floras of East and Southeast Asia, with a short note on the Cenozoic floras of Southeast Asia and China. *Geol. Palaeontol. SE Asia, 25,* 325–50.

(1987). Recent knowledge of Jurassic and Early Cretaceous floras in Japan and phytogeography of this time in East Asia. *Bull. Tokyo Gakugei Univ., 39*(4), 87–115.

(1988). Jurassic macrofloras in Japan and palaeophytogeography in East Asia. *Bull. Tokyo. Gakuge: Univ., 40,* 147–64.

Kimura, T., & Hirata, M. (1975). Early Cretaceous plants from Kochi Prefecture, Southwest Japan. *Mem. Natl. Sci. Mus., Tokyo, 8,* 67–90.

Kimura, T., Naito, G., & Ohana, T. (1986). Early Jurassic plants in Japan. Part 7: Fossil plants from the Nishinakayama Formation, Toyora Group, Yamaguchi Prefecture, Southwest Japan. *Trans. Proc. Palaeontol. Soc. Japan, N.S., 144,* 528–40.

Kimura, T. & Ohana, T. (1987a). Middle Jurassic and some Late Liassic plants from the Toyora Group, Southwest Japan (I). *Bull. Natl. Sci. Mus., Tokyo, C, 12*(2), 41–76.

(1987b). Middle Jurassic and some late Liassic plants from the Toyora Group, Southwest Japan (II). *Bull. Natl. Sci. Mus., Tokyo, C, 13*(3), 115–48.

Kimura, T., Ohana, T., & Tsujii, M. (1988). Early Jurassic plants in Japan. Part 8. Supplementary description and concluding remarks. *Trans. Proc. Palaeont. Soc. Japan, N.S., 151,* 501–22.

Kimura, T., & Tsujii, M. (1980–4). Early Jurassic plants in Japan. Parts 1–6. *Trans. Proc. Palaeontol. Soc. Japan, N.S., 119,* 339–58; *120,* 449–65; *120,* 187–207; *125,* 259–76; *129,* 35–57; *133,* 265–87.

Krassilov, V. A. (1972). Mesozoic flora from the Bureja River (Ginkgoales and Czekanowskiales) (in Russian). *Far East Geol. Inst., Far East Sci. Centre, Acad. Sci. USSR (Vladivostok).* 1–103.

Krassilov, V. A., & Schorochova, S. A. (1973). Early Jurassic flora from the Petrovka River (Primorye). In *Fossil Floras and Phytostratigraphy of the Far East* (pp. 13–27) (in Russian). Far East Geol. Instr., Far East Sci. Centre, Acad. Sci. USSR.

Lebedev, E. L. (1965). Late Jurassic flora of the Zeya River and the Jurassic/Cretaceous boundary (in Russian). *Trans. GIN, 125.*

(1973). Jurassic plants of Western Okhotsk area (in Russian). *Paleontol. J., 4,* 84–94.

(1987). Stratigraphy and age of the Okhotsk-Chukotka volcanogenic belt (in Russian). *Trans. GIN, 421.*

Markovich, E. M. (1985). New data on the flora of the South Yakutia Basin and paleobotanical grounds of intrabasian correlation of sections (in Russian). In *New Coal-bearing Areas in the South Yakutia Basin, Yakutsk* (pp. 18–41).

Miller, C. N., Jr. (1987). Land plants of the Northern Rocky Mountains before the appearance of flowering plants. *Ann. Missouri Bot. Gard., 74,* 692–706.

Oishi, S. (1940). The Mesozoic floras of Japan. *J. Fac. Sci., Hokkaido Imp. Univ., 4, 5*(2–4), 123–480.

Paraketsov, K. V. (1970). Detailed Volgian section on the Pezhenka River (Bolschio Anyui Basin) (in Russian). *Trans. SVKNII, 37,* 151–6.

Person, C. P., & Delevoryas, T. (1982). The Middle Jurassic flora of Oaxaca, Mexico. *Palaeontographica, B, 180,* 82–119.

Samylina, V. A. (1974). Early Cretaceous flora of the North-East USSR. Komarov's reading (in Russian). *Nauka, 27.*

Samylina, V. A., & Efimova, A. F. (1968). First finds of Early Jurassic flora in the Kolyma basin (in Russian). *Akad. Sci. USSR Reports, 179*(1), 166–8.

Smiley, C. J. (1970). Late Mesozoic flora from Maran, Pahang, West Malaysia. Part 2: Taxonomic considerations. *Bull. Geol. Soc. Malaysia, 3,* 69–113.

Vakhrameev, V. A. (1964). Jurassic and Early Cretaceous floras of Eurasia and the palaeofloristic provinces of this period (in Russian). *Trans. Geol. Inst. Acad. Sci. USSR, 102,* 261.

(1971). Development of the Early Cretaceous flora in Siberia. *Geophytology, 1*(1), 75–83.

(1987). Climate and the distribution of some gymnosperms in Asia during the Jurassic and Cretaceous. *Rev. Palaeobot. Palynol., 51,* 205–12.

(1988). Jurassic and Cretaceous floras and climates of the earth (in Russian). *Trans. Geol. Inst. Acad. Sci. USSR, 430,* 214.

Vakhrameev, V. A., & Doludenko, M. P. (1961). Upper Jurassic and Lower Cretaceous flora of the Bureya basin and its stratigraphic implication (in Russian). *Trans. GIN, 54,* 135.

Ward, L. F. (1905). *Status of the Mesozoic Floras of the United States.* U.S. Geological Survey, Monograph 48.

White, M. E. (1981). Revision of the Talbragar Fish Bed flora (Jurassic) of New South Wales. *Rec. Austr. Mus., 33*(15), 695–721.

Wu, S. Q., Ye, M. N., & Li, B. X. (1980). Upper Triassic and Lower and Middle Jurassic plants from the Hiangchi Group, Western Hubei (in Chinese with English abstract). *Mem. Nanjing Inst. Geol. Palaeontol., Acad. Sinica, 14,* 63–131.

Zhou, Z. Y. (1984). Early Liassic plants from Southwest Hunan, China (in Chinese with English abstract and description). *Palaeontol. Sinica, N.S., 7,* 91.

# 19 Ostracods and foraminifers of Western Interior North America

W. K. BRAUN

The paleoenvironmental assessment of this area is based on the well-documented Jurassic microfaunas of the subsurface of southern Saskatchewan (Wall 1960; Brooke and Braun 1972) and on the exposures in the northeastern Canadian Rocky Mountains (Brooke and Braun 1981). Of more general interest are shorter accounts about Jurassic ostracods or foraminifers or both groups from surface exposures in the Black Hills (South Dakota and northeastern Wyoming), southern Montana, the Little Rocky Mountains (northcentral Montana), and the southern Canadian Rocky Mountains and their foothills of southeastern British Columbia, as documented in the References section. The microfossils reached their greatest diversity and abundance in the intercratonic sequences of the Williston Basin, which straddles the Canada–United States border in the heartland of the North American continent (Figure 19.1). They are less varied, less abundant, erratically distributed, and generally poorly preserved in the deformed sedimentary pile of the Rocky Mountain miogeocline and its Western Canada Basin. Nothing has been published as yet about varied foraminiferal, radiolarian, and ostracod faunas that are known to exist in the sedimentary-volcanic Jurassic sequences of the Queen Charlotte Islands off the coast of British Columbia, representing the eugeoclinal realm.

On a generic level, most Middle Jurassic Foraminifera and Ostracoda of the Western Interior regions compare closely with western European, Russian, and other Jurassic microfaunas, but many of their species are endemic, especially among the ostracods. This imparts a distinctive character to the microfauna, setting it apart from other contemporaneous assemblages and reminding of the provinciality in ammonites and megafaunas. For this reason, the ostracods and Foraminifera are assigned to the Shoshonean Province, East Pacific Subrealm (originally a realm) of the Tethyan Realm (Taylor et al. 1984; Westermann 1984, personal communication, 1989), replacing the traditional term of "Western Interior faunas." The assignment, however, applies only to the Bajocian to Callovian microfaunas and macrofaunas – the time of major flooding of the cratonic interior regions. Once the Jurassic seas started to withdraw from the Williston Basin and became confined to the Rocky Mountain miogeocline by late Oxfordian–Kimmeridgian time, a distinctive Boreal influence became pervasive. This realignment from the East Pacific to the Boreal faunas seems to have been a direct consequence of the closing of the pathways between the miogeoclinal and eugeoclinal regions due to the first rise of the Rocky Mountains in the Columbian Orogeny (Nevadan phase), confining the sea to a narrow, eastern trough and foredeep, with its only oceanic connection in the northwest and north, until its filling-in toward the end of the Jurassic.

The compositions of the Middle and Late Jurassic microfaunas of the Western Interior regions in general, and of Western Canada in particular (Figure 19.2), reflect not only the overall evolutionary trends and faunal provincialism but equally the regional and local depositional and environmental patterns. The transgressions spread a highly diverse and relatively uniform calcareous microfauna of Ostracoda, Foraminifera, echinoderms, and other invertebrates over a wide area. Regressions, in contrast, provided initially for increased differentiation of environments and the concomitant creation of specialized niches that in turn spurred allopatric speciation and the appearance of short-lived, endemic species. The net gain in diversity, however, was soon reversed as the regressive conditions progressively ravaged the microfaunal assemblages until their final demise. Ostracods were the first to be affected and to disappear, followed by the calcareous Foraminifera, with relatively low-diversity assemblages of agglutinated-arenaceous Foraminifera lingering longest.

Too few microfossils have been recovered to date from the dominantly black shales and mudstones of the Lower Jurassic sequences of the Western Canada Basin and Rocky Mountains (assemblage I) to allow for any meaningful paleoenvironmental assessment. It was not until Bajocian time that the widespread euxinic bottom conditions had cleared up and were replaced by oxygenated conditions as the transgressions gathered momentum and spilled over into the intercratonic Williston Basin. However, a different set of restrictive conditions developed along the northern and eastern rim of this basin (southern Saskatchewan, mainly), ranging from terrestrial, freshwater to brackish water to marginal-

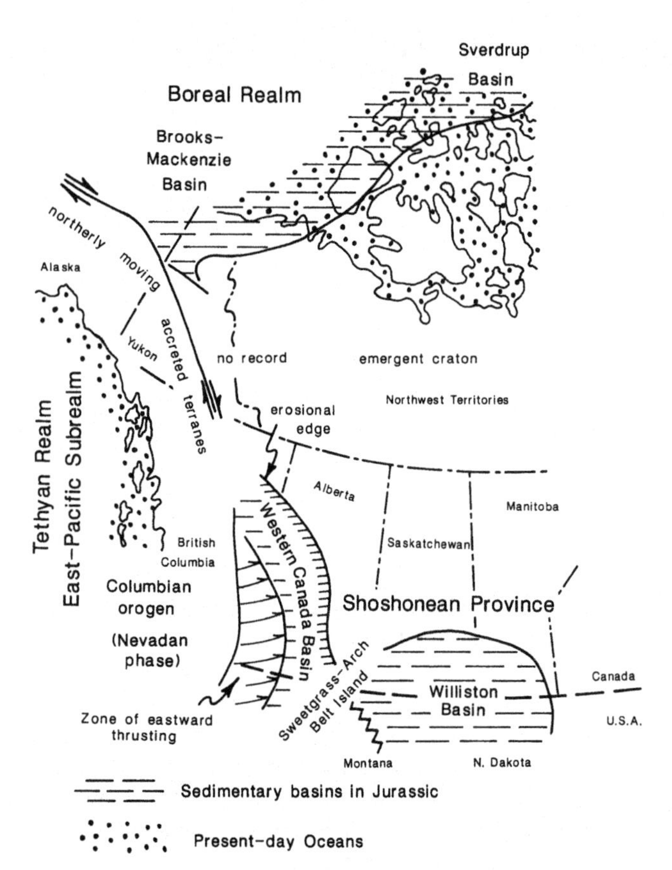


**Figure 19.1. Map of western Canada showing Jurassic sedimentary basins and major tectonic elements. (Adapted from Poulton 1984.)**

marine influences that were punctuated by short-lived and localized marine pulses. This shifting array of environments, both in the horizontal and vertical, left a facies pattern that is notoriously difficult to decipher, but in which ostracods (the Shaunavon Fauna) play an important role as environmental indicators and biostratigraphic markers (assemblages II and III). In the older subassemblage II, species of such genera as *Darwinula, Limnocythere,* and the *Metacypris–Norcanolella* groups, especially if associated with many charophytes, indicate freshwater conditions, whereas representatives of *Protocythere–Pleurocythere* and similar genera, together with crinoids and other echinoderm remains, indicate a marine setting. Brackish-water conditions are represented by a mixed fauna from which the freshwater forms are absent or diminished in number, and in which other ostracod genera such as *Aparchitocythere* and *Paracypris* and a few ostracods of still unknown origin dominate. The marginal-marine influences are indicated by mass occurrences of small, ammodiscid Foraminifera, rarely associated with a few other and slightly larger agglutinated-arenaceous forms. Equally characteristic are the many bone beds, with abundant fish teeth, scales, and phosphatic remains, that represent winnowed faunas, standstills in sedimentation, erosion, or any punctuating event.

In about early Bathonian time the situation had stabilized in the northern Williston Basin and over southern Saskatchewan, with freshwater conditions becoming all-pervasive. The freshwater ostracods flourished in great numbers (assemblage III), but to the south and in the more central parts of the basin, stable marine conditions prevailed, as indicated by the relatively diverse foraminiferal faunas, mainly calcareous forms, described from south-central Montana (Lalicker 1950).

The Bajocian–early Bathonian sequences of the Western Canada Basin and the southern Canadian Rocky Mountains have yielded thus far only impoverished and erratically distributed, restricted-marine microfaunas composed nearly exclusively of agglutinated-arenaceous and very few calcareous (pyritized) Foraminifera (Stronach 1981; Hall and Stronach 1982), with a few ostracods at the base. This combination reflects possibly deeper (and cooler?) water conditions, a higher rate of sedimentation of fine terrigenous clastics, some euxinic influences, and a combination of factors different from those in the adjoining intercratonic Williston Basin to the east.

The freshwater regime was abruptly terminated by the most forceful and far-reaching of the Jurassic transgressions that affected western North America. The event is clearly marked in the Williston Basin by the conspicuous break between the older, nonmarine ostracod fauna and the younger, fully marine and highly diversified assemblages (IV to V) of ostracods and calcareous Foraminifera – the Rierdon Fauna. The marine incursion affected the Western Canada and Williston Basins equally, and the megafaunas and microfaunas are for this reason remarkably uniform. It lasted from about middle Bathonian to Callovian, with the regression starting in middle Callovian time.

The mixture of ostracods and calcareous Foraminifera, with a liberal addition of crinoid columnals and other echinoderm fragments, indicates fully marine and peak transgressive conditions. A major, regional regression ensued, of which, unfortunately, little is preserved in the sedimentary and faunal record. The only unit that remains of this event is a wedge of sandstone in the subsurface of southwestern Saskatchewan with a distinctive marine microfauna of ostracods and calcareous Foraminifera (assemblage VI, Roseray faunule). It is dated, on general stratigraphic deductions, to be of about early middle to early late Callovian age, representing the only known fauna of that time in western Canada. All other regions were affected by a major discontinuity that involved most, if not all, of the Callovian strata, and the hiatus extends even to the latest Bathonian and earliest Oxfordian in the Rocky Mountain regions.

Unlike the Rierdon sequence and faunas with its truncated regressive portion, the Oxfordian sedimentary pile and microfaunas have fully developed regressive end members. The increase in coarse clastics during the regression was accompanied by a progressive decrease in the diversity and abundance of the microfaunas, with the calcareous elements being devastated first, leaving purely agglutinated foraminiferal faunas to the end.

In the Williston Basin, a diverse ostracod and calcareous foraminiferal fauna (assemblage VII, Swift Fauna) represents the transgressive event characterizing the lower (but not lowest) to upper Oxfordian–lowermost Kimmeridgian sequences. In overall composition, it still has a "Shoshonean" character. However, it rapidly dwindles to an impoverished, marginal-marine aggluti-

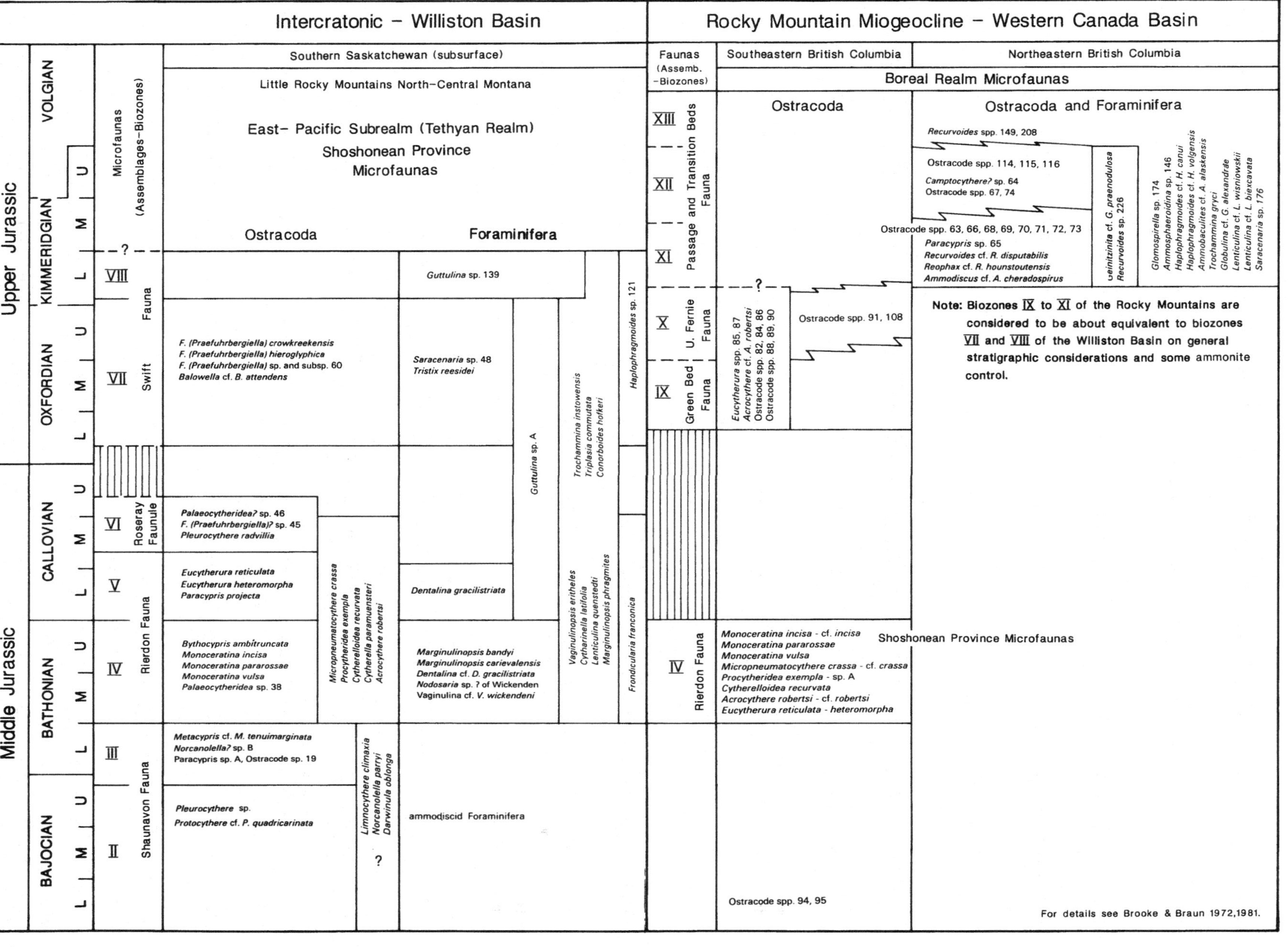


**Figure 19.2.** Key ostracods and Foraminifera within their respective assemblages (biozones) and faunal provinces. For details with respect to the precise distribution of the ostracods, see Chapter 15.

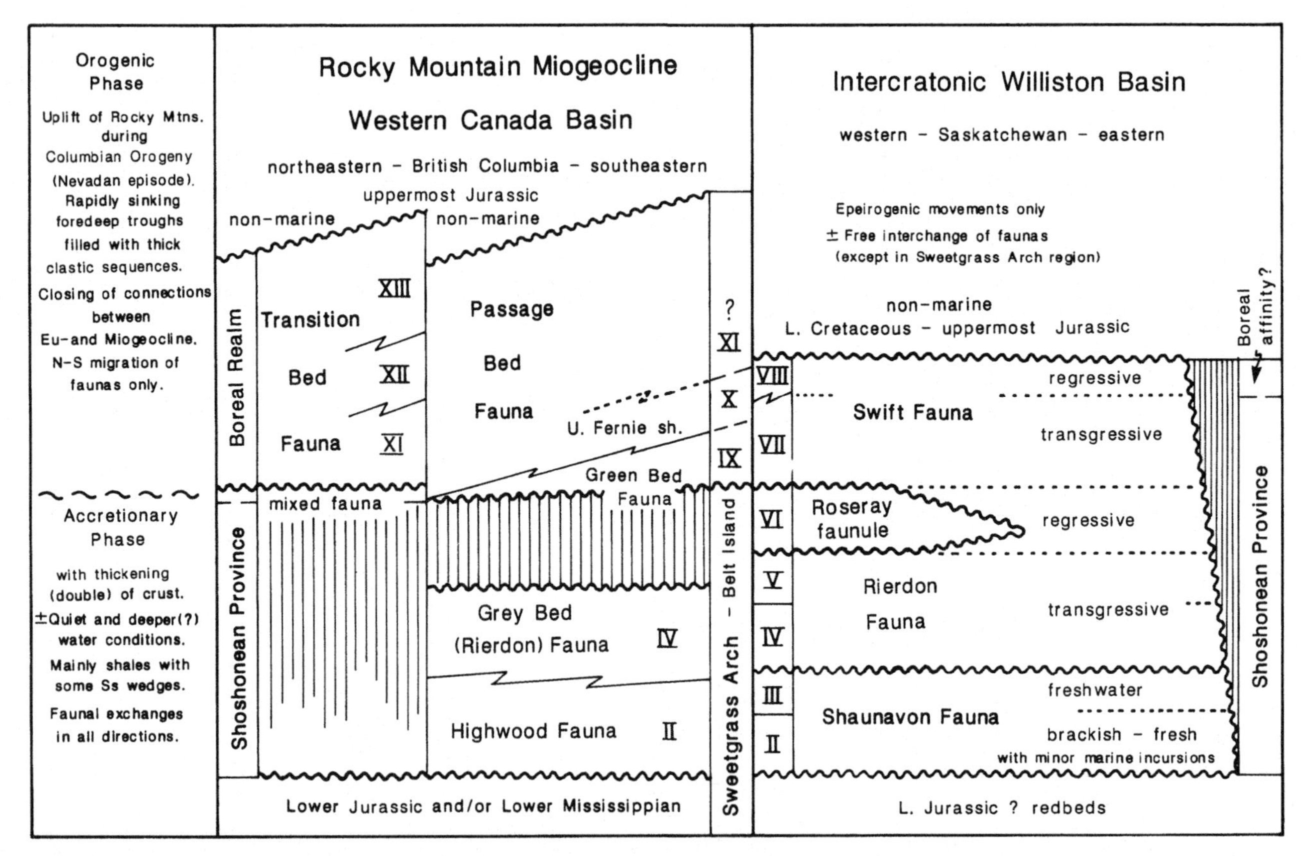


**Figure 19.3. Paleogeography, paleoenvironment, and biostratigraphy of Middle and Upper Jurassic microfaunas of Western Canada.**

nated foraminiferal fauna (assemblage VIII), with some Boreal affinities in the upper part of these units, which in turn are truncated by unfossiliferous and nonmarine Lower Cretaceous clastics in the subsurface of western Canada. The withdrawal marks the end of flooding of the Western Interior and the beginning of a long period (until early Albian time, actually) of nonmarine deposition and erosion in the southern part of the Canadian prairie provinces.

In the Rocky Mountain miogeocline and the Western Canada Basin, the major transgression of the Late Jurassic started in early Oxfordian time, ending the major Callovian hiatus, and a distinctive microfauna of calcareous Foraminifera and ostracods (assemblage IX, Green Bed Fauna) indicates relatively open-marine conditions. This fauna is different from what is considered its approximate equivalent in the Williston Basin (assemblage VII) and bears some "European" and some "Boreal" affinities. The latter strengthened progressively in the succeeding assemblage X. Assemblage IX is tentatively, and X definitely, assigned to the Boreal Realm, with major changes in the tectonic-sedimentary regimes and faunal alignments taking place throughout the Oxfordian.

Another set of microfaunas, dominated by agglutinated Foraminifera and few or no ostracods (assemblages XI to XIII, Passage and Transition Bed faunas), have been described from outcrop sections in the northern Rocky Mountains of northeastern British Columbia and adjoining foothills (Brooke and Braun 1981). They are typical Boreal in character and compare closely with faunas reported or known from arctic North America and various parts of Siberia. In most of these northern regions they extend across the Jurassic–Cretaceous boundary, but in northeastern British Columbia and the areas to the south they had vanished long before the end of Volgian time as a result of the filling-in of the seaway and the switch to nonmarine sedimentation.

The difficulties in correlating the Boreal assemblages IX to XI of the Rocky Mountains with their assumed Shoshonean counterparts (VII to VIII) of the Williston Basin are not unexpected, given their stark differences in composition. Their equivalence has to be established on independently derived stratigraphic considerations and with the help of cardioceratid ammonites. In Figure 19.3 the most influential factors are summarized that are considered responsible for the microfaunal differentiation.

## Acknowledgment

Dr. Margaret M. Brooke, now retired, has been responsible for most of the paleontological aspects of the Jurassic microfaunas of western Canada. Without her patience and dedication, this comprehensive study would have been impossible to undertake and complete.

### References

Brooke, M. M., & Braun, W. K. (1972). *Biostratigraphy and Microfaunas of the Jurassic System of Saskatchewan*. Saskatchewan Dept. Min. Res. Rept. 161.

(1981). *Jurassic Microfaunas and Biostratigraphy of North-Eastern British Columbia and Adjacent Alberta*. Geol. Surv. Canada, Bull. 283.

Hall, R. L., & Stronach, N. J. (1982). *Guidebook to the Fernie Formation of Southern Alberta and British Columbia*. Circum-Pacific Jurassic Research Group, Int. Geol. Correl. Progr. 171.

Lalicker, C. G. (1950). Foraminifera of the Ellis Group, Jurassic, at the type locality. *Univ. Kansas Paleontol. Contr., Protozoa, 2*, 3–20.

Peterson, J. A. (1954). Jurassic Ostracoda from the "Lower Sundance" and Rierdon formations, Western Interior United States. *J. Paleontol., 28*(2), 153–76.

Poulton, T. P. (1984). The Jurassic of the Canadian Western Interior, from 49°N latitude to Beaufort Sea. In D. F. Stott & D. J. Glass (eds.), *The Mesozoic of Middle North America* (pp. 15–41). Can. Soc. Petrol. Geol. Mem. 9.

Stronach, N. J. (1981). *Sedimentology and Paleoecology of a Shale Basin: The Fernie Formation of the Southern Rocky Mountains, Canada*. PhD thesis, University of Calgary.

Swain, F. M., & Peterson, J. A. (1952). *Ostracoda from the Upper Part of the Sundance Formation of South Dakota, Wyoming, and Southern Montana*. U.S. Geological Survey, Professional Paper 243-A.

Taylor, D. G., Callomon, J. H., Hall, R., Smith, P. L., Tipper, H. W., & Westermann, G. E. G. (1984). Jurassic ammonite biogeography of western North America: the tectonic implications. In G. E. G. Westermann (ed.), *Jurassic-Cretaceous Biochronology and Paleogeography of North America* (pp. 121–41). Geological Association of Canada, Special Paper 27.

Wall, J. H. (1960). *Jurassic Microfaunas from Saskatchewan*. Saskatchewan Dept. Min. Res. Rept. 53.

Weihmann, I. (1962). Jurassic microfossils from southern Alberta. In *Hermann Aldinger Festschrift, Stuttgart* (pp. 191–8).

(1964). Stratigraphy and microfauna of the Jurassic Fernie Group, Fernie Basin, southeastern British Columbia. 14th Annual Field Conference Guide Book. *Bull., Canadian Petr. Geol. 12*, 587–99.

Westermann, G. E. G. (1984). Summary of symposium papers on the Jurassic-Cretaceous biochronology and paleogeography of North America. In G. E. G. Westermann (ed.), *Jurassic-Cretaceous Biochronology and Paleogeography of North America* (pp. 307–15). Geological Association of Canada, Special Paper 27.

# 20 Ostracods of China

Y. G. WANG and M. Z. CAO

Most of the Jurassic ostracod faunas known to date from many regions of China are nonmarine. The majority of species are endemic, and the assemblages are characterized by high abundance and low diversity. Only a few meager and "marginal" marine ostracod assemblages have been reported from a few areas. The main utility of ostracods is therefore in elucidating the environment of deposition, whereas their role as time indicators is more limited.

Part of the Early Jurassic ostracod faunas are characterized by a *Naevicythere–Darwinula* assemblage, indicating estuarine conditions for regions of southwestern China and the provinces of Hunan and Guangxi. To the west and toward the northeastern Burmese border the darwinulids are more frequently associated with species questionably assigned to the genus *Gomphocythere* – a mixture indicating somewhat greater freshwater influences. This latter assemblage has been reported from the provinces of Yunnan, Sichuan, Guizhou, Hunan, Jiangxi, and Guangdong. Species of both *Naevicythere* and *Gomphocythere* are indigenous to these regions, as are most of the darwinulids. Some of the *Darwinula* species, however, seem to be related to, or even identical with, cosmopolitan species. The specific and generic compositions of both assemblages point to freshwater or brackish influences in general terms only, so that detailed environmental reconstructions have to be based on associated flora and fauna and geological considerations.

The distinction between marine and nonmarine ostracod assemblages is more pronounced in the Middle Jurassic. The *Schuleridea triebeli–Monoceratina vulsa* association of western Yunnan is undoubtedly marine. In central Yunnan, the marine horizons are often interbedded with freshwater sediments bearing characteristic and entirely different ostracod faunas. This intercalation is reminiscent of the situation along the northeastern rim of the Williston Basin in western Canada, or, in more general terms, of the nonmarine Bathonian ostracod faunas of England and parts of Russia, which are also sandwiched in between marine faunas. In contrast, a homogeneous and widespread coeval freshwater ostracod fauna, dominated by species of *Darwinula* and *Metacypris*, occurs in lacustrine-fluviatile sediments across vast regions and southern and northern China.

In the Late Jurassic, a noticeable provincialism occurred in the ostracod faunas, most likely as a result of the differentiation and proliferation of the sedimentary basins. Once again, the ostracod faunas indicate a broad range, from freshwater to brackish to marginal-marine influences.

A shallow-marine, inner neritic fauna with such cosmopolitan genera as *Protocythere, Galliaecytheridea,* and *Mandelstamia* seems to be confined to the eastern areas of Heilongjiang province of northeasternmost China. It is succeeded in the same region by a *Scabriculocypris–Mandelstamia* assemblage in which the former genus indicates brackish-water conditions, and in turn by the freshwater to brackish *Cypridea–Scabriculocypris–Vlakomia–Galliaecytheridea* fauna at the top of the Jurassic. From the faunal progression it is evident that the freshwater influences strengthened progressively throughout the Late Jurassic in these regions.

A freshwater to brackish Late Jurassic ostracod fauna is also widespread in southwestern China. The *Cetacella–Darwinula* assemblage has been reported from the base of the Upper Jurassic in the Qaidam, Quinghai, and Sichuan basins. It is followed by the *Damonella–Darwinula–Eolimnocythere* fauna mainly in the Sichuan Basin, and by the *Jinguella–Pinnocypridea–Darwinula–Cypridea* assemblage in the provinces of Sichuan, Yunnan, and Quinghai and in eastern Tibet.

In the northern inland regions, two Late Jurassic ostracod assemblages occur in the "Jehol Fauna." The older is the *Cypridea–Luangpingella–Eoparacypris* assemblage of Hebei and Gansu provinces, and the younger the *Cypridea* fauna mainly of northeastern and eastern China. This latter is also known from *Lycoptera*-bearing strata of Far East Russia. Both faunas indicate brackish-water conditions.

## References

Cao Mei-Zhen (1984). Some Early Mesozoic ostracods from South China. *Mem. Nanjing Inst. Geol. Paleontol., Acad. Sinica, 19*, 33–66.

Chengdu Institute of Geology and Mineral Resources (1983). *Paleontological Atlas of Southwest China. Microfossils, Ostracoda.* Beijing: Geol. Publ. House.

Hao Yichu, Ruan Peihua, Zhou Xiugao, Song Qishan, Yang Guo Dong, Cheng Shuwei, & Wei Zhenxin (1983). Middle Jurassic-Tertiary deposits and ostracod-Charophyta fossil assemblages of Xining and Minhe Basins. *Earth Sci. J., Wuhan Coll. Geol., 23,* 1–221.

Hou Youtang (1958). Jurassic and Cretaceous non-marine ostracods of the subfamily Cyprideinae from northwestern and northeastern regions of China. *Mem. Palaeontol. Inst., Acad. Sinica, 1,* 33–103.

Pang Qiping (1982a). Ostracoda from the middle-upper series of the Jurassic System in Yanshan Range, Hebei Province, and its stratigraphical significance. *J. Hebei Coll. Geol., 1–2,* 89–110.

(1982b). Ostracoda. In *The Mesozoic Stratigraphy and Paleontology of Guyang Coal-bearing Basin, Neimenggal Autonomous Region, China* (pp. 57–84).

(1984). Fossil Ostracoda and the boundary for terrestrial Jurassic-Cretaceous systems in Yanshan area, Hebei Province. *J. Hebei Coll. Geol., 8,* 1–16.

Su Deying, Li Yougui, et al. (1980). Ostracod fossils. In *Mesozoic Strata and Palaeontology in Shanganning Basin* (vol. 2, pp. 48–83).

Xu Maoyu (1983). Ostracods from the Mesozoic coal-bearing strata of South China. In R. F. Maddocks (ed.), *Applications of Ostracoda* (pp. 352–71). Houston: Univ. Houston Geosci.

Ye Chunhui, Gou Yunxian, & Cao Meizhen (1980). *Jurassic-Cretaceous Ostracodes from Zhejiang. Divisions and Correlations of the Non-marine Mesozoic Volcano-sedimentary Formations in the Provinces Zhejiang and Anhui* (pp. 173–210). Beijing: Sci. Press.

Ye Chunhui, Gou Yunxian, Hou Youtang, & Cao Meizhen (1977). Mesozoic and Cenozoic ostracod faunas from Yunnan. In *Mesozoic fossils from Yunnan, China* (vol. 2, pp. 153–330). Beijing: Sci. Press.

Zhang Lijun (1982). Late Jurassic to Early Cretaceous marine-brackish ostracods of eastern Heilongjiang Province, China. *Bull. Shenyang Inst. Geol. Min. Res., Chin. Acad. Geol. Sci., 5,* 201–21.

Zhong Xianchun (1964). Upper Triassic and Middle Jurassic ostracods from the Ordos Basin. *Acta Palaeontol. Sinica, 12,* 426–65.

# 21 Corals of the circum-Pacific region

L. BEAUVAIS

Circum-Pacific Jurassic corals were poorly known until recently. Only two monographs on Japan existed, by Yabe and Sugiyama (1935) on the stromatoporoids and by Eguchi (1951) on the Scleractinia. Rare and insignificant coral occurrences were reported from Indonesia by Tobler (1923) and van Bemmelen (1949).

In the past decade, numerous Jurassic corals have been collected in Canada (T. P. Poulton), western Mexico (T. E. Stump), northern Chile (P. Prinz), the Philippines, western Thailand, and Sarawak (H. Fontaine), Sakhalin and Koryakia (E. V. Krasnov), and Sumatra and Japan (L. Beauvais). Studies of these faunas have contributed to the dating of the source rocks (Lias and Dogger for Canada, Dogger for northern Chile, Dogger and Malm for western Mexico, Sumatra, Sarawak, Japan, and Sakhalin) and to paleogeography (Beauvais and Stump 1976; Krasnov 1983; Beauvais, Bernet-Rolland, and Maurin 1985, 1987).

### Lower Jurassic

Liassic corals (Figure 21.1) are very rare throughout the world (Beauvais 1981). In the circum-Pacific area they are presently known only from Canada (Vancouver Island), the lower Amur area (Russia), Argentina, and western Thailand; Argentinean scleractinian studies are in progress (by S. Morsch).

Species distributions indicate connections between western Canada and Morocco, Belgium, and Austria, as well as among Thailand, Belgium, and France. No common species occur between Canada and Thailand. It is therefore suggested (1) that Thailand was located on the northern side of the Tethyan basin, (2) that no direct coral communications occurred across Panthalassa, but the Protoatlantic began to open as early as the Liassic, and (3) that paleomagnetic and paleontological data (Stone 1980; Stanley (1986) indicate low-latitude warm water for the Late Triassic North American corals, situated on the Wrangellia Terrane, most likely in the Southern Hemisphere [based on this hypothesis, Beauvais and Poulton (1980) have suggested that Lower Jurassic sediments from the Yukon Territory and southern Alaska yielding corals had also been transported from far southern latitudes toward their present northern positions], and (4) no Liassic true reefs seem to have grown in autochthonous Canada, Russia, or Thailand. In these areas, corals were solitary or small in corallum size, few in number, and always scattered in the sediments.

### Middle Jurassic

As during the Lias, very few coral-bearing formations occur around the Pacific (Figure 21.2). During the Dogger, scleractinians principally developed in western Europe and along the western margin of the Mediterranean. It seems that conditions for reef growth were most favorable in Europe. Elsewhere, particularly in the circum-Pacific areas, corals did not develop true reefs. In the Philippines and Thailand, the microfacies of the coral-bearing sediments show mud-mound character; corals grew on these muddy mounds of bacterial origin, but did not contribute to their growth (Beauvais et al. 1985). Similar environments probably existed in North and South America, where very few corals have been collected and where no coral reef has been recorded (Imlay 1965). In western Canada, six species have been identified (Beauvais and Poulton 1980; Beauvais 1982); in California, the "simple" genus *Latomeandra* was recorded (Wells 1942; Imlay 1965); in Wyoming, only *Actinastraea hyatti* Wells is known; in Idaho, two species have been collected recently (G. D. Stanley, Jr., personal communication); in Bolivia, four genera and six species are known (Gerth 1926); in northern Chile, five species have been recognized (Prinz 1986). In western Canada, Beauvais and Poulton (1980) identified six species, of which four are new, and one species is in common with Madagascar and one with France, Switzerland, and Morocco. The species cited by Prinz (1986) from northern Chile are shared by France, Switzerland, Germany, and Morocco; only one species is known to be in common between Chile and Canada, whereas Asian species are unknown from Canada and South America. The majority of the Philippine Dogger corals are endemic, but three species are in common with France, two with England, two with Switzerland, one with Mo-

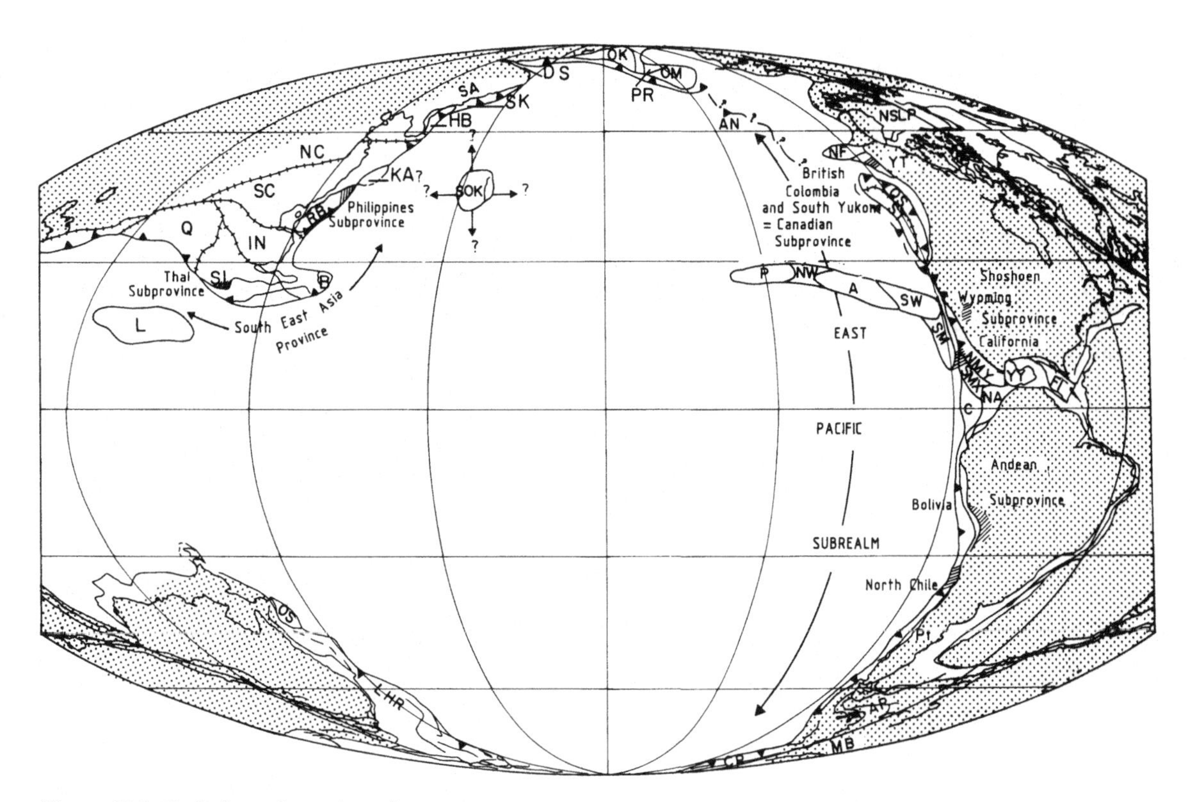


**Figure 21.1. Early Jurassic coral provinces.**

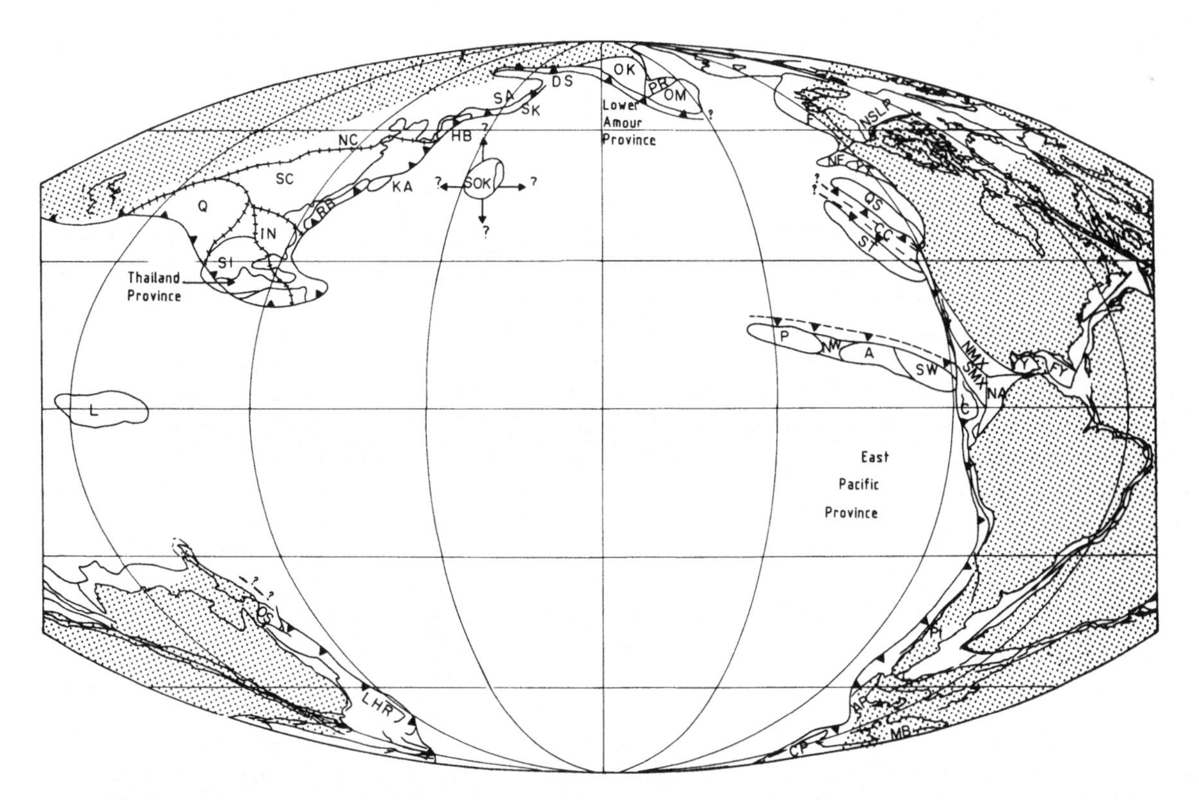


**Figure 21.2. Middle Jurassic coral provinces.**

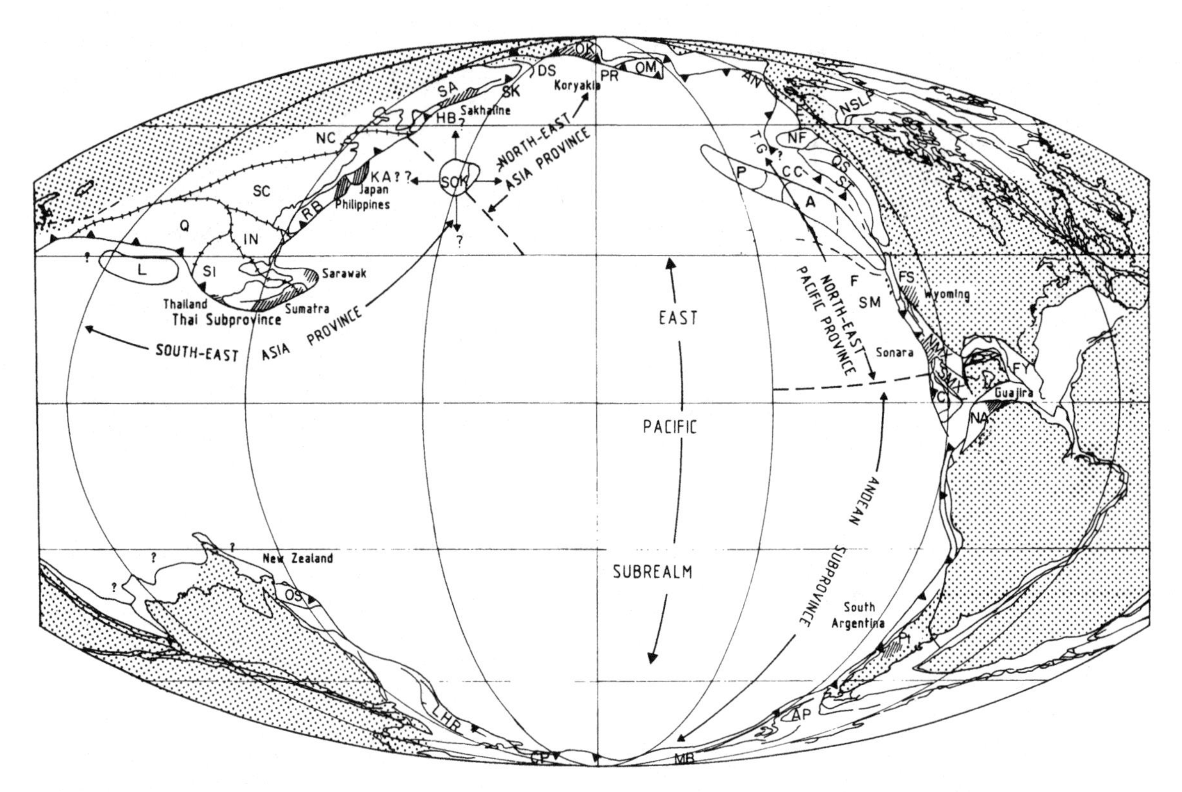


**Figure 21.3. Late Jurassic coral provinces.**

rocco, one with India, and one with Saudi Arabia. Four coral species collected in Thailand have previously been found in France, three in Switzerland, three in England, one in Morocco, one in Madagascar, and one in Iran.

It is therefore concluded that (1) seaways connected Pacific Canada, South America, Europe, and southern Tethys, but no connections existed between America and Southeast Asia, these being separated by the Panthalassa Ocean, which did not allow dispersion of coral planulas; (2) western Alaska, the Western Interior United States, and Chile–Bolivia constituted three coral subprovinces, joined by the sea but separated by paleogeographic barriers (currents, estuaries, etc.); (3) the Philippines and Thailand communicated widely with the northern Tethyan basin, but little with the southern Tethyan basin, with migrations probably passing through Europe; (4) important faunal relationships existed among India, Madagascar, and Iran, and between these and Europe and North Africa; it seems, therefore, that during Middle Jurassic time Thailand was located at the southern margin of the Eurasian craton, while India was situated on the Gondwana continental margin; (5) no common species are known between the Philippines and Thailand, and a local geographic barrier probably was responsible.

## Upper Jurassic

As elsewhere in the world, the circum-Pacific Malm is rich in development of corals and coral reefs (Figure 21.3). In Europe, true coral reefs expanded from the middle Oxfordian to the Tithonian, both geographically and in number of species; 150 genera and 300 species are known, of which 51 genera and 103 species have been identified from numerous countries in the circum-Pacific areas (i.e., Sakhalin, Koryakia, Japan, the Philippines, Thailand, Burma, Malaysia, Sumatra, Borneo, New Zealand, southern Argentina, Colombia, Mexico, Wyoming, Montana). However, although coral beds are widespread, true reefal structures seem to be extremely rare, such as in Sakhalin, where both reef-building and reefophilic corals are known (Krasnov 1983). The majority of Late Jurassic corals in Japan occur in mud mounds of bacterial origin (Beauvais et al. 1987). No Jurassic coral-reef structures of any size have been reported from North America, where isolated occurrences of both solitary and colonial forms have been found only in the Gulf of Mexico region (Burckhardt 1930; Wells 1942; Beauvais and Stump 1976). In southeastern Asia (the Philippines, Sumatra, Thailand, Sarawak), corals were reported as floating in a limy microbial mud matrix (Beauvais et al. 1985). In Burma, some colonies in growth position are reminiscent of true stroms (Fontaine et al. 1986).

Generally, Late Jurassic coral species, as in the Dogger, have a wide geographic distributions, indicating worldwide climatic uniformity (Beauvais 1973, 1981, 1986). Yet more or less connected provinces are distinguished. Among the eight species identified by Krasnov (1983) from northern Russia, five are endemic, one is known from Japan, and two from Europe, whereas none is in common with southern Tethys. From Japan, Eguchi (1951) described 58 Upper Jurassic species collected by Beauvais from the Torinosu Limestone in the Sakawa and Soma districts. Six are endemic; 1 occurs also in Sakhalin, 14 on the northern side of Tethys, and 20 in Southeast Asia, but none in the south of the Tethyan basin. Of seven species identified from the Philippines, six have been found also at the northern margin of Tethys, and one in South Yemen (southern Tethys). Of 23 species identified from Sarawak (Beauvais 1989a), 1 is endemic, 12 are known from northern Tethys, 1 from southern Tethys, 1 from South America, and 14 from other regions of Southeast Asia. Of 47 species found in Sumatra, 29 are common with northern Tethys, and 3 with southern Tethys; 1 was found in South America, and 16 in other parts of Southeast Asia. In Thailand and Burma, 37 species have been collected, 10 of which are endemic, 23 known from northern Tethys, 2 from southern Tethys, and 5 from other parts of Southeast Asia. Geyer (1968a,b) reported 10 species from Colombia (5 endemic, 5 common with northern Tethys, and 2 with southern Tethys). In Sonora (Mexico), 8 species have been identified, 2 of which are endemic, 5 common with northern Tethys and 1 with southern Tethys. No species in common occur in Colombia and Sonora.

These data may be summarized as follows: (1) Koryakia and Sakhalin constituted the Northeast Asia Coral Province, widely communicating with the northern margin of Tethys; that province may be considered as a large barrier-reef system extending as a discontinuous but rather regular coral arch and dividing the marginal Pacific geoclines and epicontinental basins of inner Japan, which differ in tectonic nature. Occurrences of corals at these high latitudes have been explained by larval dispersal in warm currents, abyssal heat flow along deep faults, and different positions of the poles or of the equatorial plane from the present (Krasnov 1983). (2) Japan, the Philippines, Sarawak, and Sumatra seem to compromise a second coral province communicating with Japan toward the north and with the northern margin of Tethys toward the west. Direct connections with southern Tethys probably were absent (the few species in common probably having crossed through the west end of Tethys). (3) Thailand seems to be a third distinct subprovince within this Asian Subrealm. A second, East Pacific Coral Subrealm includes Colombia and northwestern Mexico, but these two regions were mutually isolated, comprising two distinct provinces that communicated with southern and northern Tethys, respectively.

## References


Beauvais, L. (1973). Upper Jurassic hermatypic corals. In A. Hallam (ed.), *Atlas of Palaeogeography* (pp. 317–28). Amsterdam: Elsevier.

(1978). Révision des topotypes de Madréporaires bathoniens de Cutch (Inde). Collection Gregory. British Museum de Londres. *Ann. Paleontol. Invert., 64,* 47–77.

(1981). Données actuelles sur la Paléobiogéographie des Madréporaires mésozoiques. *C. R. Soc. Biogeogr., 57*(2), 51–64.

(1982). Etude de quelques Coelentérés infra-mésozoiques de Nord-Ouest du Canada. *Can. J. Earth Sci., 19,* 1953–73.

(1983). Jurassic Cnidaria from the Philippines and Sumatra. In *Jurassic of South-East Asia* (pp. 39–76). CCOP Tech. Bull. 16.

(1986). Evolution paléobiogéographique des formations à Scléractiniaires due bassin téthysien au cours de Mésozoique. *Bull. Soc. Geol. France, 8*(2–3), 499–509.

(1989a). Corals of the Baulimestone of Sarawak. *Rep. 25th CCOP Sess., Bagno City, Philippines.*

(1989b). Jurassic corals from the Circum-Pacific area. *5th Int. Symp. Foss. Cnidaria, Brisbane. Mem. Assoc. Austral. Palaeont., 8,* 291–302.

Beauvais, L., Bernet-Rolland, M. C., & Maurin, A. F. (1985). Reinterpretation of Pretertiary classical reefs from Indopacific examples. *Proc. 5th Internat. Congr. Coral Reefs, Tahiti, 6,* 581–6.

(1987). A microbial origin for the Torinosu limestone of Japan. *EUG IV, Strasbourg, Abstracts,* 206.

Beauvais, L., Fontaine, H., Poumot, C., & Vachard, D. (1979). Données nouvelles sur le Mésozoique des Philippines. *C. R. Somm. Soc. Géol. France, 3*(11), 117–21.

Beauvais, L., & Poulton, T. P. (1980). Quelques Coraux de Trias et du Jurassique du Canada. *Curr. Res. Geol. Surv. Can., 80,* 95–101.

Beauvais, L., & Stump, T. E. (1976). Late Jurassic corals and molluscs from Cerro Pozo Serna (Sonora, Mexico). *Palaeogeogr. Palaeoclimatol. Palaeoecol., 19,* 275–301.

Bemmelen, N. R. W. van. (1949). *The Geology of Indonesia.* The Hague.

Burckhardt, C. (1930). Remarques sur le problème du climat jurassique. In *Etude Synthetique sur le Mèsozoique mexicain, II.* Mem. Soc. Paleontol. Suisse 49.

Crickmay, C. H. (1932). Mount Jura investigations. *Bull. Geol. Soc. Am., 44,* 903.

Eguchi, M. (1951). Mesozoic hexacorals from Japan. *Sci. Rep. Tohoku Univ., Sendai, Japan, 24*(2), 1–96.

Flugel, E. (1966). Mitteljurassische Korallen vom Ostrand der Grossen Salzwuste (Shotori-Kett, Iran). *N. Jb. Geol. Palaontol. Abh., 126,* 46–91.

Fontaine, H., et al. (1985). The Pretertiary fossils of Sumatra and their environments. *Rep. 22nd CCOP Session, Guangzhou (China).*

(1986). The Upper Paleozoic and Mesozic fossils of West Thailand. *Rep. 23rd CCOP Session, Madang ( Papua-New Guinea).*

Gerth, H. (1926). Anthozoa. In E. Jaworski (ed.), *La fauna del Lias y Dogger de la Cordillera Argentina. Actas Acad. Nac. Cienc. Cordoba, 9,* 142.

(1928). Beiträge zur Kenntnis der mesozoischen Korallenfaunen von Südamerica. *Leiden Geol. Meded., 3,* 1–15.

Geyer, O. F. (1968a). Ueber den Jura der Halbinsel La Guajira (Kolumbien). *Mitt. Inst. Colombo-Aleman Invest. Cient. (Santa Marta), 2,* 67–80.

(1968b). Nota sobre la posición estratigráfica y la fauna de Corales del Jurásico superior en la península de la Guajira (Colombia). *Biol. Geol. (Bucaramanga), 24,* 9–22.

Gregory, J. W. (1900). Jurassic fauna of Cutch. The corals. *Palaeontol. Indica, 9*(2), 1–195.

Imlay, R. W. (1965). Jurassic marine faunal differentiation in North America. *J. Paleontol., 39,* 1023–38.

Krasnov, E. V. (1970). On the paleobiogeography of Late Mesozoic Scleractinia. In *Mesozoic Corals of the USSR* (pp. 49–59). Vladivostok: Akad. Nauk. SSSR, Inst. Geol.

(1983). *Coraux et formations récifales du Mésozoique d'URSS* (in Russian). Vladivostok: Akad. Nauk CCCP, Inst. Geol. Vladivostok.

Prinz, P. (1986). Mitteljurassische Korallen aus Nordchile. *N. Jb. Geol. Palaontol. Mh., 12,* 736–50.

Stanley, G. D., Jr. (1986). Late Triassic coelenterate faunas of western Idaho and northeastern Oregon: implications for biostratigraphy and palaeogeography. In T. L. Vallier & H. C. Brooks (eds.), *Geology of the Blue Mountains Region of Oregon, Idaho and Washington* (pp. 23–36). U.S. Geological Survey Professional Paper 1435.

Stone, D. B. (1980). Collision tectonics, paleomagnetism and the origin of Alaska. In *26ieme Congres Géol. Internatl. Paris, Res., 1,* 395.

Tobler, A. (1923). Unsere palaeontologische Kenntnis von Sumatra. *Eclogae Geol. Helvet., 13,* 313–42.

Wells, J. W. (1942). A new species of coral from the Jurassic of Wyoming. *Am. Mus. Nov., 1161,* 1–3.

Yabe, H., & Sugiyama, T. (1935). Jurassic stromatoporoids from Japan. *Sci. Rep., Tohuku Imp. Univ., Japan, 14*(2B), 135–91.

# 22 Brachiopods of the circum-Pacific region

M. O. MANCEÑIDO and A. S. DAGYS

Jurassic brachiopods of the circum-Pacific are poorly known. Several northern Pacific species were described from Japan (Tokuyama 1957, 1958a,b, 1959), North-East Russia (Dagys 1968a,b), eastern China (Sun 1983), Thailand (Alméras 1988), Canada (Crickmay 1933; Ager and Westermann 1963), the western United States (Crickmay 1933; Ager 1968; Perry 1979), Mexico (Alencaster and Buitrón, 1965; Alencaster 1977; Cooper 1983; Rivera-Carranco, Hernandez, and Buitrón 1984; Boullier and Michaud 1987), and Borneo (Yanagida and Lau 1978).

Similarly, in the southern Pacific, isolated descriptions and illustrations are known for Indonesia (Boehm 1907–12; Krumbeck 1922, 1923; Wandel 1936; Wanner and Knipscheer 1951; Manceñido 1978), New Zealand (Trechmann 1923; Marwick 1953; Speden and Keyes 1981), and Antarctica (Quilty 1972, 1982), and several for Argentina (e.g., Jaworski 1915, 1925, 1926; Weaver 1931; Feruglio 1934; Wahnish 1942), Chile (e.g., Thiele 1964; Pérez 1982; Cooper 1983; Manceñido 1988), and Peru (e.g., Douglas 1921; Steinmann 1929; Willard 1966; Rangel 1978). Additional earlier references for the Andean region can be found in a study by Westermann and Riccardi (1990). Yet most modern studies are concentrated mainly on Argentina (Manceñido 1978, 1981, 1983, 1990; Riccardi et al. 1991) or New Zealand–New Caledonia (MacFarlan 1985, 1990), with many results still to be published.

Though this information does not suffice to elucidate the detailed relationships of brachiopods from different Jurassic marine basins, it is enough to outline the general tendencies of their biogeography. In the following treatment, A.S.D. has been responsible for most of the northern Pacific data and interpretations, and M. O. M. for those of the southern Pacific and the overall synthesis.

Representative brachiopods are illustrated in Plates 109–114.

## Early Jurassic

The Hettangian "*Piarorhynchia*" *pomeyroli* Drot, originally described from New Caledonia (but in need of revision), and the Sinemurian *Ochotorhynchia omolonensis* Dagys occur in Siberia. The apparent Hettangian relation to the southwestern Pacific is supported by the common occurrence of bivalves such as *Otapiria* spp. The endemic genus *Ochotorhynchia* (Plate 109, Fig. 1), which resembles Tethyan forms (especially the Dimerellidae), extended from the Okhotsk coast to the northern Verkhoyansk Mountains.

European species of *Tetrarhynchia, Furcirhynchia,* and *Rimirhynchia* were described from the Sinemurian of British Columbia (Ager and Westermann 1963) (Plate 114, Figs. 4, 5). Sinemurian European species of *Quadratirhynchia* and *Lobothyris* occur in Nevada, and *Sulcirostra paronai* (Böse) in Oregon (Ager, cited by Hallam 1965) (Plate 114, Figs. 3, 6). *Furcirhynchia* is also recorded from the Hettangian–Sinemurian of Japan (Hayami 1959), from Hettangian to Toarcian beds of New Zealand (Trechmann 1923; MacFarlan 1985, 1990), from the middle Liassic of Seram (Wanner and Knipscheer 1951; revised by Manceñido 1978, 1984) (Plate 114, Fig. 7), and from the Pliensbachian of Argentina (Manceñido 1978, 1990) (Plate 111, Fig. 5).

Presumably, European species migrated through the Tethyan seas and possibly crossed the Pacific Ocean by "island hopping" at low latitudes, as envisaged by Ager (1986) for Triassic brachiopods for a much "contracted" Mesozoic Earth. Transarctic connections between the European Boreal and the Pacific basins appear less probable, considering the small-scale early Liassic transgressions in the Arctic. Yet this has been invoked, in coincidence with the opening of the North Atlantic, by Ager and Sun (1988) to explain *Furcirhynchia* distribution.

From the scarce Hettangian–Sinemurian Andean brachiopod faunules (Steinmann 1929; Manceñido 1990; also in Westermann and Riccardi 1990 and in Riccardi et al. 1991) (Plate 111, Figs. 1, 2) one may underline the occurrence of *Gibbirhynchia* from Argentina to Peru, and of a *Spiriferina* closely reminiscent of *S. ongleyi* Marw. (typical of New Zealand Aratauran beds). East–West austral links across the Pacific could be accounted for as migration along suitable neritic habitats fringing the Gondwana margin.

By the early Pliensbachian, Andean brachipods increased in diversity and, at the generic (and even specific) level, denote close relationships with European taxa, especially from Gresten-like facies, such as *Rhynchonelloidea, Tetrarhynchia, Rudirhynchia, Spiriferina, Zeilleria* (*Z.*), *Z.* (*Cincta*), *Z.* (*Pirotella*) (cf. Manceñido 1990) (Plate 111). Most conspicuously, representatives of *Squamiplana* (*Cuersithyris*) are spread from the Iberian Peninsula, through Provence, to the Carpatho-Balkanids in the Carixian (cf. Sucic-Protic 1971; Alméras and Moulan 1982; Fauré 1985). This faunal exchange may have taken place through a proto-Atlantic shallow-marine connection, also called Hispanic Corridor, as postulated for the bivalve *Weyla* (Damborenea and Mancenido 1979, 1988).

Some elements with broad European ancestry persisted into the late Pliensbachian of the Andean region (Plates 111–112); otherwise, a certain degree of endemism is apparent in this area. *Peristerothyris* appears as an East Pacific genus, described from Argentina and almost certainly shared with California (Mancenido 1983, 1990), yet further research on western Pacific terebratulaceans is still needed. Affinities with New Zealand are also recognizable, among spiriferinaceans, for instance (Damborenea and Mancenido, in press), whereas the incoming of species of *Cirpa* and *Fissirhynchia* reveals somewhat "delayed" Tethyan Mediterranean or Himalayan–Indonesian influences. These are more obvious for the western Pacific, where *Prionorhynchia* cooccurs with the last two genera in Seram (Wanner and Knipscheer 1951; revised by Mancenido 1978, 1984; Ager and Sun 1988). Moreover, *Hesperithyris*, which belongs to an unmistakable group of ribbed terebratulids characteristic of the "Grauen-Kalke" (or "Calcari Grigi") facies of the Alps, Morocco, and Portugal, is abundant in Timor (Krumbeck 1923; revised by Mancenido 1978, 1984) (Plate 114, Figs. 9, 10).

In middle Liassic times, especially late Pliensbachian, both European (*Rudirhynchia, Rimirhynchia, Furcirhynchia, Zeilleria, Spiriferina*) and endemic (*Peregrinelloidea, Orlovirhynchia, Viligothyris*) genera (Plate 109) thrived in the northwestern Pacific. Worthy of notice is the seemingly amphi-Pacific nature of the *Peregrinelloidea–Anarhynchia* couplet. The Siberian *Peregrinelloidea malkovi* Dagys is very similar to the Californian *Anarhynchia gabbi* Ager, perhaps even synonymous (if differences in internal structures are attributable to defective preservation of the latter, as Dagys suspects). Furthermore, *A. gabbi* comes from tectonized deposits of uncertain age in the Santa Ana Mountains, which were supposed to be Callovian (Ager 1968), but may be as old as Early Jurassic, according to circumstantial evidence (Mancenido 1978). Such a close relationship is further attested by undescribed material of proven Sinemurian age from the western United States (Ager, cited by Hallam 1965) (Plate 114, Figs. 1, 2). This stock is likely to be present also in the early Jurassic of Argentina (Mancenido 1978) (Plate 112, Fig. 7).

The migration routes between the northwestern Pacific and the European basins usually were along the northern shoreline of Tethys, although, as among Pliensbachian ammonoids (Dagys 1976), transarctic migrations were also possible. Arctic basin migration would explain some apparent paradoxes in the stratigraphic distribution of several genera of Liassic brachiopods. In British Columbia and Japan, *Rudirhynchia* and *Furcirhynchia* are known from the early Sinemurian, whereas in Europe and the Andes their distribution was limited to the lower Pliensbachian, and in Siberia they are found in the upper Pliensbachian. Possibly these genera occurred originally in the Tethys and lower latitudes of the Pacific, then migrated into European Boreal basins, and eventually from there, through the Arctic basin, to the northwestern Pacific. This heterochronous distribution of some Liassic genera, however, may be only apparent and due to incompleteness of the geological record and/or detailed studies. For instance, *Aulacothyris* is locally first recorded in the Aratauran (Hettangian?–Sinemurian) of New Zealand and the early Pliensbachian of western Europe (Delance 1974), but not until the Aalenian in Argentina (unpublished data) – a pattern that defies a meaningful and straightforward explanation.

Information on late Liassic brachipods is very scanty for the northern Pacific. Few European genera are known from low latitudes (i.e., *Tetrarhynchia* for Sikhote-Alin and *Pseudogibbirhynchia* for Oregon and for western Thailand). The first Boreiothyrididae (i.e., *Omolonothyris* of Arctic affinities) appeared in the Toarcian of Siberia (Plate 109, Fig. 9).

For the South Pacific, New Zealand Toarcian faunas show a typical influx of core Tethyan elements (wellerellids, norellids, certain spiriferinaceans and terebratulaceans), with no obvious Andean links. Conversely, Toarcian assemblages of Argentina (Mancenido 1990) (Plate 112) have basically European and North African affinities. Thus, in the early Toarcian, local species of *Rhynchonelloidea* and *Piarorhynchia* occur together with polymorphic *Telothyris* closely comparable to those from coeval "Spanish facies" (or Iberian province), which extended from southern France, through the Iberian Peninsula, to Morocco-Algeria (cf. Alméras and Elmi 1984; Fauré 1985). Similar influences are still perceivable in members of the somewhat impoverished late Toarcian Argentinian faunas (*Sphenorhynchia*?, *Tetrarhynchia, Lobothyris, Rhynchonelloidea*), while *Flabellirhynchia*(?) seems to foreshadow a genus more common in the Aalenian and Bajocian of Britain, France, North America, Antarctica, and China.

**Middle Jurassic**

Brachiopods are very rare in the northwestern Pacific. The supposed *Gigantothyris* (Plate 110, Fig. 1), from the Aalenian of the Okhotsk Sea region (Dagys 1968a,b), belongs perhaps to a new genus based on different brachidium morphology of the Siberian species (Cooper 1983). In Chukotka (Koryak region) this supposedly allochthonous terrane yields the endemic *Inversithyris* (Plate 110, Fig. 2) and the mainly Arctic *Ptilorhynchia*. In low latitudes of the western Pacific occur essentially different assemblages (cf. Tokuyama 1957, 1958a; Khudolej and Prozorovskaya 1985; Alméras 1988), with *Globirhynchia, Kallirhynchia, Burmirhynchia, Zeilleria*, and others, originally described from Europe and Southeast Asia, and with endemic *Naradanithyris* (possibly derived from *Peristerothyris*, according to Mancenido 1983).

Apart from *Ptilorhynchia*, the assemblages of the northeastern Pacific contain only European genera: *Globirhynchia*,

*Kallirhynchia, Flabellirhynchia, Rhactorhynchia, Capillirhynchia, Loboidothyris, Euidothyris,* and *Ornithella* (Crickmay 1933).

Aalenian and Bajocian faunas from the Andes display generic affinities mainly with western Europe: *Kallirhynchia, Cymatorhynchia, Rhactorhynchia(?), Loboidothyris, Zeilleria, Aulacothyris,* and others (Plate 113, Figs. 1–7). Rhynchonellids of the "*R.*" *moerickei–manflasensis* group may represent an endemic development, whereas the occurrence in Chile of *Plectoidothyris*(?), a marginally plicate terebratulid of distinctive Tethyan affiliation (Ager and Walley 1977), is most interesting.

Most genera recognized in the Callovian (and Bathonian?) of Argentina, such as *Loboidothyris, Torquirhynchia,* and *Rhynchonelloidella,* are widespread from England, across Europe, to the Crimea and the Caucasus and thus are currently allocated to the northern shore of the Tethys (Ager and Sun 1988; cf. Khudolej and Prozorovskaya 1985). *Lophrothyris* shows a similar distribution pattern, with a further extension into Indo-Ethiopian areas (Cutch, Somalia, Madagascar, Morocco, etc.), that is, areas of the southern Tethys (Plate 113, Figs. 8–11).

*Cryptorhynchia,* initially described from Cutch (India), according to Ager (1986; Ager and Sun 1988) also reached Madagascar and Tibet, and from there crossed the Pacific at a low latitude to southern Idaho (Perry 1979). Conversely, alleged *Cryptorhynchia* from Bathonian–Callovian beds of New Zealand should be disregarded, "*C.*" *kawhiana* Trechmanmn being in fact an Acanthothyridinae, as believed by Manceñido (1978, p. 141) and vindicated by MacFarlan (1985, 1990).

Alméras and Gupta (1986) have convincingly argued that in Bathonian–Callovian times the *Holcothyris* stock was common to both shores of the Tethys and that the genus *Kutchithyris* represents an East Tethyan offshoot thereof, being recorded from India, Somalia, Madagascar, southern China, and New Zealand (Trechmann 1923; Marwick 1953; Speden and Keyes 1981).

## Late Jurassic

The greatest geographic differentiation in the northern Pacific brachiopods occurred in the Late Jurassic. In the northwestern Pacific there was a typically Arctic brachiopod fauna. Oxfordian to early Volgian of the lower Kolyma area yielded an assemblage with predominant Boreiothyrididae: *Boreiothyris* and *Taimyrothyris* (Dagys 1968a,b) (Plate 110), plus *Uralella, Pinaxiothyris,* and *Ptilorhynchia.*

In low latitudes, brachiopods retained Tethyan affinities. The Torinosu faunas of Japan and Sarawak contain species of genera that are known from Europe and northeastern Asia (Tokuyama 1957, 1958b, 1959), such as *Burmirhynchia, Parvirhynchia, Neumayrithyris,* and probably *Disculina* (represented by "*Terebratulina*" *nishiyamensis* Tokuyama, according to Manceñido). Likewise, in the eastern Pacific, Crickmay (1933) recorded *Kallirhynchia, Terebratulina,* and *Argyrotheca* from the Tithonian of California. Cooper (1983) described the endemic genera *Animonithyris* and *Mexicaria* (Plate 114, Figs. 11, 12) from the Oxfordian of Mexico, and the ribbing pattern displayed by the latter is characteristic of Jurassic Tethyan terebratulids (Ager and Walley 1977). Representatives of *Xestosina, Septaliphoria,* and *Rhynchonella* s.s. are recorded in the Mexican Kimmeridgian and Tithonian (Alencaster and Buitrón 1965; Alencaster 1977; Rivera-Carranco et al. 1984; Boullier and Michaud 1987) (Plate 114, Fig. 13). Such connection at generic level between Europe and Mexico may be related to a Gulf-Stream-type circulation, as advocated by Ager (1986; Ager and Sun 1988; Sandy 1990).

As for the Andes, the finding of *Thurmannella* in the Oxfordian of west-central Argentina (Plate 113, Fig. 12) is a salient feature. Also reported from easternmost China (Sun 1983) and likely to be represented in Indonesia by Krumbeck's (1922, 1923) *R. tooica,* the genus is regarded as common to Tethyan and eastern European areas (Khudolej and Prozorovskaya 1985) or typical of the northern shore of the Tethys (Ager and Sun 1988). This last remark also applies to the genus *Lacunosella* probably present in the Tithonian of the Andean foothills of Argentina (cf. Burckhardt 1900; Riccardi 1983).

## Conclusions

In spite of the uneven investigation of Jurassic brachiopods in the circum-Pacific, some biogeographic trends can be recognized. In high latitudes of the northwestern Pacific (North-East Russia), and probably also on the northeastern Pacific side, the brachiopods show affinities to Arctic faunas; in low latitudes of the northern Pacific, links to Tethyan and northwest European faunas become stronger instead (cf. Khudolej and Prozorovskaya 1985).

Minimum and maximum brachiopod provincialism between faunas of different latitudes occurred in the Lias and Malm, respectively. Pliensbachian brachiopods of North-East Russia contain several European genera, but only endemic species; Late Jurassic brachiopods from this region are endemic at the genus level.

There are biogeographic differences between North-East Russian and low-latitude northern Pacific faunas: (1) The late Pliensbachian *Orlovirhynchia* and *Peregrinelloidea* from Siberia are among the largest rhynchonellids ever found, and Siberian species of *Rimirhynchia* and *Rudirhynchia* were larger than European and eastern Pacific species. This was enhanced toward the Late Jurassic, as most Boreiothyrididae reached a shell length of 60–90 mm. (2) Entirely ribbed rhynchonellids (*Orlovirhynchia, Rudirhynchia*) and biplicate terebratulids (*Viligothyris*) occurred in North-East Russia only in the Lias. For the Dogger and Malm, all rhynchonellids are only marginally plicate, and terebratulids are sulcate and rectimarginate (i.e., feebly ornamented or smooth shells). In low latitudes of the northern Pacific (including British Columbia), ribbed rhynchonellids and plicate or even ribbed (*Mexicaria*) terebratulids existed during the whole Jurassic. Low- and high-latitude faunas were thus quite distinctive in external morphology, and ocean temperature may have been the main reason for that. Differences between eastern and western northern Pacific brachiopod faunas, however, have not been recognized yet.

In the southern Pacific, the Andean region has played a hinging role between high-latitude faunas and those from Europe (celtoswabian) or the western end of the Tethys. A proto-Atlantic passageway (or Hispanic Corridor) seems to have favored connection with the latter, perhaps intermittently, with important exchange events in early Pliensbachian, Toarcian, and again in Middle Jurassic times. Endemism appears low during the early Lias and fairly strong for the late Pliensbachian. Biplicate terebratulids are known already from the Liassic (*Peristerothyris, Exceptothyris*?, *Telothyris*), as well as from the Middle and Upper Jurassic (*Loboidothyris*), whereas marginally plicate ones have been detected only in the Bajocian of Chile, so far. Toward the end of the Middle Jurassic (Bathonian–Callovian), an influx of Indo–Ethiopian elements is noticeable; otherwise, Andean Middle and Late Jurassic brachiopods essentially show prevailing affinities to European–north Tethyan faunas.

New Zealand–New Caledonia, lying at higher paleolatitudes, maintained a considerable degree of endemism (the so-called Maorian Province), though sharing some elements in common with the Andes during the late Triassic, the late Sinemurian and the Pliensbachian, a likely consequence of peri-Gondwanian migration. By middle Liassic times, rhynchonellid-spiriferinid-dominated assemblages of Seram are more closely related to certain Ururoan faunas from "distant" New Zealand than to those with predominant strongly plicate terebratulids of western Timor, in spite of being nearer (and even somewhat northward of the latter) on a present-day map. This may bear on the question of the peculiar role played by the Wallacea Line for the detailed tracing of various boundary lines of neontological biogeographic interest. This phenomenon was analyzed on the basis of marine bivalves by Hayami (1984, 1987) in relation to the suturing between what in the Mesozoic used to be opposite shores of the Tethys. From the Toarcian onward, Tethyan influences became increasingly apparent in Australasia, as duly recognized by New Zealand authors (e.g., Grant-Mackie 1985; Stevens 1990), and strengthened by the discovery of thecideaceans (Maxwell 1987) and nucleatids (Manceñido in press).

It is more than likely that the number of known endemic genera will rise after all the faunas have been thoroughly monographed. Yet provinciality is expected to remain low in comparison with the Triassic, when the Pacific was a strict barrier for the majority of Tethyan brachiopods (Dagys 1974).

Jurassic provinciality patterns probably were determined by an interplay of factors resulting chiefly from global changes in sea level plus regional (or even local) geotectonic conditions.

## References


Ager, D. V. (1968). The supposed ubiquitous Tethyan brachiopod *Halorella* and its relations. *J. Paleontol. Soc. India, 5–9,* 54–70.

(1986). Migrating fossils, moving plates and an expanding Earth. *Modern Geology, 10,* 377–90.

Ager, D. V., & Sun, D.-L. (1988). Distribution of Mesozoic brachiopods on the northern and southern shores of Tethys. *Palaeontologia Cathayana, 4,* 23–51.

Ager, D. V., & Walley, C. D. (1977). Mesozoic brachiopod migrations and the opening of the North Atlantic *Palaeogeogr. Palaeoclimatol. Palaeoecol., 21*(2), 85–99.

Ager, D. V., & Westermann, G. E. G. (1963). New Mesozoic brachiopods from Canada. *J. Paleontol., 37*(3), 595–610.

Alencaster, G. (1977). Moluscos y braquiópodos del Jurásico superior de Chiapas. *Rev. Univ. Nac. Auton. México, Inst. Geol., 1*(2), 151–66.

Alencaster, G., & Buitrón, B. (1965). Estratigrafía y paleontología del Jurásico Superior de la parte centromeridional del Estado de Puebla. Parte II. Fauna del Jurásico Superior de la Región de Petlalcingo, Estado de Puebla. *Paleontología Mexicana, 21,* 1–53.

Alméras, Y. (1988). Jurassic brachiopods from the Klo Tho–Mae Sot area. In H. Fontaine & V. Suteethorn (eds.), *The Upper Palaeozoic and Mesozoic Fossils of West Thailand* (pp. 211–7). C.C.O.P. Techn. Bull. 20.

Alméras, Y., & Elmi, S. (1984). Fluctuations des peuplements d'ammonites et de brachiopodes en liaison avec les variations bathymétriques pendant le Jurassique inférieur et moyen en Méditeranée Occidentale. *Boll. Soc. Paleontol. Ital., 21*(2–3), 169–87.

Alméras, Y., & Gupta, V. J. (1986). Importance des Terébratulidés dans les reconstitutions paléobiogéographiques: exemple du Bathonien téthysien. In P. Racheboeuf & C. Emig (eds.), *Les Brachiopodes Fossiles et Actuels. Biostratigr. Paleoz., 4,* 419–29.

Alméras, Y., & Moulan, G. (1982). Les Térébratulidés liasiques de Provence (Paléontologie – biostratigraphie – paléoécologie – phylogénie). *Doc. Lab. Geol. Fac. Sci. Lyon, 86,* 1–365.

Boehm, G. (1907–12). Beiträge zur Geologie von Niederländisch-Indien. I Abteilung: Die Südküsten der Sula-Inseln Taliabu und Mangoli. *Palaeontographica, Suppl. 4,* 9–179.

Boullier, A., & Michaud, F. (1987). Térébratulidés (Brachiopodes) nouveaux du Jurassique superieur du Chiapas (Sud-Est du Mexique). *Revue Paleobiol. (Genéve), 6*(2), 279–88.

Burckhardt, C. (1900). Coupe géologique de la Cordillere entre Las Lajas et Curacautín. *An. Mus. La Plata, Sec. Geol. Min. 3,* 1–100.

Cooper, G. A. (1983). The Terebratulacea (Brachiopoda), Triassic to Recent: a study of the brachidia (loops). *Smithson. Contrib. Paleobiol., 50,* i–ix, 1–445.

Crickmay, C. H. (1933). Attempt to zone the North American Jurassic on the basis of its brachiopods. *Bull. Geol. Soc. Am., 44*(5), 871–93.

Dagys, A. A. (1976). *Late Pliensbachian Ammonites of North Siberia* (in Russian). Novosibirsk: Nauka.

Dagys, A. S. (1968a). Brakhiopody. In A. F. Efimova et al. (eds.), *Polevoi atlas yurskoj fauny i flory severo-vostoka SSSR* (Field Atlas of Jurassic fauna and flora of northeastern USSR) (pp. 23–8). Magadan.

(1968b). Yurskie i rannemelovye brakhiopody Severa Sibiri. *Tr. Inst. Geol. Geofiz., 41,* 1–170.

(1974). *Triasovye brakhiopody (Morfologiya, sistema, filogeniya, stratigraficheskoe znachenie i biogeografiya).* Novosibirsk: Nauka.

Damborenea, S. E., & Manceñido, M. O. (1979). On the palaeogeographical distribution of the pectinid genus *Weyla* (Bivalvia, Lower Jurassic). *Palaeogeogr. Palaeoclimatol. Palaeoecol., 27*(1–2), 85–102.

(1988). *Weyla:* semblanza de un bivalvo jurásico andino. *Actas 5, Congr. Geol. Chileno, 2,* C13–25.

(in press). A comparison of Jurassic marine benthonic faunas from South America and New Zealand. *J. Roy. Soc. N.Z.*

Delance, J. H. (1974). Zeilleridés du Lias d'Europe Occidentale (Brachiopodes). Systématique des populations. Phylogénie. Biostratigraphie. *Mem. Geol. Univ. Dijon, 2,* 1–408.

Douglas, J. A. (1921). Geological sections through the Andes of Peru and Bolivia. III: From Port of Callao to the River Perene. *Q. J. Geol. Soc. London, 77*(3), 246–84.

Fauré, P. (1985). Le Lias de la partie centro-orientale des Pyrénées espagnoles (Provinces de Huesca, Lérida et Barcelona). *Bull. Soc. Hist. Nat. Toulouse, 121,* 23–37.

Feruglio, E. (1934). Fossili Liassici della valle del Rio Genua (Patagonia), *Giorn. Geol., Ann. R. Mus. Geol. Bologna, ser. 2, 9,* 1–64.
Grant-Mackie, J. A. (1985). New Zealand-New Caledonian Permian-Jurassic faunas, biogeography and terranes. *N. Z. Geol. Surv. Rec., 9,* 50–2.
Hallam, A. (1965). Observations on marine Lower Jurassic stratigraphy of North America, with special reference to United States. *Bull. Am. Assoc. Petrol. Geol., 49*(9), 1485–501.
Hayami, I. (1959). Lower Liassic lamellibranch fauna of Higashinagano Formation in West Japan. *J. Fac. Sci., Univ. Tokyo, 12*(1), 31–84.
(1984). Jurassic marine bivalve faunas and biogeography in Southeast Asia. *Geol. Palaeontol. SE Asia, 25,* 229–37.
(1987). Geohistorical background of Wallace's line and Jurassic marine biogeography. In A. Taira & M. Tashiro (eds.), *Biogeography and Plate Tectonic Evolution of Japan and Eastern Asia* (pp. 111–33). Tokyo.
Jaworski, E. (1915). Beiträge zur Kenntnis des Jura in Süd-Amerika. Teil II: Spezieller, paläontologischer Teil. *N. Jb. Min. Geol. Palaontol. Beil., 40,* 364–456.
(1925). *Contribución a la Paleontología del Jurásico sudamericano.* Dir. Gral. Min. Geol. Hidrol., Sec. Geol. (Buenos Aires). Publ. 4.
(1926). La fauna del Lias y Dogger de la Cordillera Argentina en la parte meridional de la provincia de Mendoza. *Actas Acad. Nac. Cienc., 9*(3–4), 137–316.
Khudolej, K. M., & Prozorovskaya, E. L. (1985). Osnovnye cherty paleobiogeografii Evrazii v yurskom periode. *Sovietskaya Geologiya, 9,* 65–76.
Krumbeck, L. (1922). Zur Kenntnis des Juras der Insel Rotti. *Jaarb. v.h. Mijnewezen Nederl. Oost-indië, 3,* 107–219.
(1923). Zur Kenntnis des Juras der Insel Timor sowie des Aucellen-Horizontes von Seran und Buru. *Paläontologie von Timor (Stuttgart), 12*(20), 1–120.
MacFarlan, D. A. B. (1985). *Triassic and Jurassic Rhynchonellacea (Brachiopoda) from New Zealand and New Caledonia.* PhD thesis, University of Otago, Dunedin.
(1990). Triassic and Jurassic Rhynchonellacea (Brachiopoda) from New Zealand and New Caledonia. In *Abstracts of the 2nd International Brachiopod Congress, Dunedin* (p. 63).
Manceñido, M. O. (1978). *Studies of Early Jurassic Brachipoda and Their Distribution, with Special Reference to Argentina.* PhD thesis, University of Wales, Swansea.
(1981). A revision of Early Jurassic Spiriferinidae (Brachiopoda, Spiriferida) from Argentina. In W. Volkheimer & E. Musacchio (eds.), *Cuencas sedimentarias del Jurásico y Cretácico de América del Sur* (vol. 2, pp. 625–59). Buenos Aires.
(1983). A new terebratulid genus from western Argentina and its homoeomorphs (Brachiopoda, Early Jurassic). *Ameghiniana, 20*(3–4), 347–65.
(1984). *Early Jurassic Brachiopod Faunas from Timor and Seram, Indonesia.* Circum-Pacific Jurassic Research Group, IGCP Project 171, Report 2, p. 73.
(1988). On the validity of "*Terebratula*" *inca* Forbes, 1846 (Brachiopoda, Terebratulida). In *Actas 5, Congr. Geol. Chileno* (vol. 2, pp. C1–11). Santiago.
(1990). The succession of Early Jurassic brachiopod faunas from Argentina: correlation and affinities. In D. I. MacKinnon, D. E. Lee & J. D. Campbell (eds.), *Brachiopods through Time* (pp. 397–404). *Proc. 2nd International Brachiopod Congress, Dunedin.* Rotterdam: A. Balkema.
(in press). First record of Jurassic nucleatid brachipods from the SW Pacific, with comments on the global distribution of the group. *Palaeogeogr. Palaeoclimatol. Palaeoecol.*
Manceñido, M. O., & Damborenea, S. E. (1990). Corallophilous micromorphic brachiopods from the Lower Jurassic of West Central Argentina. In *Proceedings of the 2nd International Brachiopod Congress, Dunedin* (pp. 89–96). Rotterdam: Balkema.
Marwick, J. (1953). *Divisions and Faunas of the Hokonui System (Triassic and Jurassic).* N.Z. Geol. Surv., Paleontol. Bull. 21.
Maxwell, P. A., (1987). Square snails, button-shells and jewell-box brachiopods – an unusual faunule from the North Canterbury Torlesse. In *Programs and Abstracts, Geological Society of New Zealand.* Miscellaneous Publication 37A. Dunedin.
Pérez, E. (1982). Bioestratigrafía del Jurásico de Quebrada Asientos, Norte de Potrerillos, Región de Atacama. *Bol. Serv. Nac. Geol. Min. Chile, 37,* 1–149.
Perry, D. G. (1979). A Jurassic brachiopod-oyster association, Twin Creek Limestone, southeastern Idaho. *J. Paleontol., 53*(4), 997–1004.
Quilty, P. G. (1972). Middle Jurassic brachiopods from Ellsworth Land, Antarctica. *N.Z. J. Geol. Geophys., 15*(1), 140–7.
(1982). Tectonic and other implications of Middle–Late Jurassic rocks and marine faunas from Ellsworth Land, Antarctica. In C. Craddock, (ed.), *Antarctic Geoscience* (pp. 669–78). *Proc. Symp. Antarct. Geol. Geophys., Madison, Wisconsin, 1977.*
Rangel, C. (1978). Fósiles de Lircay – Uruto. *Bol. Inst. Geol. Min. Perú, Ser. D, 6,* 1–35.
Riccardi, A. C. (1983). The Jurassic of Argentina and Chile. In M. Moullade & A. Nairn (eds.), *The Phanerozoic Geology of the World II. The Mesozoic, B.* (pp. 201–63). Amsterdam: Elsevier.
Riccardi, A. C., Damborenea, S. E., Manceñido, M. O., & Ballent, S. C. (1991). Hettangian and Sinemurian biostratigraphy of Argentina. *J. South. Am. Earth Sci., 4*(3), 159–70.
Rivera-Carranco, E., Hernandez, A. & Buitrón, B. E. (1984). *Septaliphoria potosina* n.sp. (Brachiopoda – Rhynchonellida) del Jurásico tardío de la Sierra de Catorce, San Luis de Potosí, Mexico. In *Mem. III Congr. Latinoam. Paleontol.* (pp. 216–24).
Sandy, M. R. (1990). Biogeographic affinities of some Jurassic–Cretaceous brachiopod faunas from the Americas and their relation to tectonic and paleoceanographic events. In D. I. MacKinnon, D. E. Lee, & J. D. Campbell (eds.), *Brachiopods through Time* (pp. 415–422). *Proc. 2nd International Brachiopod Congress, Dunedin.* Rotterdam: A. Balkema.
Speden, I., & Keyes, I. (1981). Illustrations of New Zealand Fossils. *N.Z. D.S.I.R. Inf. Ser., 150,* 1–109.
Steinmann, G. (1929). *Geologie von Peru.* Heidelberg: C. Winters Univ.
Stevens, G. R. (1990). The influences of palaeogeography, tectonism and eustasy on faunal development in the Jurassic of New Zealand. In G. Pallini et al. (ed.), *Atti II Conv. Int. Fossili, Evoluzione, Ambiente, Pergola 1987* (pp. 441–57).
Sucic-Protic, Z. (1971). *Mesozoic Brachiopoda of Yugoslavia. Middle Liassic Brachiopoda of the Yugoslav Carpatho-Balkanids (Part II).* Univ. Belgrade Monogr. 5.
Sun, D.-L. (1983). New rhynchonellid brachiopods from Upper Jurassic of Hulin County, eastern Heilongjiang Province (in Chinese). In *Fossils from the Middle–Upper Jurassic and Lower Cretaceous in Eastern Heilongjiang Province, China* (pp. 73–86). Heilongjiang Sci. Technol. Publ. House.
Thiele, R. (1964). Reconocimiento geológico de la Alta Cordillera de Elqui. *Publ. Fac. Cienc. Fis. y Matem., Inst. Geol., Univ. de Chile, 27,* 131–97.
Tokuyama, A. (1957). On some Jurassic rhynchonellids from Shikoku, Japan. *Trans. Proc. Paleontol. Soc. Japan, N.S., 28,* 128–36.
(1958a). On some terebratuloids from the Middle Jurassic Naradani Formation in Shikoku, Japan. *Japan. J. Geol. Geogr., 29*(1–3), 1–10.
(1958b). On some terebratuloids from the Late Jurassic Torinosu Series in Shikoku, Japan. *Japan. J. Geol. Geogr., 29*(1–3), 119–31.
(1959). Bemerkungen über die Brachiopodenfazies der oberjurassischen Torinosuserie Südwestjapans, mit Beschreibungen einiger Formen. *Japan. J. Geol. Geogr., 30,* 183–94.
Trechmann, C. T. (1923). The Jurassic rocks of New Zealand. *Q. J. Geol. Soc. London, 79*(3), 246–86, 309–12.

Wahnish, E. (1942). Observaciones geológicas en el oeste del Chubut. Estratigrafía y fauna del Liásico en los alrededores del Río Genua. *Bol. Dir. Min. Geol. (Buenos Aires)*, *51*, 1–73.

Wandel, G. (1936). Beiträge zur Paläontologie des Ostindischen Archipels XIII. *N. Jb. Min. Geol. Palaontol., Beil.*, *75*(B), 447–56.

Wanner, J., & Knipscheer, H. C. G. (1951). Der Lias der Niefschlucht in Ost-Seran (Molukken). *Eclogae Geol. Helvet.*, *44*(1), 1–28.

Weaver, C. E. (1931). Paleontology of the Jurassic and Cretaceous of West Central Argentina. *Mem. Univ. Washington*, *1*, 1–469.

Westermann, G. E., & Riccardi, A. C. (eds.). (1990). Jurassic Taxa Ranges and Correlation Charts for the Circum-Pacific. 3: South America and Antarctic Peninsula. *Newsl. Stratigr.*, *21*(2), i–vi, 75–147.

Willard, B. (1966). *The Harvey Bassler Collection of Peruvian Fossils.* Bethlehem: Lehigh Univ.

Yanagida, J., & Lau, J (1078). The Upper Jurassic and Middle Cretaceous Terebratulidae from the Bau Limestone Formation in West Sarawak, Malaysia. *Geol. Palaeontol. SE Asia*, *19*, 35–47.

# 23 Belemnites of the circum-Pacific region

A. B. CHALLINOR, P. DOYLE, P. J. HOWLETT, and T. I. NAL'NYAEVA

Several authors have studied the biogeography of the Jurassic Belemnitida (e.g., Stevens 1967, 1973; Saks and Nal'nyaeva 1970, 1972, 1975; Mutterlose 1986; Doyle 1987), but none has discussed circum-Pacific reconstructions. Boreal and Tethyan marine realms are well established by the Belemnitina and Belemnopseina, respectively. No austral belemnite realm is recognized for the Jurassic, but endemicity at the provincial level exists in the Indo–Pacific. The Boreal fauna is less sharply delineated than in Central Europe, no doubt a function of the Pacific Ocean.

In the following pages, belemnite distributions are dealt with in four major regions. No discussion of the causes of provinciality is given here, but that may be found elsewhere: Stevens (1967, 1973), Mutterlose (1986), Doyle (1987) and Doyle and Howlett (1989).

Representative specimens are illustrated in Plates 128–132.

## SOUTHWEST PACIFIC[1]

### Indonesia

Jurassic belemnites are known from the eastern Indonesian regions of Misool, Sula Islands, Roti, Timor, Jamdena, and Irianjaya (Skwarko and Yusef 1982). Early works of importance include those of Boehm (1907, 1912), Kruizinga (1920), Stolley (1929, 1935) and Stevens (1963a,b; 1964a,b). Stevens (1965) reviewed what remained of the early collections. Sula Islands belemnites were revised and described by Challinor and Skwarko (1982), and Misool belemnites by Challinor (1989b).

*Dicoelites* appears in the late Bajocian of Sula and is present at about middle Callovian and middle Oxfordian. *Conodicoelites* occurs in the middle Bathonian. *Belemnopsis* first appears in the late Bathonian, but is then unrecorded until latest Oxfordian, when it becomes abundant through to the late Tithonian. Occasional *Hibolithes* occur in the late Tithonian (Challinor and Skwarko 1982). The belemnites are dated on the basis of ammonites and bivalves, but the Bathonian–Callovian data of Sato et al. (1978) have been revised by Westermann and Callomon (1988). Early records of the ?pre-Kimmeridgian belemnites of Sula include *Hibolithes* (Boehm 1907; Kruizingas 1920), *Conodicoelites* and *Dicoelites* (Boehm 1912), and *Belemnopsis* (Kruizinga 1920). The ages of these and their relations to taxa described by Challinor and Skwarko (1982) are not clear. The Kimmeridgian–Tithonian belemnite record of Sula is thought to be reasonably complete, but the pre-Kimmeridgian record is poor.

Rich belemnite assemblages are present in the early Callovian–latest Tithonian of Misool. *Belemnopsis* and *Conodicoelites* are present in the Callovian, and *Dicoelites* in the Callovian and early Oxfordian; *Belemnopsis*, diverse *Hibolithes*, and a new genus based on "*Hibolithes weberi*" are present in the Oxfordian, and the Kimmeridgian–Tithonian is dominated by *Belemnopsis*. Occasional *Hibolithes* are present in the Tithonian. Except for the Tithonian *Hibolithes*, material is abundant, with individual species represented in some instances by hundreds of specimens (Challinor 1989b). The earliest species of *Dicoelites* and *Conodicoelites* to appear in Misool differ from their representatives in the Sula Islands, and it is thought that the earliest Callovian *Dicoelites* and *Conodicoelites* of Misool are younger than the Callovian belemnites of Sula. Correlation between Misool and Sula, based on three species of an evolving *Belemnopsis* lineage (Challinor 1989c), is excellent in the Oxfordian–Tithonian. Only belemnites of the suborder Belemnopseina are known from the Indonesian region.

### Papua New Guinea

Jurassic belemnites are known from many localities in the central highlands of Papua New Guinea. Moderately large collections, mostly from the Blücher Range, have recently been studied (Challinor 1990). *Conodicoelites* (Bathonian), *Belemnopsis* (?Oxfordian–late Tithonian), and *Hibolithes* (?Callovian–Oxfordian and Tithonian), identical at the species level with those of eastern Indonesia, are present.

[1] By A. B. Challinor.

### Australia

Middle Jurassic *Belemnopsis* are known from the Perth Basin near Geraldton (Whitehouse 1924), and Late Jurassic *Belemnopsis* from the Canning Basin near Broome (Teichert 1940; Brunnschweiler 1960). Whitehouse's *Belemnopsis* may be better placed in *Dicoelites* and may represent an incursion of the Indonesian Middle Jurassic *Conodicoelites–Dicoelites* fauna into Western Australia. The Late Jurassic *Belemnopsis* from Broome are similar to or identical with Indonesian Tithonian *Belemnopsis.*

### New Zealand

Jurassic belemnites of New Zealand are found mostly in the Murihiku Supergroup (Suggate, Stevens, and Te Punga 1978). Those few known from the Torlesse Supergroup (Suggate et al. 1978) are possibly identical with those of Murihiku rocks (Stevens 1965). Belemnites are sparse in the Toarcian (Early Jurassic) and Middle Jurassic and moderately abundant in the Late Jurassic. Their distribution is discussed here in terms of New Zealand stages (Marwick 1953); correlation of New Zealand and international Late Jurassic stages is currently under review (Chapter 17).

*Cylindroteuthis*(?) is known from the Ururoan stage (Toarcian), and *Brachybelus, Cylindroteuthis*(?), *Belemnopsis,* and *Hibolithes* from the Temaikan (Middle Jurassic) (Stevens 1965; Challinor 1977; Hudson 1983). *Conodicoelites* is present close to the Middle–Late Jurassic boundary (Hudson 1983). *Belemnopsis, Dicoelites,* and *Hibolithes* are present in the Heterian, Ohauan, and Puaroan (Late Jurassic) (Stevens 1965; Challinor 1975, 1979a,b, 1980). Belemnitina are present in the Toarcian, Belemnitina and Belemnopseina in the Middle Jurassic, and Belemnopseina in the Late Jurassic.

### New Caledonia

Similarities between Jurassic faunas of New Zealand and New Caledonia have led to the use of New Zealand stage nomenclature in the latter region (Pharo 1967). New Caledonian belemnites (Avias 1953; Stevens 1965) are now known to be nearly all of Temaikan (Middle Jurassic) age (Challinor and Grant-Mackie 1986), Belemnitina being represented in basal Temaikan beds by a new genus, and Belemnopseina by *Dicoelites* (early Temaikan), *Belemnopsis* (middle to late Temaikan), and scattered *Hibolithes* (late Temaikan). *Hibolithes* is also recorded from the Late Jurassic, but on very limited material.

### Regional assemblages

Two regional Jurassic belemnite assemblages, those of Indonesia–Papua New Guinea and New Zealand, are present in the southwestern Pacific. Indonesia, Papua New Guinea, and possibly parts of northwestern Australia form a coherent group, with identical species found throughout, but the New Zealand and New Caledonian assemblages differ from that of Indonesia in several ways.

*Cylindroteuthis*(?) and *Brachybelus* (Belemnitina) appear in New Zealand but are not known from the Indonesian region. *Dicoelites* and *Conodicoelites* are of major importance in the Bajocian–early Oxfordian of Indonesia. In New Zealand, *Conodicoelites* appears in the Kimmeridgian (according to Stevens 1980) or Callovian–Oxfordian (according to Hudson, Grant-Mackie, and Helby 1987), and *Dicoelites* occur briefly (Challinor 1980) in beds dated as Kimmeridgian by Stevens (1980) or Tithonian by Helby, Wilson, and Grant-Mackie (1988). *Belemnopsis* is present in New Zealand in the Bajocian, Kimmeridgian, and Tithonian, although it is not a continuous record. It appears in Indonesia in late Bathonian time and is abundant and consistently present from the Callovian to the end of the Tithonian. *Hibolithes* occurs first in the Bajocian–Callovian of New Zealand and reappears in the Kimmeridgian to Tithonian (Stevens 1980) or in the middle to late Tithonian (Helby et al. 1988) and is of major importance during this interval. In contrast, *Hibolithes* is a minor element in the Kimmeridgian–Tithonian of Indonesia, but important in the late Callovian–Oxfordian.

Strong similarities between New Zealand and Indonesian Belemnitida have been claimed, and identical species recognized in the two regions (Stevens 1963a, 1965; Challinor and Skwarko 1982), but this is not accepted here. The *uhligi* complex, a postulated closely interrelated widespread group recognized by Stevens (1965) from Indonesia, New Zealand, northern India, Arabia, and East Africa, and accepted by later workers (e.g., Jeletzky 1983; Mutterlose 1986), appears to have little general validity. While the Late Jurassic *Belemnopsis* of Indonesia certainly are an evolving group (Challinor 1989c), they are not closely related to New Zealand *Belemnopsis. Belemnopsis uhligi* Stevens, from which the *uhligi* complex is named, is a Himalayan species (Stevens 1963b) and is itself not known from Indonesia (Challinor and Skwarko 1982). At present there is no evidence of identical *Belemnopsis* species from northern India and Indonesia, but the two groups are similar and probably closely related.

The presence of identical species in New Zealand and Indonesia has recently been refuted (Challinor 1979a,b, 1989b; Challinor and Skwarko 1982), and there are no species common to both regions. In view of the disjunct stratigraphic distribution of genera (Chapter 17), this is not surprising. Broad within-genera differences are also present. Late Jurassic *Belemnopsis* of Indonesia (all members of a clade) are large, very strongly grooved forms (Challinor 1989c). Those of New Zealand are smaller and are relatively less strongly grooved (Stevens 1965). New Zealand Late Jurassic *Hibolithes* are all compressed, with long ventral alveolar grooves; most Indonesian Late Jurassic species are short-grooved, and some are depressed in cross section. The many dissimilarities discussed indicate that the two groups are not closely related, and there was little genetic interchange between them.

The affinities of the New Caledonian assemblage are not clear. Only Middle Jurassic taxa are known in any detail, and New Zealand Middle Jurassic forms are not well known. The presence of Belemnitina in New Caledonia and similar unusual morphology in New Zealand and New Caledonian *Belemnopsis* (Challinor and Grant-Mackie 1986) suggest that they are part of the same

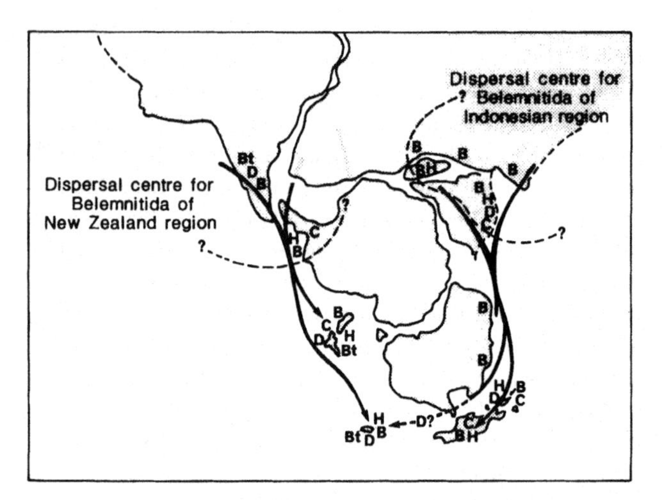


**Figure 23.1. Suggested dispersal centers and migration paths for belemnites of the Indonesian region and New Zealand. Continental positions for Tithonian time; continental shelves omitted. Abbreviations: B, *Belemnopsis;* Bt, *Belemnitina;* C, *Conodicoelites;* D, *Dicoelites;* H, *Hibolithes.* (adapted from Smith et al. 1981, map 36.)**

assemblage. However, no species are identical in both regions, and the strong development of *Dicoelites* in the New Caledonian Middle Jurassic suggests affinities with Indonesia.

### Paleobiogeographic interpretation

Recognition of two belemnite paleoprovinces in the southern and western Pacific raises the question of their origins and migration pathways into the region. These problems are discussed against the Tithonian paleogeographic construction of Gondwana by Smith, Hurley, and Briden (1981).

Stevens (1965) has suggested an origin for Indonesian Late Jurassic belemnites in the Madagascar, northern India, and East African regions, with subsequent migration to, and modification in, the southwest Pacific. Uhlig (1910) figured large, strongly grooved *Belemnopsis* from northern India (Spiti and Jandu), similar in general morphology to Indonesian forms. Late Jurassic *Belemnopsis* of southern Tibet are also similar (Yang and Wu 1964), and a migration pathway from northern India–southern Tibet into Indonesia is accepted here (Figure 23.1).

Distinctive, truly conical *Conodicoelites* differing in morphology from other members of the genus are known from the Tethyan Himalayas (Stevens 1964a), Papua New Guinea, and Indonesia (Challinor 1990; Challinor and Skwarko 1982), suggesting this pathway was operative in Bathonian–Callovian time. The evidence for origin of Indonesian Jurassic *Hibolithes* is not clear, although there is weak evidence for an origin in the Malagascan region (Challinor 1991). Other distinctive taxa migrated between central and western Europe, Madagascar, Mozambique, and Indonesia during the Cretaceous (Challinor 1989a, 1991), but are not discussed here.

Belemnites of New Zealand are considered here to have migrated from a dispersal center somewhere in the Antarctic Peninsula–Patagonia–Falkland Plateau region via a pathway marginal to western Antarctica (Figure 23.1). This is suggested by the presence of *Cylindroteuthis, Brachybelus, Dicoelites,* and *Belemnopsis* in South America (Stevens 1965), *Conodicoelites* similar to those of New Zealand in western Antarctica (Stevens 1967) and *Belemnopsis* similar to those of New Zealand found on the Falkland Plateau (Jeletzky 1983) at the relevant times. Furthermore, Late Jurassic *Belemnopsis* and *Hibolithes* of western Antarctica (Mutterlose 1986) appear almost identical with those of New Zealand.

Mutterlose (1986) suggests an origin for New Zealand forms via a circum-Gondwana pathway from Indonesia through Papua New Guinea and New Caledonia, but that hypothesis is not accepted here.

In the Middle and Late Jurassic, two provinces of the Tethyan Realm are recognized. (1) A Himalayan Province characterized mostly by *Belemnopsis, Dicoelites,* and *Conodicoelites* in the Middle Jurassic, and *Belemnopsis* and *Hibolithes* in the Late Jurassic, extended along the Tethyan coast from northern India to Papua New Guinea. (2) A South Pacific Province characterized by both Belemnopseina and Belemnitina in the Middle Jurassic, and by *Belemnopsis, Hibolithes,* and scattered Dicoelitidae in the Late Jurassic (all with intrageneric differences from those of the Himalayan Province), extended along the Pacific Coast from South America to New Zealand and possibly New Caledonia.

## ANTARCTICA AND SOUTH AMERICA[2]

### Antarctica

Jurassic sediments are known from only a few localities in Antarctica, all within the Antarctic Peninsula region (Thomson 1983). Belemnites have been found at most of these localities, but their abundances are variable. Only a single phragmocone is known from the Lower Jurassic (?Sinemurian age) (Thomson and Tranter 1986), and no Middle Jurassic belemnites have been described, though Quilty (1970) stated that belemnites were found with Middle Jurassic ammonites in Ellsworth Land. Late Jurassic belemnites are common. The most diverse and best-known faunas are from the Fossil Bluff Group on Alexander Island (Willey 1973; Crame and Howlett 1988) and the Latady Formation along the Orville Coast (Mutterlose 1986). Other known localities include the Lyon Nunataks, where the stratigraphically important *Conodicoelites* has been described (Stevens 1967). This appears to be comparable with New Zealand forms and represents the earliest Late Jurassic Antarctic belemnites, of probable Kimmeridgian age. *Belemnopsis* is common in the Kimmeridgian–Tithonian (Willey 1973; Mutterlose 1986; Crame and Howlett 1988), and *Hibolithes* also appears in the Tithonian (Mutterlose 1986; Crame and Howlett 1988). These belemnopseids are largely comparable with the faunas of New Zealand and, to a lesser degree, of South America (Crame and Howlett 1988; Doyle and Howlett 1989). The duvaliid genus *Produvalia* is also known from Antarctica, though

[2] By P. J. Howlett.

Mutterlose (1986) regards some occurrences to be Late Tithonian, whereas others appear to be Kimmeridgian–early Tithonian (Crame and Howlett 1988). These duvaliids show some similarities with Madagascan forms (Mutterlose 1986).

### South America

Though Jurassic rocks are present in many parts of South America (Chapter 6), belemnites are poorly known, and only from Argentina and Chile. Whereas recent work has been done on the Cretaceous belemnites (e.g., Riccardi 1977; Aguirre-Urreta and Suarez 1985), the most recent extensive review of the Jurassic faunas was carried out by Stevens (1965).

The earliest known belemnoid is from the Pliensbachian–Toarcian of Chile, where the xiphoteuthid (Aulacocerida) genus *Atractites* is common (Hallam 1983; Doyle 1988). "True" belemnites (Belemnitida) are, however, unknown in the earliest Jurassic of South America (Doyle 1988). A number of genera of "true" belemnites (Belemnitida) are known from the Toarcian–Middle Jurassic of Argentina and Chile, including *Megateuthis* (Möricke 1895; Tornquist 1898–1901), *Passaloteuthis* (Möricke 1895), *Dactyloteuthis*, and *Brachybelus* (Stevens 1965). These belemnites are mostly from the Andean Basin, from which Stevens (1965) also identified the belemnopseinid genera *Dicoelites* and *Belemnopsis*. According to Stevens (1965, p. 158), the *Dicoelites* show strong Indonesian affinities, whereas the *Belemnopsis* resemble more closely New Zealand species. The supposed *Cylindroteuthis* identified from Caracoles (Bülow-Trummer 1920) is now considered dubious, on biogeographic grounds (P. Doyle personal communication 1988).

The Late Jurassic belemnite faunas are dominated by the Belemnopseina, with *Dicoelites*, *Belemnopsis*, and *Hibolithes* (Feruglio 1936; Stevens 1965). Steven (1965) again considered that the *Dicoelites* showed strong Indonesian affinities. The majority of the *Belemnopsis* are referred to the endemic South American species *B. patagoniensis* Favre. This species, which shows Madagascan rather than Indo-Pacific affinities (Doyle and Howlett 1989), however, appears to be exclusively Early Cretaceous (Riccardi 1977). Stevens (1965, 1973) also considered that some of the other *Belemnopsis* from Argentina showed strong Indo-Pacific affinities. Furthermore, the few Jurassic *Hibolithes* known from Argentina (Feruglio 1936) show some similarities with Indonesian and Antarctic forms (Stevens 1965; Doyle and Howlett 1989).

## PACIFIC COAST OF NORTH AMERICA[3]

There have been relatively few belemnites recorded from North America. However, some papers detail occurrences of belemnites in the Pacific coast region of North America (e.g., Anderson 1945; Stevens 1965; Jeletzky 1980), and new data have been gleaned from museum-based material.

Belemnites first appeared in the earliest Jurassic of Europe (Stevens 1973), and although isolated examples of this age are known from elsewhere in the world (Doyle 1987), true belemnites are unknown in North America prior to the Toarcian, similar to the situation in South America (although the earliest record is that of a Pliensbachian belemnite there) and the Arctic Basin. Aulacocerids of the Xiphoteuthididae are known from the Americas prior to the Toarcian, descended from a rich fauna already established in the Cordilleran region in the Triassic (Doyle 1988). *Atractites* similar to species from the Tethyan region are known from Canadian Lower Jurassic in the Hettangian (Frebold and Little 1962; "belemnites" recorded by them are xiphoteuthids, J. A. Jeletzky personal communication 1987) and Pliensbachian [Geological Survey of Canada (GSC) collections] of British Columbia (Frebold 1969; Doyle 1990).

The Xiphoteuthididae were replaced in the Toarcian by true belemnites, which spread to most areas (Stevens 1973; Doyle 1987). Toarcian belemnites are known from the Poker Chip Shale (Fernie Formation) of Alberta (Hall 1984). These specimens, currently in the GSC collections, are of European affinity. The genus *Acrocoelites* is dominant, and the specimens resemble closely the species *A. strictus*, first described from France. *Acrocoelites* has also been identified from the Toarcian of British Columbia (Jeletzky, cited by Frebold 1969). These belemnites form part of a cosmopolitan fauna of Belemnitidae that was prevalent in the later Early Jurassic (Doyle 1987).

Although close links were maintained with Europe, an endemic Boreal fauna was developed in the Toarcian. Within the Arctic Basin (Siberia, the Arctic islands, and Arctic Canada), a distinct Arctic belemnite province was developed (Saks and Nal'nyaeva 1975; Doyle 1987). Jeletzky (1980) has recorded typical Arctic Pseudodicoelitidae (*Lenobelus* and *Pseudodicoelites*) from western British Columbia and Arctic Canada, indicating the close affinities of this part of North America with the Arctic Province. These taxa continued into the early Middle Jurassic in North America and the Arctic Basin. In Far East Russia the Pseudodicoelitidae were replaced by *Holcobelus* (discussed later), characterizing a Boreal-Pacific Province at that time. *Holcobelus* is absent from the Aalenian of North America, recorded only from the Toarcian of the Alberta Fernie Basin (Jeletzky 1980). Other Arctic taxa, such as *Paramegateuthis* and Cylindroteuthididae, are also recorded from the Early and Middle Jurassic of northwestern British Columbia (Jeletzky 1974) and Alaska (Stevens 1965). No data are available for farther south on the Pacific coast. Jeletzky (1980) has also recorded the anomalous occurrence of specimens of Dicoelitidae (previously thought to be purely Tethyan in distribution) in northwestern Canada. He suggested that this was a center of origin for the family, although in the light of contemporaneous occurrences of these taxa in southern Tibet (Doyle 1987, p. 241), that seems unlikely.

In the Late Jurassic, the Cylindroteuthididae were the dominant belemnites in the Boreal Realm, and an Arctic Province can be distinguished using endemic taxa such as the subgenus *Arctoteuthis* (Saks and Nal'nyaeva 1972; Doyle 1987). Within North America, cylindroteuthids probably were widespread, species such as *Pachyteuthis densa* being well known in central North America (Stevens 1965). Sporadic references indicate that

[3] By P. Doyle.

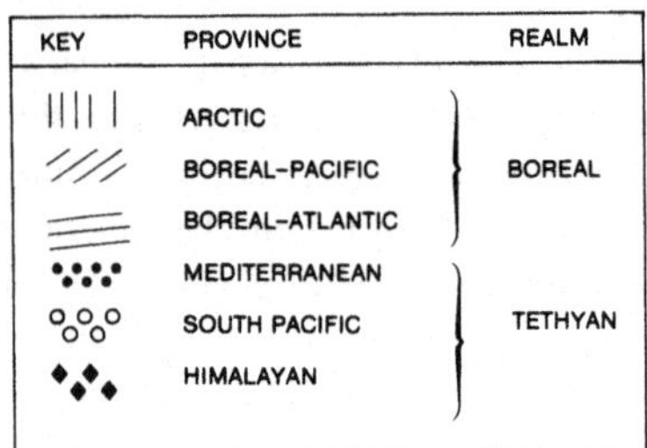


MIDDLE JURASSIC

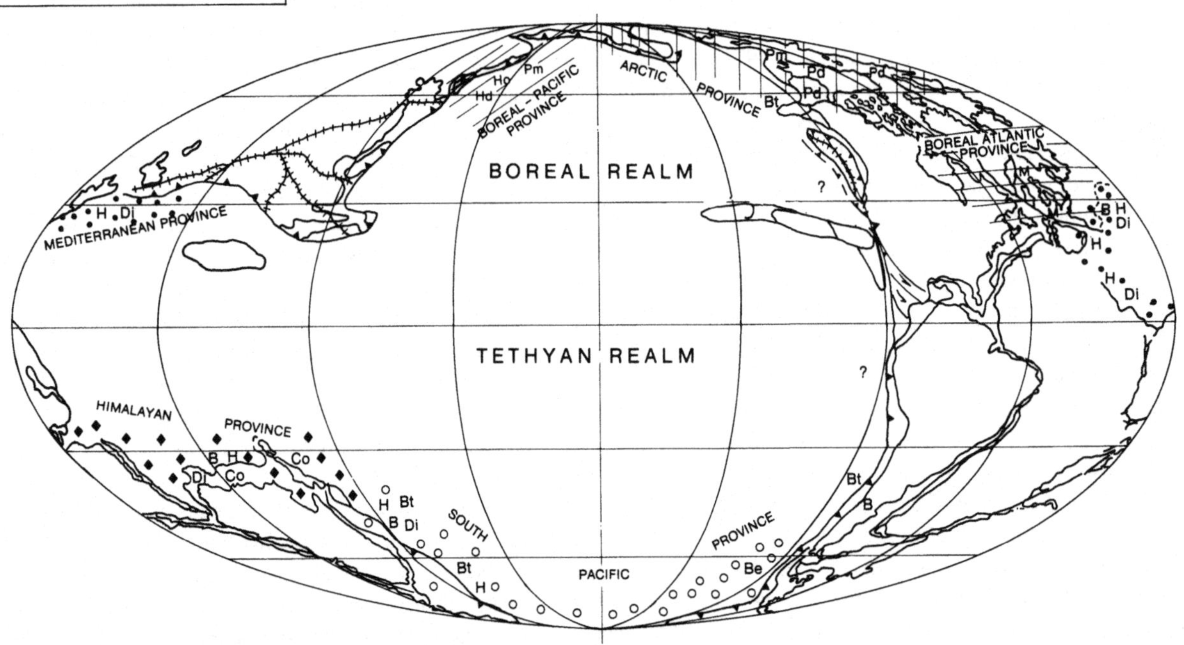


**Figure 23.2. Circum-Pacific belemnite biogeography for Middle Jurassic time. Abbreviations: B, *Belemnopsis;* Be, undifferentiated Belemnopseina; C, *Conodicoelites;* Di, *Dicoelites;* H, *Hibolithes;* Hd, Hastitidae; Ho, *Holcobelus;* Pd, Pseudodicoelitidae; Pm, *Paramegateuthis.***

*Cylindroteuthis* was also common (e.g., Anderson 1945; Imlay 1955; sources cited by Stevens 1965). In the northern part of the region, the subgenus *Arctoteuthis* is recorded from Alaska (White 1886; Stanton 1895; Imlay 1955), again indicating close affinities with the Arctic Province. Anderson (1945) described several new belemnite species from the Tithonian of California and Oregon. Among Anderson's specimens (housed in the Californian Academy of Sciences), examples of the genera and subgenera *Cylindroteuthis, Arctoteuthis,* and *Hibolithes* can be recognized (Saks and Nal'nyaeva 1972; Doyle 1987). Buitrón (1984) has recognized ?*Cylindroteuthis* as far south as central Zacatecas, Mexico. Doyle (1987) has shown that the Arctic Province of the Boreal Realm "split" down the Pacific coast of North America from the Arctic Basin, where a mingling of Tethyan and Boreal elements took place. Thus, in California, Tethyan ammonites and belemnites (*Hibolithes*) coexisted with Arctic Boreal taxa such as *Arctoteuthis* and the bivalve *Buchia,* suggesting only a subordinate role for climatic control of Jurassic provincial boundaries.

## PACIFIC COAST OF RUSSIA[4]

Belemnites from the Pacific coast of Russia have been considered in papers that have described Boreal belemnites of Russia (Saks and Nal'nyaeva 1970, 1975). At the Pacific coast, belemnites have been most adequately studied from the Lower Jurassic of the Omolon Massif. The available data suggest that belemnites make up an appreciable share of the marine fauna in areas near the Pacific coast from the Toarcian on.

The Pliensbachian has not yielded any belemnites, and only rare rostra (*Passaloteuthis* aff. *westhainensis*) have been recorded in the basal Toarcian *Tiltoniceras propinquum* Zone. Beginning with the *Harpoceras falciferum* Zone, a rather diverse assemblage was established, represented by *Acrocoelites, Mesoteuthis, Passaloteuthis,* and *Catateuthis,* which by the second half of the early Toarcian became even more diverse. At that time, in the open-sea environment of North-East Russia, *Acrocoelites* (*A. pergracilis*) was predominant, and *Brachybelus* was common, and the first *Holcobelus* (*H. kinasovi*) appeared.

The early Toarcian is marked by an extensive dispersal of Belemnitida in the northern Pacific. *Nannobelus* and *Clastoteuthis* are dispersed in the Far East (northern coast of the Sea of Okhotsk), while *Passaloteuthis tolli* is reported from the Upper Amur area. Belemnites also colonized the Japanese seas. On Honshu Island, the Toyora Group (Utano Formation) contains belemnites together with late Toarcian ammonites, but the systematics of the belemnites are unknown (Matsumoto 1956).

The Toarcian belemnite assemblage of the northern Pacific is

[4] By T. I. Nal'nyaeva.

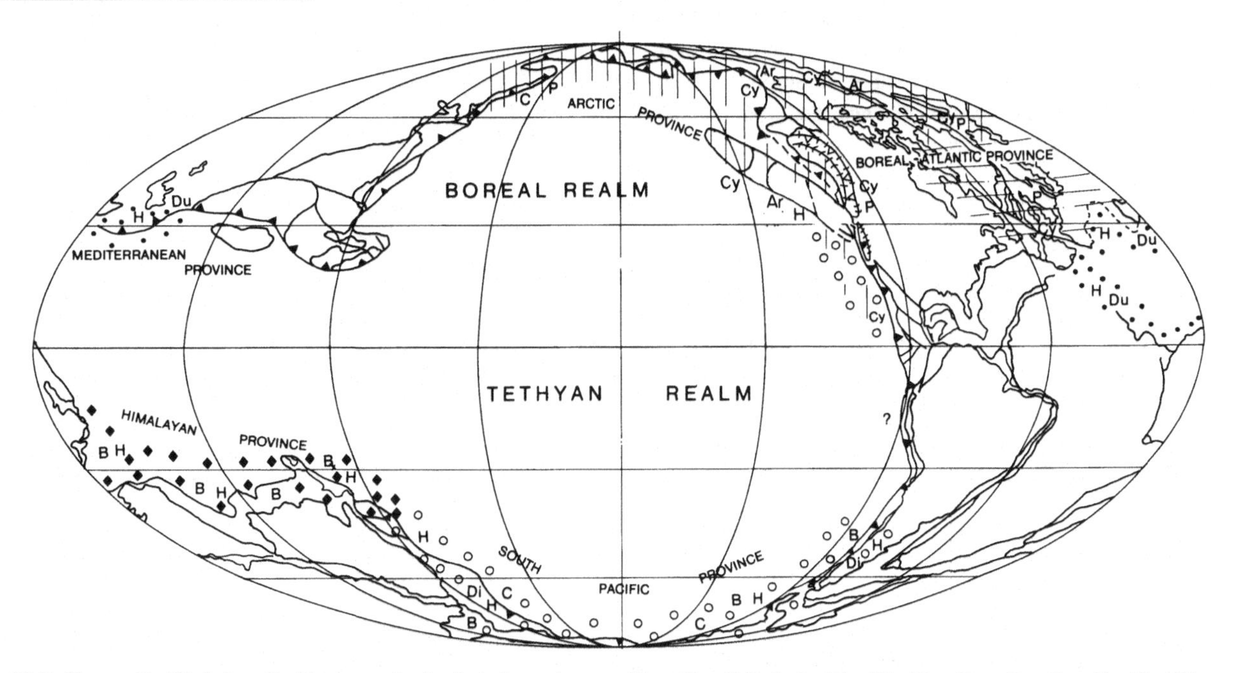


**Figure 23.3. Circum-Pacific belemnite biogeography for Late Jurassic time. Abbreviations: Ar, *Arctoteuthis;* B, *Belemnopsis;* C, *Conodicoelites;* Cy, *Cylindroteuthis,* Di, *Dicoelites;* Du, *Duvalia;* H, *Hibolithes;* P. *Pachyteuthis.***

similar to that of northern Siberia. This assemblage inhabited seas of the Arctic Province of the Boreal Realm. Presumably, the roots of this fauna are in western Europe, wherefrom it migrated via the Arctic Basin into northeastern Asia. The upper Toarcian of the Omolon Massif contains an assemblage dominated by *Passaloteuthis ignota, Mesoteuthis inornata,* and *Holcobelus* spp.

Aalenian deposits around the Pacific are poor in belemnites, but a small assemblage of *Nannobelus krimholzi, Hastites clavatiformis,* and *Holcobelus* sp. occurs on the coast of the Sea of Okhotsk. Pronounced provinciality between belemnite assemblages of the northern Pacific and northern Siberia existed. The Pacific coastal seas lacked *Sachsibelus,* as well as many *Pseudodicoelites* and *Lenobelus* species; they are dominated by species of *Holcobelus,* which are rare in Arctic seas of northern Siberia. In the Far East (western coast of the Okhotsk Sea), *Holcobelus* also plays a significant role in Aalenian belemnite assemblages, whereas the Pseudodicoelitinae and Hastitidae, abundant in northern Siberian assemblages, have not been reliably recorded here. A Boreal-Pacific Province may be distinguished in the late Aalenian.

Data on belemnites from Bajocian deposits are scarce; however, beginning with the late Bajocian, an abrupt change in faunal composition is recorded. Toarcian–Aalenian genera disappear, being replaced by an impoverished assemblage with *Paramegateuthis,* which also extends into the Bathonian. In the Bureya Basin, beds with *Lissoceras* cf. *psilodiscus* have yielded *Paramegateuthis timanensis,* and beds with *Mytiloceramus porrectus* have yielded the first Cylindroteuthididae, *C.* (*C.*) *confessa.* Beginning with late Bajocian, representatives of this family occur ubiquitously throughout Boreal basins.

The Callovian has yielded only isolated finds of *Cylindroteuthis* cf. *tornatilis* and *C.* cf. *strigata* (Tugur Bay). The uppermost Jurassic of the Pacific coast of Russia contains only few belemnites: Volgian *Pachyteuthis* (*Simobelus*) *mamillaris* (western Okhotsk area, Uda Trough).

## SYNTHESIS[5]

Belemnitina were the first belemnites to appear on Pacific coasts, replacing the Aulacocerida in the late Early Jurassic. Belemnitina remain dominant in the Russia–North America regions throughout the Jurassic, but are known farther south (southern South America–New Zealand–New Caledonia) only up to Middle Jurassic. Belemnopseina appeared in late Early to Middle Jurassic with the Pseudodicoelitidae in Canada and Russia, and with the Belemnopseidae in South America, New Zealand, New Caledonia, Australia, and Indonesia. Belemnopseidae remained

[5] By A. B. Challinor and P. Doyle.

dominant in these regions (and in Antarctica) throughout the rest of the Jurassic.

Belemnitina and Belemnopseina define, respectively, the Boreal and Tethyan Realms. The Boreal Realm can be subdivided into a European, Subboreal or Boreal-Atlantic Province (Figures 23.2 and 23.3), with Arctic and Boreal-Pacific (in the Middle Jurassic) Provinces to the north and west. Significant differences exist at the specific level between the southern and western Pacific, and the European Jurassic Belemnopseina, to subdivide the Tethyan Realm into a Himalayan Province, bordered to the west and south by the Ethiopian Province of Stevens (1973); and a South Pacific Province may be distinguished from the Mediterranean Province in the Jurassic.

## References


Aguirre-Urreta, M. B., & Suarez, M. (1985). Belemnites de una Secuencia Turbidítica Volcanoclástica de la Formación Yahgan – Titoniano – Cretácico Inferior del extremo Sur de Chile. *Actas IV Congreso Geol. Chileno, 1,* 1–16.

Anderson, F. M. (1945). Knoxville Series in California Mesozoic. *Bull. Geol. Soc. Am., 56,* 909–1014.

Avias, J. V. (1953). Contribution à l'Etude Stratigraphique et Paléontologique de la Nouvelle Caledonie Centrale. *Science Terre, 1*(1–2), 1–276.

Boehm, G. (1907). Beiträge zur Geologie von Niederlandisch-Indien. Die Südküsten der Sula-Inseln Taliabu und Mangoli. Part 2: Der Fundpunkt am Oberen Lagoi auf Taliabu. Part 3: Oxford des Wai Galo. *Palaeontographica, Suppl. 4, Abt. 1,* Abs. 2–3.

(1912). Beiträge zur Geologie von Niederlandsich-Indien. Die Südküsten der Sula-Inseln Taliabu und Mangoli. Part 4: Unteres Callovien. *Palaeontographica, Suppl. 4, Abt. 1, Abs. 4.*

Brunnschweiler, R. O. (1960). *Marine Fossils from the Upper Jurassic and the Lower Cretaceous of Dampier Peninsula, Western Australia.* Bureau of Mineral Resources of Australia, Bulletin 59.

Buitrón, B. E. (1984). Late Jurassic bivalves and gastropods from Northern Zacatecas, Mexico, and their biogeographic significance. In G. E. G. Westermann (ed.), *Jurassic–Cretaceous Biochronology and Palaeogeography of North America.* Geological Association of Canada Special Papers, *27,* 89–98.

Bülow-Trummer, E. von (1920). *Fossilium Catalogus. I. Animalia Pars Cephalopoda Dibranchiata.* Berlin: Junk.

Challinor, A. B. (1975). New Upper Jurassic belemnite from Southwest Auckland, New Zealand. *N.Z. J. Geol. Geophys., 18*(3), 361–71.

(1977). New Lower or Middle Jurassic belemnite from Southwest Auckland, New Zealand. *N.Z. J. Geol. Geophys., 20*(2), 249–62.

(1979a). The succession of *Belemnopsis* in the Heterian stratotype, Kawhia Harbour, New Zealand. *N.Z. J. Geol. Geophys., 22*(1), 105–23.

(1979b). Recognition and distribution of Heterian *Belemnopsis* in Southwest Auckland. *N.Z. J. Geol. Geophys., 22*(2), 267–75.

(1980). Two new belemnites from the Lower Ohauan (?Middle Kimmeridgian) State, Kawhia Harbour, New Zealand. *N.Z. J. Geol. Geophys., 23*(2), 257–65.

(1985). Evolution in Upper Jurassic *Belemnopsis*–phyletic gradualism or punctuated equilibrium? *N.Z. Geol. Surv. Rec., 9,* 29–30.

(1989a). Early Cretaceous belemnites from the central Birds Head, Irian Jaya, Indonesia. *Geol. Res. Devel. Centre, Indonesia, Paleont. Spec. Publ. Ser. 5.*

(1989b). Jurassic and Cretaceous Belemnitida of Misool Archipelago, Irian Jaya, Indonesia. *Geol. Res. Devel. Centre, Indonesia, Paleont. Spec. Publ. Ser. 5.*

(1989c). The succession of *Belemnopsis* in the Late Jurassic of eastern Indonesia. *Palaeontology, 32,* 571–96.

(1990). A belemnite biozonation for the Jurassic–Cretaceous of Papua New Guinea and a faunal comparison with eastern Indonesia. *BMR J. Austral. Geol. Geophys., 11*(4), 429–47.

(1991). Belemnite successions and faunal provinces in the Southwest Pacific and the belemnites of Gondwana. *BMR J. Austral. Geol. Geophys. 12*(4), 15–32.

Challinor, A. B., & Grant-Mackie, J. A. (1986). The New Caledonian Jurassic coleoid fauna (abstract). *Geol. Soc. N.Z. Misc. Publ., 35A,* 35.

Challinor, A. B., & Skwarko, S. K. (1982). Jurassic belemnites from Sula Islands, Moluccas, Indonesia. *Geol. Res. Devel. Centre, Indonesia, Paleont. Ser., 3,* 1–89.

Crame, J. A., & Howlett, P. J. (1988). Late Jurassic and Early Cretaceous biostratigraphy of the Fossil Bluff Formation, Alexander Island. *Br. Antarctic Surv. Bull., 78,* 1–35.

Doyle, P. (1987). Lower Jurassic–Lower Cretaceous belemnite biogeography and the development of the Mesozoic Boreal Realm. *Palaeogeogr. Palaeoclimatol. Palaeoecol., 61,* 237–54.

(1990). The biogeography of the Aulacocerida (Coleoidea). In G. Pallini, F. Cella, S. Cresta, & M. Santantonio (eds.), *Fossili, Evoluzione, Ambiente. Atti II Convegno Internazionale Pergola, Editore Comitato Centenario Raffaele Piccinini* (pp. 263–71), Pergola.

Doyle, P., & Howlett, P. J. (1989). Antarctic belemnite biogeography and the break-up of Gondwana. In J. A. Crame (ed.), *Origins and Evolution of the Antarctic Biota.* Geol. Soc. Spec. Publ. (London).

Feruglio, E. (1936). Palaeontographia Patagónica, Pt. 1. *Mem. 1st Geol. (R) Univ. Padova, 11,* 1–90.

Frebold, H. (1969). Subdivision and facies of Lower Jurassic rocks in the Southern Canadian Rocky Mountains and Foothills. *Proc. Geol. Assoc. Canada, 20,* 77–89.

Frebold, H., & Little H. (1962). Paleontology, stratigraphy and structure of the Jurassic rocks in Salmo map area, British Columbia. *Bull. Geol. Surv. Canada, 81,* 1–31.

Hall, R. L. (1984). Lithostratigraphy and biostratigraphy of the Fernie Formation (Jurassic) in the Southern Canadian Rocky Mountains. *Mem. Can. Soc. Petrol. Geol., 9,* 233–47.

Hallam, A. (1983). Early and mid-Jurassic molluscan biogeography and the establishment of the central Atlantic seaway. *Palaeogeogr. Palaeoclimatol. Palaeoecol., 43,* 181–94.

Helby, R., Wilson, G. J., & Grant-Mackie, J. A. (1988). A preliminary study of dinoflagellate cysts in Middle and Late Jurassic strata of Kawhia, New Zealand. *Mem. Ass. Australasian Palaeontols., 5,* 125–66.

Hudson, N. (1983). *Stratigraphy of the Ururuoan, Temaikan and Heterian Stages, Kawhia Harbour to Awakino Gorge, Southwest Auckland.* Unpublished thesis, University of Auckland.

Hudson, N., Grant-Mackie, J. A., & Helby, R. (1987). Closure of the New Zealand "Middle Jurassic Hiatus"? *Search, 18*(3), 146–8.

Imlay, R. W. (1955). *Characteristic Jurassic Mollusca from Northern Alaska* (pp. 69–96). U.S. Geological Survey, Professional Paper 274–D.

Jeletzky, J. A. (1966). Comparative morphology, phylogeny and classification of fossil Coleoidea. *Univ. Kansas Paleont. Contr., Mollusca, 6.*

(1974). Contribution to the Jurassic and Cretaceous geology of northern Yukon Territory and district of Mackenzie, Northwest Territories. *Papers Geol. Surv. Canada, 74–10.*

(1980). Dicoelitid belemnites from the Toarcian–Middle Bajocian of Western and Arctic Canada. *Bull. Geol. Surv. Canada, 338,* 1–71.

(1983). Macroinvertebrate paleontology, biochronology and paleoenvironments of Lower Cretaceous and Upper Jurassic rocks, deep

sea drilling hole 511, eastern Falkland Plateau. *Init. Rep. Deep Sea Drilling Proj.*, *71*, 9571–975. U.S. Govt. Printing Office.
Kruizinga, P. (1920). Die Belemniten uit de Jurassische afzettingen van de Soela-Eilanden. *Jaarbock Mijnwezen Nederl. Oost-Indië*, *49*(2), 161–89.
Marwick, J. (1953). Divisions and faunas of the Hokonui System (Triassic and Jurassic). *N.Z. Geol. Surv. Palaeont. Bull.*, *21*.
Matsumoto, T. (1956). Jurassic system. In: *Geology and Mineral Resources of Japan*. Tokyo.
Möricke, W. (1895). Versteinerun des Lias und Unteroolith von Chile. *N. Jb. Miner. Geol. Paläontol. Beil. Bd.*, *9*, 1–100.
Mutterlose, J. (1986). Upper Jurassic belemnites from the Orville Coast, Western Antarctica, and their palaeobiogeographic significance. *Br. Antarctic Surv. Bull.*, *70*, 1–22.
Pharo, C. H. (1967). *The Geology of Some Islands in Baie de Pritbuer, New Caledonia*. Unpublished thesis, University of Auckland.
Quilty, P. G. (1970). Jurassic ammonites from Ellsworth Land, Antarctica. *J. Paleontol.*, *44*, 110–16.
Riccardi, A. C. (1977). Berriasian invertebrate fauna from the Springhill Formation of southern Patagonia. *N. Jb. Geol. Palaontol. Abhandl.*, *155*, 216–52.
Saks, V. N., & Nal'nyaeva, T. I. (1970). *Early and Middle Jurassic belemnites of northern USSR. Nannobelinae, Passaloteuthinae and Hastitidae* (in Russian). Leningrad: Nauka.
(1972). The Berriasian marine fauna; Belemnitida. In. V. N. Saks (ed.), *The Jurassic-Cretaceous Boundary and the Berriasian Stage in the Boreal Realm* (in Russian). Leningrad: Nauka. Translation in *Israel Prog. Sci. Transl. (Jerusalem)*, 1975, pp. 216–29.
(1975). *Early and Middle Jurassic Belemnites of Northern USSR. Megateuthinae, Pseudodicoelitinae* (in Russian). Moscow: Nauka.
Sato, T., Westermann, G. E. G., Skwarko, S. K., & Hasibuan, F. (1978). Jurassic biostratigraphy of the Sula Islands, Indonesia. *Bull. Geol. Surv. Indonesia*, *4*(1), 1–28.
Skwarko, S. K., & Yusef, G. (1982). Bibliography of the invertebrate macrofossils of Indonesia (with cross references). *Geol. Res. Devel. Centre, Indonesia, Spec. Publ.*, *3*, 66.
Smith, A. G., Hurley, A. M., & Briden, J. C. (1981). *Phanerozoic Paleocontinental World Maps*. Cambridge University Press.
Stanton, T. W. (1895). *The Fauna of the Knoxville Beds. Contributions to the Cretaceous Paleontology of the Pacific Coast*. U.S. Geological Survey, Bulletin 133.
Steinmann, G. (1881). Zur Kenntnis Jura- und Kreideformation von Caracoles (Bolivia). *N. Jb. Miner. Geol. Paläontol. Beil. Bd.*, *1*, 293–301.
Stevens, G. R. (1963a). Designation of a lectotype for the nominal species *Belemnopsis alfurica* (Boehm) 1907. *Trans. R. Soc. N.Z., Geol. Ser.*, *2*(6), 101–4.
(1963b). The systematic status of Oppel's specimens of *Belemnites gerardi*. *Palaeontology*, *6*(4), 690–8.
(1964a). The genera *Dicoelites* Boehm 1906 and *Prodicoelites* Stolley 1927. *Palaeontology*, *7*(4), 606–20.
(1964b). A new belemnite from the Upper Jurassic of Indonesia. *Palaeontology*, *7*(4), 621–9.
(1965). *The Jurassic and Cretaceous Belemnites of New Zealand and a Review of the Jurassic and Cretaceous Belemnites of the Indo-Pacific Region*. New Zealand Geological Survey Paleontological Bulletin 36.
(1967). Upper Jurassic fossils from Ellsworth Land, West Antarctica, and notes on Upper Jurassic biogeography of the Southwest Pacific region. *N.Z. J. Geol. Geophys.*, *10*(2), 345–93.
(1973). Jurassic belemnites. In A. Hallam (ed.), *Atlas of Palaeobiogeography* (pp. 259–74). Amsterdam: Elsevier.
(ed.). (1980). *Geological Time Scale*. N.Z. Geol. Soc. Publ.
Stolley, E. (1929). Ueber ostindische Jura-Belemniten. *Paläontologie von Timor*, *16*, 91–213.
(1935). Zur Kenntnis des Jura und der Unterkreide von Misol. Palaontologischer Teil. *N. Jb. Miner. Geol. Paläontol. Beil. Bd.*, *73B*, 42–69.
Suggate, R. P., Stevens, G. R., & Te Punga, M. T. (eds.). (1978). *The Geology of New Zealand*. N.Z. Geol. Surv. Publ.
Teichert, C. (1940). Marine Jurassic of East Indian affinities at Broome, northwestern Australia. *J. R. Soc. West Australia*, *26*, 103–19.
Thomson, M. R. A. (1983). Antarctica. In M. Moullade & A. E. M. Nairn (eds.), *The Phanerozoic Geology of the World. II: The Mesozoic, B* (pp. 391–422). Amsterdam: Elsevier.
Thomson, M. R. A., & Tranter, T. H. (1986). Early Jurassic fossils from central Alexander Island and their geological setting. *Br. Antarctic Surv. Bull.*, *70*, 23–39.
Tornquist, A. (1898–1901). Der Dogger am Espinazito-Pass, nebst einer Zusammenstellung der jetzigen Kenntnisse von der Argentinischen Jura-formation. *Palaeontol. Abh.*, *8*, 135–204.
Uhlig, V. (1910). The fauna of the Spiti Shales (Cephalopoda). *Palaeontol. Indica, Ser. 15*, *4*, 133–306.
Westermann, G. E. G., & Callomon, J. (1988). The Macrocephalitinae and associated Early Callovian (Jurassic) ammonoids of Sula Island and New Guinea. *Palaeontographica, A207,R 1–90.*
White, C. (1886). *On Mesozoic Fossils*. U.S. Geological Survey, Bulletin 4.
Whitehouse, F. W. (1924). Some Jurassic fossils from Western Australia. *J. R. Soc. West Australia*, *11*(1), 1–13.
Willey, L. E. (1973). Belemnites from south-eastern Alexander Island. II: The occurrence of the family Belemnopseidae in the Upper Jurassic and Lower Cretaceous. *Br. Antarctic Surv. Bull.*, *36*, 33–59.
Yang, T., & Wu, S. (1964). Late Jurassic–Early Cretaceous belemnites from South Tibet, China. *Acta Palaeontol. Sinica*, *12*(2), 187–216.

# 24 Ammonites of the circum-Pacific region

A. VON HILLEBRANDT, G. E. G. WESTERMANN, J. H. CALLOMON, AND R. L. DETTERMAN

## LOWER JURASSIC[1] (Figure 24.1)

### Hettangian

Hettangian ammonites are found on both sides of the Pacific, from northeastern Asia to New Zealand and from the Arctic to central Chile. Localities, especially those with diverse assemblages, are normally less frequent than for the rest of the Lower Jurassic. Assemblages with *Badouxia, Paracaloceras, Pseudaetomoceras,* and some species of *Vermiceras* or *Metophioceras* are included that in part may be latest Hettangian (Bloos 1983, 1988) or even earliest Sinemurian (Guex and Taylor 1976; Guex 1987).

The systematics of Hettangian ammonites include genera whose synonymity remains ambiguous. Some Hettangian ammonites found in the circum-Pacific area are very similar in test morphology to genera described from the northeastern Alps, but the circum-Pacific species frequently show much simpler septal sutures. The same difference exists between northwestern Europe and the northeastern Alps. The significance of this phenomenon is controversial (G. Bloos personal communication; Guex 1982), but a possible reason could be the repeated immigration of genera and species from the open sea to shelf environments.

Faunal differentiation in the Hettangian was still weak. The richest assemblages are known from North America (mainly Nevada), Peru, and northern Chile. They all have strong Tethyan affinities. A proper Boreal Realm did not then exist, because those genera found in the northernmost part of the Pacific also occur in the Tethyan Realm and must be designated as pandemic. Differentiation between the Tethyan and Boreal Realms is achieved mainly because of richer assemblages in genera and species in the first region. One genus, *Badouxia* (Guex and Taylor 1976), seems to be restricted to the Pacific, where it was described from many localities of North America (Frebold 1967; Imlay 1981) and South America (Hillebrandt 1988), and also cited from northeastern Asia (Repin 1984). Some genera (*Eolytoceras, Fergusonites, Mullerites*) have so far been found only in the Athabascan Province of the East Pacific subrealm (Taylor et al. 1984). *Primapsiloceras* was described from northeastern Asia, but its generic and biostratigraphic position is controversial (Guex 1984; Repin 1984).

Some Hettangian species are known only from North and South America, but Hettangian ammonites are not yet known sufficiently well for exact biogeographic comparison. The Upper Hettangian genus *Sunrisites* (Guex 1980) is also found in the northeastern Alps, but seems to be more frequent in North and South America. The Upper Hettangian of northwestern Europe is dominated by *Schlotheimia,* contrary to North and South America, where different Alsatitinae are often more frequent.

Contrary to the case of northwestern Europe, but similar to that of the Mediterranean region, Phylloceratacea and Lytoceratacea are present in the Pacific area at different localities, but they are not restricted to open-water sediments, as supposedly they are in the Mediterranean.

The beginning of the Hettangian is characterized by the pandemic genus *Psiloceras.* During that stage a rapid diversification took place, and genera evolved that were confined to, or more frequent in, the Tethyan Realm.

### Sinemurian

Sinemurian ammonites are distributed more widely within the Pacific area than in the Hettangian. Pandemic genera greatly predominate, and some species have worldwide distributions and are found on both sides of the Pacific (e.g., *Arnioceras ceratitoides*). A stronger Tethyan influence is mainly seen at the species level and in the preponderance of some genera that are more frequent in the Mediterranean. The early and middle Sinemurian were characterized by Arietitinae, Arnioceratinae, and Asteroceratinae. *Arnioceras,* the most common genus, experienced a regional evolution in North America and Mexico with the genera *Arniotites, Burckhardticeras,* and *Melanhippites* (Schlatter and Schmidt-Effing 1984). The North American genus *Arctoastero-*

[1] By A. von Hillebrandt.

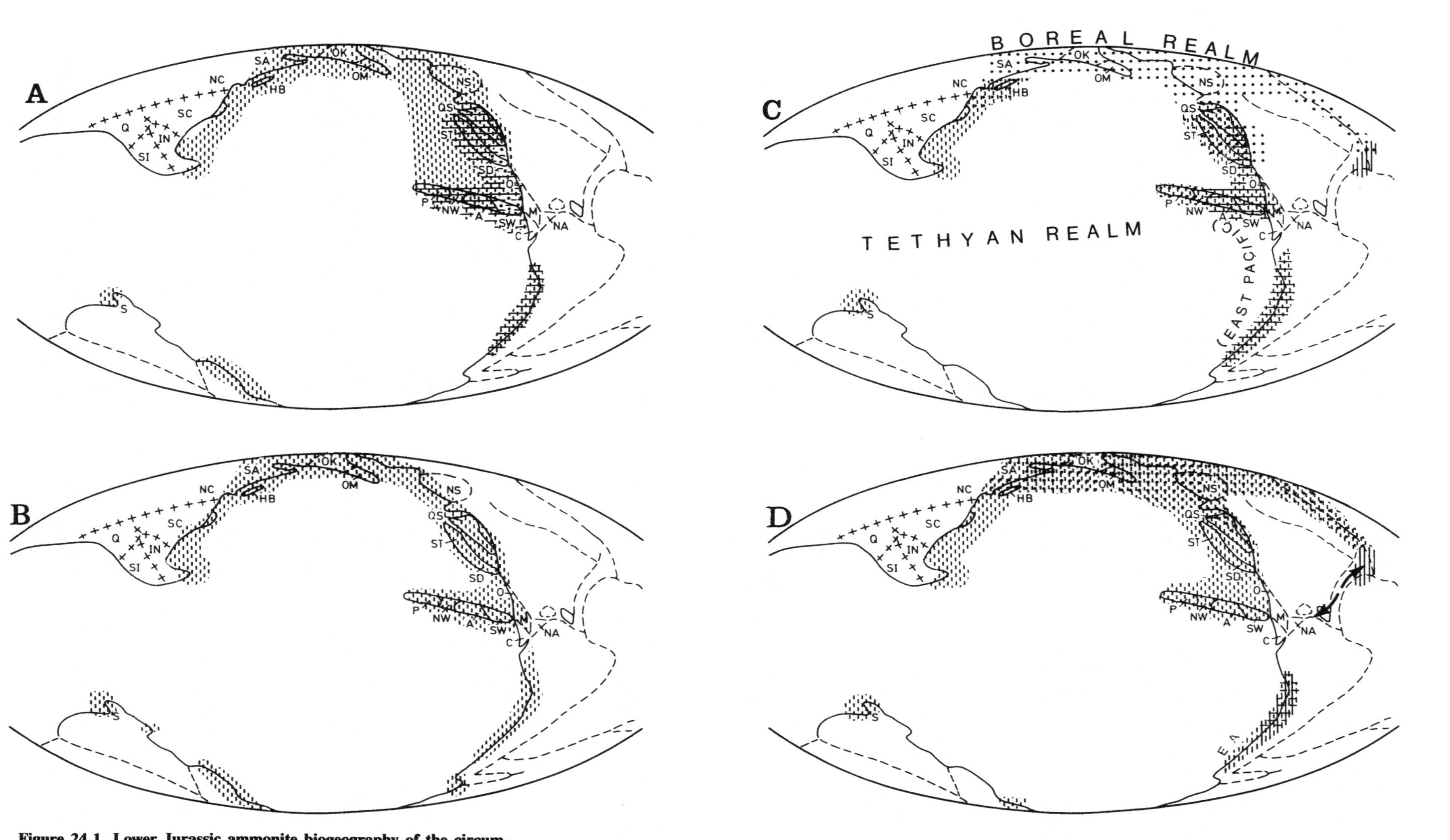


Figure 24.1. Lower Jurassic ammonite biogeography of the circum-Pacific area: A, Hettangian; B, Sinemurian–early Pliensbachian; C, late Pliensbachian; D, early–middle Toarcian. Same symbols as for Middle Jurassic (see p. 346).

*ceras* (Frebold 1960) may be synonymous with the northwestern European genus *Aegasteroceras* (Hallam 1965) of probable Boreal origin. Schlotheimiidae are mostly less frequent. *Angulaticeras kolymicum* belongs to this family and characterizes the Upper Sinemurian of northeastern Asia (Repin 1984). Echioceratidae are typical for the middle and upper parts of the Upper Sinemurian, but the northwestern European genus *Echioceras* s. str. is either absent or rare. Oxynoticeratidae are always rare, but important for biostratigraphic correlations. Eoderoceratidae are found only sparsely. *Xipheroceras* is said to occur in Vietnam (Vu Khuc 1984), and *Microderoceras* was described from Mexico (Schmidt-Effing 1980), Peru (Prinz 1985), Chile (Quinzio 1987), and the island of Rotti (Arkell 1956, p. 441).

Phylloceratacea and Lytoceratacea occur only sporadically and are never as frequent as at some localities in the Mediterranean.

### Pliensbachian

Pliensbachian ammonites occur on both sides of the Pacific. The most complete faunal successions are described from North America (Smith et al. 1988) and South America (Hillebrandt 1981, 1987). The western side of the Pacific has yielded mainly Late Pliensbachian ammonites, in northeastern Asia (Repin 1984), Japan (Hirano 1971), Vietnam (Vu Khuc 1984), and some Indonesian islands (Arkell 1956). This stage had the greatest endemism during the Early Jurassic. Faunal realms developed in the early Pliensbachian and became most accentuated in the late Pliensbachian, when Boreal and Tethyan Realms and East Pacific Subrealm are distinguishable. In North America, ammonites of all three units are found together at various places (Taylor et al. 1984).

Eoderoceratidae and Acanthopleuroceratinae are the most frequent ammonites of the Lower Pliensbachian. The basal Pliensbachian of North and South America is characterized by *Miltoceras* and *Pseudoskirroceras*, which are more typical for the Mediterranean region. The pandemic genera *Tropidoceras* and *Acanthopleuroceras* are very important for exact correlations. European species are found in North and South America. In Europe, the genus *Metaderoceras* is predominantly Mediterranean, and the genus *Dubariceras* is exclusively Mediterranean. Both genera are found in North and South America, in part with the same species. *Eoamaltheus* has been described only from South America and is linked with *Dubariceras*. The first genus gave rise to *Fanninoceras*, which is the most characteristic ammonite genus of the Upper Pliensbachian of North and South America and is restricted to the East Pacific Subrealm.

The Lower Pliensbachian of northeastern Asia is practically devoid of ammonites, and only rare *Polymorphites* are found. Early Pliensbachian ammonites are unknown from the Arctic (Frebold 1975). The genera *Phricodoceras* and *Liparoceras (Becheiceras)* are more common and longer-ranging in the Tethyan region than in northwestern Europe; they have been described from Indonesia (Rotti) (Krumbeck 1922), but are rare in North America (Taylor et al. 1984), and one specimen of *L. (Becheiceras)* was found in South America (Pérez 1982).

Different Pliensbachian genera of the Dactylioceratidae *sensu* Dommergue (1986) (*Reynesocoeloceras, Aveyroniceras, Prodactylioceras, Reynesoceras*) were reported from North America (Smith et al. 1988), and one genus from South America (Hillebrandt 1981). Some genera of this family are more typical for the Mediterranean region than for northwestern Europe.

Predominantly Tethyan genera of Upper Pliensbachian Hildoceratidae occur in North and South America, as well as Vietnam, Japan, and Far East Russia. Most frequent are the genera *Protogrammoceras, Arieticeras, Fuciniceras, Fontanelliceras, Paltarpites*, and *Leptaleoceras*. The hildoceratid genus *Tiltoniceras* appears in North America in the last zone of the Upper Pliensbachian (Smith et al. 1988).

Boreal faunal assemblages with Upper Pliensbachian Amaltheidae are found in northeastern Asia, and very rarely in Japan (Hirano 1971) and in North America (Taylor et al. 1984), where they occur either alone (southern Canadian Rockies, northern Alaska, Yukon, and Canadian Arctic) or together with Tethyan hildoceratids and *Fanninoceras* (allochthonous cordilleran terranes). The northwestern European species *Amaltheus stokesi* and *A. margaritatus* are known from North America (Smith et al. 1988) and northeastern Asia (Repin 1984). The subgenus *Nordamaltheus* and *A. (Nordamaltheus) viligaensis* are restricted to these regions, and various other species of *Amaltheus* have been found only in northeastern Asia.

The uppermost Pliensbachian amaltheid genus *Pleuroceras* is not yet known with certainty from the Pacific region, and other genera (e.g., *Beaniceras, Androgynoceras*, and *Oistocerase*) are also absent, although they are frequent in northwestern Europe.

Phylloceratacea and Lytoceratacea are mostly rare. The Tethyan genus *Juraphyllites* was described from Indonesia (Rotti) and Argentina.

### Toarcian

Toarcian ammonites are known from many localities around the Pacific Ocean. The most complete successions have been described from northeastern Asia and South America. Faunal differentiation was less pronounced than in the Pliensbachian, and, particularly in the early and middle Toarcian, most genera and some species were widely distributed. Faunal exchange was better than in the Pliensbachian, probably owing to a eustatic transgression. Some species of *Dactylioceras* s.l. have a worldwide distribution, whereas others are restricted to the Pacific region (i.e., *D. helianthoides*, first described from Japan, but later also found in North and South America). *Dactylioceras (Dacytlioceras)* (e.g., *D. commune*) and *Zugodactylites* seem to be restricted to the northern part of the Pacific (northeastern Asia, Canadian Arctic, Alberta). Some species of *Peronoceras* (= *Porpoceras*) known from Europe occur also in South America or resemble those of Europe. Thus, *Peronoceras spinatum* and *P. polare* are typical of the Canadian Arctic (Frebold 1975) and northeastern Asia (Repin 1984). *P. spinatum* was also illustrated from South America (Hillebrandt 1987), and *P. polare* was first described from Spitzbergen. The dactylioceratid genus *Kedonoceras* (Dagis 1968) has been found only in northeastern Asia, but is said to be synonymous with *Dactylioceras (Orthodactylites)* (Howarth 1973).

Different hildoceratid genera and some species are pandemic. *Tiltoniceras propinquum* was described from North America and northeastern Asia. In northeastern Asia this species appears later than in North America and occurs together with European species of this genus (Repin 1984). Species of *Eleganticeras* were figured from northeastern Asia (Dagis 1974), and this genus also exists in South America. Some European species of *Hildaites, Harpoceratoides,* and *Harpoceras,* or species very similar to European ones, were described from both sides of the Pacific. *Harpoceras chrysanthemum* was first described from Japan (Hirano 1973), but also occurs in northeastern Asia (Dagis 1974) and South America (Hillebrandt 1987). In South America, the genus *Harpoceras* is not restricted to the Lower Toarcian, as found in northwestern Europe, but extends into the Middle Toarcian, as in the Mediterranean region. *Bouleiceras* is a typical genus for the Arabo-Madagascan province, but is also found in the Mediterranean and South America (Hillebrandt 1973). The mainly Mediterranean genus *Frechiella* was figured from different localities of South America, and the almost exclusively Mediterranean genus *Leukadiella* occurs very rarely in South America (Hillebrandt and Schmidt-Effing 1981, p. 23).

The Middle Toarcian genus *Hildoceras* is very important for European biostratigraphy and is frequent there, but it has not yet been reported from the circum-Pacific. Other hildoceratid genera are endemic in the Pacific region. *Arctomercaticeras* is restricted to the Lower Toarcian of northeastern Asia (Repin 1984), and the genera *Atacamiceras* and *Hildaitoides* (Hillebrandt 1987) are typical for the Middle Toarcian of South America. In the late Toarcian, Pacific faunal differentiation increased. The genera *Phymatoceras* s.l., *Phlyseogrammoceras,* and *Pleydellia* are frequent in South America (Hillebrandt 1987). Endemic species occur together with species known from Europe. *Sphaerocoeloceras* (Jaworski 1926) is known only from the Upper Toarcian and Lower Aalenian of South America (Hillebrandt 1987). In North America the predominantly Mediterranean genus *Phymatoceras* occurs together with the more Boreal genus *Haugia* (Cameron and Tipper 1985). *Grammoceras* and *Pseudogrammoceras* are frequent in Europe, but unknown from South America, and they are found in North America (Imlay 1968, 1981; Frebold and Tipper 1970) and on the western side of the Pacific (Japan, Vietnam, Sula Islands) (Arkell 1956; Hirano 1973; Vu Khuc 1984). Different species of the Upper Toarcian genus *Hammatoceras* were described from the circum-Pacific (mainly South America and Indonesia), and some species are very similar to European ones. The Mediterranean genus *Catulloceras* has been found in North and South America. In northeastern Asia, northern Alaska, and the Canadian Arctic the Upper Toarcian is dominated by species of *Pseudolioceras* that are in part similar to species described from northwestern Europe. At this time the Bering Province was established within the Boreal Subrealm (Taylor et al. 1984).

MIDDLE JURASSIC[2] (Figure 24.2)

The Middle Jurassic probably is the most interesting part of the Jurassic from the point of view of paleobiogeography, because it provides very good examples of biogeographic changes, some of which can be explained in terms of unifying plate-tectonic theory. The biogeography of the ammonites in particular has been a subject of several recent reviews and is now rather well known (e.g., Westermann 1981; Callomon 1984; Taylor et al. 1984). Note that only minimal references are cited here, and the reader is referred to the regional documentation in Part III.

**New biogeographic units**

Within the Andean Province, the Antofagasta and Neuquén Subprovinces are here distinguished. The Antofagasta Subprovince, in the north, is typically developed in northern Chile and southwestern Mexico, beginning in the late Bajocian, and is characterized by high-diversity, low-latitude faunas with Mediterranean/West Tethyan elements. Endemics include the genera *Lupherites* and *Duashnoceras* (Bajocian).

The Neuquén Subprovince, in the south, is typically developed in the Neuquén Basin of west-central Argentina and has a low-diversity, midlatitude fauna, without Mediterranean/West Tethyan elements; *Stehnocephalites* (Bathonian) is endemic.

The subprovinces remain differentiated well into the Cretaceous (R. C. Riccardi personal communication).

Within the Indo–SW Pacific Subrealm (or Province), the Sula–New Guinean Province (or Subprovince) is here distinguished. It was well developed in Eastern Indonesia and Papua New Guinea during Bajocian–Bathonian time and is characterized by the endemic genera *Irianites, Satoceras, Praetulites,* and probably other sphaeroceratids (Westermann and Getty 1970, Pl. 54–5; cf. Westermann and Callomon 1988). For the middle Lower Bajocian, the "Pacific" *Pseudotoites* and late representatives of the cosmopolitan *Fontannesia* (including *F. kiliani* and probably the subgenus *Newmarracaroceras* in western Australia) are associated.

Coeval ammonite-bearing beds are, however, poorly represented or unknown from areas adjacent to the Sula Block and the island of New Guinea, that is, the Tethyan Himalayas to the west and eastern Australasia (including New Caledonia and New Zealand) in the south, so that the lateral extent of the province/subprovince remains unknown. It is noteworthy, however, that the very small southern Tibet fauna of the Tethyan Himalayas includes *Fontannesia kiliani* (Westermann and Wang 1988).

The choice of rank (province or subprovince) depends on the biogeographic classification of the southeastern Tethys and of the contiguous biogeographic units in the southwestern and southern Pacific. The extents and names of the biogeographic units in that area have changed greatly in the light of contemporary plate-tectonic reconstructions, in particular the recognition that faunas formerly termed "Indo-Pacific" all belong to the southern margin of eastern Tethys and the southwestern (and temporarily southern)

[2] By G. E. G. Westermann.

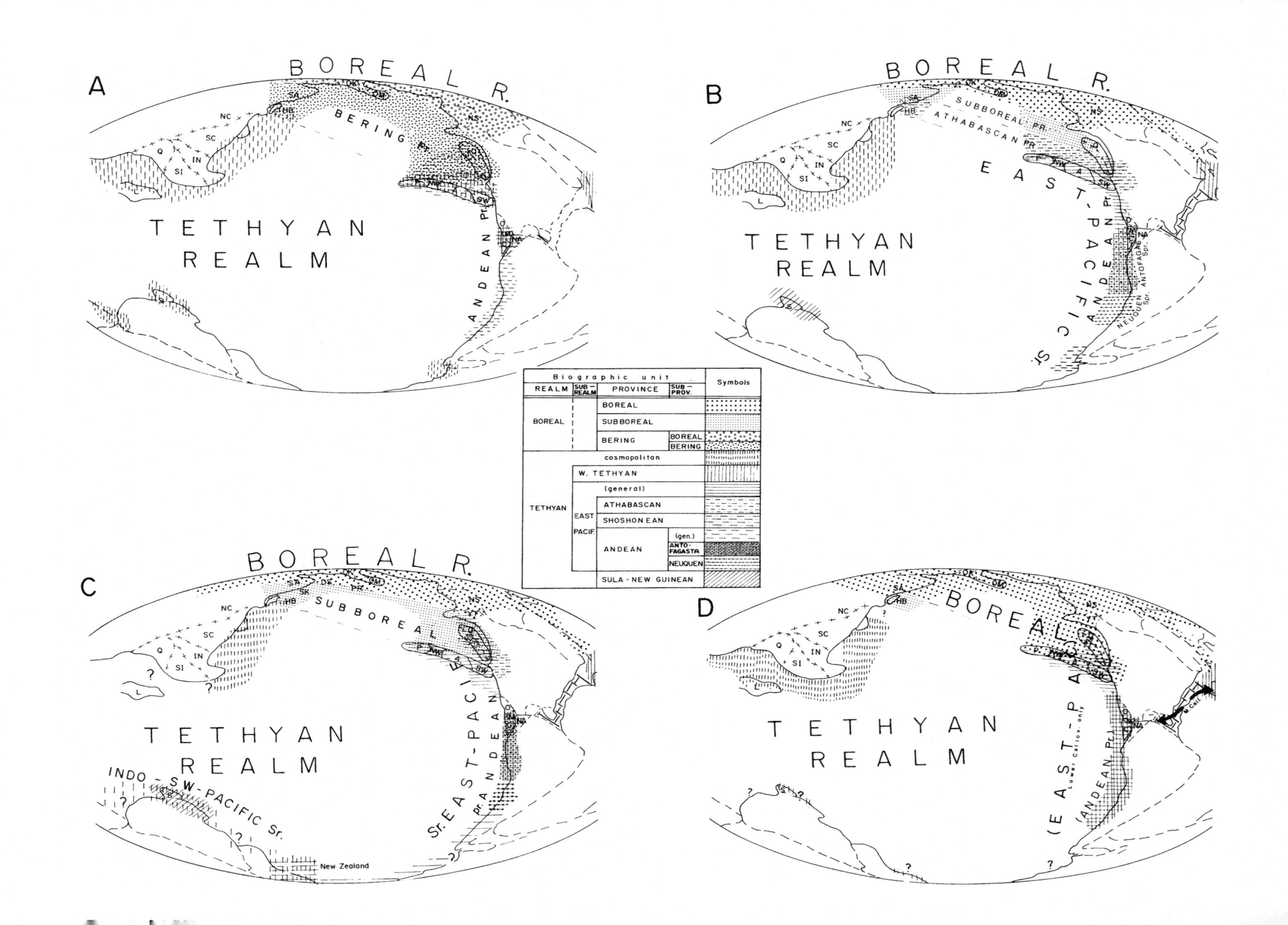
A
BOREAL R.
BERING PR.
TETHYAN REALM
ANDEAN PR.
B
BOREAL R.
SUBBOREAL PR.
ATHABASCAN PR.
EAST-PACIFIC Sr.
ANDEAN PR.
NEUQUEN Spr.
ANTOFAGAS. Spr.
TETHYAN REALM
C
BOREAL R.
SUBBOREAL
TETHYAN REALM
Sr. EAST-PACIFIC
ANDEAN Pr.
INDO-SW-PACIFIC Sr.
New Zealand
D
BOREAL
TETHYAN REALM
(EAST-PA
(ANDEAN Pr.)
Lower Callov. only
Biographic unit
REALM
SUB-REALM
PROVINCE
SUB-PROV.
Symbols
BOREAL
SUBBOREAL
BERING
BOREAL
BERING
cosmopolitan
W. TETHYAN
(general)
ATHABASCAN
SHOSHONEAN
ANDEAN
(gen.)
ANTO-FAGASTA
NEUQUEN
SULA-NEW GUINEAN
TETHYAN
EAST PACIF.

Pacific (i.e., peripheral to eastern Gondwana). In fact, it turns out that Uhlig's (1911) "Himalayan Realm" (for "Reich"), reaching from the "Aethiopian Province" and "Indian Province" to western Australia and the "Maorian Province" of New Zealand, would have been an acceptable term; but he also created a homologous "Himalayan Province" (mostly Tethyan Himalaya), which, together with the "Indian Province," is now believed to be synonymous with the Ethiopian Province (J. Krishna personal communication). This enlarged bioprovince has therefore recently been named the Indo–East African or Indo–Madagascan (Indo–Malagasy) Province, encompassing East Africa, Madagascar, and the western and northern margin of the Indian subcontinent. The best substitute name for Uhlig's "Himalayanisches Reich" is Enay's (1980) Indo–SW Pacific Subrealm (Verma and Westermann 1984). [Note that the same authors discarded the "Pacific Realm" of Arkell (1956).] The category of subrealm, rather than province, follows from the classification of all western Tethyan provinces into the West Tethyan Subrealm and the distinction of the East Pacific Subrealm. It is noted, however, that Mediterranean and/or Submediterranean faunas appear to continue along the northeastern margin of Tethys, at least as far as Thailand and, in the Aalenian, as far as southeastern Japan, which never differed more than provincially from the West Tethyan Subrealm. This subrealm thus extends along the entire southern margin of Eurasia, and consequently also needs renaming.

Beginning in the Oxfordian, the Indo–SE Pacific Subrealm becomes more uniform (mayaitids are now known also from New Zealand and the Antarctic Peninsula) and appears as a single(?) extended ammonite province in the Tithonian (Stevens 1990).

### Aalenian

The circum-Pacific part of the Boreal Realm, including a Subboreal Bering Province, extended from North-East Russia to northern Alaska and southward to allochthonous southern Alaska and British Columbia. The ammonite fauna was typically of low diversity, with the Boreal *Pseudolioceras* accompanied by rare *Erycitoides* and pandemics along the Arctic slope and most of North-East Russia. Somewhat farther south, in Far East Russia, the Peninsular Terrane of southern Alaska, and several terranes of British Columbia, *Erycitoides* became abundant and is one of the principal faunal elements defining the Bering Province.

The endemic Aalenian–early Bajocian *Pseudolioceras* is a previously geographically restricted, relict pan-Boreal Toarcian clade. Isolation from the European–Tethyan Hildocerataceae occurred presumably by the closing of the Greenland–Norway cratonic seaway. The Bering Middle Jurassic pseudolioceratids, comprising mainly the subgenus *Tugurites*, may have been ecologic equivalents of the morphologically similar graphoceratids (Sey, Kalacheva, and Westermann 1986). The restricted Boreal Sea encompassing the Bering Province was an epicontinental sea bordering the Pacific; faunas decrease in diversity and abundance with distance from the Pacific (Sey and Kalacheva 1988).

On the Asian continent, the Bering Province reaches to Sikhote-Alin in Far East Russia. The Kitakami Terrane of Japan still has *Pseudolioceras* as an ancillary element in an otherwise Subtethyan assemblage (Sato and Westermann 1991). The only Aalenian faunas known from southeastern Asia are from Thailand, not bordering the Pacific (Westermann 1981).

Along the northeastern Pacific south of Alaska, Aalenian faunas are known only from allochthonous terranes. Mixed Bering–West Tethyan–cosmopolitan assemblages occur in the Wrangellia and Stikinia Terranes, as far south as the Canada–United States border. Very rare graphoceratids (Leioceratinae) are here associated with the *Pseudolioceras–Erycitoides* assemblage. In Oregon, the faunas are typically Subtethyan, with abundant hammatoceratids (Taylor 1988); however, earlier reports of rare *Erycitoides* and the Japanese *Hossoureites* by Taylor et al. (1984) have not been confirmed (Figure 24.2A).

No Aalenian faunas have been reported between Oregon and Peru (Westermann et al. 1980). The best Aalenian Andean faunas occur in northern Chile (Hillebrandt and Westermann 1985), where a rich Tethyan hammatoceratid fauna includes the Andean endemic genera *Sphaerocoeloceras*, *"Zurcheria"* (gr. *groeberi* Westermann & Riccardi), *Puchenquia*, and *Podagrosiceras*, as well as the typically Mediterranean *Bredyia*. Even rare leioceratines appear to be present. Latest Aalenian hammatoceratids are best known from the Neuquén Basin, Argentina, where the cosmopolitan (pan-Tethyan) *Planammatoceras (Pseudaptetoceras) klimakomphalum*, species of *Eudmetoceras*, and *Tmetoceras scissum* accompany the endemic *Puchenquia (Gerthiceras)*. The latter may have been ancestral to the almost cosmopolitan Sonniniidae, which would have spread eastward through the Hispanic Corridor (Westermann and Riccardi 1985).

In conclusion, the Pacific in Aalenian time was bordered in the north by the Boreal Bering Province, which merged in the south into a distinct and broad transitional zone with Tethyan faunas. Its width today is determined by the post-Jurassic northerly terrane movement, including the southern Alaskan and British Columbian terranes. In the United States, the John Day Terrane of Oregon has a Tethyan fauna with strong European-Mediterranean affinities. The Andean Province has similar faunas, associated with plentiful endemics, indicating that the Hispanic Corridor [see Smith (1983) for the early, rifting phase of the "Proto-Atlantic"] may have been open and a two-way street for ammonite dispersal (Westermann and Riccardi 1985). Aalenian is so far unknown from Australasia and poorly known from Tethyan southeastern Asia.

### Early Bajocian

The substages are discussed separately because of major biogeographic changes between them (Figure 24.2A). The Boreal ammonite distribution resembles that of the late Aalenian. On the North Slope of eastern Asia and North America, the Boreal Province, the last species of the endemic *Pseudolioceras* is followed by similarly provincially endemic *Arkelloceras*. In North-East Russia, rare *Bradfordia* and *Holcopylloceras* occur in proximity to the Pacific. The Subboreal Bering Province is present throughout the Far East Russia. Farther south, early Bajocian ammonites are known only from the Kitakami Terrane of eastern Honshu, Japan, where the cosmopolitan (pan-Tethyan) *Sonninia, Stephanoceras,* and *Emileia* are accompanied by rare Tethyan *Strigoceras;* Boreal forms have so far not been found.

The ammonite fauna is more diverse in the Peninsular Terrane of southern Alaska and in the Stikinia Terrane of British Columbia, where Tethyan faunas of the Athabascan Province are increasingly common (Hall and Westermann 1980; Taylor et al. 1984; Taylor 1988). The well-known latest Aalenian and earliest Bajocian faunas of the Peninsular Terrane (see Chapter 4) have affinities with the Oregon faunas of the John Day Terrane. On the craton, the Boreal Province reaches southward to the central Yukon, whereas in southwestern Alberta the cosmopolitan sonniniids and stephanoceratids are accompanied by rare Athabascan *Zemistephanus* and Boreal *Arkelloceras;* Tethyan forms are absent. The ambi-Pacific *Pseudotoites* occurs rather scarcely in the Alaskan Peninsular Terrane, in the John Day Terrane of Oregon, and in New Guinea, and abundantly in the Andes and Western Australia. The Otoitinae fauna of the John Day Terrane has yielded also the Andean subgenus *Emileia (Chondromileia)* (Taylor et al. 1984).

Meso-American faunas are known mainly from the Mixteca Terrane (or eastern Guerrero Block) in Oaxaca, southwestern Mexico (Sandoval and Westermann 1986). Most of the Humphriesianum Zone is represented by cosmopolitan as well as by North Cordilleran (Athabascan) and northwestern European *Stephanoceras* species; in fact, this zone has the most cosmopolitan (pan-Tethyan) fauna of the Pacific Middle Jurassic. A distinct faunistic change with biogeographic significance occurs in the upper part of the zone, that is, the appearance of the stephanoceratid genus *Duashnoceras,* which is also found in northern Chile (see the section on the late Bajocian). Poorly preserved, cosmopolitan *Stephanoceras* [*S. (Itinsaites)* ex gr. *latansatus–itinsae*)] have also been found in a thin bed within the thick terrigenous, poorly fossiliferous sequence of Honduras (identification confirmed by Westermann; see Gordon, Chapter 5, on the Chortis Block/Terrane).

In the northern Andes, cosmopolitan *Stephanoceras (Skirroceras)* and *Emileia*(?) are associated with submarine basaltic lavaflows in Venezuela, indicating early drifting (Bartok, Renz, and Westermann 1985). The faunas of the Peruvian Andes (Westermann et al. 1980), and especially those of the Andes of northern Chile and west-central Argentina (Westermann and Riccardi 1972, 1979), consist of cosmopolitan sonniniid, sphaeroceratid, and stephanoceratid genera, the single endemic subgenus *Emileia (Chondromileia),* also known from Oregon, and the Pacific genus *Pseudotoites,* as well as of a mixture of West Tethyan and endemic species. The assemblage of the Humphriesianum Zone is close to that of Tethys and northwestern Europe, partly down to the species level, so that even its subzones can be identified. The Andean Province is therefore briefly not differentiated. Similar faunas have been recorded even for the Antarctic Peninsula (Quilty 1970), but probably are early Upper Bajocian, as discussed later. Whereas sphaeroceratids may have originated in the eastern Pacific and then migrated eastward via the Hispanic Corridor, stephanoceratids probably arrived by westward migration from the Mediterranean–northwestern European area (Westermann and Riccardi 1985; J. H. Callomon personal communication).

In the southwestern Pacific, eastern Indonesia has yielded the quasi-pandemic *Fontannesia,* including *F. kiliani* (Kruiz.), which is identical with, or very close to, forms of the Tethyan Himalaya of Tibet (Westermann and Wang 1988), but not with forms from Western Australia (Hall 1989). Stephanoceratids and sphaeroceratids occur together with the Pacific *Pseudotoites* (Westermann and Getty 1970).

In the southern Pacific, New Zealand has recently yielded the first known Bajocian fauna, consisting of *Sphaeroceras, Stephanoceras,* and a new endemic genus of the Sphaeroceratinae that formerly (Stevens 1978) was identified as an early Callovian *Macrocephalites (G. E. G. Westermann and N. Hudson unpublished data).* This appears to be another occurrence of the cosmopolitan Humphriesianum Zone fauna.

In conclusion, Boreal connections with Europe across the Arctic Ocean remained closed to Pacific ammonites, whereas the eastern Pacific was well connected with the Tethys toward the end of the early Bajocian, presumably through the Hispanic Corridor. The central western Pacific had good exchange routes with Europe and the Mediterranean, presumably along the northern margin of Tethys; Australasian faunas spread along the southern margin of Tethys.

### Late Bajocian

At the very beginning of this substage (Figure 24.2B), the East Pacific Subrealm (formerly classed as a realm, cf. Westermann 1981) of the Tethyan Realm became well established. The Boreal Realm became pan-Arctic and was characterized by the Arctocephalitinae, which evolved from the northeastern Pacific sphaeroceratid *Sphaeroceras (''Chondroceras''/Defonticeras)* (Callomon 1984, 1985). In North America, the Boreal Province reached only as far south as cratonic central Yukon, and it occupied North-East Russia, whereas the Subboreal Province reached into southern Alaska (Taylor et al. 1984; Hall 1989).

Southward along the western Pacific margin, in Primorye of Far East Russia, the *Megasphaeroceras [''Umaltites'']–Lyroxyites–Megaphylloceras–?Epizigzagiceras* assemblage and *Cobbanites (= ?Leptosphinctes),* previously known from the Athabascan Province (East Pacific Subrealm), have recently been recognized (Sey and Kalacheva 1988). This fauna closely resembles those present in the allochthonous Alaska Peninsula, Stikinia, and John Day Terranes, as well as in the cratonic Western Interior United States (Taylor et al. 1984). *Megaphylloceras, Spiroceras,* and *Leptosphinctes* have recently also been found in southwestern Alberta (Hall 1989). The Far East fauna is now included in the temporarily ambi-Pacific Athabascan Province, which soon contracted again to become part of the East Pacific Subrealm in the latest Bajocian.

South of Far East Russia, in the allochthonous Kitakami Terrane of eastern Honshu, a small Tethyan and cosmopolitan fauna occurs, with *Cadomites, ?Bigotites,* and *Leptosphinctes* (Westermann and Callomon 1988; Sato and Westermann 1991), although Sato, Kasahara, and Watika (1985) believe that a rare specimen from the Mino Belt belongs to the Boreal *Kepplerites*.

Along the eastern Pacific rim, south and east of the Athabascan Province, lies the Shoshonean Province (or Subprovince of the Athabascan Province) on the cratonic, marine embayment of the Western Interior, yielding an endemic *Eocephalites–Parachondroceras* fauna (Callomon 1984; Taylor et al. 1984).

More southerly, late Bajocian faunas belong already to the Andean Province, extending from southwestern Mexico through the Andes to the Antarctic Peninsula. In Oaxaca (Mixteca Terrane) the early Upper Bajocian ammonite fauna, according to Sandoval and Westermann (1986), consists of equal thirds of West Tethyan (Mediterranean–NW European), East Pacific, and endemic species. The large proportion of West Tethyan species is exceptional for the eastern Pacific at that time and strongly suggests the proximity of the western gate of the open, cratonic Hispanic Corridor. Curiously, the otherwise ubiquitous *Megasphaeroceras* assemblage is absent from this coeval Oaxacan ("Mixtecan") fauna.

In central Peru, northern Chile, west-central Argentina, and even the Antarctic Peninsula, the ambi-Pacific *Megasphaeroceras* assemblage includes cosmopolitan elements, that is, the possibly planktonic *Spiroceras*, which in northern Chile is gigantic (Hillebrandt 1970; Westermann 1981, 1989), nektobenthic leptosphinctines, and late stephanoceratines, as well as Tethyan cadomitines. In northern Chile occur also the Athabascan *Lupherites* and the southwestern Mexican *Duashnoceras*. The northern part of the Andean Ammonite Province, including the Mixteca Terrane of Oaxaca, is here distinguished as the Antofagasta Subprovince (from Antofagasta Province in northern Chile). Farther south, in the Neuquén Subprovince, these low-latitude elements are absent (see the section on the Bathonian). A small *Megasphaeroceras* assemblage has been found in the Antarctic Peninsula (Quilty 1970).

In Australasia, late Bajocian ammonites have been identified only from the Sula Islands and New Guinea. They are cosmopolitan leptosphinctines, Tethyan *Cadomites*, and the endemic *Irianites* and *Praetulites* (or early Bathonian) (Westermann and Getty 1970; Westermann and Callomon 1988).

In conclusion, late Bajocian ammonite faunas indicate that the Boreal Pacific was connected with the West Tethys (northwestern Europe) across the Arctic Sea, although the ammonites remained strictly segregated. The Athabascan Province of the East Pacific Subrealm extended briefly to Far East Russia, represented by the *Megasphaeroceras* assemblage. That genus is absent from the southwestern Mexican (Mixtecan) fauna that had strong West Tethyan affinities, indicating proximity of the Hispanic Corridor. Southwestern Pacific ammonite faunas probably were already partly endemic, indicating the beginning of the Sula–New Guinean Province.

#### Bathonian

During this period (Figure 24.2C) the Boreal Realm advanced southward. In North America the Boreal Province, still characterized by the Arctocephalitinae, lies on the North Slope, including cratonic northern Yukon. Southward, in the allochthonous Stikinia Terrane and the cratonic Western Interior (northern Shoshonean Province), early *Kepplerites* of the Subboreal Province overlap broadly with the East Pacific Subrealm, which is characterized by Eurycephalitinae. What appear to be middle Bathonian early eurycephalitines (i.e., *Iniskinites*) have now also been reported from as far north as the cratonic Yukon Territories (Poulton 1987). In the latest Bathonian, the *Cadoceras* fauna of the Boreal Province withdrew northward somewhat, whereas Subboreal *Kepplerites* then occurred in the allochthonous Peninsular Terrane, in the Yukon, and as far south as the Western Interior in Montana. The Peninsular fauna includes East Pacific Eurycephalitinae (Callomon 1984; Taylor et al. 1984), and that of the northern Yukon includes rather abundant phylloceratids, some *Oxycerites*, and possibly even rare *Cadomites*, characteristic of warmer, shelf seas (Poulton 1987). The presence at the North Slope of deep-water, oceanic phylloceratids would suggest that this part of the Arctic was not far from the true ocean (Westermann 1971, 1989; Callomon 1985). The Boreal Province extends far westward into North East Russia and onto the Siberian Platform (Meledina 1988); *Arctocephalites* is also recorded from Sikhote-Alin in the Far East (Khudoley, cited by Sey and Kalacheva 1988). The Subboreal Province extends to eastern Greenland and just south of the Shetlands (Callomon 1979).

Bathonian is unknown from farther south in eastern Asia, except for the Hida Block of western Japan. The Hida Plateau of Honshu has yielded a Tethyan–cosmopolitan assemblage of *Prohecticoceras* and *Bullatimorphites*, followed by what have appeared to be true Andean *Neuqueniceras*, followed by Subboreal *Kepplerites* related to basal Callovian Athabascan species (Sato 1962; Westermann and Callomon 1988; Sato and Westermann 1991). The "*Neuqueniceras*," however, has now been identified as the new genus *Pseudoneuqueniceras* of the ?Morphoceratidae (Riccardi and Westermann 1991). The Kitakami Terrane of eastern Honshu, on the other hand, has yielded a small Tethyan and Sula–New Guinean fauna of *Cadomites* and *Satoceras*, again suggesting a southerly origin of this terrane.

In the East Pacific Subrealm, the only known Bathonian of Meso-America (Mexico and Central America) occurs in Oaxaca and Guerrero (Mixteca Terrane) of southwestern Mexico (Sandoval, Westermann, and Marshall 1990), but only the Upper Bathonian is in marine facies. The small *Epistrenoceras* association, with Mediterranean affinity, is overlain by the rather rich Andean *Neuqueniceras* association, in turn followed by the basal Callovian, also Andean, *Frickites* association. Both latter associations yield abundant cosmopolitan *Choffatia*. This sequence and the majority of species are in common with those of the Central and Southern Andes; fewer species are Athabascan or West Tethyan. Southwestern Mexico and northern Chile also include in their faunas the low-latitude, presumably warm-water genera *Epistrenoceras*, *Hemigarantia*, and *Prohecticoceras*, and in Chile also *Strigoceras* (Hillebrandt 1970; Sandoval et al. 1990). All are missing from the coeval faunas of the Neuquén Basin farther south

(Riccardi, Westermann and Elmi 1989). The presence of the Mediterranean elements now distinguishes the Antofagasta Subprovince from the Neuquén Subprovince.

The East Pacific Eurycephalitinae, which abounded in the late Bathonian to early Callovian, sent offshoots into the Shoshonean Province (Western Interior United States) (i.e., the endemic *Paracephalites*) and into the Neuquén Subprovince (i.e., *Stehnocephalites*, which is endemic to this subprovince of the Andean Province).

Both *Xenocephalites* and *Lilloettia* have recently become known also from New Zealand (Westermann and Hudson 1991), extending the Andean Province and the East Pacific Subrealm during the late Bathonian. Rare *Macrocephalites* with affinities to the Bathonian *M. bifurcatus* from New Guinea and the Sula Islands occur slightly below. New Zealand was therefore in the overlap area of the East Pacific and Tethyan Subrealms. A single fragment of the New Zealand species *Xenocephalites grantmackiei* was also found among the rich Macrocephalitinae fauna of New Guinea ("*X.* cf. aff. *neuquensis*" of Westermann and Callomon 1988).

The rich Sphaeroceratidae fauna of eastern Indonesia, formerly dated as Callovian, has now been shown to be mostly Bathonian (Westermann and Callomon 1988). This includes the alleged eucycloceratines (Westermann and Getty 1970), which have turned out to be early *Macrocephalites* and are associated with the middle Bathonian *Bullatimorphites ymir* and *Cadomites* cf. *rectelobatus* of the *C. bremeri* Group. The Macrocephalitinae probably originated in what is here called the Sula–New Guinean Province, from the early Bathonian endemic sphaeroceratine *Satoceras* ("*?Bullatimorphites*" of Westermann and Getty 1970). *Satoceras* possibly descended via *Praetulites* from the Bajocian *Sphaeroceras (Chondroceras)*. Evolution of Macrocephalitinae from the East Pacific Eurycephalitinae, as accepted previously, is less likely. (The *Xenocephalites–Lilloettia* assemblage of New Zealand postdates early *Macrocephalites.*) From the Sula–New Guinea Province, the macrocephalitines spread westward, sporadically during the late Bathonian and, mainly, in the earliest Callovian, along the southern margin of the Tethys ocean into the Mediterranean and connected cratonic seas.

In summary, the Boreal Province extended along the North Slope of Asia southward into Far East Russia and North America. The Subboreal Province of Stikinia and cratonic Western Interior of North America overlapped broadly with the East Pacific Subrealm (Tethyan Realm), which extended along most of the eastern Pacific margin and, at least briefly, even to New Zealand and appears at that time not to have communicated with the West Tethys through the Hispanic Corridor. The low-latitude Antofagasta Subprovince of the Andean Province continued to exists, from allochthonous southwestern Mexico to cratonic northern Chile, and was characterized by high-diversity, low-latitude faunas. Following southward was the Neuquén Subprovince, which had low-diversity, midlatitude faunas and included the endemic *Stehnocephalites*. The Sula–New Guinean Province of the Tethyan Subrealm existed early in the stage along the southeasternmost Tethys Ocean, including several endemic genera. Late in the Bathonian, the Tethyan and East Pacific Subrealms overlapped in the New Zealand area (Westermann and Hudson 1991), and there is no indication of a cold-water "Antiboreal"/"Austral" Province (Stevens 1990).

### Callovian

The Boreal Realm and Province spread far south at the beginning of the stage (Figure 24.2D), with the *Cadoceras* fauna becoming dominant in the Peninsular Terrane and common in the Stikinia and Wrangellia Terranes, and possibly reaching as far south as the Northern Sierra Terrane of California (Taylor et al. 1984). The Subboreal Province is represented by *Kepplerites*, overlapping in the Peninsular Terrane and typically developed in parts of the Wrangellia Terrane (Queen Charlotte Islands). In most northeastern Pacific terranes these Boreal taxa are associated with late representatives of the East Pacific Eurycephalitinae, especially *Lilloettia*,. Marine Callovian is poorly developed in northeastern Asia. *Cadoceras* in North-East Russia and *?Longaeviceras* in Far East Russia indicate that all of eastern Russia was Boreal (Meledina 1988).

Callovian ammonites are unknown from eastern and continental southeastern Asia, except for the Hida Block of western Honshu, Japan, where Subboreal *Kepplerites* closely resemble Wrangellia species (Westermann and Callomon 1988). In the Kitakami Terrane of eastern Honshu, on the other hand, the Callovian, according to Sato (personal communication), has yielded the Tethyan genera *Euaspidoceras, Poculisphinctes, Horioceras,* and others, again indicating the southern origin of this terrane.

Along the eastern Pacific, south of the broad Boreal–East Pacific faunistic overlap, early Callovian is known from Middle America only from the Mixteca Terrane (Guerrero and Oaxaca) of southwestern Mexico (Sandoval et al. 1990), where the East Pacific Subrealm persisted. The Andean *Frickites* association includes also East Pacific eurycephalitines and cosmopolitan *Choffatia*. It is directly overlain by beds also yielding the Tethyan reineckeiid genus *Rehmannia,* which appears to have evolved from the subgenus *Neuqueniceras (Frickites)*. Very similar *Rehmannia* comprise the first, cryptogenic representatives of the Reineckeiinae in the Mediterranean, so that the ancestor appears to have immigrated from the persisting Andean Province (Cariou 1984; Westermann 1984; Westermann and Riccardi 1985; Dommergue et al. 1989).

In the Mixteca Terrane, as well as in the presumably autochthonous Sierra Madre Occidental of eastern Mexico, the Middle (and ?Upper) Callovian yields poorly known faunas of what appears to be the typically "coronate" *Reineckeia* of West Tethys. This signals the end of the East Pacific Subrealm by increased faunal exchange with West Tethys, via the new central Atlantic Ocean and, in part, via the deepening Gulf of Mexico (Jansa 1986).

In the Andes, relatively complete Callovian is developed in northern Chile (Hillebrandt 1970) and the Neuquén Basin (Riccardi et al. 1989), although Upper Callovian remains difficult to establish (?peltoceratids in Chile, as in southwestern Mexico; A. von Hillebrandt personal communication). The East-Pacific Eu-

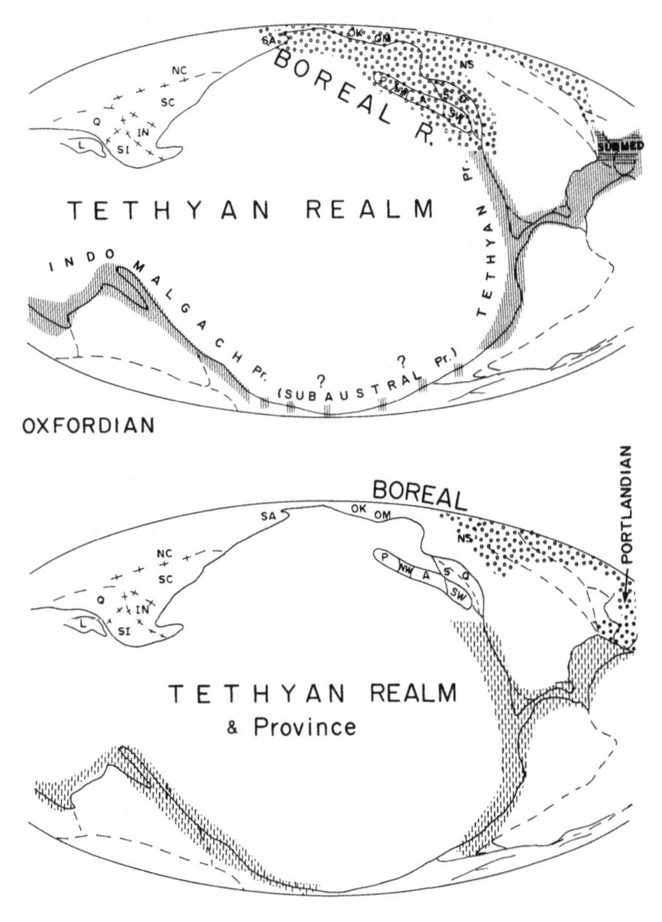


**Figure 24.3. Late Jurassic ammonite biogeography of the circum-Pacific area. For key to symbols, see fig. 24.2.**

rycephalitinae and Neuqueniceratinae are followed by the Tethyan *Rehmannia* and *Reineckeia* faunas, which usually extend only into the Middle Callovian (Riccardi and Westermann 1991). The Mediterranean *Phlycticeras* is also present. As in southwestern Mexico, the Eurycephalitinae sequence ends with late Lower Callovian beds yielding rare Mediterranean *Hecticoceras* s. str. and *H. (Chanasia)* (Riccardi et al. 1989) and is overlain by beds with *Rehmannia* and *?Reineckeia*. A southward trend of diminishing low-latitude (Mediterranean) taxa is again indicated (Westermann and Riccardi 1985), marking the Antofagasta and Neuquén Subprovinces. Curiously, the Eurycephalitinae appear to recur in the early Middle Oxfordian of the Neuquén-Mendoza Basin, and even in New Zealand (G. E. G. Westermann and N. Hudson unpublished data), with the *Lilloettia* homeomorph *Araucanites*, formerly placed in *Mayaites* (Stipanicic, Westermann, and Riccardi 1975).

In the southwestern Pacific, the Callovian remains poorly known. Only the latest *Macrocephalites* fauna of Indonesia (i.e., the *M. keeuwensis* Association) is presently considered to belong to this stage, with the associated *Oxycerites* and *Choffatia* species suggesting an early Callovian rather than late Bathonian age (Westermann and Callomon 1988). The New Guinean forms previously identified with the early Middle Callovian Eucycloceratinae have now been recognized as Bathonian *Macrocephalites* (see the section on the Bathonian). Most of the supposed *Macrocephalites* from New Zealand, however, have turned out to be a new endemic genus of Bajocian Sphaeroceratidae (see the section on the early Bajocian).

In summary, the Boreal Realm and Province spread far southward, broadly overlapping with the East Pacific Subrealm, and remained in contact with western Tethys across the Arctic; late in the early Callovian, connections between the eastern Pacific and West Tethys were renewed by the new(?), true (drifted) Central Atlantic Ocean (Westermann in press); the western Pacific was open to the Tethys Ocean.

## UPPER JURASSIC[3] (Figure 24.3)

A discussion of ammonite bioprovincialism in the circum-Pacific Upper Jurassic is severely constrained by the patchiness of the record. What appear to be locally developed faunas are known from only a few localities, and at any one locality, ammonites usually are known from only a few, widely separated levels. There are therefore great problems in establishing even the primary age relationships that are a prerequisite for a discussion of dynamic biogeographic relationships. Finally, few of the faunas that are abundant have been subjected to a critical modern taxonomic revision. It is difficult to discuss the geographic extent of evolving genera if the generic identifications are in doubt.

The main biogeographic event affecting the Upper Jurassic ammonites worldwide undoubtedly was the final opening of the direct connection between the eastern Pacific and the western Tethys through the young Central Atlantic Ocean. As detailed in earlier sections, there is much circumstantial evidence for a Hispanic Corridor open to ammonite migrations since Early Jurassic times, but the connection must have been sporadic at best and was always restricted. Besides those groups whose distribution patterns strongly suggest a direct connection, such as broad groups like Hammatoceratidae and Sonniniidae in the early Middle Jurassic, and exotica like *Vacekia* [M]/*Asthenoceras* [m] (Aalenian–Bajocian, Oregon–Spain–England) and *Subcollina* [M]/ *Parastrenoceras* [m] (late Bajocian, Mexico–Spain–England), there were always others present at the same time that did not avail themselves of the connection, such as the Graphoceratinae in the Aalenian (western Tethys), the early Stephanoceratinae (Tethys), or the Macrocephalitinae s.s. (Callovian, Tethys). The migrations appear almost always to have been from west to east, not vice versa. This changed in the Oxfordian, in which the migrations clearly could go both ways. The evidence for this is clearest in Mexico, whose ammonites show an almost unbroken succession of diverse assemblages with close European affinities from the Oxfordian well into the Neocomian, including exotica that could only have migrated westward, such as *Gregoryceras* in the middle Oxfordian and *Simoceras* cf./aff. *volanense* in the lower Tithonian. Ammonite distributions became increasingly ho-

[3] By J. H. Callomon.

mogeneous in a broad circum-equatorial belt from the Kimmeridgian upward, in a pattern that formed the basis of Neumayr's original definition of the Tethyan Realm. At the same time, the largely epicontinental shelf seas of the circum-polar northern hemisphere, steadily widening in extent through the general eustatic rise of sea level throughout the Jurassic, gave rise to the strongly segregated Boreal Realm with which we are familiar, also as originally defined. So a single broad division of the world's ammonite faunas into Boreal and Tethyan Realms, broadly latitudinally determined, remains at its most meaningful in the Upper Jurassic and Neocomian. In the Boreal Realm, numerous additional provincial complications arose, such as those making it useful even today to distinguish between Portlandian and Volgian Stages in the terminal Jurassic; but these impinge little on the Pacific and need not concern us further here.

### Oxfordian

The faunas of the Boreal Realm encroached on the margins of the northern Pacific in a wide belt. They are represented by the Cardioceratidae, found in Far East Russia, both Peninsular and North Slope Alaska, all of Wrangellia and Stikinia, and extending as dominant element on the craton as far south as Wyoming. Overlap with the Tethyan faunas farther south, as found in California and Mexico, has been lost tectonically.

The Oxfordian faunas of Mexico consist of Perisphinctidae, Aspidoceratinae, and Oppeliinae very similar to those of Cuba and the western Tethys. The record becomes clear only from the Middle Oxfordian upward. In northern Chile there is evidence for Lower Oxfordian in the form of *Peltoceras (Parawedekindia)*, a group restricted in Europe to the Mariae and Cordatum Zones of the Lower Oxfordian. Similar forms are well represented in Kutch, Madagascar, and the Sula Islands (i.e., along the whole southern margin of the Tethys), although independent evidence of their age is lacking there. The Peltoceratinae are usually classified as "Tethyan," but they appear to have been latitudinally widely tolerant. They are as common in the Oxfordian of Skye in Scotland (paleolatitude 45°N), where they are associated almost exclusively with Boreal *Cardioceras*, as they are in Sicily (20°N), Kutch, and Madagascar (27–33°S). In North America they got as far north as about paleolatitude 50°N on the craton (Montana–Alberta), but no farther. They appear, therefore, to provide one of the clearest examples of latitudinally controlled circum-equatorial pandemism.

The Middle and Upper Oxfordian ammonites of the Andes bear similarly close resemblances to those of Europe and the western Tethys, including the unmistakable late peltoceratids of the genus *Gregoryceras*, one of whose species, *G. transversarium*, is a standard zonal index. These forms have a much narrower latitudinal distribution than *Peltoceras*, being restricted in the western Tethys predominantly to the Tethyan Province itself (south of paleolatitude 30°N). They have also a restricted longitudinal distribution, extending eastward only as far as Madagascar. They are unknown in Kutch, the Himalayas, and Indonesia, where other faunas of the same age do occur. Their presence in the Andes therefore seems a clear-cut example of migration westward from the western Tethys, via the Hispanic Corridor. At the same time, another Tethyan group that might have been expected to follow this route, the Taramelliceratinae, appear to have gone no farther than Cuba. Their absence from the Andes is stressed by Gröschke and Hillebrandt (1985). Instead, there is a locally widespread specialty in the genus *Araucanites* Westermann and Riccardi (Stipanicic, Westermann and Riccardi, 1975) resembling *Mayaites* and originally tentatively placed in the Mayaitinae. If correct, this would indicate a link westward to the Indo-Malgach Province, whose Mayaitinae are common in the Oxfordian as far eastward as the southern portal of the Tethys into the Pacific in Indonesia and New Guinea, and which may be represented in New Zealand and even the Antarctic Peninsula. More recently, however, Westermann and Riccardi (1985) have placed *Araucanites* in the East Pacific Eurycephalitinae.[4]

One other remarkable feature of the Oxfordian of the Andes is the almost total absence of the "Leiostraca," the Phylloceratidae and Lytoceratidae, whose presence might have been expected in these latitudes. These groups were traditionally regarded as the hallmark of the Tethyan Realm and Province, for they largely replaced the more conventional Ammonitina in the pelagic [*sic*] sediments of the classic localities of the European Tethys. The implication was that their distribution was largely climatically controlled, being characteristic of the tropical belt. In the light of modern knowledge, however, their distribution patterns are much more readily explained as characteristic of pelagic oceanic habitats, and their food chains governed by oceanic currents, rather than by latitude. In the Oxfordian of the Alaskan Peninsula, for instance, they can make up 50% of the fauna, the rest consisting almost exclusively of *Cardioceras* – hardly a Tethyan assemblage, whatever the paleolatitude of that part of Alaska may have been. In the Oxfordian of Stikinia, now the intermontane belt of British Columbia, a similar admixture is found. Elsewhere in the Jurassic, the Leiostraca can at some horizons be predominant as far distant from the paleoequator as Far East Russia and New Zealand. Their absence from the Precordilleran basins of northern Chile, now only some 100 km from the Pacific, is most simply explained as being due to the isolation of these basins from oceanic influences, either through being sheltered behind the emergent volcanic chains of the coastal ranges or by virtue merely of distance across shallow seas on broad shelves since lost by subduction into the Andean Trench.

### Kimmeridgian

The evidence in the northern circum-Pacific becomes sparse, because ammonites are largely replaced by benthic *Buchia*. But Boreal *Amoeboceras*, the last of the Cardioceratidae,

[4] Editor's note: *Araucanites* has most recently been found in New Zealand, by N. Hudson and G. E. G. Westermann (unpublished data), possibly in association with *Mayaites*. The microconch of *Araunanites* found here clearly places the genus in the Eurycephalitinae.

extends as far south as the northern Sierra Nevada of California. *Buchia* remains widespread into Mexico.

The Tethyan ammonites of Mexico retain close relationships with those of the western Tethys. Another Mediterranean exoticum that arrived in Mexico was the unmistakable genus *Sutneria*. The general bioprovincial classification is Submediterranean.

Almost nothing is yet known about the sparse Kimmeridgian ammonites of the Andes.

### Tithonian

The worldwide distribution of the Tithonian ammonites was reviewed by Enay (1972). No Boreal Tithonian (i.e., Volgian) ammonites are known from the northern Pacific, although that changed rapidly and dramatically in the early Neocomian, when rich and diverse faunas of Tolliinae and Polyptychitinae reached northern California.

The Tethyan Tithonian ammonites left extensive records along the eastern margins of the Pacific. The broad relationships at the family level with the rest of the world remain clear. There are Haploceratacea (*Taramelliceras, Streblites, Haploceras*), Aspidoceratidae (*Aspidoceras, Physodoceras*, and, significantly, *Hybonoticeras*), and several branches of the Perisphinctaceae: Lithacoceratinae and Virgatosphinctinae; Berriasel-liinae; Idoceratinae; rare Simoceratidae and Himalayitidae; and Spiticeratidae. The compositions change latitudinally much as in the Tethys, so that locally a distinction between Tethyan and Submediterranean Provinces could be made. The pattern southward along the Andes is, however, the mirror image of that in the more familiar northern Tethys and its Submediterranean shelf seas in Europe. A more appropriate name for the climatically more temperate assemblages could therefore be Subaustral. For a proper Austral equivalent of the Boreal Province there is thus far little evidence. Going toward the South Pole, the faunas simply become poorer and less diverse, but their genera are little different (e.g., *Kossmatia* in New Zealand and the Antarctic Peninsula). The reasons were, however, probably physiographic. All the most southerly Jurassic faunas known occur in narrow perioceanic shelf-sea deposits, presumably of equable climates tempered by the proximity of oceanic currents. Diversities of all faunal elements are low, and pelagic, oceanic ammonite "Leiostraca" are relatively common. In contrast, in the Tithonian of the wide Mendoza-Neuquén shelf basins of Argentina, the "Leiostraca" are as scarce as they were in the Oxfordian of northern Chile, plausibly for the same reasons – effective distance from the ocean.

Westward, the Tithonian faunas of Argentina show equally strong resemblances to those of the southern Tethys, as epitomized by the Himalayan Spiti Shales, and to those of the scattered intermediate localities, as found in Sula and New Guinea. Not enough is yet known of the stratigraphy of the Tithonian, either of Indonesia, Papua New Guinea, or of the Himalayas, to draw any further conclusions.

## JURASSIC AMMONITE BIOGEOGRAPHY AND SOUTHERN ALASKAN ALLOCHTHONOUS TERRANES[5]

Fossil data, particularly based on ammonoids, used in conjunction with paleomagnetic data, have been of great help in the interpretation of the tectonic history of southern Alaska. This section follows the lead of Taylor et al. (1984) and uses their terminology, but includes a vast amount of recent data that were not available to them, especially from the Alaska Peninsula, whereas identification and interpretation of the Jurassic ammonids was mainly by Ralph Imlay, Gerd Westermann, and John Callomon (see Chapter 4). Chronostratigraphically important bivalves, mainly *Buchia* species, from the Upper Jurassic were identified by John W. Miller.

### Lower Jurassic

Lower Jurassic strata have been identified faunally in most of the southern Alaska terranes (Table 24.1). The faunas are mainly pandemic, but forms associated with the Tethyan Realm are present in all terranes, suggesting depositional sites at low latitude. Elements of the East Pacific Subrealm in the Sinemurian and Pliensbachian are important also, as they indicate a paleolongitude in the eastern Pacific.

Hettangian ammonites are known only from the Peninsular Terrane, mostly from Puale Bay. Pandemic genera are *Psiloceras, Schlotheimia*, and *Waehneroceras. Discamphiceras* represents the Tethyan Realm. Sinemurian ammonites are most widespread, and include pandemic *Arnioceras, Coroniceras, Crucilobiceras*, and *Paltechioceras*, as well as the Tethyan *Discamphiceras* and the bivalve *Weyla*. The last is the only form found in the Togiak Terrane and is the only common form in the Flysch Terrane. *Badouxia* and *Arctoasteroceras*, characteristic of the Athabaskan Province and eastern Siberia, are restricted to the Peninsular and Chulitna Terranes.

Pliensbachian strata are restricted to the northeastern Peninsular Terrane (Grantz 1960) and adjoining Wrangellia and are represented by the pandemic ammonites *Apoderoceras, Leptaleoceras, Paltarpites, Prodactylioceras*, and *Tropidoceras* in the late Pliensbachian and *Uptonia* in the early Pliensbachian. Tethyan *Arieticeras, Fontanelliceras*, and *Protogrammoceras* occur in the late Pliensbachian. The Boreal genus *Amaltheus* in the Peninsular Terrane suggests that the terrane had started northward by the Pliensbachian age; it is rare in the southern Alaska terranes. *Fanninoceras*, restricted to the eastern Pacific, is associated with *Amaltheus, Arieticeras*, and *Fontanelliceras* in the Talkeetna Mountains (Imlay 1981).

Toarcian strata are known only from the Peninsular Terrane, by *Dactylioceras, Grammoceras, Harpoceras, Peronoceras*, and *Phymatoceras*.

[5] By R. L. Detterman.

Table 24.1. *Ammonite genera from southern Alaska terranes*

| Age<br>Realm/Subrealm<br>Fossil | Peninsular | Wrangellia | Flysch | Chulitna |
|---|---|---|---|---|
| Hettangian | | | | |
| Tethyan | | | | |
| *Discamphiceras* | 1–7[a] | | | |
| Pandemic | | | | |
| *Psiloceras* | 2–7 | | | |
| *Scholtheimia* | 1–2 | | | |
| *Waehneroceras* | 2–16 | | | |
| Sinemurian | | | | |
| Tethyan | | | | |
| *Discamphiceras* | 1–1 | 1–1 | | |
| *Paracaloceras* | 1–6 | | 1–1 | |
| Pandemic | | | | |
| *Arnioceras* | 1–8 | 1–3 | 1–5 | 1–5 |
| *Coroniceras* | 2–5 | 1–1 | 1–1 | |
| *Crucilobiceras* | | 2–6 | | |
| *Paltechioceras* | 1–1 | 2–4 | 1–1 | |
| *Psiloceras* | 2–1 | | | 1–1 |
| East Pacific | | | | |
| *Arctoasteroceras* | 1–1 | 1–1 | 1–1 | |
| *Badouxia* | 2–4 | | 2–3 | |
| Pliensbachian | | | | |
| Tethyan | | | | |
| *Arieticeras* | 1–3 | 2–4 | | |
| *Fontanelliceras* | 1–3 | | | |
| *Protogrammoceras* | 2–5 | | | |
| Pandemic | | | | |
| *Apoderoceras* | 1–2 | | | |
| *Leptaleoceras* | 1–1 | | | |
| *Peltarpites* | 1–1 | | | |
| *Prodactilloceras* | | 2–5 | | |
| *Tropidoceras* | | 1–1 | | |
| *Uptonia* | 1–1 | 3–6 | | |
| East Pacific | | | | |
| *Fanninoceras* | 2–5 | | | |
| Boreal | | | | |
| *Amaltheus* | 1–1 | | | |
| Toarcian | | | | |
| Pandemic | | | | |
| *Dactylioceras* | 2–3 | | | |
| *Grammoceras* | 1–1 | | | |
| *Harpoceras* | 1–1 | | | |
| *Haugia* | 3–9 | | | |
| *Peronoceras* | 1–1 | | | |
| *Phymatoceras* | 1–2 | | | |
| Aalenian | | | | |
| Tethyan | | | | |
| *Abbasites* | 2–6 | | | |
| *Asthenoceras* | 2–2 | | | |
| *Eudmetoceras* | 6–21 | | | |
| *Fontannesia* | 2–2 | | | |
| *Planammatoceras* | 1–3 | | | |
| Pandemic | | | | |
| *Tmetoceras* | 4–15 | | | |
| Boreal | | | | |
| *Erycitoides* | 9–67 | | | |
| *Pseudolioceras* | 4–20 | | | |
| Early Bajocian | | | | |
| Tethyan | | | | |
| *Bradfordia* | 3–19 | | | |
| *Docidoceras* | 4–25 | | | |
| *Eudmetoceras* | 3–8 | | | |
| *Planammatoceras* | 2–5 | | | |
| Pandemic | | | | |
| *Chondroceras* | 4–23 | | | |
| *Dorsetensia* | 1–2 | | | |
| *Emileia* | 1–11 | | | |
| *Sonninia* (Euhoploceras) | 4–13 | | | |
| *Stephanoceras* | 5–18 | | | |
| *Teloceras* | 2–10 | | | |
| *Witchellia* | 2–8 | | | |
| East Pacific | | | | |
| *Alaskinia* | 1–3 | | | |
| *Parabigotites* | 2–34 | | | |
| *Zemistephanus* | 2–6 | | | |
| Boreal | | | | |
| *Arkelloceras* | 1–1 | | | |
| *Pseudolioceras* | 1–2 | | | |
| Late Bajocian | | | | |
| Pandemic | | | | |
| *Leptosphinctes* | 2–3 | | | |
| *Lissoceras* | 2–2 | | | |
| *Normannites* | 3–28 | | | |
| *Oppelia* | 2–2 | | | |
| *Stephanoceras* | 2–8 | | | |
| East Pacific | | | | |
| *Megasphaeroceras* | 1–12 | | | |
| Early Bathonian | | | | |
| Pandemic | | | | |
| *Cobbanites* | 3–14 | 2–7 | | |
| East Pacific | | | | |
| *Iniskinites* | 1–1 | | | |
| *Parareineckeia* | 2–2 | 2–16 | | |
| *Tuxednites* | 1–12 | 1–2 | | |
| Boreal | | | | |
| *Arctocephalites* | 1–4 | | | |
| *Cranocephalites* | 6–67 | 3–16 | | |
| Subboreal | | | | |
| *Kepplerites* | 5–10 | | | |
| Late Bathonian | | | | |
| East Pacific | | | | |
| *Iniskinites* | 2–8 | | | |
| Boreal | | | | |
| *Cadoceras* | 2–6 | | | |
| Subboreal | | | | |
| *Kepplerites* | 2–5 | | | |
| Callovian | | | | |
| East Pacific | | | | |
| *Lilloettia* | 5–29 | | | |
| Boreal | | | | |
| *Cadoceras* | 8–133 | | | |
| *Paracadoceras* | 4–20 | | | |
| *Pseudocadoceras* | 5–70 | | | |
| *Stenocadoceras* | 4–67 | | | |
| Subboreal | | | | |
| *Kepplerites* | 8–34 | | | |
| Oxford–Kimmeridgian | | | | |
| Boreal | | | | |
| *Amoeboceras* | | 1–3 | | |
| *Cardioceras* | 5–11 | | | |
| Pandemic | | | | |
| *Phylloceras* | 2–38 | 1–3 | | |
| *Lytoceras* | 1–8 | | | |

[a]Indicates 1 species, 7 localities.

### Middle Jurassic

This was a time of great faunal productivity and diversity, as represented in the strata exposed mainly in the southern Peninsular Terrane. Elements of several realms are commonly intermingled at the stage level (Figures 24.2 and 24.4), but a very specific pattern is developing. Tethyan Realm fauna, ubiquitous in the Triassic and common in the Lower Jurassic, continue into the early Bajocian and then completely disappear from southern Alaska terranes. At the same time, elements of the Boreal Realm, both Boreal and Bering Provinces, rapidly increase in relative abundance. The Bering Province is important for defining ancient longitudes as well as latitudes, as this fauna is restricted to northeastern Pacific areas. The Athabaskan Province of the East Pacific Subrealm becomes more important during the Middle Jurassic, and its fauna furnish a direct tie to cratonic North America.

The Aalenian ammonites are dominated by *Erycitoides* and *Pseudolioceras* of the Bering Province (Westermann 1964; Imlay 1984). Tethyan fauna, including *Abbasites, Eudmetoceras,* and *Planammatoceras,* are subordinate. The abundant Bajocian ammonite faunas were dominated by genera such as *Stephanoceras, Sonninia, Chondroceras,* and *Witchellia. Docidoceras* and *Bradfordia* were moderately abundant, *Pseudolioceras* common locally, and *Arkelloceras* rare. During the late Bajocian, East Pacific elements such as *Megasphaeroceras* and *Parareineckeia* were associated with cosmopolitan *Stephanoceras, Oppelia,* and *Leptosphinctes.* During the latest Bojacian and Bathonian, Boreal *Arctocephalites* and *Kepplerites* occur in the Peninsular Terrane, as do East Pacific *Iniskinites, Cranocephalites,* and *Parareineckeia.* In the late Bathonian, Boreal *Cadoceras* is the most abundant genus, further suggesting that southern Alaska was near its present position. Several subgenera of *Cadoceras* occur in the Callovian rocks of the Peninsular Terrane, as do a few specimens of *Kepplerites* and the East Pacific genus *Lilloettia.*

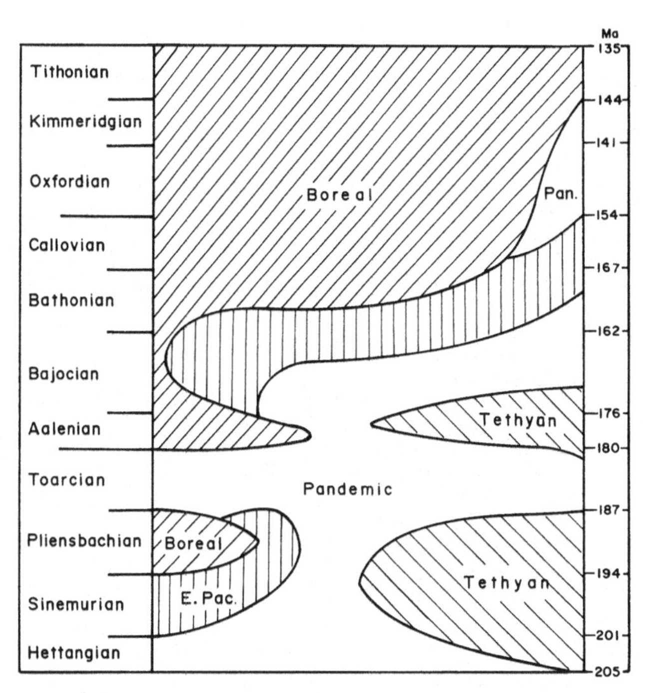


**Figure 24.4. Diagram showing relative abundances of Jurassic ammonite genera, indicating bioprovinces, for the Peninsular Terrane.**

### Upper Jurassic

Cosmopolitan bivalves, especially *Buchia,* dominate most of the Upper Jurassic, but the Boreal ammonites *Cardioceras* and *Amoeboceras* are present. This suggests that southern Alaska was north of 50°N during that time.

### Conclusions

Ammonite biogeography is consistent with, and confirms, the northward displacement of the Peninsular Terrane presented by Rowley (Chapter 3) for the Jurassic.

## AMMONITE BIOGEOGRAPHY AND ECOLOGY MODIFY MESO-AMERICAN RECONSTRUCTION[6]

Jurassic ammonoids have for a long time been used biogeographically to infer past positions of oceans and continents. Middle Jurassic ammonite biogeography in the Pacific area was reviewed and discussed earlier in this chapter, whereas the ecology of Jurassic and Cretaceous ammonoids has recently been reviewed elsewhere (Westermann 1990). Phylloceratina, Lytoceratina, and some homeomorphic Ammonitina (together known as "leiostracans") typically inhabited oceans, mostly continental slopes, whereas the majority of Ammonitina (the "trachyostracans") were neritic, inhabiting the epicontinental seas ("heteromorphs" were insignificant during the Jurassic). The duration of the presumably epiplanktonic "pseudolarva," however, remains unknown, possibly varying from a few days in typical "trachyostracans" to several weeks in some "leiostracans." Thus, the former had their dispersal essentially restricted to epicontinental seas; they tended to be more restricted paleogeographically, often delineating bioprovinces. But species (and genus) distributions were much more complex, and endemism (versus pandemism/cosmopolitanism) cannot be predicted from ammonite shell morphology or structure. Like the Present *Nautilus,* most ammonoids presumably could swim up to several kilometers per day and thus also had the potential for active dispersal.

Another, poorly known factor in ammonoid paleobiogeography is postmortem shell drift. The shells of deep-water species probably tended to sink after death, but once rare shells had reached the surface, they could have drifted for a long time. The shells of shallow-water species may have reached the surface more commonly, but then became waterlogged more rapidly. Thus, as a rule, "trachyostracan" shells could drift a few tens of kilometers onto the adjacent shelf, whereas "leiostracans" tended to stay in their habitat area, but in rare cases may have drifted up to several hundred kilometers.

Biogeographic evidence for the relative positions of, and links (seaways or land bridges) between, areas is by homeotaxis. Meso-

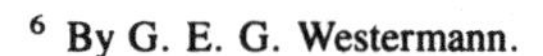
[6] By G. E. G. Westermann.

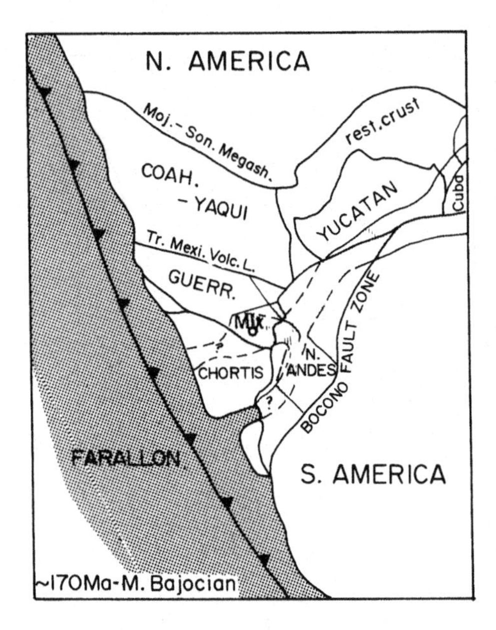


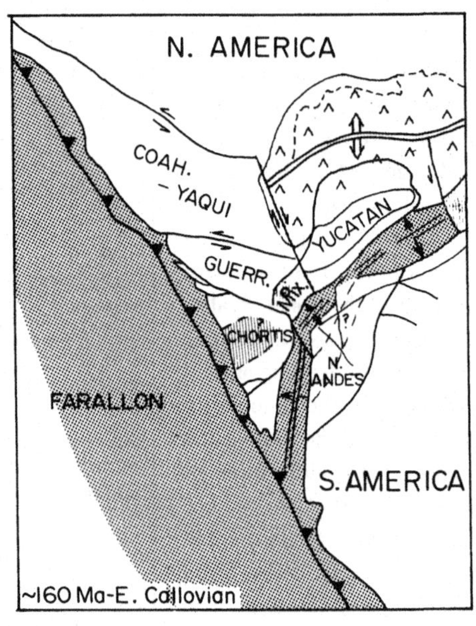


**Figure 24.5. Middle Jurassic tectonic model #1 for Middle America, according to Ross and Scotese (1988), modified according to Jansa (1986), with ocean (gray) and epicontinental seas (ruled) added, and Mixteca for southeastern part of "Guerrero Block." The early Callovian reconstruction uses a hybrid between Ross and Scotese's Oxfordian and Tithonian reconstructions; the "Mojave–Sonora Megashear" has been eliminated.**

America during Early Jurassic time was located at the western port of the epicontinental Hispanic Corridor, at some time (late?) during that epoch becoming the nascent Central Atlantic Ocean, and both connecting the eastern Pacific Ocean with the western Tethys (and the North Atlantic). In the case of ammonoids, epicontinental seaways should filter out deep-water and oceanic taxa ("leiostracans"), and the degree of endemism in the "basins" should vary with the seaway constriction, changing through time and hence reflecting the global eustatic cycle.

Jurassic palinspastic reconstructions of Meso-America and the nascent Central Atlantic (Proto-Atlantic) have recently been proposed, especially by Pindell (1985), Ross and Scotese (1988), and Jansa (1986) (Figures 24.5 and 24.6). Reconstructions are basically of two types: (1) Most reconstructions, including that of Ross and Scotese, as well as the ones by Rowley used as the base map for the Pacific in Chapter 3, envisage Meso-America of the earlier Jurassic as an integral part of Pangea, i.e., a contiguous assemblage of crustal blocks that during the later Jurassic separation of North and South America remained mutually adjacent and delimited only by "megashears" (transform faults). (The existence of an extensive "Mojave–Sonora megashear," however, has recently been refuted, based on the uninterrupted trend of Paleozoic sediments across this alleged structure; W. Hamilton personal communication.) The general hypothesis entails the consistent juxtaposition of the Southwest Mexico Blocks/Terranes (Guerrero, Mixteca, Oaxaca, Acatlan, etc.) and the Chortis Block/Terrane, so that the former would lie well inland. (2) Pindell, on the other hand, derives the Southwest Mexico and Chortis Blocks/Terranes from the northeastern Pacific, arriving together at the craton during the Jurassic, so that both major blocks/terranes were adjacent to the Pacific Ocean.

Middle Jurassic ammonite biogeography is used to test the past position of the Mixteca Block/Terrane, presently comprising eastern Guerrero and western Oaxaca states (and possibly part of Pueblo). Mixteca is the only area in Middle America known to have significant ammonite occurrences at this time, and it is tested with respect to (1) distance from the ocean ("leiostracans" versus "trachyostracans") and (2) marine access to the northeastern (North Cordilleran) and southeastern (Andean) Pacific margins, as well as to western Tethys (and the North Atlantic).

The faunistic connection of the eastern Pacific with the western Tethys (and North Atlantic) via the epicontinental Hispanic Corridor began in the Pliensbachian, perhaps already in the Sinemurian, when rifting is evident in the Huayacocotla Graben of eastern Mexico (Schmidt-Effing 1980). The corridor acted as a decreasingly restrictive, two-way dispersal route for "trachyostracans" until the end of the early Bajocian. Sudden closure early in the late Bajocian caused isolation, which lasted for two stages and resulted in almost total endemism of eastern Pacific ammonites (the East Pacific Subrealm). This marked change in the rate of endemism is correlated with the global eustatic sea-level change (Westermann in press).

#### Mid-Bajocian

The "Mixtecan" ammonite fauna combines North Cordilleran (Athabascan), Andean, and West Tethyan (Eurafrican) taxa in almost equal taxic proportions, and endemism is high (Table 24.2). All species are shallow-water "trachyostracans" (Sandoval and Westermann 1986). Scarce elements of this same fauna also constitute the only known indubitably Bajocian ammonites of Honduras, where they occur in a restricted marine facies bearing mostly plants (see Chapter 5). The tectonic reconstructions of Meso-America are all compatible with the mid-Bajocian ammonite biogeography and ecology: The shallow-marine "Guerrero embayment" may have reached all the way across Honduras (model 1, Figure 24.5), or it may have been smaller, with a second embayment existing on Chortis (model 2, Figure 24.6). Both models require an easterly marine connection with the Hispanic Corridor. The Pindell (#2) model, however, appears to be more appropriate, because the more northerly located Mixteca and Chortis Blocks/Terranes imply better access to the North Cordil-

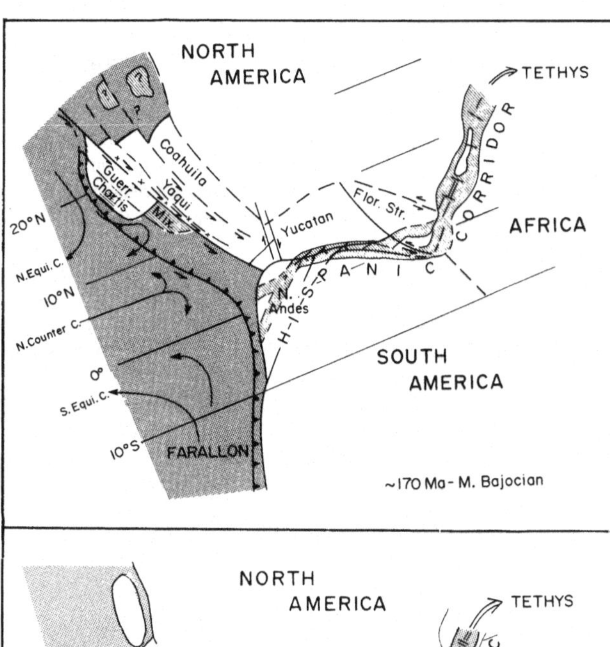


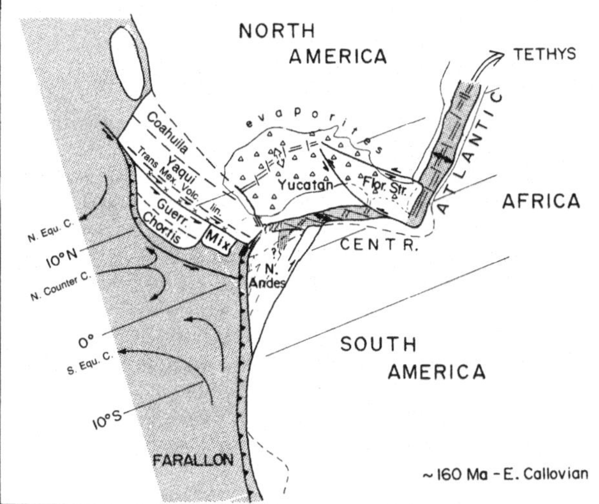


**Figure 24.6. Middle Jurassic tectonic model #2 for Middle America, according to Pindell (1985), modified only to separate Mixteca (Mix., with main fossil localities) from Guerrero Block/Terrane and with ocean (gray, with currents) and epicontinental seas added.**

leran faunas, which would not have to cross the equatorial-current systems (reconstructed in Figure 24.6).

### Latest Bathonian–early Callovian

Ammonite biogeography and ecology had changed markedly since the Bajocian (Sandoval et al. 1990) (Table 24.2). North Cordilleran elements had disappeared, Andean elements had become dominant, and endemic, oceanic "leiostracan" species had become abundant. This fauna is also compatible with the Pindell (#2) model: Mixteca has moved far southward, into the equatorial zone, and remains exposed to the Pacific Ocean. The more widely accepted model 1 (or group of tectonic models), however, has Mixteca situated inland of Chortis, where Bathonian (M. D. Gordon personal communication) and Callovian beds bear only shallow-water, rare ammonites. Yet the required modifications to these reconstructions involve essentially only a somewhat earlier, late Middle Jurassic, instead of early Late Jurassic, drifting between Chortis and the North Andean Blocks. A Bathonian start of

Table 24.2. *Biogeographic affinities of the Middle Jurassic ammonites of Guerrero and Oaxaca states (Mixteca Block/Terrane), percentages of species in common*

| | Mid-Bajocian (%) | Upper Bathonian–Lower Callovian (%) |
|---|---|---|
| Endemic | 32 | 22 |
| Mediterranean | 32 | 21 |
| Mediterranean + Andean | 0 | 8 |
| Andean | 14 | 33 |
| North Cordilleran | 18 | 0 |
| East Pacific | 0 | 6 |
| Pandemic | 5 | 0 |

the drifting here would mirror the Central Atlantic and Caribbean developments.

### References


Arkell, W. J. (1956). *Jurassic Geology of the World.* London: Oliver & Boyd.

Bartok, P., Renz, Q., & Westermann, G. E. G. (1985). The Siqoisique ophiolites, northern Lara State, Venezuela: a discussion of their Middle Jurassic ammonites and tectonic implications. *Geol. Soc. Am. Bull., 96,* 105–55.

Bloos, G (1983). The zone of *Schlotheimia marmorea* (Lower Lias) – Hettangian or Sinemurian? *Newsl. Stratigr., 12*(3), 123–31.

(1988). On the stage boundary Hettangian/Sinemurian in Northwest Europe and in Eastern Alps. In *2nd International Symposium on Jurassic Stratigraphy, Lisboa, 1987* (pp. 71–83).

Burckhardt, C. (1927). *Cefalopodeos de Jurásico Medio de Oaxaca y Guerrero.* Inst. Geol. Mex. Bol., 47.

Callomon, J. H. (1979). Marine boreal Bathonian fossils from the northern North Sea and their palaeogeographic significance. *Proc. Geol. Assoc. (London), 90,* 163–9.

(1984). A review of the biostratigraphy of the post-Lower Bajocian ammonites of western and northern North America. In G. E. G. Westermann (ed.), *Jurassic–Cretaceous Biochronology and Paleogeography of North America* (pp. 143–74). Geological Association of Canada, Special Paper 27.

(1985). The evolution of the Jurassic ammonite family Cardioceratidae. *Spec. Pap. Palaeontol., 33,* 49–90.

Cameron, B. E. B., & Tipper, H. W. (1985). *Jurassic Stratigraphy of the Queen Charlotte Islands, British Columbia.* Geol. Surv. Can., Bull. 365.

Cariou, E. (1984). Structure, origin et paléobiogéographie de la famille des Reineckeiidae, Ammonitina du Jurassique moyen. *C. R. Acad. Sci. (Paris), 298*(2), 245–8.

Cariou, E., Dietl, G., & Niederhofer, A. (1989). The ammonite faunal horizons at the Bathonian–Callovian boundary in the Swabian Jurassic and their correlation with those of Western France and England. *Stuttgarter Beitr. Naturkd., B, 148,* 1–13.

Dagis, A. A. (1968). *Toarskie ammonity (Dactylioceratidae) severa Sibiri.* Moscow: Trudy Inst. Geol. Geophys. (Akad. Nauk sibirskoe otdel.), 40.

(1974). *Toarski ammonity (Hildoceratidae) severa Sibiri.* Moscow: Trudy Inst. Geol. Geophys. (Akad. Nauk sibirskoe otdel.), 99.

Dommergue, J.-L. (1986). Les Dactylioceratidae du Domerien basal, un groupe monophyletique. Les Reynesocoeloceratinae nov. subfam. *Bull. Sci. Bourg., 39,* 1–26.

Dommergue, J.-L., Cariou, E., Contini, D., Hantzpergue, P., Marchand, D., Meister, C., & Thierry, J. (1989). Homéomorphies et canalisations évolutive; le rôle de l'ontogenèse. Quelques exemples pris chez les ammonites du Jurassique. *Geobios, 22,* 5–48.

Enay, R. (1982). Paléobiogéographie des ammonites du Jurassique terminal (Tithonique/Volgien/Portlandien s.l.) et mobilité continentale. *Géobios, 5,* 355–407.

(1980). Paléogéographie des ammonites jurassiques: "rythmes fauniques" et variations du niveau marine; voies d'échanges, migrations et domaines biogéographiques. *Livre jub. Soc. Géol. France, Mem., 10,* 261–81.

Frebold, H. (1960). *The Jurassic Faunas of the Canadian Arctic. Lower and Lowermost Middle Jurassic Ammonites.* Geol. Surv. Can. Bull., 59.

(1967). *Hettangian Ammonite Faunas of the Taseko Lakes Area.* Geol. Surv. Can., Bull. 158.

(1975). *The Jurassic Faunas of the Canadian Arctic. Lower Jurassic Ammonites, Biostratigraphy and Correlations.* Geol. Surv. Can., Bull. 243.

Frebold, H., & Tipper, H. W. (1970). Status of the Jurassic in the Canadian Cordillera of British Columbia, Alberta, and southern Yukon. *Can. J. Earth Sci., 7,* 1–21.

Grantz, A. (1960). *Geologic Map of Talkeetna A-1 Quadrangle, Alaska, and Contiguous Area to North and Northwest.* U.S. Geol. Surv. Misc. Geol. Invest. Map I-313.

Gröschke, M., & Hillebrandt, A. von (1985). Trias und Jura in der mittleren Cordillera Domeyko von Chile (23°30'–24°30'). *N. Jb. Geol. Palaontol. Abhandl., 170,* 129–66.

Guex, J. (1980). Remarques préliminaires sur la distribution stratigraphique des ammonites hettangiennes du New York Canyon (Gabbs Valley Range, Nevada). *Bull. Géol. Lausanne, 250,* and *Bull. Soc. Vaud. Sc. Nat., 75.2*(no. 358), 127–40.

(1982). Relation entre le genre Psiloceras et les Phylloceratida au voisingage de la limite Trias-Jurassique. *Bull. Géol. Lausanne, 260,* and *Bull. Soc. Vaud. Sc. Nat., 76.2,* 47–51.

(1987). Sur la phylogenèse des ammonites de Lias inferieur. *Bull. Géol. Lausanne, 292,* and *Bull. Soc. Vaud. Sc. Nat., 78.4,* 455–569.

Guex, J., & Taylor D. (1976). La limite Hettangien–Sinemurien des Préalpes romandes au Nevada. *Eclogae Geol. Helvet., 69*(2), 521–6.

Hall, R. L. (1989). New Bathonian (Middle Jurassic) ammonite faunas from the Fernie Formation, Southern Alberta. *Can. J. Earth Sci., 26,* 16–22.

Hall, R. L., & Stronach, N. H. (1981). The first record of Late Bajocian (Jurassic) ammonites in the Fernie Formation, Alberta. *Can. J. Earth Sci., 18,* 919–25.

Hall, R. L., & Westermann, G. E. G. (1980). Lower Bajocian (Jurassic) cephalopod faunas from western Canada and proposed assemblage zones for the Lower Bajocian of North America. *Paleontogr. Amer., 9*(52), 1–93.

Hallam, A. (1965). Observation on marine Lower Jurassic stratigraphy of North America, with special reference to United States. *Bull. Am. Assoc. Petrol. Geol., 49,* 1485–501.

Hillebrandt, A. von (1970). Zur Biostratigraphie und Ammoniten-Fauna des südamerikanischen Jura (insbes. Chile). *N. Jb. Geol. Paläont. Abh., 136,* 166–211.

(1973). Die Ammoniten-Gattungen Bouleiceras und Frechiella im Jura von Chile und Argentinien. *Eclogae Geol. Helvet., 66,* 351–63.

(1981). Faunas de amonites del liásico inferior y medio (Hettangiano hasta Pliensbachiano) de América del Sur (excluyendo Argentina). In W. Volkheimer (ed.), *Cuencas Sedimentarias del Jurásico y Cretácico de América del Sur* (vol. 2, pp. 499–538). Buenos Aires: Com. Sudam. Juras. Cretac.

(1987). Liassic ammonite zones of South America and correlations with other provinces – description of new genera and species of ammonites. In W. Volkheimer (ed.), *Bioestratigrafia de los Sistemas Regionales del Jurásico y Cretácico en América del Sur* (pp. 111–57). Mendoza: Com. Sudam. Juras. Cretac.

(1988). Ammonite biostratigraphy of the South American Hettangian – description of two new species of *Psiloceras.* In R. B. Rocha & A. F. Soares (eds.), *2nd Int. Symp. Jurassic Strat., Lisboa, 1987* (pp. 13–27). Lisboa: Cen. Geocien. Univ. Coinbra.

Hillebrandt, A. von, & Schmidt-Effing, R. (1981). Ammoniten au dem Toarcium (Jura) von Chile (Sudamerika). Die Arten der Gattungen Dactylioceras, Nodicoeloceras, Peronoceras und Collina. *Zittelania,* 6.

Hillebrandt, A. von, & Westermann, G. E. G. (1985). Aalenian (Jurassic) ammonite faunas and zones of the Southern Andes. *Zittelania, 12,* 3–55.

Hirano, H. (1971). Biostratigraphic study of the Jurassic Toyora Group. Part I. *Mem. Facul. Sci. Kyushu Univ., Ser. D, Geol., 21,* 93–128.

(1973). Biostratigraphic study of the Jurassic Toyora Group. Part III. *Trans. Proc. Palaeontol. Soc. Japan, N.S., 90,* 45–71.

Howarth, M. K. (1973). The stratigraphy and ammonite fauna of the Upper Liassic Grey Shales of the Yorkshire Coast. *Bull. Bri. Mus. Nat. Hist. Geol., 24,* 235–77.

Imlay, R. W. (1968). *Lower Jurassic (Pliensbachian and Toarcian) Ammonites from Eastern Oregon and California.* U.S. Geological Survey Professional Paper 593-C.

(1973). *Middle Jurassic (Bajocian) Ammonites from Eastern Oregon.* U.S. Geological Survey, Professional Paper 756.

(1981). *Early Jurassic Ammonites from Alaska.* U.S. Geological Survey, Professional Paper 1148.

(1984). *Early and Middle Bajocian (Middle Jurassic) Ammonites from Southern Alaska.* U.S. Geological Survey, Professional Paper 1322.

Jansa, L. F. (1986). Paleoceanography and evolution of the North Atlantic basin during the Jurassic. In P. R. Vogt & B. E. Tucholke (eds.), *The Geology of North America. Vol. M: The Western North Atlantic Region* (pp. 603–16). Geol. Soc. Am.

Jaworski, E. (1926). La fauna del Lias y Dogger de la Cordillera Argentina en la parte meridional de la Provincia de Mendoza. *Actas Acad. Nac. Cienc. (Córdoba),* 9, 137–318.

Kalacheva, E. D. (1988). Toarcian. In G. Y. Krymholts et al. (eds.), *The Jurassic Ammonite Zones of the Soviet Union* (pp. 1–116). Geol. Soc. Am. Spec. Pap. 223.

Komalarjun, P., & Sato, T. (1964). Aalenian (Jurassic) ammonites from Mae Sot, northwestern Thailand. *Japan J. Geol. Geogr., 35*(2–4), 149–61.

Krishna, J., & Westermann, G. E. G. (1987). The faunal associations of the Middle Jurassic ammonite genus *Macrocephalites* in Kachchh, western India. *Can. J. Earth Sci., 24,* 1570–82.

Krumbeck, L. (1922). Zur Kenntnis des Jura der Insel Rotti. *Jb. Mijnewezen Ned. Oost-Indie, 1920, Verh., 3,* 107–219.

Krymholts, G. Y., Mesezhnikov, L. M. S., & Westermann, G. E. G. (eds.) (1988). *The Jurassic Ammonite Zones of the Soviet Union.* Geol. Soc. America, Spec. Pap. 223.

Meledina, S. V. (1988). Bathonian, Callovian. In G. Y. Krymholts et al. (eds.), *The Jurassic Ammonite Zones of the Soviet Union* (pp. 29–38). Geol. Soc. America, Spec. Pap. 223.

Pérez, E. (1982). *Bioestratigrafía del Jurásico de Quebrada Asientos, norte de Potrerillos, Región de Atacama.* Serv. Nac. Geol. Min. Chile, Bol. 37.

Pindell, J. L. (1985). Alleghenian reconstruction and the subsequent evolution of the Gulf of Mexico, Bahamas and Proto-Caribbean Sea. *Tectonics, 4,* 133–56.

Poulton, T. P. (1987). *Zonation and Correlation of Middle Boreal Bathonian to Lower Callovian (Jurassic) Ammonites, Salmon Cache Canyon, Porcupine River, Northern Yukon.* Geol. Surv. Can., Bull. 358.

Prinz, P. (1985). Stratigraphie und Ammonitenfauna der Pucara-Gruppe (Obertrias-Unterjura) von Nord-Peru. *Palaeontographica, A., 188,* 153–97.

Quilty, P. G. (1970). Jurassic ammonites from Ellsworth Land, Antarctica. *J. Paleontol.*, *44*, 110–16.

Quinzio, S. L. A. (1987). Stratigraphische Untersuchungen im Unterjura des Sudteils der Provinz Antofagasta in Nord-Chile. *Berliner Geowiss. Abh.*, *A87*.

Repin, Y. S. (1984). *Lower Jurassic Ammonite Standard Zones and Zoogeography in North-East Asia.* IGCP Project 171: Circum-Pacific Jurassic, Report 2, Spec. Pap. 1.

Riccardi, A. C., & Westermann, G. E. G. (1991). The Middle Jurassic ammonoid fauna and biochronology of the Argentine–Chilean Andes. III: Eurycephalitinae, Stephanocerataceae. IV: Bathonian-Callovian Reineckeiidae. *Palaeontographica, A216*, 111–70.

Riccardi, A. C., Westermann, G. E. G., & Elmi, S. (1989). The Middle Jurassic Bathonian–Callovian ammonite zones of the Argentine–Chilean Andes. *Geobios, 22*, 553–97.

Ross, M. I., & Scotese, C. R. (1988). A hierarchical tectonic model of the Gulf of Mexico and Caribbean region. *Tectonophysics, 155*, 139–68.

Sandoval, J., & Westermann, G. E. G. (1986). The Bajocian (Jurassic) ammonite fauna of Oaxaca, Mexico. *J. Paleontol.*, *60*, 1220–71.

Sandoval, J., Westermann, G. E. G., & Marshall, M. C. (1990). Ammonite fauna, stratigraphy and ecology of the Bathonian–Callovian (Jurassic) Tecocoyunca Group, south Mexico. *Palaeontographica, A210*, 93–149.

Sato, T. (1962). Etudes biostratigraphiques des ammonites du Jurassique du Japon. *Mem. Soc. Geol. France, 94*, 5–122.

Sato, T., Kasahara, Y., & Watika, K. (1985). Discovery of a Middle Jurassic *Kepplerites* from the Mino Belt, Central Japan. *Trans. Proc. Paleont. Soc. Japan, N.S.*, *139*, 218–21.

Sato, T., & Westermann, G. E. G. (1991). Japan and South-East Asia. In G. E. G. Westermann & A. C. Riccardi (eds.), *Jurassic Taxa Ranges and Correlation Charts for the Circum-Pacific. Newsl. Strat.*, *24*, 81–108.

Schlatter, R., & Schmidt-Effing, R. (1984). Bioestratigrafía y fauna de amonites del Jurásico inferior (Sinemuriano del Area de Tenengo de Doria (Estado de Hidalgo, México). In *III Congr. Latinamericano de Paleont., Mexico.*

Schmidt-Effing, R. (1980). The Huayacocotla Aulacogen in Mexico (Lower Jurassic) and the Origin of the Gulf of Mexico. In R. H. Pilger, Jr. (ed.), *The Origin of the Gulf of Mexico and the Early Opening of the Central North Atlantic Ocean* pp. 79–86). Baton Rouge: Louisiana Geol. Surv.

Sey, I. I., & Kalacheva, E. D. (1988). Ammonites and bivalves of the Far East, Soviet Union. In G. E. G. Westermann & A. C. Riccardi (eds.), *Jurassic Taxa Ranges and Correlation Charts for the Circum-Pacific. Newsl. Stratigr.*, *19*, 35–60.

Sey, I. I., Kalacheva, D. E., & Westermann, G. E. G. (1986). The Jurassic ammonite *Pseudolioceras* (*Tugurites*) of the Bering Province. *Can. J. Earth Sci.*, *23*, 1042–5.

Smith, P. L. (1983). The Pliensbachian ammonite *Dayiceras dayiceroides* and Early Jurassic paleogeography. *Can. J. Earth Sci.*, *20*, 86–91.

Smith, P. L., Tipper, H. W., Taylor, D. G., & Guex, J. (1988). An ammonite zonation for the Lower Jurassic of Canada and the United States: the Pliensbachian. *Can. J. Earth Sci.*, *25*, 1503–23.

Stevens, G. R. (1978). Jurassic, Paleontology. In R. P. Suggate (ed.), *The Geology of New Zealand* (pp. 215–28). N.Z. Geol. Surv., D.S.I.R. Publ. Wellington: N.Z. Govt. Printer.

(1990). The influences of paleogeography, tectonism and eustasy on faunal development in the Jurassic of New Zealand (pp. 1–17). In G. Pallini et al. (eds.), *Fossili, Evoluzione, Ambiente. Proc. II Pergola Symp., 1987.* Roma: Com. Cent. R. Piccinini.

Stipanicic, P. N., Westermann, G. E. G, & Riccardi, A. C. (1975). The Indo-Pacific ammonite *Mayaites* in the Oxfordian of the Southern Andes. *Ameghiniana, 7*, 57–78.

Taylor, D. G. (1988). Middle Jurassic (Late Aalenian and Early Bajocian) ammonite biochronology of the Snowshoe Formation, Oregon. *Oregon Geol.*, *50*, 123–38.

Taylor, D. G., Callomon, J. H., Hall, R., Smith, P. L., Tipper, H. W., & Westermann, G. E. G. (1984). Jurassic ammonite biogeography of western North America: the tectonic implications. In G. E. G. Westermann (ed.), *Jurassic–Cretaceous Biochronology and Paleogeography of North America* (pp. 121–42). Geol. Assoc. Canada, Spec. Pap. 27.

Uhlig, V. (1911). Die marinen Reiche des Jura und der Unterkreise. Mitt. *Geol. Ges. Wien, IV Jahrg.*, *3*, 329–448.

Verma, H. M., & Westermann, G. E. G. (1984). The ammonoid fauna of the Kimmeridgian–Tithonian boundary beds of Mombasa, Kenya. *R. Ontario Mus. Life Sci. Contrib.*, *135*, 1–124.

Vu Khuc (1984). *Jurassic of Vietnam*, IGCP Project 171: Circum-Pacific Jurassic, Spec. Pap. 7.

Westermann, G. E. G. (1964). The ammonite fauna of the Kialagvik Formation at Wide Bay, Alaska Peninsula. Part I. Lower Bajocian (Aalenian). *Bull. Am. Paleont.*, *47*, 325–503.

(1971). Form, structure and function of shell and siphuncle in coiled Mesozoic ammonoids. *R. Ontario Mus. Life Sci. Contrib.*, *78*, 1–39.

(1981). Ammonite biochronology and biogeography of the circum-Pacific Middle Jurassic. In M. R. House & J. R. Senior (eds.), *The Ammonoidea* (pp. 459–98). System. Assoc. Spec. Pap. 18.

(1984). *Middle Jurassic Ammonite Evolution in the Andean Province.* IGCP Project 171: Circum-Pacific Jurassic, Rep. 2, pp. 69–71.

(1990). New developments in ecology of Jurassic–Cretaceous ammonoids. In G. Pallini (ed.), *Fossili, Evoluzione, Ambiente. Proc. II Pergola Symp. 1987.* Roma: Com. Cent. R. Piccinini.

(in press). Global bio-events in mid-Jurassic ammonites controlled by seaways. In M. R. House et al. (eds.), *The Ammonoidea, Evolution and Environmental Change.* Syst. Assoc. Symp., London 1991.

Westermann, G. E. G., & Callomon, J. H. (1988). The Macrocephalitinae and associated Bathonian and Early Callovian (Jurassic) ammonoids of the Sula Islands and New Guinea. *Palaeontographica, A203*, 1–90.

Westermann, G. E. G., & Getty, T. A. (1970). New Middle Jurassic Ammonitina from New Guinea. *Bull. Am. Paeontol.*, *57*(256), 227–321.

Westermann, G. E. G., & Hudson, N. (1991). Occurrence of Bathonian Eurycephalitinae in New Zealand and implication for palaeobiogeography. *J. Paleontol.*, *65*, 267–83.

Westermann, G. E. G., & Riccardi, A. C. (1972). Middle Jurassic ammonoid fauna and biochronology of the Argentine-Chilean Andes. I: Hildocerataceae. *Palaeontolographica, A140*, 1–116.

(1975). Palaeontology. In P. N. Stipanicic, G. E. G. Westermann, & A. C. Riccardi, (eds.), *The Indo-Pacific Ammonite Mayaites in the Oxfordian of the Southern Andes. Ameghiniana, 12*, 281–305.

(1979). Middle Jurassic ammonoid fauna and biochronology of the Argentine–Chilean Andes II: Bajocian Stephanocerataceae. *Palaeontographica, A164*, 85–188.

(1985). *Middle Jurassic Ammonite evolution in the Andean Province and emigration to Tethys. Lecture Notes Earth Sci.*, *1*, 6–34. Berlin: Springer-Verlag.

Westermann, G. E. G., Riccardi, A. C., Palacios, D., & Rangel, C. (1980). *Jurásico Medio en el Perú.* Geol. Min. Metal. Bol. 9, Ser. D.

Westermann, G. E. G., & Wang, Y. (1988). Middle Jurassic ammonites of Tibet and the age of the lower Spiti Shales. *Paleontology, 31*, 295–339.

# 25 Fishes of the circum-Pacific region

A. L. CIONE

The circum-Pacific Jurassic fish record is very poor in comparison with the contemporaneous European assemblages (e.g., France, England, Germany). Fishes have been reported from continental and marine sediments. Most recent descriptions are from North America (Schaeffer and Patterson 1984) and Chile (Arratia 1987). The majority of the recognized taxa from other areas need modern systematic revision (Schaeffer and Patterson 1984; Arratia 1987; Cione and Pereira 1987, 1990; Cione et al. 1987).

The record includes typical assemblages whose low diversities are mostly due to poor sampling. Marine facies in Argentina, Chile, Mexico, the United States, Canada, and Japan include a few hybodontoids and holocephalans, halecostomes *incertae sedis,* semionotids, a few pycnodontiforms, caturids, amiids, pholidophorids, pachycormids, aspidorhynchids, ichthyodectiform-osteoglossomorphs *incertae sedis,* and several genera identified as Teleostei *incertae sedis.* Freshwater facies in Argentina, the United States, China, Australia, and Antarctica include coccolepids, redfieldiids, semionotids, ichthyodectiformosteoglossomorphs *incertae sedis,* coelacanths, and dipnoans. Chondrichthyans and pycnodontiforms, which are fairly common in Europe, are very rare, and macrosemmids, ionoscopids, and oligopleurids have not been reported. Chondrosteans and coelacanths are restricted to continental facies.

The record of some supposedly well known genera such as *Hybodus, Pholidophorus,* or *Leptolepis* in different areas may be highly misleading, as Schaeffer and Patterson (1984) indicated. These temporally and geographically wide ranging taxa are non-monophyletic according to cladistic methodology. This fact, together with the provisional systematics of the remaining fishes, makes paleobiogeography or correlation highly speculative.

### References

Arratia, G. (1987). Jurassic fishes from Chile and critical comments. In W. Volkheimer (ed.), *Bioestratigrafía de los Sistemas Regionales del Jurásico y Cretácico de América del Sur,* (vol. 1, pp. 258–86). Mendoza: Com. Sudam. Juras. Cretac.

Cione, A. L., Gasparini, Z., Leanza, H., & Zeiss, A. (1987). Marine oberjurassische Plattenkalke in Argentinien (ein erster Forschungsbericht). *Archaeopteryx, 5,* 13–22.

Cione, A. L., & Pereira, S. M. (1987). Los peces del Jurásico de Argentina. El Jurásico anterior a los movimientos intermálmicos. In W. Volkheimer (ed.), *Bioestratigrafía de los Sistemas Regionales del Jurásico y Cretácico de América del Sur* (vol. 1, pp. 287–98). Mendoza: Com. Sudam. Juras. Cretac.

(1990). Los peces del Jurásico de Argentina. El Jurásico posterior a los movimientos intermálmicos y el Cretácico superior. In W. Volkheimer and E. Musacchio (eds.), *Bioestratigrafía de los Sistemas Regionales del Jurásico y Cretácico de América del Sur* (vol. 2, pp. 385–402). Mendoza: Com. Sudam. Juras. Cretac.

Schaeffer, B., & Patterson, C. (1984). Jurassic fishes from the Western United States, with comments on Jurassic fish distributions. *Am. Mus. Novit., 2796,* 1–86.

# 26 Marine reptiles of the circum-Pacific region

Z. GASPARINI

## The record

Remains of Jurassic marine reptiles are scarce in the circum-Pacific area and have been found only in Australia, Thailand, Laos, China, Canada, the United States, Cuba, Mexico, Chile, and Argentina (Figure 26.1). The most diverse fauna comes from the Andean Basin (Aubouin et al. 1973), extending from northern Chile to west-central Argentina. It includes the earliest teleosaurids, ichthyosaurs, and plesiosaurs of the Pacific area, as well as excellently preserved Tithonian plesiosaurs, ichthyosaurs, turtles, and marine crocodiles. Our knowledge of the marine reptiles of the circum-Pacific area is limited by the poor quality of material, often consisting of isolated vertebrae and indeterminable postcranial fragments on which the new taxa are based. Thus, many of the generic, specific, and even family names are invalid (Chong Díaz and Gasparini 1976; Gasparini 1979, 1985; Gasparini and Goñi 1990). Paleobiogeographic interpretations must therefore be treated with caution, particularly because the ages sometimes are unknown.

Representative specimens are illustrated in Plate 133.

*Ichthyosauria.* The oldest finds are from the Hettangian of northern Chile (Chong Díaz 1977; Gasparini 1979, 1985). Other, indeterminable fragments have been found also in the Callovian, Oxfordian, and Kimmeridgian (Chong Díaz and Gasparini 1976). Two Bajocian skulls from central Chile (Jensen and Vicente 1976) have tentatively been referred to *Ichthyosaurus* sp.

Isolated vertebrae from the Early Jurassic of the Neuquén Basin, Argentina, were the basis of the alleged new species *Ichthyosaurus*(?) *sanjuanensis* and *Ancanamunia*(?) *espinacitensis* (Rusconi 1948a, 1949), but both are "nomina vana" (Gasparini 1985). The Bajocian *Stenopterygius grandis* Cabrera (1939) and the Callovian *Ichthyosaurus "immanis"* and *I. leucopetraeus* Philippi (1895) were also based on insufficient material (McGowan 1976; Gasparini 1985) . Remains of *Ophthalmosaurus* sp. and a new longirostrate form were recently found in Bajocian rocks of southern Neuquén province. The most complete South American ichthyosaurs come from the Lower and Middle Tithonian of Mendoza and Neuquén (Dames 1893; Rusconi 1938, 1940, 1942, 1948a,b; McGowan 1972; Gasparini 1985, 1988; Gasparini and Goñi 1990), including *Ophthalmosaurus monocharactus* Appleby (1956) (Plate 133, Fig. 1) and *Ophthalmosaurus* sp.

The most ancient ichthyosaurs of North America are *Ichthyosaurus* sp. from the Liassic of Alberta and, probably, the Pliensbachian of Oregon (McGowan 1978). Late Jurassic *Ichthyosaurus* have come from Oregon (Camp and Koch 1966; Dupras 1988) and California. Preliminary examination of the California specimens indicates their similarity to Late Jurassic ichthyosaurs of Europe and Early Cretaceous species of Australia (Camp 1942; Dupras 1988). However, most North American ichthyosaurs come from the Oxfordian of Wyoming, Montana, and Dakota (Gilmore 1904–6a,b), including *Baptanodon discus* (or *Ophthalmosaurus*) (cf. Andrews 1910; Appleby 1956; McGowan 1976, 1978; Mazin 1982).

*Crocodylia.* Marine crocodiles are known from Chile, Argentina, Mexico, and the United States. The oldest is a Sinemurian teleosaurid (Plate 133, Fig. 2) from northern Chile (Chong Díaz and Gasparini 1972). Other Liassic crocodiles, indeterminable at the family level, come from Chile and Argentina (Burmeister and Giebel 1861; Huene 1927; Casamiquela 1970; Gasparini 1980, 1981, 1985). The Metriorhynchidae, first known from the European Tethys, occur in the Callovian of northern Chile: *Metriorhynchus casamiquelai* Gasparini & Chong Díaz (1977) (Plate 133, Fig. 3) and *M. westermanni* Gasparini (1980). *M casamiquelai* is closely affiliated with the European *M. brachyrhynchus* Deslongchamps, but not synonymous (Adams-Tresman 1987). The Tithonian of the Neuquén Basin has yielded *Metriorhynchus potens* (Rusconi) (1948), *M.* aff. *durobrivensis,* and *Geosaurus araucanensis* Gasparini & Dellapé (1976) (Plate 133, Fig. 4). *Geosaurus* probably occurs also in central Chile.

The first marine crocodile of Mexico has recently been found in the basal Callovian of the Mixteca Terrane in Guerrero (Plate 133, Fig. 5). The vertebrae belong to the Thalattosuchia, but are

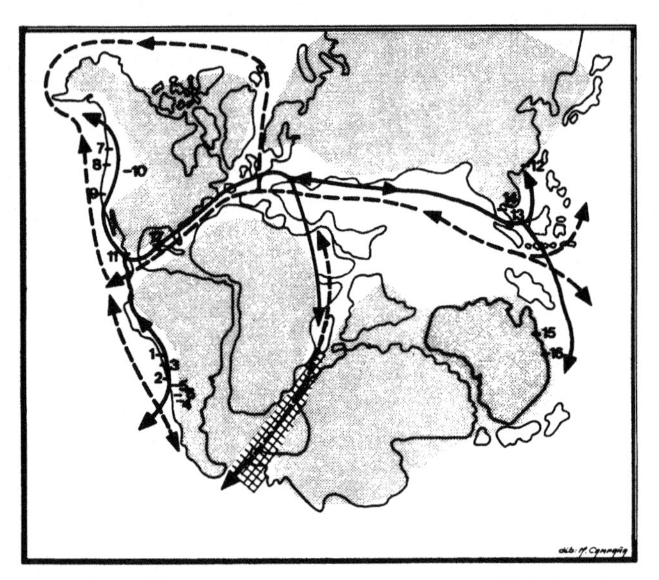

**Figure 26.1. Principal localities with Jurassic marine reptiles and possible dispersal routes: 1, northern Chile (Ichthyosauria; Plesiosauria; Teleosauridae; *Metriorhynchus casamiquelai; M. westermanni*); 2, central Chile (Ichthyosauria; Plesiosauria); 3, San Juan province (Ichthyosauria); 4, Chacaico-Sur, Neuquén province (*Ophthalmosaurus* sp.; *Simolestes* sp.; Elasmosauridae; Cryptoclididae); 5, Mendoza province (*Ophthalmosaurus* sp.; Ichthyosauria; *Metriorhynchus* aff. *durobrivensis;* Teleosauridae); 6, Neuquén province (*Ophthalmosaurus monocharactus; Metriorhynchus* sp.; *Geosaurus araucanensis; Stretosaurus* sp.; *?Eurysternum neuquinum; Notoemys laticentralis*); 7, Alberta (*Ichthyosaurus* sp.); 8, Oregon (*Ichthyosaurus* sp.; Teleosauridae); 9, California (Ichthyosauria; Plesiosauroidea); 10, Wyoming, Montana, Dakotas (*?Baptanodon;* Plesiosauroidea); 11, Cualac, Mexico (Thalattosuchia); 12, Sichuan province (*Bishanopliosaurus youngi; ?Sinopliosaurus; Yuzhoupliosaurus chengjiangensis*); 13, Ko-Kut, Thailand (Plesiosauria); 14, Laos (Plesiosauria); 15, Kolane, Australia (Plesiosauridae); 16, Mount Morgan, Australia (*?Leptocleidus* sp.); 17, Viñales, Cuba (*?Cryptoclidus* sp.). Solid lines indicate possible dispersal routes during the early and Middle Jurassic; broken lines, Late Jurassic; cross-hatching, barrier during the Early and Middle Jurassic. Paleogeography from Hallam (1983).**

indeterminable at family level. In the United States, Bajocian teleosaurids were found in central Oregon (Buffetaut 1979).

*Chelonii.* The only marine turtles known from the circum-Pacific come from Neuquén province. *Notoemys laticentralis* (Wood and Freiberg 1977; de la Fuente and Fernández 1989) is a pleurodiran that is more specialized than European Jurassic forms (Plate 133, Fig. 6); *?Eurysternum neuquinum* Fernandez & Fuente is a cryptodiran with another adaptation to marine life (Plate 133, Fig. 7). Other species occur in Tithonian rocks.

*Plesiosauria.* Early and Middle Jurassic plesiosaurs of the circum-Pacific area belong mainly to the Rhomaleosauridae and Pliosauridae.

In China, Sichuan province, occurs the late Liassic rhomaleosaurid *Bishanopliosarus youngi.* Dong (1980) suggests that the medium-size specimen entered from the sea into Sichuan Lake. From the same formation, Young (1942, 1944) earlier described the pliosaurid *Sinopliosaurus weiyuanensis,* which may be synonymous with *Pliosaurus andrewsi* (Tarlo 1960), a Callovian common species of the Oxford Clay. Another rhomaleosaurid from the Sichuan Basin, *Yuzhoupliosaurus chengjiangensis* Zhang (1985), came from the Middle Jurassic. A probable plesiosaur was found in the Lias of eastern Thailand (Buffetaut and Ingavat 1983). It was associated with hybodontid and holostean fishes, indicating estuarine or coastal habitat. A similar Liassic association of a plesiosaur tooth with *Acrodus* and *Lepidotus* is known from Laos (Hoffet and Le Maitre 1939).

Fragments of probable Plesiosauridae occur also in the Liassic of Queensland (Thulborn and Warren 1980), which appears to be partly marine. A short-necked plesiosaur of the Rhomaleosauridae was originally referred to *Leptocleidus* sp. and dated as Late Jurassic or Early Cretaceous (Bartholomai 1966), but the generic assignment is doubtful, and the age may be Liassic (Molnar 1982).

In South America, the oldest recorded plesiosaurs are from the Sinemurian of northern Chile (Chong Díaz and Gasparini 1976; Gasparini 1985), based on vertebrae and indeterminable limbs. Recently a new species of *Simolestes* was found in Bajocian rocks of Neuquén province. *S. vorax* Andrews is known from the Callovian Oxford Clay of England and is believed to be the oldest species of *Simolestes.* Remains of Callovian Elasmosauridae and Cryptoclididae came from the same area, perhaps the oldest Plesiosauroidea of the eastern Pacific. The pliosaurid *Stretosaurus* (Plate 133, Fig. 8) (Gasparini, Gonĩ, and Molina 1982) also occurs in the Lower Tithonian of Neuquén. Other Tithonian–Berriasian pliosauroids from the same province have not yet been studied.

Oxfordian plesiosauroids of western Cuba, referred to *Cryptoclidus*(?) *cuervoi* de la Torre & Rojas (1949), need taxonomic revision. Plesiosaur fragments from the Oxfordian of Wyoming were referred to *Muraenosaurus*(?) *reedi* Mehl, *"Plesiosaurus" shirleynsis* Knight, and *Pantosaurus striatus* Marsh (cf. Pearsson 1964), but the validity of these names is doubtful. Similarly, a single incomplete vertebra from the "Portlandian" of California was referred to *"Plesiosaurus" hesternus* Welles (Pearsson 1964).

## Biogeography

Our limited knowledge, reviewed earlier, of Jurassic marine reptiles from the Pacific area permits only preliminary conclusions about their biogeography. Note that the term "dispersal route" does not imply direction. The area of reference for the circum-Pacific reptiles is always the European Tethys, with the most varied known fauna of Jurassic marine reptiles, but this does not imply their ancestry. Although the phylogenies of the families here considered are poorly known (Mlynarzki 1976; Brown 1981; Buffetaut 1982; Mazin 1982), the close affinities between the eastern Pacific and western Tethys marine reptiles is obvious at the generic and even specific level, throughout the Jurassic. Western Pacific reptiles, on the other hand, are related to those from the Mediterranean only at the familiar level, although the absence of lower category affinities could be due to our poor western Pacific data.

The most ancient teleosaurid crocodiles are probably from the Sinemurian of Chile, because the longirostrians from the Kota Formation of India do not appear to be true teleosaurids (Buffetaut

1982). Most of the Liassic teleosaurids are known from the Mediterranean, with only isolated remains found in the Andes of Chile and Argentina (Gasparini 1985) and in northwestern Madagascar (Buffetaut, Termier, and Termier 1981). This distribution and the fact that evidence for a fully marine connection between western Africa and eastern South America is missing for the Jurassic support the hypothesis of the Hispanic Corridor (Bonaparte 1981; Gasparini 1985).

Liassic ichthyosaurs were functionally best adapted to marine life, so that they potentially could have dispersed along any marine route with warm to temperate waters. The poor ichthyosaurian record from the circum-Pacific area (Chile, Argentina, United States), contrasting with that of western Tethys, appears to be due to collecting failure. The genus *Ichthyosaurus* is frequent in the European Lias and is also present in North America (Alberta and Oregon).

The Liassic plesiosaurs of the circum-Pacific area are the rhomaleosaurids from southern China, a rhomaleosaurid and a probable plesiosaurid from Western Australia, and indeterminable material from Chile. Both families are well represented in the European Lias, indicating a high degree of cosmopolitanism and hence great dispersal. Rhomaleosaurids occurred also in brackish and river water, probably along the Tethian equatorial belt.

A comparison of the Bajocian-Callovian marine reptiles of the eastern Pacific with those of the Tethys shows that *Ophthalmosaurus, Simolestes,* and *Metriorhynchus,* as well as elasmosaurids and cryptoclidids, are common to both areas, but belong to different species. This indicates that the Hispanic Corridor was open, although the faunas may differ in age. Buffetaut (1979) proposed the hypothesis of a Caribbean corridor to explain the presence of Bajocian teleosaurids in Oregon, which were compared with the Callovian crocodile *Thalattosichia* of Oaxaca, Mexico.

Late Jurassic Pacific marine reptiles are known only from the eastern Pacific and its cratonic basins.

The metryorhynchid crocodiles from the Tithonian of Neuquén (i.e., *Metryorhynchus* and *Geosaurus*) are congeneric with those from France and Germany (Gasparini and Dellapé 1976). Affinities are also present in the plesiosaurid *Stretosaurus* that is associated with *Geosaurus araucanensis* in the Tithonian; the same species, *S. macomerus* (Phillips), occurs in the Callovian-Kimmeridgian of England (Tarlo 1960). Similarly, the ichthyosaurid *Ophthalmosaurus monocharactus* occurs in the Upper Jurassic of Europe, Argentina, and Arctic Canada during the early Cretaceous (McGowan 1978; Appleby 1979; Gasparini 1988). Other marine reptiles in the eastern Pacific, however, were well differentiated from Tethyan relatives. Examples are the Oxfordian ichthyosaur *?Baptanodon* from Wyoming and the Dakotas and, especially, the turtles *?Eurysternum* and the pleurodiran *Notoemys laticentralis.*

These taxonomic relationships between the Tethyan and eastern Pacific marine reptiles indicate that the Hispanic Corridor was wide open from the Late Jurassic onward and that it was the principal migration route. This is supported by the Oxfordian plesiosaurier from Cuba (*?Crypotoclidus*). Other dispersals may possibly have occurred through the seaway between Africa and South America and via the North Sea (Hallam 1983)

In conclusion, marine reptiles were widely distributed through the Early and Middle Jurassic, especially in Tethyan latitudes, and were able to disperse along the Hispanic Corridor, which connected the eastern Pacific with the western Tethys. In the Late Jurassic the corridor was followed by the nascent Central Atlantic, together with other dispersal routes to Europe (Figure 26.1).

**Acknowledgments**

I am indebted to Dr. G. E. G. Westermann, McMaster University, Canada, who edited the text. Also thanks to Dr. R. Molnar, Queensland Museum, Australia; Dr. Li Jinling, Institute of Vertebrate Paleontology and Paleoanthropology, China; Dr. J. Stewart, Los Angeles County Museum of Natural History, United States; and Mr. A. Carlini, Museo de La Plata, Argentina. The collaboration of Mr. Leonardo Gasparini is appreciated. The author is a member of the Consejo Nacional de Investigaciones Científicas y Técnicas (CONICET), Argentina. This work was supported by CONICET grant PID 390450285 and National Geographic Society grant 4265-90.

**References**

Adams-Tresman, S. (1987). The Callovian (Middle Jurassic) marine crocodile *Metriorhynchus* from central England. *Palaeontology, 30*(1), 179–94.

Andrews, C. (1910). A descriptive catalogue of the marine reptiles of the Oxford Clay. Part I. *Br. Mus. (Nat. Hist.)*, 1–207.

Appleby, R. (1956). The osteology and taxonomy of the fossil reptile *Ophthalmosaurus. Proc. Zool. Soc. London, 126,* 403–77.

(1979). The affinities of Liassic and later ichthyosaurs. *Palaeontology, 22*(4), 921–46.

Aubouin, J., Borello, A., Cecioni, G., Charrier, R., Chotin, P., Frutos, J., Thiele, R., & Vicente, J.-C. (1973). Esquisse paléogéographique et structurale des Andes Méridionales. *Rev. Géog. Phys. Geol. Dyn., 2*(15), 11–72.

Bartholomai, A. (1966). The discovery of plesiosaurian remains in freshwater sediments in Queensland. *Austral. J. Sci. 28*(11), 437.

Bonaparte, J. (1981). Inventario de los vertebrados jurásicos de América del Sur. In W. Volkheimer & E. Musacchio (eds.), *Cuencas Sedimentarias del Jurásico y Cretácico de América del Sur* (vol. 2, pp. 661–82). Buenos Aires: Com. Sudam. Juras. Cretac.

Brown, D. (1981). The English upper Jurassic Plesiosauroidea (Reptilia) and a review of the phylogeny and classification of the Plesiosauria. *Bull. Br. Mus. (Nat. Hist.)*, Geol. *Ser., 35*(4), 253–347.

Buffetaut, E. (1979). Jurassic marine crocodilians (Mesosuchia: Teleosauridae) from central Oregon: first record in North America. *J. Paleontol., 53*(1), 211–15.

(1982). Radiation évolutive, paléoécologie et biogéographie des crocodiliens mésosuchiens. *Mém. Soc. Géol. France (N.S.), Mém. 142, 60,* 1–88.

Buffetaut, E., & Ingavat, R. (1983). Vertebrates from the continental Jurassic of Thailand. *United Nations ESCAP CCOP Tech. Bull., 16,* 68–75.

Buffetaut, E., Termier, G., & Termier, H. (1981). A teleosaurid (Crocodylia, Mesosuchia) from the Toarcian of Madagascar and its paleobiogeographical significance. *Paläontol. Z., 55*(3/4), 313–19.

Burmeister, H., & Giebel, G. (1861). Die Versteinnerungen von Juntas in Tala des Rio Copiapo. *Abh. Naturhist. Gesell., 6.*

Cabrera, A. (1939). Sobre un nuevo ictiosaurio del Neuquén. *Notas Museo La Plata, 95,* 1–17.

Camp, C. (1942). Ichthyosaur rostra from central California. *J. Paleontol., 16*(3), 362–71.

Camp, C., & Koch, J. (1966). Late Jurassic ichthyosaurs from coastal Oregon. *J. Paleontol.*, *40*(1), 204–5.

Casamiquela, R. (1970). Los vertebrados jurásicos de la Argentina y de Chile. In *Actas IV Congr. Latinoamer. Zool., Caracas, 1968* (vol. 2, pp. 873–90).

Chong Díaz, G. (1977). Contribution to the knowledge of the Domeyko Range in the Andes of Northern Chile. *Geol. Rundsch.*, *66*(2), 374–404.

Chong Díaz, G., & Gasparini, Z. (1972). Presencia de Crocodilia marinos en el jurásico de Chile. *Asoc. Geol. Argent. Rev.*, *27*(4), 406–9.

(1976). Los vertebrados mesozoicos de Chile y su aporte geo-paleontológico. In *Actas VI Cong. Geol. Argent. Bahía Blanca, 1975* (vol. 1, pp. 45–67). Buenos Aires: Asoc. Geol. Argent.

Dames, W. (1893). Ueber das Vorkomen von Ichthyopterygien im Tithon Argentiniens. *Z. Deustch. Geol. Gessell.*, *45*, 23–33.

de la Fuente, M., & Fernández, M. (1989). *Notoemys laticentralis* Cattoi & Freiberg, 1961 from the upper Jurassic of Argentina: a member of the Infraorder Pleurodira (Cope, 1868). *Studia Palaeocheloniologica*, *3*(2), 25–32.

de la Torre, R., & Rojas, L. (1949). Una nueva especie y dos subespecies de Ichthyosauria del Jurásico de Viñales, Cuba. *Mem. Soc. Cubana Hist. Nat.*, *19*(2), 197–200.

Dong, Z. (1980). A new Plesiosauria from the Lias of Sichuan Basin. *Vertebrata PalAsiatica*, *18*(3), 191–7.

Dupras, D. (1988). Ichthyosaurs of California, Nevada, and Oregon. *Calif. Geol.*, May, 99–107.

Fernández, M., & de la Fuente, M. (1988). Una nueva tortuga (Cryptodira: Thalassemydidae) de la Formación Vaca Muerta (Jurásico, Tithoniano) de la provincia del Neuquén. *Ameghiniana*, *25*(2), 129–38.

Gasparini, Z. (1978). Consideraciones sobre los Metriorhynchidae (Crocodylia, Mesosuchia): su origen, taxonomía y distribución geográfica. *Obra Centen. Museo La Plata, Paleont. 5*, 1–9.

(1979). Comentarios críticos sobre los vertebrados mesozoicos de Chile. In *Actas II Cong. Geol. Chileno, Arica, 1979* (vol. 3, pp. H16–32). Santiago: Inst. Invest. Geol.

(1980). Un nuevo cocodrilo marino (Crocodylia, Metriorhynchidae) del caloviano del norte de Chile. *Ameghiniana*, *17*(2), 97–103.

(1981). Los Crocodylia fósiles de la Argentina. *Ameghiniana*, *18*(3/4), 177–205.

(1985). Los reptiles marinos jurásicos de América del Sur. *Ameghiniana*, *22*(1/2), 23–34.

(1988). *Ophthalmosaurus monocharactus* Appleby (Reptilia, Ichthyopterygia), en las calizas litográficas titonianas del área Los Catutos, Neuquén, Argentina. *Ameghiniana*, *25*(1), 3–16.

Gasparini, Z., & Chong Díaz, G. (1977). *Metriorhynchus casamiquelai* n.sp. (Crocodylia, Thalattosuchia), a marine crocodile from the Jurassic (Callovian) of Chile, South America. *N. Jb. Geol. Paläontol.*, *153*(3), 341–60.

Gasparini, Z., & Dellapé, D. (1976). Un nuevo cocodrilo marino (Thalattosuchia, Metriorhynchidae) de la Formación Vaca Muerta (Jurásico, Tithoniano) de la provincia de Neuquén. In *Actas I Cong. Geol. Chileno, Santiago, 1976* (vol. 1, pp. C1–21). Santiago: Dept. Geol., Univ. Chile.

Gasparini, Z., & Goñi, R. (1990). Los ictiosaurios jurásico-cretácicos de la Argentina. In W. Volkheimer (ed.), *Bioestratigrafía de los Sistemas Regionales del Jurásico y Cretácico de América del Sur* (vol. 2, pp. 299–312). Mendoza: Com. Sudam. Juras. Cretac.

Gasparini, Z., Goñi, R., & Molina, O. (1982). Un plesiosaurio (Reptilia) tithoniano en Cerro Lotena, Neuquén, Argentina. In *Actas V Cong. Latinoamer. Geol., Buenos Aires, 1982* (vol. 5, pp. 33–47). Buenos Aires: Serv. Geol. Argent.

Gilmore, C. (1904–6a). Osteology of *Baptanodon* (Marsh). *Mem. Carnegie Museum*, *2*(2), 77–129.

(1904–6b). Notes on osteology of *Baptanodon* with a description of a new species. *Mem. Carnegie Museum*, *2*(9), 325–42.

Hallam, A. (1983). Biogeographic evidence bearing on the creation of Atlantic seaway in the Jurassic. *Palaeogeogr. Palaeoclimatol. Palaeoecol.*, *43*, 181–93.

Hoffet, J., & Le Maitre, D. (1939). Sur la stratigraphie et la paléontologie du Lias des environs de Tchépone (Bas-Laos). *C. R. Acad. Sci. (Paris)*, *209*, 104–16.

Huene, F. von. (1927). Beitrag zur Kenntnis mariner mesozoicher Wirbeltiere in Argentinien. *Centralbl. Mineral. Geol. Paläontol.*, *B1*, 22–9.

Jensen, O., & Vicente, J.-C. (1976). Estudio geológico del área de "Las Juntas" del río Copiapó, provincia de Atacama, Chile. *Asoc. Geol. Argent. Rev.*, *31*(3), 145–73.

McGowan, C. (1972). Evolutionary trends in longipinnate ichthyosaurs with particular reference to the skull and fore fin. *Life Sci. Contrib. R. Ontario Mus.*, *83*, 1–38.

(1974). A revision of the latipinnate ichthyosaurs of the Lower Jurassic of England (Reptilia: Ichthyosauria). *Life Sci. Contrib. R. Ontario Mus.*, *100*, 1–30.

(1976). The description and phenetic relationships of a new ichthyosaur genus from the upper Jurassic of England. *Can. J. Earth Sci.*, *13*, 668–83.

(1978). Further evidence for the wide geographical distribution of ichthyosaur taxa (Reptilia: Ichthyosauria). *J. Paleontol.*, *52*(5), 1155–62.

Mazin, J.-M. (1982). Affinités et phylogénie des Ichthyopterygia. *Géobios, mem. spéc.*, *6*, 85–98.

Mlynarski, M. (1976). Testudines. *Handb. Paläoherpetol.*, *7*, 1–130.

Molnar, R. (1982). A catalogue of fossil amphibians and reptiles in Queensland. *Mem. Queensland Mus.*, *20*(3), 613–33.

Pearsson, O. (1964). A revision of the classification of the Plesiosauria with a synopsis of the distribution of the group. *Lunds Univ. Arsskrift*, *2*(59), 1–60.

Philippi, R. (1895). *Ichthyosaurus immanis* Ph. nueva especie sud-americana de este género. *An. Univ. Chile*, 1–7.

Rusconi, C. (1938). Restos de ictiosaurios del Jurásico superior de Mendoza. *Bol. Paleontol., Buenos Aires*, *10*, 1–4.

(1940). Nueva especie de ictiosaurio en el Jurásico de Mendoza. *Bol. Paleontol., Buenos Aires*, *11*, 1–4.

(1942). Nuevo género de ictiosaurio argentino. *Bol. Paleontol., Buenos Aires*, *13*, 1–2.

(1948a) Ictiosaurios del Jurásico de Mendoza. *Rev. Museo Hist. Nat. Mendoza*, *2*, 17–160.

(1948b). Plesiosaurios del Jurásico de Mendoza. *An. Soc. Cient. Argent.*, *146*(5), 327–51.

(1949). Presencia de ictiosaurios en el liásico de San Juan. *Rev. Museo Hist. Nat. Mendoza*, *3*(2), 89–94.

Tarlo, L. (1960). A review of upper Jurassic pliosaurs. *Bull. Br. Mus. (Nat. Hist.), Geol.*, *4*(5), 147–89.

Thulborn, R., & Warren, A. (1980). Early Jurassic plesiosaurs from Australia. *Nature*, *285*(5762), 224–5.

Wood, R., & Freiberg, M. (1977). Redescription of *Notoemys laticentralis* the oldest fossil turtle from South America. *Acta Geol. Lilloana*, *13*(6), 187–204.

Young, C. (1942). Fossil vertebrates from Kuangyuan, N. Szechuan, China. *Bull. Geol. Soc. China*, *22*, 293–309.

(1944). On the reptilian remains from Weiyuan, Szechuan, China. *Bull. Geol. Soc. China*, *24*(3/4), 187–210.

Zhang, Y. (1985). A new plesiosaur from Middle Jurassic of Sichuan Basin. *Vertebrata PalAsiatica*, *23*(3), 235–40.

# Part VI: Climatology and oceanography

# 27 Jurassic climate and oceanography of the Pacific region

J. T. PARRISH

## Introduction

Global paleobiogeographic patterns are principally dependent on climate and continental drift. This chapter presents conceptual models of geographically controlled climatic patterns and ocean-surface currents for two stages in the Jurassic period, in order to provide a climatic framework for the discussions of Jurassic biogeographic patterns presented elsewhere in this volume. Some of the climatic and biogeographic changes observed in the Jurassic circum-Pacific region are summarized here along with speculations on the causes of climatic change in the Jurassic.

The Jurassic was a time of climatic transition. The breakup of the supercontinent Pangea into Gondwana and Laurasia resulted in changes in continental climates in response to the breakdown of the northern Pangean monsoonal circulation (Parrish, Ziegler, and Scotese 1982; Parrish and Doyle 1984). This climatic change might have triggered changes in ocean circulation as well. Accordingly, the following discussion will emphasize monsoonal circulation.

## Paleogeography and climate

On a planet with a homogeneous surface, climatic patterns would be controlled by the meridional thermal gradient and the rotation of the planet. The patterns would be zonal, and on Earth the zonal circulation consists of easterlies at the equator, westerlies in low midlatitudes, and easterlies in high midlatitudes. This circulation is well expressed today over the Pacific Ocean, which is wide enough to permit full expression of the zonal circulation. Land–sea contrast, which is the result of the lower thermal inertia of land compared with water, disrupts the zonal circulation by introducing an additional thermal component to the system. Paleogeography is therefore the fundamental boundary condition for the distribution of paleoclimatic patterns (Parrish 1982). The paleogeographic features that have the strongest effects on climate are the area of exposed land, latitudinal distribution of the continents, and topography.

Proximity of land to ocean reduces the thermal response of the land to temperature change. This is why daily and seasonal temperature fluctuations are less in coastal regions than in continental interiors. The farther a parcel of land is from the ocean, the more responsive will be its climate to temperature change. Therefore, large continents have more variable climates than do small continents.

The magnitude of temperature change is also dependent on latitude. Although daily fluctuations on a large equatorial continent will be large, seasonal fluctuations will be relatively small. By contrast, both daily and seasonal temperature fluctuations will be great on a continent that lies in midlatitudes.

Mountains influence climate by acting as barriers and by changing the vertical thermal gradient in the atmosphere. The barrier effect of mountains may be expressed as channeling of winds or, more familiarly, as rain shadows. The best example of channeling is along the South American coast, where the surface winds associated with the eastern limb of the southern Pacific subtropical high-pressure cell are limited to the coastal region by the Andes. The rain-shadow effect is well expressed in the Americas, where the cordillera is perpendicular to the zonal circulation and creates rain shadows in Argentina, western equatorial South America, and central North America. In addition, mountains may affect global climate by influencing the positions of planetary waves (Ruddiman and Raymo 1988). The effects of mountains on the vertical thermal gradient are discussed later.

## Monsoonal circulation and the Pangean monsoon

The term "monsoon" refers to seasonally alternating high- and low-pressure cells and their climatic consequences, and the most dramatic example is the Asian monsoon. Most climatologic studies have concentrated on the intense summer monsoon because of its effects on society and agriculture (Fein and Stephens 1987). The summer monsoonal circulation in Asia is controlled by four factors: (1) heating of the Asian land surface, (2) thermal contrast between the Indian Ocean and Asia, (3) influx of moist air from the Indian Ocean and the consequent release of latent heat over the continent, and (4) effect of the Tibetan Plateau as a high-altitude heat source.

### *Monsoonal Circulation*

*Cross-equatorial contrast.* A strong monsoonal circulation should have existed for Pangea because of the great size of the continental area in midlatitudes, exceeding even the size of Eurasia, and the presence of such land areas in both hemispheres and their positions opposite one another across Tethys (Hansen, cited by Parrish, Hansen, and Ziegler 1979; Parrish 1982; Parrish and Curtis 1982). The importance of the shape of Pangea is that the seasonally alternating circulation over the large land areas would have occurred in both hemispheres, so that the high-pressure cell in the winter hemisphere would have "faced" a low-pressure cell in the summer hemisphere across Tethys. This maximal temperature and pressure contrast would have been semiannual and would have caused air to flow across the equator in opposite directions during the year.

The importance of this cross-equatorial contrast can be illustrated by comparison with the modern Asian monsoonal circulation. Described in its simplest terms, the circulation is the result of differential heating of the ocean and adjacent land (Webster 1987). The differential heating in the Asian monsoon occurs because the land and ocean receive different solar influx, and the thermal inertia of water is greater than that of land. The strongest cross-equatorial contrast today is during the northern summer. The Asian continent heats up as the water in the southern Indian Ocean cools. The differential barometric pressure between the Asian landmass and the southern Indian Ocean is substantial, greater than 30 mb. In the Pangean monsoon, differential heating between Tethys and the summer hemisphere would have been similar to that occurring during the summer Asian monsoon. However the interhemispheric thermal contrast would have been augmented by the strong cooling of the largely terrestrial winter hemisphere (Crowley, Hyde, and Short 1989). A possible analog for the pressure contrast developed in the Pangean monsoon is the seasonal pressure contrast over Asia, which is greater than 36 mb (Espenshade and Morrison 1978).

A consequence of the cross-equatorial flow today is that equatorial eastern Africa is much drier than similarly situated regions (equatorial South America, eastern East Indies). In these regions, the equatorial easterlies carry warm, moist air over the land, where the moisture is released as abundant rainfall. Because the equatorial easterlies in the Indian Ocean are diverted by the monsoon, eastern equatorial Africa does not receive much rainfall. Indeed, the greatest rainfall in equatorial Africa is in the west, and this might also be attributable to the summer monsoon over Asia. Flow in the vicinity of western equatorial Africa is west to east, especially during the northern summer, the reverse of "normal" equatorial circulation. The thermal low over the Sahara might be regarded as an extension of the Asian low, and it is intense enough to reverse equatorial flow. An excellent discussion of the African monsoon has been provided by Das (1986).

*Latent-heat release.* In the absence of water vapor, the heating of air at the ground surface over the continent (sensible heating) forces the air to rise, creating a weak inflow at ground level, and the energy of the system depends on continued heating at the ground surface. If the air contains abundant water vapor, however, the convection is self-propagating. Moisture is important to the strength of the Asian monsoon, particularly the summer monsoon. Although the summer monsoon is initiated by sensible heating in the spring, once the air starts to rise and precipitation forms, the energy of the system is increased further by the release of latent heat. The source of moisture for the Asian monsoon is the Indian Ocean. Air flowing into the monsoon at the surface originates over the southern Indian Ocean and picks up additional moisture as it continues northward over the ocean and becomes warmer.

Although the air flowing into the summer lows over Pangea is predicted to have originated over land, the expanse of warm, equatorial ocean that the air would have had to cross (Tethys) is comparable to that transversed by the modern air mass. Thus, the latent energy in the modern Asian system and that in the Pangean system might be expected to be comparable. However, the distribution of latent and sensible heat also is an important feature of the Asian monsoon, and it seems to be controlled at least partly by the Tibetan Plateau.

*High-altitude heat source.* The Tibetan Plateau enhances the summer monsoonal circulation by functioning as a high-altitude heat source. Simply put, the ground surface of the plateau increases the altitude of warm isotherms, so that the temperature on the plateau is much warmer than the temperature of air at the same altitude above the lowlands. The effect is the same as a rising column of warm air – the relative pressure is low.

The importance of the Tibetan Plateau to the strength of the Asian monsoon is well established (Flohn 1968; Hahn and Manabe 1975; Das 1986; Murakami 1987), although the consensus seems to be that the cross-equatorial thermal contrast is the primary cause (Murakami 1987; Webster 1987; Young 1987). The model simulations of Hahn and Manabe (1975) showed that the pressure gradient into the Asian interior is lower without mountains than with mountains. Partly because of this low pressure gradient, latent-heat release and rainfall did not penetrate to the continental interior in the simulation without mountains; by contrast, the simulation with mountains successfully simulated the distribution of latent heat release and rainfall of the observed monsoonal system. Summer and winter rainfall is thus predicted to have been confined to the coastal regions, and desert-like conditions are predicted for the interior. This pattern occurs in Australia today and has been predicted for Pangea (Parrish et al. 1982).

### *Monsoonal climate patterns on Pangea*

At least four distinct types of climatic patterns are expected to have been consequences of the Pangean monsoon: (1) strong seasonality, (2) an equatorial region that is dry on the east and,

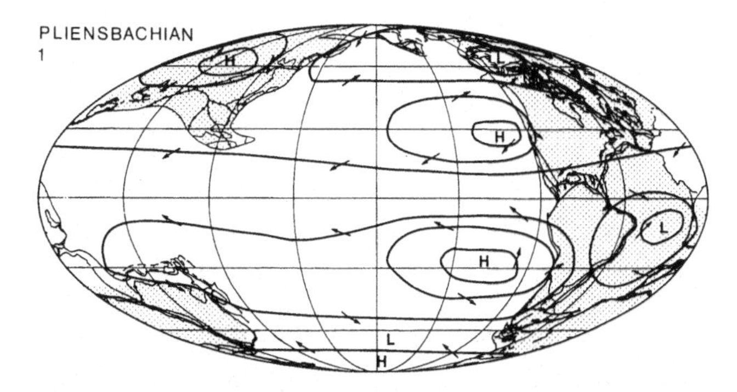


**Figure 27.1. Atmospheric circulation over the circum-Pacific region in the Pliensbachian (Early Jurassic), northern winter: H, relative high pressure; L, relative low pressure. Adapted from Parrish & Curtis (1982). In this and subsequent figures, the paleogeography is generalized; refer to maps in Ziegler et al. (1983) for details of the reconstructions.**

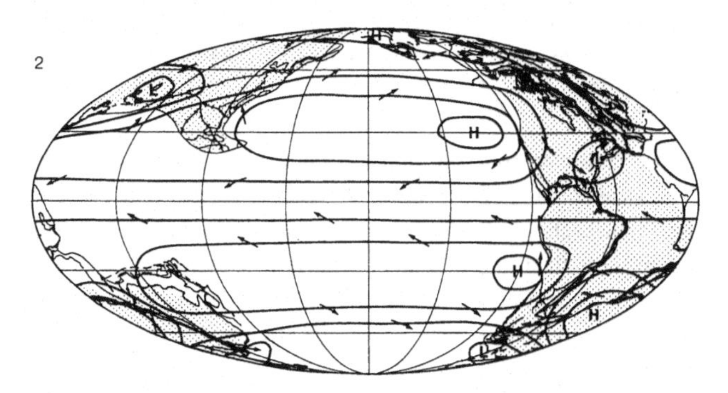


**Figure 27.2. As in Figure 27.1, northern summer. In comparing Figures 27.1 and 27.2, note strongly alternating high and low pressures in the low-latitude regions of Pangaea.**

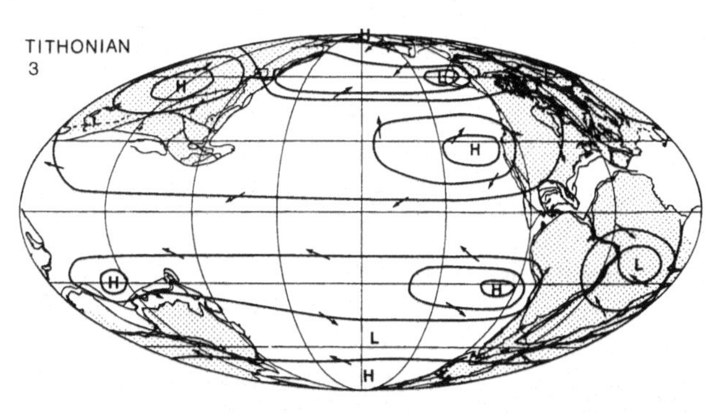


**Figure 27.3. Atmospheric circulation over the circum-Pacific region in the Tithonian (Late Jurassic), northern winter. Symbols as in Figure 27.1. (Adapted from Parrish & Curtis 1982).**

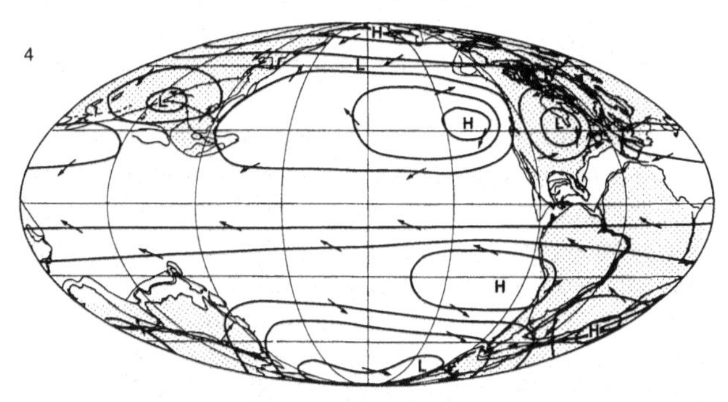


**Figure 27.4. As in Figure 27.3, northern summer. Although relative low and high pressure alternate in the low-latitude continental regions (compare Figs. 27.3 and 27.4), this seasonal fluctuation was unlikely to have been as intense in the Late Jurassic as in the Early Jurassic, although Gondwana may have been large enough to force monsoonal circulation.**

possibly, humid on the west, (3) absence of latitude-parallel climatic belts, and (4) alternating oceanic current flow, including upwelling, in western Tethys. Evidence of seasonality would be widespread. The eastern equatorial regions of Pangea would be dry compared with the high-latitude regions of the continent or with the equatorial region at others times in the planet's history. At its greatest intensity, the monsoon might have been strong enough to reverse equatorial flow on the western side of Pangea, so that the western equatorial region of the continent would be relatively humid. Pangean climatic belts would not be zonal. Although some asymmetry exists on even the smallest continents, particularly if isolatitudinal coastal regions are compared (Parrish and Ziegler 1980), the proposed monsoonal circulation over Pangea would have completely disrupted the zonal circulation. Rainfall would have been concentrated in belts paralleling the northern and southern coasts of Tethys and along the western high midlatitude coasts (i.e., in a position equivalent to southern British Columbia). Terrestrial organisms, especially plants, should reflect the loss of zonality. Finally, circulation in Tethys would alternate seasonally. Seasonally alternating upwelling zones off the low-latitude coastlines of eastern Pangea would be expected, corresponding to the Somalian upwelling that occurs today during the summer Asian monsoon (Parrish 1982).

## Predictions of Jurassic climate and ocean currents

### *Conceptual circulation models*

Maps of atmospheric circulation over the circum-Pacific region in the Jurassic for the periods before and after breakup (Pliensbachian and Tithonian stages, respectively) are presented in Figures 27.1–27.4, and relative precipitation on the continents and ocean currents is shown in Figures 27.5 and 27.6. The atmospheric circulation maps are based on the conceptual models of Parrish and Curtis (1982), modified to take into account the slightly revised continental positions (see Chapter 2). Similarly, the predictions of relative precipitation are modified from those of Parrish et al. (1982). Ocean currents are predicted from the atmospheric circulation patterns after the methods described by Ziegler et al. (1981).

Monsoonal conditions in Pangea would be expected to have reached a maximum sometime during the Triassic (Parrish, Parrish, and Ziegler 1986; Parrish and Peterson 1988). The hypothesis of Pangean monsoonal circulation is dependent on the size of Pangea and the cross-equatorial contrast between its northern and southern halves, and that contrast would be expected to be greatest when the contrasting continental areas are about equal in size. The strongest period of monsoonal circulation probably would

Table 27.1. *Area of exposed land ($10^6$ $km^2$), Carboniferous to Cretaceous*

| Time | Northern Hemisphere | Southern Hemisphere | Total |
|---|---|---|---|
| Early Carboniferous | 32.5 | 79.0 | 111.5 |
| Late Carboniferous | 37.0 | 91.5 | 128.5 |
| Late Permian | 47.4 | 84.5 | 131.9 |
| Early Triassic | 63.4 | 78.3 | 141.7 |
| Early Jurassic | 75.7 | 71.2 | 146.9 |
| Late Jurassic | 64.9 | 72.8 | 137.7 |
| Late Cretaceous | 64.7 | 62.7 | 127.4 |

*Source:* Data from Parrish (1985); paleogeographic maps from which the figures are derived are for specific geologic stages; see Parrish (1985) for more information.

have been in the later part of the Triassic, when the land area was divided in half by the equator (Table 27.1) (Parrish 1985; Parrish et al. 1986). Thus, the monsoon still would have been strong in the earliest Jurassic, and thereafter, as Pangea moved north and eventually broke up, the monsoon would have waned (Figures 27.1–27.4). The major change during the period would have been with the separation of Gondwana and Laurasia. This coincided with the relative rise of sea level in Laurasia, and the monsoon would have broken down first in the north because of the change in the thermal characteristics of Laurasia resulting from flooding of the continent. Monsoonal circulation over Gondwana might have persisted through the Jurassic, until the breakup of the continent in the Early Cretaceous.

Paleotemperature determinations for the Jurassic are relatively inadequate compared with those for the Cretaceous and Tertiary. Most data are from organisms that lived in epicontinental seas, whose temperature and salinity were likely to have been different from those in the open Pacific Ocean (Stevens 1971; Stevens and Clayton 1971; Hallam 1975; Frakes 1979; Yasamanov 1981). The probable locations of important thermal boundaries can be predicted with conceptual models, which is helpful in regions where data might be too sparse to fix the boundaries empirically. Temperature fields in the ocean are largely controlled by the latitudinal temperature gradient and by currents. The circulation in the Jurassic Pacific would be expected to resemble circulation in the modern Pacific, so for convenience, the predicted paleocurrents in Figures 27.5 and 27.6 are referred to by the names of their modern counterparts, with the prefix "paleo-," with the exception of the South Pacific current, whose closest modern counterpart is a quite different current known as the West Wind Drift, and the Antarctic Peninsula and Lord Howe currents, which have no modern counterparts. These paleocurrent names are intended to be informal.

*Pliensbachian age.* Pangea had not yet begun to break up in the Early Jurassic. Thus, monsoonal climate, that is, strong seasonality and a dry equatorial region, is predicted for this age (Figure

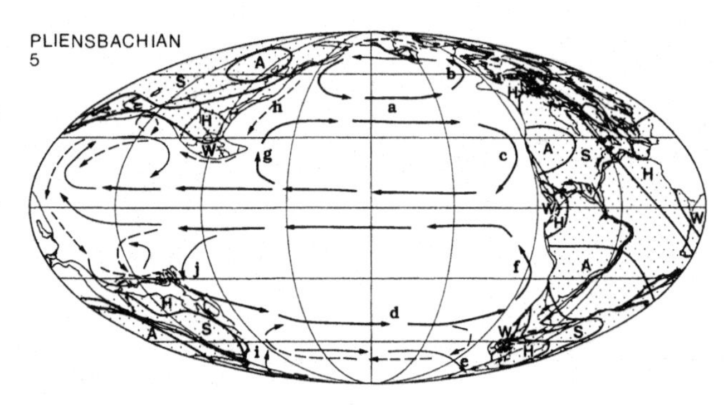


**Figure 27.5. Predicted ocean currents and relative precipitation in the Pliensbachian (Early Jurassic).** ***Oceans:*** **Thin dashed arrows indicate northern summer currents, thin solid arrows indicate northern winter currents; year-round currents are indicated by thick solid arrows. Currents: *a*, paleo-North Pacific current; *b*, paleo-Alaska current; *c*, paleo-California current; *d*, South Pacific current; *e*, Antarctic Peninsula current; *f*, paleo-Peru current; *g*, paleo-Kuroshio current; *h*, paleo-Oyoshio current; *i*, Lord Howe current; *j*, paleo-East Australia current.** ***Continents:*** **A, arid; S, semiarid; H, subhumid; W, humid (adapted from Parrish et al. 1982).**

27.5). However, the monsoonal circulation is expected to have been weaker than in the Late Triassic–earliest Jurassic because by the Pliensbachian, more exposed land area was in the Northern Hemisphere than in the Southern (Parrish 1985) (Table 27.1).

Oceanic circulation is expected to have been influenced by the monsoon only in the northwestern Pacific and in Tethys (Figure 27.5). Australia and the southeastern Pacific are predicted to have been out of the main influence of the southern monsoon, which would have been concentrated over what is now Africa. However, the northwestern Pacific, in the vicinity of what is now South China and the East Indies, might have experienced a southward-flowing current during the winter, which would have created a stronger seasonal temperature change than might have been the case without the monsoon.

On the eastern side of the paleo-Pacific, the latitudinal temperature gradient between about 70° north and south would have been low. In the north, the paleo-North Pacific current, centered on about 45° north, would have divided into a northward-flowing paleo-Alaska current and a southward-flowing paleo-California current along the northwest coast of Pangea. The counterparts in the Southern Hemisphere would have been the South Pacific current, centered on about 45° south, which would have divided into the southward-flowing Antarctic Peninsula current and the northward-flowing paleo-Peru current. The paleo-Alaska and Antarctic Peninsula currents would have carried relatively warm water poleward, whereas upwelling would have kept the paleo-California and paleo-Peru currents cool. Although the eastern equatorial Pacific probably was cooler than the west at this time, as it is today, the magnitude of the east–west gradient would depend on the strength of the equatorial upwelling and the equatorial countercurrent. Because Tethys was closed on the west, it is possible that the equatorial countercurrent was strong, and thus the east–west thermal gradient weak. Some of these factors will be discussed further in a later section.

By contrast, two distinct thermal boundaries would be expected on the western side of the Jurassic Pacific. In the north, this boundary would have been off what is now South China in the summer, where the northward-flowing paleo-Kuroshio and southward-flowing paleo-Oyoshio currents joined. In the winter, winds associated with the monsoon would have extended the southward range of the paleo-Oyoshio, and the thermal boundary would have been in what is now Southeast Asia, at about the entrance to Tethys. In the Southern Hemisphere, the thermal boundary would have been off northeast (present coordinates) Australia, at the confluence of the northward-flowing Lord Howe and southward-flowing paleo-East Australia currents. The shape of southern Pangea would have kept the winter monsoon center relatively farther from the coast than in the Northern Hemisphere, so seasonal variations in this region probably were not as great as in the north.

Tethyan circulation is predicted to have been influenced in much the same way that Indian Ocean circulation is affected by the Asian monsoon today. The warm equatorial current would have been directed northward during the northern summer and southward during the southern summer. Counter-gyres might have formed west of Australia and Southeast Asia/Tibet. In addition, seasonal (summer) upwelling zones are predicted for the eastern margin of Tethys just north and just south of the equator (Parrish and Curtis 1982).

Tethys would be expected to have been warm and virtually isothermal, owing to the long fetch of the equatorial current and complete recirculation of this water around the margins (Berggren and Hollister 1977; Ziegler et al. 1981). Salinity, rather than temperature, was likely to have been the predominate control on the distribution of marine organisms, and salinity probably was higher near the equator and along the winter coastline, and possibly quite low along the summer coastline.

*Tithonian age.* The combination of the breakup of Pangea and the flooding of Laurasia meant that the Pangean monsoonal circulation would have ceased (Figures 27.3 and 27.4) by the Tithonian. Although about the same amount of continental crust was in each hemisphere as earlier in the Jurassic, the trend toward more exposed land in the Northern Hemisphere was reversed temporarily by flooding of the Laurasian continent (Table 27.1). Continent-scale movements of some sort, either sagging of Laurasia (and no eustatic sea-level change) or uplift of Gondwana (during eustatic rise), were operating at that time, but review of those mechanisms is outside the scope of this chapter. However, the amount of exposed land in Gondwana and its latitudinal position mean that weak monsoonal circulation might have persisted there.

Because the Gondwanan monsoonal circulation would have been relatively weak, the predicted alternating circulation in Tethys illustrated in Figure 27.6 also would have been weak, although some alternation could have been forced by the Gondwanan monsoon. In addition, the passage between Gondwana and Laurasia would have permitted some throughflow of the equatorial current. Seasonal upwelling would have been possible along the southwestern Tethyan margin during the southern summer, and zonal upwelling would be predicted for the northwestern African margin (present coordinates) year-round (Parrish and Curtis 1982). Upwelling also might have occurred over the shelf region along the equator, in what is now northernmost South America or southernmost Mexico.

The temperature field in the Tithonian Pacific would have been similar to that in the Pliensbachian, except that the thermal boundaries would have shifted slightly as the continents drifted across zonal boundaries. The major difference would have been in Tethys. Once western Tethys opened, a circumequatorial current was established. This would have permitted the establishment of an east–west thermal gradient in Tethys, although some cross-equatorial mixing might still have been forced by the Gondwanan monsoon and by the diversion of the equatorial current northwestward around northern Africa. Effects of the establishment of the circumequatorial current on global climate are discussed further in a later section.

### Generalizations about Jurassic climate and ocean circulation in the circum-Pacific region

Except in the vicinity of Tethys, the conceptual models predict that climate and ocean-surface circulation did not change very much during the Jurassic. This is because the paleogeographic changes during the period were relatively subtle. Although the Pacific Ocean became slightly narrower as the central Atlantic widened, this change would have had little effect on the general patterns of circulation except in Tethys. Major changes in surface circulation patterns in the Pacific would not be expected, for two reasons. First, the change in area was small relative to the total area of the ocean. Second, the movement of the continents was mostly longitudinal, so that the same latitudinal expanse of ocean and the same distribution of land–sea temperature contrasts and meridional ocean currents were maintained throughout the period.

Because of the relatively small geographic changes in the circum-Pacific region, several features of circulation in the Jurassic are predicted to have been relatively consistent throughout the period (Figures 27.1–27.4) (Parrish and Curtis 1982). The subtropical high-pressure cells are regarded to be a relatively constant and predictable feature of global circulation, and predictions of these features in the Jurassic Pacific might be regarded as robust. Except in the southern Indian Ocean, which is strongly influenced by the monsoon, modern subtropical high-pressure cells drive upwelling currents at midlatitudes on the eastern sides of the ocean basins, off northwest and southwest Africa, Peru–Chile, and California.

Thus, counterparts to the modern subtropical upwelling zones would be expected in the Jurassic Pacific, along the western margins of the Jurassic continents. The positions of these upwelling zones would have depended on the latitudinal positions of the continents. In the Early Jurassic, upwelling would have been centered on what is now the California–Oregon margin, extending into British Columbia during the summer and Baja California and northwestern Mexico in the winter. In other words, the Early Jurassic equivalent to the modern California Current would have

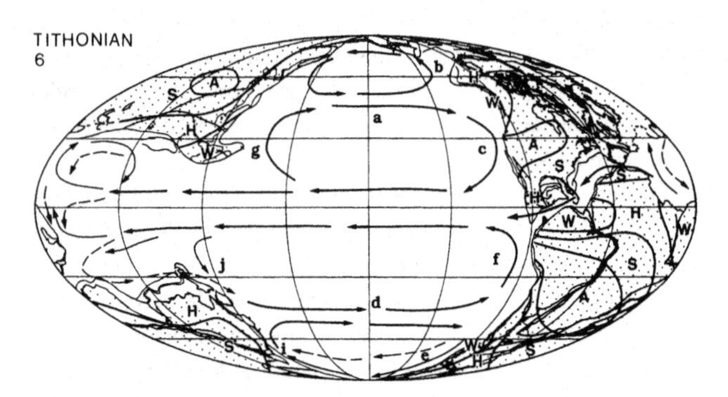


**Figure 27.6. Predicted ocean currents and relative precipitation in the Tithonian (Late Jurassic). Symbols as in Figure 27.5. Alternating circulation in western Tethys is not likely to have been as strong as in the Early Jurassic; its strength would be dependent on the strength of the Gondwanan monsoon.**

been only slightly farther north (present coordinates) than it is today, by an amount corresponding to the northward movement of North America since the Early Jurassic. Upwelling also is predicted for the southeastern Pacific in the Early Jurassic, at almost exactly its present position.

By the latest Jurassic, Gondwana had moved about 5° south and Laurasia about 5° north, and the upwelling zones would have shifted accordingly, relative to the continents. The northeastern Pacific upwelling zone would have been centered on southern California–northwestern Mexico and Baja California (in their paleogeographic positions), whereas the southeast Pacific upwelling zone would have been centered on northern Chile–southern Peru.

In addition to the coastal upwelling associated with the eastern limbs of the subtropical high-pressure cells, upwelling would have occurred in oceanic areas along the equator and about 60° north and south. Where these divergences impinged on the continental shelves (as in northernmost South America in the Tithonian), the effects of high productivity might be reflected in the sediments and shallow-water fauna.

The thermal effects of the upwelling depend partly on the vertical temperature stratification of the ocean. However, horizontal temperature gradients also are created by advection, and upwelling enhances the cool-temperature conditions along the west coasts (e.g., Philander and Pacanowski 1981). The general horizontal thermal pattern throughout the Jurassic would have included the following: weak, warm, poleward-flowing currents in low latitudes and midlatitudes on the western side of the basin; a warm, westward-flowing equatorial current; cold, eastward-flowing currents at midlatitudes; and equatorward-flowing cool currents in low latitudes and midlatitudes on the eastern side of the basin (Figures 27.5 and 27.6).

Continental climates are predicted to have been relatively wet, warm, and seasonal on the western side of the Pacific; wet and cool above about 40° north and south on the eastern side and above 60° north and south on the western side of the Pacific; and warm and dry in low latitudes on the eastern side (Parrish et al. 1982). Because of the consistent configuration of the continents, these general patterns would have existed throughout the period, although the boundaries and absolute temperature and precipitation might have fluctuated with changes in the global temperature gradient. Any major changes in the continental climates of the circum-Pacific region probably were related to the breakdown of the monsoon, even though that would have been more important in the circum-Tethyan region.

### Observed Jurassic paleoceanographic and paleoclimatic changes

*Terrestrial environment.* The major change in continental climates in the Jurassic was the breakdown of the monsoonal circulation, particularly in the north. Earliest Jurassic eolian sandstones of the Colorado Plateau record northwesterly winds, consistent with the predicted summer monsoonal circulation (Parrish and Peterson 1988). Later in the Early Jurassic, the sandstones recorded a divergence of northerly winds to the southeast and southwest, suggesting a boundary between the summer monsoonal circulation to the east and the "normal" summer subtropical high-pressure cell to the west. Parrish and Peterson (1988) interpreted the change during the Early Jurassic to be consistent with weakening summer monsoonal circulation. Data from Middle Jurassic eolian sandstones show no consistent wind field on the Colorado Plateau, and by the Late Jurassic the zonal westerlies are well expressed in the eolian sandstones. Disorganized vectors in the Middle Jurassic might have recorded a transitional climatic state, from monsoonal to zonal (Parrish and Peterson 1988).

Changes in the distribution of coals, evaporites, redbeds, and laterites during the Jurassic indicate a dramatic shift in climate sometime during the late Middle or early Late Jurassic in southern Laurasia (Vakhrameyev and Doludenko 1977; Parrish et al. 1982; Vakhrameyev 1982; Hallam 1984). In the Pliensbachian, coals were widespread throughout eastern Laurasia, even in the "dry" belt south of 45° north (Parrish et al. 1982). These coal deposits are interpreted to have resulted from a favorable combination of positive annual moisture balance and/or high groundwater table resulting from the monsoon rains and subsiding foreland basins (McCabe 1984; Ziegler et al. 1987; Parrish and McCabe 1988). [It should be pointed out that this coal deposition may have itself replaced apparently drier deposition earlier. This shift has not been documented for Laurasia as a whole, and it is therefore premature to speculate on its cause, although a possibility implied by Parrish et al. (1982) is that evaporation was reduced because of cooler temperatures brought about by the more northerly position of Laurasia in the Pliensbachian. If global temperature was dropping as well, the shift could be explained simply by changing the precipitation–evaporation balance with a change in temperature. Once the monsoon was broken up, however, the mechanism of moisture transport to those latitudes would have ceased to exist.] By the Tithonian, coal deposition was replaced by evaporite deposition between 20° and 40° north, consistent with zonal circulation and the breakdown of the northern monsoon (Parrish et al. 1982; Parrish 1988). The data from Hallam (1984) indicate that the Pliensbachian pattern persisted into the Middle Jurassic. The shift to drier climate in southern Laurasia is also reflected in changes in the distribution of *Classopollis*, the pollen of the conifer family Cheirolepidiaceae, which is thought to have been drought-tolerant (Alvin, Spicer, and Watson 1978; Alvin, Fraser, and Spicer 1981; Doyle and Parrish 1984). This shift has been documented with more resolution than the sediments and appears to have taken place during late Bathonian–Callovian time (Vakhrameyev 1982).

Climatic changes in the Southern Hemisphere roughly paralleled those of the Northern Hemisphere. This indicates that the Gondwanan monsoonal circulation probably was much weaker than the Pangean circulation, in that a strong monsoonal climate was not sustained once Pangea broke up. Redbeds, with local occurrences of evaporites and eolian sandstones, were widespread in South America in the Triassic, suggesting seasonally wet climates (e.g., Van Houten 1982), consistent with the predicted monsoonal circulation (Volkheimer 1967; Rocha-Campos 1971; Parrish and Peterson 1988). In the Early Jurassic, abundant fossil floras and widespread peat deposition occurred, and humid conditions continued into the Middle Jurassic (Volkheimer 1967). In the Late Jurassic, arid conditions were established in midlatitudes, resulting in extensive eolian sand deposition.

*Marine environment.* The major biogeographic change in the Jurassic Pacific Ocean was the establishment and disappearance of the East Pacific Subrealm in the latest Middle Jurassic (Westermann 1981, 1984) (see Chapter 5). Prior to the appearance of this subrealm, provincialism was low. Although Gondwana and Laurasia had not yet separated, short-term influxes of taxa from Tethys to the eastern Pacific and vice versa took place, probably as a result of sea-level fluctuations that periodically flooded central Pangea through the Hispanic Corridor (Westermann 1981, 1984; Hallam 1983; Brandt 1986). These fluctuations were mirrored by expected changes in biogeography, with provincialism decreasing as exchange increased.

These biogeographic changes may have been brought about by sea-level changes rather than by climate directly. Faunal interchange, which usually is noted in shallow-water faunas and is often used as a proxy indicator of geographic change (e.g., Reyment and Tait 1972; see discussion in Parrish 1987), may take place through shallow seaways during high sea-level stands prior to actual continental fragmentation, and this has led to some conclusions about the timing of rifting events that contradict geological evidence (Parrish 1987). If the tectonic events are great enough to force climatic change as well (which might be reflected in the fauna), the picture may become even more confusing, not only because the faunal exchange will precede the geographically forced climatic change but also because the faunal exchange might trigger other biogeographic changes (through diversification or competition and extinction) that are unrelated to climate.

The appearance of the East Pacific Subrealm roughly coincides with a drop in sea level on Hallam's curve (1978, 1981). However, the distinctness of the subrealm in the Middle Jurassic, compared with earlier times when the connection between western Tethys and the Pacific was closed, suggests that some additional cause operated in concert with the paleogeographic change. The disappearance of the subrealm may be attributed to the opening of western Tethys, which was accomplished by both a rise in sea level (Westermann 1984) and rifting between Gondwana and Laurasia (Dickinson and Coney 1980) (see Chapter 3). In the later part of the Late Jurassic, provinciality may have increased again (Westermann 1984), with no apparent paleogeographic cause.

Shifts in the boundary between the Boreal and Tethyan biogeographic realms have been mapped for the northwest Pacific by Sey and Kalacheva (1985) (see Chapter 11) and for the northeast Pacific by Taylor et al. (1984). Sey and Kalacheva (1985) outlined changes in the position of the boundary between the two realms. They concluded that Tethyan forms penetrated northward, to about 62° north (roughly 85° north paleolatitude) throughout the Early Jurassic and in the Callovian, after which the boundary fluctuated, but did not move farther north than 55° north (about 80° north paleolatitude). In the Middle Jurassic, the boundary was an unspecified distance farther south. Sey and Kalacheva (1985) used a fixist base map for their reconstructions. When their units are plotted on paleogeographic reconstructions, their boreal faunas are truly polar. According to some recent authors, the northern localities for the Tethyan faunas of Sey and Kalacheva (1985) are on suspected allochthonous terranes (see Chapters 3, 11), blocks that probably were islands and lay farther south relative to the continent in the Jurassic than they do now. These terranes have juxtaposed palynofloras of dramatically different character in Japan (Doyle and Parrish 1984; Kimura and Tsujii 1984). It is therefore difficult to evaluate the Soviet data, because the precise localities are not given and thus cannot be plotted accurately with respect to terrane boundaries.

In the following discussion, all latitudes referred to are paleolatitudes. Sey and Kalacheva (Chapter 11) suggest that the Boreal–Tethyan ammonite boundary during most of the Jurassic (excluding Hettangian) was at about Vladivostok (about 60° north) or slightly farther south, a few degrees farther north in the Pliensbachian and Tithonian, and off the northern tip of Sakhalin (about 75° north) in the late Bajocian to Bathonian.

On the eastern side of the Pacific basin, a boundary per se is not easily drawn for much of the Jurassic because of gaps in the cratonic record. As pointed out by Taylor et al. (1984), the paleobiogeographic picture has been obscured by the presence of faunas on allochthonous terranes; they confirmed that many of these terranes probably had significant northward movement. In addition, the East Pacific Subrealm does not have an equivalent to the west. Its presence might be taken as a priori evidence that the climatic and oceanographic conditions were different on the eastern side of the basin compared with those on the west. In particular, the fact that it was widespread and extended across the equator is evidence for the predicted relatively low thermal gradient. Nevertheless, it is useful to compare the biogeographic changes of the cratonic faunas. The following is taken from Taylor et al. (1984) and summarized in Figure 27.7 (see Chapter 12).

The Boreal Realm did not appear until the Pliensbachian, where it penetrated to about the modern United States–Canada border (about 35° north); the Tethyan Realm was represented in faunas at

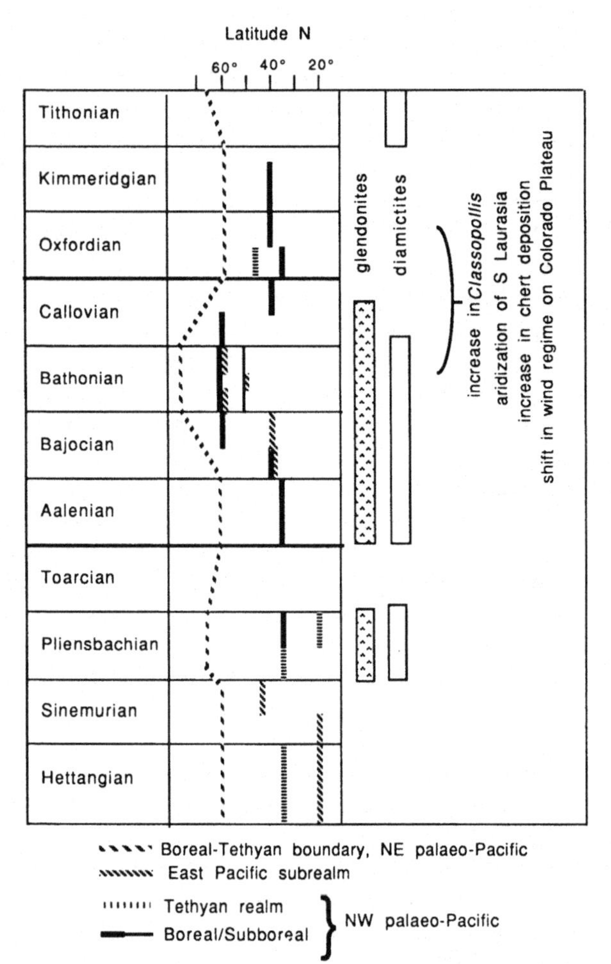


**Figure 27.7. Summary of evidence of biogeographic and climate change during the Jurassic.**

about 20° north. In the late Sinemurian, East Pacific faunas, which had previously been found at 20° north, occurred as far north as 50° north before disappearing. The next appearance was in the Bajocian, where the boundary between the Boreal Realm and East Pacific Subrealm is clearly fixed in southern Alberta, at about 40° north. In the early late Bajocian, Boreal faunas were not found farther south than about 60° north, having been replaced in southern Alberta by the East Pacific Subrealm, and at the Bajocian–Bathonian boundary the East Pacific Subrealm faunas ranged as far north as 60°. In the early and middle Bathonian, the East Pacific Subrealm retreated south again, to about 50° north, before reappearing at 60° north at the end of the Bathonian. During the Bathonian, however, faunas of the Boreal Realm, Subboreal Province, ranged as far south as 50° north. During the Callovian, the Boreal Realm/Province faunas spread southward and reached a maximum southward extent of 35° north during the early and middle Oxfordian. During that time, Tethyan Realm faunas were absent from the craton north of present-day Mexico, except in the early and middle Oxfordian, when they overlapped the Boreal Realm/Province faunas and extended as far north as 45° north. To summarize, the Boreal–Tethyan boundary was in low latitudes during the Early Jurassic, with the exception of the late Sinemurian, when there might have been a brief movement northward. During the Middle Jurassic, the boundary moved northward 15° or 20°, depending on how the Subboreal faunas are treated, and then moved southward again at the very end of the Middle Jurassic. Provinciality was maximal during the late Middle Jurassic (Westermann 1984).

The Boreal Realm appears to have been farther south in the east throughout the Jurassic than it was on the west. This is the pattern that would be expected if upwelling had been important along the northwestern Pangean margin. At the highest latitudes, however, diversity was much higher on the eastern side, which is consistent with warmer temperatures of the paleo-Alaska current compared with the paleo-Oyoshio current.

The most interesting event is the major northward shift of the boundary between the Boreal and Tethyan Realms in the Middle Jurassic (late Bajocian–Bathonian), which occurred on both sides of the ocean. This shift was coeval with the major drop in sea level that closed the connection between western Tethys and the Pacific and permitted the strongest differentiation of the East Pacific Subrealm (Westermann 1984). The poleward shift of low-latitude faunas was mirrored in the Southern Hemisphere (Stevens 1980). That time also saw the strongest evidence for cold-climate conditions in northern Pangea, with deposition of glacial-marine sediments, diamictites ("pebbly argillites"), and glendonites (Brandt 1986).

Although glaciation in the Mesozoic is an idea that has attracted a number of workers for a variety of reasons (e.g., Williams 1975; Epshteyn 1978; Sheldon 1981; Brandt 1986; Frakes and Francis 1988), the evidence has never been as convincing as for the other glacial epochs (e.g., Frakes 1979). Unlike any of the other episodes of continental glaciation, the evidence is limited to relatively unextensive deposits of dropstones and diamictites (Epshteyn 1978; Frakes and Francis 1988). Whether the Jurassic diamictites of northern Siberia are glacial in origin is highly controversial, even among Russian geologists (Chumakov 1981). As pointed out by many authors (e.g., Schermerhorn 1974; Hambrey and Harland 1981), these types of deposits may be formed in a variety of ways and thus do not constitute proof of continental glaciation. Moreover, even accepting that glaciation could occur, the continental geology of the circum-Arctic, especially for the Jurassic and Cretaceous, is very well known, and sites where glaciers could have formed were few and were restricted to highlands of limited geographic extent. Thus, it seems unlikely that glaciation of a magnitude sufficient to affect sea level to any great extent could have occurred in the Northern Hemisphere.

Although significant glaciers might not have been present, lack of glaciation does not mean that the high-latitude climate was warm. The data of Yasamanov (1981) do suggest a mild (roughly 3°C) cooling event in the Middle Jurassic in Russia, although overall the isotopic temperatures were the highest recorded for the Mesozoic. In addition, Beauvais (1973) reported hermatypic corals from Sakhalin, which would have been at about 70° north paleolatitude, but they probably are from an allochthonous terrane (see Chapter 3). Although the vegetation of Russia in the Jurassic has been described variously as "warm" or at least "extra-tropical," it was, from descriptions in the literature (Epshteyn 1978; Brandt 1986), quite similar to that of the Late Cretaceous of the North Slope of Alaska, with the exception that angiosperms had not yet appeared. The Alaskan floras, which grew within 10° of the North Pole, clearly show that the climate was cool to cold-temperate (Spicer and Parrish 1986; Parrish and Spicer 1988a,b). These floras were diverse and consisted of conifer forest with an understory of ferns, cycadophytes, ginkgophytes, and so forth, as well as angiosperms. The Jurassic flora was also dominated by conifers, bennettitalians (related to cycads), cycadophytes, and ginkgophytes. The conifers included the genus *Podozamites,* whose representatives on the North Slope were deciduous, as were the ginkgophytes and the cycadophyte genus *Nilssonia.* (It should be noted, however, that *Podozamites* may be a taxonomically diverse generic name, owing to difficulty of identification, so that the Jurassic and Cretaceous representatives may not be closely related.) Thus, as implied by Epshteyn (1978), the vegetation does not necessarily signify warm climates. The most compelling evidence that conditions at high latitudes were cold is the presence of glendonite (calcite after ikaite), which forms only in freezing seawater (Kaplan 1978, cited by Brandt 1986; Kemper 1987; Kennedy, Hopkins, and Pickthorn 1987). The presence of glendonite in Jurassic sediments in the north polar regions suggests that sea ice may have been present; sea ice, however, would not be expected to have a large effect on sea level.

In summary, the major changes appear to have been the increase in provinciality during the Middle Jurassic (Westermann 1984), accompanied by a dramatic poleward shift of the biogeographic unit boundaries (Stevens 1980; Taylor et al. 1984) coeval formation of glendonites in the polar regions, poorly organized atmospheric circulation patterns over the Colorado Plateau, and aridization of southern Laurasia between the Middle and Late Jurassic, including a dramatic increase in *Classopollis* at the Middle–Late Jurassic boundary. Another dramatic change that occurred near the Middle–Late Jurassic boundary was the increase in the deposition of radiolarian cherts in western Tethys, signifying an increase in productivity (Grunau 1965; Jenkyns and Winterer 1982; Hein and Parrish 1987). Although this increase might be partly explained by the changes in the locations of coastal upwelling resulting from the separation of Laurasia and Gondwana (Parrish and Curtis 1982), it is clear that silica production increased worldwide at that time and that most of the increase was in Tethys (Jenkyns and Winterer 1982; Hein and Parrish 1987). This will be discussed later.

## Speculations on the causes of Jurassic climatic changes in the circum-Pacific region

The major change in Jurassic paleogeography was separation of Gondwana and Laurasia. This resulted in a reorganization of equatorial circulation, possible injection of saline waters to the eastern Pacific, and the breakdown of the northern monsoon. How might these factors combine to give rise to climatic change in the Jurassic?

### *Tethyan throughflow and equatorial circulation*

Although the paleogeographic changes were relatively small – a few degrees latitudinal drift of the continents – the establishment of throughflow from Tethys and the eastern Pacific might be implicated in climatic change (e.g., Berggren and Hollister 1977). With a barrier in place across western Tethys, temperature and salinity gradients from the eastern Pacific to western Tethys would have developed, most likely with warmer, more saline water on the eastern side of the barrier, as is the case in the Caribbean and Panamanian basins today (Keigwin 1982).

The effect of the uplift of the isthmus of Panama on climate and circulation in the Atlantic and Pacific has been studied by several workers and thus may help guide interpretation of the effects of opening and closure of the connection between Tethys and the Pacific in the Jurassic. Keigwin (1982) suggested that uplift of the isthmus of Panama, at about 4 Ma, might have helped trigger a major climatic change through the intensification of the Gulf Stream by increasing its effectiveness as a moisture source for glaciation, the importance of which was emphasized by Ruddiman and McIntyre (1981) for the Quaternary. However, the Gulf Stream underwent intensification events (Kaneps 1979) before and after as well as during the uplift, and the major isotopic shift attributed to Northern Hemisphere glaciation is later, at about 2.4–2.6 Ma (Shackleton, Hall, and Boersma 1984; Prell 1985). Moreover, the presence of a Gulf Stream equivalent at any time during the Jurassic is questionable (Parrish and Curtis 1982). Closure of the connection between the Pacific and western Tethys may have only sent the waters of the equatorial current back along the equator through Tethys, not into high latitudes (Berggren and Hollister 1977).

Productivity decreased in the equatorial Pacific after about 4 Ma, although this decrease was local, not oceanwide (Pisias and Prell 1985). Pisias and Prell (1985) pointed out that decreased productivity would suggest weaker winds, as confirmed by Pisias and Rea (1988); in other words, the presence or absence of throughflow of the equatorial current would not a priori affect productivity. However, several changes did occur about the time of the final uplift of the isthmus [the barrier probably was intermittently in existence through parts of the earlier Cenozoic (Pindell and Dewey 1982)] that also might have occurred with the closing of western Tethys including a rise in the salinity of surface waters in the Caribbean, the beginning of biogeographic differentiation of the marine faunas, and terrestrial faunal interchange, as reviewed by Keigwin (1982).

It is interesting to speculate what effect the barrier between western Tethys and the Pacific would have had on the equatorial countercurrent. It seems likely that if the equatorial easterlies were comparable to strong periods today, the sea-surface topographic difference between the eastern equatorial Pacific and Tethys would have been as great or greater than today. However, the closed Tethys–Pacific geography has no modern analog. In the Atlantic, the equatorial current is largely shunted northward by the Brazilian "nose" and at least partly into the Caribbean, where it exits northeastward as the Gulf Stream. Although a strengthening of the Gulf Stream did accompany the closure of the Panamanian seaway (Kaneps 1979), strengthening also accompanied the cold bottom-water event at about 3.2 Ma (Prell 1985) and the onset of Northern Hemisphere glaciation at 2.4–2.6 Ma (Kaneps 1979; Shackleton et al. 1984; Prell 1985). In the Pacific, the equatorial current enters the Indian Ocean (Murray and Arief 1988). Godfrey and Golding (1981) speculated on the effects of shutting off the throughflow on the Indian Ocean, but, to my knowledge, similar speculations on the effects of cessation of throughflow on the Pacific have not been published.

The best analog to the Jurassic changes may be in the late Miocene, when throughflow from the Pacific to the Indian Ocean apparently did cease for a time (Kennett et al. 1985; Ziegler et al. 1983). The result was a spread of tropical foraminiferal faunas eastward and the elimination of the east–west biogeographic differentiation that had existed in the early Miocene and that was resumed again later (Kennett, Keller, and Srinivasan 1985). Kennett et al. (1985) suggested that blockage of the equatorial current resulted in a stronger equatorial countercurrent, carrying warm water eastward. Unfortunately, they did not perform the same analyses on middle Miocene faunas. During the middle Miocene, circulation was unusually vigorous in the northern Pacific (e.g., Vincent and Berger 1985), resulting in widespread high biologic productivity and abundant silica deposition. Thus, the westward spread of warm waters once this event ceased might have been established by a rebound effect in response to the weakening of the easterlies. However, this sort of response is short-term (Rasmusson 1985), and the fact that the tropical faunas penetrated all the way to the eastern rim of the basin supports the Kennett et al. (1985) hypothesis as an explanation for the longer-term pattern.

These results may suggest that when no connection between western Tethys and the Pacific existed, the equatorial countercurrent was strong. Thus, the possibility is raised that the countercurrent could have completely overcome the effects of the equatorial easterlies, suppressing even the coastal upwelling zones, and keeping the Jurassic Pacific Ocean in a permanent El Niño-like state, as discussed later. Once the barrier was breached, circulation would have assumed a state closer to the modern Pacific circulation, with variations dependent on the stability of the easterlies. However, countercurrents and undercurrents at the equator today are generated dynamically, not geographically (Philander and Pacanowski 1980), so this is purely speculative.

The effect of throughflow to the eastern Pacific in the Jurassic might also have depended on the density of the water. If the water was of normal salinity, it would have stayed on the surface and possibly suppressed the upwelling of cold water along the equator by augmenting the thermocline, although the Panamanian throughflow apparently did not have such an effect (Pisias and Prell 1985). Whether the throughflow from Tethys would have at least significantly changed the latitudinal temperature gradient in the eastern Pacific by introducing warm water into the equatorial regions depends on the effectiveness of the processes that would keep the equatorial region cool, that is, strong easterlies and resulting upwelling. Strong easterlies pile water up on the western side of the Pacific; the typical east–west difference in sea level is 0.5 m (Wyrtki 1975, cited by Rasmusson 1985). Weakening of the easterlies allows this slope to relax, whereupon warm surface waters spread eastward from the western Pacific (Rasmusson 1985).

If sea level dropped because of increased continental glaciation, the latitudinal isotherms would be expected to shift equatorward (CLIMAP 1976), and biogeographic boundaries should respond accordingly (e.g., Ingle 1981; cf. Keller and Barron 1981), although care must be taken in interpreting biogeographic data. Embayments, for example, can shelter warm-water faunas well within an otherwise cold-water biogeographic unit (Addicott 1970; Wells 1987). In addition, wind strength would have increased (Sarnthein et al. 1981, 1982; Vincent and Berger 1985), increasing upwelling (Pisias and Rea 1988), and forcing the cool isotherms even farther equatorward (Ingle 1981; Sarnthein et al. 1982; Vincent and Berger 1985). However, the observed shift in the Jurassic, coincident with a drop in sea level, was poleward, not equatorward.

On the other hand, if the sea-level drop were not eustatic, but owing rather to uplift of the region prior to rifting, equatorward compression of isotherms would not be expected. In that scenario, warming of the whole Pacific basin might have been accomplished, with the return of warm water from Tethys in a strong equatorial countercurrent [as apparently occurred in the Miocene (Kennett et al. 1985)], which would have suppressed the equatorial and, possibly, the coastal upwelling. The consequence of such a change would be a basinwide poleward shift of the biogeographic boundaries; this would not necessarily preclude cooling in the paleo-Arctic Ocean, as suggested by the glendonites.

With the opening of the connection between western Tethys and the Pacific and establishment of throughflow, upwelling might have resumed, and the warm equatorial current would simply circumnavigate the globe, with minimal mixing into higher latitudes. The result would be apparent cooling, as indicated by an equatorward shift of the biogeographic boundaries, even though the purported glacial sedimentation ceased and glendonites disappeared.

### *Warm, saline bottom waters*

Once the connection between western Tethys and the Pacific was established, if the throughflowing water was highly saline as well as warm, it would have sunk below the surface, either going all the way to the bottom, as has been suggested for the Cretaceous, or being entrained in midlevel currents flowing across the basin in the same way that the Mediterranean outflow is transported across the North Atlantic at present (Brass et al. 1982a,b). Such an occurrence immediately after the breach seems likely, since the equatorial current would have to cross the basin in which the Louann and related salt deposits were formed while the barrier was closed. However, it should be noted that saline waters generated in western Tethys were likely to have reached the Pacific Ocean throughout the Jurassic through eastern Tethys, so a breach in the barrier would not have created a major change in this sense.

Our understanding of the generation of bottom waters and the effects of changing flow is still primitive, and most recent modeling (Toggweiler and Sarmiento 1985; Wenk and Siegenthaler

1985; Volk and Liu 1988) has concentrated on changes in $O_2$ and $CO_2$, in which salinity plays a relatively minor role. However, these models do contain elements that suggest the mechanisms of Jurassic climatic change, although even qualitative application of the results of these studies to the distant past is highly speculative.

The dramatic increase in the deposition of radiolarian cherts worldwide, but especially in Tethys, may suggest a link between the events on land and in the oceans in the Jurassic. Silica accumulation is used extensively as an indicator of productivity (e.g., Lisitsyn 1972, 1977; Ramsay 1973; Leinen 1979; Brewster 1980; DeMaster 1981) and as a nutrient tracer (Miskell 1983; Miskell, Brass, and Harrison 1985). Miskell (1983) and Miskell et al. (1985) recognized several changes in global silica distribution, when the accumulation of silica switched from the Atlantic to the Pacific or vice versa, between the Late Cretaceous and the present. For example, during the Late Cretaceous, silica accumulation was greatest in the North Atlantic, suggesting that the basin received older, nutrient-rich bottom waters that were then upwelled, resulting in high productivity. Miskell et al. (1985) suggested that the bottom water was generated in saline marginal basins in low latitudes and at high latitudes.

From the standpoint of tracing thermohaline circulation and nutrients, the Jurassic is an interesting time in that the world ocean was much more continuous, and a freer exchange of water was possible. This might not have changed significantly during the period. Figure 27.8 shows very schematically a postulated thermohaline circulation for presence and absence of a connection between western Tethys and the Pacific. Bottom-water generation was likely to have occurred in the marginal basins surrounding Tethys, where evaporites were forming throughout the Jurassic (Parrish et al. 1982; Hallam 1984). In both cases, the water would have started out completely depleted in nutrients (e.g., Sarmiento, Herbert, and Toggweiler 1988). It would seem from these diagrams that the changes in circulation of the saline bottom waters would have little effect on productivity. If anything, the water entering Tethys on the east from the Pacific would be younger and less nutrient-rich in the Late Jurassic, and total silica production would be expected to be less.

In modeling anoxia, Sarmiento et al. (1988) have suggested that changes in thermohaline circulation, although important in determining productivity, cannot alone explain the fluxes required to drive the ocean to anoxia, and that convective overturn also is important. This type of circulation is occurring off Antarctica today, where Atlantic deep water is upwelled and immediately downwelled again as Antarctic bottom and intermediate water before the nutrients are fully depleted (Sverdrup, Johnson, and Fleming 1942, Fig. 164). Although Sarmiento et al. (1988) mentioned convective overturn in low latitudes in passing, the implication from their work and that of Miskell et al. (1985) is that high-latitude bottom-water generation is important to maintaining high productivity, because only in high latitudes are nutrient-rich waters likely to be downwelled. Thus, one possible change from the Middle to the Late Jurassic is onset of bottom-water generation at high latitudes (Figure 27.8). This would seem to counter the evidence from glendonites, which were more abundant during the Middle

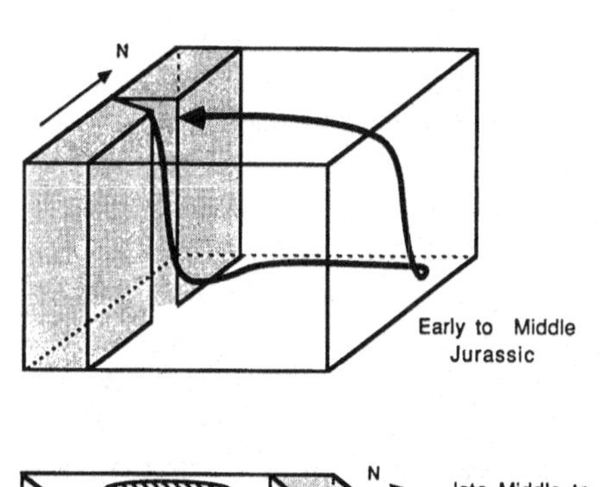


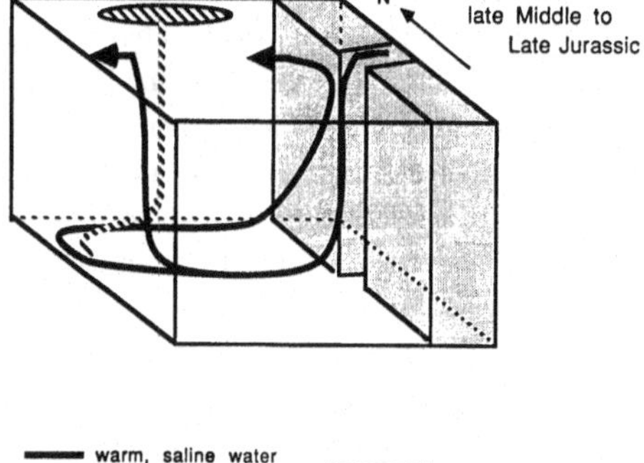

**Figure 27.8. Schematic, speculative diagrams of deep-water generation and circulation in the Jurassic; note that the circulation does not account for the positions of the spreading ridges, which are unknown. Top: Early–Middle Jurassic. Warm, saline bottom water is generated in Tethys, but not at high latitudes. Bottom: Bottom water generated both in Tethys (whence it could flow either east or west) and at high latitudes.**

Jurassic and then ceased forming until the Tithonian (Brandt 1986), suggesting warming. The answer might lie in breakdown of the monsoon.

One of the key features of the Brass et al. (1982a) model is an increase in poleward transport of latent heat (as water vapor) and an implied freshening of high-latitude waters to suppress the generation of cold bottom waters. Manabe and Wetherald (1980) pointed out that large-scale monsoonal circulation can contribute substantially to poleward latent-heat transport (i.e., moisture). Thus, under conditions of generation of warm, saline bottom water and monsoonal circulation, the high-latitude regions might be expected to be wet. As discussed earlier, northeastern Laurasia in the Early and Middle Jurassic was characterized by widespread coal deposition, even in relatively low latitudes. The rivers supplying these coal basins were likely to have drained north and west. That they did so, and that fresh water was abundant, is supported not only by the coals, but possibly by the light isotopic results of Yasamanov (1981). He interpreted these results to mean that temperatures were warm (up to 20°C in the northern basins), which is inconsistent with the presence of glendonites. Therefore, although Yasamanov (1981) claimed to have controlled for salinity, the high isotopic temperatures may nevertheless reflect freshwater input. At present, isotopic values of precipitation at high latitudes (above 60° north) are light ($-12\delta^{18}O$ to $-22\delta^{18}O$; Yurtesever 1975, cited by Anderson and Arthur 1983), so freshwater influx might change the istopic composition considerably. In addition, Hallam (1971) suggested that Jurassic biogeographic differentiation was more related to sedimentary facies and salinity than to temperature and that the Boreal Realm is a low-salinity, rather than low-temperature, unit. He supported this suggestion by pointing out that Boreal Realm faunas are characterized by low diversity and that hypersaline communities of the Purbeck are Boreal in character, showing that the organisms were euryhaline. Organisms with broad salinity tolerances might also have broad temperature tolerances, and so the low diversity and euryhaline character do not rule out temperature as a controlling factor for the Boreal Realm. Although few workers seem to have accepted the low-salinity explanation of the Boreal Realm, and Hallam himself seems to have discarded the idea (e.g., Hallam 1975, 1983), it nevertheless may have merit for explaining the sharp biogeographic boundary in Europe (Hallam 1971), and thus would be consistent with an influx of fresh water from the coal basins to the east.

Thus, high-latitude waters might have been cooler, on the evidence of the glendonites (Brandt 1986), and fresher, fresh enough to halt the generation of cold bottom water (Brass et al. 1982a; Broecker et al. 1988) and possibly enough to affect the fauna. Increased productivity in Tethys after the late Middle Jurassic, then, might have resulted from establishment of bottom-water generation at high latitudes, after the monsoonal circulation and resulting poleward transport of moisture ceased; opening of the connection between western Tethys and the Pacific and establishment of equatorial throughflow might have had only an indirect effect. It should be noted, however, that changes in productivity would not explain the loss of endemism on the west coast of Pangea (Westermann 1984). Silica production was continuous there (Hein and Parrish 1987), as well as in Japan (Iijima and Utada 1983), throughout the Jurassic, and it increased only slightly compared with Tethys (Jenkyns and Winterer 1982), so the biogeographic changes still must be attributed to the paleogeographic changes.

## Acknowledgments

The author gratefully acknowledges discussions with L. Ely and E. Wohl on recent research on the El Niño/Southern Oscillation events and suggestions by T. M. Quinn for improvement of the manuscript. This work was partially supported by National Science Foundation grant EAR-8903549.

## References

Addicott, W. O. (1970). Latitudinal gradients in Tertiary molluscan faunas of the Pacific coast. *Palaeogeogr. Palaeoclimatol. Palaeoecol., 8,* 287–312.

Alvin, K. L., Fraser, C. J., & Spicer, R. A. (1981). Anatomy and palaeoecology of *Pseudofrenelopsis* and associated conifers in the English Wealden. *Palaeontology, 24,* 759–78.

Alvin, K. L., Spicer, R. A., & Watson, J. (1978). A *Classopollis*-containing male cone associated with *Pseudofrenelopsis. Palaeontology, 21,* 847–56.

Anderson, T. F., & Arthur, M. A. (1983). Stable isotopes of oxygen and carbon and their application to sedimentologic and paleoenvironmental problems. In M. A. Arthur, (ed.), *Stable Isotopes in Sedimentary Geology.* Soc. Econ. Paleont. Mineral. Short Course 10, 1-1-151. Tulsa: SEPM.

Beauvais, L. (1973). Upper Jurassic hermatypic corals. In A. Hallam (ed.), *Atlas of Palaeobiogeography* (pp. 317–28). Amsterdam: Elsevier.

Berggren, W. A., & Hollister, C. D. (1977). Plate tectonics and paleocirculation – commotion in the ocean. *Tectonophysics, 38,* 11–48.

Brandt, K. (1986). Glacieustatic cycles in the Early Jurassic? *N. Jb. Geol. Paläontol. Mh., 5,* 257–74.

Brass, G. W., Saltzman, E., Sloan, J. L., II, Southam, J. R., Hay, W. W., Holser, W. T., & Peterson, W. H. (1982a). Ocean circulation, plate tectonics, and climate. In *Climate in Earth History* (pp. 83–9). U.S. National Research Council, Geophysics Study Committee.

Brass, G. W., Southam, J. R., & Peterson, W. H. (1982b). Warm saline bottom water in the ancient ocean. *Nature, 296,* 620–3.

Brewster, N. A. (1980). Cenozoic biogenic silica sedimentation in the Antarctic Ocean. *Geol. Soc. Am. Bull. 91,* 337–47.

Broecker, W. S., Andree, M., Wolfli, W., Oeschger, H., Bonani, G., Kennett, J., & Peteet, D. (1988). The chronology of the last deglaciation: implications to the cause of the Younger Dryas event. *Paleoceanography, 3,* 1–19.

Chumakov, N. M. (1981). Scattered stones in Mesozoic deposits of North Siberia, U.S.S.R. In M. J. Hambrey & W. B. Harland (eds.), *Earth's Pre-Pleistocene Glacial Record* (p. 264). Cambridge University Press.

CLIMAP (1976). The surface of the ice-age earth. *Science, 191,* 1131–44.

Crowley, T. J., Hyde, W. T., & Short, D. A. (1989). Seasonal cycle variations on the supercontinent of Pangaea: implications for Early Permian vertebrate extinctions. *Geology, 17,* 457–60.

Das, P. K. (1986). *Monsoons.* Fifth IMO Lecture No. 613, World Meteorological Organization.

DeMaster, D. J. (1981). The supply and accumulation of silica in the marine environments. *Geochim. Cosmochim. Acta, 45,* 1715–32.

Dickinson, W. R., & Coney, P. J. (1980). Plate tectonic constraints on the origin of the Gulf of Mexico. In R. H. Pilger, Jr. (ed.), *The Origin of the Gulf of Mexico and the Early Opening of the Central North Atlantic Ocean* (pp. 27–36). Baton Rouge: Louisiana State University.

Doyle, J. A., & Parrish, J. T. (1984). Jurassic–Early Cretaceous plant distributions and paleoclimatic models (abstract). *Abstracts, Int. Org. Paleobot. Conf.* London.

Epshteyn, O. G. (1978). Mesozoic-Cenozoic climates of northern Asia and glacial-marine deposits. *Int. Geol. Rev., 20,* 49–58.

Espenshade, E. E. B., & Morrison, J. L. (eds.). (1978). *Goode's World Atlas,* 15th ed. Chicago: Rand McNally.

Fein, J. S., & Stephens, P. L. (eds.). (1987). *Monsoons.* New York: Wiley.

Flohn, H. (1968). *Contributions to a Meteorology of the Tibetan Highlands.* Atmos. Sci. Paper 130, Dept. of Atmospheric Science, Colorado State University, Fort Collins.

Frakes, L. A. (1979). *Climates Throughout Geologic Time.* New York: Elsevier.

Frakes, L. A., & Francis, J. E. (1988). A guide to Phanerozoic cold polar climates from high-latitude ice-rafting in the Cretaceous. *Nature, 333,* 547–9.

Godfrey, J. S., & Golding, T. J. (1981). The Sverdrup relation in the Indian Ocean, and the effect of Pacific–Indian Ocean throughflow on Indian Ocean circulation and on the East Australian Current. *J. Phys. Oceanogr. 11,* 771–9.

Grunau, H. R. (1965). Radiolarian cherts and associated rocks in space and time. *Eclogae Geol. Helvet., 58,* 157–208.

Hahn, D. G., & Manabe, S. (1975). The role of mountains in the South Asian monsoon circulation. *J. Atmos. Sci.*, *32*, 1515–41.

Hallam, A. (1971). Provinciality in Jurassic faunas in relation to facies and paleogeography. Faunal provinces in space and time. *Geol. J. Spec. Iss.*, *4*, 129–52.

(1975). *Jurassic Environments.* Cambridge University Press.

(1978). Eustatic cycles in the Jurassic. *Palaeogeogr. Palaeoclimatol. Palaeoecol.*, *23*, 1–32.

(1981). A revised sea-level curve for the early Jurassic. *J. Geol. Soc. London*, *138*, 735–43.

(1983). Early and mid-Jurassic molluscan biogeography and the establishment of the central Atlantic seaway. *Palaeogeogr. Palaeoclimatol. Palaeoecol.*, *43*, 181–93.

(1984). Continental humid and arid zones during the Jurassic and Cretaceous. *Palaeogeogr. Palaeoclimatol. Palaeoecol.*, *47*, 195–223.

Hambrey, M. J., & Harland, W. B. (1981). Criteria for the identification of glacigenic deposits. In M. J. Hambrey & W. B. Harland (eds.), *Earth's Pre-Pleistocene Glacial Record* (pp. 14–17). Cambridge University Press.

Hein, J. R., & Parrish, J. T. (1987). Distribution of siliceous deposits in space and time. In J. R. Hein (ed.), *Siliceous Sedimentary Rock-hosted Ores and Petroleum.* (pp. 10–57). New York: Van Nostrand Reinhold.

Iijima, A., & Utada, M. (1983). Recent developments in sedimentology of siliceous deposits in Japan. In A. Iijima, J. R. Hein, & Siever (eds.), *Siliceous Deposits in the Pacific Region. Devel. Sedimentol.*, *36*, 45–64. Amsterdam: Elsevier.

Ingle, J. C. (1981). Origin of Neogene diatomites around the north Pacific rim. In R. E. Garrison & R. G. Douglas (eds.), *The Monterey Formation and Related Siliceous Rocks of California* (pp. 159–79). Los Angeles: Pacific Section, SEPM.

Jenkyns, H. C., & Winterer, E. L. (1982). Paleoceanography of Mesozoic ribbon radiolarites. *Earth Planet. Sci. Lett.*, *60*, 351–75.

Kaneps, A. G. (1979). Gulf Stream: velocity fluctuations during the Late Cenozoic. *Science*, *204*, 297–301.

Kaplan, M. E. (1978). Kal'citovye psevdomorfozy v jurskich; nizneme-ovych otlozenijach severa vostocnoj Sibiri. *Geolog. Geofiz.*, *12*, 62–70.

Keigwin, L. (1982). Isotopic paleoceanography of the Caribbean and East Pacific: role of Panama uplift in late Neogene time. *Science*, *217*, 350–3.

Keller, G., & Barron, J. A. (1981). Integrated planktic foraminiferal and diatom biochronology for the Northeast Pacific and Monterey Formation. In R. E. Garrison & R. G. Douglas (eds.), *The Monterey Formation and Related Siliceous Rocks of California* (pp. 43–54). Los Angeles: Pacific Section, SEPM.

Kemper, E. (1987). Das Klima der Kreide-Zeit. *Geol. Jahr.*, *A96*, 5–185.

Kennedy, G. L., Hopkins, D. M., & Pickthorn, W. J. (1987). Ikaite, the glendonite precursor, in estuarine sediments at Barrow, Arctic Alaska (abstract). *Geol. Soc. Am. Abstr. Progr.*, *19*, 725.

Kennett, J. P., Keller, G., & Srinivasan, M. S. (1985). Miocene planktonic foraminiferal biogeography and paleoceanographic development of the Indo-Pacific region. In J. P. Kennett (ed.), *The Miocene Ocean: Paleoceanography and Biogeography* (pp. 197–236). Geol. Soc. Am. Mem. 163.

Kimura, T., & Tsujii, M. (1984). Late Jurassic (Oxfordian) flora of Northeast Japan (abstract). *Abstracts, Int. Org. Paleobot. Conf.*, London.

Leinen, M. (1979). Biogenic silica accumulation in the central equatorial Pacific and its implications for Cenozoic paleoceanography: summary. *Geol. Soc. Am. Bull.*, *90*, 801–3.

Lisitsyn, A. P. (1972). *Sedimentation in the World Ocean.* Soc. Econ. Paleontol. Mineral. Spec. Pub. 17.

(1977). Biogenic sedimentation in the oceans and zonation. *Lith. Min. Res.*, *1*, 3–24.

McCabe, P. J. (1984). Depositional environments of coal and coal-bearing strata. *Spec. Publ. Inte. Ass. Sedimentol.*, *7*, 13–42.

Manabe, S., & Wetherald, R. T. (1980). On the distribution of climate change resulting from an increase in $CO_2$ content of the atmosphere. *J. Atmos. Sci.*, *37*, 99–118.

Miskell, K. J. (1983). *Accumulation of Opal in Deep Sea Sediments from the Mid-Cretaceous to the Miocene: A Paleocirculation Indicator.* M. S. Thesis, University of Miami.

Miskell, K. J., Brass, G. W., & Harrison, C. G. A. (1985). Global patterns in opal deposition from Late Cretaceous to Late Miocene. *Am. Assoc. Petrol. Geol. Bull.*, *69*, 996–1012.

Murakami, T. (1987). Orography and monsoons. In J. S. Fein & P. L. Stephens (eds.), *Monsoons* (pp. 331–64). New York: Wiley.

Murray, S. P., & Arief, D. (1988). Throughflow into the Indian Ocean through Lombok Strait, January 1985–January 1986. *Nature*, *333*, 444–7.

Parrish, J. M., Parrish, J. T., & Ziegler, A. M. (1986). Permian-Triassic paleogeography and paleoclimatology and implications for therapsid distributions. In N. H. Hotton III, P. D. MacLean, J. J. Roth, & E. C. Roth (eds.), *The Ecology and Biology of Mammal-like Reptiles* (pp. 109–32). Washington, D.C.: Smithsonian Press.

Parrish, J. T. (1982). Upwelling and petroleum source beds, with reference to the Paleozoic. *Am. Assoc. Petrol. Geol. Bull.*, *66*, 750–74.

(1985). *Latitudinal Distribution of Land and Shelf and Absorbed Solar Radiation during the Phanerozoic.* U.S. Geological Survey Open-File Rep. 85–31.

(1987). Global paleogeography and paleoclimate of the Late Cretaceous and Early Tertiary. In E. M. Friis, W. G. Chaloner, & P. R. Crane (eds.), *The Origins of Angiosperms and Their Biological Consequences* (pp. 51–73). Cambridge University Press.

(1988). Pangaean paleoclimates (abstract). *Eos*, *69*, 1060.

Parrish, J. T., & Curtis, R. L. (1982). Atmospheric circulation, upwelling, and organic-rich rocks in the Mesozoic and Cenozoic Eras. *Palaeogeogr. Palaeoclimatol. Palaeoecol.*, *40*, 31–66.

Parrish, J. T., & Doyle, J. A. (1984). Predicted evolution of global climate in Late Jurassic–Cretaceous time (abstract). *Abstracts, Int. Org. Paleobot. Conf.*, London.

Parrish, J. T., Hansen, K. S., & Ziegler, A. M. (1979). Atmospheric circulation and upwelling in the Paleozoic, with reference to petroleum source beds (abstract). *Am. Assoc. Petrol. Geol. Bull.*, *63*, 507–8.

Parrish, J. T., & McCabe, P. J. (1988). Controls on the distribution of Cretaceous coals – an overview (abstract). *Geol. Soc. Am. Abstr. Progr.*, *20*, A27.

Parrish, J. T., & Peterson, F. (1988). Wind directions predicted from global circulation models and wind directions determined from eolian sandstones of the western United States – a comparison. *Sed. Geol.*, *56*, 261–82.

Parrish, J. T., & Spicer, R. A. (1988a). Late Cretaceous terrestrial vegetation: a near-polar temperature curve. *Geology*, *16*, 22–5.

(1988b). Middle Cretaceous wood from the Nanushuk Group, central North Slope, Alaska. *Palaeontology*, *31*, 19–34.

Parrish, J. T., & Ziegler, A. M. (1980). Climate asymmetry and biogeographic distributions (abstract). *Am. Assoc. Petrol. Geol. Bull.*, *64*, 763.

Parrish, J. T., Ziegler, A. M., & Scotese, C. R. (1982). Rainfall patterns and the distribution of coals and evaporites in the Mesozoic and Cenozoic. *Palaeogeogr. Palaeoclimatol. Palaeoecol.*, *40*, 67–101.

Philander, S. G. H., & Pacanowski, R. C. (1980). The generation of equatorial currents. *J. Geophys. Res.*, *85*, 1123–36.

(1981). The oceanic response to cross-equatorial winds (with application to coastal upwelling in low latitudes). *Tellus*, *33*, 201–10.

Pindell, J., & Dewey, J. F. (1982). Permo-Triassic reconstruction of western Pangaea and the evolution of the Gulf of Mexico/Caribbean region. *Tectonics*, *1*, 179–211.

Pisias, N. G., & Prell, W. L. (1985). Changes in calcium carbonate accumulation in the equatorial Pacific during the Late Cenozoic: evidence from HPC Site 572. In E. T. Sundquist, & W. S. Broecker (eds.), *The Carbon Cycle and Atmospheric $CO_2$: Natural Variations Archean to Present* (pp. 443–54). Geophys. Monogr. 32, Amer. Geophysics Union, Washington, D.C.

Pisias, N. G., & Rea, D. K. (1988). Late Pleistocene paleoclimatology of the central equatorial Pacific: sea surface response to the Southeast Trade Winds. *Paleoceanography, 3,* 21–37.

Prell, W. L. (1985). Pliocene stable isotope and carbonate stratigraphy (holes 572C and 573A): paleoceanographic data bearing on the question of Pliocene glaciation. In L. Mayer, F. Theyer, et al. (eds.), *Init. Rep., Deep-Sea Drilling Project, 85,* pp. 723–34.

Ramsay, A. T. S. (1973). A history of organic siliceous sediment in oceans. In N. F. Hughes (ed.), *Organisms and Continents through Time* (pp. 199–234). Spec. Pap. Paleontol.

Rasmusson, E. M. (1985). El Niño and variations in climate. *Amer. Sci., 73,* 168–77.

Reyment, R. A., & Tait, E. A. (1972). Faunal evidence for the origin of the South Atlantic. *24th Int. Geol. Congr., Section 7,* 316–23.

Rocha-Campos, A. C. (1971). Upper Paleozoic and Lower Mesozoic paleogeography, and paleoclimatological and tectonic events in South America. In A. Logan & L. V. Hills (eds.), *The Permian and Triassic Systems and Their Mutual Boundary* (pp. 398–424). Can. Soc. Petrol. Geol. Mem. 2.

Ruddiman, W. F., & McIntyre, A. (1981). Oceanic mechanisms for amplification of the 23,000-year ice-volume cycle. *Science, 212,* 617–27.

Ruddiman, W. F., & Raymo, M. E. (1988). Northern Hemisphere climate régimes during the past 3 Ma: possible tectonic connections. *Philos. Trans. R. Soc. Lond., B318,* 411–30.

Sarmiento, J. L., Herbert, T. D., & Toggweiler, J. R. (1988). Causes of anoxia in the World Ocean. *Global Biogeochem. Cycles, 2,* 115–28.

Sarnthein, M., Tetzlaff, G., Koopmann, B., Wolter, K., & Pflaumann, U. (1981). Glacial and interglacial wind regimes over the eastern subtropical Atlantic and North-West Africa. *Nature, 293,* 193–6.

Sarnthein, M., Thiede, J., Pflaumann, U., Erlenkeuser, H., Fütterer, D., Koopmann, B., Lange, H., & Seibold, E. (1982). Atmospheric and oceanic circulation patterns off Northwest Africa during the past 25 million years. In U. von Rad, K. Hinz, M. Sarnthein, & E. Seibold (eds.), *Geology of the Northwest African Continental Margin* (pp. 545–604). Berlin: Springer-Verlag.

Schermerhorn, L. J. G. (1974). Late Precambrian mixtites: Glacial and/or nonglacial? *Am. J. Sci., 274,* 673–824.

Sey, I. I., & Kalacheva, E. D. (1985). Invasions of Tethyan ammonites into the Late Jurassic Boreal basins of East U.S.S.R. In G. E. G. Westermann (ed.), *Jurassic Biogeography and Stratigraphy of East U.S.S.R.* (pp. 14–17). IGCP Project 171: Circum-Pacific Jurassic, Spec. Pap. 10.

Shackleton, N. J., Hall, M. A., & Boersma, A. (1984). Oxygen and carbon isotope data from Leg 74 foraminifers. In T. C. Moore, Jr., P. D. Rabinowitz, et al. (eds.), *Init. Rep., Deep-Sea Drilling Project 74.*

Sheldon, R. P. (1981). Ancient marine phosphorites. *Ann. Rev. Earth Planet. Sci., 9,* 151–84.

Spicer, R. A., & Parrish, J. T. (1986). Paleobotanical evidence for cool North Polar climates in middle Cretaceous (Albian-Cenomanian) time. *Geology, 14,* 703–6.

Stevens, G. R. (1971). Relationship of isotopic temperatures and faunal realms to Jurassic and Cretaceous palaeogeography, particularly of the Southwest Pacific. *J. Roy. Soc. N.Z., 1,* 145–58.

(1980). Southwest Pacific faunal palaeobiogeography in Mesozoic and Cenozoic times: a review. *Palaeogeogr. Palaeoclimatol. Palaeoecol., 31,* 153–96.

Stevens, G. R., & Clayton, R. N. (1971). Oxygen isotope studies on Jurassic and Cretaceous belemnites from New Zealand and their biogeographic significance. *N.Z. J. Geol. Geophys., 14,* 829–97.

Sverdrup, H. U., Johnson, M. W., & Fleming, R. H. (1942). *The Oceans: Their Physics, Chemistry, and General Biology.* Englewood Cliffs, N.J.: Prentice-Hall.

Taylor, D. G., Callomon, J. H., Hall, R., Smith, P. L., Tipper, H. W., & Westermann, G. E. G. (1984). Jurassic ammonite biogeography of western North America: the tectonic implications. In G. E. G. Westermann (ed.), *Jurassic Cretaceous Biochronology and Paleogeography of North America* (pp. 121–41). IGCP Project 171: Geol. Assoc. Can. Spec. Pap. 27.

Toggweiler, J. R., & Sarmiento, J. L. (1985). Glacial to interglacial changes in atmospheric carbon dioxide: the critical role of ocean surface water in high latitudes. In E. T. Sundquist & W. S. Broecker (eds.), *The Carbon Cycle and Atmospheric $CO_2$: Natural Variations Archean to Present* (pp. 163–84). Geophys. Monogr. 32, Amer. Geophysics Union, Washington, D.C.

Vakhrameyev, V. A. (1982). *Classopollis* pollen as an indicator of Jurassic and Cretaceous climate. *Int. Geol. Rev., 24,* 1190–6.

Vakhrameyev, V. A., & Doludenko, M. P. (1977). The Middle–Late Jurassic boundary, an important threshold in the development of climate and vegetation of the Northern Hemisphere. *Int. Geol. Rev., 19,* 621–32.

Van Houten, F. B. (1982). Redbeds. In *McGraw-Hill Encycl. Sci. Technol., 5/e,* pp. 441–2.

Vincent, E., & Berger, W. H. (1985). Carbon dioxide and polar cooling in the Miocene: the Monterey hypothesis. In E. T. Sundquist & W. S. Broecker (eds.), *The Carbon Cycle and Atmospheric $CO_2$: Natural Variations Archean to Present* (pp. 455–68). Geophys. Monogr. 32, Amer. Geophysics Union, Washington, D.C.

Volk, T., & Liu, Z. (1988). Controls of $CO_2$ sources and sinks in the Earth scale surface ocean: temperature and nutrients. *Global Biogeochem. Cycles, 2,* 73–89.

Volkheimer, W. (1967). La paleoclimatología y los climas del Mesozoico argentino. *Rev. Minera. Soc. Argent. Min. Geol., 28*(3), 41–8.

Webster, P. J. (1987). The elementary monsoon. In J. S. Fein & P. L. Stephens (eds.), *Monsoons* (pp. 3–32). New York: Wiley.

Wells, L. E. (1987). An alluvial record of El Niño events from northern coastal Peru. *J. Geophys. Res., 92,* 14463–70.

Wenk, T., & Siegenthaler, U. (1985). The high-latitude ocean as a control of atmospheric $CO_2$. In E. T. Sundquist & W. S. Broecker (eds.), *The Carbon Cycle and Atmospheric $CO_2$: Natural Variations Archean to Present* (pp. 185–94). Geophys. Monogr. 32, Amer. Geophysics Union, Washington, D.C.

Westermann, G. E. G. (1981). Ammonite biochronology and biogeography of the circum-Pacific Middle Jurassic. In M. R. House & J. R. Senior (eds.), *The Ammonoidea.* Syst. Assoc. Spec. Vol. 18 (pp. 459–98). London: Academic Press.

(1984). Summary of symposium papers on the Jurassic-Cretaceous biochronology and paleogeography of North America. In G. E. G. Westermann (ed.), *Jurassic-Cretaceous Biochronology and Paleogeography of North America* (pp. 307–15). Geol. Assoc. Can. Spec. Pap. 27.

(1985). Comments on the biogeographic papers: a plate-tectonic alternative. In G. E. G. Westermann (ed.), *Jurassic Biogeography and Stratigraphy of East U.S.S.R.* (pp. 28–9). IGCP Project 171: Circum-Pacific Jurassic, Spec. Pap. 10.

Williams, G. E. (1975). Late Precambrian glacial climate and the earth's obliquity. *Geol. Mag., 112,* 441–544.

Wyrtki, K. (1975). El Niño – the dynamic response of the equatorial Pacific Ocean to atmospheric forcing. *J. Phys. Oceanogr., 5,* 572–84.

Yasamanov, N. A. (1981). Paleothermometry of Jurassic, Cretaceous, and Paleogene periods of some regions of the USSR. *Int. Geol. Rev., 23,* 700–5.

Young, J. A. (1987). Physics of monsoons: the current view. In J. S. Fein & P. L. Stephens (eds.), *Monsoons* (pp. 211–43). New York: Wiley.
Yurtesever, Y. (1975). *Worldwide Survey of Stable Isotopes in Precipitation*. Rept. Sect. Isotope Hydrol., IAEA, November.
Ziegler, A. M., Bambach, R. K., Parrish, J. T., Barrett, S. F., Gierlowski, E. H., Parker, W. C., Raymond, A., & Sepkoski, J. J., Jr. (1981). Paleozoic biogeography and climatology. In K. J. Niklas (ed.), *Paleobotany, Paleoecology, and Evolution* (vol. 2, pp. 231–66). New York: Praeger.
Ziegler, A. M., Raymond, A. L., Gierlowski, T. C., Horrell, M. A., Rowley, D. B., & Lottes, A. L. (1987). Coal, climate and terrestrial productivity: the Present and early Cretaceous compared. In A. C. Scott (ed.), *Coal and Coal-bearing Strata: Recent Advances* (pp. 25–49). Geol. Soc. Lond. Spec. Publ. 32.
Ziegler, A. M., Scotese, C. R., & Barrett, S. F. (1983). Mesozoic and Cenozoic paleogeographic maps. In P. Brosche & J. Sündermann (eds.), *Tidal Friction and the Earth's Rotation* (vol. 2, pp. 240–52). Berlin: Springer-Verlag.

# Appendix: Biochronology and atlas with index and guide fossils

**Contents**



**Repository abbreviations**

| | |
|---|---|
| BAS | British Antarctic Survey, Cambridge, England |
| BM (BWSM) | Burke Memorial Washington State Museum, Seattle |
| C | Geology Department, University of Auckland, New Zealand |
| CAS | California Academy of Sciences, San Francisco |
| CNIGR | Central Scientific Research Geological Museum, St. Petersburg, Russia |
| CPBA | Faculdad de Ciencias Exactas y Naturales, Buenos Aires, Argentina |
| CPC | Paleontology Collections, Bureau of Mineral Resources, Canberra, Australia |
| DNGM | Servicio Geológico Nacional, Buenos Aires, Argentina |
| GLD | Geological Laboratory, Delft, The Netherlands |
| GSC | Geological Survey of Canada, Ottawa, Ontario |
| IGG | Institut for Geology and Geophysics, Akademgorokok Univ. Ave. 3, 630090 Novosibirsk, Russia. |
| IGM | Paleontological Museum of the Institute of Geology, Autonomous University of Mexico, Mexico, D.F. |
| IIG (SNGM) | Instituto de Investigaciones Geológicas de Chile, Santiago |
| IMC | Indonesian Geological Survey, Bandung |
| McM | Department of Geology, McMaster University, Hamilton, Ontario, Canada |
| MGL | Museum of Geology, Lausanne, Switzerland |
| MLP | Museo Ciencias Naturales, La Plata, Argentina |
| NHMB | Natural History Museum, Basel, Switzerland |
| NWMNH | Northwest Museum of Natural History, Portland, Oregon |
| RGML | Rijksmuseum van Geologie en Mineralogie, Leiden, The Netherlands. |
| SGO-PI | Museo de Historia Natural de Santiago, Chile |
| TMP | Department of Geology and Geophysics, University of Calgary, Calgary, Alberta T2N 1N4, Canada |
| TUB | Institut für Geologie und Paläontologie Technische Universität, Berlin, Germany |
| TsGM | Central Geological Museum, St. Petersburg, Russia |
| UBC | University of British Columbia, Vancouver |
| UCS | University College of Swansea, Wales, Great Britain |
| USNM | United States National Museum, Washington, D.C. |
| UU | Department of Paleontology, University of Utrecht, Utrecht, The Netherlands. |
| UW | University of Washington, Seattle |

PLATE 1

NORTH AMERICA, HETTANGIAN

**Fig. 1.** *Psiloceras* cf. *planorbis* (Sowerby) (USNM 247950); Lower Hettangian, Alaska (Imlay 1981b).
**Fig. 2.** *Franziceras* cf. *ruidum* Buckman (USNM 247953); Middle Hettangian, Alaska (Imlay 1981b).
**Fig. 3.** *Waehneroceras* cf. *portlocki* (Wright) (USNM 247969); Middle Hettangian, Alaska (Imlay 1981b).
**Fig. 4.** *Schlotheimia* sp. (GSC 95560); Middle Hettangian, British Columbia (Tipper et al. 1990).
**Fig. 5.** *Franziceras* sp. (GSC 95562); Middle Hettangian, British Columbia (Tipper et al. 1990).
**Fig. 6.** *Nevadaphyllites compressus* Guex, holotype (MGL 46011); Lower Hettangian, Nevada (Guex 1980).
**Fig. 7.** *Discamphiceras silberlingi* Guex, holotype (MGL 46004); Lower Hettangian, Nevada (Guex, 1980).
**Fig. 8.** *Mullerites pleuroacanthites* Guex, holotype (MGL 46010); Middle Hettangian, Nevada (Guex 1980).
**Fig. 9.** *Sunrisites sunrisense* Guex, holotype (MGL 46015); Middle and Upper Hettangian, Nevada (Guex 1980).
**Fig. 10.** *Fergusonites striatus* Guex (GSC 95564); Lower Hettangian, British Columbia (Tipper et al. 1990).
(Compiled by Smith.)

1
a
2
b
3
4
5
6
7
8
9
10

PLATE 2

NORTH AMERICA, SINEMURIAN

**Fig. 1.** *Badouxia canadensis* (Frebold) (USNM 247975); Canadensis Zone, Alaska (Imlay 1981b).
**Fig. 2.** *Crucilobiceras* cf. *crucilobatum* Buckman (USNM 248004); Upper Sinemurian, Alaska (Imlay 1981b).
**Fig. 3.** *Arnioceras arnouldi* (Dumortier) (UBC 08); Lower Sinemurian, British Columbia (Tipper et al. 1990).
**Fig. 4.** *Metophioceras(?) rursicostatum* (Frebold) (USNM 247991); Canadensis Zone, Alaska (Imlay 1981b).
**Fig. 5.** *Paltechioceras harbledownense* (Crickmay) (GSC 95571); Upper Sinemurian, British Columbia (Tipper et al. 1990).
**Fig. 6.** *Arctoasteroceras jeletzkyi* Frebold (USNM 247999); Lower Sinemurian, Alaska (Imlay 1981b).
(Compiled by Smith.)

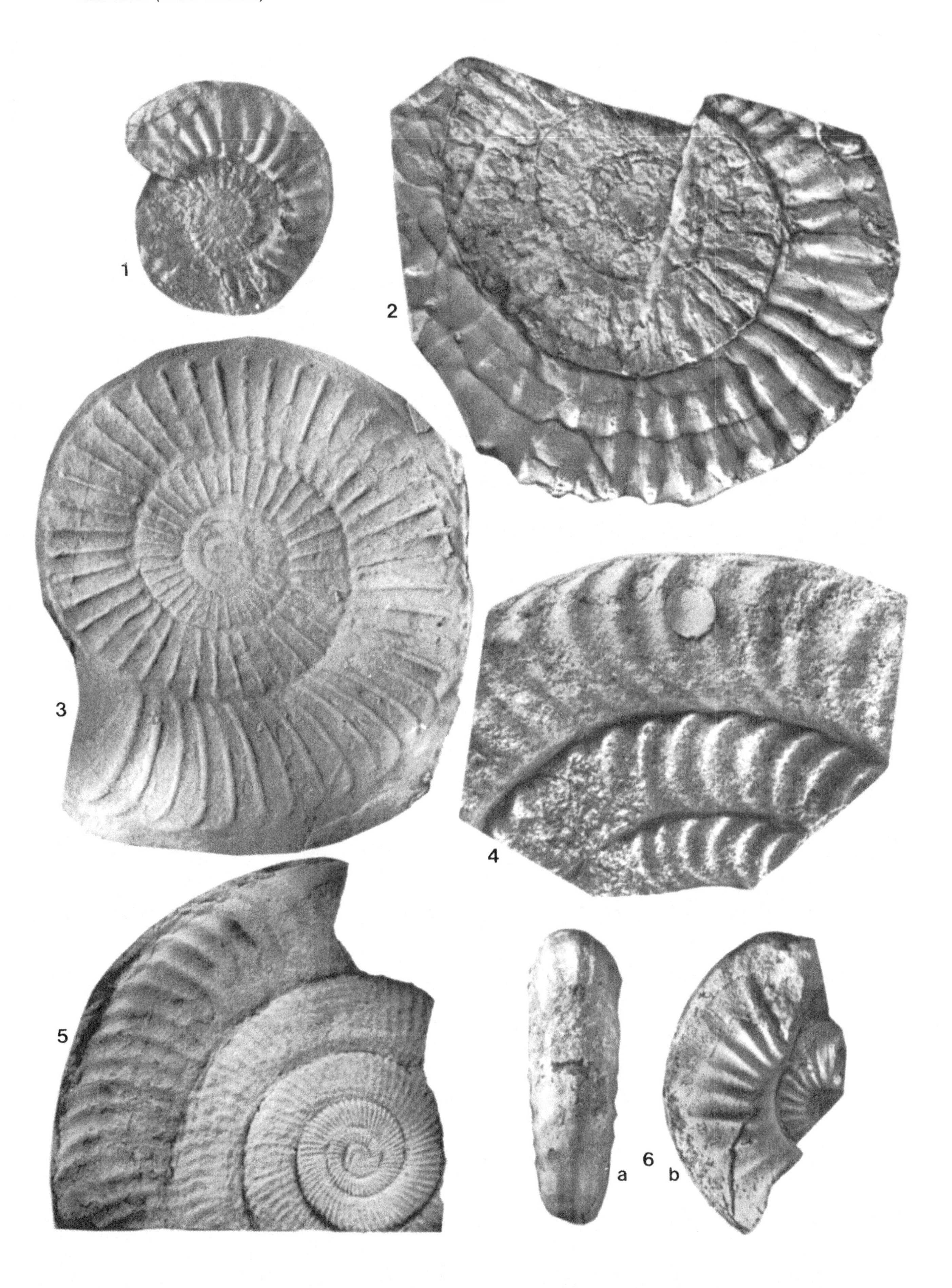
1
2
3
4
5
6
a
b

PLATE 3

NORTH AMERICA, PLIENSBACHIAN

**Fig. 1.** *Pseudoskirroceras imlayi* Smith & Tipper, holotype (GSC 87783); Imlayi Zone, British Columbia (Smith et al. 1988).
**Fig. 2.** *Acanthopleuroceras whiteavesi* Smith & Tipper, holotype (GSC 87790); Whiteavesi Zone, British Columbia (Smith et al. 1988).
**Fig. 3.** *Dubariceras freboldi* Dommergues, Mouterde, & Rivas (USNM 248015); Freboldi Zone, Alaska (Imlay 1981b).
**Fig. 4.** *Uptonia* sp. (×2) (USNM 248019); Lower Pliensbachian, Alaska (Imlay 1981b).
**Fig. 5.** *Tropidoceras actaeon* (d'Orbigny) (USNM 248020); Imlayi and Whiteavesi Zones, Alaska (Imlay 1981b).
**Fig. 6.** *Leptaleoceras* cf. *pseudoradians* (Reynes) (USNM 248040); Kunae Zone, Alaska (Imlay 1981b).
**Fig. 7.** *Aveyroniceras colubriforme* (Bettoni) (UBC 02); Freboldi to Kunae Zones, Nevada (Smith & Tipper 1986).
**Fig. 8.** *Reynesoceras ragazzonii* (Hauer) (GSC 87800); Kunae Zone, British Columbia (Smith et al. 1988).
**Fig. 9.** *Fanninoceras carlottense* McLearn, holotype (GSC 4878); Carlottense Zone, British Columbia (McLearn 1932).
**Fig. 10.** *Metaderoceras mouterdei* (Frebold) (GSC 87797); Whiteavesi and Kunae Zones, British Columbia (Smith et al. 1988).
**Fig. 11.** *Amaltheus stokesi* (Sowerby) (GSC 87803); Kunae Zone, British Columbia (Smith et al. 1988).
**Fig. 12.** *Fanninoceras kunae* McLearn, holotype (GSC 4876c); Kunae Zone, British Columbia (McLearn 1932).
(Compiled by Smith.)

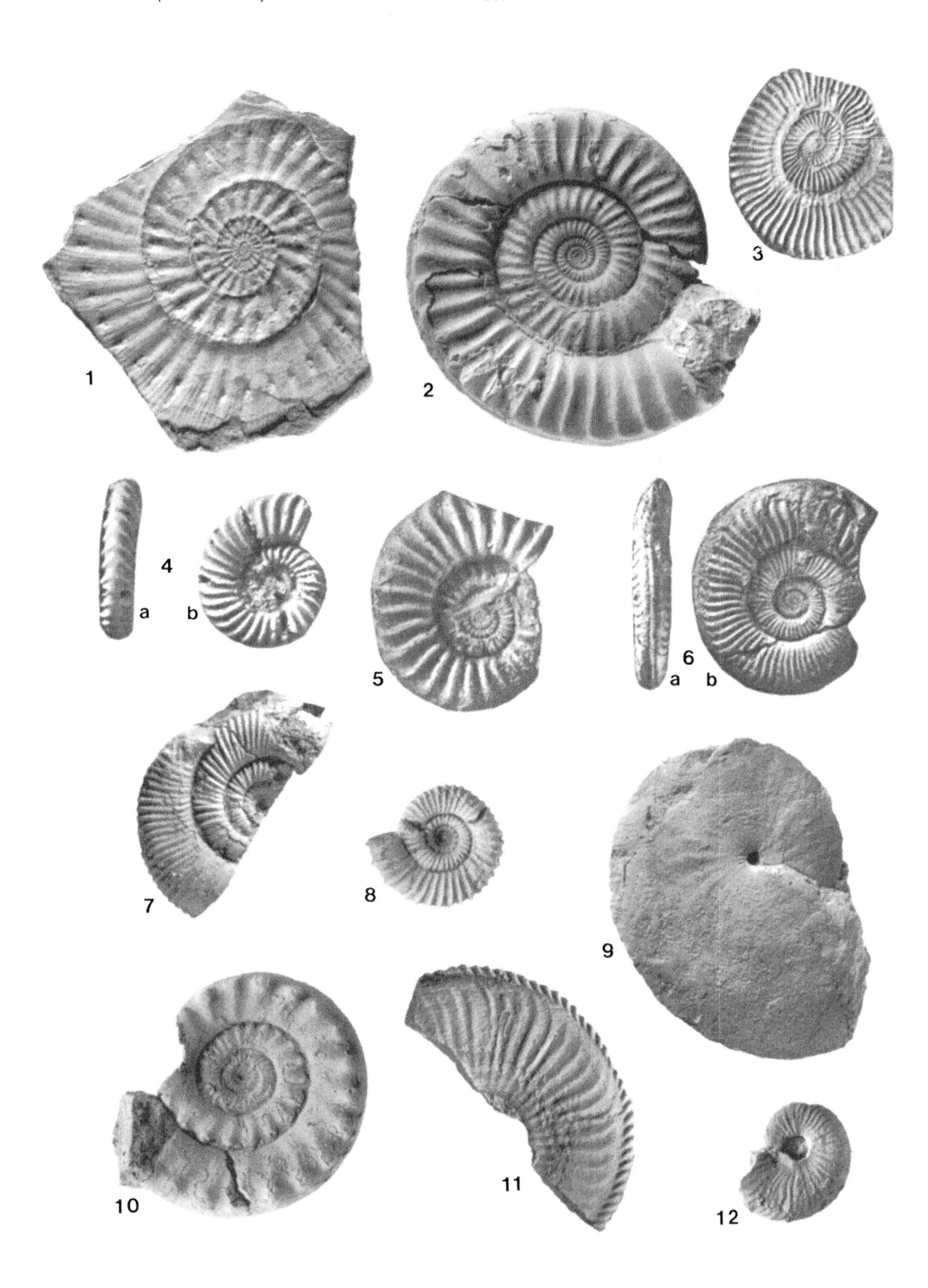
1
2
3
4
a
b
5
6
a
b
7
8
9
10
11
12

PLATE 4

NORTH AMERICA, TOARCIAN

**Fig. 1.** *Hammatoceras speciosum* (Janensch) (GSC 95583); Upper Toarcian, British Columbia (Tipper et al. 1990).
**Fig. 2.** *Sphaerocoeloceras brochiiforme* (Jaworski) (GSC 95584); Upper Toarcian, British Columbia (Tipper et al. 1990).
**Fig. 3.** *Dactylioceras kanense* McLearn, holotype (GSC 9051); Lower Toarcian, British Columbia (McLearn 1932).
**Fig. 4.** *Tiltoniceras propinquum* (Whiteaves), holotype (GSC 4877); Carlottense Zone (Pliensbachian) to Lower Toarcian, British Columbia (McLearn 1932).
**Fig. 5.** *Hildaites* cf. *chrysanthemum* (Yokoyama) (GSC 95581); Lower Toarcian, British Columbia (Tipper et al. 1990).
**Fig. 6.** *Dactylioceras* cf. *commune* (Sowerby) (USNM 248047); Lower Toarcian, Alaska (Imlay 1981b).
**Fig. 7.** *Dumortieria(?) pusilla* (Jaworski) (USNM 153888); Upper Toarcian, Oregon (Imlay 1968).
**Fig. 8.** *Haugia* cf. *compressa* Buckman (USNM 248060); Toarcian, Alaska (Imlay 1981b).
(Compiled by Smith.)

1
2
3
4
5
6
7
8

# PLATE 5

## NORTHWESTERN CANADA

### HETTANGIAN

**Fig. 1.** *Psiloceras*(?) sp., whorl fragment (GSC 92480); from lower Planorbis Zone near Bonnet Lake, northern Yukon Territory.

**Fig. 2.** *Psiloceras* (*Caloceras*) cf. *johnstoni* (Sowerby), whorl fragment (GSC 45691); from upper Planorbis Zone near Bonnet Lake, northern Yukon Territory.

### SINEMURIAN

**Fig. 3a,b.** *Arnioceras* cf. *douvillei* (Bayle), inner whorls (GSC 92486); from Semicostatum Zone, northern Yukon Territory.

**Fig. 4a,b.** *Gleviceras* sp., inner whorls (GSC 92513); from Oxynotum Zone, northern Richardson Mountains, Northwest Territories.

**Fig. 5a,b.** *Aegasteroceras* (*Arctoasteroceras?*) sp., whorl fragment (GSC 92496); from Oxynotum Zone, northern Richardson Mountains, Northwest Territories.

**Fig. 6a,b.** *Aegasteroceras* sp., whorl fragment (GSC 92501); from Oxynotum Zone, northern Richardson Mountains, Northwest Territories.

**Fig. 7.** *Echioceras aklavikense* Frebold, nearly complete specimen (GSC 83525); from Raricostatum Zone, northern Richardson Mountains, Northwest Territories.

### PLIENSBACHIAN

**Fig. 8.** *Amaltheus bifurcus* Howarth, whorl fragment (GSC 92625); from Margaritatus Zone, Loney Creek, northern Yukon Territory.

**Fig. 9.** *Amaltheus stokesi* (Sowerby), nearly complete specimen (GSC 92617); from Margaritatus Zone, Loney Creek, northern Yukon Territory.

### TOARCIAN

**Fig. 10a,b.** *Dactylioceras commune* (Simpson), whorl fragment (GSC 92714); from ?Bifrons Zone, near Bonnet Lake, northern Yukon Territory.

**Fig. 11.** *Collina*(?) aff. *simplex* (Fucini), partial specimen (GSC 92721); from ?Tenuicostatum Zone, Loney Creek, northern Yukon Territory.

**Fig. 12.** *Pseudolioceras lectum* (Simpson), small specimen (GSC 92711); from middle or late Toarcian, Johnson Creek, northern Yukon Territory.

**Fig. 13.** *Ovaticeras* cf. *ovatum* (Young & Bird), nearly complete specimen (GSC 92707); from ?Tenuicostatum Zone, Firth River, northern Yukon Territory.

(Compiled by Poulton.)

1
2
a 3 b
a 4 b
a 5 b
a 6 b
7
8
9
a 10
b
11
12
13

PLATE 6

WESTERN INTERIOR CANADA

HETTANGIAN

**Fig. 1.** *Psiloceras (Paraphylloceras) calliphyllum* (Neumayr) (GSC 69149); from Planorbis Zone of northeastern British Columbia.

SINEMURIAN

**Fig. 2.** *Epophioceras* cf. *breoni* (Reynes) (TMP 86.137.4); from Obtusum Zone of west-central Alberta.

**Fig. 3.** *Asteroceras* cf. *stellare* (J. Sowerby) (TMP 86.137.1); from the Obtusum Zone of west-central Alberta.

PLIENSBACHIAN

**Fig. 4.** *Amaltheus* cf. *stokesi* (J. Sowerby) (TMP 86.138.1); from the Margaritatus Zone of west-central Alberta.

**Fig. 5.** *Protogrammoceras* sp. indet. (TMP 86.142.3); from the Margaritatus Zone of west-central Alberta.

**Fig. 6.** *Peronoceras* cf. *subarmatum* Young & Bird (GSC 12878); from the Bifrons Zone of west-central Alberta.

(Compiled by Hall.)

1
2
3
4
5
6
X

PLATE 7

WESTERN INTERIOR CANADA, TOARCIAN

**Fig. 1.** *Harpoceras* cf. *falciferum* (J. Sowerby) (TMP 86.139.16); from the Falciferum Zone of west-central Alberta.
**Fig. 2.** *Polyplectus* cf. *subplanatus* (Oppel) (TMP 86.139.1); from the Falciferum Zone of west-central Alberta.
**Fig. 3.** *Dactylioceras* cf. *athleticum* (Simpson) (TMP 86.140.4); from the Falciferum Zone of west-central Alberta.
**Fig. 4.** *Hildaites* cf. *serpentiniformis* Buckman (GSC 41660); from the Falciferum Zone of west-central Alberta.
**Fig. 5.** ?*Haugia* sp. indet. (TMP 86.236.5); from the ?Variabilis Zone of southwestern Alberta.
(Compiled by Poulton & Hall.)

1
2
3
4
5

PLATE 8

PERU AND CHILE, HETTANGIAN

**Fig. 1.** *Psiloceras tilmanni* Lange, holotype; ↑ beginning of body chamber (TUB-GPiBo 3); from *tilmanni* Zone of Chilingote, N Peru.
**Fig. 2.** *Psiloceras primocostatum* v. Hillebrandt, holotype; ↑ beginning of body chamber (TUB-Hi 860320/3); from *primocostatum* Zone of S Portezuelo Minillas, N Chile.
**Fig. 3.** *Psiloceras rectocostatum* v. Hillebrandt, holotype; ↑ begininng of body chamber (TUB-Hi 790308/2); from *rectocostatum* Zone of S Portezuelo Minillas, N Chile.
**Fig. 4a–c.** *Curviceras armanense* (Repin); ?♂; ?completely septate (TUB-Hi 841005/6); from "*Curviceres*" spp. Zone of W Cerro Bayo, N Chile.
**Fig. 5.** *Curviceras armanense* (Repin); ?♀; last whorl body chamber (TUB-Hi 841005/6); from "*Curviceras*" spp. Zone of W Cerro Bayo, N. Chile.
**Fig. 6.** *Caloceras peruvianum* (Lange); ?completely septate (TUB-GPiBo 5); from Angulata Zone of Chilingote, N Peru.
**Fig. 7a,b.** *Discamphiceras reissi* (Tilmann), holotype, phragmocone (TUB-GPiBo 6); from *reissi* Zone of Suta (Utcubamba), N Peru.
**Fig. 8.** *Badouxia* cf. *canadensis* (Frebold), phragmocone with part of body chamber (TUB-Ch 14–060672); from Angulata Zone of Aguada de Varas, N Chile.
**Figs. 9 and 10a,b.** *Schlotheimia* cf. *montana* (Wähner), silicified specimens (TUB-Hi 871114/4); from Angulata Zone of Cerros de Cuevitas, N Chile.
(Compiled by von Hillebrandt.)

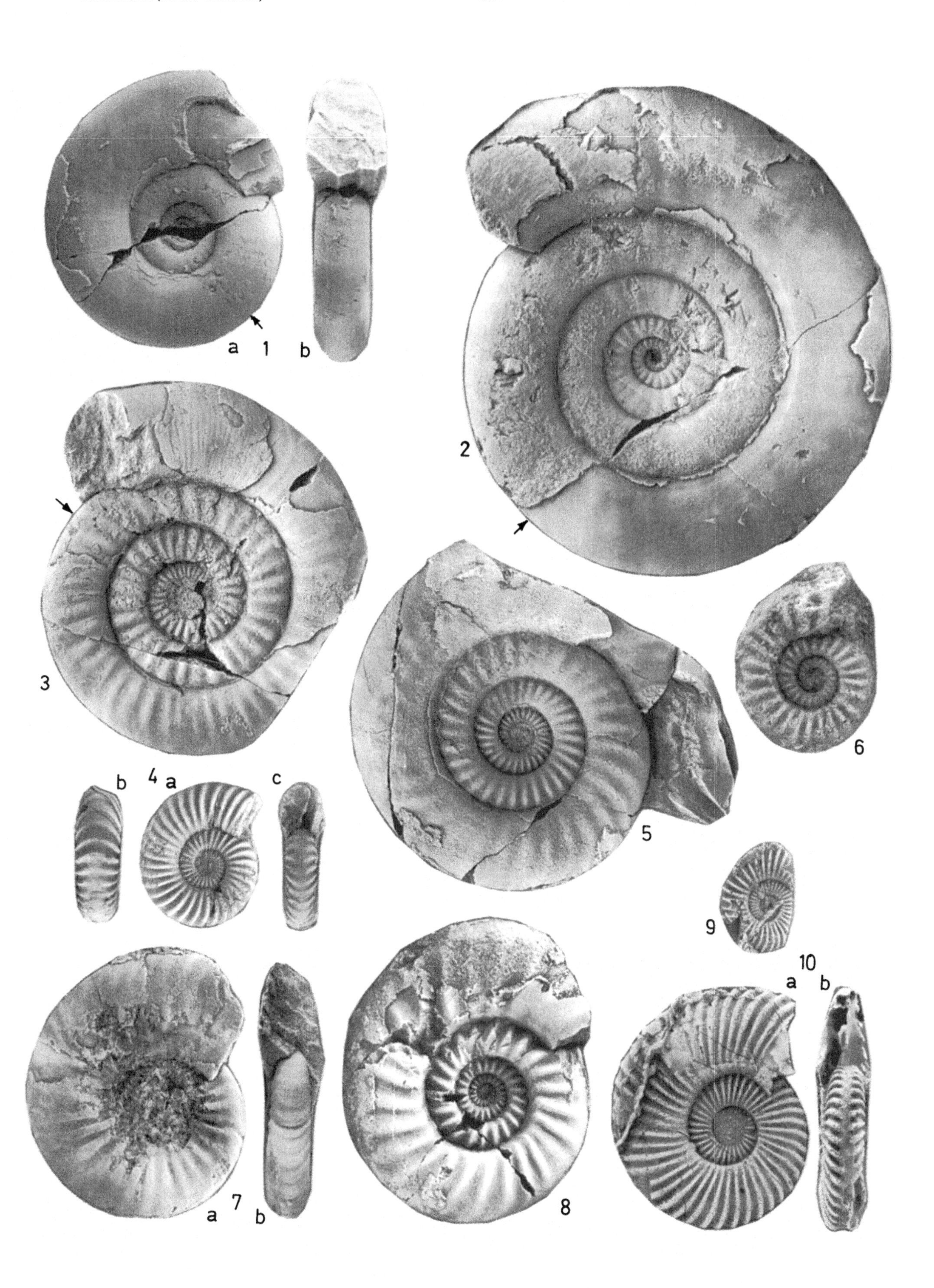
a 1 b
2
3
b 4 a c
5
6
9
10
a b
a 7 b
8

PLATE 9

NORTHERN CHILE, SINEMURIAN

**Fig. 1.** *Coroniceras* (*Metophioceras*) cf. *gracile* Spath, compressed mold (TUB-AJD-18/1); from Bucklandi Zone of Aguada Jacinto Diaz, N Chile.
**Fig. 2.** *Coroniceras* (*Coroniceras*) cf. *rotiforme* (Sowerby), latex cast of compressed outer mold (TUB-Yum-84/1); from Bucklandi Zone of Cerro Yumbes, N Chile.
**Fig. 3.** *Arnioceras semicostatum* (Young & Bird), phragmocone (Cf-72/1), from Semicostatum Zone of Posada de los Hidalgos, N Chile.
**Fig. 4.** *Arnioceras* cf. *ceratitoides* (Quenstedt), compressed outer mold (Cf-34/3); from Semicostatum Zone of Posada de los Hidalgos, N Chile.
**Fig. 5** *Epophioceras* cf. *cognitum* (Guerin-Franiatte), latex cast of outer mold (TUB-SA-33/3); from Obtusum Zone of Sierra de Argomeda/Aspera, N Chile.
**Fig. 6a,b.** *Asteroceras* cf. *stellare* (Sowerby); ↑ beginning of body chamber (TUB-AA-41/1); from Obtusum Zone of Quebrada Pan de Azucar, N Chile.
(Compiled by von Hillebrandt.)

1
2
3
4
5
6
a
b

PLATE 10

PERU AND CHILE, SINEMURIAN

**Fig. 1a,b.** *Eparietites* cf. *undaries* (Quenstedt), phragmocone (TUB-Hi 671206/4); from Obtusum Zone of Quebrada Incaguasi, N Chile.

**Figs. 2 and 3.** *Cheltonia* cf. *retentum* (Simpson), almost complete (TUB-Hi 711210/4); from Oxynotum Zone of Quebrada Yerbas Buenas, N Chile.

**Fig. 4.** *Oxynoticeras* cf. *lymense* (Wright), phragmocone (TUB-Hi 711210/4); from Oxynotum Zone of Quebrada Yerbas Buenas, N Chile.

**Fig. 5a,b.** *Plesechioceras arcticum* (Frebold), last whorl body chamber (TUB-Hi 811028/3); from Raricostatum Zone of NE Cerro Paisaje, N Chile.

**Fig. 6a,b.** *Paltechioceras* sp., phragmocone with part of body chamber (GSM 14); from Raricostatum Zone of Lagunillas (Puno), S Peru.

**Fig. 7a–c.** *Pseudoskirroceras wiedenmayeri* v. Hillebrandt, body chamber 1.5 whorls (TUB-Hi 711208/3); from Raricostatum Zone of Quebrada Vaca Muerta, N Chile.

(Compiled by von Hillebrandt.)

1
b
a
5
b
a
6
a
b
2
3
7
b
a
7c
4

PLATE 11

CHILE AND ARGENTINA, PLIENSBACHIAN

**Fig. 1.** *Apoderoceras (Miltoceras)* cf. *sellae* (Gemmellaro), phragmocone (TUB-Hi 790212/2); from "*Apoderoceras/Eoderoceras*" Zone of Arroyo Las Chircas (N Río Atuel), Argentina.
**Fig. 2.** *Tropidoceras flandrini* cf. *obtusum* (Futterer), latex cast of outer mold (TUB-Hi 790209/3b); from "*Tropidoceras*" Zone of Puesto Araya (Río Atuel), Argentina.
**Fig. 3.** *Acanthopleuroceras* cf. *whiteavesi* (Smith & Tipper); ↑ beginning of body chamber (TUB-Hi 831206/3); from "*Tropidoceras*" Zone of Puesto Araya (Río Atuel), Argentina.
**Fig. 4a,b.** *Eoamaltheus meridianus* v. Hillebrandt; ↑ beginning of body chamber (TUB-Hi 841204/4b); from *meridianus* Zone of Arroyo Blanco (Río Atuel), Argentina.
**Fig. 5a,b.** *Dubariceras* cf. *silviesi* (Hertlein), phragmocone (TUB-Hi 790209/6); from *meridianus* Zone of Puesto Araya (Río Atuel), Argentina.
**Fig. 6.** *Eoamaltheus* sp. (transitional between *Eoamaltheus* and *Fanninoceras*), mostly compressed mold (TUB-Hi 711128/3); from *meridianus* Zone of Puesto Araya (Río Atuel), Argentina.
**Fig. 7.** *Fanninoceras behrendseni* (Jaworski); ?♂; latex cast of outer mold (TUB-Hi 790216/6); from *behrendseni* Zone of Portezuelo Ancho (Valle de las Leñas), Argentina.
**Fig. 8.** *Fanninoceras behrendseni* (Jaworski); ?♀; phragmocone (TUB-Hi 711125/1); from *behrendseni* Zone of Arroyo Maihuem (Sierra Chacai-Có), Argentina.
**Fig. 9.** *Fanninoceras fannini* McLearn; ↑ beginning of body chamber (TUB-Hi 660708/1); from *fannini* Zone of Quebrada El Asiento, N Chile.
**Fig. 10a,b.** *Fanninoceras disciforme* v. Hillebrandt, phragmocone (TUB-Hi 670220/3); from *disciforme* Zone of Quebrada Noria, N Chile.
**Fig. 11.** *Fuciniceras* sp.; at least half of last whorl body chamber (TUB-Hi 811120/1); from *disciforme* Zone between Quebrada de Ceballos and Quebrada del Chacón, N Chile.
**Fig. 12.** *Paltarpites argutus* (Buckman), phragmocone; from *disciforme* Zone of Quebrada El Asiento, N Chile.
(Compiled by von Hillebrandt.)

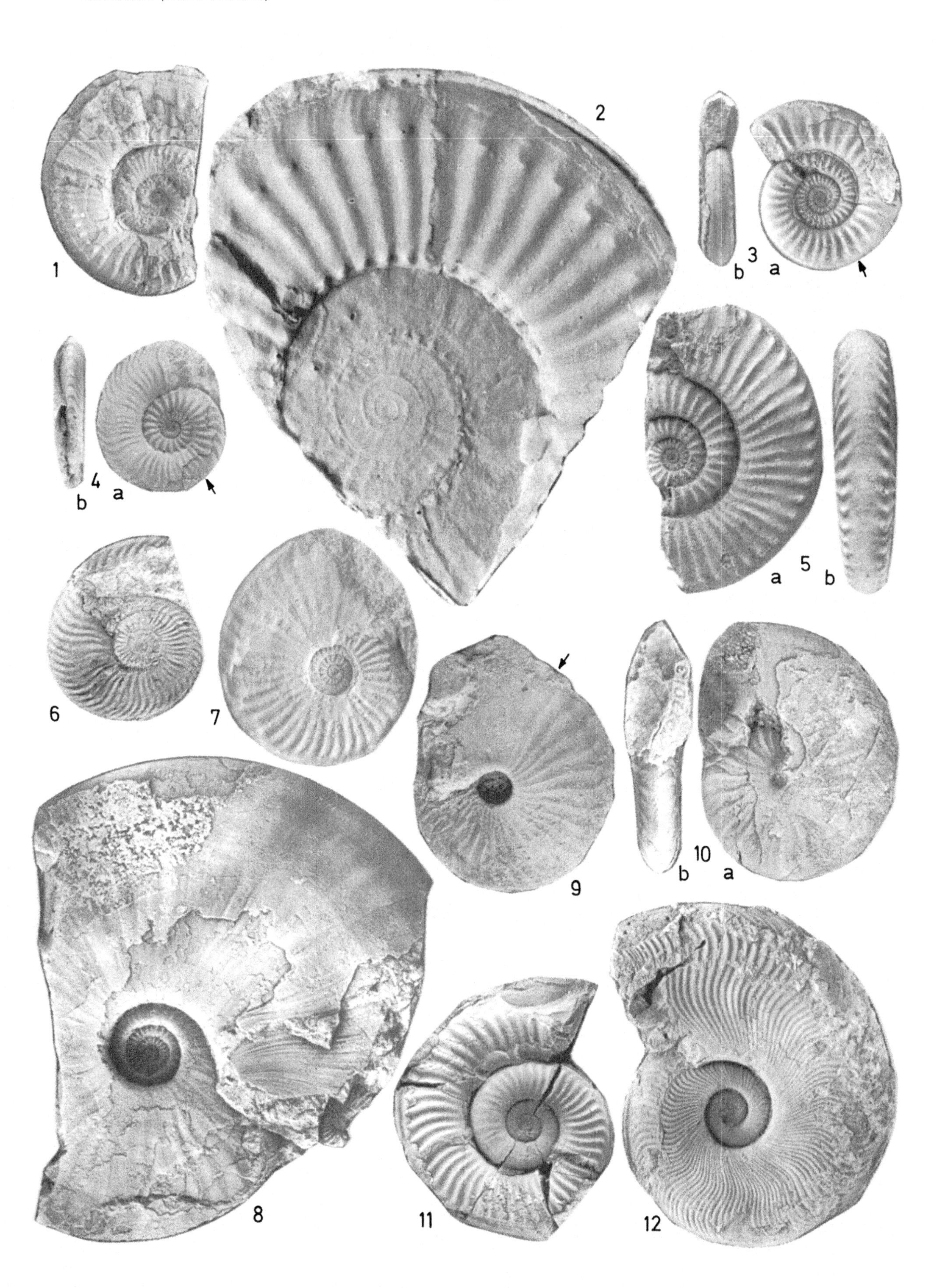
1
2
3
a
b
4
a
b
5
a
b
6
7
9
10
a
b
8
11
12

PLATE 12

NORTHERN CHILE, TOARCIAN

**Fig. 1.** *Dactylioceras* (*Eodactylites*) *simplex* Fucini, phragmocone with part of body chamber (TUB-B.St.M. 1978 II 4); from *simplex* Subzone of Quebrada Chanchoquin, N Chile.
**Fig. 2.** *Dactylioceras* (*Orthodactylites*) *tenuicostatum chilense* Schmidt-Effing, holotype; last whorl body chamber (TUB-B.St.M. 1978 II 61); from Tenuicostatum Subzone of Rio Manflas (Cerro Salto del Toro), N Chile.
**Fig. 3.** *Dactylioceras* (*Orthodactylites*) *hoelderi* Schmidt-Effing, holotype; last whorl body chamber (TUB-B.St.M. 1978 II 53); from *hoelderi* Zone of Quebrada Yerbas Buenas, N Chile.
**Fig. 4a,b.** *Peronoceras largaense* v. Hillebrandt, holotype, phragmocone with part of body chamber (TUB-B.St.M. 1978 II 106); from *largaense* Zone of Quebrada Larga, N Chile.
**Fig. 5.** *Peronoceras pacificum* v. Hillebrandt, holotype, phragmocone with part of body chamber (TUB-B.St. M. 1978 II 123); from *pacificum* Zone of Quebrada Potrerillos, N Chile.
**Fig. 6.** *Collina chilensis* v. Hillebrandt, holotype; last whorl body chamber (TUB-B.St.M. 1978 II 176); from *chilensis* Zone of Quebrada El Bolito, N Chile.
**Fig. 7.** *Peronoceras bolitoense* v. Hillebrandt, holotype; last whorl body chamber (TUB-B.St.M. 1978 II 146); from *bolitoense* Subzone of Quebrada El Bolito, N Chile.
(Compiled by von Hillebrandt.)

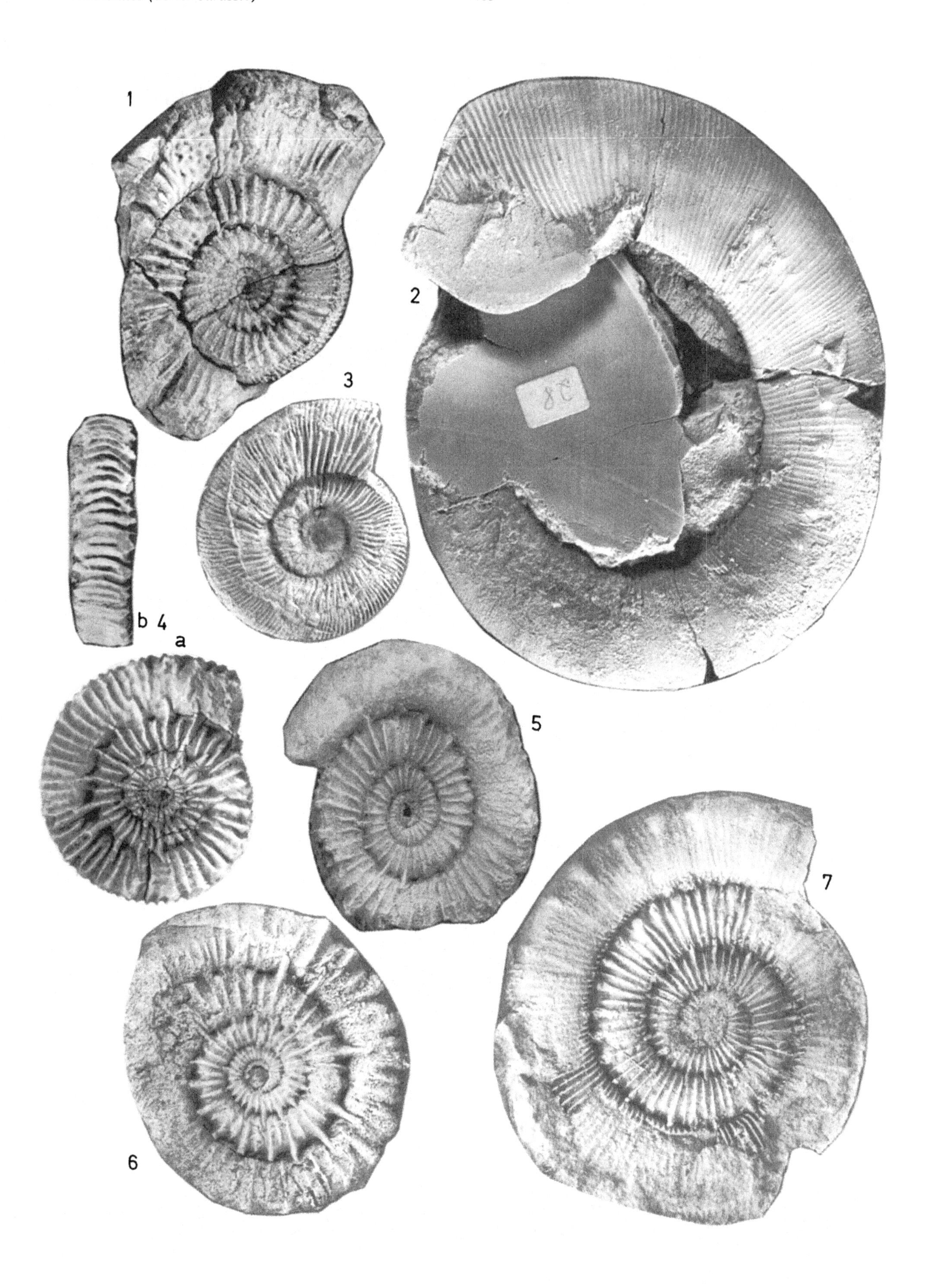
1
2
3
b 4
a
5
6
7

PLATE 13

NORTHERN CHILE AND ARGENTINA, TOARCIAN

**Fig. 1.** *Peronoceras moerickei* v. Hillebrandt, holotype; body chamber more than 1⅓ whorls long (TUB-B.St.M. 1978 II 161) (×0.5); from *moerickei* Subzone of Quebrada El Asiento, N Chile.

**Fig. 2.** *Phymatoceras toroense* v. Hillebrandt, holotype; ↑ beginning of body chamber (TUB-Hi 670109/14); from *toroense* Zone of Río del Toro (Río Manflas), N Chile.

**Fig. 3a,b.** *Phymatoceras copiapense* (Möricke), paratype, phragmocone (TUB-SMNS); from *copiapense* Zone of side valley of Quebrada Paipote, N Chile.

**Fig. 4.** *Phlyseogrammoceras*(?) *tenuicostatum* (Jaworski); ↑ beginning of body chamber (TUB-Hi 790222/2); from *tenuicostatum* Zone of Arroyo Negro, Argentina.

**Fig. 5.** *Pleydellia* cf. *lotharingica* (Branco), phragmocone (TUB-Hi 680109/10); from "lotharingica" Zone of Arroyo Honda, Argentina.

**Fig. 6.** *Pleydellia* cf. *fluitans* (Dumortier); ↑ beginning of body chamber (TUB-Hi 680109/12); from "*fluitans*" Zone of Arroyo Honda, Argentina.

**Fig. 7a,b.** *Pleydellia* cf. *fluitans* (Dumortier); ↑ beginning of body chamber (TUB-Hi 670222/10); from "*fluitans*" Zone of Quebrada El Bolito, N Chile.

(Compiled by von Hillebrandt.)

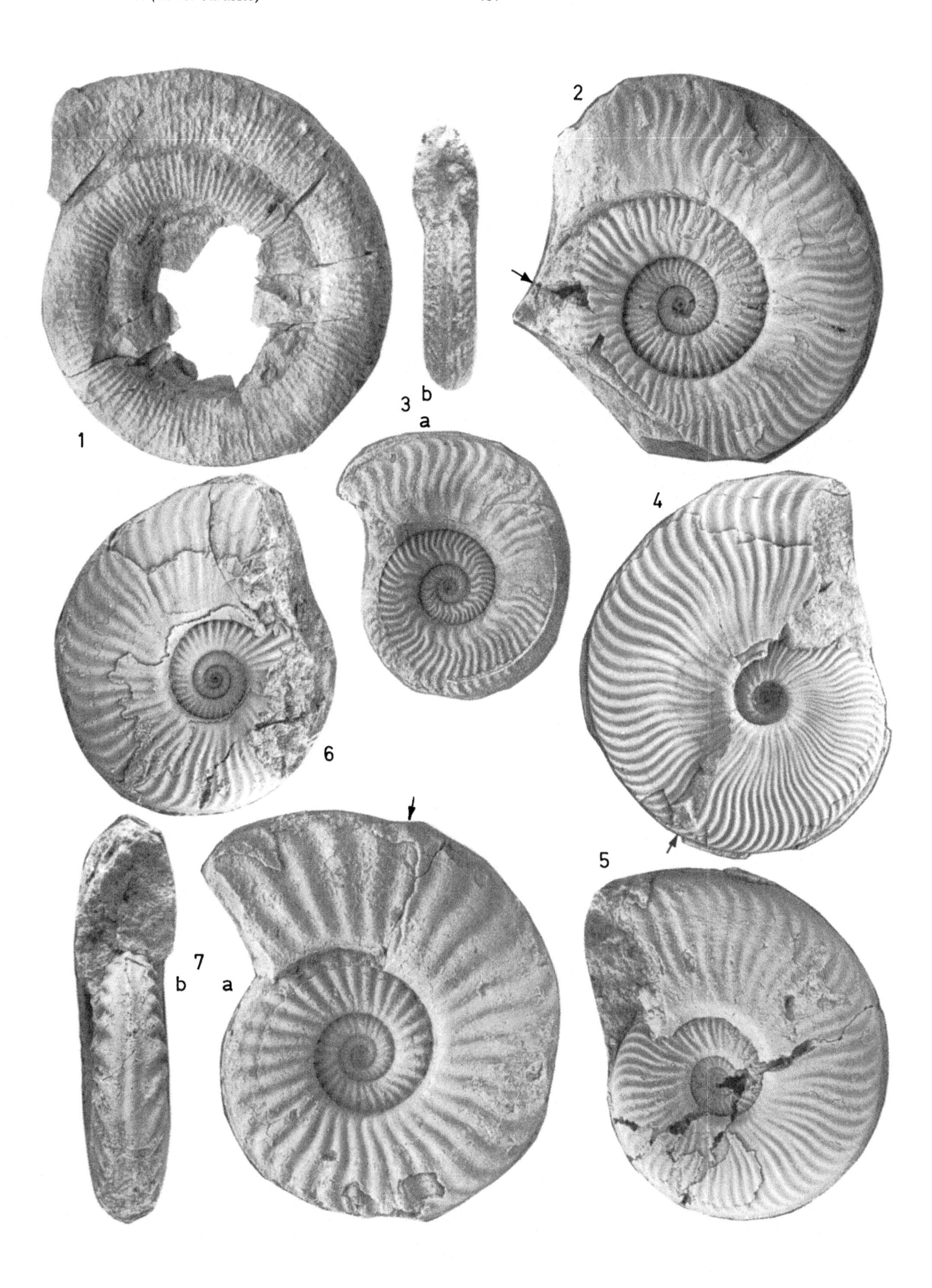
1
2
3
b
a
4
5
6
7
b
a

PLATE 14

JAPAN, VIETNAM, AND THE PHILIPPINES, SINEMURIAN–AALENIAN

**Fig. 1.** *Arnioceras yokoyamai* (Matsumoto), phragmocone, Lower Sinemurian; from Nakazai, Shizukawa area, South Kitakami, NE Japan.

**Fig. 2.** *Amaltheus* sp. phragmocone, upper Pliensbachian; from Teradani, Kuruma area, central Japan.

**Fig. 3.** *Fontanelliceras fontanellense* (Gemmellaro), phragmocone, *Fontanelliceras fontanellense* Assemblage Zone; from Sakuraguchi-dani, Toyora area, W Japan.

**Fig. 4.** *Protogrammoceras nipponicum* (Matsumoto), phragmocone, *Protogrammoceras nipponicum* Assemblage Zone; from Sakuraguchi-dani, Toyora area, W Japan.

**Fig. 5.** *Dactylioceras helianthoides* Yokoyama, phragmocone, *Dactylioceras helianthoides* Assemblage Zone; from Nishinakayama, Toyora area, W Japan.

**Fig. 6.** *Hosoureites ikianus* (Yokoyama), phragmocone, *Hosoureites ikianus* Assemblage Zone; from Hosoura, Shizukawa area, NE Japan.

**Fig. 7.** *Planammatoceras planinsigne* (Vacek), complete (×3/5), *Planammatoceras planinsigne* Assemblage Zone; from Hosoura, Shizukawa area, NE Japan.

**Fig. 8.** *Dumortieria lantenoisi* (Mansuy), phragmocone, *Dumortieria lantenoisi* Zone; from Lo-Duc, near Bienhoa, Vietnam.

**Fig. 9.** *Pseudogrammoceras*(?) *loducensis* Sato, external cast of a phragmocone, *Dumortieria lantenoisi* Assemblage Zone; from Lo-Duc, near Bienhoa, Vietnam.

**Fig. 10.** *Tmetoceras scissum* (Ben.), phragmocone (×2), Aalenian; from Mae Sot, Thailand.

**Fig. 11 a,b.** *Graphoceras concavum* (Sow.), phragmocone, Aalenian; from Mae Sot, Thailand.

(Compiled by Sato.)

1
2
3
4
5
6
7
8
9
10
11
a
b

PLATE 15

EASTERN RUSSIA, HETTANGIAN

**Figs. 1, 2, 6.** *Psiloceras viligense* Chud. & Polub. (1, complete; 2, phragmocone; 6, nucleus); from Planorbis Zone, North-East Russia, Viliga River (coll. Polubotko & Repin).

**Fig. 3.** *Waehneroceras angustum* A. Dagis, complete; from Liasicus Zone, North-East Russia, Arman River.

**Fig. 4.** *Waehneroceras frigga* (Waehner), phragmocone?; from Liasicus Zone, North-East Russia, Yana River basin (coll. Repin).

**Fig. 5.** *Schlotheimia* ex gr. *angulata* (Schloth.), phragmocone?; from Angulata Zone, North-East Russia, Omolon River basin (coll. Polubotko & Repin).

**Fig. 7.** *Psiloceras olenekense* (Kiparisova), phragmocone?; from Planorbis Zone, North-East Russia, Olenek River basin (coll. Knjazev).

**Fig. 8.** *Primapsiloceras primulum* (Repin), phragmocone?; from *P. primulum* Zone, North-East Russia, Omolon River basin (coll. Polubotko & Repin).

**Fig. 9.** *Alsatites*(?) sp. indet., complete?; from Liasicus Zone, North-East Russia, Omolon River basin (coll. Polubotko & Repin).

**Fig. 10.** *Waehneroceras armanense* Repin, complete; from Liasicus Zone, North-East Russia, Arman River.

**Fig. 11.** *Waehneroceras portlocki* (Wright), phragmocone?; from Liasicus Zone, North-East Russia, Omolon River basin. These specimens are kept in Central Scientific Research Geological Museum (CNIGR Museum), St. Petersburg, and in the Geological Museum of the Geological Survey of North-East Russia, Magadan.

(Compiled by Repin, Sey, & Kalacheva.)

a
1
2
3
a
5
4
6
a
b
7
8
9
10
a
b
11

PLATE 16

EASTERN RUSSIA, SINEMURIAN

**Figs. 1, 2.** *Arietites libratus* Repin (1, complete; 2, juv.); from Bucklandi Zone, North-East Russia, Omolon River basin (coll. Polubotko & Repin).
**Figs. 3–6.** *Angulaticeras* (*Gydanoceras*) *ochoticum* Repin (3, complete; 4, body chamber; 5, juv. (×2); 6, phragmocone); from *A. kolymicum* Zone, North-East Russia, Ghyzhiga River basin, Buynda River basin, (coll. Polubotko & Repin).
**Fig. 7.** *Coroniceras (Paracoroniceras) siverti* (Tuchkov), phragmocone (×0.4); from *C. siverti* Zone, North-East Russia, Omolon River basin.
**Fig. 8.** *Coroniceras* (*Primarietites*) *reynesi* (Spath), complete; from *C. siverti* Zone, North-East Russia, Omolon River basin (coll. Polubotko & Repin).
**Fig. 9.** *Angulaticeras (Gydanoceras)* cf. *ochoticum* Repin; from Upper Sinemurian, Far East Russia, southern Primorye, Trudny Peninsula (coll. Sey & Kalacheva).
**Fig. 10.** *Angulaticeras* (*Gydanoceras*) *kolymicum* Repin, complete; from *A. kolymicum* Zone, North-East Russia, Korkodon River basin (coll. Polubotko & Repin).
(Compiled by Repin, Sey, & Kalacheva.)

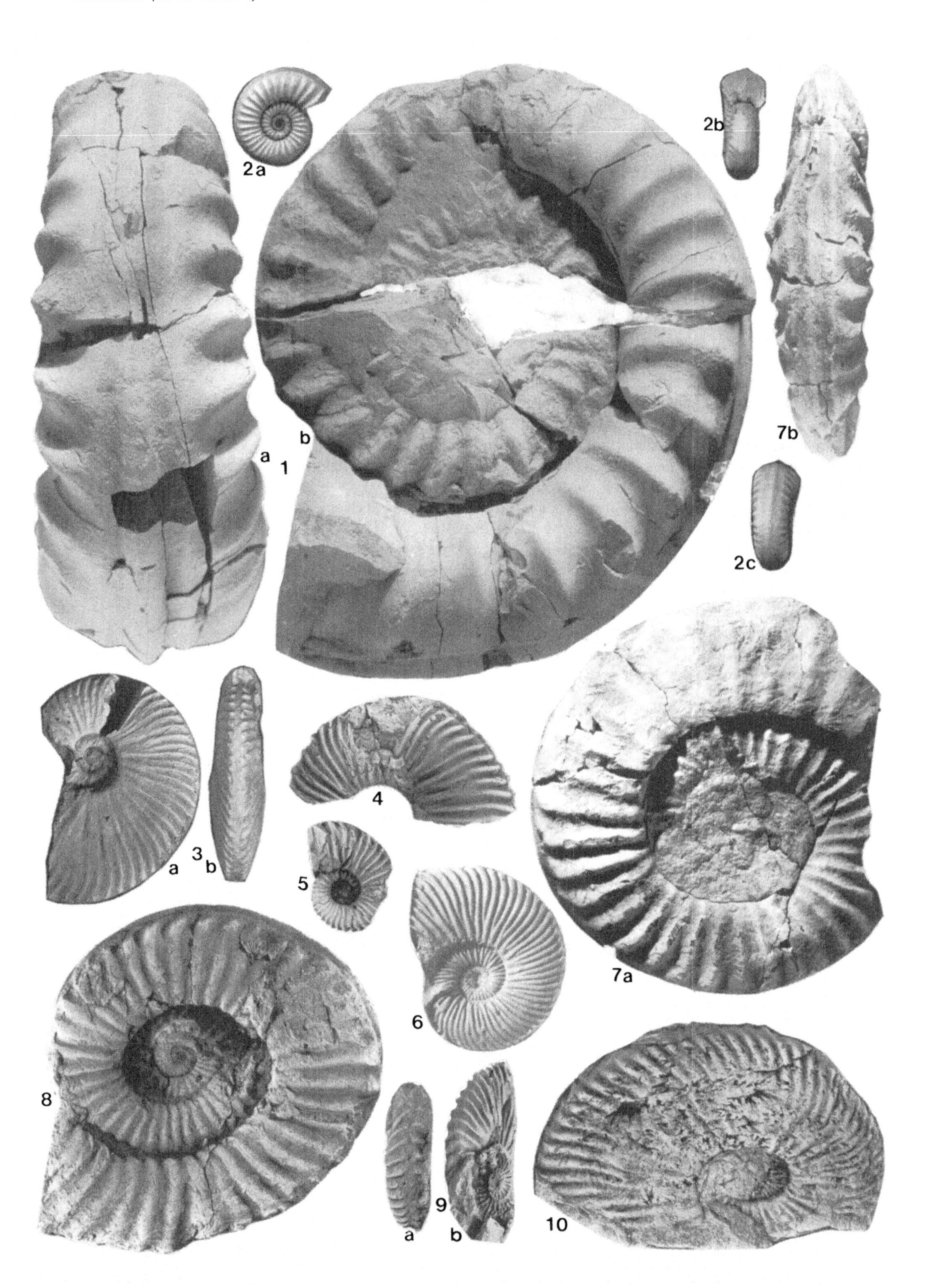
2a
2b
7b
b
a 1
2c
3
a
b
4
5
7a
6
8
9
a
b
10

PLATE 17

EASTERN RUSSIA, UPPER PLIENSBACHIAN

**Fig. 1.** *Amaltheus (Amaltheus) stokesi* (Sow.), complete; from Stokesi Zone, North-East Russia, Sededema River basin.
**Fig. 2.** *A. (Amaltheus) subbifurcus* Repin, complete; from Stokesi Zone, North-East Russia, Omolon River basin.
**Figs. 3, 6.** *A. (Amaltheus) margaritatus* (Montf.); from Margaritatus Zone [3, North-East Russia, Lena River (coll. Repin); 6, Far East Russia, Bureya River basin] (coll. Sey & Kalacheva).
**Fig. 4.** *A. (Amaltheus) talrosei* Repin, complete; from *A. talrosei* Zone, North-East Russia, Korkodon River basin (coll. Repin).
**Figs. 5, 7.** *Amaltheus (Nordamaltheus) viligaensis* (Tuchkov); from *A. (N.) viligaensis* Zone, North-East Russia, Viliga River (coll. Repin).
**Fig. 8.** *A. (Amaltheus) extremus* Repin, phragmocone; from *A. (N.) viligaensis* Zone, North-East Russia, Omolon River basin (coll. Repin).
**Fig. 9.** *Protogrammoceras* cf. *serotinum* (Bettoni); from upper Pliensbachian, Far East Russia, Southern Sikhote-Alin, Izvilinka River basin (coll. Sey & Kalacheva).
**Fig. 10.** *Fontanelliceras* cf. *fontanellense* (Gemm.), complete; from upper Pliensbachian, Far East Russia, Sikhote-Alin (coll. Sey & Kalacheva).
**Figs. 11, 12, 14.** ''*Dactylioceras*'' *simplex* Fucini; from upper Pliensbachian, Far East Russia, Southern Sikhote-Alin (coll. Sey & Kalacheva).
**Fig. 13.** ''*Dactylioceras*'' *polymorphum* Fucini, phragmocone?; from Upper Pliensbachian, Far East Russia, Southern Sikhote-Alin (coll. Sey & Kalacheva).
**Figs. 15, 16.** *Arieticeras japonicum* Matsumoto, complete; from upper Pliensbachian, Far East Russia, Southern Sikhote-Alin (coll. Sey & Kalacheva).
**Fig. 17.** *Protogrammoceras* sp. indet.; from upper Pleinsbachian, Far East Russia, Southern Sikhote-Alin (coll. Sey & Kalacheva) [= *Paltarpites* sp. indet.].
(Compiled by Repin, Sey, & Kalacheva.)

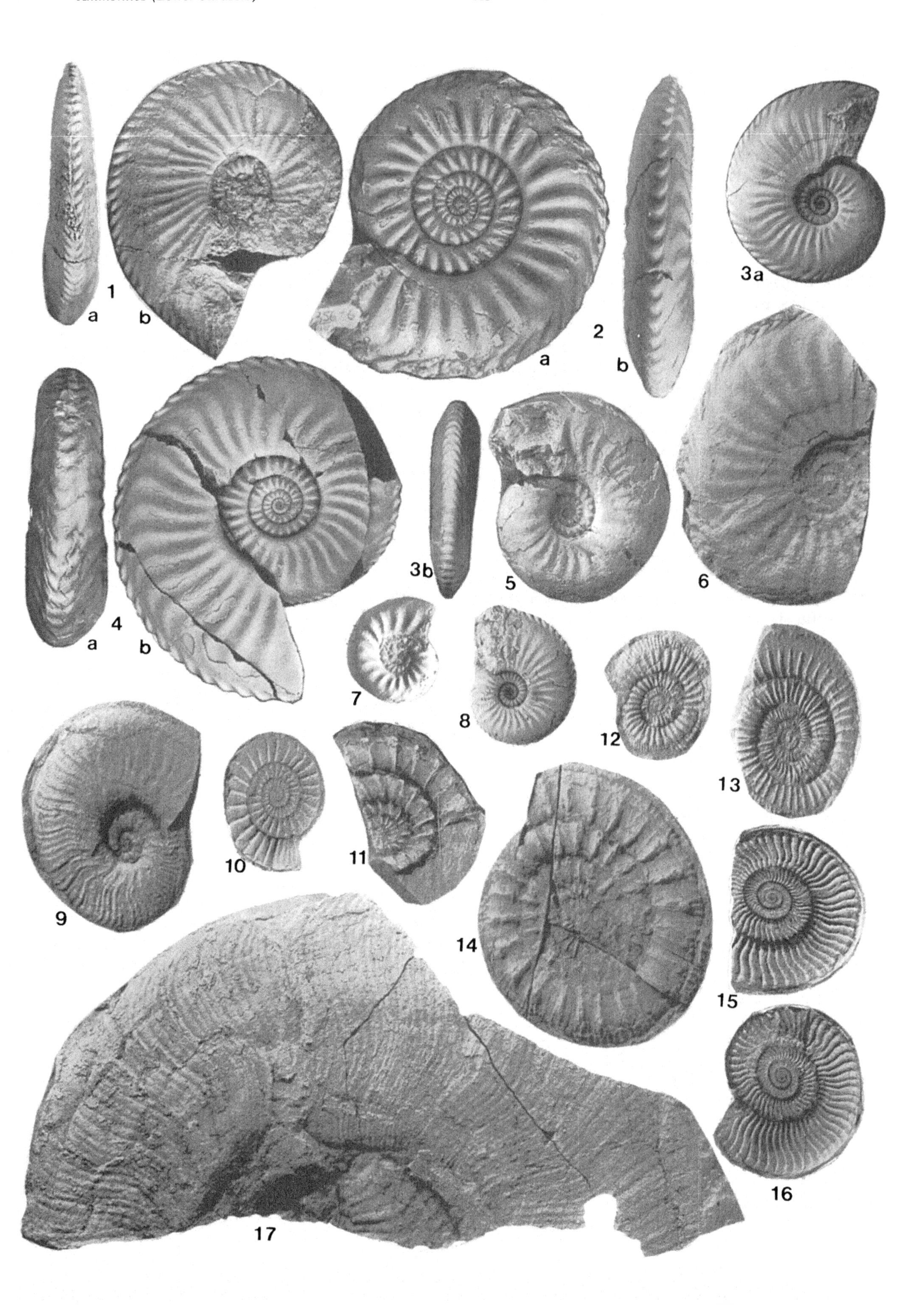
1
a
b
2
a
b
3a
3b
4
a
b
5
6
7
8
9
10
11
12
13
14
15
16
17

PLATE 18

EASTERN RUSSIA, TOARCIAN

**Figs. 1, 3.** *Tiltoniceras propinquum* (Whit.) (1, phragmocone; 3, complete); from *T. propinquum* Zone, North-East Russia, Omolon River basin.
**Fig. 2.** *Arctomercaticeras costatum* Repin, complete; from *T. propinquum* Zone, North-East Russia, Omolon River basin.
**Fig. 4.** *Harpoceras exaratum* (Young & Bird), complete; from Falciferum Zone, North-East Russia, Omolon River basin (coll. Polubotko & Repin).
**Fig. 5.** *Eleganticeras alajaense* (Repin), phragmocone?; from *E. elegantulum* Zone, North-East Russia, Omolon River basin (coll. Polubotko & Repin).
**Figs. 6, 8.** *Eleganticeras elegantulum* (Y. & B.), complete; from *E. elegantulum* Zone, North-East Russia, Omolon River basin (coll. Repin).
**Fig. 7.** *Hildaites chrysanthemum* (Yok.), complete; from *D. athleticum* Zone, North-East Russia, Omolon River basin (coll. Repin).
**Fig. 9.** *Eleganticeras planus* Repin, complete; from *E. elegantulum* Zone, North-East Russia, Omolon River basin.
**Figs. 10, 13.** *Pseudolioceras kedonense* Repin, complete; from *Z. monestieri* Zone; 10, Far East Russia, Tugur Bay (coll. Sey & Kalacheva); 13, North-East Russia, Omolon River basin (coll. Repin).
**Fig. 11.** *Harpoceras falciferum* (Sow.), complete; from Falciferum Zone, North-East Russia, Omolon River basin (coll. Repin).
**Fig. 12.** *Dactylioceras commune* (Sow.), complete; from *D. athleticum* Zone, North-East Russia, Omolon River basin (coll. Polubotko & Repin).
**Fig. 14.** *Dactylioceras athleticum* (Simps.), complete; from *D. athleticum* Zone, North-East Russia, Omolon River basin.
(Compiled by Repin, Sey, & Kalacheva.)

1
a
b
2
a
b
3
a
b
4
5
6
a
b
7
8
9
10
11
12
a
b
13
a
b
14

PLATE 19

EASTERN RUSSIA, TOARCIAN

**Fig. 1.** *Porpoceras spinatum* (Frebold), complete; from *P. spinatum* Zone, North-East Russia, Korkodon River basin.

**Figs. 2, 5.** *Pseudolioceras lythense* (Y. & B.); 2, complete (×0.64); 5, phragmocone?; from *Z. monestieri* Zone; 2, North-East Russia, Omolon River basin (coll. Polubotko & Repin); 5, Far East Russia, Tugur Bay (coll. Sey & Kalacheva).

**Figs. 3, 9, 15.** *Zugodactylites braunianus* (Orb.); 3, complete; 9, 15, phragmocone; from *Z. monestieri* Zone; 3, North-East Russia, Korkodon River basin (coll. Repin); 9, 15, Far East Russia, Tugur Bay (coll. Sey & Kalacheva).

**Fig. 4.** *Pseudolioceras gradatum* Buckman, complete; from *P. spinatum* Zone, North-East Russia, Omolon River basin.

**Figs. 6, 16.** *Pseudolioceras rosenkrantzi* A. Dagis, complete; from *P. rosenkrantzi* Zone, North-East Russia, Omolon River basin (coll. Repin).

**Fig. 7.** *Zugodactylites exilis* A. Dagis, phragmocone; from *Z. monestieri* Zone, North-East Russia, Korkodon River basin (coll. Repin).

**Fig. 8.** *Kolymoceras* aff. *viluiense* Krimholz; from *Z. monestieri* Zone, North-East Russia, Omolon River basin (coll. Polubotko & Repin).

**Fig. 12.** *Porpoceras polare* (Freb.), phragmocone; from *P. spinatum* Zone, North-East Russia, Paren River basin (coll. Repin).

**Fig. 13.** *Collina* aff. *orientalis* A. Dagis, complete; from *P. spinatum* Zone, North-East Russia, Paren River basin (coll. Repin).

**Fig. 14.** *Pseudolioceras replicatum* Buckman, complete; from *P. rosenkrantzi* Zone-*P. beyrichi* Zone, North-East Russia, Olenek River basin (coll. Polubotko & Repin).

(Compiled by Repin, Sey, & Kalacheva.)

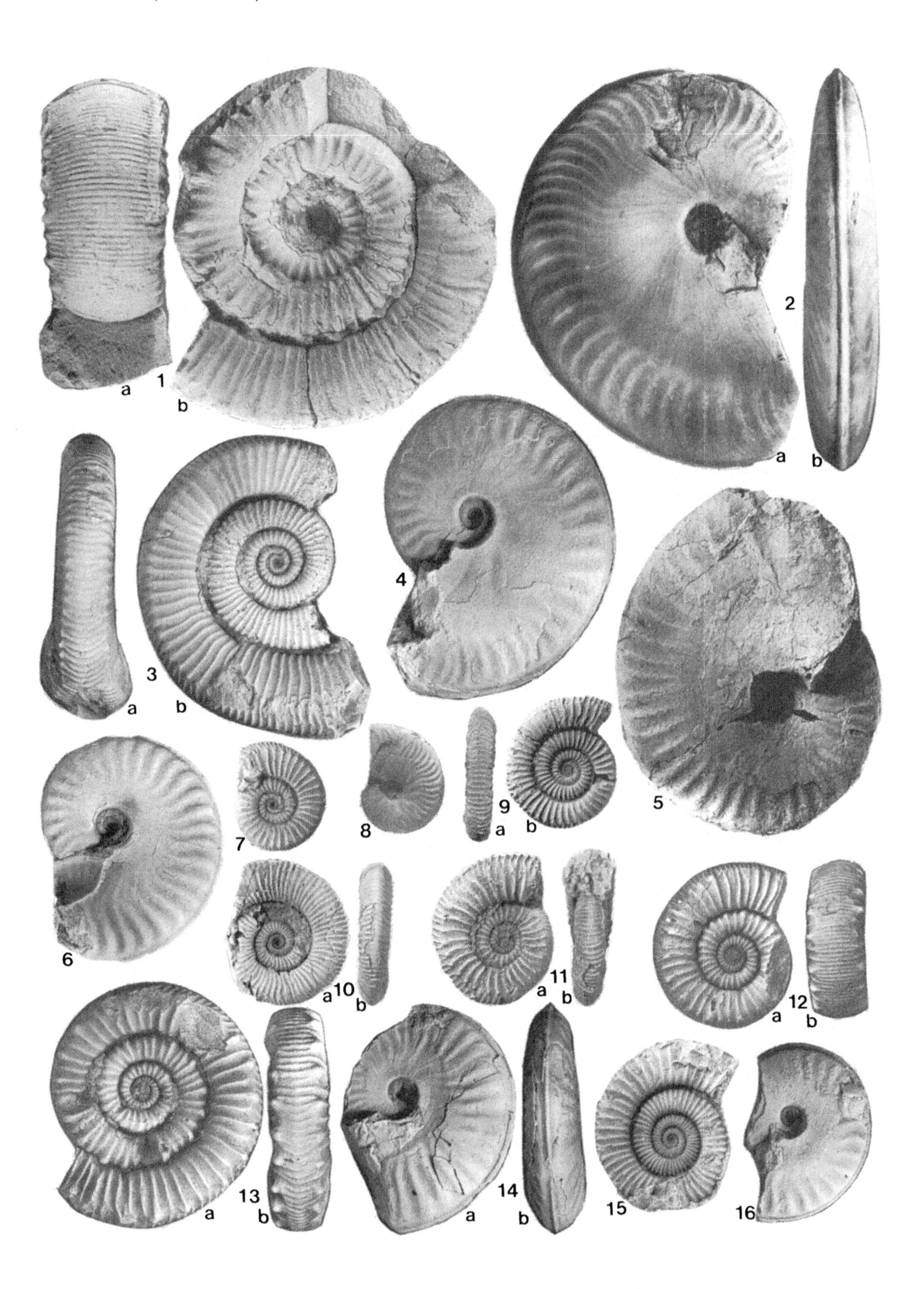
a 1
b
2
a b
3
a b
4
5
6
7
8
9
a b
a10
b
11
a b
12
a b
13
a
b
14
a
b
15
16

## PLATE 20

## NORTHWESTERN CANADA

AALENIAN

**Fig. 1.** *Leioceras* aff. *opalinum* (Reinecke), fragmentary specimen (GSC 92778); from ?Opalinum Zone, Johnson Creek, northern Yukon Territory.

**Fig. 2.** *Pseudolioceras* aff. *whiteavesi* (White), whorl fragment (GSC 92774); from Howelli Zone, near Trout Lake, northern Yukon Territory.

**Fig. 3.** *Pseudolioceras* aff. *whiteavesi* (White), inner whorls (GSC 92762); from ?Opalinum Zone, Murray Ridge, northern Yukon Territory.

**Fig. 4a,b.** *Erycitoides howelli* (White), nearly complete specimen (GSC 92745); from Howelli Zone, Johnson Creek, northern Yukon Territory.

**Fig. 5.** *Planammatoceras* sp., whorl fragment (GSC 92779); from Howelli Zone, Johnson Creek, northern Yukon Territory.

BATHONIAN

**Fig. 6a,b.** *Arctocephalites spathi* Poulton, phragmocone and partial body chamber (GSC 68266); from *A. spathi* Zone, Porcupine River, northern Yukon Territory.

**Fig. 7.** *Arctocephalites callomoni* Frebold, nearly complete specimen (GSC 68285); from *A. porcupinensis* Zone, Porcupine River, northern Yukon Territory.

**Fig. 8.** *Cadomites* sp., fractured, nearly complete specimen (GSC 68417); from *A. harlandi* Zone, Porcupine River, northern Yukon Territory.

**Fig. 9.** *Arctocephalites porcupinensis* Poulton, complete specimen (GSC 68277); from *A. porcupinensis* Zone, Porcupine River, northern Yukon Territory.

**Fig. 10.** *Oxycerites*(?) sp., juvenile whorl fragment (GSC 68433); from ?Ishmae Zone, Porcupine River, northern Yukon Territory. See also Plate 14, Figs. 6, 7, 10, and 11.

(Compiled by Poulton.)

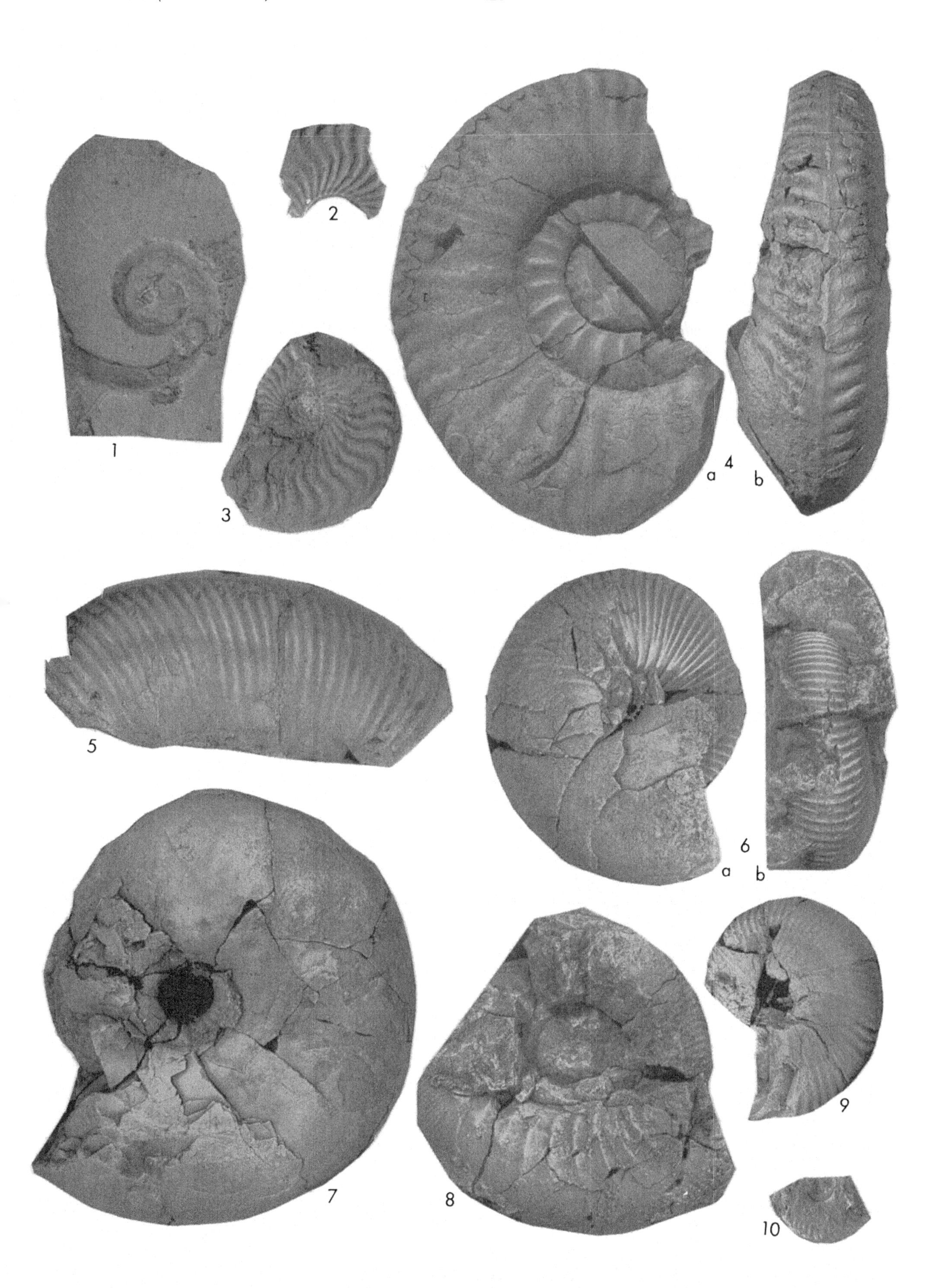
1
2
3
4
a
b
5
6
a
b
7
8
9
10

PLATE 21

NORTHWESTERN CANADA, BATHONIAN

**Fig. 1.** *Arcticoceras harlandi* Rawson, whorl fragment (GSC 68346); from *A. harlandi* Zone, Porcupine River, northern Yukon Territory.
**Fig. 2.** *Arcticoceras ishmae* (Keyserling), complete specimen (GSC 68353); from Ishmae Zone, Porcupine River, northern Yukon Territory.
**Fig. 3.** *Parareineckeia* sp., whorl fragments (GSC 68438); from Ishmae Zone, Porcupine River, northern Yukon Territory.
**Fig. 4.** *Oxycerites* (*Paroecotraustes*)(?) sp., small specimen (GSC 68431); from (?)*A. spathi* Zone, Porcupine River, northern Yukon Territory.
**Fig. 5.** *Oxycerites birkelundi* Poulton, nearly complete specimen (GSC 68437); from Ishmae Zone, Porcupine River, northern Yukon Territory.
**Fig. 6.** *Loucheuxia bartletti* Poulton, phragmocone (GSC 68369); from Ishmae Zone, Porcupine River, northern Yukon Territory.
(Compiled by Poulton.)

1
2
3
4
5
6

## PLATE 22

## NORTHWESTERN CANADA

BATHONIAN

**Figs. 1, 2.** *Cadoceras barnstoni* (Meek); 1, complete specimen (GSC 68390); 2, cross section showing flattened venter of body chamber (GSC 68394); from *C. barnstoni* Zone, Porcupine River, northern Yukon Territory.

**Fig. 3.** *Iniskinites* sp., partial specimen (GSC 68414); from *C. barnstoni* Zone, Porcupine River, northern Yukon Territory.

**Fig. 5a,b.** *Kepplerites* aff. *K. rosenkrantzi* Spath, partial phragmocone (GSC 68419), showing flattened juvenile venter; from *C. barnstoni* Zone, Porcupine River, northern Yukon territory.

CALLOVIAN

**Fig. 4.** *Cadoceras bodylevskyi* Frebold, complete specimen (GSC 68402); from *C. bodylevskyi* Zone, Porcupine River, northern Yukon Territory.

OXFORDIAN

**Fig. 6.** *Cardioceras* (*Scarburgiceras*) aff. *scarburgense* (Young & Bird), partial specimen (GSC 83526); from Mariae Zone, northern Richardson Mountains, northern Yukon Territory.

(Compiled by Poulton.)

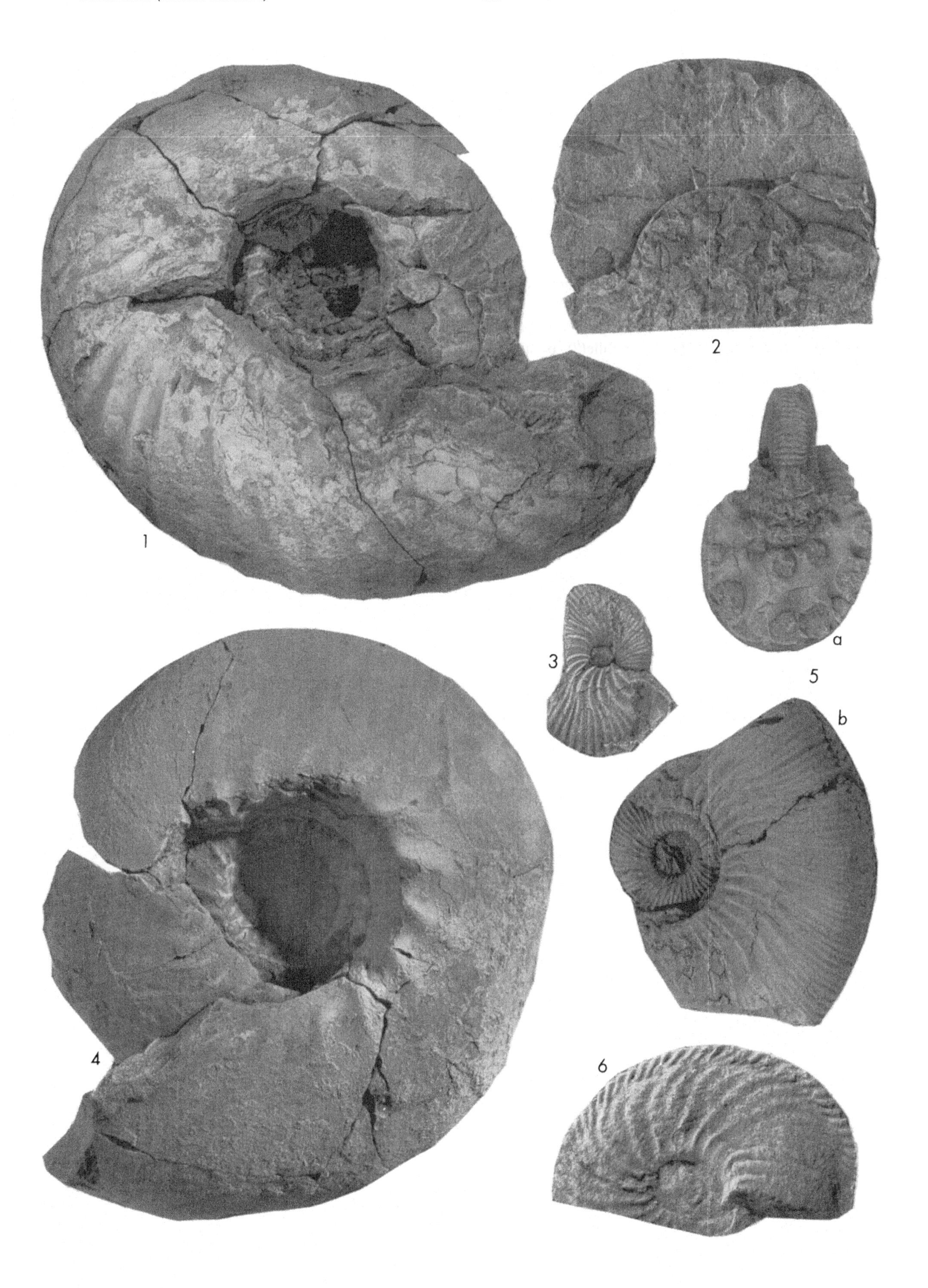
1
2
3
4
5
a
b
6

PLATE 23

SOUTHERN ALASKA, AALENIAN. HOWELLI ZONE (AND BELOW)

**Figs. 1, 2.** *Eudmetoceras nucleospinosum* West.; 1, (UW 16635); 2a,b, holotype; from Wide Bay.

**Fig. 3.** *Tmetoceras scissum* (Ben.) ♀, about complete (USNM 132047); from ca. 100 m below Howelli Zone at Wide Bay.

**Fig. 4.** *Tmetoceras flexicostatum* West., ♀, holotype; from Wide Bay.

**Fig. 5a,b.** *Tmetoceras kirki* West., holotype; from middle Howelli Zone of Wide Bay.

**Fig. 6a–c.** *Tmetoceras tenue* West., ♂ (''*Tmetoites*''), holotype, complete, (×0.3); from *tenue-flexicostatum* faunule of Wide Bay.

**Figs. 7, 8.** *Pseudolioceras (Tugurites) whiteavesi* (White); 7, ♀, phragmocone, lectotype, from ''Wrangel Bay''; 8, juvenile(?), paralectotype (USNM 132044); both near Wide Bay.

**Fig. 9.** *Pseudolioceras (Tugurites) tugurense* Kal. & Sey (CAS), ♀, from Wide Bay.

(Compiled by Westermann.)

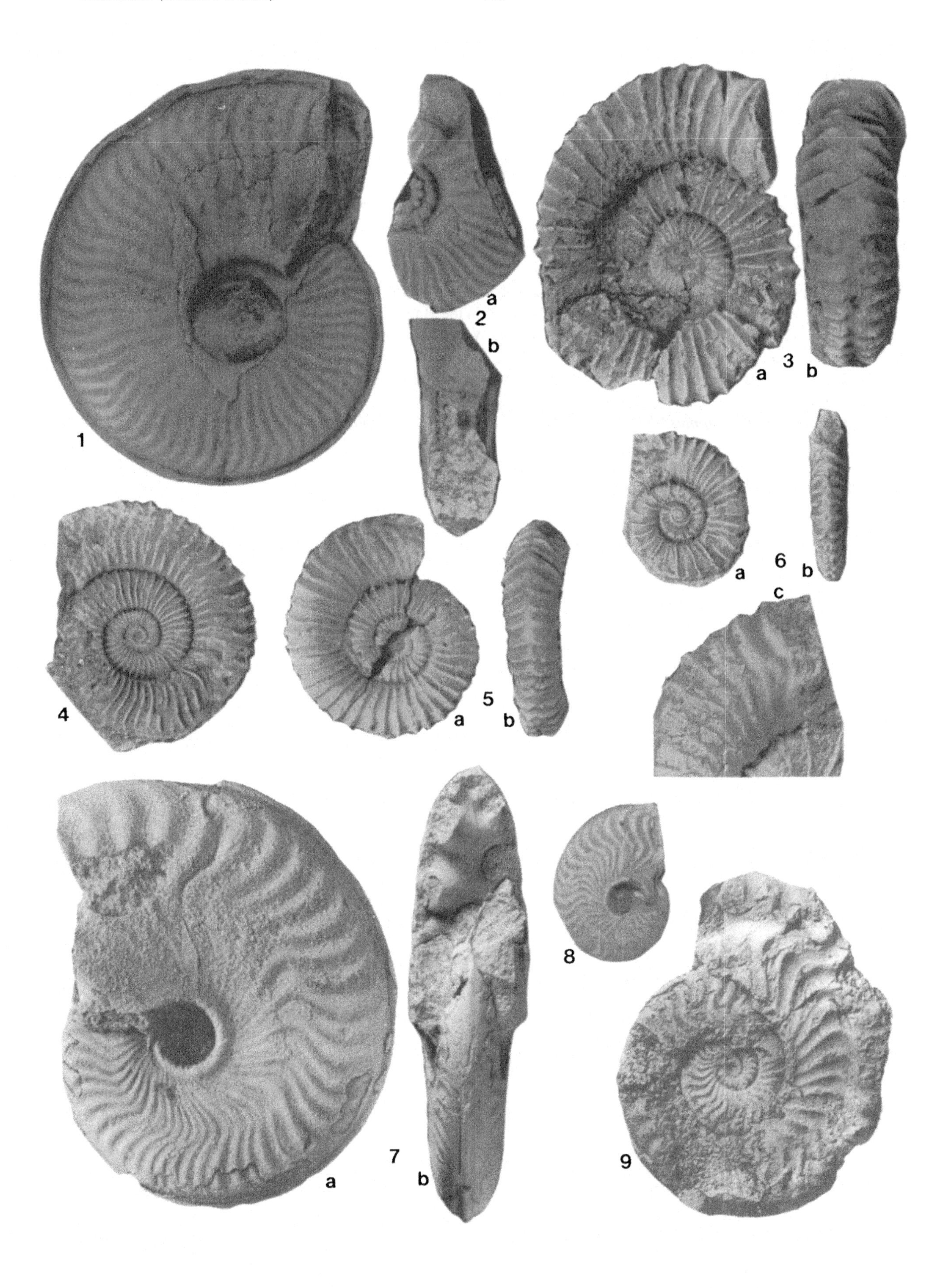
1
a
2
b
a
3
b
4
a
5
b
a
6
b
c
a
7
b
8
9

## PLATE 24

## SOUTHERN ALASKA, AALENIAN. HOWELLI ZONE

**Fig. 1a,b.** *Erycitoides howelli* (White), ♀, phragmocone (UW 16600); from upper Howelli Zone of Wide Bay.
**Fig. 2.** *Erycitoides profundus* West., ♀, phragmocone (UW 16608); from *profundus* horizon of Wide Bay.
**Fig. 3a,b.** *Erycitoides teres* West., ♀, phragmocone (USNM 132028) (holotype of subsp. *compressus* West.); from upper Howelli Zone of Wide Bay.
**Fig. 4a,b.** *Erycitoides howelli* (White), ♂, [''*Kialagvikes kialakvikensis* (White)], complete (UW 16615); from *tenue-flexicostatum* faunule of Wide Bay.
**Fig. 5.** *Erycitoides levis* West., ♂, almost complete (UW 16628); from (?)*profundus* horizon at Wide Bay.
**Fig. 6a,b.** *Erycitoides spinatus* West., ♂, holotype, almost complete; from upper Howelli Zone of Wide Bay.
**Fig. 7a,b.** *Abbasites platystomus* West., ♀, holotype, phragmocone; from *howelli* zonule of Wide Bay.
**Fig. 8 a,b.** *Erycites imlayi* West., ♀, complete; from Wide Bay.
(Compiled by Westermann.)

a
1
b
2
a
3
b
a
5
b
a
4
b
a
6
b
a
7
b
a
8
b

PLATE 25

SOUTHERN ALASKA, LOWER BAJOCIAN. WIDEBAYENSE ZONE

**Fig. 1a,b.** *Eudmetoceras amplectens* (Buck.), ♀, phragmocone (McM-J 952); from *amplectens* bed at Wide Bay.
**Fig. 2a,b.** *Hebetoxyites* aff. *hebes* Buck., ♂, complete (McM-J 1043); from *amplectens* bed at Wide Bay.
**Fig. 3a,b.** *Planammatoceras (Pseudaptetoceras) klimakomphalum discoidale* West., ♀, mostly septate (USNM 160928); from Camachoi Subzone of Wide Bay.
**Fig. 4a,b.** *Praeoppelia oppeliiformis* West., ♀, phragmocone; a, lateral view (McM-J 1040); b, venter of holotype; from Wide Bay.
**Fig. 5a,b.** *Asthenoceras* aff. *nannodes* (Buck.), ♀, almost complete (McM-J 1035b); from Wide Bay.
**Fig. 6a,b.** *Pseudolioceras (Tugurites) fastigatum* West., ♀, holotype, mostly septate; from Camachoi Subzone of Wide Bay.
**Fig. 7a–c.** *Pseudolioceras (Tugurites) costistriatum* West.; a, b, ♀, holotype, from *sudneroides* Subzone; c, juv. or ♂, showing typical striae (USNM 160295); from Wide Bay.
(Compiled by Westermann.)

1
a
b
2
a
b
3
a
b
4
a
b
5
a
b
6
a
b
7
a
b
7c

PLATE 26

SOUTHERN ALASKA, LOWER BAJOCIAN. WIDEBAYENSE ZONE

**Figs. 1, 2.** *Euhoploceras bifurcatum* West., from Camachoi Subzone of Wide Bay; 1a,b, ♀, holotype, with beginning of body chamber; 2a,b, inner whorls of spinose variant (USNM 160238).

**Fig. 3a,b.** *Witchellia sudneroides* West., ♀, holotype, phragmocone; from *sudneroides* Subzone of Wide Bay.

**Figs. 4, 5.** *Witchellia* (''*Pelekodites*'') aff. *spatians* (Buck.), ♂, both specimens with lappets (McM-J 1028a, McM-J 939); from Wide Bay.

**Fig. 6a,b.** *Alaskinia alaskensis* (West.), ♀, phragmocone (McM-J 1021); from lower Widenbayense Zone of Wide Bay.

**Fig. 7a,b.** ''*Docidoceras*'' aff. *longalvum* (Vacek), ♀, inner whorls (McM-J 937); from lower Widebayense Zone of Wide Bay.

(Compiled by Westermann.)

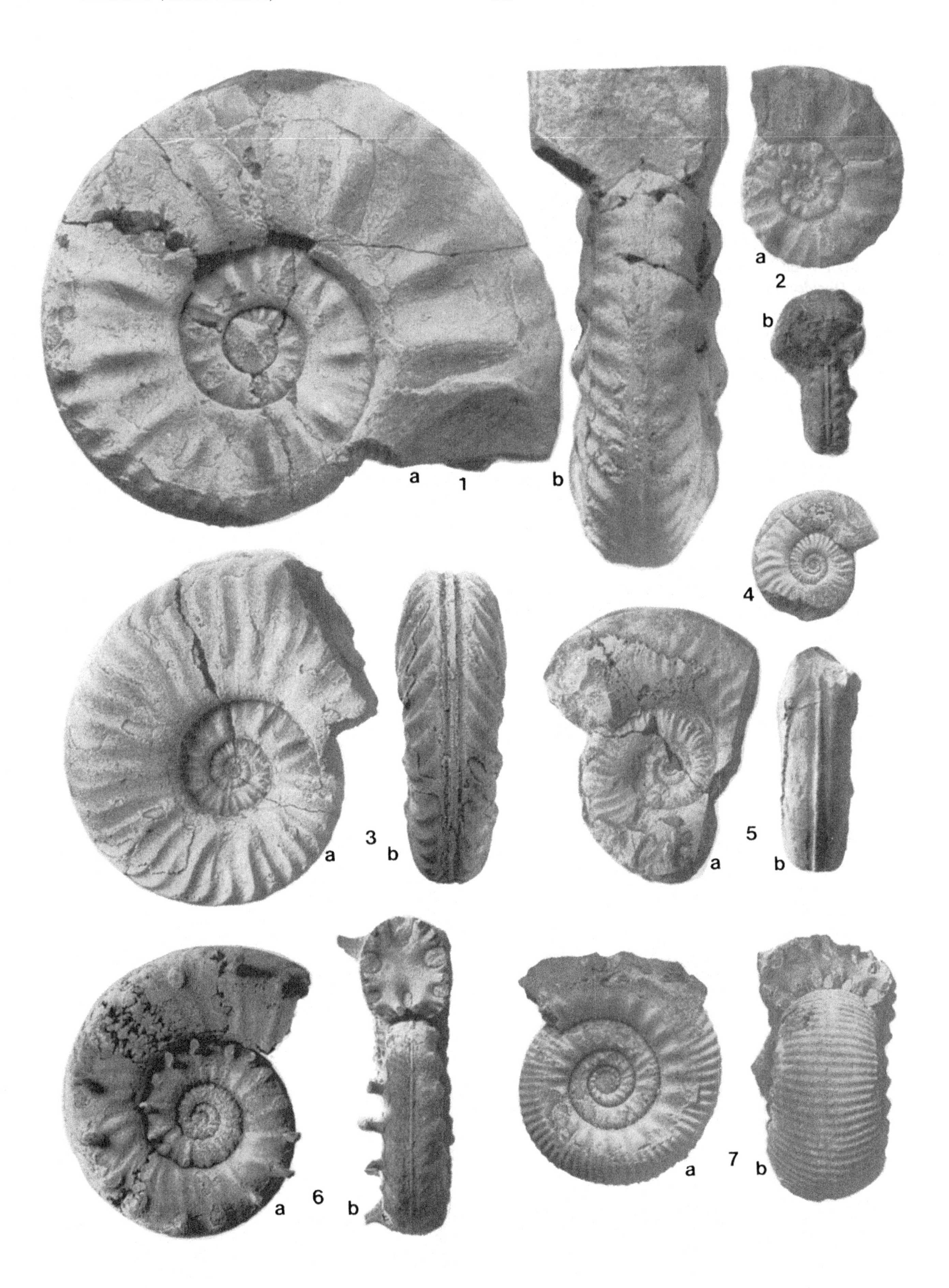
a
1
b
a
2
b
4
3
a
b
5
a
b
6
a
b
7
a
b

PLATE 27

SOUTHERN ALASKA, LOWER BAJOCIAN. WIDEBAYENSE ZONE

**Figs. 1–3.** *Docidoceras (Pseudocidoceras) widebayense* West.; 1a–c, ♀, holotype, complete with inner whorls (c); from Camachoi Subzone of Wide Bay; 2a,b, ♀, inner whorls (McM-J 897); 3, ♂, complete (J 932); Camachoi Subzone of Wide Bay.

**Figs. 4, 5.** *Docidoceras (Pseudocidoceras) camachoi* West.; 4a,b, ♀, phragmocone (USNM 160258); 5a,b, septate inner whorls or ♂ (J 961); both from Camachoi Subzone of Wide Bay.

**Fig. 6.** *Docidoceras*(?) *paucinodosum* West., ♀, holotype, complete; from *Eudmetoceras amplectens* beds of Wide Bay.

**Fig. 7.** *Pseudotoites kialagvikensis* West. & Ricc., ♀, holotype, almost complete; from middle Widebayense Zone of Wide Bay.

(Compiled by Westermann.)

a 1 b
1c
a 2 b
a
5
b
3
4
a
b
6
7

PLATE 28

SOUTHERN ALASKA, COOK INLET, LOWER BAJOCIAN, CRASSICOSTATUS ZONE

**Fig. 1a,b.** *Bradfordia costidensa* Imlay, ♀, holotype, approximately complete (USNM 131396).
**Fig. 2a,b.** *Labyrinthoceras glabrum* Imlay, ♀, holotype, with incomplete body chamber (USNM 131389).
**Fig. 3a–c.** *Arkelloceras* sp. juv. indet. (×2) (USNM 132014).
**Fig. 4a,b.** *Emileia constricta* Imlay, ♀, holotype with incomplete body chamber (USNM 131884). All reproduced from Imlay (1964).
(Compiled by Westermann.)

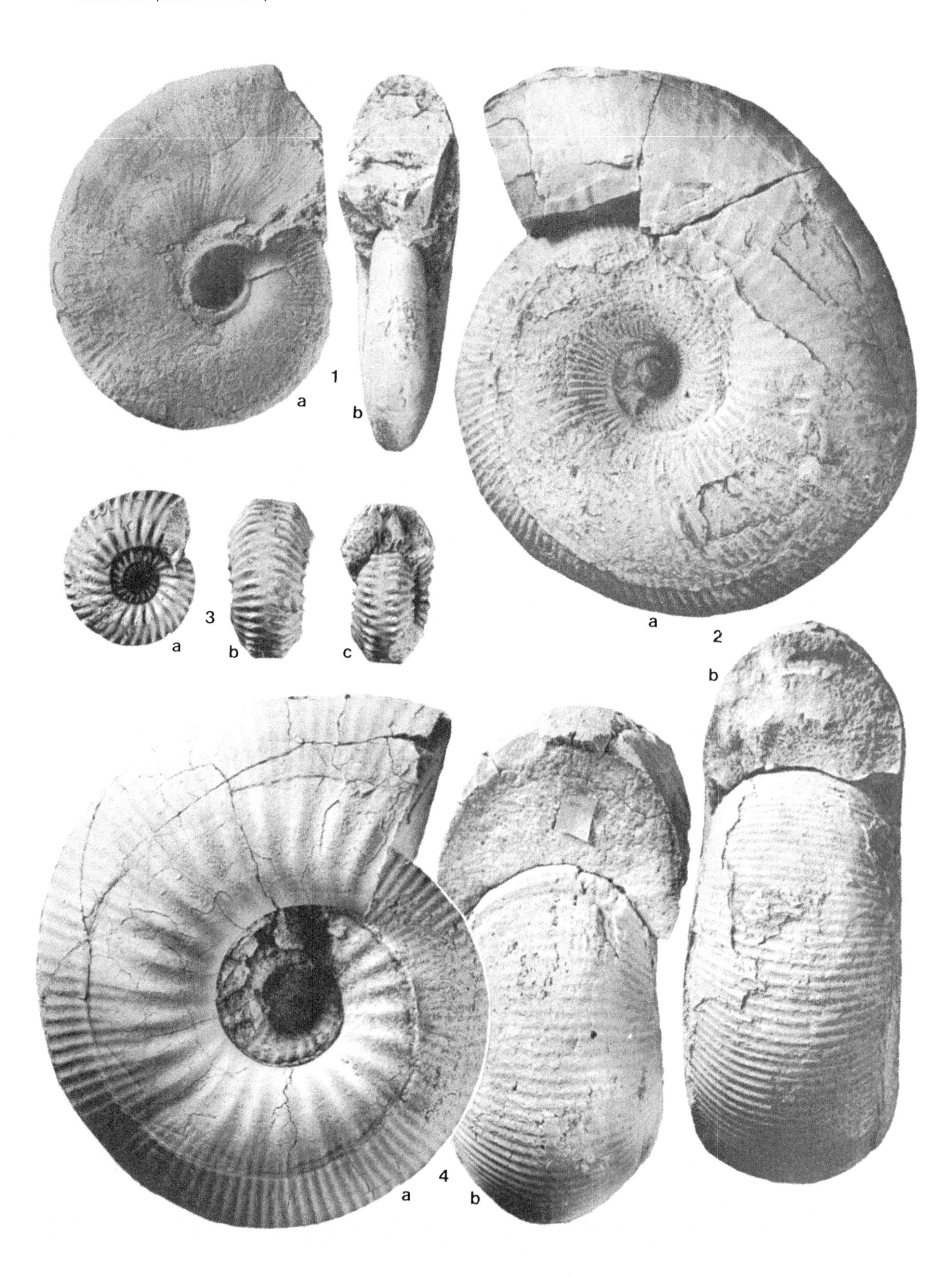
1
a
b
3
a
b
c
a
2
b
4
a
b

PLATE 29

SOUTHERN ALASKA, COOK INLET, LOWER BAJOCIAN

CRASSICOSTATUS ZONE

**Figs. 1, 2.** *Parabigotites crassicostatus* Imlay, ♀ & ♂; 1a,b, holotype almost complete macroconch (USNM 130909); 2a–c, allotype (here), complete microconch (holotype of "*Normannites kialagvikensis*" Imlay) (USNM 131401).

**Figs. 3, 4.** *Stephanoceras kirschneri* Imlay, ♀; 3, large holotype with part of body chamber (×0.35) (USNM 131421); 4a,b, inner whorls (USNM 131423).

OBLATUM ZONE

**Fig. 5a,b.** *Sphaeroceras* (*Chondroceras*) *allani* (McLearn), complete (USNM 131399).
All reproduced from Imlay (1964).

(Compiled by Westermann.)

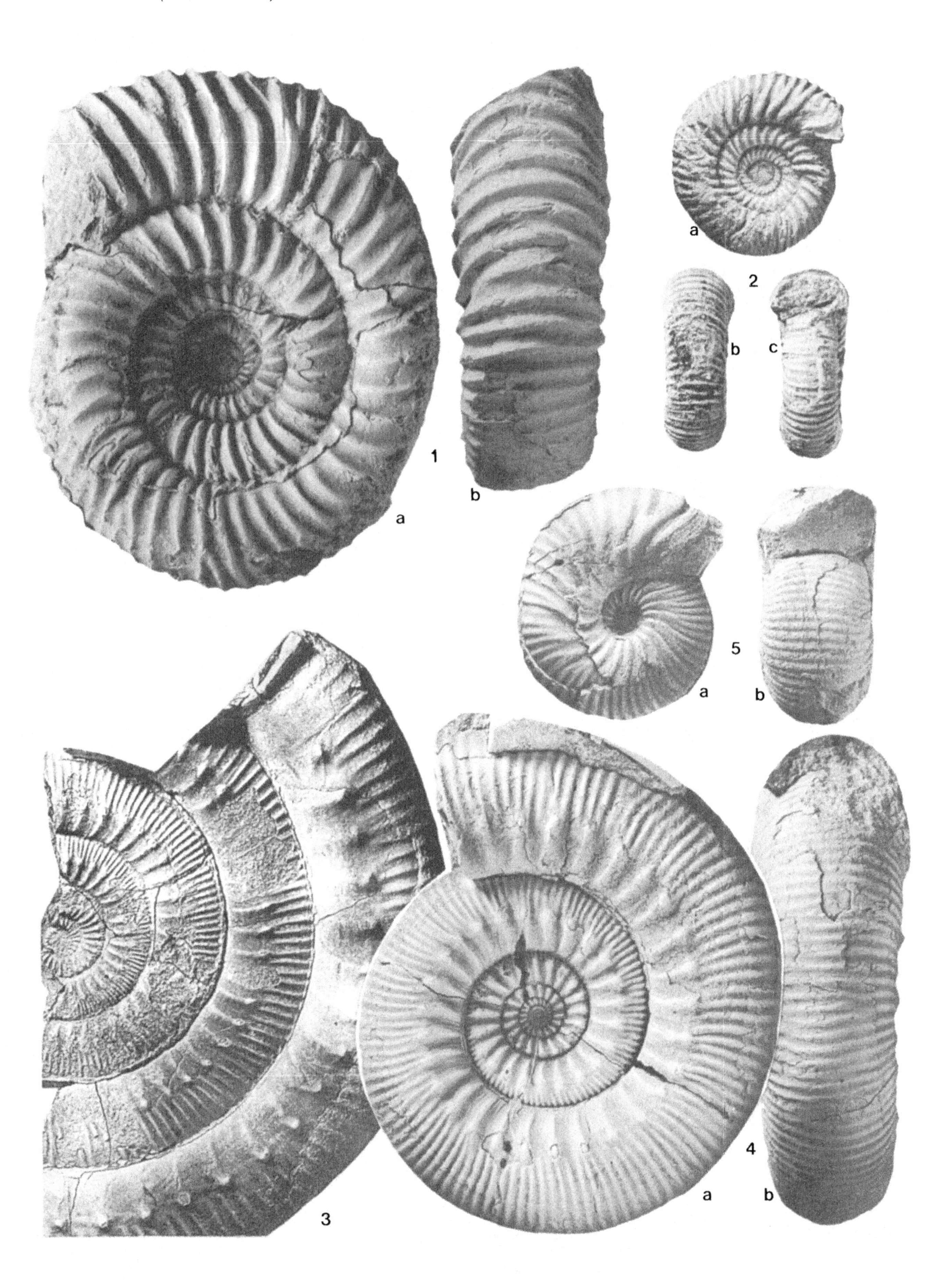
1
a
b
a
2
b
c
5
a
b
3
4
a
b

## PLATE 30

## SOUTHERN ALASKA, COOK INLET, LOWER BAJOCIAN

KIRSCHNERI ZONE

**Fig. 1a,b.** *Zemistephanus alaskensis* Hall & Westermann, ♀, holotype, complete (USNM 131434).

ROTUNDUM ZONE

**Figs. 2, 3.** *Megasphaeroceras rotundum* Imlay; 2, complete topotype (USNM 170899a); 3a,b, holotype, incomplete (USNM 130898).

**Figs. 4, 5.** *Strigoceras* (*Liroxyites*) *kellumi* Imlay; 4, holotype, complete but deformed (USNM 130886); 5a,b, phragmocone (USNM 130889b). All reproduced from Imlay (1964).

(Compiled by Westermann.)

1
a
b
2
3
a
b
4
5
a
b

PLATE 31

SOUTHERN ALASKA, COOK INLET, UPPER BAJOCIAN,
ROTUNDUM ZONE

**Fig. 1.** *Stephanoceras vigorosum* (Imlay), ♂, ("*Dettermanites*"), holotype, complete (USNM 130895).

**Fig. 2a,b.** *Sphaeroceras* (*S.*) *talkeetnanum* Imlay, holotype, complete (USNM 130902).

**Fig. 3a,b.** *Lissoceras bakeri* Imlay, ♀, holotype, fragment of phragmocone (USNM 130891).

**Fig. 4.** *Macrophylloceras grossimontanum* Imlay, phragmocone (USNM 130903).

**Fig. 5.** *Leptosphinctes cliffensis* Imlay, holotype, damaged but probably almost complete (USNM 130901).

**Fig. 6.** *Leptosphinctes delicatus* Imlay, holotype, damaged but with almost complete body chamber (USNM 130902).

All reproduced from Imlay (1964).

(Compiled by Westermann.)

1
2
a
b
3
a
b
4
5
6

PLATE 32

SOUTHERN ALASKA

BATHONIAN

**Fig. 1a,b.** *Iniskinites intermedius* (Imlay), ♀, holotype, complete (USNM 108225); from Upper Bathonian of Cook Inlet.

CALLOVIAN

**Fig. 2a,b.** *Kepplerites loganianus* (Whiteaves), ♀, complete (USNM 108127); from Lower Callovian of Cook Inlet.

**Fig. 3a,b.** *Cadoceras* (*Paracadoceras*) *tonniense* Imlay, ♀, holotype, damaged, with part of body chamber (USNM 108088); from Lower Callovian of Cook Inlet.

All reproduced from Imlay (1953).

(Compiled by Westermann.)

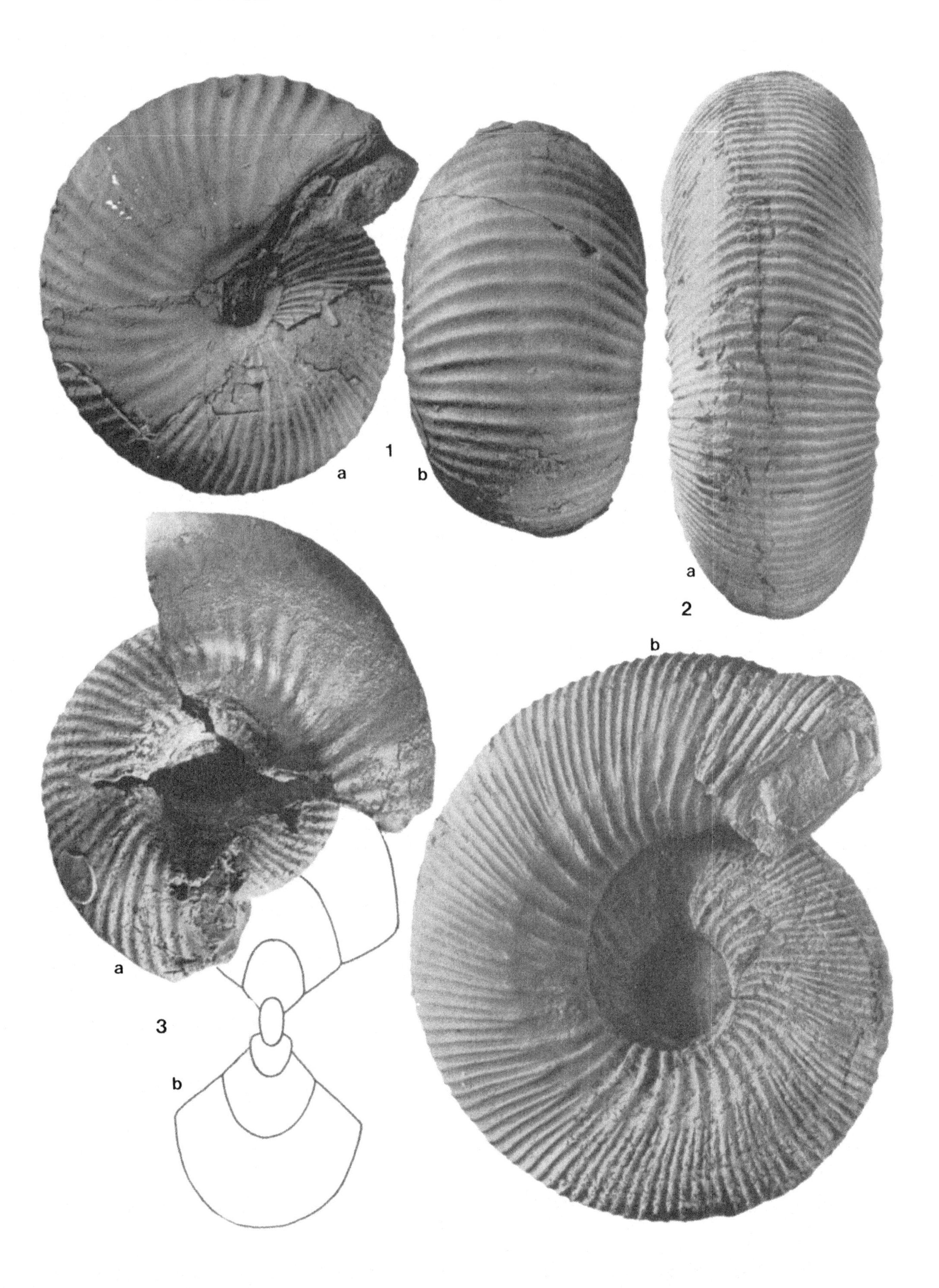
1
a
b
a
2
b
a
3
b

PLATE 33

SOUTHERN ALASKA, CALLOVIAN

**Fig. 1a,b.** *Lilloettia lilloetensis* (Crickmay), ♀, phragmocone with fragment of body chamber (USNM 108031); from Lower Callovian of Cook Inlet.

**Fig. 2a,b.** *Xenocephalites vicarius* Imlay, ♂, holotype with complete body chamber (USNM 108041); from Lower Callovian of Cook Inlet.

**Fig. 3a,b.** *Cadoceras coma* Imlay, ♀, holotype, incomplete (USNM 108057); from Lower Callovian of Cook Inlet.

**Fig. 4a,b.** *Cadoceras (Stenocadoceras) stenoloboide* Pompeckj, ♀, phragmocone (USNM 108092) (holotype of *C. multicostata* Imlay); from Middle Callovian of Cook Inlet.

**Fig. 5a–c.** *Cadoceras (Pseudocadoceras) petilini* Pompeckj, ♂, almost complete (USNM 108095b) (= *C. stenoloboide*?); from Middle Callovian of Alaska Peninsula.

**Fig. 6a,b.** *Longaeviceras pomeroyense* (Imlay), ♀, holotype with beginning of body chamber (USNM 108110); from Upper Callovian of Cook Inlet.

**Fig. 7a,b.** *Pseudocadoceras crassicostatum* Imlay, ♂, probably almost complete (USNM 108119); from Upper Callovian of Cook Inlet.

All reproduced from Imlay (1953).

(Compiled by Westermann.)

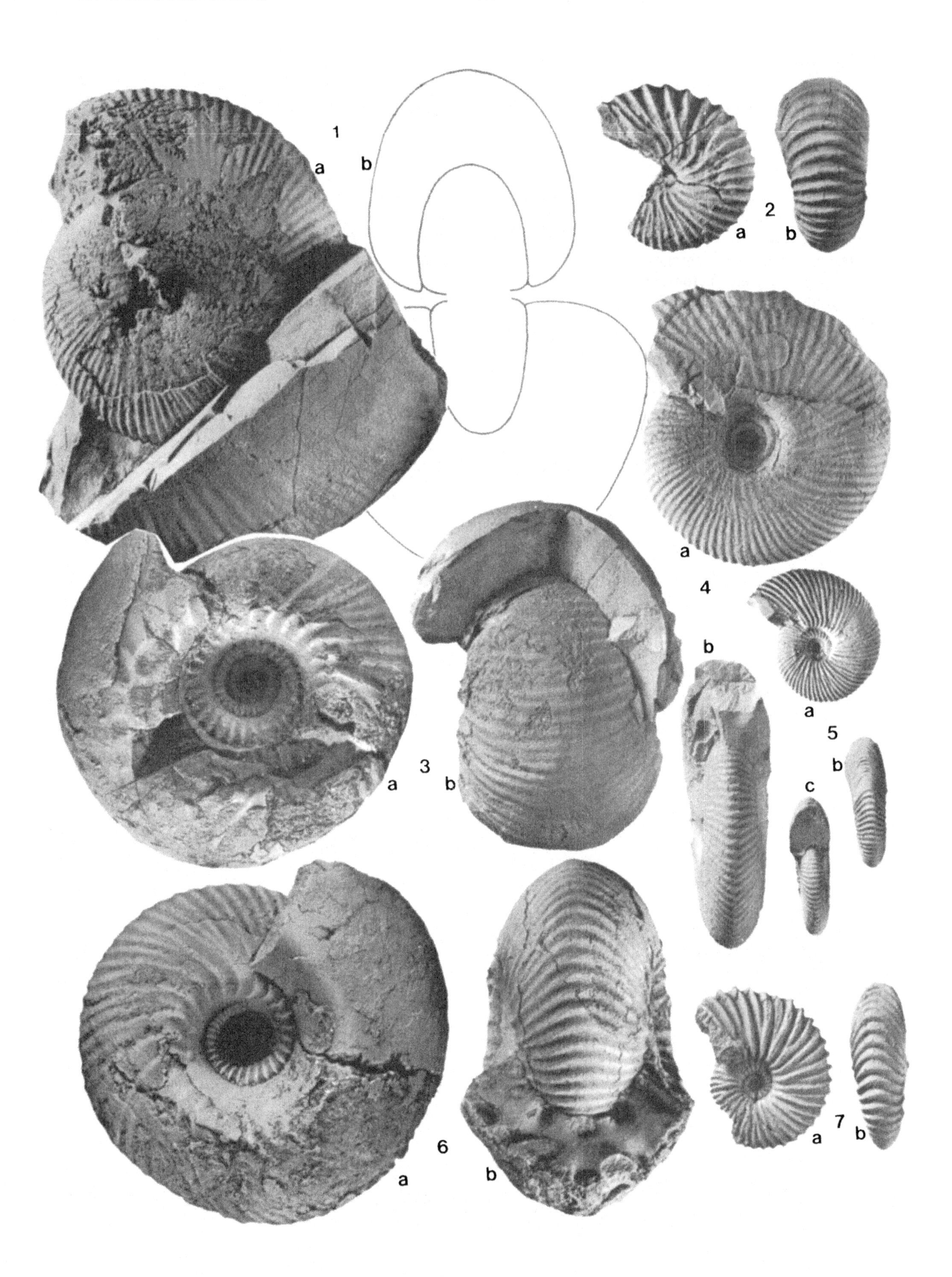
1
a
b
2
a
b
a
4
b
a
5
b
c
3
a
b
6
a
b
7
a
b

PLATE 34

BRITISH COLUMBIA, QUEEN CHARLOTTE ISLANDS, BAJOCIAN

**Figs. 1–3.** *Zemistephanus richardsoni* (Whiteaves), ♀ & ♂; 1, complete macroconch (McM J 1747a); 2a,b, macroconch phragmocone (McM-J 1797b); 3, microconch, allotype, complete (McM-J 1796a) (''*Kanastephanus*''); from *Z. richardsoni* Subzone, *S. kirschneri* Zone.

**Figs. 4, 5.** *Stephanoceras* (*Stephanoceras*) *skidegatensis* (Whiteaves), ♀ & ♂; 4, macroconch, fragment of phragmocone and body chamber (×0.7) (McM-J 1878); 5a,b, allotype, complete microconch (McM J 1802b); from Yakoun Formation, bed d.

All reproduced from Hall & Westermann (1980).

(Compiled by Westermann.)

1
2
a
b
3
4
5
a
b

PLATE 35

BRITISH COLUMBIA, SMITHERS AREA

BAJOCIAN

**Fig. 1.** *Epizigzagiceras evolutum* Frebold, holotype, incomplete (GSC 31583); from Upper Bajocian.

**Fig. 2.** *Epizigzagiceras crassicostatum* Frebold, holotype, damaged (GSC 31589); from Upper Bajocian.

BATHONIAN

**Fig. 3.** *Iniskinites tenacensis* Frebold, ♀, holotype, almost complete (GSC 54005); from ''*Iniskinites* fauna,'' late middle–early Late Bathonian.

**Fig. 4.** *Iniskinites robustus* Frebold, ♀, holotype, almost complete (GSC 53995); from ''*Iniskinites* fauna,'' late middle–early Late Bathonian.

All reproduced from Frebold & Tipper (1973) and Frebold (1978).

(Compiled by Westermann.)

1
2
3
4

PLATE 36

WESTERN INTERIOR CANADA, BAJOCIAN

**Fig. 1.** *Sonninia (Euhoploceras) modesta* Buckman (GSC 41673); from the Lower Bajocian of southwestern Alberta.
**Fig. 2a,b.** *Chondroceras oblatum* (Whiteaves), ♀, macroconch (McM-J 1836b); from the zone of *Chondroceras oblatum* (= Middle and Upper Humphriesianum Zone) of southwestern Alberta.
**Fig. 3.** *Alaskinia gracilis* (Whiteaves) (GSC 4809); from Lower Bajocian of southwestern Alberta.
**Fig. 4a,b.** *Stephanoceras itinsae* (McLearn), ♀, macroconch (McM-J 1880); from the zone of *Chondroceras oblatum* (= Middle and Upper Humphriesianum Zone) of southwestern Alberta.
**Fig. 5a,b.** *Stephanoceras itinsae* (McLearn), ♂, microconch (McM-J 1838); from the zone of *Chondroceras oblatum* (= Middle and Upper Humphriesianum Zone) of southwestern Alberta.
(Compiled by Hall.)

a
2
b
1
3
a
4
b
5a
5b

PLATE 37

WESTERN INTERIOR CANADA

BAJOCIAN(?)

**Fig. 7.** *Cranocephalites costidensus* Imlay (TMP 82.48.17); probably from Upper Bajocian, southeastern British Columbia.

BATHONIAN

**Fig. 1.** *Procerites engleri* Frebold, holotype (GSC 12906); from the Lower Bathonian of southwestern Alberta.

**Fig. 2.** *Parareineckeia* cf. *shelikofana* Imlay (×2) (TMP 82.48.13); from the Lower Bathonian of southeastern British Columbia.

**Fig. 3.** *Cobbanites* cf. *talkeetnanus* Imlay (×2) (TMP 82.48.1); from the Lower Bathonian of southeastern British Columbia.

**Fig. 4.** *Kepplerites* cf. *tychonis* Ravn (TMP 82.48.12); from the Boreal Variabile Zone of southwestern Alberta.

**Fig. 5a,b.** *"Imlayoceras" miettense* Frebold (GSC 14707); probably Upper Bathonian from central-western Alberta.

**Fig. 6.** *Xenocephalites vicarius* Imlay (TMP 87.155.6); from the Upper Bathonian of southwestern Alberta.

(Compiled by Hall.)

1
2
3
4
5b
6
7

PLATE 38

WESTERN INTERIOR CANADA, BATHONIAN

**Fig. 1a,b.** *Lilloettia imlayi* Frebold (GSC 12897); from Lower or Middle Bathonian, southwestern Alberta.

**Fig. 2.** *Kepplerites* cf. *costidensus* (Imlay) (TMP 87.155.2); from Boreal Cranocephaloides Zone of southwestern Alberta.

**Figs. 3, 5.** *Paracephalites glabrescens* Buckman; 3, (×0.8) (GSC 14695); 5 (GSC 14705); from the Lower Bathonian of southwestern Alberta.

**Fig. 4.** *Kepplerites subitus* (Imlay) (TMP 87.155.3); from the Upper Bathonian of southwestern Alberta. See also Plate 78, Fig. 1.

(Compiled by Hall.)

a
1
b
2
3
4
5

PLATE 39

WESTERN INTERIOR UNITED STATES

BAJOCIAN

**Fig. 1a,b.** *Eocephalites primus* Imlay, holotype (USNM 132938); from the Rotundum Zone of Wyoming.

**Fig. 2a,b.** *Megasphaeroceras* cf. *rotundum* Imlay (USNM 132939); from the Rotundum Zone of Utah.

**Fig. 3.** *Sohlites spinosus* Imlay, holotype (USNM 132788); from the Upper Bajocian of Montana.

BATHONIAN

**Fig. 4a,b.** *Xenocephalites saypoensis* (Imlay), holotype (USNM 104149); from the ?Lower Bathonian of Montana.

**Fig. 5a,b.** *Parachondroceras andrewsi* Imlay, holotype (USNM 132811); from the ?Lower Bathonian of Montana.

**Figs. 6, 7a,b.** *Paracephalites henryi* (Imlay); 6, holotype (USNM 314), from the Lower Bathonian of South Dakota; 7a,b (USNM 108258f), from the Lower Bathonian of South Dakota.

(Compiled by Hall.)

1
a
2
a
b
3
4
a
b
5
a
b
6
7a
7b

PLATE 40

WESTERN INTERIOR UNITED STATES, BATHONIAN

**Fig. 1a,b.** *Paracephalites codyensis* (Imlay), holotype (USNM 104132); from the ?Lower Bathonian of Wyoming.
**Fig. 2.** *Paracephalites muelleri* (Imlay) (USNM 108263); from the ?Upper Bathonian of Montana.
**Fig. 3a,b.** *Xenocephalites shoshonense* (Imlay), holotype (USNM 104143); from the ?Upper Bathonian of Montana.
**Fig. 4a,b.** *Procerites (Siemiradzkia) warmdonensis* (Imlay), holotype (USNM 108309); from the ?Upper Bathonian of Montana.
**Fig. 5a,b.** *Kepplerites costidensus* (Imlay), holotype (USNM 108297); from the Upper Bathonian of Montana.
**Fig. 6a,b.** *Kepplerites subitus* (Imlay), holotype (USNM 104155); from the Upper Bathonian of Montana.
**Fig. 7a,b.** *Kepplerites tychonis* Ravn (USNM 104173); from the Upper Bathonian of Montana.
(Compiled by Hall.)

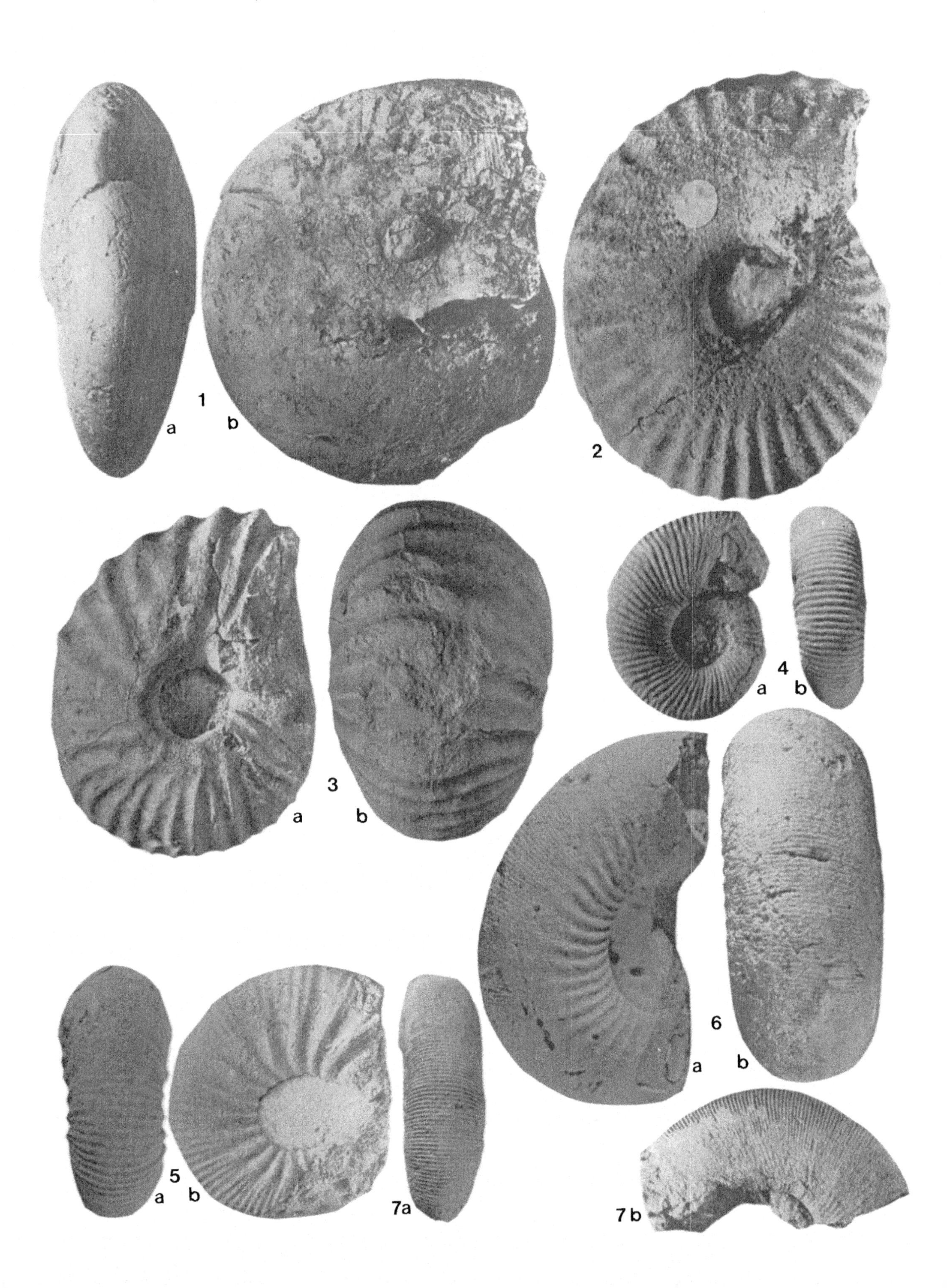
1
a
b
2
3
a
b
4
a
b
6
a
b
5
a
b
7a
7 b

PLATE 41

WESTERN INTERIOR UNITED STATES, BATHONIAN

**Fig. 1a,b.** *Xenocephalites crassicostatus* (Imlay), holotype (USNM 108250); from the ?Lower Bathonian of Wyoming.
**Fig. 2.** *Xenocephalites bearpawensis* Imlay, holotype (USNM 108240); from the Upper Bathonian of Montana.
**Fig. 3.** *Paracephalites muelleri* (Imlay), holotype (×0.4) (USNM 108262); from the ?Upper Bathonian of Montana.
**Fig. 4a,b.** *Kepplerites (Torricellites) knechteli* (Imlay), holotype (USNM 104152); from the Boreal Calyx Zone of Montana.
**Fig. 5a,b.** *Xenocephalites phillipsi* Imlay, holotype (USNM 108238); from the Boreal Calyx Zone of Montana.
**Fig. 6.** *Kepplerites mclearni* Imlay, holotype (×0.33) (USNM 104125); from the Boreal Calyx Zone of Montana.
For Callovian, see Plate 79, Figs. 1 and 2.
(Compiled by Hall.)

1
a
b
2
3
4
a
b
5
a
b
6

PLATE 42

OREGON, AALENIAN

**Fig. 1a,b.** *Docidoceras schnabelei* Taylor, holotype, phragmocone with incomplete body chamber (NWMNH 25019); from Packardi Zone, Snowshoe Formation, Oregon.

**Fig. 2a,b.** *Fontannesia involuta* Taylor, holotype, phragmocone with incomplete body chamber (38158); from Packardi Zone, Snowshoe Formation, Oregon.

**Fig. 3a,b.** *Planammatoceras*? *robertsoni* Taylor, holotype (×0.67), with beginning of body chamber (NWMNH 25005); from Packardi Zone, Snowshoe Formation, Oregon.

**Fig. 4a,b.** *Planammatoceras mowichense* Taylor, holotype, phragmocone (NWMNH 25003); from Mowichense Zone, Snowshoe Formation, Oregon.

**Fig. 5a,b.** *Planammatoceras vigrassi* Taylor, holotype, with incomplete body chamber of nearly ½ volution (NWMNH 25004); from Packardi Zone, Snowshoe Formation, Oregon.

**Fig. 6a,b.** *Strigoceras harrisense* Taylor, holotype, phragmocone (NWMNH 25015); from Sparsicostatus Zone, Snowshoe Formation, Oregon.

(Compiled by Taylor.)

1
a
b
a 2 b
3
a
b
4
a
b
5
a
b
6
a
b

PLATE 43

OREGON, AALENIAN & BAJOCIAN

**Fig. 1a,b.** *Sonninia washburnensis* Taylor, holotype (×0.5), nearly complete (NWMNH 25008); from Tuberculatum Zone, Snowshoe Formation, Oregon.
**Fig. 2a,b.** *Euhoploceras grantense* Taylor, holotype (×0.5) nearly complete (NWMNH 25010); from Tuberculatum Zone, Snowshoe Formation, Oregon.
**Fig. 3a,b.** *Euhoploceras crescenticostatum* Taylor, holotype (×0.55), with incomplete body chamber of ½ volution (NWMNH 25009); from Packardi Zone, Snowshoe Formation, Oregon.
**Fig. 4.** *Euhoploceras rursicostatum* Taylor, holotype (×0.5), probably almost complete (NWMNH 25014); from Tuberculatum Zone, Snowshoe Formation, Oregon.
**Fig. 5a,b.** *Euhoploceras ochocoense* Taylor, holotype (×0.5), with beginning of body chamber (NWMNH 25011); from Tuberculatum Zone, Snowshoe Formation, Oregon.
(Compiled by Taylor.)

1
a
b
2
a
b
3
a
b
4
5
b
a

PLATE 44

OREGON, BAJOCIAN

**Fig. 1a,b.** *Strigoceras lenticulare* Taylor, holotype, complete phragmocone (NWMNH 25016); from Tuberculatum Zone, Snowshoe Formation, Oregon.

**Fig. 2.** *Sonninia grindstonensis* Taylor, holotype, crushed and with incomplete body chamber of ½ volution (NWMNH 25007); from Crassicostatus Zone, Snowshoe Formation, Oregon.

**Fig. 3a,b.** *Sonninia burkei* Taylor, holotype, (×0.67), phragmocone with incomplete body chamber (NWMNH 25006); from Burkei Zone, Snowshoe Formation, Oregon.

**Fig. 4.** *Euhoploceras transiens* Taylor, holotype (×0.5), nearly complete (NWMNH 25012); from Tuberculatum Zone, Snowshoe Formation, Oregon.

**Fig. 5a,b.** *Euhoploceras westi* Taylor, holotype (×0.5) phragmocone with incomplete body chamber (NWMNH 25013); from Tuberculatum Zone, Snowshoe Formation, Oregon.

(Compiled by Taylor.)

1
a
b
2
3
a
b
4
5
a
b

PLATE 45

OREGON, BAJOCIAN

**Fig. 1a,b.** *Docidoceras amundsoni* Taylor, holotype (×0.5), almost complete (NWMNH 25018); from Tuberculatum Zone, Snowshoe Formation, Oregon.
**Fig. 2a,b.** *Docidoceras striatum* Taylor, holotype (×0.5), with incomplete body chamber less than one volution (NWMNH 25020); from Tuberculatum Zone, Snowshoe Formation, Oregon.
**Fig. 3a,b.** *Holcophylloceras supleense* Taylor, holotype, complete (NWMNH 25001); from Burkei Zone, Snowshoe Formation, Oregon.
**Fig. 4a,b.** *Hebetoxyites snowshoensis* Taylor, holotype (×0.67), complete (NWMNH 25017); from Tuberculatum Zone, Snowshoe Formation, Oregon.
**Fig. 5a,b.** *Ptychophylloceras compressum* Taylor, holotype, complete phragmocone (NWMNH 25002); from Tuberculatum Zone, Snowshoe Formation, Oregon.
(Compiled by Taylor.)

1
a
b
3
a
b
2
a
b
4
a
b
5
a
b

## PLATE 46

## EASTERN OREGON

### BAJOCIAN

#### CRASSICOSTATUS ZONE

**Fig. 1.** *Emileia (Chondromileia) buddenhageni* Imlay, ♀, holotype cast of damaged mold (USNM 168562).

**Fig. 2.** *Emileia* cf. *contracta* (J. Sowerby), ♂ (''*Otoites*''), cast of incomplete mold (USNM 168569).

**Fig. 3.** *Lissoceras hydei* Imlay, ♀, holotype, cast of damaged but almost complete mold (CAS 13475).

**Figs. 4, 5.** *Phaulostephanus movichensis* (Imlay), ♀, casts of complete molds; 4, holotype (USNM 168594); 5 (USNM 168595).

#### ROTUNDUM ZONE

**Figs. 6–8.** *Lupherites senecaensis* Imlay, ♀ & ♂; 6, holotype, damaged microconch (CAS 13526); 7, incomplete (CAS 13527); 8, fragmentary macroconch (USNM 168597).

### BATHONIAN

**Fig. 9.** *Iniskinites acuticostatum* Imlay, ♀, holotype, almost complete (USNM 252645); from late Middle–early Upper Bathonian.

All reproduced from Imlay (1964).

(Compiled by Westermann.)

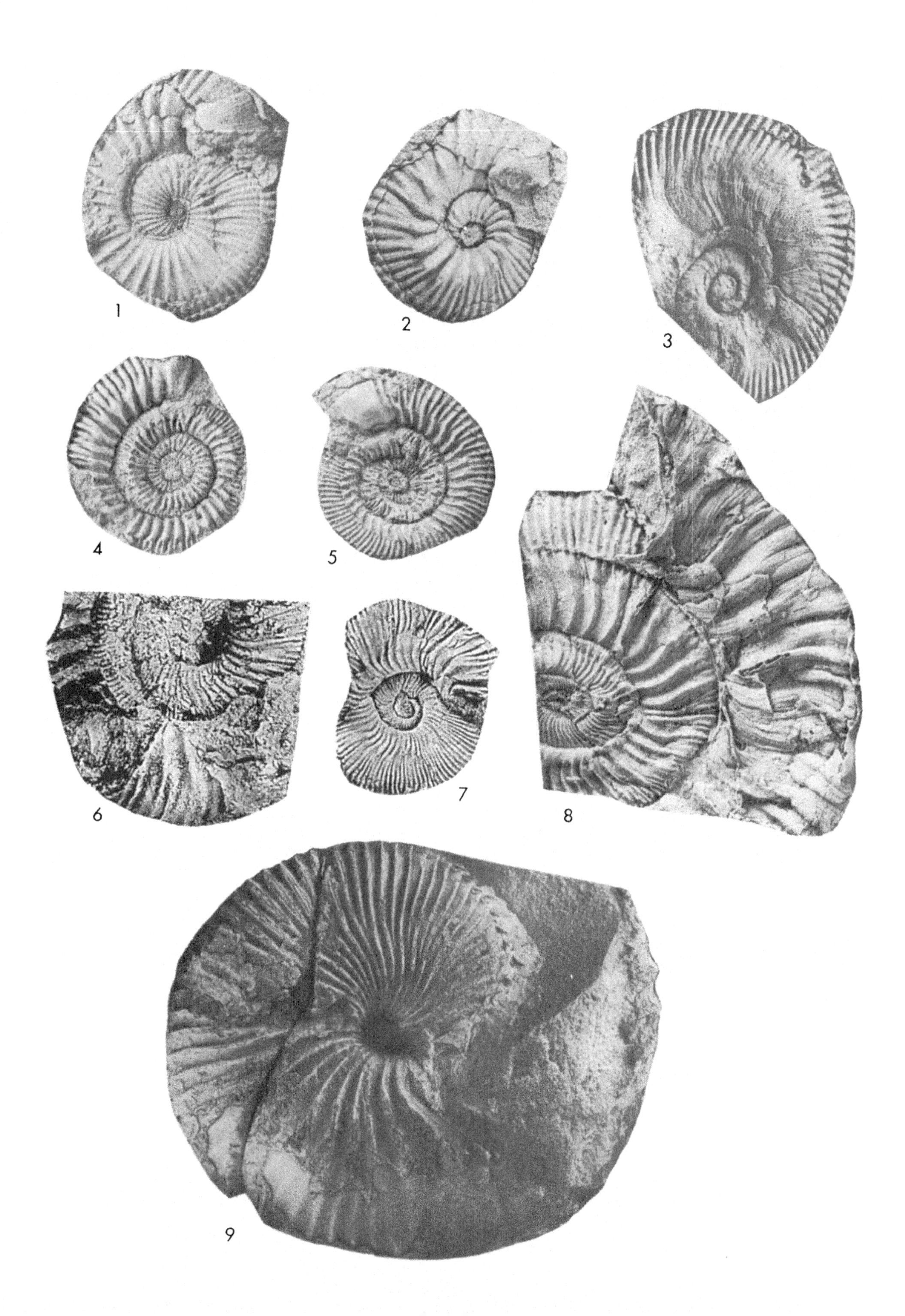
1
2
3
4
5
6
7
8
9

PLATE 47

SOUTHWESTERN MEXICO, OAXACA STATE, BAJOCIAN

**Fig. 1.** *Stephanoceras (S.) boulderense* (Imlay), ♂, almost complete (IGM 3904); from *Duashnoceras floresi* Zone of Mixtepec.
**Figs. 2, 3.** *Duashnoceras floresi* (Burck.); 2, ♂, almost complete (IGM 3913); 3, ♀, almost complete (J 2129); from *D. floresi* Zone of Mixtepec.
**Fig. 4a,b.** *Duashnoceras paucicostatum* (Felix), ♀, neotype, with beginning of body chamber (IGM 3917); from *D. floresi* Zone of Mixtepec.
**Fig. 5.** *Duashnoceras andinense* (Hilleb.), ♂, complete (IGM 3922); from *D. floresi* Zone of Mixtepec.
**Fig. 6a,b.** *Duashnoceras undulatum* (Burckh.), ♂, complete (IGM 3922); from *D. floresi* Zone of Mixtepec.
**Fig. 7a,b.** *Phaulostephanus burckhardti* Sand. & West., complete (IGM 3932); from *D. floresi* Zone of Mixtepec.
**Fig. 8a,b.** *Subcollina lucretia* (d'Orb.), ♂, almost complete (IGM 3936a); from *D. floresi* Zone of Mixtepec.
**Fig. 9.** *Parastrenoceras tlaxiacense* Ochot., ♀, probably almost complete (IGM 3943); from *P. zapotecum* Zone of Tlaxiaco.
**Fig. 10a,b.** *Parastrenoceras zapotecum* Ochot., ♂, fragment of phragmocone and body chamber (IGM 3942a); from *P. zapotecum* Zone of Tlaxiaco.
**Figs. 11, 12.** *Stephanosphinctes buitroni* Sand. & West.; 11, ?♂, (IGM 3947c); 12, holotype, ?♀, almost complete (IGM 3946); from *D. floresi* Zone of Mixtepec.
**Fig. 13.** *Leptosphinctes tabernai* West., septate (IGM 3945); from *P. zapotecum* Zone of Mixtepec.
(Compiled by Westermann.)

1
2
3
4
a
b
5
6
a
b
7
a
b
8
a
b
9
10a
b
11
12
13

PLATE 48

SOUTHWESTERN MEXICO, GUERRERO STATE, UPPER BATHONIAN

**Fig. 1a,b.** *Oxycerites* (*Alcidellus*) *obsoletoides* Ricc. et al., ♀, body chamber (IGM 4206); from Vergarensis Zone of Cualac.
**Fig. 2.** *Oxycerites* (*Alcidellus*) cf. *tenuistriatus* (Gorss.), ♀, phragmocone (IGM 4203); from Steinmanni Zone of Cualac.
**Fig. 3.** *Epistrenoceras histricoides* (Rollier), ♀, almost complete (IGM 4251); from Steinmanni Zone of Cualac.
**Fig. 4a,b.** *Prohecticoceras blanazense* Elmi, ♀, incomplete phragmocone (IGM 4256); from Steinmanni Zone of Cualac.
**Fig. 5a,b.** *Lilloettia steinmanni* (Spath), ♀, complete (IGM 4228); from Steinmanni Zone of Cualac.
**Fig. 6a,b.** *Bullatimorphites* (*Kheraiceras*) *bullatum* (d'Orb.), ♀, almost complete (IGM 4247); from Steinmanni Zone, *histricoides* Association, of Cualac.
**Figs. 7, 8.** *Neuqueniceras* (*N.*) *plicatum* (Burckh.), ♀; 7a,b, with complete body chamber (McM-J 2119a), purchase; 8a,b, outer whorl (IGM 4260); from Steinmanni Zone of Cualac, *Lilloettia* Association.
(Compiled by Westermann.)

a
1
b
2
3
a
4
b
a
5
b
6
a
b
a
7
b
8 a
b

PLATE 49

SOUTHWESTERN MEXICO, GUERRERO STATE, CALLOVIAN

**Fig. 1a,b.** *Eurycephalites vergarensis* (Burckh.), ♀, complete (IGM 4229); from Vergarensis Zone of Cualac.
**Fig. 2.** *Xenocephalites nikitini* (Burckh.), ♂, body chamber, (×2) (IGM 4227); from Bodenbenderi Zone, *inflatum* Association, of Cualac.
**Fig. 3.** *Jeanneticeras malbosei* Elmi, ♂, complete (×2) (IGM 4210); from Bodenbenderi Zone, *inflatum* Association, of Cualac.
**Fig. 4, 5.** *Guerrericeras inflatum* (West.), ♂; 4a,b, holotype; 5a,b, fragment; from Bodenbenderi Zone, *inflatum* Association, of Cualac.
**Fig. 6a–d.** *Parapatoceras distans* (Baugier & Sauze), fragments (IGM 4222–4); from Bodenbenderi Zone, *inflatum* Association, of Cualac.
**Fig. 7a,b.** *Neuqueniceras (Frickites)* cf. *bodenbenderi* (Torn.), ♀, phragmocone; from Bodenbenderi Zone of Cualac.
**Fig. 8a,b.** *Reineckeia* gr. *franconica* (Que.), ♂, phragmocone; from Bodenbenderi Zone, *rehmanni* Association, of Cualac.
**Fig. 9a,b.** *Rehmannia* gr. *rehmanni* (Oppel), ♂, fragments of phragmocone and body chamber (IGM 4284); from Bodenbenderi Zone, *rehmanni* Association, of Cualac.
**Fig. 10a,b.** *Reineckeia latesellata* Burckh., ♀, holotype, almost complete (×0.75); from Middle–Upper Callovian of El Consuelo, Oaxaca.
(Compiled by Westermann.)

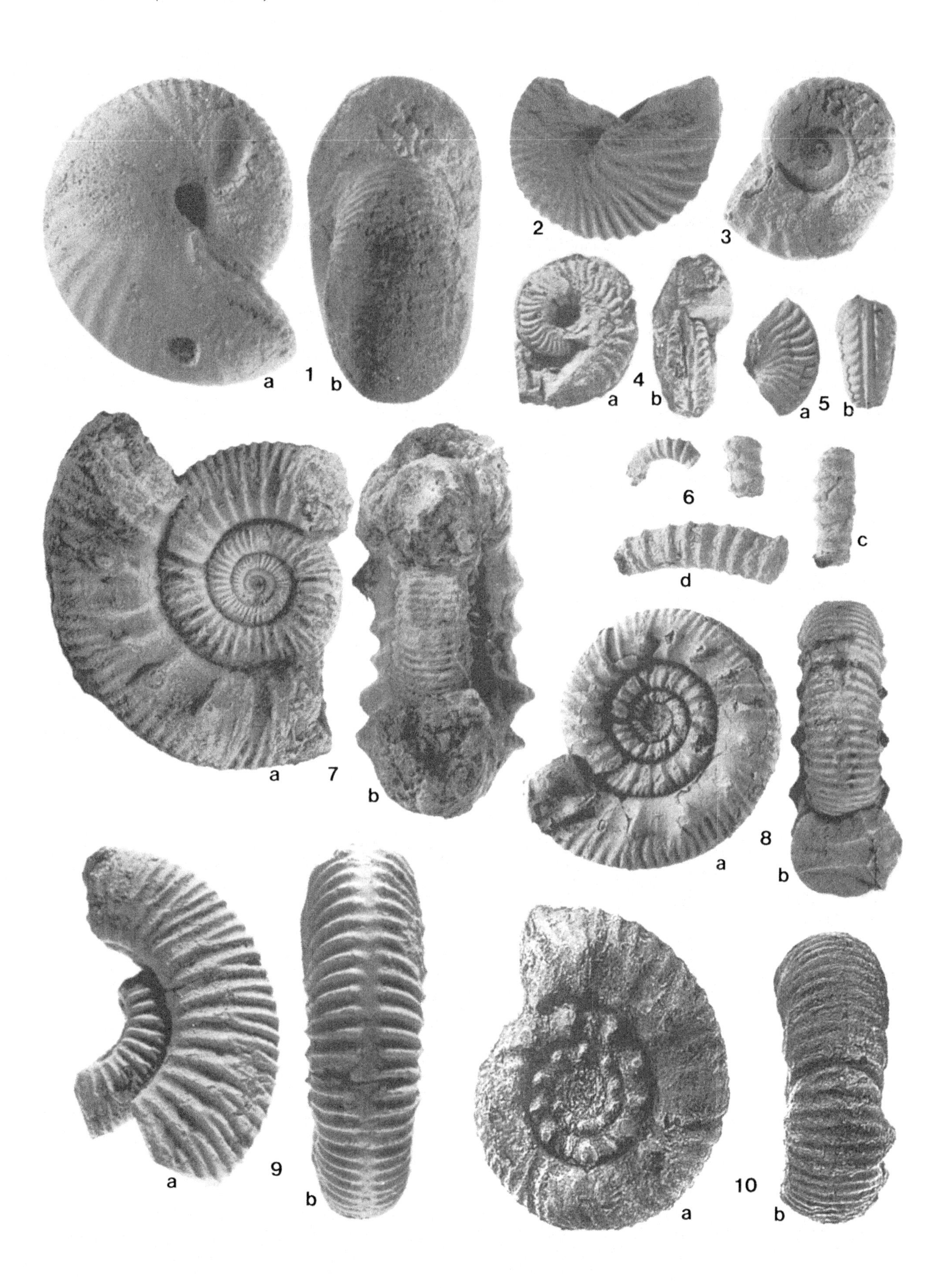

1 a b
2
3
4 a b
5 a b
6 c d
7 a b
8 a b
9 a b
10 a b

PLATE 50

NORTHERN CHILE, AALENIAN

MANFLASENSIS ZONE

**Fig. 1a,b.** *Sphaerocoeloceras brochiiforme* Jaw., complete (TUB-B.St.M. 1983V13); from Manflas, Atacama.

**Fig. 2a,b.** *Bredyia manflasensis* West., ♀, holotype, phragmocone; from Manflas.

**Fig. 3a,b.** *Bredyia delicata* West., ♀, holotype, phragmocone; from Manflas.

GROEBERI ZONE

**Fig. 4a,b.** *Parammatoceras jenseni* West., ♀, holotype, phragmocone; from Quebrada el Asiento, Atacama.

(Compiled by Westermann.)

a
1
b
2
a
b
b
3
a
b
4 a

# PLATE 51

## NORTHERN CHILE AND ARGENTINA, AALENIAN

MANFLASENSIS ZONE

**Fig. 1a,b.** *Ancolioceras*(?) *chilense* (West.), ♀, holotype, phragmocone; from lower Manflasensis Zone of Manflas, Atacama.

**Figs. 2, 3.** *Podagrosiceras athleticum* Maub. & Lamb.; 2a,b, ♂, almost complete; 3a,b, *P.* cf. *athleticum,* ♀, phragmocone, with end of body chamber (MLP 12817); from Sierra de Reyes, Mendoza.

MALARGUENSIS ZONE

**Fig. 4a–c.** *Podagrosiceras maubeugei* West. & Ricc., ♀, holotype, damaged complete specimen, inner whorls (c) (×2); from Cerro, Loteno, Neuquén.

**Fig. 5** *Planammatoceras* (*Pseudaptetoceras*) *moerickei* (Jaw.), ♀, holotype (plaster cast), phragmocone; from Arroyo Blanco, Mendoza.

**Fig. 6a,b.** *Fontannesia*(?) *austroamericana* Jaw., holotype, complete; from Cerro Tricolor, Mendoza.

GROEBERI ZONE

**Fig. 7a,b.** "*Zurcheria*" *groeberi* West. & Ricc., holotype, with body chamber; from Bardas Blancas, Mendoza.

(Compiled by Westermann.)

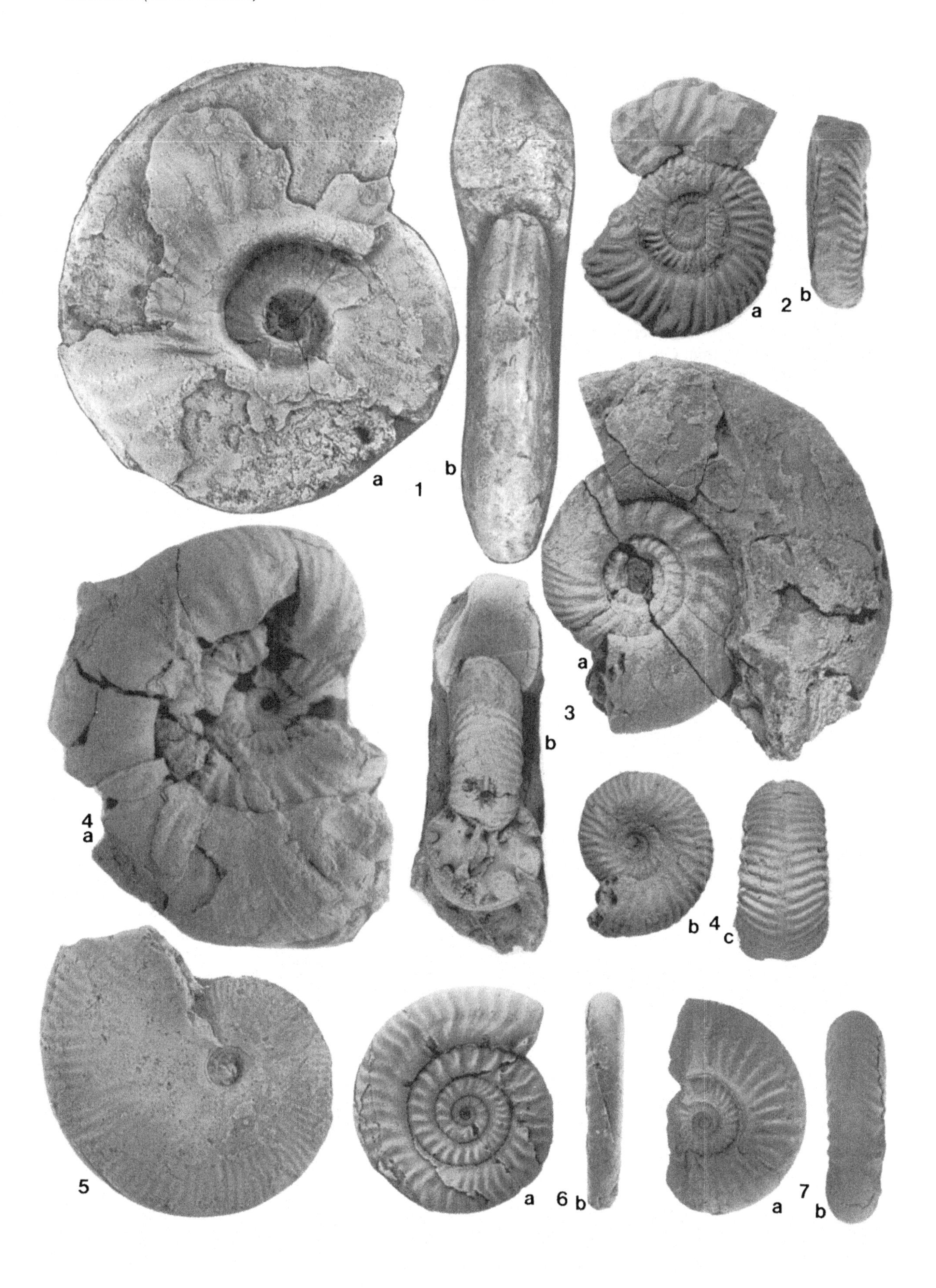
a
1
b
a
2
b
a
3
b
4
a
b
4
c
5
a
6
b
a
7
b

PLATE 52

ARGENTINA, AALENIAN–BASAL BAJOCIAN

MALARGUENSIS ZONE

**Fig. 1a,b.** *Planammatoceras (Pseudaptetoceras) tricolore* West. & Ricc., ♀, holotype, phragmocone with beginning of body chamber; from *P. moerickei* assemblage of Cerro Tricolor, Mendoza.

**Fig. 2a,b.** *Puchenquia (Gerthiceras) compressa* West. & Ricc., ♀, holotype, almost complete; from Río Potimalal, Mendoza.

**Fig. 3.** *Praeleptosphinctes jaworskii* West., holotype, phragmocone; from Cerro Tricolor, Mendoza.

**Fig. 4a,b.** *Eudmetoceras (Euaptetoceras) amplectens* (Buck.), ♀, damaged phragmocone (MLP 14497); from Cerro Tricolor.

**Fig. 5.** *Planammatoceras (P.) gerthi* (Jaw.), lectotype, phragmocone; from Cerro Tricolor.

GROEBERI ZONE

**Fig. 6.** *Tmetoceras scissum* (Ben.), phragmocone and body-chamber fragments; from Arroyo Blanco, Mendoza.

(Compiled by Westermann.)

1 a
b
2 a
b
a 6
b
3
a 4 b
5

PLATE 53

ARGENTINA, LOWER BAJOCIAN

**Figs. 1–3.** *Puchenquia malarguensis* (Burck.); 1, ♀, body chamber (McM-J 1300:38); 2a,b, ♀, lectotype, phragmocone; 3a,b, ♂, allotype ("*H. puchense*"); from Malarguensis Zone of Cerro Puchénque, Mendoza.
**Figs. 4–6.** *Sonninia (Euhoploceras) amosi* West. & Ricc.; 4a,b, ♀, phragmocone (McM-J 1748:29); 5a,b, inner whorls of spinose variant (McM-J 1300:32); 6a,b, ♂, allotype, complete; from Malarguensis Zone of Cerro Puchénque.
**Figs. 7, 8.** *Sonninia (S.?) alsatica* (Haug); 7a,b, ♀, phragmocone (McM-J 1300:36); 8a,b, ♂, complete (McM-J 1717:5); from Giebeli Zone of Charahuilla, Neuquén.
**Figs. 9, 10.** *Sonninia (Fissilobiceras) zitteli* (Gott.), ♀; 9a,b, phragmocone of involute variant near type; 10a,b, juvenile, modal form (MLP 11422); from Singularis Zone of Arroyo Blanco, Mendoza.
(Compiled by Westermann.)

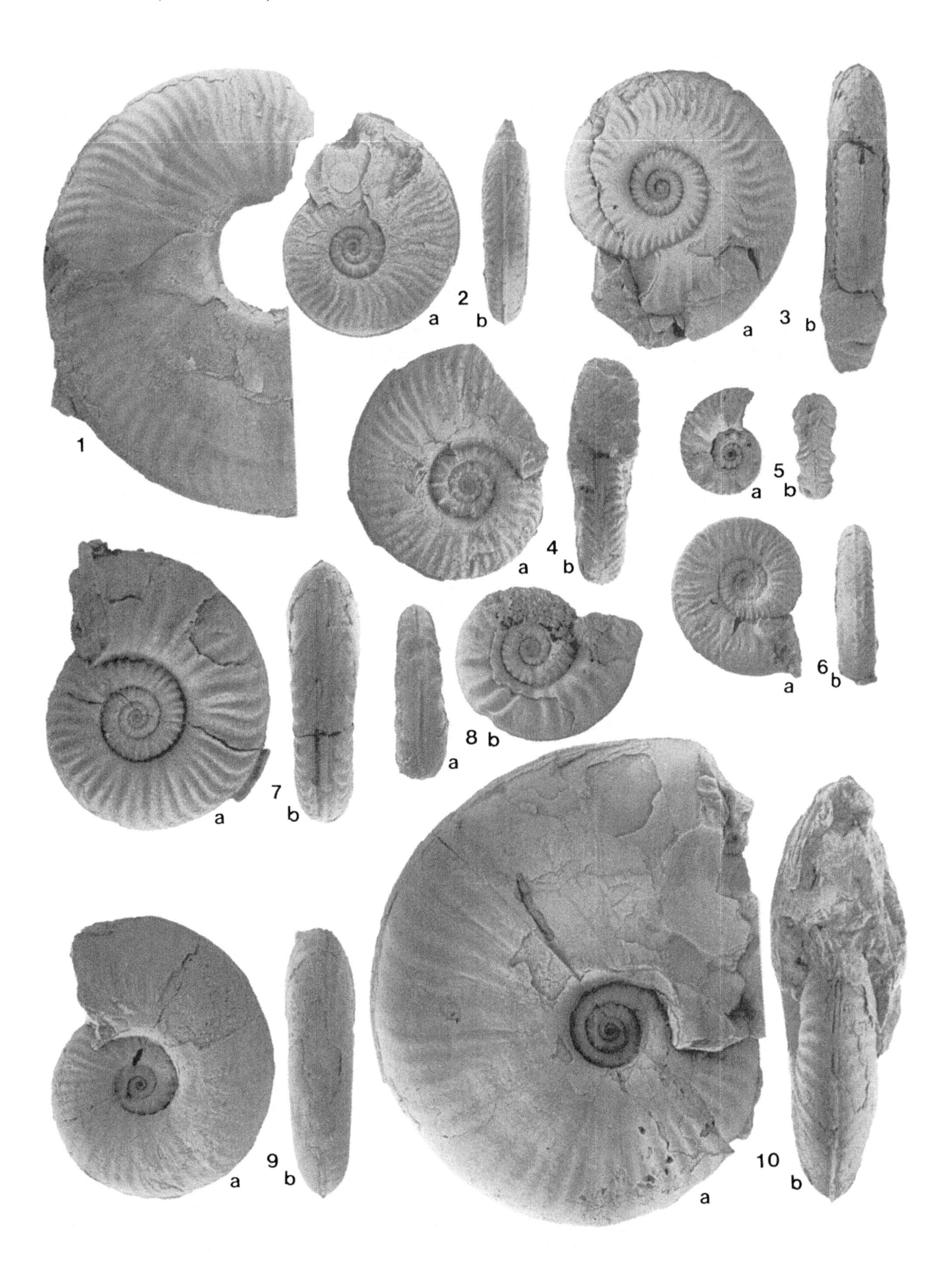
1
2
a
b
3
a
b
4
a
b
5
a
b
6
a
b
7
a
b
8
a
b
9
a
b
10
a
b

PLATE 54

ARGENTINA, LOWER BAJOCIAN. GIEBELI ZONE

**Figs. 1, 2.** *Sonninia* (*Papilliceras*) *espinazitensis* Tornq., ♀; 1, mature phragmocone (MLP), from Paso del Espinacito, San Juan; 2a,b, inner whorls (McM-J 1311:4), from Charahuilla, Neuquén.

**Figs. 3, 4.** *Sonninia* (*Papilliceras*) cf. *espinazitensis* Tornq., ♂; 3a,b, syntype of "*S. subdeltafalcata*" Tornq.; 4a,b, holotype of "*S. bodenbenderi*" Tornq.; from Paso del Espinacito, San Juan.

**Fig. 5.** *Dorsetensia mendozai* West. & Ricc., ♀, holotype, almost complete; from upper Giebeli Zone of Arroyo Blanco, Mendoza.

**Fig. 6a,b.** *Dorsetensia*(?) cf. *mendozai* West. & Ricc.; ♂, complete (McM-J 1324:8); from Arroyo Blanco.

**Figs. 7, 8.** *Dorsetensia blancoensis* West. & Ricc.; 7a,b, ♀, holotype; 8a,b, ♂, damaged, but with aperture (McM-J 1324:4); from upper Giebeli Zone of Arroyo Blanco, Mendoza.

(Compiled by Westermann.)

1
2
a
b
3
a
b
4
a
b
5
6
a
b
7
a
b
8
a
b

PLATE 55

CHILE AND ARGENTINA, LOWER BAJOCIAN

HUMPHRIESIANUM ZONE

**Figs. 1, 2.** *Dorsetensia romani* (Oppel), ♀, phragmocone and (♂?) fragment (TUB 66017/6,6a); from Salar de Pedernales, N Chile.

**Fig. 3.** *Dorsetensia tecta* Buck., ♀, phragmocone (TUB 660614/6b); from Salar del Pedernales.

GIEBELI ZONE

**Figs. 4–6.** *Emileia* (*E.*) *multiformis* (Gott.); 4, ♀, phragmocone and end of body chamber (MLP 12519), from Los Molles, Neuquén; 5a,b, ♀, phragmocone (McM-J 15554), from Carro Quebrado, Neuquén; 6, ♂, complete (MLP 12517), from Paso de la Guardia, San Juan.

**Fig. 7a,b.** *Emileia* (*Chondromileia*) *giebeli* (Gott.) s.s., ♀, complete (MLP 12548); from Charahuilla, Neuquén.

(Compiled by Westermann.)

1
2
3
4
5
a
b
6
7
a
b

PLATE 56

ARGENTINA, LOWER BAJOCIAN

GIEBELI ZONE

**Figs. 1, 2.** *Emileia* (*Chondromileia*) *giebeli submicrostoma* (Gott.), ♀; 1a,b, complete (MLP 12564), from Charahuilla, Neuquén; 2a,b, phragmocone (MLP 12586), from Carro Quebrado, Neuquén.

**Fig. 3a,b.** *Emileia* (*Chondromileia*) *giebeli* s.s. (Gott.), ♂, complete (MLP 12530); from Charahuilla.

SINGULARIS ZONE

**Figs. 4, 5.** *Pseudotoites sphaeroceroides* (Tornq.); 4a,b, ♀, phragmocone (MLP 12616); 5a,b, ♂, allotype, complete; from Carro Quebrado, Neuquén.

**Fig. 6a,b.** *Pseudotoites transatlanticus* (Tornq.), ♀, phragmocone and part of body chamber (MLP 12620); from Carro Quebrado.

(Compiled by Westermann.)

1
a
b
2
a
b
3
a
b
4
a
b
5
a
b
6
a
b

PLATE 57

CHILE AND ARGENTINA, LOWER BAJOCIAN

SINGULARIS ZONE

**Fig. 1a,b.** *Pseudotoites crassus* West. & Ricc., ♀, holotype, almost complete; from Paso del Espinacito, San Juan.

**Figs. 2, 3.** *Pseudotoites singularis* (Gott.); 2a,b, ♀, almost complete (MLP), from Cerro Puchénque; 3a,b, ♂ [''*Latotoites evolutus* (Torn.)''], complete, from Carro Quebrado, Neuquén.

GIEBELI ZONE

**Figs. 4, 5.** *Sphaeroceras rectecostatum* (West. & Ricc.); 4a,b, ♀, holotype, almost complete; 5a,b, ♂, complete, allotype; from Carro Quebrado, Neuquén.

**Fig. 6.** *Stephanoceras* (*Skirroceras*) cf. *macrum* (Qu.), ♀, fragment of last two whorls (MLP 18106); from Chacaico, Neuquén.

**Fig. 7.** *Stephanoceras* (*Skirroceras*) cf. *julei* Imlay, ♀, septate (MLP 6645); from Chacaico, Neuquén.

HUMPHRIESIANUM ZONE

**Fig. 8.** *Stephanoceras* (*S.*) *pyritosum* (Qu.), ♀, phragmocone with body-chamber fragment (TUB-GA 1035b); from Salar de Pedernales, N Chile.

**Fig. 9.** *Sphaeroceras* (*Chondroceras*/*Defonticeras*) cf. *defontii* (McLearn), ♀, phragmocone and part of aperture (McM-J 1190a); from Manflas, N Chile.

(Compiled by Westermann.)

1
a
b
2
a
b
9
4
a
b
3
a
b
6
5
a
b
7
8

PLATE 58

CHILE AND ARGENTINA, UPPER BAJOCIAN, ROTUNDUM ZONE

**Fig. 1a,b.** *Teloceras crickmayi chacayi* West. & Ricc., ♀, holotype, almost complete, but body chamber largely cut away (×0.7) (MLP 13204); from Chacay Melehue, Neuquén.

**Fig. 2.** *Duashnoceras paucicostatum chilense* (Hill.), ♀, holotype, phragmocone; from Quebrada Juncal, Antofagasta.

**Figs. 3, 4.** *Lupherites dehmi* (Hill.); 3a,b, ♂?, incomplete phragmocone (B.St.M. 1976XV121); 4, ♀, phragmocone with beginning of body chamber (XV121); from Cordillera Domeyko, N Chile.

**Fig. 5.** *Cadomites* sp. nov. *A.* aff. *deslongchampsii* (d'Orb.), ♀, parts of phragmocone and body chamber (MLP); loose from Arroyo Blanco, Mendoza.

(Compiled by Westermann.)

a
1
b
2
3
a
b
4
5

PLATE 59

CHILE AND ARGENTINA, UPPER BAJOCIAN

**Fig. 1a,b.** *Megasphaeroceras magnum* Ricc. & West., holotype, almost complete; from Rotundum Zone.
**Fig. 2.** *Megasphaeroceras magnum spissum* Ricc. & West., almost complete (TUB 30–280873); from N Chile.
**Figs. 3, 4.** *Leptosphinctes* aff. *coronarius* Buck.; 3a,b, inner whorls; 4a,b, damaged mature specimen (MLP); from Rotundum Zone, Chacay Melehue, Neuquén.
**Fig. 5.** ?*Strenoceras* (or ?*Parastrenoceras*) sp. juv. (MLP); from Rotundum Zone, Chacay Melehue.
(Compiled by Westermann.)

1
a
b
2
3
a
b
a
4
b
5

PLATE 60

CHILE AND ARGENTINA, BATHONIAN

**Fig. 1.** *Cadomites* ex gr. *rectelobatus* (Hauer), ♀, latex mold of damaged impression, with part of aperture; from *Cadomites*–Tulitidae mixed assemblage of Chacay Melehue, Neuquén.

STEINMANNI ZONE

**Fig. 2a,b.** *Lilloettia steinmanni* (Spath), ♀, lectotype, phragmocone with beginning of body chamber; from Caracoles, N Chile.

**Fig. 3a,b.** *Xenocephalites neuquensis* (Stehn), ♂, holotype; from Chacay Melehue.

**Fig. 4a,b.** *Xenocephalites neuquensis chilensis* Ricc. & West., ♂, holotype, complete; from Antofagasta province, Chile.

**Fig. 5a,b.** *Stehnocephalites gerthi* (Spath), complete topotype; from *gerthi* horizon, Chacay Melehue.

**Fig. 6.** *Choffatia jupiter* (Stein.), ♀, fragment with incomplete body chamber; from *jupiter* horizon, Caracoles.

(Compiled by Westermann.)

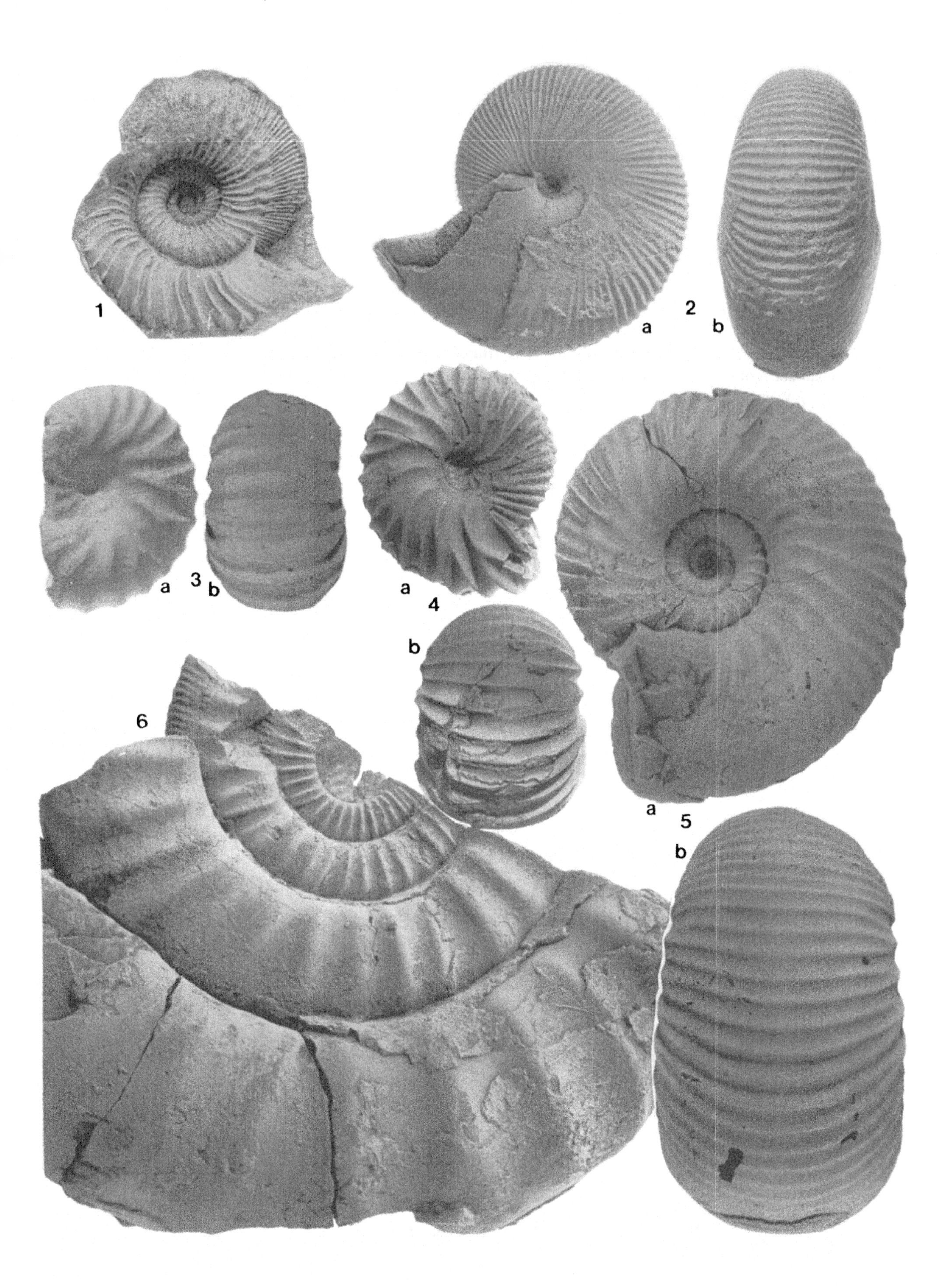
1
a
2
b
a
3
b
a
4
b
6
a
5
b

## PLATE 61

## CHILE AND ARGENTINA, TOP BAJOCIAN–BATHONIAN

**Fig. 1.** *Lobosphinctes intersertus* Buck., ♀, almost complete (×0.5); from uppermost *Megasphaeroceras* range zone of Chacay Melehue, Neuquén.

STEINMANNI ZONE

**Fig. 2a–c.** *Hecticoceras* (*Prohecticoceras*) *blanazense* Elmi, ♀, phragmocone; Caracoles, Antofagasta.

**Fig. 3a,b.** *Eohecticoceras* sp., with part of body chamber; from Caracoles.

**Fig. 4a–d.** *Oxycerites* (*Paroxycerites*) *exoticus* (Stein.), ♀, almost complete; c–d, inner whorls; from Caracoles.

**Fig. 5a,b.** *Bullatimorphites* (*Kheraiceras*) *v-costatus* (Burck.), ♀, complete, part of body chamber removed in b to show venter; from Caracoles.

(Compiled by Westermann.)

1
a
2 b
c
3
a
b
4
a
b
c
d
5
a
b

PLATE 62

ARGENTINA, NEUQUEN PROVINCE, UPPER BATHONIAN.
STEINMANNI ZONE

**Fig. 1a,b.** *Neuqueniceras steinmanni* (Stehn), ♀, almost complete topotype; from Chacay Melehue.
**Fig. 2a,b.** *Neuqueniceras biscissum* (Stehn), ♀, almost complete topotype (MLP 23957); from Chacay Melehue.
**Fig. 3.** *Iniskinites crassus* Ricc. & West., ♀, almost complete (MLP 12727); from Chacay Melehue.
**Fig. 4.** *Iniskinites gulisanoi* Ricc. & West., ♀, holotype; from Chacay Melehue.
(Compiled by Westermann.)

a
1
b
a
2
b
3
4

PLATE 63

ARGENTINA, LOWER CALLOVIAN

VERGARENSIS ZONE

**Fig. 1a,b.** *Eurycephalites vergarensis* (Burckh.), ♀, complete (MLP 12658); from Chacay Melehue, Neuquén.

**Fig. 2a,b.** *Xenocephalites gottschei* (Torn.), ♂, almost complete body chamber; from Chacay Melehue.

BODENBENDERI ZONE

**Fig. 3a,b.** *Eurycephalites rotundus* (Torn.), ♀, complete (MLP 12694); from Chacay Melehue.

**Fig. 4a,b.** *Xenocephaltes stipanicici* Ricc. et al., ♂, holotype, complete; from Vega de Veranda, Neuquén, (or ?Proximum Zone).

**Fig. 5.** *Neuqueniceras (Frickites) bodenbenderi* (Torn.), ♀, lectotype, damaged but almost complete (×0.5); from Paso del Espinacito, San Juan.

(Compiled by Westermann.)

1
a
b
4a
a
2
b
3
a
b
5
4b

PLATE 64

CHILE AND ARGENTINA, LOWER CALLOVIAN

BODENBENDERI ZONE

**Fig. 1a,b.** *Oxycerites* (*O.*) cf. *oppeli* Elmi, ♀, phragmocone; from Caracoles, Antofagasta.

**Fig. 2a,b.** *Oxycerites* (*Alcidellus*) *obsoletoides* Ricc. et al., ♀, holotype; from Caracoles.

PROXIMUM ZONE

**Fig. 3a,b.** *Hecticoceras* (*H.*) *proximum* Elmi, ♀, fragment with complete body chamber; from Chacay Melehue, Neuquén.

**Fig. 4a,b.** *Hecticoceras* (*Chanasia*) *ardenscicum* Elmi, ♀, almost complete body chamber; from Chacay Melehue.

**Fig. 5.** *Eulunulites lunula* (Ziet.), ♂, almost complete; from Chacay Melehue.

**Fig. 6a,b.** *Rehmannia brancoi* (Stein.), ♂, incomplete (TUB 790316); N Chile.

**Fig. 7a,b.** *Rehmannia stehni* (Zeiss), incomplete (TUB 1-050576); from N Chile.

(Compiled by Westermann.)

1
a
b
2
a
b
3
a
b
4
a
b
5
a
b
6
a
b
7
a
b

PLATE 65

ARGENTINA, MENDOZA PROVINCE

MIDDLE CALLOVIAN

**Fig. 1a,b.** *Rehmannia* (*Loczyceras*) *pathagoniensis* (Weaver), phragmocone (MLP 12841); from *pathagoniensis* horizon at Picún Lefú.

MIDDLE OXFORDIAN

**Figs. 2, 3.** *Araucanites stipanicici* West. & Ricc.; 2a,b, holotype, phragmocone with body-chamber fragment; 3a,b, inner whorls; from ?Plicatilis Zone of Sierra de Reyes.

(Compiled by Westermann.)

2
a
b
3
a
b
1
a
b

PLATE 66

EASTERN INDONESIA, BAJOCIAN

**Fig. 1a,b.** *Fontannesia kiliani* (Kruiz.), ♀, complete phragmocone (RGML-St 126185); loose from Kemaboe Valley.
**Fig. 2a,b.** *"Docidoceras" longalvum* (Vacek) cf. *limatum* (Pomp.), ♀, complete (RGML-St 126189); loose from Kemaboe Valley.
**Figs. 3, 4.** *Stephanoceras* (*Stemmatoceras*?) *etheridgei* (Gerth); 3a,b, holotype, ♀ (RGML-St 1272); loose from Vogelkop Peninsula; 4a,b, ♂ or juv. (RGML-St 126192); from Kemaboe Valley.
**Fig. 5.** *Stephanoceras* (*S.*) ex gr. *humphriesianum* (Sow.), ♀ (RGML-St 126191); loose from Kemaboe Valley.
**Fig. 6a,b.** *Leptosphinctes* (*Caumontisphinctes*?) aff. *engleri* (Frebold), ♀, probably almost complete (RGML-St 126191); loose from Kemaboe Valley.
(Compiled by Westermann.)

1
a
b
2
a
b
3
a
b
4
a
b
5
6
a
b

PLATE 67

EASTERN INDONESIA, UPPER BAJOCIAN–LOWER BATHONIAN

**Figs. 1, 2.** *Irianites moermanni* (Kruiz.); 1a–c, ♀, with ¾-whorl body chamber; c (×1.5) (RGML-St 126221); 2a–c, ♂, complete, with lappet; loose from Kemaboe Valley, Irianjaya.

**Fig. 3.** *Praetulites kruizingai* West., holotype, complete (IMC 457); loose from Taliabu, Sula Islands.

**Fig. 4.** *Satoceras satoi* West. & Call., holotype, complete (Tokyo Univ. 010701–6); loose from Kemaboe Valley, Irianjaya.

**Fig. 5a–c.** *Satoceras*? *costidensum* West. & Getty; a,b, holotype, almost complete (RGML 126203); c, constricted aperture, fragment; loose from Kemaboe Valley, Irianjaya.

(Compiled by Westermann.)

1c
b
1
a
b
c
a
2
3
4
5
a
b
c

PLATE 68

EASTERN INDONESIA, SULA ISLANDS. MIDDLE–UPPER(?) BATHONIAN

**Fig. 1.** *Cadomites* cf. *rectelobatus* (Hauer), ♀, almost complete, but distorted (UU 654); from ''Keeuw'' on Taliabu.

**Fig. 2a,b.** *Tulites* (*Rugiferites*) *godohensis* (Boehm), ♀, holotype, with beginning of body chamber; from ''Keeuw.''

**Fig. 3.** *Tulites*? (*Rugiferites*?) *sofanus* (Boehm), ♀, lectotype, complete, but damaged; from ''Keeuw.''

**Fig. 4a,b.** *Bullatimorphites* (*B.*) *ymir* (Oppel), ♀, complete phragmocone and end of body chamber (IMC 465); from *M. bifurcatus* Zone of Taliabu.

**Fig. 5a,b.** *Macrocephaltes bifurcatus intermedius* (Spath), ♂, neotype, complete (IMC 421); from *M. b. intermedius* Subzone of Taliabu.

**Fig. 6a,b.** *Macrocephalites bifurcatus bifurcatus* Boehm, ♂, lectotype, complete (UU 640); from ''Keeuw.''

(Compiled by Westermann.)

1
2
a
b
3
4
a
b
5
a
b
6
a
b

## PLATE 69

## EASTERN INDONESIA AND PAPUA NEW GUINEA, MIDDLE–UPPER BATHONIAN

**Fig. 1a,b.** *Macrocephalites bifurcatus intermedius* (Spath), ♀, complete (CPC 22632); from *M. b. intermedius* Subzone of Strickland River, Papua New Guinea.
**Fig. 2a,b.** *Macrocephalites bifurcatus* s.s. Boehm, ♀, inner whorls of compressed variant (IMC 432); from *M. bifurcatus* Zone of Taliabu, Sula Islands.
**Fig. 3a,b.** *Macrocephalites apertus* (Spath), ♀, holotype, complete; from "Keeuw," Taliabu.
(Compiled by Westermann.)

1
a
b
a
2
b
3
a
b

PLATE 70

INDONESIA AND PAPUA NEW GUINEA, UPPER BATHONIAN–LOWER CALLOVIAN

**Fig. 1a,b.** *Macrocephalites mantataranus* Boehm, ♀, holotype, almost complete (Utrecht coll.); from "Keeuw," Taliabu, Sula Islands, *M. apertus* Zone.
**Fig. 2a,b.** *Macrocephalites keeuwensis* Boehm, ♂, lectotype, almost complete (Utrecht coll.); from "Keeuw," *M. keeuwensis* Zone.
**Fig. 3a,b.** *Macrocephalites etheridgei* (Spath), ♂, quasi-topotype, complete (McM-J 1974R); loose from Telefomin, Papua New Guinea.
(Compiled by Westermann.)

×
1
a
b
2
a
b
a
3
b

PLATE 71

EASTERN INDONESIA, SULA ISLANDS, UPPER BATHONIAN–LOWER CALLOVIAN

**Figs. 1–3.** *Oxycerites* (*O.*) *sulaensis* West. & Call.; 1a,b, ♀, holotype, complete phragmocone (IMC 409); 2a,b, ♂, allotype, complete (IMC 411); 3a,b, inner whorls (IMC 408); from *Macrocephalites keeuwensis* Association of Taliabu.

**Fig. 4.** *Oxycerites* (*Alcidellus*) cf. *tenuistriatus* (Gross.), ♀, phragmocone (IMC 412); from *M. apertus* Zone of Taliabu.

**Figs. 5, 6.** *Choffatia* (*Homoeplanulites*) aff./cf. *furcula* (Neum.), ♀; from *M. keeuwensis* Association of Taliabu.

**Fig. 7.** *Prohecticoceras* cf. *haugi* (Pop.-Hatz.), ♀, phragmocone (UU 490, 690–1905); from "Keeuw."

(Compiled by Westermann.)

1
a
b
2
a
b
3
a
b
4
a
b
5
a
b
6
7

PLATE 72

JAPAN AND THE PHILIPPINES

**Fig. 1.** *Sonninia* gr. *sowerbyi* (Sowerby), phragmocone; from *Sonninia* Assemblage Zone, Tsunakizaka, in the Karakuwa area, South Kitakami, NE Japan.
**Fig. 2.** *Hecticoceras* (*Sublunuloceras*) cf. *quthei* (Noetling), phragmocone; from Lower to Middle Callovian, Amaga River, Mindoro, Philippines.
**Fig. 3.** *Leptosphinctes* gr. *davidsoni-coronarius* Buckman, phragmocone; from *Leptosphinctes* gr. *davidsoni* Zone, Niida, in the Shizukawa area, southern Kitakami, NE Japan.
**Fig. 4.** *Pseudoneuqueniceras yokoyamai* (Kobayashi & Fukada), phragmocone; from *P. yokoyamai* Zone, Kuzuryu, central Japan.
**Fig. 5.** *Kepplerites japonicus* Kobayashi, phragmocone; from *K. japonicus* Zone, Kuzuryu, central Japan.
**Fig. 6.** *Oxycerites* cf. *sulaensis* Westernmann & Callomon, phragmocone; from *O.* cf. *sulaensis* Assemblage Zone, Kuzuryu, central Japan.
(Compiled by Sato.)

1
2
4
3
5
6

PLATE 73

EASTERN RUSSIA, AALENIAN

**Figs. 1, 6.** *Pseudolioceras* (*Pseudolioceras*) *beyrichi* (Schloenbach), phragmocone; from *P.* (*P.*) *beyrichi* Zone; 1, Far East Russia, Tugur Bay (coll. Sey & Kalacheva); 6, North-East Russia, Sededema River basin (coll. Repin).

**Figs. 2, 3, 17.** *Pseudolioceras* (*Tugurites*) *whiteavesi* (White); 2, 3, phragmocone; 17, complete; from *P.* (*Tugurites*) *tugurense* Zone; 2, North-East Russia, Paren River basin (coll. Repin); 3, 17, Far East Russia, Tugur Bay (coll. Sey & Kalacheva).

**Figs. 4, 15, 16.** *Pseudolioceras* (*Tugurites*) *maclintocki* (Haughton), phragmocone; from *P.* (*T.*) *maclintocki* Zone; 4, 16, Far East Russia, Bureya River basin, Tugur Bay (coll. Sey & Kalacheva); 15, North-East Russia, Kegali River basin (coll. Repin).

**Fig. 5.** *Erycitoides* sp., phragmocone (×2.5); from *P.* (*T.*) *tugurense* Zone, North-East Russia, Penzhina River basin (coll. Repin).

**Fig. 7.** *Erycitoides howelli* (White), phragmocone; from *P.* (*T.*) *turgurense* Zone, Far East Russia, Tugur Bay (coll. Sey).

**Figs. 8–10.** *P.* (*Tugurites*) *tugurense* Kalach. & Sey, phragmocone; from *P.* (*T.*) *tugurense* Zone, Far East Russia, Tugur Bay (coll. Sey & Kalacheva).

**Figs. 11–13.** *P.* (*Tugurites*) sp. (= "*Grammoceras*" sp. indet.); from *P.* (*T.*) *maclintocki* Zone; 11, North-East Russia, Sugoy River basin (coll. Repin); 12, 13, Far East Russia, Tugur Bay (coll. Sey).

**Fig. 14.** *Erycitoides* (*Kialagvikes*) *spinatum* Wester., complete; from *P.* (*T.*) *tugurense* Zone, Far-East Russia, Tugur Bay (coll. Sey & Kalacheva).

(Compiled by Repin, Sey, & Kalacheva.)

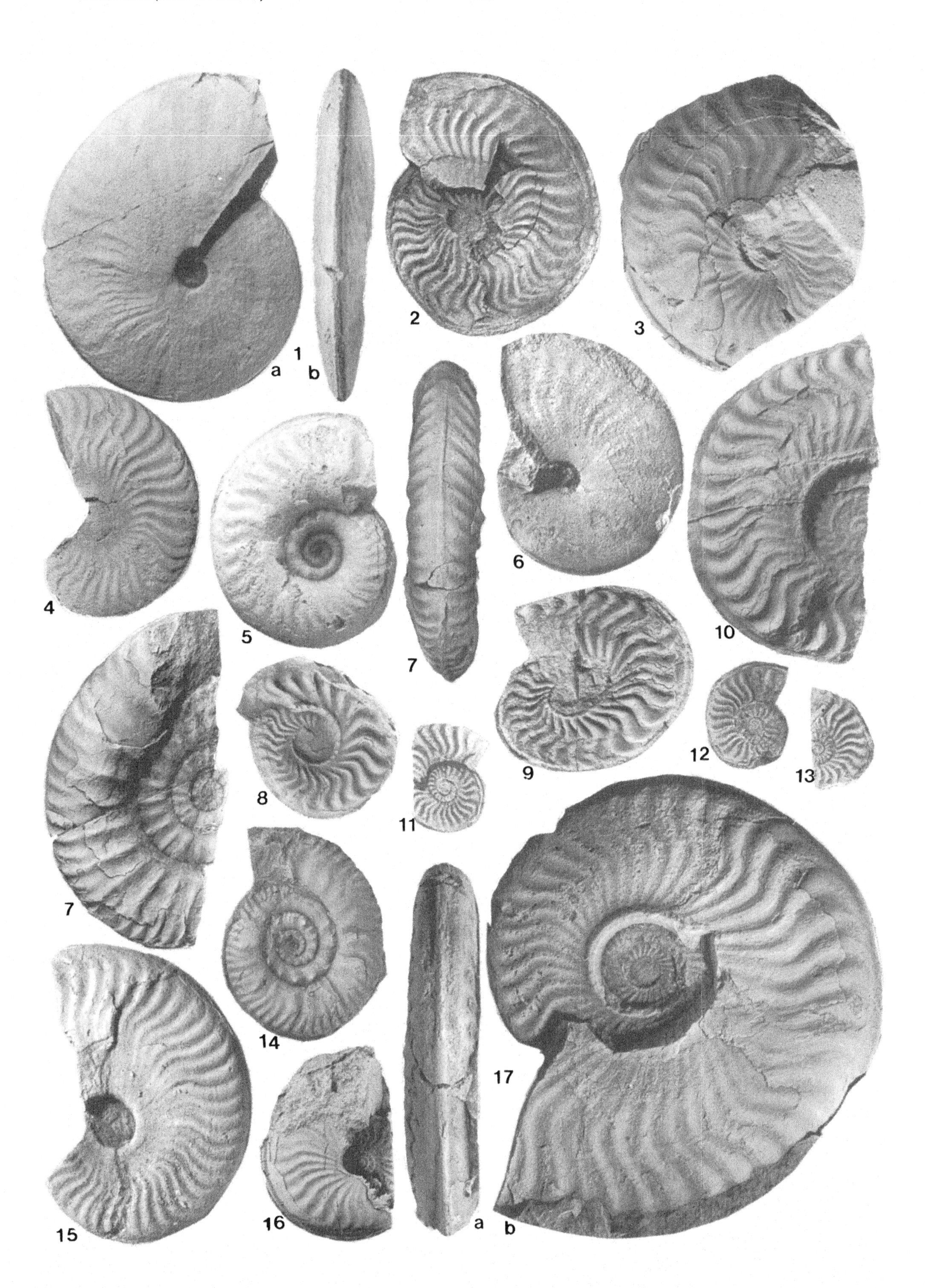
1
a
b
2
3
4
5
6
7
8
9
10
11
12
13
7
14
15
16
17
a
b

PLATE 74

EASTERN RUSSIA, BAJOCIAN

**Figs. 1, 2, 5.** *Pseudolioceras (Tugurites) fastigatum* Wester., phragmocone; from *P. (T.) fastigatum* Zone; 1,2, Far East Russia, Bureya River basin (coll. Sey & Kalacheva); 5, North-East Russia, Kegali River basin (coll. Repin).

**Figs. 3, 4.** *P. (Tugurites) costistriatum* Wester., phragmocone?; from *P. (T.) fastigatum* Zone; 3, Far East Russia, Bureya River basin (coll. Sey); 4, North-East Russia, Paren River basin (coll. Repin).

**Fig. 6.** *Arkelloceras tozeri* Frebold, phragmocone?; from *A. tozeri* Zone, Far East Russia, Bureya River basin (coll. Sey & Kalacheva).

**Figs. 7, 13.** *Arkelloceras elegans* Frebold, phragmocone, complete; from *A. tozeri* Zone; 7, Far East Russia, Bureya River basin (coll. Sey & Kalacheva); 13, North-East Russia, Anadyr River basin (coll. Repin).

**Fig. 8.** *Oppelia (Liroxyites)* cf. *kellumi* Imlay; from Upper Bajocian, Far East Russia, (coll. Sey & Kalacheva).

**Fig. 9.** *Bradfordia alaseica* Repin, complete; from *A. tozeri* Zone, North-East Russia, Sededema River basin (coll. Repin).

**Fig. 10.** *Arkelloceras* cf. *maclearni* Freb.; from *A. tozeri* Zone; Far East Russia, Shilka River basin (coll. Sey).

**Fig. 11.** *Chinitnites*(?) sp. (= *Umaltites era* [m]), complete; from Upper Bajocian, *Umaltites era* Beds, Far East Russia, Bureya River basin (coll. Sey).

**Figs. 12, 14, 15.** *Umaltites* (= *Megasphaeroceras*) *era* (Krimholz); 12,14, complete; 15, phragmocone; from Upper Bajocian, *Umaltites era* Beds, Far East Russia, Bureya River basin (coll. Sey). [*Editor's comment: Umaltites* closely resembles the coeval genus *Megasphaeroceras,* in which it should be included as a subgenus, if not synonym.]

**Fig. 16a–c.** *Chondroceras* cf. *marshalli* (McLearn), complete?; a (×3); b,c (×2); from Lower Bajocian, *C. marshalli* Beds, North-East Russia, Anadyr River basin (coll Repin).

(Compiled by Repin, Sey, & Kalacheva.)

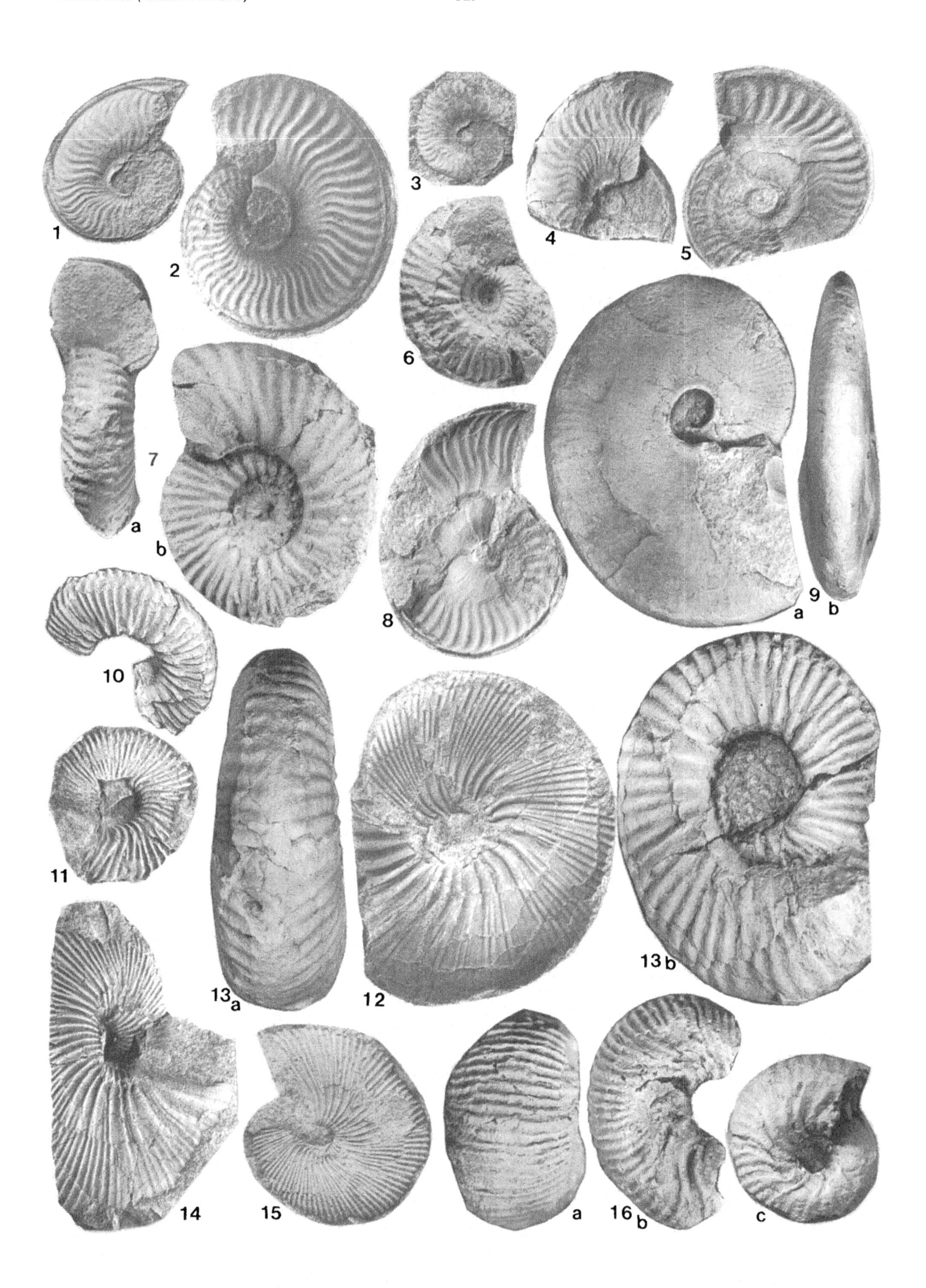
1
2
3
4
5
6
7
a
b
8
9
a
b
10
11
12
13a
13b
14
15
16
a
b
c

PLATE 75

EASTERN RUSSIA, BAJOCIAN–BATHONIAN

**Fig. 1.** *Itinsaites*(?) sp.; from Bajocian, North-East Russia, Yana River basin (coll. Repin).
**Fig. 2.** *"Oxycerites" jugatus* Ersch. & Meledina, phragmocone (×2); from Lower Bathonian, *A. elegans* Zone, North-East Russia, Takhtoyama River basin (coll. Polubotko & Repin).
**Fig. 3.** *Cadoceras*(?) sp.; from Upper Bathonian, North-East Russia, Bolshoi Anyui River basin (coll. Repin).
**Fig. 4.** *"Cobbanites" talkeetnanus* Imlay, complete; from Upper Bajocian, Far East Russia, southern Primorye (coll. Sey & Kalacheva).
**Fig. 5.** *Cranocephalites furcatus* Spath, complete?; from *C. vulgaris* Zone, North-East Russia, Takhtoyama River basin (coll. Polubotko & Repin).
**Fig. 6.** *Costacadoceras*(?) cf. *bluethgeni* Rawson; from Upper Bathonian, North-East Russia, Okhotsk Sea, Babushkin Bay (coll. Paraketzov).
**Fig. 7.** *Iniskinites* cf. *magniformis* (Imlay); from Upper Bathonian, North-East Russia, Okhotsk Sea, Babushkin Bay (coll. Paraketzov).
**Fig. 8.** *Umaltites*(?) sp. (= *Megasphaeroceras*),[2] complete?; from Upper Bajocian, North-East Russia, Indigirka River basin (coll. Nikonov).
**Figs. 9, 10.** *Arctocephalites elegans* Spath, phragmocone; from Upper Bathonian, *A. elegans* Zone, North-East Russia, Viliga River.
**Fig. 11.** *Cranocephalites vulgaris* Spath; from Upper Bajocian, *C. vulgaris* Zone, North-East Russia, Takhtoyama River basin (coll. Polubotko & Repin).
(Compiled by Repin, Sey, & Kalacheva.)

[2] See caption to Plate 74.

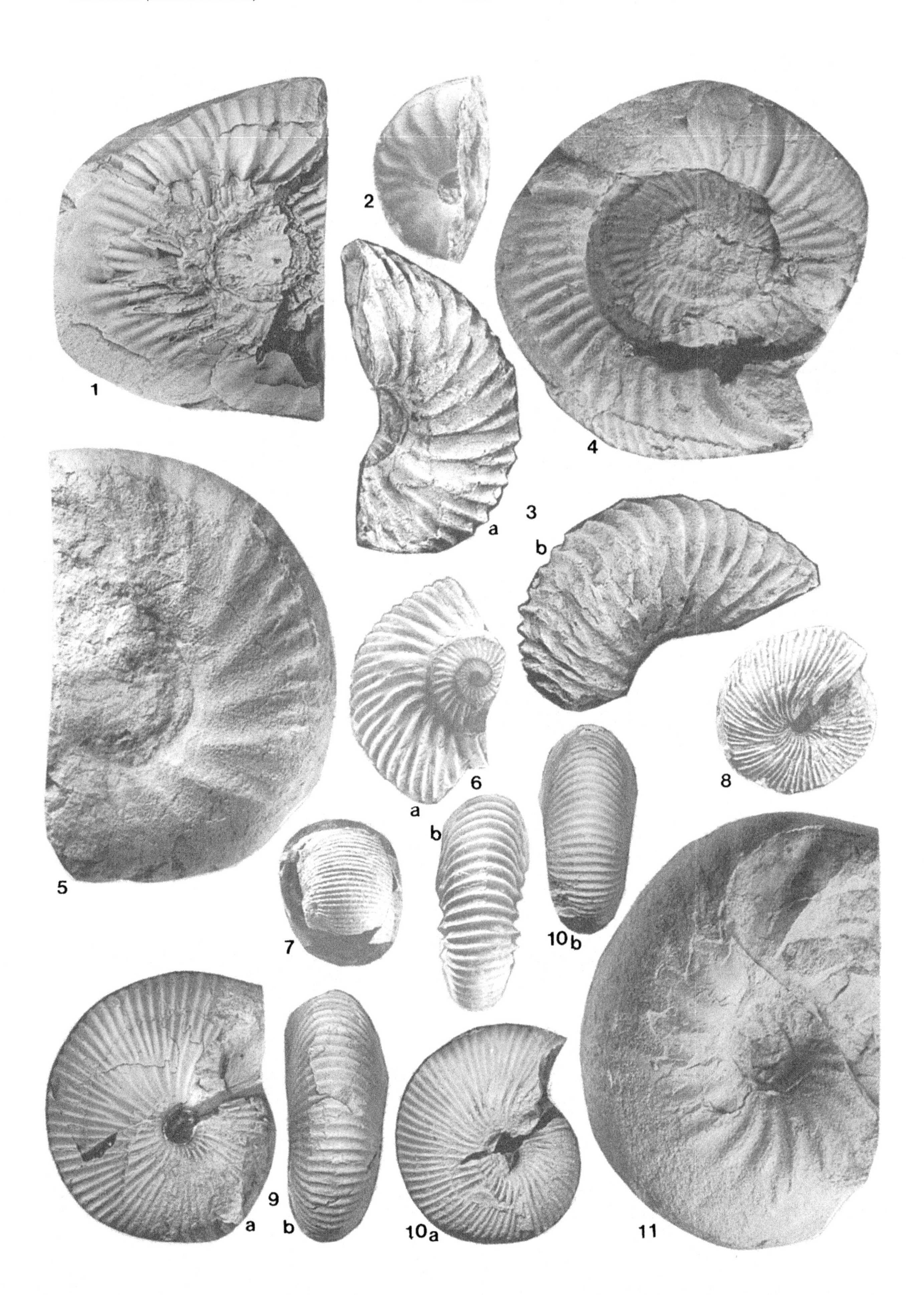
1
2
3
a
b
4
5
6
a
b
7
8
9
a
b
10a
10b
11

PLATE 76

EASTERN RUSSIA, CALLOVIAN

**Figs. 1–3.** *Longaeviceras* cf. *keyserlingi* (Sok.); from Upper Callovian; 1, North-East Russia, Yana River basin (coll. Repin); 2,3, Far East Russia, Tugur Bay (coll. Sey & Kalacheva).
**Fig. 4.** *Lunuloceras*(?) sp. indet.; from Callovian, North-East Russia, Koryak Upland (coll. Terekhova).
**Figs. 5, 6.** *Choffatia* cf. *leptonota* Spath; sense 6, fragment of body chamber; from Callovian, North-East Russia, Koryak Upland (coll. Terekhova).
**Figs. 7, 8.** *Zieteniceras* sp. indet.; from Callovian, North-East Russia, Koryak Upland (coll. Terekhova) (coll. Sey & Kalacheva).
**Fig. 9.** *Cadoceras* (*Paracadoceras*) cf. *anabarense* Bodyl., phragmocone; from Lower Callovian, North-East Russia, Korkodon River basin.
(Compiled by Repin, Sey, & Kalacheva.)

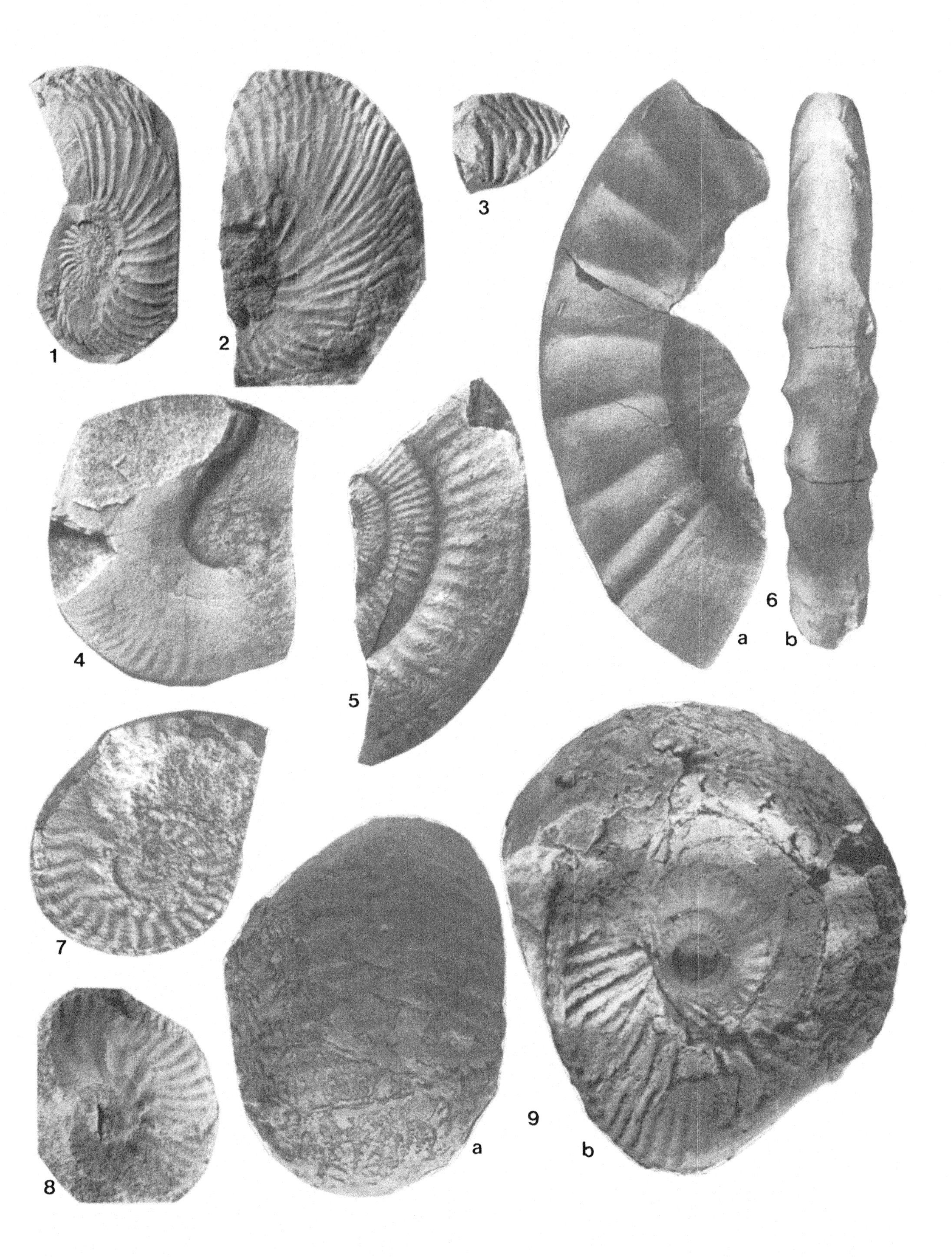
1
2
3
4
5
6
a
b
7
8
9
a
b

PLATE 77

SOUTHERN ALASKA, OXFORDIAN

**Figs. 1, 2.** *Cardioceras martini* Reeside, ♀ & ♂; 1a,b, macroconch (USNM 256926); 2a–c, microconch (USNM 32317); from Lower Oxfordian of Cook Inlet.
**Fig. 3.** *Cardioceras spiniforme* Reeside, ♀, holotype, fragment with part of body chamber (USNM 32348); from Lower Oxfordian of Cook Inlet.
**Fig. 4a,b.** *Cardioceras whiteavesi* Reeside, ♂, fragment (USNM 256934); from Middle Oxfordian of Alaska Peninsula.
**Figs. 5–8.** *Amoeboceras* aff. *transitorium* Spath, fragments (USNM 256938/44–46); from Upper Oxfordian of Wrangell Mountains.
All reproduced from Imlay (1981c).
(Compiled by Westermann.)

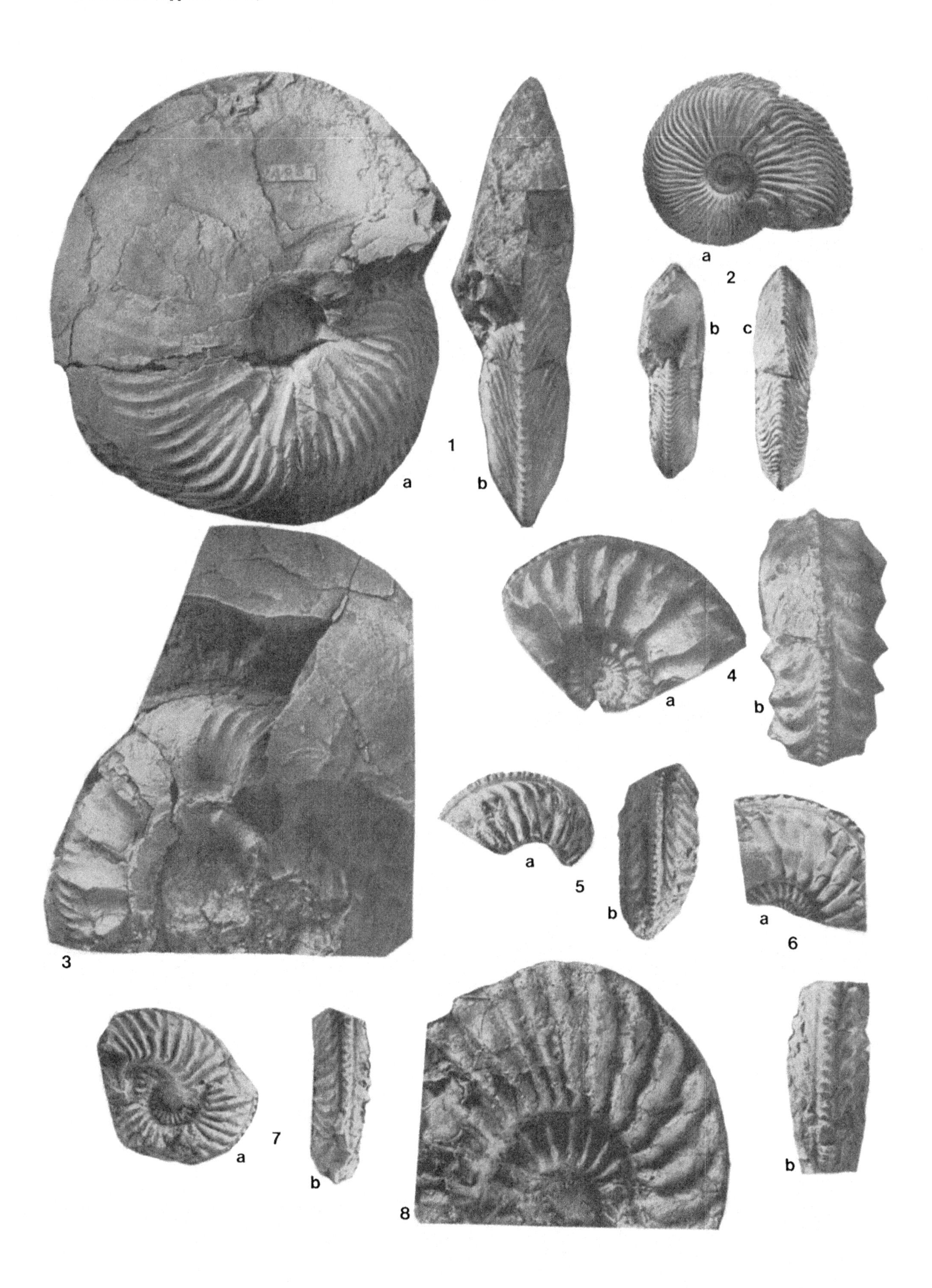
1
a
b
2
a
b
c
3
4
a
b
5
a
b
6
a
7
a
b
8
b

PLATE 78

WESTERN INTERIOR CANADA

BATHONIAN

**Fig. 1a,b.** *Paracephalites henryi* (Meek & Hayden) (×0.5) (GSC 14688); from the Middle Bathonian of southwestern Alberta.

OXFORDIAN

**Fig. 2a,b.** *Cardioceras canadense* (Whiteaves), holotype (GSC 7437); from the Densiplicatum Zone of southeastern British Columbia.

**Fig. 3a,b.** *Cardioceras mountjoyi* Frebold, holotype (GSC 13895); from the Cordatum Zone of west-central Alberta.

**Fig. 4.** "*Titanites*" *occidentalis* Frebold, "holo-plasto-type" (McM-J 1132) (×0.06); ?Tithonian, from southeastern British Columbia.

For Oxfordian, see also Plate 22, Fig. 6.

(Compiled by Hall.)

1
2
a
b
a
3
b
4

PLATE 79

WESTERN INTERIOR UNITED STATES

CALLOVIAN

**Fig. 1a,b.** *Perisphinctes* (*Prososphinctes*) sp. indet. (×2) (USNM 303722); from the Lamberti Zone of Montana.

**Figs. 2, 3a,b.** *Quenstedtoceras collieri* Reeside; 2 (USNM 104137); 3a,b (×2) (USNM 303605); from the Lamberti Zone of Montana.

OXFORDIAN

**Figs. 4a,b, 5.** *Cardioceras distans* (Whitfield); 4a,b, holotype (USNM 32332); 5 (USNM 303643); from the Cordatum Zone of Wyoming.

**Fig. 6a,b.** *Cardioceras canadense* Whiteaves (USNM 303705); from the Densiplicatum Zone of Montana.

**Fig. 7a,b.** *Cardioceras cordiforme* (Meek & Hayden), holotype (USNM 203); from the Mariae Zone of South Dakota.

**Fig. 8.** *Cardioceras sundancense* Reeside, holotype (USNM 29314); from the Tenuiserratum Zone of Wyoming.

(Compiled by Hall.)

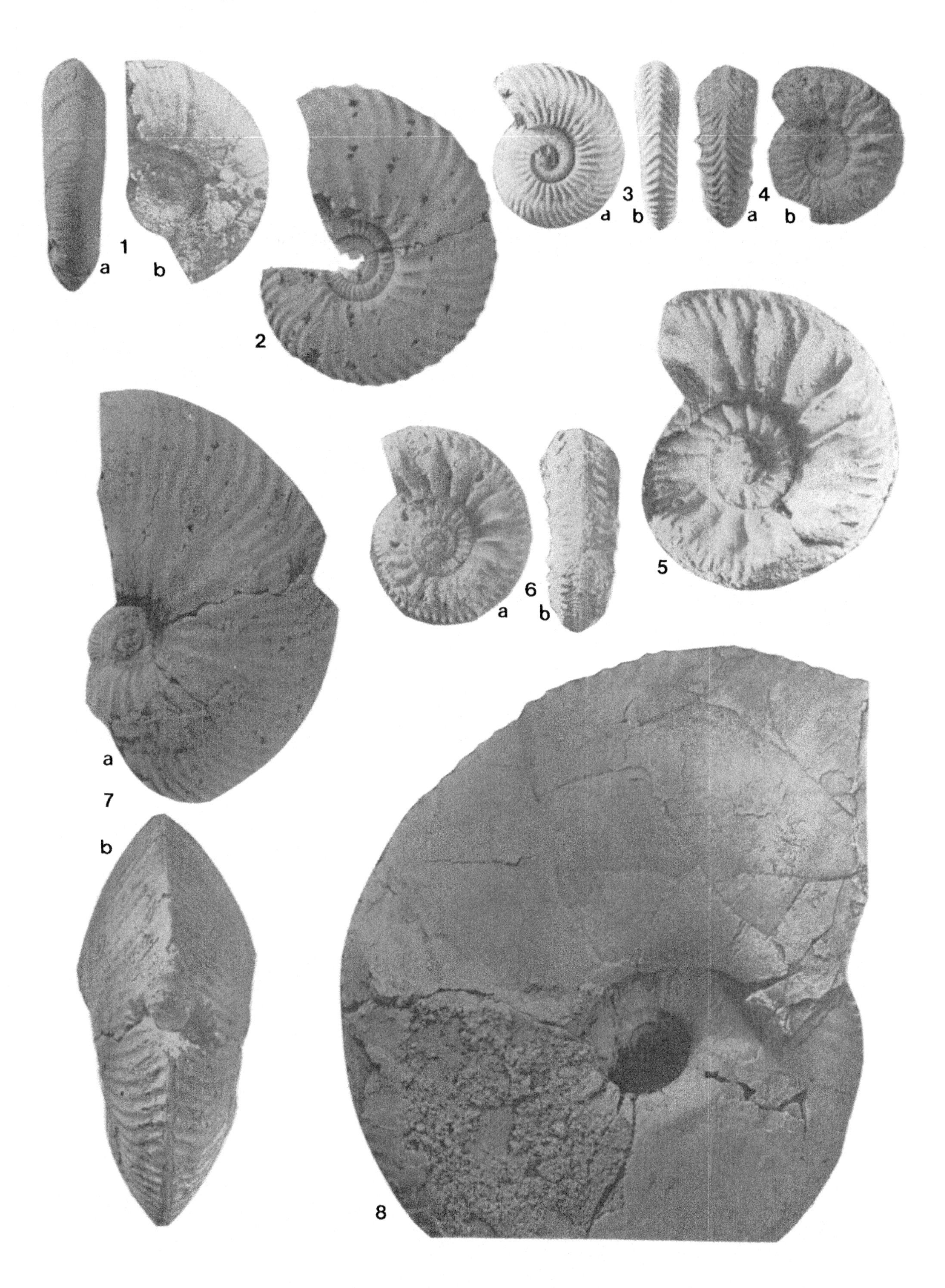
1
a
b
2
3
a
b
4
a
b
5
6
a
b
7
a
b
8

PLATE 80

ARGENTINA, TITHONIAN, *Virgatosphinctes medozanus* ZONE

**Fig. 1a,b.** *Pseudinvoluticeras douvillei* Spath, phragmocone (MLP 8210); from Picún Leufú, Neuquén province.
**Fig. 2a,b.** *Virgatosphinctes evolutus* H. Leanza, phragmocone (MLP 3405); from Casa Pincheira, Mendoza province.
**Fig. 3a,b.** *Choicensphinctes choicensis* (Burckhardt), phragmocone (MLP 6342); from Cerro Lotena, Neuquén province.
(Compiled by Riccardi.)

1
a
b
2
a
b
3
a
b

PLATE 81

ARGENTINA, TITHONIAN

*Aulacosphinctes proximus* ZONE

**Fig. 1a–c.** *Aspidoceras andinum* Steuer, phragmocone (MLP 3568); from Río Diamante, Mendoza province.

**Fig. 2a,b.** *Aulacosphinctes proximus* (Steuer), phragmocone and body chamber (MLP 14629); from Mina La Eloisa, Mendoza province.

**Fig. 3a,b.** *Pseudhimalayites steinmanni* (Haupt), phragmocone (MLP 900); from Cerro Lotena, Neuquén province.

*Corongoceras alternans* ZONE

**Fig. 4a,b.** *Corongoceras lotenoense* Spath, phragmocone and body chamber (MLP 3355); from Río Diamante, Mendoza province.

*Substeueroceras koeneni* ZONE

**Fig. 5a,b.** *Substeueroceras koeneni* (Steuer), phragmocone; from Mina de Rafaelita, Mendoza province.

(Compiled by Riccardi.)

1
a
b
c
2
a
b
4
a
b
3
a
b
5
a
b

## PLATE 82

## ARGENTINA, TITHONIAN

*Windhauseniceras interspinosum* ZONE

**Fig. 1a,b.** *Hemispiticeras* aff. *steinmanni* (Steuer), phragmocone (MLP 14941); from Cañada de Leiva, Mendoza province.

*Substeueroceras koeneni* ZONE

**Fig. 2a,b.** *Paradontoceras calistonides* (Behrendsen), complete (MLP 9924); from Tril West, Neuquén province.

**Fig. 3a,b.** *Substeueroceras koeneni* (Steuer), phragmocone and body chamber (MLP 6768); from Cienenguita, Mendoza province.

See also Plate 65, Figs. 2 and 3.

(Compiled by Riccardi.)

1
a
b
2
a
b
3
a
b

PLATE 83

EASTERN INDONESIA, SULA ISLANDS, LOWER OXFORDIAN

**Fig. 1a,b.** *Peltoceratoides* (*Peltomorphites*?) *tjapalului* (Boehm), ♀, lectotype, almost complete (×0.7).
**Fig. 2a,b.** *Peltoceratoides* (*Parawedekindia*) cf. *arduennensis* (d'Orb.), ♂ (×0.7).
**Fig. 3.** *Retroceramus galoi* (Boehm) (×0.8).
All reproduced from Boehm (1907).
(Compiled by Westermann.)

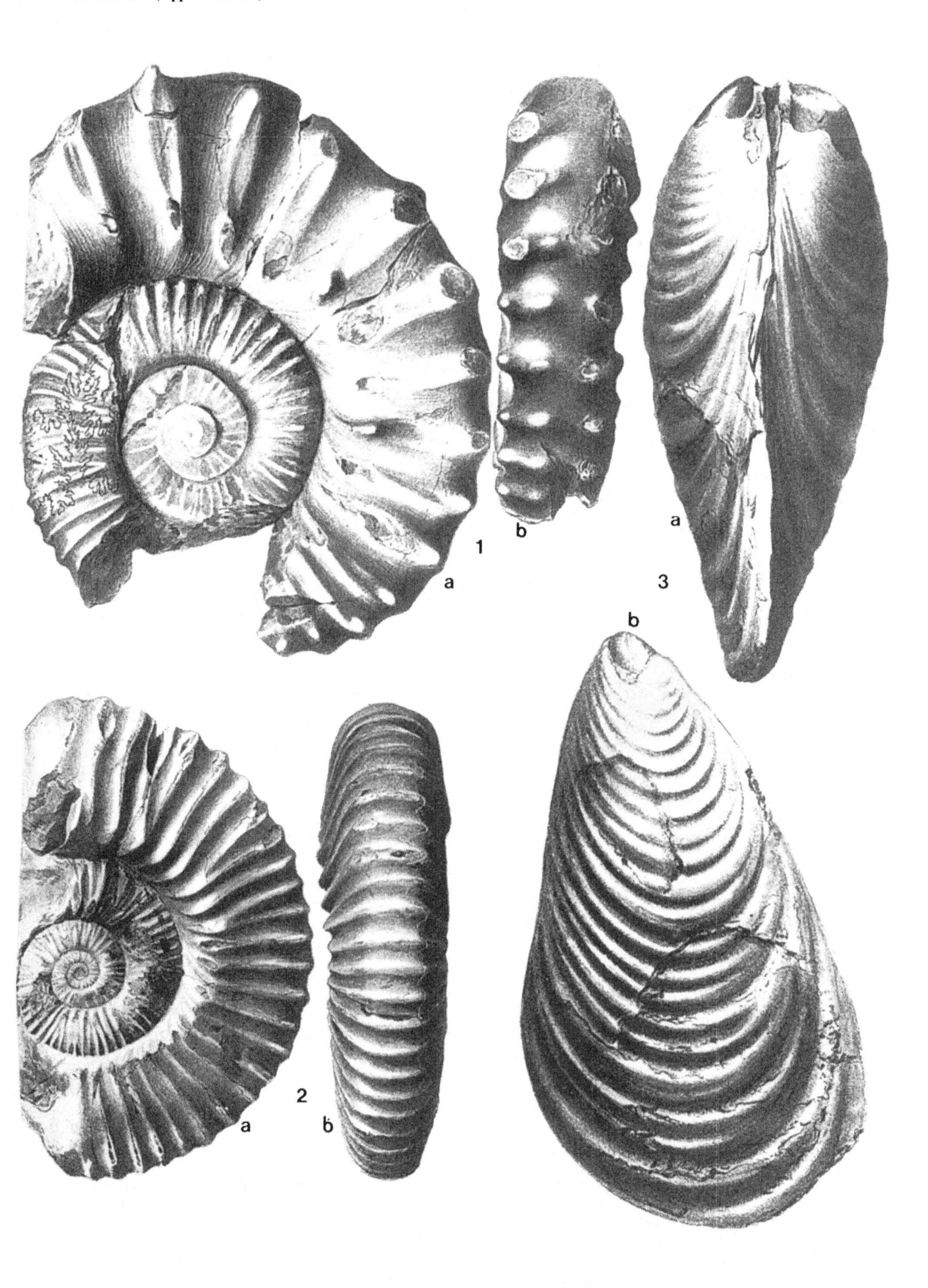
1
a
b
2
a
b
3
a
b

PLATE 84

EASTERN INDONESIA, SULA ISLANDS, LOWER–MIDDLE OXFORDIAN

**Fig. 1a,b.** *Mayaites (Epimayaites) rotangi* (Boehm), ♀, lectotype, phragmocone (×0.7); from Wai Galo.
**Fig. 2a,b.** *Mayaites (Epimayaites) sublemoini* Spath, ♀, holotype, phragmocone (×0.8); from Wai Galo.
**Fig. 3a,b.** *Mayaites (Epimayaites) palmarum* (Boehm), ♀, holotype, with part of body chamber? (×0.8); from Wai Galo.
All reproduced from Boehm (1907).
(Compiled by Westermann.)

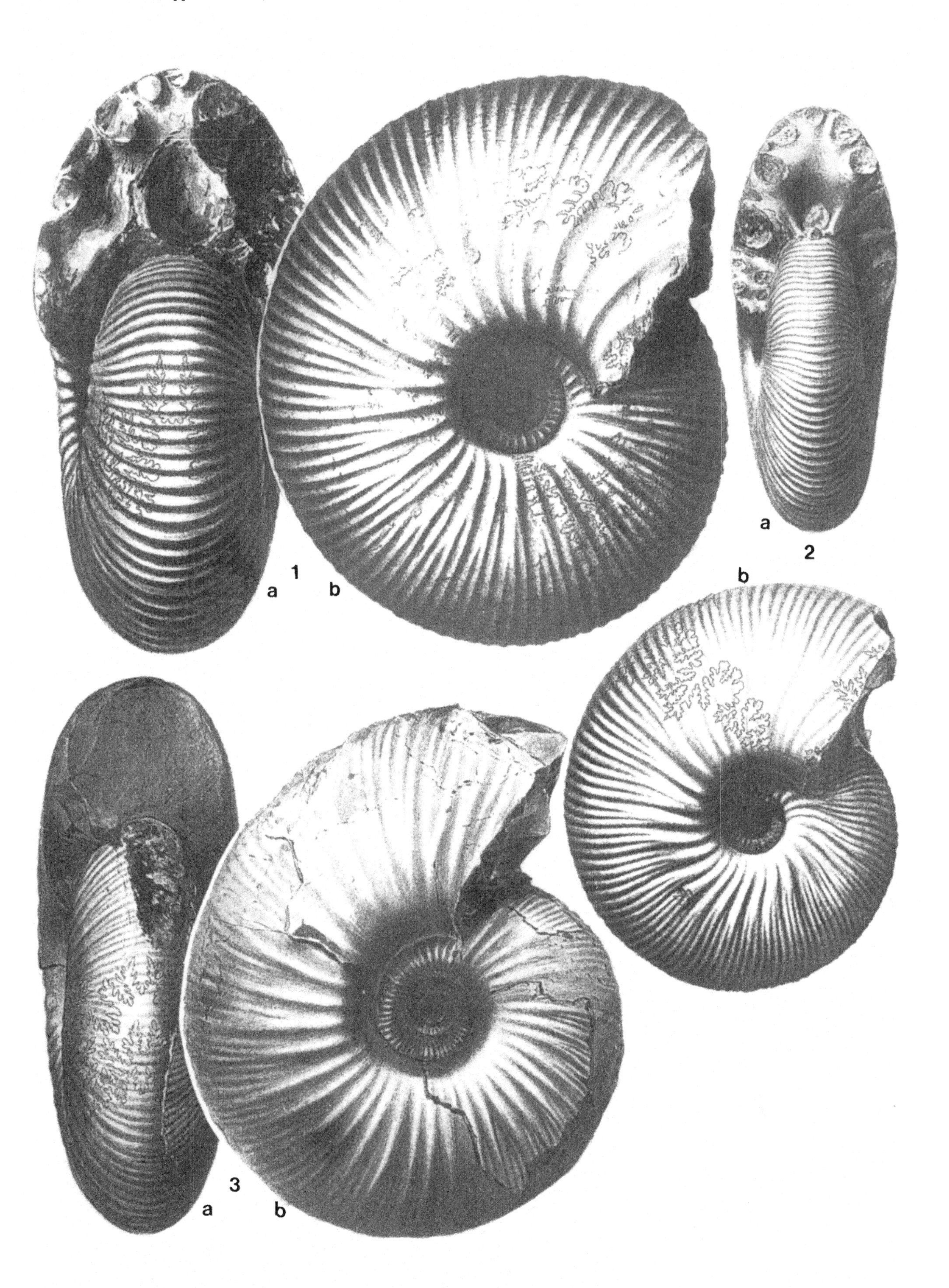
1
a
b
2
a
b
3
a
b

PLATE 85

EASTERN INDONESIA, SULA ISLANDS, LOWER–MIDDLE OXFORDIAN

**Fig. 1.** *Epimayaites* (*Paryphoceras*) *alfuricus* (Boehm), ♂, holotype, almost complete (×0.8).
**Fig. 2a,b.** *Epimayaites* (*Paryphoceras*) *sinuatus* Spath, ♂, holotype, complete (×0.8).
**Fig. 3.** *Epimayaites* (*Paryphoceras*) *cocosi* (Boehm), ♂, lectotype, approximately complete (×0.8).
**Fig. 4.** *Perisphinctes* (*Kranaosphinctes?*) *galoi* Boehm, ♀, phragmocone (×0.8).
All reproduced from Boehm (1907).
(Compiled by Westermann.)

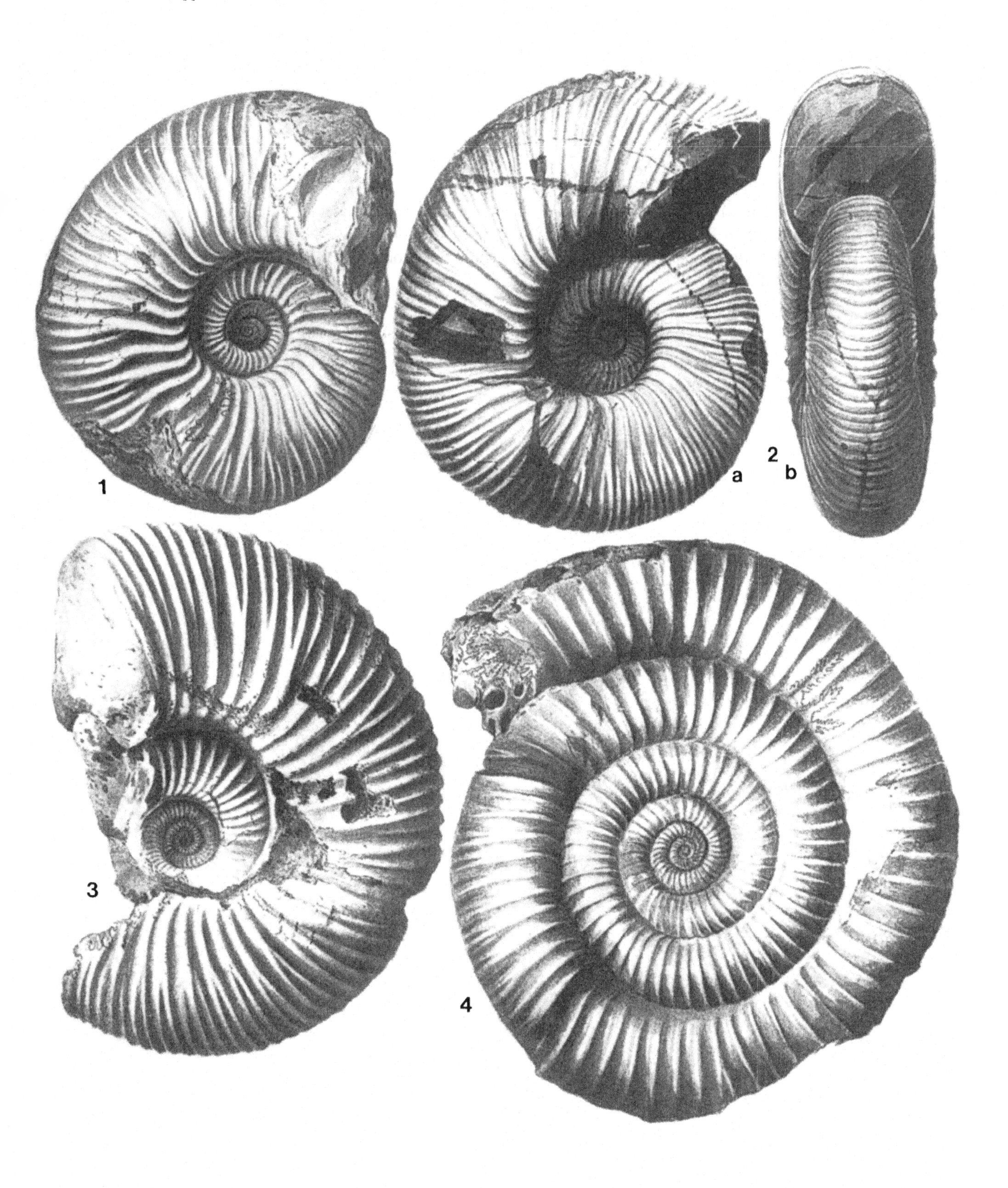
1
a
2
b
3
4

PLATE 86

EASTERN INDONESIA, SULA ISLANDS, UPPER TITHONIAN

**Fig. 1a,b.** *Haplophylloceras strigile* (Blanford); from Taliabu.
**Fig. 2a,b.** *Blanfordiceras* cf. *wallichi* (Gray) (×0.8); from Taliabu.
**Fig. 3a,b.** *B(?) rosenbloomi* (Boehm), holotype, almost complete (×0.65); from Taliabu.
**Fig. 4a,b.** *Himalayites treubi* Boehm, holotype, phragmocone (×0.9); from Taliabu.
**Fig. 5a–c.** *Himalayites nederburghi* Boehm, holotype, phragmocone; from Taliabu. All reproduced from Boehm (1904).
(Compiled by Westermann.)

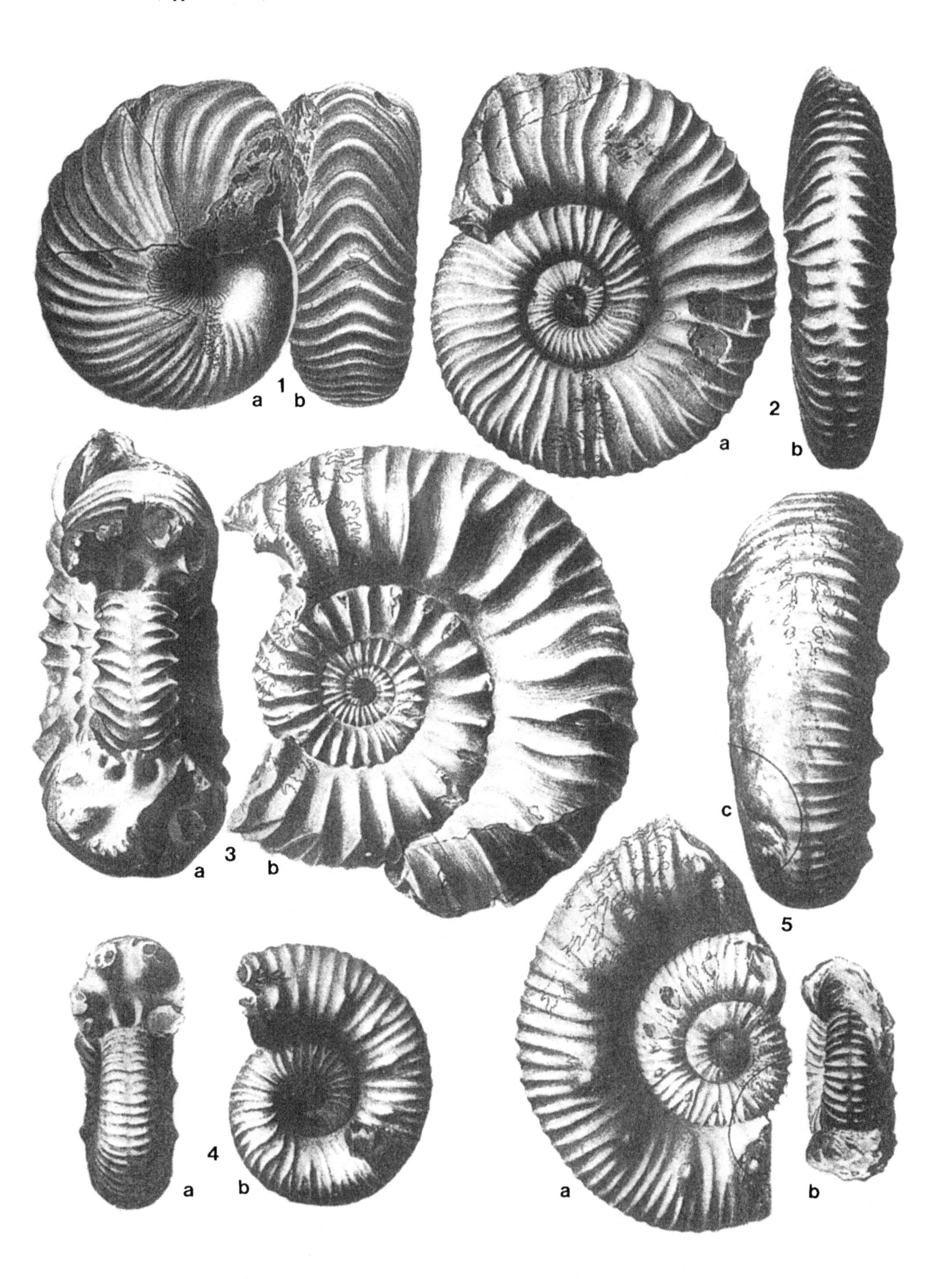
a
1
b
2
a
b
a
3
b
c
5
4
a
b
a
b

PLATE 87

EASTERN INDONESIA, IRIANJAYA, TITHONIAN

**Fig. 1a,b.** *Kossmatia desmidoptycha* Uhlig; loose from Telefomin (coll. B. Crouch).
**Fig. 2a,b.** *Kossmatia tenuistriata* (Gray); loose from Telefomin (coll. B. Crouch).
**Fig. 3a,b.** *Paraboliceras mutilis* (Oppel); loose from Telefomin (coll. B. Crouch).
**Fig. 4a,b.** *Paraboliceras* cf. *sabineanus* (Oppel), phragmocone; loose from Telefomin (coll. B. Crouch).
(Compiled by Westermann.)

1a
b
2a
b
3
a
b
4
b
a

PLATE 88

JAPAN AND THE PHILIPPINES

**Fig. 1.** *Kranaosphinctes matsushimai* (Yokoyama), phragmocone; from *K. matsushimai* Assemblage Zone, Nagano, in the Kuzuryu area, central Japan.
**Fig. 2.** *Ataxioceras kurisakense* Kobayashi & Fukada, phragmocone; from *A. kurisakense* Assemblage Zone, Kurisaka, in Shikoku, SW Japan.
**Fig. 3.** *Aulacosphinctoides* cf. *steigeri* (Shimizu), phragmocone (×1.5); from Lower Tithonian, Sakawa, in Shikoku, SW Japan.
**Fig. 4a,b.** *Parawedekindia arduennensis* (d'Orbigny), phragmocone; from *P. arduennensis* Assemblage Zone, Parucpoc Hill, Mindoro, Philippines.
**Fig. 5a,b.** *Euaspidoceras* cf. *hypselum* (Oppel), phragmocone; from *E.* cf. *hypselum* Assemblage Zone, Amaga River, Mindoro, Philippines.
(Compiled by Sato.)

1
2
3
4
a
b
5
a
b

PLATE 89

EASTERN RUSSIA, OXFORDIAN–KIMMERIDGIAN

**Figs. 1, 3.** *Cardioceras* (*Scarburgiceras*) *praecordatum* R. Douville; from lower Oxfordian, *Scarburgiceras* Beds, Far East Russia, Tugur Bay (coll. Sey & Kalacheva).
**Fig. 2.** *C.* (*Scarburgiceras*) cf. *gloriosum* Arkell; from lower Oxfordian, *Scarburgiceras* Beds, Far East Russia, Tugur Bay (coll. Sey & Kalacheva).
**Fig. 4.** *Cardioceras* (*Maltoniceras*) aff. *schellwieni* Boden, complete?; from Middle Oxfordian, Far East Russia, Tugur Bay (coll. Sey & Kalacheva).
**Figs. 5–7.** *Ochetoceras elgense* Chudoley & Kalacheva; from Upper Kimmeridgian–?Lower Tithonian, Far East Russia, Uda River basin (coll. Sey & Kalacheva).
**Figs. 8–10.** *Amoeboceras* (*Amoebites*) ex gr. *kitchini* (Salfeld); 9, (×2); from Lower Kimmeridgian, *A.* (*A.*) ex gr. *kitchini* Beds; 8,9, Far East Russia, Tugur Bay (coll. Sey & Kalacheva); 10, North-East Russia, Korkodon River basin.
**Fig. 11.** *Perisphinctes* (*Dichotomosphinctes*) cf. *mühlbachi* Hyatt, from Middle Oxfordian, Far East Russia, Tugur Bay (coll. Sey & Kalacheva).
(Compiled by Repin, Sey & Kalacheva.)

1
2
3
4
5
6
7
8
9
10
11

PLATE 90

EASTERN RUSSIA, TITHONIAN

**Fig. 1.** *Subplanitoides* aff. *tithonicum* Zeiss, phragmocone?; from middle Tithonian, *Pseudolissoceras zitteli* Zone, Far East Russia, southern Primorye, Putiatin Island (coll. Sey & Kalacheva).

**Figs. 2, 17.** *Subplanitoides* ex gr. *altegyratum* Zeiss; from middle Tithonian, *P. zitteli* Zone, Far East Russia, Putiatin Island (coll. Sey & Kalacheva).

**Fig. 3.** *Subplanitoides putiatinensis* (Chudoley); from Middle Tithonian, Far East Russia, Putiatin Island.

**Fig. 4.** *Haploceras* cf. *elimatum* (Oppel), complete; from middle Tithonian, *P. zitteli* Zone, Putiatin Island (coll. Sey & Kalacheva).

**Figs. 5, 6, 10–12, 15.** *Glochiceras jollyi* (Oppel); from *P. zitteli* Zone, Putiatin Island (coll. Sey & Kalacheva).

**Figs. 7–9, 13, 14.** *Pseudolissoceras* ex gr. *zitteli* (Burckhardt); from *P. zitteli* Zone, Putiatin Island (coll. Sey & Kalacheva).

**Fig. 16.** *Virgatosphinctes* cf. *mexicanus* (Burckh.), phragmocone [= *Subplanites* (*Parapallasiceras*) *contiguus* (Zittel)]; from Lower Tithonian, Far East Russian, southern Primorye.

**Figs. 18, 19.** *Subplanitoides*(?) sp.; from *P. zitteli* Zone, Putiatin Island (coll. Sey & Kalacheva).

**Fig. 20.** *Sublithacoceras*(?) sp.; from *Aulacosphinctes proximus* Zone, Putiatin Island (coll. Sey & Kalacheva).

(Compiled by Repin, Sey, & Kalacheva.)

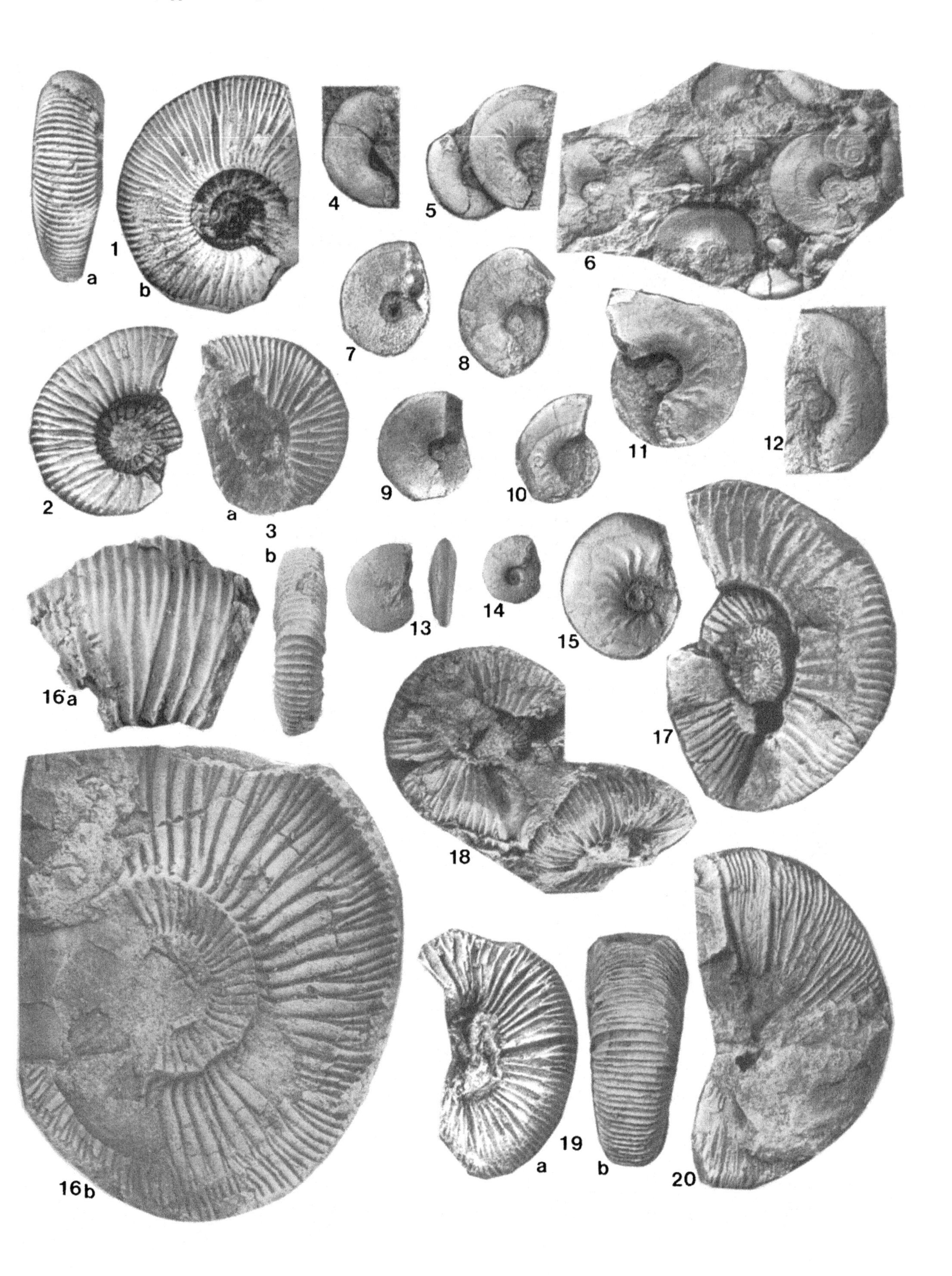
1
a
b
2
3
a
b
4
5
6
7
8
9
10
11
12
13
14
15
16a
16b
17
18
19
a
b
20

PLATE 91

EASTERN RUSSIA, TITHONIAN/VOLGIAN

**Fig. 1.** *Lemencia* aff. *adeps* (Schneid); from middle Tithonian, *Aulacosphinctes proximus* Zone, Far East Russia, southern Primorye, Putiatin Island (coll. Sey & Kalacheva).
**Fig. 2, 3, 7.** *Aulacosphinctes proximus* (Steuer); from *A. proximus* Zone, Putiatin Island (coll. Sey & Kalacheva).
**Fig. 4.** *Lemencia* sp.; from *A. proximus* Zone, Putiatin Island (coll. Sey & Kalacheva).
**Fig. 5.** *Subplanitoides* ex gr. *subpraecox* (Donze & Enay); from *Pseudolissoceras zitteli* Zone, Putiatin Island (coll. Sey & Kalacheva).
**Fig. 6.** *Sublithacoceras* sp. juv.; from *A. proximus* Zone, Putiatin Island (coll. Sey & Kalacheva).
**Fig. 8.** *Parapallasiceras* sp., fragment of body chamber; from *P. zitteli* Zone, Putiatin Island (coll. Sey & Kalacheva).
**Fig. 9.** *Subplanitoides* sp., phragmocone; from *A. proximus* Zone, Putiatin Island (coll. Sey & Kalacheva).
**Fig. 10.** *Durangites* sp. indet.; from Middle Volgian, Far East Russia, Uda River basin.
**Fig. 11.** *Chetaites*(?) sp. indet.; from Upper Volgian, North-East Russia, Bolshoi Anyui River basin.
**Fig. 12.** *Sublithacoceras* cf. *penicillatum* (Schneid), complete?; from middle Tithonian, *A. proximus* Zone, Far East Russia, Putiatin Island (coll. Sey & Kalacheva).
**Fig. 13.** *Dorsoplanites* cf. *transitorius* Spath, phragmocone; from Middle Volgian, North-East Russia, Bolshoi Anyui River basin (coll. Paraketzov).
(Compiled by Repin, Sey, & Kalacheva.)

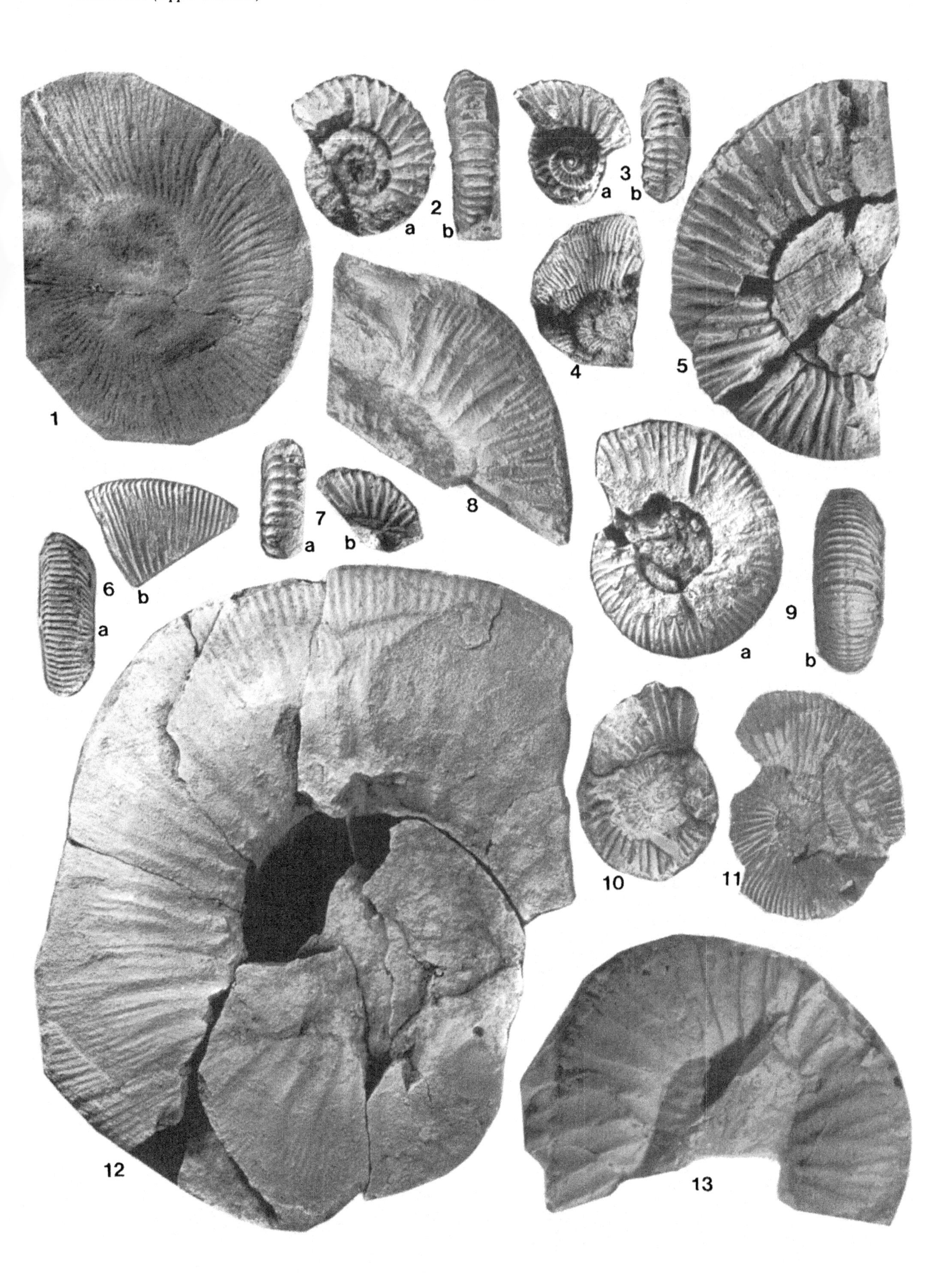
1
2
a
b
3
a
b
4
5
6
a
b
7
a
b
8
9
a
b
10
11
12
13

PLATE 92

AUSTRALIA

*Callialasporites dampieri* SUPERZONE
**Fig. 1.** *Polycingulatisporites crenulatus.*
**Fig. 2.** *Crassitudisporites problematicus.*
**Fig. 3.** *Callialasporites turbatus.*
**Fig. 4.** *Perinopollenites elatoides.*
**Fig. 5.** *Corollina torosa.*
**Fig. 6.** *Retitriletes circolumenus.*
**Fig. 7.** *Murospora florida.*
**Fig. 8.** *Contignisporites cooksoniae.*
**Fig. 9.** *Dictyosporites complex.*
**Fig. 10.** *Callialasporites dampierii.*
**Fig. 11.** *Gleicheniidites* sp.
**Fig. 12.** *Microcachryidites antarcticus.*
**Fig. 13.** *Exesipollenties tumulus.*
**Fig. 14.** *Concayissimisporites varieverrucatus.*
**Fig. 15.** *Contignisporites* sp.
**Fig. 16.** *Matonisporites* sp.
**Fig. 17.** *Aequitriradites* sp.

*Microcachryidites* SUPERZONE
**Figs. 18, 19.** *Retitriletes facetus.*
**Fig. 20.** *Pilosisporites notensis.*
**Fig. 21.** *Retitriletes watherooensis.*
**Figs. 22, 23.** *Clavatipollenites hughesii.*
**Fig. 24.** *Aequitritadites hispidus.*
**Fig. 25.** *Cicatricosporites australiensis.*
**Fig. 26.** *Crybelosporites stylosus.*
**Fig. 27.** *Foraminisporis asymmetricus.*
**Fig. 28.** *Coronatispora perforata.*
**Fig. 29.** *Foraminisporis wonthaggiensis.*
**Fig. 30.** *Aequitriradites acusus/spinulosus.*
**Fig. 31.** *Ceratosporites equalis.*
**Fig. 32.** *Biretisporites eneabbaensis.*
**Fig. 33.** *Cicatricosisporites hughesii.*
**Fig. 34.** *Cyclosporites hughesii.*
**Figs. 35, 36.** *Microcachryidites antarcticus.*
**Fig. 37.** *Trilobosporites trireticulosus.*
(Compiled by Sarjeant.)

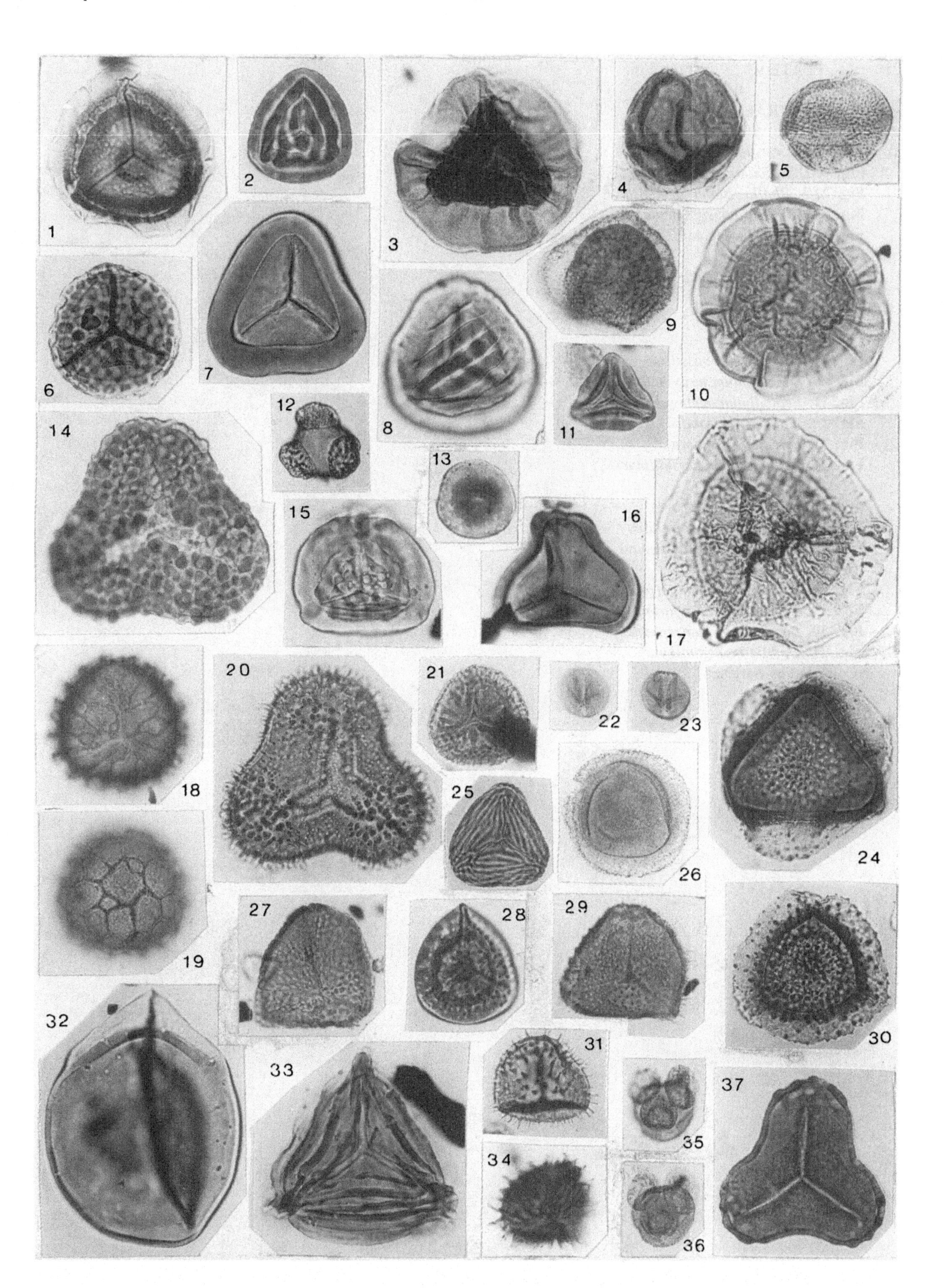
1
2
3
4
5
6
7
8
9
10
11
12
13
14
15
16
17
18
19
20
21
22
23
24
25
26
27
28
29
30
31
32
33
34
35
36
37

PLATE 93

AUSTRALIA

*Pareodinia ceratophora* SUPERZONE
**Fig. 1.** *Dissiliodinium caddaense.*
**Fig. 2, 3.** *Phallocysta erregulensis.*
**Fig. 4.** *Nannoceratopsis spiculata.*
**Fig. 5.** *Energlynia indotata.*
**Fig. 6.** *Ternia balmei.*
**Fig. 7.** *Rigaudella aemula.*
**Fig. 8.** *Wuroia capnosa.*

*Pyxidiella* SUPERZONE
**Fig. 9.** *Tubotuberella missilis.*
**Fig. 10.** *Dingodinium jurassicum.*
**Fig. 11.** *Leptodinium ambiguum.*
**Fig. 12.** *Wanaea spectabilis.*
**Fig. 13.** *Pyxidiella pandora.*
**Fig. 14.** *Scriniodinium crystallinum.*

*Fromea cylindrica* SUPERZONE
**Fig. 15.** *Peridictyocysta mirabilis.*
**Fig. 16.** *Scriniodinium? irregulare.*
**Fig. 17.** *Fromea cylindrica.*
**Fig. 18.** *Cribroperidinium perforans.*
**Fig. 19.** *Herendeenia pisciformis.*
**Fig. 20.** *Leptodinium eumorphum.*
**Fig. 21.** *Omatia montgomeryi.*
(Compiled by Sarjeant.)

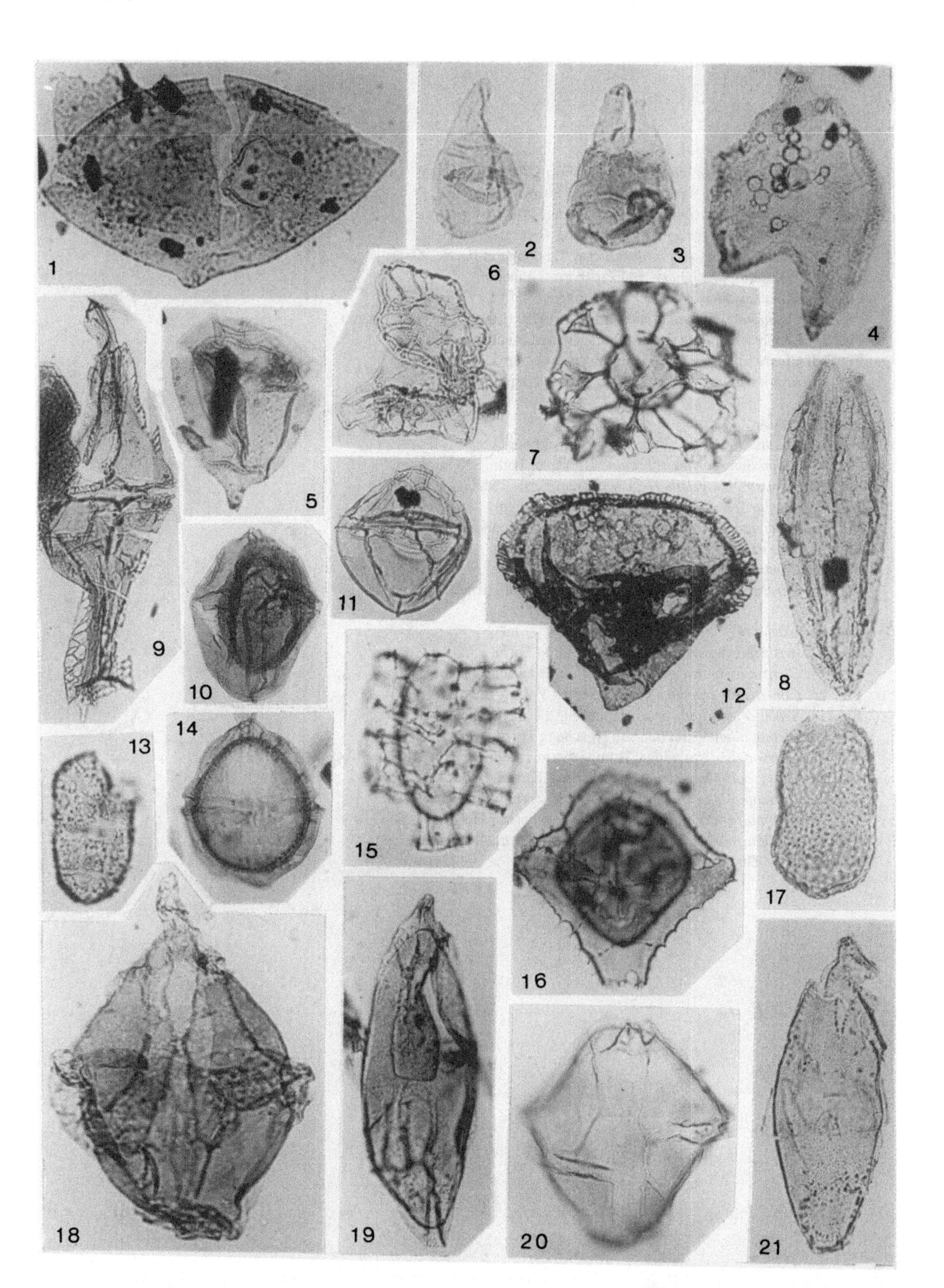
1
2
3
4
5
6
7
8
9
10
11
12
13
14
15
16
17
18
19
20
21

PLATE 94

CHINA

LOWER JURASSIC

**Fig. 2.** *Dictyophyllidites mortoni,* Fuxian Form., Neimongol.
**Fig. 3.** *Osmundacidites wellmanii,* Beipiao Form., Liaoning.
**Fig. 6.** *Klukisporites variegatus,* Tandonggou Form., Gansu.
**Fig. 8.** *Lycopodiacidites rugulatus,* Ziliujing Form., Sichuan.
**Fig. 10.** *Crassitudisporites problematicus,* Tandonggou Form., Gensu.
**Fig. 11.** *Cibotiumspora juncta,* Beipiao Form., Liaoning.
**Fig. 12.** *Marattisporites scabratus,* Tandonggou Form., Gensu.
**Fig. 19.** *Cycadopites subganulatus,* Fuxian Form. (Lower Jurassic), Shanxi.
**Fig. 20.** *Chasmatosporites minor,* Daling Form., Guangxi.
**Fig. 22.** *Classopollis parvus,* Daling Form., Guangxi.
**Fig. 29.** *Protoconiferus flavus,* Fuxian Form., Neimongol.

MIDDLE JURASSIC

**Fig. 1.** *Cayathidites minor,* Yanan Form., Neimongol.
**Fig. 4.** *Lycopodiumsporites subrotundus,* Yanan Form., Neimongol.
**Fig. 5.** *Neoraistrickia gristhorpensis,* Yanan Form., Shanxi.
**Fig. 7.** *Todisporites major,* Yanan Form., Neimongol.
**Fig. 13.** *Laevigatosporites ovatus,* Yanan Form., Neimongol.
**Fig. 16.** *Leptolepidites major,* Lanqi Form., Liaoning.
**Fig. 21.** *Caytonipollenites pallidus,* Yanan Form., Neimongol.
**Fig. 26.** *Cerebropollenites carlylensis,* Xingtiangou Form., Sichuan.

UPPER JURASSIC

**Fig. 9.** *Densoisporites microrugulatus,* Anding Form., Shanxi.
**Fig. 14.** *Cicatricosisporites minutaestriatus,* Houcheng Form., Hebei.
**Fig. 15.** *Concavissimisporites minor,* Anding Form., Shanxi.
**Fig. 17.** *Schizaeoisporites certus,* Houcheng Form., Hebei.
**Fig. 18.** *Pirinopollenites undulatus,* Houcheng Form., Hebei.
**Fig. 23.** *Classopollis classoides,* Anning Form., Yunnan.
**Fig. 24.** *Classopollis monotriatus,* Houcheng Form., Hebei.
**Fig. 25.** *Classopollis qiyangensis,* Houcheng Form., Hebei.
**Fig. 27.** *Callialasporites dampieri,* Houcheng Form., Hebei.
**Fig. 28.** *Callialasporites trilobatus,* Houcheng Form., Hebei.
(Compiled by Zhang.)

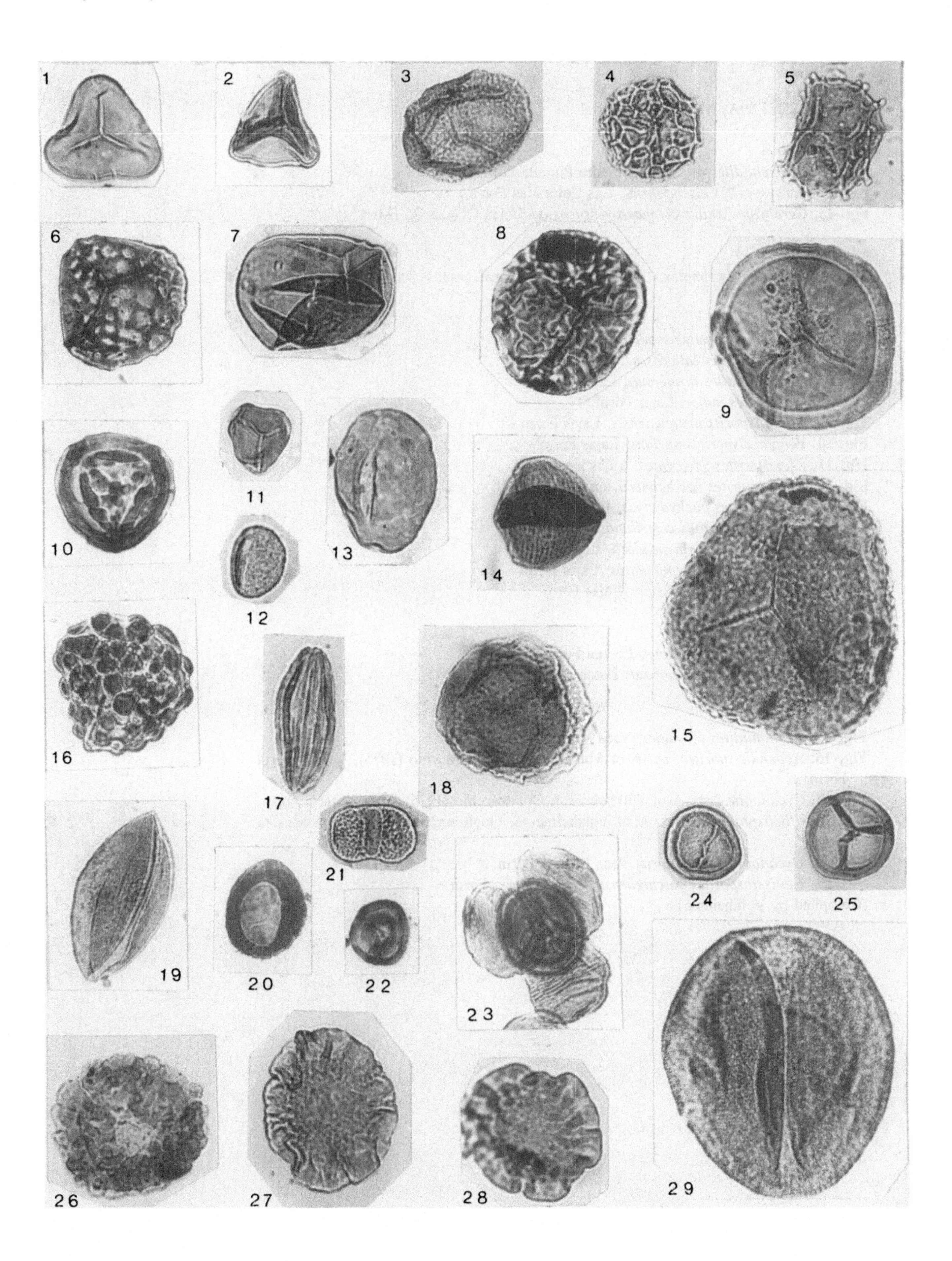
1
2
3
4
5
6
7
8
9
10
11
12
13
14
15
16
17
18
19
20
21
22
23
24
25
26
27
28
29

# PLATE 95

## ARGENTINA, NEUQUÉN BASIN

PLIENSBACHIAN

**Fig. 1.** *Gleicheniidites argentinus*, Piedra Pintada Form.
**Fig. 2.** *Skarbysporites elsendoornii*, Las Coloradas Form.
**Fig. 15.** *Cerebropollenites* cf. *macroverrucosus*, Sierra Chacia Có Form.

BAJOCIAN

**Fig. 21.** *Classopollis simplex*, "Arcillas Negras" Form.

EARLY CALLOVIAN

**Fig. 4.** *Lycopodiumsporites semimurus*, Lajas Form.
**Fig. 5.** *Concavisporites laticrassus*, Lajas Form.
**Fig. 6.** *Osmundacidites araucanus*, Lajas Form.
**Fig. 7.** *Todisporites major*, Lajas Form.
**Fig. 9.** *Rugulatisporits neuquenensis*, Lajas Form.
**Fig. 10.** *Verrucosisporites varians*, Lajas Form.
**Fig. 11.** *Klukisporites variegatus*, Lajas Form.
**Fig. 13.** *Ischyosporites volkheimerii*, Lajas Form.
**Fig. 14.** *Klukisporites pachydictyus*, Lajas Form.
**Fig. 17.** *Microcachryidites castellanosii*, Lajas Form.
**Fig. 19.** *Equisetosporites menendezii*, Lajas Form.
**Fig. 22.** *Classopollis intrareticulatus*, Lajas Form.
**Fig. 23.** *Callialasporites turbatus*, Lajas Form.

MIDDLE CALLOVIAN

**Fig. 8.** *Staplinisporites caminus*, Lotena Form.
**Fig. 12.** *Callialasporites trilobatus* Lotena Form.

EARLY TITHONIAN

**Fig. 3.** *Eucommiidites* cf. *minor*, Vaca Muerta Form.
**Fig. 16.** *Appendicisporites*, sp. A of Volkheimer & Quattrocchio (1975), Vaca Muerta Form.
**Fig. 18.** *Cycadopites*, sp. A of Volkheimer & Quattrocchio (1975), Vaca Muerta Form.
**Fig. 20.** *Phrixipollenites*, sp. A of Volkheimer & Quattrocchio (1975), Vaca Muerta Form.
**Fig. 24.** *Cycadopites follicularis*, Vaca Muerta Form.
**Fig. 25.** *Equisetosporites caichiguensis*, Vaca Muerta Form.
(Compiled by Volkheimer.)

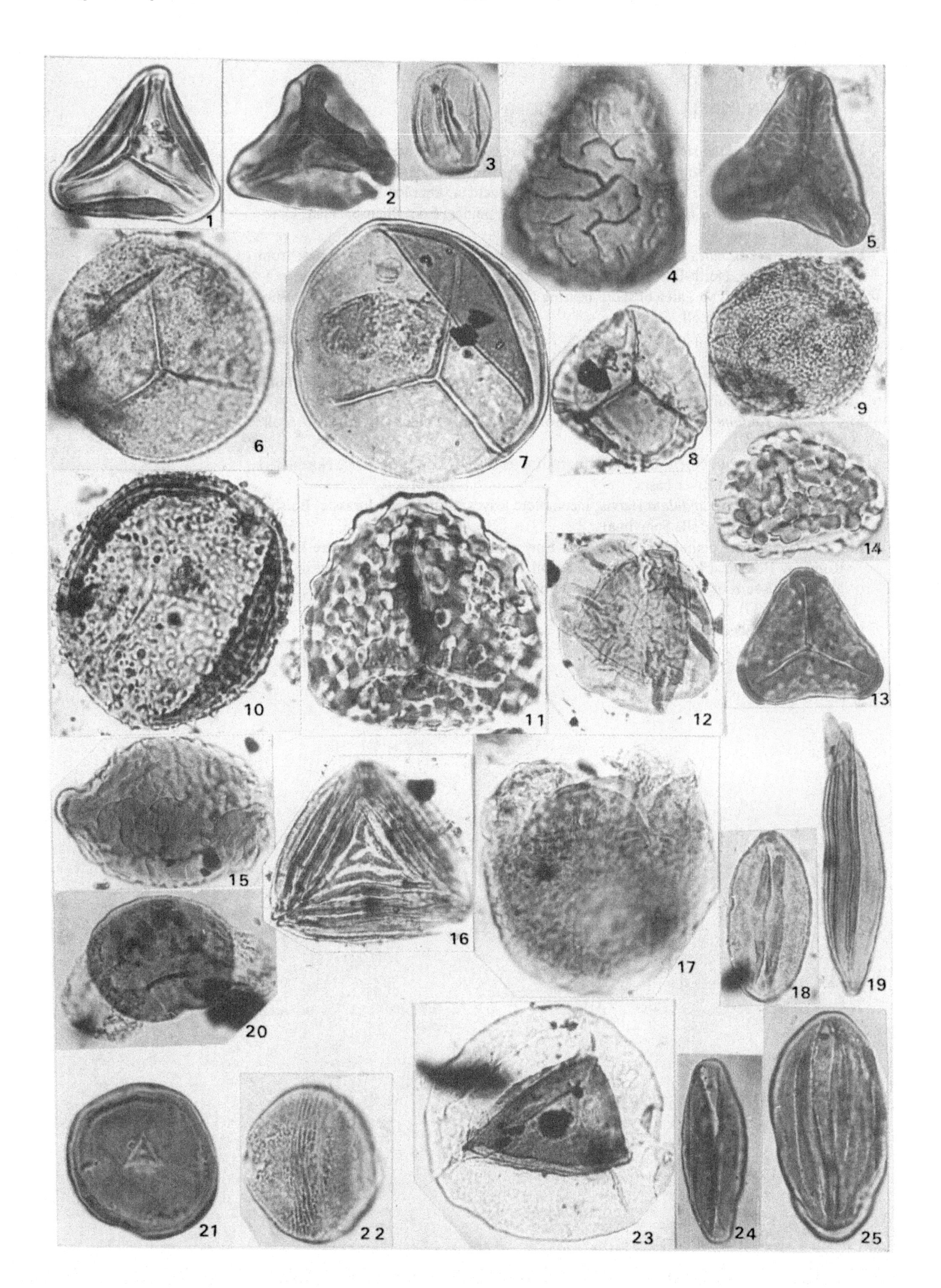
1
2
3
4
5
6
7
8
9
10
11
12
13
14
15
16
17
18
19
20
21
22
23
24
25

PLATE 96

EASTERN RUSSIA, LOWER JURASSIC

**Figs. 1–3.** *Ptiophyllum sibiricum* Samyl.; 1,2, leaves; 3 (×100), area of epiderm of the lower surface of leaf segment; from Toarcian, North-East Russia, Omolon River basin.
**Fig. 4.** *Taeniopteris* sp., leaf fragment; from Pliensbachian, Omolon River basin.
**Figs. 5, 6.** *Sagenopteris nilssoniana* (Brongn.) Ward, pinnae; from Pliensbachian, Omolon River basin.
**Figs. 7, 8.** *Neocalamites carrerei* (Zeil.) Halle, shoot fragments with internodes; from Lower Jurassic, North-East Russia, Bolshoi Anyui River basin (coll. Samylina).
**Fig. 9.** *Marrattiopsis* sp., area of spore-bearing leaf, (×3); from Lower Jurassic, Bolshoi Anyui River basin (coll. Samylina).
**Fig. 10.** *Nilssonia* sp., part of a leaf; from Lower Jurassic, Omolon River basin (coll. Samylina).
**Fig. 11.** *Cladophlebis* aff. *denticulatus* Kiritch, fragment of pinnule; from Lower Jurassic, Bolshoi Anyui River basin (coll. Samylina).
**Fig. 12.** *Dicroidium* sp., leaf fragment; from Toarcian, Omolon River basin (coll. Samylina).
**Fig. 13.** *Thaumatopteris schenkii* Nath., part of a leaf; from Lower Jurassic, North-East Russia, Korkodon River basin.
**Fig. 14.** *Nilssonia* aff. *undulata* Harris, incomplete leaves; from Lower Jurassic, Bolshoi Anyui River basin (coll. Samylina).
**Fig. 15.** *Ctenis* sp., pinnula segment; from Lower Jurassic. All specimens are kept in Komarov Botanical Institute, Leningrad (coll. 508, 509, 514, 515, 531) and in Geological Museum of the Geological Survey of North-East Russia, Magadan (coll. 38, 375, 452, 635).
(Compiled by Lebedev et al.)

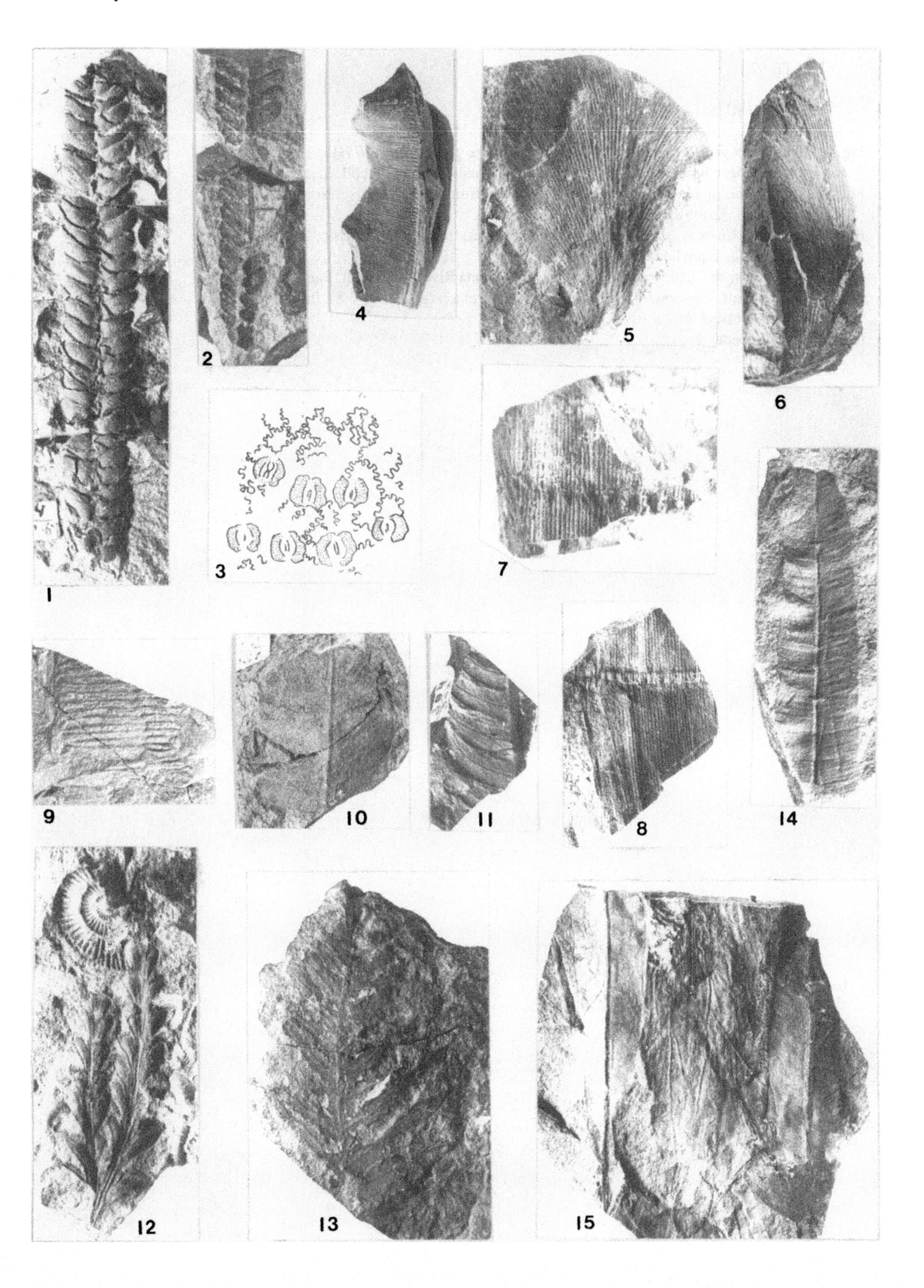
1
2
3
4
5
6
7
8
9
10
11
12
13
14
15

PLATE 97

EASTERN RUSSIA, UPPER JURASSIC

**Fig. 1a.** *Ginkgo* ex gr. *sibirica* Heer; 1b, *Pityophyllum* ex gr. *staratschinii* (Heer) Nath.; leaves; from Upper Jurassic, North-East Russia, Kolyma River basin (coll. Samylina).
**Fig. 2.** *Cladophlebis aldanensis* Vachr., frond fragment; from Upper Jurassic, North-East Russia, Bolshoi Anyui River basin (coll. Paraketzov).
**Fig. 3.** *Czekanowskia* ex gr. *setacea* Heer, bundle of leaves; from Upper Jurassic, Kolyma River basin (coll. Samylina).
**Fig. 4.** *Ginkgodium* sp., leaf; from Upper Jurassic, Kolyma River basin (coll. Samylina).
**Fig. 5.** *Heilungia* aff. *amurensis* (Novopokr.) Pryn, part of a large leaf (×0.8); from Upper Jurassic, Bolshoi Anyui River basin.
(Compiled by Lebedev et al.)

1
2
3
4
5

PLATE 98

EASTERN RUSSIA, UPPER JURASSIC

**Fig. 1.** *Ctenis* aff. *borealis* (Dawson) Bell, upper part of a leaf; from Upper Jurassic, North-East Russia, Bolshoi Anyui River basin (coll. Philippova).
**Fig. 2.** *Podozamites* ex gr. *eichwaldii* Schimp., leaf-bearing shoots; from Upper Jurassic, North-East Russia, Kolyma River basin (coll. Samylina).
**Fig. 3.** *Equisetites tschetschumensis* Vassilevsk, fragments of shoots with leaf whorl; from Upper Jurassic, Kolyma River basin (coll. Samylina).
**Fig. 4.** *Ctenis anyuensis* Philipp., part of a leaf (×0.8); from Upper Jurassic, Bolshoi Anyui River basin (coll. Philippova).
**Fig. 5.** *Raphaelia diamensis* Sew., part of a leaf; from Upper Jurassic, North-East Russia, Ulbeja River (coll. Samylina).
(Compiled by Lebedev et al.)

1
2
3
4
5

PLATE 99

MOSTLY NORTH AMERICA

HETTANGIAN

**Figs. 4, 8, 13, 18.** *Canoptum merum* Pessagno & Whalen, holotype (USNM 307206); from Hettangian of Queen Charlotte Islands, Zone 05; scale = 85, 48, 48, 30 μm.

PLIENSBACHIAN

**Figs. 1, 2, 9.** *Noritus lilihomensis* Pessagno & Whalen; 1, paratype (Pessagno coll.); 2,9, holotype (USNM 307202); from Lower Pliensbachian of Queen Charlotte Islands, Zone 02; scale = 99, 99, 42 μm.

**Fig. 3.** *Paracanoptum anulatum* (Pessagno & Poisson), holotype (USNM 264011); from Upper Pliensbachian of Turkey; known from California Coast Ranges, Izee Terrane, east-central Oregon, and Queen Charlotte Islands; scale = 150 μm.

**Figs. 5, 14, 19.** *Canutus hainaensis* Pessagno & Whalen; from Lower Pliensbachian of Queen Charlotte Islands, Zone 02; scale = 150, 75, 75 μm.

**Figs. 6, 10, 11, 15.** *Katroma* sp.; from Lower Pliensbachian of Queen Charlotte Islands, Zone 01; scale = 150, 75, 99, 99 μm.

**Figs. 7, 12, 16, 17, 20, 21.** *Katroma neagui* Pessagno & Poisson, holotype (USNM 264015); from Gumuslu Allochthon, Upper Pliensbachian, Turkey; scale = 199.8, 75, 66, 75, 42, 75 μm.

(Compiled by Pessagno.)

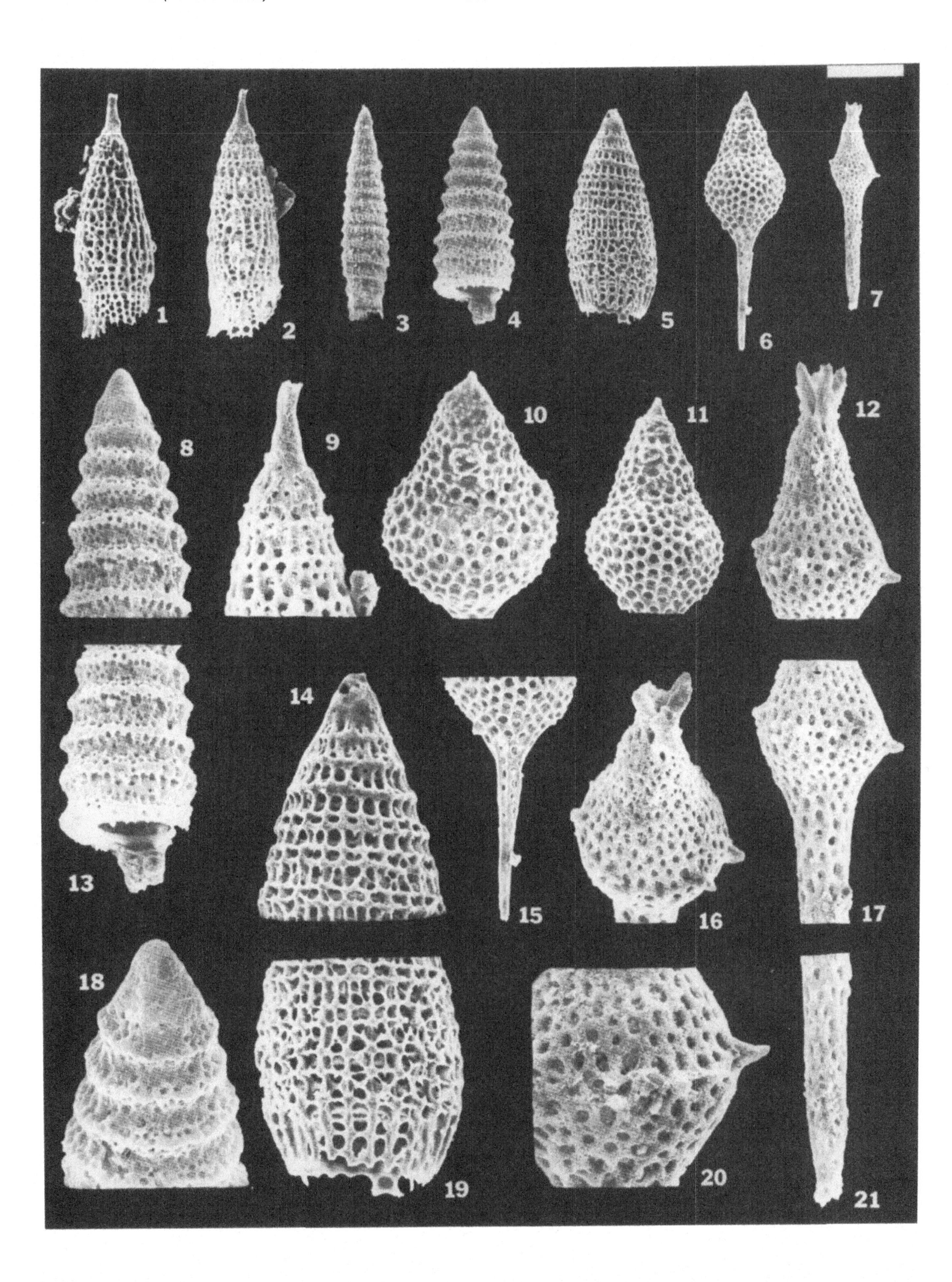
1
2
3
4
5
6
7
8
9
10
11
12
13
14
15
16
17
18
19
20
21

PLATE 100

NORTH AMERICA

BAJOCIAN

**Figs. 1, 11, 12.** *Xiphostylus halli* Pessagno & Yang, holotype (USNM 424174); from Lower Bajocian of Izee terrane, Oregon; scale = 120, 60, 48, 48 μm.

**Figs. 2, 18.** *Xiphostylus fragilis* Pessagno & Yang, holotype (USNM 424170); from Lower Bajocian of Izee Terrane, Oregon; scale = 120 μm.

**Figs. 10, 15.** *Parastatumalis* spp., from Pliensbachian–Lower Bajocian of Queen Charlotte Islands, Zone 02, Lower Pliensbachian; scale = 150, 150 μm.

BATHONIAN (CALLOVIAN–OXFORDIAN)

**Fig. 7.** *Turanta capsensis* Pessagno & Blome, holotype (USNM 319133); from Upper Bathonian of Izee Terrane, east-central Oregon; Superzone 1, Zone1F; scale = 240 μm.

**Figs. 8, 9.** *Hilarisirex inflatus* Pessagno, Whalen, & Yeh; from Upper Bathonian of Izee Terrane, east-central Oregon; Superzone 1, Zone 1F; scale = 99, 99 μm.

**Figs. 5, 14, 16.** *Tetraditryma praeplena* Baumgartner; from ?uppermost Callovian and Oxfordian of Superzone 1, Zone J of Pesssagno, Blome, & Six (in press); scale = 150, 30, 60 μm.

KIMMERIDGIAN

**Figs. 6, 13, 17.** *Eucytridiellum ptyctum* (Riedel & Sanfilippo); from Lower–Upper Kimmeridgian of Alamo Creek, San Luis Obispo County, California; Zone 2, Subzone 2 beta; scale = 48, 48, 30 μm.

OXFORDIAN

**Fig. 19.** *Acanthocirus suboblongus* (Yao); from Oxfordian of Klamath Mountains, northwestern California; Zone 2, Subzone 2 gamma; scale = 85.8 μm.

TITHONIAN

**Figs. 3, 4.** *Acanthocirus dicranocanthos* Squinabol; from Upper Tithonian of east-central Mexico; Zone 4, Subzone 4 beta; scale = 300, 199.8 μm.

(Compiled by Pessagno.)

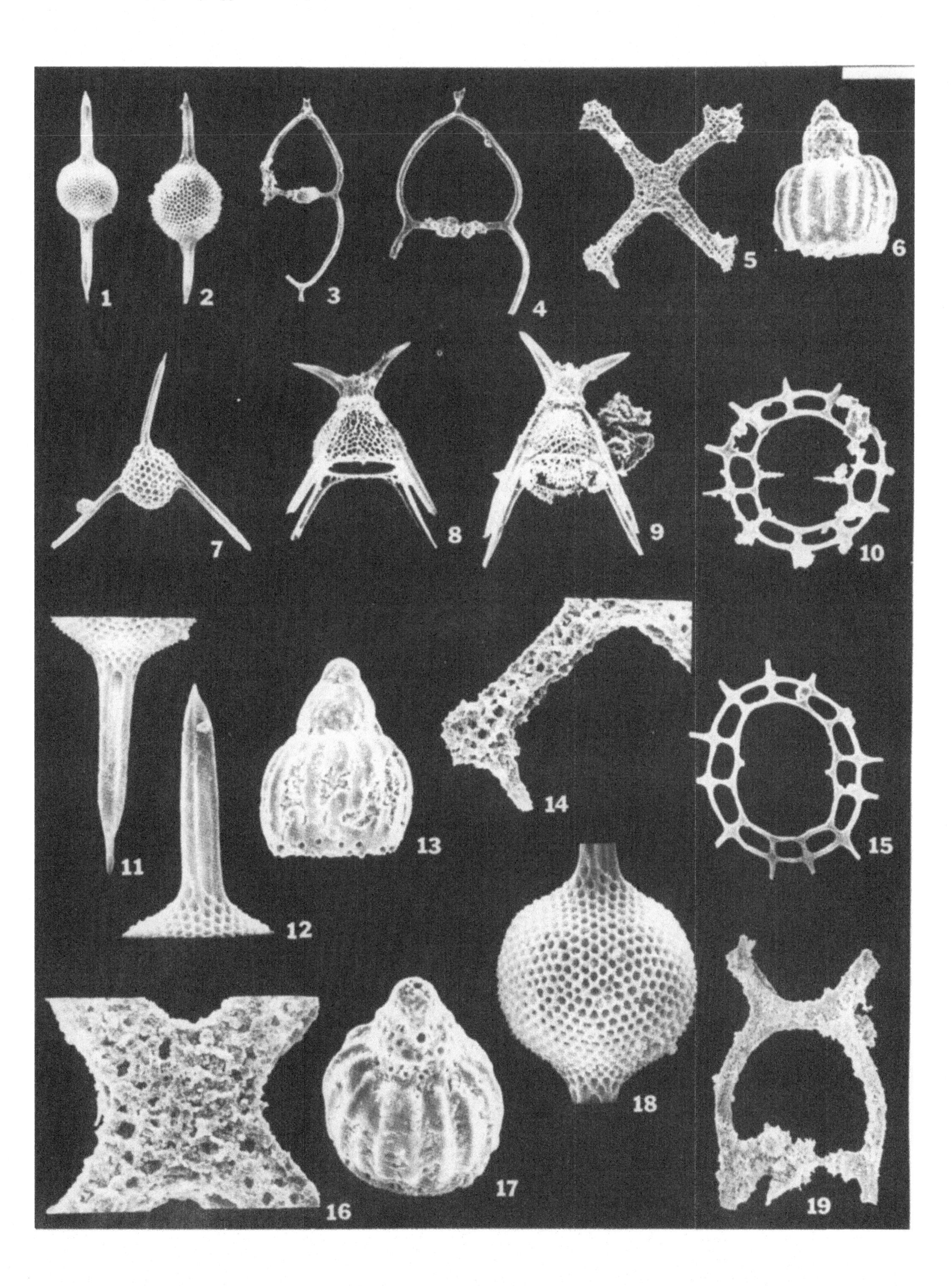
1
2
3
4
5
6
7
8
9
10
11
12
13
14
15
16
17
18
19

PLATE 101

NORTH AMERICA

BATHONIAN

**Figs. 9, 10.** *Perispyridium nitidum* Pessagno & Blome, Bathonian paratypes (Pessagno coll.); from Upper Bathonian of Izee Terrane, east-central Oregon; Superzone 1, Zone 1F; scale = 120 μm.

TITHONIAN

**Fig. 1.** *Ristola altissima* Rust; from Lower Tithonian of Santa Barbara County, California; Zone 3, Subzone 3 beta; scale = 300 μm.

**Figs. 2, 3, 13, 17.** *Parvicingula* s.s.; from Upper Tithonian of San Luis Obispo County, California; Zone 4, Subzone 4 alpha; scale = 150, 85.8, 60, 60 μm.

**Figs. 4, 16.** *Pseudodictyomitra*(?) *simplex* Matsuoka & Yao; from Upper Tithonian of east-central Mexico; Zone 4, Subzone 4 beta; scale = 66.6, 60 μm.

**Figs. 5, 11, 15, 19.** *Pseudodictyomitra*(?) sp.; from California Coast Ranges; Zone 8; scale = 150 μm.

**Figs. 6, 14, 18.** *Pseudodictyomitra pentacolaensis* Pessagno, from California Coast Ranges; Zone 8; scale = 120, 48, 48 μm.

**Fig. 7.** *Mirifusus baileyi* Pessagno; from Lower Tithonian of Santa Barbara County, California; Zone 3, Subzone 3 beta; scale = 240 μm.

**Figs. 8, 12.** *Mirifusus guadalupensis* Pessagno; from Upper Kimmeridgian/Lower Tithonian of Santa Barbara County, California; Zone 2, Subzone 2 alpha; scale = 199.8, 120 μm.

(Compiled by Pessagno.)

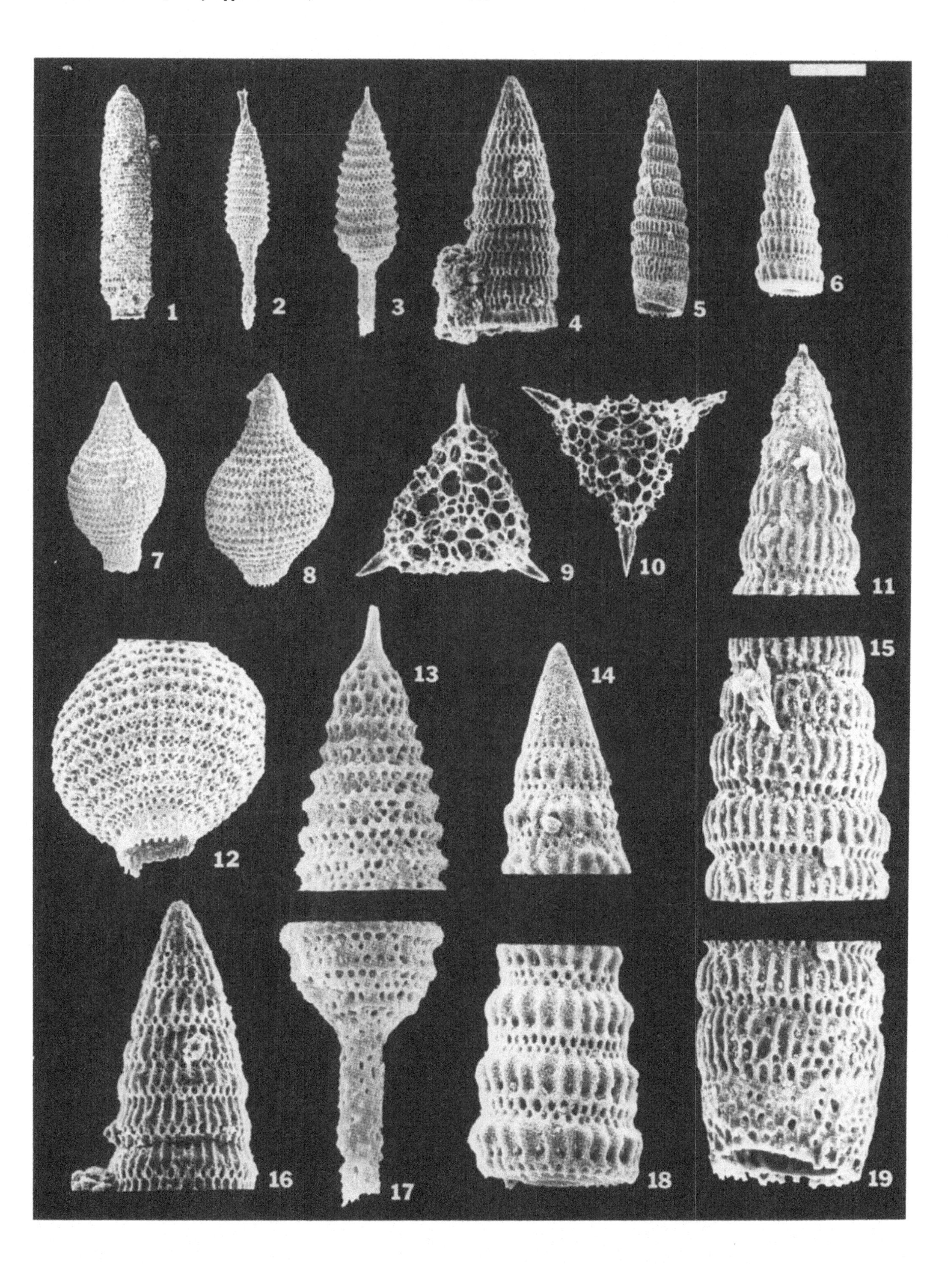
1
2
3
4
5
6
7
8
9
10
11
12
13
14
15
16
17
18
19

PLATE 102

NORTH AMERICA, TITHONIAN

**Figs. 1, 5.** *Vallupus hopsoni* Pessagno & Blome; from Zone 4, Subzone 4 beta; scale = 61.2, 61.2 μm.

**Figs. 2, 3, 12, 13, 20.** *Bivallupus longoriai* Pessagno & MacLeod; Zone 4, Subzone 4 beta; 2,3,12,20, paratypes (Pessagno coll.); scale = 85.7, 61.2, 61.2, 31.5 μm; 13, holotype (USNM 401060); scale = 61.2 μm; note in Fig. 12 that CC1 and CC2 = two cortical collars characterizing *Bivallupus* Pessagno & MacLeod.

**Figs. 4, 10, 11, 19.** *Bivallupus mexicanus* Pessagno & MacLeod; Zone 4, Subzone 4 beta; 4,19, holotype (USNM 401063); scale = 61.2, 31.5 μm; 10,11, paratypes (Pessagno coll.); scale = 61.2, 61.2 μm.

**Figs. 6, 7.** *Mesovallupus guadalupensis* Pessagno & MacLeod; Zone 4, Subzone 4 beta; 6, holotype (USNM 401065); scale = 61.2 μm; 7, paratype (Pessagno coll.); scale = 61.2μm.

**Figs. 8, 16.** *Vallupus zeissi* Pessagno & MacLeod; Zone 4, Subzone 4 beta; holotype (USNM 401075); scale = 61.2, 30,0 μm.

**Fig. 9.** *Vallupus* sp. aff. *V. nodosus* Pessagno & MacLeod; Zone 4, Subzone 4 beta; scale = 61.2 μm.

**Figs. 14, 15, 18.** *Vallupus nodosus* Pessagno & MacLeod; Zone 4, Subzone 4 beta; 14,18, holotype (USNM 401073); scale = 61.2, 37.5 μm; 15, paratype (Pessagno coll.); scale = 61.2 μm.

**Fig. 17.** *Vallupus* sp. aff. *V hopsoni* Pessagno & Blome; Zone 4, Subzone 4 beta; scale = 37.5 μm. All from Upper Tithonian of San Luis Potosí, east-central Mexico (Pessagno, Longoria, MacLeod, & Six 1987a).

(Compiled by Pessagno.)

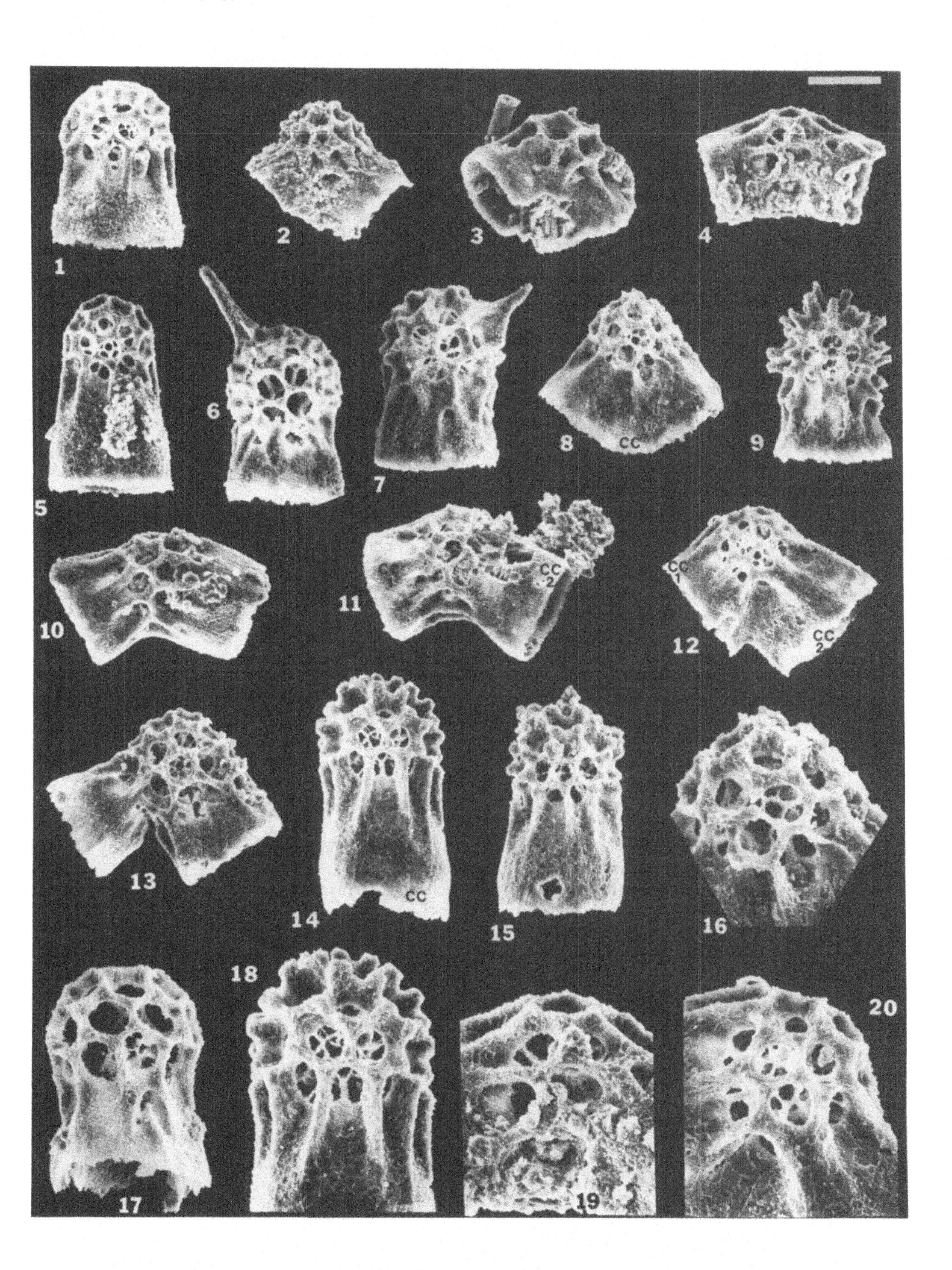
1
2
3
4
5
6
7
8
CC
9
10
11
CC
2
12
CC
1
CC
2
13
14
CC
15
16
17
18
19
20

PLATE 103

ARGENTINA, NEUQUÉN PROVINCE, MIDDLE JURASSIC

**Fig. 1.** *Reophax* cf. *tener* Seibold & Seibold (MLP-Mi 646) (×96); from early Bajocian *Emileia giebeli* Zone, Charahuilla.
**Fig. 2.** *Ammobaculites* cf. *subcretaceus* Cushman & Alexander (MLP-Mi 647) (×102); from early Bajocian *Emileia Geibeli* Zone, Paso del Carro Quebrado (northern limb of the anticline).
**Fig. 3.** *Ammobaculites* cf. *alaskensis* Tappan (MLP-Mi 630) (×55); from Middle–Late Callovian *Reineckeia* Zone, María Rosa Curicó.
**Fig. 4.** *Nodosaria mutabilis* Terquem (MLP-Mi 642) (×143); from early Bajocian *Puchenquia malarguensis* Zone, Picún Leufú.
**Fig. 5.** *Citharina heteropleura* (Terquem) (MLP-Mi 648) (×80); from early Bajocian *Emileia giebeli* Zone, Charahuilla.
**Fig. 6.** *Citharina heteropleura* (Terquem) (MLP-Mi 631) (×80); from Middle–Late Callovian *Reineckeia* Zone, María Rosa Curicó.
**Fig. 7.** *Citharina serratocostata* (Gümbel) MLP-Mi 633) (×65); from Middle–Late Callovian *Reineckeia* Zone, María Rosa Curicó.
**Fig. 8.** *Frondicularia involuta* Terquem (MLP-Mi 611) (×150); from early Bajocian *Puchenquia malarguensis* Zone, Picún Leufú.
**Fig. 9.** *Frondicularia involute* Terquem (MLP-Mi 612) (×150); from early Bajocian *Pseudotoites singularis* Zone, Paso del Carro Quebrado.
**Fig. 10.** *Lenticulina muensteri* (Roemer) (MLP-Mi 547) (×86); from early Bajocian *Emileia giebeli* Zone, Paso del Carro Quebrado.
**Fig. 11.** *Lenticulina* (*Astacolus*) *d'orbignyi* (Roemer) (MLP-Mi 649) (×100); from late Aalenian *Puchenquia malarguensis* Zone, Picún Leufú.
**Fig. 12.** *Lenticulina* (*Astacolus*) *d'orbignyi* (Roemer) (MLP-Mi 650) (×165); from early Bajocian *Emileia giebeli* Zone, Paso del Carro Quebrado.
**Fig. 13.** *Lenticulina* (*Astacolus*) *violetae* Ballent (MLP-Mi 643/1) (×160); from early Bajocian *Puchenquia malarguensis* Zone, Picún Leufú.
**Fig. 14.** *Lenticulina varians suturaliscostata* (Franke) (MLP-Mi 593) (×86); from early Bajocian *Emileia giebeli* Zone, Paso del Carro Quebrado.
**Fig. 15.** *Planularia beierana* (Gümbel) (MLP-Mi 644) (×200); from early Bajocian *Puchenquia malarguensis* Zone, Picún Leufú.
**Fig. 16.** *Planularia beierana* (Gümbel) (MLP-Mi 637) (×160); from Middle–Late Callovian *Reineckeia* Zone, María Rosa Curicó.
**Fig. 17.** *Vaginulina* cf. *flabelloides* (Terquem) (MLP-Mi 638) (×43); from Middle–Late Callovian *Reineckeia* Zone, María Rosa Curicó.
**Fig. 18.** *Lingulina nodosaria* Reuss (MLP-Mi 641) (×120); from Middle–Late Callovian *Reineckeia* Zone, María Rosa Curicó.
**Fig. 19.** *Vaginulinopsis epicharis* Loeblich & Tappan (MLP-Mi 640) (×100); from Middle–Late Callovian *Reineckeia* Zone, María Rosa Curicó.
(Compiled by Ballent.)

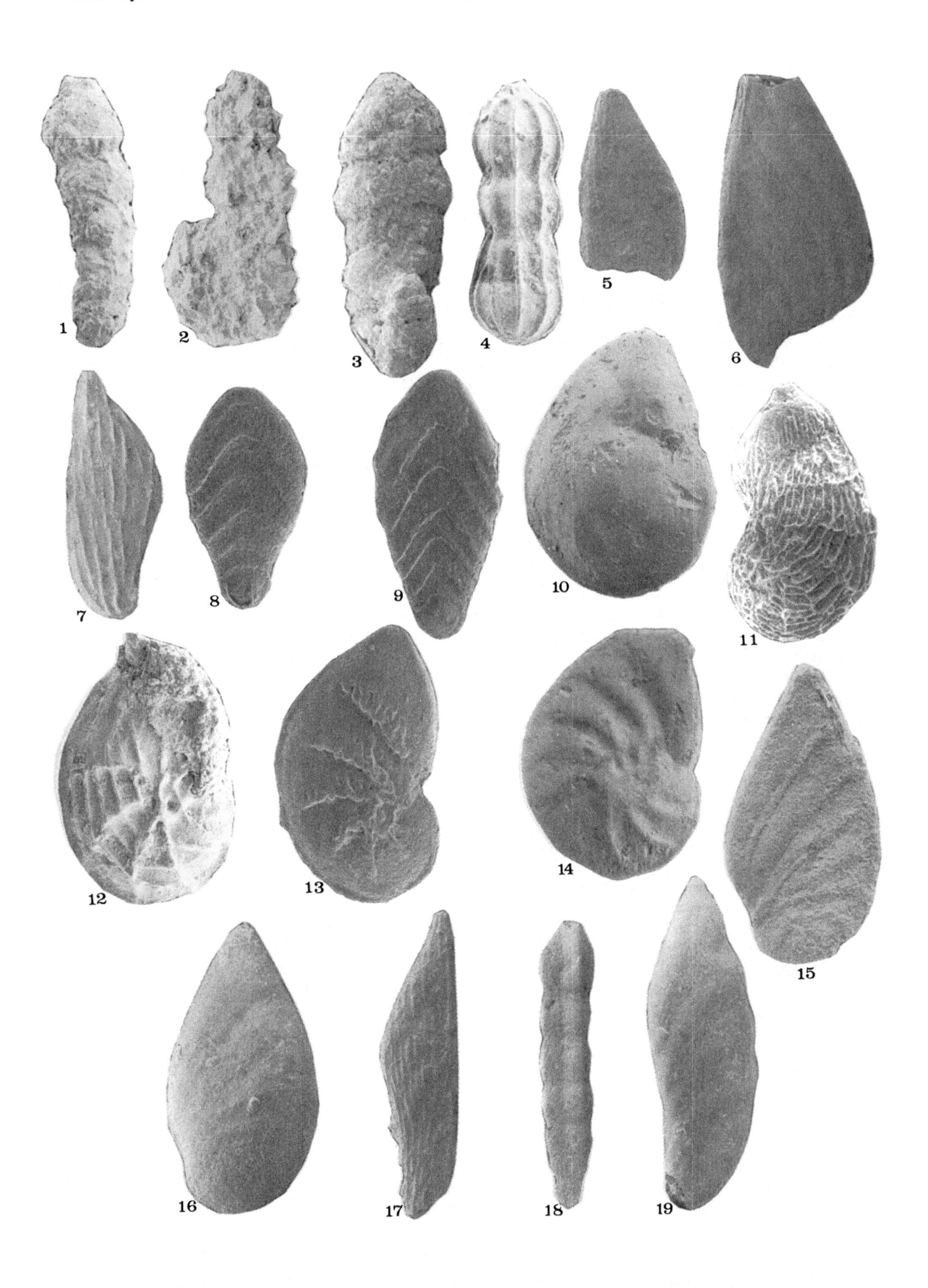
1
2
3
4
5
6
7
8
9
10
11
12
13
14
15
16
17
18
19

PLATE 104

ARGENTINA, NEUQUÉN PROVINCE, LATE PLIENSBACHIAN (AND LATE TOARCIAN)

**Fig. 1.** *Nodosaria* cf. *apheilolocula* Tappan (MLP-Mi 501) (×160); from *Fanninoceras* Zone of Estancia Santa Isabel.
**Fig. 2.** *Nodosaria kuhni* Franke (MLP-Mi 503) (×47); from *Fanninoceras* Zone of Picún Leufú.
**Fig. 3.** *Astacolus matutinus* (d'Orbigny) MLP-Mi 506) (×95); from *Fanninoceras* Zone of Picún Leufú.
**Fig. 4.** *Astacolus quadricostatus* (Terquem) (MLP-Mi 507) (×117); from *Fanninoceras* Zone of Picún Leufú.
**Fig. 5.** *Dentalina pseudocommunis* Franke (MLP-Mi 509) (×47); from *Fanninoceras* Zone of Picún Leufú.
**Fig. 6.** *Frondicularia terquemi bicostata* d'Orbigny (MLP-Mi 512) (×100); from *Fanninoceras* Zone, Cerro Trapial Mahuida.
**Fig. 7.** *Frondicularia terquemi sulcata* Bornemann (MLP-Mi 514) (×165); from *Fanninoceras* Zone of Picún Leufú.
**Fig. 8.** *Lenticulina gottingensis* (Bornemann) (MLP-Mi 519) (×117); from *Fanninoceras* Zone of Picún Leufú.
**Fig. 9.** *Lenticulina varians* (Bornemann) (MLP-Mi 520) (×82); from *Fanninoceras* Zone of Picún Leufú.
**Fig. 10.** *Lenticulina polygonata* (Franke) (MLP-Mi 589) (×115); from *Fanninoceras* Zone of Picún Leufú.
**Fig. 11.** *Lenticulina varians suturaliscostata* (Franke) (MLP-Mi 516) (×90); from late Toarcian *Dumortieria* faunule, 3 km north of Cerro Granito.
**Fig. 12.** *Marginulina prima prima* d'Orbigny (MLP-Mi 523) (×65); from *Fanninoceras* Zone of Picún Leufú.
**Fig. 13.** *Planularia protracta* (Bornemann) (MLP-Mi 527) (×100); from *Fanninoceras* Zone of Picún Leufú.
**Fig. 14.** *Pseudonodosaria oviformis* (Terquem) (MLP-Mi 530) (×176); from *Fanninoceras* Zone of Picún Leufú.
**Fig. 15.** *Pseudonodosaria vulgata* (Bornemann) (MLP-Mi 531) (×90); from *Fanninoceras* Zone of Estancia Santa Isabel.
**Fig. 16.** *Eoguttulina liassica* (Strickland) (MLP-Mi 608) (×100); from *Fanninoceras* Zone, hill south of Cerro Roth.
**Fig. 17.** *Lingulina tenera tenera* Bornemann (MLP-Mi 537) (×180); from *Fanninoceras* Zone of Picún Leufú.
**Fig. 18.** *Conicospirillina trochoides* (Berthelin) (MLP-Mi 541) (×90); from *Fanninoceras* Zone of Estancia Santa Isabel.
**Fig. 19.** *Isobythocypris* sp. (MLP-Mi 559) (×80); from *Fanninoceras* Zone of Picún Leufú.
**Fig. 20.** *Liasina*(?) sp. (MLP-Mi 565) (×160); from *Fanninoceras* Zone of Picún Leufú.
**Fig. 21.** *Eucytherura isabelensis* Ballent (MLP-Mi 576/5) (×170); from *Fanninoceras* Zone of Estancia Santa Isabel.
**Fig. 22.** *Ogmoconcha* sp. (MLP-Mi 584) (×143); from *Fanninoceras* Zone of Picún Leufú.

(Compiled by Ballent.)

1
2
3
4
5
6
7
8
9
10
12
13
11
14
15
16
17
18
19
20
21
22

PLATE 105

ARGENTINA, NEUQUÉN PROVINCE, MIDDLE JURASSIC

**Fig. 1.** *Cytherelloidea* sp. A (MLP-Mi 562) (×102); from early Bajocian, *Emileia giebeli* Zone of Paso del Carro Quebrado.
**Fig. 2.** *Cytherelloidea* sp. B (MLP-Mi 653) (×125); from early Bajocian, *E. giebeli* Zone of Paso del Carro Quebrado.
**Fig. 3.** *Eucytherura argentina* Ballent (MLP-Mi 654) (×175); from early Bajocian, *Puchenquia malarguensis* Zone of Picún Leufú.
**Fig. 4.** *Procytherura* cf. *euglyphea* Ainsworth (MLP-Mi 623) (×235); from early Bajocian, *P. malarguensis* Zone of Picún Leufú.
**Fig. 5.** *Procytherura* cf. *celtica* Ainsworth (MLP-Mi 625) (×175); from early Bajocian, *P. malarguensis* Zone of Picún Leufú.
**Fig. 6.** *Procytherure bispinata* Ballent (MLP-Mi 624) (×265); from early Bajocian, *P. malarguensis* Zone of Picún Leufú.
**Fig. 7.** *Hemiparacytheridea* sp. (MLP-Mi 655) (×150); from early Bajocian, *P. malarguensis* Zone of Picún Leufú.
**Fig. 8.** *Rutlandella* cf. *transversiplicata* Bate & Coleman (MLP-Mi 656) (×175); from early Bajocian, *P. malarguensis* Zone of Picún Leufú.
**Fig. 9.** *Rutlandella* sp. A (MLP-Mi 622) (×175); from early Bajocian, *P. malarguensis* Zone of Picún Leufú.
**Fig. 10.** *Acrocythere pichia* Ballent (MLP-Mi 645/2) (×235); from early Bajocian, *P. malarguensis* Zone of Picún Leufú.
**Fig. 11.** *Etkyphocythere australis* Ballent (MLP-Mi 628) (×100); from early Bajocian, *P. malarguensis* Zone of Picún Leufú.
**Fig. 12.** *Ektyphocythere australis* Ballent (MLP-Mi 500/2) (×115); from early Bajocian, *E. giebeli* Zone of Paso del Carro Quebrado.
**Fig. 13.** *Progonocythere neuguenensis* Muscchio (MLP-Mi 620) (×80); from middle–late Callovian, Reineckeia Zone of María Rosa Curicó.
**Fig. 14.** *Sondagella* sp. (MLP-Mi 621) (×130); from Callovian–Oxfordian boundary, Mallín de la Cueva.
(Compiled by Ballent.)

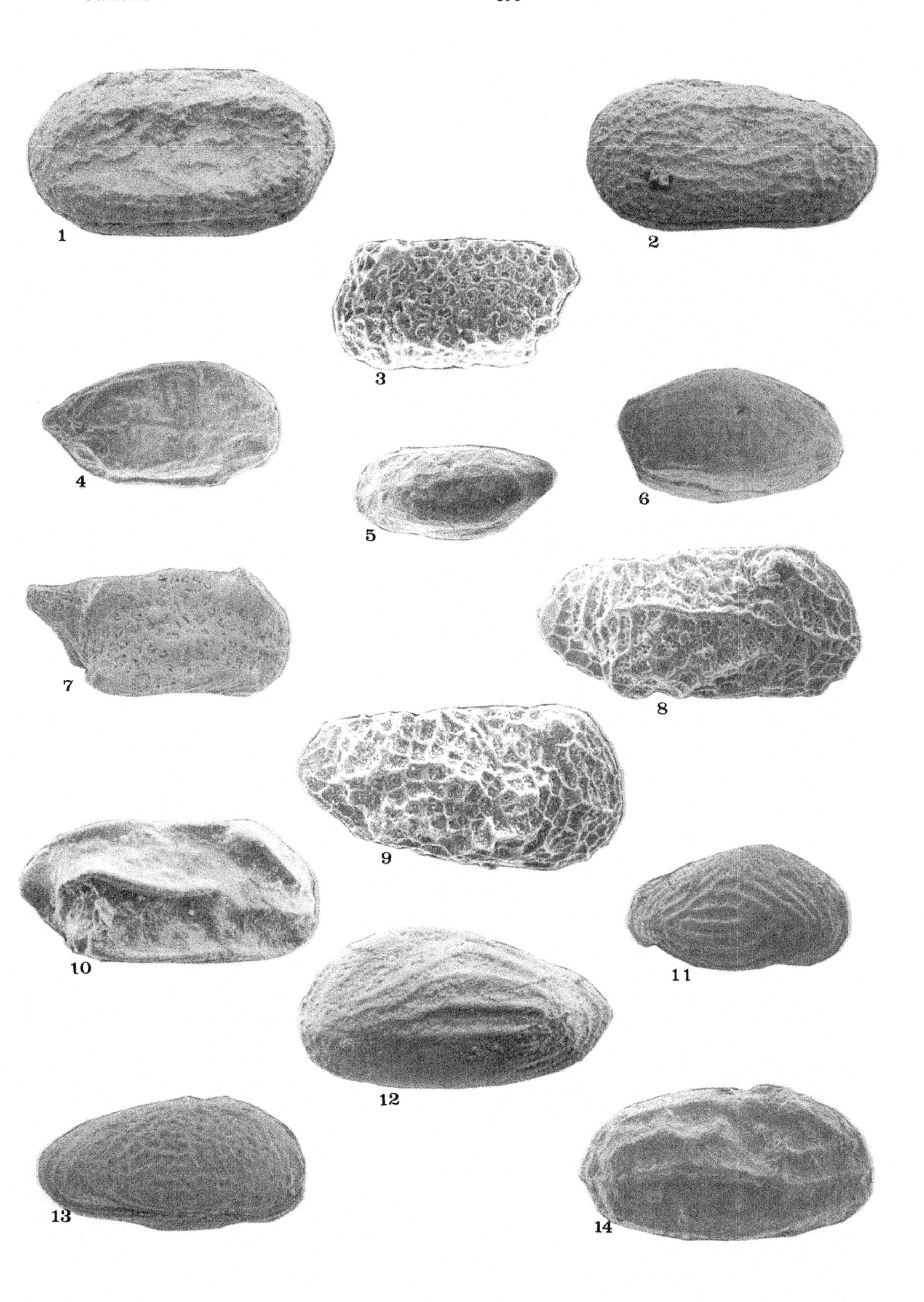
1
2
3
4
5
6
7
8
9
10
11
12
13
14

PLATE 106

WESTERN CANADA, BAJOCIAN–LOWER BATHONIAN (all magnified about ×50; numbers for one or several specimens). SHAUNAVON FAUNA, BIOZONES II AND III

**Fig. 1.** *Darwinula oblonga* (Roemer); note extreme variations in size; larger forms dominate in Zone II.
**Fig. 2.** *Norcanollela parryi* Loranger.
**Fig. 3.** ?*Aparchitocythere typica* Swain & Peterson.
**Fig. 4.** *Aparchitocythere* cf. *compressa* Peterson.
**Fig. 5.** *Aparchitocythere*(?) sp. 59 (Brooke & Braun).
**Fig. 6.** *Limnocythere climaxia* (Loranger).
**Fig. 7.** *Limnocythere pustulosa* Wall.
**Fig. 8.** Genus ind. sp. 19 (Brooke & Braun).
**Fig. 9.** *Mandelstamia*(?) *elongata* (Peterson).
**Fig. 10.** *Metacypris* cf. *tenuimarginata* Bernard, Bizon, & Oertli.
**Fig. 11.** *Pseudoperissocytheridea*(?) *anoda* (Peterson).
**Fig. 12.** *Lophocythere* (*Neurocythere*) aff. *minuta* (Peterson).
**Fig. 13.** *Protocythere* cf. *guadricarinata* Swain & Peterson.
**Fig. 14.** *Acrocythere rushlakensis* (Wall).
**Fig. 15.** *Procytheridea*(?) sp. 57 (Brooke & Braun).
**Fig. 16.** *Paracypris*? sp. A Wall.

All specimens in Figs. 1–10 are from the subsurface of southern Saskatchewan, representing most of the fresh- to brackish-water ostracod taxa of the northern part of the Williston Basin. Most of the marine indicators, from a few scattered intercalations, are shown in Figs. 11–16.

(Compiled by Braun & Brooke.)

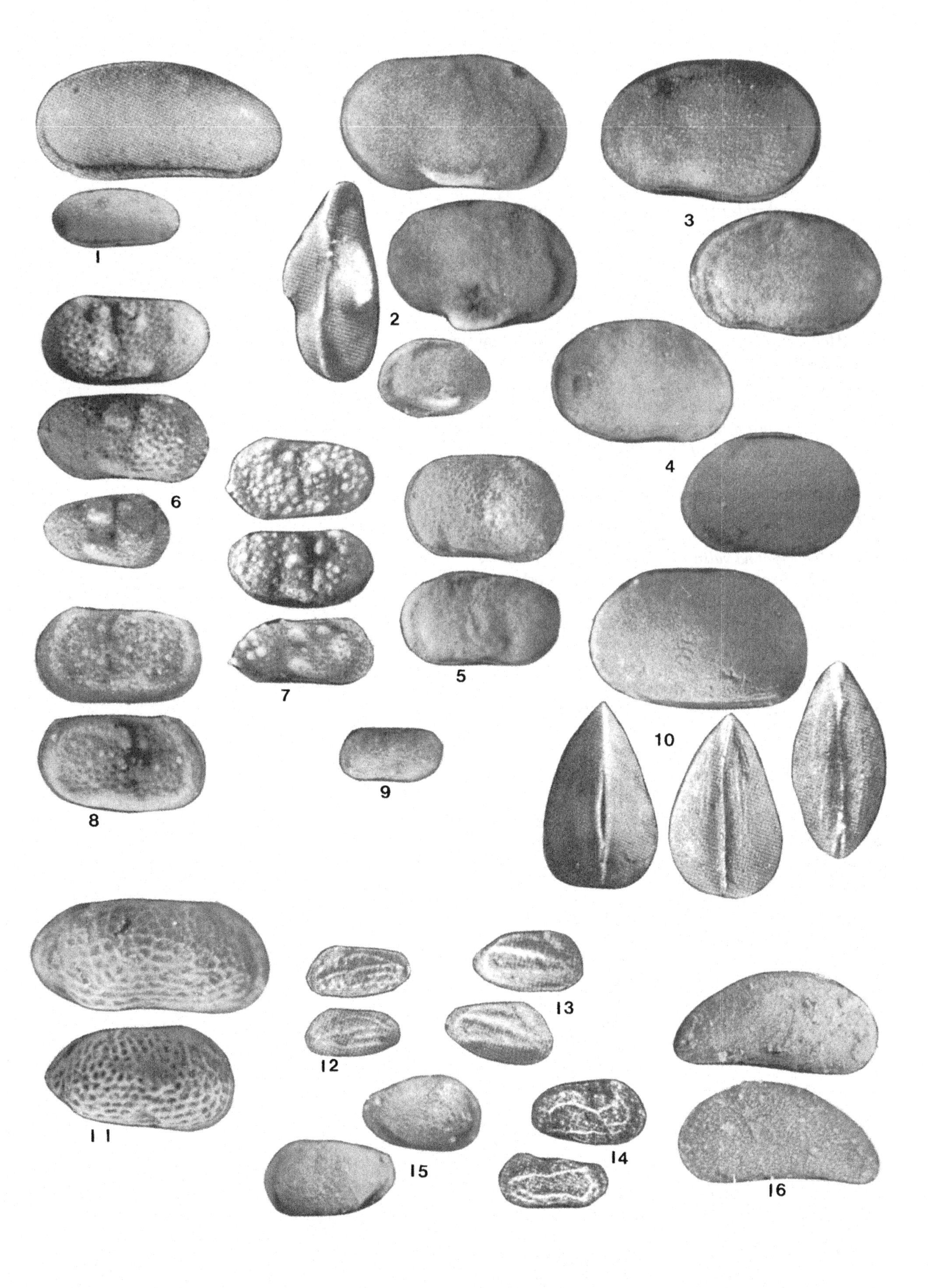
1
2
3
4
5
6
7
8
9
10
11
12
13
14
15
16

## PLATE 107

## WESTERN CANADA, MIDDLE BATHONIAN–MIDDLE CALLOVIAN

COSMOPOLITAN RIERDON FAUNA, BIOZONES IV AND V (all magnified about ×50; numbers for one or several specimens)

**Fig. 1.** *Procytheridea exempla* s.l. Peterson.
**Fig. 2.** *Palaeocytheridea* sp. 38 (Brooke & Braun).
**Fig. 3.** *Micropneumatocythere crassa* (Peterson).
**Fig. 4.** *Micropneumatocythere* cf. *crassa* of Grey Beds, Rocky Mountains.
**Fig. 5.** *Acrocythere robertsi* (Wall).
**Fig. 6.** *Acrocythere robertsi* subsp. 55 (Brooke & Braun).
**Fig. 7.** *Monoceratina incisa* Peterson.
**Fig. 8.** *Monoceratina* cf. *incisa* of Grey Beds, Rocky Mountains.
**Fig. 9.** *Orthonotacythere* cf. *dorsoconvexa* Peterson.
**Fig. 10.** *Paracytheropteron*(?) cf. *polyornata* (Peterson).
**Fig. 11.** *Eucytherura reticulata* (Peterson).
**Fig. 12.** *Eucytherura heteromorpha* (Peterson).
**Fig. 13.** *Monoceratina* cf. *sundancensis* Swain & Peterson.
**Fig. 14.** *Cytherelloidea recurvata* Peterson.
**Fig. 15.** *Cytherella* cf. *catenulata* Jones & Sherborn.
**Fig. 16.** *Cytherella paramuensteri* Swain & Peterson.
**Fig. 17.** *Monoceratina pararossae* Peterson.
**Fig. 18.** *Monoceratina vulsa* (Jones & Sherborn).
**Fig. 19.** *Monoceratina* cf. *vulsa* of Grey Beds, Rocky Mountains.
**Fig. 20.** *Monoceratina* cf. *obtusifrons* Triebel & Bartenstein.
**Fig. 21.** *Monoceratina* cf. *striata* Triebel & Bartenstein.
**Fig. 22.** *Cytheropteron* (*Eocytheropteron*?) *lanceolata* Peterson.
**Fig. 23.** *Bythocypris ambitruncata* Peterson.
**Fig. 24.** *Paracypris projecta* Peterson.
**Fig. 25.** *Macrocypris*(?) *minutus* Swain & Peterson.

ROSERAY FAUNULE, BIOZONE VI, SUBSURFACE OF SW SASKATCHEWAN ONLY

**Fig. 26.** *Pleurocythere radvillia* (Loranger).
**Fig. 27.** *Fuhrbergiella* (*Praefuhrbergiella*?) sp. 45 (Brooke & Braun).
**Fig. 28.** *Palaeocytheridea*(?) sp. 46 (Brooke & Braun).
**Fig. 29.** *Protocythere quadricarinata* Swain & Peterson.
**Fig. 30.** *Fuhrbergiella* (*Praefuhrbergiella*) sp. 60 (Brooke & Braun).
(Compiled by Braun & Brooke.)

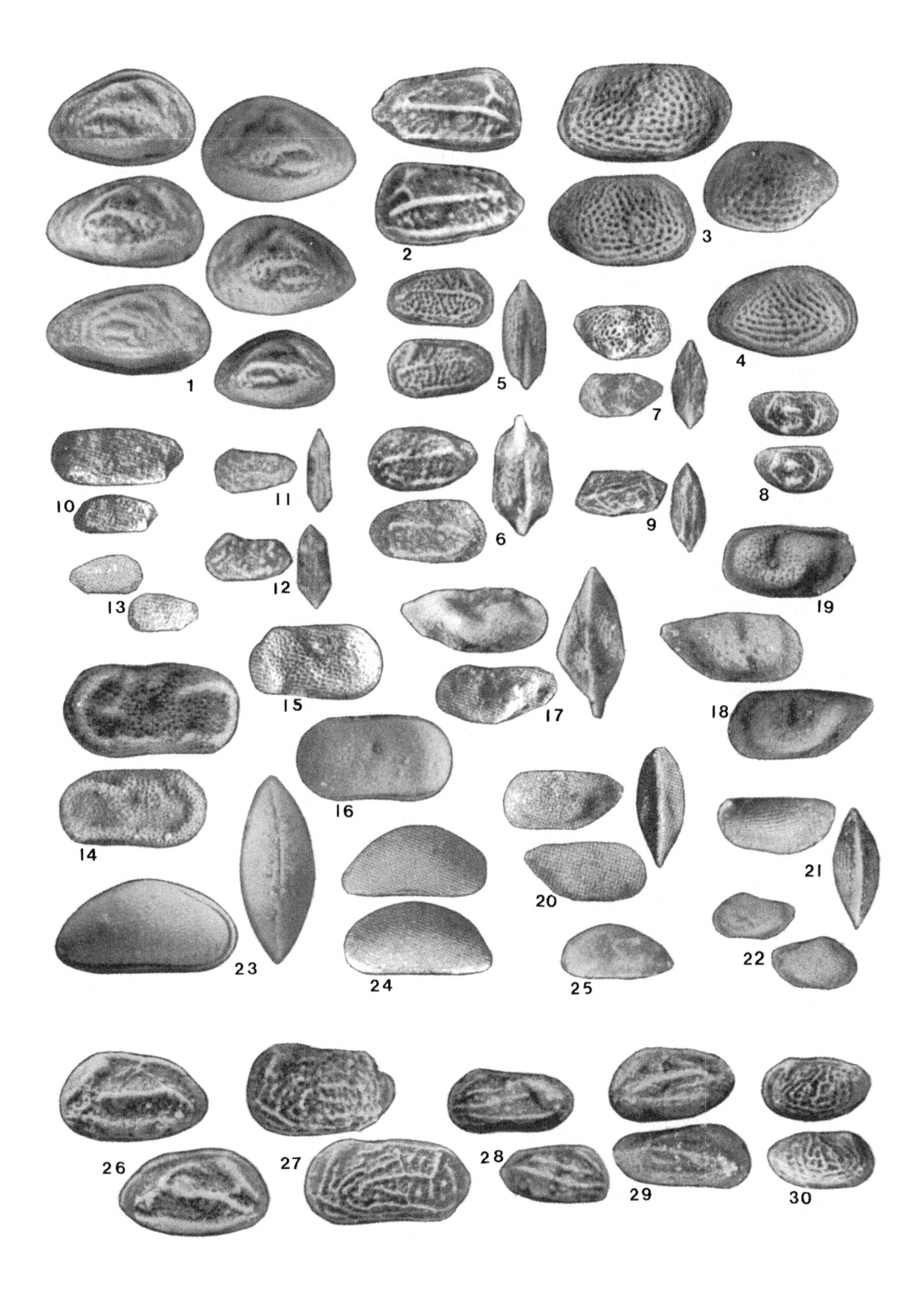
1
2
3
4
5
6
7
8
9
10
11
12
13
14
15
16
17
18
19
20
21
22
23
24
25
26
27
28
29
30

PLATE 108

WESTERN CANADA, LATE JURASSIC

OXFORDIAN SWIFT FAUNA OF WILLISTON BASIN, BIOZONE VII (all magnified about ×50; numbers for one or several specimens)

**Fig. 1.** *Fuhrbergiella (Praefuhrbergiella) crowcreekensis* (Swain & Peterson).
**Fig. 2.** *Fuhrbergiella (Praefuhrbergiella) hieroglyphica* (Swain & Peterson).
**Fig. 3.** *Balowella* cf. *attendens* (Ljubimova).
**Fig. 4.** ?*Aparchitocythere loeblichorum* Swain & Peterson.
**Fig. 5.** *Camptocythere*(?) sp. 44 (Brooke & Braun).
**Fig. 6.** *Cytherella*(?) sp. 3 (Brooke & Braun).

OXFORDIAN–VOLGIAN OF ROCKY MOUNTIANS, BIOZONES IX–XII

**Fig. 7.** Genus indet. sp. 88.
**Fig. 8.** *Eucytherura* (*Vesticytherura*) sp. 87.
**Fig. 9.** *Eucytherura* (*Vesticytherura*) sp. 85.
**Fig. 10.** Genus indet. sp. 70 (Brooke & Braun).
**Fig. 11.** Genus indet. sp. 91.
**Fig. 12.** Genus indet. sp. 63 (Brooke & Braun).
**Fig. 13.** *Camptocythere*(?) sp. 64 (Brooke & Braun).
**Fig. 14.** Genus indet. sp. 69 (Brooke & Braun).
**Fig. 15.** Genus indet. sp. 68 (Brooke & Braun).
**Fig. 16.** Genus indet. sp. 89.
**Fig. 17.** Genus indet. sp. 84.
**Fig. 18.** Genus indet. sp. 90.
**Fig. 19.** *Paracypris* sp. 65 (Brooke & Braun).
**Fig. 20.** Genus indet. sp. 66 (Brooke & Braun).
**Fig. 21.** Genus indet. sp. 116.
**Fig. 22.** Genus indet. sp. 73 (Brooke & Braun).
**Fig. 23.** Genus indet. sp. 74 (Brooke & Braun).
**Fig. 24.** Genus indet. sp. 67 (Brooke & Braun).
**Fig. 25.** Genus indet. sp. 72 (Brooke & Braun).
**Fig. 26.** Genus indet. sp. 71 (Brooke & Braun).

*Because of the poor state of preservation, lack of single valves, and low abundance, most ostracods could not be identified to generic rank. The species numbers assigned are those used for data recording. Some of the species were discussed earlier (Brooke & Braun 1981); the rest are new forms still to be described.

(Compiled by Braun & Brooke.)

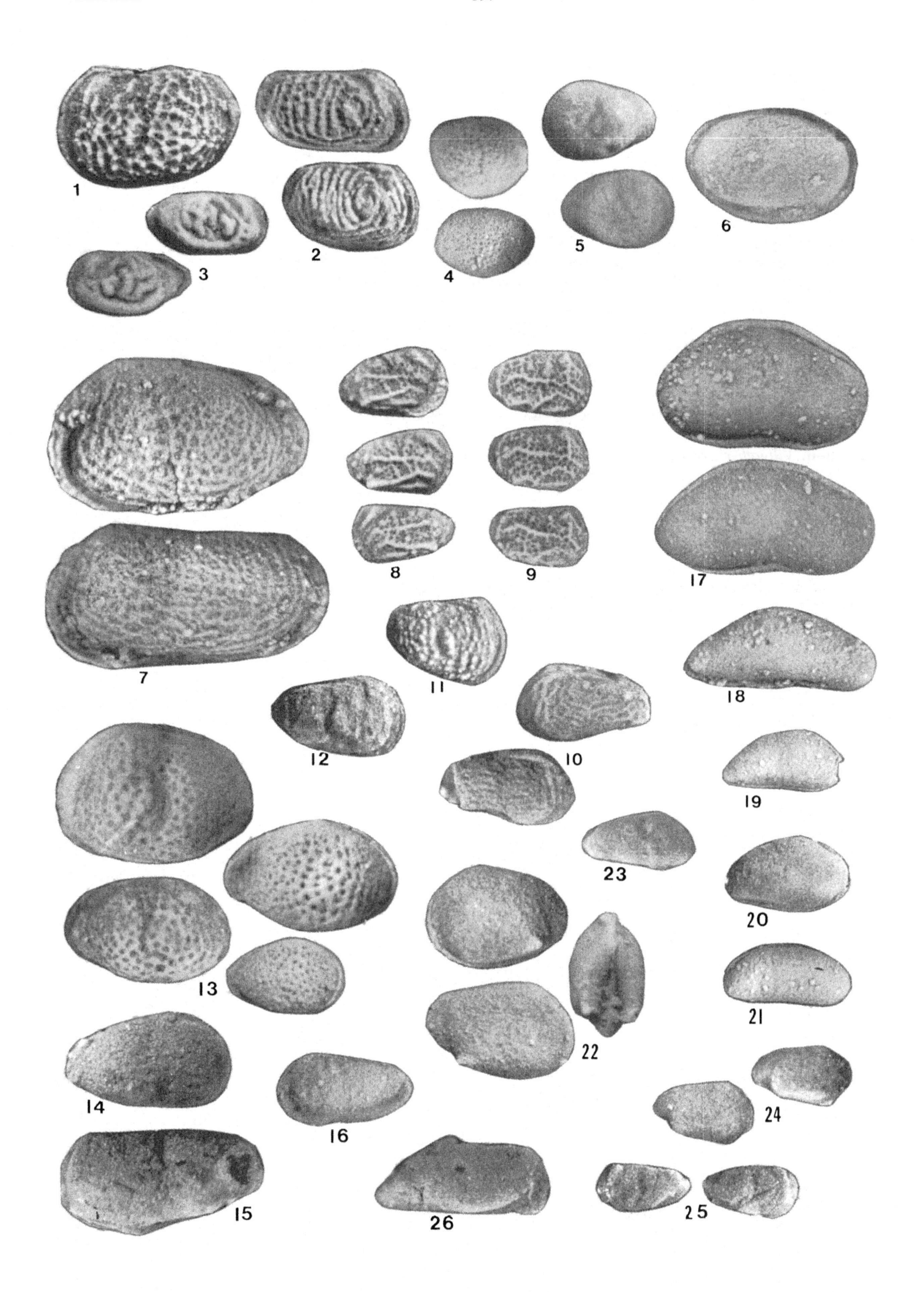
1
2
3
4
5
6
7
8
9
10
11
12
13
14
15
16
17
18
19
20
21
22
23
24
25
26

PLATE 109

NORTH-EAST RUSSIA

SINEMURIAN

**Fig. 1a–d.** *Ochotorhynchia omolonensis* Dagys, holotype (IGG 230/78)(×2); from Omolon Massif.

UPPER PLIENSBACHIAN

**Figs. 2a, b, 3a–c, 4.** *Peregrinelloidea malkovi* Dagys; 2, paratype (IGG 223/78); 3, holotype of *P. tenuicostata* (IGG 222/78); 4, paratype (IGG 221/78); from Kolyma River basin.

**Fig. 5a–d.** *Rimirhynchia maltanensis* Dagys, holotype (IGG 362/78); from Kolyma River basin.

**Fig. 6a–c.** *Orlovirhynchia viligaensis* (Moisseiev), whole shell (IGG 315/78); from Viliga River, Okhotsk Sea coast.

**Fig. 7a–c.** *Rudirhynchia najahaensis* (Moisseiev), whole shell (IGG 335/78); from Viliga River, Okhotsk Sea coast.

**Fig. 8a–d.** *Viligothyris orientalis* (Dagys), holotype (IGG 135/78); from Viliga River, Okhotsk Sea coast.

TOARCIAN

**Fig. 9a–d.** *Omolonothyris inopinatus* Dagys, holotype (IGG 92/78); from Omolon Massif.

(Compiled by Dagys.)

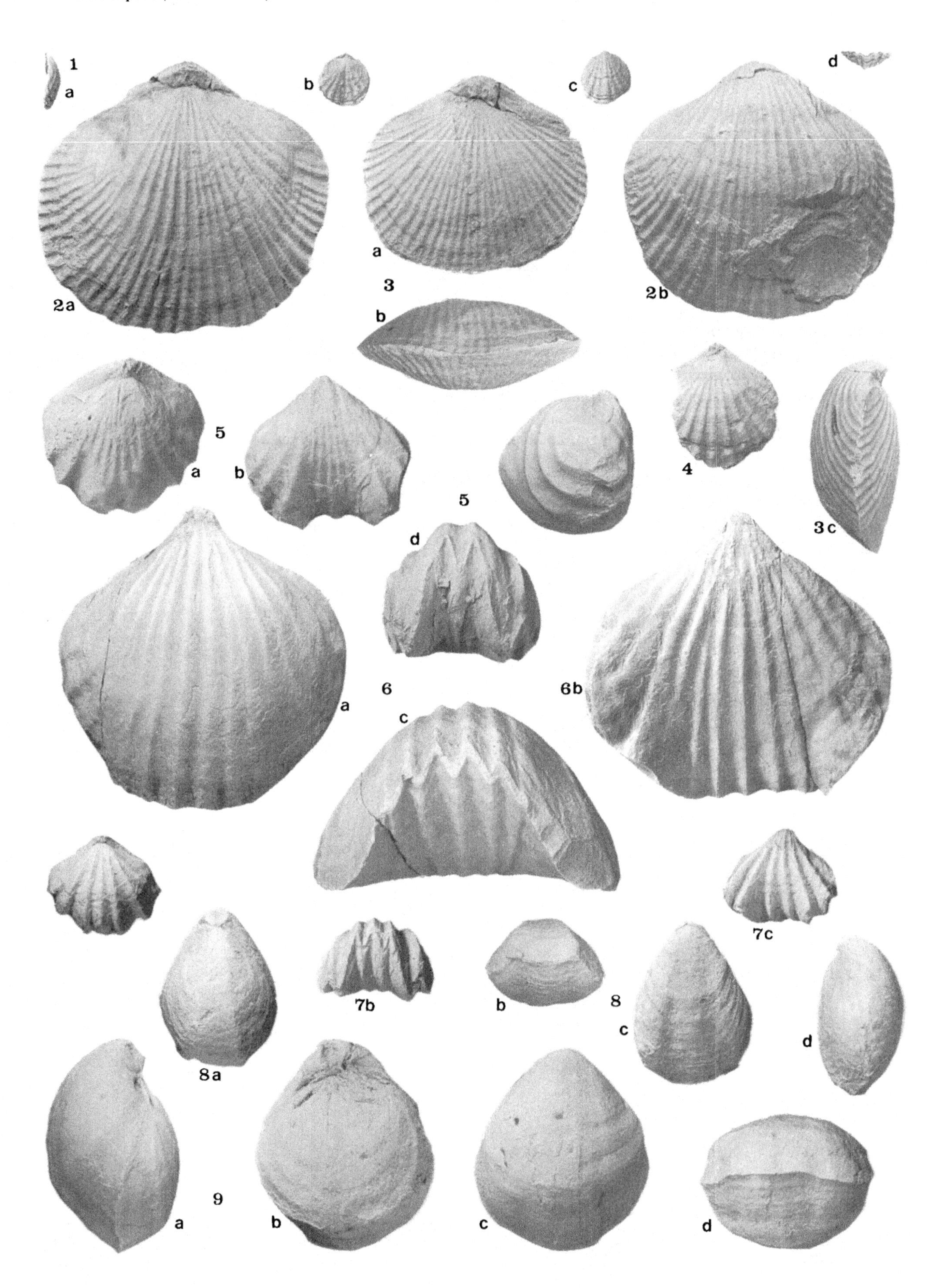
1
a
b
c
d
a
3
2a
2b
b
5
a
b
4
5
3c
d
6
a
6b
c
7c
7b
b
8
c
d
8a
9
a
b
c
d

PLATE 110

NORTH-EAST RUSSIA

AALENIAN

**Fig. 1a–d.** *Gigantothyris*(?) *ochoticus* Dagys, holotype (IGG 117/78); from Viliga River, Okhosk Sea coast.

BATHONIAN–CALLOVIAN

**Fig. 2a–d.** *Inversithyris rhomboidalis* Dagys, holotype (IGG 173/78); from Anadyr River basin.

OXFORDIAN(?)–LOWER VOLGIAN

**Fig. 3a,b.** *Boreiothyris lamutkensis* (Moisseiev), holotype (TsGM 1/5586); from Zyrianka River, Kolyma River basin.

**Fig. 4.** *Boreiothyris pelecypodaeformis* (Moisseiev), holotype (TsGM); from Zyrianka River, Kolyma River basin.

**Fig. 5a–c.** *Boreiothyris simkini* (Moisseiev), holotype (TsGM 29/5586); from Zyrianka River, Kolyma River basin.

**Fig. 6a,b.** *Boreiothyris goliensis* (Moisseiev), holotype (TsGM 27/5586); from Zyrianka River, Kolyma River basin.

**Fig. 7a–c.** *Taimyrothyris kropotkini* (Moisseiev), holotype (TsGM 16/5586); from Zyrianka River, Kolyma River basin.

(Compiled by Dagys.)

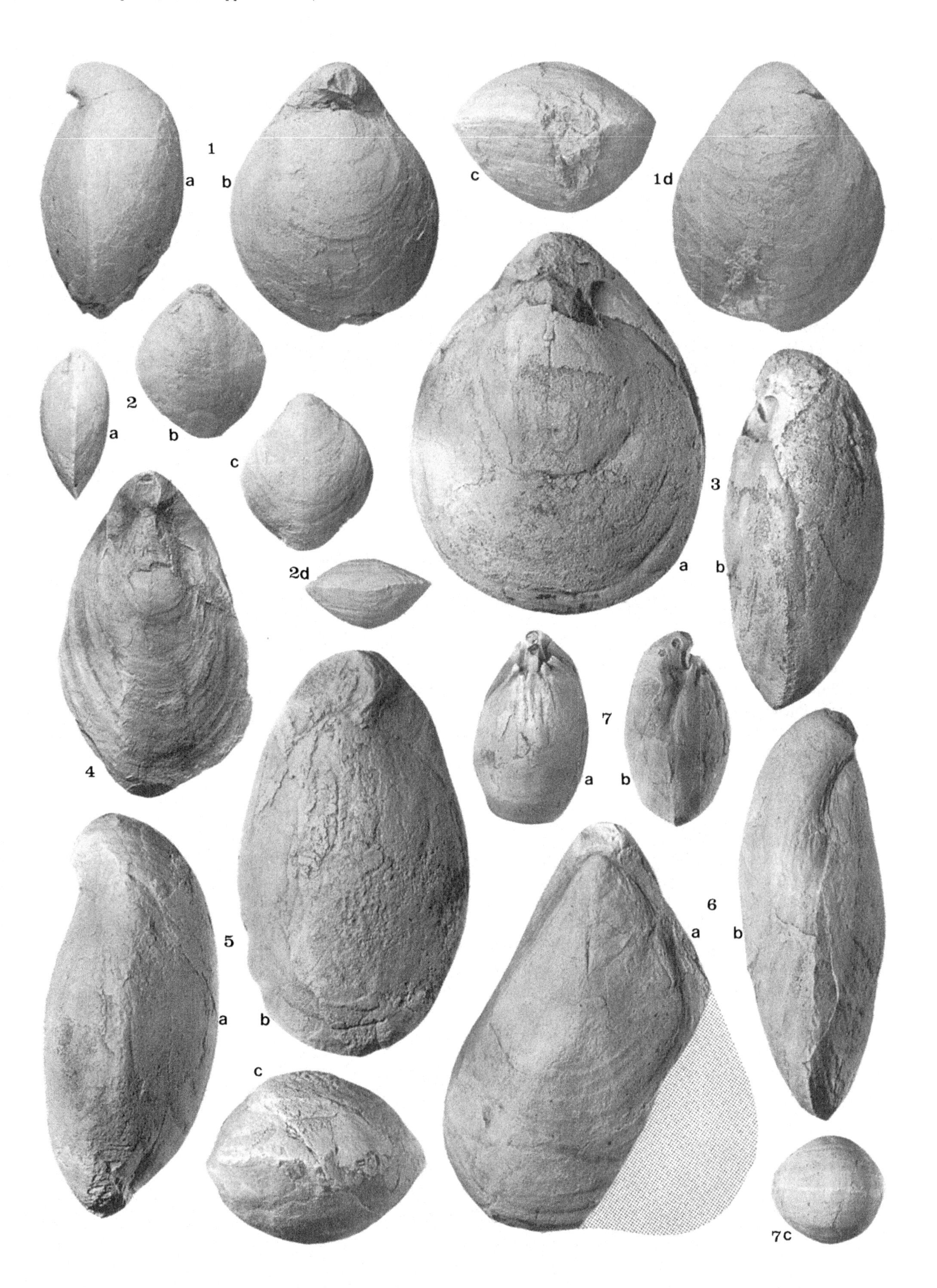
1
a
b
c
1d
2
a
b
c
2d
3
a
b
4
5
a
b
c
6
a
b
7
a
b
7c

PLATE 111

ARGENTINA

HETTANGIAN

**Fig. 1.** *Lingula* cf. *metensis* Terquem, both valves shifted (MLP 22258) (×4); from *B. canadensis* Zone, Río Atuel area, Mendoza.

SINEMURIAN (BRACHIOPOD ASSEMBLAGE LJ1)

**Fig. 2a–c.** *Gibbirhynchia* sp. nov., whole shell (MLP 24413) (×2); from Río Atuel area, Mendoza.

PLIENSBACHIAN (s.l.)

**Fig. 3a,b.** *Spiriferina chilensis* (Forbes), plaster cast of BWSM S-834 (= MLP 18839); from "Middle Lias," Catàn Lil area, Neuquén.

**Fig. 6a–d.** *Spiriferina hartmanni* (Zieten), steinkern (MLP 15265); from Río Atuel area, Mendoza.

**Fig. 7.** *Acanthothyropsis*(?) sp., uncoated brachial valve (MLP 24416) (×2); from Charahuilla area, Neuquén.

**Fig. 12.** *Spiriferina tumida* (Buch), brachial internal mold (MLP 10015a); from "Middle Lias," Piedra Pintada area, Neuquén.

BRACHIOPOD ASSEMBLAGE LJ2

**Fig. 4a–c.** *Squamiplana* (*Cuersithyris*) *davidsoni* (Haime), steinkern (MLP 24414); from Río Atuel area, Mendoza.

**Fig. 5.** *Furcirhynchia* sp., brachial internal mold (MLP 24415) (×2); from Piedra Pintada area, Neuquén.

**Fig. 8a,b.** *Zeilleria* (*Zeilleria*) cf. *sarthacensis* (d'Orbigny), two fragmentary specimens from same bed (MLP 24417a,b); from Río Atuel area, Mendoza.

**Fig. 9a,b.** *Zeilleria* (*Cincta*) cf. *numismalis* (Lamarck), steinkern (MLP 24418); from Cerro Puchénque area, Mendoza.

**Fig. 10.** *Zeilleria* (*Pirotella*) *anserirostrata* Manc., steinkern (MLP 24419); from Cerro Puchénque area, Mendoza.

**Fig. 11a–c.** *Rudirhynchia* aff. *rothi* Manc., whole shell (MLP 24420) (×2); from Piedra Pintada area, Neuquén.

BRACHIOPOD ASSEMBLAGE LJ3

**Fig. 13a–c.** *Exceptothyris*(?) *bodenbenderi* Manc., steinkern (MLP 24421); from Cerro Puchénque area, Mendoza.

**Fig. 14a–d.** *Fissirhynchia piremapuana* Manc., steinkern with shell (MLP 24422) (×1.5); from Piedra Pintada area, Neuquén.

**Fig. 15a–c.** *Quadratirhynchia crassimedia* S. Buckman, steinkern (MLP 24423) (×2); from Cerro Puchénque area, Mendoza.

(Compiled by Manceñido.)

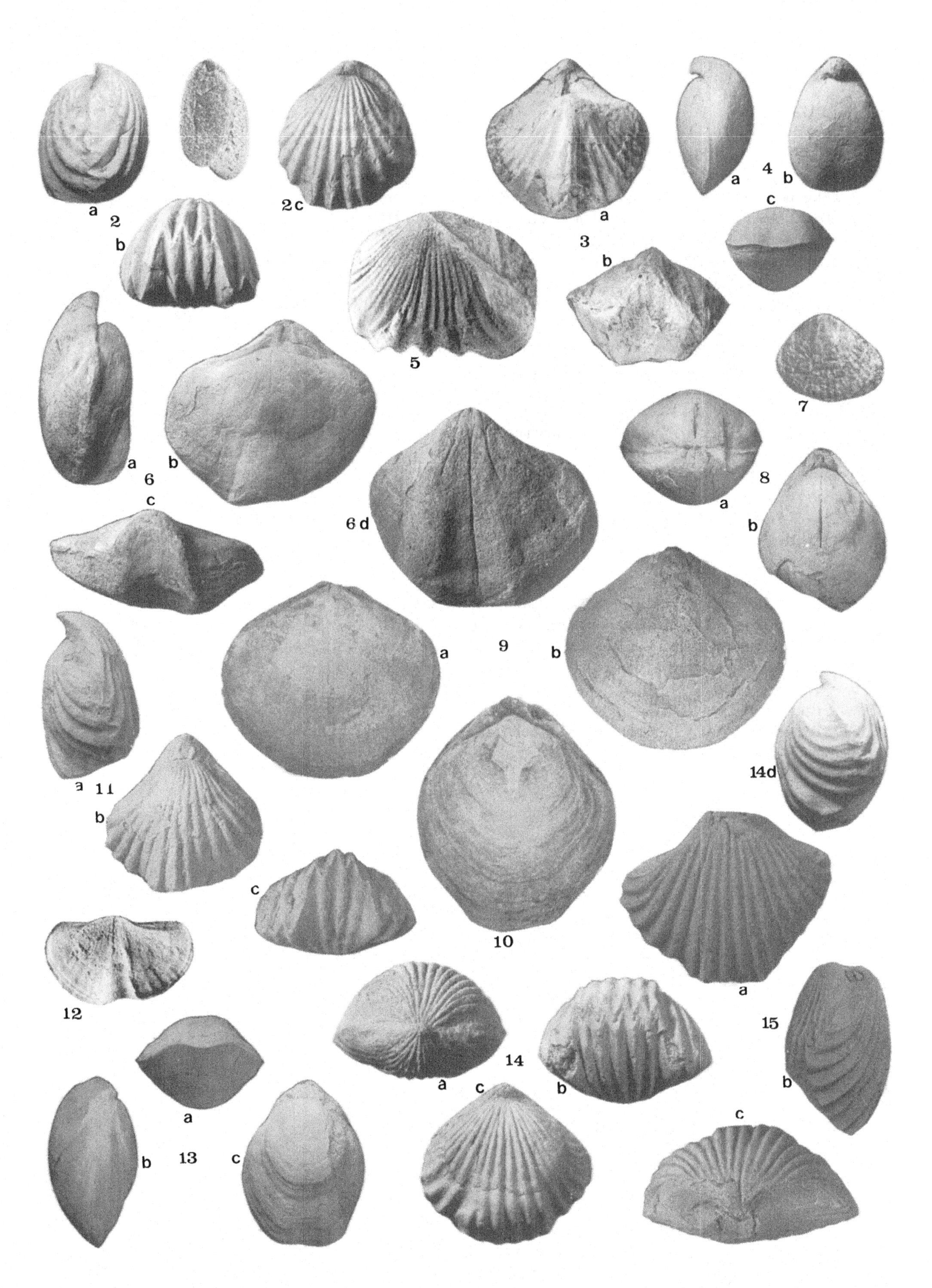

a 2
b
2c
a
3
b
a 4 b
c
5
7
a 6 b
c
6d
8
a
b
a 9 b
14d
a 11
b
c
10
a
12
15
a
14
a c b
b
a
b 13 c
c

PLATE 112

ARGENTINA

UPPER PLIENSBACHIAN–LOWER TOARCIAN. BRACHIOPOD ASSEMBLAGE LJ3

**Figs. 1a–c, 2a,b.** *Peristerothyris columbiniformis* Manceñido; 1, paratype (MLP 17726); 2, holotype (MLP 17725); both steinkerns; from Piedra Pintada area, Neuquén.

**Fig. 3a–c.** *Rhynchonelloidea cuyana* Manc., steinkern with shell (MLP 21339) (×2); from Cordón del Espinacito area, San Juan.

**Fig. 4a–c.** *Lobothyris subpunctata* (Davidson), whole shell (MLP 24424); from Piedra Pintada area, Neuquén.

**Fig. 5a,b.** *Zeilleria (Zeilleria)* ex gr. *quadrifida* (Lamarck), whole shell (MLP 24425); from upper Río Salado area, Mendoza.

**Fig 6a–d.** *Spiriferina tumida ericensis* de Gregorio, steinkern with shell (MLP 15285); from Cordón del Espinacito area, San Juan.

TOARCIAN BRACHIOPOD ASSEMBLAGE LJ4

**Fig. 8a–c.** *Telothyris* ex gr. *jauberti* (Deslongchamps), steinkern (MLP 24427); from Cerro Lotena area, Neuquén.

**Fig. 9a–c.** *Quadratirhynchia* sp. nov., whole shell (MLP 24428) (×2); from Cerro Lotena area, Neuquén.

**Fig. 10a–c.** *Piarorhynchia keideli* Manc. whole shell (MLP 24429) (×2); from Cerro Lotena area, Neuquén.

BRACHIOPOD ASSEMBLAGE LJ5

**Fig. 11a,b.** *Tetrarhynchia* ex gr. *subconcinna* (Davidson), plaster cast of BWSM S-826 (= MLP 24430) (×2); from Malargüe area, Mendoza.

**Fig. 12a–c.** *Sphenorhynchia*(?) cf. *rubrisaxensis* (Rothpletz), steinkern (MLP 14622) (×1.5); from Bardas Blancas area, Mendoza.

LOWER JURASSIC (s.l.)

**Fig. 7a–c.** *Anarhynchia*(?) sp., steinkern with shell (MLP 24426) (×1.5); from Río Atuel area, Mendoza.

(Compiled by Manceñido.)

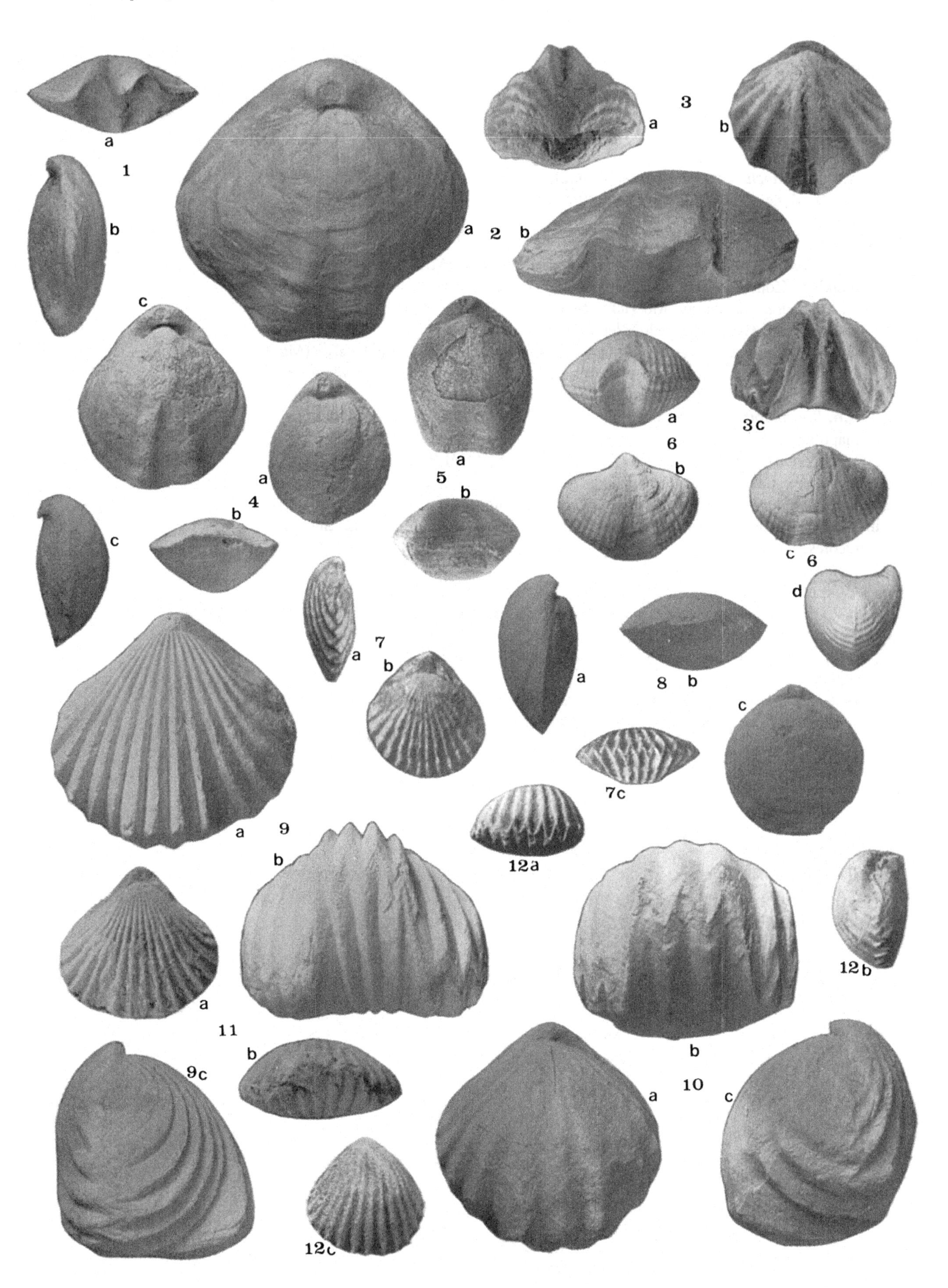
1
a
b
c
2
a
b
3
a
b
3c
4
a
b
c
5
a
b
6
a
b
c
d
7
a
b
7c
8
a
b
c
9
a
b
9c
10
a
b
c
11
a
b
12a
12b
12c

## PLATE 113

## ARGENTINA AND CHILE

AALENIAN–BAJOCIAN

**Fig. 1a,b.** *"Rhynochonella" moerickei* Tornquist, topotype (MLP 24431); from *P. malarguensis* Zone (?), Cordón del Espinacito, San Juan.

BAJOCIAN

**Fig. 2a–c.** *Kallirhynchia transatlantica* (Tornquist), topotype (MLP 24432) (×2); from *P. singularis* Zone(?), Cordón del Espinacito, San Juan.

**Fig. 3a,b.** *Lingula plagemanni* Moericke, both valves (MLP 24433a,b) (×2); from *P. singularis* Zone, Carro Quebrado, Neuquén.

**Fig. 4a,b.** *"Rhynchonella"* ex gr. *moerickei–manflasensis;* 4a, brachial valve (MLP 42434) (×1.5), from Cordón del Espinacito area, San Juan; 4b, pedicle valve (coll. Covacevich, SNGM) (×1.5), from Copiapó area, Atacama.

**Fig. 5a,b.** *Plectoidothyris*(?) sp., brachial valve (coll. Covacevich, SNGM) (×1.5); from Copiapó area, Atacama.

**Fig. 6.** *Zeilleria* (*Cincta*?) cf. *anglica* (Oppel), whole shell (MLP 24435) (×2); from *E. giebeli* Zone, Carro Quebrado, Neuquén.

**Fig. 7a–c.** *Rhactorhynchia*(?) *caracolensis* (Gottsche), uncoated plaster cast of topotype in McM (= MLP 24436) (×1.5); from Humphriesianum Zone, *S. chilense* Subzone, Caracoles, Antofagasta.

LOWER CALLOVIAN

**Fig. 8a–c.** *Lophrothyris* cf. *euryptycha* (Kitchin), whole shell (MLP 22041a) (×1.5); from Cerro Puchénque area, Mendoza.

**Fig. 9a,b.** *Torquirhynchia* sp., shell (MLP 24437) (×1.5); from Vega de la Veranada, Neuquén.

**Fig. 10a–c.** *Loboidothyris* (*Loboidothyris*) sp., whole shell (MLP 24438); from *N. bodenbenderi* Zone, Vega de la Veranda, Neuquén.

MIDDLE CALLOVIAN

**Fig. 11a–c.** *Rhynchonelloidella* sp., whole shell (MLP 24439) (×2): from *"Reineckeia"* Assemblage Zone, Picún Leufú area, Neuquén.

OXFORDIAN

**Fig. 12a–c.** *Thurmannella* sp., whole shell (MLP 24440) (×2); from *Perisphinctes–Araucanites* Association, Sierra de Reyes area, Mendoza.

(Compiled by Manceñido.)

a 1
b
2a
4 b
a
2
b
c
5 b
a
3
b
6
7
a
b
c
8
a
b
9
a
b
8c
10
a
b
c
11
b
a
c
12
a
b
c

## PLATE 114

## MISCELLANEOUS CIRCUM-PACIFIC AREAS

UNITED STATES AND CANADA, SINEMURIAN

**Figs. 1a,b, 2a–c.** *Anarhynchia* sp. nov. aff. *gabbi* Ager, two uncoated plaster casts (coll. Ager, UCS and MLP 24441); 2 (×1.5); from Adritch Mountains, Oregon.

**Fig. 3a–d.** *Sulcirostra paronai* (Boese)(?), uncoated cast (coll. Ager, UCS) (×1.5); from Aldritch Mountains, Oregon.

**Fig. 4a–c.** *Furcirhynchia striata* (Quenstedt), shell (McM J 579) (×2); from Peace River area, British Columbia.

**Fig. 5a–c.** *Tetrarhynchia dunrobinensis* (Rollier), shell (McM J 582) (×2); from Jasper area, British Columbia.

**Fig. 6a–c.** *Quadratirhynchia* aff. *quadrata* S. Buckman, uncoated plaster cast (coll. Ager, UCS and MLP 24442) (×2); from Nevada.

INDONESIA, PLIENSBACHIAN (s.l.)

**Fig. 7a,b.** *Furcirhynchia laevigata* (Quenstedt), uncoated shell (NHMB L4485) (×2); from Liassic Nief gorge, East Seram.

**Fig. 8a,b.** *Cirpa seranensis* (Wanner & Knipscheer), uncoated plaster cast of paratype (NHMB L4482 = MLP 24443) (×4); from Nief gorge, East Seram.

**Figs. 9a,b, 10a–c.** *Hesperithyris renieri timorensis* (Krumbeck), two syntypes (GLD); 9, from Fatu Kenapa; 10, from Fatu Nimassi; both western Timor.

MEXICO, OXFORDIAN

**Figs. 11a,b, 12a–c.** *Mexicaria mexicana* (Ochoterena), two topotypes (×1.5); from Tlaxiaco area, NW Oaxaca.

TITHONIAN

**Fig. 13a–c.** *Septaliphoria potosina* River et al., holotype (IGM 3588) (×1.5); from Sierra de Catorce, San Luis Potosí.

(Compiled by Manceñido.)

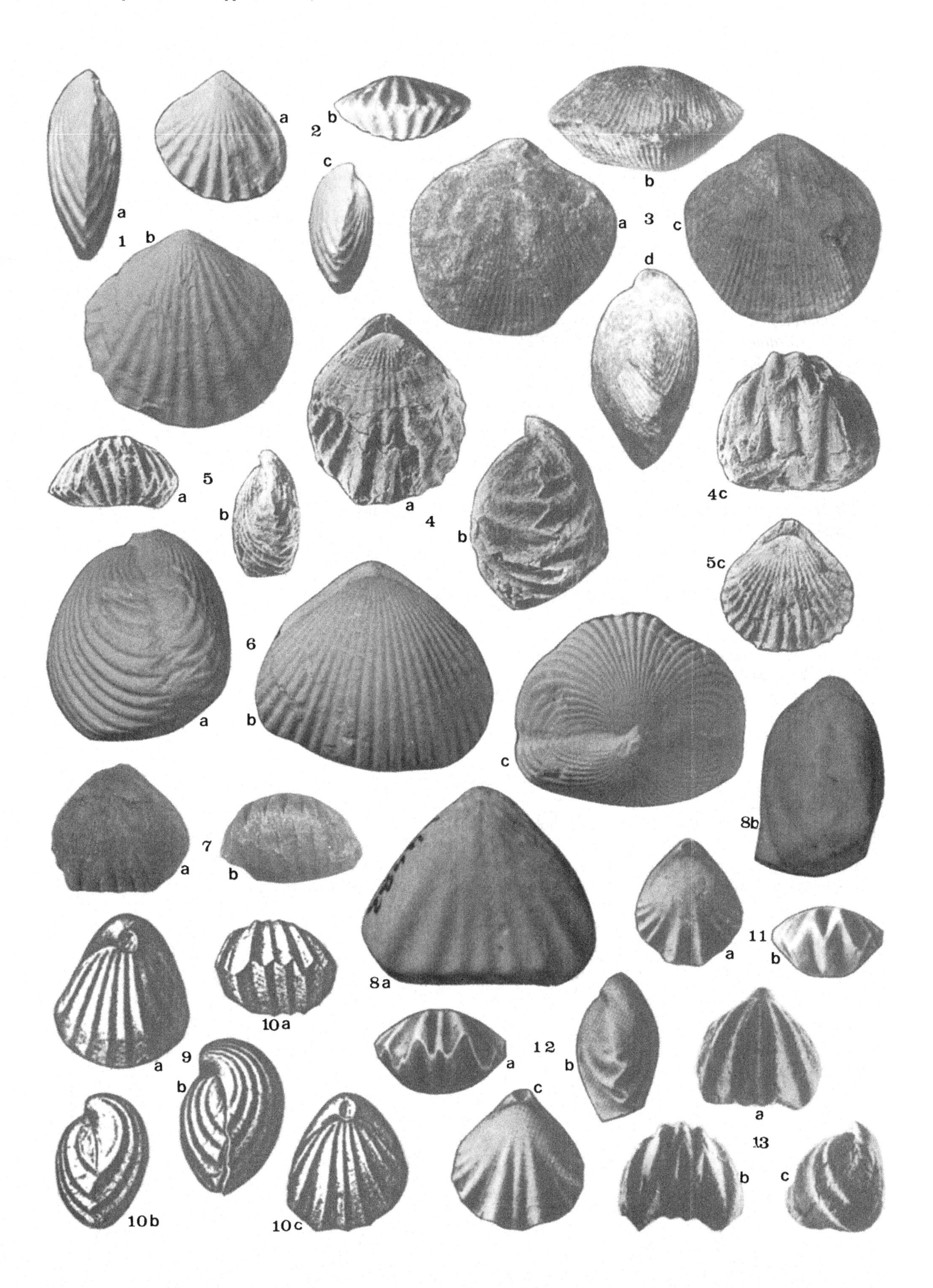
1a
1b
2a
2b
2c
3a
3b
3c
3d
4a
4b
4c
5a
5b
5c
6a
6b
6c
7a
7b
8a
8b
9a
9b
10a
10b
10c
11a
11b
12a
12b
12c
13a
13b
13c

# PLATE 115

## ARGENTINA AND CHILE

HETTANGIAN

**Fig. 1.** *Palmoxytoma* sp. (MLP 22253); from *Waehneroceras–Schlotheimia* Zone, Río Atuel, Mendoza.

**Fig. 2.** *Eopecten* cf. *velatus* (Goldfuss) (MLP 22256); from *B. canadensis* Zone, Río Atuel, Mendoza.

SINEMURIAN

**Figs. 3,4.** *Otapiria pacifica* (Covacevich & Escobar; 3, paratype (IIG 4130); 4, holotype (IIG 4148) (×1.5); from *O. pacifica* Association Zone, Vaquillas Altas, Antofagasta.

**Fig. 10.** *Cardinia* cf. *listeri* (J. Sowerby) (MLP 24315); from *Epophioceras* faunule, Río Atuel, Mendoza.

PLIENSBACHIAN *O. neuquensis* ASSOCIATION ZONE

**Figs. 5,6.** *Otapiria neuquensis* Damborenea; 5, holotype (MLP 16480); 6, paratype (MLP 16484) (×2); from Piedra Pintada, Neuquén.

**Figs. 7,8.** *Kalentera* n. sp. Damborenea; 7 (MLP 24294); 8 (MLP 24308); from Río Atuel, Mendoza.

**Fig. 9.** *Parainoceramus apollo* (Leanza), lectotype (MLP 6252); from Piedra Pintada, Neuquén.

**Fig. 13.** *Grammatodon costulatus* (Leanza), lectotype (MLP 6074) (×2); from Piedra Pintada, Neuquén.

**Fig. 17.** *Weyla (Weyla) alata* (v. Buch) (MLP 16528); from Río Atuel, Mendoza.

*R. sosneadoensis* ASSOCIATION ZONE

**Figs. 11,12.** *Kolymonectes coloradoensis* (Weaver); 11 (MLP 23688); 12 (MLP 23686) (×1.5); from Piedra Pintada, Neuquén.

**Fig. 15.** *Nuculana* cf. *ovum* (J. Sowerby) (MLP 16197) (×2); from Piedra Pintada, Neuquén.

PLIENSBACHIAN (s.l.)

**Fig. 14.** "*Myophorigonia*" *neuquensis* (Groeber), holotype (MLP 24324); from Puruvé Pehuén, Neuquén.

**Fig. 16.** *Weyla (Weyla) unca* (Philippi), holotype (SGO-PI 909); from Amolanas, Atacama.

All figures ×1 except where otherwise indicated.

(Compiled by Damborenea; photographs of Figs. 3 and 4 kindly provided by V. Covacevich.)

1
2
3
4
5
6
7
8
9
10
11
12
13
14
15
b
16
a
17

PLATE 116

ARGENTINA

PLIENSBACHIAN ( s.l.)

**Fig. 1.** *Gervillaria*(?) *pallas* (Leanza) (DNGM 8646); from Nueva Lubecka, Chubut.
**Fig. 2.** ''*Astarte*'' *keideli* Wahnish (DNGM 8689); from Nueva Lubecka, Chubut.
**Fig. 4.** *Neocrassina aureliae* (Feruglio) (DNGM 8726); from Nueva Lubecka, Chubut.
**Fig. 8.** ''*Lucina*'' *feruglioi* Wahnish (DNGM 8711); from Nueva Lubecka, Chubut.
**Fig. 15.** *Pholadomya* cf. *decorata* Hart. (DNGM 6932); from Río Atuel, Mendoza.
**Fig. 17.** *Frenguelliella chubutensis* (Feruglio); syntype (MLP 3729); from Río Genua, Chubut.

*R. sosneadoensis* ASSOCIATION ZONE

**Fig. 3.** *Aguilerella neuquensis* Damborenea, paratype (MLP 16370); from Piedra Pintada, Neuquén.
**Fig. 5.** *Weyla* (*Weyla*) *bodenbenderi* (Behrendsen) (MLP 16543); from Piedra Pintada, Neuquén.
**Figs. 6,7.** *Radulonectites sosneadoensis* (Weaver); 6, holotype (BM SA 1136/153), from Río Atuel, Mendoza; 7 (MLP 6033); from Piedra Pintada, Neuquén.
**Figs. 9,10.** *Camptochlamys wunschae* (Marwick); 9 (MLP 23660); 10 (MLP 23658); Arroyo Ñireco, Neuquén.
**Fig 11.** *Pulvinites* (*Hypotrema*) *liasicus* Damborenea, holotype (MLP 16471); from Cerro Puchénque, Mendoza.
**Fig. 12.** *Myophorella araucana* (Leanza) (MLP 13008); from Piedra Pintada, Neuquén.
**Figs. 13,14.** *Plicatula rapa* Bayle & Coquand; 13 (MLP 16512), from Arroyo Lonqueo, Neuquén; 14 (MLP 16511), from Vuta Picún Leufú, Neuquén (×2).
**Fig. 16.** *Frenguelliella inexspectata* (Jaworski) (MLP 13008); from Piedra Pintada, Neuquén.
**Fig. 18.** *Frenguelliella tapiai* (Lambert) (MLP 21076); from Arroyo Ñireco, Neuquén.
All figures ×1 except where otherwise indicated.
(Compiled by Damborenea.)

a
1
b
2
3
4
a
5
b
6
7
8
9
10
11
12
13
14
15
16
17
18

PLATE 117

ARGENTINA

PLIENSBACHIAN. *R. sosneadoensis* ASSOCIATION ZONE

**Fig. 1.** *Cardinia andium* (Giebel) and *Atreta* sp. (MLP 24316); from Piedra Pintada, Neuquén.

**Fig. 2.** *Cardinia densestriata* Jaworski (MLP 24317); from Piedra Pintada, Neuquén.

**Fig. 3.** *Falcimytilus*(?) *gigantoides* (Leanza) (MLP 16319); from Piedra Pintada, Neuquén.

**Fig. 4.** *Chlamys textoria* (Schlotheim) (MLP 23607); from Portezuelo Ancho, Mendoza.

**Fig. 5.** *Parallelodon riccardii* Damborenea, holotype (MLP 16251); from Piedra Pintada, Neuquén.

**Fig. 6.** *Modiolus* cf. *thiollierei* (Dumortier) (MLP 16341); from Piedra Pintada, Neuquén.

**Fig. 7.** *Cucullaea rothi* Leanza (MLP 16293); from Piedra Pintada, Neuquén.

TOARCIAN *P. cancellata* ASSOCIATION ZONE

**Figs. 9–11.** *Posidonotis cancellata* (Leanza); 9 (MLP 15761) (×2), from Arroyo Lapa, Neuquén; 10 (MLP 19746) (×2); from Mallín de Ibáñez, Neuquén; 11 (MLP 15761), from Arroyo Lapa, Neuquén.

**Fig. 12.** *Weyla* (*Weyla*) *alata angustecostata* (Philippi) (DNGM 12463); from Sierra de Agnia, Chubut, b (×0.5).

*P. pumilus* ASSOCIATION ZONE

**Fig. 8.** *Modiolus gerthi* Damborenea, holotype (MLP 19702); from Arroyo Serrucho, Mendoza.

**Figs. 13–15.** *Propeamussium* (*Variamussium*) *pumilum* (Lamarck); 13 (MLP 17477), from Río Salado, Mendoza; 14 (MLP 15448), from Cordillera del Viento, Neuquén; 15 (MLP 15539), (×2), from Cordillera del Viento, Neuquén.

All figures ×1 except where otherwise indicated.

(Compiled by Damborenea.)

1
2
3
4
5
6
7
8
9
10
11
12 a
12 b
13
14
15

PLATE 118

ARGENTINA

BAJOCIAN. SINGULARIS ZONE

**Fig. 3.** *Coelastarte jurensis* (Tornquist) (MLP 24318); from Carro Quebrado, Neuquén.

GIEBELI ZONE

**Fig. 2.** *Lycettia* cf. *lunularis* (Lycett) (MLP 19091); from Carro Quebrado, Neuquén.
**Fig. 4.** *Neuquenitrigonia hunickeni* (Leanza & Garate) (MLP 24320); from Carro Quebrado, Neuquén.
**Fig. 5.** *Propeamussium andium* (Tornquist) (MLP 24323) (×2); from Carro Quebrado, Neuquén.
**Fig. 6.** *Cercomya peruviana* Cox (MLP 24321); from Carro Quebrado, Neuquén.
**Fig. 7.** *Neocrassina andium* (Gottsche) (MLP 24319); from Carro Quebrado, Neuquén.

HUMPHRIESIANUM ZONE

**Fig. 1.** *Parainoceramus* n. sp. Damborenea (MLP 23362); from Chacay Melehue, Neuquén.

ROTUNDUM ZONE

**Fig. 8.** *Retroceramus* cf. *marwicki* (Speden) (MLP 23371); from Chacay Melehue, Neuquén.

BAJOCIAN (s.l.)

**Fig. 10.** *Iotrigonia radixscripta* (Lambert) (MLP 6710); from Chacai-co, Neuquén.
**Fig. 11.** *Scaphorella leanzai* (Lambert) holotype (MLP 6305); Chacai-co, Neuquén.
**Fig. 12.** *Trigonia stelzneri* Gottsche (MLP 6723); from Chacai-co, Neuquén.

BATHONIAN

**Fig. 9.** *Retroceramus patagonicus* (Philippi) (MLP 23388); from Chacay Melehue, Neuquén.

CALLOVIAN

**Fig. 13.** *Retroceramus stehni* Damborenea (CPBA 7392); from Chacay Melehue, Neuquén.

All figures ×1 except where otherwise indicated.
(Compiled by Damborenea.)

1
2
3
4
5
6
7
8
9
10
11
12
13

PLATE 119

ARGENTINA AND CHILE

CALLOVIAN

**Fig. 1.** *Myoconcha mollensis* (Weaver), holotype (BM SA 966/192); from Picún Leufú, Neuquén.

**Fig. 2.** *Ctenostreon neuquense* Weaver, syntype (BM SA 1001/183); from Picún Leufú, Neuquén.

OXFORDIAN

**Fig. 3.** *Gryphaea* cf. *impressimarginata* McLearn (MLP 24322); Sierra de Reyes, Mendoza.

TITHONIAN

**Fig. 4.** *Grammatodon (Indogrammatodon) lotenoensis* (Weaver) (MLP 2267); from Cerro Lotena, Neuquén.

**Fig. 5.** *Deltoideum lotenoense* (Weaver) (MLP 2148); from Cerro Lotena, Neuquén.

**Fig. 6.** *Anditrigonia carrincurrensis* (Leanza), holotype (MLP 4129); from Carrán Curá, Neuquén.

**Fig. 7.** *Eriphyla lotenoensis* Weaver (MLP 1669); from Cerro Lotena, Neuquén.

**Fig. 8.** *"Lucina" lotenoensis* Weaver (MLP 1713); Cerro Lotena, Neuquén.

**Fig. 9.** *Anditrigonia discors* (Philippi) (IIG 108), paratype of *Trigonia eximia* var. *multicostata* Corvalán; Río Leñas, O'Higgins.

**Fig. 10.** *Myoconcha transatlantica* Burckhardt (MLP 3874); Chacai-co, Neuquén.

(Compiled by Damborenea.)

1
2
3
4
5
6
7
8
9
10

PLATE 120

EASTERN RUSSIA, HETTANGIAN–PLIENSBACHIAN

**Figs. 1–3.** *Otapiria originalis* (Kipar.); 2 (×3); 3 (×2); from Hettangian–Lower Sinemurian, North-East Russia, Indigirka River basin, northern coast of Okhotsk Sea, Viliga River (coll. Kiparisova).
**Fig. 4.** *"Pseudomytiloides" sinuosus* Polub.; from Hettangian–Lower Sinemurian, North-East Russia, right bank of Kolmya River, Korkodon River basin.
**Figs. 5,6.** *"Pseudomytiloides" rassochaensis* Polub.; 5b,6 (×2); from Sinemurian, Korkodon River basin.
**Figs. 7–9.** *Meleagrinella subolifex* Polub. (×2); from Hettangian, North-East Russia, Omolon River basin.
**Figs. 10–12.** *Otapiria omolonica* Polub.; 12a,b (×2); from Lower Sinemurian, Korkodon River basin.
**Fig. 13.** *Otapiria tailleuri* Imlay; from Sinemurian, Korkodon River basin.
**Figs. 14–16.** *Otapiria limaeformis* Zakh.; 16 (×2); from Upper Sinemurian, North-East Russia, Viliga River.
**Figs. 17–21.** *"Monotis" inopinata* Polub.; 20,21 (×2); from Upper Sinemurian, North-East Russia, right bank of Kolyma River (coll. Polubotko & Repin).
**Figs. 22,23.** *Kolymonectes kedonensis* Polub. (×2); from Lower Sinemurian, Omolon River basin (coll. Polubotko & Milova).
**Figs. 24,25.** *Kolymonectes mongkensis* Polub.; from lowermost Upper Pliensbachian, North-East Russia, northern coast of Okhotsk Sea, Viliga River (coll. Milova).
**Figs. 26,27.** *Kolymonectes staeschei* (Polub.); from Upper Sinemurian, Korkodon River basin.
**Fig. 28.** *Chlamys textoria* (Schloth.); from Sinemurian, Far East Russia, Amur River.
**Figs. 29–31.** *Chlamys tapensis* Milova (×5); from Lower Pliensbachian, Viliga River.
**Figs. 32,33.** *"Amonotis"* sp.; 32 (×4); 33 (×3); from Lower Pliensbachian, Viliga River (coll. Polubotko & Milova).
All figured specimens are kept in CNIGR (coll. 4039, 5802, 10151, 10334, 12125, 12309, 12312, 12566) and in Geological Museum of the Geological Survey of North-East Russia, Magadan (coll. 400).
(Compiled by Sey & Polubotko.)

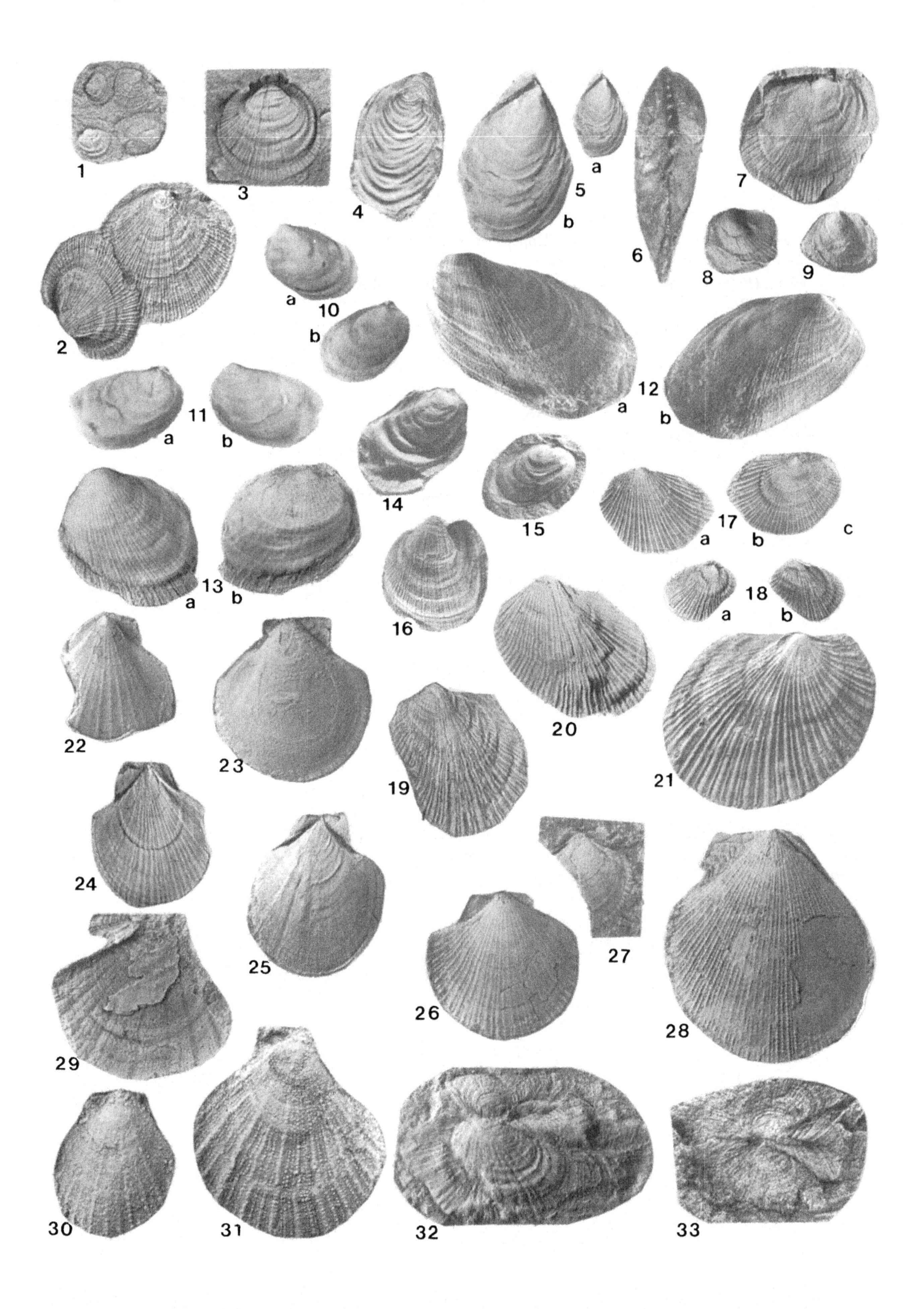
1
2
3
4
5
a
b
6
7
8
9
10
a
b
11
a
b
12
a
b
13
a
b
14
15
16
17
a
b
c
18
a
b
19
20
21
22
23
24
25
26
27
28
29
30
31
32
33

PLATE 121

EASTERN RUSSIA, UPPER PLIENSBACIAN–TOARCIAN

**Figs. 1,2.** "*Velata*" *viligaensis* Tuchk.; from lower Upper Pliensbachian, North-East Russia, Viliga River (coll. Milova).
**Figs. 3,4.** *Harpax spinosus* (Sow.); 4 (×2); from Upper Pliensbachian, Stokesi Zone, Far East Russia, Bureya River basin (coll. Sey).
**Figs. 5,6.** *Radulonectites hayamii* Polub.; from Upper Pliensbachian, North-East Russia, Omolon River basin (coll. Polubotko).
**Fig. 7.** *Oxytoma* (*Palmoxytoma*) *cygnipes* (J. & B.); from Upper Pliensbachian, Stokesi Zone, Far East Russia, S Primorye (coll. Voronetz).
**Fig. 8.** *Ochotochlamys bureiensis* Sey; from Upper Pliensbachian, Margaritatus Zone, Bureya River basin.
**Figs. 9,10.** *Amuropecten solonensis* (Sey), external and internal mold of right valves; from Upper Pliensbachian, Margaritatus Zone, Bureya River basin.
**Fig. 11.** *Kolymonectes* ex gr. *staeschei* (Polub.); from Upper Pliensbachian, Stokesi Zone, Bureya River basin (coll. Kusmin).
**Figs. 12, 13.** *Meleagrinella ansparsicosta* (Polub.); from Upper Pliensbachian, Omolon River basin.
**Fig. 14.** *Meleagrinella ptchelincevae* Polub. (×2); from Upper Pliensbachian, Omolon River basin.
**Figs. 15,16.** *Meleagrinella oxytomaeformis* Polub.; 15 (×2) (RV); 16 (LV); from Upper Pliensbachian, Omolon River basin.
**Fig. 17.** *Meleagrinella substriata* (Münst.) (×2); from Lower Toarcian, Omolon River basin.
**Figs. 18,19.** *Meleagrinella faminaestriata* Polub.; from middle part of Toarcian, North-East Russia, Korkodon River basin, Viliga River.
**Figs. 20,21.** *Propeamussium pumilum* (Lamk.); from middle part of Toarcian, Korkodon River basin.
**Fig. 22.** *Pseudomytiloides* sp.; from Toarcian, *Z. monestieri* Zone, from Korkodon River basin.
**Fig. 23.** *Galinia borsjaensis* Okun. (×2); from Lower Toarcian, Far East Russia, E Transbaikal (coll. Okuneva).
**Figs. 24,30.** *Pseudomytiloides marchaensis* (Petr.); from Upper Toarcian, Korkodon River.
**Fig. 25.** "*Pseudomy- tiloides*" aff. *amygdaloides* (Goldf.); from Lower Toarcian, Omolon River basin.
**Figs. 26,27.** *Vaugonia literata* (Y. & B.); from Toarcian, *P. polare* Zone, Omolon River basin.
**Figs. 28,29.** *Kolymonectes* sp., from Lower Toarcian *T. propinquum–E. elegantulum* Zones, Omolon River basin (coll. Polubotko & Repin).
(Compiled by Sey & Polubotko.)

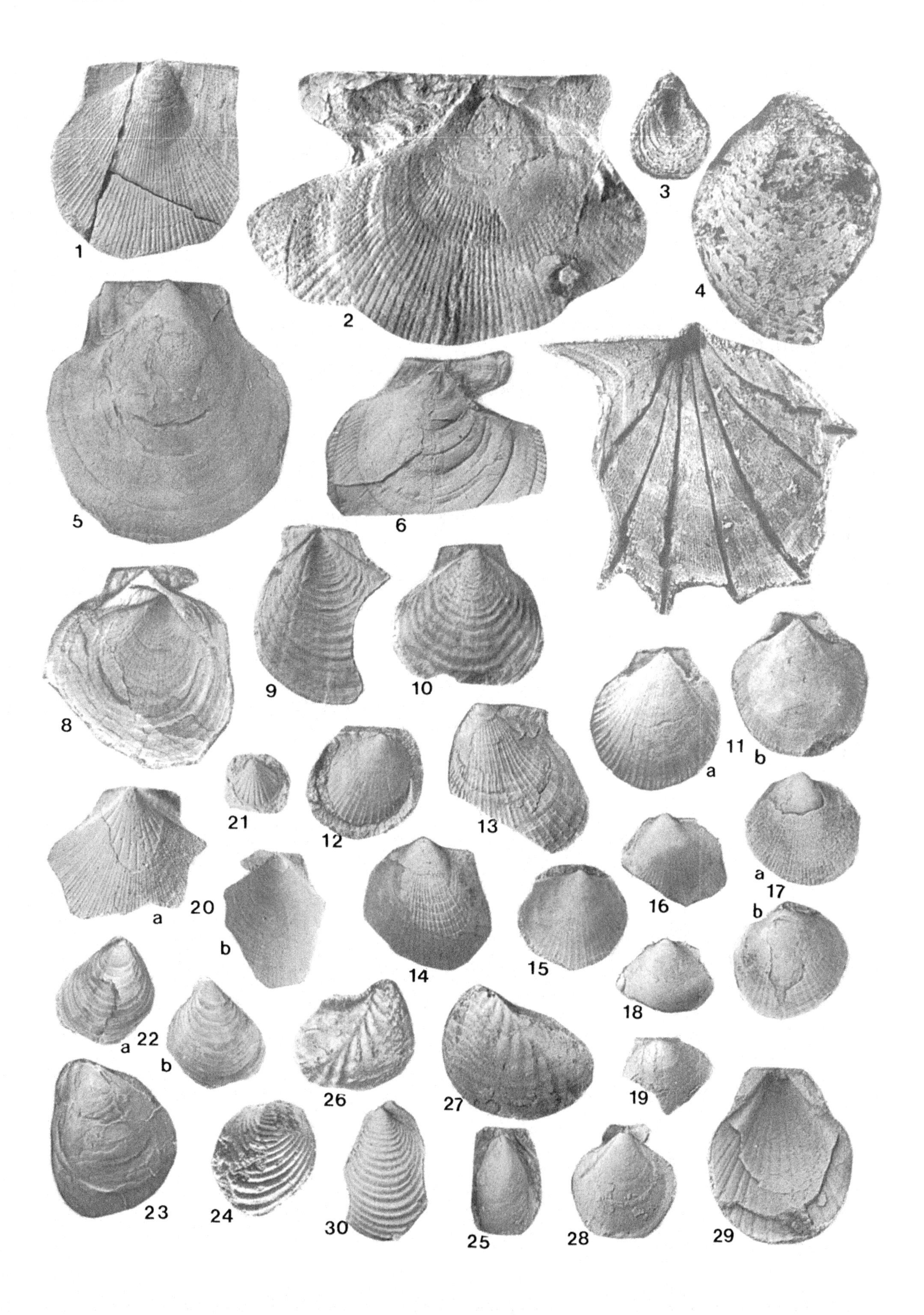
1
2
3
4
5
6
8
9
10
11
a
b
21
12
13
20
a
b
14
15
16
17
a
b
18
22
a
b
23
24
26
27
19
30
25
28
29

PLATE 122

EASTERN RUSSIA, AALENIAN–LOWER BAJOCIAN

**Figs. 1,2.** *Propeamussium olenekense* (Bodyl.); from Aalenian, North-East Russia, Viliga River basin.
**Fig. 3.** *Oxytoma* (*Oxytoma*) *jacksoni* (Pomp.); from Lower Aalenian, North-East Russia, northern coast of Okhotsk Sea, Gizhiga River basin.
**Fig. 4.** *Trigonia alta* Vor.; from Lower Aalenian, *T. alta* Beds, Viliga River (coll. Polubotko).
**Fig. 5.** *Mytiloceramus priscus* (Sey); from Lower Aalenian, *P. beyrichi* Zone, Far East Russia, Tugur Bay.
**Fig. 6.** *Mytiloceramus popovi* (Kosch.); from Upper Aalenian, *P. tugurense* Zone, North-East Russia, Gizhiga River basin (coll. Polubotko).
**Fig. 7.** *Mytiloceramus subtilis* (Sey); from Lower Aalenian, *P. beyrichi* Zone, Tugur Bay.
**Fig. 8.** *Mytiloceramus polyplocus* (Roem.); from Upper Aalenian, *P. tugurense* Zone, Tugur Bay.
**Fig. 9.** *Mytiloceramus obliquus* (Mor. & Lyc.); from Upper Aalenian, *P. tugurense* Zone, Tugur Bay.
**Fig. 10.** *Mytiloceramus anilis* (G. Pčel.); from Upper Aalenian, *P.* (*T.*) *tugurense* Zone, Tugur Bay.
**Figs. 11,12.** *Mytiloceramus jurensis* (Kosch.); from Lower Bajocian, *P. fastigatum* Zone, Bureya River basin (all coll. Sey & Kalacheva).
(Compiled by Sey & Polubotko.)

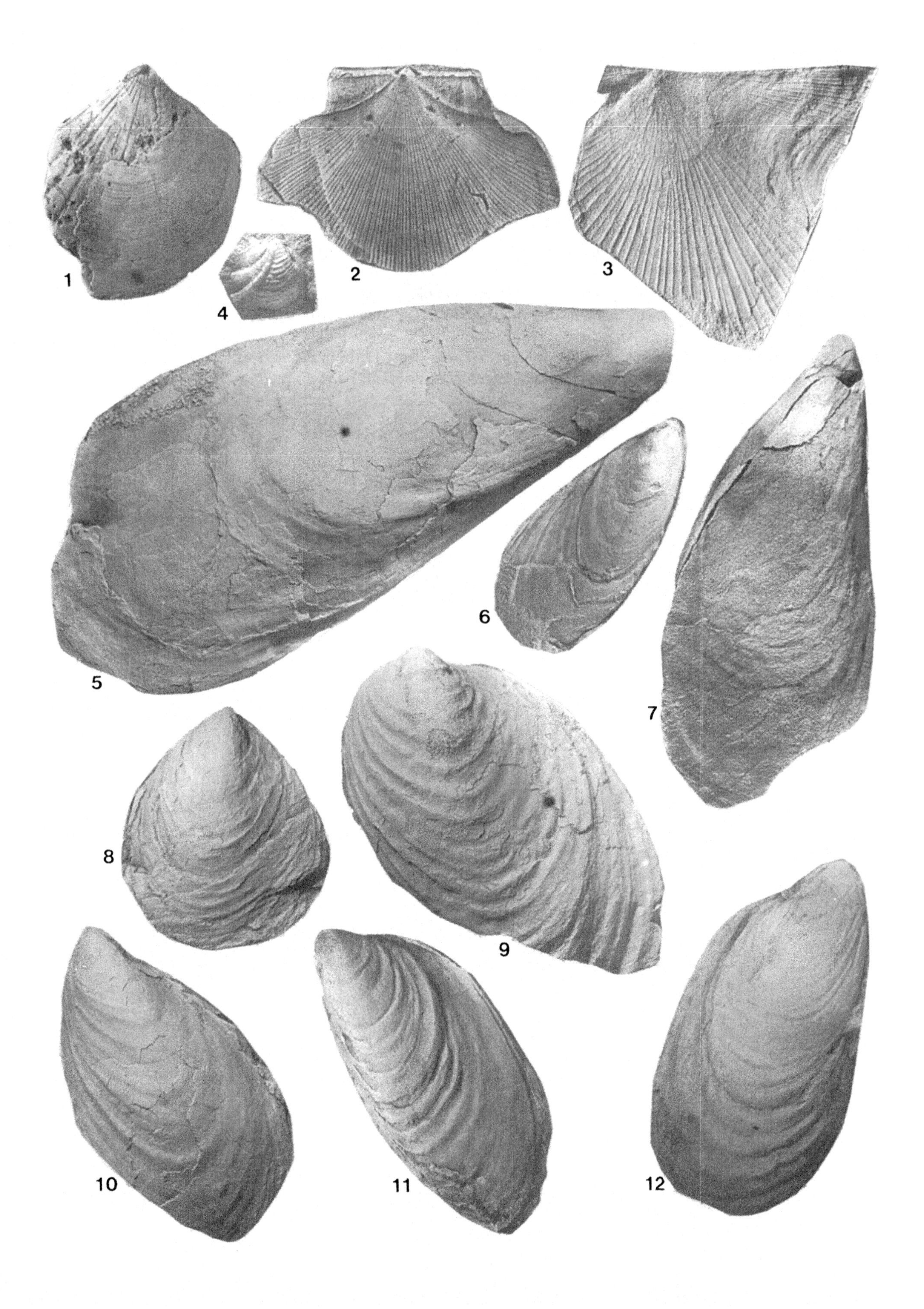
1
2
3
4
5
6
7
8
9
10
11
12

PLATE 123

EASTERN RUSSIA, LOWER BAJOCIAN

**Fig. 1.** *Mytiloceramus ussuriensis* (Vor.); from *A. tozeri* Zone, Far East Russia, Bureya River basin.
**Figs. 2,3.** *Mytiloceramus formosulus* (Vor.); from *A. tozeri* Zone, Bureya River basin.
**Figs. 4–6.** *Mytiloceramus lucifer* (Eichw.); from *A. tozeri* Zone, North-East Russia, Indigirka, Kolyma, and Alazeja river basins.
**Fig. 7.** *Mytiloceramus omolonensis* (Polub.); from *A. tozeri* Zone, North-East Russia, Omolon River basin.
**Figs. 8,9.** *Arctotis sublaevis* Bodyl.; from lower Bajocian, Omolon River basin.
**Fig. 10.** *Mytiloceramus porrectus* (Eichw.); from *M. clinatus* Beds, Bureya River basin (all coll. Sey & Kalacheva).
**Figs. 11,12.** *Mytiloceramus clinatus* (Kosch.); from *M. clinatus* beds; 11, from Far East Russia, Tugur Bay; 12, from North-East Russia, Omolon River basin.
(Compiled by Sey & Polubotko.)

1
2
3
4
5
6
7
8
9
10
11
12

PLATE 124

EASTERN RUSSIA, UPPER BAJOCIAN–LOWER BATHONIAN

**Figs. 1,4.** *Mytiloceramus retrorsus* (Keys.); from Upper Bajocian, *M. retorsus–M. electus* Beds, North-East Russia, northern coast of Okhotsk Sea.
**Fig. 2.** *Mytiloceramus kystatymensis* (Kosch.) (×0.6); from Upper Bajocian–Lower Bathonian, North-East Russia, right bank of Kolyma River.
**Fig. 3.** *Mytiloceramus tongusensis* (Lah.) (×0.7); from Upper Bajocian, *M. retrorsus–M. electus* Beds, northern coast of Okhotsk Sea.
**Fig. 5.** *Mytiloceramus electus* (Kosch.); from Upper Bajocian, *M. retrorsus–M. electus* Beds, northern coast of Okhotsk Sea (coll. Polubotko).
(Compiled by Sey & Polubotko.)

1
2
3
4
5

PLATE 125

EASTERN RUSSIA, BATHONIAN

**Fig. 1.** *Mytiloceramus polaris* (Kosch.); from *M. polaris–M. vagt* Beds, North-East Russia, northern coast of Okhotsk Sea.
**Figs. 2,3.** *Mytiloceramus vagt* (Kosch.) (×0.8); from *M. polaris–M. vagt* Beds, North-East Russia, Gizhiga and Bolshoi Anyui rivers.
**Fig. 4.** *Mytiloceramus ultimus* (Kosch.); from *M. polaris–M. vagt* Beds, North-East Russia, Gizhiga River basin (coll. Polubotko).
**Fig. 5.** *Mytiloceramus bulunensis* (Kosch.) (×0.9); from *A. elegans* Zone, North-East Russia, Viliga River.
**Fig. 6.** *Mytiloceramus tuchkovi* (Polub.); from *M. polaris–M. vagt* Beds, Gizhiga River.
(Compiled by Sey & Polubotko.)

1
2
3
4
5
6

## PLATE 126

## EASTERN RUSSIA, OXFORDIAN–MIDDLE VOLGIAN

**Fig. 1.** *Praebuchia*(?) *impressae* (Quenst.); right valve (RV) external mold (×1.2); from Lower Oxfordian, Far East Russia, Tugur Bay.

**Figs. 2–5.** *Praebuchia lata* (Traut.) [''*P. kirghisensis*'']; 2 (×1, ×2); 4,5 (×2); from Middle–Upper Oxfordian, Tugur Bay, Uda River basin; 3, from Oxfordian, North-East Russia, Eropol River basin.

**Figs. 6–11.** *Buchia concentrica* (Sow.); 6 (×2); 6–8, from Middle–Upper Oxfordian, Far East Russia, Uda River basin; 9, from Lower Kimmeridgian, Tugur Bay; 10,11, from Lower Kimmeridgian, North-East Russia, Eropol River basin, Bolshoi Anyui River.

**Figs. 12–15.** *Buchia tenuistriata* (Lah.); 12 (×1, ×2); 12,14, from Lower Kimmeridgian, Tugur Bay; 13,15, from Upper Kimmeridgian(?), Uda River basin.

**Figs. 16,17.** *Buchia ochotica* Sey; from Lower Kimmeridgian, Tugur Bay.

**Figs. 18–22.** *Buchia rugosa* (Fisch.); 18, from Upper Kimmeridgian, Tugur Bay; 19 (×2); from Lower Volgian, Uda River basin; 20 (×1, ×2); 21, from Middle Volgian, Uda River basin (coll. Sey); 22, from Lower Volgian, North-East Russia, Bolshoi Anyui River.

**Figs. 23–28.** *Buchia mosquensis* (Buch.); 23,24, from Upper Kimmeridgian–Lower Volgian(?); 26, from Lower Volgian; 27,28, from Middle Volgian, Uda River basin (coll. Sey); 25, from Lower–Middle Volgian, Bolshoi Anyui River basin (coll. Paraketzov).

**Figs. 29–38.** *Buchia russiensis* (Pavl.); 29, 33, from Lower–Middle Volgian, Bolshoi Anyui River basin (coll. Paraketzov); 30–32,34–38, from Middle Volgian, Uda River basin (coll. Sey).

(Compiled by Sey & Polubotko.)

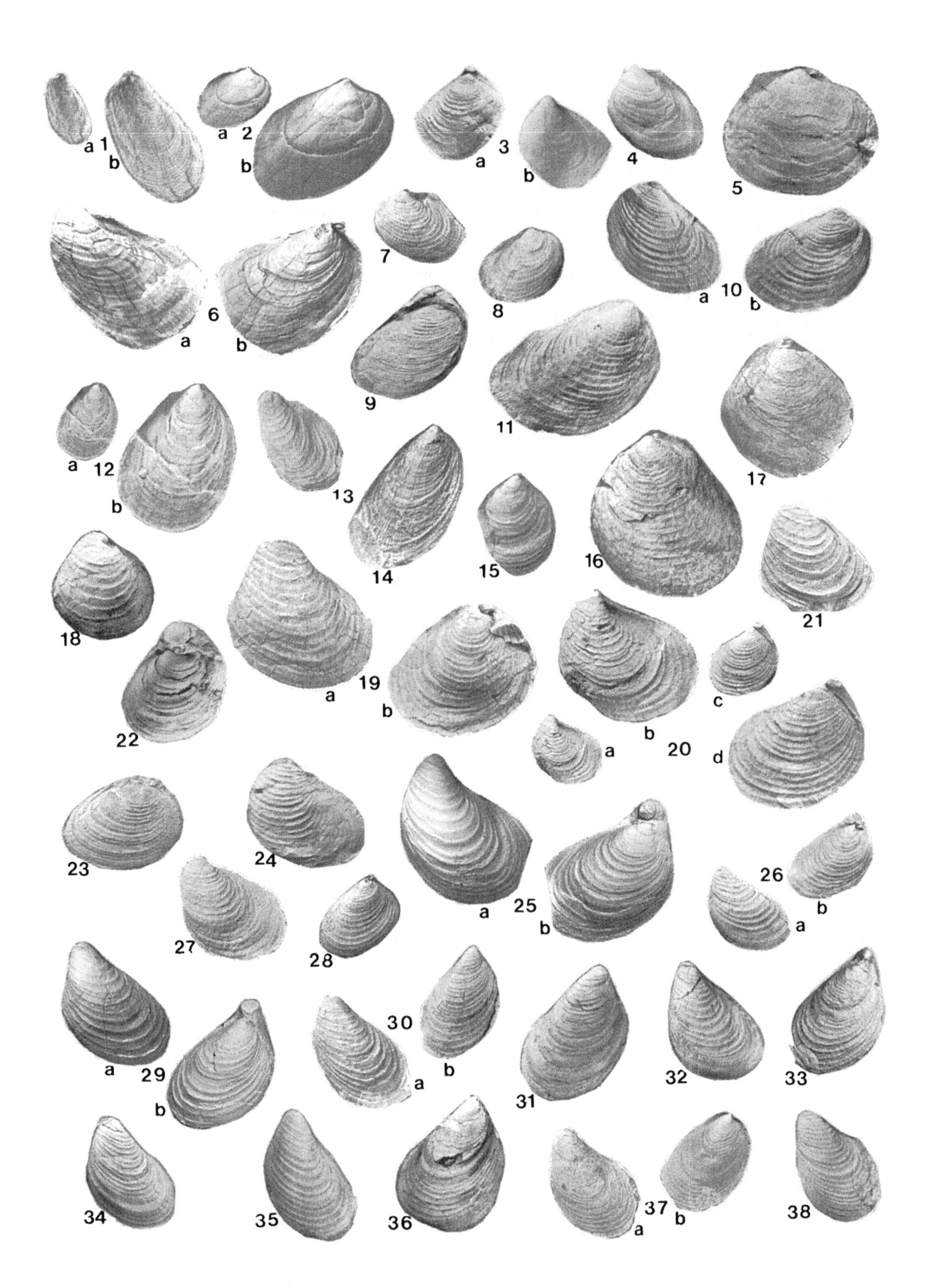
a 1
b
a 2
b
3
a
b
4
5
6
a
b
7
8
9
a 10
b
11
a 12
b
13
14
15
16
17
18
19
a
b
20
a
b
c
d
21
22
23
24
25
a
b
26
a
b
27
28
29
a
b
30
a
b
31
32
33
34
35
36
37
a
b
38

PLATE 127

EASTERN RUSSIA, MIDDLE–UPPER VOLGIAN

**Figs. 1–7.** *Buchia fischeriana* (Orb.); 1–3, from Middle Volgian; 5–7, from Upper Volgian, Far East Russia, Uda River basin (coll. Sey); 4, from Middle Volgian, North-East Russia, Bolshoi Anyui River basin (coll. Paraketzov).
**Figs. 8–12.** *Buchia trigonoides* (Lah.); 8,9,11, from Middle Volgian, 10,12, from Upper Volgian, Uda River basin (coll. Sey).
**Figs. 13–21.** *Buchia piochii* (Gabb); 13, from Upper Volgian, Bolshoi Anyui River basin (coll. Paraketzov); 14–17, from Upper Volgian, Uda River basin; 18–21, from Upper Tithonian–lowermost Berriasian, S. Primorye, Ussuri Bay (coll. Sey).
**Figs. 22–31.** *Buchia terebratuloides* (Lah.); 22–26, 31, from Upper Volgian, Uda River basin; 27–30, from Upper Tithonian–lowermost Berriasian, Ussuri Bay (coll. Sey).
**Figs. 32–41.** *Buchia* ex gr. *unschensis* (Pavl.); 32–38, from Upper Volgian, Uda River basin; 39–41, from Upper Tithonian–lowermost Berriasian, Ussuri Bay (coll. Sey).
**Figs. 42,43.** *Buchia lahuseni* (Pavl.); from Upper Volgian, Uda River basin (coll. Sey).
(Compiled by Sey & Polubotko.)

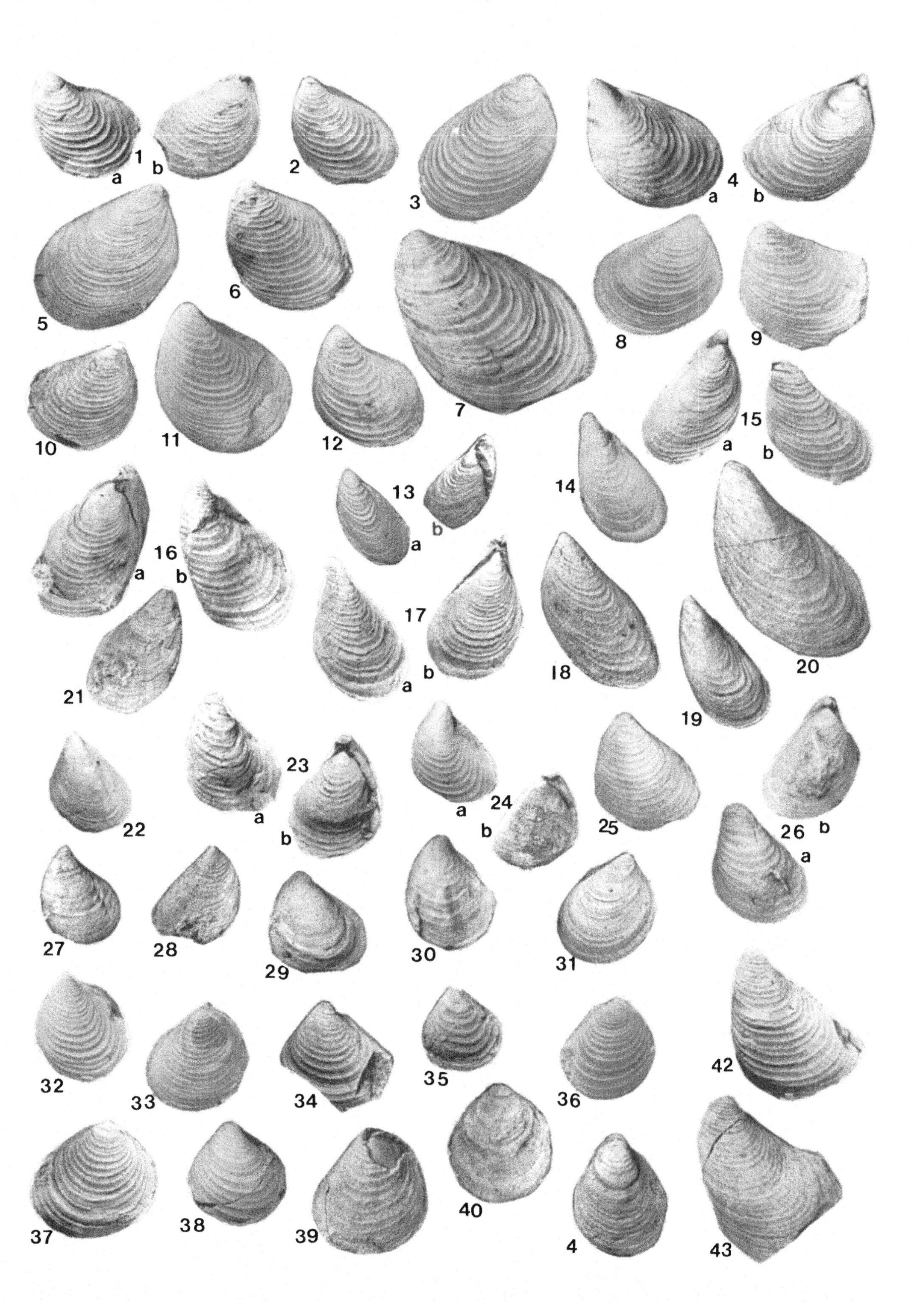

1
a
b
2
3
4
a
b
5
6
7
8
9
10
11
12
13
a
b
14
15
a
b
16
a
b
17
a
b
18
19
20
21
22
23
a
b
24
a
b
25
26
a
b
27
28
29
30
31
32
33
34
35
36
37
38
39
40
4
42
43

PLATE 128

ANTARCTICA AND NORTH AMERICA, UPPER JURASSIC

**Fig. 1.** *Cylindroteuthis* (*Cylindroteuthis*) *occidentalis* Anderson, holotype (CAS 28037.03); from Tithonian, Glenn County, California; 1a, ventral outline; 1b, left profile; 1c, transverse section.
**Fig. 2.** *Cylindroteuthis* (*Arctoteuthis*) *clavicula* Anderson, holotype (CAS 28037.07); from Tithonian, Glenn County, California; 2a, ventral outline; 2b, left profile.
**Fig. 3.** *Belemnopsis* cf. *aucklandica* (Hochstetter) (BAS KG.2910.27); from Kimmeridgian, Alexander Island, Antarctica; 3a, ventral outline; 3b, left profile.
**Fig. 4.** *Produvalia* sp. (BAS KG.2902.55); from Kimmeridgian, Alexander Island, Antarctica; 4a, ?ventral outline; 4b, left profile.
**Fig. 5.** *Hibolithes argentinus* Feruglio (BAS KG.2802.259); from Upper Tithonian, Alexander Island, Antarctica; 5a, ventral outline; 5b, left profile.
**Fig. 6.** *Hibolithes*(?) *mercurialis* (Anderson), holotype (CAS 28607.01); from Tithonian, Napa County, California; 6a, ventral outline; 6b, transverse section; 6c, left profile.
**Fig. 7.** *Hibolithes belligerundi* Willey, holotype (BAS KG.712.63a); from Lower Tithonian, Alexander Island, Antarctica; 7a, ventral outline; 7b, right profile.
(Compiled by Doyle.)

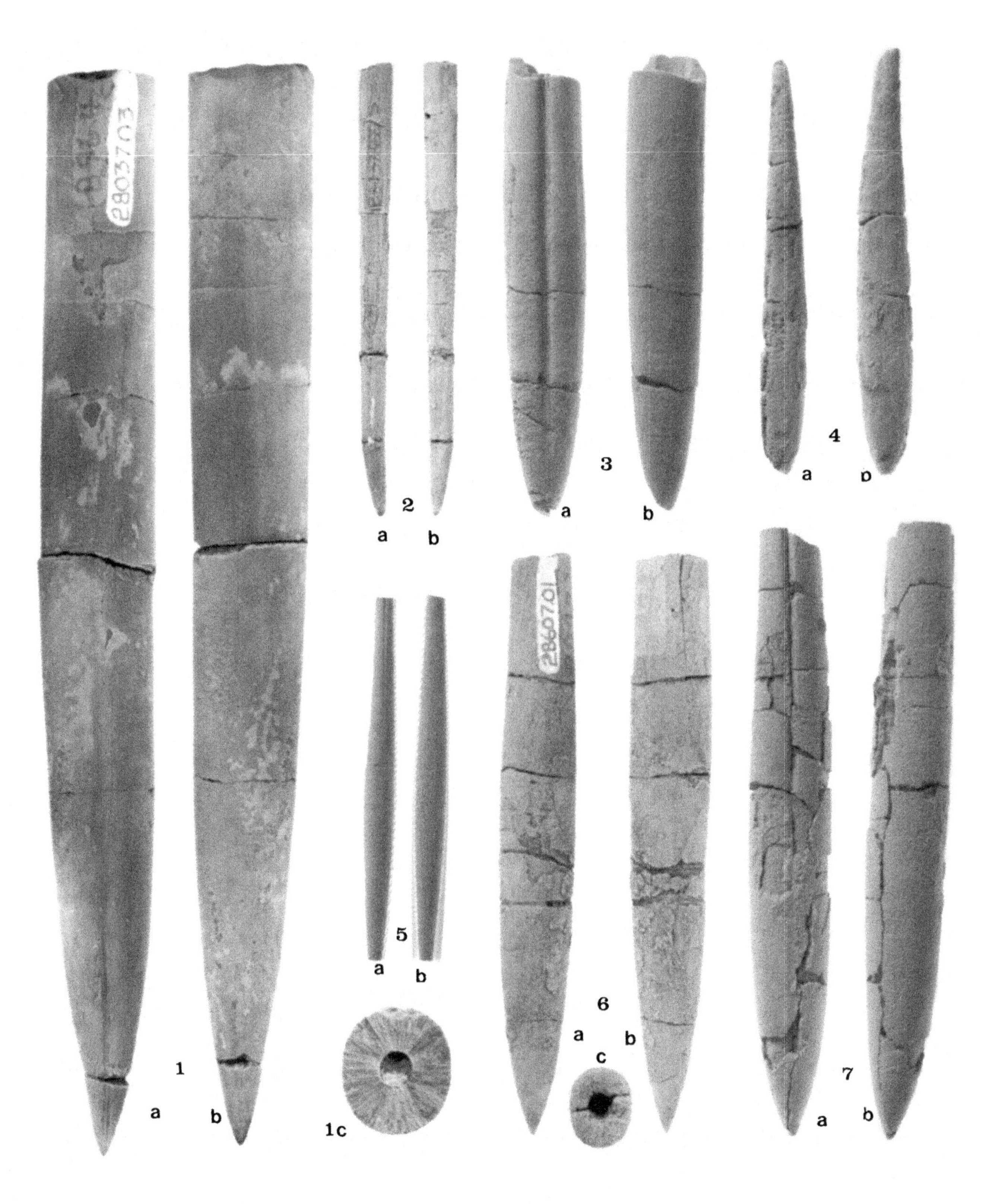
1
a
b
1c
2
a
b
3
a
b
4
a
b
5
a
b
6
a
b
c
7
a
b

PLATE 129

AUSTRALASIA, EASTERN INDONESIA, AND PAPUA NEW GUINEA, BAJOCIAN–TITHONIAN

**Fig. 1a,b.** *Dicoelites* sp. C (IMC 236); from Oxfordian, Sula Islands, Indonesia; ventral and dorsal views.
**Fig. 2a,b.** *Dicoelites* sp. A (IMC 232); from Bajocian, Sula Islands; ventral and dorsal views.
**Fig. 3a,b.** *Conocoelites kalepuensis* Challinor (CPC 27688); from Bathonian, Lagaip River, Papua New Guinea; ventral and right lateral (ventral surface facing right) views.
**Fig. 4a,b.** *Hibolites boloides* Stolley (IMC 671); from Oxfordian, Misool, Indonesia; ventral and right lateral views.
**Fig. 5a,b.** *Belemnopsis muluccana* (Boehm) (IMC 271); from Oxfordian–Kimmeridgian, Sula Islands; ventral and left lateral views.
**Fig. 6.** *Belemnopsis alfurica* (Boehm) (IMC 254); from Kimmeridgian, Sula Islands; ventral view.
**Fig. 7a,b.** *Belemnopsis stolleyi* Stevens (IMC 530); from Tithonian, Misool, Indonesia; ventral and left lateral views.
**Fig. 8.** *Belemnopsis galoi* (Boehm) (IMC 295); from Tithonian, Sula Islands; ventral view.
(Compiled by Challinor.)

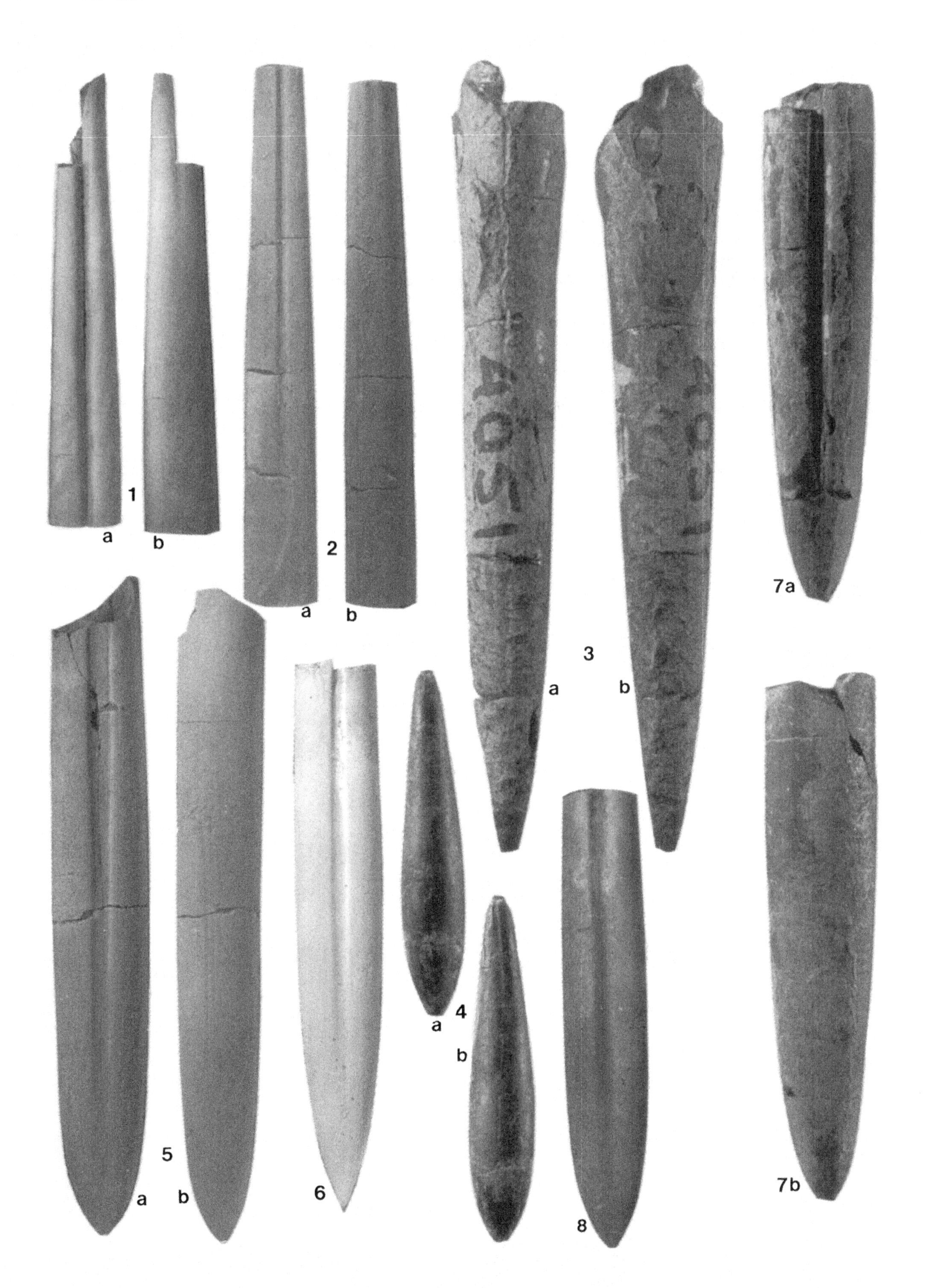
1
a
b
2
a
b
3
a
b
7a
5
a
b
6
4
a
b
8
7b

PLATE 130

AUSTRALASIA, NEW CALEDONIA, AND NEW ZEALAND

**Fig. 1a–c.** *Dicoelites aviasi* Chall. & Grant-M. (C 1356); from Temaikan, New Caledonia; ventral, left lateral, and dorsal views.
**Fig. 2a,b.** *Cylindroteuthis* sp. (C 1375); from Temaikan, Kawhia Harbour, New Zealand; ventral and left lateral views (plastic cast).
**Fig. 3a,b.** *Brachybelus* sp. (C 1377); from Temaikan, Awakino, New Zealand; ventral and right lateral views.
**Fig. 4a,b.** *Belemnopsis keari* Stevens (C 1258a); from Heterian, Kawhia Harbour, New Zealand; ventral and left lateral views.
**Fig. 5a–c.** *Conodicoelites* sp. (C 1374); from Heterian, Kawhia, New Zealand; ventral, left lateral, and dorsal views (plastic cast).
**Fig. 6a,b.** *Hibolites arkelli arkelli* Stevens (C 1376); from Puaroan, Kawhia Harbour, New Zealand; ventral and left lateral views.
**Fig. 7a,b.** *Belemnopsis aucklandica aucklandica* (Hochst.) (C 1373); from Puaroan, Port Waikato, New Zealand; ventral and left lateral views.
(Compiled by Challinor.)

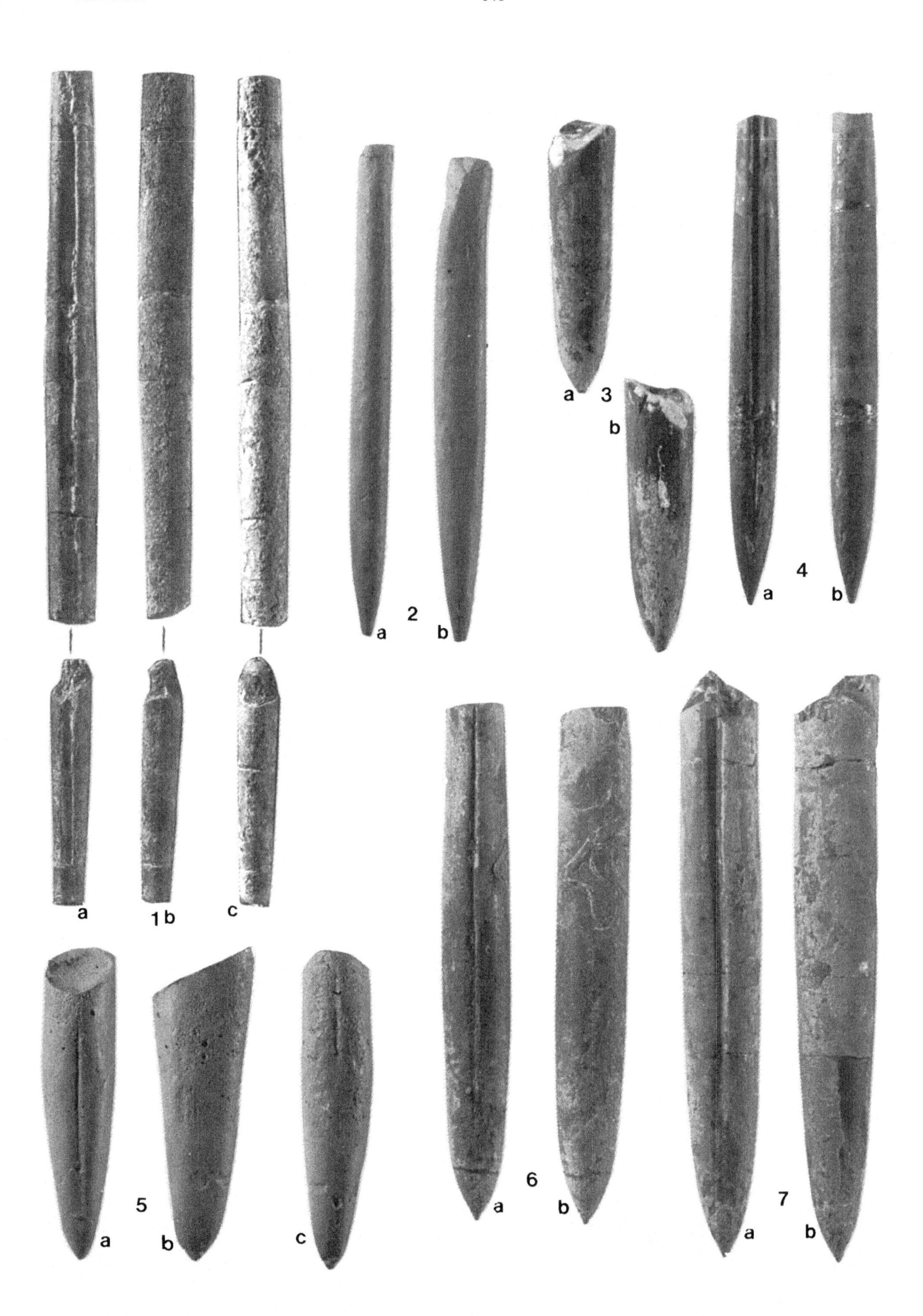
a
1 b
c
2
a
b
a
3
b
4
a
b
5
a
b
c
6
a
b
7
a
b

PLATE 131

EASTERN RUSSIA, TOARCIAN–AALENIAN

**Figs. 1,2.** *Acrocoelites pergracilis* Sachs; from Lower Toarcian, *Dactylioceras commune* Zone, North-East Russia, Omolon River basin (coll. Nalnjaeva).
**Figs. 3,4.** *Passaloteuthis ignota* Naln.; from Upper Toarcian, *Pseudolioceras rosenkrantzi* Zone, North-East Russia, Omolon River basin (coll. Nalnjaeva).
**Fig. 5.** *Holcobelus umaraensis* Tuchkov; from Lower Aalenian, North-East Russia, northern coast of Okhotsk Sea, Ola River basin (coll. Nalnjaeva).
**Figs. 6,7.** *Mesoteuthis inornata* (Phill.); from Upper Toarcian, *P. rosenkrantzi* Zone; 6, North-East Russia, Omolon River basin; 7, Siberia, Olenek River basin (coll. Nalnjaeva).
(Compiled by Nalnjaeva.)

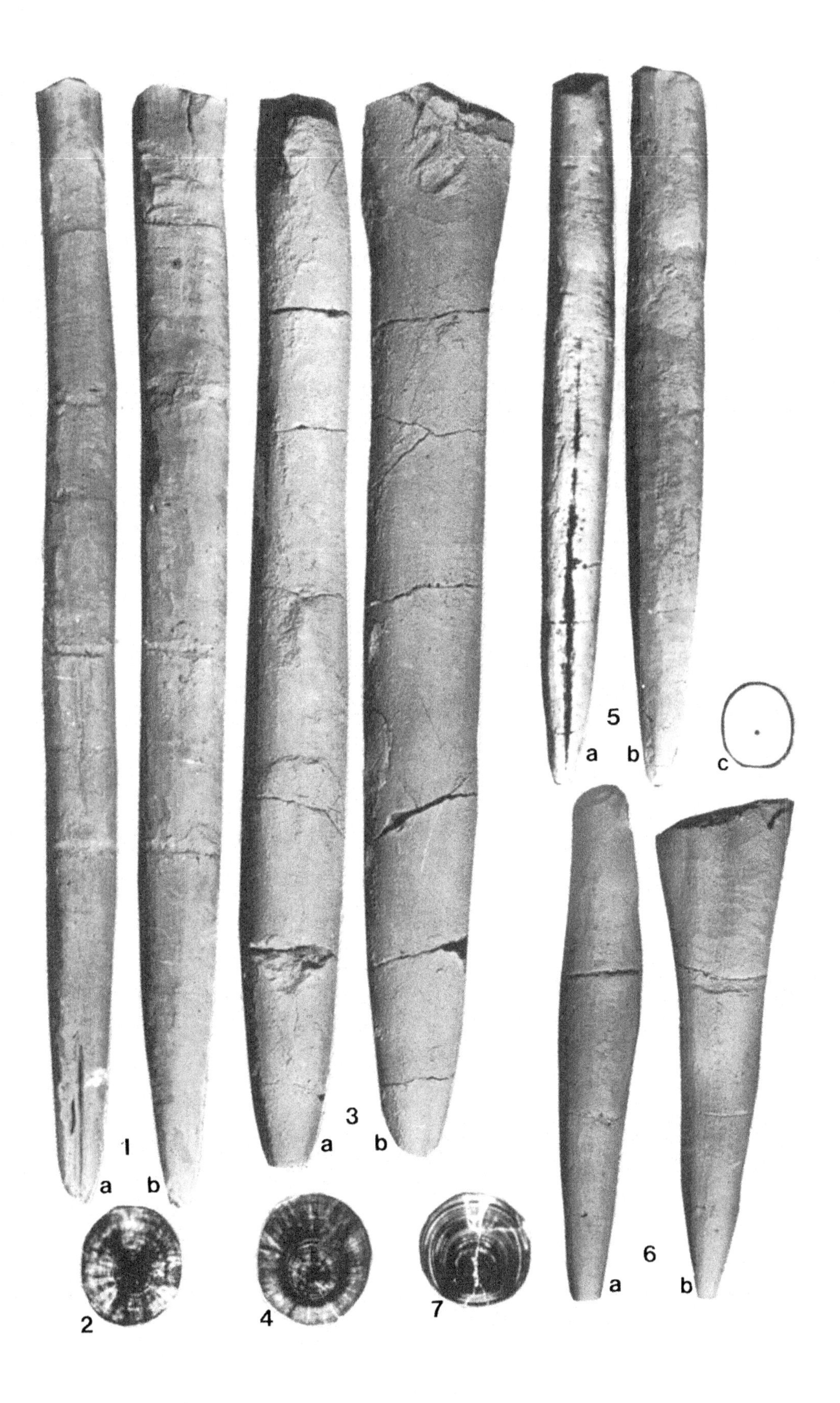
1
a
b
3
a
b
5
a
b
c
6
a
b
2
4
7

PLATE 132

EASTERN RUSSIA

**Fig. 1.** *Cylindroteuthis* (*Cylindroteuthis*) *confessa* Nalnjaeva; from Upper Lower Bajocian, *Mytiloceramus clinatus* Zone, Far East Russia, Soloni River, in Bureya River basin (coll. Nalnjaeva).

**Fig. 2.** *Cylindroteuthis* (*Cylindroteuthis*) cf. *tornatilis* Phillips; from Callovian, North-East Russia, northern coast of Okhotsk Sea (coll. Nalnjaeva).

**Fig. 3.** *Paramegateuthis timanensis* Gustomesov; from lower upper Bajocian, *Mytiloceramus kystatymensis* Beds, Far East Russia, Soloni River, in Bureya River basin (coll. Nalnjaeva).

**Fig. 4.** *Holcobelus* sp.; from lower Aalenian, *Pseudolioceras* (*Tugurites*) *maclintocki* Zone, Far East Russia, Soloni River, in Bureya River basin (coll. Nalnjaeva).

(Compiled by Nalnjaeva.)

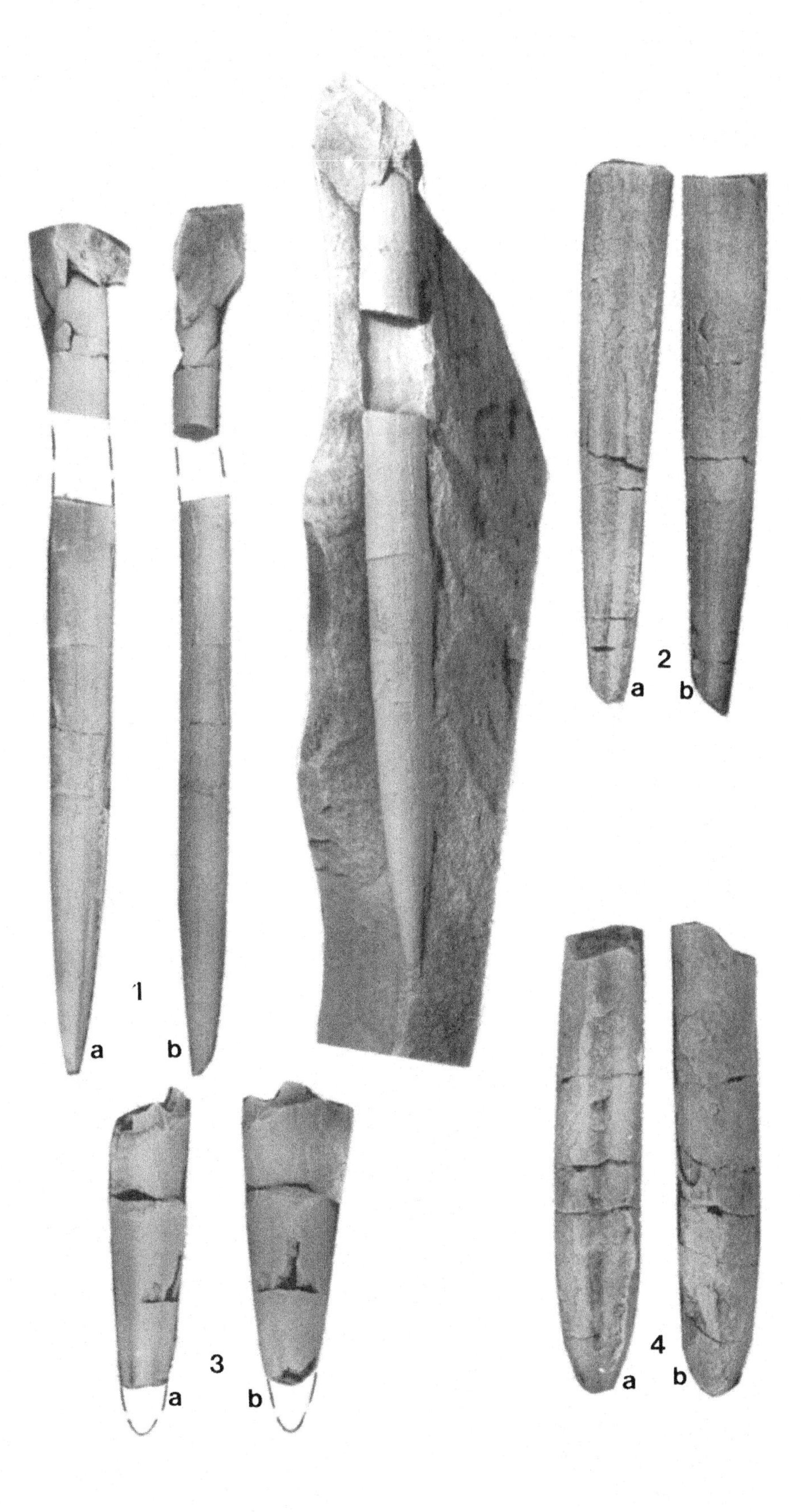
1
a
b
2
a
b
3
a
b
4
a
b

PLATE 133

MEXICO, CHILE, AND ARGENTINA

**Fig. 1.** *Ophthalmosaurus monocharactus* Appleby (×0.1); from Los Catutos, Argentina.
**Fig. 2.** Teleosauridae, metatarsals, Alto de Varas, Chile.
**Fig. 3.** *Metriorhynchus casamiquelai* Gasparini & Chong Díaz (×0.14); from Caracoles, Chile.
**Fig. 4.** *Geosaurus araucanensis* Gasparini & Dellapé (×0.16); from Cerro Lotena, Argentina.
**Fig. 5.** *Thalattosuchia,* dorsal vertebrae (×0.5); from Cualac, Mexico.
**Fig. 6.** *Notoemys laticentralis* Cattoi & Freiberg (×0.25); from Cerro Lotena, Argentina.
**Fig. 7.** ?*Eurysternum* sp. (×0.2); from Cerro Lotena, Argentina.
**Fig. 8.** *Stretosaurus* sp. (×0.4); from Cerro Lotena, Argentina.

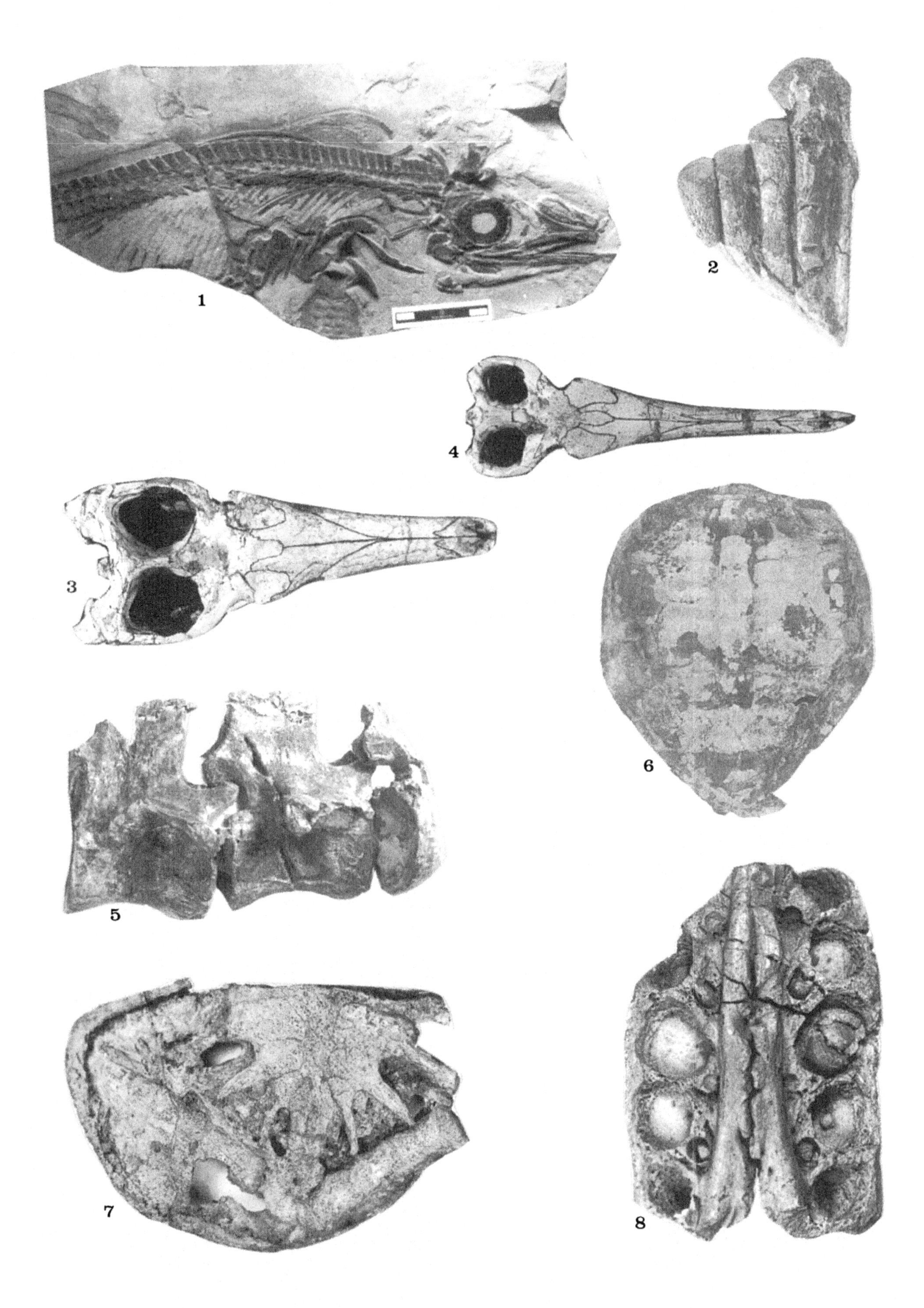
1
2
3
4
5
6
7
8

# General Index

Bold-face type is used throughout this index to indicate authors' names and pages with illustrations.

# Index of Guide- and Indexfossils